REVIEWS in M... and GEOCHEMISTRY

Volume 48 2002

PHOSPHATES:

GEOCHEMICAL, GEOBIOLOGICAL, AND MATERIALS IMPORTANCE

EDITORS:

MATTHEW J. KOHN *University of South Carolina Columbia, South Carolina*

JOHN RAKOVAN &
JOHN M. HUGHES *Miami University Oxford, Ohio*

COVER PHOTOGRAPH:

Cathodoluminescence photomicrograph of hexagonal growth hillocks on the {001} face of an apatite from the Siglo XX Mine, Llallagua, Bolivia. Luminescence (purple, activated by REEs) is homogenous among the six symmetrically equivalent pyramidal vicinal faces. Yellow luminescence (Mn^{2+}-activated) dominates flat regions of the {001} face and the terminal faces of the hillocks. [Used by permission of the Mineralogical Society of America, from Rakovan and Reeder (1994) *American Mineralogist*, Vol. 79, Fig. 7, p. 897.] *See Rakovan, Fig. 24, p. 77, this volume for further details.*

Series Editor: ***Paul H. Ribbe***
Virginia Polytechnic Institute and State University
Blacksburg, Virginia

MINERALOGICAL SOCIETY of AMERICA

Washington, DC

REVIEWS IN MINERALOGY AND GEOCHEMISTRY

(Formerly: REVIEWS IN MINERALOGY)

ISSN 1529-6466

Volume 48

Phosphates:

Geochemical, Geobiological and Materials Importance

ISBN 0-93995060-X

** This volume is the tenth in a series of review volumes published jointly under the banner of the Mineralogical Society of America and the Geochemical Society. The newly titled *Reviews in Mineralogy and Geochemistry* has been numbered contiguously with the previous series, *Reviews in Mineralogy*.

Additional copies of this volume as well as others in this series may be obtained at moderate cost from:

THE MINERALOGICAL SOCIETY OF AMERICA

1015 EIGHTEENTH STREET, NW, SUITE 601

WASHINGTON, DC 20036 U.S.A.

Dedication

Dr. William C. Luth has had a long and distinguished career in research, education and in the government. He was a leader in experimental petrology and in training graduate students at Stanford University. His efforts at Sandia National Laboratory and at the Department of Energy's headquarters resulted in the initiation and long-term support of many of the cutting edge research projects whose results form the foundations of these short courses. Bill's broad interest in understanding fundamental geochemical processes and their applications to national problems is a continuous thread through both his university and government career. He retired in 1996, but his efforts to foster excellent basic research, and to promote the development of advanced analytical capabilities gave a unique focus to the basic research portfolio in Geosciences at the Department of Energy. He has been, and continues to be, a friend and mentor to many of us. It is appropriate to celebrate his career in education and government service with this series of courses in cutting-edge geochemistry that have particular focus on Department of Energy-related science, at a time when he can still enjoy the recognition of his contributions.

— PHOSPHATES —

GEOCHEMICAL, GEOBIOLOGICAL, AND MATERIALS IMPORTANCE

FOREWORD

Several years ago, John Rakovan and John Hughes (colleagues at Miami of Ohio), and later Matt Kohn (at South Carolina), separately proposed short courses on phosphate minerals to the Council of the Mineralogical Society of America (MSA). Council suggested that they join forces. Thus this volume, *Phosphates: Geochemical, Geobiological, and Materials Importance*, was organized. It was prepared in advance of a short course of the same title, sponsored by MSA and presented at Golden, Colorado, October 25-27.

As Series Editor I thank Matt, John and John for careful selection of the topics and many authors who contributed to Volume 48 and for their subsequent management of the manuscript review and preparation process. All four of us appreciate the considerable efforts of MSA's Executive Director, Alex Speer, who manages all the Society's many programs, including short courses and publications, so calmly (well, maybe not always) and competently (without exception).

Paul H. Ribbe
Blacksburg, Virginia
September 20, 2002

PREFACE

We are pleased to present this volume entitled *Phosphates: Geochemical, Geobiological and Materials Importance*. Phosphate minerals are an integral component of geological and biological systems. They are found in virtually all rocks, are the major structural component of vertebrates, and when dissolved are critical for biological activity. This volume represents the work of many authors whose research illustrates how the unique chemical and physical behavior of phosphate minerals permits a wide range of applications that encompasses phosphate mineralogy, petrology, biomineralization, geochronology, and materials science. While diverse, these fields are all linked structurally, crystal-chemically and geochemically. As geoscientists turn their attention to the intersection of the biological, geological, and material science realms, there is no group of compounds more germane than the phosphates.

The chapters of this book are grouped into five topics: *Mineralogy and Crystal Chemistry, Petrology, Biomineralization, Geochronology*, and *Materials Applications*. In the first section, three chapters are devoted to mineralogical aspects of apatite, a phase with both inorganic and organic origins, the most abundant phosphate mineral on earth, and the main mineral phase in the human body. Monazite and xenotime are highlighted in a fourth chapter, which includes their potential use as solid-state radioactive waste repositories. The *Mineralogy and Crystal Chemistry* section concludes with a detailed examination of the crystal chemistry of 244 other naturally-occurring phosphate phases and a listing of an additional 126 minerals.

In the *Petrology* section, three chapters detail the igneous, metamorphic, and sedimentary aspects of phosphate minerals. A fourth chapter provides a close look at analyzing phosphates for major, minor, and trace elements using the electron microprobe.

A final chapter treats the global geochemical cycling of phosphate, a topic of intense, current geochemical interest.

The *Biomineralization* section begins with a summary of the current state of research on bone, dentin and enamel phosphates, a topic that crosses disciplines that include mineralogical, medical, and dental research. The following two chapters treat the stable isotope and trace element compositions of modern and fossil biogenic phosphates, with applications to paleontology, paleoclimatology, and paleoecology.

The *Geochronology* section focuses principally on apatite and monazite for U-Th-Pb, (U-Th)/He, and fission-track age determinations; it covers both classical geochronologic techniques as well as recent developments.

The final section—*Materials Applications*—highlights how phosphate phases play key roles in fields such as optics, luminescence, medical engineering and prosthetics, and engineering of radionuclide repositories. These chapters provide a glimpse of the use of natural phases in engineering and biomedical applications and illustrate fruitful areas of future research in geochemical, geobiological and materials science. We hope all chapters in this volume encourage researchers to expand their work on all aspects of natural and synthetic phosphate compounds.

This volume has benefited from the efforts of numerous colleagues. First and foremost, Paul H. Ribbe continues to produce a remarkable series, and we are deeply indebted to him for his continuing efforts on this volume. The U.S. Department of Energy generously provided funds to help defray publication costs. All the authors worked diligently to produce and revise their chapters, and the volume editors are deeply appreciative of their efforts. Chapter reviews were provided by Serena Best, Adele Boskey, John Ferry, Karl Föllmi, Gerhard Franz, John Hanchar, Tony Hurford, John Jaszczak, Bradley Joliff, Paul Koch, Barry Kohn, Lee Kump, Al Meldrum, Rick Murray, Yuangming Pan, Randy Parrish, Richard Reeder, Christian Rey, Lois Roe, George Rossman, Cathy Skinner, Jeffrey Tepper, Günther Wagner, William White, David Williams, and Peter Zeitler, in addition to several anonymous reviewers. The volume was greatly improved by their thorough and timely efforts. We trust that we have done justice to all your hard work, as well as to all phosphates.

Matthew J. Kohn

Columbia, South Carolina

John Rakovan &
John M. Hughes

Oxford, Ohio

August 17, 2002

PHOSPHATES:
Geochemical, Geobiological, and Materials Importance

TABLE of CONTENTS

1 The Crystal Structure of Apatite, $Ca_5(PO_4)_3(F,OH,Cl)$

John M. Hughes, John Rakovan

2 Compositions of the Apatite-Group Minerals: Substitution Mechanisms and Controlling Factors

Yuanming Pan, Michael E. Fleet

3 Growth and Surface Properties of Apatite

John Rakovan

4 Synthesis, Structure, and Properties of Monazite, Pretulite, and Xenotime

Lynn A. Boatner

5 The Crystal Chemistry of the Phosphate Minerals

Danielle M.C. Huminicki, Frank C. Hawthorne

6 Apatite in Igneous Systems

Philip M. Piccoli, Philip A. Candela

7 Apatite, Monazite, and Xenotime in Metamorphic Rocks

Frank S. Spear, Joseph M. Pyle

8 Electron Microprobe Analysis of REE in Apatite, Monazite and Xenotime: Protocols and Pitfalls

Joseph M. Pyle, Frank S. Spear, David A. Wark

9 Sedimentary Phosphorites—An Example: Phosphoria Formation, Southeastern Idaho, U.S.A.

Andrew C. Knudsen, Mickey E. Gunter

10 The Global Phosphorus Cycle

Gabriel M. Filippelli

11 Calcium Phosphate Biominerals

James C. Elliott

12 Stable Isotope Compositions of Biological Apatite

Matthew J. Kohn, Thure E. Cerling

13 Trace Elements in Recent and Fossil Bone Apatite

Clive N. Trueman, Noreen Tuross

14 U-Th-Pb Dating of Phosphate Minerals

T. Mark Harrison, Elizabeth J. Catlos, Jean-Marc Montel

15 (U-Th)/He Dating of Phosphates: Apatite, Monazite, and Xenotime

Kenneth A. Farley, Daniel F. Stockli

16 Fission Track Dating of Phosphate Minerals and the Thermochronology of Apatite

Andrew J.W. Gleadow, David X. Belton, Barry P. Kohn, Roderick W. Brown

17 Biomedical Application of Apatites

Karlis A. Gross, Christopher C. Berndt

18 Phosphates as Nuclear Waste Forms

Rodney C. Ewing, LuMin Wang

19 Apatite Luminescence

Glenn A. Waychunas

1 The Crystal Structure of Apatite, $Ca_5(PO_4)_3(F,OH,Cl)$

John M. Hughes and John Rakovan

Department of Geology
Miami University
Oxford, Ohio 45056

INTRODUCTION

As illustrated by the broad range of topics presented in this volume, the mineral apatite, $Ca_5(PO_4)_3(F,OH,Cl)$, is of importance in a greater variety of fields than virtually any other mineral. It is of particular significance in Earth science, life science, and material science; the foundation of this significance is the apatite atomic arrangement.

Apatite is the most abundant naturally occurring phosphate on Earth. Consequently, it is the major source of phosphorous, both as an ore and the base of the global phosphorous cycle. As the major ore mineral of phosphorous, apatite is critical for the production of huge quantities of fertilizers, detergents and phosphoric acid; the extracted phosphorous is also used in many other applications such as phosphors, rust removers, motor fuels, and insecticides to name but a few (McConnell 1973). The global biogeochemical cycling of phosphorous starts by its release from apatite at the Earth's surface and ultimately leads to the formation of other geological apatites through sedimentary processes or tectonic recycling. Along the way, however, it is an essential element to all life on Earth (Filippelli, this volume).

The structure and chemistry of apatite allow for numerous substitutions, including a multitude of metal cations (i.e., K, Na, Mn, Ni, Cu, Co, Zn, Sr, Ba, Pb, Cd, Sb, Y, REEs, U) that substitute for Ca in the structure, and anionic complexes (i.e., AsO_4^{3-}, SO_4^{2-}, CO_3^{2-}, SiO_4^{4-}, etc.) that replace PO_4^{3-} (Pan and Fleet, this volume). Indeed, apatite incorporates half the periodic chart in its atomic arrangement. These substitutions are usually in trace concentrations, but large concentrations and even complete solid solutions exist for certain substituents. This complex and variable chemistry has great implications, and is utilized in all areas of apatite research.

Geologically, apatite is a ubiquitous accessory mineral in igneous rocks, and because it is the most abundant phosphate it is essential to understanding phase equilibria of systems containing P. The presence of apatite can also strongly influence the trace element evolution of magmas (Piccoli and Candela, this volume). Many studies have shown apatite to be one of the most important minerals affecting REE trends in igneous rocks (e.g., Nash 1972; Bergman 1979; Watson and Capobianco 1981; Kovalenko et al. 1982; Gromet and Silver 1983; Watson and Harrison 1984a,b). Thus, apatite chemistry plays a critical role in understanding and modeling of igneous petrogenetic processes.

Apatite is also commonly found in metamorphic rocks (Spear, this volume), low-temperature sedimentary environments (Knudsen and Gunter, this volume), as well as a precipitate from hydrothermal solutions (Rakovan and Reeder 1994). Because of its high affinity for many trace metals, the presence of apatite in all of these environments can strongly influence their trace element signature and evolution (Bergman 1979; Watson and Capobianco 1981; McLennan 1989). Many of the elements (i.e., U, Sm, etc.) that the apatite structure can accommodate have radioisotopes that are commonly utilized as geochronometers. Hence, the dating of apatite has become an important tool for studying

1529-6466/00/0048-0001$05.00

the rocks in which it is found (Harrison, this volume; Farley and Stockli, this volume; Gleadow et al., this volume).

In the realm of biology, hydroxylapatite is the main mineral constituent of human bones, teeth, and many pathological calcifications. Indeed, except for small portions of the inner ear, all hard tissue of the human body is formed of apatite materials. Hence it is a focus of the medical fields of orthopedics, dentistry, and pathology (Elliott 1994; Elliott, this volume). It is important to recognize that although this inorganic material is part of bone tissue, skeletal apatite is by no means inert, and plays an important role in the metabolic functions of the body. For example, heavy metal sequestering by apatite is important in our understanding of the fate and transport of toxic species in the body. The presence of certain elements in trace concentrations is also known to inhibit as well as promote apatite growth in the body (e.g., McLean and Bundy 1964; Posner 1987; Hahn 1989). Because the bones and teeth of animals can persist for relatively long periods of time in geologic environments, modern to ancient biogenic apatites have been important in paleontology, archeology, and more recently in studies of palioclimates (Kohn and Cerling, this volume), animal diets and tracing environmental pollutants (Trumen, this volume).

Finally, because of their physical and chemical properties, apatite and the apatite group minerals are utilized in many materials applications. The presence of Mn, REE, U and other activators in apatite gives it luminescence properties that are of great utility in both the phosphor and laser industries (Waychunas, this volume). The fact that the apatite structure can accommodate many radionuclide species is not only of utility in geochronology but has lead to the serious evaluation of apatite as a radionuclide solid waste form for anthropogenic radioactive wastes (Ewing, this volume). Because bone and teeth are comprised of apatite it follows that surgical replacements of natural apatite precipitates are formed of the synthetic phase. Indeed, to increase biocompatibility, apatite coatings are applied to metal and ceramic prostheses (Gross and Berndt, this volume).

The foundation for understanding the complex chemistry and the many and varied uses of apatite starts with an understanding of its atomic arrangement. For this purpose we review the structure of apatite *sensu stricto* in this chapter.

THE $P6_3/m$ APATITE STRUCTURE

As will be illustrated in subsequent chapters, numerous variants of the apatite atomic arrangement exist. The variant structures, all sub-symmetries of the atomic arrangement of apatite *sensu stricto*, result principally from chemical substitutions (or, in the case of pure chlorapatite and pure hydroxylapatite, *lack* of chemical substitutions) on cation and/or anion sites. To provide a framework for discussion, we here describe the holosymmetric apatite structure, in space group $P6_3/m$ (#176).

The term apatite *sensu stricto* defines three unique minerals, *fluorapatite* [$Ca_5(PO_4)_3F$], *chlorapatite* [$Ca_5(PO_4)_3Cl$], and *hydroxylapatite* [$Ca_5(PO_4)_3(OH)$], all with $Z = 2$. As shown by numerous authors, most recently Hughes et al. (1989), the atomic arrangements of the three apatite phases differ principally in the positions of the occupants of the 0,0,*z* anion positions, i.e., fluorine, chlorine, and hydroxyl for the three end-members, respectively. Below we compare the component polyhedra of the three phases and use this comparison as a basis for subsequent discussion of other apatite group phases. Portions of the comparison are taken from Hughes et al. (1989), which also cites previous works that provide a rich history of research on mineralogical apatite. Table 1 provides the atomic parameters for well-characterized samples of fluorapatite, hydroxylapatite, and chlorapatite (Hughes et al. 1989).

Table 1. Positional parameters (*x,y,z*) and equivalent isotropic temperature factors (*B*) for fluor-, hydroxyl-, and chlorapatite. Lattice parameters are at the base of the table (from Hughes et al. 1989).

Atom	*x*	*y*	*z*	**B** (Å^2)
Ca1				
F	2/3	1/3	0.0010(1)	0.91(1)
OH	2/3	1/3	0.00144(8)	0.929(7)
Cl	2/3	1/3	0.0027(1)	0.99(1)
Ca2				
F	-0.00712(7)	0.24227(7)	1/4	0.77(1)
OH	-0.00657(5)	0.24706(5)	1/4	0.859(9)
Cl	0.00112(6)	0.25763(6)	1/4	1.14(1)
P				
F	0.36895(8)	0.39850(8)	1/4	0.57(1)
OH	0.36860(6)	0.39866(6)	1/4	0.62(1)
Cl	0.37359(7)	0.40581(7)	1/4	0.77(1)
O1				
F	0.4849(2)	0.3273(3)	1/4	0.99(4)
OH	0.4850(2)	0.3289(2)	1/4	1.00(3)
Cl	0.4902(2)	0.3403(2)	1/4	1.34(4)
O2				
F	0.4667(2)	0.5875(3)	1/4	1.19(5)
OH	0.4649(2)	0.5871(2)	1/4	1.25(3)
Cl	0.4654(2)	0.5908(2)	1/4	1.47(4)
O3				
F	0.2575(2)	0.3421(2)	0.0705(2)	1.32(3)
OH	0.2580(1)	0.3435(1)	0.0703(2)	1.57(2)
Cl	0.2655(2)	0.3522(2)	0.0684(3)	1.88(3)
X				
F	0	0	1/4	1.93(6)
OH	0	0	0.1979(6)	1.31(8)
Cl	0	0	0.4323(4)	2.68(5)
F:		*a* = 9.397(3) Å	*c* = 6.878(2) Å	
OH:		*a* = 9.417(2)	*c* = 6.875(2)	
Cl:		*a* = 9.598(2)	*c* = 6.776(4)	

Several comments are warranted about the apatite structure before its description is proffered. As noted in Table 1, the atoms lie on or near four (00*l*) planes in the atomic arrangement. Ca2, P, O1, O2, and the *X* anion (where *X* = F, OH, Cl) lie on (or are disordered about) special positions on the mirror planes at $z = {}^1/_4$ and ${}^3/_4$. Intercalated approximately halfway between these planes are Ca1 (in a special position at z = ~0, ~$^1/_2$) and O3, in the general position with *z* values of ~0.07 and ~0.57. On the basis of the layer structure of the atomic arrangement, O'Keeffe and Hyde (1985) noted the similarity between the cation positions in the apatite structure and the Mn_5Si_3 intermetallic phase, and offered a description of apatite as a *cation*-closest-packed atomic arrangement. Dai et al. (1991) and Dai and Harlow (1991) elucidated that intriguing cation-closest packing relationship in an examination of arsenate apatites, and their work is worthy of detailed review.

The atomic arrangement of apatite *sensu stricto* is formed of three cation polyhedra, and the structural variations among the three anion end-members is perhaps best understood by examining variations that occur in the three polyhedra concomitant with substitution of the three column anions. The three polyhedra as they occur in fluorapatite, hydroxylapatite, and chlorapatite are compared and superimposed below.

The PO_4 tetrahedron

Phosphorous occurs in apatite *sensu stricto* in tetrahedral coordination, with the central P atom in the *6h* special position (Table 1). Typical of such rigid polyhedra, the tetrahedron is essentially invariant in fluor-, chlor-, and hydroxylapatite. Figure 1 displays the superposition of PO_4 tetrahedra for fluorapatite, hydroxylapatite, and chlorapatite, and illustrates the invariance of the polyhedron among the three end-members.

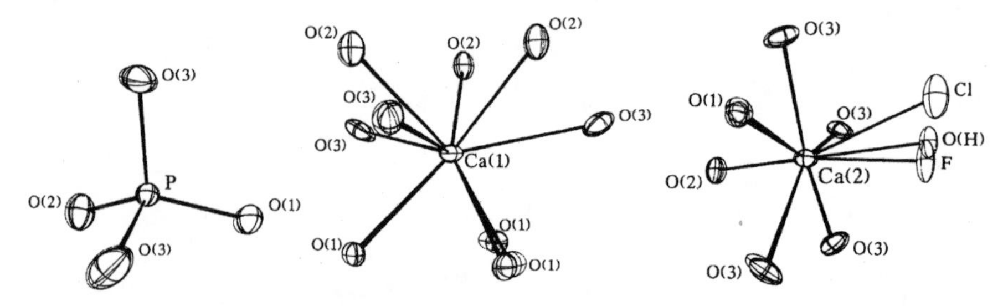

Figure 1. Drawings of PO_4 tetrahedron (left) and Ca1 and Ca2 polyhedra for the three apatite structures. Each overlay is a superposition of the analogous polyhedra from the fluorapatite, hydroxylapatite, and chlorapatite end-members, drawn to the same scale and with coincident central cations. From Hughes et al. (1989).

The $Ca1O_9$ polyhedron

In apatite *sensu stricto*, the ten Ca ions in the unit cell exist in two polyhedra. Ca1, with the central cation in the $4f\,(^1/_3,^2/_3,z)$ position, is coordinated to nine oxygen atoms in the arrangement of a tricapped trigonal prism. Ca1, with z values near 0 and $^1/_2$, bonds to six of those oxygen atoms (3 × O1, 3 × O2) in planes $^1/_2$ unit cell above and below the central cation, forming a trigonal prism. Three more oxygen atoms (3 × O3), essentially coplanar with Ca1, are bonded through the prism faces to form the tricapped trigonal prism.

Figure 1 displays the superposition of Ca1 and its ligands for fluorapatite, hydroxylapatite, and chlorapatite. As revealed by the figure, the Ca1 polyhedron shows little response to incorporation of the different column anions in the three end-members of apatite *sensu stricto*.

The $Ca2O_6X$ polyhedron

Ca2, in the *6h* special position of space group $P6_3/m$, bonds to 6 oxygen atoms (O1, O2, 4 × O3) and one column anion (X). The major structural response to substitution of the three column anions occurs in this polyhedron. Figure 1 displays the superposition of the three Ca2 polyhedra for the pure anion end-members, and illustrates the large shifts that occur in the positions of the X anions in the [00z] anion column.

The Ca2 cations form triangles on the planes at $z = ^1/_4$ and $z = ^3/_4$ (Fig. 2). Each of the three Ca atoms at the corners of the triangles is bonded to the central anion in the [00z] column. Fluorine, the smallest of the three anions, lies on the planes at z = $^1/_4,^3/_4$ (at $0,0,^1/_4$ and $0,0,^3/_4$) with Ca2-F bonds lying in (001). The larger OH anionic complex and

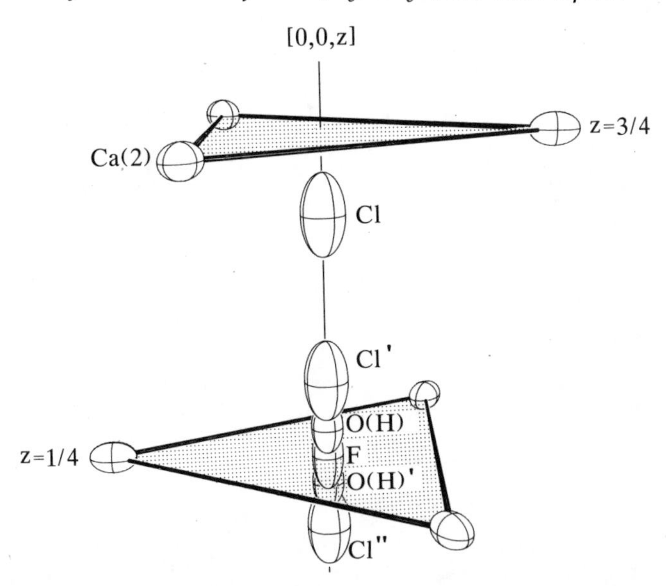

Figure 2. Depiction of possible anion positions in the hexagonal ternary apatite structure. Stippled planes represent mirror planes at $z = {}^1/_4, {}^3/_4$, each containing a triangle of Ca2 atoms (connected by "bonds"). Atom *Cl* represents a Cl atom disordered below the mirror plane at $z = {}^3/_4$, and other column anions represent five possible anion neighbors associated with the mirror plane at $z = {}^1/_4$ (*Cl'* = Cl disordered above plane; *O(H)* = OH oxygen disordered above that plane; *F* = fluorine atoms at $0,0,{}^1/_4$; *O(H)'* and *Cl''* represent OH and Cl disordered below the mirror plane at $z = {}^1/_4$). From Hughes et al. (1990).

Cl anion are too large to lie on the rigid plane defined by the Ca atoms, and the OH or Cl anion associated with the plane is displaced above *or* below the plane. Such a displacement locally destroys the $P6_3/m$ symmetry by eliminating the mirror plane, as only one of the two mirror-symmetric sites above and below the plane is occupied. However, except in rare cases discussed below, over the crystal as a whole each mirror-related site is half-occupied, thus preserving the average $P6_3/m$ symmetry.

The polyhedral components of the apatite atomic arrangement combine to form the atomic arrangement pictured in Figure 3. That [001] projection shows the packing of the three polyhedra described above to yield the atomic arrangement of apatite *sensu stricto*.

APATITE *SENSU STRICTO* STRUCTURAL VARIANTS

Numerous structural variants of the apatite atomic arrangement exist, all related to the holosymmetric structure depicted in Figure 3. Many of the substructures result from ordering of cation substituents on the Ca and P sites, and those structures will be discussed in subsequent chapters. Here we elucidate structural changes in apatite *sensu stricto* that result from anion substitutions or ordering in the anion column.

Monoclinic hydroxylapatite and chlorapatite

Although the ideal apatite atomic arrangement is described in space group $P6_3/m$, the pure hydroxylapatite and chlorapatite end-members actually crystallize in the sub-symmetric monoclinic space group $P2_1/b$. As noted above, OH and Cl are too large to fit in the triangles of Ca atoms (Fig. 2). Early structure elucidation of the two end-members demonstrated (OH: Kay et al. 1964, Elliot et al. 1973; Cl: Hounslow and Chao 1968) that the OH and Cl are disordered above *or* below the Ca triangles at $z = {}^1/_4, {}^3/_4$ at any

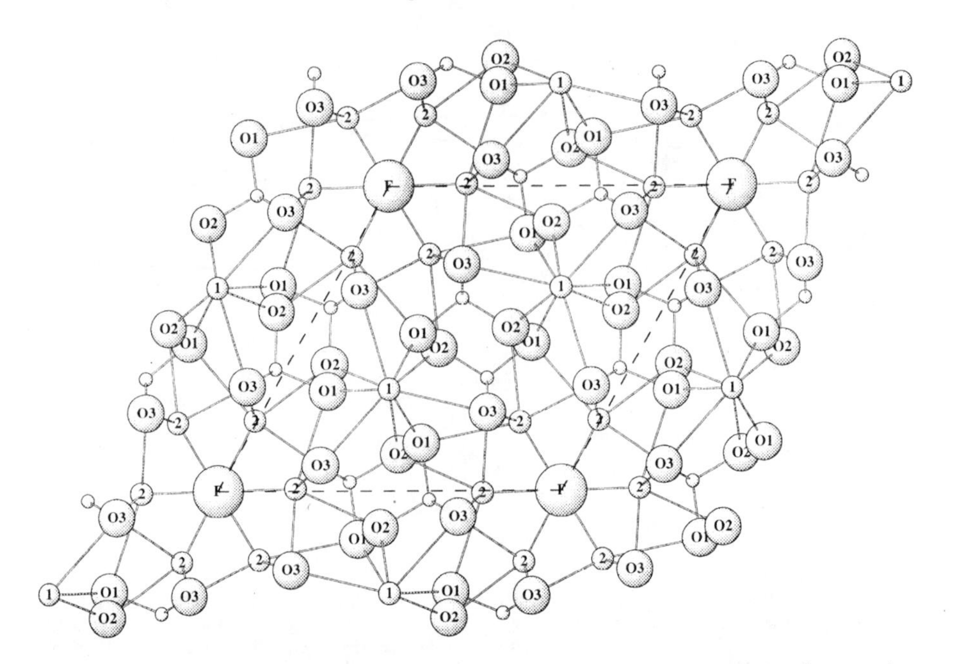

Figure 3. The atomic arrangement of fluorapatite projected on (001). Phosphorous (smallest atoms) are unlabeled, and Ca1 and Ca2 are labeled "1" and "2" respectively. The P bond to O3 represents two such bonds to overlying O3 atoms. Coordination polyhedra are explained in text.

individual anion site. In [00*z*] columns in hydroxylapatite or chlorapatite in which few vacancies or impurities occur, the OH or Cl column anions are ordered either above *or* below the mirror planes in a given anion column (Fig. 4); importantly, in any individual column in the pure phases, all anions are ordered in the same sense. The sense of ordering (e.g., anions above *or* below the plane) in any individual column is transmitted to the adjacent column along *b*. Tilting of polyhedra caused by the ordering in one column causes the adjacent column along *b* to be ordered in the opposite sense (e.g., below *or* above the plane), thereby doubling the *b* axis length. The symmetry thus degenerates to $P2_1/b$, a consequence of column anion ordering. Most natural chlorapatite and all natural hydroxylapatite, however, contain enough impurities or vacancies in the anion columns (as in Fig. 4) to destroy the ordering in the column, thus they exist in the putative $P6_3/m$ space group. Hounslow and Chow (1968) have reported natural monoclinic chlorapatite, and, in a particularly insightful study, Elliot et al. (1973) illustrated the structural details of monoclinic hydroxylapatite. Despite the monoclinic nature of hydroxylapatite and chlorapatite at room temperature, however, the phases invert to the hexagonal structure at elevated temperatures (OH: Bauer 1991; Cl: Prener 1967, Bauer and Klee 1993).

Anion compatibility

Fluor-, chlor-, and hydroxylapatite end-members exist with a single anion occupant of the anion columns. However, as solid solution among the column anion components occurs the [00*z*] columns must accommodate the variety of anions, and the means of accommodation is not straightforward. The extent of solid solution between and among the column anion constituents, and the structural response to such solution, is not well characterized (McConnell 1973). The locations of the anions in the end-members demonstrate that the structural configurations are not miscible without a structural

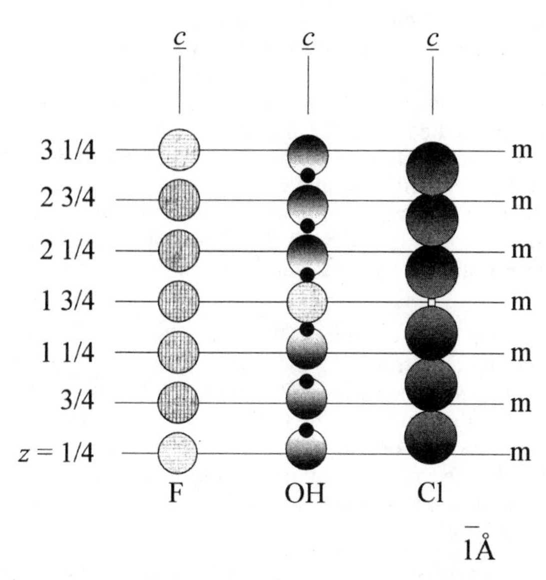

Figure 4. Anion columns in hexagonal fluorapatite, hydroxylapatite and chlorapatite depicted to scale. Column "F" depicts anions in fluorapatite located on mirror planes at $z = {}^1/_4, {}^3/_4$ in successive unit cells. Column "OH" depicts three successive hydroxyls in hydroxylapatite disordered 0.35 Å above the mirror planes and three successive hydroxyls disordered below the mirror planes, with the sense of ordering reversed by an F anion (stippled) "impurity". Column "Cl" depicts three successive Cl anions in chlorapatite disordered above the mirror planes and three successive Cl anions disordered below the mirror planes; the vacancy (□) at $z = 1^3/_4$ must exist in order to reverse the sense of ordering, as F and OH species are prohibited. Radii: F = OH = 1.4 Å; Cl = 1.8 Å. Scale of H atom (solid circles) is arbitrary. After Hughes et al. (1989).

response to solution of the binary and ternary anion columns. The structural responses to binary and ternary solid solution have not been well studied, and are a fruitful area of future research.

The apatite end-members contain column anions associated with the mirror planes at $z = {}^1/_4$ and ${}^3/_4$ in each unit cell. In fluorapatite the F atoms are located on the mirror planes, whereas in hydroxylapatite and chlorapatite the anions are disordered about each mirror plane, with OH displaced ~0.35 Å above or below the plane and Cl displaced ~1.2 Å above or below the plane. Because the anion sites in any column are separated by only $c/2$, or approximately 3.4 Å, there is extensive interaction between the occupants of the adjacent sites, and the occupants of a given site can affect the occupant of the adjacent site in the same anion column along c, creating a Markovian sequence of anion occupants.

Figure 4 depicts anion columns in fluorapatite, and near-end-member hydroxylapatite and chlorapatite, and illustrates the nature of interactions in the anion columns in binary and ternary apatites. The spherical atoms are drawn to scale and illustrate that in fluorapatite, with F atoms lying on the mirror planes, the hard-sphere model allows a fluorine atom at each successive anion site.

In hydroxylapatite the hydroxyls are located ~0.35 Å above or below each mirror plane. To achieve the average disordered $P6_3/m$ structure, half the hydroxyls in any column must be located above the particular mirror plane, and half below; this arrangement requires "reversal sites" in which the sense of the ordering (above *vs.* below) is reversed. Figure 4 illustrates such a reversal site, with a fluorine "impurity" allowing reversal of the sense of the ordering; a vacancy would also allow reversal. Without such impurities, however, the hydroxyls at a reversal site would be separated by ~2.7 Å, and the hydrogen atoms associated with the hydroxyls would be ~0.8 Å distant, clearly impossible. Thus, with the addition of sufficient F or □ reversal sites, reversal of the ordering of the hydroxyls is facilitated and hydroxylapatite inverts from $P2_1/b$ to $P6_3/m$ symmetry. Sudarsanan and Young (1969) have shown that the well-characterized Holly Springs hydroxylapatite is hexagonal with 8% substitution of F for OH in the anion columns; their paper should be regarded as the seminal paper on the structure of natural hydroxylapatite.

Incorporation of the large Cl anion (1.81 Å radius) in the anion columns adds additional constraints to the structure, and hence chemistry, of the anion columns. Because the larger Cl anion is displaced ~1.2 Å above or below the plane, the interaction with the occupant of the adjacent site along the anion column is profound. As depicted in Figure 4, the reversal from a Cl ordered above the mirror plane to a Cl below the plane can only take place with a vacancy, *assuming no adjustment in anion sites*. Thus, for pure chlorapatite, the hexagonal phase must have stoichiometric vacancies to effect anion reversals in the anion columns.

The interaction of the column anions suggests that binary and ternary calcium apatite phosphates are not mere additions of the end-member structures, but that structural adjustments must occur to accommodate mixed-anion columns. Table 2 depicts the distance between anion occupants of adjacent planes, and illustrates which adjacent neighbors are allowed with mixtures of the end-member structures.

Table 2.

Anion Associated with Mirror Plane at z = 1/4 \ Anion Associated with Mirror Plane at z = 3/4	OH_a	OH_b	Cl_a	Cl_b
OH_a	3.41	2.70	4.30	1.81
OH_b	4.12	3.41	5.01	2.52
Cl_a	2.52	1.81	3.41	0.92
Cl_b	5.01	4.30	5.90	3.41

Solution in the anion columns

The discussion above illustrates that various anion column occupants are incompatible as nearest neighbors, at least given the end-member structures. Yet, such solutions do indeed occur; solid solution among all three anions is not uncommon. Clearly, structural adjustments must occur to accommodate anion mixtures in the columns; for example, without such accommodations, Cl could not coexist in the columns with other anions. The situation is illustrated in Figure 2, which depicts a Cl ordered below a plane at $z = {}^3/_4$, and the five possible anion occupants at the adjacent mirror plane at $z = {}^1/_4$. As depicted in Figure 2, possible column-anion neighbors associated with the adjacent mirror plane at $z = {}^1/_4$ are (1) a Cl located above the $z = {}^1/_4$ plane (Cl-Cl′ distance 0.92 Å), (2) an OH disordered above the mirror plane (Cl-O(H) distance = 1.81 Å), (3) an F ion in the plane of the mirror (Cl-F distance = 2.16 Å), (4) an OH disordered below the mirror plane (Cl-O(H)′ distance = 2.51 Å), and (5) a Cl ion disordered below the $z = {}^1/_4$ mirror plane (Cl-Cl″ distance = $c/2$ = 3.4 Å). On the basis of anion positions in the hexagonal end-member structures, all of these anion neighbors except the Cl ion below the adjacent plane (= Cl″) are prohibited on the basis of anion-anion distances. It is thus not immediately apparent what structural accommodations must occur to achieve solid solution among the anions.

In a particularly insightful study, however, Sudarsanan and Young (1978) demonstrated that in Cl-bearing apatite solid solutions, a new Cl site can be created in the column with attendant shifts in the positions of the Ca atoms in the surrounding triangle. The new site allows solid solution to occur among the three-anion occupants by creation of favorable sequences in the anion column. Hughes et al. (1990) examined two natural F-, OH-, and Cl-bearing apatites ("ternary apatites"), and found that the structural adjustments proffered by Sudarsanan and Young (1978) operated in hexagonal apatites to effect solid solution among the anions. In addition to the hexagonal phase, however, Hughes et al. (1990) described a monoclinic ternary apatite in which ordering of the ternary column anions yielded a $P2_1/b$ structure. The two phases will be used to elucidate the solution of the three anions in calcium phosphate apatite.

Hexagonal ternary apatite

Hughes et al. (1990) refined the atomic arrangement of a hexagonal ternary apatite with the column anion composition of $(F_{0.39}Cl_{0.33}OH_{0.28})$, and much of the following is taken contextually or verbatim from their description of the phase.

As noted previously, the anion positions in the three end-member apatite *sensu stricto* phases are not compatible in solid solution, as demonstrated in Figure 2. As shown there, a Cl atom disordered below the $z = {}^3/_4$ plane could only be followed along $-c$ by another Cl atom disordered below the mirror plane at $z = {}^1/_4$, ad infinitum; on the basis of the anion positions in the hexagonal end-member atomic arrangements, all of these anion neighbors for the Cl anion except Cl″ (Fig. 2) are prohibited on the basis of anion-anion distance.

Hughes et al. (1990) noted that in "end-member" hexagonal chlorapatite, the Cl atom is unfettered by adjacent non-Cl atoms in the [0,0,z] anion columns, and the Cl half-occupies a position at (0,0,0.5677) about the $z = {}^3/_4$ mirror plane. In hexagonal *ternary* apatite, however, the Cl atom position shifts in response to the presence of other anion neighbors, as first demonstrated in a perceptive study by Sudarsanan and Young (1978). In mixed-anion columns, a portion of the Cl atoms shift 0.4 Å closer to the $z = {}^3/_4$ mirror plane, to a half-occupied (0,0,0.632) position (continuing the example of a Cl atom disordered below the $z = {}^3/_4$ plane). The shift in Cl position allows accommodation of an OH neighbor at the OH position disordered below the adjacent plane at $z = {}^1/_4$ [O(H)′ in Fig. 2]. The Cl-O distance is now an allowable 2.95 Å, in contrast to 2.52 Å calculated with anion positions in the hexagonal end-member chlor- and hydroxylapatite atomic arrangements.

The shift of anion positions in apatite solid solutions is not without concomitant shifts of the coordinating cation positions. The shift of the Cl position causes a splitting of the single Ca2 site into two sites. In a detailed structure refinement of a ternary apatite, Sudarsanan and Young (1969) demonstrated that the shift of the Cl atoms closer to the Ca2 triangles causes a portion of the Ca2 atoms (proportional to the number of Cl atoms in the new position) to shift to a new position that enables the ~2.70 Å Ca-Cl bond distance to be maintained. Thus, in ternary apatites, two Ca2 positions exist: $Ca2_A$, in which Ca2 atoms bond to F, OH, or unshifted Cl atoms, and $Ca2_B$, which bond to the shifted Cl atoms.

Miscibility of F, Cl and OH in hexagonal ternary apatites thus results from a Markovian sequence of anions in the [00*z*] anion columns, i.e., a sequence in which the occupant of a given position is dependent on the occupant of the adjacent position(s). For example, to change the anion sequence in a given column from a Cl_A atom disordered below a plane to a Cl_A atom disordered above a plane (Fig. 2), the following sequence can be postulated. Moving down the anion column to successive mirror planes at $z = {}^3/_4, {}^1/_4$, etc., along $-c$, a change from a Cl_A atom disordered *below* the plane to a Cl_A atom disordered *above* the plane can be effected, without vacancies, by the following Markovian sequence of anions at each successive site: $Cl_{A,\text{below plane}}$ –(~2.95 Å)– $Cl_{B,\text{below plane}}$ –(~2.95 Å)– $OH_{\text{below plane}}$ –(~3.08 Å)– $F_{\text{on plane}}$ –(~3.08 Å)– $OH_{\text{above plane}}$ –(~2.95 Å)– $Cl_{B,\text{above plane}}$ –(~2.95 Å)– $Cl_{A,\text{above plane}}$. Thus, the creation of new anion sites in response to solid solution enables accommodation of the three anions in a single column without vacancies.

Monoclinic ternary apatite

In a study of ternary apatites *sensu stricto*, Hughes et al. (1990) revealed a monoclinic variant of ternary apatite that illustrates the unaddressed complexities of apatite crystal chemistry. In the $P2_1/b$ variety, the positions of the non-column atoms are similar to those in $P6_3/m$ ternary apatite, but the reduction in symmetry results from ordering in the anion columns, similar to the cause of monoclinicity in the chlorapatite and hydroxylapatite end-members. The positions of the column anions in hexagonal and monoclinic ternary apatite are depicted in Figure 5. The petrologic implications of monoclinic ternary apatite remain unexplored, and it is unknown how common the phase is.

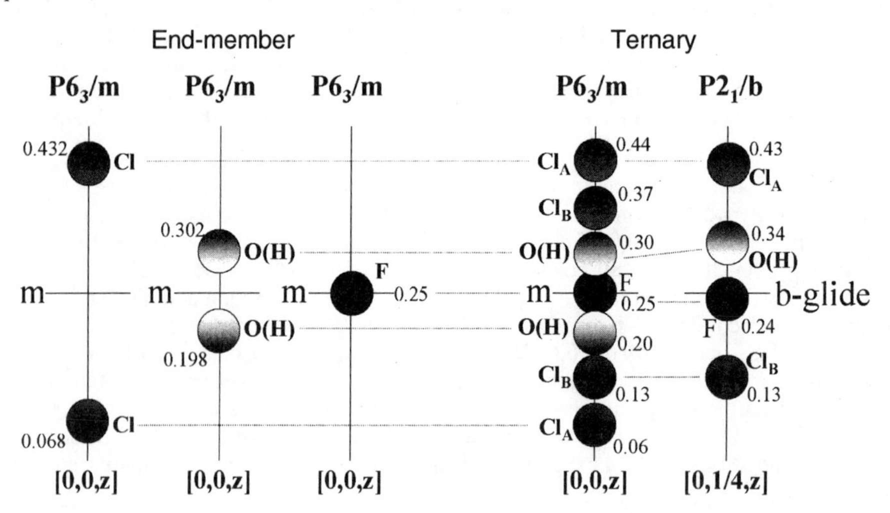

Figure 5. Correlation of anion sites in end-member and $P6_3/m$ and $P2_1/b$ ternary apatites. Atoms represent possible anion sites associated with the mirror plane ($P6_3/m$) or the *b*-glide ($P2_1/b$) at $z = {}^1/_4$. Atom size is arbitrary, and the *z* positional parameter is given. Dashed lines connect correlative positions in different structures. The shift of the O(H) position between the ternary structures is largely a result of the difference in Cl content. Modified from a drawing by Maryellen Cameron.

It is perhaps illustrative of the unexplored complexities of apatite crystal chemistry that a study of apatite that begins with simple film studies of a few crystals quickly uncovers a new structural variant of ternary apatite (Hughes et al. 1990). Researchers are urged to explore in detail the crystal chemistry of the apatite *sensu stricto* group of minerals, the minerals that collectively comprise the tenth most abundant mineral on Earth and perhaps the premier example of biomineralization, the biomineral that is essential to human life.

SUMMARY

The apatite *sensu stricto* group of minerals is formed of three phases that are among the most abundant minerals on Earth, and are essential to human life. As the attention of mineralogists turns to biomineralization, it is fitting that research on apatite re-enters the forefront of mineralogical research to address the many unanswered questions on apatite crystal chemistry.

Such fundamental parameters as the extent of solid solution between the anion end-members and among the ternary anion components remain unanswered by modern studies. Indeed, the outline of column anion interactions given above does not provide a viable mechanism for binary solution of fluorapatite and chlorapatite, and the study of mineral solution along the join is of fundamental interest in petrology, as constraints on incorporation of volatile components in apatite are important. Mackie and Young (1974) addressed the incompatibility of F and Cl in the apatite anion columns, and suggested that solid solution may be possible with creation of new F sites in the anion column. However, the bond-valence sums on their proposed new F site occupant are unreasonably low, and questions remain as to the extent and mechanism of solid solution. Hughes et al. (1990) suggested that in natural apatites, OH may be an *essential* component in stabilizing fluorapatite-chlorapatite solid solution, but this observation has not been explored. In addition to X-ray diffraction studies that average larger volumes of unit cells, microbeam and atomic-scale imaging methods are of interest in elucidating local atomic constraints in the apatite *sensu stricto* phases. Such studies are of fundamental importance to the fields of petrology, medicine and material science, and are a fruitful area for mineralogical researchers.

ACKNOWLEDGMENTS

We thank Yuanming Pan for constructive reviews of the manuscript and helpful suggestions for its improvement, and the National Science Foundation for support of many of the projects cited herein.

REFERENCES

Bauer M (1991) Röntgenographische und Dielektrische Untersuchungen an Apatiten. Dissertation, Fakultät für Physik, Universität Karlsruhe

Bauer M, Klee WE (1993) The monoclinic-hexagonal phase transition in chlorapatite. Eur J Mineral 5: 307-316

Bergman, SC (1979) The significance of accessory apatite in the REE modelling of magma genesis. EOS Trans Am Geophys Union 60:412

Dai Y, Hughes JM, Moore PB (1991) The crystal structures of mimetite and clinomimetite, $Pb_5(AsO_4)_3Cl$. Can Mineral 29:369-376

Dai Y, Harlow GE (1991) Structural relationships of arsenate apatites with their anion-devoid intermetallic phase Ca_5As_3. Geol Soc Am Ann Meet, Progr Abstr 23:219

Elliot JC, Mackie PE, Young RA (1973) Monoclinic hydroxylapatite. Science 180:1055-1057

Elliott JC (1994) Structure and Chemistry of the Apatites and other Calcium Orthophosphates. Elsevier, Amsterdam

Gromet LP, Silver LT (1983) Rare earth element distributions among minerals in a granodiorite and their petrologic implications. Geochim et Cosmochim Acta 47:925-939

Hahn TJ (1989) Aluminum-related disorders of bone and mineral metabolism. *In* Bone and Mineral Research. Vol. 6. Peck WA (ed) Elsevier, New York

Hounslow AW, Chao GY (1968) Monoclinic chlorapatite from Ontario. Can Mineral 10:252-259

Hughes JM, Cameron M, Crowley KD (1989) Structural variations in natural F, OH and Cl apatites. Am Mineral 74:870-876

Hughes JM, Cameron M, Crowley KD (1990) Crystal structures of natural ternary apatites: Solid solution in the $Ca_5(PO_4)_3X$ (X = F, OH, Cl) system. Am Mineral 75:295-304

Kay MI, Young RA, Posner AS (1964) Crystal structure of hydroxylapatite. Nature 204:1050-1052

Kovalenko VI, Antipin VS, Vladykin NV, Smirnova YV, Balashov YA (1982) Rare-earth distribution coefficients in apatite and behavior in magmatic processes. Geochem Intl 19:174-183

Mackie PE, Young RA (1974) Fluorine-chlorine interaction in fluor-chlorapatite. J Solid State Chem 11:319-329

McConnell D (1973) Apatite: its crystal chemistry, mineralogy, utilization, and geologic and biologic occurrences. Springer-Verlag, New York

McLean FC, Bundy AM (1964) Radiation, Isotopes, and Bone. Academic Press, New York.

McLennan SM (1989) Rare earth elements in sedimentary rocks: Influence of provenance and sedimentary processes. Rev Mineral 21:169-200

Nash WP (1972) Apatite chemistry and phosphorus fugacity in a differentiated igneous intrusion. Am Mineral 57:877-886

O'Keeffe M, Hyde BG (1985) An alternative approach to non-molecular crystal structures, with emphasis on the arrangements of cations. Struct Bond 16:77-144

Posner AS (1987) Bone mineral and the mineralization process. *In* Bone and Mineral Research Vol. 5. Peck WA (ed) Elsevier, New York.

Prener JS (1967) The growth and crystallographic properties of calcium fluor- and chlorapatite crystals. J Electrochem Soc 114:77-83

Rakovan J, Reeder RJ (1994) Differential incorporation of trace elements and dissymmetrization in apatite: The role of surface structure during growth. Am Mineral 79:892-903

Sudarsanan K, Young RA (1969) Significant precision in crystal structure details: Holly Springs hydroxyapatite. Acta Crystallogr B25:1534-1543

Sudarsanan K, Young RA (1978) Structural interactions of F, Cl and OH in apatites. Acta Crystallogr B34:1401-1407

Watson, EB, Capobianco, CJ (1981) Phosphorus and rare-earth elements in felsic magmas: An assessment of the role of apatite. Geochim Cosmochim Acta 45:2349-2358

Watson EB, Harrison TM (1984a) Accessory minerals and the geochemical evolution of crustal magmatic systems: A summary and prospectus of experimental approaches. Phys Earth Planet Inter 5:19-30

Watson EB, Harrison TM (1984b) What can accessory minerals tell us about felsic magma evolution? A framework for experimental study. Proc 27th Intl Geol Cong 11:503-520

2 Compositions of the Apatite-Group Minerals: Substitution Mechanisms and Controlling Factors

Yuanming Pan

Department of Geological Sciences
University of Saskatchewan
Saskatoon, Saskatchewan S7N 5E2, Canada

Michael E. Fleet

Department of Earth Sciences
University of Western Ontario
London, Ontario N6A 5B7, Canada

INTRODUCTION

The apatite-group minerals of the general formula, $M_{10}(ZO_4)_6X_2$ (M = Ca, Sr, Pb, Na…, Z = P, As, Si, V…, and X = F, OH, Cl…), are remarkably tolerant to structural distortion and chemical substitution, and consequently are extremely diverse in composition (e.g., Kreidler and Hummel 1970; McConnell 1973; Roy et al. 1978; Elliott 1994). Of particular interest is that a number of important geological, environmental/paleoenvironmental, and technological applications of the apatite-group minerals are directly linked to their chemical compositions. It is therefore fundamentally important to understand the substitution mechanisms and other intrinsic and external factors that control the compositional variation in apatites.

The minerals of the apatite group are listed in Table 1, and representative compositions of selected apatite-group minerals are given in Table 2. Also, more than 100 compounds with the apatite structure have been synthesized (Table 3). Phosphate apatites, particularly fluorapatite and hydroxylapatite, are by far the most common in nature and are often synonymous with "apatite(s)". For example, fluorapatite is a ubiquitous accessory phase in igneous, metamorphic, and sedimentary rocks and a major constituent in phosphorites and certain carbonatites and anorthosites (McConnell 1973; Dymek and Owens 2001). Of particular importance in biological systems, hydroxylapatite and fluorapatite (and their carbonate-bearing varieties) are important mineral components of bones, teeth and fossils (McConnell 1973; Wright et al. 1984; Grandjean-Lécuyer et al. 1993; Elliott 1994; Wilson et al. 1999; Suetsugu et al. 2000; Ivanova et al. 2001).

Following Fleischer and Mandarino (1995), Table 1 also includes melanocerite-(Ce), tritomite-(Ce), and tritomite-(Y), the compositions of which correspond closely to synthetic rare-earth borosilicate oxyapatites [e.g., $Ce_{10}(SiO_4)_4(BO_4)_2O_2$, Ito 1968]. These minerals, however, have not been characterized adequately because they are invariably metamict. Hogarth et al. (1973) showed that tritomite-(Ce) and tritomite-(Y), after heating in air for 2 hours at 900°C, recrystallized to britholite-(Ce) and britholite-(Y), respectively, with or without CeO_2 as an additional phase (see also Portnov et al. 1969). Also, it remains unclear whether the compositionally similar melanocerite-(Ce) and tritomite-(Ce) are separate mineral species or not. Other minerals whose structures are closely related to those of apatites include ganomalite (Dunn et al. 1985a), nasonite (Giuseppetti et al. 1971), and samuelsonite (Moore and Araki 1977).

Despite a long history of heated debate and controversy (see McConnell 1973 and Elliott 1994 for reviews), carbonate-bearing apatites with lattice-bound CO_3^{2-} ions are now well established and recognized as the major minerals of phosphorites and the main

1529-6466/00/0048-0002$05.00

Table 1. Summary of the apatite-group minerals

Mineral name	*Formula*	*Space Group*	*Reference*
fluorapatite	$Ca_{10}(PO_4)_6F_2$	$P6_3/m$	Hughes et al. (1989)
hydroxylapatite	$Ca_{10}(PO_4)_6(OH)_2$	$P2_1/b$	Ikoma et al. (1999)
chlorapatite	$Ca_{10}(PO_4)_6Cl_2$	$P2_1/b$	Mackie et al. (1972)
fermorite	$Ca_{10}(PO_4)_3(AsO_4)_3(OH)_2$	$P2_1/m$	Hughes & Drexler (1991)
alforsite	$Ba_{10}(PO_4)_6Cl_2$	$P6_3/m$	Newberry et al. (1981)
pyromorphite	$Pb_{10}(PO_4)_6Cl_2$	$P6_3/m$	Dai & Hughes (1989)
strontium-apatite	$Sr_{10}(PO_4)_6(OH)_2$	$P6_3$	Pushcharovskii et al. (1987)
belovite-(La)	$Sr_6(Na_2La_2)PO_4)_6(OH)_2$	$P\bar{3}$	Pekov et al. (1996)
belovite-(Ce)	$Sr_6(Na_2Ce_2)PO_4)_6(OH)_2$	$P\bar{3}$	Rakovan & Hughes (2000)
deloneite-(Ce)	$NaCa_2SrCe(PO_4)_3F$	$P3$	Khomyakov et al. (1996)
svabite	$Ca_{10}(AsO_4)_3F_2$	$P6_3/m$	Welin (1968)
johnbaumite	$Ca_{10}(AsO_4)_6(OH)_2$	$P6_3/m$	Dunn et al. (1980)
clinomimetite	$Pb_{10}(AsO_4)_6Cl_2$	$P2_1/b$	Dai et al. (1991)
hedyphane	$Pb_6Ca_4(AsO_4)_6Cl_2$	$P6_3/m$	Rouse et al. (1984)
mimetite	$Pb_{10}(AsO_4)_3Cl_2$	$P6_3/m$	Dai et al. (1991)
morelandite	$Ba_{10}(AsO_4)_3Cl_2$	$P6_3/m$ or $P6_3$	Dunn & Rouse (1978)
turneaureite	$Ca_{10}(AsO_4)_3Cl_2$	$P6_3/m$	Dunn et al. (1985b)
britholite-(Ce)	$Ce_6Ca_4(SiO_4)_6(OH)_2$	$P6_3$	Oberti et al. (2001)
britholite-(Y)	$Y_6Ca_4(SiO_4)_6(OH)_2$	$P2_1$	Zhang et al. (1992)
chlorellestadite	$Ca_{10}(SiO_4)_3(SO_3)_3Cl_2$	$P6_3$ or $P6_3/m$	Rouse et al. (1982)
fluorellestadite	$Ca_{10}(SiO_4)_3(SO_3)_3F_2$	$P6_3/m$	Chesnokov et al. (1987)
hydroxylellestadite	$Ca_{10}(SiO_4)_3(SO_3)_3(OH)_2$	$P2_1/m$	Hughes & Drexler (1991)
mattheddleite	$Pb_{10}(SiO_4)_3(SO_4)_3Cl_2$	$P6_3/m$	Steele et al. (2000)
cesanite	$Na_6Ca_4(SO_4)_6(OH)_2$	$P6_3/m$	Deganello (1983)
caracolite	$Na_6Pb_4(SO_4)_6Cl_2$	$P2_1/m$	Schneider (1967)
vanadinite	$Pb_{10}(VO_4)_6Cl_2$	$P6_3/m$	Dai & Hughes (1989)
melanocerite-(Ce)	$Ce_5(Si,B)_3O_{12}(OH,F)\cdot nH_2O$	?	Anovitz & Hemingway (1996)
tritomite-(Ce)	$Ce_5(Si,B)_3(O,OH,F)_{13}$	?	Hogarth et al. (1973)
tritomite-(Y)	$Y_5(Si,B)_3(O,OH,F)_{13}$	?	Hogarth et al. (1973)

components of bones and teeth of the vertebrates (e.g., Wallaeys 1952; Bonel and Montel 1964; Elliott 1964; LeGeros 1965; McConnell 1973; Jahnke 1984; McArthur 1985; Elliott 1994; Wilson et al. 1999; Suetsugu et al. 2000; Ivanova et al. 2001). Bonel and Montel (1964), using the structural positions of the CO_3^{2-} ions inferred from infrared (IR) absorption spectra, classified carbonate-bearing apatites into two types: A-type with CO_3^{2-} ions in the *c*-axis anion channels and B-type with CO_3^{2-} ions substituting for tetrahedral PO_4^{3-} groups. "Francolite" and "dahlite" have been used widely in the literature to describe carbonate-bearing fluorapatite and hydroxylapatite, respectively, but are not valid mineral names because CO_3^{2-} ions are not known to be the dominant anion species substituting for tetrahedral groups (or in the *c*-axis anion channels) in natural carbonate-bearing apatites.

This chapter outlines the compositional variations of the apatite-group minerals, with emphasis on the chemical substitutions that appear to be responsible for these variations. We purposely include data from the large number of synthetic apatites, which may or may not have natural equivalents but are extremely informative in understanding the crystal chemistry of this complex group of minerals. Also, we use the uptake of rare earth elements (REEs) in fluorapatite, hydroxylapatite, and chlorapatite as examples to illustrate some of the important factors that control the compositional variation in apatites. Following the practice in much of the chemical and mineralogical literature, fluorapatite, hydroxylapatite, and chlorapatite are hereafter abbreviated as FAp, OHAp and ClAp, respectively.

CATION AND ANION SUBSTITUTIONS IN APATITES

It is convenient to discuss the chemical substitutions in apatites relative to FAp, which is the most studied mineral and compound of this group and has the ideal $P6_3/m$ structure (e.g., Sudarsanan et al. 1972; Hughes et al. 1989). The structures of many natural and synthetic apatites deviate from the $P6_3/m$ structure (e.g., Mackie et al. 1972; Hughes et al. 1990; 1992; 1993; Hughes and Drexler 1991; Huang and Sleight 1993; Takahashi et al. 1998; Ikoma et al. 1999; Fleet et al. 2000a,b; Rakovan and Hughes 2000), but these deviations are generally very small. It is reasonable, therefore, to discuss the atomic arrangements of all apatites relative to the hexagonal unit cell and the $P6_3/m$ structure to facilitate direct comparisons among them (e.g., Hughes et al. 1990; Fleet et al. 2000a,b).

Notes for Table 2 (from the next three pages).

1: fluorapatite, Corro de Mercado mine, Durango, Mexico (Young et al. 1969)
2: REE-rich fluorapatite, Pajarito, Otero County, New Mexico, USA (Roeder et al. 1987; Hughes et al. 1991b)
3: Sr-rich fluorapatite, Lovozero massif, Kola Peninsula, Russia (Rakovan & Hughes 2000)
4: carbonate-bearing fluorapatite ("francolite"), Staffel, Germany (Brophy & Nash 1968)
5: hydroxylapatite, Holly Springs, Georgia, USA (Mitchell et al. 1943; a, including 0.15 wt % insoluble)
6: carbonate-bearing hydroxylapatite ("dahllite"), Allendorf, Saxony, Germany (Brophy & Nash 1968)
7: REE-bearing hydroxylapatite in kalsilite-bearing leucitite, Grotta del Cervo, Abruzzi, Italy (Comodi et al. 1999)
8: chlorapatite, Bob's Lake, Ontario, Canada (Hounslow & Chao 1970; b, unit-cell parameters from powder XRD)
9: fermorite, Sitipar deposit, Chhinwara district, India (Smith & Prior 1911; Hughes & Drexler 1991)
10: alforsite, Big Creek, California, USA (Newberry et al. 1981)
11: strontium-apatite, Inagli massif, Aldan, Yakutia, Russia (Efimov et al. 1962)
12: belovite-(La), Mt. Kukisvumchorr, Kola Peninsula, Russia (Pekov et al. 1996)
13: belovite-(Ce), Durango, Mexico (Rakovan & Hughes 2000)
14: deloneite-(Ce), Mt. Koashva, Kola Peninsula, Russia (Khomyakov et al. 1996)
15: svabite, Jakobsberg, Sweden (Welin 1968)
16: johnbaumite, Franklin, New Jersey, USA (Dunn et al. 1980)
17: hedyphane, Långban, Sweden (Rouse et al. 1984)
18. clinomimetite, Johanngeorgenstadt, Germany (Dai et al. (1991)
19: morelandite, Jakobsberg, Sweden (Dunn & Rouse 1978)
20: turneaureite, Långban, Sweden (Dunn et al. 1985b).
21: britholite-(Ce),Vico volcanic complex, Capranica, Latium, Italy (Oberti et al. 2001; c, including 2.12 wt % UO_2)
22: britholite-(Y), Henan, China (Zhang et al. 1992)
23: fluorellestadite, Kopeysk, Chelyabinsk basin, south Ural Mountains, Russia (Chesnokov et al. 1987)
24: hydroxylellestadite, Chichibu mine, Saitama Prefecture, Japan (Harada et al. 1971; d, including 0.72 wt % H_2O^-)
25: cesanite from Cesano geothermal field, Latium, Italy (Cavaretta et al. 1981)
26: melanocerite, Burpala alkalic intrusion, north Baikal region, Russia (Portnov et al. 1969; e, total rare earth oxides; f, including 0.63 wt % TiO_2 and 0.17% UO_2; g, unit-cell parameters obtained after heating at 670°C)

Table 2. Representative compositions and unit-cell parameters of selected apatite-group minerals.

Analysis (wt %)	**1**	**2**	**3**	**4**	**5**	**6**	**7**	**8**
P_2O_5	40.78	36.55	39.15	40.33	42.05	39.39	34.22	41.20
As_2O_5	0.10	0.003						
V_2O_5	0.01							
B_2O_3								
SO_3	0.37						1.53	
SiO_2	0.34	1.71					3.11	
Al_2O_3	0.07							
Fe_2O_3	0.06							
FeO		0.169					0.04	
CaO	54.02	38.35	47.14	51.42	55.84	51.58	54.61	53.40
MgO	0.01			1.35	0.10	1.16		
MnO	0.01	0.065			0.07			
PbO								
BaO		0.200						
SrO	0.07	2.63	10.67				1. 57	
Na_2O	0.23	0.01		1.17		0.80		
K_2O	0.01	0.01		0.38				
Y_2O_3	0.097	0.306					0.86	
La_2O_3	0.493	4.48					1.58	
Ce_2O_3	0.551	8.50						
Pr_2O_3	0.094	0.93					0.48	
Nd_2O_3	0.233	3.69						
Sm_2O_3	0.035	0.50						
Eu_2O_3	0.002	<0.059						
Gd_2O_3	0.023	0.33						
Tb_2O_3	0.012	0.477						
Dy_2O_3	0.017	<0.033						
Ho_2O_3	0.003	<0.082						
Er_2O_3	0.011	<0.024						
Tm_2O_3	0.001	<0.005						
Yb_2O_3	0.006	<0.017						
Lu_2O_3	0.001	<0.015						
ThO_2	0.02							
CO_2	0.05			2.70		3.51		
H_2O	0.01			0.63	1.86	1.48	1.10	0.09
F	3.53	3.65	3.37	3.89	0.16	0.44	1.46	0.13
Cl	0.41	0.01		trace	trace	trace	0.02	6.20
-O=F,Cl	1.58		1.42	1.64	0.07	0.19	0.61	1.45
Total	99.94	101.11	98.91	100.23	100.16[a]	98.17	99.92	99.57
a (Å)	9.391(1*	9.406(3	9.416(1)	9.346	9.4166	9.419	9.4035	9.606(4[b]
b		9.405(2						
c	6.878(2	6.913(2	6.924(1)	6.887	6.8745	6.886	6.8990	6.785(3
α (°)		90.02(2						
β		89.98(3						
γ		120.00(2						

* Small number [e.g., (3] following unit cell parameters is 1σ error in the last decimal place.

Table 2 (continued)

9	10	11	12	13	14	15	16	17
20.11	22.7	30.44	28.30	27.5	30.71	0.38	1.7	0.4
25.23						51.05	52.2	28.2
		0	0.03			0.69		
	0.1	0.90	0.24	0.9	0.74			
		0.40				0.08		
		0.15						
							0.2	10.3
44.34		10.80	0.50	0.61	14.77	42.07	43.5	
		1.64				0.52	0.1	
	<0.1	0				0.26		
	0.8					3.02		58.0
	67.7	2.70	2.35	2.15	0.10			0.1
9.93	2.7	46.06	40.09	38.1	18.19			1.0
		0.64	4.09	4.19	4.45	0.56		
		0.10			0.07	0.30		
			0.01		0.02			
		0.98	13.08	6.67	8.12			
		2.02	8.15	12.47	13.14			
		0.19	0.30	3.74	1.13			
		0.50	0.30		3.81			
		0.02	0.03		0.34			
		0.004						
		0.011	0.01					
		0.002						
		0.002						
		0.005						
			0.43		0.02			
trace		0.61	0.22	0.01	0.38	0.25	1.3	
0.83	0.7	1.67	2.04	2.39	2.03	1.99	0.2	
0.08	3.6			0.03		0.12	0.1	3.3
0.35	1.2	0.70	0.86	1.01	0.85	0.87	0.1	0.7
100.17	101.7	99.74	99.31	97.86	97.18	100.42	99.2	100.6
9.594(2*	10.284(2	9.66(1	9.647(1	9.659(2	9.51(1	9.75	9.70(2	10.140(3
9.597(2								
6.975(3	7.651(3	7.19(1	7.170(1	7.182(2	7.01(1	6.92	6.93(2	7.185(4
90.03(4								
89.95(3								
119.97(1								

* Small number [e.g., (3] following unit cell parameters is 1σ error in the last decimal place.

Table 2 (concluded)

18	19	20	21	22	23	24	25	26
0.33	2.05	6.1	1.11	0.327	1.31	0.66		2.85
22.05	28.11	44.9						
								4.37
			0.35		20.75	21.56	52.6	
0.14			21.10	25.254	15.3	17.30		14.73
				0.317	1.84	trace		0.9
					1.38	0.21		1.5
	0.41			0.121	0.18			
0.00	8.85	43.8	16.80	13.707	55.0	54.51	18.9	12.6
	0.39	0.0		0		trace		0.5
		1.9	0.18	0.041		0.04		0.1
74.61	24.85	0.7						
	33.00							
			0.00			0.28	0.72	0.4
			0.00	0	0.33	0.34	23.3	
				0.086	0.10	0.07	0.21	
			1.71	44.43				
			11.23					38.89
			21.70	5.957				
			2.19					
			5.92					
			0.72					
			0.06					
			0.50					
			0.31	4.806				
			0.14	2.663				
			0.12					
			11.92	0.399				12.39
					0.66	1.65		
	trace	nd	0.13		0.30	2.76[d]	2.91	2.55
	0.00	1.2	2.12		3.60	0.28	0.25	9.10
2.58	3.69	3.2				0.91	0.44	
0.58	0.83	1.2	0.89			0.32	0.21	3.86
99.28	100.52	100.6	99.54[c]	98.11	100.76	100.25	99.12	100.6
10.189(3*	10.869(2	9.810(4	9.547(4	9.504(5	9.485(2	9.491(1	9.442(4	9.517
20.372(8				9.414(4				
7.46(1	7.315(2	6.868(4	6.991(1	6.922(2	6.916(2	6.921(1	6.903(3	6.989
119.88(3				119.71(4				

* Small number [e.g., (3] following unit cell parameters is 1σ error in the last decimal place.

Substitution for fluorine (X anions)

The X anions in the *c*-axis channels of natural apatites are dominated by F^-, OH^-, and Cl^- (Tables 1 and 2). Additional substituents in the *c*-axis anion channels include other monovalent anions (Br^-, I^-, O_2^-, O_3^-, BO_2^-, NCO^-, NO_3^-, and NO_2^-), divalent anions (O^{2-}, CO_3^{2-}, O_2^{2-}, S^{2-}, NCN^{2-}, and NO_2^{2-}), vacancy (□) and vacancy clusters, and neutral and organic molecules (McConnell 1973; Trombe and Montel 1978; Elliott 1994). Major substitutions responsible for the incorporation of these anions and vacancies into the *c*-axis channels are as follows:

$$X^- = F^- \quad (1)$$

$$\square + X^{2-} = 2\,F^- \quad (2)$$

$$\square + M^+ = F^- + Ca^{2+} \quad (3)$$

$$2\,\square + \square = 2\,F^- + Ca^{2+} \quad (4)$$

$$\square + ZO_4^{4-} = F^- + PO_4^{3-} \quad (5)$$

$$X^{2-} + M^{3+} = F^- + Ca^{2+} \quad (6)$$

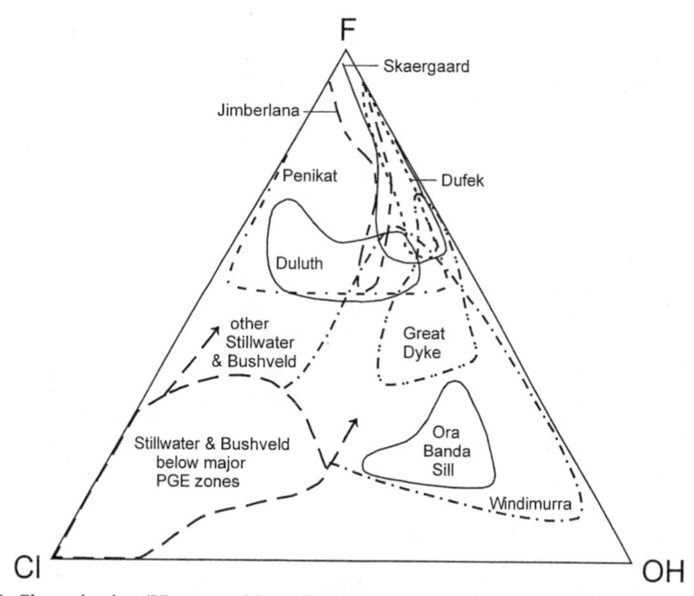

Figure 1. Channel anion (X) composition of apatites from a variety of layered intrusions showing extent of solid solution among FAp-OHAp-ClAp end-members. The molar proportion of F, OH and Cl in apatite of near ideal $Ca_{10}(PO_4)_6X_2$ composition appears to be limited only by the composition of these components in coexisting fluid/melt (after Boudreau 1995).

Monovalent anions. Extensive substitutions among F, OH, and Cl have been well documented in natural apatite-group minerals. For example, the compositions of apatites from layered intrusions span almost the complete range of F, OH, and Cl end-members (Fig. 1; Boudreau 1995 and references therein). Also, complete binary (F-OH, F-Cl, and Cl-OH) solid solutions of Ca, Sr, and Pb apatites have been synthesized (e.g., Wallaeys 1952; Prener 1967; Ruszala and Kostiner 1975). These monovalent anions are known to reside mostly in the *c*-axis anion channels but may differ greatly in their precise location owing to significant differences in ionic radius (Mackie et al. 1972; Elliott et al. 1973; Sudarsanan and Young 1980; Hughes et al. 1989, 1990; Fleet et al. 2000a,b; Rakovan and Hughes 2000). The crystal-chemical complexity related to substitutions involving these monovalent anions in calcium phosphate apatites has been discussed in Hughes and Rakovan (this volume).

Table 3. Formulas of selected synthetic compounds with the apatite structure.

Composition	*Reference*
$Ba_{10}(PO_4)_6F_2$	Mathew et al. (1979)
$Ba_{10}(PO_4)_6(OH)_2$	Engel (1973); Fowler (1974)
$Ba_{10}(PO_4)_6Cl_2$	Hata et al. (1979)
$Ba_{10}(PO_4)_6CO_3$	Mohseni-Koutchesfehani (1961)
$Ca_{10}(PO_4)_6Br_2$	Elliott et al. (1981)
$Ca_{10}(PO_4)_6CO_3$	Elliott et al. (1980)
$Ca_{10}(PO_4)_6O\square$	Wallaeys (1952)
$Ca_{10}(PO_4)_6O_2$	Trombe & Montel (1978)
$Ca_{10}(PO_4)_6(NCN)\square$	Trombe & Montel (1981)
$Ca_6Eu_2Na_2(PO_4)_6F_2$	Mayer & Cohen (1983)
$Ca_8Eu_2(PO_4)_6O_2$	Piriou et al. (1987)
$Mn_{10}(PO_4)_6Cl_2$	Klement & Haselbeck (1965)
$Sr_{10}(PO_4)_6F_2$	Kreidler & Hummel (1970)
$Sr_{10}(PO_4)_6Cl_2$	Kingsley et al. (1965)
$Sr_{10}(PO_4)_6CO_3$	Nadal et al. (1971)
$Sr_{10}(PO_4)_6O\square$	Hata et al. (1978)
$Sr_{10}(PO_4)_6O_2$	Trombe & Montel (1978)
$Sr_{9.4}Na_{0.2}\square_{0.4}(PO_4)_6BO_2$	Calvo et al. (1975)
$Cd_{10}(PO_4)_6(OH)_2$	Hata et al. (1978)
$Cd_{10}(PO_4)_6Cl_2$	Wilson et al. (1977)
$Cd_{10}(PO_4)_6Br_2$	Wilson et al. (1977)
$Cd_{10}(PO_4)_6I_2$	Sudarsanan et al. (1977)
$Pb_{10}(PO_4)_6(OH)_2$	Engel (1970)
$Pb_{10}(PO_4)_6O\square$	Wondratschek (1963)
$Pb_{10}(PO_4)_6S\square$	Trombe & Montel (1975)
$Pb_8K_2(PO_4)_6\square_2$	Mathew et al. (1980)
$\square Pb_9(PO_4)_6\square_2$	Hata et al. (1980)
$Bi_2Ca_8(PO_4)_6O_2$	Buvaneswari & Varadaraju (2000)
$La_2Ca_8(PO_4)_6O_2$	Buvaneswari & Varadaraju (2000)
$La_2Sr_8(PO_4)_6O_2$	Lacout & Mikou (1989)
$Ca_{10}(AsO_4)_6CO_3$	Roux & Bonel (1977)
$Sr_{10}(AsO_4)_6(OH)_2$	Mayer et al. (1975)
$Sr_{10}(AsO_4)_6CO_3$	Hitmi et al. (1986)
$Cd_{10}(AsO_4)_6Br_2$	Sudarsanan et al. (1977)
$Cd_{10}(AsO_4)_6I_2$	Sudarsanan et al. (1977)
$Pb_{10}(AsO_4)_6(OH)_2$	Engel (1970)
$Eu_{10}(AsO_4)_6(OH)_2$	Mayer et al. (1975)
$Ca_{10}(VO_4)_6(OH)_2$	Kutoglu (1974)
$Sr_{10}(VO_4)_6(OH)_2$	Mayer et al. (1975)
$Cd_{10}(VO_4)_6Br_2$	Sudarsanan et al. (1977)
$Cd_{10}(VO_4)_6I_2$	Sudarsanan et al. (1977)
$Pb_{10}(VO_4)_6(OH)_2$	Engel (1970)
$Bi_2Ca_8(VO_4)_6O_2$	Huang & Sleight (1993)
$Pb_{9.85}\square_{0.15}(O_4)_6I_{1.7}\square_{0.3}$	Audubert et al. (1999)

Table 3, continued

Composition	***Reference***
$Ca_{10}(CrO_4)_6(OH)_2$	Banks & Jaunarajs (1965)
$Sr_{10}(SiO_4)_3(CrO_4)_3F_2$	Schwarz (1967a)
$K_6Pb_4(CrO_4)_6F_2$	Pascher (1963)
$Na_6Pb_4(SO_4)_6F_2$	Kreidler & Hummel (1970)
$K_6Ca_4(SO_4)_6F_2$	Fayos et al. (1987)
$Na_6Cd_4(SO_4)Cl_2$	Perret & Bouillet (1975)
$Na_6Pb_4(SO_4)_6Cl_2$	Perret & Bouillet (1975)
$Pb_6K_4(PO_4)_4(SeO_4)_2\square_2$	Schwarz (1967a)
$Pb_6K_4(AsO_4)_4(SeO_4)_2\square_2$	Schwarz (1967a)
$Ca_4La_6(SiO_4)_6(OH)_2$	Ito (1968)
$Ba_4La_6(SiO_4)_6(OH)_2$	Ito (1968)
$Sr_4La_8(SiO_4)_6(OH)_2$	Ito (1968)
$Cd_4La_6(SiO_4)_6(OH)_2$	Ito (1968)
$Mg_4La_6(SiO_4)_6(OH)_2$	Ito (1968)
$Pb_4La_6(SiO_4)_6(OH)_2$	Ito (1968)
$Mn_4La_6(SiO_4)_6(OH)_2$	Ito (1968)
$Ca_2La_8(SiO_4)_6O_2$	Ito (1968)
$Ba_2La_8(SiO_4)_6O_2$	Ito (1968)
$Sr_2La_8(SiO_4)_6O_2$	Ito (1968)
$Cd_2La_8(SiO_4)_6O_2$	Ito (1968)
$Mg_2La_8(SiO_4)_6O_2$	Ito (1968)
$Pb_2La_8(SiO_4)_6O_2$	Ito (1968)
$Mn_2La_8(SiO_4)_6O_2$	Ito (1968)
$NaLa_9(SiO_4)_6O_2$	Ito (1968)
$LiY_9(SiO_4)_6O_2$	Ito (1968)
$\square CaLa_8(SiO_4)_6F_2$	Grisafe & Hummel (1970)
$\square_2La_8(SiO_4)_6\square_2$	Grisafe & Hummel (1970)
$Sm_{10}(SiO_4)_4(SiO_3N)_2O_2$	Gaudé et al. (1975)
$Cr_2Sm_8(SiO_4)_4(SiO_3N)_2O_2$	Maunaye et al. (1976)
$Y_{10}(SiO_4)_4(BO_4)_2O_2$	Ito (1968)
$La_{10}(SiO_4)_4(BO_4)_2O_2$	Mazza et al. (2000)
$Sr_{10}(SO_4)_3(GeO_4)_3F_2$	Schwarz (1967b)
$Sr_{10}(PO_4)_4(GeO_4)_2\square_2$	Schwarz (1968)
$Ca_4La_6(GeO_4)_6(OH)_2$	Cockbain & Smith (1967)
$NaLa_9(GeO_4)_6O_2$	Takahashi et al. (1998)
$Ba_{10}(ReO_5)_6Cl_2$	Besse et al. (1979)
$Ba_{10}(ReO_5)_6Br_2$	Baud et al. (1979)
$Ba_{10}(ReO_5)_6CO_3\square$	Baud et al. (1980)
$Ba_{10}(ReO_5)_6(O_2)_2$	Besse et al. (1980)
$Na_6Pb_4(BeF_4)_6F_2$	Engel (1978)

Bromine and I occur only as trace constituents in natural apatites (up to 100 ppm Br, O'Reilly and Griffin 2000; Dong and Pan 2002), although several compounds of BrAp and IAp have been synthesized (e.g., Akhavan-Niaki 1961; Sudarsanan et al. 1977; Baud et al. 1979; Elliott et al. 1981; Audubert et al. 1999). Single-crystal X-ray refinements of synthetic BrAp and IAp revealed that Br^- and I^- ions reside in the *c*-axis anion channels but, unlike F in FAp, are located at (0,0,0) (Sudarsanan et al. 1977; Wilson et al. 1977; Elliott et al. 1981; Audubert et al. 1999). These positions suggest that Br and I are incompatible for solid solution with F, OH or Cl, which partly explains the paucity of Br and I in natural apatites. Another important reason for the low Br and I contents in apatites is that these elements partition strongly into coexisting solutions/melts (e.g., Böhlke and Irwin 1992; Berndt and Seyfried 1997; Dong and Pan 2002).

Other monovalent anions, such as O_2^-, O_3^-, BO_2^-, NCO^-, NO_3^-, and NO_2^-, have been shown to occur in various synthetic apatites (Calvo et al. 1975; Dugas and Rey 1977; Tochon-Danguy et al. 1978; Trombe and Bonel 1978; Dowker and Elliott 1983; Ito et al. 1988). For example, an electron paramagnetic resonance (EPR) study by Dugas and Rey (1977) detected the presence of superoxide O_2^- ions in synthetic oxygen-rich apatites and suggested the location of the O_2^- ions away from (0,0,1/4) on the basis of an anisotropic Zeeman splitting factor (*g*). Similarly, Besse et al. (1980) reported that the O_2^- ions in the compound $Ba_{10}(ReO_5)_6(O_2)_2$ are located at (0,0,0.673). Tochon-Danguy et al. (1978) assigned an asymmetrical EPR signal from an OHAp sample excited in an atmosphere of oxygen gas at 80°C and 130 Pa to the presence of O_3^- ions in the *c*-axis anion channels. Calvo et al. (1975) showed that BO_2^- ions in the apatite anion channels have a linear configuration with the B atom at (0,0,1/2) and O atoms at (0,0,0.3278) and (0,0,0.6722) (see also Ito et al. 1988). Similarly, a polarized IR study of heated enamel by Dowker and Elliott (1983) showed that the NCO^- ion, formed from reaction between NH_4^+ and CO_2 during heating, is highly oriented in the *c*-axis direction, indicative of a location in the *c*-axis anion channels. Elliott (1994) suggested that the NCO^- ion may be located at a position similar to that of the BO_2^- ion, although the C atom must be slightly displaced from (0,0,1/2) along the *c*-axis because of the absence of a center of symmetry in the NCO^- ion. The nitrate (NO_3^-) and nitrite (NO_2^-) ions also have been detected by IR spectroscopy of A-type carbonate-bearing apatite samples heated in nitrogen monoxide and have been suggested to be located in the *c*-axis anion channels (Dugas et al. 1978).

Vacancy and vacancy clusters. Sudarsanan et al. (1977) and Wilson et al. (1977) showed that vacancies are common in the *c*-axis anion channels, and apatites with completely vacant anion channels via Substitutions (3), (4), and (5) have been synthesized [for example,

$Pb_8K_2(PO_4)_6\square_2$, Mathew et al. 1980;

$\square Pb_9(PO_4)_6\square_2$, Hata et al. 1980;

$Pb_{10}((PO_4)_4(SiO_4)_2\square_2$, Merker et al. 1970).

The anion vacancies in CaF_2-, $Ca(OH)_2$-, and $CaCl_2$-deficient apatites are most likely compensated by loss of Ca atoms [i.e., Substitution (4); see also Audubert et al. 1999; Christy et al. 2001]. For example, vacancy clusters of the type $\square_{OH}\square_{Ca2}HPO_4$ have been proposed to occur in synthetic $Ca(OH)_2$-deficient OHAp (Labarthe et al. 1973). Cho and Yesinowski (1993; 1996) detected a lack of coherence in the ···OH OH OH··· chains in OHAp by a multiple-quantum NMR dynamics study on the quasi-one-dimensional distribution of protons and interpreted it to represent OH^- deficiency. Another type of vacancy in the anion channels is related to the incorporation of divalent anions (e.g., O^{2-}, S^{2-}, and CO_3^{2-}) in the channels via Substitution (2). However, $Ca_{10}(PO_4)_6O\square$, an end-member of this substitution, is not stable (Ito 1968) and hydrates readily to oxyhydroxylapatite in air (Trombe and Montel 1978). EPR studies of FAp (Warren 1972) provided evidence for different arrangements of vacancies in the anion channels, namely: (1) $(O\square)^0$, one vacancy and one O atom replacing two F^- ions; and (2) $(\square O \square)^+$, two vacancies and one O atom substituting for three F^- ions. Wondratschek (1963) reported the loss of the 6_3 axis in $Pb_{10}(PO_4)_6O\square$ and attributed it to an ordered arrangement of vacancy and O^{2-} along the *c*-axis.

Divalent anions. Partial or complete replacement of the monovalent ions (F^-, OH^-, or Cl^-) in the *c*-axis channels by various divalent anions (e.g., O^{2-}, CO_3^{2-}, O_2^{2-}, S^{2-}, NCN^{2-}, and NO_2^{2-}) has also been well established. For example, the substitution of O^{2-} for F^- is indicated by the presence of natural oxygen-rich FAp (e.g., Young and Munson 1966; Sudarsanan and Young 1980) and has also been demonstrated by the synthesis of oxyapatites (Ito 1968; Felsche 1972; Schroeder and Mathew 1978; Azimov et al. 1981; Piriou et al. 1987; Lacout and Mikou 1989; Takahashi et al. 1998; Buvaneswari and Varadaraju 2000). Two examples of Substitution (2) are the incorporation of O^{2-} and CO_3^{2-} ions into oxyapatites and A-type carbonate-bearing apatites (see below), respectively. Examples for Substitution (6) will be given in the section on the incorporation of trivalent REEs into apatites.

In synthetic A-type carbonate apatites [e.g., $Ca_{10}(PO_4)_6CO_3\square$, $Ba_{10}(PO_4)_6CO_3\square$ and $Sr_{10}(AsO_4)_6CO_3\square$], the incorporation of CO_3^{2-} ions into the *c*-axis anion channels has been shown to be of the type:

$$CO_3^{2-} + \square = 2\,F^- \qquad (2a)$$

(Mohseni-Koutchesfehani 1961; Bonel 1972; Baran et al. 1983; Hitmi et al. 1986). Polarized IR studies (Elliott 1964) suggested that the planar CO_3^{2-} ions in the anion channels are oriented approximately parallel to the *c*-axis to minimize the steric strain related to the incorporation of this large ion (Gruner and McConnell 1937). This configuration has recently been confirmed by single-crystal X-ray structure refinement of a flux-grown, A-type carbonate apatite (Suetsugu et al. 2000).

Trombe and Montel (1978) reported the presence of the peroxide O_2^{2-} ions in OHAp and A-type carbonate-bearing apatites that have been heated in dry oxygen. They proposed a reaction between OHAp and O_2 at 900°C:

$$Ca_{10}(PO_4)_6(OH)_{2-2x}(O^{2-})_x\square_x + 1/2\, yO_2 \rightarrow Ca_{10}(PO_4)_6(OH)_{2-2x}(O^{2-})_{x-y}(O_2^{2-})_y\square_x$$

where some of the O^{2-} ions in the OHAp anion channels are oxidized to the O_2^{2-} ions without any disruption of other structural constituents. Similarly, Trombe and Montel (1981) reported the presence of the cyanamide ions, NCN^{2-}, in the *c*-axis anion channels, as indicated by the formation of $Ca_{10}(PO_4)_6(NCN)\square$ from heating of A-type carbonate apatite in an atmosphere of NH_3 at 600 to 900°C. Dowker and Elliott (1983) suggested that the NCN^{2-} ions may be located at a position similar to that of NCO^- ions. Dugas et al. (1978) suggested that the nonlinear NO_2^{2-} ions are also present in the anion channels and oriented with the O-O direction parallel to the *c*-axis.

Neutral molecules and organic molecules. Neutral molecules, including H_2O, O_2, and CO_2, have been proposed to occur in the apatite *c*-axis anion channels (e.g., Joris and Amberg 1971; Rey et al. 1978a; Tochon-Danguy et al. 1978; Ivanova et al. 2001). Similarly, Tochon-Danguy et al. (1978) suggested that Ar is most likely trapped in the *c*-axis anion channels in enamel and bone powder samples, after they were subjected to low-temperature ashing in the presence of excited Ar gas molecules. Organic molecules such as glycine (Rey et al. 1978b), acetate (Bacquet et al. 1981a), and amino-2-ethylphosphate (Bonel et al. 1988) also have been reported to occur in the *c*-axis anion channels of OHAp precipitated from aqueous solutions containing the respective organic molecules.

Substitutions for calcium (M cations)

A large number of divalent cations (Sr^{2+}, Pb^{2+}, Ba^{2+}, Mn^{2+}, etc.) have been reported to substitute for Ca in the apatite-group minerals. Similarly, many monovalent (e.g., Na^+), trivalent (REE^{3+}), tetravalent (Th^{4+} and U^{4+}), and hexavalent cations (U^{6+}; Rakovan et al. 2002) commonly occur in significant quantities in apatites and may substitute for Ca. In addition, Ca-deficiency (i.e., □) has been reported to occur in both natural and synthetic apatites, especially in biogenic apatites (e.g., Elliott 1994; Wilson et al. 1999; Suetsugu et al. 2000; Ivanova et al. 2001). There are two Ca sites in the apatite structure: nine-coordinated Ca1 and seven-coordinated Ca2 (Hughes et al. 1989). Therefore, possible cation ordering between these Ca positions is not only of intrinsic interest in structural studies of the apatite-group minerals but has important implications for substitution mechanisms.

Proposed substitutions at the Ca sites include:

$$M^{2+} = Ca^{2+} \quad (7)$$

$$\square + 2\,\square = Ca^{2+} + 2\,F^- \quad (4)$$

$$\square + 2\,ZO_4^{2-} = Ca^{2+} + 2\,PO_4^{3-} \quad (8)$$

$$M^+ + \square = Ca^{2+} + F^- \quad (3)$$

$$2\,M^+ = Ca^{2+} + \square \quad (9)$$

$$M^+ + M^{3+} = 2\,Ca^{2+} \quad (10)$$

$$M^+ + ZO_4^{2-} = Ca^{2+} + PO_4^{3-} \quad (11)$$

$$M^{3+} + X^{2-} = Ca^{2+} + F^- \quad (6)$$

$$M^{3+} + ZO_4^{4-} = Ca^{2+} + PO_4^{3-} \quad (12)$$

$$2\,M^{3+} + \square = 3Ca^{2+} \quad (13)$$

$$2\,M^{3+} + ZO_4^{5-} = 2\,Ca^{2+} + PO_4^{3-} \quad (14)$$

$$M^{4+} + \square = 2\,Ca^{2+} \quad (15)$$

Divalent cations. Strontium is one of the most common M cation substituents in apatites and forms extensive solid solution series with Ca in natural apatites (Efimov et al. 1962; Pushcharovskii et al. 1987; Hughes et al. 1991a; Rakovan and Hughes 2000; Chakhmouradian et al. 2002). In addition, there are several Sr minerals of the apatite group: belovite-(La), belovite-(Ce), deloneite-(Ce), and strontium-apatite (Table 1). Also, complete solid solution series between Ca and Sr end-members have been established experimentally for FAp, OHAp, and ClAp (Khudolozhkin et al. 1972; 1973a; Heijligers et al. 1979; Sudarsanan and Young 1980; Khattech and Jemal 1997). Despite early contradictory results on the Ca site occupancy of Sr (e.g., Khudolozhkin et al. 1972; 1973a; Heijligers et al. 1979), it is now well established that Sr almost exclusively occupies the Ca2 site (or equivalent sites) in apatite structures (Sudarsanan and Young 1974, 1980; Hughes et al. 1991a; Bigi et al. 1998; Rakovan and Hughes 2000). Sudarsanan and Young (1980) also found that the site preference of Sr for Ca2 decreases with increase in the content of Sr and that the Cl position shifts from (0,0,0.44) in pure ClAp (Mackie et al. 1972; Hughes et al. 1989) to (0,0,1/2) in SrClAp at or above 48% of the replacement of Ca by Sr. This new location of the Cl^- ions leads to the formation of the SrO_6Cl_2 polyhedron in $Sr_{10}(PO_4)_6Cl_2$ (Sudarsanan and Young 1980). Similarly, Rakovan and Hughes (2000) reported the presence of the SrO_6Cl_2 polyhedron in a Cl-bearing belovite-(Ce), on the basis of the location of Cl^- ions at (0,0,1/2).

Complete solid solutions between Pb and Ca end-members have been synthesized for OHAp and ClAp (Akhavan-Niaki 1961; Engel et al. 1975; Miyake et al. 1986), although Verbeeck et al. (1981) noted a miscibility gap in OHAp at 800°C. Single-crystal and powder XRD studies on Pb apatites synthesized at elevated temperatures revealed that the Pb^{2+} ions have a strong preference for Ca2 sites (Engel et al. 1975; Hata et al. 1980; Mathew et al. 1980; Verbeeck et al. 1981), as that in caracolite (Schneider 1967). Engel et al. (1975) attributed this strong preference to the ability of Pb^{2+} cations to form partial covalent bonds (e.g., Pb2-O2 = 2.238 Å in $Pb_8K_2(PO_4)_6\square_2$; Mathew et al. 1980). One notable exception to this general trend is the study of Miyake et al. (1986) who, on the basis of Rietveld XRD refinements of Pb^{2+} ion-exchanged OHAp, ClAp, and FAp, suggested that Pb^{2+} ions have no preference for Ca1 or Ca2 sites.

A complete solid solution between Ba and Ca end-members has been confirmed for ClAp (Table 4), whereas large miscibility gaps are known to exist in FAp and OHAp (e.g., between 6 and 61 mol % $Ba_{10}(PO_4)_6F_2$ in FAp at 900-1100°C; Akhavan-Niaki 1961; Bigi et al. 1984). Similarly, natural Cl-poor apatites contain only small amounts of Ba (up to 12.54 wt % BaO; Edgar 1989). Khudolozhkin et al. (1973a) showed that Ba^{2+} ions have preference for Ca2 sites and that this preference increases with increase in the content of Ba. Also, this site preference is more marked in the Ca-Ba apatites than in the Sr-Ba apatites (Khudolozhkin et al. 1973a).

Table 4. Solubility limits of some divalent cations in apatites $Ca_{10-n}M_n(PO_4)_6X_2$.

M	*X*	*n*	*References*
Sr	F	10	Akhavan-Niaki (1961), Khattech & Jamel (1997)
	OH	10	Heijligers et al. (1979)
	Cl	10	Akhavan-Niaki (1961)
Ba	F	0.6	Akhavan-Niaki (1961)
	Cl	10	Akhavan-Niaki (1961)
Pb	OH	10?	Engel et al. (1975), Verbeeck et al. (1981)
	Cl	10	Kreidler & Hummel (1970)
Cd	F	10	Akhavan-Niaki (1961) Kreidler & Hummel (1970)
	OH	10	Bigi et al. (1986)
	Cl	10	Klement & Haselbeck (1965)
Mg	F	0.9	Kreidler & Hummel (1970)
	OH	10?	Patel (1980); Chiranjeevirao et al. (1982)
	Cl	3.0	Klement & Haselbeck (1965)
Fe	F	1.5	Khudolozhkin et al. (1974)
Mn	F	1.37	Ercit et al. (1994)
	Cl	10	Klement & Haselbeck (1965)
Co	F	1.5	Grisafe & Hummel (1970)
	Cl	2.5	Grisafe & Hummel (1970)
Ni	F	0.75	Kreidler & Hummel (1970)
	F	1.0	Brasseur & Dallemagne (1949)
	Cl	3.0	Klement & Haselbeck (1965)
Cu	Cl	4.0	Klement & Haselbeck (1965)
Zn	F	1.0	Brasseur & Dallemagne (1949)
Sn	F	5.0	Klement & Haselbeck (1965)
	Cl	8.0	Klement & Haselbeck (1965)

Note that the reported complete solid solutions between Pb and Ca in OHAp and between Mg and Ca in OHAp are questionable (see text for discussion).

The substitution of Mn in apatites has been extensively investigated because Mn-doped FAp is used in the fluorescent-light industry, and Mn is ubiquitous in natural apatites. The maximum content of 1.37 Mn atoms per formula unit (apfu) in natural FAp (Ercit et al. 1994; see also Hughes et al. 1991a) exceeds the early proposed limit of 1 Mn apfu (Suitch et al. 1985). A more extensive solid solution between Ca and Mn may exist in ClAp, as indicated by the synthesis of pure manganese ClAp (Table 4; Klement and Haselbeck 1965). There is a general consensus that Mn^{2+} ions have a strong preference for the Ca1 sites (Ryan et al. 1972; Warren and Mazelsky 1974; Suitch et al. 1985; Hughes et al. 1991a; Pan et al. 2002a). Manganese in the most Mn-rich FAp, however, shows only a slight preference for Ca1 (64%) (Ercit et al. 1994). This disordering in the Mn-rich FAp is consistent with the EPR results of Warren (1970), who showed that the site preference of Mn for Ca1 decreases with increase in the Mn content (see also Warren and Mazelsky 1974). Ryan et al. (1972) suggested that the site preference of Mn for Ca1 is greater in CaFAp than that in SrFAp. Ohkubo (1968) reported that the EPR signal of Mn^{2+} at Ca1 (i.e., center Mn_I) decreases with increase in the Cl content in binary Cl-FAp. Warren and Mazelsky (1974) also showed that the signal from Mn^{2+} at the Ca2 site (Mn_{II}) in CaF_2-deficient FAp is very weak and is replaced by a different signal from Mn in a modified Ca2 site (i.e., Mn_{IIm}). Warren and Mazelsky (1974) noted that the signal intensity from Mn_{II} increases with increase in the Mn/Ca value, whereas the Mn_{IIm} signal intensity decreases. Warren and Mazelsky (1974) associated the Mn_{IIm} center with the $(\square O \square)^+$ defect (Warren 1972; see above). Suitch et al. (1985) suggested that incorporation of Mn into FAp results in a reduction of symmetry to $P6_3$ or $P3$. This suggestion, however, was not supported by studies of natural Mn-rich FAp (Hughes et al. 1991a; Ercit et al. 1994).

Iron occurs only as a minor to trace element and rarely exceeds 1 wt % as FeO in natural apatites (up to 2.2 wt % FeO; Fransolet and Schreyer 1981). Khudolozhkin et al. (1974) reported that the solubility limit of Fe in FAp is ~15 mol % replacement of Ca^{2+} by Fe^{2+} (Table 4). Their Mössbauer spectroscopic study suggested that Fe^{2+} is randomly distributed between the Ca1 and Ca2 sites in Fe-poor FAp (<1 mol %), but has a strong preference for Ca1 at high concentrations towards the solubility limit of Fe^{2+} in FAp. These results, however, are opposite to that of Hughes et al. (1993) who, on the basis of a single-crystal X-ray structural refinement of a natural, Fe-bearing monoclinic FAp, showed that Fe^{2+} preferentially occupies Ca2-equivalent sites.

Although Mg appears to have a limited solubility in FAp (Table 4), a complete replacement of Ca by Mg has been reported for OHAp (Patel 1980; Chiranjeevirao et al. 1982). However, an attempt by Terpstra and Driessens (1986) to confirm the results of Chiranjeevirao et al. (1982) was unsuccessful. Neuman and Mulryan (1971) showed that nearly 90% of the Mg in OHAp precipitated from Mg^{2+}-bearing solutions is readily exchangeable and, hence, is most likely located at surface positions. In synthetic $Mg_2REE_8(SiO_4)_6O_2$, Mg^{2+} appears to preferentially occupy Ca1 sites and may alternate with the REE^{3+} ions (Ito 1968).

Other divalent cations, which substitute for Ca^{2+} in apatites, include Ni^{2+}, Co^{2+}, Cu^{2+}, Zn^{2+}, Sn^{2+}, Cd^{2+} (Table 4), and Eu^{2+} (Table 3). The solubility limits for Co^{2+} in FAp and ClAp are 15 and 25 mol %, respectively, but are less than 10 mol % in the Sr analogs (Grisafe and Hummel 1970). Single-crystal X-ray refinements of Co-bearing ClAp showed that the Co^{2+} ions are located exclusively at the Ca2 sites, with the Cl^- ions shifted along the *c*-axis toward the Co^{2+} ion to maintain a reasonable Co-Cl bond distance (Anderson and Kostiner 1987). A complete replacement of Ca^{2+} by Cd^{2+} has been demonstrated by the synthesis of various Cd apatites (Sudarsanan et al. 1977; Wilson et al. 1977; Hata et al. 1978; Bigi et al. 1986; Christy et al. 2001). A cadmium K-edge EXAFS study by Sery et al. (1996) suggested that Cd occupies both Ca sites with a slight preference for Ca2. The presence of Eu^{2+} in apatites has been well established for synthetic materials (Mayer et al. 1975) and has been confirmed by a synchrotron wavelength dispersive XANES study on natural FAp (Rakovan et al. 2001). Rakovan et al. (2001) also suggested that the presence of both Eu^{2+} and Eu^{3+} is most likely responsible for the abnormal partitioning behavior of Eu, relative to other REEs, between the <001> and <011> sectors in the Llallagua FAp crystals investigated by Rakovan and Reeder (1994, 1996).

Monovalent cations. Sodium is a common minor constituent in natural calcium phosphate apatites (e.g., Roeder et al. 1987; Rønsbo 1989; Comodi et al. 1999) and becomes a major component in belovite-(La), belovite-(Ce), deloneite-(Ce), cesanite, caracolite, and many synthetic apatites (Table 3). Potassium, Li, and Rb are only trace constituents in natural apatites, but attain significant concentrations in some synthetic apatites (e.g., Schwarz 1967a; Simpson 1968; Mathew et al. 1980). These monovalent cations in apatites have been shown to have strong preference for the Ca1 sites (Calvo et al. 1975; Mathew et al. 1979; Fleet and Pan 1997a; Takahashi et al. 1998; Rakovan and Hughes 2000) and commonly involve REE^{3+} ions (see below) or other coupled substitutions (e.g., CO_3^{2-} or SO_4^{2-} for PO_4^{3-}; see below) to preserve electrostatic neutrality.

Other monovalent ions that have been proposed to substitute for Ca^{2+} in apatites include NH_4^+ and H_3O^+ (McConnell 1952; Simpson 1968; Doi et al. 1982; Vignoles et al. 1987). For example, McConnell (1952) and Simpson (1968) postulated the presence of H_3O^+ as replacement for Ca^{2+} in Ca-deficient OHAp; however, this was questioned by Elliott (1969), because this ion is generally restricted to the structure of strong acids. The NH_4^+ ion has been shown to occur (at up to 0.12 wt % N) in carbonate apatites precipitated from NH_4^+-bearing solutions (Doi et al. 1982; Vignoles et al. 1987). Doi et al. (1982) suggested that NH_4^+ may substitute for Ca^{2+} via:

$$2\,NH_4^+ = Ca^{2+} + \square \qquad (9a)$$

Ivanova et al. (2001) interpreted a slight decrease of atomic scattering at the Ca2 site in a synthetic, NH_4^+-bearing apatite to a preferential incorporation of NH_4^+ ions into this site.

Trivalent cations. Apatite-group minerals have long been known as important hosts for REEs and Y in igneous, metamorphic, and sedimentary rocks (e.g., Watson et al. 1985; Roeder et al. 1987; Rønsbo 1989; Hughes et al. 1991b; Fleet and Pan 1995a; Gaft et al. 1997' Pan and Breaks 1997) and in the biomass as well (e.g., Wright et al. 1984; Grandjean-Lécuyer et al. 1993; Holmden et al. 1996, 1998). The ability of apatites to accommodate significant amounts of REEs and Y is also demonstrated by the formation of many natural and synthetic REE apatites (Tables 1, 2, and 3; Cockbain and Smith 1967; Ito 1968; Felsche 1972; Steinbruegge et al. 1972; Mayer et al. 1974; Azimov et al. 1981; Mayer and Cohen 1983). Although early studies on REE site preference yielded contradictory results (e.g., Ca1, Urusov and Khudolozhkin, 1974; Ca2, Borisov and Klevtsova 1963), recent X-ray structure refinements of natural and synthetic apatites have all shown that REEs in FAp, OHAp, and OAp generally prefer the Ca2 site and that the site-occupancy ratios (REE-Ca2/REE-Ca1) decrease monotonically with increase in atomic number through the 4f transition series (Hughes et al. 1991b, 1992; Fleet and Pan 1994, 1995a, 1997a; Takahashi et al. 1998; Fleet et al. 2000a; Serret et al 2000). However, Fleet et al. (2000b) showed that REEs in ClAp preferentially occupy the Ca1 equivalent sites, with the exception of Nd, which has a marginal preference for the Ca2 equivalent sites [(Nd-Ca2/Nd-Ca1) = 1.11].

Four main types of charge-compensating mechanisms have been proposed for the substitution of Ca^{2+} by REE^{3+} (and Y^{3+}) in apatites (Ito 1968; Felsche 1972; Roeder et al. 1987; Rønsbo 1989; Fleet and Pan 1995a; Comodi et al. 1999; Cherniak 2000; Serret et al. 2000; Chen et al. 2002a,b):

$$REE^{3+} + X^{2-} = Ca^{2+} + F^- \qquad (6a)$$

$$REE^{3+} + M^+ = 2\,Ca^{2+} \qquad (10a)$$

$$REE^{3+} + SiO_4^{4-} = Ca^{2+} + PO_4^{3-} \qquad (12a)$$

$$2\,REE^{3+} + \square = 3\,Ca^{2+} \qquad (13a)$$

Substitution (6a), involving concomitant replacement of O^{2-} or S^{2-} ions for F^- ions in the *c*-axis anion channels, has been demonstrated by the synthesis of $Ca_8REE_2(PO_4)_6O_2$ (e.g., Ito 1968; Schroeder and Mathew 1978; Piriou et al. 1987; Serret et al. 2000) and $Ca_{10-x}Eu_x(PO_4)_6S_{1+x/2}\square_{1-x/2}$ (Suitch et al. 1986; Taïtaï and Lacout 1989). Single-crystal EPR studies of synthetic FAp containing 1.2 wt % Gd_2O_3 (Chen et al. 2002a) and 97 ppm ^{157}Gd (Chen et al. 2002c) revealed the presence of a Gd^{3+} center 'a', corresponding to occupancy of Gd^{3+} ions at Ca2 sites, and that the center 'a' has a rhombic (i.e., triclinic) local symmetry, different from a uniaxial symmetry expected for the ideal Ca2 site in pure FAp. This distortion was interpreted to be related to a replacement (and minor displacement away from z = 1/4 or 3/4; cf. Fleet al. 2000a) of O^{2-} for F^- in the *c*-axis anion channel (Chen et al. 2002a).

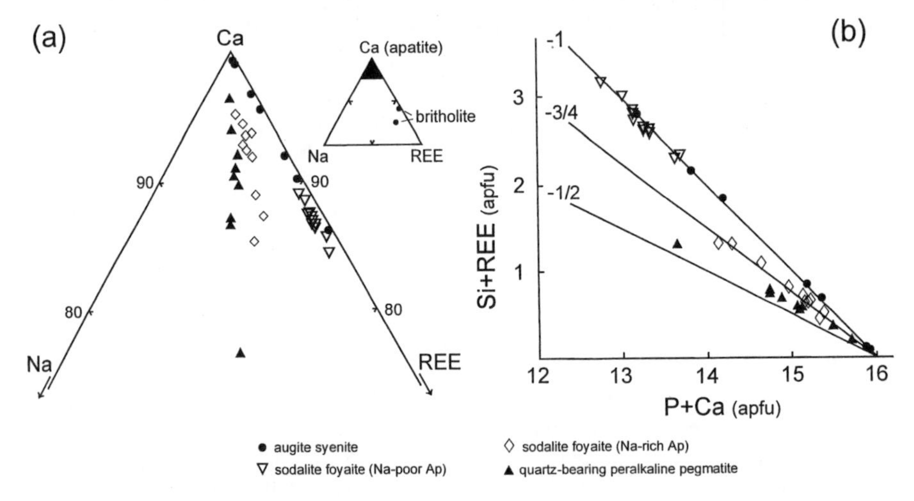

Figure 2. Substitution of REEs into apatite from the Ilímaussaq intrusion, South Greenlend: (a) molar proportion of Ca, REEs and Na; (b) (Si+REE) apfu *versus* (P+Ca) atoms per formula unit (apfu). Line of slope -1 represents the coupled substitution: $REE^{3+} + SiO_4^{4-} = Ca^{2+} + PO_4^{3-}$. Note that apatite compositions from Na-rich sodalite foyaite and quartz-bearing peralkaline pegmatite reflect progressively increasing importance of the substitution: $REE^{3+} + Na^+ = 2Ca^{2+}$ (after Rønsbo 1989).

Substitution (10a) is well established on the basis of compositional data from natural apatites (e.g., Roeder et al. 1987; Rønsbo 1989; Peng et al. 1997; Comodi et al. 1999; Fig. 2), and is largely responsible for accommodating REEs into belovites (e.g., Pekov et al. 1996; Rakovan and Hughes 2000). Although the end-member composition $Na_5REE_5(PO_4)_6F_2$ for this substitution is known to be unstable, intermediate compositions such as $Na_4REE_4Ca_2(PO_4)_6F_2$ have been synthesized (Mayer et al. 1974). Other monovalent cations that have been shown to participate in this type of coupled substitution for the incorporation of REE^{3+} into synthetic apatites include Li^+ (Ito 1968; Felsche 1972) and Ag^+ (Mayer and Swissa 1985).

Substitution (12a) is also well documented in natural REE-bearing apatites (Fig. 2; Roeder et al. 1987; Rønsbo 1989; Comodi et al. 1999), and is supported by a complete solid solution between OHAp and britholite-(Y) (Ito 1968; Khudolozhkin et al. 1973b). This substitution leads to the end-member $Ca_4REE_6(SiO_4)_6F_2$, which has been synthesized for compositions involving La, Ce, Nd, and Y (Ito 1968; Mayer et al. 1974).

Substitution (13a), involving a vacancy at Ca sites, is partly responsible for the accommodation of REEs in synthetic $\square CaREE_8(SiO_4)_6F_2$ and $\square_2REE_8(SiO_4)_6\square_2$ (Grisafe and Hummel 1970). Chen et al. (2002b), on the basis of an EPR study on flux-grown FAp crystals containing 57 ppm Gd, detected a Gd^{3+} center 'b' corresponding to occupancy of

Gd^{3+} ions at the Ca1 sites. Chen et al. (2002b) showed that the Gd^{3+} center 'b' has a highly triclinic local symmetry, different from the uniaxial symmetry of the ideal Ca1 site in pure FAp. Chen et al. (2002b) suggested that the triclinic symmetry of this Gd^{3+} center is related to the presence of a vacancy at the next-nearest-neighbor Ca2 site, resulting in a Gd^{3+} -- □--- Gd^{3+} arrangement, with the cations well separated.

Another extensively investigated trivalent cation in apatites is Sb^{3+}, because Sb-doped FAp acts as an activator in fluorescent-light tubes (e.g., Davis et al. 1971; Soules et al. 1971; Mishra et al. 1987; DeBoer et al. 1991; Moran et al. 1992). Rietveld XRD refinements of a FAp powder sample with 2.2 wt % Sb suggested that Sb^{3+} is ordered at the Ca2 site (DeBoer et al. 1991), consistent with the site occupancy deduced from excitation and emission spectra of Sb-doped ClAp and FAp (Davis et al. 1971; Soules et al. 1971). However, the same study on a different sample containing 3.1 wt % Sb did not find any evidence for substitution at the Ca2 site but suggested, on the basis of electron density maps, that the Sb^{3+} ions occupy the (1/3,2/3,1/4) and (2/3,1/3,1/4) sites (DeBoer et al. 1991). Also, the Ca2 site assignment for Sb^{3+} ions is not consistent with results from ^{121}Sb Mössbauer (Mishra et al. 1987) or ^{19}F and ^{31}P MAS NMR studies (Moran et al. 1992; see below).

Other trivalent cations that have been shown to substitute for Ca in synthetic oxyapatites include Bi^{3+} and Cr^{3+} (Table 2). The composition of $Bi_2Ca_8(PO_4)_6O_2$ (Buvaneswari and Varadaraju 2000) suggests a coupled substitution of the type:

$$Bi^{3+} + O^{2-} = Ca^{2+} + F^{-} \qquad (6b)$$

Huang and Sleight (1993) showed that Bi^{3+} ions in $Bi_2Ca_8(VO_4)_6O_2$ preferentially occupy Ca1 sites (see also Mayer and Semadja 1983), whereas Cr^{3+} ions in $Cr_2Sm_8(SiO_4)_4(SiO_3N)_2O_2$ were found to reside exclusively in Ca2 sites (Maunaye et al. 1976). Mayer and Semadja (1983) also noted that Bi tends to favor apatites with vacant anion channels.

Tetravalent cations. Elevated amounts of Th have been reported for natural apatites, particularly in the REE-rich varieties [e.g., up to 15.9 wt % ThO_2 in britholite-(Ce)] from alkaline and peralkaline igneous rocks (Hughson and Sen Gupta 1964; Arden and Halden 1999; Della-Ventura et al. 1999; Oberti et al. 2001). Uranium is also a common, minor to trace element (up to 3.4 wt % UO_2) in natural apatites (Clarke and Altschuler 1958; Arden and Halden 1999; Della-Ventura et al. 1999; Oberti et al. 2001), although the U contents in apatites from some early studies may be overestimated owing to the common occurrences of U-rich inclusions (Baumer et al. 1983). Compositional data (Hughson and Sen Gupta 1964; Baumer et al. 1983) have shown that Th^{4+} ions substitute for Ca^{2+} via:

$$Th^{4+} + \square = 2\,Ca^{2+} \qquad (15a)$$

Similarly, Clarke and Altschuler (1958) suggested that U in apatites is mainly tetravalent and that U^{4+} ions occupy the Ca sites via a similar substitution (Baumer et al. 1983):

$$U^{4+} + \square = 2Ca^{2+} \qquad (15b)$$

Substitutions for phosphate (ZO_4 group)

The PO_4^{3-} group in the apatite-group minerals is commonly replaced by a variety of other tetrahedral anion groups (e.g., AsO_4^{3-}, VO_4^{3-}, MnO_4^{3-}, CrO_4^{3-}, SO_4^{2-}, SeO_4^{2-}, CrO_4^{2-}, BeF_4^{2-}, SiO_4^{4-}, GeO_4^{4-}, SbO_3F^{4-}, SiO_3N^{5-}, and BO_4^{5-}). Another tetrahedral anion group proposed by McConnell (1973) to substitute for PO_4^{3-} is $(OH)_4^{4-}$, by analogy to that in "hydrogarnets" (Nobes et al. 2000; Armbruster et al. 2001 and references therein). But to our knowledge, no structural evidence for the $(OH)_4^{4-}$ group has been found in the apatite-group minerals or their synthetic analogs. Other polyhedral groups that have been shown to substitute for PO_4^{3-} in the apatite-group minerals include CO_3^{2-}, BO_3^{3-}, and ReO_5^{3-}. Proposed mechanisms for the replacement of the PO_4^{3-} group in the apatite-group minerals include:

$$ZO_4^{3-} = PO_4^{3-} \quad (16)$$
$$2\ ZO_4^{2-} + \square = 2\ PO_4^{3-} + Ca^{2+} \quad (8)$$
$$ZO_4^{2-} + M^{+} = PO_4^{3-} + Ca^{2+} \quad (11)$$
$$ZO_4^{2-} + ZO_4^{4-} = 2\ PO_4^{3-} \quad (17)$$
$$ZO_4^{4-} + \square = PO_4^{3-} + F^{-} \quad (5)$$
$$ZO_4^{4-} + M^{3+} = PO_4^{3-} + Ca^{2+} \quad (12)$$
$$ZO_4^{5-} + 2\ M^{3+} = PO_4^{3-} + 2\ Ca^{2+} \quad (14)$$

Trivalent anion groups. Extensive substitution of the PO_4^{3-} group by the tetrahedrally coordinated and isovalent AsO_4^{3-} ion has been well established by the existence of a complete solid solution between pyromorphite and mimetite (Kautz and Gubser 1969; Förtsch and Freiburg 1970) and by data from As-bearing FAp (Persiel et al. 2000 and references therein). One notable exception is the study of Bothe and Brown (1999), who observed no solid solution between $Ca_{10}(PO_4)_6(OH)_2$ and $Ca_{10}(AsO_4)_6(OH)_2$ at ambient temperatures. Hughes and Drexler (1991) showed that replacement of PO_4^{3-} by AsO_4^{3-} in fermorite causes little disruption in the atomic arrangement of this mineral and that there is no evidence of ordering accompanying the substitution:

$$AsO_4^{3-} = PO_4^{3-} \quad (16a)$$

Other tetrahedrally coordinated and trivalent ions that have been shown to substitute for the PO_4^{3-} group include VO_4^{3-}, MnO_4^{3-}, and CrO_4^{3-} (Banks and Jaunarajs 1965; Kingsley et al. 1965; Sudarsanan et al. 1977; Huang and Sleight 1993). Vanadinite and synthetic vanadate apatites indicate the substitution (Sudarsanan et al. 1977; Dai and Hughes 1989; Huang and Sleight 1993):

$$VO_4^{3-} = PO_4^{3-} \quad (16b)$$

and a complete solid solution series between $Ca_{10}(PO_4)_6F_2$ and $Ca_{10}(VO_4)_6F_2$ has been confirmed by Kreidler and Hummel (1970). Kingsley et al. (1965) showed the presence of MnO_4^{3-} ions as replacement for PO_4^{3-} ions in $MnCl_2$ flux-grown ClAp. Similarly, replacement of PO_4^{3-} ions by CrO_4^{3-} ions in synthetic apatites was reported by Banks and Jaunarajs (1965) and Banks et al. (1971).

Other trivalent ions that have been suggested to substitute for PO_4^{3-} include SbO_3^{3-}, BO_3^{3-}, and ReO_5^{3-} (Calvo et al. 1975; Ito et al. 1988; Mishra et al. 1987; Moran et al. 1992; Schriewer and Jeitschko 1993). For example, an ^{121}Sb Mössbauer study by Mishra et al. (1987) suggested that Sb in Sb-doped FAp occurs as the SbO_3^{3-} ion, which may be coordinated with F^- as the fourth ligand to form the SbO_3F^{4-} group. Also, the presence of SbO_3^{3-} ions in Sb-doped FAp is supported by results from ^{19}F and ^{31}P MAS NMR studies (Moran et al. 1992). Ito et al. (1988) showed, on the basis of isotopic frequency shifts (IR), the presence of BO_3^{3-} as replacement for PO_4^{3-} in ^{11}B- and ^{10}B-doped apatites. Schriewer and Jeitschko (1993) emphasized that the apex oxygen atoms of the pyramidal ReO_5^{3-} groups in $Ba_{10}(ReO_5)_6Cl_2$ all point approximately down the *c*-axis, resulting in the loss of the horizontal mirror plane (see also Besse et al. 1979).

Divalent anion groups. Extensive substitution between SO_4^{2-} and PO_4^{3-} has been demonstrated by compositional data from natural apatites (McConnell 1937; Schneider 1967; Rouse and Dunn 1982; Hughes and Drexler 1991; Liu and Comodi 1993; Peng et al. 1997; Comodi et al. 1999) and is supported by the synthesis of various sulfate apatites (e.g., Schwarz 1967a,b; Kreidler and Hummel 1970; Khorari et al. 1994). Common correlations between the Si and S contents in natural apatites (Rouse and Dunn 1982; Baumer et al. 1990; Liu and Comodi 1993; Peng et al. 1997; Comodi et al. 1999; Steele et al. 2000) indicate a charge-compensating mechanism of the type:

$$SO_4^{2-} + SiO_4^{4-} = 2\ PO_4^{3-} \quad (17a)$$

For example, Rouse and Dunn (1982) reported that SO_4^{2-} and SiO_4^{4-} in ellestadites occur in consistently equal proportions, suggesting a possible ordering of Si and S in the ellestadite structure. Sudarsanan (1980) showed that the equivalent tetrahedral sites in the $P6_3/m$ structure split into 3 non-equivalent T1, T2 and T3 sites in monoclinic hydroxylellestadite and that some degree of Si-S ordering does appear to occur. This observation has since been confirmed by Hughes and Drexler (1991), who showed that the Si^{4+} and S^{6+} ions in monoclinic hydroxylellestadite preferentially occupy the T1 and T2 sites, respectively, and that the T3 site is occupied by both ions in approximately equal proportions. Khorari et al. (1994), however, did not find any evidence for Si-S ordering in synthetic FAp. Compositional data from natural and synthetic FAp (Liu and Comodi 1993; Peng et al. 1997) also indicate a substitution of the type:

$$SO_4^{2-} + Na^{+} = PO_4^{3-} + Ca^{2+} \quad (11a)$$

This substitution, however, appears to be less important than Substitution (17a) (Peng et al. 1997).

The common occurrence of B-type carbonate-bearing apatite involving replacement of the PO_4^{3-} group by CO_3^{2-} is now well established (McConnell 1973; Elliott 1994; Ivanova et al. 2001). Polarized IR studies suggested that the orientation of the CO_3^{2-} ion lies in the position of the sloping face of the replaced PO_4^{3-} tetrahedron (Elliott 1964). A Ca *K*-edge EXAFS study of carbonate-bearing OHAp showed that the coordination of the Ca^{2+} ions by the nearest-neighbor O atoms is not notably affected by the replacement of CO_3^{2-} for PO_4^{3-}, but marked changes in the transformation occur beyond 3 Å (Harries et al. 1987). These data led Harries et al. (1987) to suggest that the O atoms of the planar CO_3^{2-} ion occupy three of the four vacant O sites left by a PO_4^{3-} ion, and that the fourth O site is directed away from the CO_3^{2-} ion. A Rietveld XRD refinement of a synthetic, B-type carbonate-bearing OHAp by Ivanova et al. (2001) showed that the CO_3^{2-} ions randomly occupy the two vertical faces of the PO_4 tetrahedron (i.e., the faces with a common edge parallel to the *c*-axis).

Six charge-compensating mechanisms have been proposed for the incorporation of CO_3^{2-} ions into the PO_4^{3-} sites:

$$CO_3OH^{3-} = PO_4^{3-} \quad (16c)$$

$$CO_3F^{3-} = PO_4^{3-} \quad (16d)$$

$$2\ CO_3^{2-} + \square = 2\ PO_4^{3-} + Ca^{2+} \quad (8a)$$

$$CO_3^{2-} + SiO_4^{4-} = 2\ PO_4^{3-} \quad (17b)$$

$$CO_3^{2-} + Na^{+} = PO_4^{3-} + Ca^{2+} \quad (11b)$$

$$CO_3^{2-} + H_3O^{+} = PO_4^{3-} + Ca^{2+} \quad (11c)$$

The two substitutions involving CO_3OH^{3-} and CO_3F^{3-} were proposed mainly on the basis of compositional data from natural and synthetic carbonate-bearing apatites (e.g., $F^- + OH^- > 2$ apfu; Borneman-Starynkevich 1938; Borneman-Starynkevich and Belov 1953; Trueman 1966; Labarthe et al. 1971; Vignoles and Bonel 1978; Sommerauer and Katz-Lehnert 1985; Binder and Troll 1989). These two substitutions are appealing in that the CO_3F^{3-} and CO_3OH^{3-} groups are tetrahedrally coordinated and are isovalent with the PO_4^{3-} group for which they substitute. However, evidence for the presence of CO_3F^{3-} and CO_3OH^{3-} groups from IR and EPR studies (e.g., Labarthe et al. 1971; Vignoles and Bonel 1978; Bacquet et al. 1980; 1981b) is ambiguous (Okazaki 1983; Elliott 1994; Regnier et al. 1994). Also, Okazaki (1983) and Jahnke (1984) found no clear positive correlation between the F^- and CO_3^{2-} contents in synthetic apatites. Indeed, no structural evidence for CO_3F^{3-} or CO_3OH^{3-} groups has been found in any of the ^{1}H, ^{19}F and ^{13}C NMR and EXAFS studies of natural and synthetic carbonate-bearing apatites (e.g., Harries et al. 1987; Rey et al. 1989; Beshah et al. 1990; Regnier et al. 1994). Also, *ab-initio* quantum mechanical calculations suggested that CO_3F^{3-} is unstable in apatites and that the excess F^- ions in carbonate-bearing FAp most likely occupy an interstitial site (Regnier et al. 1994). Another variation of these two substitutions is replacement of $CO_3H_2O^{2-}$ for PO_4^{3-}, which also suffers from the lack of structural evidence for the existence of $CO_3H_2O^{2-}$ (Beshah et al. 1990).

X-ray structural data from synthetic carbonate-bearing apatites (Suetsugu et al. 2000) showed that the charge imbalance introduced when CO_3^{2-} replaces PO_4^{3-} is compensated primarily by Ca vacancies [i.e., Substitution (8a)]. However, uncertainties remain regarding the location of the associated vacancies (e.g., Ca2 site, Wilson et al. 1999; Ca1 site, Ivanova et al. 2001). The coupled substitution involving concomitant replacement of SiO_4^{4-} for PO_4^{3-} has been well established on the basis of compositional data from high-temperature FAp and OHAp in igneous rocks (e.g., Sommerauer and Katz-Lehnert 1985; Comodi et al. 1999). The coupled substitution involving parallel replacement of Na^+ for Ca^{2+} was also proposed on the basis of compositional data from natural carbonate-bearing apatites (McConnell 1952; Ames 1959), and has been confirmed by the synthesis of Na- and carbonate-bearing apatites (e.g., Bonel et al. 1973). On the other hand, the similar substitution involving concomitant replacement of H_3O^+ for Ca^{2+} (McConnell 1952; Simpson 1968) is less certain.

Two other divalent complex ions that have been shown to substitute for PO_4^{3-} in synthetic apatites are SeO_4^{2-} and CrO_4^{2-} (Schwarz 1967a,b). The compositions of $Pb_6K_4(PO_4)_4(SeO_4)_2\Box_2$ and $Pb_6K_4(AsO_4)_4(SeO_4)_2\Box_2$ (Schwarz 1967a) suggest a coupled substitution of the type:

$$SeO_4^{2-} + K^+ = PO_4^{3-} + Ca^{2+} \qquad (11d).$$

The compositions of $K_6Pb_4(CrO_4)_6F_2$ and $Sr_{10}(SiO_4)_3(CrO_4)_3F_2$ (Schwarz 1967b) suggest the charge-compensating mechanisms are respectively:

$$CrO_4^{2-} + K^+ = PO_4^{3-} + Ca^{2+} \qquad (11e)$$

$$CrO_4^{2-} + SiO_4^{4-} = 2\ PO_4^{3-} \qquad (17c)$$

Tetravalent anion groups. A complete solid solution between OHAp and britholite-(Y), indicative of extensive substitution of PO_4^{3-} by SiO_4^{4-}, has been confirmed by data from both natural apatites and experimental synthesis (Ito 1968; Felsche 1972; Rouse and Dunn 1982; Roeder et al. 1987; Rønsbo 1989; Liu and Comodi 1993; Comodi et al. 1999). Two coupled substitutions of the type:

$$SiO_4^{4-} + REE^{3+} = PO_4^{3-} + Ca^{2+} \qquad (12a)$$

and

$$SiO_4^{4-} + (SO_4,CO_3)^{2-} = 2\ PO_4^{3-} \qquad (17a,b)$$

have been proposed for incorporating Si into apatites (Roeder et al. 1987; Rønsbo 1989; Liu and Comodi 1993; Comodi et al. 1999) and have been discussed above. Peng et al. (1997) showed that these two substitutions alone cannot account for the observed Si contents in their FAp synthesized from FMQ-buffered experiments and proposed an additional coupled substitution of the type:

$$SiO_4^{4-} + \Box = PO_4^{3-} + F^- \qquad (5a)$$

A replacement of GeO_4^{4-} ions for PO_4^{3-} ions is indicated by synthetic germanate apatites: $M_{10}(ZO_4)_3(GeO_4)_3F_2$ (M = Sr, Pb; Z = S, Cr) and $Sr_{10}(PO_4)_4(GeO_4)_2\Box_2$ (Schwarz 1967b, 1968). These compositions suggest that the charge-compensating mechanisms for the incorporation of GeO_4^{4-} ions are of the type:

$$GeO_4^{4-} + (SO_4^{2-},CrO_4^{2-}) = 2\ PO_4^{3-} \qquad (17d,e)$$

$$GeO_4^{4-} + \Box = PO_4^{3-} + F^- \qquad (5b)$$

Frondel and Ito (1957) documented the substitution of GeO_4^{4-} for AsO_4^{3-} in natural mimetite.

Pentavalent anion groups. SiO_3N^{5-} ions have been suggested to occur in synthetic silicate oxyapatites (Gaudé et al. 1975; Maunaye et al. 1976; Guyader et al. 1978; Dupree et al. 1988; Harris et al. 1989), and are apparently incorporated via a coupled substitution of the type:

$$SiO_3N^{5-} + 2\ REE^{3+} = PO_4^{3-} + 2\ Ca^{2+} \qquad (14a)$$

as indicated by the composition of $Sm_{10}(SiO_4)_4(SiO_3N)_2O_2$ (Gaudé et al. 1975).

It has been noted above that the synthetic rare-earth borosilicate oxyapatites of Ito (1968) suggest the presence of BO_4^{5-} ions (see also Mazza et al. 2000), which can be related to PO_4^{3-} in FAp via a coupled substitution of the type:

$$BO_4^{5-} + 2\ REE^{3+} = PO_4^{3-} + 2\ Ca^{2+} \qquad (14b)$$

Ito (1968) also showed that there is a complete solid solution between $Ca_2Y_8(SiO_4)_6O_2$ and $Y_{10}(SiO_4)_4(BO_4)_2O_2$, where the exchange reaction is:

$$BO_4^{5-} + Y^{3+} = SiO_4^{4-} + Ca^{2+}.$$

INTRINSIC AND EXTERNAL CONTROLS ON UPTAKE OF REEs IN APATITES

It is not possible to detail all of the factors that may influence the complex chemical variation in apatites (see also reviews by McConnell 1973; Roy et al. 1978; Elliott 1994), especially because data for many solid solutions are either incomplete or absent. Accordingly, we have selected the uptake of REEs in FAp, OHAp, and ClAp as examples to illustrate some of the important factors, both intrinsic (crystal-chemical) and external (P-T-X), that control the compositional variation in apatites.

Crystal-chemical controls

Data from natural apatites and laboratory experiments have shown that the uptake of REEs in natural apatites is highest in the range Nd-Gd, the peak in the uptake curve being near Nd for synthetic FAp (Fleet and Pan 1995b; 1997b), OHAp (Fleet et al. 2000a), and ClAp (Fleet et al. 2000b), while uptake is lowest for Lu (Figs. 3 and 4). Also, Watson and Green (1981) showed that the partition coefficient for Sm [D(Sm)] between FAp and melts of basanitic to granitic compositions is greater than those for La, Dy, and Lu. Similarly, Ayers and Watson (1993) reported that D(Gd) is greater than D(Ce) and D(Yb) for partitioning between FAp and aqueous fluids at 1.0 GPa and 1000°C. The overall consistency of this behavior for rocks, melts and fluids of widely different composition points to crystal-chemical controls on the uptake of REEs in apatites.

From systematic analysis of site occupancies and structural change in REE-substituted FAp, OHAp, and ClAp (Fleet and Pan 1994; 1995a; 1997a,b; Fleet et al. 2000a,b), Fleet et al. (2000b) noted that the crystal-chemical factors that control site preference and uptake of REEs in apatites are complex and include substitution mechanisms (Mackie and Young 1973), spatial accommodation, equalization of bond valence (Hughes et al. 1991b; Takahashi et al. 1998) and a possible crystal field contribution for Nd.

Substitution mechanisms. Fleet et al. (2000b) noted that different substitution mechanisms were responsible for the incorporation of REEs into synthetic apatites. For example, REEs in OHAp that crystallized from H_2O-bearing phosphate melts were charge-compensated by concomitant replacement of PO_4^{3-} by SiO_4^{4-}, i.e.,

$$REE^{3+} + SiO_4^{4-} = Ca^{2+} + PO_4^{3-} \qquad (12a)$$

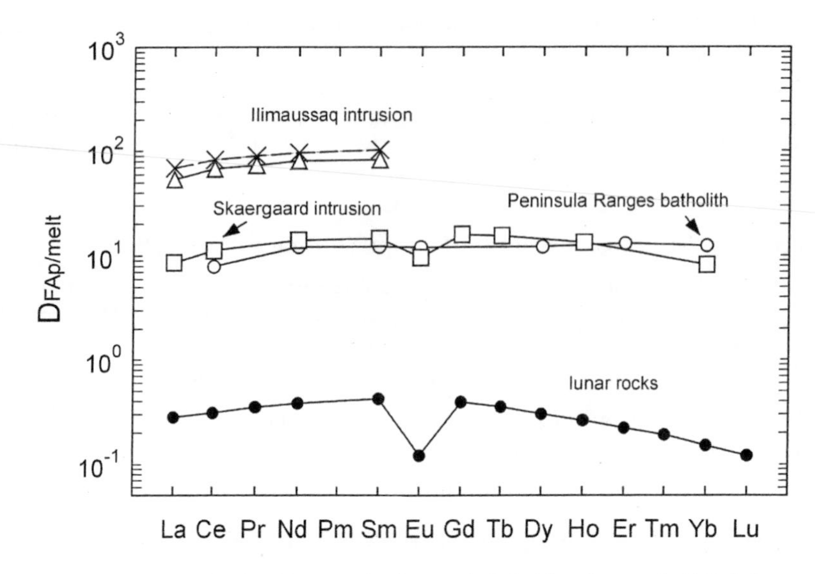

Figure 3. Melt-normalized REE patterns for fluorapatite (FAp) from lunar rocks (full circles; Jolliff et al. 1993), Skaergaard layered series, Greenland (upper zone, □; Paster et al. 1974), granodiorite from the eastern Peninsular Ranges batholith, southern California (apatite/whole rock data, ○; Gromet and Silver 1983), and the Ilímaussaq intrusion, South Greenland (sodalite foyaite; Δ and ✕ are ranges in REE content of FAp; Rønsbo, 1989; Larsen 1979) (from Fleet and Pan 1997b, with permission).

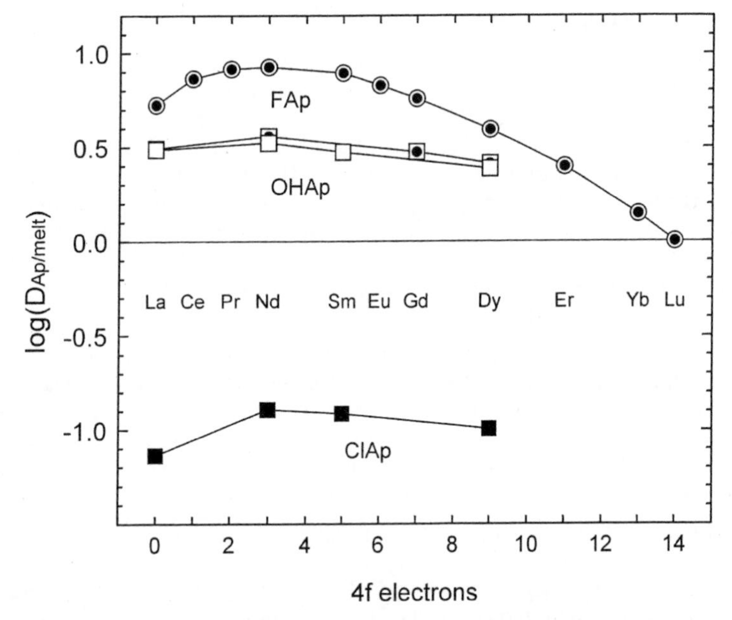

Figure 4. Melt-normalized REE contents of apatites synthesized from H_2O-bearing phosphate melts: REE-ClAp (Fleet et al. 2000a; ■); REE-FAp (Fleet and Pan 1995a; □ with central dot; Fleet and Pan 1997b, minor contents of REE; ○ with central dot; REE-OHAp (Fleet et al. 2000b; □) (from Fleet et al. 2000a, with permission).

(Fleet et al. 2000a), whereas these elements in ClAp that crystallized from H_2O-bearing, Na-rich phosphate-chloride melts were charge-compensated by parallel substitution of Na for Ca, i.e.,

$$REE^{3+} + Na^{+} = 2Ca^{2+} \quad (10a)$$

(Fleet et al. 2000b). The REEs in FAp that crystallized from H_2O-bearing phosphate-fluoride melts were charge-compensated by concomitant substitutions of both Si for P and Na for Ca (Fleet and Pan 1997b). Fleet et al. (2000b) noted that the site occupancy ratio of REE (i.e., REE-Ca2/REE-Ca1) in these synthetic apatites broadly correlated with substitution mechanism: e.g., for La-doped apatites, this ratio is 11 in OHAp (with charge compensated by Si), 4 in FAp (with charge compensated by both Si and Na), and 0.71 in ClAp (with charge compensated by Na). These observations support the suggestion of Mackie and Young (1973), who found that Nd^{3+} ions substitute for Ca in both Ca1 and Ca2 positions at approximately equal atomic proportion in NdF_3-doped FAp, but exclusively in Ca2 in Nd_2O_3-doped FAp. However, Chen et al. (2002b and Pan et al. (2002a) showed that Gd^{3+} ions in flux-grown FAp crystals containing 0.8 and 57 ppm Gd, prepared under similar conditions to those of Mackie and Young (1973), occupy both Ca1 and Ca2 sites and have Gd-Ca2/Gd-Ca1 values of 0.13 and 0.20, respectively. This marked preference of Gd for the Ca1 site in these crystals, opposite to those observed in Gd-ricl FAp (e.g., Gd-Ca2/Gd-Ca1 = 2.0 for a FAp crystal containing 10.36 wt % Gd_2O_3; Fleet and Pan 1995a), is most likely attributable to the availability of intrinsic Ca^{2+} vacancies in the *c*-axis channels at ppm concentrations [i.e., via Substitution (13a); Pan et al. 2002b]. Cherniak (2000) also attributed the differences in the diffusion rates of REE in FAp among three sets of experiments (i.e., ion-implantation, in-diffusion, and out-diffusion) to differences in substitution mechanisms.

Spatial accommodation. The effect of spatial accommodation on the uptake of REE by apatites has been investigated by Fleet and Pan (1995a; 1997a) and Fleet et al. (2000a,b). It was noted that the effective size of the Ca2 site, which preferentially incorporates light REEs in FAp and OHAp, depends on the volatile anion component. Fleet et al. (2000b) showed that plots of the REE site occupancy ratio versus change in the cell volume relative to end-member structures converge toward (REE-Ca2/REE-Ca1) = 1 and $\Delta V_{unit\ cell} = 0$ for both FAp and OHAp (Fig. 5). This relationship suggests that minimization of volume strain is an important factor in the strong preference of light REEs for the Ca2 position. Ca2 is a fairly open site and readily accommodates substituents, as is evident from plots of polyhedral volume versus unit-cell volume (Fig. 6). Conversely, the individual Ca2-O distances and Ca1 polyhedral volume do not change homogeneously with change in unit-cell volume. Fleet et al. (2000b) suggested that the Ca1 polyhedron, a distorted trigonal prism, does not readily accommodate cations that are either appreciably larger or smaller than Ca2. Thus, the REE site-occupancy ratio decreases monotonically for FAp through the range of 4*f* series in response to progressive minimization of volume strain. A similar behavior is observed for OHAp, except that somewhere beyond Sm, the REE^{3+} cations become too small for strain-free substitution into Ca1 and preferentially enter Ca2. Also, uptake is optimized for $Nd^{3+} \rightarrow Gd^{3+}$, because these cations fit most readily into the Ca positions of FAp and OHAp, although this explanation appears to be more quantitative for OHAp than for FAp. The unusual site preference of REEs in ClAp also has been attributed to the large increase in size (6-8%) and distortion of the $Ca2O_6X$ polyhedron on substitution of Cl for (F,OH) (Fleet et al. 2000b).

Equalization of bond valence. Bond valence is a measure of the bonding power of an atom and is calculated using empirical bond strength-bond length correlations (Brown 1981). It has been used extensively to understand the site preference of REEs in apatites and calc-silicate minerals (Hughes et al. 1991b; Fleet and Pan 1995a,b; Pan and Fleet 1996a; Fleet et al. 2000a; Rakovan and Hughes 2000). Ca2 is the only Ca position coordinated to the volatile

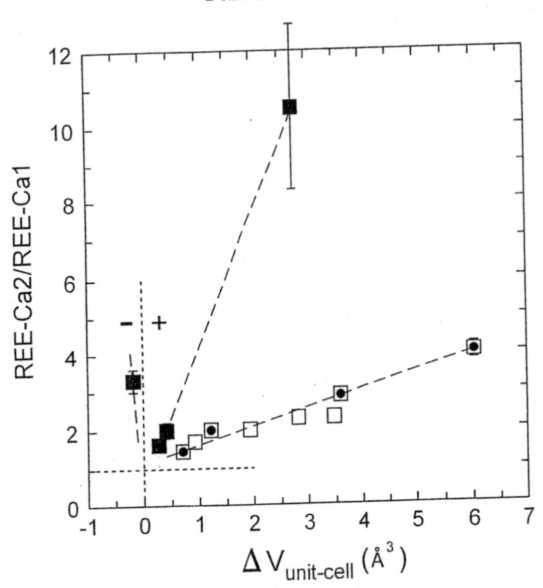

Figure 5. REE site-occupancy ratio (REE-Ca2/REE-Ca1) of REE-OHAp (Fleet et al. 2000b; ■) and REE-FAp (Fleet and Pan 1995a, □ with central dot; Fleet and Pan 1997a, □) compared with change in unit-cell volume relative to end-member OHAp and (F,OH)Ap solid solution, respectively. Note that plots reveal no site preference for REE at $\Delta V_{unit\ cell} = 0$ (from Fleet et al. 2000b, with permission).

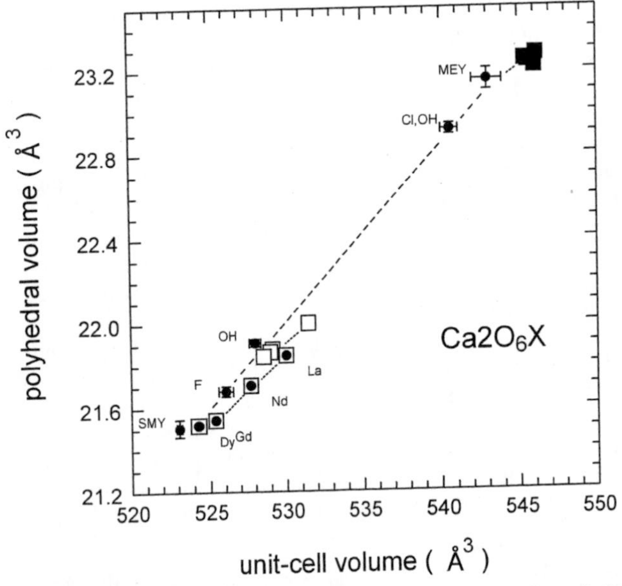

Figure 6. Variation in volume of $Ca2O_6X$ polyhedron with unit-cell volume for REE-ClAp (Fleet et al. 2000a; ■), REE-FAp (Fleet and Pan 1995a, □ with central dot); REE-OHAp (Fleet et al. 2000b, □), and synthetic (Sudarsanan et al. 1972, SMY; Mackie et al. 1972, MEY) and natural (Hughes et al. 1989) FAp, OHAp and (Cl,OH)Ap (●): trend lines have been fitted visually (from Fleet et al. 2000a, with permission).

anion component (F, OH, and Cl) and is underbonded in FAp. Therefore, Fleet and Pan (1995b) suggested that minor amounts of trivalent REEs in FAp should favor Ca2 over Ca1 to increase the bond valences of both Ca2 and F. Hughes et al. (1991b) used calculated bond valence to show that HREEs (Gd to Lu) are underbonded in both Ca positions, whereas La, Ce and Pr are slightly overbonded in the Ca1 position and therefore should prefer Ca2. Promethium and Sm should favor Ca1, and Nd should readily substitute into either Ca1 or Ca2. These results suggested that bond valence might influence both REE site preference and selectivity of apatites for REEs. However, X-ray structure refinements (Fleet and Pan 1995a; 1997a; Fleet et al. 2000a,b) revealed an apparent monotonic decrease in REE site-occupancy ratio (REE-Ca2/REE-Ca1) through the 4*f* transition-metal series, with the bond valences of Ca1 and Ca2 remaining more or less equal. In contrast, the melt-normalized REE patterns for synthetic FAp peaked near Nd-Sm (Fig. 2). These observations led Fleet and Pan (1997a) to suggest that the overall site preference for REEs is determined by equalization of bond valence but that the effective size of the Ca positions (as discussed above) exerts greater control on the selectivity of apatites for REEs. Rakovan and Hughes (2000) also extended the bond-valence requirement to explain the observed site occupancies of Sr and REEs in belovite-(Ce). They suggested that Sr, which is overbonded in the Ca1 site of FAp by as much as 0.97 valence units (Hughes et al. 1991a), competes for the Ca2 site and preferentially occupies a Ca2 site, forcing the REEs into a Ca1 equivalent site.

Crystal field contribution. Fleet et al. (2000b) noted that progressive change in individual Ca-O bond distances and O-Ca-O bond angles with substitution of REE for Ca in apatites is not continuous but tends to hinge at Nd. For example, the changes in the Ca2-O1 distance in REE-doped FAp, OHAp and ClAp all show anomalous contraction at Nd (Fig. 7). In particular, incorporation of Nd into ClAp resulted in a marked decrease in the Ca2-O1 distance, even though this element was present in greater abundance than neighboring REE cations (La and Sm). Clearly, the anomalies at Nd are not consistent with the incorporation of spherical, hard-shell cations of progressively increasing (or decreasing) radius. Fleet et al. (2000b) suggested that Nd imposes a local Jahn-Teller distortion on the Ca2 position. The 4*f* crystal field effect should be stronger for the Ca2 position because of its asymmetrical crystal field, more pronounced for Pr and Nd than adjacent REEs and heavy REEs, and absent for La, Gd and Lu (e.g., Morss 1976). This interpretation is analogous to that of Hughes et al. (1993), who suggested that the strong preference of Fe^{2+} for the Ca2 equivalent positions of a monoclinic, Fe-bearing FAp resulted from a contribution by the 3*d* crystal field stabilization energy (CFSE; cf. Burns 1993). It is noteworthy that Co^{2+} and Cr^{3+} ions have large CFSEs in asymmetrical crystal fields (Burns 1993) and therefore are expected to preferentially occupy the distorted Ca2 sites, consistent with results from X-ray structure refinements of synthetic Co and Cr-bearing apatites (Maunaye et al. 1976; Anderson and Kostiner 1987). Mn^{2+} ions of the high-spin configuration have zero CFSE, and, therefore, are controlled largely by geometric factors, and hence prefer the Ca1 position (Hughes et al. 1991a).

External (P-T-X) controls

Watson and Green (1981) reported that the partition coefficients of REEs [D(REE)] between FAp and silicate melts increase systematically with decrease in temperature and with increase in the SiO_2 content of the melt. Also, Khudolozhkin et al. (1973b) showed that the site preference of REEs for Ca2 in the FAp-britholite series decreases with increase in both temperature and content of Si, approaching zero as P is completely replaced. Similarly, Chen et al. (2002b) noted that the Gd site-occupancy ratios (Gd-Ca2/Gd-Ca1) of FAp grown in CaF_2-rich melts at 1220°C and atmospheric pressure are significantly lower than that of a Gd-rich FAp synthesized hydrothermally at 700°C and 0.12 GPa (Fleet and Pan 1995a). These results point to possible geothermometric applications using the intracrystalline partitioning of REEs in apatites. However, further studies are needed to quantitatively isolate the effect of temperature from those of other factors (e.g., substitution mechanisms) on the REE site-occupancy ratios in apatites.

Chen et al. (2002b) noted that the Gd site-occupancy ratios (Gd-Ca2/Gd-Ca1) of FAp grown in CaF_2-rich melts at 1220°C and atmospheric pressure are significantly lower than that of a Gd-rich FAp synthesized hydrothermally at 700°C and 0.12 GPa (Fleet and Pan 1995a). These results point to possible geothermometric applications using the intracrystalline partitioning of REEs in apatites. However, further studies are needed to quantitatively isolate the effect of temperature from those of other factors (e.g., substitution mechanisms) on the REE site-occupancy ratios in apatites.

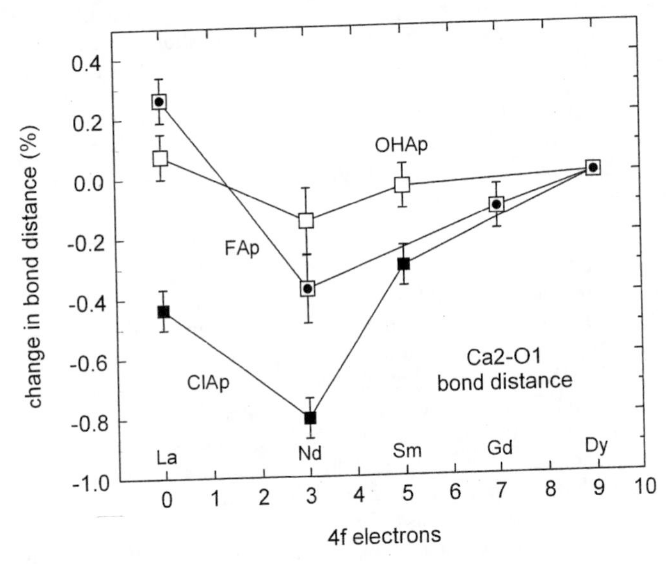

Figure 7. Change in Ca2-O1 bond distance of REE-substituted apatites with substitution of REE relative to Dy-Ap, showing anomalous decrease at Nd attributed to a *4f* crystal field contribution (from Fleet et al. 2000a, with permission).

Jolliff et al. (1993) attributed the very low apparent values of D(REE-FAp/melt) in whitlockite-bearing lunar rocks to the high temperatures expected for lunar magmas and somewhat low SiO_2 and Na_2O contents. They also noted that other melt components could be contributing factors as well, citing the positive correlation between REE abundances and content of Cl reported by Murrell et al. (1984) for FAp coexisting with whitlockite. A control on the REE uptake in FAp by melt/fluid compositions is evident in Pan and Breaks (1997), who reported marked discontinuities at Nd and Er on chondrite-normalized REE patterns of FAp from rare-element mineralized pegmatites. These discontinuities are attributed to depletion of these two elements relative to neighboring REEs in the residual melts, probably related to extreme fractionation involving monazite and garnet.

The dependence of D(REE) on SiO_2 content is generally attributed to a decrease in the number of melt sites suitable for REE^{3+} cations with increase in the degree of polymerization (Watson 1976; Ellison and Hess 1989; Gaetani and Grove 1995). Also, an increase in SiO_2 content in the melt is expected to promote the coupled substitution:

$$REE^{3+} + SiO_4^{4-} = Ca^{2+} + PO_4^{3-} \qquad (12a)$$

further enhancing D(REE) values. Similarly, activities of other impurities (e.g., Na) in the melt/fluid are also expected to affect the uptake of REEs by apatites because they participate directly in coupled substitutions for the incorporation of REEs into apatites (see above).

Oxygen fugacity is expected to exert direct controls on the uptake of Ce and Eu in apatites, because these two REEs commonly occur in mixed valences (e.g., Rakovan et al. 2001). Similarly, many other elements are known to be multi-valent in apatites (e.g., Cr, +3, +5 and +6; Mn, +2, +5 and +6; and S, -2 and +6) and, therefore, are expected to be influenced by oxygen fugacity. For example, the experimental data of Peng et al. (1997) revealed a correlation between the SO_3 content of FAp and fo_2, signifying a strong control on the uptake of S by fo_2.

SUMMARY AND SUGGESTIONS FOR FUTURE RESEARCH

Available data from natural occurrences and synthetic materials have shown that apatites are capable of accommodating a large number of elements and molecules because of the remarkable tolerance of these phases to structural distortion and chemical substitution. The chemistry of apatites is further complicated by nonstoichiometry, order-disorder in all of the *c*-axis anion channel, tetrahedral and Ca sites, and the presence of elements with multiple valences (e.g., Cr, Eu, Mn, and S). The example on the uptake of REEs in FAp, OHAp, and ClAp showed that the complex compositional variation in apatites is controlled by both crystal-chemical and external factors.

The diverse compositions of apatites have contributed to such important applications as petrogenetic modelling in crustal-mantle studies (e.g., Watson et al. 1985; Brenan 1993; Boudreau 1995; Pan and Fleet 1996b; Pan and Breaks 1997), paleoenvironmental reconstruction (Holmden et al. 1996, 1998; Pan and Stauffer 2000), immobilization of heavy metals (Chen et al. 1997; Arey et al. 1999) and radioactive wastes (Wronkiewicz et al. 1996), agriculture (Nriagu 1984), medical sciences (Harris et al. 2000; Kato et al. 2001) and material sciences (Steinbruegge et al. 1972; Mishra et al. 1987; Rakovan and Hughes 2000). In particular, many of these applications make direct use of specific compositional characteristics of apatites. For example, apatites as major hosts of REEs have been shown to be important in geochemical models for crustal anatexis, magma evolution and mantle compositions (Watson et al. 1985). Also, apatites with elevated Sr, REE, U, and Th contents have been widely used in radiogenic isotope analysis. In particular, the $^{87}Sr/^{86}Sr$ values of Ca apatites have long been used as a proxy of initial $^{87}Sr/^{86}Sr$ values for tracing the source and evolution of magmas and fluids, because Ca apatites typically have very low Rb/Sr values and hence relatively small amounts of radiogenic Sr from the decay of ^{87}Rb (e.g., Creaser and Gray 1992). Similarly, the compositions of biogenic apatites from fossils and sedimentary rocks have allowed a wide range of stable and radiogenic isotope (e.g., O, S, C, Sr, Nd, and U-Th-Pb) analyses with applications for environmental studies and paleoenvironmental reconstruction (Holmden et al. 1996, 1998, and references therein).

The combined geological, environmental, medical and economic importance has made apatites some of the most extensively researched minerals in the past. We have every reason to believe that these same applications and potentially new ones will attract continuing and future research on these minerals. Many applications of apatites require better understanding of their chemistry; in particular, data are generally limited or even absent for biogenic apatites due to their very fine grain size. For example, little is known about the mechanisms or rates of the uptake of REEs into biogenic apatites. Therefore, future research on factors that control on the compositions of biogenic apatites should be rewarding, because of their important applications in medical sciences, environmental studies, and paleoenvironmental reconstruction (e.g., Holmden et al. 1996, 1998; Kato et al. 2001). Similarly, the great versatility in both structure and chemistry of apatites makes them good candidates in the quest for new and better materials, including biomaterials. For example, carbonate-bearing OHAp has been investigated extensively for its nucleation and interactions with organic molecules (e.g., proteins) in connection with its use in artificial bones (e.g., Harris et al. 2000; Kato et al. 2001; Vali et al 2001). Rakovan and Hughes (2000) suggested that it may be possible to tailor the emission characteristics of apatite hosts by controlling the distribution of activating lanthanides between the two Ca sites with specific Sr codoping. Also, Huang and Sleight (1993) showed that synthetic $Bi_2Ca_8(VO_4)_6O_2$ compound does not have a center of symmetry and therefore is a candidate for ferroelectricity.

ACKNOWLEDGMENTS

We thank John Hughes and Bradley Jolliff for constructive reviews of the manuscript and helpful suggestions for its improvement and NSERC for financial support of our research.

REFERENCES

Akhavan-Niaki AN (1961) Contribution a l'étude des substitutions dans les apatites. Ann Chim 6:51-79

Ames LLJr (1959) The genesis of carbonate apatites. Econ Geol 54:829-841

Anderson JB, Kostiner E (1987) The crystal structure of cobalt-substituted calcium chlorapatite. J Solid State Chem 66:343-349

Anovitz LM, Hemingway BS (1996) Thermodynamics of boron minerals: Summary of structural, volumetric and thermochemical data. Rev Mineral 33:181-261

Arden KM, Halden NM (1999) Crystallization and alteration history of britholite in rare-earth-element-enriched pegmatitic segregations associated with the Eden Lake Complex, Manitoba, Canada. Can Mineral 37: 1239-1253

Arey JS, Seaman JC, Bertsch PM (1999) Immobilization of uranium in contaminated sediments by hydroxyapatite addition. Environ Sci Tech 33:337-342

Armbruster T, Kohler T, Libowiitzky E, Friedrich A, Miletich R, Kunz M, Medenbach O, Gutzmer J (2001) Structure, compressibility, hydrogen bonding, and dehydration of the tetragonal Mn^{3+} hydrogarnet, henritermiertite. Am Mineral 86:147-158

Audubert F, Savariault JM, Lacout JL (1999) Pentaleadtris(vanadate)iodide, a defect vanadinite-type compound. Acta Crystallogr C55:271-273

Ayers JC, Watson EB (1993) Apatite/fluid partitioning of rare earth elements and strontium: Experimental results at 1.0 GPa and 1000°C and application to models of fluid-rock interaction. Chem Geol 110:299-314

Azimov SY, Ismatov AA, Fedorov NF (1981) Synthetic silicophosphates, silicovanadates and silicoarsenates with the apatite structure. Inorg Mater 17:1384-1387

Bacquet G, Vo QT, Bonel G, Vignoles M (1980) Résonance paramagnétique élétronique du centre F^+ dans les fluorapatites carbonatées de type B. J Solid State Chem 33:189-195

Bacquet G, Vo QT, Vignoles M, Bonel G (1981a) EPR detection of acetate ions trapping in B-type carbonated fluorapatites. J Solid State Chem 39:148-153

Bacquet G, Vo QT, Vignoles M, Bonel G (1981b) ESR of the F^+ centre in B-type carbonated hydroxyapatite. Phys Status Solidi 68:71-74

Banks E, Greenblatt M, McGarvey BR (1971) Electron spin resonance of CrO_4^{3-} in chlorapatite $Ca_5(PO_4)_3Cl$. J Solid State Chem 3:308-313

Banks E, Jaunarajs KL (1965) Chromium analogs of apatite and spodiosite. Inorg Chem 4:78-83

Baran EJ, Baud G, Besse J-P (1983) Vibrational spectra of some rhenium-apatites containing ReO_5-groups. Spectrochim Acta 39A:383-386

Baud G, Besse J-P, Sueur G, Chevalier R (1979) Structure de nouvelles apatites au rhenium contenant des anions volumineux: $Ba_{10}(ReO_5)_6X_2$ (X = Br,I). Mater Res Bull 14:675-682

Baud G, Besse J-P, Capestan M, Sueur G, Chevalier R (1980) Étude comparative d'apatites contenant l'ion $(ReO_5)^{3-}$. Structure des fluoro et carbonatoapatites. Ann Chim Sci Mater 5:575-583

Baumer A, Caruba R, Bizzouard H, Peckett A (1983) Chlorapatite de synthésis: substitution et inclusions de Mn, Ce, U et Th traces. Can Mineral 21:567-573

Baumer A, Caruba R, Ganteaume M (1990) Carbonate-fluorapatite: Mise en évidence de la substitution 2 $PO_4^{3-} \rightarrow SiO_4^{4-} + SO_4^{2-}$ par spectrométrie infrarouge. Eur J Mineral 2:297-304

Berndt ME, Seyfried WEJr (1997) Calibration of Br/Cl fractionation during phase separation of seawater: Possible halite at 9 to 10°N East Pacific Rise. Geochim Cosmochim Acta 61:2849-2854

Beshah K, Rey C, Glimcher MJ, Schimizu M, Griffin RG (1990) Solid state carbon-13 and proton NMR studies of carbonate-containing calcium phosphates and enamel. J Solid State Chem 84:71-81

Besse J-P, Baud G, Levasseur G, Chevallier R (1979) Structure crystalline de $Ba_5(ReO_5)_3Cl$: une nouvelle apatite contenant l'ion $(ReO_5)^{3-}$. Acta Crystallogr B35:1756-1759

Besse J-P, Baud G, Chevallier R, Zarembowitch J (1980) Mise en évidence de l'ion O_2^- dans l'apatite au rhénium $Ba_5(ReO_5)_3O_2$. Mater Res Bull 15:1255-1261

Bigi A, Foresti E, Marchetti F, Ripamonti A, Roveri N (1984) Barium calcium hydroxyapatite solid solutions. J Chem Soc Dalton Trans, p 1091-1093

Bigi A, Gazzano M, Ripamonti A, Foresti E, Roveri N (1986) Thermal stability of cadmium-calcium hydroxyapatite solid solutions. J Chem Soc Dalton Trans, p 241-244

Bigi A, Falini G, Gazzano M, Roveri N, Tedesco E (1998) Structural refinements of strontium substituted hydroxyapatites. Mater Sci Forum 278:814-819

Binder G, Troll G (1989) Coupled anion substitution in natural carbon-bearing apatites. Contrib Mineral Petrol 101:394-401

Böhlke JK, Irwin JJ (1992) Laser microprobe analysis of Cl, Br, I and K in fluid inclusions: Implications for sources of salinity in some hydrothermal fluids. Geochim Cosmochim Acta 56:203-225

Bonel G (1972) Contribution a l'étude de la carbonatation des apatites: I. Ann Chim 7:65-87

Bonel G, Montel G (1964) Sur une nouvelle apatite carbonatée synthétique. Compt Rend Acad Sci 258:923-926

Bonel G, Labarthe JC, Vignoles C (1973) Contribution a l'étude structurale des apatites carbonatees de type B. Colloq Intern CNRS 230:117-125

Bonel G, Heughebaert J-C, Hughebaert M, Lacout JL, Lebugle A (1988) Apatitic calcium orthophosphates and related compounds for biomaterials preparation. Ann New York Acad Sci 523:115-130

Borisov SV, Klevtsova RF (1963) The crystal structure of REE-Sr apatite. Zh Struct Khim 4:629-631

Borneman-Starynkevich ID (1938) On some isomorphic substitutions in the apatite group. Dokl Akad Nauk SSSR 19:253-255

Borneman-Starynkevich ID, Belov NV (1953) Carbonate-apatites. Doklad Akad Nauk SSSR 90:89-92

Bothe JV Jr, Brown PW (1999) Arsenic immobilization by calcium arsenate formation. Environ Sci Tech 33:3806-3811

Boudreau AE (1995) Fluid evolution in layered intrusions: Evidence from the chemistry of halogen-bearing minerals. *In* Magmas, Fluids, and Ore Deposits. Thompson JFH (ed) Mineral Assoc Can Short Course Series 23:25-45

Brasseur H, Dallmagne MJ (1949) Stnthèse de l'apatites. Bull Soc Chim France 135-137

Brenan JM (1993) Partitioning of fluorine and chlorine between apatite and aqueous fluids at high pressure and temperature: Implications for the F and Cl contents of high P-T fluids. Earth Planet Sci Lett 117:251-263

Brophy GP, Nash JT (1968) Compositional, infrared, and X-ray analysis of fossil bone. Am Mineral 53: 445-454

Brown ID (1981) The bond-valence method: An empirical approach to chemical structure and bonding. *In* Structure and Bonding in Crystals. O'Kieffe M, Navrotsky A (eds) Academic Press, p 1-30

Burns RG (1993) Mineralogical Applications of Crystal Field Theory (2nd edn). Cambridge University Press, Cambridge, UK

Buvaneswari G, Varadaraju UV (2000) Synthesis and characterization of new apatite-related phosphates. J Solid State Chem 149:133-136

Calvo C, Faggiani R, Krishnamachari N (1975) Crystal structure of $Sr_{9.402}Na_{0.209}(PO_4)_6B_{0.996}O_2$—a deviant apatite. Acta Crystallogr B31:188-192

Cavaretta G, Mottana A, Tecce E (1981) Cesanite, $Ca_2Na_3[(OH)(SO_4)_3]$, a sulfate isotypic to apatite from the Cesano geothermal field (Latium Italy). Mineral Mag 44:269-273

Chakhmouradian A, Reguir E, Mitchell R (2002) Strontium-apatite: New occurrences and the extent of Ca-Sr substitution in apatite-group minerals. Can Mineral 40:121-136

Chen N, Pan Y, Weil JA (2002a) Electron paramagnetic resonance spectroscopic study of synthetic fluorapatite: Part I. Local structural environment and substitution mechanism of Gd at the Ca2 site. Am Mineral 87: 37-46

Chen N, Pan Y, Weil JA, Nilges MJ (2002b) Electron paramagnetic resonance spectroscopic study of synthetic fluorapatite: Part II. Gd^{3+} at the Ca1 site, with a neighboring Ca2 vacancy. Am Mineral 87:47-55

Chen N, Pan Y, Weil JA, Nigles MJ (2002c) EPR study of ^{157}Gd-doped fluorapatite: Hyperfine and nuclear quadrupole splitting anisotropy. Geol Assoc Can Mineral Assoc Can Abstr 27:19-20

Chen XB, Wright J, Conca I, Peurrung LM (1997) Evaluation of heavy metal remediation using mineral apatite. Water Air Soil Pollut 98:57-78

Cherniak DJ (2000) Rare earth element diffusion in apatite. Geochim Cosmochim Acta 64:3871-3885

Chesnokov BV, Bazhenova LF, Bushmakin AF (1987) Fluorellestadite $Ca_{10}[(SO_4),(SiO_4)]_6F_2$—a new mineral. Zap Vser Mineral Obsh 1167:743-746

Chiranjeevirao SV, Hemmerle J, Voegel JC, Frank RM (1982) A method of preparation and characterization of magnesium-apatites. Inorg Chim Acta 67:183-187

Cho G, Yesinowski JP (1993) Multiple-quantum NMR dynamics in the quasi-one-dimensional distribution of protons in hydroxyapatite. Chem Phys Lett 205:1-5

Cho G, Yesinowski JP (1996) ^1H and ^{19}F multiple-quantum NMR dynamics in quasi-one-dimensional spin clusters in apatites. J Chem Phys 100:15716-15725

Christy AG, Alberius-Henning P, Lidin SA (2001) Computer modelling and description of nonstoichiometric apatites $Cd_{5-/2}(VO_4)_3I_{1-}$ and $Cd_{5-/2}(PO_4)_3Br_{1-}$ as modified chimney-ladder structures with ladder-ladder and chimney-ladder coupling. J Solid State Chem 156:88-100

Clarke RS, Altschuler ZS (1958) Determination of the oxidation state of uranium in apatite and phosphorite deposits. Geochim Cosmochim Acta 13:127-142

Cockbain AG, Smith GV (1967) Alkaline-earth – rare-earth silicate and germanate apatites. Mineral Mag 36:411-421

Comodi P, Liu Y, Stoppa F, Woolley AR (1999) A multi-method analysis of Si-, S- and REE-rich apatite from a new kind of kalsilite-bearing leucitite (Abruzzi, Italy). Mineral Mag 63:661-672

Creaser RA, Gray CM (1992) Preserved initial $^{87}Sr/^{86}Sr$ in apatites from altered felsic igneous rocks: A case study from the Middle Proterozoic of South Australia. Geochim Cosmochim Acta 56:2789-2795

Dai Y, Hughes JM (1989) Crystal structure refinements of vanadinite and pyromorphite. Can Mineral 27: 189-192

Dai Y, Hughes JM, Moore PB (1991) The crystal structure of mimetite and clinomimetite, $Pb_5(AsO_4)_3Cl$. Can Mineral 29:369-376

Davis TS, Kreidler ER, Parodi JA, Soules TF (1971) The luminescent properties of antimony in calcium halophosphates, J Lumines 4:48-62

DeBoer BG, Sakthivel A, Cagle JR, Young RA (1991) Determination of the antimony substitution site in calcium fluorapatite from powder X-ray diffraction data. Acta Crystallogr B47:683-692

Deganello S (1983) The crystal structure of ceanite at 21 and 236°C. N Jahrb Mineral Monatsh 305-313

Della-Ventura G, Williams CT, Cabella R, Oberti R, Caprilli E, Bellatreccia F (1999) Britholite-hellandite intergrowths and associated REE-minerals from he alkali-syenitic ejecta of the Vico volcanic complex (Latium, Italy): petrological implications bearing on REE mobility in volcanic systems. Eur J Mineral 11:843-854

Doi Y, Moriwaki Y, Aoba T, Takahashi J, Joshin K (1982) ESR and IR studies of carbonate-containing hydroxyapatites. Calcif Tissue Intl 34:178-181

Dong P, Pan Y (2002) F-Cl-Br partitioning between apatites and halide-rich melts: Experimental studies and applications. Geol Ass Can Mineral Ass Can Abstr 27:29

Dowker SEP, Elliott JC (1983) Infrared study of the formation, loss and location of cyanate and cyanamide in thermally treated apatites. J Solid State Chem 49: 334-340

Dugas J, Rey C (1977) Electron spin resonance characterization of superoxide ions in some oxygenated apatites. J Phys Chem 81:1417-1419

Dugas J, Bejjaji B, Sayah D, Trombe JC (1978) Etude par RPE de l'ion NO_2^{2-} dans une apatite nitrée. J Solid State Chem 24:143-151

Dunn PJ, Rouse RC (1978) Morelandite, a new barium arsenate chloride member of the apatite group. Can Mineral 16:601-604

Dunn PJ, Peacor DR, Newberry N (1980) Johnbaumite, a new member of the apatite group from Franklin, New Jersey. Am Mineral 65:1143-1145

Dunn PJ, Peacor DR, Valley JW, Randall CA (1985a) Ganomalite from Franklin, New Jersey, and Jakobsberg, Sweden; new chemical and crystallographic data. Mineral Mag 49:579-582

Dunn PJ, Petersen EU, Peacor DR (1985b) Turneaureite, a new member of the apatite group from Franklin, New Jersey, Balmant, New York and Laangban, Sweden. Can Mineral 23:251-254

Dupree R, Lewis MH, Smith ME (1988) High resolution silicon-29 nuclear magnetic resonance in the Y-Si-O-N system. J Am Chem Soc 110:1083-1087

Dymek RF, Owens BE (2001) Petrogenesis of apatite-rich rocks (nelsonites and oxide-apatite gabbronorites) associated with massif anorthosites. Econ Geol 96:797-815

Edgar AD (1989) Barium- and strontium-enriched apatites in lamproites from West Kimberley, Western Australia. Am Mineral 74:889-895

Efimov AS, Kravchenko SM, Vasil'eva ZV (1962) Strontium-apatite a new mineral. Dokl Akad Nauk SSSR 142:439-442

Elliott JC (1964) The crystallographic structure of dental enamel and related apatites. PhD dissertation, University of London, London, UK
Elliott JC (1969) Recent progress in the chemistry, crystal chemistry and structure of the apatites. Calcif Tissue Res 3:293-307
Elliott JC (1994) Structure and Chemistry of the Apatites and Other Calcium Orthophosphates. Elsevier, Amsterdam
Elliott JC, Mackie PE, Young RA (1973) Monoclinic hydroxyapatite. Science 180:1055-1057
Elliott JC, Bonel G, Trombe JC (1980) Space group and lattice constants of $Ca_{10}(PO_4)_6CO_3$. J Appl Crystallogr 13:618-621
Elliott JC, Dykes E, Mackie PE (1981) Structure of bromapatite, $Ca_5(PO_4)_3Br$, and the radius of the bromide ion. Acta Crystallogr B37:435-438
Ellison AJG, Hess PC (1989) Solution properties of rare earth elements: Inference from immiscible liquids. Geochim Cosmochim Acta 53:1965-1974
Engel G (1970) Hydrothermal synthese von bleihyoxylapatiten $Pb_5(XO_4)_3OH$ mit X = P, As, V. Naturwissenschaften 57:355
Engel G (1973) Infrarotspektroskopische und röntgenographische untersuchungen von bleihydorxylapatit, bleioxyapatit und bleialkaliapatit. J Solid State Chem 6:286-292
Engel G (1978) Fluoroberyllate mit apatitstruktur und ihre beziehungen zu sulfaten und silicaten. Mater Res Bull 13:43-48
Engel G, Kreig F, Reif G (1975) Mischekristallbildung und kationeordnung im system bleihydroxylapatit-calciumhydroxylapatit. J Solid State Chem 15:117-126
Ercit TS, Cerny P, Groat LA (1994) The crystal structure of Mn-rich fluorapatite and the role of Mn in the apatite structure. Geol Assoc Can Mineral Ass Can Abstr 19:A34
Fayos J, Watkin DJ, Pérez-Méndez M (1987) Crystal structure of the apatite-like compound $K_3Ca_2(SO_4)_3F$. Am Mineral 72:209-212
Felsche F (1972) Rare earth silicates with apatite structure. J Solid State Chem 5:266-275
Fleet ME, Pan Y (1994) Site preference of Nd in fluorapatite $[Ca_{10}(PO_4)_6F_2]$. J Solid State Chem 111:78-81
Fleet ME, Pan Y (1995a) Site preference of rare earth elements in fluorapatite. Am Mineral 80:329-335
Fleet ME, Pan Y (1995b) Crystal chemistry of rare earth elements in fluorapatite and calc-silicate minerals. Eur J Mineral 7:591-605
Fleet ME, Pan Y (1997a) Site preference of rare earth elements in fluorapatite: Binary (LREE+HREE)-substituted crystals. Am Mineral 82:870-877
Fleet ME, Pan Y (1997b) Rare earth elements in apatite: Uptake from H_2O-bearing phosphate-fluoride melts and the role of volatile components. Geochim Cosmochim Acta 61:4745-4760
Fleet ME, Liu X, Pan Y (2000a) Rare earth elements in chlorapatite $[Ca_{10}(PO_4)_6Cl_2]$: Uptake, site preference and degradation of monoclinic structure. Am Mineral 85:1437-1446
Fleet ME, Liu X, Pan Y (2000b). Site preference of rare earth elements in hydroxyapatite $[Ca_{10}(PO_4)_6(OH)_2]$. J Solid State Chem 149:391-398
Fleischer M, Altschuler ZS (1986) The lanthanides and yttrium in minerals of the apatite group – an analysis of the available data. N Jahrb Mineral Monatsh 467-480
Fleischer M, Mandarino JA (1995) Glossary of Mineral Species (7th ed). Mineral Record, Tucson, Arizona.
Förtsch E, Freiburg IB (1970) Untersuchungen as mineralien der pyromorphit gruppe. N Jahrb Mineral Abh 113:219-250
Fowler BO (1974) Infrared studies of apatites. II. Preparation of normal and isotopically substituted calcium, strontium, and barium hydroxyapatite and spectra-structure-composition-correlations. Inorg Chem 13: 207-214
Fransolet A-M, Schreyer W (1981) Unusual, iron-bearing apatite from a garnetiferous pegmatoid, Northampton Block, Western Australia. N Jahrb Mineral Monatsh 317-327
Frondel C, Ito J (1957) Geochemistry of germanium in the oxidized zone of the Tsumeb Mine, South-West Africa. Am Mineral 42:743-753
Gaetani GA, Grove TL (1995) Partitioning of rare earth elements between clinopyroxene and silicate melt: Crystal-chemical controls. Geochim Cosmochim Acta 59:1951-1962
Gaft M, Reisfeld DR, Ranczer G, Shoval S, Champagnon B, Boulon G (1997) Eu^{3+} luminescence in high-symmetry sites of natural apatite. J Lumines 72-74:572-574
Gaudé J, L'Haridon P, Hamond C, Marchand R, Laurent Y (1975) Composés à structure apatite. I. Structure de l'oxynitrure $Sm_{10}Si_6N_2O_{24}$. Bull Soc fr Minéral Cristal 98:214-217
Giuseppetti G, Rossi G, Tadini C (1971) The crystal structure of nasonite. Am Mineral 56:1174-1179
Grandjean-Lécuyer P, Feist R, Albarède F (1993) Rare earth elements in old biogenic apatites. Geochim Cosmochim Acta 57:2507-2514
Grisafe DA, Hummel FA (1970) Crystal chemistry and color in apatites containing cobalt, nickel and rare-earth ions. Am Mineral 55:1131-1145

Gromet L P, Silver LT (1983) Rare earth element distributions among minerals in a granodiorite and their petrogenic implications. Geochim Cosmochim Acta 47:925-939

Gruner JW, McConnell D (1937) The problem of the carbonate-apatites. Z Krist 97A:208-215

Guyader J, Grekov FF, Marchand R, Lang J (1978) Nouvelles séties de silicoapatites enrichies en azote. Rev Chim Minéral 15:431-438

Harada K, Nagashima K, Nakao K, Kato A (1971) Hydroxylellestadite, a new apatite from Chichibu mine, Saitama Prefecture, Japan. Am Mineral 56:1507-1518

Harries JE, Hasnain SS, Shah JS (1987) EXAFS study of structural disorder in carbonate-containing hydroxyapatite. Calcif Tissue Intl 41:346-350

Harris NL, Rattray KR, Tye CE, Underhill TM, Somerman MJ, D'Errico JA, Chambers AF, Hunter GK, Goldberg HA (2000) Functional analysis of bone sialoproteIn Identification of the hydroxyapatite-nucleating and cell-bindingdomains by recombinant peptide expression and site-directedmutagenesis. Bone 27:795-802

Harris RK, Leach MJ, Thompson DP (1989) Silicon-29 magic-angle spinning nuclear magnetic resonance study of some lanthanum and yttrium silicon oxynitride phases. Chem Material 1:336-338

Hata M, Marumo F, Iwai S, Aoki H (1979) Structure of barium chlorapatite. Acta Crystallogr B35:2382-2384

Hata M, Marumo F, Iwai S, Aoki H (1980) Structure of a lead apatite $Pb_9(PO_4)_6$. Acta Crystallogr B36: 2128-2130

Hata M, Okada K, Iwai S, Akao M, Aoki H (1978) Cadmium hydroxyapatite. Acta Crystallogr B34:3062-3064

Heijligers HJM, Verbeeck RMH, Driessens FCM (1979) Cation distribution in calcium- strontium-hydroxyapatites. J Inorg Nuclear Chem 41:763-764

Hitmi N, LaCabanne C, Bonel G, Roux P, Young RA (1986) Dipole co-operative motions in an A-type carbonated apatite, $Sr_{10}(AsO_4)_6CO_3$. J Phys Chem Solids 47:507-515

Hogarth DD, Staecy HR, Semenov EI, Proshchenko EG, Kazakova ME, Kataeva ZT (1973) New occurrences and data of spencite. Can Mineral 12:66-71

Holmden C, Creaser RA, Muehlenbachs K, Leslie SA, Bergstrom SM (1996) Isotopic and elemental systematics of Sr and Nd in 454 Ma biogenic apatites: Implications for paleoseawater studies. Earth Planet Sci Lett 142:425-437

Holmden C, Creaser RA, Muehlenbachs K, Leslie SA, Bergstrom SM (1998) Isotopic evidence for geochemical decoupling between epeiric seas and bordering oceans: Implications for secular curves. Geology 26:567-570

Hounslow AW, Chao GY (1970) Monoclinic chlorapatite from Ontario. Can Mineral 10:252-259

Huang J, Sleight AW (1993) The apatite structure without an inversion center in a new bismuth calcium vanadium oxide $BiCa_4V_3O_{13}$. J Solid State Chem 104:52-58

Hughes JM, Cameron M, Crowley KD (1989) Structural variations in natural F, OH, and Cl apatites. Am Mineral 74:870-876

Hughes JM, Cameron M, Crowley KD (1990) Crystal structures of natural ternary apatites: solid solutions in the $Ca_5(PO_4)_3X$ (X=F, OH, Cl) system. Am Mineral 75:295-304

Hughes JM, Cameron M, Crowley KD (1991a) Ordering of divalent cations in apatite structure: crystal structure refinements of natural Mn- and Sr-bearing apatites. Am Mineral 76:1857-1862

Hughes JM, Cameron M, Mariano AN (1991b) Rare-earth element ordering and structural variations in natural rare-earth-bearing apatite. Am Mineral 76:1165-1173

Hughes JM, Drexler JW (1991) Cation substitution in the apatite tetrahedral site: crystal structures of type hydroxylellstadite and type fermorite. N Jahrb Mineral Monatsh 327-336

Hughes JM, Fransolet A-M, Schreyer W (1993) The atomic arrangement of iron-bearing apatite. N Jahrb Mineral Monatsh 504-510

Hughes JM, Mariano AN, Drexler JW (1992) Crystal structures of synthetic Na-REE-Si oxyapatites, synthetic monoclinic britholite. N Jahrb Mineral Monatsh 311-319

Hughson MR, Sen Gupta JG (1964) A thorian intermediatemember of the britholite-apatiteseries. Am Mineral 49:937-951

Ikoma T, Yamazaki A, Nakamura S, Akao M (1999) Preparation and structure refinement of monoclinic hydroxyapatite. J Solid State Chem 144:272-276

Ito J (1968) Silicate apatites and oxyapatites. Am Mineral 53:890-907

Ito A, Aoki H, Akao M, Miura N, Otsuka R, Tsutsumi S (1988) Structure of borate groups in borate-containing apatite. J Ceram Soc Japan 96:695-697

Ivanova TI, Frank-Kamenetskaya OV, Kol'tsov V, Ugolkov L (2001) Crystal structure of calcium-deficient carbonated hydroxyapatite. Thermal decomposition. J Solid State Chem 160:340-349

Jahnke RA (1984) The synthesis and solubility of carbonate fluorapatite. Am J Sci 284:58-78

Jolliff BL, Haskin LA, Colson RO, Wadhwa M (1993) Partitioning in REE-saturating minerals: Theory, experiment and modelling of whitlockite, apatite, and evolution of lunar residual magmas. Geochim Cosmochim Acta 57:4069-4094

Joris SJ, Amberg CH (1971) The nature of deficiency in nonstoichiometric hydroxyapatite. II. Spectroscopic studies of calcium and strontium hydroxyapatite. J Phys Chem 75:3172-3178

Kato H, Nishiguchi S, Furukawa T, Neo M, Kawanabe K, Saito K, Nakamura T (2001) Bone bonding in sintered hydroxyapatite combined with a new synthesized agent, TAK-778. J Biomed Mater Res 54:619-629

Kautz K, Gubser R (1969) Untersuchungen mit der elektronen-mikrosonde an zonargebauten mineralen der pyromorphit gruppe. Contrib Mineral Petrol 20:298-305

Khattech I, Jemal M (1997) Thermochemistry of phosphate products. II. Standard enthalpies of formation and mixing of calcium and strontium fluorapatites. Thermochim Acta 298:23-30

Khomyakov AP, Lisitsyn DV, Kulikova IM, Rastsvetayeva RK (1996) Deloneite-(Ce), $NaCa_2SrCe(PO_4)_3F$, a new mineral with a belovite-like structure. Zap Vser Mineral Obsh 125:83-94

Khorari S, Cahay R, Rulmont A, Tarte P (1994) The coupled isomorphic substitution $2(PO_4)^{3-} = (SO_4)^{2-} + (SiO_4)^{4-}$ in synthetic apatite $Ca_{10}(PO_4)_6F_2$: a study by X-ray diffraction and vibrational spectroscopy. Eur J Solid State Inorg Chem 31:921-934

Khudolozhkin VO, Urusov VS, Kurash VV (1974) Mössbauer study of the ordering of Fe^{2+} in the fluor-apatite structure. Geochem Intl 11:748-750

Khudolozhkin VO, Urusov VS, Tobelko KI (1972) Ordering of Ca and Sr in cation positions in the hydroxylapatite-belovite isomorphous series. Geochem Intl 9:827-833

Khudolozhkin VO, Urusov VS, Tobelko KI (1973a) Distribution of cations between sites in the structure of Ca, Sr, Ba - apatites. Geochem Intl 10:266-269

Khudolozhkin VO, Urusov VS, Tobelko KI, Vernadskiy VI (1973b) Dependence of structural ordering of rare earth atoms in the isomorphous series apatite-britholite (abukumalite) on composition and temperature. Geochem Intl 10:1171-1177

Kingsley JD, Prener JS, Segall B (1965) Spectroscopy of $(MnO_4)^{3-}$ in calcium halophosphates. Phys Rev A137:189-202

Klement R, Haselbeck H (1965) Apatite and wagnerite zweiwertige metalle. Z Anorg Allgem Chem 336: 113-128

Kreidler ER, Hummel FA (1970) The crystal chemistry of apatite: structure fields of fluor- and chlorapatite. Am Mineral 55:170-184

Kutoglu A von (1974) Structure refinement of the apatite $Ca_5(VO_4)_3(OH)$. N Jahrb Mineral Monatsh 210-218

Labarthe J-C, Bonel G, Montel G (1971) Sur la localisation des carbonate dans le réseau des apatites calciques. Compt Rend Acad Sci 273:349-351

Labarthe J-C, Therasse M, Bonel G, Montel G (1973) Sur la structure des apatites phosphocalciques carbonatées de type B. Compt Rend Acad Sci 276:1175-1178

Lacout JL, Mikou M (1989) Sur les dioxyapatites phosphostrontiques contenant deux ions de terres rares. Ann Chim 14:9-14

Larsen L M (1979) Distribution of REE and other trace elements between phenocrysts and peralkaline undersaturated magmas, exemplified by rocks from the Gardar igneous province, south Greenland. Lithos 12:303-315

LeGeros RZ (1965) Effect of carbonate on the lattice parameters of apatite. Nature 206:403-404

Liu Y, Comodi P (1993) Some aspects of the crystal chemistry of apatites. Mineral Mag 57:709-719

Mackie PE, Elliott JC, Young RA (1972) Monoclinic structure of synthetic $Ca_5(PO_4)_3Cl$ chlorapatite. Acta Crystallogr A28:1840-1848

Mackie PE, Young RA (1973) Location of Nd dopant in fluorapatite, $Ca_5(PO_4)_3F$:Nd. J Appl Crystallogr 6: 26-31

Mathew M, Mayer I, Dickens B, Schroeder LW (1979) Substitution in barium-fluoride apatite: the crystal structures of $Ba_{10}(PO_4)_6F_2$, $Ba_6La_2Na_2(PO_4)_6F_2$ and $Ba_4La_3Na_3(PO_4)_6F_2$. J Solid State Chem 28:79-95

Mathew M, Brown WE, Austin M, Negas T (1980) Lead alkali apatites without hexad anion: the crystal structure of $Pb_8K_2(PO_4)_6$. J Solid State Chem 35:69-76

Maunaye M, Hamon C, L'Haridon P, Laurent Y (1976) Composés á structure apatite. IV. Étude structurale de l'oxynitrure $Sm_8Cr_2Si_6N_2O_{24}$. Bull Soc fr Minéral Cristal 99:203-205

Mayer I, Cohen S (1983) The crystal structure of $Ca_6Eu_2Na_2(PO_4)_6F_2$. J Solid State Chem 48:17-20

Mayer I, Roth RS, Brown WE (1974) Rare earth substituted fluoride-phosphate apatites. J Solid State Chem 11:33-37

Mayer I, Fischbein E, Cohen S (1975) Apatites of divalent europium. J Solid State Chem 14:307-312

Mayer I, Semadja A (1983) Bismuth-substituted calcium, strontium, and lead apatites. J Solid State Chem 46:363-366

Mayer I, Swissa S (1985) Lead and strontium phosphate apatites substituted by rare earth and silver ions. J Less Common Metals 110:411-414

Mazza D, Tribaudino M, Delmstro A, Lebech B (2000) Synthesis and neutron structure of $La_5Si_2BO_{13}$, an analogue of the apatite mineral. J Solid State Chem 155:389-393

McArthur MJ (1985) Francolite geochemistry-compositional controls during formation, diagenesis, metamorphism, and weathering. Geochim Cosmochim Acta 49:23-35

McConnell D (1937) The substitution of SiO_4 and SO_4 for PO_4 groups in the apatite structure; ellestadite, the end member. Am Mineral 22:977-986

McConnell D (1952) The problem of the carbonate apatites. IV. Structural substitutions involving CO_3 and OH. Bull Soc Fr Mineral Cristall 75:428-445

McConnell D (1973) Apatite. Its Crystal Chemistry, Mineralogy, Utilization, and Geologic and Biologic Occurrences. Springer, New York

Merker L, Engel G, Wondratschek H, Ito J (1970) Lead ions and empty halide sites in apatites. Am Mineral 55:1435-1436

Mishra KC, Patton RJ, Dale EA, Das TP (1987) Location of antimony in a halophosphate phosphor. Phys Rev B35:1512-1520

Mitchell L, Faust GT, Hendricks SB, Reynolds DS (1943) The mineralogyand genesis of hydroxylapatite. Am Mineral 28:356-371

Miyake M, Ishigaki K, Suzuki T (1986) Structure refinements of Pb^{2+} ion-exchanged apatites by X-ray powder pattern-fitting. J Solid State Chem 61:230-235

Mohseni-Koutchesfehani S (1961) Contribution à l'étude des apatites barytiques. Ann Chim: 463-479

Moore PB, Araki T (1977) Samuelsonite: its crystal structure and relation to apatite and octacalcium phosphate. Am Mineral 62:229-245

Moran LB, Berkowitz JK, Yeinowski JP (1992) ^{19}F and ^{31}P magic-angle spinning nuclear magnetic resonance of antimony(III)-doped fluorapatite phosphors: Dopant sites and spin diffusion. Phys Rev B45:5347-5360

Morss LS (1976) Thermochemical properties of yttrium, lanthanum, and lanthanide elements and ions. Chem Rev 76:827-841

Murrell MT, Brandriss M, Woolum DS, Burnett DS (1984) Pu-REE-Y partitioning between apatite and whitlockite. Lunar Planet Sci XV:579-580

Nadal M, LeGeros RZ, Bonel G, Montel G (1971) Mise en évidence d'un phénomène d'order-désordre dans le réseau des carbonate-apatites strontique. Compt Rend Acad Sci 272:45-48

Neuman WF, Mulryan BJ (1971) Synthetic hydroxyapatite crystals IV. Magnesium incorporation. Calcif Tissue Intl 7:133-138

Newberry NG, Essene EJ, Peacor DR (1981) Alforsite, a new member of the apatite group: The barium analogue of chlorapatite. Am Mineral 66:1050-1053

Nobes RH, Akhmatskaya EV, Milman V, White JA, Winkler B, Pickard CJ (2000) An *ab initio* study of hydrogarnets. Am Mineral 85:1706-1715

Nriagu JO (1984) Formation and stability of base metal phosphates in soils and sediments. *In* Phosphate Minerals. Nriagu JO, Moore PB (eds) Springer-Verlag, New York, p. 318-329

Oberti R, Ottolini L, Della Ventura G, Pardon GC (2001) On the symmetry and crystal chemistry of britholite: New structural and microanalytical data. Am Mineral 86:1066-1075

Ohkubo Y (1968) EPR spectra of manganese(II)ions in synthetic calcium chloride fluoride phosphates. J Appl Phys 39:5344-5345

Okazaki M (1983) $F^-CO_3^{2-}$ interactions in IR spectra of fluoridated CO_3-apatites. Calcif Tissue Intl 35:78-81

O'Reilly SY, Griffin WL (2000) Apatite in the mantle: implications for metasomatic processes and high heat production in Phanerozoic mantle. Lithos 53 217-232

Pan Y, Breaks FW (1997) Rare earth elements in fluorapatite, Separation Lake area, Ontario: Evidence for S-type granite - rare-element pegmatite linkage. Can Mineral 35:659-671

Pan Y, Chen N, Weil JA, Nilges MJ (2002a) Electron paramagnetic resonance spectroscopicstudy of synthetic fluorapatite: Part III. Structural characterization of sub-ppm-level Gd and Mn in minerals at W-band frequency. Am Mineral (in press)

Pan Y, Fleet MA, Chen N, Weil JA (2002b) Site preference of Gd in synthetic fluorapatite by single-crystal W-band EPR and X-ray refinement of structure: A comparative study. Can Mineral 40 (in press)

Pan Y, Fleet ME (1996a) Intrinsic and external controls on rare earth elements in calc-silicate minerals. Can Mineral 34:147-159

Pan Y, Fleet ME (1996b) Rare-earth element mobility during prograde granulite-facies metamorphism: significance of fluorine. Contrib Mineral Petrol 123:251-262

Pan Y, Stauffer MR (2000) Cerium anomaly and Th/U fractionation in the 1.85-Ga Flin Flon paleosol: Clues from accessory minerals and implications for paleoatmospheric reconstruction. Am Mineral 85:898-911

Pascher F (1963) Untersuchungen über die Austausch-barkeit des phosphors durch chrom, selen, molybdän, wolfram, und titan im apatitgitter. Tech Wiss Abhandl Osram 8:67-77

Paster TP, Schauwecker DS, Haskin LA (1974) The behaviour of some trace elements during solidification of the Skaergaard layered intrusion. Geochim Cosmochim Acta 38:1549-1577

Patel PN (1980) Magnesium calcium hydroxylapatite solid solutions. J Inorg Nuclear Chem 42:1129-1132

Pekov IV, Kulikova IM, Kabalov YuK, Yeletskaya OV, Chukanov NV, Men'shikov YuP, Khomyakov AP (1996) Belovite-(La), $Sr_3Na(La,Ce)(PO_4)_3(F,OH)$, a new rare earth mineral in the apatite group. Zap Vser Mineral Obsh 125:101-109

Peng GY, Juhr JF, McGee JJ (1997) Factors controlling sulfur concentrations in volcanic apatite. Am Mineral 82:1210-1224

Perret R, Bouillet AM (1975) The sulfate apatites $Na_3Cd_2(SO_4)_3Cl$ and $Na_3Pb_2(SO_4)_3Cl$. Bull Soc fr Minéral Cristallogr 98:254-255

Persiel E-A, Blanc P, Ohnenstetter D (2000) As-bearing fluorapatite in manganiferous deposits from St. Marcel-Praborna, Val d'Aosta, Italy. Can Mineral 38:101-117

Piriou B, Fahmi D, Dexpert-Ghys J, Taïtaï A, Lacout JL (1987) Unusual fluorescent properties of Eu^{3+} in oxyapatites. J Lumines 39:97-103

Portnov AM, Sidorenko GA, Dubinchuk VT, Kunetsova NN, Ziborova TA (1969) Melanocerite from the northern Baikal region. Doklad Akad Nauk SSSR, 185:901-904

Prener JS (1967) The growth and crystallographic properties of calcium fluor- and chlorapatite crystals. J Electrochem Soc 114:77-83

Pushcharovskii DYu, Nadezhina TN, Khomyakov AP (1987) Crystal structure of strontium apatite from Khibiny. Soviet Phys Crystallogr 32:524-526

Rakovan J, Reeder RJ (1994) Differential incorporation of trace elements and dissymmetrization in apatite: The role of surface structure during growth. Am Mineral 79:892-903

Rakovan J, Reeder RJ (1996) Intracrystalline rare earth element distributions in apatite: surface structural influences on incorporation during growth. Geochim Cosmochim Acta 60:4435-4445

Rakovan J, Hughes JM (2000) Strontium in the apatite structure: strontian fluorapatite and belovite-(Ce). Can Mineral 38:839-845

Rakovan J, Reeder RJ, Elzinga EJ, Cherniak DJ, Tait CD, Morris DE (2002) Structural characterization of U(VI) in apatite by X-ray absorption spectroscopy. Environ Sci Technol 36:3114-3117

Rakovan J, Newville M, Sutton S (2001) Evidence for heterovalent europium in zoned Llallagua apatite using wavelength dispersive XANES. Am Mineral 86:697-700

Regnier P, Lasaga AC, Berner RA, Han OH, Zilm KW (1994) Mechanism of CO_3^{2-} substitution in carbonate-fluorapatite: evidence from FTIR spectroscopy, ^{13}C NMR and quantum mechanical calculations. Am Mineral 79:809-818

Rey C, Trombe J-C, Montel G (1978a) Some features of the incorporation of oxygen in different oxidation states in the apatite lattice: III. Synthesis and properties of some oxygenated apatites. J Inorg Nuclear Chem 40:27-30

Rey C, Trombe J-C, Montel G (1978b) Sur la fixation de la glycine dans le réseau des phosphates à structure d'apatite. J Chem Res 188:2401-2416

Rey C, Collins B, Goehl T, Dickson IR, Glimcher MJ (1989) The carbonate environment in bone mineral: A resolution-enhanced Fourier transform infrared spectroscopy study. Calcif Tissue Intl 45:157-164

Roeder PL, MacArthur D, Ma XP, Palmer GR (1987) Cathodoluminescence and microprobe study of rare-earth elements in apatites. Am Mineral 72:801-811

Rønsbo JG (1989) Coupled substitution involving REEs and Na and Si in apatites in alkaline rocks from the Illimaussaq intrusions, South Greenland, and the petrological implications. Am Mineral 74:896-901

Rouse RC, Dunn PJ (1982) A contribution to the crystal chemistry of ellestadite and the silicate sulfate apatites. Am Mineral 67:90-96

Rouse RC, Dunn PJ, Peacor DR (1984) Hedyphane from Franklin, New Jersey and Laangban, Sweden; cation ordering in an arsenate apatite. Am Mineral 69:920-927

Roux P, Bonel G (1977) Sur la preparation de l'apatite carbonatée de type A, à haute température par évolution, sous pression de gaz carbonique, des arséniates tricalcique et tristrontique. Ann Chim, p 159-165

Roy DM, Drafall LE, Roy R (1978) Crystal chemistry, crystal growth, and phase equilibria of apatites. *In* Phase Diagrams, Material Sciences and Technology 6-V. Alper AM (ed) Academic Press, New York, p 186-239

Ruszala F, Kostiner E (1975) Preparation and characterization of single crystals in the apatite system $Ca_{10}(PO_4)_6(Cl,OH)_2$. J Crystal Growth 30:93-95

Ryan FM, Hopkins RH, Warren RW (1972) The optical properties of divalent manganese in strontium fluorophosphate: a comparison with calcium fluorophosphate. J Lumines 5:313-333

Schneider W (1967) Caracolit, das $Na_3Pb_2(SO_4)_3Cl$ mit apatitstruktur. N Jahrb Mineral Monatsh 284-289

Schriewer MS, Jeitschko W (1993) Preparation and crystal structure of the isotypic orthorhombic strontium perrhenate halides $Sr_5(ReO_5)_3X$ (X = chloride, bromide, iodide) and structure refinement of the related hexagonal apatite-like compound barium perrhenate chloride ($Ba_5(ReO_5)_3Cl$). J Solid State Chem 107:1-17

Schroeder LW, Mathew M (1978) Cation ordering in $Ca_2La_8(SiO_4)_6O_2$. J Solid State Chem 26:383-387

Schwarz H (1967a) Apatite des typs $Pb_6K_4(X^VO_4)_4(X^{VI}O_4)_2$ (X^V = P,As; X^{VI} = S, Se). Z Anorg Allgem Chem 356:29-35

Schwarz H (1967b) Apatite des typs $M^{II}_{10}(X^{VI}O_4)_3(X^{IV}O_4)_3$ (M^{II} = Sr,Pb; X^{VI} = S, Cr; X^{IV} = Si ,Ge). Z Anorg Allgem Chem 356:36-45
Schwarz H (1968) Strontiumapatite des typs $Sr_{10}(PO_4)_4(X^{IV}O_4)_2$ (X^{IV} = Si ,Ge). Z Anorg Allgem Chem 357: 43-53
Serret A, Cabañas MV, Vallet-Regí M (2000) Stabilization of calcium oxyapatites with lanthanum(III)-created anionic vacancies. Chem Mater 12:3836-3841
Sery A, Manceau A, Greaves GN (1996) Chemical state of Cd in apatite phosphate ores as determined by EXAFS spectroscopy. Am Mineral 81:864-873
Simpson DR (1968) Substitutions in apatite: I. Potassium-bearing apatite. Am Mineral 53:432-444
Smith GFH, Prior GT (1911) On fermorite, a new arsenate and phosphate of lime and strontia, and tilasite, from the manganese ore deposit of India. Mineral Mag 16:84-96
Sommerauer T, Katz-Lehnert K (1985) A new partial substitution mechanism of CO_3^{2-}/CO_3OH^{3-} and SiO_4^{4-} for PO_4^{3-} group in hydroxy-apatite from the Kaiserstuhl alkaline complex. Contrib Mineral Petrol 91:360-368
Soules TF, Davis TS, Kreidler ER (1971) Molecular orbital model for antimony luminescent centers in fluorophosphates. J Chem Phys 55:1056-1064
Steele IM, Pluth JJ, Livingstone A (2000) Crystal structure of mattheddleite, a Pb, S, Si phase with the apatite structure. Mineral Mag 64:915-921
Steinbruegge KB, Henningsen T, Hopkins RH, Mazelsky R, Melamed NT, Riedel EP, Roland DW (1972) Laser properties of Nd^{+3} and Ho^{+3} doped crystals with the apatite structure. Appl Optics 11:999-1012
Sudarsanan K (1980) Structure of hydroxylellestadite. Acta Crystallogr B36:1636-1639
Sudarsanan K, Mackie PE, Young RA (1972) Comparison of synthetic and mineral fluorapatite, $Ca_5(PO_4)_3F$, in crystallographic detail. Mater Res Bull 7:1331-1338
Sudarsanan K, Young RA (1974) Structure refinement and random error analysis for strontium "chlorapatite", $Sr_5(PO_4)_3Cl$. Acta Crystallogr B30:1381-1386
Sudarsanan K, Young RA (1980) Structure of partially substituted chlorapatite $(Ca,Sr)_5(PO_4)_3Cl$. Acta Crystallogr B36:1525-1530
Sudarsanan K, Young RA, Wilson AJC (1977) The structures of some cadmium "apatites" $Cd_5(MO_4)_3X$: I. Determination of the structures of $Cd_5(VO_4)_3I$, $Cd_5(PO_4)_3Br$, $Cd_5(AsO_4)_3Br$ and $Cd_5(VO_4)_3Br$. Acta Crystallogr B33:3136-3142
Suetsugu Y, Takahashi Y, Okamura FP, Tanaka J (2000) Structure analysis of A-type carbonate apatite by a single-crystal X-ray diffraction method. J Solid State Chem 155:292-297
Suitch PR, Lacout JL, Hewat A, Young RA (1985) The structural location and role of Mn^{2+} partially substituted for Ca^{2+} in fluorapatite. Acta Crystallogr B41:173-179
Suitch PR, Taïtaï A, Lacout JL, Young RA (1986) Structural consequences of the coupled substitution of Eu, S, in calcium sulfoapatite. J Solid State Chem 63:267-277
Taïtaï A, Lacout JL (1989) On the coupled introduction of Eu^{3+} and S^{2+} ions into strontium apatites. J Phys Chem Solids 50:851-855
Takahashi M, Uematsu K, Ye ZG, Sato M (1998) Single-crystal growth and structure determination of a new oxide apatite, $NaLa_9(GeO_4)_6O_2$. J Solid State Chem 139:304-309
Terpstra RA, Driessens FCM (1986) Magnesium in tooth enamel and synthetic apatites. Calcif Tissue Intl 39:348-354
Tochon-Danguy HT, Very JM, Geoffroy M, Baud CA (1978) Paramagnetic and crystallographic effects of low temperature ashing on human bones and tooth enamel. Calcif Tissue Intl 25:99-104
Trombe J-C, Montel G (1975) Sur les conditions de préparation d'une nouvelle apatite contenant des ions sulfure. Compt Rend Acad Sci 280:567-570
Trombe J-C, Montel G (1978) Some features of the incorporation of oxygen in different oxidation states in the apatite lattice-II. On the synthesis and properties of calcium and strontium peroxiapatites. J Inorg Nuclear Chem 40:23-26
Trombe J-C, Montel G (1981) On the existence of bivalent ions in the apatite channels. A new example-phosphocalcium cyanamido-apatite. J Solid State Chem 40:152-160
Trueman NA (1966) Substitutions for phosphate ions in apatite. Nature 210:937-938
Urusov VS, Khudolozhkin VO (1974) An energy analysis of cation ordering in apatite. Geochem Intl 11:1048-1053
Vali H, McKee MD, Ciftcioglu N, Sears SK, Plows FL, Chevet E, Ghiabi P, Plavsic M, Kalander EO, Zare RN (2001) Nanoforms; a new type of protein-associated mineralization. Geochim Cosmochim Acta 65: 63-74
Verbeeck RMH, Lassuyt CJ, Heijligers HJM, Driessens FCM, Vrolijk JWGA (1981) Lattice parameters and cation distribution of solid solutions of calcium and lead hydroxyapatites. Calcif Tissue Intl 33:243-247
Vignoles M, Bonel G (1978) Sur la localisation des ions fluorure dans les carbonate-apatites de type B. Compt Rend Acad Sci 287:321-324

Vignoles M, Bonel G, Young RA (1987) Occurrence of nitrogenous species in precipitated B-type carbonated hydroxyapatites. Calcif Tissue Intl 40:64-70

Wallaeys R (1952) Contribution à l'étude des apatites phosphocalciques. Ann Chim 7:808-848

Warren RW (1970) EPR of Mn^{+2} in calcium fluorophosphate: I. The Ca(II) site. Phys Rev B2:4383-4388

Warren RW (1972) Defect centres in calcium fluorophosphate. Phys Rev B6:4679-4689

Warren RW, Mazelsky R (1974) EPR of Mn^{+2} in calcium fluorophosphate. II. Modified Ca(II) site. Phys Rev B10:19-25

Watson EB (1976) Two-liquid partition coefficients: Experimental data and geochemical implications. Contribut Mineral Petrol 56:119-134

Watson EB, Green TH (1981) Apatite/liquid partition coefficients for the rare earth elements and strontium. Earth Planet Sci Lett 56:405-421

Watson EB, Harrison TM, Ryerson FJ (1985) Diffusion of Sm, Sr, and Pb in fluorapatite. Geochim Cosmochim Acta 49:1813-1823

Welin E (1968) X-ray powder data for minerals from Långban and the related mineral deposits of Central Sweden. Arkiv Mineral Geol 4:499-541

Wilson AJC, Sudarsanan K, Young RA (1977) The structures of some cadmium "apatites" $Cd_5(MO_4)_3X$: II. The distribution of the halogen atoms in $Cd_5(VO_4)_3I$, $Cd_5(PO_4)_3Br$, $Cd_5(AsO_4)_3Br$ and $Cd_5(VO_4)_3Br$. Acta Crystallogr B33:3142-3154

Wilson RM, Elliott JC, Dowker SEP (1999) Rietveld refinement of the crystallographic structure of human dental enamel apatites. Am Mineral 84:1406-1414

Wondratschek H (1963) Untersuchungen zur kristallchmie der blei-apatite (pyromorphite). N Jahrb Mineral Abh 99:113-160

Wright J, Seymour RS, Shaw HF (1984) REE and neodymium isotopes in conodont apatite: Variations with geological age and depositional environment. Geol Soc Am Spec Pap 196:325-340

Wronkiewicz DJ, Wolf SF, Disanto TS (1996) Apatite- and monazite-bearing glass-crystal compositions for the immobilization of low-level nuclear and hazardous wastes. Mater Res Soc 412:345-352

Young EJ, Munson EL (1966) Fluor-chlor-oxy-apatite and sphene from Crystal Lode pegmatite near Eagle, Colorado. Am Mineral 51:1476-1493

Young EJ, Myers AT, Munson EL, Conklin NM (1969) Mineralogy and geochemistry of fluorapatite from Cerro de Mercado, Durango, Mexico. U S Geol Surv Paper 650-D:84-93

Zhang JH, Fang Z, Liao LB (1992) A study of crystal structure of britholite-Y. Acta Mineral Sinica 12: 132-142 (in Chinese).

3 Growth and Surface Properties of Apatite

John Rakovan

Department of Geology
Miami University
Oxford, Ohio 45056

INTRODUCTION

The surface defines the interface between a mineral and its surroundings. In a dynamic context, reactions that occur between apatites and the environments in which they exist take place at or through their surfaces. This includes, but is not restricted to, crystal growth, dissolution, and surface-mediated reactions such as sorption, surface complexation, and catalysis (Hochella and White 1990). Because this interface is partly defined by the nature of the environment around the crystal, the properties, structure and chemistry of the crystal surface are always different than those of the bulk, and can be quite varied depending on the environment. For example, the crystal surface of apatite may have very different characteristics in contact with an aqueous solution as opposed to a polymerized silicate melt.

Understanding the behavior of apatite surfaces is important on both the local and global scale. Apatite-water interactions can influence global biogeochemical systems such as the phosphorus cycle and the fate and transport of trace elements in Earth surface environments. The rapidly growing integration of surface science with the geological sciences is due to the realization that mineral surfaces are of great importance in a wide variety of natural processes. In order to understand the role of mineral surfaces on a global scale we must first understand the details of their structure and the interactions that take place on those surfaces at the atomic scale. This chapter provides a review of the nature of the apatite surface, particularly as it exists in contact with aqueous solutions, and the processes involved in apatite crystal growth.

CRYSTAL MORPHOLOGY AND SURFACE MICROTOPOGRAPHY

Morphology

In igneous, metamorphic and hydrothermal systems apatite is most often found as euhedral to subhedral hexagonal crystals. The habit is usually prismatic, elongate along [001], or tabular, although equant crystals are not uncommon (Fig. 1). The most common forms are {100}, {001}, {101}, and {110}. However, the combination of many other forms can lead to complex morphologies. Goldschmidt (1913) provides 13 pages of crystal drawings exhibiting more than 17 distinct forms. Palache et al. (1951) list 30 forms that have been observed on apatite. Although many apatite crystals are only bound by two forms, {001} and {100}, some crystals are very complex, with six or more forms on a single crystal (Fig. 2). Crystals are also found as anhedral massive, granular or globular. Sedimentary apatites are most often found as massive, cryptocrystalline crusts, concretions, stalactites and masses of radiating fibers, collectively known as collophane.

Much is known about the causes of morphologic variations in crystals (Kostov and Kostov 1999, Sunagawa 1987a); however, little has been done to investigate the factors that control apatite crystal morphology specifically. In a study of the affect of fluorine concentration in solution on the morphology and growth rate of fluorapatite, Deutsch and Sarig (1977) found that at high fluorine concentrations (F concentrations in excess of 4× the stoichiometric amount for fluorapatite) the crystal habit was found to be

1529-6466/00/0049-0003$05.00

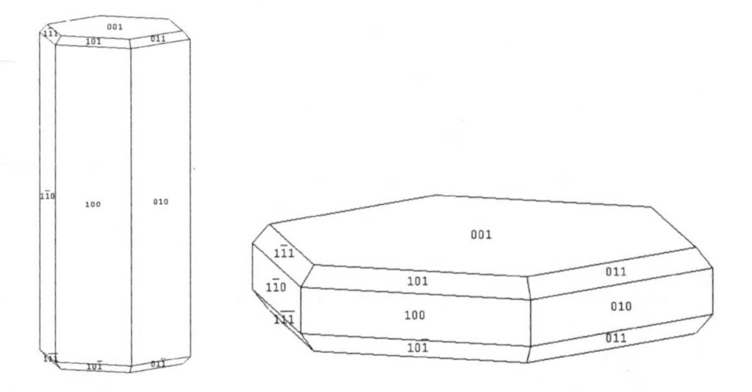

Figure 1. Depictions of prismatic and tabular apatite crystals.

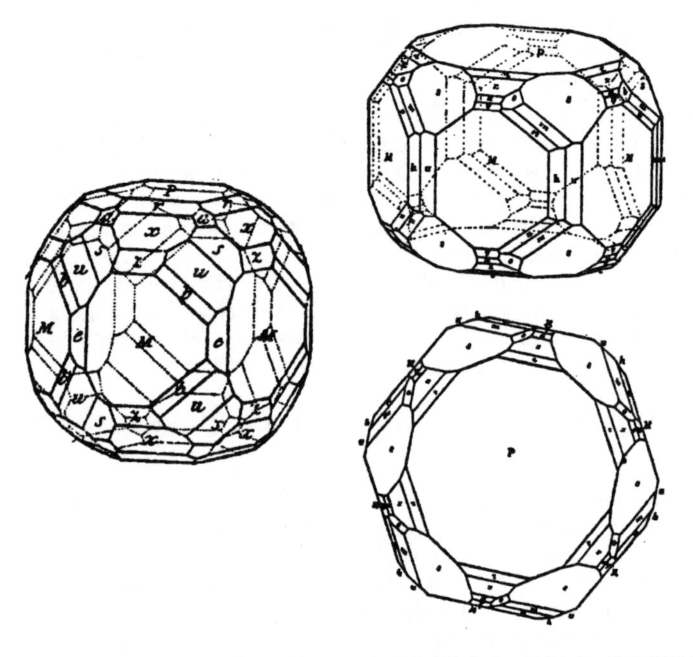

Figure 2. Depictions of complex apatite crystals. [Modified after Goldschmidt (1913).]

tabular, whereas at low fluorine concentrations (F concentrations less than 4× the stoichiometric amount for fluorapatite) the habit was needle-like.

Biological apatite in bone and teeth often has a distinct morphology, unlike any inorganic apatites. Individual crystals are usually platy or ribbon-like (Nylen et al. 1963, Christoffersen and Landis 1991), with rectangular cross sections (Fig. 3). It has been suggested that these form as pseudomorphs of a precursor phase such as octacalcium phosphate, OCP (Brown 1966). OCP forms thin, bladed crystals that are dominated by the {100} form and elongate in [001] (Terpstra and Bennema 1987). Some biological apatites, however, show a hexagonal morphology (Fig. 4; Nylen et al. 1963). The existence of precursor phases in the growth of apatite is discussed in detail later.

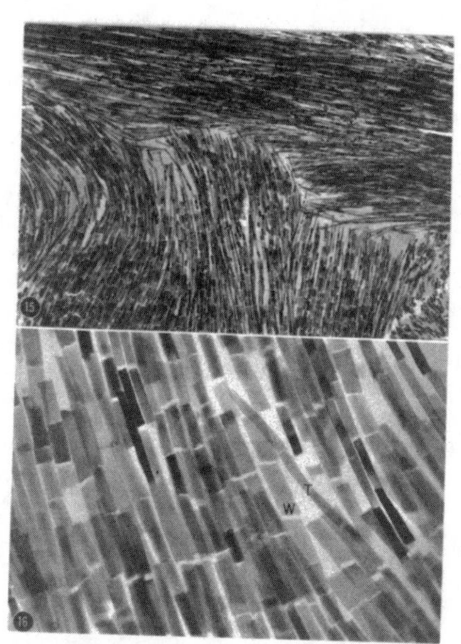

Figure 3. TEM photomicrographs of hydroxylapatite from rat bone. The direction of crystal elongation is [001]. Magnification is 35,000× and 150,000× in the upper and lower images respectively. [Used by permission of The Rockefeller University Press, from Nylen et al. (1963) *Journal of Cell Biology*, Vol. 18, Figs. 15 & 16, p. 119.]

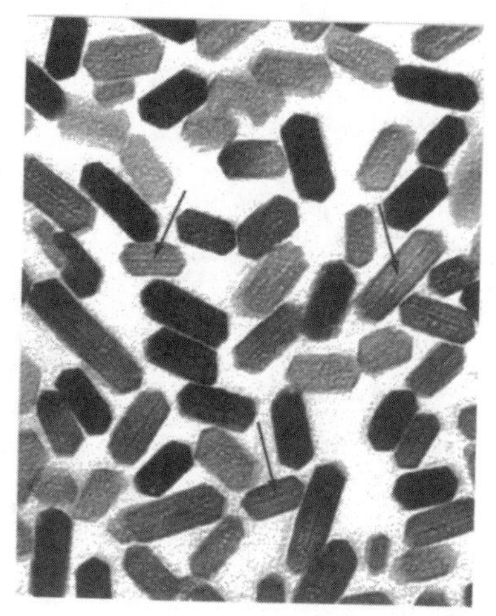

Figure 4. TEM photomicrograph of hydroxylapatite from rat bone. Crystals are oriented with {001} parallel to the image plane. Magnification is 300,000x. [Used by permission of The Rockefeller University Press, from Nylen et al. (1963) Jour. Cell Biology, Vol. 18, Fig. 13, p. 117.]

Apatites range in size from submicron crystals, like those found in biological systems, to giant metamorphic crystals weighing many tons. Velthuizen (1992) and Hogarth (1974) summarize reports of large crystals from Grenville province metasediments including "a solid but irregular mass of green crystalline apatite, 15 feet long and 9 feet wide" and a single euhedral crystal from the Aetna mine measuring 2.1 × 1.2 m with an estimated weight of 6 tons.

Surface microtopography

The microtopography of apatite crystal surfaces can vary greatly depending on the crystal face observed, conditions of crystal growth and growth mechanism, the degree and type of physical abrasion, cleavage or fracture, and the degree and mechanism of dissolution. The scale of microtopographic features on crystal surfaces ranges from atomic dimensions to dimensions of the entire crystal. Surface microtopographic features observed *ex situ* are often important clues to the conditions in, and mechanisms by, which crystals have formed, as well as post-growth reactions with the environment (Sunagawa 1987b). Thus, surface microtopography can be powerful for the interpretation of the growth and weathering history of apatite.

Growth. For crystal growth processes, one microtopographic feature that is unambiguously indicative of growth mechanism is a growth spiral (Burton et al. 1951, Sunagawa 1984, Sunagawa 1987b). Under conducive growth conditions (temperatures below the roughening transition and relatively low supersaturation) spirals may be highly polygonized. Amelinckx (1952a) observed growth spirals of several different morphologies on the {100} faces of apatite crystals from Mexico. The exact location is not given, but from the description of the samples it is likely that they were from the deposits of Cerro del Mercado, Durango. The spirals indicate that growth was by the spiral mechanism (see section on growth mechanisms). Growth steps on the observed spirals were measured by reflected light techniques to be 10 ± 2 Å, which is, within error, the unit cell dimension normal to {100}. Crystals from Sulzbachtal, Austria also exhibit well-developed polygonized spirals on {100} faces, indicating growth by the spiral mechanism (Fig. 5) (Amelinckx 1952b). Apatites from both locations most likely grew from aqueous solutions (Amelinckx 1952b).

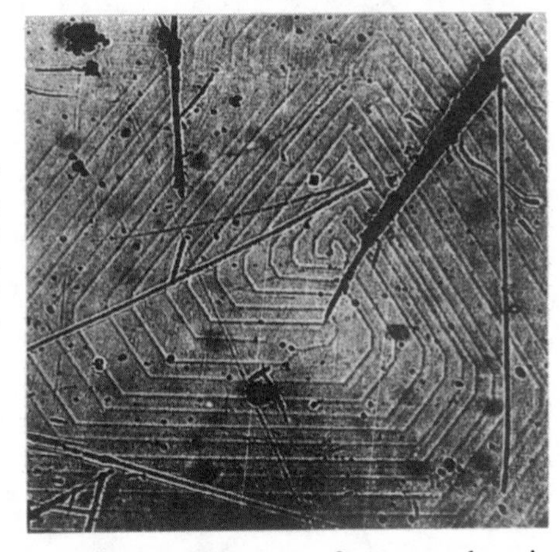

Figure 5. Reflected light photomicrograph of a complex growth spiral on a {100} face of an apatite crystal from Sulzbachtal, Austria. The horizontal steps run in the [001] direction. [Used by permission of Macmillan Publishers Ltd., from Amelinckx (1952) *Nature*, Vol. 170, Fig. 1, p. 760.]

Growth spirals can cause low mounds or growth hillocks to form on otherwise macroscopically flat crystal faces (Sunagawa 1987b, Reeder and Rakovan 1999, Paquette and Reeder 1995, Teng et al. 1999, Pina et al. 1998). If the growth spiral is polygonized, well-defined vicinal faces may be observed on the hillock (Müller-Krumbhaar et al. 1977, Sunagawa and Bennema 1982). Vicinal faces are shallow (usually only hundredths of one degree in inclination from the crystal faces on which they form), non-rational surfaces that are formed from multiple steps of like orientation that propagate in the same direction as adatoms or growth units are added to them during crystal growth. Growth hillocks have been observed on multiple forms of apatite. Rakovan and Reeder (1994) reported growth hillocks on {100} and {001} faces of hydrothermally precipitated apatites from numerous locations. They observed three-sided hillocks on {100} faces (Fig. 6a). The vicinal faces of these hillocks are made up steps that strike parallel to [001], [011], and [011]. The face symmetry of {100} is a mirror plane perpendicular to [001]. This mirror equates [011] and [$01\bar{1}$] steps, however the [001] steps of the third vicinal face are symmetrically nonequivalent (Fig. 6b). These step orientations are parallel to the orientations of the dominant steps of growth spirals reported by Amelinckx (1952b). The majority of {100} faces on well-developed apatite crystals that form from aqueous solutions are striated with steps that strike along [001] only. However, three-sided hillocks on {100} faces, similar to those in Figure 6, are not uncommon and have been observed on apatites from pegmatite pockets, veins and other hydrothermal deposits from dozens of locations (personal observation). The {001} faces of apatites often exhibit

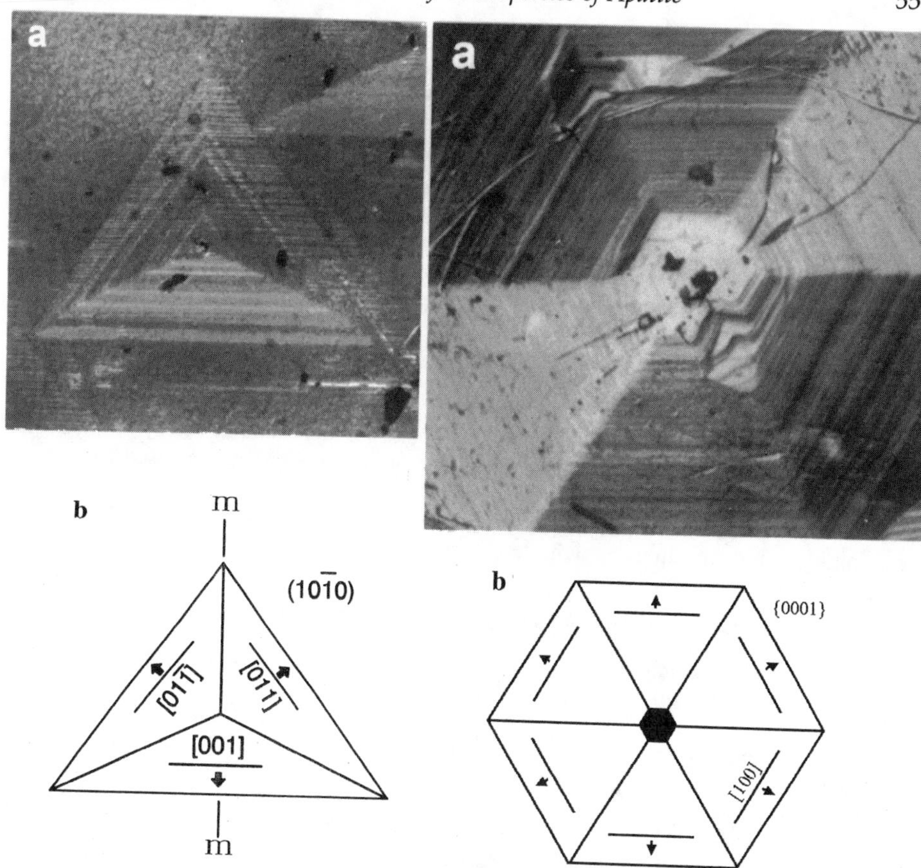

Figure 6 (left). (a) Differential interference contrast (DIC) photomicrograph of a trigonal growth hillock on a {100} face of apatite from the Golconda Mine, Minas Gerais Brazil. The three vicinal faces of the hillock exhibit macrosteps. The horizontal steps run in the [001] direction. The three step orientations are parallel to the three dominant step orientations of the spiral in Figure 5. Image is approximately 666 μm across. (b) Schematic of the face symmetry with respect to the hillock in a. Steps on the basal vicinal face parallel [001]. Lines within each vicinal face represent the orientations of growth steps. Arrows indicate the directions of advancement of steps during growth. [Modified after Rakovan and Reeder (1994)].

Figure 7 (right). (a) DIC photomicrograph of a hexagonal growth hillock on a {001} face of an apatite from the Siglo XX Mine, Llallagua, Bolivia. Steps that form the six vicinal faces run in the <100> directions. Image is approximately 1.2 mm across. (b) Schematic of the face symmetry with respect to the hillock in a. A six-fold axis of rotation perpendicular to {001} symmetrically equates steps of the six vicinal faces. Lines within each vicinal face represent the orientations of growth steps. Arrows indicate the directions of advancement of steps during growth. [Modified after Rakovan and Reeder (1994)].

well-developed hexagonal growth hillocks with six pyramidal vicinal faces (Fig. 7a). The vicinal faces of these hillocks are composed of steps that strike parallel to <100>, all of which are symmetrically related by a six-fold axis of rotation perpendicular to {001} (Fig. 7b). Akizuki et al. (1994) report similar hillocks on the {001} faces of apatites from the Asio Mine, Japan. The {001} faces of hydrothermal apatites commonly exhibit six-sided hillocks such as those in Figure 7. Symmetry and structural nonequivalence of growth steps on different crystal faces or on a single crystal face have important implications for growth kinetics and trace element incorporation (see sections below on growth rates and surface structural controls on trace element incorporation during growth).

Dissolution. As with crystal growth, dissolution of apatite can lead to surface microtopographic features that are indicative of the dissolution mechanism and also yield information about structural defects in the crystals. In a study of the dissolution of large fluorapatite crystals from Cerro de Mercado, Mexico, in hydrochloric and citric acids, Thirioux et al. (1990) observed the formation of well defined, polygonized etch pits on {100} and {001} surfaces. Regular hexagonal etch pits form on the {001} and uniformly increase in size as dissolution progresses. Etch pits with a trapezohedral geometry form on the {100} surfaces and anisotropically expand along [001] as dissolution continues, resulting in elongate etch troughs. Natural elongate etch pits of a similar morphology were reported on the {100} faces of apatites from Minas Gerais, Brazil (Rakovan and Reeder 1994). The formation of such microtopography suggests the presence and role of dislocations in the dissolution mechanism. Phakey and Leonard (1970) studied the structural defects in Mexican apatites (probably from Cerro de Mercado based on sample description) using X-ray diffraction topography. The dislocation density of the crystals was found to be low, with the majority of dislocations aligned with [001]. Preferential etching of such dislocations explain the morphology of etch pits observed on {001} and {100} surfaces.

APATITE CRYSTAL GROWTH FROM SOLUTION

Geologically, apatite precipitates from melts, concentrated hydrothermal brines, low-temperature aqueous solutions and possibly from the vapor phase (McConnell 1973). Biologically the phase precipitates from intra- and extra-cellular fluids, influenced by a variety of organic molecular species (Skinner 1987, Christoffersen and Landis 1991, McConnell 1973, Elliott 1994). In the laboratory, apatite synthesis is carried out by an array of techniques that includes melt, solution and plasma growth (Gross and Brendt, this volume; Elliott 1994 and references therein). With such a wide variety of environments of formation, it is not unexpected that the kinetics and mechanisms of growth also vary widely. In this section the growth of apatite in aqueous fluids is reviewed. It is in solution growth where the structure and chemistry of the apatite surface is found to exert the greatest influence on the growth process. Most studies of apatite growth from solution have focused on hydroxylapatite. This has been driven by motivation to understand apatite formation in biological systems. Solution growth of geologic apatite is important in surface water, sedimentary and hydrothermal environments. Chlorapatite growth kinetics are rarely addressed.

Precipitation or the formation of apatite begins with nucleation and continues through growth. There is an obvious continuum between nucleation and growth; however, the processes are different in their energetics and mechanisms, which are dependent on many variables, from the type of growth environment to subtle differences within a given environment. Several of the steps involved in precipitation are associated with all mechanisms in all environments. Nucleation must occur and a critical size must be reached by the crystal nucleus so that growth of a stable crystal can continue. Mineral constituents must move from the surrounding environment to the crystal surface. Attachment of these constituents to the surface may be followed by surface diffusion, binding at surface sites of incorporation (protosites) and dissociation of complexing ions from solution. Eventually, burial leads to incorporation and increases in crystal size and volume. [General aspects of nucleation and growth have been reviewed by Nancollas (1979), Nielsen (1984), Nancollas (1984), and Boistelle and Astier (1988).]

NUCLEATION

The driving force for nucleation and crystal growth from solution can be expressed as the positive difference between the chemical potential of a species in a supersaturated solution and that of a saturated solution. Once a solution becomes supersaturated with respect to a particular mineral, 3-D nucleation may occur. For the spontaneous formation of an apatite nucleus (for the sake of discussion we will refer to fluorapatite) in solution the change in the Gibbs free energy (ΔG) of the reaction

$$5\,Ca_{(aq)} + 3\,(PO_4)_{(aq)} + F_{(aq)} \rightarrow Ca_5(PO_4)_3F_{(s)} \tag{1}$$

must be negative. The change in the Gibbs free energy of this crystallization reaction (ΔG_c) can be expressed as

$$-\Delta G_c = R\,T\,\ln S \tag{2}$$

where R is the universal gas constant, T is the absolute temperature, and S is the saturation ratio (Nielsen 1984) defined as

$$\boxed{X}\, S = \left(\frac{[Ca]^5 [PO_4]^3 [F]}{_{eq}[Ca]^5\, _{eq}[PO_4]^3\, _{eq}[F]} \right)^{\frac{1}{v}} = \left(\frac{IAP}{K_s} \right)^{\frac{1}{v}} \tag{3}$$

where [x] is the activity of species x in solution, $_{eq}[x]$ is the activity of species x in solution at equilibrium, ν = the number of ions in the formula unit of apatite [$\nu = 9$ for fluorapatite as expressed in Equation (1)], IAP is the ion activity product and K_s is the equilibrium constant for Reaction (1), which is equivalent to the solubility product. Saturation, S, can be defined in different ways. Throughout the text S is defined as in Equation (3) unless otherwise noted.

The free energy of the formation of a crystal has two components

$$\Delta G_c = \Delta G_{bulk} + \Delta G_{surf} \tag{4}$$

due to the energy gained by the formation of bonds (ΔG_{bulk}) and the work required to increase surface area (ΔG_{surf}). For a supersaturated solution ΔG_{bulk} is negative and ΔG_{surf} is always positive. The ΔG_{surf} has a geometric component that describes the surface area and that depends on the shape of the nucleus. For a spherical nucleus

$$\Delta G_{surf} = 4\pi r^2 \gamma \tag{5}$$

where γ is the interfacial or surface free energy, which is assumed to be independent of nucleus/crystal size. During nucleation the surface free energy makes up a large part of the total free energy of the nucleus (Boistelle and Astier 1988). This positive surface contribution to the free energy acts as an energy barrier to nucleation, and is why supersaturation is necessary for a nucleus to form (Fig. 8). The interfacial energy, γ, of apatites has been studied by a variety of different methods including crystal growth and dissolution as well as surface wetting experiments. The obtained values vary widely depending on the method of determination.

Surface free energies of hydroxylapatite and fluorapatite were determined by Busscher et al. (1987) from contact angle measurements. They found that in vacuum γranges between 72 and 95 mJ/m^2. In the presence of a saturated vapor, γ was found to be between 28 and 48 mJ/m^2. The difference in γobtained by the two methods was explained as resulting from the presence of an adsorbed film originating from the liquid droplets employed in the measuring procedure. Face-specific measurements revealed for the {001} crystal face $\gamma = 28 \pm 3$ mJ/m^2, and $\gamma = 48 \pm 7$ mJ/m^2 on {100} for fluorapatite, and for hydroxylapatite the {001} crystal face $\gamma = 39 \pm 11$ mJ/m^2, and $\gamma = 30 \pm 3$ mJ/m^2 on

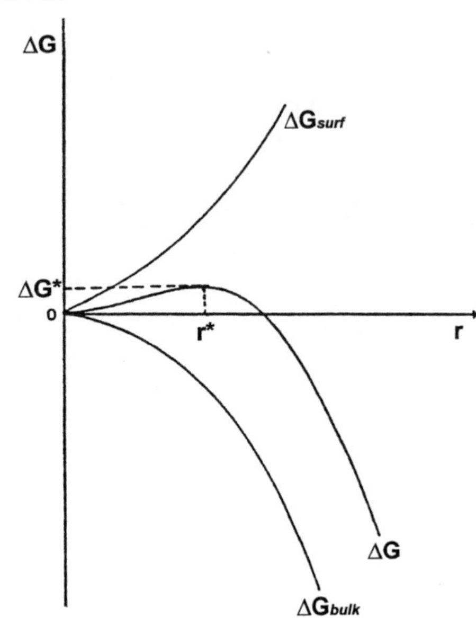

Figure 8. A plot of the change in Gibbs free energy, ΔG, with size of a crystal, r. The activation energy ΔG* must be reached for the creation of a nucleus of critical radius r*.

{100}. For hydroxylapatite, Arends et al. (1987) calculated from growth rate data $\gamma = 87$ mJ/m^2 and 80 mJ/m^2 at pH = 7 and 6, respectively. On a theoretical basis, Christoffersen et al. (1991) calculated the surface free energy of fluorapatite to be 250 mJ/m^2. This is most closely approximated by the results of Vancappellen and Berner (1991), who conducted seeded fluorapatite growth rate experiments in a carbonate-free simulated seawater solution. From analysis of their kinetic data they calculated the $\gamma = 289$ mJ/m^2. Christoffersen and Christoffersen (1992a) developed a revised version of the rate model of Hilig (1966) for the growth of crystals by a nucleation and spread mechanism, which included the relationship of surface free energy and solubility. The model was applied to the rate of growth of hydroxylapatite, and γ was calculated to be 80 mJ/m^2. Christoffersen and Christoffersen (1992b) found the same for supersaturations between 3.25 and 7.53. They suggested that the deviation of this value from the value expected from theory, 250 mJ/m^2, is the result of imperfections in the crystal surfaces. Liu and Nancollas (1996) calculated an interfacial free energy $\gamma = 30$ mJ/m^2 from their growth kinetic data, and they determined $\gamma = 18.5$ mJ/m^2 from contact angle measurements. The surface free energy of fluorapatite was found to be 120 mJ/m^2 from the growth results and 42 mJ/m^2 from the dissolution results in kinetic experiments by Christoffersen et al. (1996a). For hydroxylapatite, Christoffersen et al. (1998b) determined $\gamma = 100$ mJ/m^2 (from growth kinetic data) and $\gamma = 40$ mJ/m^2 (from dissolution kinetic data). This difference in γ values has been explained by Christoffersen et al. (1998b) and Christoffersen and Christoffersen (1992a) by the presence of surface imperfections that they conclude will affect the rate of dissolution to a greater extent than the rate of growth.

Interfacial free energy is of critical importance in determining the thermodynamics and kinetics of nucleation and crystal growth (Stumm and Morgan 1996). It is particularly important in understanding controls on crystal morphology, rate-limiting steps in the growth process, and the interactions with adsorbed species from solution. It is obvious from the review above, that there is considerable uncertainty in the value of the surface free energy of apatite. Although values vary widely, those determined from growth data are generally higher than those determined from dissolution kinetics, and γ values for fluorapatite are higher than for hydroxylapatite.

In homogeneous nucleation a crystal nucleus forms free from contact with another solid. One mechanism by which the energy barrier to nucleation, ΔG^* in Figure 8, can be decreased is by heterogeneous nucleation, where the surface of another material (i.e., another crystal) acts as a substrate on which nucleation occurs. The interaction of the substrate surface with adsorbates decreases the surface free energy of a nucleus (Boistelle and Astier 1988). For polynuclear growth (see growth mechanisms below) of hydroxylapatite Christoffersen and Christoffersen (1992a) calculated the critical radius of the surface nuclei as $2.8 < r^* < 5.1$ Å from growth kinetic data. In the simplest case of heterogeneous nucleation, the influence of the substrate is only on the energy of nucleation and no specific orientational relationship occurs between the substrate and the overgrowth. However, in many cases the structure of a crystalline substrate may also have an influence on the orientation of the nucleus and crystal overgrowth. When the substrate dictates the orientation of an overgrowth, it is known as epitaxy. Epitaxy occurs when one or more of the structural dimensions on the surface of the substrate matches, or is very close to, one or more of the structural dimensions in the overgrowth. In such a case the lowest energy configuration is often one in which the directions of similar structural dimension in the two phases are in register. Hence, there is a control on the orientation of the overgrowth as it forms. Homoepitaxy, the growth of a mineral on a preexisting substrate of the same mineral, is essentially crystal growth by a layer mechanism and is the basis for the use of seeds in growth experiments.

GROWTH MECHANISMS

Once a stable nucleus has formed, a crystal can continue to grow by several different mechanisms. All mechanisms of crystal growth involve diffusion of adatoms or growth units from the bulk of the growth medium through a diffusion boundary layer to the crystal surface and adsorption thereon. In the case of growth from solutions of extremely high supersaturation, the crystal surface can be atomically rough. In this situation, adsorption and incorporation of growth units may occur with equal probability at all exposed attachment sites with little or no surface diffusion after attachment. Such attachment and incorporation leads to macroscopically rough or anhedral crystals, and is called continuous growth. Most apatites, however, do not grow in a regime of such high supersaturation. In the majority of cases, growth occurs by a layer mechanism, where addition of growth units results in the development of planer crystal faces.

One mechanism of layer growth is two-dimensional nucleation (Fig. 9a). Analogous to heterogeneous nucleation, growth units must organize on the crystal face to reach the critical diameter of a surface nucleus, with an associated activation energy barrier. Once this has occurred, the addition of growth units is energetically favored at the edge (step) of this layer nucleus. Kinks in the layer edge are even more favorable for attachment and incorporation. By addition of growth units to edge and kink sites, the new layer spreads over the preexisting face (Fig. 9b). Normal growth of the crystal face proceeds by repeated two-dimensional nucleation and spread. Because of the activation barrier to nucleation, the rate-limiting step in this type of growth is often considered to be the initial two-dimensional nucleation of each progressive layer. Two-dimensional nucleation and growth, also called surface nucleation, birth-and-spread, and nucleated layer growth, can be divided into two categories that will manifest themselves in different reaction rates and rate orders. In the mononuclear case each surface nucleus will spread over the entire crystal face on which it is formed. The polynuclear mechanism involves the formation of multiple surface nuclei. These will interfere with one another before any one can cover the entire surface.

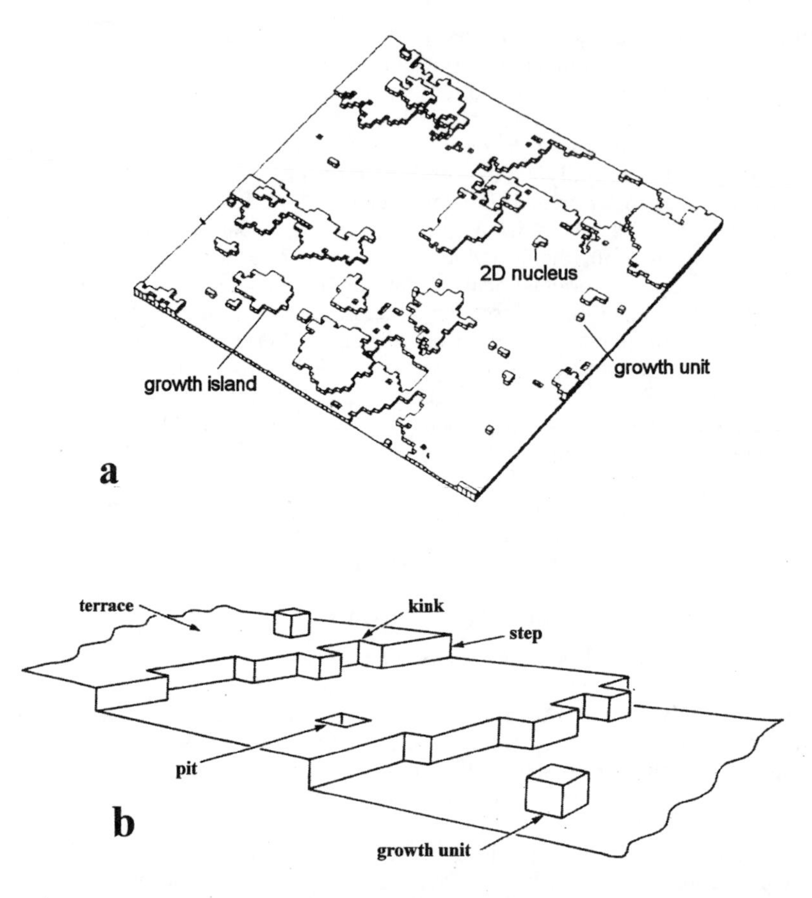

Figure 9. (a) Schematic results of a computer simulation of a crystal surface growing by the birth and spread mechanism. Numerous growth islands form the surface microtopography. [Modified from Gilmer and Jackson (1977).] (b) Schematic of the microtopography on a growing crystal face and the different types of surface sites.

A second type of layer growth is the spiral growth mechanism (Burton et al. 1951). The potential for spiral growth occurs when a screw dislocation or other extended defect in the crystal intersects a crystal face. The displacement of the structure around the dislocation creates a step on the face where the dislocation emerges. Again, for energetic reasons (Burton et al. 1951, Sunagawa 1984), incorporation of adatoms or growth units on the crystal face during growth takes place preferentially along the step, and more specifically at kink sites within the step (Fig. 9b). Because kink sites are the most likely surface sites for eventual incorporation they have been distinguished as proto-bulk sites or simply "protosites" (Nakamura 1973, Dowty 1976). A protosite is the surface representation of a bulk crystallographic site for which the occupant of the site is not yet emplaced or is not fully coordinated. As growth units are added at the step, the step spreads laterally. Because the step terminates at the dislocation, the step will spiral during growth, hence the name spiral growth. One common manifestation of spiral growth is the formation of a surface microtopographic spiral (Figs. 10 and 5). The presence of a growth spiral on a crystal face is thus an indication that the face has grown by the spiral mechanism.

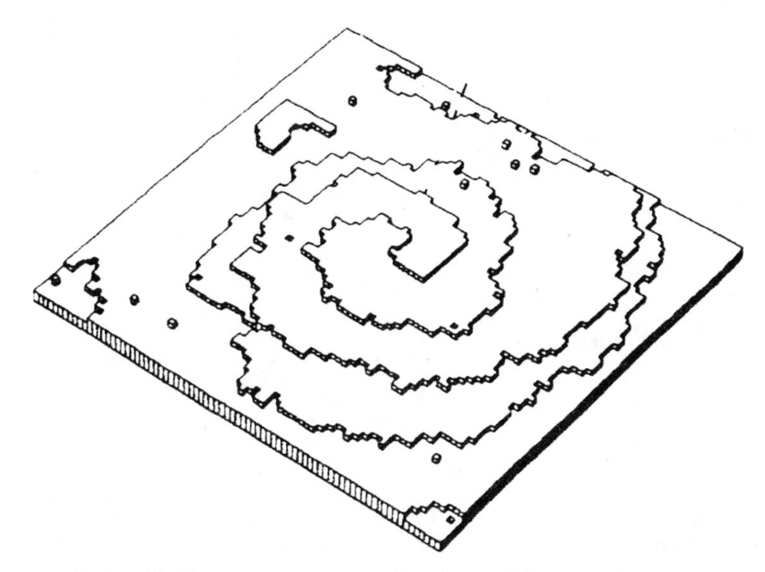

Figure 10. Results of a computer simulation of a crystal surface growing by the spiral growth mechanism. A growth spiral forms the surface microtopography. [Used by permission of Elsevier Science Ltd., from Gilmer (1976) *Journal of Crystal Growth*, Vol. 35, Fig. 4d, p. 21.]

In part, the type of growth mechanism that can operate on a given crystal face is dictated by the structure of that face. Periodic bond chain (PBC) theory predicts the possible growth mechanisms of a given face by analysis of the atomic structure of a surface slice parallel to that face (Hartman and Perdok 1955a, Hartman and Perdok 1955b). A PBC is a structural sequence or chain of uninterrupted strong bonds within a structure. Morphologic features such as crystal faces and microtopographic steps on faces are often dictated by the presence and direction of PBCs. A crystal face can be categorized by the number of non-parallel PBCs within a slice parallel to that face. F faces contain two or more PBCs, S faces contain only one PBC, and K faces have no PBCs. In Hartman and Perdok's PBC theory, only F faces should grow by a spiral mechanism. The growth rate of F faces is predicted to be slower than that of S and K faces, hence F faces should dominate the crystal habit.

In an analysis of the apatite structure Terpstra et al. (1986) identified nine different PBCs: $\langle 001\rangle$, $\langle 100\rangle$, $\langle 101\rangle$, $\langle 1\bar{1}0\rangle$, $\langle 120\rangle$, $\langle 121\rangle$, $\langle 122\rangle$, $\langle \bar{1}20\rangle$, and $\langle \bar{2}10\rangle$. Twelve different forms, including the {100} and {001}, were identified as F type (Terpstra et al. 1986), and thus by PBC theory can grow by a spiral mechanism (Hartman and Perdok 1955a). In a {100} slice (a monolayer of thickness d_{hkl} deposited on a given face during layer growth) there are PBCs that parallel the directions of the growth spiral steps reported by Amelinckx (1952b) (Fig. 5) and the vicinal face steps on growth hillocks reported by Rakovan and Reeder (1994) (Fig. 6). Growth step directions of hexagonal hillocks on the {001} face of apatite (Rakovan and Reeder 1994, Akizuki et al. 1994) (Fig. 7) were also identified as PBCs in the {001} slice.

The growth mechanism of a crystal will vary depending on factors that include the surface and step free energies of the crystal faces, the nature and density of defects within the crystal, and most importantly the degree of supersaturation, S, of solution. Figure 11 shows the relationship between S and growth mechanism. In this figure theoretical

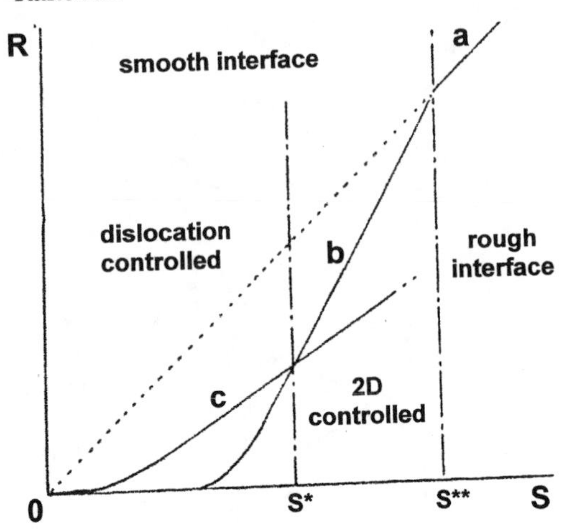

Figure 11. Illustration of the relationship between growth rate, R, and supersaturation, S. Curve a for continuous growth; curve b for a birth and spread growth mechanism; and curve c for a spiral growth mechanism. The vertical lines at S* and S** indicate the transitions from growth dominated by the spiral mechanism and the birth-and-spread mechanism, respectively. [Modified after Sunagawa (1984).]

growth rate curves are plotted for spiral growth (c), 2-D nucleation and spread (b) and continuous growth (a). The values of S* and S** are system dependant and represent the saturation states where the dominant growth mechanism changes from spiral growth to 2-D nucleation and spread, and from 2-D nucleation and spread to continuous growth, respectively. The mechanism by which apatite grows has been extensively studied (discussed later). Most of this work is based on inferences made from the reaction orders from growth kinetic data. Several studies, however, have used direct observation of growth *in situ* or the surface microtopography to determine the mechanism of growth on specific crystal faces.

Solubility, equilibrium, and precursor phases in apatite growth

The hydroxylapatite-solution reactions of growth and dissolution depend on the solubility of apatite and other phases in the $Ca(OH)_2$-H_3PO_4-H_2O system. In this system, hydroxylapatite is the thermodynamically most stable calcium orthophosphate phase at normal Earth-surface conditions and in most biological systems. Figure 12 (from Elliott 1994) shows an example of calculated solubility isotherms of the orthophosphates at 37°C. The exact positions of the isotherms depend on the ionic strength of the solutions and the thermodynamic constants used to calculate them (solubility products and stability constants). Solubility data for apatites vary widely and identification of ideal solubility constants has been difficult. This is particularly true of hydroxylapatites. Table 1 gives a sampling of published solubility products, K_s, for fluorapatite and hydroxylapatite. Chander and Fuerstenau (1984) review solubility measurements of hydroxylapatite made through 1978 and suggest that there is good agreement that the solubility product of stoichiometrically pure, well-crystallized hydroxylapatite is $1 \times 10^{-57.5}$. Apatite can incorporate large amounts of impurities and structural defects that lead to varying degrees of non-stoichiometry, which in part has lead to varying solubility measurements. At low pH values dicalcium phosphates become more stable than hydroxylapatite (Fig. 12). Nancollas (1984) suggests that in hydroxylapatite dissolution experiments conducted at low pHs, apatite may form a surface coating of a more acid phosphate. Determination of solubility will then reflect the surface phase and not apatite. Reaction kinetics can be slow for the calcium orthophosphates and make determination of equilibrium difficult. It has also been shown that kinetic factors may play a more important role in growth and dissolution reactions than has been traditionally accounted for. All of these factors may in part account for the discrepant solubility data for apatites in the literature (Table 1).

Table 1. A sampling of published solubility products, K_s, for fluorapatite, FAP [$Ca_5(PO_4)_3F$] and hydroxylapatite, HAP [$Ca_5(PO_4)_3OH$].

Reference	*Phase*	K_s	*Temp. (°C)*
Jaynes et al. (1999)	HAP	$1 \times 10^{-56.02}$	20-25
Valsami-Jones et al. (1998)	HAP	1×10^{-58}	25
Elliott (1994): *calculated*	HAP	2.57×10^{-63}	25
Chander & Fuerstenau (1984)	HAP	$1 \times 10^{-57.5}$	25
Nancollas (1982)	HAP	4.7×10^{-59}	25
Stumm & Morgan (1981)	HAP	$1 \times 10^{-57.0}$	25
Snoeyink & Jenkins (1980)	HAP	$1 \times 10^{-55.9}$	25
*Fawzi et al. (1978)	HAP	1×10^{-61}	30
*McDowell et al. (1977)	HAP	$1 \times 10^{-88.5}$	25
*Brown et al. (1977)	HAP	1×10^{-57}	?
*Wu et al. (1976)	HAP	$1 \times 10^{-62.5}$	30
*Smith et al. (1976)	HAP	$1 \times 10^{-58.5}$	20
*Avnimelech et al. (1973)	HAP	$1 \times 10^{-58.2}$	25
*Saleeb & deBruyn (1972)	HAP	$1 \times 10^{-57.5}$	?
*Chien (1972)	HAP	$1 \times 10^{-60.5}$	25
*Weir (1971)	HAP	$1 \times 10^{-58.5}$	25
Jaynes et al. (1999)	FAP	$1 \times 10^{-58.13}$	20-25
Valsami-Jones et al. (1998)	FAP	$\sim 1 \times 10^{-70}$	25
Elliott (1994): *calculated*	FAP	$2.50 \times 10^{-68.5}$	25
Chin & Nancollas (1991)	FAP	9×10^{-61}	37
Driessens (1982)	FAP	$1 \times 10^{-60.6}$	25
Amjad et al. (1981)	FAP	$1 \times 10^{-60.15}$	37
Stumm & Morgan (1996)	FAP	$1 \times 10^{-59.0}$	25
Lindsay (1979)	FAP	$1 \times 10^{-58.89}$	

* Taken from tabulated values in Chander & Fuerstenau (1984).

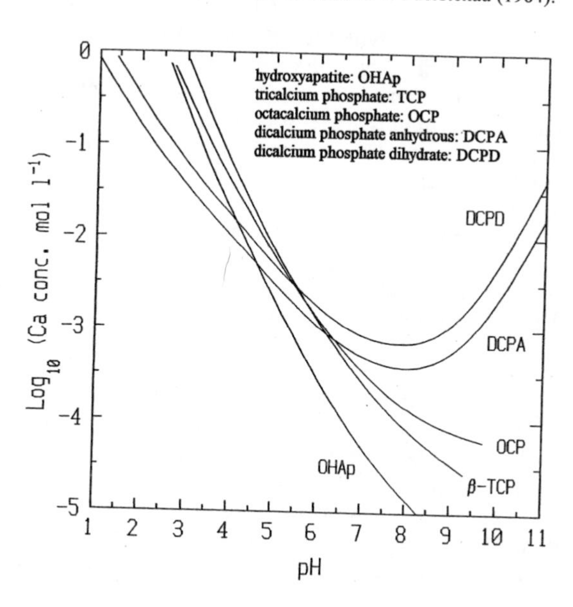

Figure 12. Solubility isotherms of calcium orthophosphate phases in the system $Ca(OH)_2$-H_3PO_4-H_2O at 37°C (equimolar calcium and phosphate). [Modified after Elliott (1994).]

Many calcium phosphate solutions prepared for hydroxylapatite growth experiments and possibly some natural solutions are supersaturated with respect to other calcium orthophosphates such as tricalcium phosphate, TCP [$Ca_3(PO_4)_2$], octacalcium phosphate, OCP [$Ca_4H(PO_4)_3$], dicalcium phosphate anhydrous, DCPA ($CaHPO_4$), and dicalcium phosphate dihydrate, DCPD ($CaHPO_4 \cdot 2H_2O$). Slightly acidic to basic solutions that are supersaturated with TCP, OCP, or DCP will also be supersaturated, to a higher degree, with hydroxylapatite (Fig. 12). The activation barrier to nucleation of a phase is directly related to the free energy of a surface, or more appropriately described as the surface-solution interfacial free energy. The activation barrier to nucleation, and hence the degree of supersaturation necessary to overcome this barrier and form a nucleus, generally increases with increasing surface free energy. Thus it is possible for a thermodynamically less stable phase to nucleate and grow if its surface free energy is significantly less than that of a more stable phase. This is often times seen in the precipitation of calcium phosphates. In many cases hydroxylapatite formation is preceded by the precipitation of another, less stable calcium phosphate, which may ultimately transform to apatite. Numerous studies have noted that a precursor phase to the formation of hydroxylapatite can be an amorphous calcium phosphate phase, ACP (Eanes et al. 1966, Brecevic and Furedi-Milhofer 1972, Termine et al. 1970). Francis and Web (1971) proposed DCPD as a precursor. Eanes et al. (1966) suggested TCP as a precursor and Brown (1966) suggested OCP. Many studies have proposed various mixtures of hydroxylapatite and other precursor phases, especially OCP, to explain the nonstoichiometry of some synthetic and natural bio-hydroxylapatites (Brown 1966, Young and Brown 1982 and references therein). The ribbon-like habit of biological apatites (Fig. 3) has also been used to indicate that OCP was a precursor (Brown 1966). More recent studies that involve hydroxylapatite precipitation experiments have shown that different precursor phases can form depending on solution conditions, particularly depending on saturation state with respect to hydroxylapatite and other calcium orthophosphates.

Boskey and Posner (1973) synthesized ACP and studied the kinetics of its transformation to hydroxylapatite as a function of pH at 26°C and as a function of temperature at a pH of 8. The rate of transformation increases with increasing pH. The mechanism of transformation is dissolution of the ACP and reprecipitation of hydroxylapatite, with the energetically least favorable and rate-limiting step in the process being the nucleation of hydroxylapatite. The activation energy of this conversion was determined as 33 kcal/mol, which includes nucleation and the energy associated with the dissolution of ACP. In seeding experiments there was no measurable induction period, in contrast to the unseeded experiments. This indicates that the activation barrier to ACP dissolution is relatively small and that the 33 kcal/mol is dominated by the barrier to hydroxylapatite nucleation. The rate of conversion increases proportionally with the amount of hydroxylapatite already formed (hence the apatite surface area), suggesting that the process of conversion was autocatalytic. Boskey and Posner (1976) homogenously precipitated hydroxylapatite in solutions with supersaturations (defined in this study as $S = [IAP - K_{sp}/K_{sp}]$, see discussion of growth rates) from 22 to 47, pH = 7.4, temperature = 26.5°C, and ionic strengths 5.0-0.005 M. Hydroxylapatite was found to precipitate without the formation of ACP as a precursor. This is in contrast to high supersaturations ($S = [IAP - K_{sp}/K_{sp}] = 500$) and pH = 6.8 where ACP did form first (Boskey and Posner 1973). Nancollas and Tomazic (1974), working in the same range of solution conditions as Boskey and Posner (1976), found direct hydroxylapatite precipitation in experiments with large seeds, and they found OCP directly precipitated on the seed crystals at higher supersaturations than in Boskey and Posner (1976). They conclude from their work and the work of others that pH, Ca and PO_4 concentrations and ratios, ionic strength, temperature and the presence of seeds are all critical in determining the initial precipitate.

Feenstra and Bruyn (1979) studied the dependence of the induction period between ACP and hydroxylapatite on the degree of S and pH, in high supersaturations, where precursor phases form before hydroxylapatite (26°C, ionic strength of 0.15 mol/L, and pHs 6.7, 7.4, and 8.5). They found the first step was nucleation of ACP followed by OCP as an intermediate phase in the conversion of ACP to hydroxylapatite. They pose the possibility that ACP acts as a substrate for the heterogeneous nucleation of OCP and the OCP acts as a substrate for the heterogeneous nucleation and epitaxial overgrowth of hydroxylapatite. Koutsoukos et al. (1980) studied the crystallization of hydroxylapatite using a constant composition method. In experiments with differing amounts of seed crystals, varying Ca/P ratios, pH from 6-8.5, 37°C, and constant ionic strength in solutions supersaturated with hydroxylapatite only, they found that hydroxylapatite was precipitated on the seed crystals without the formation of a precursor phase. In similar solutions that were also supersaturated with TCP, TCP was never convincingly identified. Amjad et al. (1981) conducted experiments similar to those of Koutsoukos et al. (1980) for fluorapatite and found that no precursor phases were observed in this range of supersaturations.

Christoffersen et al. (1989) investigated the formation of calcium-deficient hydroxylapatite. In their precipitation experiments they found that ACP and in some cases DCPD (at low temperature, 15°C) were the initial precipitates. The original ACP (ACP1) transformed to a second ACP form (ACP2), which in turn transformed to OCP and finally to hydroxylapatite. These transformations are thought to be solution-mediated where one phase dissolved as the following phase precipitated. Transformation rates increased with temperature. Lundager-Madsen and Christensson (1991) studied calcium phosphate precipitation in the pH range of 5-7.5 at 40°C with Ca and PO_4 concentrations up to 0.04M. ACP was the first solid phase to precipitate in most of their experiments. At low pH brushite ($CaHPO_4 \cdot 2H_2O$) is the only precipitate, and at low concentrations OCP dominates. ACP ultimately transforms to brushite or hydroxylapatite.

It is evident from growth experiments and observations of natural calcium phosphates that precursor phases such as ACP, TCP, OCP, or DCP can be involved in the precipitation of hydroxylapatites at low temperature conditions. The presence of these phases, mixed with apatite, has, in part lead to the difficulty in making accurate solubility measurements of apatites.

The formation of precursor phases is in accordance with Oswald's rule of stages for multiple phase precipitation of sparingly soluble salts. In a system like the orthocalcium phosphates, the phase with the highest solubility and lowest surface free energy, which means that it will be at the lowest supersaturation at a given set of conditions, is preferentially formed due to more favorable kinetics. Ultimately, thermodynamic factors will drive the system to formation of a more stable phase. Thus, ACP, OCP, etc. may initially form in supersaturated solutions but these phases will ultimately transform to apatite.

Growth rates

The growth rate for many sparingly soluble salts, $M_{\nu+}A_{\nu-}$, can be expressed as:

$$\frac{d(M_{\nu+}A_{\nu-})}{dt} = -ks[([M]^{\nu+}[A]^{\nu-})^{1/\nu}] - [([M]_{eq}^{\nu+}[A]_{eq}^{\nu-})^{1/\nu}]^{n} \tag{6}$$

where [x] is activity and $[x]_{eq}$ is the activity at equilibrium, k is the precipitation rate constant, s is proportional to the number of available growth sites on a seed or crystal (reactive surface area), $\nu = (\nu+ + \nu-)$, and n is the reaction order (Inskeep and Silvertooth

1988). Thus for fluorapatite

$$5\ Ca_{(aq)} + 3\ (PO_4)_{(aq)} + F_{(aq)} \rightarrow Ca_5(PO_4)_3F_{(s)} \quad (1)$$

$$\frac{d(Ca_5(PO_4)_3F)}{dt} = -ks[([Ca]^5[PO_4]^3[F])^{1/9}] - [([Ca]^5_{eq}[PO_4]^3_{eq}[F]_{eq})^{1/9}]^n \quad (7)$$

From growth kinetic theory, distinction between different growth mechanisms and the rate-limiting step during growth is often inferred on the basis of reaction order. If the rate-limiting step in the growth process is diffusion of growth units through the solution, $n = 1$. Rates controlled by spiral growth will have $1 < n \leq 2$, and for surface nucleation and spread mechanisms n is assumed to be greater than 3.

Determination of growth mechanism from rate data. Most determinations of the growth mechanisms of apatites precipitated in solution experiments are based on fitting kinetic data (usually changes in solution chemistry with time) to empirical rate equations in pure systems. From crystal growth theory the order of these rate equations depend on the growth mechanism and hence can be used to infer the dominant mechanism active in the experiment. Most apatite growth rate experiments have involved precipitation of hydroxylapatite. Nancollas (1979) used the semi-empirical equation $R = d[Mv + Av-]/dt = -ks[(IAP)^{1/v} - Ksp^{1/v}]^n$ and found for hydroxylapatite grown by the constant composition method at pH 5.0-6.5 and supersaturations, S, between 2 and 7, that the overall growth rate was proportional to (S-1). From this relationship they suggested that growth occurred by the spiral mechanism. Koutsoukos et al. (1980) fit their rate data with the same rate equation and determined the reaction order n = 1.25 for low hydroxylapatite supersaturations; again, indicating growth by the spiral mechanism. They also found that chemistry of the precipitate was independent of the Ca/P ratio of the solution. Moreno (1981) studied the dependence of the kinetics of hydroxylapatite growth on the concentration of seed crystals and the degree of supersaturation, S. Seeded growth experiments were conducted in solutions at 37°C, pH = 7.4, S_{HA} = 15.9, 11.9 and 5.32, and Ca:PO_4 of 5:3 and 3:5 at varying seed concentrations. They found that the rates of precipitation were highly dependent on the amount of seeds added, hence total surface area. The ratio of Ca:PO_4 at the same S_{HA} did not affect the rate of hydroxylapatite formation but may have influenced the precursor formation of OCP. Diffusion-limiting and 2-D nucleation models did not fit their data, although a spiral growth model did. Inskeep and Silvertooth (1988) studied seeded hydroxylapatite precipitation at S = 7.7-21.4, pH 7.4-8.4 (geochemically relevant in earth surface conditions), and using the model of Nielsen and Toft (1984) they predicted growth by the spiral mechanism. A rate equation was developed on the basis of the reaction orders determined with respect to solution species: $R = k_f s \gamma_2 \gamma_3 [Ca^{2+}][PO_4^{3-}]$, where k_f= rate constant for the forward reaction, s = surface area, and γ_2 and γ_3 are the divalent and trivalent species, Ca^{2+} and PO_4^{3-}, with ion activity coefficients calculated from the Davies equation (Davies 1962). This equation is first order in $[Ca^{2+}]$, $[PO_4^{3-}]$ and s. The average rate constant k_f was determined to be 173 ± 11 L^2 mol^{-1} m^{-2} sec^{-1}. The apparent Arrhenius activation energy was determined over a temperature range of 10 to 40°C to be 186 ± 15 kJ mol^{-1}. Inskeep and Silvertooth (1988) speculated that the rate-limiting step is a surface process, specifically that it is the surface diffusion of Ca and PO_4 ions and dehydration commensurate with binding at surface sites of incorporation. Using a constant-composition method and seeded solutions with S_{HA} = 3.05-5.58 Koutsopoulos (2001) found hydroxylapatite formed without a precursor phase. The spiral growth model was found to best fit their kinetic data.

In contrast to the above studies where the growth mechanism was inferred to be spiral growth, Arends et al. (1987), using the constant composition technique of Hohl et al. (1982) studied the rate of crystal growth of hydroxylapatite at 37 °C, pHs of 6.0 and 7.0 and supersaturations S = 2 - 6. Reaction orders n = 3.2 and 2.9 for experiments at pH = 7 and 6 respectively were determined. It was concluded that hydroxylapatite grew by the polynuclear mechanism in these experiments. The rate of growth differed by a factor of two between these two pH values. Christoffersen and Christoffersen (1992a) developed a revised version of the model of (Hilig 1966) for the growth of crystals by a nucleation and spread mechanism. In their model growth rates are expressed in activities of solutes rather than concentrations, and the relationship of surface free energy and solubility is built into the rate expression. The model was then applied to the rate of growth of hydroxylapatite (Christoffersen and Christoffersen 1992b). Rate data from their experiments at supersaturations between 3.25 and 7.53 are consistent with growth controlled by a surface process. Two models, the spiral growth model and the polynuclear model, were found to be consistent with the results. Using their revised growth model, the frequency of ion integration was found to be 9×10^4 s^{-1} for the polynuclear model, close to the 1.6×10^5 s^{-1} value expected if the rate-determining step is partial dehydration of calcium ions. If similar assumptions to those used for the application of the polynuclear theory are applied to the spiral theory, similar values of the ion integration frequency are obtained, making it difficult to distinguish the rate-controlling mechanism. Christoffersen et al. (1998a) used the revised mechanistic model of polynuclear growth, developed by Christoffersen et al. (1996b), to fit the growth of hydroxylapatite. In this study they make the assumption that the incorporation of OH ions strongly influences the growth of hydroxylapatite. They propose that H_2O is initially incorporated into the column anion position and that OH formation takes place by deprotonation after burial. It is suggested that this will then slow the further incorporation of Ca ions at kink sites on the surface. Christoffersen et al. (1998b) examined the role of varying Ca:PO_4 ratios (from 0.1-20) in solution on the kinetics of hydroxylapatite growth at pHs of 6 and 7.2, and supersaturations S = 3.03-6.86. Again, they modeled their kinetic results with a polynuclear growth mechanism in which nucleation is expressed as a function of mean-ion activity and OH formation by H_2O dissociation in the growth layer is taken into account.

Although the number of studies of fluorapatite growth kinetics is smaller than for hydroxylapatite, they have particular relevance to sedimentary apatite formation. Amjad et al. (1981) repeated the experiments of Koutsoukos et al. (1980) (pH = 7.40, T = 37°C, S = 10-30) for fluorapatite and similarly found that the rate equation held for FAP with an apparent reaction order of 1.25 indicating a spiral growth mechanism. Van Cappellen and Berner (1991) conducted seeded fluorapatite growth rate experiments in a carbonate-free simulated seawater solution at different degrees of saturation, S. They fit their growth rate data with an empirical rate law of the form

$$r = K_a(S-1)^n \tag{8}$$

where r is the growth rate (in units of mass of fluorapatite per unit time and unit seed surface area), and K_a is an apparent rate constant. It was found that for supersaturations $S < 37$ the reaction order $n = 1.8 \pm 0.3$, and for $S > 37$ $n = 4.9 \pm 0.4$ (Fig. 13). The slope of the line fit to plots of log (J/m_O) vs. log (1-S) is equal to the reaction order, n, for growth rates that are normalized to surface area. They interpreted this to indicate a transition from growth by the spiral growth mechanism to growth dominated by a polynuclear or two-dimensional nucleation and spread mechanism; thus the saturation value of 37 = S*, where the dominant growth mechanism changes from spiral to two-dimensional nucleation and spread (Fig. 11). Christoffersen et al. (1996a) conducted seeded growth and dissolution experiments with fluorapatite microcrystals in solutions with Ca:P:F = 10:6:2, and 5.0 < pH < 6.5. In contrast to Van Cappellen and Berner (1991) they found that both growth and dissolution appear to be controlled by a polynuclear surface mechanism in the supersaturation range studied: 2 < S < 7 for growth and 0.1 < S < 0.7 for dissolution. They note that the highest supersaturation in Amjad et al. (1981) was in the lower range of those used in Van Cappellen and Berner (1991), yet the rate of growth determined in Van Cappellen and Berner (1991) was approximately 300 times slower than that measured in Amjad et al. (1981). Figure 14 is a plot of the kinetic data from Amjad et al. (1981), Van Cappellen and Berner (1991) and Christoffersen et al. (1996a). The results of Amjad et al. (1981) fit well with those of Christoffersen et al. (1996a).

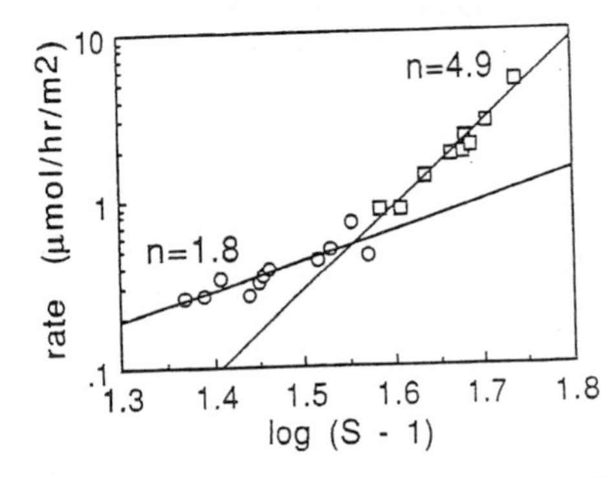

Figure 13. Relationship between growth rate and supersaturation for fluorapatite plotted on a log-log scale. The change in the empirical order of the growth rate, n , indicates a change in the dominant growth mechanism from spiral to birth-and-spread. [Modified after Van Cappellen and Berner (1991).]

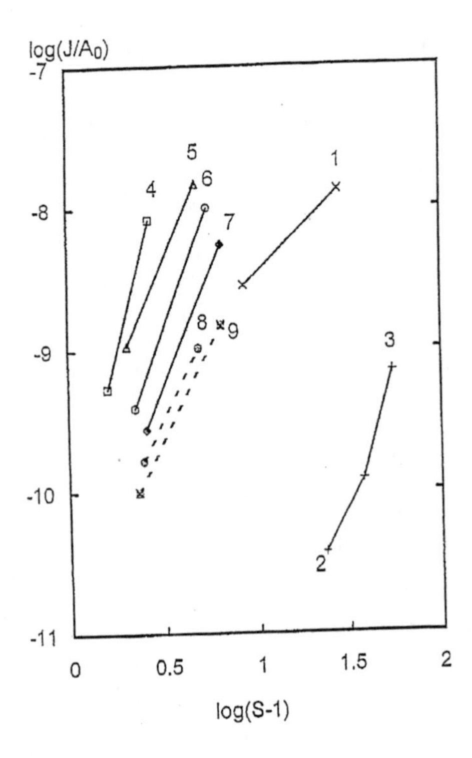

Figure 14. Relationship between growth rate and supersaturation, plotted as the log of the quotient of the overall growth rate, J, and the initial surface area of the seed crystals, A_O vs. the log of the supersaturation S – 1, for apatite from different studies:

Line No. *(Reference)*

1 (Amajad et al. 1981);

2-3 (Van Cappellen & Berner 1991);

4-7 (Christoffersen et al. 1996);

8-9 (Christoffersen & Christoffersen 1992b).

Lines 1-7 are for fluorapatite and 8-9 for hydroxylapatite. [Used by permission of Elsevier Science Ltd., from Christoffersen et al. (1996) *Journal of Crystal Growth*, Vol. 163, Fig. 5, p. 301.]

The influence of impurities on growth kinetics. The presence of impurities in solution, including foreign ions, small molecules and polymers, can have a marked effect on the kinetics of apatite crystal growth (Wu and Nancollas 1998). Chin and Nancollas (1991) investigated the effect of Zn and Mg on the dissolution kinetics of fluorapatite. Dissolution was retarded in the presence of both Zn and Mg ions at pH 5.0; however, Zn was much more effective in inhibiting the dissolution. The effective adsorption affinity, calculated from the decrease in rates of dissolution, was 2.5×10^5 L/mol. Vancappellen and Berner (1991) investigated the affect of Mg^{2+} on fluorapatite growth over the dissolved composition range of 0-60 mM. They found that Mg^{2+} inhibited growth, and in the Mg^{2+} concentration range typically found in marine pore waters (40-60 nM) the rate of growth was slowed by a factor of 15 to 20 compared to solutions free of Mg^{2+}. The inhibition of fluorapatite growth was explained by blocking of active growth sites by adsorbed Mg^{2+}. Kanzaki et al. (2000) examined the role of Mg and Zn on the growth kinetics of hydroxylapatite. Both elements inhibited the lateral growth rate of 2-D nuclei on the {001} surface, with Zn affecting the growth rate approximatly 1000 times more than Mg. Supersaturation of hydroxylapatite was 22 in these solutions and Mg and Zn concentrations ranged from 0-1.5 mM and 0-7.5 μM respectively. It was concluded from AFM measurements and kinetic data fitted to adsorption models that the inhibitory affect of both Mg and Zn results from blocking of kink sites along steps of growth islands. Kanzaki et al. (2001) examined the affect of Mg and Zn sorption on the edge free energy and the rate of formation of 2-D nuclei. They found that it was negligible, although these impurities did block the lateral growth of 2-D nuclei. Brown (1981) found that the presence of both HCO_3 and Mg significantly decreased the rate of growth of hydroxylapatite. In these experiments the P-removal rate (used to indicate apatite growth rate) was approximately seven times faster at 1 mM HCO_3 than at 6 mM. In contrast, a 1 mM increase in Mg concentration in solution slowed P-removal by more than two orders of magnitude. Strontium was also found to inhibit the rates of both dissolution and growth of hydroxylapatite (Christoffersen et al. 1997). Two main mechanisms for the inhibition of apatite growth by impurity ions, exemplified in the previous studies, have been proposed: (1) impediment of the spreading of growth steps, and (2) poisoning of active growth sites or protosites. For cations a general trend of increasing inhibition with increasing charge has been observed (Mullin 1993). Wu and Nancollas (1998) derive expressions describing the influence of impurities on crystal growth kinetics based on their affect on surface free energy. Their model predicts increasing surface free energy with increasing charge of adsorbed foreign cations, with a commensurate increase in the growth inhibition with charge. It is interesting to note that a given impurity ion may have distinctly different effects on different phases. For example, while Mg strongly diminishes the growth rate of apatite it has a much smaller effect on the growth of OCP and DCPD (Nancollas and Zhang 1994). In general the kinetics of thermodynamically more stable phases are more sensitive to impurities in solution.

The influence of molecular species and extended polymers on apatite growth kinetics has also been investigated. Zieba et al. (1996) studied the kinetics of crystal growth of hydroxylapatite in the presence of seven different phosphonate additives. Their results indicate that traces of some phosphonates (less than or equal to 10^{-6} mol L^{-1}) are extremely effective in inhibiting crystal growth. Amjad and Reddy (1998) investigated the influence of seven humic compounds on the growth of hydroxylapatite. They found that fluvic acid, tannic acid, benzene hexacarboxylic acid, and poly(acrylic acid) had an inhibitory effect on growth, but salicyclic acid did not inhibit growth under the same experimental conditions (concentrations in solution between 0.25 and 5 ppm).

One of the major problems with using bulk solution data to determine growth mechanism is that the presence of impurities in solution, even at concentrations as low as

1 ppm, can significantly affect growth kinetics and have a profound affect on the processes involved in crystallization (Davey 1976, Wu and Nancollas 1998). Thus, interpretations based on growth rate theory of ideal systems can be in error. More accurate determination of growth mechanism is made by direct observations, either *in situ* or by inference based on certain types of surface microtopographic features.

DIRECT OBSERVATIONS OF GROWTH AT SURFACES

A second drawback with bulk solution studies of growth kinetics is that they yield an average of the kinetics of all the crystal faces. This leads to uncertainties in the estimation of surface properties determined from growth rate data, e.g., surface free energy. Indeed, on the basis of theory as well as limited direct measurements, we would expect that surface properties of symmetrically and structurally nonequivalent faces on the same crystal would be different. Because of symmetry constraints, apatite crystals will always be bound by at least two symmetrically nonequivalent forms, and as discussed previously, they may have many forms on a single crystal. Growth rate and mechanism may vary among different faces on the same crystal. To determine the mechanism of growth on any one face direct observations of growth or post-growth surface microtopography are needed.

Several studies have observed the growth mechanism *in situ* on specific crystal faces using atomic force microscopy (AFM). Using *in situ* AFM observations of growth on the {100} faces of synthetic hydroxylapatite, Onuma et al. (1995a) reported growth by step flow combined with two-dimensional nucleation. The rate-limiting step in growth was interpreted to be—based on step velocities—incorporation of growth units at the step fronts. They estimated the normal growth rate of the {100} face in their experiments to be on the order of 10^{-4} nm/s, on the basis of step advancement rates. In a similar study Onuma et al. (1995b) observed a hillock on {100} that advanced by a layer-spreading mechanism. The hillocks were anhedral and it was undetermined if they originated from a dislocation or 2-D nucleation. Using phase-shift interferometry and atomic force microscopy, Kanzaki et al. (1998) studied growth on the {001} face of a synthetic hydroxylapatite crystal in a simulated body fluid. Under the conditions of their experiments they found that growth took place by the formation of rounded islands of precipitate on the apatite surface, indicating a two-dimensional nucleation growth mode. They also found that the growth rate gradually decreased and eventually stopped, even though the solution chemistry was kept constant. The original growth rate was 1 to 2 orders of magnitude faster than that of on {100} determined by Onuma et al. (1995a). Onuma et al. (1998) investigated the growth rate dependence on supersaturation ($0.85 < S < 22.0$) by the same methods, and they estimate the edge free energy of a step on the {100} face to be 3.3 kT (k = Boltzmann constant). They also suggest that the critical supersaturation for growth to be 0.6. It is uncertain, however, if this is for spiral growth or two-dimensional nucleated growth.

Other studies have used *ex situ* observations of the surface microtopography of apatite to infer the growth mechanism in geologic samples. Growth spirals (Fig. 5) and hillocks (Figs. 6 and 7) on the {100} and {001} faces indicate that growth was by the spiral mechanism (Amelinckx 1952a,b; Rakovan and Reeder 1994, 1996). It is unclear, however, if polygonized growth hillocks can form by mechanisms other than spiral growth (Rakovan and Jaszczak 2002). For further discussion of these features, see the previous section on *Surface microtopography*.

IMPURITY INCORPORATION DURING GROWTH

The structure and composition of apatite lends itself to many compositional substitutions; indeed, apatite incorporates over half the periodic table in its atomic arrangement (Hughes and Rakovan, Ch. 1, Pan and Fleet, Ch. 2 —both in this volume). These substitutions are usually in trace concentrations; however, complete solid solutions exist for certain substituents. This complex and variable chemistry has immense implications, and is utilized in all areas of apatite research, exemplified in many chapters in this volume.

Apatite can strongly influence the trace element evolution of magmas (e.g., Belousova et al. 2002). Many studies have shown apatite to be one of the most important minerals affecting rare earth element (REE) trends in igneous rocks (e.g., Nash 1972, Bergman 1979, Watson and Capobianco 1981, Kovalenko et al. 1982, Gromet and Silver 1983, Watson and Harrison 1984a,b), and volatile evolution in magmas (O'Reilly and Griffin 2000, Boudreau et al. 1986, Boudreau and McCallum 1989, Picolli and Candella, this volume). Thus, apatite chemistry can play a critical role in understanding and modeling of igneous petrogenetic processes (e.g., Papike et al. 1984, Sha and Chappell 1999, Spear and Pyle, Ch. 7, this volume; Picolli and Candella, Ch. 6, this volume). The formation of apatite in low-temperature sedimentary environments and from hydrothermal solutions can also strongly influence their trace element signature and evolution (e.g., Graf 1977, Bergman 1979, Watson and Capobianco 1981, McLennan 1989, Knudsen and Gunter, Ch. 9, this volume). Heavy metal sequestering by hydroxylapatite is important in our understanding of the fate and transport of toxic species in the body. The presence of certain elements in trace concentrations is also known to inhibit as well as promote apatite growth in the body (e.g., McLean and Bundy 1964, Posner 1987, Hahn 1989). Because of its affinity for many environmentally important elemental species, apatite is being used as an engineered contaminant barrier (i.e., Jeanjean et al. 1995, Chen et al. 1997, Arey et al. 1999, Seaman et al. 2001) for heavy metals and being investigated as a solid waste form for many radionuclides (Ewing 2001; Ewing, this volume). The presence of Mn and REEs in apatite gives it luminescence properties that are of great utility in both the phosphor and laser industries (e.g., McConnell 1973; Waychunas, this volume). Hence, knowledge of the variables that govern the partitioning of trace elements into apatite is of great importance.

Many complex variables affect the partitioning of trace elements into minerals during their formation. Trace element charge and size (Goldschmidt 1937), electronegativity (Ringwood 1955), crystal field effects (Burns 1970), and crystal growth rate (Burton et al. 1953), as well as state variables such as temperature, pressure, and composition, have all been identified and studied as factors affecting trace element partitioning. More recently, the role of surface structure on the partitioning of elements into crystals has been demonstrated (Nakamura 1973, Dowty 1976, Reeder and Grams 1987, Paquette and Reeder 1990, Rakovan and Reeder 1994, Reeder and Rakovan 1999).

Surface structural controls on trace element incorporation during growth

Compositional zoning. Studies of the incorporation of trace elements into apatite crystals that precipitated from hydrothermal solutions exemplify the role of crystal surface structure on heterogeneous reactivity. Rakovan and Reeder (1994) present cathodoluminescence (CL) and synchrotron X-ray fluorescence (SXRFMA) data that show heterogeneous incorporation of Y, Sr, Mn and REE into coeval regions of symmetrically different crystal faces (sectoral zoning). Differences in CL and photoluminescence (Waychunas, this volume) between different crystal faces can reveal sectoral zoning when luminescence-activating ions are differentially incorporated. Figure 15 shows the photoluminescence of an apatite crystal from Llallagua, Bolivia. The difference in luminescence color between {100}, orange, and {001}, lavender, indicates that the luminescence activators are segregated between these sets of crystal faces and

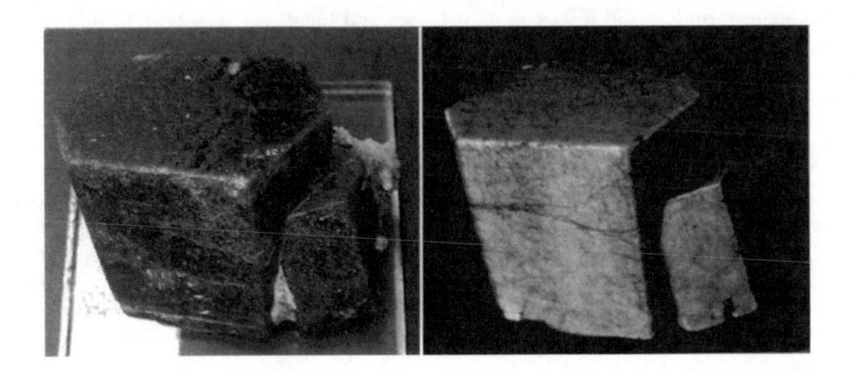

Figure 15. (*left*) Fluorapatite from the Siglo XX Mine, Llallagua, Bolivia. The larger crystal measures 3 × 3 × 2 cm. (*right*) Same, in long-wave ultraviolet light. Photoluminescence is activated by REEs. Differential luminescence between {001} faces, violet, and {100} faces, orange, indicates sectoral zoning REEs. See **Color Plate 1**, opposite page 718.

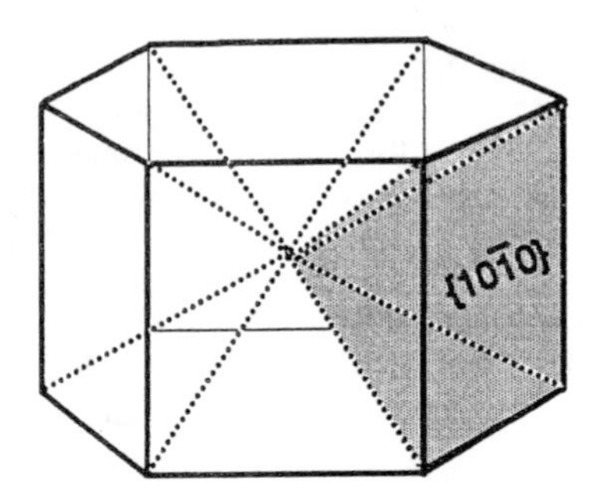

Figure 16. Diagram of a hexagonal apatite crystal showing the distribution of {100} and {001} sectors. One of the {100} prism sectors is shaded. [Modified after Rakovan and Reeder (1996).]

their associated sectors (Fig. 16) in the bulk crystal. Quantitative measurements, by isotope dilution mass spectrometry, of REE concentrations in coeval portions of {100} and {001} sectors of a Llallagua apatite (Fig. 17) show sectoral differences in concentration from roughly 1.4 for Yb to almost an order of magnitude (9.5) for La (Rakovan et al. 1997). One of the striking results of sectoral zoning of the REE, seen in Figure 17, is the significantly different REE patterns from different regions of the same crystal that grew simultaneously from the same fluid. Sectoral zoning of REE in the Llallagua apatite was used by Rakovan et al. (1997) for the development of an intracrystalline isochron dating technique (Fig. 18).

The {100} and {001} faces of the sectorally zoned apatites from Llallagua and numerous other locations exhibit polygonized growth hillocks (Figs. 6 and 7) that indicate a layer growth mechanism; most likely by spiral growth. In spiral growth, incorporation of constituent or substituent atoms takes place dominantly at kink sites along growth steps. Because of the symmetrical nonequivalence of {100} and {001}, the atomic structure of steps on those forms is different. Rakovan and Reeder (1994) suggested that this difference in the structure of surface sites of incorporation leads to a differential affinity for different trace elements and differential incorporation thereof, leading to sectoral zoning (Rakovan and Waychunas 1996, Rakovan et al. 1997).

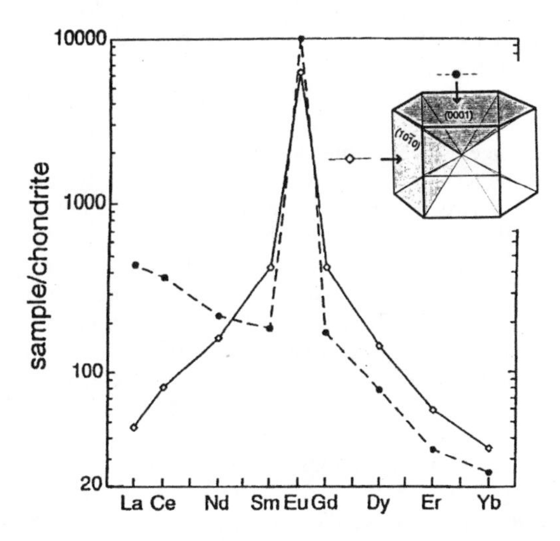

Figure 17. Chondrite-normalized REE patterns for coeval portions of {100} and {001} sectors of a single fluorapatite crystal from the Siglo XX Mine, Llallagua, Bolivia. Sectoral difference in the concentration of La is almost an order of magnitude. [Modified after Rakovan et al. (1997).]

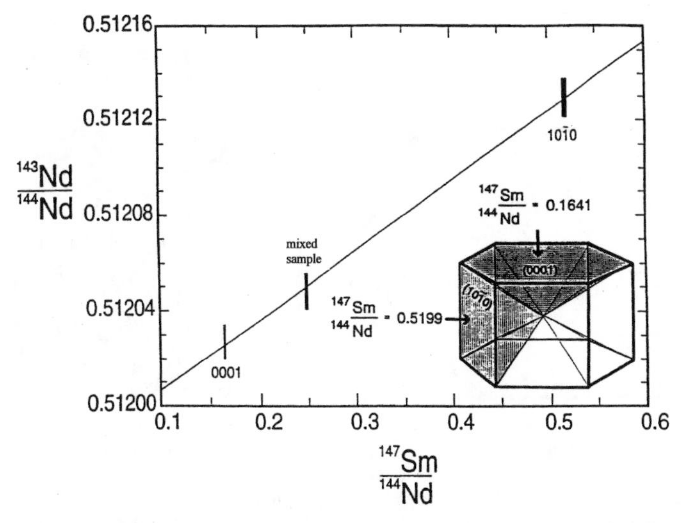

Figure 18. Sm-Nd isochron plot from coeval samples from different sectors of the sectorally zoned apatite in Figure 17. The isochron yields an age of crystallization of 43.8 ± 4.7 Ma. [Modified after Rakovan et al. 1997.]

Like sectoral zoning, intrasectoral zoning of trace elements in apatite also results from differential incorporation among structurally nonequivalent growth steps. In this case, however, the different steps reside on the same crystal face rather than different faces, and make up symmetrically nonequivalent vicinal faces of growth hillocks on the crystal face. Intrasectoral zoning, compositional differences between coeval portions of symmetrically different subsectors [regions of the bulk crystal that grew by incorporation into a specific vicinal face of a growth hillock (Paquette and Reeder 1990)] was identified by Rakovan and Reeder (1994, 1996). Figure 19 shows the ideal subsector distribution for trigonal growth hillocks on {100} apatite faces. Intrasectoral differences in the distribution of luminescence-activating trace elements are clearly seen in cathodoluminescence (Figs. 20 and 21). Quantitative determination of REE concentration differences between the symmetrically nonequivalent subsectors by SXRFMA indicates that all of the REE, except for Eu, are intrasectorally zoned (e.g., Fig. 22). The partitioning behavior between subsectors associated with the nonequivalent growth steps was found to be correlated with the size of the REE ion relative to Ca^{2+}, the ion for which it substitutes in the apatite structure (Pan and Fleet, this volume); this indicates that substituent size was one of the main factors causing differential incorporation between nonequivalent surface sites. REEs with an ionic radius larger than Ca^{2+} (e.g., La^{3+}) are enriched in the [001] subsector relative to the <011> subsectors. All of the REEs analyzed that are smaller than Ca^{2+}, except Eu, are depleted in the [001] subsector relative to the <011>. A similar size affect is observed in sectoral zoning of REE between the {100} and {001} sectors in Llallagua apatites (Rakovan et al. 1997). The anomalous incorporation behavior of Eu can be explained by the fact that Eu is heterovalent in the Llallagua apatites (Fig. 23) (Rakovan et al. 2001a,b). In apatite, the ionic radius of Eu^{2+} is larger than Ca^{2+}, whereas Eu^{3+} is smaller. If Eu^{2+} follows the same partitioning behavior as the trivalent REEs that are larger than Ca, it will be enriched in the [001] subsector; the opposite is true of Eu^{3+}. Thus, roughly equal and opposite partitioning of the two ions between the nonequivalent sectors would yield the observed lack of intrasectoral zoning of total Eu.

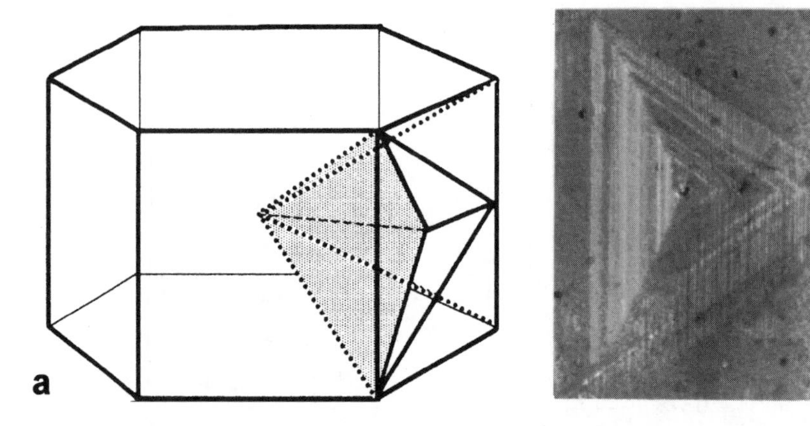

Figure 19. (a) Schematic of an apatite crystal with a three-sided growth hillock on a {100} face. The dotted lines indicate the sector associated with the prism face and the subsector of the basal vicinal face of the hillock is shaded. (b) A DIC photomicrograph of an actual hillock on the {100} face of a fluorapatite from the Golconda Mine, Minas Gerais, Brazil. Image is ~666 μm vertical. [Modified after Rakovan and Reeder (1996).]

The {001} faces of the Llallagua apatites studied by Rakovan and Reeder (1994) also exhibit growth hillocks, and associated subsectors display intrasectoral zoning. Six pyramidal vicinal faces and a terminating surface parallel to {001} form the hexagonal growth hillocks on the {001} faces. The face symmetry of {001} is a six-fold axis of rotation that symmetrically equates the <100> steps of all six pyramidal vicinal faces. CL of the six pyramidal vicinal faces of hillocks on the {001} of Llallagua apatite crystals is homogenous, as expected on the basis of their symmetry equivalence. Differential incorporation, however, does take place between the pyramidal vicinal faces and the {001} terminations and regions of the {001} surface where no hillocks are present (Fig. 24, and cover). The exact mechanism of this intrasectoral zoning is uncertain (Rakovan and Reeder 1994).

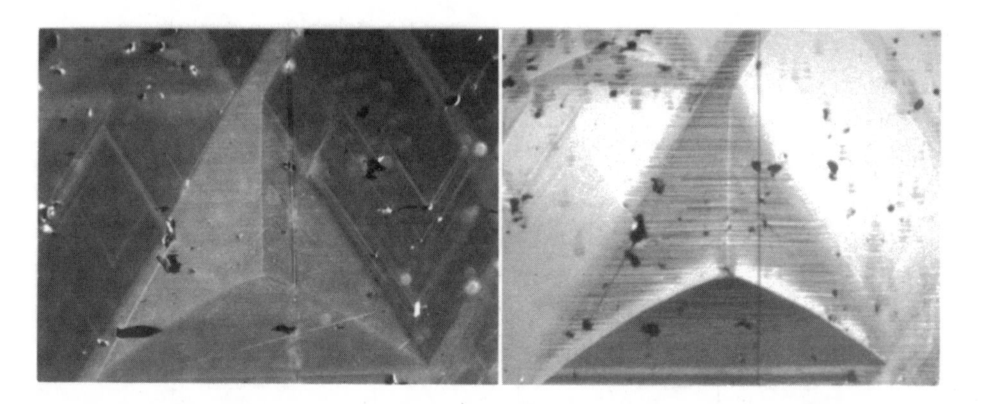

Figure 20. (*left*) Cathodoluminescence (REE activated) photomicrograph of the growth hillock in right image. Luminescence is homogenous between the two symmetrically equivalent vicinal faces (color is orange). Differential luminescence exists between these and the symmetrically nonequivalent vicinal face, whose luminescence is blue. See **Color Plate 2**, opposite page 718. (*right*) DIC photomicrograph of a trigonal growth hillock on the {100} face of a fluorapatite from the Golconda Mine, Minas Gerais, Brazil. Image is ~1 mm across. [Used by permission of the Mineralogical Society of America, from Rakovan and Reeder (1994) *American Mineralogist*, Vol. 79, Fig. 6, p. 897.]

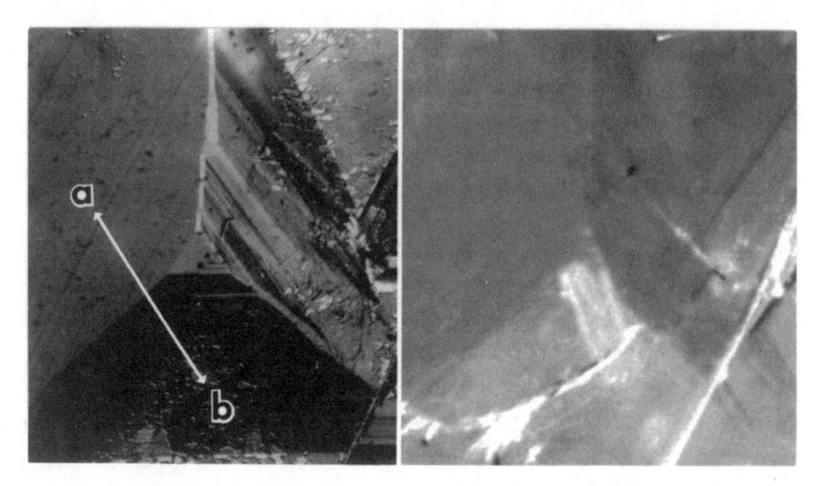

Figure 21. (*left*) DIC photomicrograph of a trigonal growth hillock on the {100} face of a fluorapatite from the Siglo XX Mine, Llallagua, Bolivia. Steps of [$01\bar{1}$] orientation comprise vicinal face **a**, and steps of [001] orientation comprise vicinal face **b**. Image is ~2.4 mm across. (*right*) Cathodoluminescence photomicrograph of the hillock in left image. Luminescence is homogenous between the two symmetrically equivalent vicinal faces (yellow luminescence activated by Mn^{2+}). Differential luminescence exists between these and the symmetrically nonequivalent vicinal face, whose luminescence is blue (activated by REE). See **Color Plate 3**, opposite page 718. [Modified after Rakovan and Reeder (1996).]

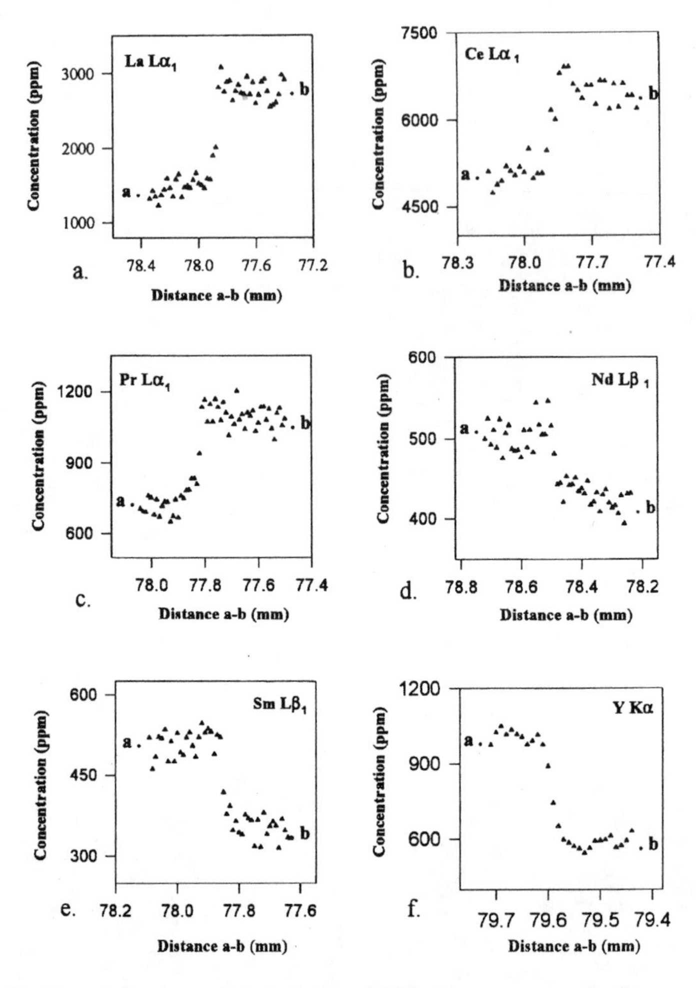

Figure 22. Plots of the concentration of selected REEs between symmetrically nonequivalent vicinal faces, and their associated subsectors, along the line *a-b* shown in Figure 21a. [Used by permission of Elsevier Science Ltd., from Rakovan and Reeder (1996) *Geochimica et Cosmochimica Acta*, Vol. 60, Fig. 3, p. 4439.]

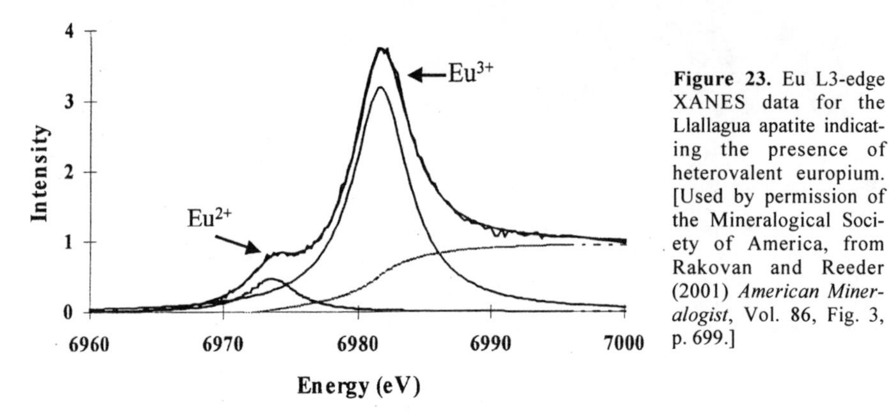

Figure 23. Eu L3-edge XANES data for the Llallagua apatite indicating the presence of heterovalent europium. [Used by permission of the Mineralogical Society of America, from Rakovan and Reeder (2001) *American Mineralogist*, Vol. 86, Fig. 3, p. 699.]

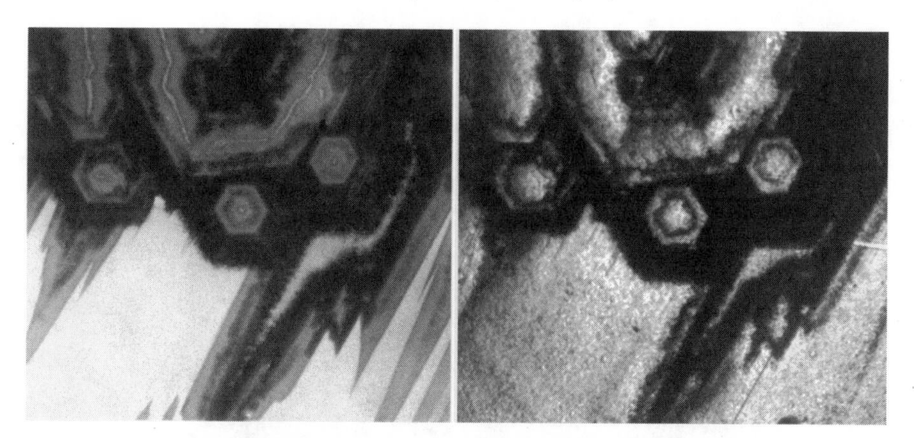

Figure 24. (*left*) Cathodoluminescence photomicrograph of hexagonal growth hillocks on the {001} face of an apatite from the Siglo XX Mine, Llallagua, Bolivia. Luminescence (purple, activated by REEs) is homogenous among the six symmetrically equivalent pyramidal vicinal faces. Yellow luminescence (Mn^{2+} activated) dominates flat regions of the {001} face and the terminal faces of the hillocks. See cover or **Color Plate 4**, opposite page 718. (*right*) DIC photomicrograph showing the microtopography of the {001} apatite face in left image. Image is ~1.2 mm across. [Used by permission of the Mineralogical Society of America, from Rakovan and Reeder (1994) *American Mineralogist*, Vol. 79, Fig. 7, p. 897.]

Superposition of sectoral, intrasectoral, and concentric zoning can lead to complex trace element distribution patterns within single crystals that may be difficult to interpret in randomly-oriented crystal sections. Several authors (i.e., Marshall 1988, Jolliff et al. 1989) have noted a clustering or patchy distribution of REE within individual crystals of apatite. Several explanations have been given, but it is quite possible that such anomalous distributions are due to intrasectoral or sectoral zoning. Kempe and Gotze (2002) identified concentric and sectoral zoning in oriented sections of apatite crystals (Figs. 25 and 26). The polygonized domains of different luminescence intensity within single sectors in Figure 25 may be subsectors indicating intrasectoral zoning as well.

Figure 25. Cathodoluminescence photomicrograph of a section cut through an apatite from Ehrenfriedersdorf, Germany, exhibiting concentric, sectoral and intrasectoral zoning. Yellow luminescence activated by Mn^{2+}. [Used by permission of the Mineralogical Society, from Kempe and Götze (2002) *Mineralogical Magazine*, Vol. 66, Fig. 2e, p. 156.]

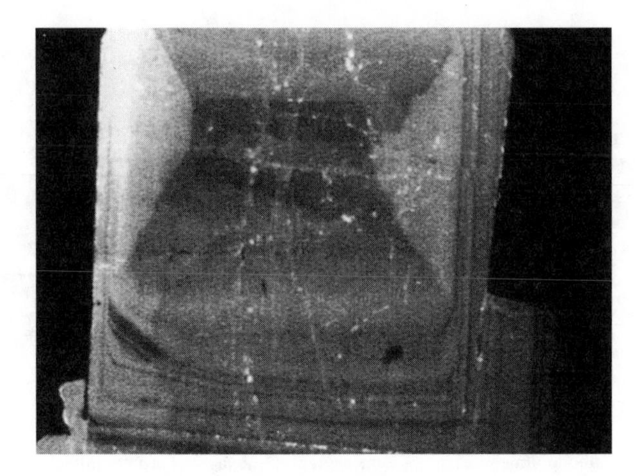

Figure 26. Cathodoluminescence photomicrograph of a section cut through an apatite from Ehrenfriedersdorf, Germany exhibiting oscillatory and sectoral zoning. Yellow luminescence activated by Mn^{2+}. [Used by permission of the Mineralogical Society, from Kempe and Götze (2002) *Mineralogical Magazine*, Vol. 66, Fig. 2f, p. 156].

Surface-induced dissymmetrization. Akizuki et al. (1994) and Rakovan and Reeder (1994) reported optical anomalies in apatite crystals from the Asio Mine, Japan, and Llallagua, Bolivia, respectively. In both cases the apatite crystals exhibit concentric, sectoral and intrasectoral domains with different optical orientation and character. The correlation of these domains with surface features, such as specific crystal faces, indicates that the optical anomalies were created during growth of the crystals. There are several possible causes for the optically anomalous behavior. Certain anomalies may be the result of inhomogeneous mismatch strain from impurities. It has also been shown that the type of surface structural control on element incorporation that leads to compositional sectoral and intrasectoral zoning can also cause ordering and a lowering of crystal symmetry (dissymmetrization) during growth (Bulka et al. 1980, Akizuki 1987, Akizuki and Sunagawa 1978, Shtukenberg et al. 2001).

In the apatites from Japan, Akizuki et al. (1994) found biaxial optical character with domains of different optical orientation and $2V$ (Fig. 27). Although the optical properties indicate a monoclinic or triclinic structure, electron and X-ray diffraction data showed no reflections inconsistent with $P6_3/m$ symmetry or that indicate a superstructure. They postulated that the dissymmetrization is due to ordering of OH and F (Hughes and Rakovan, this volume) along growth steps during incorporation.

In the Llallagua apatites, {001} sectors with two different optical characters have been found from different parts of the mine. The {001} sectors are typically uniaxial negative, as would be expected for $P6_3/m$ apatite. However, in some crystals (i.e., those in Figs 7 and 15) the {001} sector is optically biaxial and sub-sectoral domains associated with the pyramidal vicinal faces of hexagonal hillocks on the {001} faces show different optical orientation (Fig. 28). In all of the Llallagua samples studied, {101}, {111} and {100} sectors show biaxial character with distinct concentric and sectoral zoning of extinction orientation and $2V$. Coeval portions of a given sector, as indicated by a single continuous concentric zone, show subsector differences in $2V$ from 2° to as much as 15° (Fig. 29). Rakovan and Reeder (1994) also postulated that the optical anomalies in these crystals were due to ordering during growth, potentially of F and OH ions. Less complex, but distinctly different, are optically anomalous apatites reported by Richards and Rakovan (2000). These crystals show single biaxial domains that correlate with the six {100} sectors of tabular crystals (Fig. 30).

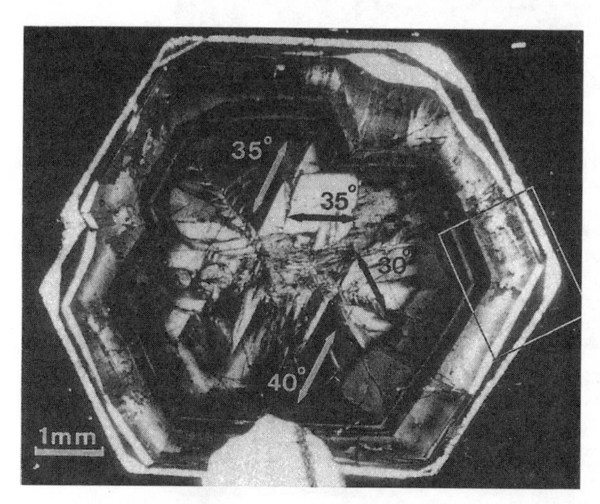

Figure 27. Optical photomicrograph in crossed polarized transmitted light of a thin section (parallel to {001}) through an apatite from the Asio Mine, Shimotsuke Province, Japan. Optically distinct regions correspond to different sectors and subsectors throughout the crystal. [Used by permission of the Mineralogical Society, from Akizuki et al. (1994) *Mineralogical Magazine*, Vol. 58, Fig. 4, p. 311.]

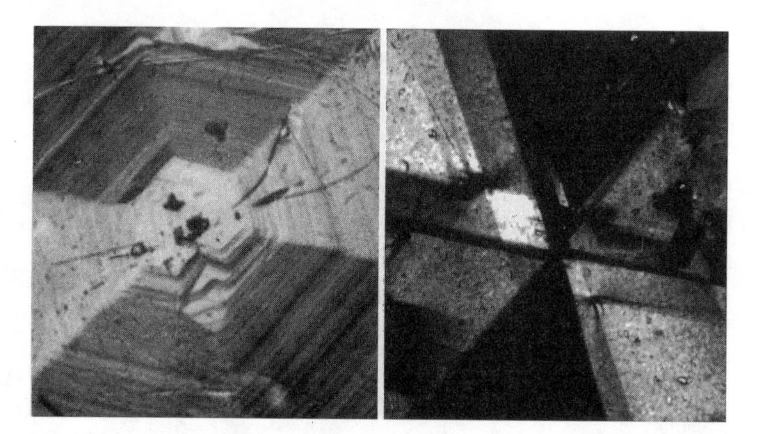

Figure 28. (*left*) DIC photomicrograph of a hexagonal growth hillock on a {001} face of an apatite from the Siglo XX Mine, Llallagua, Bolivia (see Fig. 7). Image is ~1.2 mm across. (*right*) Optical photomicrograph of a thin section (parallel to {001} within a {001} sector) in crossed polarized transmitted light. Optically-distinct regions correspond to the different subsectors beneath the vicinal faces of the growth hillock in a.

Studies of surface structural controls on trace element incorporation into apatite and other minerals during growth exemplify that a bulk crystallographic site may have multiple, distinctly different surface representations (surface sites). Differences between these surface sites may lead to heterogeneous reactivity at the mineral-water interface (Reeder and Rakovan 1999). Sectoral and intrasectoral zoning in apatite demonstrates nonequilibrium incorporation and that factors such as growth mechanism and surface structure are important in trace element partitioning. Furthermore, these zoning types

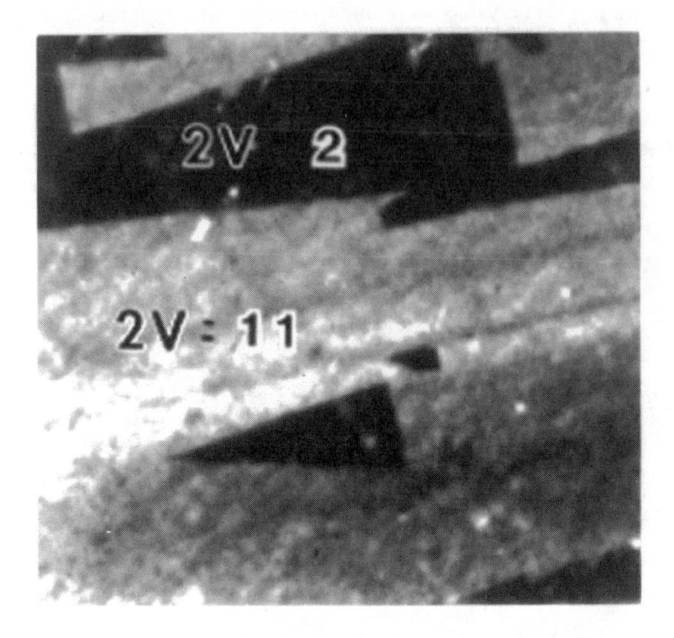

Figure 29. Optical photomicrograph of a thin section (parallel to {001} within a {100} sector) in crossed polarized transmitted light; from the same apatite crystal in Figure 28. Optically-distinct regions correspond to the different concentric zones and subsectors within the {100} sector. Image is ~500 μm across.

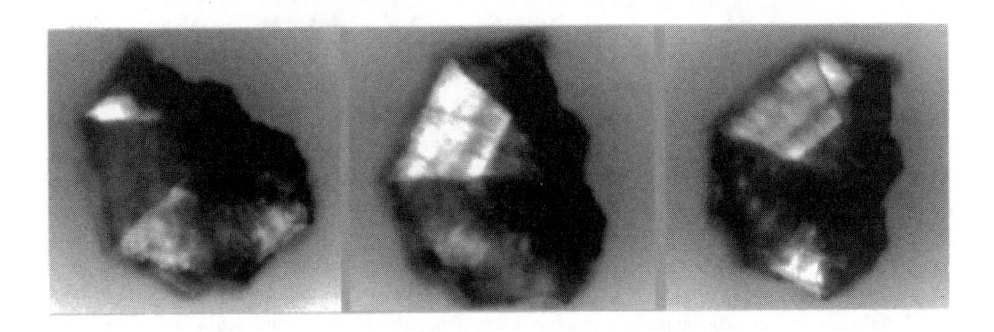

Figure 30. Optical photomicrographs in crossed polarized transmitted light of an apatite in three different orientations from Huron County, Ohio, United States. The optical path is parallel to [001]. Optically distinct regions correspond to the different {100} sectors. The crystal is ~2 mm across.

suggest that partition coefficients, K_d, for a given mineral-fluid system are not unique; rather, the K_d may differ for structurally distinct regions of a crystal surface. Ultimately, this bears on the use of trace element concentrations in minerals as indicators of petrogenetic processes in all types of rock forming environments where apatite is found.Surface structure not only plays an important role in differential partitioning of trace elements into apatite during growth but also influences ordering of atoms, leading to dissymmetrization.

CONCLUSIONS

The nature of the apatite surface, particularly as it exists in contact with complex aqueous solutions, and the processes involved in apatite crystal growth and trace element incorporation have been studied extensively. However the apatite surface and mechanisms of growth are very complex and many questions about the structure of active surface sites of adsorption and incorporation, the free energies of different crystal faces and growth steps, controlling factors on morphology and surface microtopography, mechanisms of impurity affects on growth, and the relationships between differential incorporation leading to compositional zoning and dissymmetrization remain outstanding.

In studies of crystals growth, microtopographic features such as growth spirals are well understood and can be used as indicators of growth mechanisms and to some extent growth conditions. Other commonly observed microtopographic features are not so well understood. Likewise, controls on the morphology and specific face development of apatite are poorly understood. A better understanding of the controls on morphologic and surface microtopographic features, and variations there of, will allow scientists to more widely use these features on natural samples to unravel information about the crystallization and post crystallization histories of biologic and geologic apatites.

The formation of precursor phases, the strong influence of impurity species, and the face specific and step specific incorporation of trace elements all point to kinetic and mechanistic factors being particularly important in the crystal growth of apatite from solution. Surface structural and chemical affects clearly may be more important than thermodynamics in driving crystallization processes of apatite at low temperatures. This is especially true in biological and geological systems where complex environments contain many different impurities, from single ions to complex organic molecules and polymers. The role of precursor phases in the formation of apatite in low temperature geologic environments is poorly understood. This is of particular significance in understanding the role of apatite on the fate of environmentally significant species such as heavy metals and radionuclides.

With the rapid advances in nanotechnology, including new methods for direct observation and analysis of surfaces at the nanometer and atomic scales, more insights pertaining to these and other unanswered questions will be made.

ACKNOWLEDGMENTS

I thank Rich Reeder, John Jaszczak and John Hughes for their reviews of this chapter. I also thank Chris Thomas for his help in editing this chapter and many others in this volume.

REFERENCES

Akizuki M, Sunagawa I. (1978) Study of the sector structure in adularia by means of optical microscopy, infra-red absorption, and electron microscopy. Mineral Mag 42:453-462

Akizuki M (1987) Crystal symmetry and order-disorder structure of brewsterite. Am Mineral 72:645-648

Akizuki M, Nisidoh H, Kudoh Y, Watanabe T, Kurata K (1994) Sector growth and symmetry of (F,OH) apatite from the Asio mine, Japan. Mineral Mag 58:307-314

Amelinckx S (1952a) Spiral growth patterns on apatite crystals. Nature 169:841-842

Amelinckx S (1952b) Growth spirals and their relation to crystal habit as illustrated by apatite. Nature 170:760-761

Amjad Z, Koutsoukos PG, Nancollas GH (1981) The crystallization of fluorapatite: A constant-composition study. J Colloid Inter Sci 82:394-400

Amjad Z, Reddy MM (1998) Influence of humic compounds on the crystal growth of hydroxylapatite. *In* Water Soluble Polymers. Amjad Z (ed) Plenum Press, New York

Arends J, Christoffersen J, Christoffersen MR, Eckert H, Fowler BO, Heughebaert JC, Nancollas GH, Yesinowski JP, Zawacki SJ (1987) A calcium hydroxylapatite precipitated from an aqueous-solution—an international multimethod analysis. J Cryst Growth 84:515-532
Arey JS, Seaman JC, Bertsch PM (1999) Immobilization of uranium in contaminated sediments by hydroxyapatite addition. Environ Sci Tech 33:337-342
Belousova EA, Griffin WL, O'Reilly SY, Fisher NI. (2002) Apatite as an indicator mineral for mineral exploration: Trace-element compositions and their relationship to host rock type. J Geochem Explor 76:45-69
Bergman SC (1979) The significance of accessory apatite in the REE modeling of magma genesis. EOS Trans Am Geophys Union 60:412
Boistelle R, Astier JP (1988) Crystallization mechanisms in solution. J Cryst Growth 90:14-30
Boskey AL, Posner AS (1973) Conversion of amorphous calcium phosphate to microcrystalline hydroxylapatite. A pH-dependent, solution-mediated, solid-solid conversion. J Phys Chem 77: 2313-2317
Boskey AL, Posner AS (1976) Formation of hydroxylapatite at low supersaturation. J Phys Chem 80:40-45
Boudreau AE, Mathez EA, Mccallum IS (1986) Halogen geochemistry of the Stillwater and Bushveld complexes—Evidence for transport of the platinum-group elements by Cl-rich fluids. J Petrol 27: 967-986
Boudreau AE, McCallum IS (1989) Investigations of the Stillwater Complex; Part 5, Apatites as indicators of evolving fluid composition. Contrib Mineral Petrol 102:138-153
Brecevic LJ, Furedi-Milhofer H (1972) Precipitation of calcium phosphates from electrolyte solutions II: The formation and transformation of the precipitates. Calc Tissue Res 10:82-90
Brown WE (1966) Crystal Growth of Bone Mineral. Clin Orthopaed 44:205-220
Brown JL (1981) Calcium phosphate precipitation: Effects of common and foreign ions on hydroxyapatite crystal growth. Soil Sci Soc J 45:482-486
Bulka GR, Vinokurov VM, Nizamutdinov NM, Hasanova NM (1980) Dissymmetrization of crystals: Theory and experiment. Phys Chem Minerals 6:283-293
Burns RG (1970) Mineralogical Applications of Crystal Field Theory. Cambridge University Press, Cambridge
Burton JA, Prim RC, Slichter WP (1953) The distribution of solute in crystals grown from the melt. Part I: Theoretical. J Chem Phys 21:1987-1991
Burton WK, Cabrera N, Frank FC (1951) The Growth of Crystals and the Equilibrium Structure of their Surfaces. Phil Trans 243:299-358
Busscher HJ, De Jong HP, Arends J (1987) Surface free energies of hydroxylapatite, fluorapatite and calcium fluoride. Mater Chem Phys 17:553-558
Chander S, Fuerstenau DW (1984) Solubility and interfacial properties of hydroxylapatite: A review. *In* Adsorption On and Surface Chemistry of Hydroxyapatite. Misra DN (ed) Plenum Press, New York
Chen XB, Wright JV, Conca JL, Peurrung LM (1997) Evaluation of heavy metal remediation using mineral apatite. Water Air Soil Poll 98:57-78
Chin KOA, Nancollas GH (1991) Dissolution of fluorapatite—A constant-composition kinetics study. Langmuir 7:2175-2179
Christoffersen J, Christoffersen MR, Kibalczyc W, Andersen FA (1989) A contribution to the understanding of the formation of calcium phosphates. J Cryst Growth 94:767-777
Christoffersen J, Landis WJ (1991) A contribution with review to the description of mineralization of bone and other calcified tissues *in vivo*. Anatom Rec 230:435-450
Christoffersen J, Rostrup E, Christoffersen MR (1991) Relation between interfacial surface-tension of electrolyte crystals in aqueous suspension and their solubility—A simple derivation based on surface nucleation. J Cryst Growth 113:599-605
Christoffersen J, Christoffersen, MR (1992a) A revised theory for the growth of crystals by surface nucleation. J Cryst Growth 121:608-616
Christoffersen MR, Christoffersen J (1992b) Possible mechanisms for the growth of the biomaterial, calcium hydroxylapatite microcrystals. J Cryst Growth 121:617-630
Christoffersen J, Christoffersen MR, Johansen T (1996a) Kinetics of growth and dissolution of fluorapatite. J Cryst Growth 163:295-303
Christoffersen J, Christoffersen MR, Johansen T (1996b) Some new aspects of surface nucleation applied to the growth and dissolution of fluorapatite and hydroxylapatite. J Cryst Growth 163:304-310
Christoffersen J, Christoffersen MR, Kolthoff N, Barenholdt O (1997) Effects of strontium ions on growth and dissolution of hydroxyapatite and on bone mineral detection. Bone 20:47-54
Christoffersen J, Dohrup J, Christoffersen MR (1998a) The importance of formation of hydroxyl ions by dissociation of trapped water molecules for growth of calcium hydroxylapatite crystals. J Cryst Growth 186:275-282

Christoffersen MR, Dohrup J, Christoffersen J (1998b) Kinetics of growth and dissolution of calcium hydroxylapatite in suspensions with variable calcium to phosphate ratio. J Cryst Growth 186:283-290
Davey RJ (1976) The effect of impurity adsorption on the kinetics of crystal growth from solution. J Cryst Growth 34:109-19
Davies CW (1962) Ion Association. Butterworths, London
Deutsch Y, Sarig S (1977) The effect of fluoride ion concentration on apatite formation and on its crystal habit. J Cryst Growth 42:234-237
Dowty E. (1976) Crystal structure and crystal growth II: Sector zoning in minerals. Am Mineral 61:460-469
Driessens FCM (1982) Mineral Aspects of Dentistry. S. Karger, Basel
Eanes ED, Gillessen IH, Posner AS (1966) Mechanism of conversion of non-crystalline calcium phosphate to crystalline hydroxylapatite. *In* Crystal Growth. Peiser HS (ed) Pergamon Press, Oxford
Elliott JC (1994) Structure and Chemistry of the Apatites and Other Calcium Orthophosphates. Elsevier, Amsterdam
Ewing RC (2001) The design and evaluation of nuclear-waste forms: clues from mineralogy. Can Mineral 39:697-715
Feenstra TP, Bruyn PLD (1979) Formation of calcium phosphates in moderately supersaturated solutions. J Phys Chem 83:475-479
Francis MD, Web NC (1971) Hydroxylapatite formation from hydrated calcium monohydrogen phosphate precursor. Calc Tissue Res 6:335-342
Gilmer GH (1976) Growth on imperfect crystal faces I. Monte-Carlo growth rates. J Cryst Growth 35: 15-28
Gilmer GH, Jackson KA (1977) Computer simulation of crystal growth. *In* Current Topics in Materials Science. Vol 2. Kaldis E, Scheel HJ (eds) North-Holland Pub, Amsterdam
Goldschmidt VM (1913) Atlas der Kristallformen. Heidelberg
Goldschmidt VM (1937) The principles of distribution of chemical elements in minerals and rocks. J Chem Soc London, p 655-673
Graf JL (1977) Rare earth elements as hydrothermal tracers during the formation of massive sulfide deposits in volcanic rocks Appears. Econ Geol 72:527-548
Gromet, L.P., and Silver L.T. (1983) Rare earth element distributions among minerals in a granodiorite and their petrologic implications. Geochim Cosmochim Acta 47:952-938
Hahn TJ (1989) Aluminum-related disorders of bone and mineral metabolism. *In* Bone and Mineral Research Vol. 6. Peck WA (ed) Elsevier, New York
Hartman P, Perdok WG (1955a) On the relations between structure and morphology of crystals. I. Acta Crystallogr 8:49-52
Hartman P, Perdok WG (1955b) On the relations between structure and morphology of crystals. II. Acta Crystallogr 8:521-529
Hillig WB (1966) A derivation of classical two-dimensional nucleation kinetics and the associated crystal growth laws. Acta Metall 14:1868-1969
Hochella MF, White AF (eds) (1990) Mineral-Water Interface Geochemistry—an Overview. Rev Mineral 23. Mineralogical Society of America, Washington, DC
Hogarth DD (1974) The discovery of apatite on the Lièvre River, Quebec. Mineral Rec 5:178-182
Hohl H, Koutsoukos PG, Nancollas GH (1982) The crystallization of hydroxylapatite and dicalcium phosphate dihydrate; representation of growth curves. J Cryst Growth 57:325-35
Inskeep WP, and Silvertooth JC (1988) Kinetics of hydroxylapatite precipitation at pH 7.4 to 8.4. Geochim Cosmochim Acta 52:1883-1893
Jaynes WF, Moore PA Jr., Miller DM (1999) Solubility and ion activity products of calcium phosphate minerals. J Environ Qual 28:530-536
Jeanjean J, Rouchaud JC, Tran L, Fedoroff M (1995) Sorption of uranium and other heavy metals on hydroxyapatite. J Radioanal Nuclear Chem Lett 201:529-539
Jolliff BL, Papike JJ, Shearer CK, Shimizu N (1989) Inter- and intra-crystal REE variations in apatite from the Bob Ingersoll pegmatite, Black Hills, South Dakota. Geochim Cosmochim Acta 53:429-441
Kanzaki N, Onuma K, Ito A, Teraoka, K, Tateishi T, Tsutsumi S (1998) Direct growth rate measurement of hydroxylapatite single crystal by Moire phase shift interferometry. J Phys Chem B 102:6471-6476
Kanzaki N, Onuma K, Treboux G, Tsutsumi S, Ito A (2000) Inhibitory effect of magnesium and zinc on crystallization kinetics of hydroxylapatite (0001) face. J Phys Chem B 104:4189-4194
Kanzaki N, Onuma K, Treboux G, Tsutsumi S, Ito A (2001) Effect of impurity on two-dimensional nucleation kinetics: Case studies of magnesium and zinc on hydroxylapatite (0001) face. J Phys Chem B 105:1991-1994
Kempe U, Gotze J (2002) Cathodoluminescence (CL) behaviour and crystal chemistry of apatite from rare-metal deposits. Mineral Mag 66:151-172

Kostov I, Kostov RI (1999) Crystal Habits of Minerals. Pensoft Publisher, Sofia
Koutsopoulos S (2001) Kinetic study on the crystal growth of hydroxylapatite. Langmuir 17:8092-8097
Koutsoukos P, Amjad Z, Tomson MB, Nancollas GH (1980) Crystallization of calcium phosphates—Constant-composition study. J Am Chem Soc 102:1553-1557
Kovalenko VI, Antipin VS, Vladykin NV, Smirnova YV, Balashov YA (1982) Rare-earth distribution coefficients in apatite and behavior in magmatic processes. Geokhimiya 2:230-242
Lindsay (1979) Chemical Equilibria in Soils. Wiley Interscience, New York
Liu Y, Nancollas GH (1996) Fluorapatite growth kinetics and the influence of solution composition. J Cryst Growth 165:116-123
Lundager-Madsen HE, Christensson F (1991) Precipitation of calcium phosphate at 40 degrees C from neutral solution. J Cryst Growth 114:613-18
Marshall DJ (1988) Cathodoluminescence of Geological Materials. Unwin Hyman, Boston
McLean FC, Bundy AM (1964) Radiation, Isotopes, and Bone. Academic Press, New York
McLennan SM (1989) Rare earth elements in sedimentary rocks: Influence of provenance and sedimentary processes. Rev Mineral 21:169-200
McConnell D (1973) Apatite: Its Crystal Chemistry, Mineralogy, Utilization and Geologic and Biologic Occurrences. Springer Verlag, New York
Moreno EC, Varughese K (1981) Crystal growth of calcium apatites from dilute solutions. J Cryst Growth 53:20-30
Müller-Krumbhaar H, Burkhard TW, Kroll D (1977) A generalized kinetic equation for crystal growth. J Cryst Growth 38:13-22
Mullin JW (1993) Crystallization. Butterworth-Heinemann Ltd., London
Nakamura Y (1973) Origin of sector zoning in igneous clinopyroxenes. Am Mineral 58:986-990
Nancollas GH, Tomazic B (1974) Growth of calcium phosphate on hydroxylapatite crystals. Effect of supersaturation and ionic medium. J Phys Chem 78:2218-2225
Nancollas GH (1979) The growth of crystals in solution. Adv Colloid Inter Sci 10:215-52
Nancollas GH (1982) Phase transformation during precipitation of calcium salts. *In* Biological Mineralization and Demineralization. Nancollas GH (ed) Springer-Verlag, Berlin, p 79-99
Nancollas GH (1984) The nucleation and growth of phosphate minerals. *In* Phosphate Minerals: Their Properties and General Modes of Occurrence. Nriagu JO, Moore PB (eds) Springer-Verlag, Berlin
Nancollas GH, Zhang J (1994) Formation and dissolution mechanisms of calcium phosphates in aqueous systems. *In* Hydroxylapatite and Related Materials. CRC Press, Boca Raton, Florida
Nash WP (1972) Apatite chemistry and phosphorus fugacity in a differentiated igneous intrusion. Am Mineral 57:877-886
Nielsen AE (1984) Electrolyte crystal growth mechanisms. J Cryst Growth 67:289-310
Nielsen AE, Toft JM (1984) Electrolyte crystal growth kinetics. J Cryst Growth 67:278-288
Nylen MU, Eanes ED, Omnell KA (1963) Crystal Growth in Rat Enamel. J Cell Biol 18:109-123
Onuma K, Ito A, Tateishi T, Kameyama T (1995a) Growth kinetics of hydroxylapatite crystal revealed by atomic force microscopy. J Cryst Growth 154:118-125
Onuma K, Ito A, Tateishi T, Kameyama T (1995b) Surface observations of synthetic hydroxylapatite single crystal by atomic force microscopy. J Cryst Growth 148:201-206
Onuma K, Kanzaki N, Ito A, Tateishi T (1998) Growth kinetics of the hydroxylapatite (0001) face revealed by phase shift interferometry and atomic force microscopy. J Phys Chem B 102:7833-7838
O'Reilly SY, Griffin WL (2000)Apatite in the mantle: Implications for metasomatic processes and high heat production in Phanerozoic mantle. Lithos 53:217-232
Palache C, Berman H, Frondel C (1951) The system of mineralogy of James Dwight Dana and Edward Salisbury Dana. Wiley and Sons, London
Papike JJ, Jensen M, Laul JC, Shearer CK, Simon SB, Walker RJ (1984) Apatite as a recorder of pegmatite petrogenesis. Geol Soc Am Abstr Progr 16:617
Paquette J, Reeder RJ (1990) A new type of compositional zoning in calcite: Insights into crystal-growth mechanisms. Geology 18:1244-1247
Paquette J, Reeder RJ (1995) Relationship between surface structure, growth mechanism, and trace element incorporation in calcite. Geochim Cosmochim Acta 59:735-751
Phakey PP, Leonard JR (1970) Dislocations and fault surfaces in natural apatite. J Appl Crystallogr 3:38-44
Pina C, Becker U, Risthaus P, Bosbach D, Putnis A (1998) Molecular-scale mechanisms of crystal growth in barite. Nature 395:483-486
Posner A.S. (1987) Bone Mineral and the Mineralization Process. Bone Mineral Res Vol. 5. Peck WA (ed) Elsevier, New York
Rakovan J, Reeder RJ (1994) Differential incorporation of trace-elements and dissymmetrization in apatite—The role of surface-structure during growth. Am Mineral 79:892-903

Rakovan J, Reeder RJ (1996) Intracrystalline rare earth element distributions in apatite: Surface structural influences on incorporation during growth. Geochim Cosmochim Acta 60:4435-4445
Rakovan J, Waychunas G (1996) Luminescence in Minerals. Mineral Rec 27:7-19
Rakovan J, McDaniel DK, Reeder RJ (1997) Use of surface-controlled REE sectoral zoning in apatite from Llallagua, Bolivia, to determine a single-crystal Sm-Nd age. Earth Planet Sci Lett 146:329-336
Rakovan J, Newville M, Sutton S (2001a) Evidence of heterovalent europium in zoned Llallagua apatite using wavelength dispersive XANES. Am Mineral 86:697-700
Rakovan J, Sutton S, Newville M (2001b) Evaluation of europium oxidation state and anomalous partitioning behavior in intrasectorally zoned apatite using wavelength dispersive micro-XANES. Advanced Photon Source, 2001 Activity Report
Rakovan J, Jaszczak JA (2002) Multiple length scale growth spirals on metamorphic graphite {001} surfaces studied by atomic force microscopy. Am Mineral 87:17-24
Reeder RJ, Grams JC (1987) Sector zoning in calcite cement crystals: Implications for trace element distributions in carbonates. Geochim Cosmochim Acta 51:187-194
Reeder RJ, Rakovan J (1999) Surface structural controls on trace element incorporation during crystal growth. *In* Growth, Dissolution and Pattern Formation in Geosystems. Jamtveit B, Meakin P (eds) Kluwer, Boston
Richards RP, Rakovan J (2000) The first occurrence of apatite crystals in Ohio. Rocks Mineral 75:255
Ringwood AE (1955) The principles governing trace element distribution during crystallization. Part I: The influence of electronegativity. Geochim Cosmochim Acta 7:189-202
Seaman JC, Arey JS, Bertsch PM (2001) Immobilization of nickel and other metals in contaminated sediments by hydroxyapatite addition. J Environ Qual 30:460-469
Sha LK, Chappell BW (1999) Apatite chemical composition, determined by electron microprobe and laser-ablation inductively coupled plasma mass spectrometry, as a probe into granite petrogenesis. Geochim Cosmochim Acta 63:3861-3881
Shtukenberg AG, Punin YO, Haegele E, Klapper H (2001) On the origin of inhomogeneity of anomalous birefringence in mixed crystals: An example of alums. Phys Chem Mineral 28:665-674
Skinner HCW (1987) Bone: Mineralization. *In* The Scientific Basis of Orthopaedics. Albright JA, Brand RA (eds) Appleton and Lange, Norwalk, Connecticut
Snoeyink VL, Jenkins D (1980) Water Chemistry. Wiley and Sons, New York
Stumm W, Morgan JJ (1996) Aquatic Chemistry. Wiley Interscience, New York
Sunagawa I, Bennema P (1982) Morphology of growth spirals: Theoretical and experimental. *In* Preparation and Properties of Solid State Materials. Wilcox WR (ed) Marcel Dekker, New York
Sunagawa I. (1984) Growth of crystals in nature. *In* Materials Science of the Earth's Interior. Sunagawa I (ed) Terra Scientific Publishing, Tokyo
Sunagawa I (1987a) Morphology of minerals. *In* Morphology of Crystals. Sunagawa I (ed) Terra Scientific Publishing, Tokyo
Sunagawa I (1987b) Surface microtopography of crystal faces. *In* Morphology of Crystals. Sunagawa I (ed) Terra Scientific Publishing, Tokyo
Teng HH, Dove PM, De Yoreo JJ (1999) Reversed calcite morphologies induced by microscopic growth kinetics: Insight into biomineralization. Geochim Cosmochim Acta 63:2507-2512
Termine JD, Peckauskas RA, Posner AS (1970) Calcium phosphate formation *in vitro*. II. Effect of environment on amorphous-crystalline transformation. Arch Biochem Biophys 140:318-325
Terpstra RA, Bennema P (1987) Crystal morphology of octacalcium phosphate: Theory and observation. J Cryst Growth 82:416-26
Terpstra RA, Bennema P, Hartman P, Woensdregt CF, Perdok WG, Senechal ML (1986) F faces of apatite and its morphology: Theory and observation. J Cryst Growth 78:468-78
Thirioux L, Baillif P, Ildefonse JP, Touray JC (1990) Surface reactions during fluorapatite dissolution-recrystallization in acid media (hydrochloric and citric acids). Geochim Cosmochim Acta 54: 1969-1977
Valsami-Jones, E., Ragnarsdottir, K.V., Putnis, A., Bosbach, D., Kemp, A.J., andCressey, G. (1998) The dissolution of apatite in the presence of aqueous metal cations at pH 2-7. Chem Geol 151: 215-233
Van Cappellen P, Berner RA (1991) Fluorapatite crystal-growth from modified seawater solutions. Geochim Cosmochim Acta 55:1219-1234
Velthuizen JV (1992) Giant fluorapatite crystals: a question of locality. Mineral Rec 23:459-463
Watson EB, Capobianco CJ (1981) Phosphorus and rare-earth elements in felsic magmas: An assessment of the role of apatite. Geochim Cosmochim Acta 45:2349-2358
Watson EB, Harrison TM (1984a) Accessory minerals and the geochemical evolution of crustal magmatic systems: A summary and prospectus of experimental approaches. Phys Earth Planet Inter 5:19-30
Watson EB, Harrison TM (1984b) What can accessory minerals tell us about felsic magma evolution? A framework for experimental study. Proc 27th Intl Geol Congr 11:503-520

Wu W, Nancollas GH (1998) The influence of additives and impurities on crystallization kinetics: An interfacial tension approach. *In* Water Soluble Polymers. Amjad Z (ed) Plenum Press, New York
Young RA, Brown WE (1982) Structures of biological minerals. *In* Biological Mineralization and Demineralization. Nancollas GH (ed) Springer-Verlag, Berlin
Zieba A, Sethuraman G, Perez F, Nancollas GH, Cameron D (1996) Influence of organic phosphonates on hydroxyapatite crystal growth kinetics. Langmuir 12:2853-2858

4

Synthesis, Structure, and Properties of Monazite, Pretulite, and Xenotime

Lynn A. Boatner

Solid State Division
Oak Ridge National Laboratory
Oak Ridge, Tennessee 37831

PROPERTIES OF RARE-EARTH-, Sc-, AND Y-ORTHOPHOSPHATES

General characteristics

Orthophosphate compounds of the type $A(PO_4)$ include the minerals monazite [$RE(PO_4)$ with RE = the light rare-earth ions, e.g., La, Ce, Nd...], xenotime [$Y(PO_4)$, and also incorporating the heavy RE ions], and pretulite [$Sc(PO_4)$]. The name of the mineral monazite is derived from the Greek word *monazein*, "to be solitary," whereas that of xenotime is derived from the Greek words *xenos*, meaning "foreign" and *time*, meaning "honor." The mineral pretulite is named for Pretul Mountain, located in the Fischbacher Alps in Styria, Austria where the mineral was discovered (Bernhard et al. 1998). Pretulite is only the sixth mineral to have been found in which scandium is a major constituent.

The anhydrous rare-earth (RE) orthophosphates can be structurally divided between the light RE-element compounds with the monoclinic monazite structure (space group $P2_1/n$, $Z = 4$) and the heavier RE compounds with the tetragonal (zircon-type) xenotime structure (space group $I4_1/amd$, $Z = 4$). In terms of the crystal chemistry of these two groups of compounds, a primary distinguishing structural feature is the coordination of the RE, Y, or Sc ions. In the monoclinic monazite structure, the RE ion is located in a polyhedron in which it is coordinated with nine oxygen ions. In the tetragonal xenotime structure, the heavier REs, Y, or Sc are located in a polyhedron in which they are coordinated with eight oxygen ions. Scandium orthophosphate is the basic constituent of the recently identified mineral, pretulite, which belongs to the orthophosphate group with the tetragonal xenotime structure (Bernhard et al. 1998).

Both monazite and xenotime are relatively widely distributed as microcrystalline (or small crystalline) accessory inclusions in granitic rocks, rhyolites, pegmatites, and gneisses. Monazite also occurs in carbonatites, charnockites, migmatites, and quartz veins (Rapp and Watson 1986) as well as in alluvial deposits, including beach sands (e.g., the black beach sands of Kerala, India and beach sands in San Mateo County, California). These alluvial deposits and beach sands are produced by the weathering of peraluminous granites, granitic pegmatites, and other host rock types. The occurrence of the alluvial deposits of monazite and their derivation from the weathering of granitic rocks attest to the ability of the mineral to survive metamorphic and sedimentary cycles that can extend over several hundred million years (Rapp et al. 1987).

Chemical composition

Natural monazite contains not only cations of the light rare-earth elements, but it also incorporates uranium and thorium. Monazite is the principal ore for the commercial extraction of thorium, and it has also been used as a secondary source of uranium. Monazite deposits are located in the United States, Australia, South Africa, Sri Lanka, Brazil, India, Malagasy, and Canada. As shown in Table 1, monazite is capable of incorporating significant amounts of both uranium and thorium (Boatner and Sales 1988, Houk 1943). Because of the chemical durability of monazite, relatively harsh chemical

1529-6466/00/0048-0004$05.00

treatments at elevated temperatures are required for the extraction of thorium or uranium. One method of extraction consists of treatment in 45% sodium hydroxide at 138°C, and a second commercial extraction process consists of subjecting the ore to a 93% solution of sulfuric acid at 210°C.

Table 1. Composition of natural monazites.

Mineral source	*UO_2 (wt %)*	*ThO_2 (wt %)*	*Combined lanthanide oxides (wt %)*	*Other oxides (wt%)*	*P_2O_5 (wt %)*
Piona, Italy[a]	15.64	11.34	35.24	6.76	31.02
Ratunapura, Sri Lanka[b]	0.10	14.32	53.51	5.03	26.84
Burke County,[C] N.C.	0	6.49	62.26	1.97	29.28
Brazil[d(1)]	N.D,	10.05	58.13	6.98	25.51
Malay[d(2)]	N.D.	8.38	60.98	6.43	23.92
Australia[d(3)]	N.D.	3.80	65.40	3.94	26.89
India[e]	N.D.	10.22	60.36	3.23	26.82

[a] Gramaccioli & Segalstad (1978)
[b] Kato T (1958)
[c] Penfield SC (1882)
[d]Houk LG (1943) Monazite Sand. *Information Circular IC 7233*, U S Dept of the Interior-Bureau of Mines
[d(1)] monazite from the river sands of Rio Paraguassir in Bahia, Banderiro do Mello.
[d(2)] concentrated monazite from the Sempang Tin Co., Pahang, Malay Peninsula.
[d(3)] monazite from Cooglegong and Moolyella, Western Australia.
[e] Johnstone SJ (1914)

The extensive and detailed analytical results given in Table 1 of Förster (1998a) illustrate the extreme compositional diversity exhibited by natural monazites. In fact, Förster has even reported the existence of a monazite–group mineral that is intermediate between monazite and huttonite. Förster's analysis indicated that complete miscibility exists between common monazite-(Ce) and the phosphate mineral brabanite [Ca,Th,U$(PO_4)_2$]. The monazite/xenotime compositional systematics have also been examined by Heinrich et al. (1997). Additional data related to the compositional diversity of monazites can be found in the work of Bea (1996), Bea et al. (1994), and Hinton and Patterson (1994).

The large compositional variability that can be exhibited by natural xenotime is illustrated by the results of Förster (1998b), who analyzed xenotime specimens from various peraluminous granites found in Erzgebirge, Germany. This work showed that, with respect to the heavy rare earths as well as Y, U, and Th, a wide range of xenotime compositions occurs. Specifically, xenotime grains normally obtained from granites worldwide are characterized by compositions consisting of 70 to 80 mol % $Y(PO_4)$ + 16 to 25 mol % heavy rare-earth phosphates. Förster (1998b) found specimens of xenotime with as much as 45 mol % heavy rare-earth ions replacing yttrium.

The new mineral pretulite represents a scandium-dominant analog of xenotime-(Y). Bernhard et al. (1998) determined an empirical formula of [$Sc_{0.98}Y_{0.02}(PO_4)$] for pretulite obtained from the Styrian lazulites. They reported a range of Y content replacing Sc [i.e., Y/(Y + Sc)] in the range of 0.5 to 3.2 mol %; traces of the heavy rare-earth elements Yb, Er, and Dy were also found. The composition of pretulite was reported to be inhomogeneous on a micron scale due to variations in yttrium content of 0.39 to 2.49 wt %.

Monazite, xenotime, and pretulite, in addition to minerals such as purpurite [$Mn(PO_4)$] and lithiophylite [$LiMn(PO_4)$], represent members of a relatively small group of phosphate minerals that are classified as anhydrous and as "lacking foreign anions." Hydrated modifications of these substances, including rhabdophane-type phosphates [$RE(PO_4)\cdot nH_2O$, with RE = La to Tb], weinschenkite-type phosphates [$RE(PO_4)\cdot 2H_2O$ with RE = Dy, Y, Er, Yb, or Lu], the orthorhombic compound $Dy(PO_4)\cdot 1.5H_2O$, and other hydrated phosphates such as the pseudo-hexagonal, orthorhombic mineral ningyoite [$(Ca,U,RE)(PO_4)\cdot xH_2O$; Muto et al. 1959) can be synthesized at reduced temperatures and are well known. Recently, the yttrium-based rhabdophane-type compounds $Y(PO_4)\cdot 0.8H_2O$ and $Er(PO_4)\cdot 0.9H_2O$ have also been synthesized and investigated by Hikichi et al. 1989. A hydrated form of $Sc(PO_4)$ occurs in nature as kolbeckite [$Sc(PO_4)\cdot 2H_2O$] – a mineral of supergene or hydrothermal derivation. An irreversible structural transformation of rhabdophane-to-monazite or rhabdophane-to-xenotime can be achieved by heating the hydrated phosphate compounds. The transformation involves an initial dehydration in the range of ~100 to 400°C to produce a hexagonal form of the $A(PO_4)$ compound, followed by a structural transformation that occurs exothermically in the range of 500-900°C – depending on the specific cation (Jonasson and Vance 1986, Hikichi et al. 1988, 1996). Hikichi et al. (1978) have noted that the dehydrated hexagonal form of the RE orthophosphates will readily re-hydrate when exposed to air at ambient temperature. The mechanochemical preparation (i.e., by grinding in various media) and the conversion of various types of hydrated $RE(PO_4)$ and $Y(PO_4)$ compounds have been described in detail by Hikichi et al. (1989, 1991, 1993, 1995, 1996).

Chemical durability

The anhydrous RE, Y, and Sc orthophosphates are compounds that are extremely insoluble and chemically durable in an aqueous environment over a relatively wide pH range (Sales et al. 1983). In particular, the chemical durability of monazite is indicated by the extreme treatments, noted in the previous section, that are required to extract thorium from natural ores. Some alteration of monazite does occur, however, under hydrothermal conditions, and the responsible mechanisms and geochemical implications of these alteration processes have been studied in detail by Poitrasson et al. 1996. Subsequent investigations of the behavior of monazite under hydrothermal conditions and the relevance of this behavior to U-Pb isotope systematics and the associated geochronological implications have been carried out by Teufel and Heinrich (1997). These workers reported only minor surface dissolution effects in the case of monazite grains, but more significant alterations of monazite powders were observed. Specifically, Teufel and Heinrich found a temperature-dependent loss of lead from monazite powders exposed to hydrothermal conditions. Andrehs and Heinrich (1998) and Gratz and Heinrich (1997) have hydrothermally synthesized and investigated (RE+Y)-orthophosphate solid solutions. They produced both monazite- and xenotime-structure phases and demonstrated the utility of the temperature-dependent partitioning of the rare earths between monazite and xenotime as a method of geothermometry. Additional hydrothermal investigations have been carried out by Podor and Cuny (1997). Surface reactions (including replacement reactions) of synthetic monazite- and xenotime-structure RE-orthophosphates in various environments have been also investigated by Jonasson et al. (1988).

Radiation damage effects

As previously noted, monazite usually contains significant amounts of thorium as well as uranium, and xenotime also contains uranium. Naturally radioactive minerals are, therefore, exposed to displacive radiation damage events over geological time scales. Accordingly, such minerals are frequently found in the metamict state, i.e. they can exhibit external faceting and the appearance of a crystal, but they have been rendered structurally amorphous by cumulative radiation damage effects. The issue of the metamictization of natural monazite has been examined in the early structural work of Ghouse (1968), who examined the effects of heating on monazites from Kerala beach sands. Ghouse found only a small difference in the structure of the natural monazite specimens that he examined following heat treatments up to 1130°C. More recently, Ewing (1975) and Ewing and Haaker (1980) have noted that monazite is, in fact, generally always found in a highly crystalline (i.e., not fully metamict) form despite being subjected to displacive radiation damage over hundreds of millions to billions of years. This represents an important issue for mineral phases such as monazite that are considered as host waste forms for the disposal of actinides and other types of high-level radioactive waste. The concern is that, in such an application, structural alterations leading to amorphization will render the waste form less chemically durable and more susceptible to the loss of radioactive ions by dissolution or leaching in aqueous media (Eyal and Kaufman 1982, Eyal and Fleischer 1985, Roy and Vance 1981). The subject of radiation damage effects and their impact on the performance of radioactive waste forms is thoroughly reviewed by Ewing and Wang (this volume), and the subject is very well referenced in their chapter. In brief, however, the original idea that monazite was simply very resistant to radiation damage and thus did not experience metamictization has been found to be incorrect. The material is easily amorphized by displacive radiation events but it also recovers its ordered structure by annealing processes that occur at relatively low temperatures. This issue has recently been well resolved and placed on solid ground by the work of Meldrum et al. (1997a,b,c, 1998, 2000). One additional interesting point is that, in the case of the rare-earth orthophosphates, Sales et al. (1983) found that even in the radiation-damaged phase of a monazite matrix the dissolution rate remained quite low.

Thermophysical and thermochemical properties

Monazite, xenotime, and (presumably) pretulite in their stable anhydrous forms are refractory materials with reported melting points in excess of 2000°C (Bondar 1976, Rouanet et al. 1981, Hikichi and Nomura 1987). In the case of natural monazite from Nogisawa-mura, Japan, Hikichi and Nomura (1987) used a solar furnace to measure a melting temperature of 2057(±40)°C. These orthophosphate compounds remain relatively stable chemically until near or at their melting point, where some thermal decomposition is initiated that is apparently associated with the loss of phosphorous-containing species. Table 2 lists the melting point of various phase-pure RE orthophosphates and synthetic xenotime as determined by Bondar (1976) and Hikichi and Nomura (1987). It should be noted that there are significant differences between the earlier values for the melting points reported by Bondar and the more recent results of Hikichi and Nomura.

The thermochemical properties of the rare-earth orthophosphates [plus $Sc(PO_4)$ and $Y(PO_4)$] have recently been investigated in detail by Ushakov et al. (2001). These workers obtained the formation enthalpies of 14 orthophosphates by using calorimetric techniques and found an almost linear dependence between the enthalpies of formation and the rare-earth radius, from $La(PO_4)$ (-321.4 kJ/mol) to $Lu(PO_4)$ (-236.9 kJ/mol); xenotime and pretulite were found to be consistent with this behavior as well. The structural transition from the xenotime structure to the monazite structure was not manifested in a significant discontinuity in the relatively linear trend in the enthalpies of formation. The complete results of these detailed thermochemical studies are tabulated in Ushakov et al. (2001).

Table 2. Melting points of rare-earth- and Y-orthophosphates.

Compound	*Melting point (°C)*	*Liquid spectral emissivity* [a] *(at 0.65 μm)*
$La(PO_4)$	2072 (20)[a]	1.00
	2300 [b]	
$Ce(PO_4)$	2045 (20)[a]	0.99
$Pr(PO_4)$	1938 (20)[a]	0.82
$Nd(PO_4)$	1975 (20)[a]	0.85
	2250 [b]	
$Sm(PO_4)$	1916 (20)[a]	0.90
$Eu(PO_4)$	2200 [b]	
$Gd(PO_4)$	2200 [b]	
$Tb(PO_4)$	2150 [b]	
$Dy(PO_4)$	2150 [b]	
$Er(PO_4)$	1896 (20)[a]	1.00
	2150 [b]	
$Yb(PO_4)$	2150 [b]	
$Y(PO_4)$	1995 (20)[a]	1.00
	2150 [b]	

(a) Hikichi and Nomura (1987)
(b) Bondar et al. (1976)

CRYSTAL GROWTH OF MONAZITE, PRETULITE, AND XENOTIME

To date, pretulite has only been found in lazulite and quartz-rich domains (Bernhard et al. 1998), and euhedral pretulite crystals up to a maximum size of ~200 μm were observed. Large crystals of monazite weighing up to ~25 kg have been reported; however, natural xenotime crystals are more limited in size relative to monazite. One of the larger known xenotime crystals, as represented by a specimen from a pegmatite (Minas Gerais, Brazil), is ~4.2 cm tall and is presently housed in the Natural History Museum in Los Angeles, California. Regardless of their size, natural crystal specimens of these minerals are generally characterized by compositional variations (monazite, of course, being a mixed rare-earth orthophosphate by origin), by structural defects, and radiation damage effects (because most monazites contain thorium and uranium, and xenotime frequently contains uranium.) Additionally, the sample histories of these minerals in the geological environment and the alteration conditions to which they have been subjected are not known with certainty. Because of the limitations of natural mineral specimens in terms of purity, compositional variation, crystal perfection, and the availability of appropriate well-characterized naturally occurring specimens, synthetic high-purity (or controlled purity) single crystals of monazite, xenotime, and pretulite are essential for fundamental studies of the structural characteristics, solid state chemistry, thermochemistry, and physical properties of the pure orthophosphate phases. In many cases, synthetic single crystals are also necessary for use in the practical applications outlined elsewhere in this chapter. The following sections detail studies on the synthesis of such specimens.

Flux growth of RE-, Sc-, and Y-orthophosphates

A crystal growth technique similar to that originally described by Feigelson (1964) has been used to grow phase-pure single crystals of all of the RE-orthophosphates [except for the radioactive member $Pm(PO_4)$]—including the end members, $La(PO_4)$ and $Lu(PO_4)$, as well as $Y(PO_4)$ and $Sc(PO_4)$. This growth method has been applied to the preparation of orthophosphate-host single crystals doped with controlled amounts of iron-group, other RE, actinide, and other (e.g., Zr and Hf) impurities. Additionally, the technique has been used to grow mixed "alloy" rare-earth orthophosphate single crystals for use in x-ray diffraction structural refinement studies, the results of which are described elsewhere in this chapter.

The growth of the subject orthophosphate single crystals is carried out by means of a high-temperature solution (or flux) growth process using lead pyrophosphate $Pb_2P_2O_7$ as the high-temperature solvent (Wickham 1963, Wanklyn 1972). The flux is prepared by first chemically precipitating lead hydrogen phosphate from a lead nitrate solution through the addition of phosphoric acid. The resulting $PbH(PO_4)$ precipitate is subsequently oven dried in air. The lead hydrogen phosphate is then combined with the appropriate RE oxide plus any desired dopant oxides, and the mixture is placed in a platinum crucible and covered with a loose-fitting Pt lid. In a typical small laboratory-scale growth (e.g. in a 50 ml Pt crucible), 60 g of $PbHPO_4$ would be combined with 3.5 g of RE_2O_3. The growth crucible is placed in a temperature-controlled and programmable resistance-heated furnace and heated to 1360°C. During the initial heating phase, the lead hydrogen phosphate decomposes to form $Pb_2P_2O_7$, which then reacts at elevated temperatures to form the RE, Y, or Sc orthophosphate. The unreacted excess Pb pyrophosphate then serves as a high-temperature solvent in which the orthophosphate is dissolved. Following a 1360°C "soaking period," the duration of which can be varied from several hours to several days, the furnace is cooled at a linear rate of 0.5 to 2.0°C per hour to ~900°C. Below this temperature, the crystal growth does not proceed further, and thus the furnace is rapidly cooled to room temperature. Figure 1 illustrates the Pt crucible and contents following the crystal growth of synthetic pretulite, $Sc(PO_4)$. The faceted $Sc(PO_4)$ crystals have nucleated on and are located around the periphery of the crucible and have formed on the surface of the white $Pb_2P_2O_7$ flux that occupies the remainder of the solidified surface as shown in Figure 1. In this case, the long dimension of the largest $Sc(PO_4)$ crystal is ~2.0 cm.

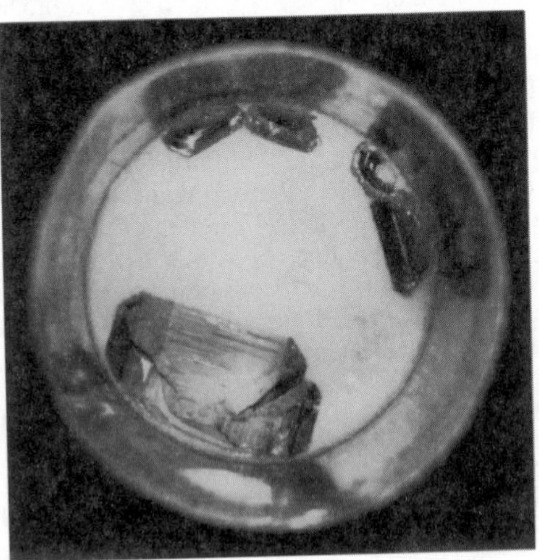

Figure 1. Single crystals of synthetic pretulite [pure $Sc(PO_4)$] are shown inside a 50 ml Pt crucible following a high-temperature-solution (flux) growth run. The large white central area is solidified $Pb_2P_2O_7$ flux, and the $ScPO_4$ crystals have nucleated and grown around the periphery of the Pt crucible. A normal flux-growth run takes ~30 days, and an additional ~30 days is required to chemically remove the $Sc(PO_4)$ crystals from the solidified flux. The outer diameter of the Pt crucible is ~3.9 cm.

As illustrated in Figure 1, the orthophosphate single crystals that are grown by means of the spontaneous-nucleation, flux-growth technique are initially entrained in the solidified $Pb_2P_2O_7$ flux upon cooling to room temperature. This entraining can be minimized by an elevated-temperature decanting of the molten flux prior to the final cooling step. In some cases, however, (e.g., when one of the tetragonal crystals grows completely across the diameter of the crucible), this subjects the phosphate crystals to thermomechanical stresses that may induce fracture. An example of a large (originally ~3 cm long) $Y(PO_4)$ crystal that fractured on removal from the flux is illustrated in Figure 2.

Figure 2. A flux-grown single crystal of synthetic xenotime [pure $Y(PO_4)$] is shown following its removal from a 50 ml Pt crucible. This "bar-like" form is typical of the growth habit of the tetragonal rare-earth and Y orthophosphates. The four-fold symmetry axis of the tetragonal structure lies along the long dimension of the "bar." Growth striations that frequently occur on the surface of the flux-grown tetragonal-structure orthophosphate crystals are evident. The crystal has fractured on cooling due to thermomechanical stresses. The horizontal dimension is ~3.5 cm.

This figure also illustrates the surface striations that are often observed, the long dimension of which lies along the tetragonal *c* axis. A more benign but significantly slower method for removing the solidified flux is to place the crucible and its contents in boiling nitric acid. Even under this somewhat extreme condition, the solubility of the solidified $Pb_2P_2O_7$ is sufficiently low that a period of four-to-five weeks may be required to completely free the phosphate crystals and remove any residual pyrophosphate flux (including flux inclusions with open porosity exposure.) Accordingly, in those cases where the maximum size crystals are not required for a given experimental investigation, the flux-decanting method provides a more expedient means of obtaining crystal samples. In this regard, a special technique and apparatus for decanting the flux from a number of Pt growth crucibles simultaneously has been described by Smith and Wanklyn (1974). This apparatus had a relatively large thermal mass so that the crucibles cooled more slowly than in the case of simply removing an individual Pt crucible with tongs and decanting the flux into a hollow firebrick or a nickel boat filled with alumina powder. This feature undoubtedly accounts for the observation of Smith and Wanklyn (1974) that by using the hot-pouring device that they describe, the orthophosphate (or orthovanadate) crystals could be recovered without any evidence for thermal shock or mechanical damage. Tanner and Smith (1975) have examined the perfection of single crystals of $Tb(PO_4)$, $Dy(PO_4)$, $Ho(PO_4)$, and $Yb(PO_4)$ that were grown by slow cooling (i.e., spontaneous nucleation) using a $Pb_2P_2O_7$ flux. These workers found that, in the case of optically-clear samples, the average dislocation density was ~10^4 cm^{-2}, whereas large dislocation-free areas were found in the case of crystals that formed as thin plates. These workers employed a hot pouring (straining) technique to recover the crystals, and they observed cracking and dislocation loops originating from the edges of the crystals as well as evidence for plastic deformation that had apparently occurred after the growth was complete.

RE-, Sc-, and Y-orthophosphate crystal morphology

In the case of the monoclinic rare-earth orthophosphates grown using the method outlined above, the crystals generally grow with two habits that are morphologically quite different. Rappaz et al. (1981a,b) have determined the relationship between the growth faces of the monoclinic crystals and the *b* axis using precession methods. Figure 3 illustrates the two monoclinic crystal morphological forms as exemplified by $La(PO_4)$ and $Eu(PO_4)$ and the relation of the observed crystal faces to the *b*-axis. The monoclinic orthophosphates, $Ce(PO_4)$, $Pr(PO_4)$, and $La(PO_4)$, are more frequently characterized by the growth habit illustrated on the left of Figure 3, whereas $Nd(PO_4)$, $Sm(PO_4)$, and $Pr(PO_4)$ are frequently characterized by the morphology shown at the right of Figure 3. Occasionally, a third monoclinic growth habit that (not shown in Fig. 3) is observed in which the crystals consist of thin, flat {010} plates, the edges of which form the monoclinic angle of the crystal structure.

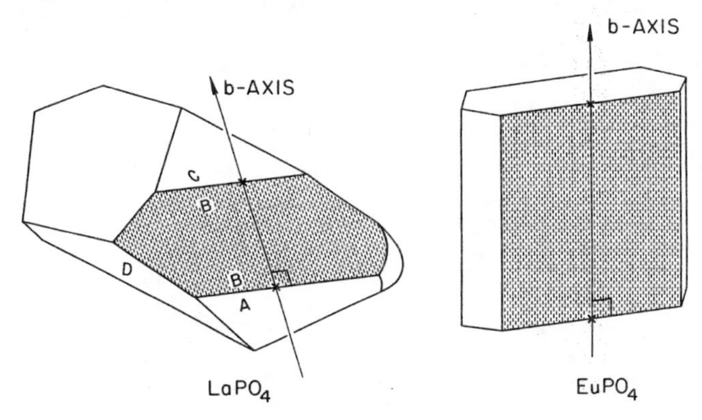

Figure 3. Single-crystal habits for monoclinic rare-earth orthophosphates grown from a $Pb_2P_2O_7$ flux. The relationship between the monoclinic crystal *b*-axis and the crystal faces for the two morphologies is indicated. In some cases, a third habit (not shown) occurs that consists of thin, diamond-shaped plates whose angles are determined by the monoclinic angle β (after Rappaz et al. 1981a).

With the exception of $Sc(PO_4)$, all of the tetragonal orthophosphates grow with a rectangular parallelepiped habit like that shown in Figure 2 for $Y(PO_4)$. The long dimension of the crystals lies along the tetragonal *c*-axis. The orthophosphate crystals generally nucleate and grow as elongated {001} plates on the top surface of the $Pb_2P_2O_7$ flux. Smaller crystals frequently exhibit tetragonal prism forms and pyramid-capped ends, but the larger crystals usually grow faster parallel to the $Pb_2P_2O_7$ nutrient liquid surface so that the dimension (thickness) of the face in the direction perpendicular to the flux surface is smaller. The growth habit of flux-grown single crystals of $Sc(PO_4)$ is generally not in the form of elongated bars but is more compact as exemplified by the large faceted scandium orthophosphate crystal shown in Figure 1.

Impurities and dopant incorporation

In the case of growth in a $Pb_2P_2O_7$ solvent, some level of contamination by lead is anticipated, and does in fact occur –particularly in the case of $Ce(PO_4)$. Low-level trivalent lead impurities were observed in $Y(PO_4)$ by Abraham et al. (1980a) and investigated using electron paramagnetic resonance (EPR) spectroscopy. Figure 4 illustrates the EPR spectrum due the Pb^{3+} "even" isotopes, along with EPR lines due to Fe^{3+} and Gd^{3+} impurities. More recently, the nature and dependence of the Pb content that is present in orthophosphate crystals grown using a $Pb_2P_2O_7$ flux has been studied quantitatively by Donovan et al. (2002a,b). These studies were undertaken as a result of the currently widespread use of flux-grown RE, Y, and Sc orthophosphate single crystals as microprobe analytical standards for rare-earth, Y, and Sc analysis (Jarosewich and Boatner 1991). The analytical results of Donovan et al. (2002a,b) showed that none of the xenotime-structure orthophosphates [$Y(PO_4)$, $Sc(PO_4)$ or Gd-Lu(PO_4)] contained appreciable amounts of Pb. Only the monoclinic-structure compounds $Ce(PO_4)$ [and possibly $La(PO_4)$ and $Sm(PO_4)$] were found to contain Pb in a concentration sufficient to result in a 2 to 4% variance in composition.

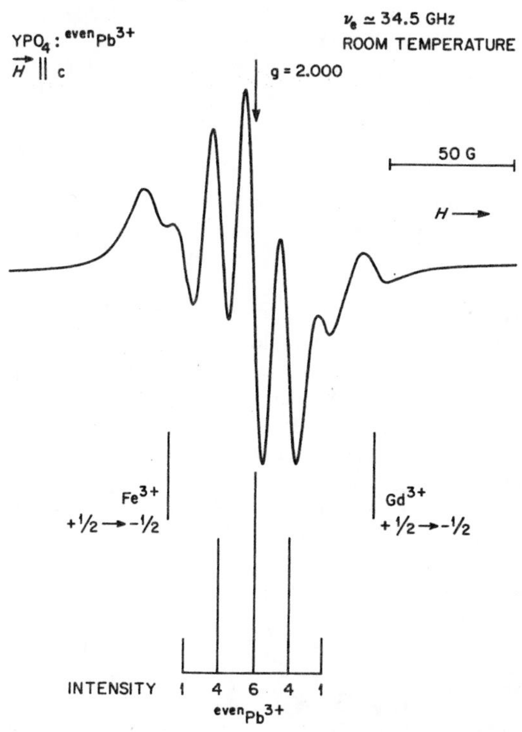

Figure 4. The electron paramagnetic resonance (EPR) spectrum of $Y(PO_4)$ (obtained with the applied magnetic field along the crystal *c*-axis) shows signals from the "even" isotopes of Pb that exhibit a superhyperfine structure due to interactions with four equivalent ^{31}P ($I = 1/2$) neighbors. Also shown are EPR lines due to impurities of Fe^{3+} and Gd^{3+}. EPR lines due to ^{207}Pb are also observed in the full spectrum, but are not shown here. Low levels of Pb impurities are incorporated in orthophosphate single crystals grown from a $Pb_2P_2O_7$ flux, but the concentration levels are significant mainly in the case of $Ce(PO_4)$ and not in the case of the tetragonal-structure orthophosphates (after Abraham et al. 1980a).

The tetragonal RE, Y, and Sc orthophosphates in particular have been widely used as host media for a variety of solid state chemical, spectroscopic, magnetic resonance, neutron and other studies of rare-earth and actinide impurities. These materials have proved to be ideal hosts for the incorporation of other rare-earth dopants (e.g., Er-doped $Lu(PO_4)$ for microlaser studies). Doped orthophosphates with desired levels of dopants are desirable for both basic investigations and applications. Unfortunately, there are apparently no available quantitative data on the segregation coefficients for the rare earths in the tetragonal orthophosphates.

Experience has shown that the lighter rare earths may be rejected to a significant degree during the flux growth of the heavier RE orthophosphates such as $Lu(PO_4)$. In practice, it has proven difficult, for example, to incorporate more than 1.0% Ce in $Lu(PO_4)$. In the case of doped orthophosphate crystals, the actual dopant concentration is usually determined by quantitative analysis subsequent to crystal growth by using techniques such as glow-discharge mass spectrometry. In the case of the growth of orthophosphate single crystals that are doped with radioactive actinide elements, special glove-box containment facilities are used because of the alpha activity of these materials. In previous work, the method described here for orthophosphate single-crystal growth by utilizing a high-temperature solvent has been applied to the preparation of U-, Np-, Pu-, Am-, Cm-, and Cf-doped specimens. These unique materials have been the subject of a wide range of fundamental solid state chemical studies (Murdoch et al. 1996, Sytsma et al. 1995, Kot et al. 1993a,b; Abraham et al. 1980a,b, 1982, 1985, 1987; Liu et al. 1997, 1998; Kelly et al. 1981, Huray et al. 1982).

Orthophosphate crystal size and alternate growth methods

The single crystal of synthetic xenotime shown in Figure 2 represents the approximate practical size limit achieved during numerous laboratory-scale spontaneously-nucleated flux crystal growths carried out over a period of ~20 years. The flux crystal growth approach is extremely useful for producing research specimens with various doping levels, compositions, etc., because eight or more growth runs in 50 ml Pt crucibles can be carried out simultaneously in a laboratory-size box furnace. The relatively recent identification of a variety of practical applications for the tetragonal-symmetry orthophosphates, however, has created a need for the controlled growth of larger single crystals of these materials. Unfortunately, attempts to grow large crystals of, for example, $Lu(PO_4)$ from the melt in iridium crucibles have not been successful because these materials do not melt congruently and exhibit some decomposition at the melting temperature. This leaves, as alternative growth methods, either hydrothermal growth, like that used to commercially grow single crystals of berlinite [$Al(PO_4)$], or top-seeded solution growth—a more sophisticated variant of high-temperature solution growth. Using the latter, Eigermann et al. (1978) have previously determined the solubility curves for $Lu(PO_4)$, $Yb(PO_4)$, $Tm(PO_4)$, $Er(PO_4)$, $Ho(PO_4)$, $Dy(PO_4)$, $Tb(PO_4)$, and $Yb(PO_4)$ in $Pb_2P_2O_7$. These results are illustrated in Figure 5. Additionally, these workers determined the saturation temperatures of various concentrations of $Y(PO_4)$ in $Pb_2P_2O_7$ to which varying amounts of PbO were added and found that the solubility increased with increasing PbO content in the flux. The growth of rare-earth orthophosphate crystals from a lead-phosphate flux consisting of various mixtures of PbO and P_2O_5 has also been investigated by Wanklyn (1978). The data of Eigermann et al. (1978) and Wanklyn (1978) are particularly valuable in regard to the application of top-seeded-solution growth to the tetragonal rare-earth and yttrium orthophosphates. Top-seeded-solution crystal-growth studies of the tetragonal-symmetry orthophosphates are currently ongoing in at least two laboratories. Finally, Hintzman and Müller-Vogt (1969) have described a variation of the flux growth technique in which a relatively small oscillating temperature that is superimposed on the normal linear cooling rate used during standard flux growth is employed. As indicated by the size of the crystals obtained by means of the oscillating temperature method, as shown in the work of Hintzman and Müller-Vogt, there is no apparent advantage to this approach. Finally, a method for the transfer and growth of single crystals of $Ho(PO_4)$ has been described by Orlovskii et al. (1975). In this relatively unusual approach, the authors found that the crystal-growth habit strongly depended on the crystallization temperature, and they were able to grow $Ho(PO_4)$ crystals as a dipyramid, dipyramid with prism face, columnar, and blade-like crystals.

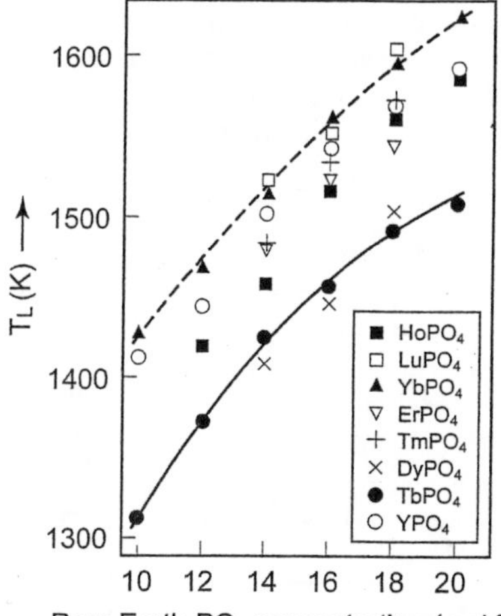

Figure 5. The solubility curves of the rare-earth orthophosphates in $Pb_2P_2O_7$ are shown [plotted from the data of Eigermann et al. (1978)]. The solubility data are plotted for $Lu(PO_4)$, $Yb(PO_4)$, $Tm(PO_4)$, $Er(PO_4)$, $Ho(PO_4)$, $Dy(PO_4)$, $Tb(PO_4)$, and $Y(PO_4)$. The solubility shows a decrease with increasing atomic number of the rare-earth elements, although the differences in solubility between neighboring elements are relatively small.

CHEMICAL SYNTHESIS OF MONAZITE, PRETULITE, AND XENOTIME

Urea precipitation

A method has been described by Abraham et al. (1980b) that can be used to synthesize all of the rare-earth orthophosphates, plus $Y(PO_4)$ and $Sc(PO_4)$, in powder form and that exercises a relatively high degree of control over the particulate size. This process has proven to be extremely useful and versatile in the preparation of high-purity, stoichiometric orthophosphate powders for use in a wide range of investigations, including studies of sintering and compaction, x-ray diffraction determinations of orthophosphate thermal expansion coefficients, and the preparation of doped cathodoluminescent phosphors.

In the urea synthesis and precipitation process illustrated schematically in Figure 6, the RE oxides are converted into the orthophosphate form by first dissolving a given oxide in a hot nitric acid solution. Through the addition of ammonium dihydrogen phosphate to the solution, a metathesis reaction is initiated that forms the RE, Y, or Sc orthophosphate. To carry out the controlled precipitation process that yields a uniform particle size, urea $[(NH_2)_2CO]$ is added in a granular form, and the resulting mixture is heated at ~180°C until the orthophosphate precipitate begins to form. Upon further heating in a fume hood to ~400°C, ammonia, nitrogen, hydrogen, water vapor, and CO_2 are evolved. At this point, the material is transferred to an alumina crucible and a final calcination is carried out between 800 and 850°C. Larger amounts of urea correspond to the formation of smaller orthophosphate particles in the resulting powder. Abraham et al. (1980b) employed up to a 720:1 mole ratio of urea to orthophosphate and were able to produce relatively monodispersed, sub-micron $Ce(PO_4)$ powders. As indicated in the flow chart shown in Figure 6, the powders produced by controlled precipitation from urea can be used to form high-density orthophosphate ceramic bodies by hot pressing or by standard cold pressing and sintering methods. Using powder formed by this approach, Abraham et al. (1980b) were able to produce orthophosphate ceramics with up to 97% of the theoretical density by hot pressing at 1100°C at 280 bar (4000 psi) for one hour. Floran et al. (1981a,b) also investigated the use of orthophosphate powders synthesized by urea precipitation to form synthetic monazite ceramics by employing both hot pressing and cold pressing followed by sintering.

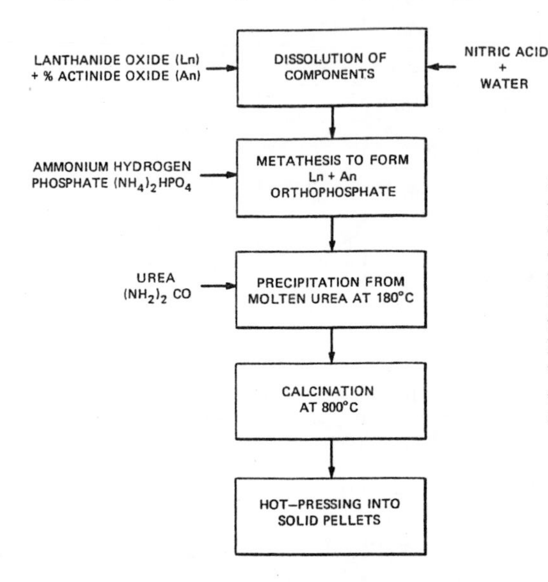

Figure 6. Flow diagram for the preparation of rare-earth, Y, and Sc orthophosphate powders with controlled particle size by means of precipitation from molten urea. Powders synthesized by this technique have been compacted by hot pressing to form ceramic bodies with >97% of the theoretical density of the compound (after Abraham et al. 1980b).

In the case of urea-precipitation-synthesized $Ce(PO_4)$ powder, Floran et al. (1981a) produced a ceramic microstructure consisting of large uniaxially-oriented cryptocrystalline grains by hot pressing at 340 bar for 1 hour at 1000°C. The urea precipitation process was also employed by Petek et al. (1982) to synthesize mixed rare-earth orthophosphates that contained a diverse suite of iron-group and other cations (i.e., either cations found in simulated light water reactor wastes or in Savannah River defense waste). By using the urea precipitation process, Petek et al. were able to synthesize $La(PO_4)$ powders that contained up to 10 wt % of chemically simulated light water reactor waste. These powders were subsequently compacted using both cold pressing and sintering or hot pressing (4000 psi, 1050°C, 1 h), and in both cases, ceramic bodies with a density of 96% of theoretical were obtained, and no phases other than the monoclinic monazite phase were observed. Additionally, in their studies of the chemical durability of monazite loaded with Cs, Sr, or U, Sales et al. (1983) employed the urea precipitation process to synthesize doped $La(PO_4)$ precursor powders that were subsequently compacted using hot pressing to produce standard high-density samples for leach testing in aqueous media. The preparation of chemically diverse light rare-earth-element orthophosphates by means of urea precipitation and studies of the subsequent formation of high-density ceramic bodies by various compaction methods have been reviewed and summarized previously in some detail by Boatner and Sales (1988).

Metathesis reactions with BPO_4

Rare-earth orthophosphates and the actinide phosphates $Pu(PO_4)$ and PuP_2O_7 have been synthesized using a process developed by Bamberger (1982) and Bamberger et al. (1984) that consists of a metathesis reaction with boron phosphate $B(PO_4)$. Bamberger first synthesized $B(PO_4)$ by heating mixtures of $H_3(PO_4)$ and H_3BO_3 to 1000°C in air.

Subsequently, the $B(PO_4)$ was mixed with the rare-earth oxide and heated in a Pt boat at temperatures up to 1000°C. In the case of the formation of $Ce(PO_4)$, the metathesis reaction proceeded according to the relation:

$$CeO_2 + B(PO_4) \rightarrow Ce(PO_4) + 1/2\,(B_2O_3) + 1/4\,(O_2)\,.$$

The evolution of oxygen in this reaction was reported to occur in the range of 550 to 600°C. A metathesis reaction using $B(PO_4)$ was also shown to be effective in producing rare-earth orthophosphates starting with the trifluoride form of the rare earth, and this reaction proceeded according to the relation:

$$NdF_3 + B(PO_4) \rightarrow Nd(PO_4) + BF_3\,.$$

Bamberger also applied a similar metathesis reaction process to the synthesis of phosphates of uranium and neptunium, thus this represents a versatile and alternative synthesis method for the conversion of rare-earth oxides or halides to the phosphate form.

Solid state reactions

Hikichi (1991) and Hikichi et al. (1978, 1980) have reported the solid state synthesis of the orthophosphates of La, Ce, Pr, Nd, Sm, Gd, Y, Dy, Er, and Yb by heating the rare-earth trichloride $RECl_3$, oxide RE_2O_3, CeO_2 or $Ce(NO_3)_3$ with either $(NH_4)_2H(PO_4)$, $K_2H(PO_4)$, $Na_2H(PO_4)$, or $H_3(PO_4)$, with an atomic ratio of one phosphorous per rare earth. In this process, the RE and phosphorous-containing compounds were mixed and initially dried for one day at 100°C. The material was subsequently heated in a Pt crucible from 200 to 1000°C for an additional day. The crucibles were then removed from the electric furnace and quenched to room temperature. A 1.3 N HNO_3 aqueous solution followed by rinsing in distilled water was used to clean the material of soluble unreacted compounds.

Other orthophosphate synthesis methods

A process for synthesizing monazite ceramics that contain a variety of cations (including both thorium and uranium) has been described by McCarthy et al. (1978). This synthesis employed nitrate solutions of the REs, calcium, thorium, UO_2, $NH_4H_2(PO_4)$, and colloidal SiO_2 that were initially heated to form a paste. The paste was subsequently calcined at 600°C for two hours. The resulting material was ground and cold-pressed into pellets that were fired in the range 1050 to 1200°C for four or 48 hours. Different charge-compensation mechanisms that involve the addition of calcium or silicon were investigated. In the case of the highest concentrations of uranium, phase-pure monazite was not obtained, and the phases $U_3(PO_4)_4$ and $U_3P_2O_{10}$ were observed along with the dominant monazite component.

A direct precipitation technique for the preparation of the RE orthophosphates has been described by Hikichi et al. (1978) in which a 0.05 mol/l solution of the rare-earth trichloride was added to a dilute, stirred $H_3(PO_4)$ solution. The mixed solutions were then maintained at 20, 50, and 90°C for 1 to 900 days and were titrated to maintain a constant pH by adding $H_3(PO_4)$ or other phosphates. Both the hydrated low-temperature hexagonal forms of the rare-earth phosphates (as discussed in the previous section on Chemical composition) and the anhydrous RE orthophosphates were produced by Hikichi et al. (1978).

Hydrothermal methods have been used by Gratz and Heinrich (1997) to synthesize solid solutions of $(Ce,Y)PO_4$ at temperatures in the range of 300 to 1000°C and at pressures between two and 15 kbar. These workers reported the formation of two immiscible phases (monazite and xenotime) over a wide range of compositions. Complete solid solutions between $La(PO_4)$ and $(Ca_{0.5}Th_{0.5})PO_4$ were also hydrothermally synthesized at 780°C and 200 MPa by Podor and Cuney (1997). Their results show: (1) that the substitution of thorium in monazite is not limited by the temperature and pressure conditions experienced by granitic magmas, and (2) that radiogenic lead is stable in natural monazites.

APPLICATIONS OF RE-, SC -, AND Y-ORTHOPHOSPHATES

Orthophosphate waste forms for actinides and high-level radioactive wastes

The application of the RE orthophosphates to the disposal or long-term storage of actinides and high-level radioactive wastes is thoroughly treated by Ewing and Wang in this volume. This concept was advanced by Boatner et al. (1980, 1981a, 1983) and McCarthy et al. (1978), and the use of monazite as a radioactive waste form was subsequently reviewed by Boatner and Sales (1988). In brief, this concept was initially based on four known and established primary characteristics of monazite that are clearly advantageous for its use as a radioactive waste form. These characteristics are:

(1) The established chemical stability of monazite in the earth's crust over geological time scales (Floran et al. 1981b, Pasteels 1970).

(2) The ability of monazite to incorporate large amounts of thorium and uranium (see Table 1) and the structural compatibility of monazite with $Pu(PO_4)$ (Bamberger et al. 1984).

(3) The ability of monazite to recover from displacive radiation damage and thereby retain its crystallinity and chemical integrity [Ewing and Wang, this volume; Meldrum et al. (1997a,b,c; 1998, 2000)].

(4) Monazite exhibits a negative temperature coefficient of solubility and, therefore, will actually be less soluble during the highly radioactive early "thermal" period of a high-level radioactive waste form (Marinova and Yaglov 1976).

In the case of the disposal of high-level light water reactor waste, these wastes intrinsically contain up to 35 wt % rare-earth oxides that could be directly converted to form part of the monazite host matrix. Because Ewing and Wang (this volume) have provided a comparison of the properties of monazite-based waste forms to those of other crystalline host media and nuclear waste phosphate glasses, no further details regarding this application of the rare-earth orthophosphates will be given here.

Gamma- and X-ray scintillator and phosphor applications

The heavy zircon-structure rare-earth orthophosphate $Lu(PO_4)$ (or LOP), when activated with cerium, has been found to be a fast, efficient, and dense scintillator for the detection of gamma rays (Lempicki et al. 1993, Wojtowicz et al. 1994, 1995; Moses et al. 1997, 1998). This scintillator was found to have a 24 nsec decay time constant and a light output that is over double that of bismuth germanium oxide. Another Lu-based, but non-phosphate efficient and fast scintillator, $Lu_2O(SiO_4)$:Ce (or LSO) is currently finding widespread application in medical diagnostic equipment such as positron emission tomography systems. This lutetium silicate has a very important advantage over LOP in that the compound melts congruently, and large single crystals can be extracted from the melt (i.e., Czochralski growth). As previously noted, top-seeded-solution or hydrothermal techniques will have to be developed in order to grow large single crystals of LOP, and this has yet to be accomplished. More recently, Nd-activated $Lu(PO_4)$ and $Y(PO_4)$ have been found to be interesting VUV scintillators whose short wavelength output (~190 nm) can be matched to appropriately activated proportional counters, thereby eliminating the need for photomultiplier tubes and their associated electronics (Wisniewski et al. 2002a,b). Other candidate orthophosphate scintillators and phosphors have been identified by Moses et al. (1997, 1998). Specifically, Eu- and Sm-doped $Lu(PO_4)$, $Sc(PO_4)$, and $Y(PO_4)$ were found to have a high light output in the wavelength range of 600 to 900 nm. In the case of $Lu(PO_4)$:20%Eu, Moses et al. reported a value of 123,171 photons per MeV in the 600-900 nm range. This light-output region matches the high quantum efficiency region of silicon photodiode detectors. Orthophosphates doped with Tb, Dy, Er, Pr, and Tm were also reported by Moses et al. to have a significant light output when excited by x-rays. Although the decay times of some of these scintillators may be too long for them to find applications in some medical imaging devices, they may be appropriate for other applications, including imaging screens and displays.

Allison et al. (1995) have shown that $Lu(PO_4)$:(1% Dy, 2% Eu) can be used as a thermophosphor that can be calibrated to carry out high-temperature measurements on moving parts in a remote, non-contact mode. Similarly, both $Y(PO_4)$ and $Sc(PO_4)$ doped with Dy and Eu can be used for remote, non-contact, high-temperature measurements, e.g., on pistons operating in a reciprocating engine, on moving turbine blades, or in environments where it is not practical to use metallic thermocouples - such as in microwave sintering furnaces (Allison et al. 1998, 1999).

Orthophosphates as weak interfaces in ceramic composites

Morgan and Marshall (1993, 1995, 1996, 1997) and Marshall et al. (1997, 1998) have introduced the concept of using $La(PO_4)$, xenotime, and other rare-earth orthophosphates as weak interfaces in ceramic composites. In this concept, the fiber component of a ceramic composite is coated with a material (e.g., a rare-earth orthophosphate) that will debond from the fiber. This allows fiber slippage or "pull out" to occur as a mechanism for arresting crack propagation by deflecting the crack along the interface, thereby avoiding failure by brittle fracture. The demands on a coating material of this type are severe. Morgan and Marshall (1995) have identified the following required characteristics for a weak-interface material: (1) chemical and morphological compatibility, (2) refractory behavior, (3) stability in oxidizing (and limited reducing) conditions, (4) stability in H_2O vapor and CO_2 environments, and (5) weak interfaces that exhibit the desired debonding behavior.

The application of weak-interface RE and Y orthophosphate coatings to both Al_2O_3 fibers and ZrO_2-based ceramics has been investigated. Marshall et al. (1997) have studied composites formed by alternating layers of monazite and various types of ZrO_2 (e.g., Y-ZrO_2 and Ce-ZrO_2). They found that no reaction occurred between $La(PO_4)$ and Y-ZrO_2 at temperatures as high as 1600°C. Similar results were obtained by Marshall et al. (1998) for the case of the Al_2O_3/$La(PO_4)$ system, and as long as the ratio of lanthanum to phosphorous was near 1:1, the $La(PO_4)$/Al_2O_3 interface remained sufficiently weak to exhibit debonding and deflect cracks. Reactions occurred, however, if excess La or P was present, and other phases [e.g. berlinite, $Al(PO_4)$] were formed on heating to 1600°C. Research on structural ceramic applications of this type is continuing at several laboratories, and clearly, the development of applications of this nature would create a significantly increased demand for the rare-earth elements.

Other applications of RE-, Sc -, and Y-orthophosphates

The rare-earth orthophosphates plus $Y(PO_4)$ and $Sc(PO_4)$ represent the basis for a set of reference samples for rare-earth analysis, developed by Jarosewich and Boatner (1991). The sixteen orthophosphates that are included in this microprobe reference sample set were initially synthesized in single-crystal form by using the high-temperature-solvent process described in a previous section this chapter. High-purity oxides and unused Pt growth crucibles were utilized in all of the single-crystal growth operations. These orthophosphates were determined to be homogeneous on a micron scale, and they were found to be stable under exposure to an electron beam for extended periods of time. In terms of contamination from other REs, neutron activation analysis revealed only traces of RE elements other than the principal RE(PO_4) component. As discussed previously, Donovan et al. (2002a,b) have recently re-examined these reference samples in terms of potential lead contamination arising from the use of the lead pyrophosphate flux in the crystal growth process. These microprobe analytical reference materials are now widely distributed on an international scale, and they have proven to be useful for the rare-earth analysis of minerals and a wide range of other materials.

The heavier rare-earth phosphates are excellent hosts for the incorporation of other rare-earth dopants, and accordingly, host systems such as Lu(PO_4) have been used in investigations of new materials for potential applications as micro lasers (Rapaport et al. 1999a,b). The lasing properties of crystals of Lu(PO_4) doped with Nd were studied by Rapaport et al. (1999a), who reported that 88% of the pump power (at 804.4 nm) was absorbed in a 1.0 mm-thick crystal. The Lu(PO_4):Er system has also been investigated by Rapaport et al. (1999b), but the prospects for using this system as a laser were not encouraging because of the difficulty in achieving a population inversion as a result of a low absorption of the 804 nm pump wavelength and up-conversion effects that serve to deplete the population of the upper Er level.

STRUCTURAL PROPERTIES OF RE-, SC-, AND Y-ORTHOPHOSPHATES

Early crystallographic studies

The structural properties of the rare-earth and related orthophosphates have been treated for over 60 years with varying degrees of detail and increasing levels of sophistication by several workers—beginning, it appears, with the work of Parish (1939) and Gliszczynski (1939). Structural information was obtained on La(PO_4), Ce(PO_4), Pr(PO_4), and Nd(PO_4) by Rose Mooney in 1944 at the Metallurgical Laboratory of the University of Chicago in conjunction with the Manhattan Project, and the results were later published in the open literature (Mooney 1948). The structural results for the hexagonal form of Ce(PO_4) and related phosphates were described in more detail in a later publication (Mooney 1950). The relationship between the ionic radii and the structures of RE phosphates, vanadates, and arsenates was examined by Carron et al. (1958). Unit cell data for the monazite-type RE orthophosphates [including data for $PmPO_4$)] were determined by Weigel et al. (1965), and data for the zircon-type rare-earth phosphates, including Sc(PO_4), and for several rare-earth arsenates and vanadates were obtained by Schwarz (1963). Crystallographic data have been obtained by Kizilyalli and Welch (1976), who prepared crystalline powders by means of a direct precipitation and calcining technique and by Pepin and Vance (1981) who also prepared and investigated a series of precipitated rare-earth orthophosphate powders. The crystal structure of natural monazite was studied by Ueda (1967) using a sample from Ishikawa-yama, Fukushima Pref. in Japan, and refinements of the structure of heat-treated monazite from India have been carried out by Ghouse (1968), whose results were described previously in the section on radiation damage effects.

Crystal chemical background and characteristics

The anhydrous 4*f* transition-series (rare-earth) elements undergo a change in ionic radius in accordance with the well-known lanthanide contraction, in which the lanthanide ions exhibit a decreasing ionic radius as the atomic number increases. This change in ionic radius is accompanied by a change in the RE orthophosphate crystal structure as one goes across the series from the light end member La(PO_4) to the heavy end-member Lu(PO_4). The orthophosphates from La(PO_4) to Eu(PO_4) only exist in the monoclinic monazite structure, space group $P2_1/n$, Z = 4, whereas the orthophosphates from Ho to Lu are only found with the tetragonal zircon structure, space group $I4_1/amd$, $Z = 4$. The orthophosphates $Gd(PO_4)$, $Tb(PO_4)$, and $Dy(PO_4)$ have been synthesized in both the monoclinic and tetragonal forms (Ushakov et al. 2001, Bondar et al. 1976) with the tetragonal structure representing the high-temperature structural form. Xenotime, $Y(PO_4)$, and pretulite, $Sc(PO_4)$, have the tetragonal zircon structure. These two different orthophosphate structural types are also common to most of the rare-earth vanadates and arsenates. The xenotime structure, however, only accepts the larger rare-earth ions as the metal-oxide tetrahedra increase in size in going from PO_4 to AsO_4 to VO_4. Ushakov et al. (2001) have noted that in the rare-earth vanadate series, only $La(VO_4)$ exhibits the monoclinic monazite structure—although the vanadates from Ce to Nd have been reported to have a low-temperature monazite polymorph. These workers also noted that, in the case of the actinide-series elements, $Am(VO_4)$ has the tetragonal zircon structure, whereas $Pu(PO_4)$ and $Am(PO_4)$ exhibit the monoclinic monazite structure. In the case of the orthosilicates, either the tetragonal structure of $Th(SiO_4)$ (thorite) or the monoclinic monazite structure (huttonite) can be formed – depending on the formation temperature—with the tetragonal zircon structural form being stable at higher temperatures. An extremely useful compilation that illustrates the structural relationships between the phosphates, silicates, arsenates, and vanadates of the rare earths and some actinides is shown in Figure 7 (after Ushakov et al. 2001).

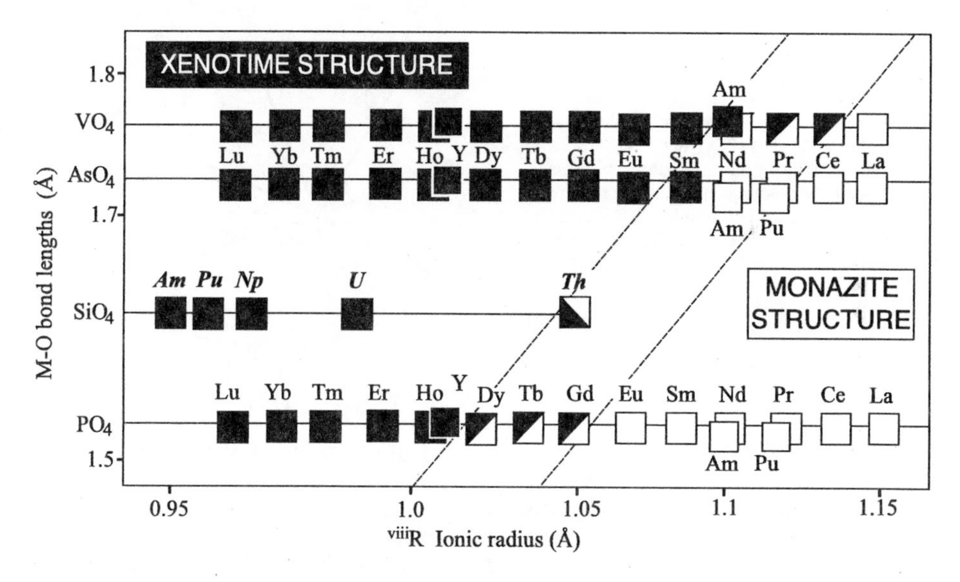

Figure 7. Illustration of the relationship between phosphates, vanadates, arsenates and silicates with the xenotime or monazite structure for compounds of the type $R(MO_4)$. Here, R represents: RE^{3+}, An^{3+}; An^{4+}; and M represents P, Si, As, or V. After Ushakov et al. (2001).

A major distinguishing crystal-chemical feature between the monoclinic monazite-structure compounds and those with the tetragonal zircon or xenotime structure lies in the oxygen coordination of the rare-earth ion in the RE-oxygen polyhedra. In the monazite structure, the RE ion is coordinated with nine oxygen ions—one of which, however, has a somewhat longer bond length (2.78 Å) than the others (~2.53 Å). A view of the REO_9 polyhedron with one attached PO_4 tetrahedron is offered in Figure 8 (Beall et al. 1981). In the tetragonal zircon structure, the RE, Y, or Sc ions are coordinated with 8 surrounding oxygen ions, and this type of coordination is illustrated in Figure 9 (Milligan et al. 1982).

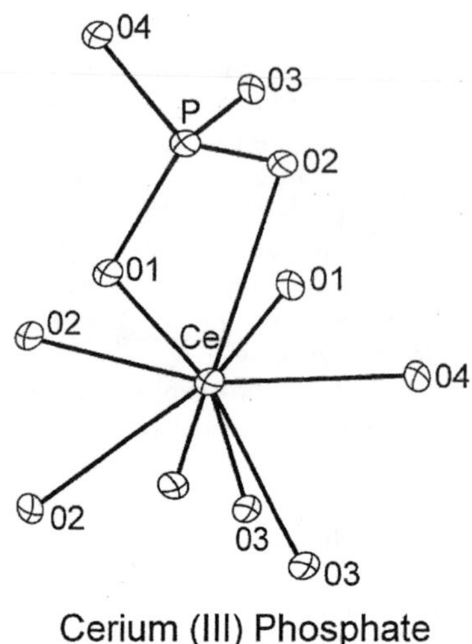

Figure 8. The nature of the nine-fold coordination characteristic of the monoclinic monazite-structure ($P2_1/n$) orthophosphates is illustrated in a view of Ce(PO_4) that also shows the linkage to one of the surrounding PO_4 structural units (after Beall et al. 1981).

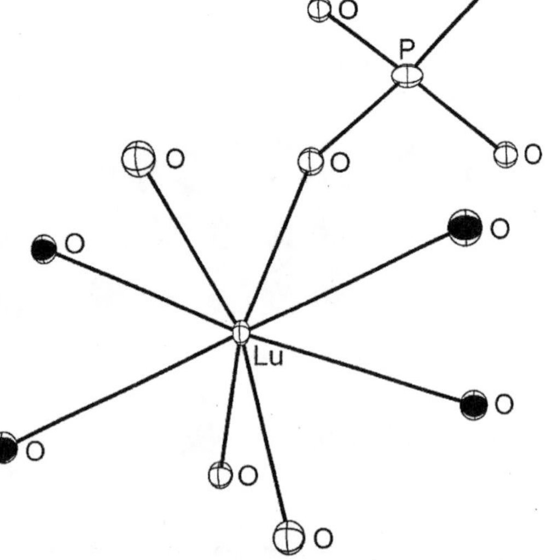

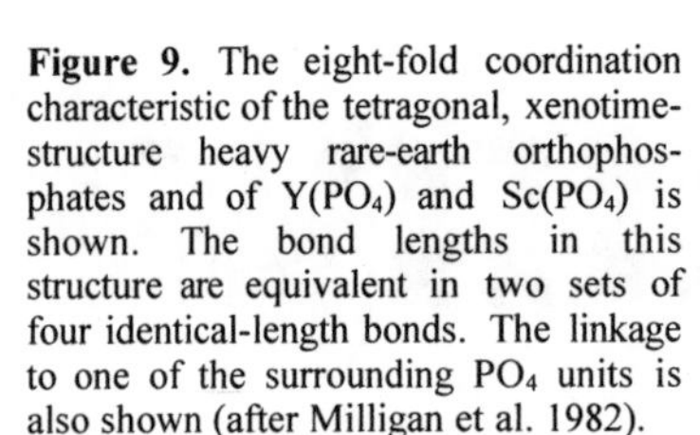

Figure 9. The eight-fold coordination characteristic of the tetragonal, xenotime-structure heavy rare-earth orthophosphates and of Y(PO_4) and Sc(PO_4) is shown. The bond lengths in this structure are equivalent in two sets of four identical-length bonds. The linkage to one of the surrounding PO_4 units is also shown (after Milligan et al. 1982).

The application of monoclinic monazites to the chemical sequestration of radioactive wastes has been noted in a prior section of the present chapter, and it is reviewed in detail by Ewing and Wang (this volume). From the crystal-chemical point of view, it has been previously suggested (Beall et al. 1981) that the observed ability of monazites to incorporate the diverse set of cations found in most radioactive wastes may be due to a relaxation of chemical constraints associated with the irregular 9-fold coordination.

Single-crystal monazite and xenotime structural refinements

Single crystals of the anhydrous monazite and xenotime structure RE orthophosphates plus Y(PO_4) and Sc(PO_4) that were grown by the $Pb_2P_2O_7$ flux technique described previously have been used in carrying out a series of single-crystal structural refinements by various workers. The structure of La(PO_4) was determined by Mullica et al. (1984) and that of Ce(PO_4) by Beall et al. (1981). Structural refinements of Pr(PO_4) and Nd(PO_4) were reported by Mullica et al. (1985a), of Sm(PO_4), Eu(PO_4), and Gd(PO_4) by Mullica et al. (1985b), of Tb(PO_4), Dy(PO_4) and Ho(PO_4) by Milligan et al.

(1983a), and the structures of $Er(PO_4)$, $Tm(PO_4)$, and $Yb(PO_4)$ by Milligan et al. (1983b). The structure of synthetic xenotime, pretulite and the RE series end member $Lu(PO_4)$ were described by Milligan et al. (1982). The structure of the mixed RE/actinide crystal $Ce_{0.9}U_{0.1}(PO_4)$ was refined and reported by Mullica et al. (1989). More recently the same suite of synthetic flux-grown orthophosphate single crystals [with the exception of $Sc(PO_4)$] was re-examined, and the crystal chemistry of these materials was described by Ni et al. (1995). These latter workers also carried out refinements for natural monazite and natural xenotime samples. The crystal data and the results of the single-crystal structural refinements based on the recent work of Ni et al. (1995) are shown in Table 3 for the monazite-structure materials and in Table 4 for the xenotime-structure phases. The data for $Sc(PO_4)$ and the available structural data for $Pm(PO_4)$ are given in Table 5.

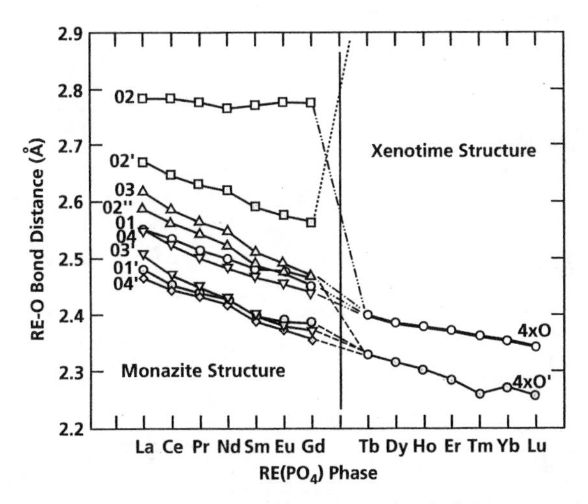

Figure 10. The variation of the rare-earth-to-oxygen bond distances are shown for a number of rare-earth orthophosphate compounds with the monoclinic monazite (nine-fold oxygen coordinated) and tetragonal xenotime (eight-fold oxygen coordinated) structures (after Ni et al. 1995).

For the monoclinic structure, there are nine distinct bond distances between the RE and oxygen ions (versus two sets of four equivalent distances for the xenotime structures). The RE-oxygen bond distances have been plotted for the various monazite and xenotime structure RE phases by Ni et al. (1995), and their results are shown in Figure 10. These workers point out the regular variation of the RE cation-to-oxygen bond distances that is typical of both the monazite and xenotime structures. The nine-fold geometric coordination of the light RE, monoclinic orthophosphates and a view of the resulting coordination between the RE ions and the orientation of the surrounding PO_4 tetrahedra is shown in Figure 11. This figure represents a view that is slightly tilted from looking directly along the *c* axis of the structure. The single-crystal refinement data were carefully examined by Mullica et al. (1984) for the case of $La(PO_4)$, and these workers determined that the proper description of this system was that of a pentagonal interpenetrating tetrahedral polyhedron (PITP). Figure 12(a) shows a typical $La(PO_4)$ structural unit that is reoriented in Figure 12(b) to illustrate the PITP polyhedron and the C_s symmetry plane. Figure 12(c) illustrates the idealized PITP and C_s plane of symmetry. The bidentate bonding to two of the PO_4 tetrahedra is illustrated in Figure 13 (top), and Figure 13 (bottom) shows a view illustrating the pentagonal arrangement of the RE-O bonds to the five PO_4 tetrahedral units that lie in the pentagonal plane.

TABLE 3. Crystal data and results of structure refinements for monazite structure phases

Phases	Monazite*	La(PO$_4$)	Ce(PO$_4$)	Pr(PO$_4$)	Nd(PO$_4$)	Sm(PO$_4$)	Eu(PO$_4$)	Gd(PO$_4$)
Unit cells by least squares (unconstrained)								
a (Å)	6.7902(10)	6.8313(10)	6.7880(10)	6.7596(8)	6.7352(10)	6.6818(12)	6.6613(10)	6.6435(9)
b (Å)	7.0203(6)	7.0705(9)	7.0163(8)	6.9812(10)	6.9500(9)	6.8877(9)	6.8618(9)	6.8414(10)
c (Å)	6.4674(7)	6.5034(9)	6.4650(7)	6.4344(9)	6.4049(8)	6.3653(9)	6.3491(8)	6.3281(6)
α (°)	90.007(8)	89.993(11)	89.997(9)	89.997(11)	90.004(10)	89.982(11)	90.005(11)	90.013(9)
β (°)	103.38(1)	103.27(1)	103.43(1)	103.53(1)	103.68(1)	103.86(1)	103.96(1)	103.976(9)
γ (°)	89.989(9)	89.981(11)	90.011(10)	89.979(10)	89.967(11)	89.979(12)	89.975(12)	89.983(11)
No. collected	3541	1945	1910	1884	1860	1812	1797	1778
No. unique	945	891	943	861	851	827	819	811
No. > $3\sigma_I$	774	790	717	753	763	732	712	710
R_{merge}	0.012	0.014	0.011	0.012	0.010	0.013	0.012	0.011
R	0.015	0.017	0.014	0.016	0.015	0.016	0.016	0.016
R_w	0.023	0.026	0.019	0.021	0.017	0.019	0.019	0.019
Largest peaks on difference maps (e/Å^3)								
(+)	0.741	0.897	0.653	0.971	0.841	0.885	1.039	1.073
(−)	0.704	1.023	0.632	1.276	0.995	1.236	0.884	1.265
Atomic positions								
RE x	0.28152(4)	0.28154(2)	0.28182(2)	0.28177(2)	0.28178(3)	0.28153(3)	0.28152(3)	0.28150(3)
y	0.15929(4)	0.16033(2)	0.15914(2)	0.15862(3)	0.15806(3)	0.15638(3)	0.15595(3)	0.15529(3)
z	0.10006(4)	0.10068(2)	0.10008(2)	0.09988(2)	0.09950(3)	0.09813(3)	0.09757(3)	0.09695(3)
B (Å^2)	0.329(4)	0.294(3)	0.431(3)	0.351(3)	0.280(3)	0.228(3)	0.193(3)	0.306(3)
P x	0.3048(2)	0.3047(1)	0.3047(1)	0.3040(1)	0.3037(1)	0.3034(2)	0.3029(1)	0.3031(2)
y	0.1630(2)	0.1639(1)	0.1635(1)	0.1630(1)	0.1626(1)	0.1618(2)	0.1615(1)	0.1612(2)
z	0.6121(2)	0.6121(1)	0.6124(1)	0.6127(1)	0.6127(1)	0.6130(2)	0.6130(1)	0.6131(2)
B (Å^2)	0.31(2)	0.33(1)	0.51(1)	0.36(1)	0.30(1)	0.25(2)	0.24(1)	0.38(2)
O1 x	0.2501(5)	0.2503(4)	0.2508(4)	0.2498(4)	0.2502(4)	0.2499(5)	0.2513(4)	0.2539(5)
y	0.0068(5)	0.0077(4)	0.0055(4)	0.0051(4)	0.0046(4)	0.0020(5)	0.0012(4)	0.0013(5)
z	0.4450(5)	0.4477(4)	0.4458(4)	0.4441(4)	0.4430(4)	0.4405(5)	0.4409(4)	0.4385(5)
B (Å^2)	0.63(6)	0.78(4)	0.93(4)	0.69(4)	0.67(5)	0.57(6)	0.68(5)	0.80(6)
O2 x	0.3814(5)	0.3799(4)	0.3811(4)	0.3816(4)	0.3815(4)	0.3822(5)	0.3833(4)	0.3837(5)
y	0.3307(6)	0.3315(3)	0.3320(3)	0.3327(4)	0.3331(4)	0.3341(5)	0.3350(4)	0.3346(5)
z	0.4975(6)	0.4964(4)	0.4982(4)	0.4990(4)	0.4987(4)	0.5008(5)	0.5017(4)	0.5024(5)
B (Å^2)	0.69(6)	0.61(4)	0.80(4)	0.67(4)	0.61(4)	0.57(5)	0.52(5)	0.66(5)
O3 x	0.4742(6)	0.4748(4)	0.4745(4)	0.4744(4)	0.4747(4)	0.4748(5)	0.4743(4)	0.4729(5)
y	0.1070(6)	0.1071(4)	0.1054(4)	0.1046(4)	0.1040(4)	0.1025(5)	0.1022(4)	0.1016(5)
z	0.8037(6)	0.8018(4)	0.8042(4)	0.8057(4)	0.8073(4)	0.8102(5)	0.8116(4)	0.8126(5)
B (Å^2)	0.73(6)	0.63(4)	0.84(4)	0.63(4)	0.62(5)	0.56(5)	0.53(5)	0.62(5)
O4 x	0.1274(5)	0.1277(3)	0.1268(3)	0.1260(4)	0.1249(4)	0.1217(5)	0.1204(4)	0.1187(5)
y	0.2153(5)	0.2168(3)	0.2164(4)	0.2150(4)	0.2153(4)	0.2125(5)	0.2135(4)	0.2138(5)
z	0.7104(6)	0.7101(4)	0.7108(4)	0.7120(4)	0.7127(4)	0.7113(5)	0.7119(5)	0.7131(5)
B (Å^2)	0.63(6)	0.66(4)	0.77(4)	0.62(4)	0.52(4)	0.57(5)	0.48(4)	0.67(5)

* Natural sample.

TABLE 4. Crystal data and results of structure refinements for xenotime structure phases

Phases	Xenotime*	$Tb(PO_4)$	$Dy(PO_4)$	$Ho(PO_4)$	$Er(PO_4)$	$Tm(PO_4)$	$Yb(PO_4)$	$Lu(PO_4)$
Unit cells by least squares (unconstrained)								
a (Å)	6.8951(6)	6.9319(12)	6.9046(12)	6.8772(8)	6.8510(13)	6.8297(7)	6.8083(9)	6.7848(14)
b (Å)	6.8943(5)	6.9299(11)	6.9057(9)	6.8773(12)	6.8505(16)	6.8290(10)	6.8103(6)	6.7807(10)
c (Å)	6.0276(6)	6.0606(11)	6.0384(6)	6.0176(8)	5.9968(10)	5.9798(10)	5.9639(5)	5.9467(6)
α (°)	89.993(7)	90.016(14)	90.013(9)	90.004(13)	89.986(16)	90.009(12)	89.987(7)	89.969(10)
β (°)	90.012(8)	89.997(14)	90.019(12)	89.991(10)	89.963(14)	89.967(11)	89.997(9)	90.008(12)
γ (°)	90.018(7)	89.997(13)	90.006(13)	89.988(13)	90.026(17)	89.996(10)	90.019(9)	90.010(14)
No. collected	919	924	922	903	891	884	890	1046
No. unique	145	135	134	132	130	129	128	151
No. > $3\sigma_I$	109	113	112	111	108	107	97	119
R_{merge}	0.018	0.010	0.011	0.014	0.014	0.020	0.013	0.016
R	0.016	0.009	0.008	0.015	0.012	0.017	0.013	0.009
R_w	0.028	0.019	0.019	0.021	0.022	0.026	0.033	0.012
Largest peaks on difference maps (e/Å^3)								
(+)	0.624	0.464	0.496	0.669	1.105	0.934	0.817	0.978
(−)	0.766	0.498	0.537	1.283	0.922	1.187	0.734	0.802
Atomic positions								
RE x	0	0	0	0	0	0	0	0
y	0.75	0.75	0.75	0.75	0.75	0.75	0.75	0.75
z	0.125	0.125	0.125	0.125	0.125	0.125	0.125	0.125
B (Å^2)	0.455(7)	0.350(4)	0.346(4)	0.049(6)	0.406(7)	0.104(7)	0.28(1)	0.245(2)
P x	0	0	0	0	0	0	0	0
y	0.25	0.25	0.25	0.25	0.25	0.25	0.25	0.25
z	0.375	0.375	0.375	0.375	0.375	0.375	0.375	0.375
B (Å^2)	0.62(2)	0.43(2)	0.43(2)	0.26(4)	0.48(4)	0.14(4)	0.50(7)	0.31(2)
O x	0	0	0	0	0	0	0	0
y	0.0753(6)	0.0764(5)	0.0760(6)	0.0757(8)	0.0743(9)	0.072(1)	0.074(1)	0.0735(4)
z	0.2158(6)	0.2175(5)	0.2162(6)	0.2165(8)	0.216(1)	0.213(1)	0.215(1)	0.2138(4)
B (Å^2)	0.94(6)	0.60(5)	0.61(5)	0.38(8)	0.65(9)	0.3(1)	0.5(1)	0.52(4)

* Natural sample.

Table 5. Sc(PO_4) (pretulite) and Pm(PO_4) crystallographic data.

	Summary data		***$ScPO_4$ bond and contact distances (Å) & bond angles (°)***	
	$ScPO_4$	***$PmPO_4$***		
a (Å)	6.574(1)	6.72	M-O	2.153(1)
b (Å)	-	6.89		2.260(1)
c (Å)	5.791(1)	6.37	Avg.	2.206
β	-	104° 17′	P-O	1.534(1)
D_c (g cm^{-3})	3.71	5.62	**Phosphate Group**	
V ($Å^3$)	250.27	-	O(1)-O(2)	2.374(3)
			O(3)-O(4)	
			O(1)-O(3), O(4)	2.569(2)
			O(2)-O(3), O(4)	
			O(1)-P-O(2)	101.39(2)
			O(3)-P-O(4)	
			O(1)-P-O(3), O(4)	113.72(2)
			O(2)-P-O(3), O(4)	
			Avg. (O-P-O)	109.61

Sc(PO_4) data after Milligan et al. (1982); Pm(PO_4) data after Kizilyalli & Welch (1976)

$ScPO_4$ atomic positions and thermal parameters (Å × 10^2)[a]

Atom	*x*	*y*	*z*	U_1	U_{22}	U_{33}	U_1	U_{13}	U_{23}
Sc	0	3/4	1/8	0.43(1)	-	0.34(2)	0	0	0
P	0	1/4	3/8	0.39(2)	-	0.24(3)	0	0	0
O	0	0.4305(2)	0.2071(2)	0.89(5)	0.52(4)	0.38(3)	0	0	0.07(3)

[a] The anisotropic temperature factors are of the form:
$T = \exp[-2\pi^2(U_{11}h^2a^{*2}+U_{22}k^2b^{*2}+U_{33}l^2c^{*2}+2U_{12}hka^*b^*\cos\gamma^*+2U_{13}hla^*c^*\cos\beta^*+2U_{23}klb^*c^*\cos\alpha^*)]$
where U_{ij} values are the thermal parameters denoted in terms of mean-squared amplitudes of vibration.

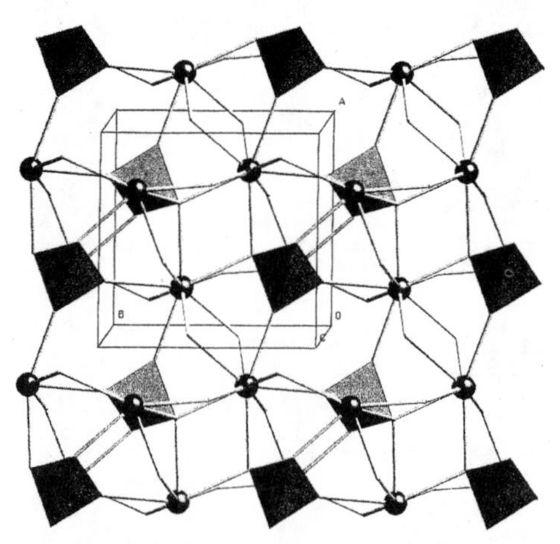

Figure 11. The nine-fold coordination of the rare-earth ion in the monoclinic monazite structure is illustrated along with the linkage to various surrounding PO_4 tetrahedral units. This view is slightly off of "looking down" the *c*-axis of the structure.

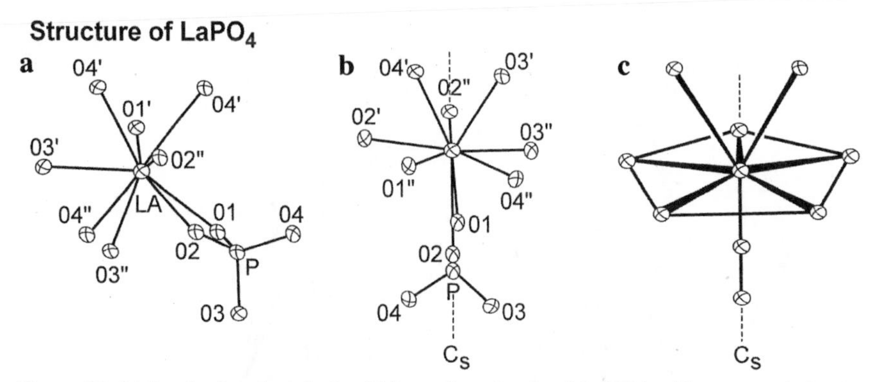

Figure 12. (a) Randomly oriented nine-fold coordinated unit of La(PO_4) with one attached PO_4 tetrahedron, (b) the same structural unit as shown in (a) but reoriented to the reveal thecoordination described in the text and the C_s symmetry plane, (c) idealized pentagonal rare-earth-to-oxygen bond arrangement (after Mullica et al. 1984).

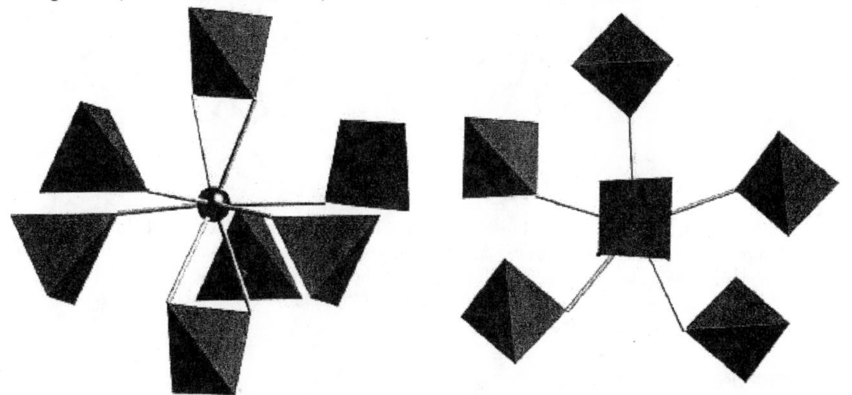

Figure 13. (Left) The bidentate bonding to two of the PO_4 tetrahedral units in the monoclinic monazite structure is illustrated. (Right) The pentagonal arrangement of the rare-earth-to-oxygen bonds on five surrounding PO_4 tetrahedra is illustrated in a projected view.

The structure of the monoclinic materials consists of chain-like interlocked strands of nine-fold coordinated polyhedra that are linked together by the PO_4 tetrahedral units (Mullica et al. 1985a,b), sometimes referred to as "polyhedron-tetrahedron" chains (Ni et al. 1995). An apical linking of the light rare-earth atoms by the distorted tetrahedral groups forms the chain-like monoclinic monazite structure. A stereoscopic view of the interlocking mechanism of the chain-like units for the case of the monoclinic RE orthophosphate structure is shown in Figure 14 (Mullica et al. 1985a,b).

In the xenotime structure, there are two sets of four equivalent bond distances (Ni et al. 1995). This coordination is illustrated in Figure 15 (left; after Milligan et al. 1982). These two sets of four equivalent bond distances are oriented orthogonal to each other, forming two unique and orthogonal interpenetrating tetrahedra. This results in a RE-O configuration (Fig. 15, right) that has been described by Milligan et al. (1983c) as a distorted dodecahedron D_{2d} (bibisphenoid) derived from a distorted cube. The relationship between the eight-fold-coordinated RE ion and the surrounding PO_4 tetrahedral units is illustrated in Figure 16. This figure shows the details of the molecular packing in the unit cell of the tetragonal RE orthophosphate structure in a stereo view and clearly illustrates how the RE-PO_4 polyhedral units fit together. Ni et al. (1995) pointed out that the RE-P

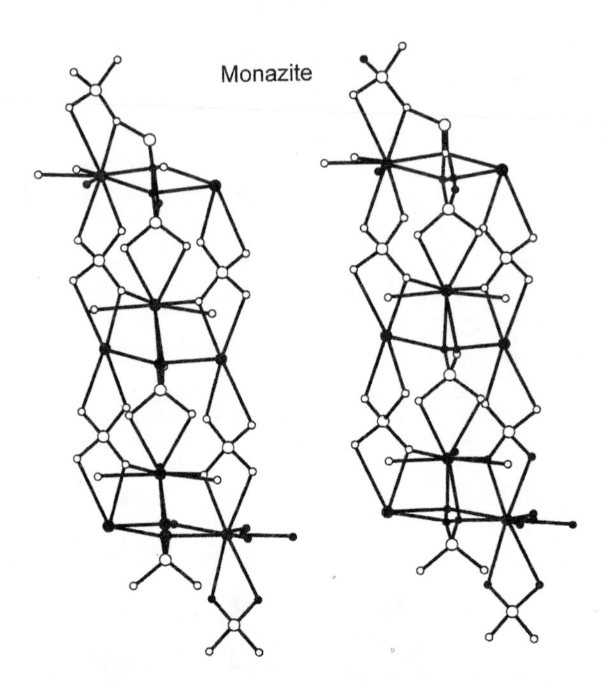

Figure 14. The chain-like nature and interlocking mechanism of the monoclinic monazite structure illustrates the apical linking that forms the chains, shown in a stereo view. The large open circles represent phosphorous while the large solid circles represent the rare-earth ions. The small open circles represent oxygen (after Mullica et al. 1985a,b).

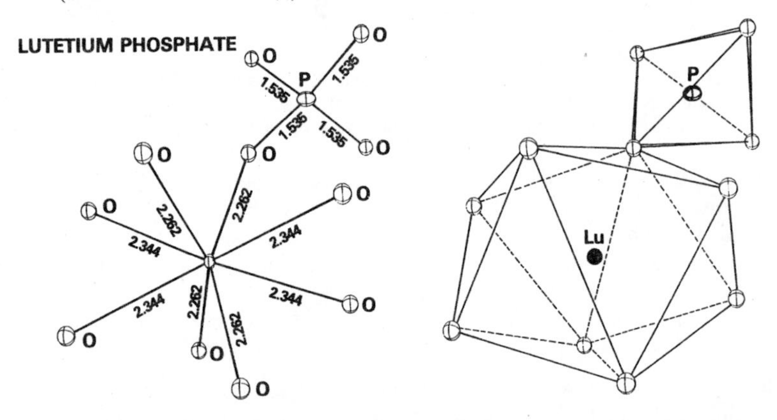

Figure 15. (Left) The eight-fold coordination of lutetium in the tetragonal-symmetry structure of $Lu(PO_4)$ is illustrated. As shown, there are four Lu-O bonds with a length of 2.344 Å and four Lu-O bonds with a length of 2.262 Å. (Right) The distorted dodecahedron D_{2d} described by Milligan et al. (1983).

inter-chain distances exhibit a regular variation with the rare-earth Z value. Two different RE-phosphorous distances occur in the case of the monazite structure and only one distance occurs for the xenotime structure. The variation of the RE-phosphorous distances with the RE element radius is shown in Figure 17 (after Ni et al. 1995).

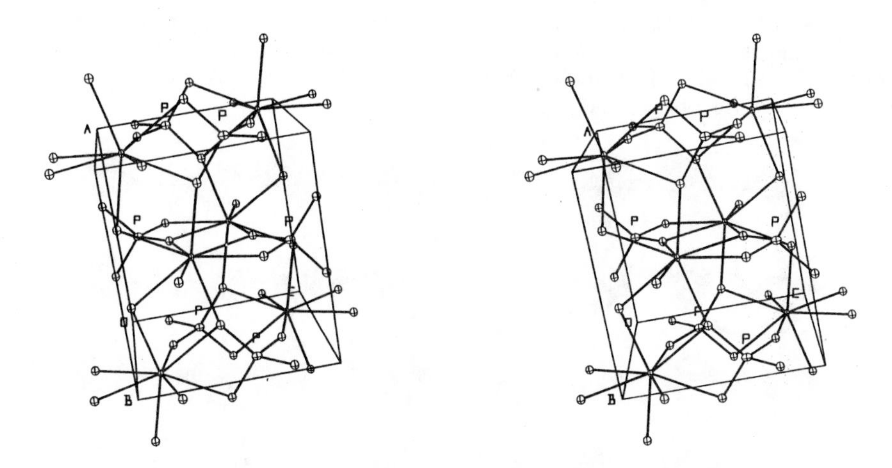

Figure 16. The details of the molecular packing in a tetragonal-structure orthophosphate are shown in a stereo view that illustrates the nature of the linkage between RE-O structural units (after Milligan et al. 1982).

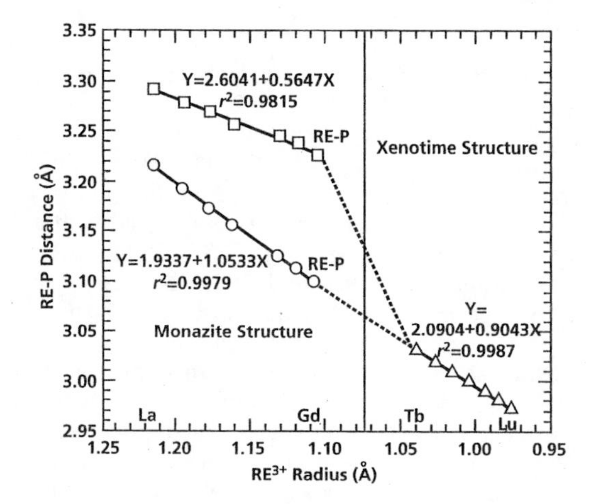

Figure 17. The variation of the RE-to-phosphate distances is shown as a function of the RE-ion radius for both the monazite- and xenotime-structure orthophosphate compounds. As shown in this figure, the shorter RE-to-phosphorous distances in the monoclinic structure compounds vary linearly with the trivalent RE ion radius with a slope that is close to 1. A similar variation is evident for the RE-P distances in the tetragonal xenotime-structure compounds—a trend that supports the comparison of the [001] polyhedron-PO_4 tetrahedron "chain arrangement" in the two structural types (after Ni et al. 1995).

A chain-like structure similar to that exhibited by the monoclinic orthophosphates is also found for the tetragonal zircon type materials. Each of these structure types has four chains in each unit cell, with the principal difference between the xenotime and monazite structural types residing in the difference in the coordination number. The linking of the chains occurs laterally through edge-sharing of adjacent polyhedra (Ni et al. 1995). One view of how the RE ions link to the PO_4 tetrahedral units in the xenotime structure is provided by the stereo view of $Y(PO_4)$ looking down the *c* axis of the structure shown in

Figure 18. The relationship between the monazite and xenotime structures has been described in detail by Ni et al. (1995). These workers point out that the two structures are related by a shift of the (100) planes along [010] by 2.23 Å and 1/2 a cosβ = 0.79 Å along [001], plus a small rotation of the tetrahedron about [001].

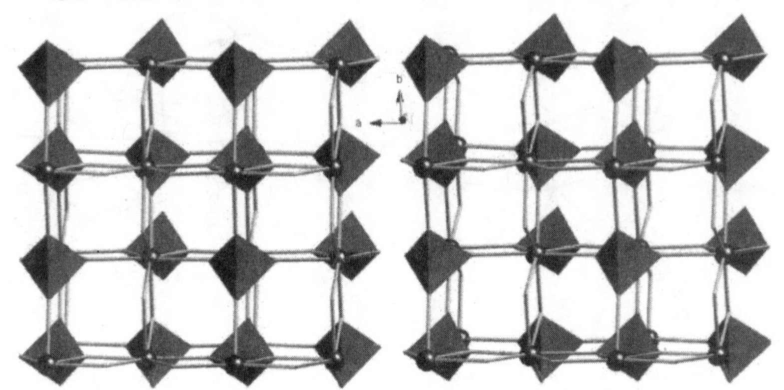

Figure 18. A stereo view looking down the c-axis is shown for the xenotime structure, i.e., of $Y(PO_4)$. This illustrates the three-dimensional nature of the vertical and lateral linkage between the structural units.

The structural properties of several mixed RE-orthophosphates were also investigated by Mullica et al. (1986, 1990, 1992, 1996). In the case of 1:1 Gd/Yb orthophosphate (Mullica et al. 1986), the structure was found to be of the tetragonal zircon structure type with a = 6.865(2) Å, c = 6.004(2) Å, and unit cell volume = 283.0(3) $Å^3$. Structural refinements were also carried out for Gd/Er-, Gd/Y-, and Gd/Yb-orthophosphates with a 1:1 RE ratio and for $(Gd/Yb)PO_4$ and with a 75:25 ratio of Gd toYb (Mullica et al. 1990). The mixed systems 1:1 $(Gd/Tb)PO_4$, 3:1$(Gd/Tb)PO_4$ and 9:1 $(Lu/Tb)PO_4$ were also investigated (Mullica et al. 1992) as were seven mixed compounds in the (Ce/Tb), (Nd/Tb), and $(Sm/Tb)PO_4$ families of orthophosphates (Mullica et al. 1996). All of the mixed orthophosphates investigated by Mullica et al. were found to crystallize with the tetragonal zircon structure. At this time, a considerable body of structural and crystal chemical data exists for the synthetic RE, Y, and Sc orthophosphates as evidenced by the previously cited references of Mullica et al., Milligan et al., and Ni et al. to which the reader is referred for details. Additionally the related structural investigations of Chakoumakos et al. (1994) should be noted. In these studies, structural refinements of the zircon-type RE-, Y-, and Sc-orthovanadates are reported.

PHYSICAL, OPTICAL, AND SOLID-STATE CHEMICAL PROPERTIES

The solid-state chemical, optical, and physical properties of the RE, Y, and Sc orthophosphates have been extensively investigated by means of numerous techniques. Such studies include optical spectroscopy (Trukhin and Boatner 1997), x-ray absorption (Shuh et al. 1994), electron paramagnetic resonance (EPR) spectroscopy (Abraham et al. 1981, Boatner et al. 1981b), Mössbauer (Huray et al. 1982), Rutherford backscattering (Sales et al. 1983), and other techniques. Additionally, scanning ellipsometry has been used by Jellison and Boatner (2000) to determine the spectroscopic refractive indices of the xenotime–structure RE orthophosphates. The extensive range of studies of these orthophosphates was motivated initially by the potential application of the ortho-phosphates to radioactive waste disposal and subsequently by the other applications discussed previously in this chapter. In addition to the motivation provided by existing and potential applications, the RE, Y, and Sc orthophosphates have proven to be versatile hosts (or actual subjects in their own right in the pure form) for numerous basic scientific studies of rare-earth electronic, phonon, and magnetic properties.

Numerous studies of the solid-state chemical properties of the RE-, Y-, and Sc-orthophosphates have been performed using EPR techniques. These investigations have shed considerable light on the ability of the orthophosphates to incorporate a diverse range of cation impurities encompassing other rare-earth ions, iron-group ions, actinide impurities, and other impurities such as Hf and Zr. EPR studies of a variety of rare-earth impurities in the tetragonal hosts $Lu(PO_4)$, $Y(PO_4)$, and $Sc(PO_4)$ were carried out by Abraham et al. (1983). A representative EPR spectrum showing the EPR signals from isotopically enriched $^{145}Nd^{3+}$ as well as EPR lines due to Gd^{3+} and Er^{3+} is shown in Figure 19. Results such as these illustrate the solid-state chemical capability of incorporating many combinations of RE impurities in various orthophosphate hosts. The EPR method is applicable to the study of the solid-state chemistry of paramagnetic species in both orthophosphate single crystals and powders or ceramics. Figure 20 shows an example of the ERP spectra of a Gd^{3+} impurity in both single crystals and powders of synthetic xenotime (Rappaz et al. 1980).

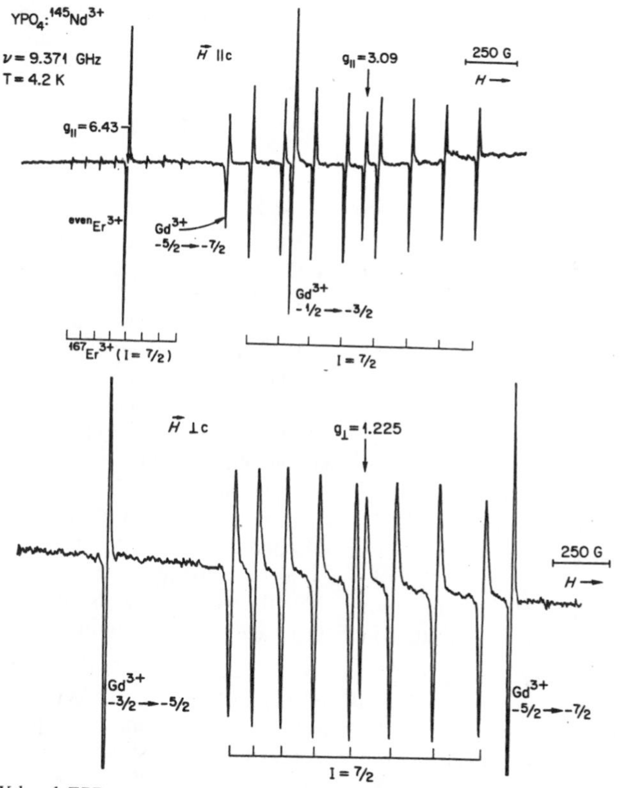

Figure 19. X-band EPR spectra (at T = 4.2 K) showing the hyperfine structure due to Nd^{3+} (isotopically enriched with ^{145}Nd). The spectra are shown for the applied magnetic field oriented parallel and perpendicular to the *c*-axis of the $Y(PO_4)$ host single crystal. The EPR spectrum of Er^{3+} and lines from the spectrum of Gd^{3+} are also present. This type of magnetic resonance spectroscopy has proven to be very useful in the study of the solid-state chemical properties of the rare-earth-, Y-, and Sc-orthophosphates (after Abraham et al. 1983).

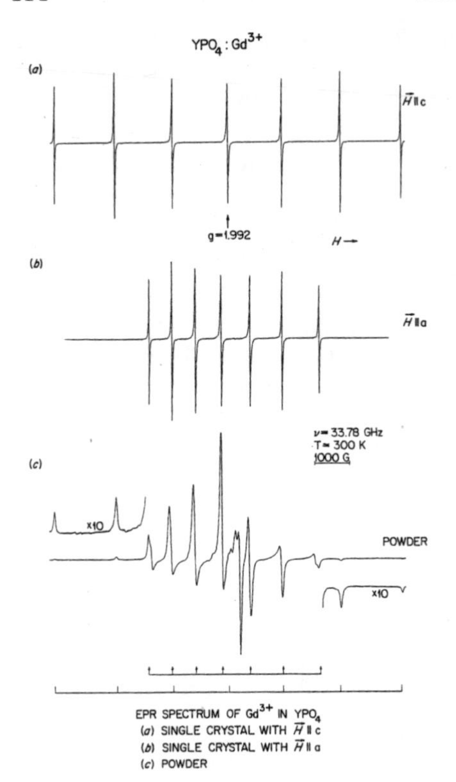

Figure 20. X-band EPR spectra of both single crystals and powders of $Y(PO_4)$ doped with Gd^{3+} are shown. The single-crystal Gd^{3+} spectra are shown for both the parallel and perpendicular orientations of the applied magnetic field relative to the c-axis of the $Y(PO_4)$ crystal. The powder spectrum of Gd^{3+} shown at the bottom of the figure represents an effective geometrical average of the EPR transitions over all possible orientations of the numerous $Y(PO_4)$ powder grains (after Rappaz et al. 1980).

One of the more intriguing aspects of the solid-state chemical characteristics of synthetic xenotime, pretulite, and $Lu(PO_4)$ was revealed by EPR techniques. In the work of Abraham et al. (1984, 1985, 1986), EPR spectroscopy was used to identify the unusual 3^+ valence state of Zr, Ti, and Hf impurities incorporated in the noted orthophosphate hosts. In the case of Zr and Ti, the trivalent state was observed as a stable valence state that existed in some as-grown crystals (i.e. stable at room temperature and above). In the case of Hf^{3+} it was necessary to gamma irradiate the Hf-doped crystals at 77 K and to maintain the samples at that temperature during the EPR observations. Figure 21 shows the room temperature EPR spectrum of a single crystal of $Sc(PO_4)$ doped with isotopically enriched $^{91}Zr^{3+}$. Trivalent iron-group impurities other than Ti^{3+} can be readily incorporated in the subject orthophosphates as evidenced by the study of Fe^{3+} in $Y(PO_4)$, $Sc(PO_4)$, and $Lu(PO_4)$ reported by Rappaz et al. (1982). Divalent iron group impurities can also be incorporated in the Re, Y, and Sc orthophosphates as illustrated by case of Mn^{2+} impurities in these tetragonal hosts (Boldu et al. 1985).

Several basic investigations of the spectroscopic properties of REs, either in the form of the pure compound or as impurities doped into hosts [$Y(PO_4)$, $Sc(PO_4)$, and $Lu(PO_4)$ in particular], have been performed. Raman spectra for the orthophosphates were reported by Begun et al. (1981), and other more-detailed investigations employing Raman spectroscopy were carried out by Becker et al. (1984, 1985, 1986, 1992) and by Williams et al. (1989a,b,c). Optical spectroscopic studies of Pr, Nd, and Er have been reported by Hayhurst et al. (1981, 1982), and studies of Ce^{3+} in $Lu(PO_4)$ and $Y(PO_4)$ by Sytsma et al. (1993). Magnetic transitions in $Tb(PO_4)$ have been investigated by Liu et al. (1994). Neutron scattering and diffraction techniques have been applied to studies of the ground-state properties of pure rare-earth orthophosphate compounds by Loong et al.

(1993a,b,c; 1994, 1999), and the lattice dynamics of $Lu(PO_4)$ were investigated by Nipko et al. (1997a,b). Anomalous temperature-induced variations in the lattice parameters of $Ho(PO_4)$ and $Ho(VO_4)$ were reported by Skanthakumar et al. (1995).

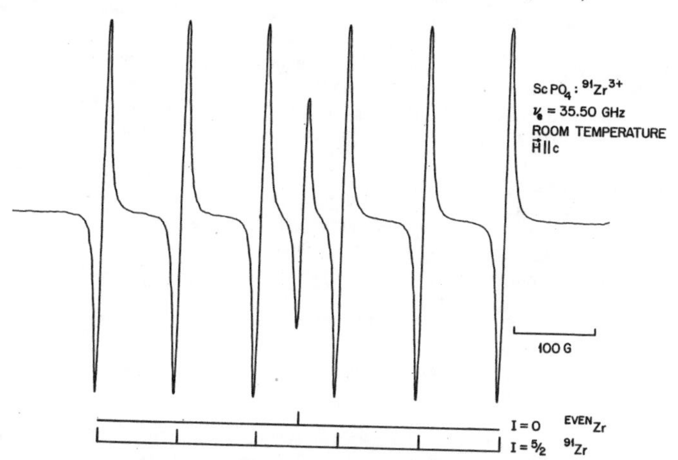

Figure 21. Room temperature EPR spectrum of a $^{91}Zr^{3+}$ (isotopically enriched) impurity in synthetic pretulite. The observation of unusual valence states like Zr^{3+} and Hf^{3+}, some of which are stable in the as-grown single crystals, represents an unusual feature of the solid state chemical properties of the xenotime-structure orthophosphates (after Abraham et al. 1984).

SUMMARY

The properties and applications of orthophosphates formed from the light and heavy rare-earth elements and from yttrium and scandium continue to represent important research topics in the fields of geology, geochronology, radioactive waste disposal, ceramic composites, gamma-ray scintillators, and a variety of optical devices. As discussed in the preceding sections, these are remarkable materials that are characterized by an unusual combination of chemical, physical, radiation-response, mineralogical, and other properties. Interest in these intriguing materials continues to grow, and undoubtedly, the future will witness a wide range of new investigations that will both further enlighten us in regard to the fundamental aspects of these materials and lead to the identification of new applications and to the development of new devices.

ACKNOWLEDGMENTS

This chapter is dedicated to Donald F. Mullica—a loyal colleague—who was professionally committed to the science of crystallography and personally committed to the welfare of his students and his friends. The able and tireless assistance of Joanne Ramey in many aspects of this work is gratefully acknowledged, as is the assistance of J. Matt Farmer for his help in the preparation of a number of the crystal-structure illustrations. This research was sponsored by the Division of Materials Sciences, Office of Basic Energy Sciences, United States Department of Energy under contract DE-AC05-00OR22725. Oak Ridge National Laboratory is managed and operated by UT-Battelle, LLC.

REFERENCES

Abraham MM, Boatner LA (1982) Electron-paramagnetic-resonance investigations of $^{243}Cm^{3+}$ in $LuPO_4$ single crystals. Phys Rev B 26:1434-1437

Abraham MM, Boatner LA, Rappaz M (1980a) Novel measurement of hyperfine interactions in solids: $^{207}Pb^{3+}$ in YPO_4 and $LuPO_4$. Phys Rev Letters 45:839-842

Abraham MM, Boatner LA, Quinby TC, Thomas DK, Rappaz M (1980b) Preparation and compaction of synthetic monazite powders. Radioactive Waste Management 1:181-191

Abraham MM, Boatner LA, Rappaz M (1981) EPR investigations of impurities in the lanthanide orthophosphates. *In* Nuclear and Electron Resonance Spectroscopies Applied to Materials Science. Kaufmann EN, Shenoy GK (ed) Elsevier North-Holland, p 475-480

Abraham MM, Boatner LA, Ramey JO, Rappaz M (1983) An EPR study of rare-earth impurities in single crystals of the zircon-structure orthophosphates $ScPO_4$, YPO_4, and $LuPO_4$. J Chem Phys 78:3-10

Abraham MM, Boatner LA, Ramey JO, Rappaz M (1984) The occurrence and stability of trivalent zirconium in orthophosphate single crystals. J Chem Phys 81:5362-5366

Abraham MM, Boatner LA, Ramey JO, (1985) The observation by EPR of trivalent hafnium in $LuPO_4$, YPO_4, and $ScPO_4$. J Chem Phys 83:2754-2758

Abraham MM, Boatner LA, Aronson MA (1986) EPR observations of trivalent titanium in orthophosphate single crystals. J Chem Phys 85:1-6

Abraham MM, Boatner LA, Finch CB, Kot W, Conway JG, Shalimoff GV, Edelstein NM, (1987) EPR and optical-absorption studies of Cm^{3+} in YPO_4 and $LuPO_4$ single crystals. Phys Rev B – Condens Matter 35:3057-3061

Allison SW, Boatner LA, Gillies GT (1995) Characterization of high-temperature thermographic phosphors: spectral properties of $LuPO_4$:Dy(1%), Eu(2%). Appl Opt 34:5624-5627

Allison SW, Cates MR, Gillies GT, Boatner LA (1998) High temperature thermometric phosphors for use in a temperature sensor. U.S. Patent 5,730,528

Allison SW, Cates MR, Boatner LA, Gillies GT (1999) High temperature thermometric phosphors. U.S. patent 5,885,484

Andrehs G and Heinrich W (1998) Experimental determination of REE distributions between monazite and xenotime: potential for temperature-calibrated geochronology. Chemical Geology 149:83-96

Bamberger CE (1982) Preparation of metal phosphates by metathesis reaction with BPO_4. J Am Ceram Soc 65:C107-C108

Bamberger CE, Haire RG, Hellwege HE, and Begun GM (1984) Synthesis and characterization of crystalline phosphates of plutonium(III) and plutonium(IV). J Less-Common Metals 97:349-356

Bea F, Periera MD, Corretge LG, Fershtater GB (1994) Differentiation of strongly peraluminous perphosphorous granites: The Pedrobernardo pluton, central Spain. Geochim Cosmochim Acta 58:2609-2627

Bea F (1996) Residence of REE, Y, Th and U in granites and crystal protoliths: Implications for the chemistry of crystal melts. J Pet 37:521-552

Beall GW, Boatner LA, Mullica DF, Milligan WO (1981) The structure of cerium orthophosphate, a synthetic analog of monazite. J Inorg Nucl Chem 43:101-105

Becker P, Hayhurst T, Shalimoff G, Conway JG, Edelstein N, Boatner LA, Abraham MM (1984) Crystal field analysis of Tm^{3+} and Yb^{3+} in YPO_4 and $LuPO_4$. J Chem Phys 81:2872-2878

Becker PC, Williams GM, Edelstein N, Bucher JJ, Russo RE, Koningstein JA, Boatner LA, Abraham MM (1985) Intensities and asymmetries of electronic Raman scattering in $ErPO_4$ and $TmPO_4$. Phys Rev B 31:8102-8110

Becker PC, Williams GM, Edelstein N, Koningstein JA, Boatner LA, Abraham MM (1986) Resonance electronic Raman scattering in erbium phosphate crystals. Optics Lett 11:282-284.

Becker PC, Williams GM, Edelstein NM, Koningstein JA, Boatner LA, Abraham MM (1992) Observation of strong electron-phonon coupling effects in $YbPO_4$. Phys Rev B - Condens Matter 45:5027-5030

Begun GM, Beall GW, Boatner LA, Gregor WT (1981) Raman spectra of the rare earth orthophosphates. J Raman Spectrosc 11:273-278

Bernhard F, Walter F, Ettinger K, Taucher J, Mereiter K (1998) Pretulite, $ScPO_4$, a new scandium mineral from the Styrian and lower Austrian lazulite occurrences, Austria. Am Mineral 83:625-630

Boatner LA, Sales BC (1988) Monazite. *In* Radioactive Waste Forms for the Future. Lutze W, Ewing RC (eds) Elsevier North-Holland, Amsterdam, Ch 8

Boatner LA, Beall GW, Abraham MM, Finch CB, Huray PG, Rappaz M (1980) Monazite and other lanthanide orthophosphates as alternate actinide waste forms. *In* Scientific Basis for Nuclear Waste Management. Northrup CJ (ed) Plenum Publishing, New York, p 289-296

Boatner LA, Beall GW, Abraham MM, Finch CB, Floran RJ, Huray PG, Rappaz M (1981a) Lanthanide orthophosphates for the primary immobilization of actinide wastes. *In* Management of Alpha-Contaminated Wastes. Intl Atomic Energy Agency, Vienna (IAEA-SM246/73), p 411-422

Boatner LA, Abraham MM, Rappaz M (1981b) The characterization of nuclear waste forms by EPR spectroscopy. *In* Scientific Basis for Nuclear Waste Management. Moore JG (ed) Plenum Publishing, New York, p 181-188

Boatner LA, Abraham MM, Sales BC (1983) Lanthanide orthophosphate ceramics for the disposal of actinide-contaminated nuclear wastes. Inorganica Chimica Acta 94(E23):146

Boldú JL, Muñoz E, Abraham MM, Boatner LA (1985) EPR and thermoluminescence investigations of Mn^{2+} in $LuPO_4$, YPO_4, and $ScPO_4$. J Chem Phys 83:6113-6120

Bondar IA, Domanskii AI, Mezentseva LP, Degen MG, Kalinina NE (1976) A physicochemical study of lanthanide orthophosphates. Russ J Inorg Chem (Engl. Transl.) 21:1126-28

Carron MK, Mrose ME, Murata KJ (1958) Relation of ionic radius to structures of rare-earth phosphates, arsenates, and vanadates. Am Mineral 43:985-989

Chakoumakos BC, Abraham MM, Boatner LA (1994) Crystal structure refinements of zircon-type MVO_4 (M = Sc, Y, Ce, Pr, Nd, Tb, Ho, Er, Tm, Yb, Lu). J Solid State Chem 109:197-202

Donovan JJ, Hanchar JM, Picolli PM, Schrier MD, Boatner LA, Jarosewich E (2002a) A re-examination of the rare-earth element orthophosphate reference samples for electron microprobe analysis. Can Mineral, submitted for publication.

Donovan JJ, Hanchar JM, Picolli P, Schrier MD, Boatner, LA (2002b) Contamination in the rare-earth-element orthophosphate reference samples. NIST J Res (submitted)

Eigermann W, Müller-Vogt G, Wendl W (1978) Solubility curves in high-temperature melts for the growth of single crystals of rare earth vanadates and phosphates. Phys Stat Sol (a) 49:145-148

Ewing RC (1975) The crystal chemistry of complex niobium and tantalum oxides. IV. The metamict state. Am Mineral 60:728-733

Ewing RC, Haaker RF (1980) The metamict state: Implications for radiation damage in crystalline waste forms. Nucl Chem Waste Management 1:51-57

Eyal Y and Kaufman A (1982) Alpha-recoil damage in monazite: preferential dissolution of the radiogenic actinide isotopes. Nuclear Technology 58:77-83

Eyal Y and Fleischer RL, (1985) Preferential leaching and the age of radiation damage from alpha decay in minerals. Geochim Cosmochim Acta 48:1155-1164

Feigelson RS (1964) Synthesis and single crystal growth of rare-earth orthophosphates. J Am Ceram Soc 47:257-258

Floran RJ, Abraham MM, Boatner LA, Rappaz M (1981a) Geologic stability of monazite and its bearing on the immobilization of actinide wastes. *In* Scientific Basis for Nuclear Waste Management. Moore JG (ed) Plenum Publishing, New York, p 507-514

Floran RJ, Rappaz M, Abraham MM, Boatner LA (1981b) Hot and cold pressing of $(La,Ce)PO_4$-based nuclear waste forms. *In* Alternate Nuclear Waste Forms and Interactions in Geologic Media. Boatner LA, Battle GC (ed) USDOE CONF-8005107 p 185-193

Förster H-J (1998a) The chemical composition of REE-Y-Th-U-rich minerals in peraluminous granites of the Erzgebirge-Fichtelgebirge region, Germany, Part I: The monazite-(Ce)-brabanite solid solution series. Am Mineral 83:259-272

Förster H-J (1998b) The chemical composition of REE-Y-Th-U-rich accessory minerals in peraluminous granites of the Erzgebirge-Fichtelgebirge region, Germany. Part II: Xenotime. Am Mineral 83: 1302-1315

Ghouse KM (1968) Refinement of the crystal structure of heat-treated monazite crystal. Indian J Pure Appl Phys 6:265-268

Von Gliszcynski S (1939) Beitrag zur isomorphis von monazit und krokoit. Z Kristallogr 101:1-16

Gramaccioli CM, Segalstad TU (1978) A uranium- and thorium-rich monazite from a south-alpine pegmatite at Piona, Italy. Am. Mineral. 63:757-761

Gratz R, Heinrich W (1997) Monazite-xenotime thermobarometry: Experimental calibration of the miscibility gap in the binary system $CePO_4$-YPO_4. Am Mineral 82:772-780

Hayhurst T, Shalimoff G, Edelstein N, Boatner LA, Abraham MM (1981) Optical spectra and Zeeman effect for Er^{3+} in $LuPO_4$ and $HfSiO_4$. J Chem Phys 74:5449-5452

Hayhurst T, Shalimoff G, Conway JG, Edelstein N, Boatner LA, Abraham (1982) Optical Spectra and Zeeman effect for Pr^{3+} and Nd^{3+} in $LuPO_4$ and YPO_4. J Chem Phys 76:3960-3966

Heinrich W, Andrehs G. and Franz G (1997) Monazite-xenotime miscibility gap thermometer: I. An empirical calibration. J Metamorph Geol 15:3-16

Hikichi Y (1991) Synthesis of monazite (RPO_4, R = La, Ce, Nd, or Sm) by solid state reaction. Mineral J 15:268-275

Hikichi Y, Nomura T (1987) Melting temperatures of monazite and xenotime. J Am Ceram Soc 70: C-252–C-253

Hikichi Y, Hukuo K, Shiokawa J (1978) Synthesis of rare earth orthophosphates. Bull Chem Soc Jpn 51:3645-3646

Hikichi Y, Hukuo K, Shiokawa J (1980) Solid state reactions between rare earth orthophosphate and oxide. Bul. Chem Soc Jpn 53:1455-1456

Hikichi Y, Sasaki T, Suzuki S, Murayama K (1988) Thermal reactions of hydrated hexagonal $RPO_4 \cdot nH_2O$ (R = Tb or Dy, n = 0.5 to 1). J Am Ceram Soc 71:C-354–C-355

Hikichi Y, Sasaki T, Murayama K, Nomura T (1989) Mechanochemical changes of weinschenkite-type $RPO_4 \cdot 2H_2O$ [R = Dy, Y, Er, or Yb] by grinding and thermal reactions of the ground specimens. J Am Ceram Soc 72:1073-1076

Hikichi Y, Yu CF, Miyamoto M, Okada S (1991) Mechanical conversion of rhabdophane type $RPO_4 \cdot nH_2O$ (r = La, Ce, Pr, Nd or Sm, n~1/2) to the monazite type analogues. Mineral J 15:349-355

Hikichi Y, Yu CF, Miyamoto M, Okada S (1993) Mechanochemical changes in hydrated rare earth orthophosphate minerals by grinding. J Alloys Comp 192:102-104

Hikichi Y, Yogi K, Ota T (1995) Preparation of rhabdophane-type $RPO_4 \cdot nH_2O$ (R = Y or Er, n = 0.7-0.8) by pot-milling churchite-type $RPO_4 \cdot 2H_2O$ at 20-25°C in air. J Alloys Comp 224:L1-L3

Hikichi Y, Ota T, Hattori T Imaeda T (1996) Synthesis and thermal reactions of rhabdophane = (Y). Mineral J 18:87-96

Hinton RW and Paterson BA (1994) Crystallization history of granitic magma: Evidence from trace element zoning. Mineral Mag 58A:416-417

Hintzmann W, Müller-Vogt G (1969) Crystal growth and lattice parameters of rare-earth doped yttrium phosphate, arsenate and vanadate prepared by the oscillating temperature flux technique. J Cryst Growth 5:274-278

Houk LG (1943) Monazite Sand. Information Circular 7233, U S Department of the Interior – Bureau of Mines

Huray PG, Spaar MT, Nave SE, Legan JM, Boatner LA, Abraham MM (1982) The Application of ^{57}Fe Mössbauer spectroscopy to the characterization of nuclear waste forms. *In* Scientific Basis for Nuclear Waste Management. Topp SV (ed) Elsevier, North Holland, p 59-66

Jarosewich E and Boatner LA (1991) Rare-earth element reference samples for electron microprobe analysis. Geostandards Newsletter 15:397-399

Jellison GE, Boatner LA (2000) Spectroscopic refractive indices of metal-orthophosphates with the zircon type structure. Opt Mater 15:103-109

Johnstone SJ (1914) Monazites from some new localities. J Chem Industry 33:55-59

Jonasson RG and Vance ER (1986) DTA study of the Rhabdophane to monazite transformation in rare earth (La–Dy phosphates. Thermochemica Acta 108:65-72

Jonasson RG, Bancroft GM, Boatner LA (1988) Surface reactions of synthetic, end-member analogues of monazite, xenotime and rhabdophane, and evolution of natural waters. Geochim Cosmochim Acta 52:767-770

Kato T (1958) A study on monazite from the Ebisu mine, Gifu prefecture. Mineral J (Japan) 2:224

Kelly KL, Beall GW, Young JP, Boatner LA (1981) Valence states of actinides in synthetic monazites. *In* Scientific Basis for Nuclear Waste Management. Moore JG (ed) Plenum Publishing, New York, p 189-195

Kizilyalli M and Welch AJE (1976) Crystal data for lanthanide orthophosphates. J Appl Crystallogr 9: 413-414

Kot WK, Edelstein NM, Abraham MM, Boatner LA (1993a) Zero-field splitting of Cm^{3+} in $LuPO_4$ single crystals. Phys Rev B 48:12704-12712

Kot WK, Edelstein NM, Abraham MM, Boatner LA (1993b) Electron paramagnetic resonance of Pu^{3+} and Cf^{3+} in single crystals of $LuPO_4$. Phys Rev B 47:3412-3414

Lempicki A, Berman E, Wojtowicz AJ, Balcerzyk M, Boatner LA (1993) Cerium-doped orthophosphates: new promising scintillators. IEEE Trans Nucl Sci 40:384-387

Liu GK, Loong CK, Trouw F, Abraham MM, Boatner LA (1994) Spectroscopic studies of magnetic transitions in $TbPO_4$. J Appl Phys 75:7030-7032

Liu GK, Li ST, Zhorin VV, Loong CK, Abraham MM, Boatner LA (1998) Crystal-field splitting, magnetic interaction, and vibronic excitations of $^{244}Cm^{3+}$ in YPO_4 and $LuPO_4$. J Chem Phys 109: 6800-6808

Loong CK, Soderholm AL, Abraham MM, Boatner LA, Edelstein NM (1993a) Crystal-field excitations and magnetic properties of $TmPO_4$. J Chem Phys 98:4214-4222

Loong CK, Soderholm AL, Hammonds JP, Abraham MM, Boatner LA (1993b) Neutron study of crystal-field transitions in $ErPO_4$. J Appl Phys 73:6069-6071

Loong CK, Soderholm L, Goodman GL, Abraham MM, Boatner LA (1993c) Ground-state wave functions of Tb^{3+} ions in paramagnetic $TbPO_4$: a neutron scattering study. Phys Rev B 48:6124-6131

Loong CK, Soderholm L, Xue JS, Abraham MM, Boatner LA (1994) Rare earth energy levels and magnetic properties of $DyPO_4$. J Alloys Comp 207/208:165-169

Loong CK, Loewenhaupt M, Nipko JC, Braden M, Reichardt W, Boatner LA (1999) Dynamic coupling of crystal-field and phonon states in $YbPO_4$. Phys Rev B 60:R12549-R12552
Marinova LA and Yaglov VN (1976) Thermodynamic characteristic of lanthanide phosphates. Russ J Physical Chem 50:477 (Translated from: Zh Fizicheskoi Khimi 50:802-803)
Marshall DB, Morgan PED, Housley RM (1997) Debonding in multilayered composites of zirconia and $LaPO_4$. J Am Ceram Soc 80:1677-1683
Marshall DB, Morgan PED, Housley RM, Cheung JT (1998) High-temperature stability of the Al_2O_3-$LaPO_4$ system. J. Am. Ceram. Soc. 81:951-956
McCarthy GJ, White WB, Pfoertsch DE, (1978) Synthesis of nuclear waste monazites, ideal actinide hosts for geological disposal. Mater Res Bull 13:1239-1245
Meldrum A, Boatner LA, Ewing RC (1997a) Displacive radiation effects in the monazite- and zircon-structure orthophosphates. Phys Rev B 56:13805-13814
Meldrum A, Boatner LA, Ewing RC (1997b) Electron-irradiation-induced nucleation and growth in amorphous $LaPO_4$ and $ScPO_4$, and zircon. J Mater Res 12:1816-1827
Meldrum A, Boatner LA, Wang LM, Ewing RC (1997c) Ion-beam-induced amorphization of $LaPO_4$ and $ScPO_4$. Nucl Instrum Methods Phys Res B 127/128:160-165
Meldrum A, Boatner LA, Weber WJ, Ewing RC (1998) Radiation damage in zircon and monazite. Geochim Cosmochim Acta 62:2509-2520
Meldrum A, Boatner LA, Ewing RC (2000) A comparison of radiation effects in crystalline ABO_4-type phosphates and silicates. Mineral Mag 64:185-194
Milligan WO, Mullica DF, Beall GW, Boatner LA (1982) Structural investigations of YPO_4, $ScPO_4$, and $LuPO_4$. Inorganica Chimica Acta 60: 39-43
Milligan WO, Mullica DF, Beall GW, Boatner LA (1983a) The structures of three lanthanide orthophosphates. Inorganica Chimica Acta 70:133-136
Milligan WO, Mullica DF, Beall GW, Boatner LA (1983b) Structures of $ErPO_4$, $TmPO_4$, and $YbPO_4$. Acta Crystallogr C 39:23-24
Milligan WO, Mullica DF, Perkins HO, Beall GW, Boatner LA (1983c) Crystal data for lanthanide orthophosphates with the zircon-type structure. Inorganica Chimica Acta 77:L23-L25
Mooney RCL (1948) Crystal structures of a series of rare earth phosphates. J Chem Phys 16:1003
Mooney RCL (1950) X-ray diffraction study of cerous phosphate and related crystals. I. Hexagonal modification. Acta Crystallogr 3:337-340
Morgan PED and Marshall DB (1993) Functional interfaces for oxide/oxide composites. Mater Sci Engin A162:15-25
Morgan PED and Marshall DB (1995) Ceramic composites of monazite and alumina. J Am Ceram Soc 78:1553-1563
Morgan PED and Marshall DB (1996) Ceramic composites having a weak bond material selected from monazites and xenotimes. U.S. patent 5,514,474
Morgan PED and Marshall DB (1997) Fibrous composites including monazites and xenotimes. U.S. Patent 5,665,463
Moses WW, Weber MJ, Derenzo SE, Perry D, Berdahl P, Schwarz L, Sasum U, Boatner LA (1997) Recent results in a search for inorganic scintillators for X- and gamma-ray detection. Proc Intl Conf Inorganic Scintillators and Their Applications. Chinese Academy of Sciences Press, Shanghai, China p 358-361
Moses WW, Weber MJ, Derenzo SE, Perry D, Berdahl P, and Boatner LA (1998) Prospects for dense, infrared emitting scintillators. IEEE Trans on Nucl Sci 45:462-466
Mullica DF, Milligan WO, Grossie DA, Beall GW, Boatner LA (1984) Nine-fold coordination in $LaPO_4$: pentagonal interpenetrating tetrahedral polyhedron. Inorganica Chimica Acta 95:231-236
Mullica DF, Grossie DA, Boatner LA (1985a) Structural refinements of praseodymium and neodymium orthophosphate. J Solid State Chem 58:71-77
Mullica DF, Grossie DA, Boatner LA (1985b) Coordination geometry and structural determinations of $SmPO_4$, $EuPO_4$, and $GdPO_4$. Inorganica Chimica Acta 109:105-110
Mullica DF, Grossie DA, Boatner LA (1986) Crystal structure of 1:1 gadolinium/ytterbium orthophosphate. Inorganica Chimica Acta 118:173-176
Mullica DF, Sappenfield EL, Wilson GA, Boatner LA (1989) The crystal structure of $Ce_{0.9}U_{0.1}PO_4$. Lanthanide Actinide Res 3:51-61
Mullica DF, Sappenfield EL, Boatner LA (1990) A structural investigation of several mixed lanthanide orthophosphates. Inorganica Chemica Acta 174:155-159
Mullica DF, Sappenfield EL, Boatner LA (1992) Single-crystal analysis of mixed ($Ln/TbPO_4$) orthophosphates. J Solid State Chem 99:313-318
Mullica DF, Sappenfield EL, Boatner LA (1996) Monazite-and zircon-type structures of seven mixed $(Ln/Ln)PO_4$ compounds. Inorganica Chimica Acta 244:247-252

Murdoch KM, Edelstein NM, Boatner LA, Abraham MM (1996) Excited state absorption and fluorescence line narrowing studies of Cm^{3+} in $LuPO_4$. J Chem Phys 105:2539-2546

Muto T, Meyrowitz R. Pommer AM, Murano T (1959) Ningyoite, a new uranous phosphate mineral from Japan. Am Mineral 44:633-650

Ni Y, Hughes JM, and Mariano AN (1995) Crystal chemistry of the monazite and xenotime structures. Am Mineral 80:21-26

Nipko JC, Loong CK, Loewenhaupt M, Braden M, Reichart W, Boatner LA (1997a) Lattice dynamics of xenotime: The phonon dispersion relations and density of states of $LuPO_4$. Phys Rev B 56: 11584-11592

Nipko JC, Loong CK, Loewenhaupt M, Reichardt W, Braden M, Boatner LA (1997b) Lattice dynamics of $LuPO_4$. J Alloys Comp 250:573-576

Orlovskii VP, Khalikov BS, Bugakov VI, and Kurbanov KhM (1975) Conditions for transfer and formation of single crystals of $HoPO_4$. Izvestiya Akademii Nauk SSSR, Neorganicheskie Materialy 11:494-497

Parish W (1939) Unit cell and space group of monazite $(LaCe:Y)PO_4$. Am Mineral 24:651-652

Pasteels P (1970) Uranium-lead radioactive ages of monazite and zircon from the Vire-Caroles granite (Normandy): A case of zircon-monazite discrepancy. Eclogae Geol Helv 63:231-237

Penfield SC (1882) Sand from Brindletown district, Burke County, NC. Am J. Sci. 24:252

Pepin JG and Vance ER (1981) Crystal data for rare-earth orthophosphates of the monazite structure type. J Inorg Nucl Chem 43:2807-2809

Petek M, Abraham MM, Boatner LA (1982) Lanthanide orthophosphates as a matrix for solidified radioactive defense and reactor wastes. *In* Scientific Basis for Nuclear Waste Management. Topp SV (ed) Elsevier, North Holland, p 181-186

Podor R and Cuney M (1997) Experimental study of Th-bearing $LaPO_4$ (780°C, 200 MPa):Implications for monazite and actinide orthophosphate stability. Am Mineral 82:765-771

Poitrasson F, Chenery S, Bland D (1996) Contrasted monazite hydrothermal alteration mechanisms and their geochemical implications. Earth Planetary Sci Letters 145:79-96

Rapaport A, Moteau O, Bass M, Boatner L, Deka C (1999a) Optical spectroscopy and lasing properties of neodymium-doped lutetium orthophosphate. J Opt Soc Am B 16:911-916

Rapaport A, David V, Bass M, Deka C, Boatner LA (1999b) Optical spectroscopy of erbium-doped lutetium orthophosphate. J Lumin 85:155-161

Rapp RP and Watson EB (1986) Monazite solubility and dissolution kinetics: Implications for the thorium and light rare earth chemistry of felsic magmas. Contrib Mineral Petrol 94:304-316

Rapp RP, Ryerson FJ, Miller CF (1987) Experimental evidence bearing on the stability of monazite during crystal anatexis. Geophys. Res Letters 14:307-310

Rappaz M, Boatner LA, Abraham MM (1980) EPR investigations of Gd^{3+} in single crystals and powders of the zircon-structure orthophosphates YPO_4, $ScPO_4$, and $LuPO_4$. J Chem Phys 73:1095-1103

Rappaz M, Abraham MM, Ramey JO, Boatner LA (1981a) EPR spectroscopic characterization of Gd^{3+} in the monazite-type rare-earth orthophosphates: $LaPO_4$, $CePO_4$, $PrPO_4$, $NdPO_4$, $SmPO_4$, and $EuPO_4$. Phys Rev B 23:1012-1030

Rappaz M, Boatner LA, Abraham MM (1981b) The application of EPR spectroscopy to the characterization of crystalline nuclear waste forms. *In* Alternate Nuclear Waste Forms and Interactions in Geologic Media. Boatner LA, Battle GC (ed) USDOE CONF-8005107, p 185-193

Rappaz M, Ramey JO, Boatner LA, Abraham MM (1982) EPR investigations of Fe^{3+} in single crystals and powders of the zircon-structure orthophosphates $LuPO_4$, YPO_4, and $ScPO_4$. J Chem Phys 76: 40-45

Rouanet A, Serra JJ, Allaf K, Orlovskii VP (1981) Rare earth orthophosphates at high temperatures. Izv. Akad. Nauk SSSR, Neorg Mater 17:76-81

Roy R and Vance ER (1981) Irradiated and metamict materials: relevance to radioactive waste science J. Mater Sci 16:1187-1190

Sales BC, White CW, Boatner LA (1983) A comparison of the corrosion characteristics of synthetic monazite and borosilicate glass containing simulated nuclear waste glass. Nucl Chem Waste Management 4: 281-289

von Schwarz H (1963) Die Phosphate, Arsenate und Vanadate der seltenen Erden. Z Anorgan Allgemeine Chem 323:44-56

Shuh DK, Terminello LJ, Boatner LA, Abraham MM, Perry D (1994) Characterization of Ce-doped $LaPO_4$ by x-ray absorption spectroscopy. Mater Res Soc Symp Proc Vol. 329, Materials Research Society, p. 91-96

Skanthakumar S, Loong CK, Soderholm L, Nipko J, Richardson JW Jr, Abraham, MM, Boatner LA (1995) Anomalous temperature dependence of the lattice parameters in $HoPO_4$ and $HoVO_4$: rare earth quadrupolar effects. J Alloys Compd 225:595-598

Smith SH and Wanklyn BM (1974) Flux growth of rare earth vanadates and phosphates. J Cryst Growth 21:23-28

Sytsma J, Piehler D, Edelstein NM, Boatner LA, Abraham MM (1993) Two-photon excitation of the $4f^1 \rightarrow 5d^1$ transitions of Ce^{3+} in $LuPO_4$ and YPO_4. Phys Rev B 47:14786-14794

Sytsma J, Murdoch K, Elelstein NM, Boatner LA, Abraham MM (1995) Spectroscopic studies and crystal-field analysis of Cm^{3+} and Gd^{3+} in $LuPO_4$. Phys Rev B 52:12668-12676

Tanner BK and Smith SH (1975) X-ray topographic study of the perfection of flux-grown rare earth phosphates. J Cryst Growth 30:323-326

Teufel S and Heinrich W (1997) Partial resetting of the U-Pb isotope system in monazite through hydrothermal experiments: An SEM and U-Pb isotope study. Chem Geol 137:273-281

Trukhin AN, Boatner LA, (1997) Electronic structure of $ScPO_4$ single crystals: optical and photoelectric properties. Proc 13th Intl Conf Defects in Insulating Crystals. Mater Sci Forum, Transactions Technical Publications, p 239-241

Ueda T (1967) Re-examination of the crystal structure of monazite. J Jpn Assoc Mineral Petrol Econ Geol. 58:170-179

Ushakov SV, Helean KB, Navrotsky A, Boatner LA (2001) The thermochemistry of rare-earth orthophosphates. J Mater Res 16:2623-2633

Wanklyn BM (1972) Flux growth of some complex oxide materials. J Mater Science 7:813-821

Wanklyn BM (1978) Effects of modifying starting compositions for flux growth. J Cryst Growth 43: 336-344

Weigel F. Scherer V, Henschel H (1965) Unit cells of the monazite-type rare-earth phosphates. J Am Ceram Soc 48:383-384

Wickham DG (1963) Use of lead pyrophosphate as a flux for crystal growth. J Appl Phys 33:3597-98

Williams GM, Becker PC, Conway JG, Edelstein N, Boatner LA, Abraham MM (1989a) Intensities of electronic Raman scattering between crystal-field levels of Ce^{3+} in $LuPO_4$: Nonresonant and near-resonant excitation. Phys Rev B 40:4132-4142

Williams GM, Becker PC, Edelstein N, Boatner LA, Abraham MM (1989b) Excitation profiles of resonance electronic Raman scattering in $ErPO_4$ crystals. Phys Rev B 40:1288-1296

Williams GM, Edelstein N, Boatner LA, Abraham MM (1989c) Anomalously small $4f$-$5d$ oscillator strengths and $4f$-$4f$ electronic Raman scattering cross sections for Ce^{3+} in crystals of $LuPO_4$. Phys Rev B 40:4143-4152

Wisniewski D, Tavernier S, Dorenbos P, Wisniewska M, Wojtowicz AJ, Bruyndonckx P, Loef EV, Eijk CWE, Boatner LA (2002a) VUV Scintillation of $LuPO_4$:Nd and YPO_4:Nd. IEEE Trans Nucl Sci (submitted)

Wisniewski D, Tavernier S, Dorenbos P, Wisniewska M, Wojtowicz AJ, Bruyndonckx P, Loef EV, Eijk CWE, Boatner LA (2002b) $LuPO_4$:Nd and YPO_4:Nd—new promising VUV scintillation materials. Nucl Instrum Methods A, (submitted)

Wojtowicz AJ, Lempicki A, Wisniewski D, Boatner LA (1994) Cerium-doped orthophosphate scintillators. Mat Res Soc Symp Proc, Materials Research Society 348:123-129

Wojtowicz AJ, Wisniewski D, Lempicki A, Boatner LA (1995) Scintillation mechanisms in rare earth orthophosphates. *In* Radiation Effects and Defects in Solids, Vol. 135. Biersack (ed) Overseas Publishers Assoc., Amsterdam B.V., p. 305-310

5 The Crystal Chemistry of the Phosphate Minerals

Danielle M.C. Huminicki and Frank C. Hawthorne

Department of Geological Sciences
University of Manitoba
Winnipeg, Manitoba, Canada R3T 2N2

INTRODUCTION

Phosphorus was discovered in 1669 by Hennig Brand. The word *phosphorus* originates from the two Greek words *phôs*, meaning light, and *phoros*, meaning bearer, due to the phosphorescent nature of white phosphorus. Phosphorus is the tenth most abundant element on Earth and tends to be concentrated in igneous rocks. It is an incompatible element in common rock-forming minerals, and hence is susceptible to concentration via fractionation in geochemical processes. It reaches its highest abundance in sedimentary rocks: the major constituents of phosphorite are the minerals of the apatite group. Phosphorus is the second most abundant inorganic element in our bodies (after Ca); it makes up about 1% of our body weight, occurring primarily in bones and teeth. Phosphorus (atomic number 15) is a non-metal in group VA of the periodic table, and has the ground-state electronic configuration $1s^2\ 2s^2\ 2p^6\ 3s^2\ 3p_x^{\ 1}\ 3p_y^{\ 1}\ 3p_z^{\ 1}$ or $[Ne]3s^2 3p^3$. There are three orbitals occupied with only one electron each in the third energy level (the *M* shell). Phosphorus participates in essentially covalent bonds; electron gain to form P^{3-} from P requires considerable energy (on the order of 1450 kJ mol^{-1}). Loss of electrons is also difficult due to the high ionization potentials of P (the sum of the first three ionization potentials is 60.4 eV).

CHEMICAL BONDING

Here we use bond-valence theory (Brown 1981) and its developments (Hawthorne 1985a, 1994, 1997) to consider structure topology and hierarchical classification of crystal structures, and we point out that bond-valence theory can be considered as a simple form of molecular-orbital theory (Burdett and Hawthorne 1993; Hawthorne 1994, 1997).

STEREOCHEMISTRY OF ($P\phi_4$) POLYHEDRA IN MINERALS

The variation of P-ϕ (ϕ: O^{2-}, OH^-) distances and ϕ-P-ϕ angles is of great interest for several reasons:

(1) mean bond-length and empirical cation and anion radii play an important role in systematizing chemical and physical properties of crystals;

(2) variations in individual bond-lengths give insight into the stereochemical behavior of structures, particularly with regard to the factors affecting structure stability;

(3) there is a range of stereochemical variation beyond which a specific oxyanion or cation-coordination polyhedron is not stable; it is obviously useful to know this range, both for assessing the stability of hypothetical structures (calculated by DLS [Distance Least-Squares] refinement, Dempsey and Strens 1976; Baur 1977) and for assessing the accuracy of experimentally determined structures.

Here, we examine the variation in P-ϕ distances in minerals and review previous work on polyhedral distortions in ($P\phi_4$) tetrahedra. Data for 408 ($P\phi_4$) tetrahedra were taken from 244 refined crystal structures with $R \leq 6.5\%$ and standard deviations of $\leq$0.005 Å on P-ϕ bond-lengths; structural references are given in Appendix A.

1529-6466/00/0048-0005$10.00

Variation in <P-ϕ> distances

The variation in <P-ϕ_4> distances (< > denotes a mean value; in this case, of P in tetrahedral coordination) is shown in Figure 1. The grand <P-ϕ> distance (i.e., the mean value of the <P-ϕ> distances) is 1.537 Å, in agreement with the value of 1.537 Å given by Baur (1974). The minimum and maximum <P-ϕ> distances are 1.459 and 1.602 Å, respectively (the larger values in Fig. 1 are considered unreliable), and the range of variation is 0.143 Å. Shannon (1976) lists the radius of $^{[4]}$P as 0.17 Å; assuming a mean anion-coordination number of 3.25 and taking the appropriate O/OH ratio, the sum of the constituent radii is 0.17 + 1.360 = 1.53 Å, in accord with the grand <P-ϕ> distance of 1.537 Å. Brown and Shannon (1973) showed that variation in <M-O> distance correlates with bond-length distortion Δ (= Σ[l(o) – l(m)] / l(m); l(o) = observed bond-length, l(m) = mean bond-length) when the bond-valence curve of the constituent species shows a strong curvature, and when the range of distortion is large. There is no significant correlation between <P-ϕ> and Δ; this is in accord with the bond-valence curve for P-O given by Brown (1981).

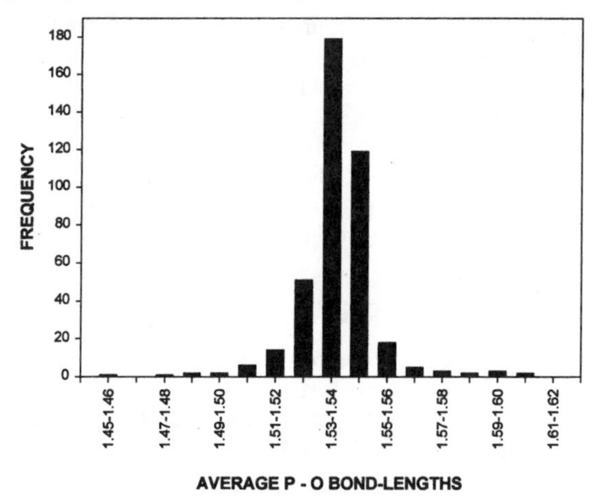

Figure 1. Variation in average P-O distance in minerals containing (Pϕ_4) tetrahedra.

Variation in P-ϕ distances

The variation in individual P-ϕ distances is shown in Figure 2; the grand mean P-ϕ distance is 1.537 Å, in close agreement with the value of 1.537 Å found by Baur (1974). The minimum and maximum observed P-ϕ distances are 1.439 and 1.625 Å, respectively, and the range of variation is 0.186 Å; the distribution is a skewed Gaussian. According to the bond-valence curve for P (the universal curve for second-row elements) from Brown (1981), the range of variation in P-ϕ bond-valence is 1.05-1.67 vu (valence units).

General polyhedral distortion in P-bearing minerals

Baur (1974) considered geometrical distortion in (PO_4) tetrahedra in great detail. In particular, he examined the variation in <P-ϕ> distance as a function of the mean coordination number of the simple anions of the phosphate group, and as a function of the dispersion of P-O distances, O-P-O angles and O-O angles from their respective mean values (described as distortion parameters). Baur (1970) found a correlation between <P-O> (corrected for dependence on the dispersion of the individual P-O distances from their mean value in the [PO_4] group) and the mean coordination numbers (including hydrogen bonds) of the constituent simple anions: <P-O> = 1.514(2) + 0.0059(7) CN, r = 0.49; the amount of variation explained is 24% in a sample size of 211. The correlation is shown in Figure 3, together with the regression line.

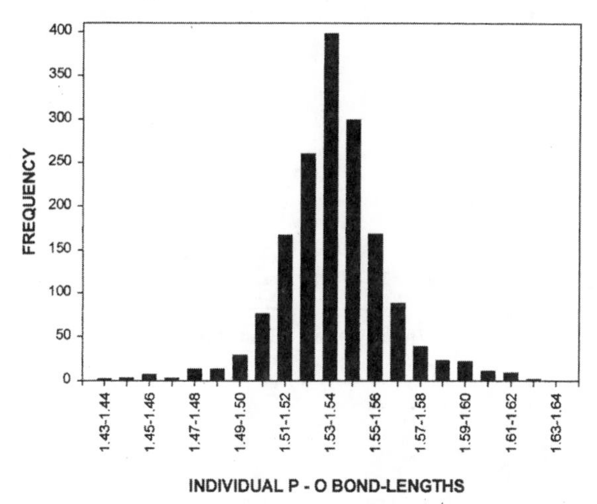

Figure 2. Variation in individual P-O distance in minerals containing ($P\phi_4$) tetrahedra.

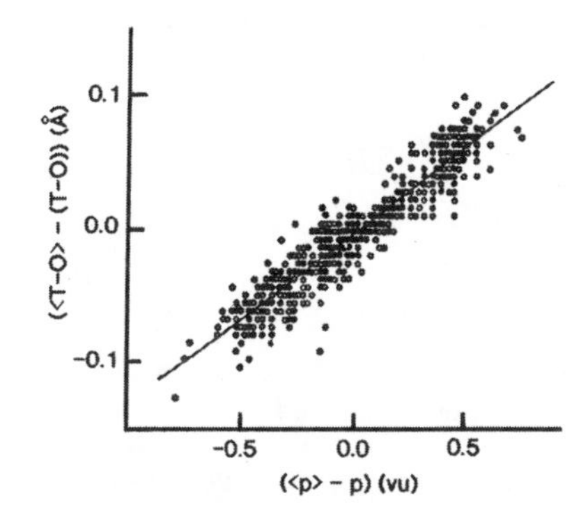

Figure 3. Variation in <T-O>-(T-O) for (PO_4) groups as a function of <p>-p, the deviation of the anion from exact agreement with Pauling's second rule; after Baur (1974).

Baur (1974) and Griffen and Ribbe (1979) have considered polyhedral distortions in general. Baur (1974) identified three types of distortion: (1) bond-length distortion; (2) bond-angle distortion; (3) polyhedron edge-length distortion, whereas Griffen and Ribbe (1979) considered two of these three distortion parameters, omitting bond-angle distortion. Baur (1974) showed that the variation in P-O distances (expressed as deviation from the <P-O> distance) correlates well with the deviation from ideal agreement with Pauling's second rule (Pauling 1929, 1960) (Fig. 4). This correlation is in accord with the general concepts of bond-valence theory (Brown and Shannon 1973; Brown 1981). Baur (1974) also showed that the dispersion (distortion) of P-O distances is much higher than the dispersion in corresponding O-O distances, and stated that the (PO_4) group can be "viewed, to a first approximation, as a rigid regular arrangement of O atoms, with the P atom displaced from their centroid."

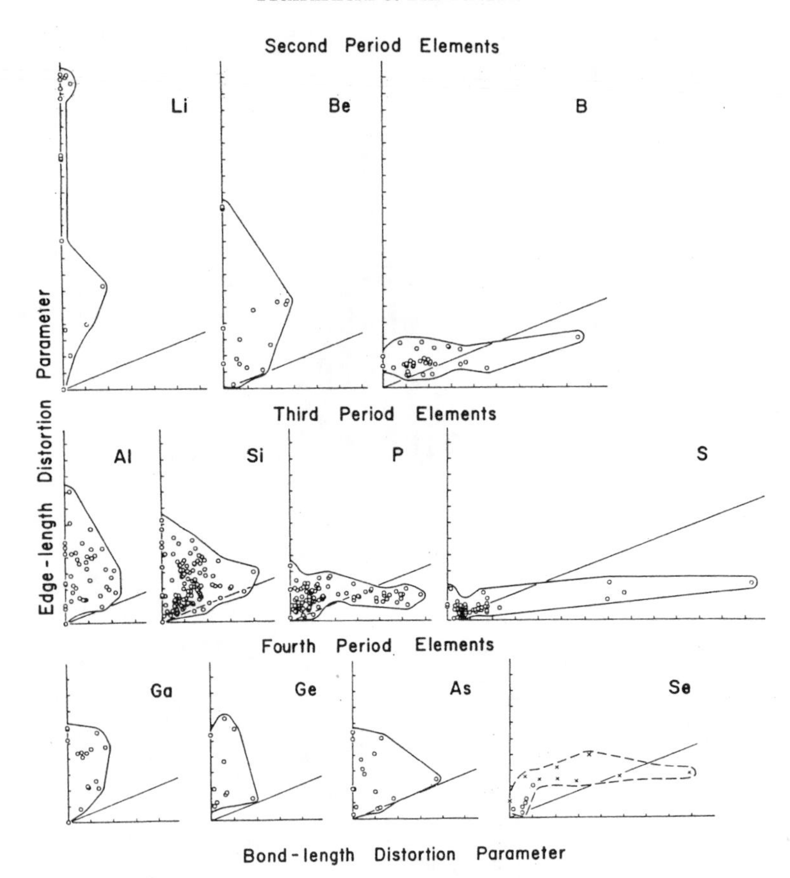

Figure 4. Variation in BLDP (Bond-Length Distortion Parameter) and ELDP (Edge-Length Distortion Parameter) for second-, third- and fourth-period non-transition elements in tetrahedral coordination by oxygen. Used by permission of E. Schweizerbart'sche Verlagsbuchhandlung, from Griffen and Ribbe (1979), *Neues Jarhrbuch für Mineralogie Abhandlungen*, Vol. 137, Fig. 2, p. 59.

Griffen and Ribbe (1979) considered two ways in which polyhedra may distort (i.e., depart from their holosymmetric geometry): (1) the central cation may displace from its central position [bond-length distortion]; (2) the anions may displace from their ideal positions [edge-length distortion]; these authors designate these as BLDP (Bond-Length Distortion Parameter) and ELDP (Edge-Length Distortion Parameter), respectively. Figure 4 shows the variation in both these parameters for the second-, third- and fourth-period (non-transition) elements in tetrahedral coordination. Some very general features of interest (Griffen and Ribbe 1979) are apparent from Figure 4:

(1) A BLDP value of zero only occurs for an ELDP value of zero; presuming that ELDP is a measure also of the O-T-O angle variation, this is in accord with the idea that variation in orbital hybridization (associated with variation in O-T-O angles) must accompany variation in bond-length.

(2) Large values of BLDP are associated with small values of ELDP, and vice versa. The variation in mean ELDP correlates very highly with the grand mean tetrahedral-edge length for each period (Fig. 5).

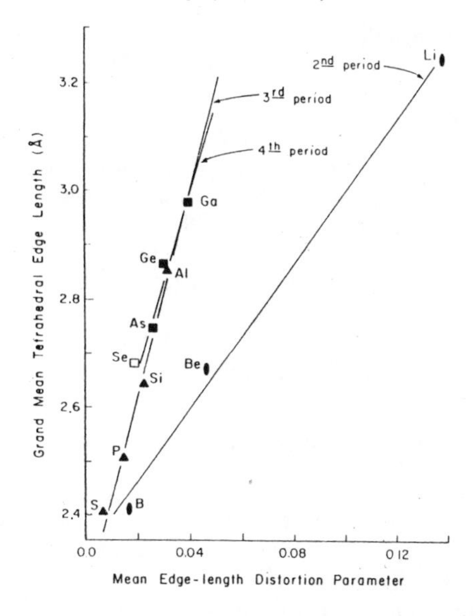

Figure 5. Variation in grand mean tetrahedral edge-length with mean ELDP for the second-, third- and fourth-period elements of the periodic table. Used by permission of E. Schweizerbart'sche Verlagsbuchhandlung, from Griffen and Ribbe (1979), *Neues Jarhrbuch für Mineralogie Abhandlungen*, Vol. 137, Fig. 6a, p. 65.

Griffen and Ribbe (1979) suggested that the smaller the tetrahedrally-coordinated cation, the more the tetrahedron of anions resists edge-length distortion because the anions are in contact, whereas the intrinsic size of the interstice is larger than the cation which can easily vary its cation-oxygen distances by 'rattling' within the tetrahedron. This general conclusion is in accord with the conclusion of Baur (1974) for the (PO_4) group.

HIERARCHICAL ORGANIZATION OF CRYSTAL STRUCTURES

The most fundamental characteristic of a mineral is its crystal structure, a complete description of which involves the identities, amounts and arrangement of atoms that constitute the mineral. The physical, chemical and paragenetic characteristics of a mineral arise as natural consequences of its crystal structure and the interaction of that structure with the environment in which it occurs. A structural hierarchy is an arrangement of crystal structures that reflects the systematic change in the character of their bond topologies. As the bond topology is a representation of the energetic characteristics of a structure (Hawthorne 1994, 1997), an adequate structural hierarchy of minerals should provide an epistemological basis for the interpretation of the role of minerals in Earth processes. This is not yet the case for any major class of minerals, but significant advances have been made. Bragg (1930) classified the major rock-forming silicate minerals according to the type of polymerization of $(Si,Al)O_4$ tetrahedra, and this scheme was extended by Zoltai (1960) and Liebau (1985); it is notable that this scheme parallels Bowen's reaction series (Bowen 1928) for silicate minerals in igneous rocks. Much insight can be derived from such structural hierarchies, particularly with regard to controls on bond topology (Hawthorne 1983a, 1994), mineral chemistry (Schindler and

Hawthorne 2001a,b; Schindler et al. 2002) and mineral paragenesis (Moore 1965b, 1973a; Hawthorne 1984, 1998; Hawthorne et al. 1987; Schindler and Hawthorne 2001c).

Hawthorne (1983a) proposed that structures be ordered or classified according to the polymerization of those cation coordination polyhedra with higher bond-valences. Higher bond-valence polyhedra polymerize to form *homo-* or *heteropolyhedral clusters* that constitute the *fundamental building block (FBB)* of the structure. The *FBB* is repeated, often polymerized, by translational symmetry operators to form the *structural unit*, a complex (usually anionic) polyhedral array (not necessarily connected) the excess charge of which is balanced by the presence of *interstitial* species (usually large, low-valence cations) (Hawthorne 1985a). The possible modes of cluster polymerization are obviously (1) unconnected polyhedra; (2) finite clusters; (3) infinite chains; (4) infinite sheets; (5) infinite frameworks.

POLYMERIZATION OF ($P\phi_4$) AND OTHER ($T\phi_4$) TETRAHEDRA

Bond valence is a measure of the strength of a chemical bond, and, in a coordination polyhedron, can be approximated by the formal valence divided by the coordination number (the Pauling bond-strength). Thus, in a (PO_4) group, the mean bond-valence is 5/4 = 1.25 vu. The valence-sum rule (Brown 1981) states that the sum of the bond valences incident at an atom is equal to the magnitude of the formal valence of that atom. Thus any oxygen atom linked to the central P cation receives ~1.25 vu from that cation, and hence must receive ~0.75 vu from other coordinating cations. Hence an oxygen atom is unlikely to link to two P atoms as it would receive, on average, $2 \times 1.25 = 2.50$ vu and the linking oxygen atom would violate the valence-sum rule. For this reason (PO_4) groups are unlikely to polymerize in crystal structures. This conclusion is not completely followed, as there are three (very rare) minerals in which (PO_4) groups polymerize [canaphite: $Na_2Ca[P_2O_7](H_2O)_4$, wooldridgeite: $Na_2CaCu^{2+}{}_2[P_2O_7]_2(H_2O)_{10}$ and kanonerovite: $Ma_3Mn^{2+}[P_3O_{10}](H_2O)_{12}$], and polyphosphates are common among synthetic compounds (Corbridge 1985). It was commonly thought that polyphosphates would be unstable in the presence of any H-bearing species as H would attack the bridging anion, resulting in depolymerization. However, the common existence of hydrated polyphosphates (Corbridge 1985) vitiates this argument. The only attempt to consider this problem in any detail is due to Byrappa (1983) who presented very limited experimental data and concluded that polyphosphates cannot form under hydrothermal conditions with $P(H_2O) > 6$ atm. This conclusion accounts for the absence of condensed phosphates in many phosphate parageneses (e.g., granitic pegmatites). However, many phosphate minerals crystallize under surficial conditions and yet only three minerals with polymerized phosphate groups are currently known. Our lack of understanding concerning this issue is obviously an important gap in our knowledge of phosphate crystal chemistry. Suffice it to say here that polymerized (PO_4) groups are sufficiently rare in minerals that phosphates cannot be classified in an analogous way to silicates (i.e., by the polymerization characteristics of the principal oxyanion).

The simple anions of a (PO_4) group each require ~0.75 vu to satisfy their bond-valence requirements. What type of linkage with other tetrahedral oxyanions is possible with this constraint? Obviously, any tetrahedral oxyanion with a mean bond-valence of $\leq$0.75 vu, which includes (AlO_4), (BO_4), (BeO_4) and (LiO_4) groups. Moreover, P-O bonds in specific structural arrangements may have bond valences somewhat less that 1.25 vu, raising the possibility that (PO_4) groups might polymerize with (SiO_4) groups. Phosphates show all of these particular polymerizations, in accord with the valence-sum rule.

POLYMERIZATION OF ($P\phi_4$) TETRAHEDRA AND OTHER ($M\phi_N$) POLYHEDRA

In oxysalt minerals, the coordination number of oxygen is most commonly [3] or [4]. This being the case, the *average* bond-valence incident at the oxygen atom bonded to one P cation is ~0.75/3 = 0.25 vu and ~0.75/2 = 0.38 vu for the other cation-oxygen bonds. The more common non-tetrahedrally coordinated cations available in geochemical systems are [6]-coordinated divalent (e.g., Mg, Fe^{2+}, Mn^{2+}) and trivalent (e.g., Al, Fe^{3+}) cations, and [7]- and higher coordinated monovalent (e.g., Na, K) and divalent (e.g., Ca, Sr) cations. The average bond-valences involved in linkage to these cations are ${}^{[6]}M^{2+} \approx 0.33$, ${}^{[6]}M^{3+} \approx 0.50$, ${}^{[7]}M^{1+} \approx 0.14$, ${}^{[7]}M^{2+} \approx 0.29$ vu. Hence, (PO_4) groups link easily to all of these cations, particularly in hydroxy-hydrated phosphates where hydrogen bonds (0.1-0.3 vu) commonly supply additional bond-valence to the (simple) anions of the structure.

A STRUCTURAL HIERARCHY FOR PHOSPHATE MINERALS

This promiscuous polymerization suggests that we should classify the phosphates according to the types of polymerization of their principal coordination polyhedra, as suggested by Hawthorne (1983a, 1998) and discussed briefly above. The most common polymerizations in phosphate minerals are between tetrahedra and tetrahedra, between tetrahedra and octahedra, and between tetrahedra and large-cation polyhedra (i.e., [7]-coordinated and above). Hence we will divide the phosphates into the following three principal groups, involving

(1) polymerization of tetrahedra;

(2) polymerization of tetrahedra and octahedra;

(3) polymerization of tetrahedra and > [6]-coordinated polyhedra.

Large-cation polyhedra occur in group (1) and group (2) phosphates, as cations such as Al and Ca often occur together in minerals. In principle, octahedra should not occur (by definition) in group (3) phosphates; however, in some cases, the structural affinity of the mineral indicates it to be a large-cation phosphate mineral (e.g., whitlockite, $Ca_{18}Mg_2(PO_4)_{12}(PO_3\{OH\})_2$).

In accord with the above discussion, phosphate minerals are classified into three distinct groups. The first group, involving polymerization of ($T\phi_4$) tetrahedra (T = P plus Be, Zn, B, Al and Si), is fairly small, in accord with the observation that mixed oxyanion minerals are uncommon. The second group, involving polymerization of (PO_4) tetrahedra and ($M\phi_6$) octahedra, is very large. Within this group, the structures are arranged as suggested by Hawthorne (1983a), similar to the classification of the sulfate minerals given by Hawthorne et al. (2000), according to the mode of polymerization of the tetrahedra and octahedra: (1) unconnected polyhedra; (2) finite clusters of polyhedra; (3) infinite chains of polyhedra; (4) infinite sheets of polyhedra; (5) infinite frameworks of polyhedra. Within each class, structures are arranged in terms of increasing connectivity of the constituent polyhedra of the structural unit. Detailed chemical and crystallographic information, together with references, are given in Appendix A. In the following figures, (PO_4) groups are dashed-line-shaded, and cell dimensions with an arrow on one end only are (slightly) tilted to the plane of the figure.

STRUCTURES WITH POLYMERIZED ($T\phi_4$) GROUPS

As noted above, (PO_4) tetrahedra can polymerize with other (PO_4) groups, plus tetrahedrally coordinated Be, Zn, B, Al and Si. However, in minerals, only the following polymerizations are known: (PO_4)-(PO_4), (PO_4)-($Be\phi_4$), (PO_4)-(ZnO_4) and (PO_4)-(AlO_4).

Although there seems no obvious reason why (PO_4) should not polymerize with $(B\phi_4)$ or $(Si\phi_4)$ groups, this has not been observed, although there are several minerals containing both (PO_4) and $(B\phi_4)$ or $(Si\phi_4)$ groups in which the different polyhedra do not polymerize.

Table 1. Phosphate minerals* based on $(T\phi_4)$ clusters.

Mineral	*Cluster*	*Space group*	*Figure*
Canaphite	$[P_2O_7]$	*Pc*	6a,b
Wooldridgeite	$[P_2O_7]$	*Fdd2*	6c,d,e
Kanonerovite	$[P_3O_{10}]$	$P2_1/n$	7a
"Pyrocoproite"**	P_2O_7	–	–
"Pyrophosphate"**	P_2O_7	–	–
"Arnhemite"**	P_2O_7	–	–
Gainesite *	$[Be(PO_4)_4]$	$I4_1/amd$	7b
McCrillisite	$[Be(PO_4)_4]$	$I4_1/amd$	7b
Selwynite	$[Be(PO_4)_4]$	$I4_1/amd$	7b

* For isostructural minerals, the name of the group is indicated by a * in this and all following tables;
** These names are used in the literature, but have not been approved by CNMMN of IMA.

Finite clusters of tetrahedra

The minerals in this class (Table 1) are dominated by polymerized (PO_4) groups; only in gainesite do (PO_4) tetrahedra polymerize with another type of $(T\phi_4)$ group.

In **canaphite**, $Na_2Ca(H_2O)_4[P_2O_7]$, (PO_4) tetrahedra link to form $[P_2O_7]$ groups in the eclipsed configuration. When viewed down [100], the structure consists of layers of $(Na\phi_6)$ and $(Ca\phi_6)$ octahedra intercalated with intermittent layers of $[P_2O_7]$ groups (Fig. 6a). The layers of octahedra are not completely continuous (Fig. 6b); staggered α-PbO_2-like chains of $(Na\phi_6)$ octahedra extend along *a* and are linked in the *b*-direction by $(Ca\phi_6)$ octahedra to form a sheet of octahedra punctuated by dimers of vacant octahedra. Additional linkage is provided by an extensive network of hydrogen bonds involving the (H_2O) groups of the $(CaO_5\{H_2O\})$ and $(NaO_3\{H_2O\}_3)$ octahedra.

In **wooldridgeite**, $Na_2Ca(H_2O)_6[Cu^{2+}_2(P_2O_7)_2(H_2O)_2](H_2O)_2]$, $[P_2O_7]$ groups occur in the eclipsed configuration. A key part of the wooldridgeite structure is the $[Cu^{2+}(P_2O_7)(H_2O)]$ chain (Fig. 6c) in which $(Cu\phi_6)$ octahedra link by sharing one set of *trans* ligands (H_2O) to form a 7-Å chain (Moore 1970), decorated by $[P_2O_7]$ groups, that extends along [101] (and [101]). Each chain is flanked by a chain of corner-sharing $(Na\phi_6)$ octahedra in which the Na-ϕ-Na linkage is through *cis* vertices and each $(Na\phi_6)$ octahedron shares an edge with a $(Cu\phi_6)$ octahedron, and these chains are linked in the [101] direction into a sheet by $(Ca\phi_6)$ octahedra (Fig. 6d). These sheets stack along the [010] direction (Fig. 6e), with each sheet rotated 90° with respect to the adjacent sheets.

In **kanonerovite**, $Na_3Mn^{2+}[P_3O_{10}](H_2O)_{12}$, three (PO_4) tetrahedra link into a $[P_3O_{10}]$ fragment. All three (PO_4) tetrahedra of this trimeric group share one vertex with the same $(Mn^{2+}\phi_6)$ octahedron (Fig. 7a) to form an $[Mn^{2+}(H_2O)_3P_3O_{10}]$ cluster. $(Na\phi_6)$ octahedra

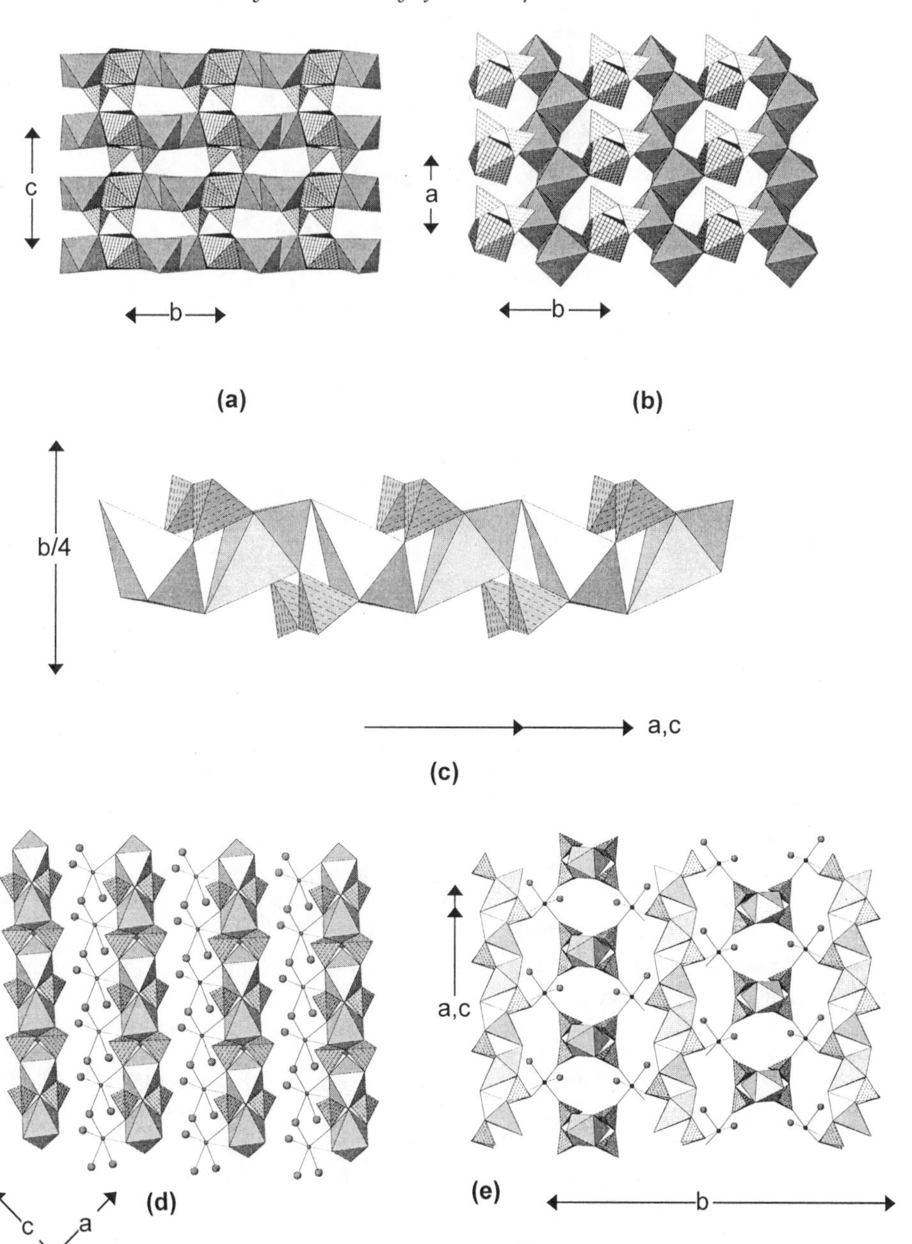

Figure 6. The crystal structures of canaphite and wooldridgeite: (a) canaphite projected onto (100); (b) canaphite projected onto (001); $(Ca\phi_6)$ octahedra are 4^4-net-shaded, $(Na\phi_6)$ octahedra are shadow-shaded; (c) a perspective view of the $[Cu^{2+}(P_2O_7)(H_2O)]$ chain in wooldridgeite; (d) wooldridgeite projected onto (010); (e) wooldridgeite, showing two orthogonal sets of $[Cu^{2+}(P_2O_7)(H_2O)]$ chains; $(Cu^{2+}\phi_6)$ octahedra are shadow-shaded, Na atoms are shown as small dark circles, (H_2O) groups are shown as large shaded circles.

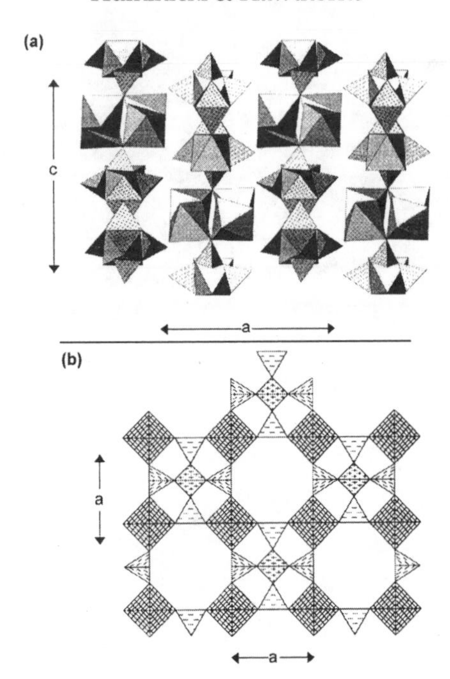

Figure 7. The crystal structures of kanonerovite and gainesite: (a) kanonerovite projected onto (010) showing the $[P_3O_{10}]$ trimers linked to $(Mn^{2+}\phi_6)$ octahedra (dot-shaded); the resulting clusters are linked by $(Na\phi_6)$ octahedra (shadow-shaded) and by hydrogen bonding involving (H_2O) groups (not shown); (b) the $[BeP_4O_{16}]$ pentamer that is the finite tetrahedron cluster in gainesite linked by (ZrO_6) octahedra (4^4-net-shaded).

link by sharing vertices to form clusters that link $[Mn^{2+}(H_2O)_3P_3O_{10}]$ clusters adjacent in the *c*-direction. All other linkage is via hydrogen bonds emanating from the (H_2O) groups of the $[Mn^{2+}(H_2O)_3P_3O_{10}]$ cluster and involving interstitial (H_2O) groups.

In **gainesite**, $Na_2Zr_2[Be(PO_4)_4]$, and the isostructural minerals **mccrillisite**, $Cs_2Zr_2[Be(PO_4)_4]$, and **selwynite**, $Na_2Zr[Be(PO_4)_4]$, a (BeO_4) tetrahedron links to four (PO_4) tetrahedra to form the pentameric cluster $[BeP_4O_{16}]$ that is topologically identical to the $[Si_5O_{16}]$ cluster in zunyite, $Al_{13}O_4[Si_5O_{16}](OH)_{18}Cl$. These clusters are linked in a continuous framework through (ZrO_6) octahedra (Fig. 7b). The Be and P sites in the gainesite structure are only half-occupied, and in the tetrahedral-octahedral framework, tetrahedral clusters alternate with cavities occupied by interstitial Na atoms.

Infinite chains of tetrahedra

The minerals in this class can be divided into two broad groups based on the (bond valence) linkage involved in the infinite chains. Minerals of this class are listed in Table 2.

Moraesite, $[Be_2(PO_4)(OH)](H_2O)_4$, contains chains (ribbons) of (PO_4) and (BeO_4) tetrahedra. The (PO_4) tetrahedra are four-connected and the (BeO_4) tetrahedra are three-connected, and the resulting $[Be_2(PO_4)(OH)]$ ribbons extend along the *c*-direction (Fig. 8a). These ribbons form a face-centered array (Fig. 8b) and are linked by hydrogen bonds involving interstitial (H_2O) groups.

Table 2. Phosphate minerals based on ($T\phi_4$) chains.

Mineral	*Chain*	*Space Group*	*Figure*
Moraesite	$[Be_2(PO_4)(OH)]$	*C2/C*	8a,b
Väyrynenite	$[Be(PO_4)(OH)]$	$P2_1/A$	8c,d
Fransoletite	$[Be_2(PO_4)_2(PO_3\{OH\})_2]$	$P2_1/A$	9a,b
Parafransoletite	$[Be_2(PO_4)_2(PO_3\{OH\})_2]$	$P\bar{1}$	–
Roscherite	$[Be_4(PO_4)_6(OH)_6]$	*C2/C*	9c,d
Zanazziite	$[Be_4(PO_4)_6(OH)_6]$	*C2/C*	9c,d
Spencerite	$[Zn(PO_4)(OH)(H_2O)]$	$P2_1/C$	9e,f

Väyrynenite, $Mn^{2+}[Be(PO_4)(OH)]$, contains chains of (PO_4) and (BeO_4) tetrahedra extending in the *a*-direction (Fig. 8c). (BeO_4) tetrahedra link by corner-sharing to form a pyroxenoid-like [TO_3] chain that is decorated on both sides by (PO_4) tetrahedra to form a ribbon in which the (BeO_4) tetrahedra are four-connected and the (PO_4) tetrahedra are two-connected. These ribbon-like chains are linked by edge-sharing pyroxene-like chains of ($Mn^{2+}O_6$) octahedra that also extend parallel to the *a*-axis. The resulting structural arrangement consists of modulated sheets of tetrahedra and octahedra (Fig. 8d).

The dimorphs, **fransoletite** and **parafransoletite**, have the composition $Ca_3[Be_2(PO_4)_2(PO_3\{OH\})_2](H_2O)_4$. The principal motif in each structure is a complex chain consisting of four-membered rings of alternating (PO_4) and (BeO_4) tetrahedra that link through common (BeO_4) tetrahedra; these chains extend in the *a*-direction (Fig. 9a). Viewed end-on (Fig. 9b), the chains form a square array and are linked by [6]- and [7]-coordinated Ca atoms that form sheets parallel to {001}; further interchain linkage occurs through H-bonding involving (H_2O) groups. The fransoletite and parafransoletite structures differ only in the relative placement of the octahedrally coordinated Ca atom and the disposition of adjacent chains along their length (Kampf 1992).

Roscherite and **zanazziite** are composed of very convoluted chains of $Be\phi_4$ and (PO_4) tetrahedra extending in the [101] direction (Fig. 9c; note that in this view, the two chains appear to join at a mirror plane parallel to their length; however, the plane in question is a glide plane and the two chains are displaced in the *c*-direction). The chain consists of four-membered rings of alternating $Be\phi_4$ and (PO_4) tetrahedra linked through (PO_4) tetrahedra that are not members of these rings (Fig. 9c). These chains are linked by $(Al,)O_6$ and $(Mg,Fe^{2+})O_6$ octahedra that form edge-sharing chains parallel to [110] and [$1\bar{1}0$]; the octahedral chains link to each other in the [001] direction by sharing *trans* vertices (Fig. 9d). The resultant octahedral-tetrahedral framework is strengthened by [7]-coordinated Ca occupying the interstices.

The structure and composition of these minerals is not completely understood. Roscherite (Slavík 1914) is the Mn^{2+}-dominant species and zanazziite (Leavens et al. 1990) is the Mg-dominant species. Lindberg (1958) also reported an Fe^{2+}-dominant species from the Sapucaia pegmatite, Minas Gerais, that is currently unnamed. The situation is complicated by the fact that the original crystal-structure determination of roscherite (Fanfani et al. 1975) was done on a crystal of what was later determined to be *zanazziite* with the ideal end-member formula $Ca_2Mg_4(Al_{0.67}G_{0.33})_2[Be_4(PO_4)_6(OH)_6]$ $(H_2O)_4$. Fanfani et al. (1977) report a triclinic structure for roscherite that is Mn^{2+}

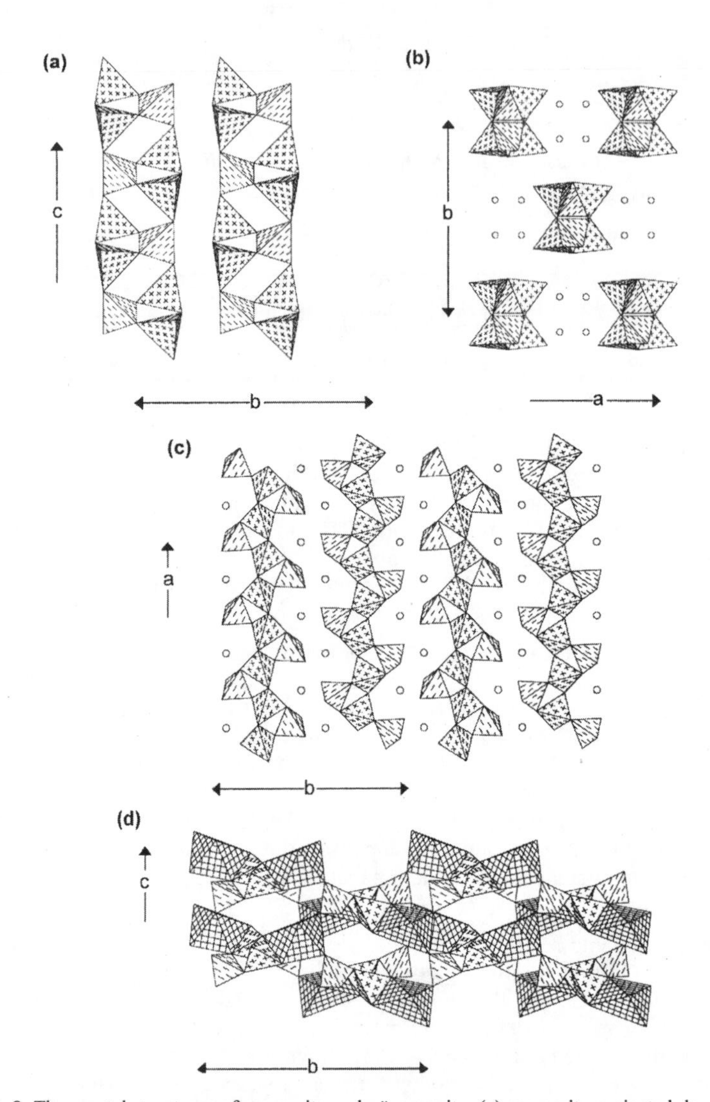

Figure 8. The crystal structures of moraesite and väyrynenite: (a) moraesite projected down the *a*-axis; (b) moraesite projected down the *c*-axis; (BeO_4) tetrahedra are cross-shaded, (H_2O) groups are shown as circles; (c) väyrynenite projected down the *c*-axis; Al atoms are shown as circles; (d) väyrynenite projected down the *a*-axis.

dominant, i.e., *roscherite* with the ideal end-member formula $Ca_2Mn^{2+}_4(Fe^{3+}_{0.67}G_{0.33})(G)$ $[Be_4(PO_4)_6(OH)_4(H_2O)_2](H_2O)_4$. Note that the trivalent-cation content ($Al_{1.33}$ vs. $Fe^{3+}_{0.67}$) and type are different in the two species, and electroneutrality is maintained by replacement of OH by H_2O via the exchange Fe^{3+} + □ (vacancy) + 3 $H_2O \rightarrow Al^{3+}_2$ + 3 OH. Whether the monoclinic → triclinic transition is caused by the $Mn^{2+} \rightarrow Mg$ replacement or by the reaction noted above is not yet known.

Spencerite, $Zn_2[Zn(OH)(H_2O)(PO_4)]_2(H_2O)$, contains simple linear chains of alternating ($Zn\phi_4$) [$\phi_4 = O_2(OH)(H_2O)$] and (PO_4) tetrahedra extending along the *c*

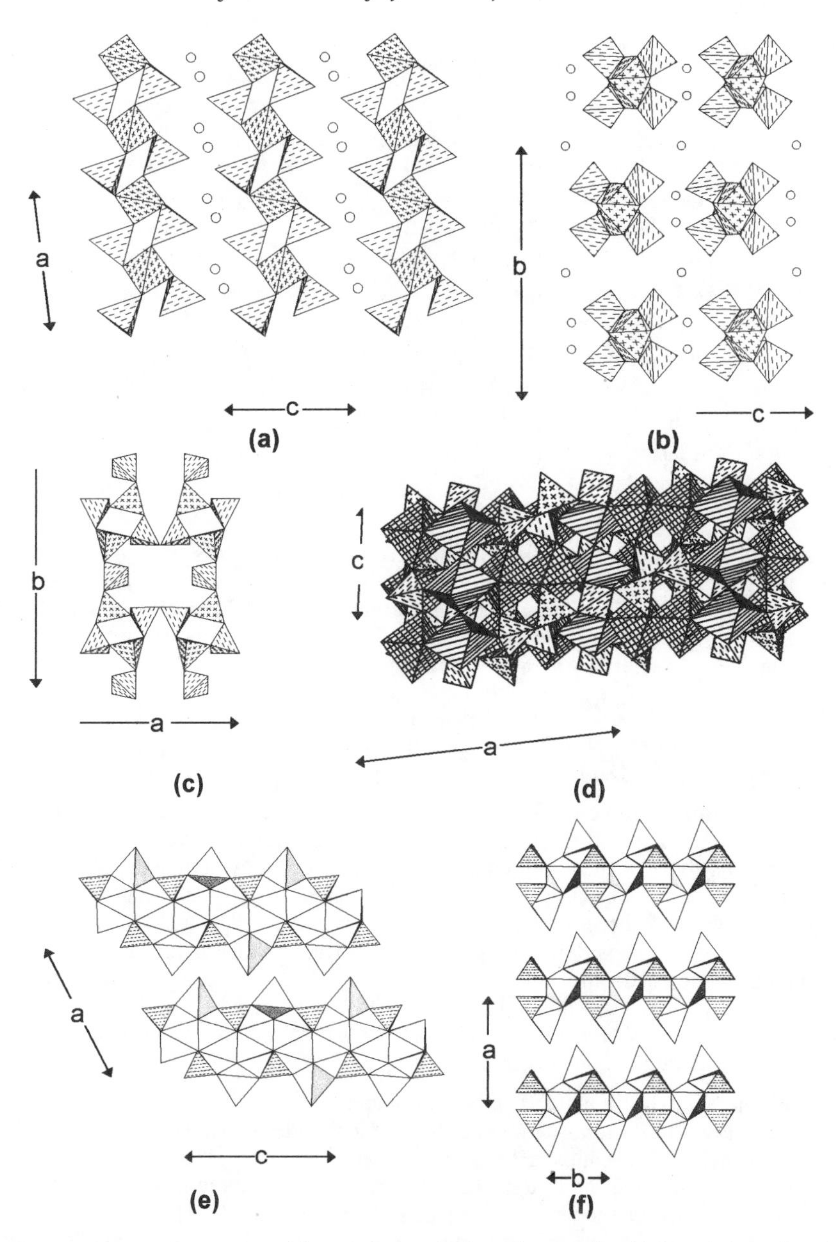

Figure 9. The crystal structures of fransoletite, roscherite and spencerite: (a) fransoletite projected onto (010); Mn^{2+} atoms are shown as circles; (b) fransoletite projected down the *a*-axis; a view in which the chains are seen end-on; (c) the structural unit in roscherite projected down the *c*-axis; note that the (PO_4) tetrahedra in the center of the figure do *not* share a common anion, but are separated in the *c*-direction; (d) roscherite projected onto (010); note that the trivalent octahedra (4^4-net-shaded) are only two-thirds occupied (by Al) and that Ca atoms are omitted for clarity; (e) spencerite projected onto (010); (f) spencerite projected onto (001); chains of (PO_4) and (ZnO_4) are seen 'end-on'.

Table 3. Phosphate minerals based on ($T\phi_4$) sheets.

Mineral	*Sheet*	*Space group*	*Figure*
Herderite	$[Be(PO_4)(OH)]$	$P2_1/a$	10a,b
Hydroxylherderite	$[Be(PO_4)(OH)]$	$P2_1/a$	10a,b
Uralolite	$[Be_4P_3O_{12}(OH)_3]$	$P2_1/n$	10c,d
Ehrleite	$[BeZn(PO_4)_2(PO_3\{OH\})]$	$P\bar{1}$	10e,f
Hopeite	$[Zn(PO_4)]$	$Pnma$	11a,b
Parahopeite	$[Zn(PO_4)]$	$P\bar{1}$	11c,d
Phosphophyllite	$[Zn(PO_4)]$	$P2_1/c$	12a,b
Veszelyite	$[Zn(PO_4)(OH)]$	$P2_1/a$	12c,d
Kipushite	$[Cu^{2+}_5Zn(PO_4)_2(OH)_6(H_2O)]$	$P2_1/c$	12e,f
Scholzite	$[Zn(PO_4)]$	$Pbc2_1$	13a,b
Parascholzite	$[Zn(PO_4)]$	$I2/c$	13c,d

direction (Fig. 9e) and cross-linked into heteropolyhedral sheets by ($Zn\phi_6$) octahedra. These sheets are also shown in Figure 9f, where it can be seen that the ($Zn\phi_6$) octahedra share all their vertices with ($Zn\phi_4$) and (PO_4) tetrahedra. The heteropolyhedral sheets link solely via hydrogen bonding that involves one (H_2O) group (not shown in Figs. 9e or 9f) held in the structure solely by hydrogen bonding.

Infinite sheets of tetrahedra

The minerals in this class (Table 3) can be divided into two groups: (PO_4)-(BeO_4) linkages, and (PO_4)-(ZnO_4) linkages.

Hydroxylherderite, $Ca[Be(PO_4)(OH)]$ and **herderite**, $Ca[Be(PO_4)F]$, are isostructural, although the structures were reported in different orientations, $P2_1/c$ and $P2_1/a$, respectively. The sheet unit consists of (PO_4) and ($Be\phi_4$) tetrahedra at the vertices of a two-dimensional net (Fig. 10a). Four-membered rings of alternating (PO_4) and ($Be\phi_4$) tetrahedra link directly by sharing vertices between (PO_4) and ($Be\phi_4$) tetrahedra; thus the sheet can be considered to be constructed from chains of four-membered rings that extend in the [110] and [1$\bar{1}$0] directions (Fig. 10a). These sheets stack in the *c*-direction (Fig. 10b) and are linked by layers of [8]-coordinated Ca atoms. Note that the structure reported by Lager and Gibbs (1974) seems to have been done on hydroxylherderite rather than herderite.

Uralolite, $Ca_2[Be_4P_3O_{12}(OH)_3](H_2O)_5$, contains ($PO_4$) and ($BeO_4$) tetrahedra linked into a sheet (Fig. 10c). Eight-membered rings of tetrahedra (P-Be-P-Be-P-Be-P-Be) link through common (PO_4) groups to form chains that extend along [101]. These chains link in the (010) plane via sharing of tetrahedral vertices, forming three-membered (Be-Be-Be and Be-Be-P) and four-membered (Be-Be-Be-P) rings. Interstitial [7]-coordinated Ca atoms lie within the eight-membered rings (in projection). The layers stack along the *b*-direction (Fig. 10d) and are linked by Ca atoms (circles) and H-bonding; in this view, the three- and four-membered rings are easily seen.

Ehrleite, $Ca_2[BeZn(PO_4)_2(PO_3\{OH\})](H_2O)_4$, has a very complicated sheet of tetrahedra, both from topological and chemical viewpoints. There is one distinct (BeO_4) tetrahedron that links to four ($P\phi_4$) groups (Fig. 10e); similarly, there is one (ZnO_4) tetrahedron and this links to four ($P\phi_4$) groups. The ($P\phi_4$) groups link only to three or two other tetrahedra. Four-membered rings of alternating (PO_4) and (BeO_4) tetrahedra link

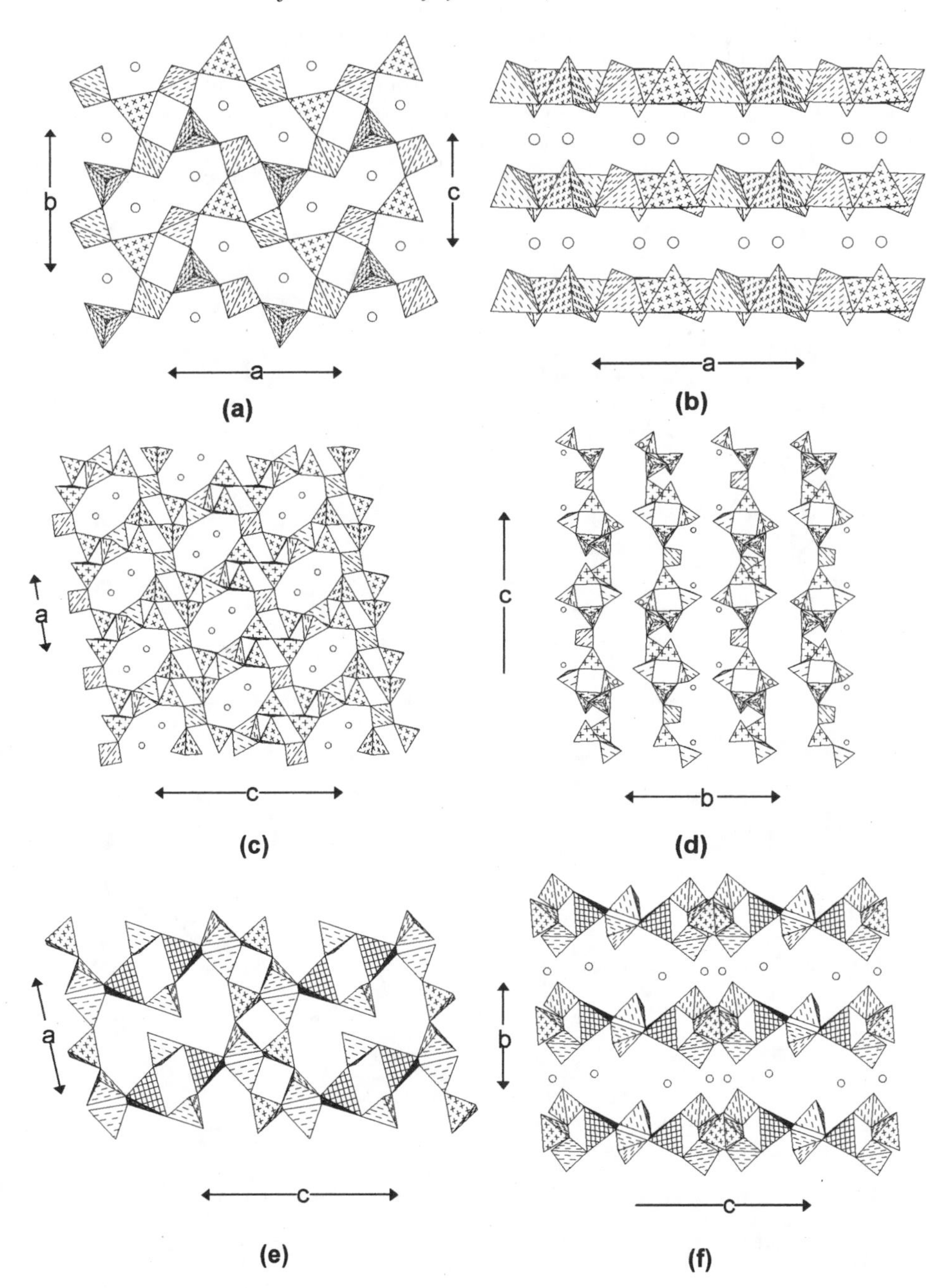

Figure 10. The crystal structures of herderite, uralolite and ehrleite: (a) herderite projected onto (001); (b) herderite projected onto (010); (c) uralolite projected onto (010); (d) uralolite projected down the *a*-axis; (e) the structural unit in ehrleite projected onto (010); (ZnO_4) tetrahedra are 4^4-net-shaded; (f) ehrleite projected down the *a*-axis. Ca atoms are shown as circles.

through common (BeO_4) tetrahedra to form chains in the *a*-direction (Fig. 10e). These chains are linked in the *c*-direction by four-membered rings of alternating (PO_4) and (ZnO_4) tetrahedra to form additional four-membered rings (Zn-P-Be-P). The result is an open sheet, parallel to (010), with buckled twelve-membered rings (Fig. 10e) into which project the H atoms of the acid-phosphate groups. These sheets stack along the *b*-direction (Fig. 10f) and are linked together by [7]-coordinated and [8]-coordinated interstitial Ca atoms.

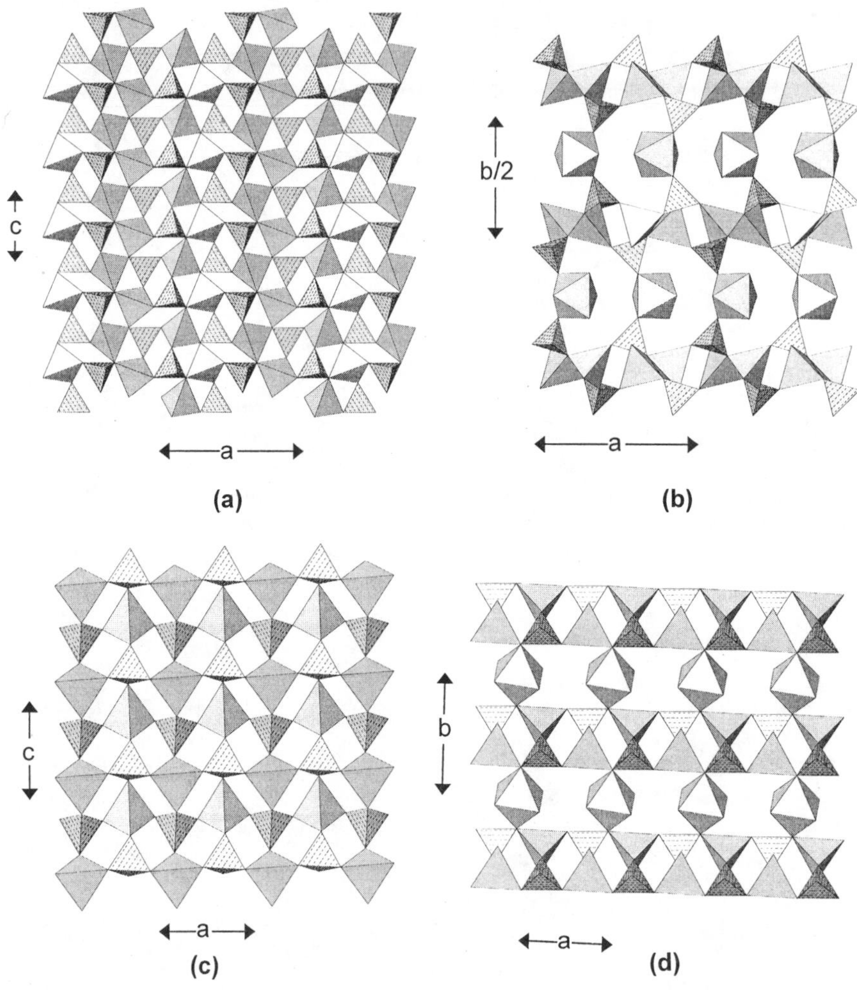

Figure 11. The crystal structures of hopeite and parahopeite: (a) hopeite projected onto (010); (b) hopeite projected onto (001); hydrogen bonds are omitted for clarity; (c) parahopeite projected onto (010); (d) parahopeite projected onto (001). ($Zn\phi_4$) tetrahedra and ($Zn\phi_6$) octahedra are shadow-shaded.

In **hopeite**, $Zn(H_2O)_4[Zn(PO_4)]_2$, kinked chains of ($Zn\phi_4$) tetrahedra extend in the *c*-direction, and adjacent chains are linked by (PO_4) tetrahedra to form a sheet parallel to (101) (Fig. 11a). The (ZnO_4) tetrahedra are four-connected, but the (PO_4) tetrahedra are only three-connected; this difference in connectivity is very important as it promotes structural linkage perpendicular to the sheet. A continuous sheet with this connectivity requires unusual coordination numbers for some of the simple anions of the sheet: for the (PO_4) group, the anion coordination numbers within the sheet are [1], [2] × 2 and [3], and for the (ZnO_4) group, the anion coordination numbers within the sheet are [2] × 2 and [3] × 2. Hence the sheet is quite corrugated, as can be seen in Figure 11b: the (ZnO_4) tetrahedra form a central layer and the (PO_4) tetrahedra form two outer (or sandwiching) layers. The sheets are linked in the *b*-direction by ($ZnO_2\{H_2O\}_4$) octahedra (Fig. 11b), the [1]-coordinated anion of the phosphate group forming a ligand of the linking $^{[6]}Zn$ cation. In **parahopeite**, $Zn(H_2O)_4[Zn(PO_4)]_2$, (PO_4) and (ZnO_4) tetrahedra lie at the vertices of a 4^4 net to form a sheet in which (PO_4) tetrahedra link only to (ZnO_4) tetrahedra, and vice versa. Thus all tetrahedra are four-connected, and all vertices (simple anions) are two-connected within the resultant sheet (Fig. 11c). These sheets are parallel to (101) and are linked in the *b*-direction by ($ZnO_2\{H_2O\}_4$) octahedra in which the *trans* O-atoms belong to adjacent sheets (Fig. 11d).

In **phosphophyllite**, $Fe^{2+}(H_2O)_4[Zn(PO_4)]_2$, ($PO_4$) and ($Zn\phi_4$) tetrahedra form a sheet (Fig. 12a) that is topologically identical to the [$Zn(PO_4)$] sheet in hopeite (Fig. 11a). These sheets are linked by ($Fe^{2+}O_2\{H_2O\}_4$) octahedra, similar to the linkage by ($ZnO_2\{H_2O\}_4$) octahedra in hopeite. However, in phosphophyllite, the O-atoms of the ($Fe^{2+}O_2\{H_2O\}_4$) octahedron are in a *trans* configuration (Fig. 12b), whereas in hopeite, the O-atoms of the ($ZnO_2\{H_2O\}_4$) octahedron are in a *cis* configuration (Fig. 11b).

In **veszeylite**, $Cu^{2+}_2(OH)_2(H_2O)_2[Zn(PO_4)(OH)]$, ($PO_4$) and ($Zn\phi_4$) tetrahedra occur at the vertices of a 4.8^2 net (Fig. 12c) in which each type of tetrahedron points both up and down relative to the plane of the sheet. Both (PO_4) and ($Zn\phi_4$) tetrahedra are three-connected within the sheet, and (PO_4) tetrahedra and ($Zn\phi_4$) tetrahedra always alternate in any path through the 4.8^2 net. In the four-membered ring, the tetrahedra point *uudd*, and in the eight-membered ring, the tetrahedra point *uuuudddd*. The ($Cu^{2+}\phi_6$) octahedra form an interrupted [$M\phi_2$] sheet (Hawthorne and Schindler 2000; Hawthorne and Sokolova 2002) in which the vacant octahedra are ordered as dimers. The sheets of tetrahedra and octahedra stack in the *c*-direction (Fig. 12d) with hydrogen bonds (not shown) providing additional linkage between octahedra and tetrahedra.

Kipushite, [$Cu^{2+}_5Zn(PO_4)_2(OH)_6(H_2O)$], contains ($PO_4$) and ($ZnO_4$) tetrahedra that are arranged at the vertices of a 4.8^2 net (as occurs in veszelyite) and link by corner-sharing (Fig. 12e, c.f. Fig. 12c). ($Cu^{2+}\phi_6$) octahedra share edges to form a sheet with ordered vacancies. It is actually a sheet of the form [$M_6\phi_{12}$] ≡ [$M\phi_2$]$_6$ with M_6 = Cu^{2+}_5 , where is a vacant octahedron; these 'vacant octahedra' share a face with a (PO_4) tetrahedron on one side of the sheet of octahedra. Two of these sheets then link by sharing the apical vertices of their (PO_4) tetrahedra with octahedron vertices of the adjacent sheet to form a thick slab (Fig. 12f). These slabs stack in the *a*-direction and are linked by the (PO_4)-(ZnO_4) sheet through sharing of vertices between tetrahedra and octahedra.

In **scholzite**, $Ca(H_2O)_2[Zn(PO_4)]_2$, (ZnO_4) tetrahedra share pairs of vertices to form simple linear chains parallel in the *c*-direction. Adjacent (ZnO_4) tetrahedra are further linked by sharing vertices with a (PO_4) tetrahedron, and the (PO_4) tetrahedra are in a staggered arrangement along the length of the chain. Chains adjacent in the *b*-direction link through (PO_4) tetrahedra to form a sheet parallel to (100) (Fig. 13a). In this sheet, the (ZnO_4) tetrahedra are four-connected and the (PO_4) tetrahedra are three-connected. The bridging anions of the chain of (ZnO_4) tetrahedra are three-connected; all other anions of the sheet are two-connected except for the one-connected anion of the (PO_4) tetrahedron. The resulting sheet (Fig. 13a) forms quite a thick slab that is linked by two crystallographically distinct octahedrally coordinated Ca atoms (Fig. 13b). In

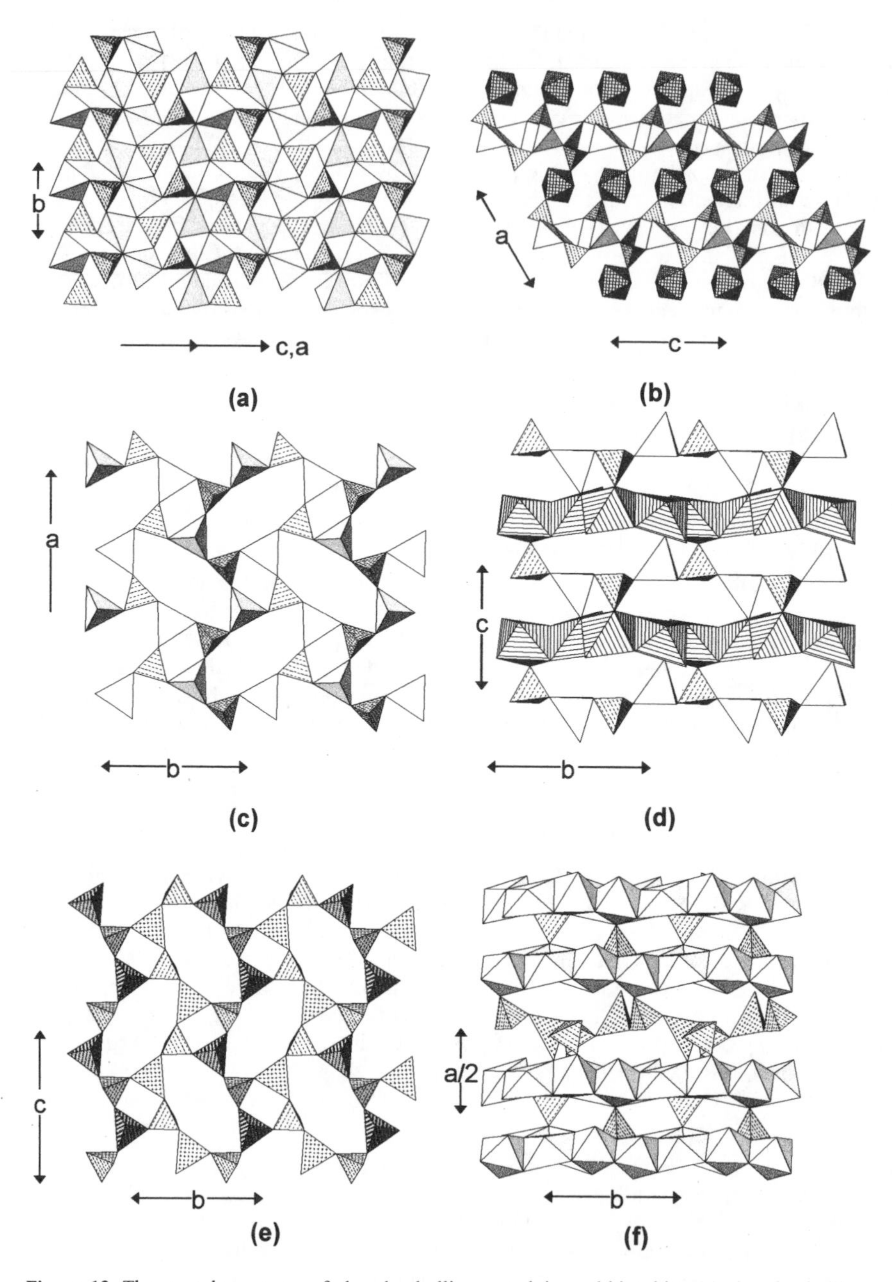

Figure 12. The crystal structures of phosphophyllite, veszelyite and kipushite: (a) phosphophyllite showing (PO_4) and ($Zn\phi_4$) tetrahedra at the vertices of a 4.8^2 net; (b) phosphophyllite projected onto (010); ($Zn\phi_4$) tetrahedra are shadow-shaded, ($Zn\phi_6$) octahedra are 4^4-net-shaded; (c) veszelyite projected onto (001); tetrahedra are arranged at the vertices of a 4.8^2 net; (d) veszelyite projected onto (100); ($Cu^{2+}\phi_6$) octahedra are line-shaded; (e) the sheet of corner-linked (PO_4) and (ZnO_4) tetrahedra in kipushite projected onto (100); (f) the structure of kipushite projected onto (001); ($Zn\phi_4$) tetrahedra are shadow-shaded.

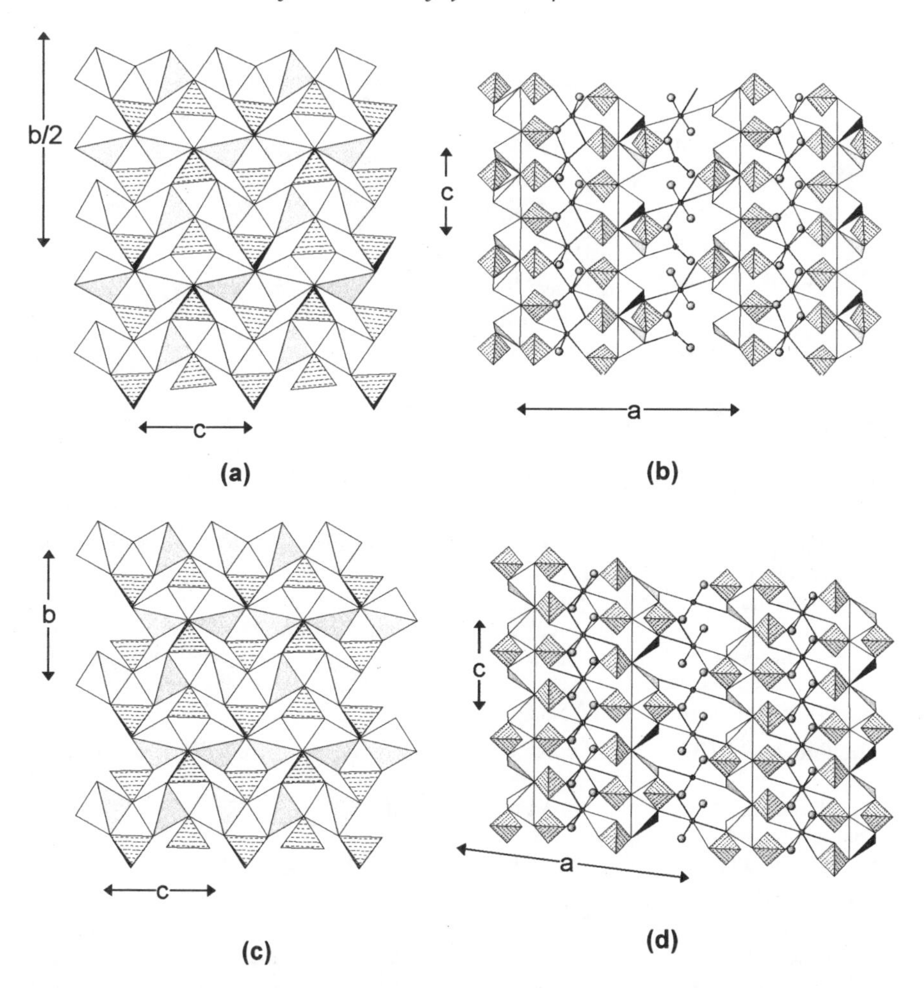

Figure 13. The crystal structures of scholzite and parascholzite: (a) scholzite projected onto (100); (b) scholzite projected onto (010); (c) parascholzite projected onto (100); (d) parascholzite projected onto (010). $(Zn\phi_4)$ tetrahedra are shadow-shaded, Ca atoms are shown by small shaded circles, (H_2O) groups are shown by large shaded circles.

parascholzite, $Ca(H_2O)_2[Zn(PO_4)]_2$, the sheet of (PO_4) and (ZnO_4) tetrahedra (Fig. 13c) is topologically identical to the analogous sheet in scholzite (Fig. 13a). Scholzite and parascholzite are dimorphs, and the difference between these two structures involves linkage of the sheets in the *a*-direction (Figs. 13b,d). The details of the coordination of the interstitial Ca atoms differ in the two structures, leading to a different arrangement of adjacent sheets that produces an orthorhombic arrangement in scholzite and a monoclinic arrangement in parascholzite.

Infinite frameworks of tetrahedra

The minerals of this class (Table 4) are also dominated by PO_4-BeO_4 linkages. Only berlinite is different, as it is the only structure with polymerized (PO_4) and AlO_4 groups.

Table 4. Phosphate minerals based on ($T\phi_4$) frameworks.

Mineral	*Framework*	*Space group*	*Figure*
Berlinite	[$AlPO_4$]	$P3_12$	–
Beryllonite	[$BePO_4$]	$P2_1/n$	14a,b
Hurlbutite	[$Be_2(PO_4)_2$]	$P2_1/a$	14c,d
Babefphite	[$Be(PO_4)F$]	$F\bar{1}$	14e,f
Tiptopite	[$Be_6(PO_4)_6$]	$P6_3$	15a,b
Weinebeneite	[$Be_3(PO_4)_2(OH)_2$]	Cc	15c,d
Pahasapaite	[$Be_{24}P_{24}O_{96}$]	$I23$	15e

Berlinite, [$AlPO_4$], is a framework structure, topologically identical to the structure of α-quartz. Both structures have the same space group, $P3_121$, but the c dimension in berlinite is twice that of α-quartz in order to incorporate two distinct types of tetrahedra, AlO_4 and PO_4.

Beryllonite, Na[$Be(PO_4)$], consists of a well-ordered framework of alternating four-connected (PO_4) and (BeO_4) tetrahedra arranged at the vertices of a 6^3 net, with (PO_4) and (BeO_4) tetrahedra pointing in opposing directions along the b-axis (Fig. 14a). This arrangement is topologically identical to the tridymite framework. These sheets stack along the b-direction and share tetrahedron corners to form four-membered and eight-membered rings (Fig. 14b). The resultant framework has large channels containing [6]- and [9]-coordinated interstitial Na.

Hurlbutite, Ca[$Be_2(PO_4)_2$], consists of an ordered array of (PO_4) and (BeO_4) tetrahedra in which all tetrahedra are four-connected and there is alternation of (PO_4) and (BeO_4) tetrahedra in the structure. Viewed down [001] (Fig. 14c), the tetrahedra are arranged at the vertices of a 4.8^2 net with [7]-coordinated Ca occupying the interstices; these sheets link along the [001] direction by vertex-sharing (Fig. 14d).

Babefphite, Ba[$Be(PO_4)F$], is a rather unusual mineral; it is an ordered framework of (PO_4) and (BeO_3F) tetrahedra. Projected down the c-direction, tetrahedra are arranged at the vertices of a 6^3 net (Fig. 14e) with the tetrahedra pointing (*uuuddd*). Projected down the a-direction, again the tetrahedra occur at the vertices of a 6^3 net (Fig. 14f) but the tetrahedra point (*uuuuuu*). Both the (PO_4) and the ($Be\phi_4$) tetrahedra are three-connected, and the F anions are the non-T-bridging species in the ($Be\phi_4$) tetrahedra. The interstices of the framework are occupied by [9]-coordinated Ba.

Tiptopite, $K_2(Li_{2.9}Na_{1.7}Ca_{0.7}G_{0.7})[Be_6(PO_4)_6](OH)_2(H_2O)_4$, is isotypic with the minerals of the cancrinite group: $Ca_2Na_6[Al_6(SiO_4)_6(CO_3)_2](H_2O)_2$ for the silicate species. The (PO_4) and (BeO_4) tetrahedra are arranged at the vertices of a two-dimensional net (Fig. 15a) such that all tetrahedra are three-connected when viewed down [001]. Prominent twelve-membered rings are arranged at the vertices of a 3^6 net such that they two-connect four-membered rings and three-connect through six-membered rings. These sheets link in the c-direction such that all tetrahedra are four-connected and, projected down the b-direction, form a two-dimensional net of four- and six-membered rings (Fig. 15b). The latter can be considered as a 6^3 net in which every third row of hexagons have a linear defect corresponding to an a-glide operation along c, i.e., double chains of hexagons extending in the c-direction and interleaved by single ladders of edge-sharing squares. Details of the rather complex relations between the interstitial species are discussed by Peacor et al. (1987).

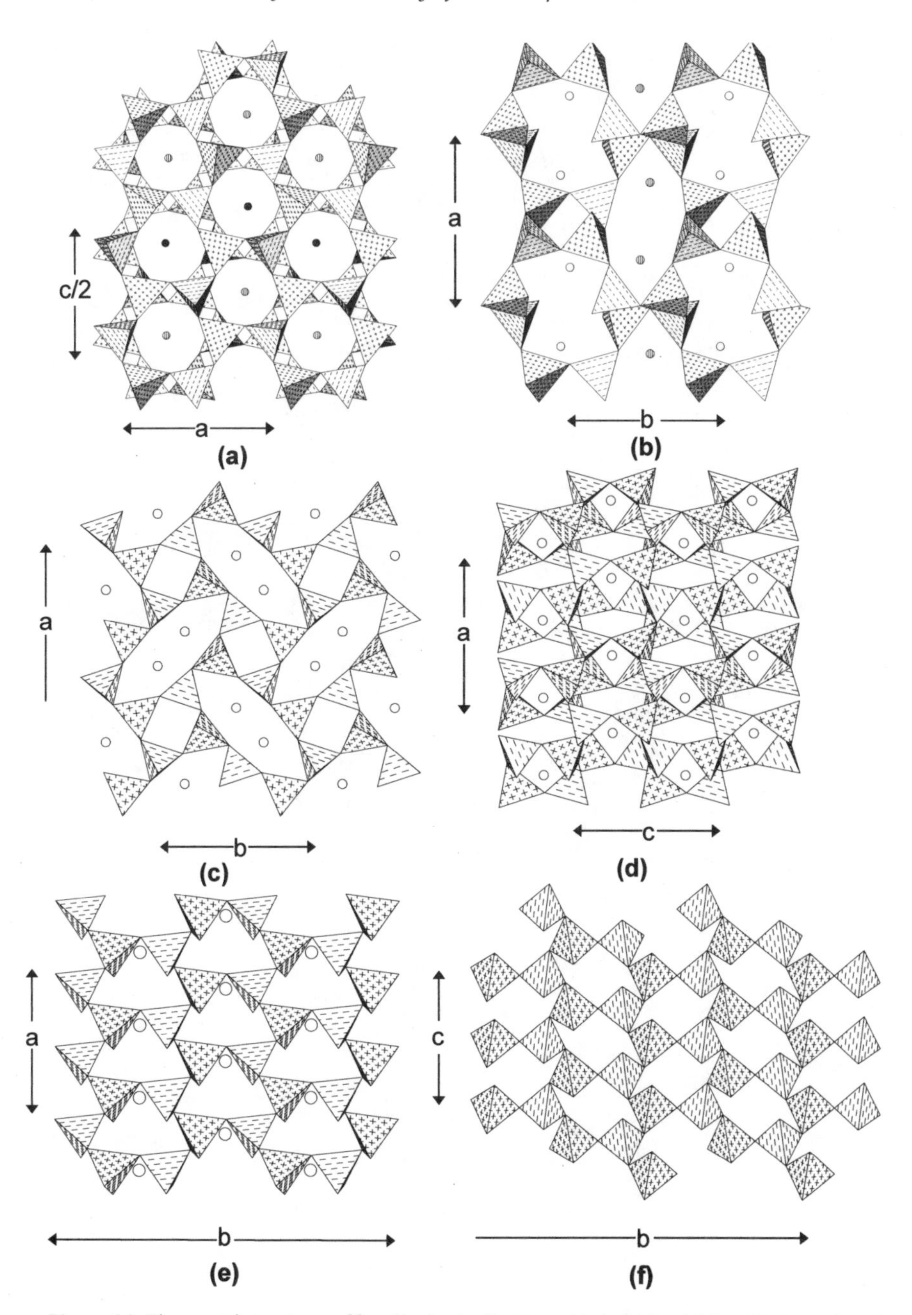

Figure 14. The crystal structures of beryllonite, hurlbutite and babefphite: (a) beryllonite projected onto (010); (b) beryllonite projected onto (001); (c) hurlbutite projected onto (001); (d) hurlbutite projected down the *c*-axis; (e) babefphite projected onto (001); tetrahedra occur at the vertices of a 6^3 net; (f) babefphite projected down the *a*-axis; tetrahedra occur at the vertices of a 6^3 net. Interstitial cations are shown as circles.

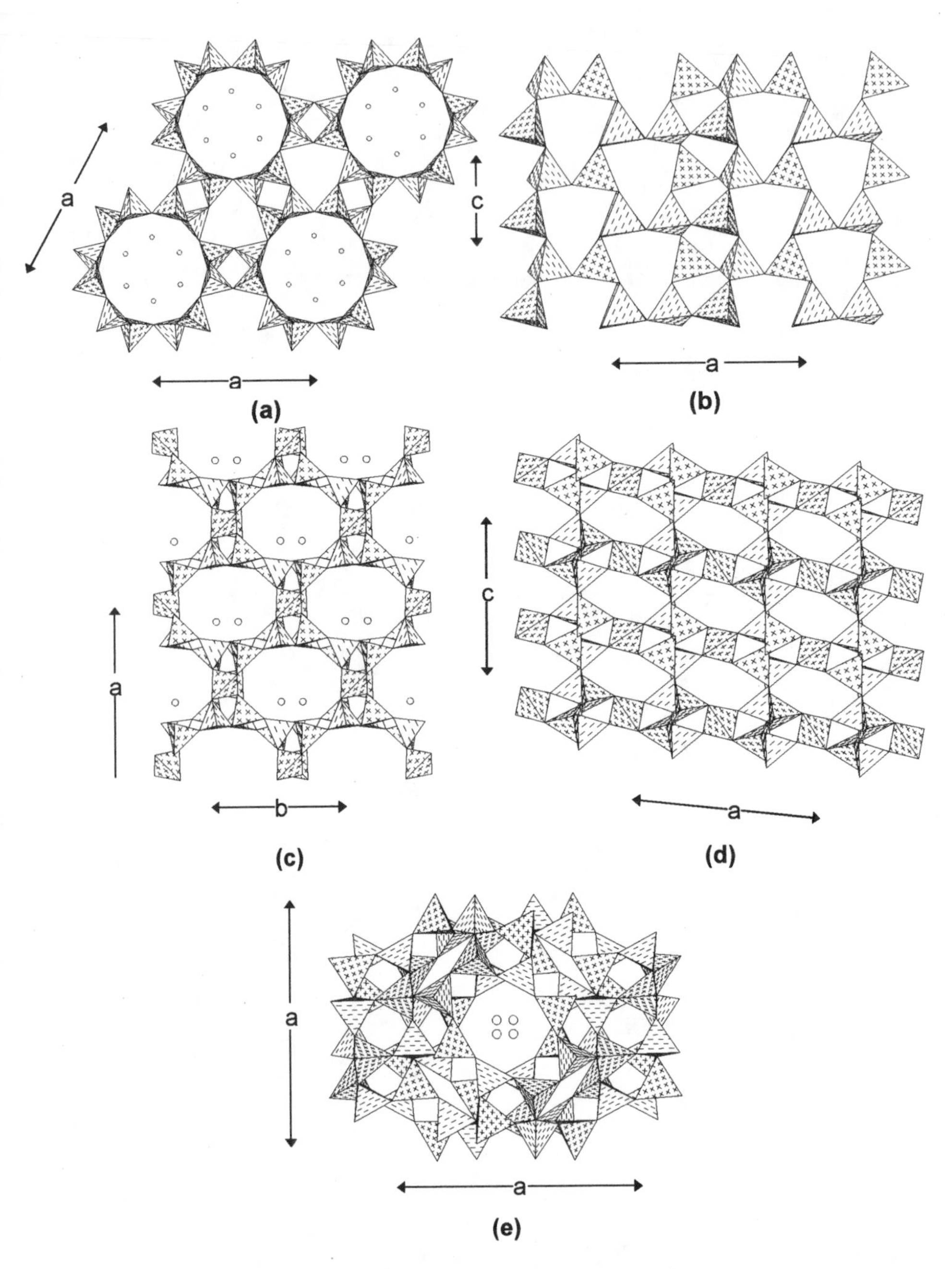

Figure 15. The crystal structures of tiptopite, weinebeneite, and pahasapaite: (a) tiptopite projected onto (001); (b) tiptopite projected onto (010); (c) weinebeneite projected down the c-axis; (d) weinebeneite projected onto (010); in both (c) and (d), 4.8^2 nets of tetrahedra link in the *a*-direction through an additional (BeO_4) group, (H_2O) groups are omitted for clarity; (e) pahasapaite projected onto (001); Li and (H_2O) are omitted for clarity. Interstitial cations are shown as circles.

Weinebeneite, $Ca[Be_3(PO_4)_2(OH)_2](H_2O)_4$, contains an ordered framework of (PO_4) and (BeO_4) tetrahedra; the (PO_4) tetrahedra connect only to (BeO_4) tetrahedra, but the (BeO_4) tetrahedra connect to both (PO_4) and (BeO_4) tetrahedra, the Be-Be linkages occurring through the (OH) groups of the framework. Viewed down [100], the structure consists of alternating (PO_4) and (BeO_4) tetrahedra at the vertices of a 4.8^2 net (view not shown). Projected onto (001) (Fig. 15c) and viewed down [010] (Fig. 15d), the 4.8^2 sheets stack in the [100] direction and link together through additional (non-sheet) (BeO_4) tetrahedra. Interstitial [7]-coordinated Ca is situated to one side of the large channels thus formed, with channel (H_2O) also bonded to the Ca.

Pahasapaite, $Ca_8Li_8[Be_{24}P_{24}O_{96}](H_2O)_{38}$, has an ordered array of (PO_4) and (BeO_4) tetrahedra arranged in a zeolite-rho framework, topologically similar to the minerals of the faujasite group and related to the synthetic aluminophosphate zeolite-like frameworks. Viewed along any crystallographic axis, the structure consists of prominent eight-membered rings of alternating (PO_4) and (BeO_4) tetrahedra (Fig. 15e) in an I-centered (F-centered in projection) array. The eight-membered rings are connected along the axial directions by linear triplets of four-membered rings, and to nearest-neighbor eight-membered rings through six-membered rings. All tetrahedra are four-connected; (PO_4) tetrahedra link only to (BeO_4) tetrahedra, and vice versa. The structure has large cages (Rouse et al. 1989) and prominent intersecting channels (Fig. 15e) that contain interstitial Li, [7]-coordinated Ca and strongly disordered (H_2O) groups.

STRUCTURES WITH $(T\phi_4)$ AND $(M\phi_6)$ GROUPS

As noted above, we classify the structures within each sub-group in terms of the connectivity of the constituent polyhedra of the structural unit. We use the nomenclature of Hawthorne (1983a) to denote the linkage: - denotes corner-sharing (e.g., *M-M*), = denotes edge-sharing (e.g., *M=M*), and ≡ denotes (triangular) face-sharing (e.g., *M* ≡ *M*).

Structures with unconnected (PO_4) groups

Phosphate minerals of this class are listed in Table 5. In these minerals, the (PO_4) groups and $(M\phi_6)$ octahedra are linked together by hydrogen bonding.

In **struvite**, $[Mg(H_2O)_6][PO_4]$, the (PO_4) tetrahedra and $(Mg\{H_2O\}_6)$ octahedra are linked solely by hydrogen bonding from the (H_2O) groups bonded to Mg directly to the anions of the (PO_4) groups, or by hydrogen bonding from the interstitial (NH_4) group

Table 5. Phosphate minerals based on isolated tetrahedra and octahedra and finite clusters of tetrahedra and octahedra.

Mineral	*Structural unit*	*Space group*	*Figure*
	Isolated polyhedra		
Strüvite	$[Mg(H_2O)_6][PO_4]$	$Pmn2_1$	16a
Phosphorrösslerite	$[Mg(H_2O)_6][PO_3(OH)]$	$C2/c$	16b
	Clusters		
Anapaite	$[Fe^{2+}(PO_4)_2(H_2O)_4]$	$P\bar{1}$	17a,b
Schertelite	$[Mg(PO_3\{OH\}_2(H_2O)_4]$	$Pbca$	17c,d
Morinite	$[Al_2(PO_4)_2F_4(OH)(H_2O)_2]$	$P2_1/m$	17e,f

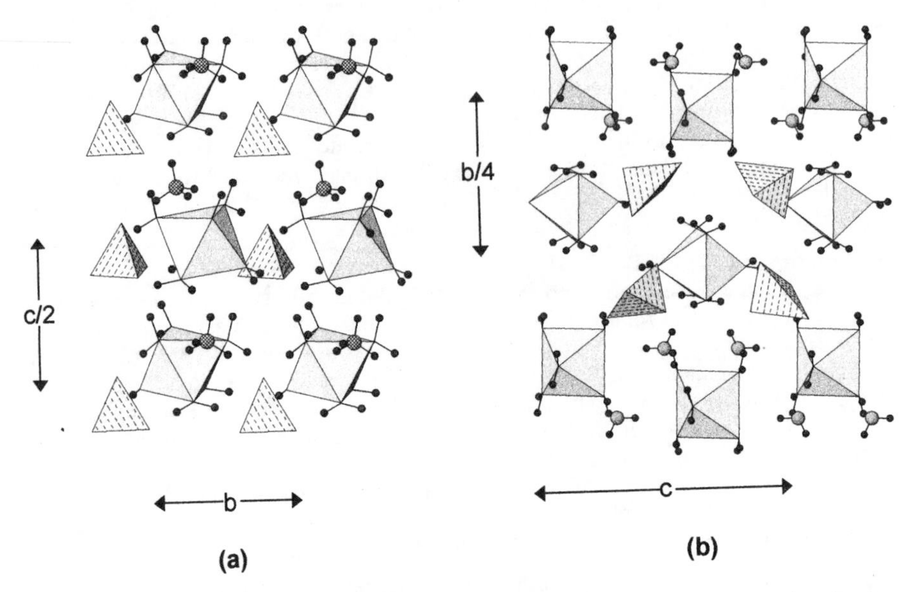

Figure 16. The crystal structures of struvite and phosphorrösslerite: (a) struvite projected onto (100); (b) phosphorrösslerite projected onto (100). $(Mg\{H_2O\}_6)$ octahedra are shadow-shaded, hydrogen atoms are shown by small shaded circles, N [as part of the (NH_4) group] is shown as cross-hatched circles, O-atoms of interstitial (H_2O) groups are shown as large shaded circles.

(Fig. 16a). In **phosphorrösslerite**, $[Mg(H_2O)_6][PO_3(OH)](H_2O)$, the phosphate group is an acid phosphate, one of the phosphate anions being an (OH) group. The $(Mg\{H_2O\}_6)$ octahedron hydrogen bonds to the $(P\phi_4)$ group, but there is also an interstitial (H_2O) group that is held in the structure solely by hydrogen bonding (Fig. 16b), acting both as a hydrogen-bond donor and as a hydrogen-bond acceptor.

Structures with finite clusters of tetrahedra and octahedra

Phosphate minerals of this class are listed in Table 5.

M-T linkage. In **anapaite**, $Ca_2[Fe^{2+}(PO_4)_2(H_2O)_4]$, two (PO_4) groups link to *trans* vertices of an $(Fe^{2+}\phi_6)$ octahedron to form an $[M(TO_4)_2\phi_4]$ cluster, where $M = Fe^{2+}$, $T = P$, and $\phi = (H_2O)$ (Fig. 17a). These clusters are arranged in open layers parallel to (001) (Fig. 17b), and these layers are linked by Ca atoms and by hydrogen bonding. The atomic arrangement in **schertelite**, $(NH_4)_2[Mg\{PO_3(OH)\}_2(H_2O)_4]$, is similar to that in anapaite (and also the sulfate minerals bloedite, $Na_2[Mg(SO_4)_2(H_2O)_4]$, and leonite, $K_2[Mn^{2+}(SO_4)_2(H_2O)_4]$, Hawthorne 1985b). The $[Mg(PO_3\{OH\})_2(H_2O)_4]$ clusters are arranged in a centered rectangular array (Fig. 17c), with the projection of the long axis of the cluster parallel to the *a*-direction. The clusters are arranged in layers parallel to (010) (Fig. 17d), and the clusters are linked by hydrogen bonding involving (H_2O) groups of the cluster and interstitial (NH_4) groups.

M-M, M-T linkage. In **morinite**, $NaCa_2[Al_2(PO_4)_2F_4(OH)(H_2O)_2]$, two $(Al\phi_6)$ octahedra link through one vertex to form a dimer, and (two pairs of) vertices from each octahedron, *cis* to their common vertex, are linked by (PO_4) groups to form a cluster of the general form $[M_2(TO_4)_2\phi_7]$. These clusters are arranged in a centered array when viewed down [001] (Fig. 17e). Adjacent clusters are linked by $^{[8]}Ca$ (Fig. 17f), $^{[5]}Na$ in triangular-bipyramidal coordination, and by hydrogen bonds. As shown by Hawthorne

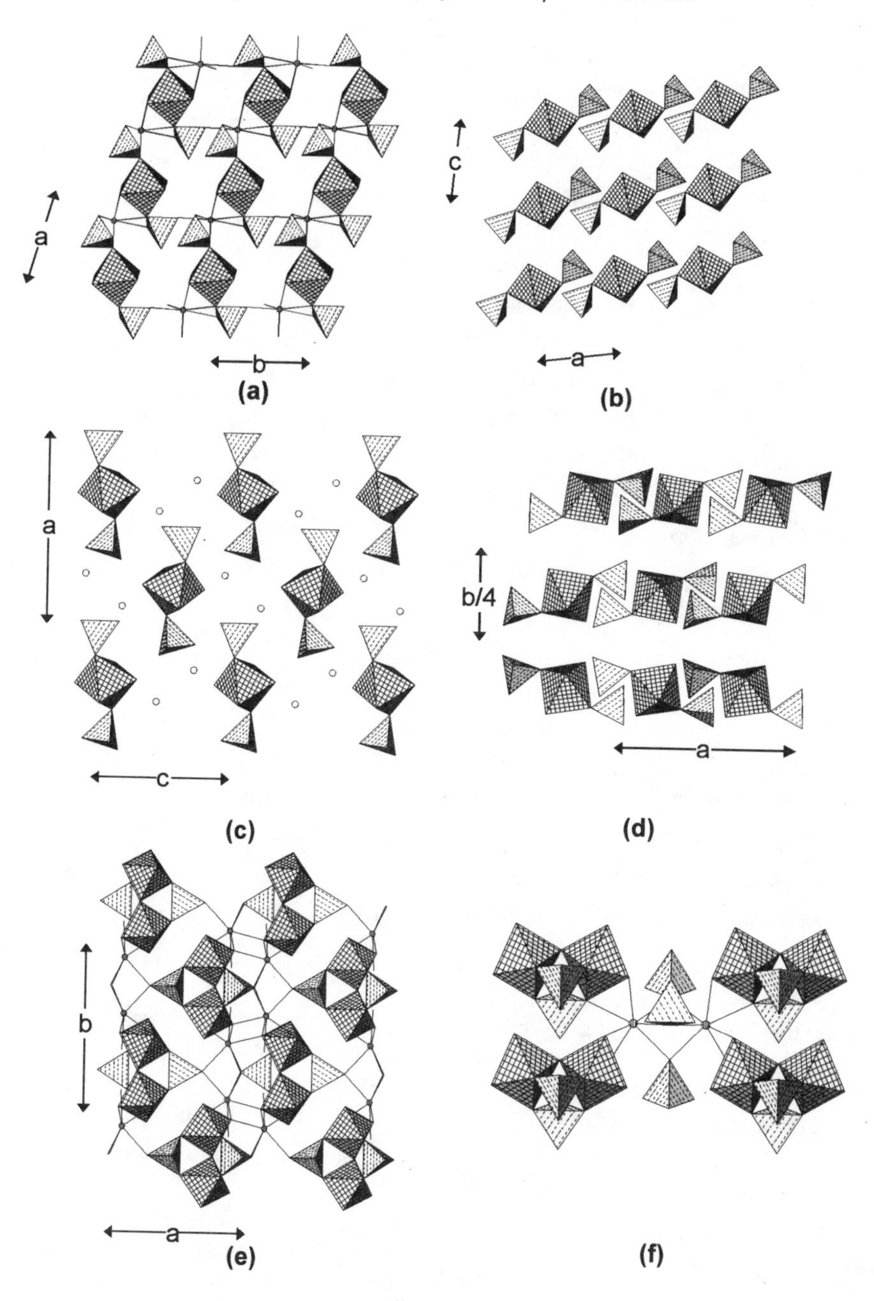

Figure 17. The crystal structures of anapaite, schertelite and morinite: (a) anapaite projected onto (001); (b) anapaite projected onto (010); (c) schertelite projected onto (010); N shown as circles; (d) schertelite projected onto (001); (e) morinite projected onto (001); Ca atoms are shown as circles, Na atoms and hydrogen bonds are not shown; (f) morinite, showing the linkage of adjacent clusters by interstitial Ca atoms.

Table 6. Phosphate minerals based on infinite chains of tetrahedra and octahedra.

Mineral	*Structural unit*	*Space group*	*Figure*
Bøggildite	$[Al_2(PO_4)F_9]$	$P2_1/c$	19a,b
Cassidyite	$[Ni(PO_4)_2(H_2O)_2]$	$P\bar{1}$	19c
Collinsite*	$[Mg(PO_4)_2(H_2O)_2]$	$P\bar{1}$	19c
Fairfieldite*	$[Mn^{2+}(PO_4)_2(H_2O)_2]$	$P\bar{1}$	19d
Messelite	$[Fe^{2+}(PO_4)_2(H_2O)_2]$	$P\bar{1}$	19d
Childrenite*	$[Al(PO_4)(OH)_2(H_2O)]$	$Bbam$	20a,b
Eosphorite	$[Al(PO_4)(OH)_2(H_2O)]$	$Bbam$	20a,b
Jahnsite*	$[Fe^{3+}(PO_4)_2(OH)]_2$	$P2/a$	20c,d
Rittmanite	$[Al(PO_4)_2(OH)]_2$	$P2/a$	20c,d
Whiteite	$[Al(PO_4)_2(OH)]_2$	$P2/a$	20c,d
Whiteite-(CaMnMg)	$[Al(PO_4)_2(OH)]_2$	$P2/a$	20c,d
Lun'okite	$[Al(PO_4)_2(OH)]_2$	$Pbca$	21a,b
Overite*	$[Al(PO_4)_2(OH)]_2$	$Pbca$	21a,b
Segelerite	$[Fe^{3+}(PO_4)_2(OH)]_2$	$Pbca$	21a,b
Wilhelmvierlingite	$[Fe^{3+}(PO_4)_2(OH)]_2$	$Pbca$	21a,b
Tancoite	$[Al(PO_4)_2(OH)]$	$Pbcb$	21c,d
Sinkankasite	$[Al(PO_3\{OH\})_2(OH)]$	$P\bar{1}$	21e,f
Bearthite	$[Al(PO_4)_2(OH)]$	$P2_1/m$	22a,b
Brackebuschite *	$[Mn^{3+}(VO_4)_2(OH)]$	$P2_1/m$	22a,b
Goedkenite	$[Al(PO_4)_2(OH)]$	$P2_1/m$	22a,b
Tsumebite	$[Cu^{2+}(PO_4)(SO_4)(OH)]$	$P2_1/m$	22a,b
Vauquelinite	$[Cu^{2+}(PO_4)(CrO_4)(OH)]$	$P2_1/n$	22c,d

(1979a), this $[M_2(TO_4)_2\phi_7]$ cluster is the basis of a short hierarchy of phosphate minerals of higher connectivity: minyulite, olmsteadite, hureaulite, phosphoferrite, kryzhanovskite, melonjosephite and whitmoreite.

Structures with infinite chains of (PO_4) tetrahedra and ($M\phi_6$) octahedra

The minerals of this class are listed in Table 6. The topologically distinct chains and their corresponding graphs are shown in Figure 18.

M-T linkage. **Bøggildite**, $Na_2Sr_2[Al_2(PO_4)F_9]$, is a rare phosphate-aluminofluoride mineral. The structural unit consists of a chain of alternating (PO_4) tetrahedra and (AlO_2F_4) octahedra that is decorated by flanking ($AlOF_5$) octahedra attached to the (PO_4) groups (Figs. 18a, 19a). The (PO_4) groups are three-connected and alternately point up and down along the length of the chain. The chain extends along the *b*-direction (Fig. 19b) and are linked by [8]- and [9]-coordinated Sr, and [7]- and [9]-coordinated Na. Bøggildite is the only phosphate-aluminofluoride mineral currently known.

The minerals of the **collinsite**, $Ca_2[Mg(PO_4)(H_2O)_2]$, and **fairfieldite**,

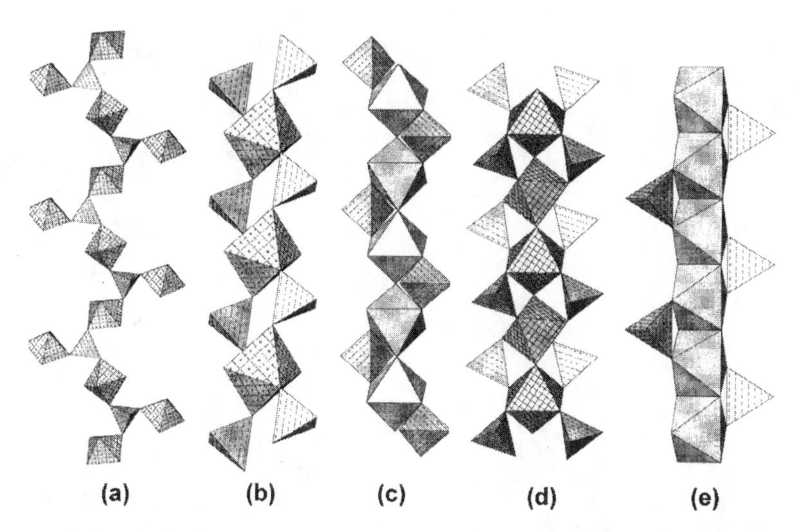

Figure 18. Topologically distinct chains of (PO_4) tetrahedra and ($M\phi_6$) octahedra, and their corresponding graphs: (a) the [$M(TO_4)\ \phi_2$] chain in bøggildite; (b) the [$M(TO_4)\ \phi_2$] chain in the minerals of the collinsite and fairfieldite groups; (c) the [$M(TO_4)\ \phi_2$] chain in the minerals of the childrenite group; (d) the [$M(TO_4)_2\phi$] chain in the minerals of the jahnsite group; (e) the [$M(TO_4)_2\phi$] chain in bearthite (and the minerals of the brackebuschite group).

$Ca_2[Mn^{2+}(PO_4)_2(H_2O)_2]$, groups are both based on a general [$M(TO_4)_2\phi_2$] chain that also occurs in the (non-phosphate) minerals of the kröhnkite, $Na_2[Cu^{2+}(SO_4)(H_2O)_2]$, group. This chain is formed of alternating ($M^{2+}O_4\{H_2O\}_2$) octahedra and pairs of (PO_4) tetrahedra (Figs. 19c,d), with the (H_2O) groups in a *trans* arrangement about the divalent cation (Fig. 18b). The repeat distance along the length of the chain is ~5.45 Å, and this is reflected in the *c*-dimensions of these minerals. The minerals of the collinsite and fairfieldite groups are often incorrectly grouped together as the fairfieldite group because they all have triclinic symmetry. However, the interaxial angles in the two groups are significantly different (see Appendix). Adjacent chains in both structures are linked by [7]-coordinated Ca atoms and by hydrogen bonding. The two structures differ in the details of their hydrogen bonding (Figs. 19c,d).

M-M, M-T linkage. **Childrenite**, $Mn^{2+}(H_2O)[Al(PO_4)(OH)_2]$, consists of [$Al\phi_5$] chains in which ($Al\phi_6$) octahedra link through pairs of *trans* vertices. The chains are decorated by (PO_4) groups that link adjacent octahedra and are arranged in a staggered fashion along the length of the chain (Fig. 18c) to give the general form [$M(TO_4)\phi_3$]. The chains extend in the *c*-direction in childrenite (Fig. 20a) and are cross-linked by [6]-coordinated Mn^{2+}, the coordination octahedra of which form an edge-sharing chain in the *c*-direction. Viewed down [001], the chains are arranged at the vertices of a primitive orthorhombic net, and four adjacent chains are linked through one ($Mn^{2+}\phi_6$) octahedra (Fig. 20b).

Jahnsite, $CaMn^{2+}Mg_2[Fe^{3+}(PO_4)_2(OH)](H_2O)_8$, consists of [$Fe^{3+}\phi_5$] chains of *trans*-corner-sharing octahedra that are decorated by bridging (PO_4) groups to give the general form [$M^{3+}(TO_4)_2\phi$] (Fig. 18d). These chains extend in the *b*-direction and have a repeat distance of ~7.1Å, leading Moore (1970) to designate these, and related, chains as the

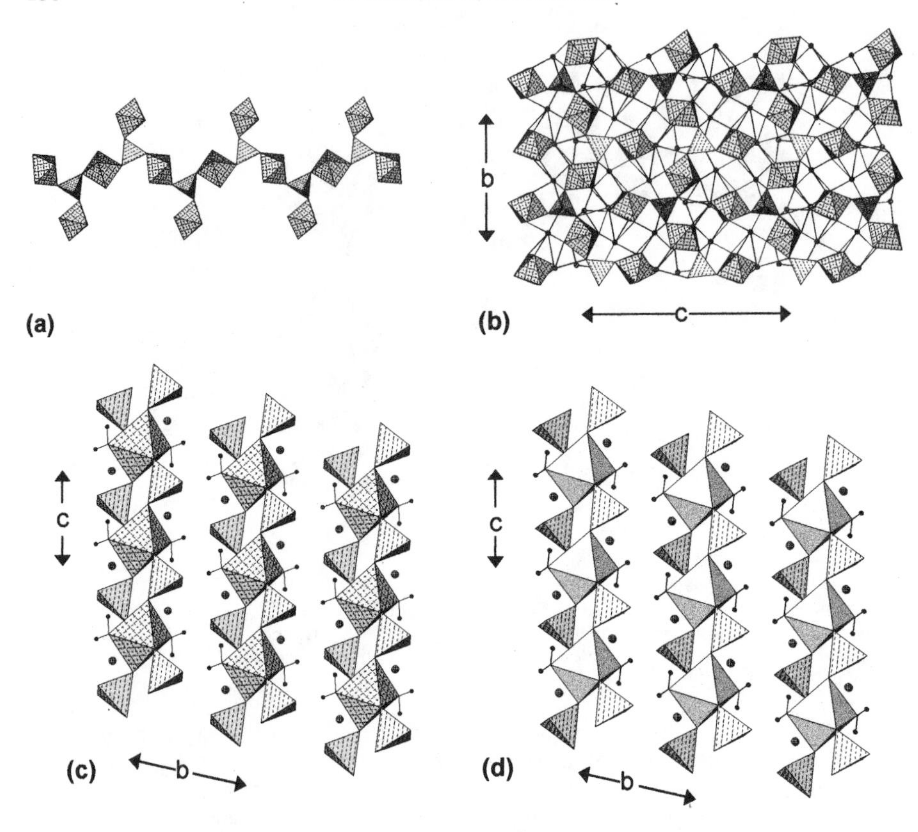

Figure 19. The crystal structures of bøggildite, collinsite and fairfieldite: (a) the $[Al_2(PO_4)F_9]$ chain in bøggildite; (b) bøggildite projected onto (100); Ca atoms are shown as dark circles, Na atoms are shown as shaded circles; (c) collinsite projected onto (100); (d) fairfieldite projected onto (100); Ca atoms are shown as shaded circles, H atoms are shown as small dark circles.

7-Å chains. These chains are linked in the *a*-direction by [6]-coordinated Ca (Fig. 20c) that form chains of edge-sharing polyhedra in the *b*-direction, forming slabs (Fig. 20d) parallel to (100) that are linked by octahedrally coordinated divalent-metal cations and by hydrogen bonding. In addition to the minerals of this group listed in Table 6, Matsubara (2000) reports the Fe^{2+} equivalent of jahnsite, ideally $CaFe^{2+}Fe^{2+}{}_2[Fe^{3+}(PO_4)_2(OH)]_2(H_2O)_8$, but this has not been approved as a valid species by the IMA.

Overite, $Ca_2Mg_2[Al(PO_4)_2(OH)]_2(H_2O)_8$, and **tancoite**, $Na_2LiH[Al(PO_4)_2(OH)]$, are both based on the $[Al(PO_4)_2(OH)]$ chain that is shown in Figure 18d, and in both structures, this chain defines the *c*-dimension, 7.11 Å in overite and 7.03 × 2 = 14.06 Å in tancoite. $(Al\phi_6)$ octahedra link through one set of *trans* vertices, corresponding to the (OH) groups, to form an $[Al\phi_5]$ chain. Adjacent octahedra are linked by pairs of (PO_4) tetrahedra that point alternately up and down the *b*-direction in overite (Fig. 21a) and the *a*-direction in tancoite (Fig. 21c). In overite, the chains are linked in the *a*-direction by [8]-coordinated Ca to form slabs parallel to (010), the Ca linking to both tetrahedra and octahedra. These slabs are linked in the *b*-direction by $(MgO_2\{H_2O\}_4)$ octahedra (Fig. 21b), and the resulting structure is strengthened by hydrogen bonds from the (H_2O) groups, all of which are bonded to the interstitial Mg cations. In tancoite, the chains are linked in the *b*-direction by [8]-coordinated Na and [5]-coordinated Li, forming slabs

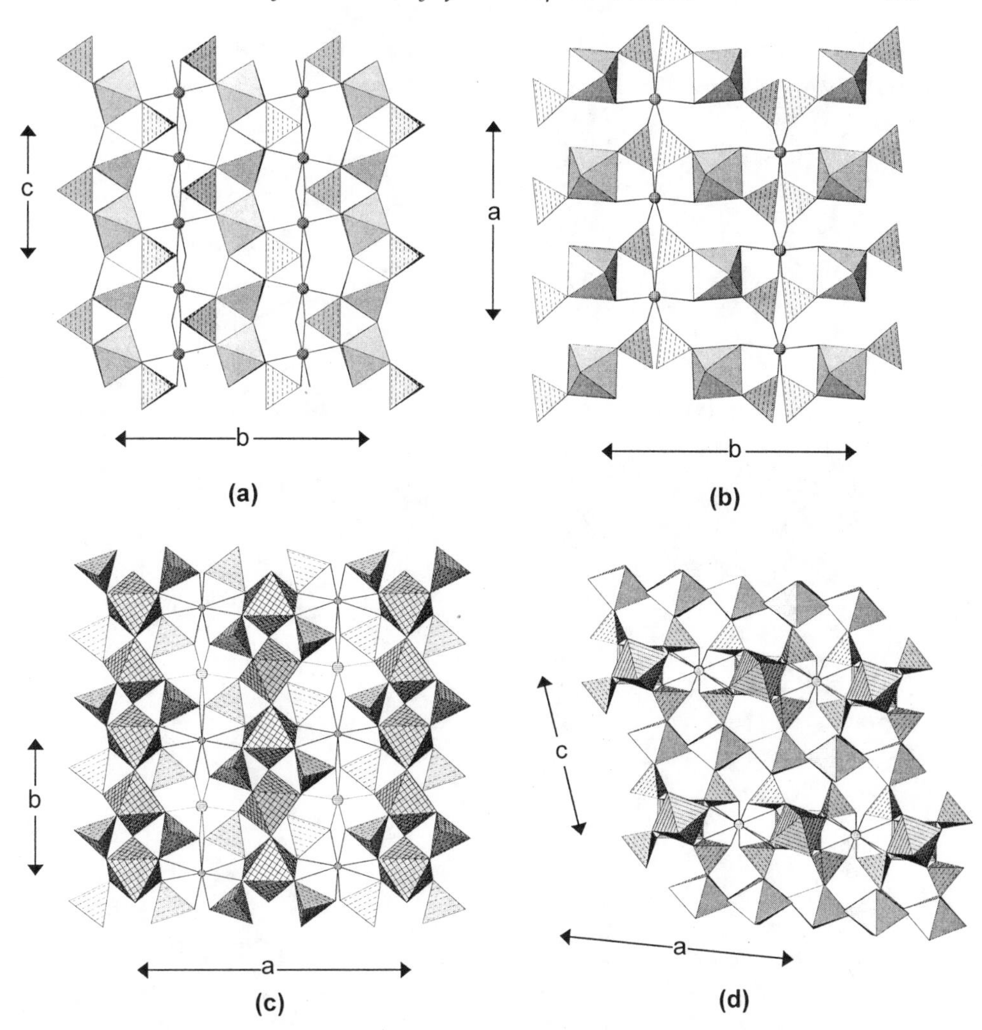

Figure 20. The crystal structures of childrenite and jahnsite: (a) childrenite projected onto (100); (b) childrenite projected onto (001); ($Al\phi_6$) octahedra are shadow-shaded, Mn^{2+} cations are shown as circles; (c) jahnsite projected onto (001); (d) jahnsite projected onto (010); ($Fe^{3+}\phi_6$) octahedra are 4^4-net-shaded, Ca atoms are shown as circles, ($Mn^{2+}\phi_6$) and ($Mg\phi_6$) octahedra are shadow-shaded.

parallel to (100) (Fig. 21c). These slabs are linked in the *a*-direction by [8]-coordinated Na (Fig. 21d). In addition, there is a symmetrical hydrogen-bond between two anions of adjacent (PO_4) groups.

Sinkankasite, $Mn^{2+}(H_2O)_4[Al(PO_3\{OH\})_2(OH)](H_2O)_2$, is also based on the $[M(T\phi_4)_2\phi]$ chain of Figure 18d, extending in the *c*-direction to give a repeat of ~7 Å. However, it is topochemically different from the analogous chain in overite and tancoite because one of the tetrahedron vertices is occupied by (OH), forming an acid-phosphate group. The chains are linked in the *b*-direction (Fig. 21e) by ($Mn^{2+}O_2\{H_2O\}_4$) octahedra to form a thick slab parallel to (100). These slabs stack in the *a*-direction (Fig. 21f) and

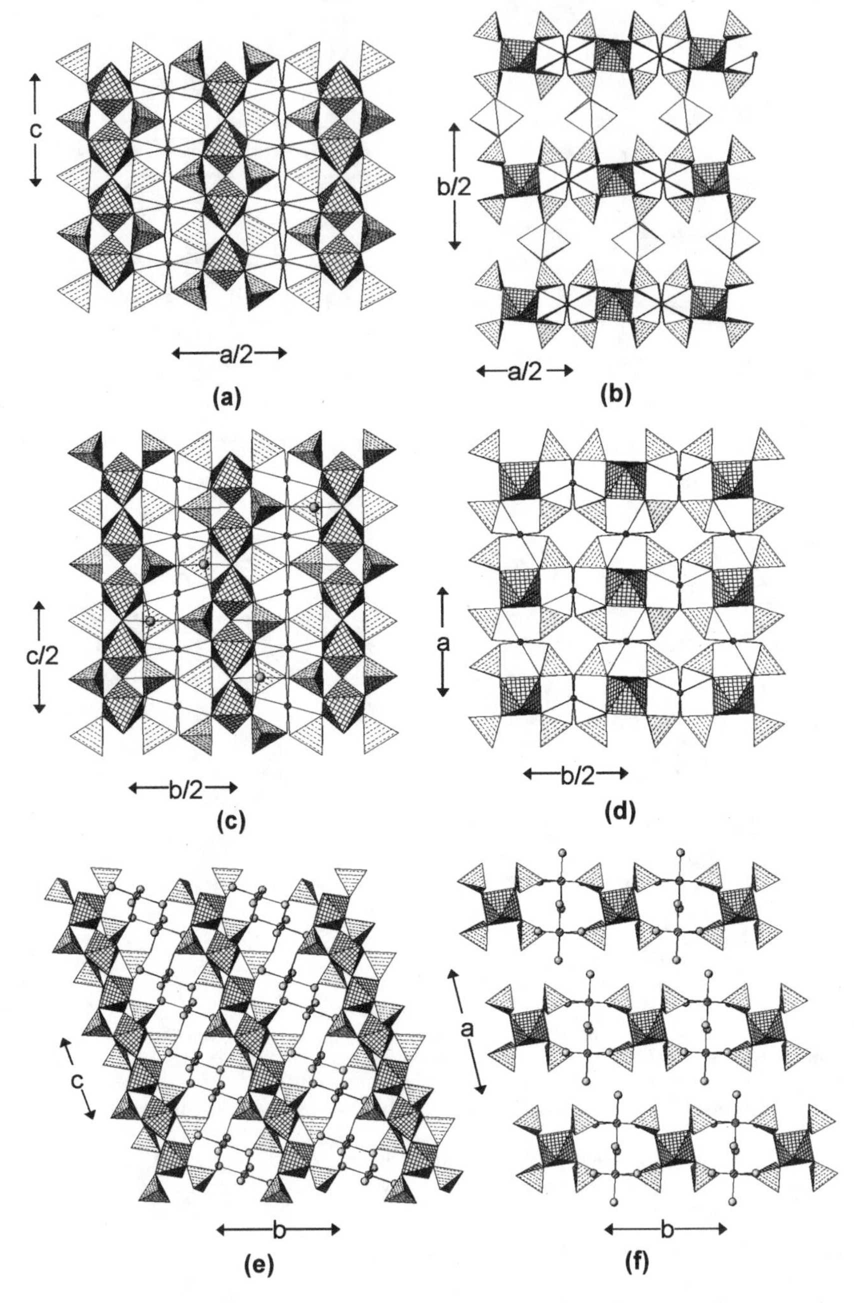

Figure 21. The crystal structures of overite, tancoite and sinkankasite: (a) overite projected onto (010); (b) overite projected onto (001); Ca atoms are shown as small shaded circles; (c) tancoite projected onto (100); (d) tancoite projected onto (001); Na atoms are shown as small shaded circles, Li atoms are shown as large shaded circles; (e) sinkankasite projected onto (100); (f) sinkankasite projected onto (001); hydrogen atoms and bonds are omitted for clarity, $(Al\phi_6)$ octahedra are 4^4-net-shaded.

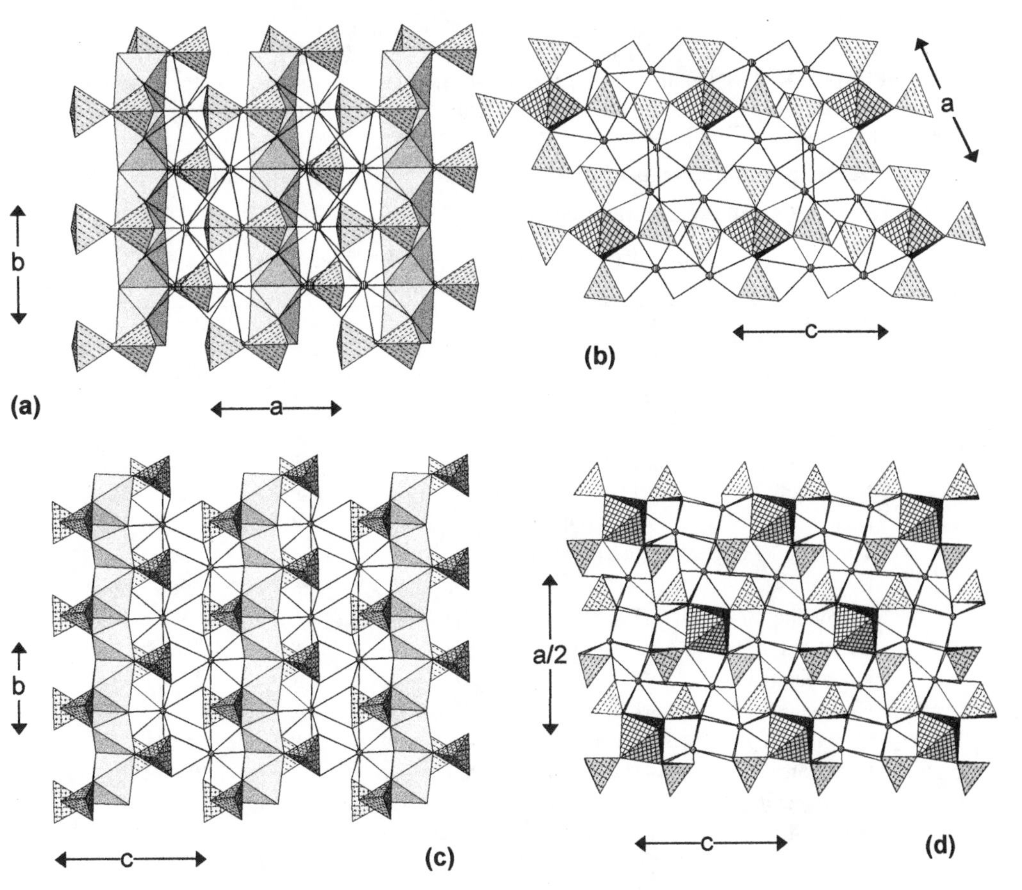

Figure 22. The crystal structures of bearthite and vauquelinite: (a) bearthite projected onto (001); (b) bearthite projected onto (010); $(Al\phi_6)$ octahedra are shadow-shaded in (a) and 4^4-net-shaded in (b), interstitial Ca is shown as shaded circles; (c) vauquelinite projected onto (100); (d) vauquelinite projected onto (010); (CrO_4) tetrahedra are square-pattern-shaded, $(Cu^{2+}\phi_6)$ octahedra are shadow-shaded in (c) and 4^4-net-shaded in (d), Pb^{2+} is shown as shaded circles.

are linked solely by hydrogen bonds involving the H atom of the acid-phosphate group, the (H_2O) groups of the interstitial $(Mn^{2+}O_2\{H_2O\}_4)$ octahedron, and interstitial (H_2O) groups not bonded directly to any cations.

M=M, M-T linkage. **Bearthite**, $Ca_2[Al(PO_4)_2(OH)]$, contains $(Al\phi_6)$ octahedra which share one set of *trans* edges with adjacent octahedra to form an $[Al\phi_4]$ chain. Adjacent octahedra are linked (bridged) by (PO_4) tetrahedra in a staggered arrangement on either side of the chain to form a decorated chain of the general form $[M(TO_4)_2\phi]$ (Fig. 18e). These chains are linked in the *a*-direction by [10]-coordinated Ca (Fig. 22a). Viewed along [010] (Fig. 22b), the chains resemble four-membered pinwheels; linkage in the *c*-direction is also provided by interstitial Ca cations. A topologically identical chain, $[M(TO_4)_2\phi]$, occurs in **vauquelinite**, $Pb^{2+}_2[Cu^{2+}(PO_4)(CrO_4)(OH)]$; however, there are two symmetrically (and chemically) distinct tetrahedra in vauquelinite, (PO_4) and (CrO_4) (Fig. 22c). In vauquelinite, $(Cu^{2+}\phi_6)$ octahedra form the $[M\phi_4]$-type chain, (PO_4)

tetrahedra bridge vertices of adjacent octahedra in the chain, and (CrO_4) tetrahedra link to one vertex of the edge shared between adjacent octahedra (Fig. 22c). The resulting $[Cu^{2+}(PO_4)(CrO_4)(OH)]$ chains extend in the *b*-direction, and are linked in the *a*-direction and *c*-direction by [9]-coordinated Pb^{2+}. When viewed end-on (Fig. 22d), the chains resemble four-membered pinwheels.

Structures with infinite sheets of (PO_4) tetrahedra and ($M\phi_6$) octahedra

The minerals of this class are listed in **Table 7**.

Table 7. Phosphate minerals based on infinite sheets of tetrahedra and octahedra.

Mineral	*Structural Unit*	*Space Group*	*Figure*
Johnwalkite	$[Nb(PO_4)_2O_2]$	$Pb2_1m$	23a.b
Olmsteadite*	$[Nb(PO_4)_2O_2]$	$Pb2_1m$	23a,b
Brianite	$[Mg(PO_4)_2]$	$P2_1/c$	23c,d
Merwinite*	$[Mg(SiO_4)_2]$	$P2_1/c$	23c,d
Newberyite	$[Mg(PO_3OH)(H_2O)_3]$	$Pbca$	24a,b
Hannayite	$[Mg_3(PO_3\{OH\})_4]$	$P\bar{1}$	24c,d
Minyulite	$[Al_2(PO_4)_2F(H_2O)_4]$	$Pba2$	25a,b
Benauite	$[Fe^{3+}_3(PO_4)(PO_3\{OH\})(OH)_6]$	$R\bar{3}m$	25c,d
Crandallite	$[Al_3(PO_4)(PO_3\{OH\})(OH)_6]$	$R\bar{3}m$	25c,d
Eylettersite	$[Al_3(PO_4)(PO_3\{OH\})(OH)_6]$	$R\bar{3}m$	25c,d
Florencite-(Ce)	$[Al_3(PO_4)(PO_3\{OH\})(OH)_6]$	$R\bar{3}m$	25c,d
Florencite-(La)	$[Al_3(PO_4)(PO_3\{OH\})(OH)_6]$	$R\bar{3}m$	25c,d
Florencite-(Nd)	$[Al_3(PO_4)(PO_3\{OH\})(OH)_6]$	$R\bar{3}m$	25c,d
Gorceixite	$[Al_3(PO_4)(PO_3\{OH\})(OH)_6]$	$R\bar{3}m$	25c,d
Plumbogummite	$[Al_3(PO_4)(PO_3\{OH\})(OH)_6]$	$R\bar{3}m$	25c,d
Waylandite	$[Al_3(PO_4)(PO_3\{OH\})(OH)_6]$	$R\bar{3}m$	25c,d
Zairite	$[Fe^{3+}_3(PO_4)_2(OH)_6]$	$R\bar{3}m$	25c,d
Gordonite	$[Al_2(PO_4)_2(OH)_2(H_2O)_2]$	$P\bar{1}$	26a,b
Laueite*	$[Fe^{3+}_2(PO_4)_2(OH)_2(H_2O)_2]$	$P\bar{1}$	26a,b
Mangangordonite	$[Al_2(PO_4)_2(OH)_2(H_2O)_2]$	$P\bar{1}$	26a,b
Paravauxite	$[Al_2(PO_4)_2(OH)_2(H_2O)_2]$	$P\bar{1}$	26a,b
Sigloite	$[Al_2(PO_4)_2(OH)_2(H_2O)_2]$	$P\bar{1}$	26a,b
Ushkovite	$[Fe^{3+}_2(PO_4)_2(OH)_2(H_2O)_2]$	$P\bar{1}$	26a,b
Curetonite	$[Al(PO_4)(OH)]$	$P2_1/n$	26c,d
Kastningite	$[Al_2(PO_4)_2(OH)_2(H_2O)_2]$	$P\bar{1}$	27a,b
Stewartite*	$[Fe^{3+}_2(PO_4)_2(OH)_2(H_2O)_2]$	$P\bar{1}$	27a,b
Pseudolaueite	$[Fe^{3+}(PO_4)(OH)(H_2O)]_2$	$P2_1/a$	27c,d
Strunzite*	$[Fe^{3+}(PO_4)(OH)(H_2O)]_2$	$P\bar{1}$	28a,b
Ferrostrunzite	$[Fe^{3+}(PO_4)(OH)(H_2O)]_2$	$P\bar{1}$	28a,b
Metavauxite	$[Al(PO_4)(OH)(H_2O)]_2$	$P2_1/c$	28c,d
Montgomeryite	$[MgAl_4(PO_4)_6(OH)_4(H_2O)]$	$C2/c$	29a,b
Mitryaevaite	$[Al_5(PO_4)_2(PO_3\{OH\})_2F_2(OH)_2(H_2O)_8]$	$P\bar{1}$	29c,d
Bonshtedite	$[Fe^{2+}(PO_4)(CO_3)]$	$P2_1/m$	29e,f
Bradleyite*	$[Mg(PO_4)(CO_3)]$	$P2_1/m$	29e,f
Sidorenkoite	$[Mn^{2+}(PO_4)(CO_3)]$	$P2_1/m$	29e,f
Bermanite*	$[Mn^{3+}(PO_4)(OH)]_2$	$P2_1$	30a,b

Ercitite	$[Mn^{3+}(PO_4)(OH)]_2$	$P2_1/n$	30a,b
Schoonerite	$[Mn^{2+}Fe^{2+}{}_2ZnFe^{3+}(PO_4)_3(OH)_2(H_2O)_7]$	$Pmab$	30c,d
Nissonite	$[Cu^{2+}Mg(PO_4)(OH)(H_2O)_2]$	$C2/c$	31a,b,c
Foggite	$[Al(PO_4)(OH)_2]$	$A2_122$	32a,b,c
Earlshannonite	$[Fe^{3+}(PO_4)(OH)]_2$	$P2_1/c$	32d,e
Whitmoreite*	$[Fe^{3+}(PO_4)(OH)]_2$	$P2_1/c$	32d,e
Mitridatite*	$[Fe^{3+}{}_3(PO_4)_3O_2]$	Aa	33a
Robertsite	$[Mn^{3+}{}_3(PO_4)_3O_2]$	Aa	33a
Arupite	$[Ni_3(PO_4)_2(H_2O)_8]$	$C2/m$	34a,b
Vivianite *	$[Fe^{2+}{}_3(PO_4)_2(H_2O)_8]$	$C2/m$	34a,b
Bobierrite	$[Mg_3(PO_4)_2(H_2O)_8]$	$C2/c$	34c,d

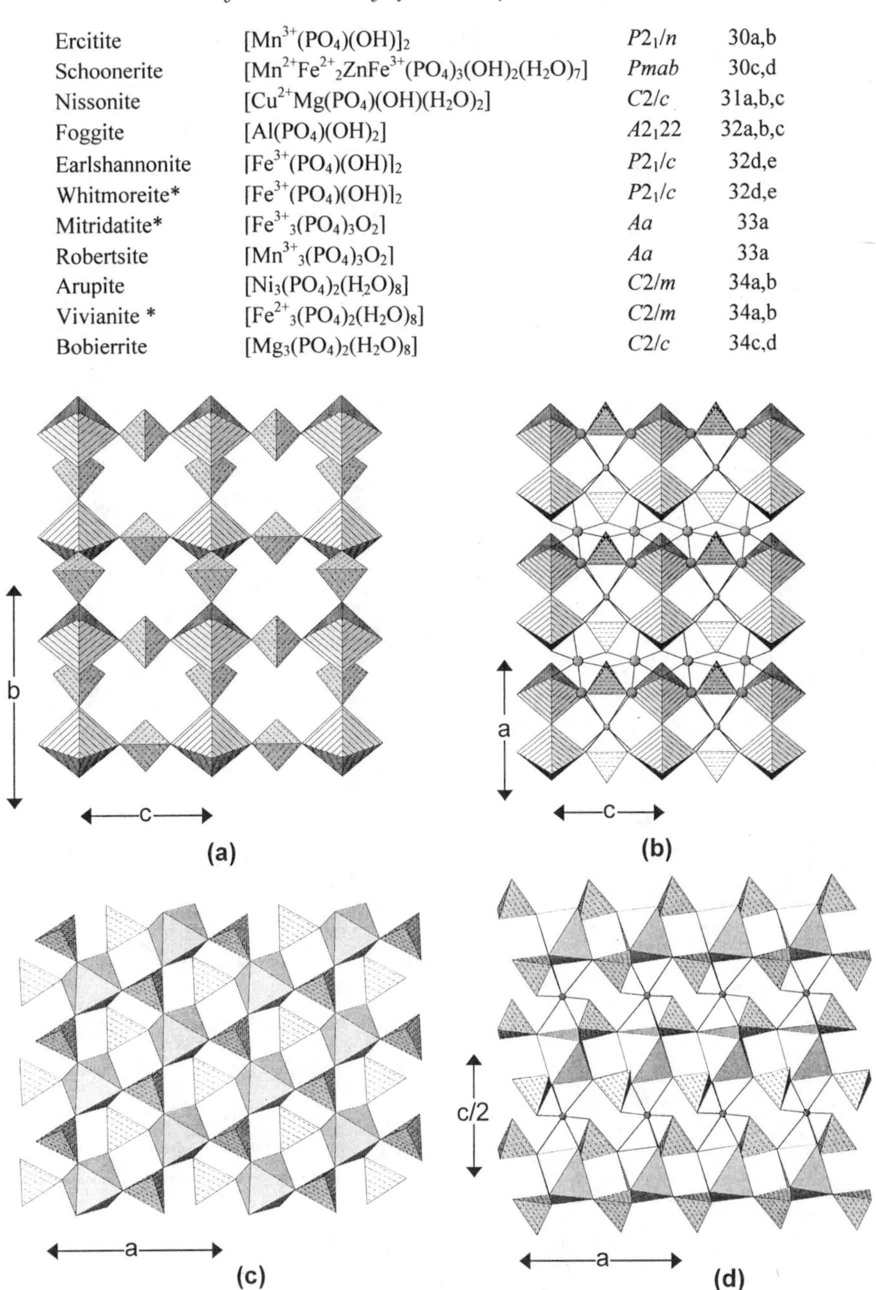

Figure 23. The crystal structures of olmsteadite and brianite: (a) olmsteadite projected onto (100); (b) olmsteadite projected onto (010); (NbO_6) octahedra are line-shaded, Fe^{2+} atoms are shown as line-shaded circles, (H_2O) groups are shown as dot-shaded circles; (c) brianite projected onto (001); (d) brianite projected onto (010); (MgO_6) octahedra are shadow-shaded, interstitial cations are shown as circles.

M-T linkage. The minerals of the **olmsteadite**, $K_2Fe^{2+}_4(H_2O)_4[Nb_2(PO_4)_4O_4]$, group consist of (PO_4) tetrahedra and (NbO_6) octahedra at the vertices of a 4^4 plane net, linked by sharing corners to form a sheet parallel to (100) (Fig. 23a, above). In the *c*-direction, the (PO_4) groups link to *trans* vertices of the (NbO_6) octahedra, but in the *b*-direction, the (PO_4) groups link to *cis* vertices of the (NbO_6) octahedra, and these *cis* vertices alternate above and below the plane of the sheet in the *b*-direction. The sheets link in pairs by sharing octahedron corners to form slabs that incorporate the interstitial [8]-coordinated K (Fig. 23b). These slabs are linked in the *a*-direction by rutile-like $[M\phi_4]$ chains of $(Fe^{2+}\phi_6)$ octahedra that extend in the *c*-direction.

Brianite, $Na_2Ca[Mg(PO_4)_2]$ is a member of the merwinite group (Table 7) and consists of (PO_4) tetrahedra and (MgO_6) octahedra at the vertices of a $(4^3)_2\ 4^6$ plane net. The (PO_4) groups link to both the upper and lower corners of the octahedra (Fig. 23c) to

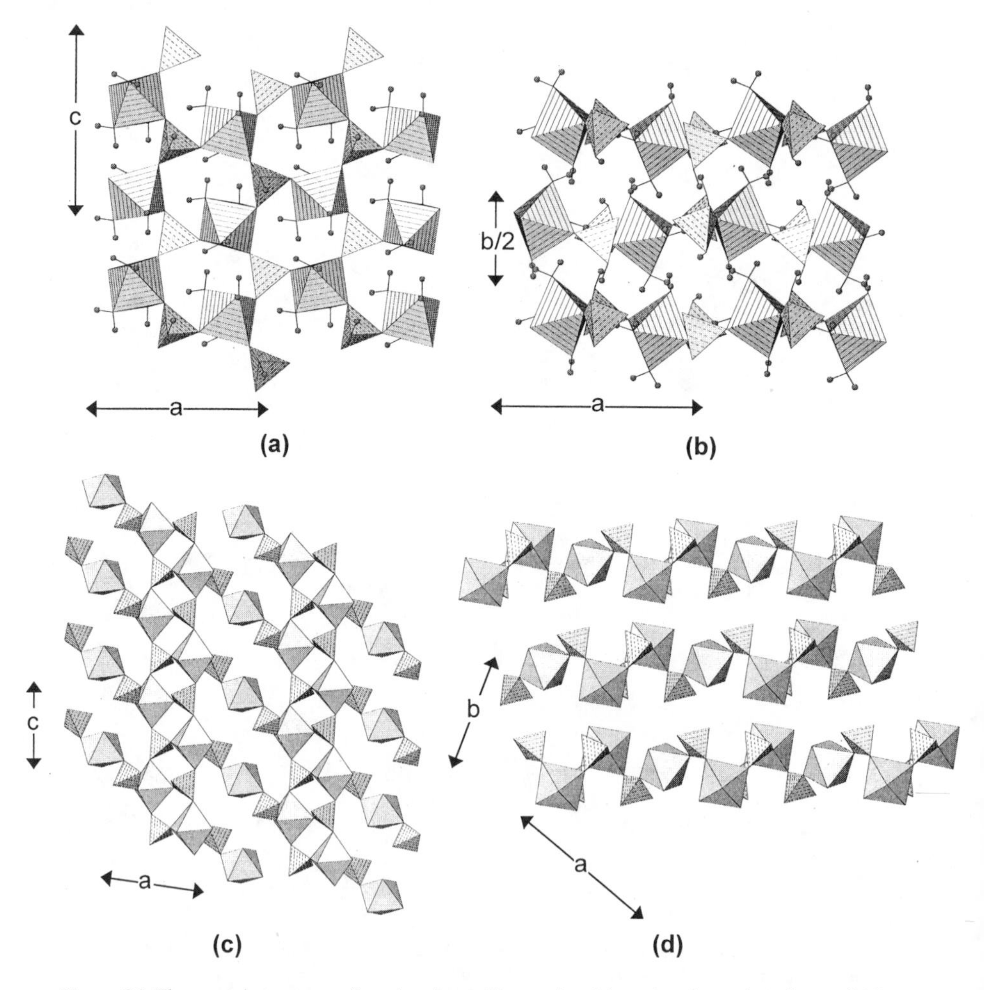

Figure 24. The crystal structures of newberyite and hannayite: (a) newberyite projected onto (010); (b) newberyite projected onto (001); $(Mg\phi_6)$ octahedra are line-shaded, H atoms are shown as small shaded circles; (c) hannayite projected onto (010); (d) hannayite projected onto (001); $(Mg\phi_6)$ octahedra are shadow-shaded.

form pinwheels (Moore 1973b) and the resulting sheet has a layer of octahedra sandwiched between two layers of tetrahedra (Fig. 23d). These sheets are linked in the *c*-direction by interstitial Na and Ca.

Newberyite, $[Mg(PO_3\{OH\})(H_2O)_3]$, consists of $(P\phi_4)$ tetrahedra and $(Mg\phi_6)$ octahedra at the vertices of a 6^3 plane net; the two different types of polyhedra alternate on any path through the net (Fig. 24a), and the (PO_4) tetrahedra point both up and down relative to the plane of the sheet. Both tetrahedra and octahedra are three-connected, and all one-connected vertices in the net are 'tied-off' by H atoms. Thus the $(P\phi_4)$ group is actually an acid-phosphate group, $(PO_3\{OH\})$, and the three one-coordinated anions of the $(Mg\phi_6)$ octahedron are (H_2O) groups. Hawthorne (1992) used newberyite as an example of the role of H atoms in controlling the dimensional character of a structural unit. The sheets in newberyite stack in the *b*-direction (Fig. 24b) and are linked solely by hydrogen bonds. Newberyite is an unusual structure in that it undergoes a low-temperature crystal-to-amorphous transition (Sales et al. 1993). When heated above 150°C (but below 600°C), newberyite becomes amorphous. With continued heating above 150°C, the amorphous phase develops chains of polymerized (PO_4) tetrahedra (up to 13 tetrahedra long), until at 600°C, crystalline $Mg_2P_2O_7$ forms. Heating under (unspecified) pressure results in a phase of the form $Mg_3(PO_3\{OH\})[P_2O_7](H_2O)_{4.5}$, the only known crystalline phosphate containing two different phosphate anions (Sales et al. 1993).

Hannayite, $(NH_4)_2[Mg_3(PO_3\{OH\})_4(H_2O)_8]$, consists of a very exotic sheet of alternating $(PO_3\{OH\})$ tetrahedra and $(Mg\phi_6)$ octahedra in a very open array. Alternating tetrahedra and octahedra fuse to form an $[M(TO_4)\phi_4]$ chain. Pairs of these chains meld by sharing corners between tetrahedra and octahedra to form ribbons of the type $[M(TO_4)\phi_3]$ that extend in the *c*-direction. These ribbons are linked in the *a*-direction by *trans* $[Mg(PO_4)_2\phi_4]$ clusters to form an open sheet parallel to (010) (Fig. 24c). These sheets stack in the *b*-direction (Fig. 24d) and are linked by hydrogen bonds directly from sheet to sheet, and by hydrogen bonds involving the interstitial (NH_4) groups.

M-M, M-T linkage. **Minyulite**, $K[Al_2(PO_4)_2F(H_2O)_4]$, contains a sheet that is made up of $[Al_2(PO_4)_2F(H_2O)_4O_2]$ clusters that are topologically identical to the $[Al_2(PO_4)_2F_4(OH)(H_2O)_2]$ clusters in morinite (Fig. 17e). These clusters occur at the vertices of a 4^4 net (Fig. 25a) and link by sharing vertices between tetrahedra and octahedra. This arrangement leads to large interstices within the sheet, and these are occupied by [10]-coordinated K atoms (Fig. 25a); the sheet is parallel to (001). When viewed in the *b*-direction, it can be seen (Fig. 25b) that each sheet consists of a layer of tetrahedra and a layer of octahedra. The interstitial K atoms actually lie completely *within* each sheet and hence do not participate in intersheet linkage. All (H_2O) groups of the structural unit occur on the underside of each sheet (Fig. 25b) and adjacent sheets are linked solely by hydrogen bonds.

The minerals of the **crandallite**, $Ca[Al_3(PO_4)(PO_3\{OH\})(OH)_6]$, group are based on an open sheet of corner-sharing $(Al\phi_6)$ octahedra that is decorated with (PO_4) and $(PO_3\{OH\})$ groups. This sheet can be envisaged as parallel $[M\phi_5]$ chains of octahedra that extend in the *a*- (plus symmetrically equivalent) direction (Fig. 25c) and are linked into a sheet by sharing corners with linking octahedra; thus all octahedra are four-connected within this sheet. There are prominent three-membered and six-membered rings of octahedra within this sheet, and the $(P\phi_4)$ tetrahedra share three vertices with octahedra of the three-membered rings (Fig. 25c). There is only one crystallographically distinct $(P\phi_4)$ group in the structure of crandallite, and hence the normal and acid phosphate group must be disordered. The resultant $[M^{3+}{}_3(T\phi_4)\phi_6]$ sheets (Fig. 25c) stack in the *c*-direction (Fig. 25d) and are linked by hydrogen bonds between the $(OH)_6$ anions

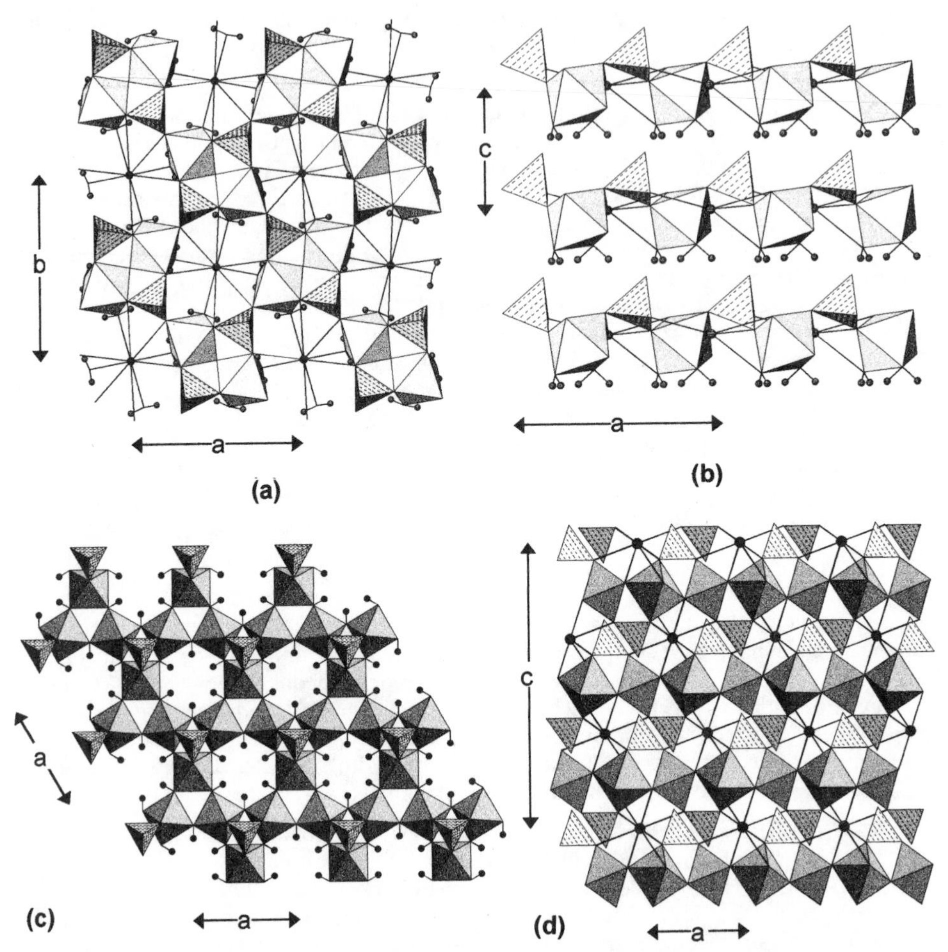

Figure 25. The crystal structures of minyulite and crandallite: (a) minyulite projected onto (001); (b) minyulite projected onto (010); (c) crandallite projected onto (001); (d) crandallite projected onto (001). ($Al\phi_6$) octahedra are shadow-shaded, H atoms are shown as small shaded circles, K atoms are shown as large shaded circles, Ca atoms are shown as dark circles.

of the octahedra and the 'free' vertex of the ($P\phi_4$) group in the adjacent sheet, and by [10]-coordinated Ca. **Viséite** is a poorly crystalline mineral that McConnell (1952, 1990) has proposed as an aluminophosphate isotype of analcime. However, Kim and Kirkpatrick (1996) showed that viséite is a mixture of several phases; the dominant phase has a structure similar to that of crandallite, plus admixed phases that include an unidentified aluminophosphate, opal and a zeolitic framework aluminosilicate.

A prominent feature in **laueite**, $Mn^{2+}(H_2O)_4[Fe^{3+}{}_2(PO_4)_2(OH)_2(H_2O)_2](H_2O)_2$, and the minerals of the laueite group (Table 7) is the 7-Å chain shown in Figure 18c. ($Fe^{3+}\phi_6$) octahedra link by sharing *trans* vertices to form an [$M\phi_5$] chain that is decorated by flanking (PO_4) groups, and the resulting chains extend in the *c*-direction, giving a *c*-repeat of ~7.1 Å (see Appendix). These chains meld in the *a*-direction by sharing one quarter of the flanking (PO_4) vertices with octahedra of adjacent chains to form an [$Fe^{3+}{}_2(PO_4)_2(OH)_2(H_2O)_2$] sheet (Fig. 26a); note that the sheet is written with two

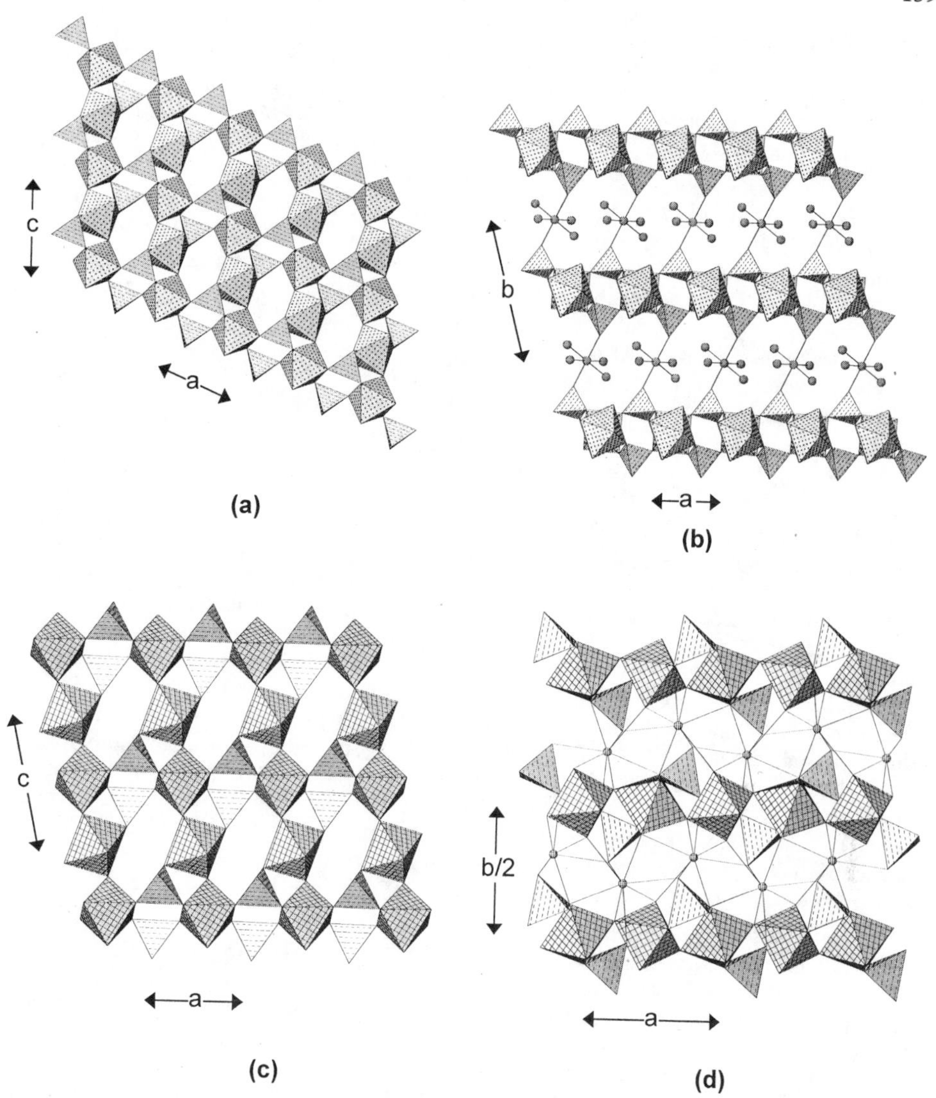

Figure 26. The crystal structures of laueite and curetonite: (a) laueite projected onto (010); (b) laueite projected onto (001); (H_2O) groups not bonded to any cations are not shown; (c) curetonite projected onto (010); (d) curetonite projected onto (001). ($Fe\phi_6$) and ($Al\phi_6$) octahedra are dot-shaded and 4^4-net-shaded, respectively; Mn^{2+} and Ba are shown as shaded circles, selected (H_2O) groups are shown as grey circles.

octahedrally coordinated cations, rather than as $[M(TO_4)\phi_2]_2$ because the two octahedra are topologically distinct. In the resulting sheet, the (PO_4) tetrahedra are three-connected. Note that there are two distinct octahedra in these sheets, one of which is six-connected within the sheet, and the other of which is only four-connected and has (H_2O) at two vertices. Another prominent feature of this sheet is the $[M(TO_4)_2\phi_2]$ chain (Fig. 18b) that extends from SE to NW in Figure 26a. Thus we could also think of the laueite sheet as composed of $[Fe^{3+}(PO_4)_2O_2]$ chains that are linked by ($Fe^{3+}O_6$) octahedra. This occurrence of two different types of chain in a more highly connected structural unit is a common feature in minerals, and reflects Nature's parsimony in designing structural arrangements in crystals. These sheets stack in the *b*-direction and are linked by $(Mn^{2+}O_2\{H_2O\}_4)$ octahedra (Fig. 26b), and by hydrogen bonds involving the interstitial (H_2O) groups bonded to Mn^{2+} *and* interstitial (H_2O) groups held in the structure solely by hydrogen bonds.

Curetonite, $Ba_2[Al_2(PO_4)_2(OH)_2F_2]$, contains an $[Al_2(PO_4)_2(OH)_2F_2]$ sheet (Fig. 26c) topologically identical to the analogous sheet in laueite (Figs. 26a). Note that the formula of curetonite has previously been written as half the formula unit given above, but that formulation ignored the fact that there are two topologically distinct $(Al\phi_6)$ octahedra in the structural unit. There is another interesting wrinkle in the chemistry of curetonite, replacement of Al by Ti and (OH) by O^{2-}, which can give local areas of titanite-like arrangement within the sheet. The sheets stack in the *b*-direction (Fig. 26d) and are linked by interstitial [10]-coordinated Ba.

Stewartite, $Mn^{2+}(H_2O)_4[Fe^{3+}_2(PO_4)_2(OH)_2(H_2O)_2](H_2O)_2$, and **pseudolaueite**, $Mn^{2+}(H_2O)_4[Fe^{3+}_2(PO_4)_2(OH)_2(H_2O)_2](H_2O)_2$, are polymorphs of laueite. Both contain $[Fe^{3+}_2(PO_4)_2(OH)_2(H_2O)O^P_2]$ chains (cf. Fig. 18c), but the way in which these chains cross-link to form a sheet is different from the analogous linkage in laueite. In stewartite, there are three symmetrically distinct $(Fe\phi_6)$ octahedra in the 7-Å chain, with coordinations $(\{OH\}_2O_2\{H_2O\}_2)$, $(\{OH\}_2O_2\{H_2O\}_2)$ and $(\{OH\}_2O_4)$ with multiplicities of 1, 1 and 2, respectively. In laueite, there are two symmetrically distinct $(Fe\phi_6)$ octahedra in the 7-Å chain, with coordinations $(\{OH\}_2O_2\{H_2O\}_2)$ and $(\{OH\}_2O_4)$ with multiplicities of 2 and 2, respectively. However, the cross-linkage of chains is different from in laueite, as is apparent from the presence of $[Fe^{3+}(PO_4)_2\phi_2]$ chains in laueite (Fig. 26a) and only fragments of this chain in stewartite (Fig. 27a). These sheets stack in the *c*-direction, linked by $(Mn^{2+}O_2\{H_2O\}_4)$ octahedra (Fig. 27b) and hydrogen bonds involving (H_2O) bonded to interstitial cations and (H_2O) held in the structure solely by hydrogen bonding. In pseudolaueite, the $[Fe^{2+}_2(PO_4)_2(OH)_2(H_2O)O^P_2]$ chains condense to form a sheet (Fig. 27c) topologically distinct from those in laueite and stewartite; Moore (1975b) discusses in detail the isomeric variation in these (and related) sheets. These sheets stack along the *c*-direction (Fig. 27d) and are linked by $(Mn^{2+}O_2\{H_2O\}_4)$ octahedra and by hydrogen bonds.

The sheets in **strunzite**, $Mn^{2+}(H_2O)_4[Fe^{3+}(PO_4)_2(OH)(H_2O)]_2$, and **metavauxite**, $Fe^{2+}(H_2O)_6[Al(PO_4)(OH)(H_2O)]_2$, are built from topologically identical $[M(TO_4)\phi_3]$ chains. In strunzite, the 7-Å chains extend in the *c*-direction and cross-link to form an $[Fe^{3+}(PO_4)(OH)(H_2O)]$ sheet (Fig. 28a) that is a graphical isomer of the $[Fe^{3+}_2(PO_4)_2(OH)_2(H_2O)_2]$ sheet in stewartite (Fig. 27a). These sheets stack in the *a*-direction (Fig. 28b) and are linked by $(Mn^{2+}O_2\{H_2O\}_4)$ octahedra and hydrogen bonds. In metavauxite, the 7-Å chains also extend in the *c*-direction, and cross-link to form an $[Al(PO_4)(OH)(H_2O)]$ sheet (Fig. 28c). These sheets stack in the *a*-direction (Fig. 28d) and are linked by hydrogen bonds emanating from the interstitial $(Fe^{2+}\{H_2O\}_6)$ groups.

Montgomeryite, $Ca_4Mg(H_2O)_{12}[Al_2(PO_4)_3(OH)_2]_2$, contains 7-Å chains of the form $[M(T\phi_4)\phi_2]$ (Fig. 18c) in which alternate octahedra are decorated by two tetrahedra that attach to *trans* vertices (Fig. 29a) to give a chain of the form $[M_2(TO_4)_4\phi_4]$ that extends in the [101] direction. These chains meld in the $[10\bar{1}]$ direction by sharing flanking (PO_4) groups to form an $[Al_2(PO_4)_3(OH)_2]$ sheet that is parallel to (010) (Fig. 29a). These sheets stack in the [010] direction (Fig. 29b). The decorating tetrahedra of the 7-Å chains project above and below the plane of the sheet, and one Ca cation occurs in the interstices created by these tetrahedra, being coordinated by four O-atoms of the sheet and four interstitial (H_2O) groups. The second Ca cation links to four anions of the sheet and

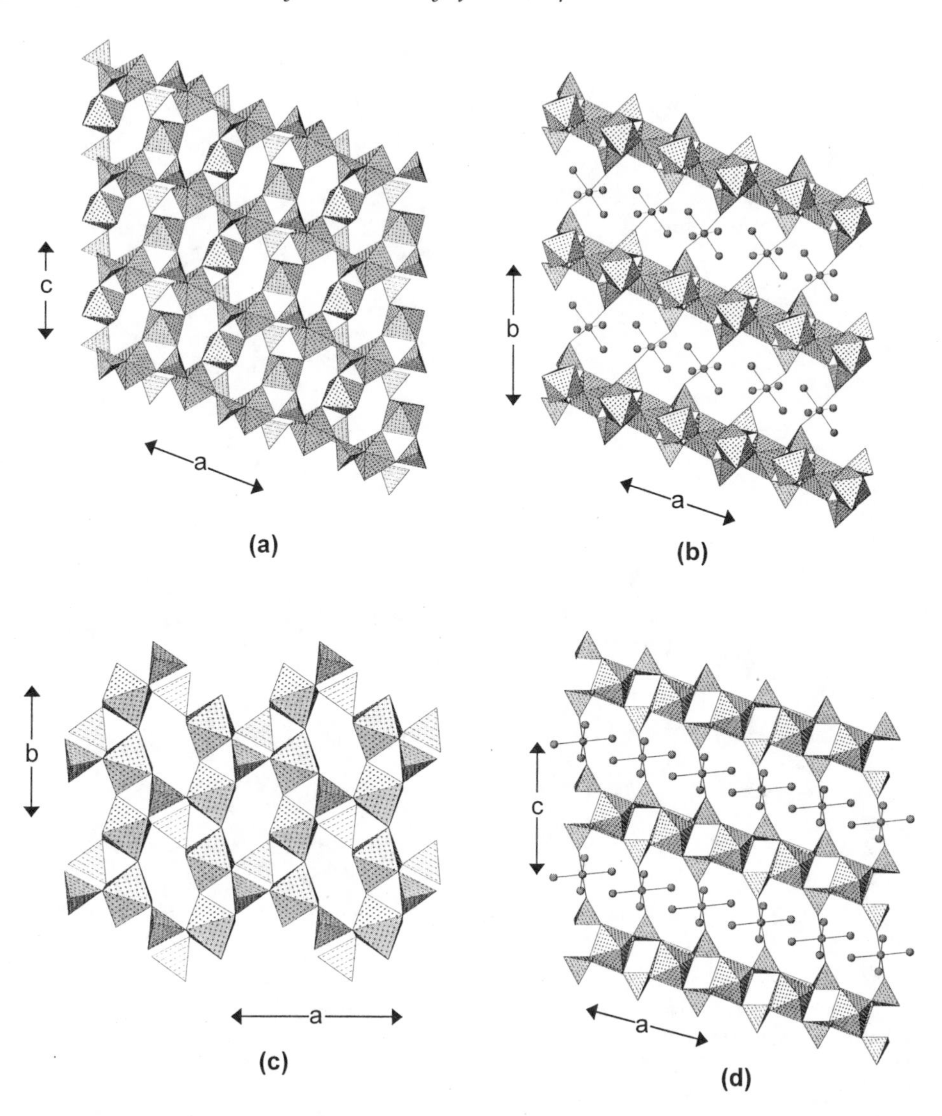

Figure 27. The crystal structures of stewartite and pseudolaueite: (a) stewartite projected onto (010); (b) stewartite projected onto (001); (c) pseudolaueite projected onto (001); (d) pseudolaueite projected onto (010). ($Fe^{3+}\phi_6$) octahedra are dot-shaded, Mn^{2+} and (H_2O) are shown as shaded circles.

shares four interstitial (H_2O) groups with an adjacent Ca that, in turn, links to the adjacent sheet. Further intersheet linkage is provided by octahedrally coordinated interstitial Mg that bonds to four interstitial (H_2O) groups.

Mitryaevaite, $[Al_5(PO_4)_2(PO_3\{OH\})_2F_2(OH)_2(H_2O)_8](H_2O)_{6.5}$, has quite a complex sheet that, nevertheless, can be related to other sheets in this group. An important motif in this sheet is an $[M_5(TO_4)_4\phi_{17}]$ fragment (Fig. 29c) of the $[M(TO_4)\phi]$ chain (Fig. 18c) that extends along ~[120]. These fragments meld in the ~$[1\bar{1}0]$ direction through tetrahedron-

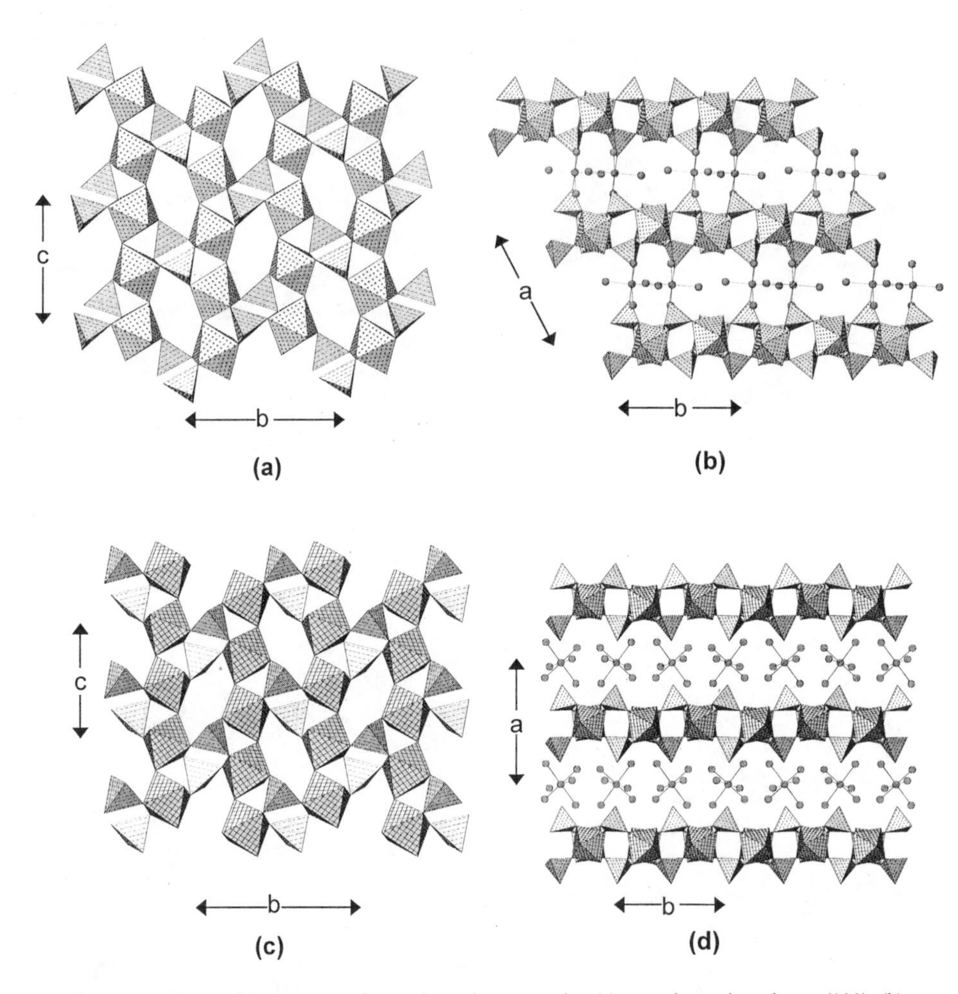

Figure 28. The crystal structures of strunzite and metavauxite: (a) strunzite projected onto (100); (b) strunzite projected onto (001); $(Fe^{3+}\phi_6)$ octahedra are dot-shaded; (c) metavauxite projected onto (100); (d) metavauxite projected onto (001); $(Al\phi_6)$ octahedra are 4^4-net-shaded, interstitial Mn^{2+} and (H_2O) groups are shown as shaded circles.

octahedron linkages to form a sheet (Fig. 29c) parallel to (110). The chain fragments are inclined to the plane of the sheet, giving it a very corrugated appearance in cross-section (Fig. 29d). These sheets stack in the *c*-direction and are linked by hydrogen bonds via inclined sheets of interstitial (H_2O) groups that do not bond to any cation.

Figure 29 (next page). The crystal structures of montgomeryite, mitryaevaite and sidorenkoite: (a) montgomeryite projected onto (010); (b) montgomeryite projected onto (001); (c) mitryaevaite projected onto (001), showing the $[Al_5(P\phi_4)_4\phi_{12}]$ sheet that is made up of $[M_5(TO_4)_4\phi_{17}]$ fragments (one is shown in black) of the 7-Å $[M(TO_4)\,\phi]$ chain; (d) mitryaevaite projected onto (100); $(Al\phi_6)$ octahedra are 4^4-net-shaded, Mg are shown as small shaded circles, (H_2O) groups are shown as large shaded circles in (b) and small unshaded circles in (d); (e) sidorenkoite projected onto (100); (f) sidorenkoite projected onto (010); $(Mn^{2+}\phi_6)$ octahedra are shadow-shaded, (CO_3) groups are lined triangles.

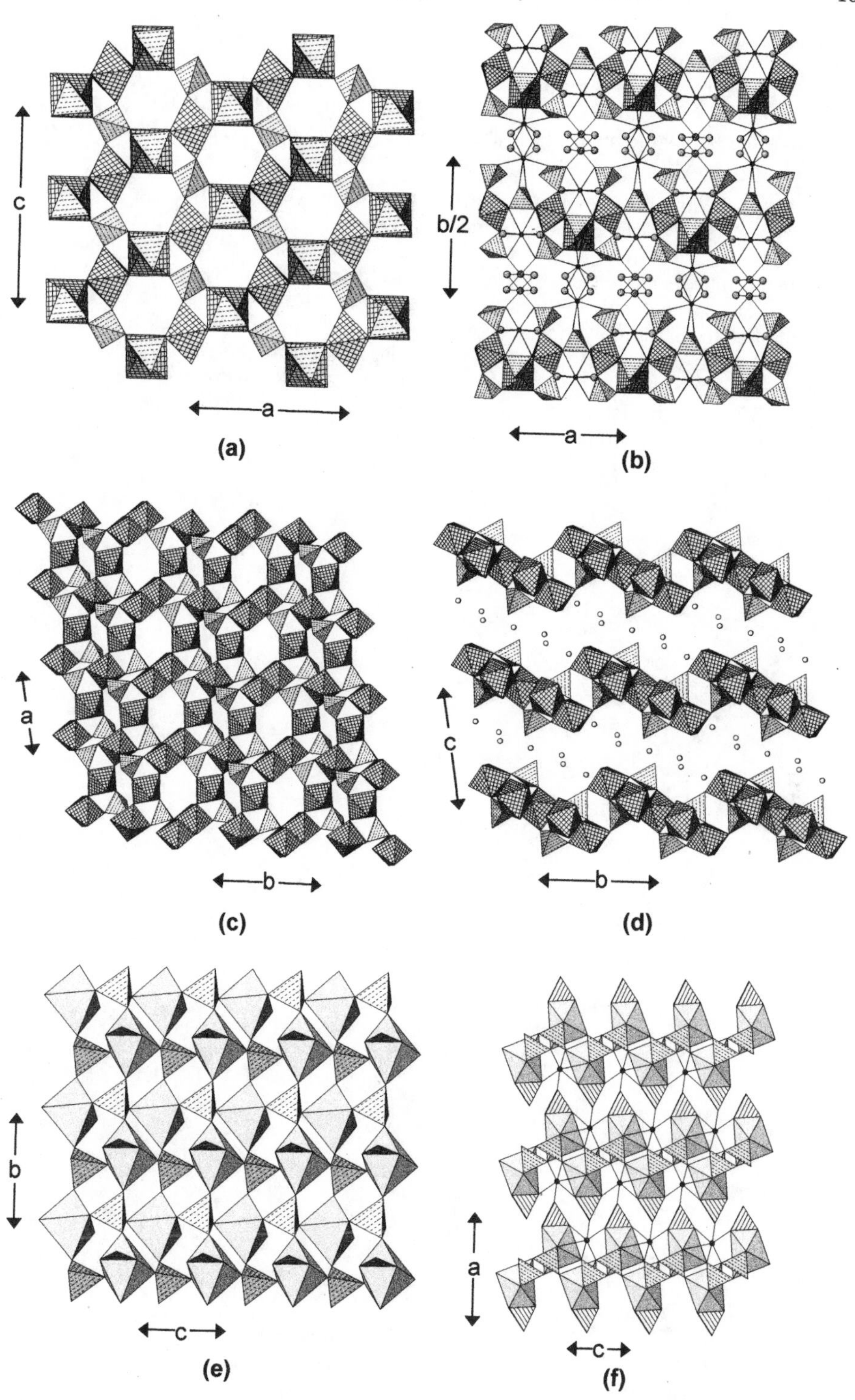
c
a
(a)
b/2
a
(b)
a
b
(c)
c
b
(d)
b
c
(e)
a
c
(f)

Sidorenkoite, $Na_3[Mn^{2+}(PO_4)(CO_3)]$, and the other minerals of the **bradleyite** group consist of (PO_4) groups and $(M^{2+}O_6)$ octahedra at the vertices of a 4^4 plane net and link by sharing corners to form a sheet parallel to (100) (Fig. 29e). This leaves two octahedron vertices that do not link to (PO_4) groups; these link to (CO_3) groups that decorate the sheet above and below the plane of the sheet (Fig. 29f). These sheets are linked in the *a*-direction by [6]- and [7]-coordinated interstitial Na cations.

M=M, M-T linkage. **Bermanite**, $Mn^{2+}(H_2O)_4[Mn^{3+}(PO_4)(OH)]_2$, and **ercitite**, $Na_2(H_2O)_4[Mn^{3+}(PO_4)(OH)]_2$, are not formally isostructural as they have different space-group symmetries, but they contain topologically and chemically identical structural units. $(M\phi_6)$ octahedra share pairs of *trans* edges to form an $[M\phi_4]$ chain decorated with flanking tetrahedra that link vertices of adjacent octahedra (Fig. 18e). These chains extend parallel to [101] and link together by sharing octahedral vertices to form an $[M(TO_4)\phi]$ sheet that is parallel to (010) in bermanite and ercitite (Fig. 30a). These sheets stack in the *b*-direction and are linked by $(Mn^{2+}O_2\{H_2O\}_4)$ octahedra and by hydrogen bonds (Fig. 30b). The interstitial linkage is somewhat different in ercitite. One Mn^{2+} atom plus one vacancy (space group $P2_1$) is replaced by two Na atoms (space group $P2_1/m$), the Mn^{2+} and being ordered in bermanite and giving rise to the non-centrosymmetric space group.

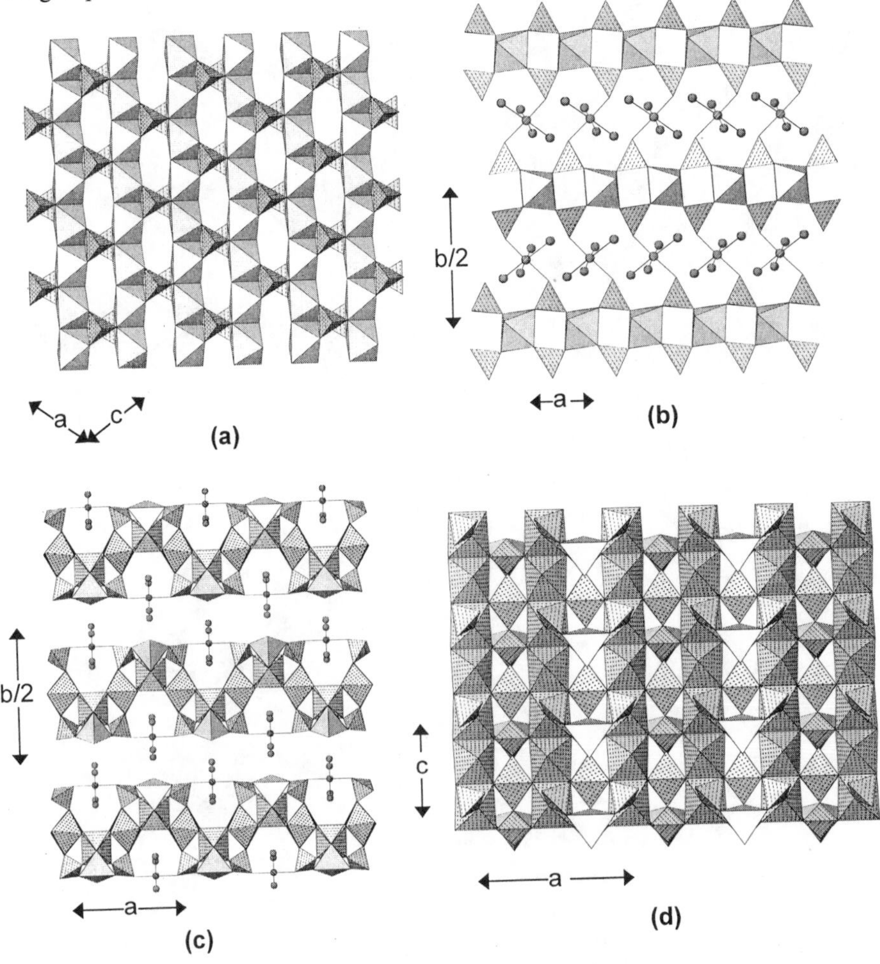

M=M, M-M, M-T linkage. **Schoonerite**, $[Mn^{2+}Fe^{2+}{}_2ZnFe^{3+}(PO_4)_3(OH)_2(H_2O)_7](H_2O)_2$, is a very complicated structure, and its assignment to a specific structural class is somewhat ambiguous. Figures 30c,d show the polyhedra and their connectivity. Inspection of Figure 30c indicates the sheet-like nature of the structure. However, the sheet includes both divalent *and* trivalent cations, and is further complicated by the fact that Zn is [5]-coordinated. There are two prominent motifs within the sheet, an $[Fe^{2+}\phi_4]$ chain of edge-sharing octahedra extending in the *c*-direction, and an $[Fe^{3+}\,{}^{[5]}Zn(PO_4)_2\phi_6]$ cluster. These link in the *a*-direction to form a continuous sheet (Fig. 30d) that is further strengthened by $(Mn^{2+}O_2\{H_2O\}_4)$ octahedra occupying dimples in the sheet. These sheets stack in the *b*-direction and are linked by hydrogen bonds. Assigning the divalent cations as interstitial species results in a finite-cluster structure, not in accord with the dense distribution of polyhedra in the sheet arrangement of Figure 30d. However, we must recognize a somewhat arbitrary aspect of the assignment here. Another aspect that suggests a sheet structure is the 7-Å chain that extends in the *a*-direction; this chain involves both Fe^{3+} and Fe^{2+}.

Nissonite, $[Cu^{2+}Mg(PO_4)(OH)(H_2O)_2]_2(H_2O)$, consists of a thick slab of polyhedra linked solely by hydrogen bonds. $(Mg\phi_6)$ octahedra and (PO_4) tetrahedra lie at the vertices of a 6^3 plane net (Fig. 31a); this layer, $[Mb(PO_4)(OH)(H_2O)_2]$, is topologically identical with the $[Mg(PO_3\{OH\})(H_2O)_3]$ sheet in newberyite (Fig. 24a). However, the tetrahedra in newberyite point alternately up and down relative to the plane of the sheet, whereas the tetrahedra in nissonite all point in the same direction; hence these sheets are topologically identical but graphically distinct, and are geometrical isomers (Hawthorne 1983a, 1985a). Edge-sharing $[Cu^{2+}{}_2O_8(OH)_2]$ dimers link by sharing corners to form the sheet shown in Figure 31b. The $[Mg(PO_4)(OH)(H_2O)_2]$ sheets sandwich the $[Cu^{2+}{}_2O_8(OH)_2]$ sheet to form a thick slab parallel to (100). These slabs link through hydrogen bonds both directly and involving interstitial (H_2O) groups not bonded to any cation (Fig. 31c).

Foggite, $Ca[Al(PO_4)(OH)_2](H_2O)$, contains $[Al\phi_4]$ α-PbO_2-like chains of edge-sharing $(Al\phi_6)$ octahedra that extend in the *c*-direction and are cross-linked into a sheet by (PO_4) tetrahedra (Fig. 32a). These sheets are parallel to (010), and are linked by [8]- and [10]-coordinated interstitial Ca (Fig. 32b) and by hydrogen bonds involving interstitial (H_2O) groups. The structure of foggite is strongly related to the pyroxene structure, specifically calcium tschermakite. Figure 32c depicts the structure of foggite projected onto (100), showing the *M*(1)-like chains and their associated tetrahedra. Moore et al. (1975b) expressed the relation as follows:

Foggite	$[CaAl_2P_2O_8(OH)_4]$-$Ca(H_2O)_2$
Px	$[CaAl_2\,{}^{[6]}T_2O_{12}]$-$Ca^{[4]}T_2$

Whitmoreite, $Fe^{2+}(H_2O)_4[Fe^{3+}(PO_4)(OH)]_2$, consists of a fairly densely packed sheet of (PO_4) tetrahedra and $(Fe^{3+}\phi_6)$ octahedra parallel to (100) (Fig. 32d). Pairs of $Fe^{3+}\phi_6$ octahedra condense to form edge-sharing $[Fe^{3+}{}_2\phi_{10}]$ dimers that occupy the vertices of a 4^4 plane net and link by sharing corners. This results in an interrupted sheet of octahedra, the interstices of which are occupied by (PO_4) tetrahedra (Fig. 32d). These sheets stack in the *a*-direction, and are linked by interstitial $(Fe^{2+}O_2\{H_2O\}_4)$ octahedra and by hydrogen bonds (Fig. 32e).

Figure 30 (opposite page). The crystal structures of bermanite and schoonerite: (a) bermanite projected onto (010); (b) bermanite projected onto (001); (c) schoonerite projected onto (001); note that Zn is [5]-coordinated; (d) schoonerite projected onto (010). $(Mn^{3+}\phi_6)$ octahedra and $(Zn\phi_4)$ tetrahedra are shadow-shaded, $(\{Mn^{2+},Fe^{2+}, Fe^{3+}\}\phi_6)$ octahedra are dot-shaded, Mn^{2+} and (H_2O) groups are shown as shaded circles.

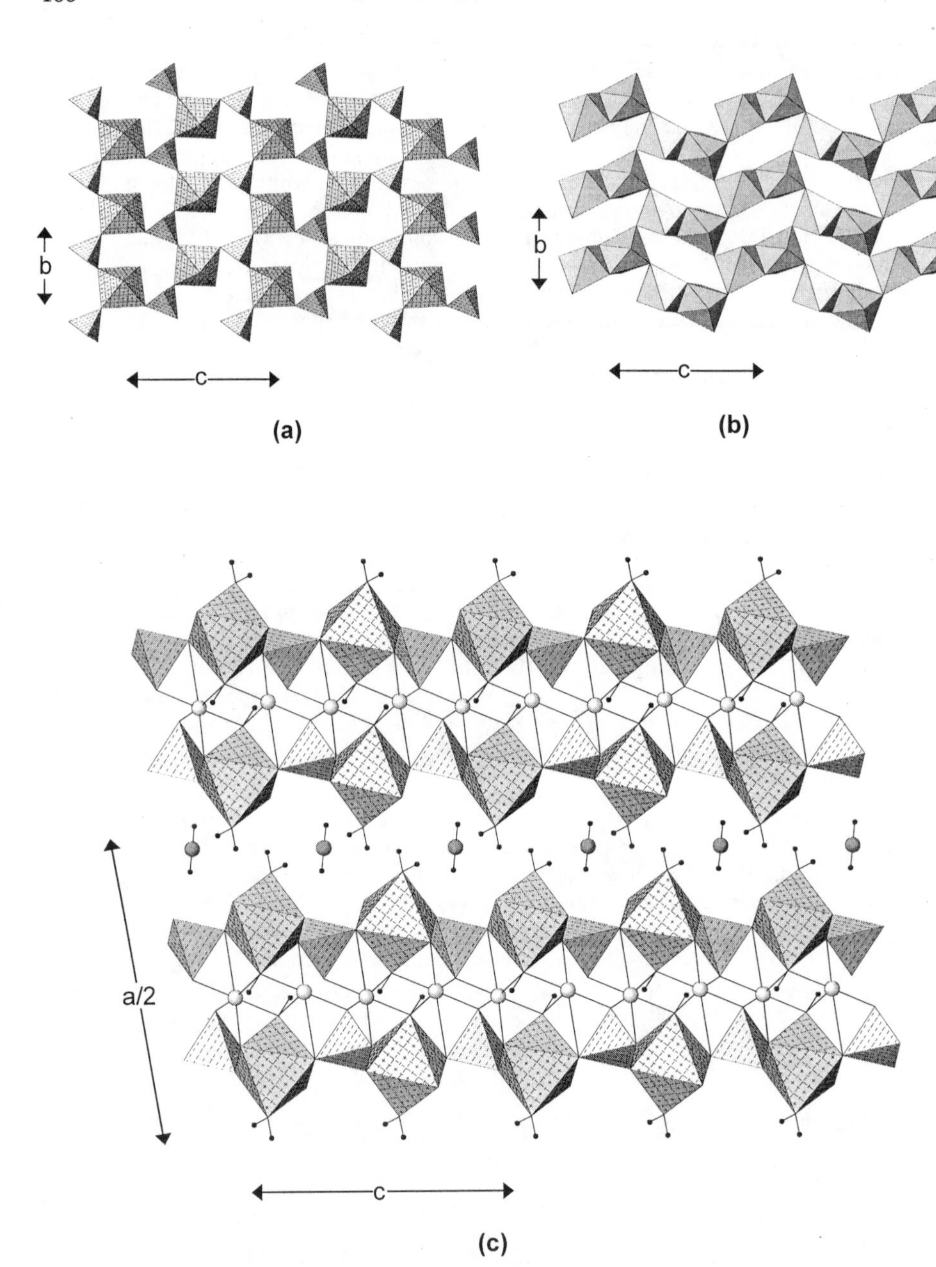

Figure 31. The crystal structure of nissonite: (a) the $[Mg(PO_4)(OH)(H_2O)_2]$ layer parallel to (100); (b) the $[Cu^{2+}{}_2O_8(OH)_2]$ layer parallel to (100); (c) a view of the $[Cu^{2+}Mg(PO_4)(OH)(H_2O)_2]$ sheet in the *b*-direction, showing the $[Cu^{2+}{}_2O_8(OH)_2]$ layer sandwiched by two $[Mg(PO_4)(OH)(H_2O)_2]$ layers. $(Mg\phi_6)$ octahedra are square-pattern-shaded, $(Cu^{2+}\phi_6)$ octahedra are shadow-shaded, Cu^{2+} cations are shown as light circles, H atoms are shown as small black circles, O atoms of interstitial (H_2O) groups are shown as shaded circles.

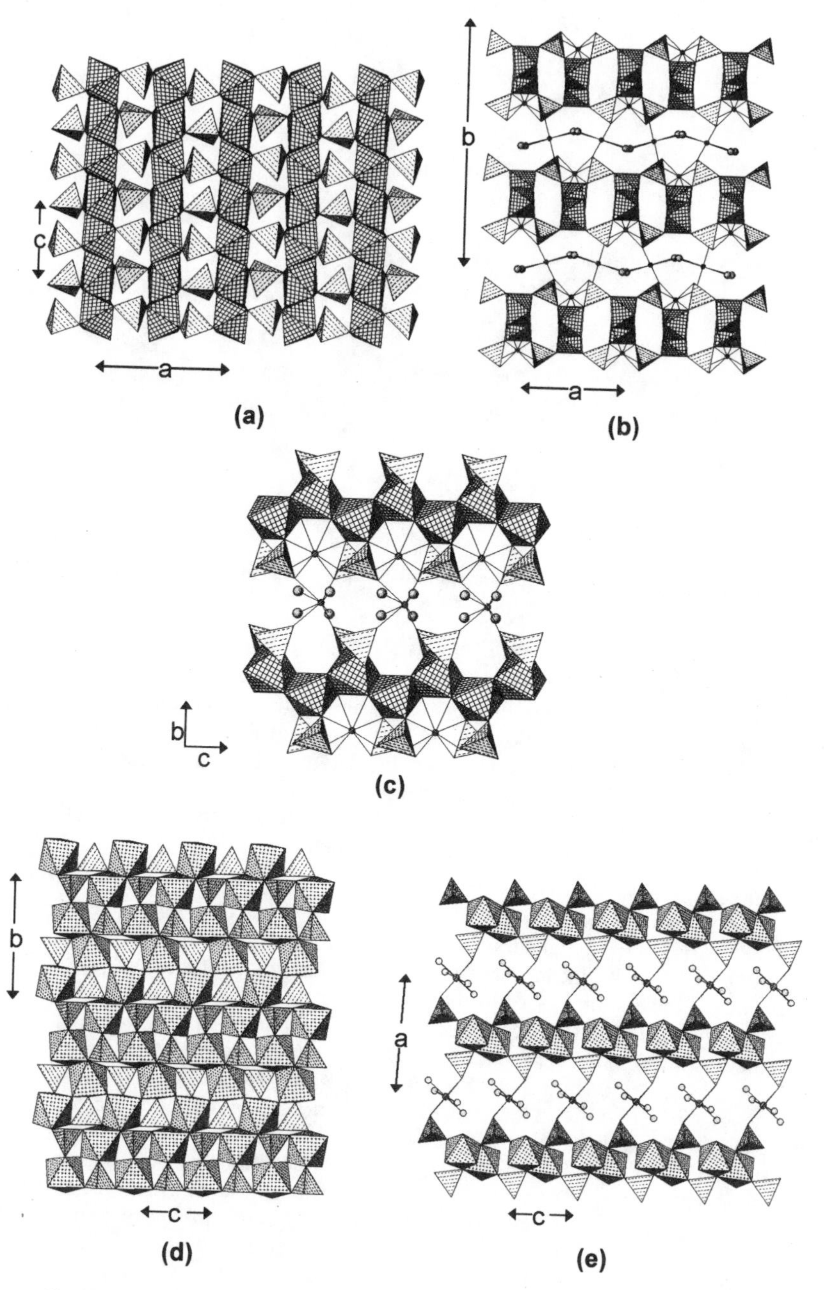

Figure 32. The crystal structures of foggite and whitmoreite: (a) foggite projected onto (010); (b) foggite projected onto (001); (c) foggite projected onto (100), showing its similarity to the monoclinic pyroxene (calcium tschermaks) structure; (d) whitmoreite projected onto (100); (e) whitmoreite projected onto (010). ($Al\phi_6$) octahedra are 4^4-net-shaded, ($Fe^{3+}\phi_6$) octahedra are dot-shaded, Ca atoms are shown as small shaded circles, (H_2O) groups are shown as large shaded circles, Fe^{2+} atoms are shown as small shaded circles in (e).

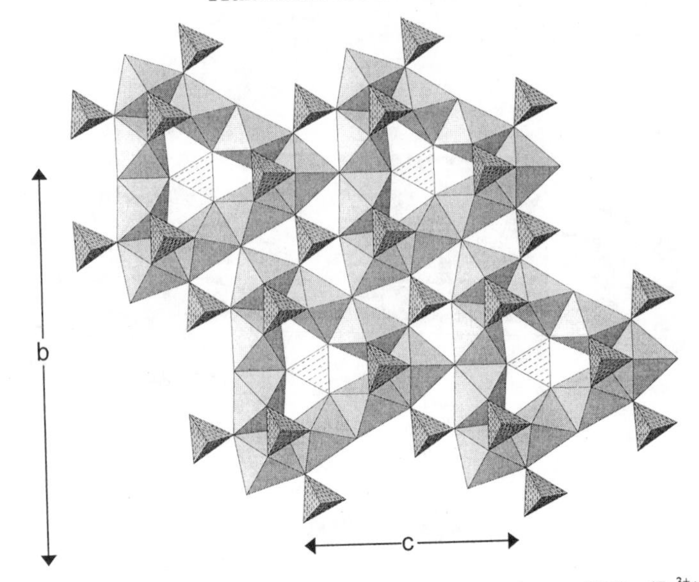

Figure 33. The crystal structure of mitridatite projected onto (100); ($Fe^{3+}\phi_6$) octahedra are shadow-shaded.

Mitridatite, $Ca_6(H_2O)_6[Fe^{3+}{}_9O_6(PO_4)_9](H_2O)_3$, has a sheet structural unit of exotic complexity. ($Fe^{3+}\phi_6$) octahedra share edges to form nonameric triangular rings that are braced by a central (PO_4) group that shares corners with six octahedra (Fig. 33). These clusters link by their corners linking to the mid-points of the edges of adjacent clusters. The resulting interstices are occupied by (PO_4) tetrahedra that point in the opposite direction to the tetrahedra occupying the centres of the clusters. These sheets are linked by [7]-coordinated interstitial Ca and by hydrogen bonds involving interstitial (H_2O) bonded to Ca and interstitial (H_2O) groups not bonded to any cation.

Vivianite, $[Fe^{2+}{}_3(PO_4)_2(H_2O)_8]$, contains two crystallographically distinct Fe^{3+} cations that are octahedrally coordinated by ($O_2\{H_2O\}_4$) and ($O_4\{H_2O\}_2$), respectively. Pairs of ($Fe^{2+}O_4\{H_2O\}_2$) octahedra share edges to form a dimer that is decorated by two (PO_4) groups that each link to corners of each octahedron, forming an $[Fe^{2+}{}_2(PO_4)_2\phi_6]$ cluster. These clusters are linked in the *c*-direction by ($Fe^{2+}O_2\{H_2O\}_4$) octahedra (Fig. 34a). These chains link in the *a*-direction (Fig. 34b) by corner-sharing between tetrahedra and octahedra to form sheets parallel to (010). The sheets are linked solely by hydrogen bonds in the *b*-direction (Fig. 34a). **Bobierrite**, $[Mg_3(PO_4)(H_2O)_8]$, has a structure very similar to that of vivianite. The sheets of octahedra and tetrahedra are topologically identical (Figs. 34c,d), but the attitude of adjacent sheets in the *b*-direction is sufficiently different such that the hydrogen-bond linkage between the sheets differs from that in vivianite. In vivianite, the hydrogen-bond linkages are at an angle to the plane of the sheet (Fig. 34a), whereas in bobierrite, the hydrogen-bond linkages are orthogonal to the plane of the sheet (Fig. 34c). These difference in the attitude of adjacent sheets is reflected in the symmetries of the two structures: *C*2/*m* (vivianite) versus *C*2/*c* (bobierrite).

Structures with infinite frameworks of (PO_4) tetrahedra and ($M\phi_6$) octahedra

The minerals of this class are listed in Table 8; graphs are not shown for framework structures because representing such a structure in two dimensions is not satisfactory and the representation can be confusing.

Table 8. Phosphate minerals based on infinite frameworks of tetrahedra and octahedra.

Mineral	*Structural unit*	*Space group*	*Figure*
Kolbeckite	$[Sc(PO_4)(H_2O)_2]$	$P2_1/n$	35a,b
Metavariscite*	$[Al(PO_4)(H_2O)_2]$	$P2_1/n$	35a,b
Phosphosiderite	$[Fe^{3+}(PO_4)(H_2O)_2]$	$P2_1/n$	35a,b
Strengite	$[Fe^{3+}(PO_4)(H_2O)_2]$	$Pbca$	35c,d
Variscite*	$[Al(PO_4)(H_2O)_2]$	$Pbca$	35c,d
Kosnarite	$[Zr_2(PO_4)_3]$	$R\bar{3}c$	35e,f
Isokite	$[Mg(PO_4)F]$	$C2/c$	36a,b
Lacroixite	$[Al(PO_4)F]$	$C2/c$	36a,b
Panasqueraite	$[Mg(PO_4)(OH)]$	$C2/c$	36a,b
Titanite *	$[Ti(SiO_4)O]$	$C2/c$	36a,b
Amblygonite *	$[Al(PO_4)F]$	$C\bar{1}$	36c,d
Montebrasite	$[Al(PO_4)(OH)]$	$C\bar{1}$	36c,d
Natromontebrasite	$[Al(PO_4)(OH)]$	–	36c,d
Tavorite	$[Fe^{3+}(PO_4)(OH)]$	–	36c,d
Cyrilovite	$[Fe^{3+}{}_3(PO_4)_2(OH)_4]$	$P4_12_12$	37a,b
Wardite*	$[Al_3(PO_4)_2(OH)_4]$	$P4_12_12$	37a,b
Fluellite	$[Al_2(PO_4)F_2(OH)]$	$Fddd$	37c,d
Wavellite	$[Al_3(PO_4)_2(OH)_3(H_2O)_2]$	$Pcmn$	37e,f
Augelite	$[Al_2(PO_4)(OH)_3]$	$C2/m$	38a,b
Jagowerite *	$[Al(PO_4)(OH)]_2$	$P\bar{1}$	38c,d
Mari_ite	$[Fe^{2+}(PO_4)]$	$Pmnb$	38e,f
Kovdorskite	$[Mg_2(PO_4)(OH)(H_2O)_3]$	$P2_1/a$	39a,b
Libethenite	$[Cu^{2+}{}_2(PO_4)(OH)]$	$Pnnm$	39c,d
Adamite*	$[Cu^{2+}{}_2(AsO_4)(OH)]$	$Pnnm$	39c,d
Tarbuttite	$[Zn_2(PO_4)(OH)]$	$P\bar{1}$	39e,f
Paradamite*	$[Zn_2(AsO_4)(OH)]$	$P\bar{1}$	39e,f
Mixite*	$[Cu^{2+}{}_6(AsO_4)_3(OH)_6]$	$P6_3/m$	40a,b
Petersite-(Y)	$[Cu^{2+}{}_6(PO_4)_3(OH)_6]$	$P6_3/m$	40a,b
Brazilianite	$[Al_3(PO_4)_2(OH)_4]$	$P2_1/n$	40c,d
Pseudomalachite	$[Cu^{2+}{}_5(PO_4)_2(OH)_2(H_2O)]$	$P2_1/c$	41a,b
Reichenbachite	$[Cu^{2+}{}_5(PO_4)_2(OH)_4(H_2O)]$	$P2_1/a$	41c,d
Ludjibaite	$[Cu^{2+}{}_5(PO_4)_2(OH)_4(H_2O)]$	$P\bar{1}$	41e,f
Magniotriplite	$[Mg_2(PO_4)F]$	$I2/a$	42a,b
Triplite *	$[Mn^{2+}{}_2(PO_4)F]$	$I2/c$ (?)	42a,b
Zweiselite	$[Fe^{2+}{}_2(PO_4)F]$	$I2/a$ (?)	42a,b
Triploidite *	$[Mn^{2+}{}_2(PO_4)(OH)]$	$P2_1/a$	42c,d
Wagnerite	$[Mg_2(PO_4)F]$	$P2_1/a$	42c,d
Wolfeite	$[Fe^{2+}{}_2(PO_4)(OH)]$	$P2_1/a$	42c,d
Alluaudite *	$[Fe^{2+}(Mn,Fe^{2+},Fe^{3+},Mg)_2(PO_4)_3]$	$I2/a$	43a,b
Hagendorfite	$[Mn^{2+}(Fe^{2+},Mg,Fe^{3+})_2(PO_4)_3]$	$I2/a$	43a,b
Maghagendorfite	$[Mn^{2+}(Mg,Fe^{2+},Fe^{3+})_2(PO_4)_3]$	–	43a,b
Quingheite			43a,b
Varulite	$[Mn^{2+}(Mn,Fe^{2+},Fe^{3+})_2(PO_4)_3]$	–	43a,b
Rosemaryite	$[Mn^{2+}Fe^{3+}Al(PO_4)_3]$	$C2/c$ (?)	43c,d
Wyllieite *	$[Mn^{2+}Fe^{2+}Al(PO_4)_3]$	$P2_1/n$	43c,d
Bobfergusonite	$[Mn^{2+}Fe^{3+}Al(PO_4)_6]$	$P2_1/n$	43e,f
Ludlamite	$[Fe^{2+}{}_3(PO_4)_2(H_2O)_4]$	$P2_1/a$	45a,b
Melonjosephite	$[(Fe^{2+},Fe^{3+})(PO_4)(OH)]$	$Pnam$	45c,d

Bertossaite *	$[Al(PO_4)(OH)]_4$	$I*aa$	45e,f
Palermoite	$[Al(PO_4)(OH)]_4$	$Imcb$	45e,f
Arrojadite*	$[Fe^{2+}{}_{14}Al(PO_4)_{12}(OH)_2]$	$C2/c$	–
Dickinsonite	$[Mn^{2+}{}_{14}Al(PO_4)_{12}(OH)_2]$	$C2/c$	–
Farringtonite	$[Mg_3(PO_4)_2]$	$P2_1/n$	46a,b
Beusite	$[Mn^{2+}{}_3(PO_4)_2]$	$P2_1/c$	46c,d
Graftonite*	$[Fe^{2+}{}_3(PO_4)_2]$	$P2_1/c$	46c,d
Bederite	$[Mn^{2+}{}_2Fe^{3+}{}_2Mn^{3+}{}_2(PO_4)_6]$	$Pcab$	47a,b,c
Wicksite	$[Fe^{2+}{}_4MgFe^{3+}(PO_4)_6]$	$Pcab$	47a,b,c
Aheylite	$[Al_6(PO_4)_4(OH)_8]$	$P\bar{1}$	47d,e
Chalcosiderite	$[Fe^{3+}{}_6(PO_4)_4(OH)_8]$	$P\bar{1}$	47d,e
Coeruleolactite	$[Al_6(PO_4)_4(OH)_8]$	$P\bar{1}$	47d,e
Faustite	$[Al_6(PO_4)_4(OH)_8]$	$P\bar{1}$	47d,e
Planerite	$[Al_6(PO_4)_2(PO_3\{OH\})_2(OH)_8]$	$P\bar{1}$	47d,e
Turquoise *	$[Al_6(PO_4)_4(OH)_8]$	$P\bar{1}$	47d,e
Leucophosphite*	$[Fe^{3+}{}_2(PO_4)_2(OH)(H_2O)]$	$P2_1/n$	48a,b,c
Tinsleyite	$[Al_2(PO_4)_2(OH)(H_2O)]$	$P2_1/n$	48a,b,c
Cacoxenite	$[Fe^{3+}{}_{25}(PO_4)_{17}O_6(OH)_{12}]$	$P6_3/m$	49a,b,c,d
Althausite	$[Mg_4(PO_4)_2(OH)F]$	$Pnma$	50a,b
Hureaulite	$[Mn^{2+}{}_5(PO_3\{OH\})_2(PO_4)_2(H_2O)_4]$	$C2/c$	50c,d
Thadeuite	$[CaMg_3(PO_4)_2(OH)_2]$	$C222_1$	50e,f
Bakhchisaraitsevite	$[Mg_5(PO_4)_4(H_2O)_5]$	$P2_1/c$	51a,b
Kryzhanovskite	$[Mn^{2+}Fe^{3+}{}_2(PO_4)_2(OH)_2(H_2O)]$	$Pbna$	51c,d
Phosphoferrite*	$[Fe^{2+}{}_3(PO_4)_2(H_2O)_3]$	$Pbna$	51c,d
Griphite	$[A_{24}Fe^{2+}{}_4Al_8(PO_4)_{24}]$	$Pa\bar{3}$	52a,b,c,d
Cornetite	$[Cu^{2+}{}_3(PO_4)(OH)_3]$	$Pbca$	52e,f
Chladniite	$[Mg_7(PO_4)_6]$	$R\bar{3}$	–
Fillowite*	$[Mn^{2+}{}_7(PO_4)_6]$	$R\bar{3}$	–
Galileiite	$[Fe^{2+}{}_7(PO_4)_6]$	$R\bar{3}$	–
Johnsomervilleite	$[Mg_7(PO_4)_6]$	$R\bar{3}$	–
Gladiusite	$[Fe^{2+}{}_4Fe^{3+}{}_2(PO_4)(OH)_{11}(H_2O)$	$P2_1/n$	53a,b,c
Lipscombite	$[Fe^{2+}Fe^{3+}{}_2(PO_4)_2(OH)_2]$	$P4_32_12$	53d,e
Burangaite	$[Fe^{2+}Al_5(PO_4)_4(OH)_6(H_2O)_2]$	$C2/c$	54a
Dufrénite	$[Fe^{2+}Fe^{3+}{}_5(PO_4)_4(OH)_6(H_2O)_2]$	$C2/c$	54a
Natrodufrénite	$[Fe^{2+}Fe^{3+}{}_5(PO_4)_4(OH)_6(H_2O)_2]$	$C2/c$	54a
Frondellite	$[Fe^{2+}Fe^{3+}{}_4(PO_4)_3(OH)_5]$	$Bbmm$	54b,c
Rockbridgeite*	$[Fe^{2+}Fe^{3+}{}_4(PO_4)_3(OH)_5]$	$Bbmm$	54b,c
Barbosalite	$[Fe^{3+}(PO_4)(OH)]_2$	$P2_1/c$	54d,e
Hentschelite	$[Fe^{3+}(PO_4)(OH)]_2$	$P2_1/c$	54d,e
Lazulite*	$[Al(PO_4)(OH)]_2$	$P2_1/c$	54d,e
Scorzalite	$[Al(PO_4)(OH)]_2$	$P2_1/c$	54d,e
Trolleite	$[Al_4(PO_4)_3(OH)_3]$	$I2/c$	55a,b
Seamanite	$[Mn^{2+}{}_3(PO_4)(B\{OH\}_4)(OH)_2]$	$Pbnm$	55c,d
Holtedahlite	$[Mg_{12}(PO_3\{OH\})(PO_4)_5(OH)_6]$	$P31m$	55e,f
Satterlyite	$[Fe^{2+}{}_4(PO_4)_2(OH)_2]$	$P\bar{3}1m$	55e,f
Triphylite*	$[Fe^{2+}(PO_4)]$	$Pbnm$	56a,b
Lithiophylite	$[Mn^{2+}(PO_4)]$	$Pbnm$	56a,b
Natrophilite	$[Mn^{2+}(PO_4)]$	$Pbnm$	56a,b
Ferrisicklerite	$[Mn^{2+},Fe^{3+}(PO_4)]$	$Pbnm$	56c,d
Sicklerite*	$[Fe^{2+},Mn^{3+}(PO_4)]$	$Pbnm$	56c,d
Heterosite*	$[Fe^{3+}(PO_4)]$	$Pmnb$	56e,f
Purpurite	$[Mn^{3+}(PO_4)]$	$Pmnb$	56e,f

Senegalite	$[Al_2(PO_4)(OH)_3(H_2O)]$	$P2_1nb$	57a,b,c
Sarcopside	$[Fe^{2+}_3(PO_4)_2]$	$P2_1/a$	58a,b
Bjarebyite*	$[Al_2(PO_4)_3(OH)_3]$	$P2_1/m$	58c,d
Kulanite	$[Al_2(PO_4)_3(OH)_3]$	$P2_1/m$	58c,d
Penikisite	$[Al_2(PO_4)_3(OH)_3]$	$P2_1/m$	58c,d
Perloffite	$[Fe^{3+}_2(PO_4)_3(OH)_3]$	$P2_1/m$	58c,d

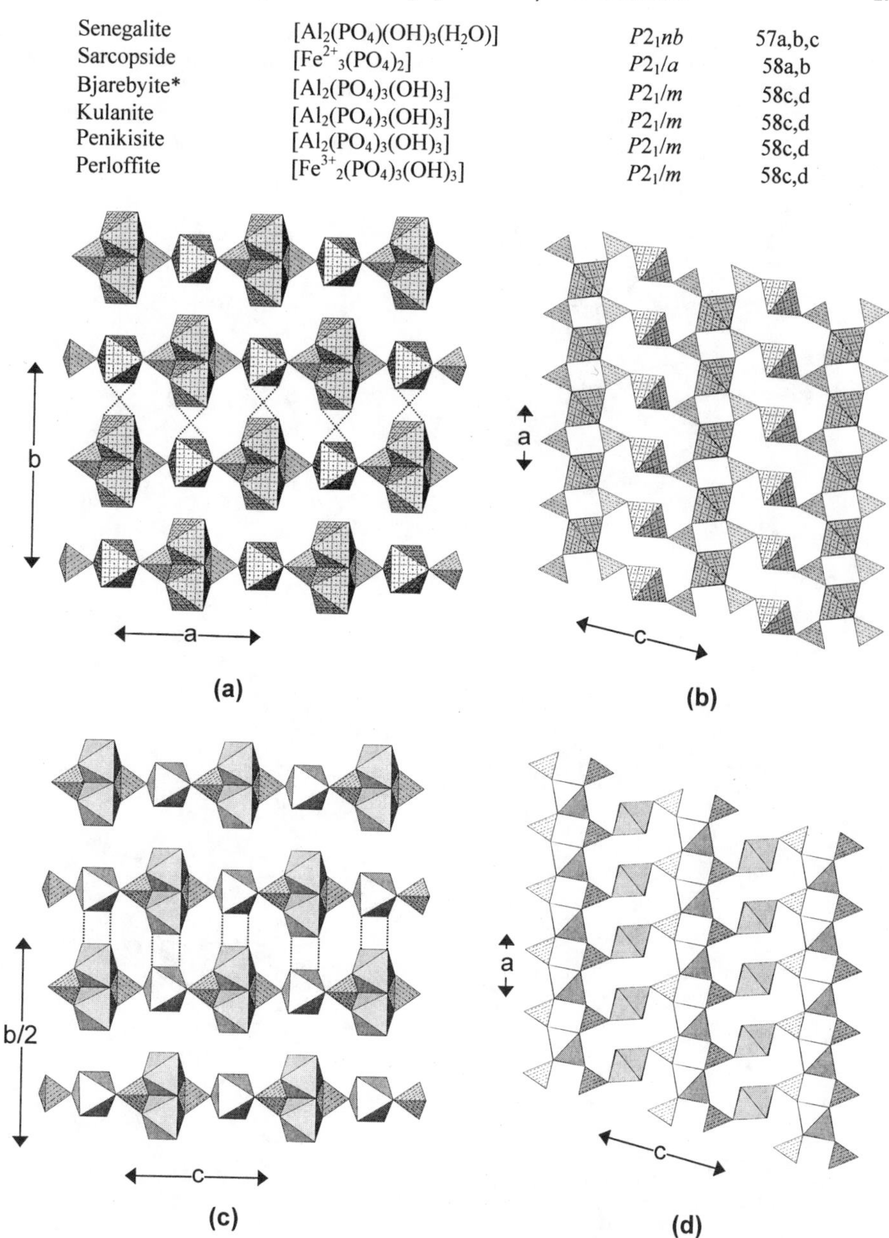

Figure 34. The crystal structures of vivianite and bobierrite: (a) vivianite projected onto (001); (b) vivianite projected onto (010); (c) bobierrite projected onto (100); (d) bobierrite projected onto (010). $(Fe^{2+}\phi_6)$ octahedra are square-pattern-shaded, $(Mg\phi_6)$ octahedra are shadow-shaded, donor-acceptor pairs for hydrogen bonds are shown by dotted lines.

M-T linkage. The minerals of the **metavariscite**, $[Al(PO_4)(H_2O)_2]$, and **variscite**, $[Al(PO_4)(H_2O)_2]$, groups consist of simple frameworks of alternating (PO_4) tetrahedra and $(Al\phi_6)$ octahedra. As there are equal numbers of tetrahedra and octahedra, both polyhedra are four-connected, and hence two vertices of the (Alϕ_6) octahedron must be one-connected. The local bond-valence requirements of the anions at these one-connected vertices require that the anions be (H_2O) groups. When viewed down the *c*-direction, octahedra and tetrahedra occupy the vertices of a 6^3 net, and Figures 35a,c show two layers of such nets. When metavariscite is viewed in the *a*-direction (Fig. 35b), the

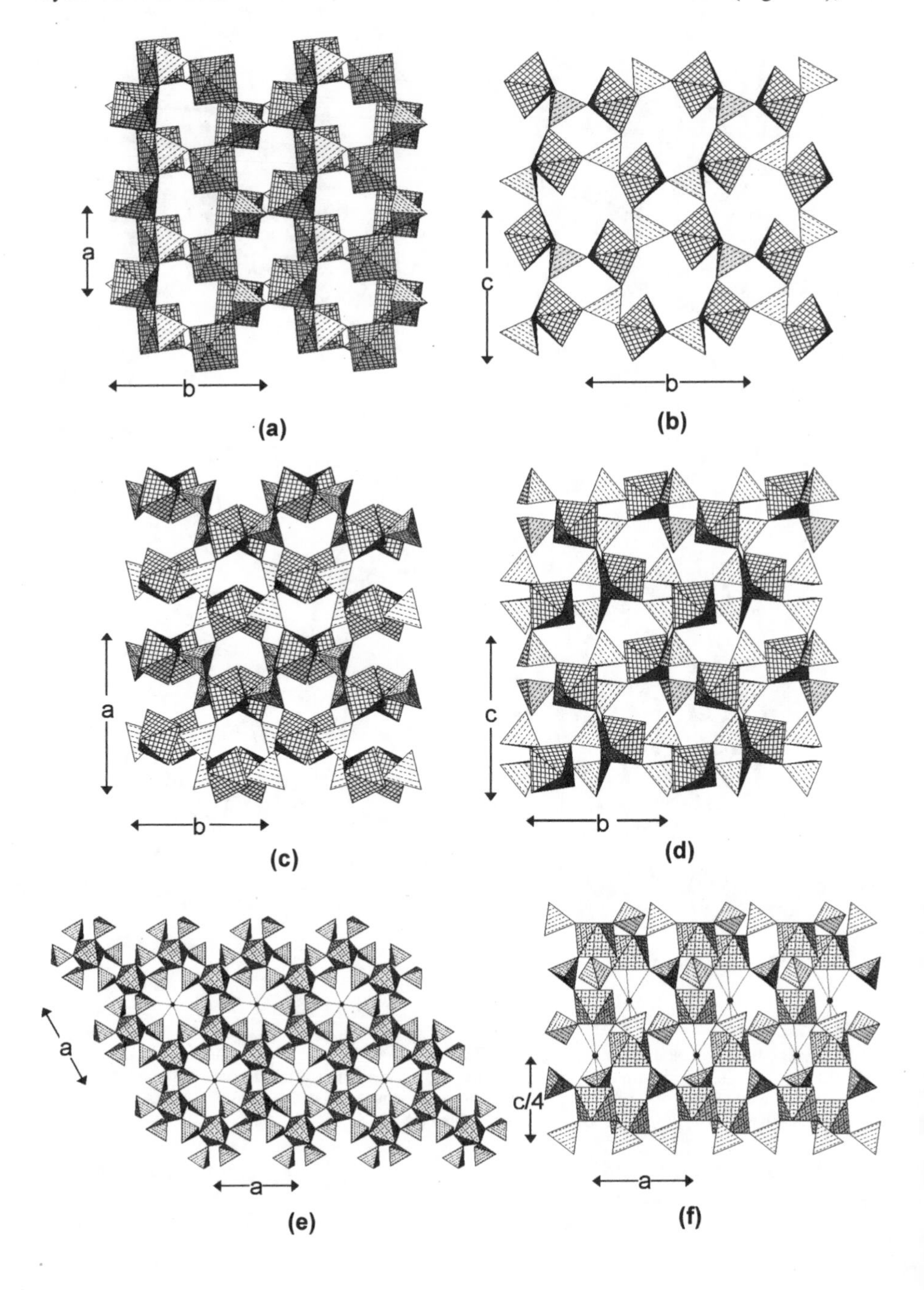

tetrahedra and octahedra occupy the vertices of a 4.8^2 plane net. Note how the one-connected vertices of the octahedra project into the large eight-membered ring, giving room for the H atoms of the (H_2O) groups at these vertices. When viewed down [100] (Fig. 35d), variscite shows alternating tetrahedra and octahedra occupying the vertices of a very puckered 8^3 net. As with metavariscite, the one-connected vertices of the octahedra project into the large cavities.

Kosnarite, X[$Zr_2(PO_4)_3$], contains octahedrally coordinated Zr. In projection down [001] (Fig. 35e), (ZrO_6) octahedra occupy the vertices of a 6^3 net, and all octahedron vertices link to (PO_4) tetrahedra, forming a slab with prominent interstices. These slabs stack in the *c*-direction and link by sharing of octahedron-tetrahedron vertices (Fig. 35f), with [6]-coordinated X cations in the interstices of the framework.

M-M, M-T linkage. The minerals of the **amblygonite**, Li[Al(PO_4)F], group and the phosphate members of the titanite group, such as **lacroixite**, Na[Al(PO_4)F], have topologically identical structural units. However, lacroixite is monoclinic, whereas the amblygonite-group minerals are triclinic; because of their topological identity, we use the unconventional space group $C\bar{1}$ to emphasize the congruity of these two structures (Table 8). A key feature of both structures is the 7-Å [$M\phi_5$] chain of corner-sharing octahedra that extends in the *c*-direction (Figs. 36a,c). This chain is decorated by staggered flanking (PO_4) groups that link the chains in both the *a*- and *b*-directions, a feature that is very apparent in an end-on view of the chains (Figs. 36b,d). The frameworks are strengthened by interstitial alkali cations Na and Li in the minerals of the amblygonite group and both Ca and Na in the minerals of the titanite group.

Cyrilovite, Na[$Fe^{3+}_3(PO_4)_2(OH)_4(H_2O)_2$] is a member of the wardite group (Table 8). The principal motif in cyrilovite is the [$Fe^{3+}\phi_5$] chain that is decorated by (PO_4) tetrahedra arranged in a staggered fashion at the periphery of the chain (the [$M(TO_4)\phi_3$] chain shown in Fig. 18c). These chains extend parallel to the *a*- and *b*-directions (note the tetragonal symmetry) to form a slab of corner-sharing octahedra and tetrahedra (Fig. 37a), tetrahedra on opposite sides of each chain pointing in opposing directions along *c*. The tetrahedral vertices that project out of the plane of the slab link to octahedra of adjacent slabs (Fig. 37a) to form a framework that consists of successive layers of octahedra and tetrahedra along the *c*-direction. [8]-coordinated Na occupies the large interstices in this framework (Fig. 37b), and hydrogen bonds strengthen the framework.

Fluellite, [$Al_2(PO_4)F_2(OH)(H_2O)_3$]($H_2O$)$_4$, is an open framework of corner-sharing (PO_4) tetrahedra and ($Al\phi_6$) octahedra. The principal motif of the framework is a 7-Å chain of the form [$M(TO_4)\phi_3$] (Fig. 18c) consisting of ($AlF_2\{OH\}(H_2O)_3$) octahedra linked through pairs of *trans* vertices (= F) and decorated by (PO_4) tetrahedra that link adjacent octahedra along the chain. These chains extend in both the *a*- and *b*-directions (Fig. 37c) by sharing (PO_4) groups between chains extending in orthogonal directions (Fig. 37d). There are large interstices within the framework that accommodate (H_2O) groups held in the structure solely by hydrogen bonds emanating from the (H_2O) groups bonded directly to the Al of the structural unit.

Wavellite, [$Al_3(PO_4)_2(OH)_3(H_2O)_4$]($H_2O$), is an open framework of corner-sharing octahedra and tetrahedra (Fig. 37e) with interstitial non-transformer (H_2O) groups held in the interstices by hydrogen bonds. ($Al\phi_6$) octahedra share one set of *trans* corners with each other to form [$M\phi_5$] chains that are decorated by (PO_4) tetrahedra bridging adjacent

Figure 35 (opposite page). The crystal structures of metavariscite, variscite and kosnarite: (a) metavariscite projected onto (001); (b) metavariscite projected onto (100); (c) variscite projected onto (001); (d) variscite projected onto (100); ($Al\phi_6$) octahedra are 4^4-net-shaded; (e) kosnarite projected onto (001); (f) kosnarite projected onto (010); ($Zr\phi_6$) octahedra are square-pattern-shaded.

octahedra (Fig. 37e) to give chains of the form [$M(TO_4)\phi$] extending in the *c*-direction (Fig. 18c). These chains cross-link in the *a*-direction by sharing octahedron-tetrahedron corners (Fig. 37f) with undecorated [$Al\phi_5$] chains (i.e., the tetrahedra linked to these chains do not bridge octahedra within the chain). The resulting framework (Figs. 37e,f) has large cavities that contain the interstitial (H_2O) groups held in the structure solely by hydrogen bonds.

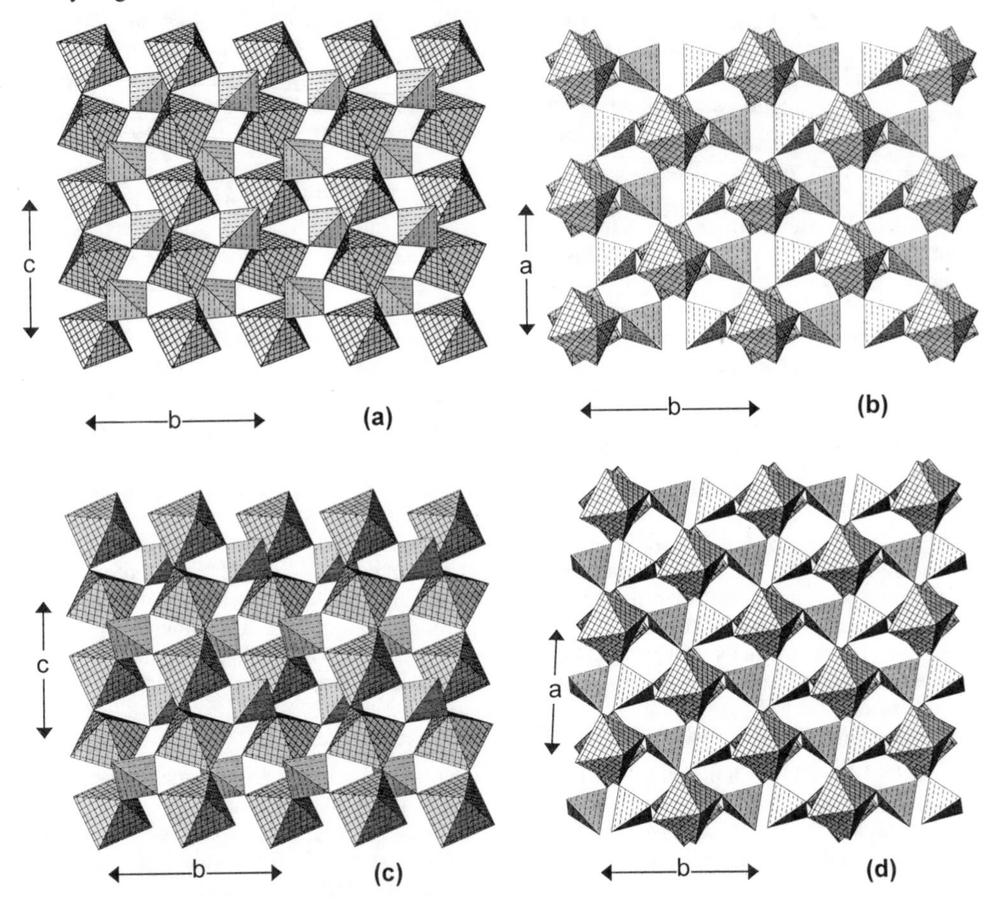

Figure 36. The crystal structures of lacroixite and amblygonite: (a) lacroixite projected onto (100); (b) lacroixite projected onto (001); (c) amblygonite projected onto (100); note the similarity with (a); (d) amblygonite projected onto (001); note the similarity with (b). ($Al\phi_6$) octahedra are 4^4-net-shaded.

M=M, M-T linkage. **Augelite**, [$Al_2(PO_4)(OH)_3$], contains Al in both octahedral and trigonal-bipyramidal coordinations. Pairs of ($Al\phi_6$) octahedra share an edge to form [$Al_2\phi_{10}$] dimers that are oriented with their long axis in the *b*-direction. The dimers are arranged at the vertices of a centered orthorhombic plane net (Fig. 38a), and dimers adjacent in the *b*-direction are linked through pairs of (PO_4) tetrahedra to form [$Al_2(PO_4)_2\phi_6$] chains. The dimers are decorated by ($Al\phi_5$) trigonal bipyramids that bridge pairs of vertices from each octahedron. These ($Al\phi_5$) groups link to (PO_4) groups of adjacent chains to link them in the *a*-direction. Viewed in the *b*-direction (Fig. 38b), the structure appears as layers of dimers linked by chains of (PO_4) and ($Al\phi_5$) groups.

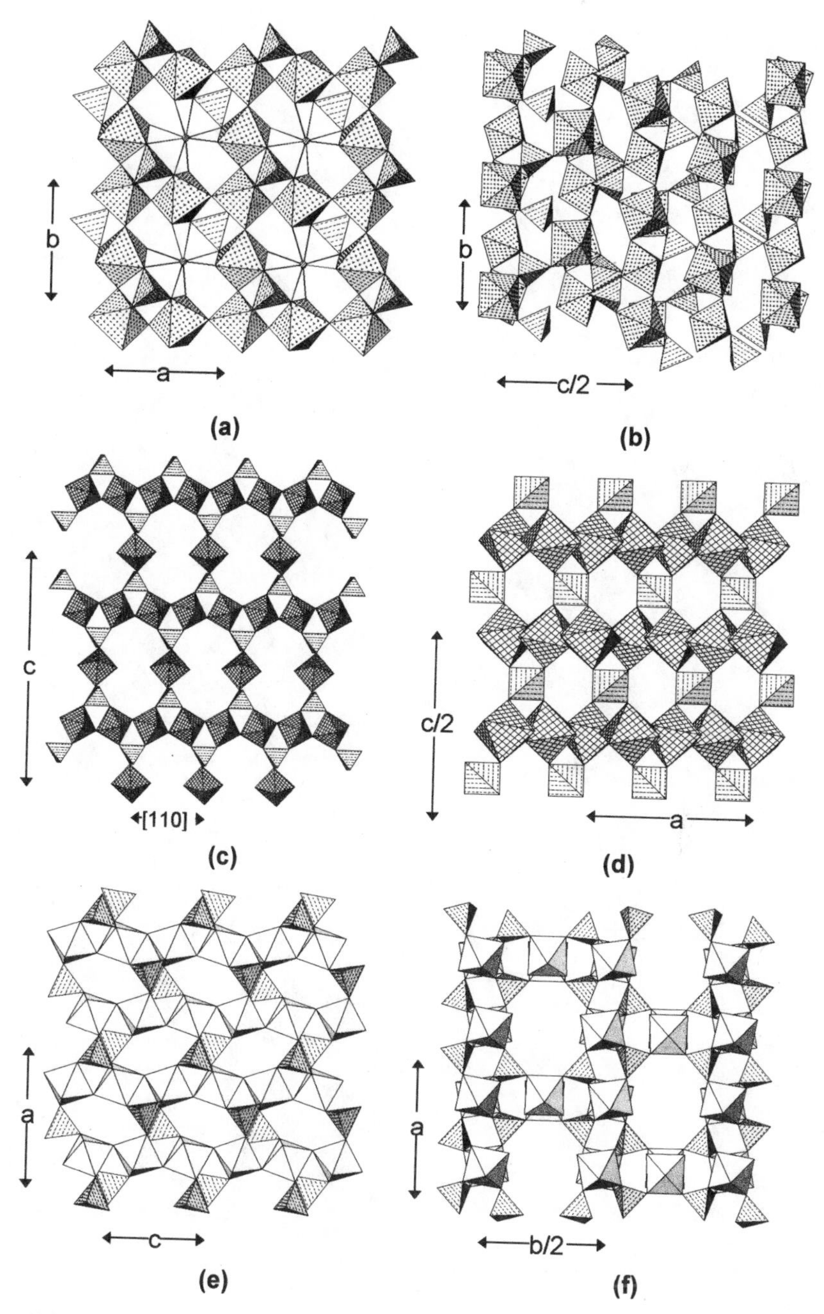

Figure 37. The crystal structures of cyrilovite, fluellite and wavellite: (a) cyrilovite projected a few degrees away from onto (001); (b) cyrilovite projected onto (100); (c) fluellite projected down [110]; (d) fluellite projected onto (010); (e) wavellite projected onto (010); (f) wavellite projected onto (001). $(Al\phi_6)$ octahedra are 4^4-net-shaded and shadow-shaded, $(Fe^{3+}\phi_6)$ octahedra are dot-shaded.

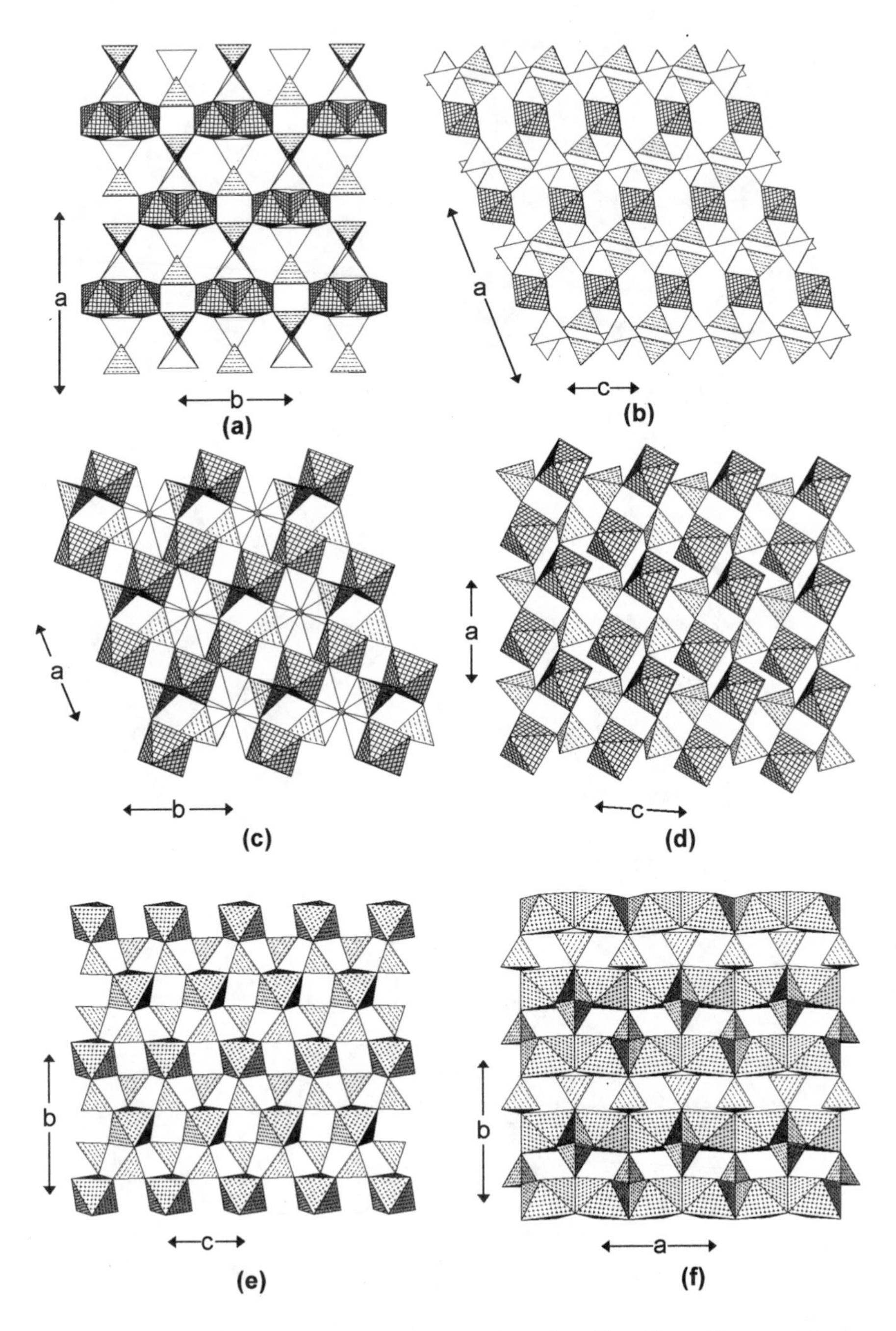

Figure 38. The crystal structures of augellite, jagowerite and mariċite: (a) augelite projected onto (001); (b) augelite projected onto (010); (c) jagowerite projected onto (001); (d) jagowerite projected a few degrees away from onto (010); (e) mariċite projected onto (100); (e) maricite projected a few degrees away from onto (001). Legend as in Figure 37.

Maričite, $Na[Fe^{2+}(PO_4)]$, is a dense-packed framework of $(Fe^{2+}O_6)$ octahedra and (PO_4) tetrahedra. Each $(Fe^{2+}O_6)$ octahedron links to six (PO_4) groups to form what Moore (1973b) calls a 'pinwheel'. The octahedra occupy the vertices of a 3^6 net, and the resulting sheet (Fig. 38e) is topologically identical to the $[Mg(PO_4)_2]$ sheet in brianite (Fig. 23c). These sheets stack along the *a*-direction with octahedra from adjacent sheets sharing edges to form $[MO_4]$-type chains extending in the *a*-direction when viewed down [001] (Fig. 38f). [10]-coordinated Na occupies interstices in the framework.

Kovdorskite, $[Mg_2(PO_4)(OH)(H_2O)_3]$, consists of two distinct $(Mg\phi_6)$ octahedra that condense to form tetramers via edge-sharing. These tetramers are decorated by pairs of (PO_4) tetrahedra to form $[Mg_4(PO_4)_2\phi_8]$ clusters. The clusters occur at the vertices of a 4^4 plane net and link together by sharing octahedron-tetrahedron vertices to form open sheets parallel to (001) (Fig. 39a). These sheets stack in the *c*-direction by sharing octahedron-tetrahedron vertices (Fig. 39b) to form a very open framework that is strengthened by extensive hydrogen bonding involving the (OH) and (H_2O) groups of the structural unit.

Libethenite, $[Cu^{2+}_2(PO_4)(OH)]$, is a member of the **adamite** group (Table 8) (Hawthorne 1976) in which Cu^{2+} is both [5]- and [6]-coordinated, triangular bipyramidal and octahedral, respectively. Chains of *trans* edge-sharing $(Cu^{2+}\phi_6)$ octahedra extend in the *c*-direction and are decorated by (PO_4) tetrahedra to give chains of the general form $[M_2(TO_4)_2\phi_4]$ (Fig. 39c). These chains link in the *a*- and *b*-directions by sharing octahedron-tetrahedron corners (Fig. 39d) to form an open framework with channels extending in the *c*-direction. These channels are clogged with dimers of edge-sharing $(Cu\phi_5)$ triangular bipyramids. Note that this is also the structure of andalusite, $[Al_2(SiO_4)O]$.

Tarbuttite, $[Zn_2(PO_4)(OH)]$, is a member of the paradamite group (Table 8) (Hawthorne 1979b). $(Zn\phi_5)$ bipyramids share edges to form a chain extending in the *b*-direction: $[Zn\phi_4]$. (PO_4) groups and $(Zn\phi_3)$ bipyramids alternate along a chain of corner-sharing polyhedra that also extends in the *b*-direction. These chains link in the *a*-direction by sharing polyhedron vertices (Fig. 39e) to form a rather thick slab parallel to (001). These slabs stack in the *c*-direction (Fig. 39f) by sharing polyhedron edges and corners. Apart from the presence of both [5]-coordinated divalent cations, there is no structural relation with the stoichiometrically similar libethnite, $Cu^{2+}_2(PO_4)(OH)$.

Petersite-(Y), $Y[Cu^{2+}_6(PO_4)_3(OH)_6](H_2O)_3$, is the only phosphate member of the mixite group (Table 8). Cu^{2+} is [5]-coordinated with a long sixth distance to (H_2O). Six-member rings of corner-sharing alternating (PO_4) tetrahedra and $(Cu^{2+}\phi_5)$ square-pyramids occur parallel to (001) and link by corner-sharing to form four-membered and twelve-membered rings of polyhedra (Fig. 40a). An alternative description is as six-member rings occupying the vertices of a 6^3 net. The layers of Figure 40a stack along the *c*-direction (Fig. 40b), and link by edge-sharing between the $(Cu^{2+}\phi_5)$ square pyramids. In the cross-linkage of the rings in the (001) plane, note how the (PO_4) groups bridge apical vertices of square pyramids adjacent along the *c*-direction (Fig. 40b). The interstitial (H_2O) groups occupy the channels of the twelve-membered rings, and interstitial Y occupies the channels generated by the six-membered rings (Fig. 40a).

Brazilianite, $Na[Al_3(PO_4)_2(OH)_4]$, contains chains of edge-sharing $(Al\phi_6)$ octahedra that extend in the [101] direction (Fig. 40c). These chains are fairly contorted as the shared edges are not in a *trans* configuration and hence a slight helical character results. The chains are decorated by (PO_4) tetrahedra which link next-nearest-neighbor octahedra, a rather unusual linkage that is promoted by the helical nature of the chains (Fig. 40c). Adjacent chains link by sharing octahedron vertices with the decorating tetrahedra (Fig. 40c,d). Interchain linkage is also promoted by [7]-coordinated interstitial Na.

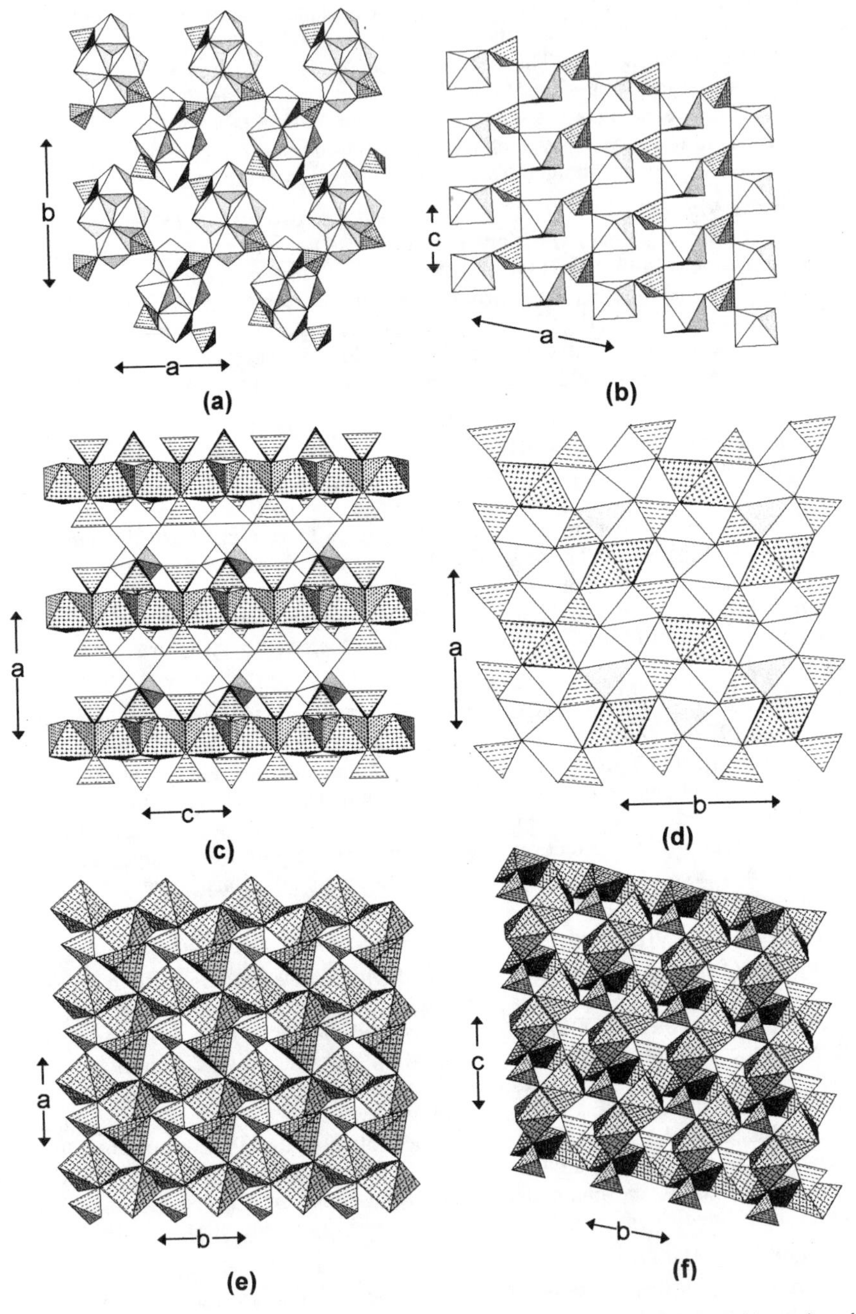

Figure 39. The crystal structures of kovdorskite, libethenite and tarbuttite: (a) kovdorskite projected onto (001); (b) kovdorskite projected onto (010); ($Mg\phi_6$) octahedra are shadow-shaded; (c) libethenite projected onto (010); (d) libethenite projected onto (001); ($Cu^{2+}\phi_6$) are dot-shaded, ($Cu^{2+}\phi_5$) are shadow-shaded; (e) tarbuttite projected onto (001); (f) tarbuttite projected onto (100); ($Zn\phi_6$) octahedra and ($Zn\phi_5$) polyhedra are square-pattern-shaded.

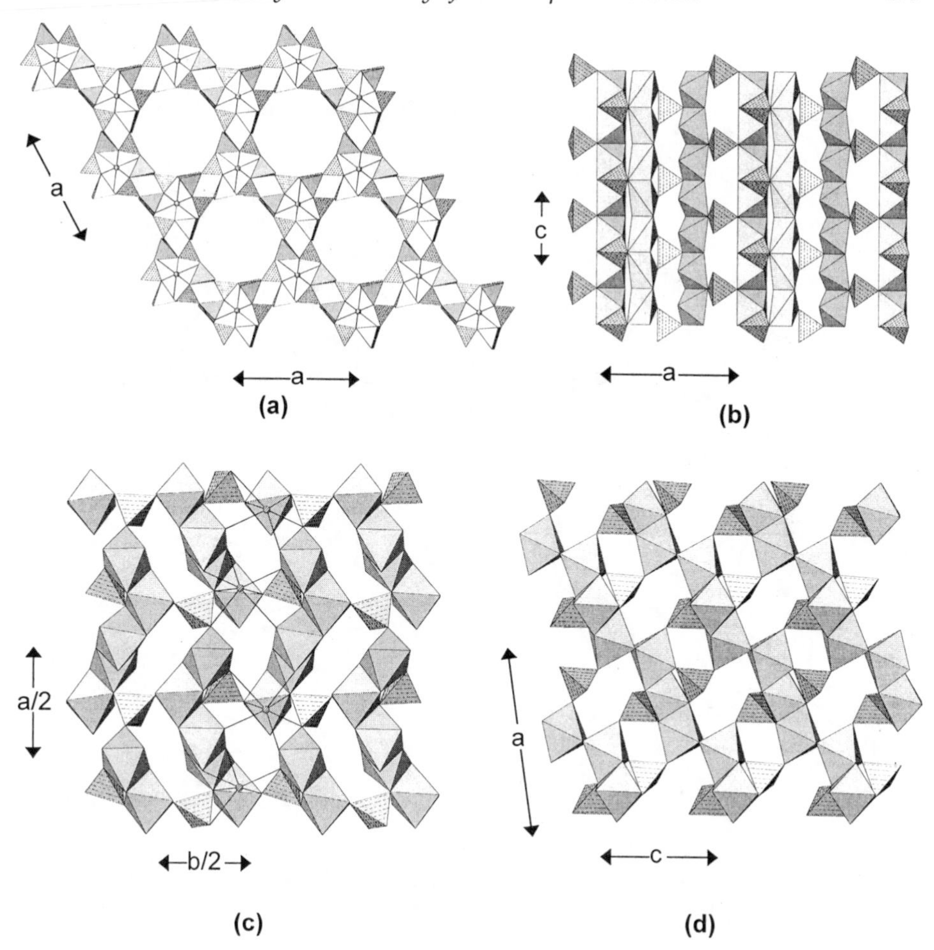

Figure 40. The crystal structures of petersite-(Y) and brazilianite: (a) petersite-(Y) projected onto (001); (b) petersite-(Y) projected onto (010); $(Cu^{2+}\phi_6)$ octahedra are shadow-shaded, Y is shown as shaded circles; (c) brazilianite projected onto (001); (d) brazilianite projected onto (010); $(Al\phi_6)$ octahedra are shadow-shaded, Na is shown as shaded circles.

M=M, M-M, M-T linkage. There are three polymorphs of $[Cu^{2+}_5(PO_4)_2(OH)_4(H_2O)]$, pseudomalachite, reichenbachite and ludjibaite, and their structures are all based on sheets of octahedra that are linked by (PO_4) tetrahedra. The sheets of octahedra are somewhat unusual in that they are not close-packed octahedra interspersed with vacancies (as is common in this type of structure). In pseudomalachite (Fig. 41a), linear $[M\phi_4]$ chains of octahedra extend in the *b*-direction at $z \approx {}^1/_4$ and ${}^3/_4$, and are linked by trimers of edge-sharing octahedra packed such that there are square interstices in the sheet. The sheets stack in the *a*-direction and are linked by (PO_4) groups that share two vertices with each sheet (Fig. 41b). In reichenbachite (Fig. 41c), the arrangement of octahedra within the sheet is fairly irregular. It can be envisioned as edge-sharing trimers of octahedra at (0 1/2 *z*) and (1/2 0 *z*) linked by edge-sharing with dimers of edge-sharing octahedra at (0 1/8 *z*) and at (5/8 3/8 *z*) (Fig. 41c). These sheets stack in the *c*-direction (Fig. 41d) and are linked by (PO_4) groups that each share two vertices with adjacent sheets. In ludjibaite (Fig. 41e), linear $[M\phi_4]$ chains extend in both the *b* and *c*-directions,

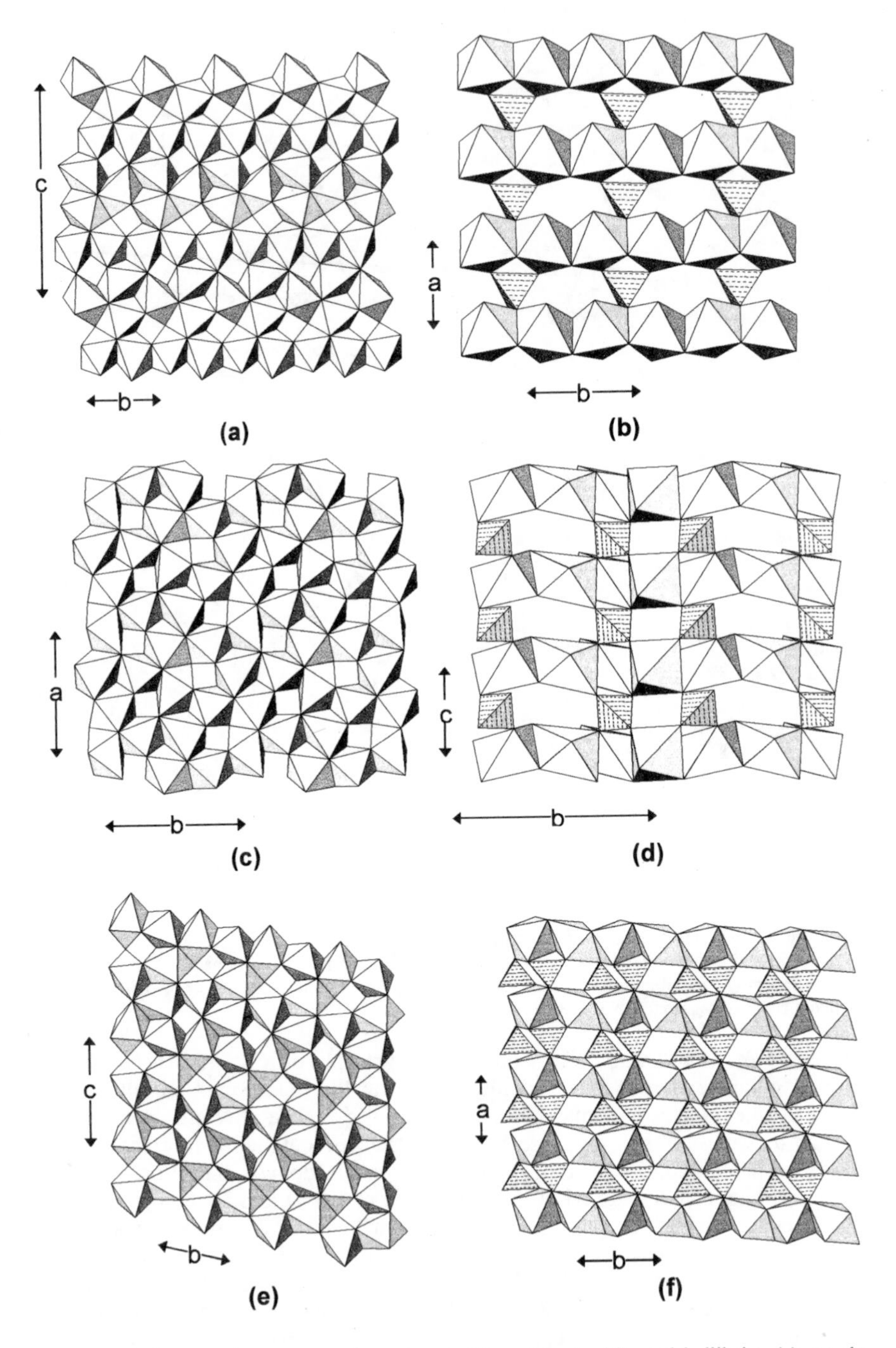

Figure 41. The crystal structures of pseudomalachite, reichenbachite and ludjibaite: (a) pseudomalachite projected onto (100); (b) pseudomalachite projected onto (001); (c) reichenbachite projected onto (001); (d) reichenbachite projected onto (100); (e) ludjibaite projected onto (100); (f) ludjibaite projected onto (001). ($Cu^{2+}\phi_6$) octahedra are shadow-shaded.

and link together by sharing edges with a [$M\phi_5$] chain of corner-sharing octahedra that extends in the *c*-direction. These sheets stack in the *a*-direction (Fig. 41f) and, as with the other two structures, are linked by (PO_4) groups that share pairs of vertices with adjacent sheets.

In the minerals of the **triplite**, [$Mn^{2+}{}_2(PO_4)F$], group (Table 8), ($M^{2+}\phi_6$) octahedra share edges to form [$M^{2+}{}_2\phi_{10}$] dimers (Fig. 42a) that share corners with (PO_4) tetrahedra to form slightly corrugated layers that are parallel to (010) (Fig. 42a). These layers link in the *b*-direction by sharing corners between tetrahedra and octahedra (Fig. 42b).

The minerals of the **triploidite**, [$Mn^{2+}{}_2(PO_4)(OH)$], group have divalent cations in both octahedral and triangular bipyramidal coordinations. Pairs of octahedra share an edge to form [$M^{2+}{}_2\phi_{10}$] dimers, and these dimers are linked by sharing corners with both

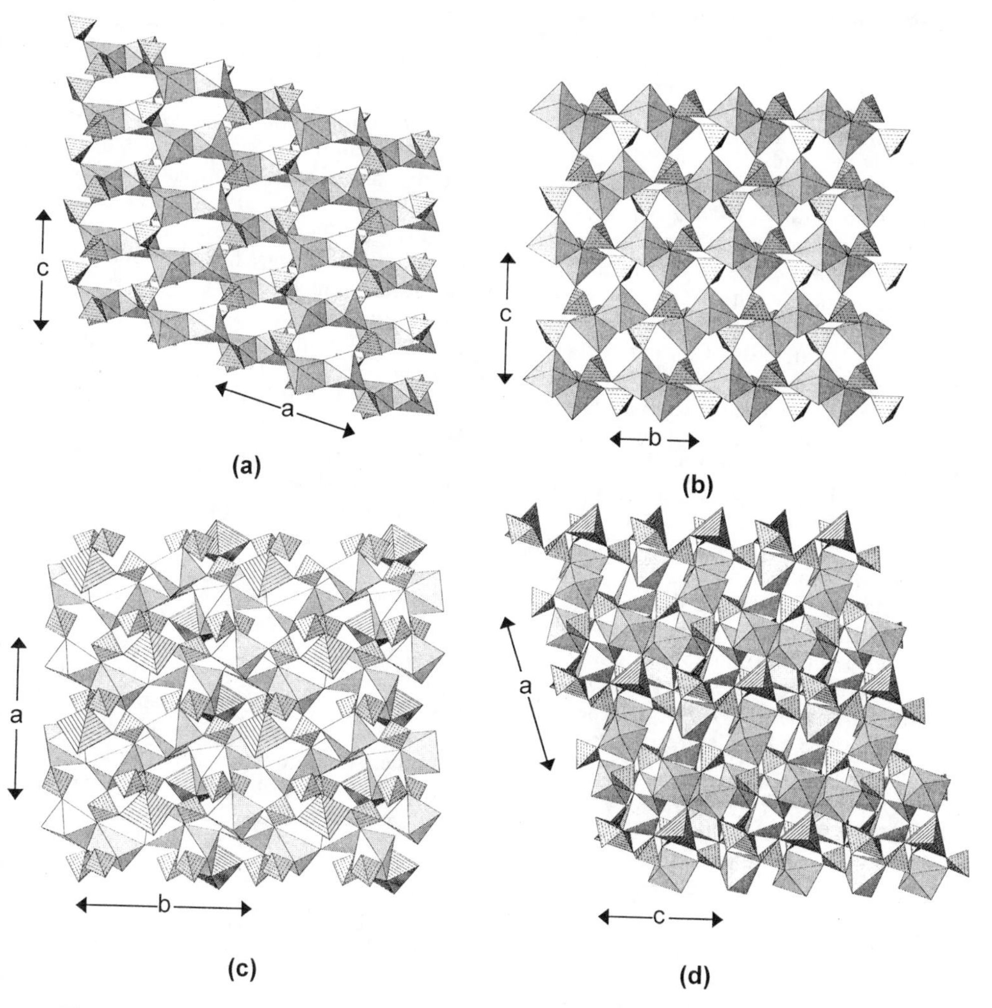

Figure 42. The crystal structures of triplite and triploidite: (a) triplite projected onto (010); (b) triplite projected onto (100); (c) triploidite projected onto (001); (d) triploidite projected onto (001). ($Mn\phi_6$) octahedra are shadow-shaded, ($Mn\phi_5$) polyhedra are line-shaded.

triangular bipyramids and (PO_4) groups, and by sharing one octahedron edge with a tetrahedron (Fig. 42c). Triangular bipyramids also form edge-sharing dimers, $[M^{2+}{}_2\phi_8]$, and chains of corner-sharing octahedra, triangular bipyramids and tetrahedra extend in the *b*-direction (Fig. 42d). It should be noted that this is an extremely complicated structure, and is not easily related to any other structure, except at a trivial level.

Table 9. Cell dimensions for the different structures of the alluaudite-group (*sensu lato*) minerals.

	a (Å)	*b*(Å)	*c* (Å)	β(°)	Sp grp	*V* (Å³)	Ref.
Alluaudite	12.004(2)	12.533(45)	6.404(1)	114.4(1)	*C*2/*c*	877.4	(1)
Wyllieite	11.868(15)	12.382(12)	6.354(9)	114.5(1)	$P2_1/n$	849.5	(2)
Bobfergusonite	12.776(2)	12.488(2)	11.035(2)	97.21(1)	$P2_1/n$	1746.7(4)	(3)

References: (1) Fisher (1965); (2) Moore and Molin-Case (1974); (3) Ercit et al. (1986a,b)

The minerals of the **alluaudite**, **wyllieite** and **bobfergusonite** groups are topologically identical but are distinguished by different cation-ordering schemes over the octahedrally coordinated cation sites in the basic structure. The cell dimensions of the principal mineral in each group are given in Table 9. The principal feature of each structure is a linear trimer of edge-sharing octahedra (Figs. 43a,c,e). These trimers link together by sharing edges to form chains of octahedra in (010) that are linked by sharing octahedron corners with (PO_4) groups to form thick sheets parallel to (010). These sheets link in the *b*-direction by sharing corners between tetrahedra and octahedra (Figs. 43b,d,f). The minerals of these three groups differ primarily in their Al content and the pattern of cation order over the trimer of edge-sharing octahedra. The minerals of the alluaudite group (sensu stricto) contain negligible Al (Al_2O_3 < 0.10 wt %), the minerals of the wyllieite group contain moderate Al (Al_2O_3 ≈ 3 wt %), and bobfergusonite contains far more Al than wyllieite (Al_2O_3 ≈ 7.5 wt %). In addition, there is a fourth (as yet undescribed) structure type with ~15 wt % Al_2O_3 (unpublished data). The differences in cation order in these three structure types are summarized in Figure 44. In alluaudite, there is no Al, and hence Al is not involved in the ordering scheme. There are only two distinct sites in the trimer in alluaudite, and the pattern of cation order can vary from complete M^{2+}-cation disorder to complete Fe^{3+}-M^{2+} order (Fig. 44a). In wyllieite, there are three distinct sites in the trimer; Al is completely ordered at one site, and the other two sites can vary from complete M^{2+}-cation disorder to complete Fe^{3+}-M^{2+} order (Fig. 44b). In bobfergusonite, there are two crystallographically distinct trimers (Fig. 44c); Al is ordered in one trimer, Fe^{3+} is ordered in the other trimer, and M^{2+} is disordered over the other sites. This picture is somewhat idealized, and each structure-type may show minor ordering characteristic of one or more of the other structure types. Moore and Ito (1979) discuss the nomenclature of the alluaudite and wyllieite groups in detail and propose a nomenclature based on suffixes, but this has not been used very extensively.

Figure 43 (opposite page). The crystal structures of alluaudite, wyllieite and bobfergusonite: (a) alluaudite projected onto (001); (b) alluaudite projected onto (100); shadow-shaded octahedra are occupied predominantly by Mn^{2+}, 4^4-net-shaded octahedra are occupied predominantly by Fe^{3+}; (c) wyllieite projected onto (010); (d) wyllieite projected onto (100); dot-shaded octahedra are occupied predominantly by Fe, 4^4-net-shaded octahedra are occupied by Al; (e) bobfergusonite projected onto (010); (f) bobfergusonite projected onto (100); shadow-shaded octahedra are occupied by Mn^{2+}, Fe^{3+} and Al.

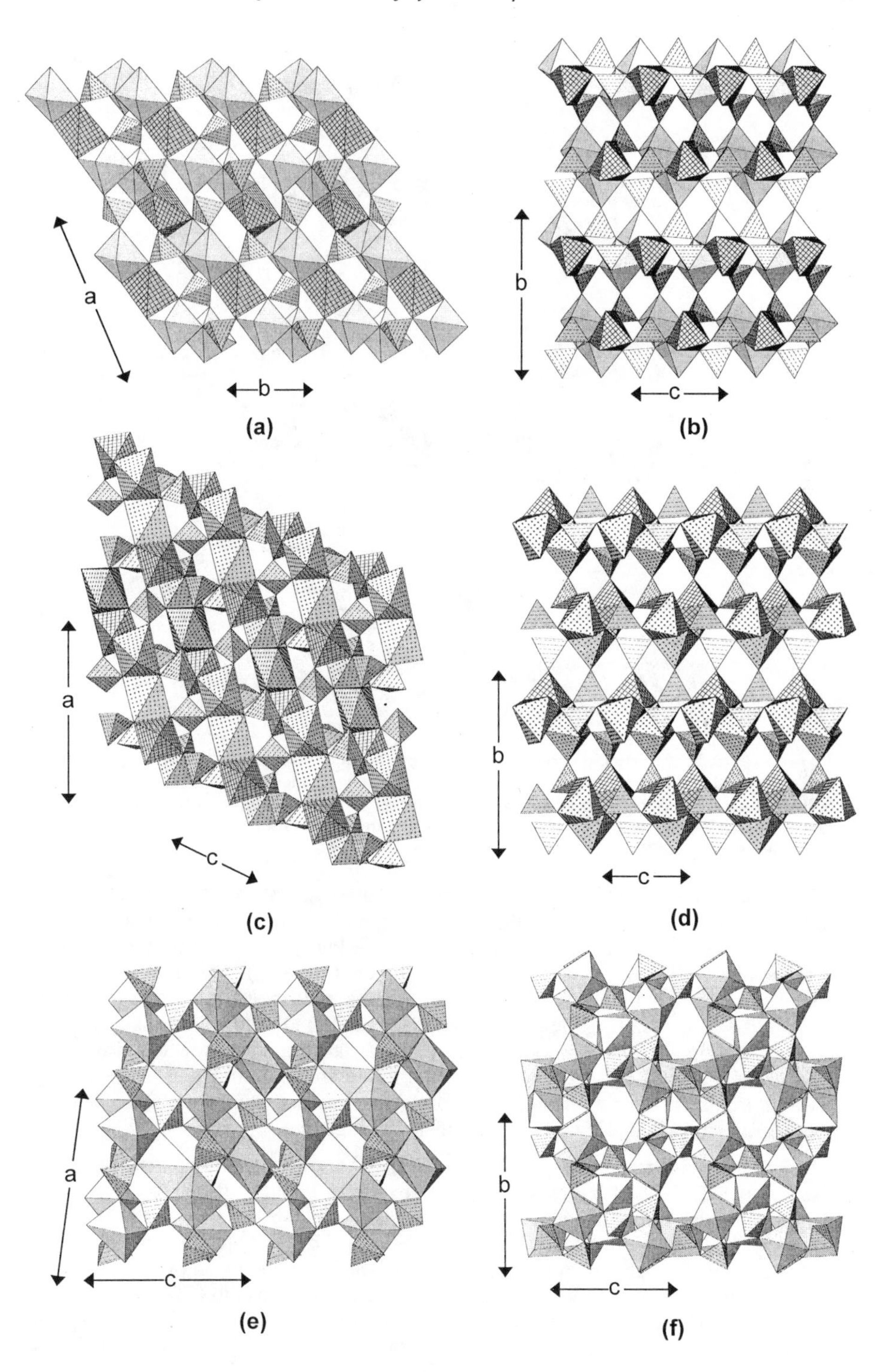
a
b
(a)
b
c
(b)
a
c
(c)
b
c
(d)
a
c
(e)
b
c
(f)

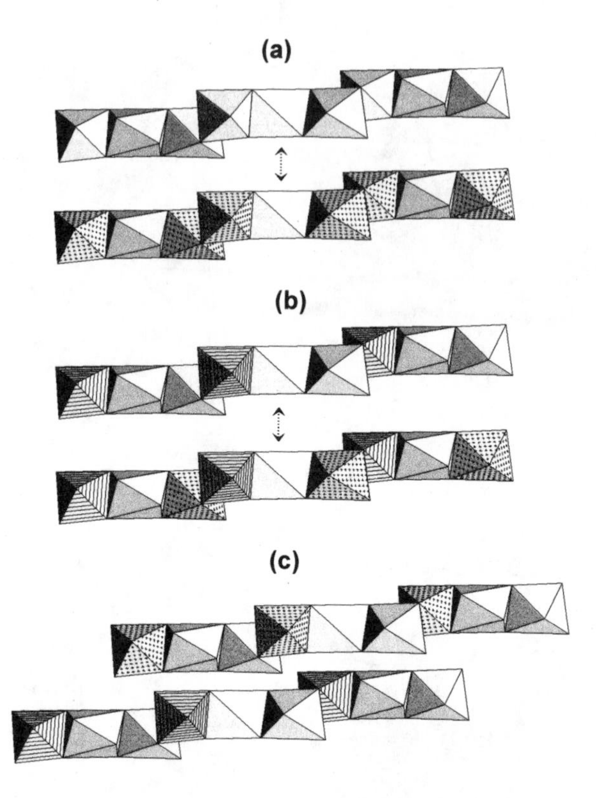

Figure 44. Octahedral-cation-ordering patterns in (a) alluaudite; (b) wyllieite; (c) bobfergusonite; shadow-shaded octahedra are occupied by any divalent M^{2+} cation, dot-shaded octahedra are occupied by Fe^{3+}, line-shaded octahedra are occupied by Al. In (a) and (b), the arrows indicate the range of possible ordering within a single chain; in (c), there are two distinct chains (shown here) in which the ordering is different (after Ercit et al. 1986a).

Ludlamite, $[Fe^{3+}_3(PO_4)_2(H_2O)_4]$ consists of $(Fe^{3+}\phi_6)$ octahedra that share edges to form $[Fe^{3+}_3\phi_{14}]$ linear trimers with (PO_4) tetrahedra bridging between adjacent octahedra in a staggered fashion on each site of the trimer: $[Fe^{3+}_3(PO_4)_2\phi_{10}]$. These trimers extend in the *c*-direction and link by sharing octahedron corners (Fig. 45a). The crankshaft chains link in the *b*-direction by sharing octahedron corners and by sharing corners between tetrahedra and octahedra (Fig. 45a) to form a sheet parallel to (100). These sheets link in the *a*-direction by sharing corners between (PO_4) tetrahedra and octahedra (Fig. 45b). Note that the chains of octahedra shown in this figure are not completely edge-sharing; for every third octahedra, the linkage is by corner-sharing, as is apparent by the change in direction of the top triangular faces of the octahedra (Fig. 45a).

In **melonjosephite**, $Ca[Fe^{2+}Fe^{3+}(PO_4)_2(OH)]$, there are two crystallographically distinct octahedra, both of which are occupied by equal amounts of Fe^{2+} and Fe^{3+}. One type of octahedron forms linear chains of edge-sharing octahedra ($[M\phi_4]$ of the rutile-type) extending in the *c*-direction. This chain is decorated by (PO_4) tetrahedra linking free vertices of adjacent octahedra in a staggered arrangement, producing an $[M(T\phi_4)\phi_2]$ chain (Fig. 18e). The other crystallographically distinct octahedron links to (PO_4) tetrahedra to form $[M(PO_4)\phi_4]$ chains. These $[M(PO_4)\phi_4]$ chains link in a pair-wise fashion by the octahedra sharing edges, and the resulting structure consists of the two types of chains,

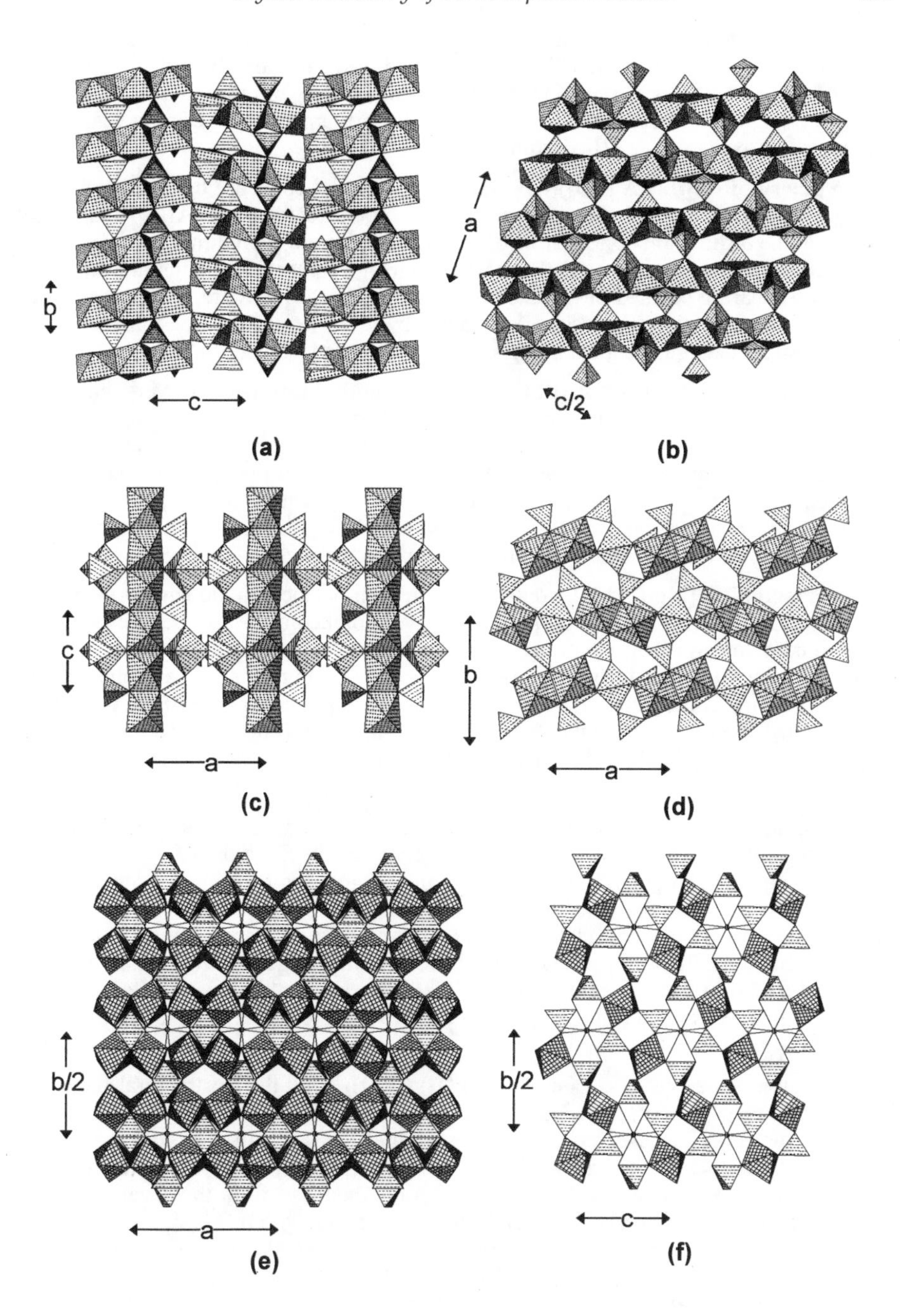

Figure 45. The crystal structures of ludlamite, melonjosephite and palermoite: (a) ludlamite projected onto (100); (b) ludlamite projected onto (010); (c) melonjosephite projected onto (010); (d) melonjosephite projected onto (001); ($\{Fe^{2+},Fe^{3+}\}\phi_6$) octahedra are dot-shaded; (e) palermoite projected onto (001); (f) palermoite projected onto (100); ($Al\phi_6$) octahedra are 4^4-net-shaded, Ca is shown as small shaded circles.

both extending in the *c*-direction and cross-linked by sharing octahedron-tetrahedron and octahedron-octahedron corners (Fig. 45c). Viewed down the length of the chains (Fig. 45d), the dimers that link the two $[M(PO_4)\phi_4]$ chains are very prominent, and the key role of the (PO_4) groups in cross-linking the chains is very apparent. The interstices of the framework are occupied by [7]-coordinated Ca.

In **palermoite**, $SrLi_2[Al(PO_4)(OH)]_4$, $(Al\phi_6)$ octahedra condense by sharing edges to form $[Al_2\phi_{10}]$ dimers, and these dimers share corners to form an $[Al_2\phi_8]$ chain that extends in the *a*-direction. One pair of octahedron vertices in each dimer is bridged by a (PO_4) tetrahedron to form an $[Al_2(PO_4)\phi_6]$ chain (Fig. 45e); these chains link in the *b*-direction by sharing octahedron-tetrahedron vertices. These chains are seen end-on when viewed in the *a*-direction (Fig. 45f), cross-linked by (PO_4) tetrahedra. The framework has large interstices that are occupied by [8]-coordinated Sr and [5]-coordinated Li.

Arrojadite, $KNa_4Ca[Fe^{2+}_{14}Al(PO_4)_{12}(OH)_2]$, and **dickinsonite**, the Mn^{2+} analogue, are infernally complex structures, each with several partly occupied cation sites, and the complete details of their structure exceeds our spatial parameters. Moore et al. (1981) describe the structure as six distinct rods (columns) of cation polyhedra decorated by (PO_4) tetrahedra and occurring at the vertices of a {6·3·6·3} and {6·6·3·3} net. Moore et al. (1981) also compare the structure of arrojadite with wyllieite, but the relation to the general alluaudite-type structures has not been explored.

Farringtonite, $[Mg_3(PO_4)_2]$, contains Mg in both octahedral and square-pyramidal coordinations. As is common with [5]-coordinated polyhedra, $[Mg\phi_5]$ square pyramids share an edge to form $[Mg_2\phi_8]$ dimers, and the terminal edges of this dimer are shared with (PO_4) tetrahedra to form a $[Mg_2(PO_4)_2\phi_4]$ cluster (Fig. 46a). These clusters are linked by sharing corners with $(Mg\phi_6)$ octahedra. When projected onto (010), prominent $[Mg(PO_4)_2\phi_2]$ chains are evident, extending in the *c*-direction (Fig. 18b). These chains are bridged in the a-direction by $[Mg_2(PO_4)_2\phi_4]$ clusters (Fig. 46b).

Beusite, $[Mn^{2+}_3(PO_4)_2]$, and **graftonite**, $[Fe^{2+}_3(PO_4)_2]$, show unusual coordination numbers for the divalent cations: ${}^{[7]}M(1)$, ${}^{[5]}M(2)$, ${}^{[6]}M(3)$. Perhaps as a result of this unusual coordination these minerals can accept considerable Ca at the *M*(1) site (Wise et al. 1990), and the latter authors report a composition for Ca-rich beusite close to $CaFe^{2+}Mn(PO_4)_2$. ${}^{[7]}M(1)$ polyhedra share an edge to form a dimer; these dimers occur at the vertices of a 4^4 net and share corners to form a sheet parallel to (100) that is strengthened by (PO_4) groups (Fig. 46c). Pyroxene-like edge-sharing chains of *M*(3) octahedra extend in the *c*-direction and are linked by chains of alternating (PO_4) groups and *M*(2) square pyramids (Fig. 46d), and these two types of sheet alternate in the [100] direction.

Wicksite, $NaCa_2[Fe^{2+}_2(Fe^{2+}Fe^{3+})Fe^{2+}_2(PO_4)_6(H_2O)_2]$, and the isostructural **bederite**, $\square Ca_2[Mn^{2+}_2Fe^{3+}_2Mn^{2+}_2(PO_4)_6(H_2O)_2]$, are complex heteropolyhedral framework structures that may be resolved into layers parallel to (001). In wicksite at $z \approx 1/4$, $(Fe^{2+}\phi_6)$ and $(Fe^{3+}\phi_6)$ octahedra share an edge to form $[M_2\phi_{10}]$ dimers that are canted to both the *a* and *b* axes, and are linked by (PO_4) tetrahedra to form the sheet shown in Figure 47a. At $z \approx 0$, two $(Fe^{2+}\phi_6)$ octahedra share edges with an $(Na\phi_6)$ octahedra to form an $[M_3\phi_{14}]$ trimer that is decorated by (PO_4) tetrahedra linking adjacent free octahedron vertices to form a cluster of the form $[M_3(PO_4)_2\phi_6]$. These clusters link by sharing of octahedron-tetrahedron vertices to form the layer shown in Figure 47b. There are two types of interstice within this layer. In the first type of interstice is the Ca site coordinated by nine anions, and in the second type of interstice are four H atoms that belong to the two peripheral (H_2O) groups (Fig. 47b). The layers of Figure 47a and 47b link by edge-sharing between the $(Fe^{2+}\phi_6)$ [= *M*(1)] octahedron of one sheet with the $(Fe^{2+}\phi_6)$ [= *M*(3) octahedron of the other sheet (Fig. 47c). The relation between wicksite

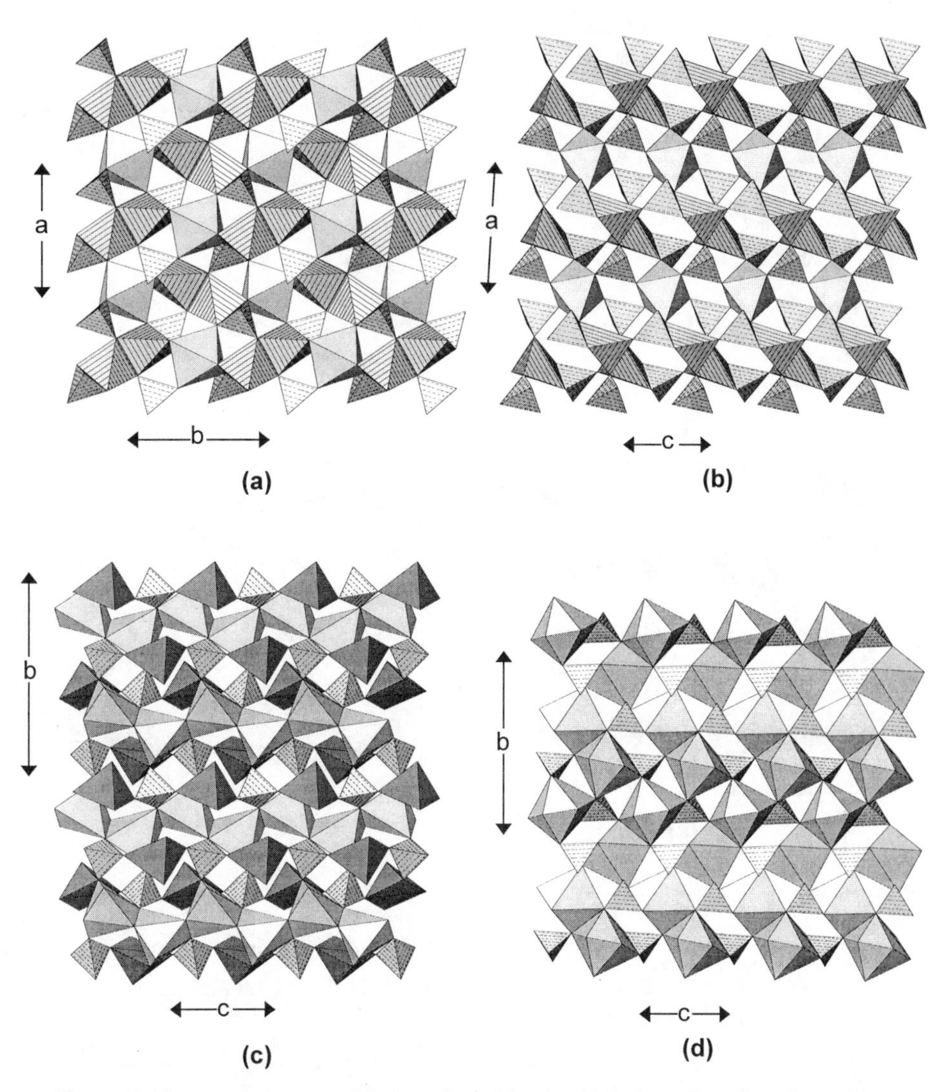

Figure 46. The crystal structures of farringtonite and beusite: (a) farringtonite projected onto (001); (b) farringtonite projected onto (010); (MgO_6) octahedra are shadow-shaded, (MgO_5) polyhedra are line-shaded; (c) a layer of the beusite structure projected onto (100); (d) another layer of the beusite structure projected onto (100); [7]- and [6]-coordinated polyhedra are shadow-shaded, [5]-coordinated polyhedra are dark-shadow-shaded.

and bederite is as follows: the Fe^{2+} = Na = Fe^{2+} triplet in wicksite (cf. Fig. 47b) is replaced by the [Mn^{2+} = = Mn^{2+}] triplet in bederite.

Chalcosiderite, $[Cu^{2+}Fe^{3+}_6(PO_4)_4(OH)_8(H_2O)_4]$, is a member of the turquoise group (Table 8). The structure contains trimers of edge-sharing octahedra, two $(Fe^{3+}\phi_6)$ and one $(Cu^{2+}\phi_6)$ octahedra that link by sharing corners with (PO_4) tetrahedra and other $(Fe^{3+}\phi_6)$ octahedra parallel to (100) (Fig. 47d). This linkage is also seen in Figure 47e, with additional linkage between trimers through corner-sharing with additional $(Fe^{3+}\phi_6)$ octahedra

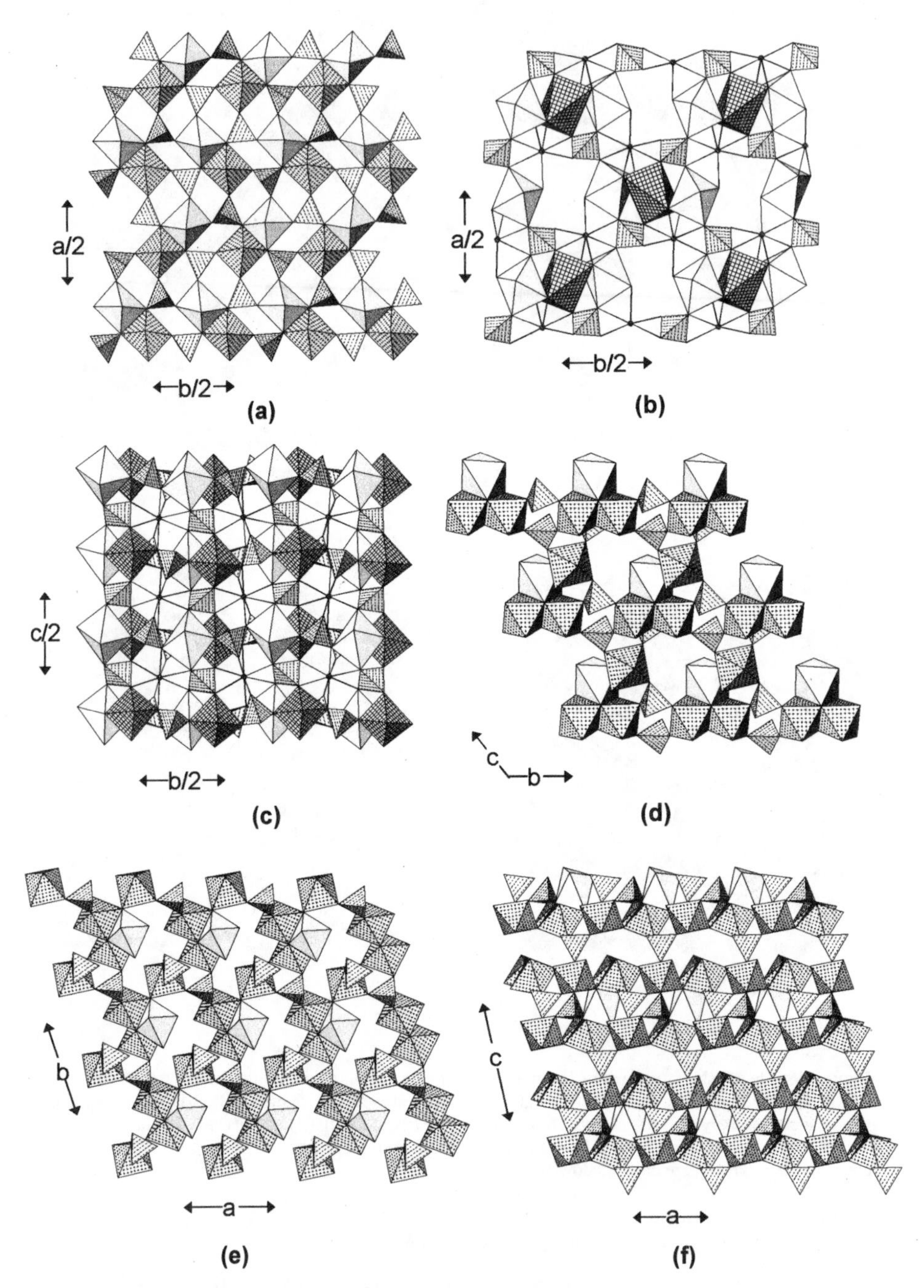

Figure 47. The crystal structures of bederite and chalcosiderite: (a) layer 1 of bederite projected onto (001); (b) layer 2 of bederite projected onto (001); (c) stacking of layers projected onto (100); ($Fe^{3+}\phi_6$) octahedra are 4^4-net-shaded, ($Fe^{2+}\phi_6$) octahedra are shadow-shaded, Ca atoms are shown as shaded circles; (d) chalcosiderite projected onto (100); (e) chalcosiderite projected onto (001); (f) chalcosiderite projected onto (010); ($Fe^{3+}\phi_6$) octahedra are dot-shaded and shadow-shaded.

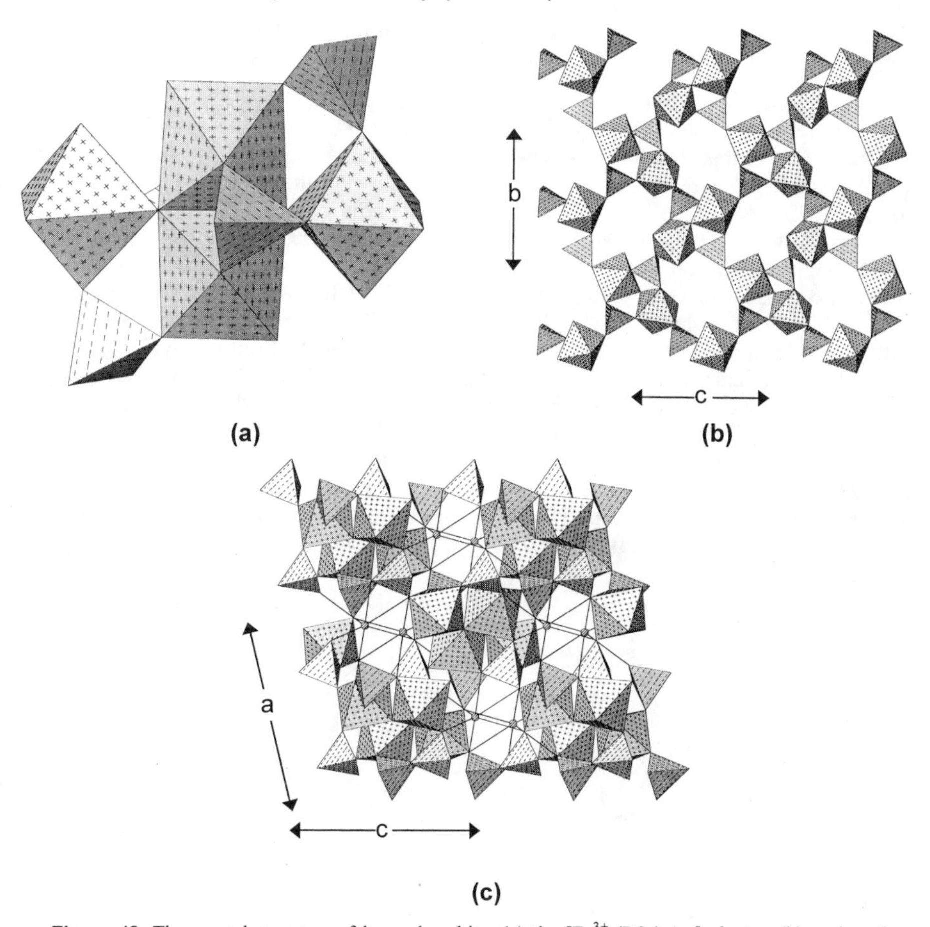

Figure 48. The crystal structure of leucophosphite: (a) the $[Fe^{3+}_4(PO_4)_4\phi_{12}]$ cluster; (b) projected onto (100); (c) projected onto (010). $(Fe^{3+}\phi_6)$ octahedra are cross/dot-shaded, K is shown as shaded circles.

and (PO_4) groups to form a thick slab parallel to (001). These slabs stack along the *c*-direction (Fig. 47f) and are linked through bridging (PO_4) tetrahedra. The structure is fairly open to accommodate the extensive hydrogen-bonding associated with the (OH) and (H_2O) groups of the structural unit.

Leucophosphite, $K_2(H_2O)[Fe^{3+}_2(PO_4)_2(OH)(H_2O)]_2(H_2O)$, and **tinsleyite**, $K_2(H_2O)$-$[Al_2(PO_4)_2(OH)(H_2O)]_2(H_2O)$, are based on a prominent tetramer of octahedra in which two $(Fe^{3+}O_6)$ octahedra share an edge and an additional $(Fe^{3+}O_6)$ octahedron links to the anions at each end of the shared edge. Moore (1972b) notes that the topologically identical cluster occurs in the sulfate mineral amarantite. This cluster is decorated by four (PO_4) tetrahedra to form an $[Fe^{3+}_4(SO_4)_4\phi_{12}]$ cluster (Fig. 48a). These clusters link by sharing vertices between octahedra and tetrahedra to form a framework (Figs. 48b,c) with K in the interstices. Inspection of Figure 48a shows that the decorated tetramer can be regarded as a condensation of two $[M_2(TO_4)_2\phi_7]$ clusters (Fig. 17e), a group that Hawthorne (1979a) showed is common as a fragment in several complex phosphate structures. In fact, when the structure is viewed down [100], it can be considered as sheets of corner-shared $[M_2(TO_4)_2\phi_3]$ clusters, similar to those in the structure of minyulite (Fig. 25a).

Cacoxenite, $[Fe^{3+}_{24}Al(PO_4)_{17}O_6(OH)_{12}(H_2O)_{24}](H_2O)_{51}$, surely has to qualify as one of the more complicated of Nature's masterpieces. Moore and Shen (1983a) identified two key *FBB*s in this structure. Pairs of $(Fe^{3+}\phi_6)$ octahedra share an edge to form dimers, and three dimers share octahedron corners to form a ring that has a (PO_4) group at its core, linking to one end of each of the shared edges in the cluster (Fig. 49a). The resulting *FBB* has the form $[Fe^{3+}_6(PO_4)\phi_{24}]$ and resembles the $[Fe^{3+}_6(PO_4)\phi_{24}]$ group in mitridatite (Fig. 33) and the central girdle of the Keggin molecule. The second *FBB* consists of one dimer of edge-sharing octahedra with two additional octahedra linked by sharing corners with each end of the shared edge of the dimer. Four (PO_4) groups each share two corners with octahedra at the periphery of the cluster, and a fifth (PO_4) group shares corners with three of the octahedra (Fig. 49b). The resulting *FBB* has the form $[Fe^{3+}_3Al(PO_4)_5\phi_9]$, and has some similarities with clusters in melonjosephite (Fig. 45c) and leucophosphate (Fig. 48a). These two *FBB*s polymerize by sharing polyhedron corners to form rings consisting of twelve *FBB*s, each type alternating around the ring. These rings are arranged at the vertices of a 3^6 net (Fig. 49c). The layer shown in Figure 49c repeats in the *c*-direction (Fig. 49d), linking by sharing polyhedron edges and corners, with the addition of some linking octahedra, to form a framework with extremely wide channels that are filled with (H_2O) groups.

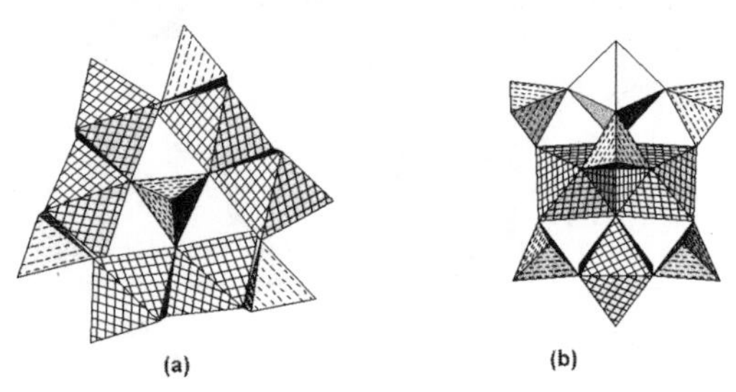

Figure 49. The structure of cacoxenite: (a) the $[Fe^{3+}_6(PO_4)\,\phi_{24}]$ cluster; (b) the $[Fe^{3+}_3Al(PO_4)_5\phi_9]$ cluster. $(Fe^{3+}\phi_6)$ octahedra are 4^4-net-shaded, $(Al\phi_6)$ octahedra are shadow-shaded.

Althausite, $[Mg_4(PO_4)_2(OH)F]$, and **satterlyite**, $[Fe^{2+}_4(PO_4)_2(OH)_2]$, have their divalent cations in both [5]- and [6]-coordination, triangular bipyramidal and octahedral. In althausite, $[M\phi_4]$ chains of *trans* edge-sharing $(Mg\phi_6)$ octahedra extend in the *b*-direction. These chains link in the *a*-direction by sharing corners between tetrahedra and octahedra (Fig. 50a) to form a sheet parallel to (001). These sheets are linked in the *c*-direction by sharing octahedron edges with $(Mg\phi_5)$ triangular bipyramids (Fig. 50b). In althausite, ~20% of the (OH) is replaced by O^{2-} and the excess charge is compensated by omission (i.e., incorporation of vacancies) of F.

In **hureaulite**, $[Mn^{2+}_5(PO_3\{OH\})_2(PO_4)_2(H_2O)_4]$, five $(Mn^{2+}\phi_6)$ octahedra share edges to form a kinked linear pentamer that extends in the *a*-direction (Fig. 50c). These pentamers occur at the vertices of a plane centered orthorhombic net and link by sharing corners (4 per pentamer) to form a sheet of octahedra parallel to (001). Adjacent pentamers are also linked through (PO_4) groups with which they share corners to form a thick slab parallel to (001). These slabs link in the *c*-direction through corner sharing between octahedra and tetrahedra (Fig. 50d). There are fairly large interstices within the resulting framework (Fig. 50c), but these are usually unoccupied. However, Moore and Araki (1973) suggest that alkalis or alkaline earths could occupy this cavity with loss of the acid character of the acid-phosphate groups.

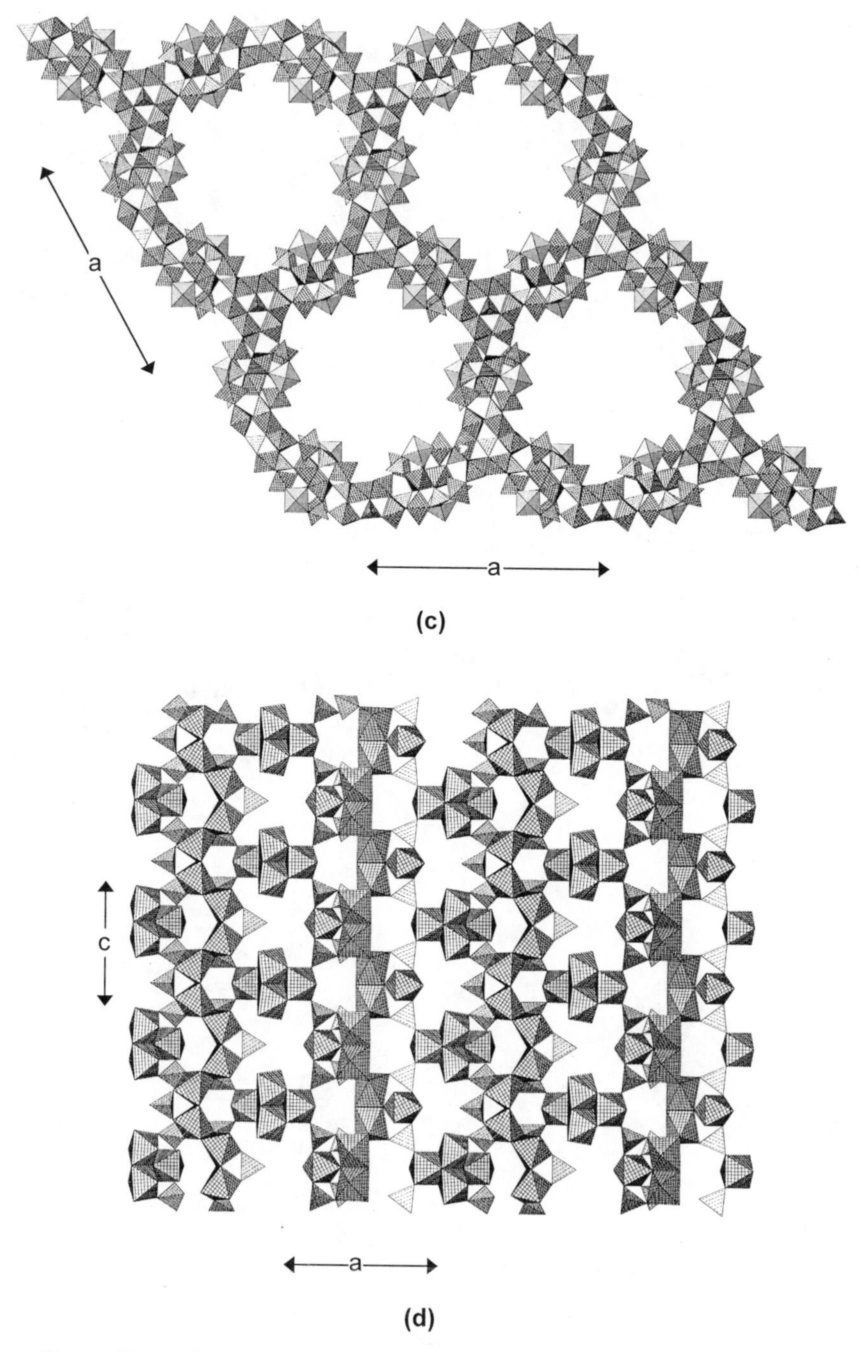

Figure 49 (continued). The structure of cacoxenite: (c) projected onto (001); (d) projected onto (010). ($Fe^{3+}\phi_6$) octahedra are 4^4-net-shaded, ($Al\phi_6$) octahedra are shadow-shaded.

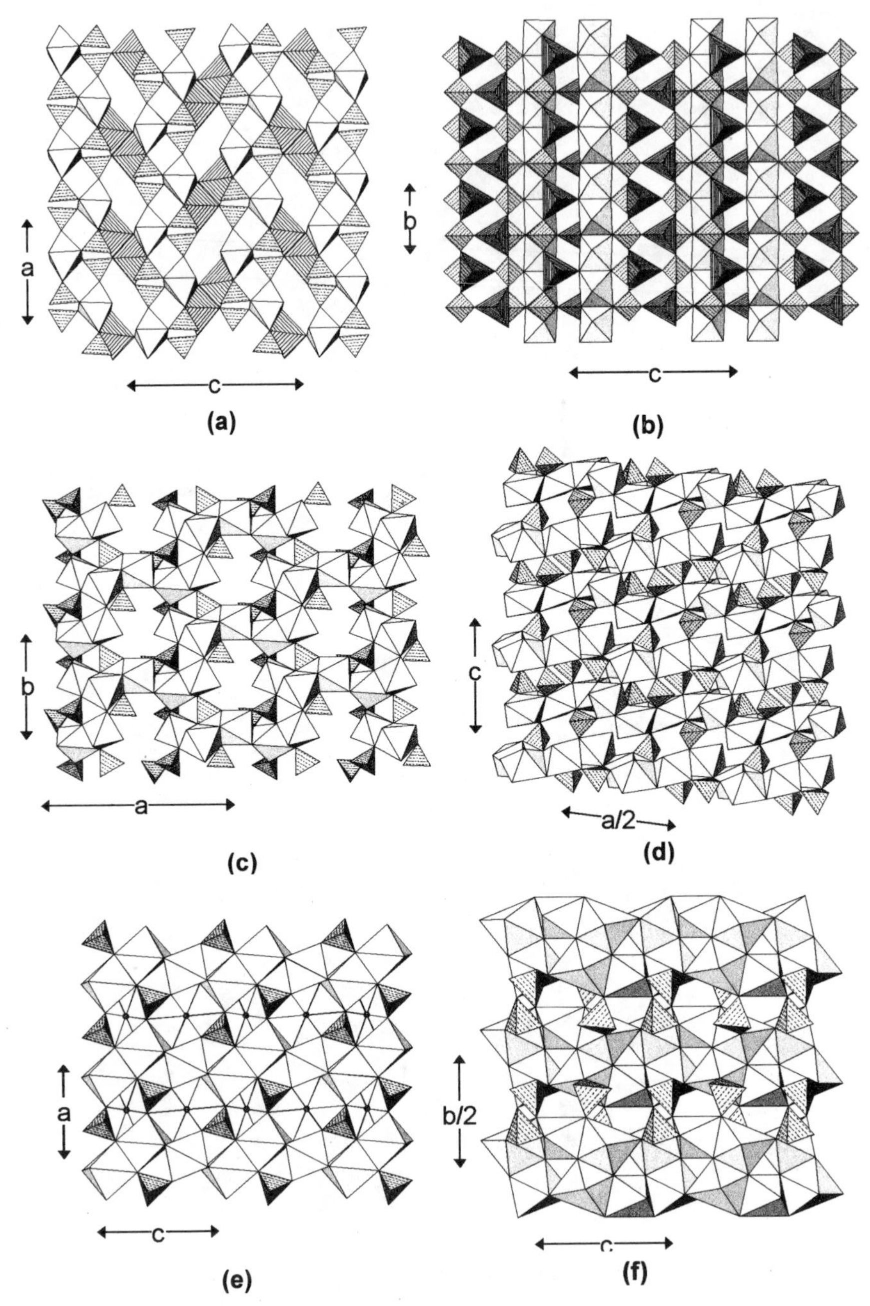

Figure 50. The crystal structures of althausite, hureaulite and thadeuite: (a) althausite projected onto (010); (b) althausite projected onto (100); ($Mg\phi_6$) octahedra are shadow-shaded, ($Mg\phi_5$) triangular bipyramids are line-shaded; (c) hureaulite projected onto (001); (d) hureaulite projected onto (010); ($Mn^{2+}\phi_6$) octahedra are shadow-shaded; (e) thadeuite projected onto (010); (f) thadeuite projected onto (100); ($Mg\phi_6$) octahedra are shadow-shaded.

Thadeuite, [$CaMg_3(PO_4)_2(OH)_2$], is a densely packed framework of (PO_4) tetrahedra and both ($Ca\phi_6$) and ($Mg\phi_6$) octahedra. ($Mg\phi_6$) octahedra share edges to form chains that extend in the *c*-direction (Fig. 50e). These chains are decorated by (PO_4) tetrahedra that link octahedra along the chain, and also link between chains in the *a*-direction. Interchain linkage also occurs by edge-sharing with ($Ca\phi_6$) octahedra (shown as ball-and-stick in Fig. 50e). The resulting layers stack in the *b*-direction (Fig. 50f) and are linked by (PO_4) groups. In this view, the more complicated nature of the chains of octahedra is apparent: two single [$M\phi_4$] chains are joined by edge-sharing between octahedra, and these two chains twist together in a helical fashion. Despite its common stoichiometry, $M_2(T\phi_4)\phi$, thadeuite shows no close structural relation with any other minerals of this stoichiometry.

Bakhchisaraitsevite, $Na_2(H_2O)[Mg_5(PO_4)_4(H_2O)_5](H_2O)$, has to be one of Nature's masterpieces of complexity. Pairs of ($Mg\phi_6$) octahedra meld to form [$Mg_2\phi_{10}$] dimers which then link by sharing edges to form zig-zag [$Mg\phi_4$] chains that extend in the *a*-direction (Fig. 51a). The vertices of the shared edge of each dimer link to (PO_4) groups which also link to the corresponding vertices of the neighboring dimer in the chain, and chains adjacent in the *b*-direction link by octahedron-tetrahedron and octahedron-octahedron corner-linkages, forming a complex sheet parallel to (001). These sheets are cross-linked in the *c*-direction by [$Mg_2\phi_{10}$] dimers, leaving large interstices between the sheets (Fig. 51b). Within these interstices are interstitial Na and (H_2O) groups: [5]- and [7]-coordinated Na each bond to one (H_2O) group and four and six O-atoms, respectively, of the structural unit, and there is one interstitial (H_2O) group not bonded to any cations, but held in the structure solely by hydrogen bonds.

The minerals of the **phosphoferrite** group have the general formula $M(1)M(2)_2(PO_4)_2X_3$, where the *M* cations may be divalent or trivalent and X = (OH), (H_2O); these minerals are isostructural, despite differences in both cation and anion charges (Moore and Araki 1976; Moore et al. 1980). The currently known species of this group are **phosphoferrite**, [$Fe^{2+}_3(PO_4)_2(H_2O)_3$], **reddingite**, [$Mn^{2+}_3(PO_4)_2(H_2O)_3$], **landesite**, [$Fe^{3+}Mn^{2+}_2(PO_4)_2(OH)(H_2O)_2$], and **kryzhanovskite**, [$Fe^{3+}_3(PO_4)_2(OH)_3$]. A prominent feature of these structures is a trimer of edge-sharing octahedra that is canted at about 20° to the *c*-axis (Fig. 51c). These trimers link by sharing octahedron edges to form chains of en-echelon trimers that extend in the *c*-direction. These chains link in the *b*-direction by sharing octahedron vertices to form a sheet of octahedra parallel to (100). The upper and lower surfaces of the sheet are decorated by (PO_4) tetrahedra, and a prominent feature of this decorated sheet is the [$M_2(TO_4)_2\phi_7$] cluster (Fig. 17e) (Hawthorne 1979a). These sheets stack in the *a*-direction, and link by sharing octahedron and tetrahedron vertices (Fig. 51d). Moore and Araki (1976) showed that single crystals of phosphoferrite can be transformed by heating (oxidation-dehydroxylation) in air to single crystals of kryzhanovskite.

Griphite, $Ca_4F_8[A_{24}Fe^{2+}_4Al_8(PO_4)_{24}]$, where $A \approx Li_2Na_4Ca_2Fe^{2+}_2Mn^{2+}_{14}$ and has triangular bipyramidal coordination, is rather complicated from both a chemical and a structural perspective, and we could not write a satisfactory end-member chemical formula; even the simplification of the above formula produces a substantial charge imbalance (~2^+). (AlO_6) octahedra share all vertices with (PO_4) tetrahedra, forming -(AlO_6)-(PO_4)-(AlO_6)-(PO_4)- chains that extend in the *a*-, *b*- and *c*-directions to form a very open framework of the form [$Al_8(PO_4)_{24}$] (Fig. 52a). The ($Fe^{2+}O_6$) octahedron links to six (PO_4) groups by sharing corners, and the resultant clusters link to a framework of corner-sharing (PO_4) groups and (CaO_8) cubes (Fig. 52b). The triangular bipyramids of the *A* cations share corners to form a very irregular sheet centered on $z \approx 0.62$ (Fig. 52c). The three sheets of Figures 52a,b,c meld to form a very complicated hetero-polyhedral framework (Fig. 52d, in which the *A* cations are shown as circles for simplicity).

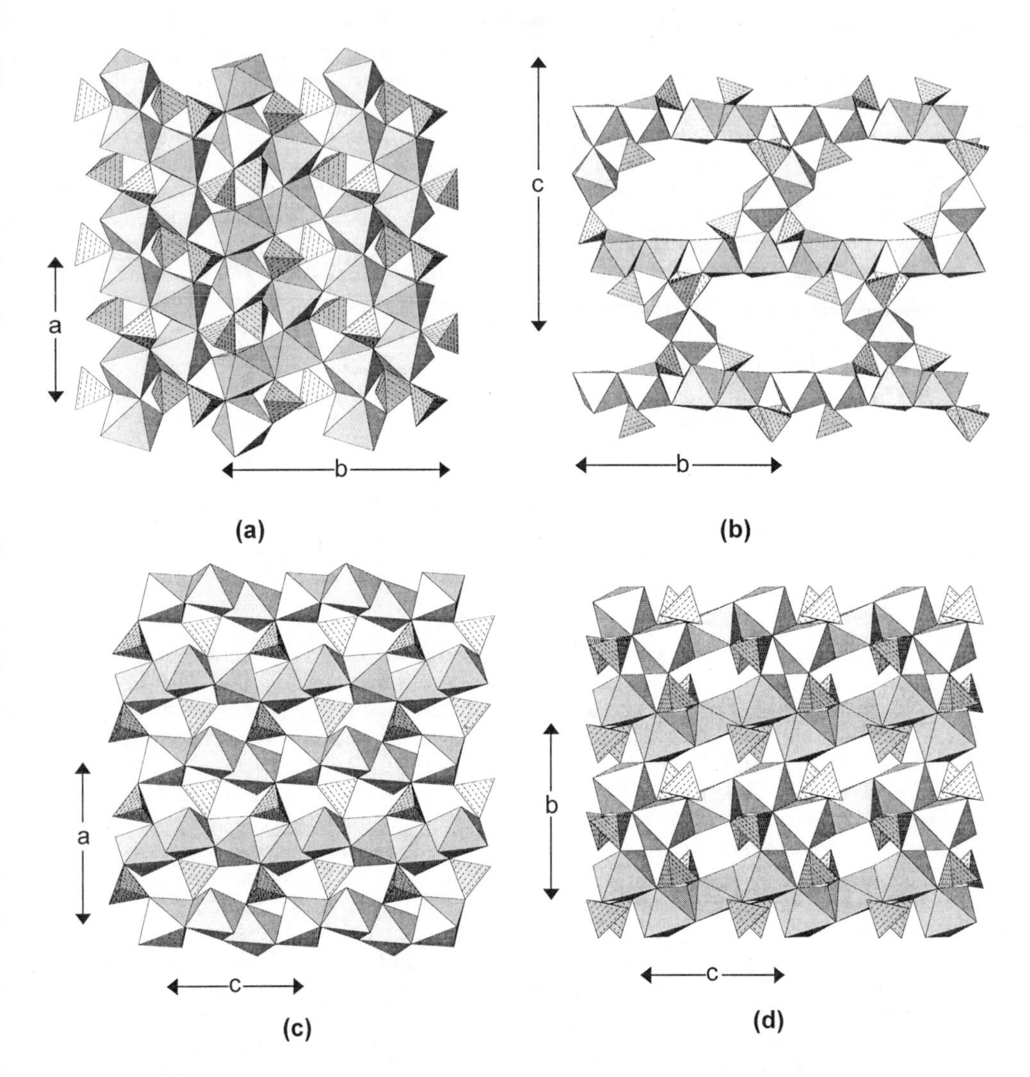

Figure 51. The crystal structures of bakhchisaraitsevite and kryzhanovskite/phosphoferrite: (a) bakhchisaraitsevite projected onto (001); (b) bakhchisaraitsevite projected onto (100); (c) kryzhanovskite/phosphoferrite projected onto (010); (d) kryzhanovskite/phosphoferrite projected onto (100). ($Fe\phi_6$) octahedra are shadow-shaded.

Cornetite, $[Cu^{2+}_3(PO_4)(OH)_3]$, contains Cu^{2+} in both octahedral and triangular bipyramidal coordinations. Pairs of $(Cu^{2+}\phi_6)$ octahedra share an edge to form $[Cu^{2+}_2\phi_{10}]$ dimers that are inclined at ~30° to the *b*-direction (Fig. 52e). Dimers adjacent in the *c*-direction show opposite cants and link by an octahedron from one dimer bridging the apical vertices of the adjacent dimer to form serrated ribbons that extend in the *c*-direction. These ribbons are linked by sharing corners with (PO_4) tetrahedra, and edges and corners with$(Cu^{2+}\phi_5)$ triangular bipyramids (Fig. 52f).

Gladiusite, $Fe^{2+}_4Fe^{3+}_2(PO_4)(OH)_{11}(H_2O)$, is an open framework structure with extensive hydrogen bonding. In the structure, $(Fe^{2+}\phi_6)$ and $(Fe^{3+}\phi_6)$ octahedra form $[M\phi_4]$

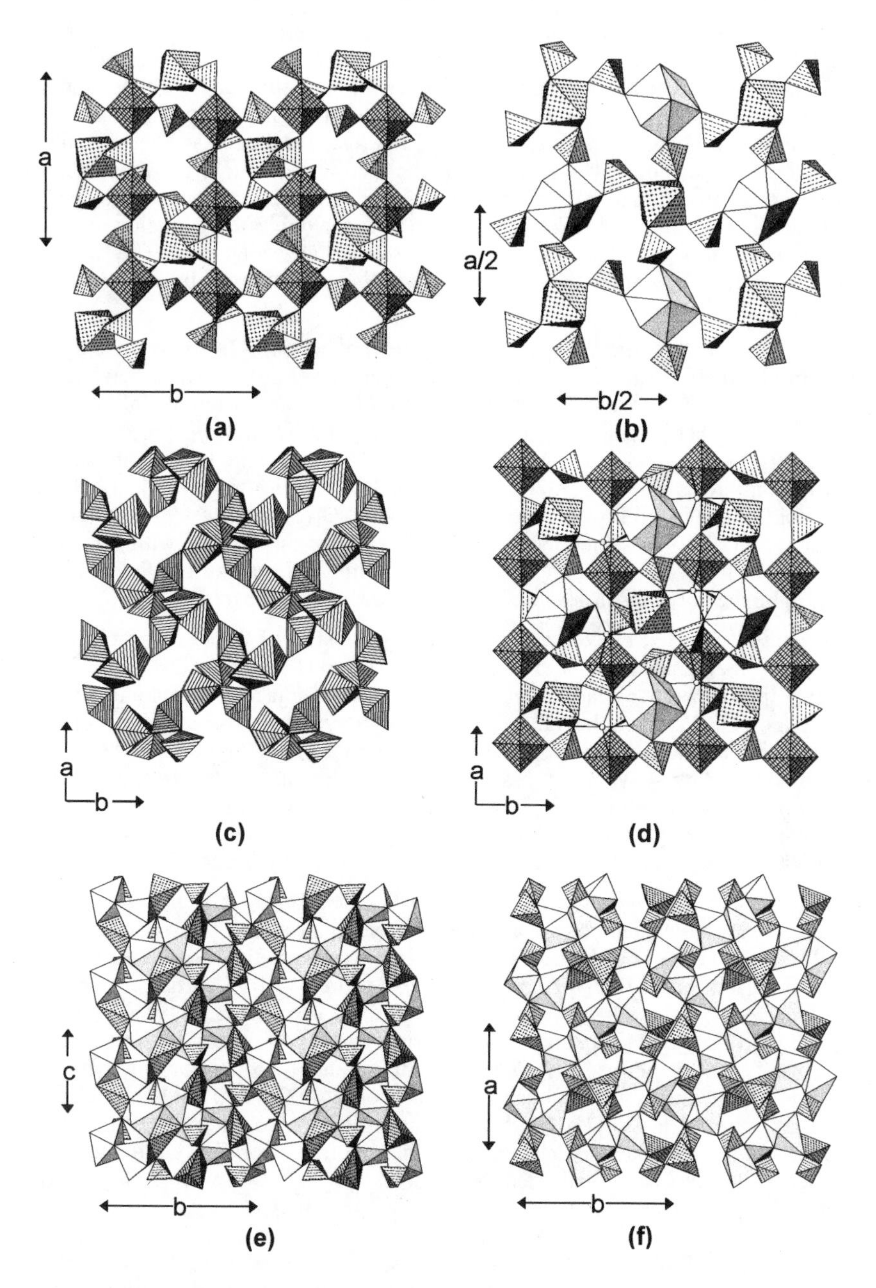

Figure 52. The crystal structures of griphite and cornetite: (a) griphite layer at $z \sim 0.25$ projected onto (001); (b) griphite layer at $z \sim 0.50$ projected onto (001); (c) griphite layer at $z \sim 0.62$ projected onto (001); (d) griphite structure projected onto (001); $(Al\phi_6)$ octahedra are 4^4-net-shaded, $(Ca\phi_8)$ cubes are shadow-shaded, $(Fe^{3+}\phi_6)$ octahedra are dot-shaded, triangular bipyramids are line-shaded; in (d), the triangular-bipyramidally coordinated cation is shown as a circle; (e) cornetite projected onto (100); (f) cornetite projected onto (001); $(Cu^{2+}\phi_6)$ octahedra are shadow-shaded, $(Cu^{2+}\phi_5)$ polyhedra are dot-shaded.

chains of edge-sharing octahedra that extend in the *c*-direction (Fig. 53a). Pairs of these chains meld by sharing edges to form ribbons, and these ribbons link in triplets by sharing corners between octahedra such that the plane of each succeeding ribbon is offset from the first (Fig. 53b). The ribbons are further linked through $(P\phi_4)$ groups that share one vertex with each ribbon (Figs. 53a,b). Figure 53c illustrates the linkage between these ribbons in three dimensions. The ribbons are inclined at ~45° to the *c*-axis and are repeated by the *b*-translation to form a row of parallel ribbons centered on $z \approx 0$. Adjacent rows centered on $z \approx {}^1/_2$ have the ribbons arranged with the opposite inclination to the *a*-axis, and adjacent ribbons link by sharing octahedron vertices (Fig. 53c). The resultant framework is very open, and the interstitial space is criss-crossed by a network of hydrogen bonds.

M≡M, M-M, M-T linkage. **Lipscombite**, $[Fe^{2+}Fe^{3+}_2(PO_4)_2(OH)_2]$, is an enigma. Katz and Lipscomb (1951) applied this name to synthetic $Fe_7(PO_4)_4(OH)_4$ with symmetry $I4_122$ and $a = 5.37$, $c = 12.81$ Å. Gheith (1953) used the name for tetragonal synthetic compounds varying between $Fe^{2+}_8(PO_4)_4(OH)_4$ and $Fe^{3+}_{3.5}(PO_4)_4(OH)_4$. More recently, Vochten and DeGrave (1981) and Vochten et al. (1983) gave the cell parameters of synthetic lipscombite as $a = b = 5.3020(5)$, $c = 12.8800(5)$ Å. However, Lindberg (1962) reported natural manganoan lipscombite with symmetry $P4_12_12$ and $a = 7.40$, $c = 12.81$ Å. Vencato et al. (1989) presented the structure of synthetic lipscombite with symmetry $P4_32_12$ and $a = 7.310(3)$, $c = 13.212(7)$ Å, in accord with the results of Lindberg (1962), who seems to be the only person who has actually characterized the mineral.

The structure reported by Vencato et al. (1989) consists of face-sharing chains of $(Fe^{2+}\phi_6)$ and $(Fe^{3+}\phi_6)$ octahedra that extend in the [110] and $[\bar{1}10]$ directions (Fig. 53d,e). These chains link by corner-sharing between octahedra of adjacent chains and also by sharing corners with (PO_4) tetrahedra. Because of the 4_3 symmetry, the structure consists of layers in which the face-sharing chains extend only in a single direction, and adjacent layers that are related by 4_3 symmetry have the chains extending in orthogonal directions. A single layer is shown in Figure 50e, in which all the chains extend along [110] and are linked within the layer by rows of bridging (PO_4) groups. Note that in the face-sharing chain, two of the three symmetrically distinct octahedra are partly occupied.

The minerals of the **burangaite**, $Na[Fe^{2+}Al_5(PO_4)_4(OH)_6(H_2O)_2]$, group contain a trimer of face-sharing octahedra that is a feature of several basic iron-phosphate minerals (Moore 1970). An $(Fe^{2+}\phi_6)$ octahedron shares two *trans* faces with $(Al\phi_6)$ octahedron to form a trimer of the form $[M_3\phi_{12}]$ (the *h* cluster of Moore 1970). This trimer is corner linked to two $(Al\phi_6)$ octahedra and two (PO_4) tetrahedra to produce a cluster of the general form $[M_5(TO_4)_2\phi_{18}]$. This cluster polymerizes in the *c*-direction to form a dense slab by corner-sharing between $(Al\phi_6)$ octahedra and by corner-sharing between octahedra and tetrahedra. This slab is oriented parallel to (100) (Fig. 54a) and adjacent slabs are weakly linked in the [100] direction by additional $(Al\phi_6)$ octahedra that share corners with both tetrahedra and octahedra. The resulting framework has large interstices that are occupied by [8]-coordinated Na that is bonded to two (H_2O) groups. Note that Moore (1970) gave the formula of the isostructural dufrénite as $Ca_{0.5}Fe^{2+}Fe^{3+}_5(PO_4)_4(OH)_6\,(H_2O)_2$, which is in accord with the requirements for an end-member composition (Hawthorne 2002). However, both Moore (1984) and Nriagu (1984) incorrectly list the formula of dufrénite as $CaFe^{3+}_6(PO_4)_4(OH)_6(H_2O)_2$; this formula has a net charge of 2^+. Van der Westhuizen et al. (1990) reported electron-microprobe analyses for dufrénite, but many of the resultant formulae are incompatible with the dufrénite structure.

The minerals of the **rockbridgeite**, $[Fe^{2+}Fe^{3+}_4(PO_4)_3(OH)_5]$, group are also based on the *h* cluster, but the mode of linkage of these clusters is very different from that in the

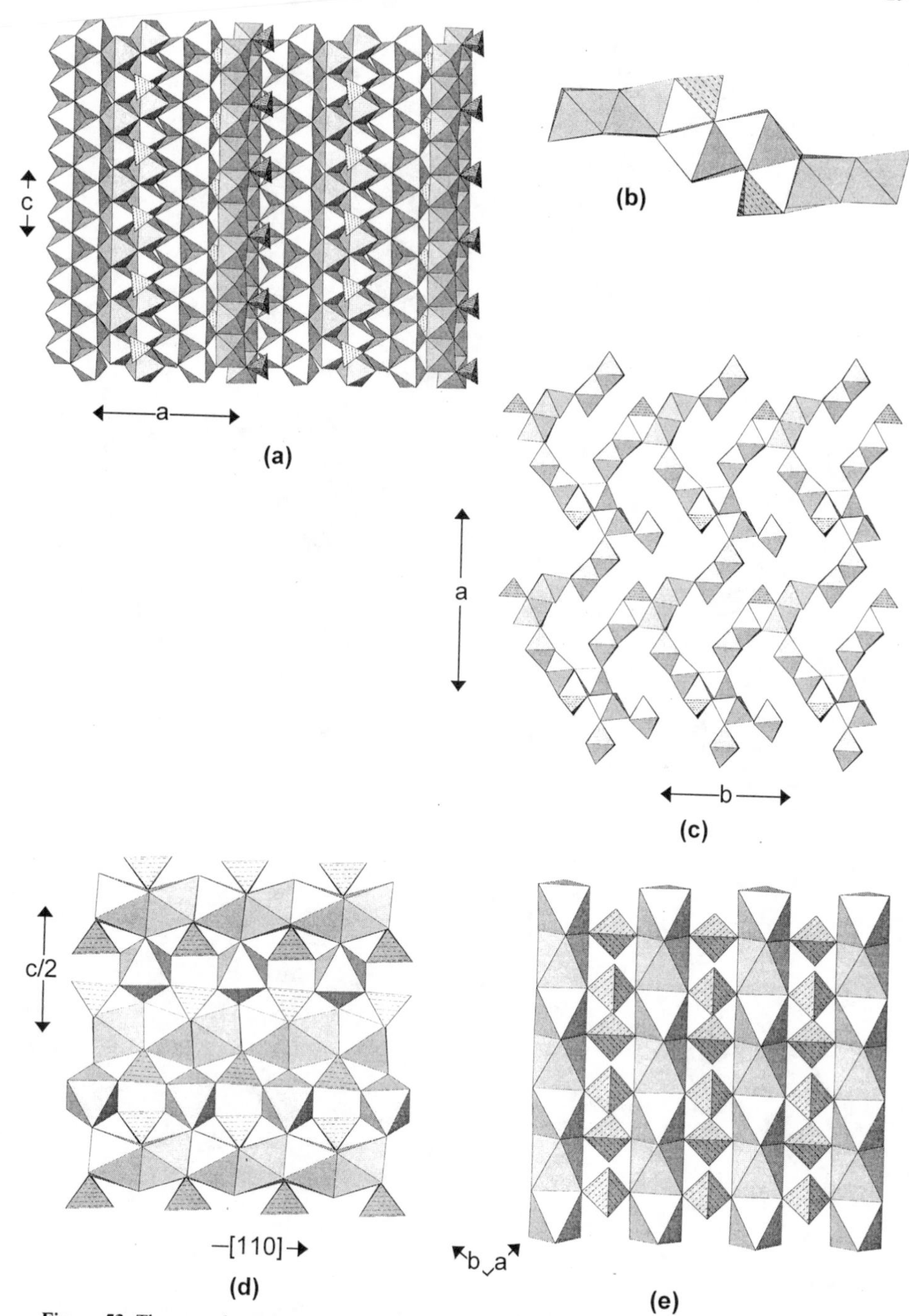

Figure 53. The crystal structures of gladiusite and lipscombite: (a) gladiusite projected onto (010); (b) in gladiusite, the linking of adjacent pairs of chains to form a triplet of offset chains with linking (PO_4) tetrahedra; (c) gladiusite projected onto (001); (d) lipscombite projected onto 100); (e) lipscombite projected onto (001). $(\{Fe^{2+},Fe^{3+}\}\phi_6)$ octahedra are shadow-shaded.

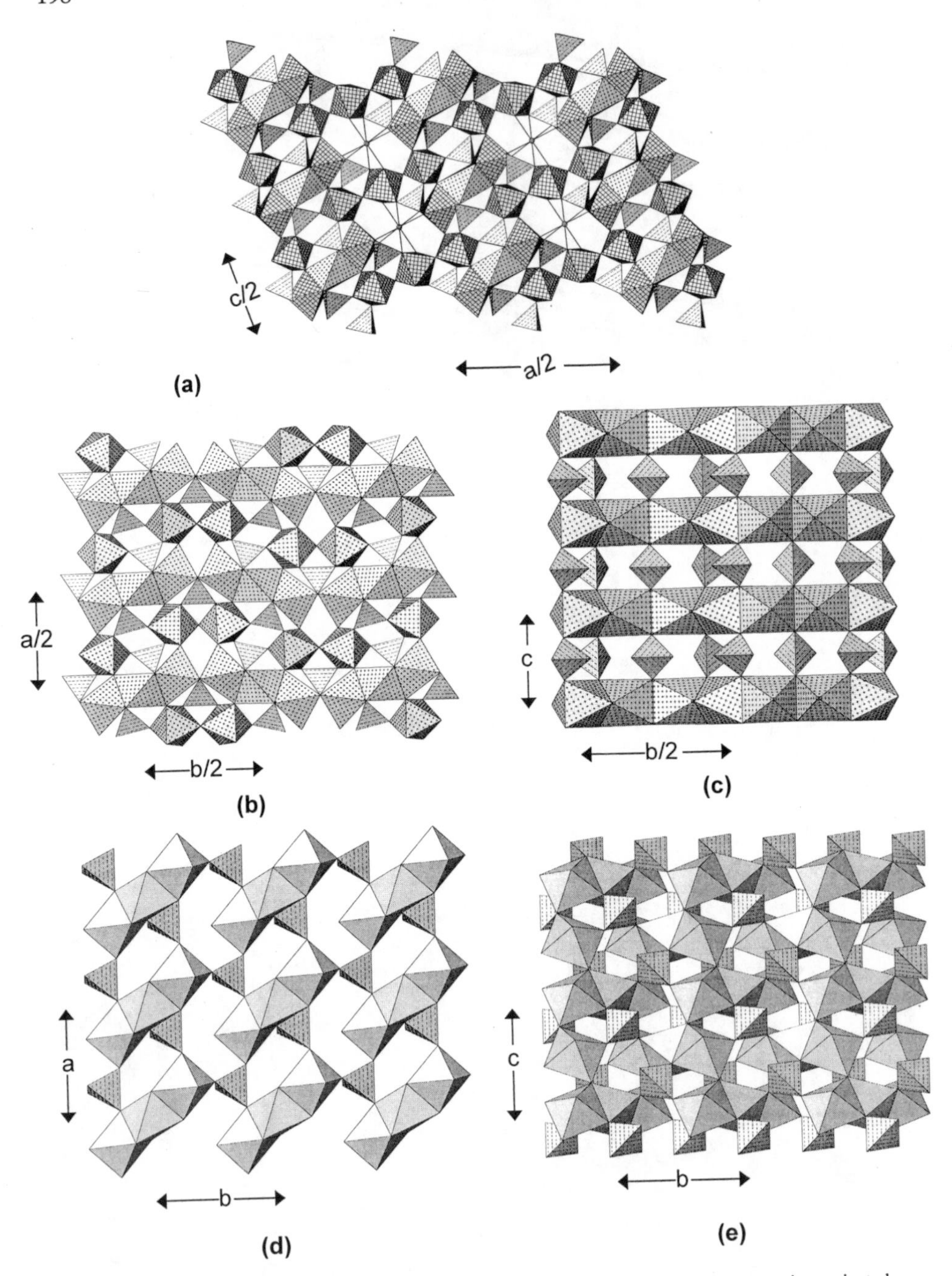

Figure 54. The crystal structures of burangaite, rockbridgeite and lazulite: (a) burangaite projected onto (010); ($Al\phi_6$) octahedra are 4^4-net-shaded, ($Fe^{2+}\phi_6$) octahedra are dot-shaded, Na is shown as shaded circles; (b) rockbridgeite projected onto (001), (c) rockbridgeite projected onto (100); ($\{Fe^{3+},Fe^{2+}\}\phi_6$) octahedra are dot-shaded; (d) lazulite projected onto (001); (e) lazulite projected onto (100); ($\{Mg,Al\}\phi_6$) octahedra are shadow-shaded.

minerals of the burangaite group The face-sharing trimers link by sharing octahedron corners to form chains of octahedra that extend in the *b*-direction (Fig. 54b). Chains adjacent in the *a*-direction are linked by [$M_2(TO_4)\phi_8$] clusters and (PO_4) groups that link to two adjacent trimers and two [$M_2(TO_4)\phi_8$] clusters, forming complex sheets parallel to (001). When viewed in the *a*-direction (Fig. 54c), the very layered aspect of the structure is apparent, layers of octahedra alternating with layers of tetrahedra.

The minerals of the **lazulite**, [$MgAl_2(PO_4)_2(OH)_2$], group contain the *h* cluster, a trimer of face-sharing octahedra, that is characteristic of several basic phosphate minerals. An ($Mg\phi_6$) octahedron is sandwiched between two ($Al\phi_6$) octahedra, and the resulting trimers are arranged at the vertices of a 4^4 net and extending in the [110] direction (Fig. 54d). Adjacent trimers are linked by sharing corners with (PO_4) groups, and when viewed down [001], the structure consists of layers of octahedra and tetrahedra. When viewed down [100] (Fig. 54e), it can be seen that the trimers of adjacent layers are canted in opposing direction, thereby promoting linkage of each (PO_4) tetrahedron to four different trimers. The resulting arrangement is quite densely packed.

Trolleite, [$Al_4(PO_4)_3(OH)_3$], is a very dense structure with some similarities to the structural arrangement of the minerals of the lazulite group (Table 8). There are two prominent chain motifs that constitute the building blocks of this structure. There is an [$Al(PO_4)\phi_3$] 7-Å chain (Fig. 18c) that extends in the *c*-direction (Fig. 55a), giving the 7.1 Å repeat along the *c*-axis. There is also an [$Al(PO_4)\phi_4$] chain that assumes a very contorted geometry (Fig. 55a) so that it has the same repeat distance along its length as the [$Al(PO_4)\phi_3$] chain to which it is attached by sharing octahedron faces. These rather complex double-chains link in the *b*-direction by sharing vertices of the tetrahedra of the [$Al(PO_4)\phi_3$] chain with the octahedra of the [$Al(PO_4)\phi_4$] chain (Fig. 55a). These slabs repeat in the *a*-direction in a very complex manner. As shown in Figure 55b, these slabs meld by sharing octahedron-tetrahedron vertices between adjacent [$Al(PO_4)\phi_3$] chains to form a thick slab: [$Al(PO_4)\phi_4$]-[$Al(PO_4)\phi_3$]-[$Al(PO_4)\phi_3$]-[$Al(PO_4)\phi_4$] that constitute one-half the cell in the *a*-direction. The thick slabs link by sharing octahedron vertices between [$Al(PO_4)\phi_4$] chains to form a very dense framework.

Seamanite, [$Mn^{2+}_3(B\{OH\}_4)(PO_4)$], is a mixed phosphate-borate mineral based on chains of ($Mn\phi_6$) octahedra that consist of free-sharing [$M_3\phi_{12}$] trimers that link by sharing octahedron edges to form an [$M_3\phi_{10}$] chain that extends in the *c*-direction (Fig. 55c). The rather unusual [$M_3\phi_{12}$] trimer is apparently stabilized by the ($B\phi_4$) group that spans the apical vertices of the edge-sharing octahedra (Moore and Ghose 1971). Additional linkage along the length of the chain is provided by (PO_4) tetrahedra that link apical vertices on neighboring ($M\phi_6$) octahedra such that the (PO_4) and ($B\phi_4$) tetrahedra adopt a staggered configuration on either side of the [$M(B\phi_4)(PO_4)\phi_6$] chain. These chains condense in pairs by sharing both octahedron-octahedron and octahedron-tetrahedron vertices to form columns, seen end-on in Figure 55d. These columns link together in the *a*- and *b*-directions by sharing vertices between tetrahedra and octahedra, with additional linkage involving hydrogen bonds.

M=M, M=T, M-T linkage. **Holtedahlite**, [$Mg_{12}(PO_3\{OH\})(PO_4)_5(OH)_6$], contains dimers of face-sharing ($Mg\phi_6$) (Fig. 55e). These dimers link by sharing edges to form ribbons that extend in the *c*-direction and contain ($PO_3\{OH\}$) tetrahedra that link to all three ribbons (Fig. 55e). These channels link in the *a*- and *b*-directions by sharing octahedron corners and by sharing octahedron corners with bridging (PO_4) tetrahedra (Figs. 55e,f).

The crystallographic and chemical details of the minerals of the **triphylite-lithiophyllite**, **sicklerite-ferrisicklerite** and **heterosite-purpurite** groups are summarized in Table 10 (for consistency, some of the axial orientations have been changed from

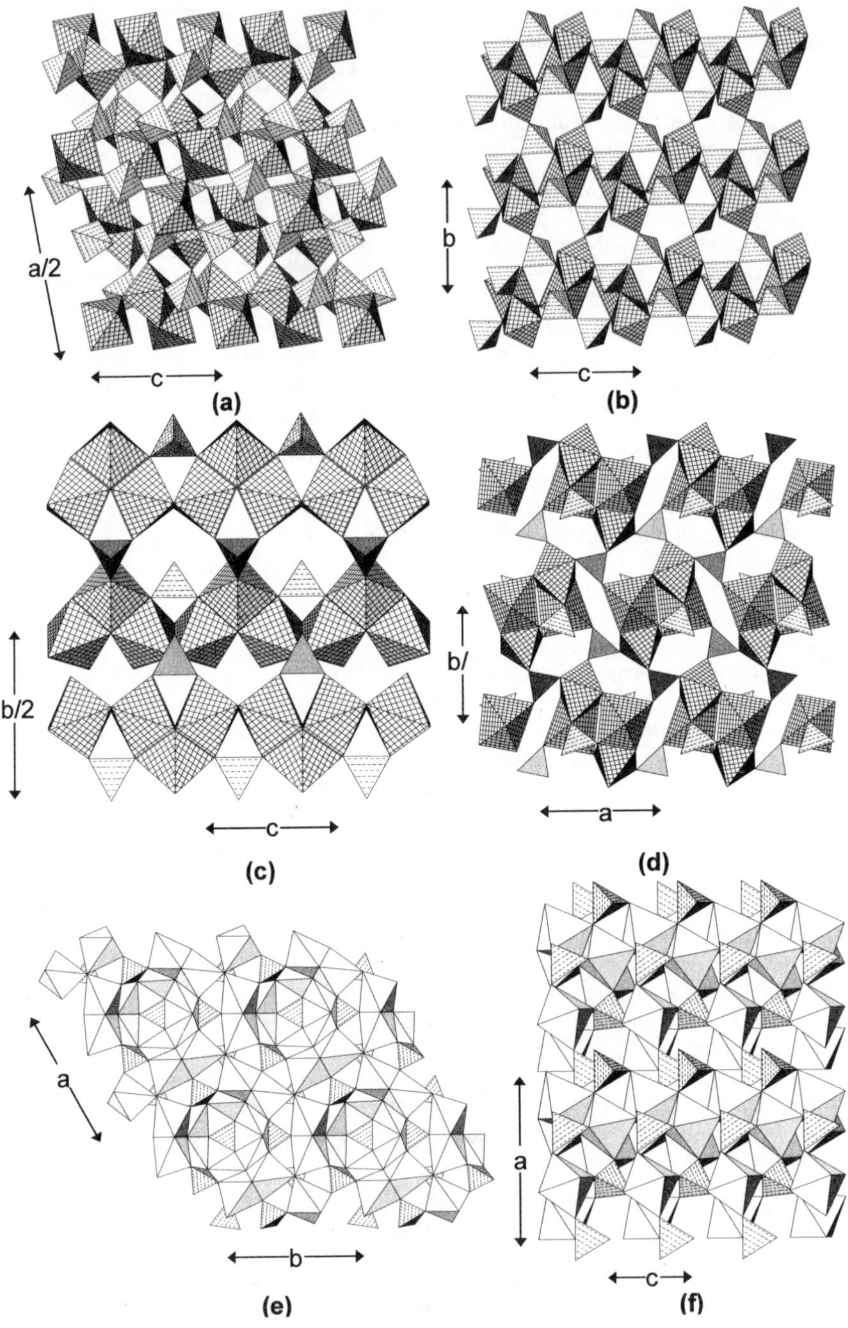

Figure 55. The crystal structures of trolleite, seamanite and holtedahlite: (a) trolleite projected onto (010); (b) trolleite projected onto (100); $(Al\phi_6)$ octahedra are 4^4-net-shaded; (c) seamanite projected onto (100); (d) seamanite projected onto (001); $(Mn^{2+}\phi_6)$ octahedra are 4^4-net-shaded; (e) holtedahlite projected onto (001); (f) holtedahlite projected onto (010); $(Mg\phi_6)$ octahedra are shadow-shaded.

Table 10. Details of the minerals of the triphylite-lithiophyllite, sicklerite-ferrisicklerite and heterosite-purpurite groups.

	a (Å)	*b*(Å)	*c* (Å)	Sp grp	Ref.
Triphylite	4.704	10.347	6.0189	*Pbnm*	(1)
Lithiophyllite	4.744(10)	10.460(30)	6.100(20)	*Pbnm*	(2)
Sicklerite	4.794	10.063	5.947	*Pbnm*	(3)
Ferrisicklerite	4.978	10.037	5.918	*Pbnm*	(4)
Heterosite	4.769(5)	9.760(10)	5.830(10)	*Pbnm*	(5)
Purpurite	4.76	9.68	5.819	*Pbnm*	(6)

References: (1) Yakubovich et al. (1977), (2) Geller and Durand (1960), (3) Blanchard (1981), (4) Alberti (1976), (5) Eventoff et al. (1972), (6) Bjoerling and Westgren (1938).

those reported in the original papers). All of these structures have the olivine arrangement. [MO_4] chains of edge-sharing (LiO_6) or (NaO_6) octahedra extend parallel to the *a*-direction (Fig. 56a,c,e) and are decorated by (Fe^{2+},$Mn^{2+}O_6$) or (□O_6) octahedra (□ = vacancy). These decorated chains are linked in the *b*-direction by sharing octahedron corners with (PO_4) groups, although such 'linkage' is not effective when the decorating octahedra are vacant (Fig. 56e); in this case, the chains link to other chains above and below the plane (Fig. 56f).

Consider the Fe end-members of each group:

triphylite	Li	Fe^{2+}	(PO_4)
ferrisicklerite	(Li,□)	(Fe^{2+},Fe^{3+})	(PO_4)
heterosite	□	Fe^{3+}	(PO_4)

As all three minerals have the same structure, the ranges in chemical composition for triphylite and heterosite are $LiFe^{2+}(PO_4)$B($Li_{0.5}$□$_{0.5}$)($Fe^{2+}_{0.5}Fe^{3+}_{0.5}$)($PO_4$) and □$_{0.5}Li_{0.5}$($Fe^{3+}_{0.5}Fe^{2+}_{0.5}$)($PO_4$)- □$Fe^{3+}$($PO_4$), respectively. Ferrisicklerite is an unnecessary name for intermediate-composition triphylite and heterosite. Similarly, sicklerite is an unnecessary name for intermediate lithiophyllite and purpurite.

Senegalite, [$Al_2(PO_4)(OH)_3(H_2O)$], contains Al in both triangular bipyramidal and octahedral coordinations. ($Al\phi_5$) and ($Al\phi_6$) polyhedra share an edge to form a dimer, and these dimers link by sharing corners to form a [$^{[5]}Al^{[6]}Al\phi_8$] chain that extends in the [101] (and [$\bar{1}$01]) direction (Fig. 57a). These chains are decorated by (PO_4) groups that link them to form a slab parallel to (010). These slabs stack in the *b*-direction, and link by sharing tetrahedron-octahedron and tetrahedron-bipyramid corners (Fig. 57b). This framework is fairly open, and the interstices are criss-crossed by a network of hydrogen bonds.

M=M, M-M, M=T, M-T linkage. **Sarcopside**, [$Fe^{2+}_3(PO_4)_2$], is chemically similar to the minerals of the graftonite group but is structurally more similar to the structures of triphylite-lithiophyllite and its derivatives (Table 8). When viewed down [100] (Fig. 58a), the structure consists of a sheet of corner-linked octahedra at the vertices of a 4^4 net, and further linked by edges and corners with (PO_4) tetrahedra. When viewed down [010], the structure consists of trimers of edge-sharing octahedra linked into a sheet by sharing corners with (PO_4) groups (Fig. 58b). Sarcopside usually contains significant Mn^{2+}, but assuming complete solid-solution between graftonite and beusite, it seems that sarcopside is a polymorph of graftonite.

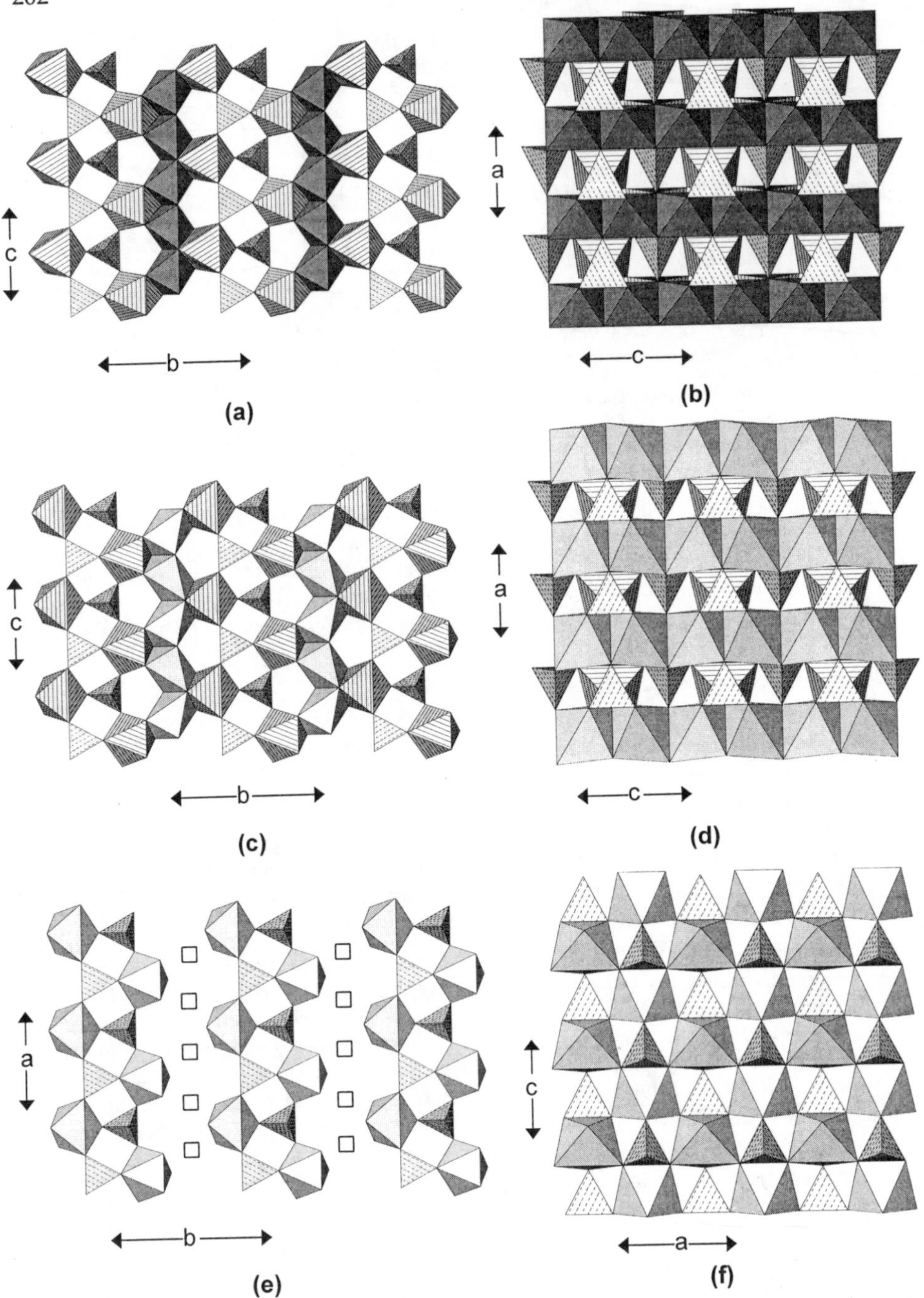

Figure 56. The crystal structures of lithiophylite, ferrisicklerite and heterosite: (a) lithiophylite projected onto (100); (b) lithiophylite projected onto (010); (c) ferrisicklerite projected onto (100); (d) ferrisicklerite projected onto (010); (e) heterosite projected onto (001); (f) heterosite projected onto (010). ($Fe^{2+}O_6$) octahedra are dark-shadow-shaded, (Fe^{2+},Fe^{3+}) octahedra are shadow-shaded, ($\{Li,\square\}O_6$) octahedra are line-shaded, vacancies are shown as squares.

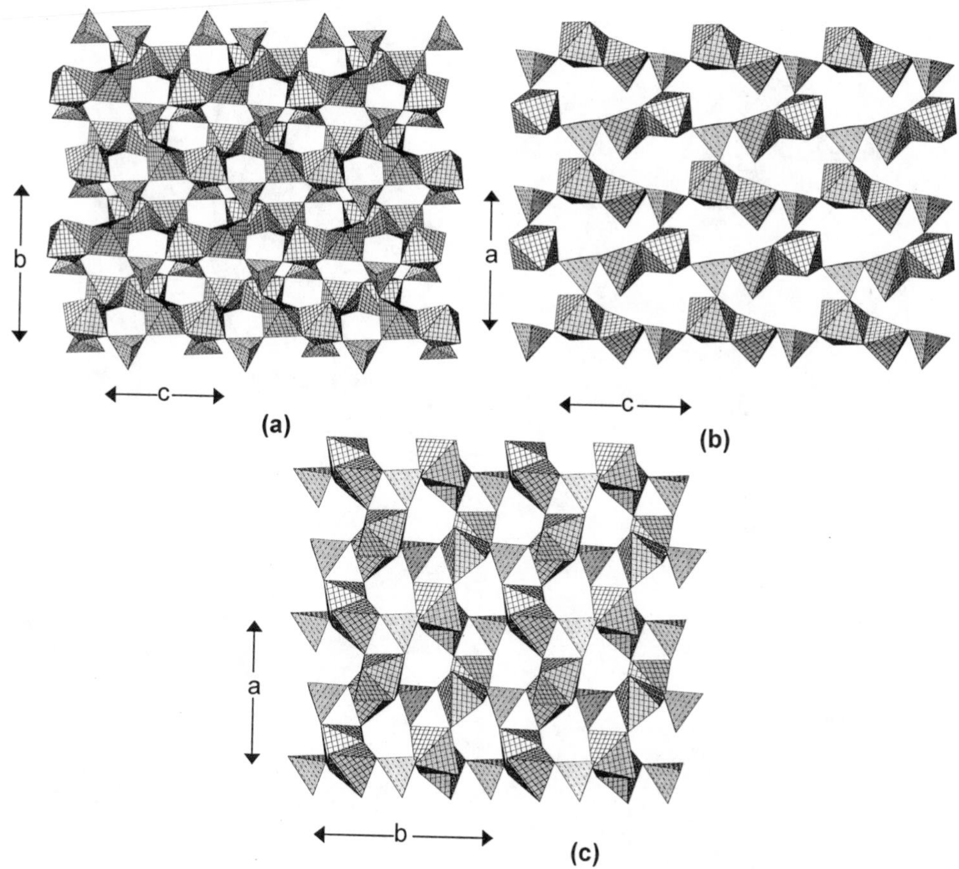

Figure 57. The crystal structure of senegalite: (a) projected onto (100); (b) projected onto (010); (c) projected onto (001). ($Al\phi_n$) polyhedra are 4^4-net-shaded.

In **bjarebyite**, $BaMn^{2+}{}_2[Al_2(PO_4)_3(OH)_3]$, and the minerals of the bjarebyite group (Table 8), pairs of ($Al\phi_6$) octahedra link by sharing edges to form dimers, and these dimers link together by sharing corners to form a chain of the form $[Al_2\phi_9]$, intermediate between the corner-sharing $[Al\phi_5]$ chain and the edge-sharing $[Al\phi_4]$ chain (i.e., $[Al_2\phi_9] \equiv 2\,[Al\phi_{4.5}]$). Octahedra that share corners are also linked by a flanking (PO_4) tetrahedron, similar to the linkage in the chain of Figure 18c. The *cis* vertex of each octahedron also links to a (PO_4) tetrahedron to give a chain of the form $[Al_2(PO_4)_3(OH)_3]$. These chains extend in the *b*-direction (Fig. 58c) and are linked in the *a*-direction by [6]-coordinated Mn^{2+} and by [11]-coordinated Ba (Fig. 58d). Kulanite and penikisite were originally reported as triclinic (Mandarino and Sturman 1976; Mandarino et al. 1977). However, Cooper and Hawthorne (1994a) showed that kulanite is monoclinic (and is isostructural with bjarebyite). It is probable that penikisite is also monoclinic.

STRUCTURES WITH ($T\phi_4$) GROUPS AND LARGE CATIONS

Xenotime-(REE), (REE)(PO_4), and **pretulite**, Sc(PO_4) (Table 11), belong to the zircon, Zr(SiO_4), group. The larger trivalent cation is coordinated by eight O-atoms in an arrangement that is known as a Siamese dodecahedron (Hawthorne and Ferguson 1975).

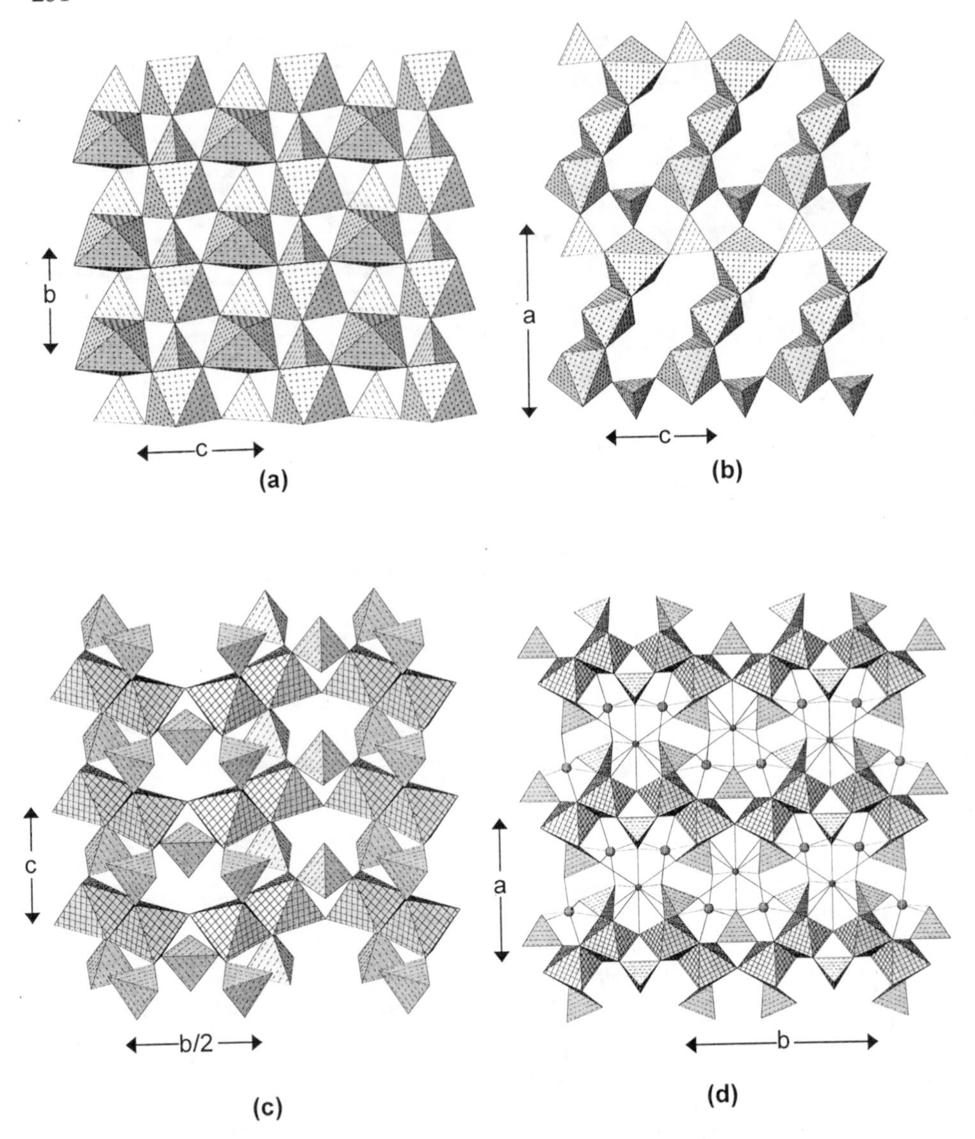

Figure 58. The crystal structures of sarcopside and bjarebyite: (a) sarcopside projected onto (100); (b) sarcopside projected onto (010); $(Fe^{2+}\phi_6)$ octahedra are dot-shaded; (c) bjarebyite projected onto (100); (d) bjarebyite projected onto (001). $(Al\phi_6)$ octahedra are 4^4-net-shaded, Mn^{2+} are shown as large shaded circles, Ba are shown as small shaded circles.

These dodecahedra link by sharing edges to form chains that extend in the *b*-direction (Fig. 59a). These chains are linked in the *c*-direction by (PO_4) tetrahedra that share an edge with a dodecahedron of one chain and a vertex with a dodecahedron of the adjacent chain, forming a layer in the (100) plane (Fig. 59a). These layers stack in the *a*-direction by edge-sharing between dodecahedra of adjacent layers to form dodecahedral chains orthogonal to the layers. Hence the zircon structure is tetragonal (Fig. 59b).

Monazite-(REE), (REE)(PO_4), is a dimorph of xenotime-(REE). In this structure, the larger trivalent cation is coordinated by nine O-atoms in a rather irregular arrange-

Table 11. Large-cation phosphate minerals.

Mineral	*Formula*	*Space group*	*Figure*
	Xenotime group		
Pretulite	$Sc(PO_4)$	$I4_1/amd$	59a,b
Xenotime-(Y)	$Y(PO_4)$	$I4_1/amd$	59a,b
Xenotime-(Yb)	$Yb(PO_4)$	$I4_1/amd$	59a,b
	Monazite group		
Brabantite	$CaTh(PO_4)_2$	$P2_1$	59c,d
Cheralite-(Ce)	$Ce(PO_4)$	$P2_1/n$	59c,d
Monazite-(Ce)	$Ce(PO_4)$	$P2_1/n$	59c,d
	Rhabdophane group		
Brockite		Aa	60a,b
Grayite	$Th(PO_4)(H_2O)$	$P6_222$	60a,b
Ningyoite	$U_2(PO_4)_2(H_2O)_{1-2}$	$P222$	60a,b
Rhabdophane	$Ce(PO_4)$	$P6_222$	60a,b
	General		
Archerite	$K(PO_3\{OH\}_2)$	$I\bar{4}2d$	60c,d
Biphosphammite	$(NH_4)(PO_2\{OH\}_2)$	$I\bar{4}2d$	60c,d
Brushite	$Ca(PO_3\{OH\})(H_2O)_2$	Ia	61a,b
Churchite-(Y)	$Y(PO_4)(H_2O)_2$	$I2/a$	61a,b
Ardealite	$Ca_2(PO_3\{OH\})(SO_4)(H_2O)_4$	Cc	61c,d,e
Dorfmanite	$Na_2(PO_3\{OH\})(H_2O)_2$	$Pbca$	62a,b,c
Monetite	$Ca(PO_3\{OH\})$	$Pmna$	62d
Nacaphite	$Na_2Ca(PO_4)F$	$P\bar{1}$	63a,b
Arctite	$(Na_5Ca)Ca_6Ba(PO_4)_6F_3$	$P\bar{1}$	63c,d
Nabaphite	$NaBa(PO_4)(H_2O)_9$	$P2_13$	–
Nastrophite	$NaSr(PO_4)(H_2O)_9$	$P2_13$	–
Lithiophosphate	$[Li_3(PO_4)]$	$Pcmn$	64a,b
Nalipoite	$NaLi_2(PO_4)$	$Pmnb$	64c,d
Nefedovite	$Na_5Ca_4(PO_4)_4F$	$I\bar{4}$	65a,b
Olgite	$NaSr(PO_4)$	$P3$	65c,d,e
Phosphammite	$(NH_4)_2(PO_3\{OH\})$	$P2_1/c$	66a,b
Vitusite-(Ce)	$Na_3Ce(PO_4)_2$	$Pca2_1$	66c
Stercorite	$Na(NH_4)(PO_3\{OH\})(H_2O)_4$	$P\bar{1}$	67a,b
Natrophosphate	$Na_7(PO_4)_2F(H_2O)_{19}$	$Fd\bar{3}c$	67c
Buchwaldite	$NaCa(PO_4)$	$Pn2_1a$	68a,b
Olympite	$LiNa_5(PO_4)_2$	$Pcmn$	–

ment. These polyhedra link by sharing edges to form chains that extend in the *b*-direction (Fig. 59c). The chains are linked in the *c*-direction by (PO_4) tetrahedra that share edges with polyhedra of adjacent chain to form a layer parallel to (100) (Fig. 59c). These layers stack in the *a*-direction by sharing edges between the $(\{REE\}O_9)$ polyhedra to form rather staggered chains that extend in the [101] direction (Fig. 59d). The monazite structure preferentially incorporates the larger light REEs whereas the xenotime structure preferentially incorporates the smaller heavy REEs (Ni et al. 1995); this is in accord with the difference in trivalent-cation coordination numbers in these two structure-types, and is also in accord with $Sc(PO_4)$ crystallizing in the xenotime (zircon) structure (as pretulite) rather than in the monazite-type structure.

Rhabdophane, $Ca(PO_4)$, contains Ce coordinated by eight O-atoms in a dodeca-

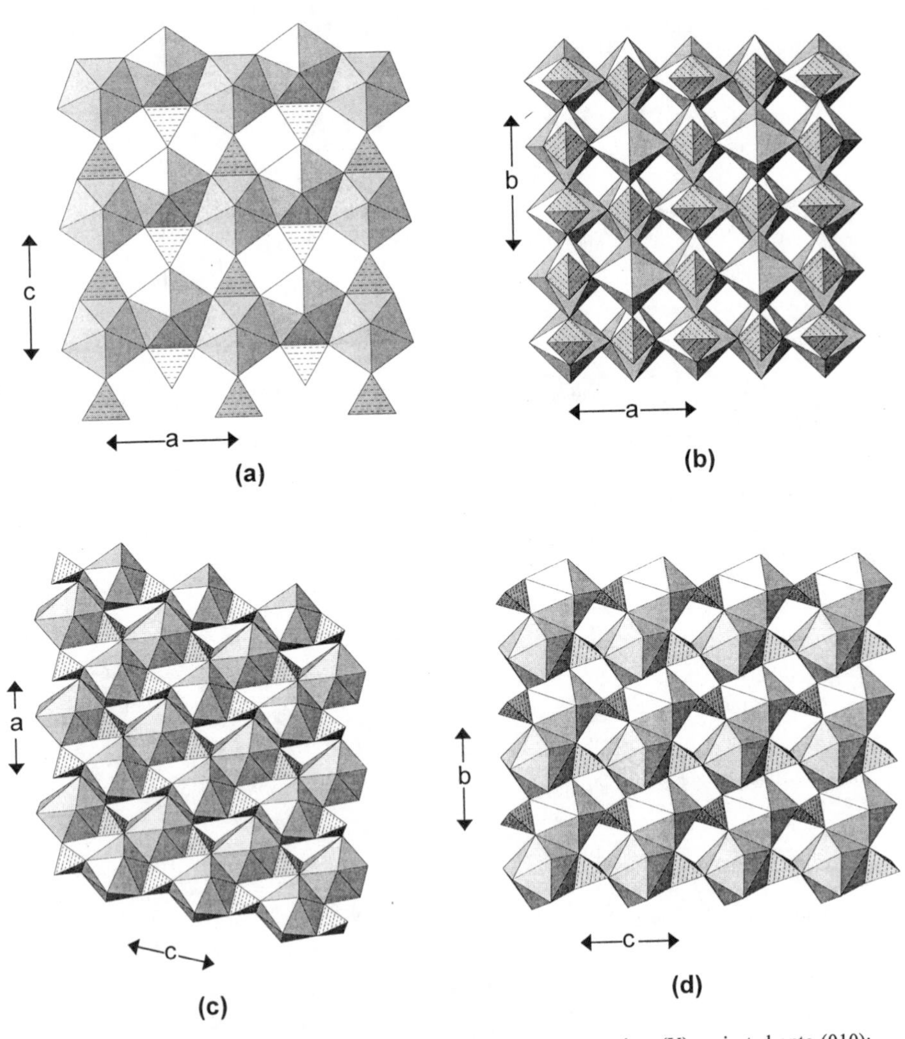

Figure 59. The crystal structures of xenotime and monazite: (a) xenotime-(Y) projected onto (010); (b) xenotime-(Y) projected onto (001); (c) monazite-(Ce) projected onto (010); (d) monazite-(Ce) projected onto (100). $(REE\phi_7)$ polyhedra are shadow-shaded.

hedral arrangement. These dodecahedra polymerize by sharing edges to form chains that extend in the *a*- and *b*-directions (Fig. 60a). These chains link in the *c*-direction by sharing edges with (PO_4) tetrahedra to form sheets parallel to (100) and (010) (Fig. 60a). These sheets interpenetrate in the *a*- and *b*-directions (Fig. 60b) to form a framework with large hexagonal channels extending parallel to the *c*-axis.

Archerite, $K(PO_2\{OH\}_2)$, and **biphosphammite**, $(NH_4)(PO_2\{OH\}_2)$, are isostructural. In archerite, K is [8]-coordinated by O-atoms that are arranged at the vertices of a Siamese dodecahedron. These dodecahedra share edges to form a chain in the *b*- (and *a*-) directions, and adjacent chains link by sharing edges with (PO_4) tetrahedra (Fig. 60c) to form layers parallel to (011) and (101). These layers meld by sharing edges (i.e., mutually intersecting) to form a framework (Fig. 60d) that is topologically identical to the

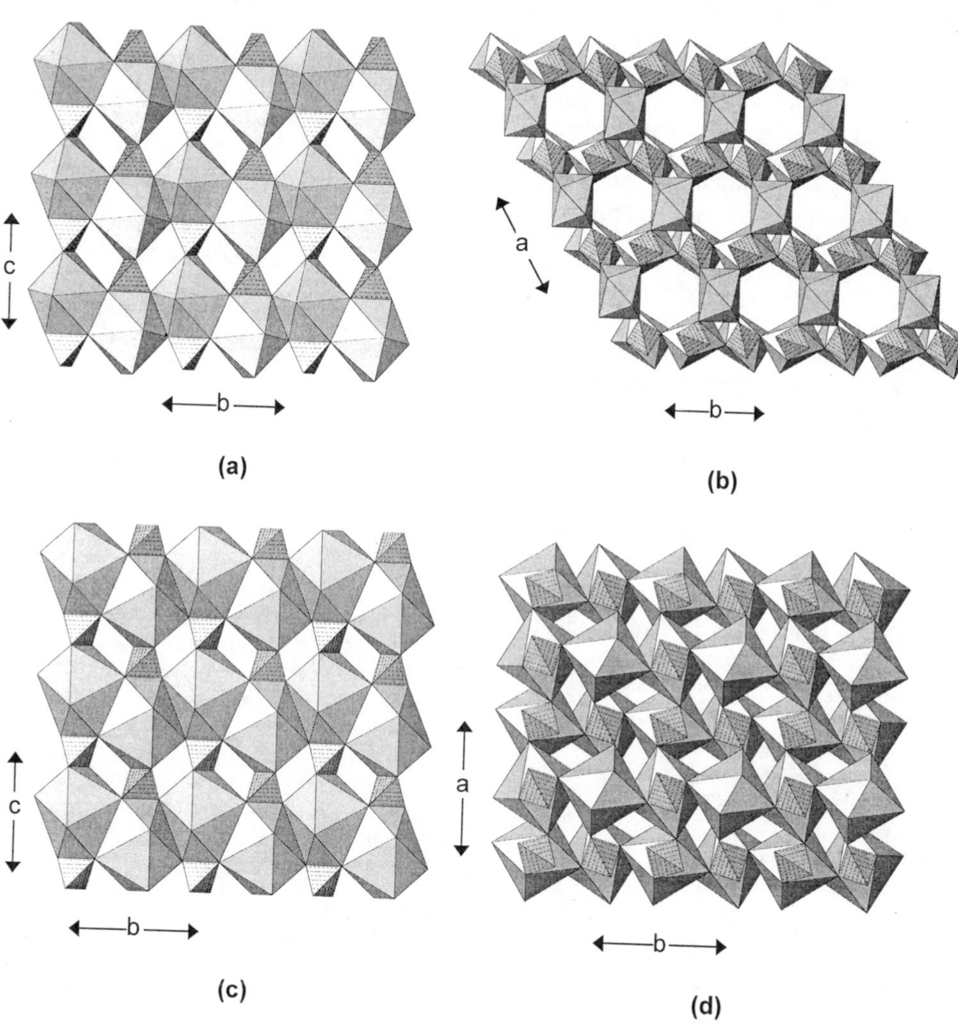

Figure 60. The crystal structure of rhabdophane and archerite: (a) rhabdophane projected onto (010); (b) rhabdophane projected onto (001); $(REE\phi_7)$ polyhedra are shadow-shaded; (c) archerite projected onto (100); (d) archerite projected onto (001); $(K\phi_8)$ polyhedra are shadow-shaded.

framework in xenotime (Figs. 59a,b), although geometrical distortions result in a lower symmetry arrangement in archerite ($I\bar{4}2d$) as compared with xenotime ($I4_1/amd$).

Brushite, $Ca(H_2O)_2(PO_3\{OH\})$, and **churchite**, $Y(H_2O)_2(PO_4)$, are essentially isostructural, although Curry and Jones (1971) report the space group *Ia* for brushite and Kohlmann et al. (1994) report *I*2/*a* for churchite. The large cation is [8]-coordinated with the bonded anions in a dodecahedral arrangement. The dodecahedra share edges with the $(P\phi_4)$ tetrahedra to form chains that extend in the [101] direction (Fig. 61a); these chains are a common feature of large-cation structures, and occur in gypsum and other Ca-sulfate minerals. These chains link in the [$10\bar{1}$] direction by sharing edges between $(Ca\phi_8)$ polyhedra of adjacent chains, and by sharing of vertices between $(P\phi_4)$ tetrahedra and $(Ca\phi_8)$ dodecahedra, forming a dense sheet parallel to (101) (Fig. 61a). These sheets stack in the *b*-direction (Fig. 61b) and are linked solely by hydrogen bonds (not shown in Fig. 61b).

Ardealite, $Ca_2(H_2O)_4(PO_3\{OH\})(SO_4)$, is an intriguing structure in that P and S seem to be disordered over the two symmetrically distinct tetrahedrally coordinated sites (Sakae et al. 1978); presumably, the acid H atom is locally associated with the $(P\phi_4)$ tetrahedra and hence shows analogous disorder. In the synthetic analogue, each of the two Ca atoms is coordinated by six O-atoms and two (H_2O) groups in a dodecahedral arrangement (as is the case in brushite). Chains of $(Ca\phi_8)$ and $(\{P,S\}\phi_4)$ polyhedra are formed by edge-sharing between the two types of polyhedra, chains that are topologically identical to the corresponding chains in brushite (Fig. 61a). These chains extend along [110] (Fig. 61c) and [001] (Fig. 61d), forming thick slabs that resemble the slabs in brushite (cf. Figs. 61b,e). Intercalated between these slabs are sheets of [8]-coordinated Ca and tetrahedra (Fig. 61e), and the structure is held together by a network of hydrogen bonds, the details of which are not known.

Dorfmanite, $Na_2(PO_3\{OH\})(H_2O)_2$, contains Na in both [5]- and [6]-coordination. $(Na\phi_5)$ polyhedra occur at the vertices of a 6^3 plane net and link by sharing corners (Fig. 62a) to form a sheet that is decorated by $(P\phi_4)$ tetrahedra. $(Na\phi_6)$ octahedra share edges to form chains that extend in the *c*-direction and are linked in the *b*-direction by the $(P\phi_4)$ groups (Fig. 62b) that decorate the underlying sheet of Figure 62a. These two sheets stack in the *a*-direction (Fig. 62c) and link through the $(P\phi_4)$ groups.

Monetite, $CaH(PO_4)$, contains Ca in both [7]- and [8]-coordination, and the polyhedra share edges to form chains that extend in the *a*-direction. These chains are linked in the *b*-direction by sharing edges and vertices of the (CaO_n) polyhedra with $(P\phi_4)$ groups (Fig. 62d). These sheets link in the *c*-direction via corner-sharing between (CaO_n) polyhedra and $(P\phi_4)$ groups. Catti et al. (1977a) have carefully examined the evidence for a symmetrical hydrogen-bond in monetite. In space group $P\bar{1}$, one of the three symmetrically distinct H-atom sites lies on, or disordered off, a centre of symmetry, and another H-atom is statistically distributed between two centrosymmetric positions. In space group *P*1, the first H-atom is displaced slightly off the pseudo-centre of symmetry, and another H-atom is either ordered or disordered. Catti et al. (1977a) propose that the crystal they examined is a mixture of domains of both $P\bar{1}$ and *P*1 structure.

Nacaphite, $Na_2Ca(PO_4)F$, contains six octahedrally coordinated sites, four of which are each half-occupied by Ca and Na, and two of which are occupied solely by Na. Two $\{(Ca_2Na_2)O_6]$ and one (NaO_6) octahedra link to form an $[M_3\phi_{11}]$ trimer, and there are two such symmetrically distinct trimers in this structure. The trimers link in the (100) plane by sharing corners (Fig. 63a), and the resultant sheet is braced by (PO_4) tetrahedra that link three adjacent trimers. The sheets stack in the *a*-direction (Fig. 63b), and are linked by edge-and face-sharing between trimers and by corner-sharing between (PO_4) groups and trimers of adjacent layers. This nacaphite structure is related to the structures of

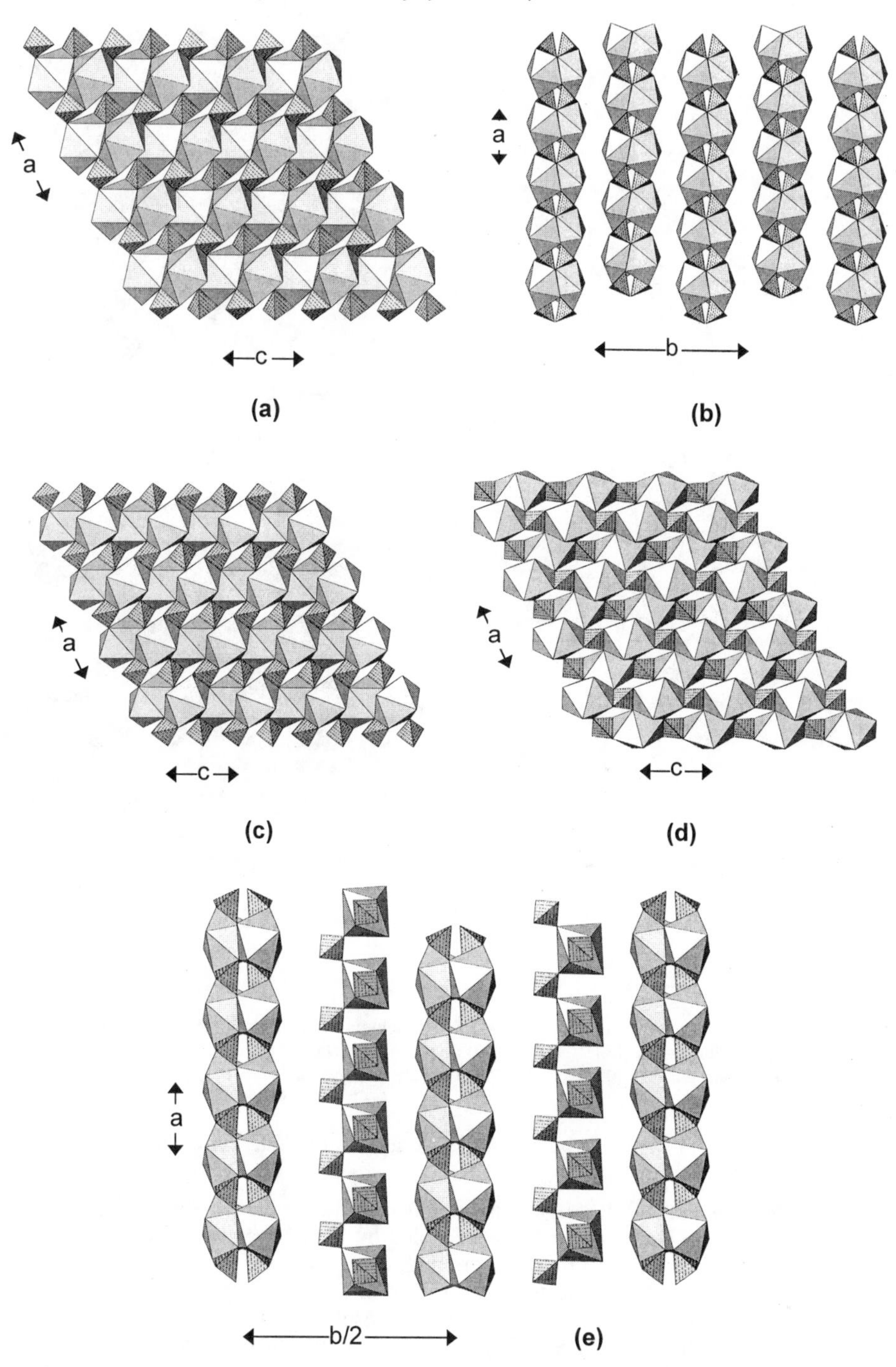

Figure 61. The crystal structures of brushite and ardealite: (a) brushite projected onto (010); (b) brushite projected onto (001); (c) ardealite: one layer projected onto (010); (d) ardealite: the next layer projected onto (010); (e) ardealite projected onto (001). ($Ca\phi_8$) polyhedra are shadow-shaded.

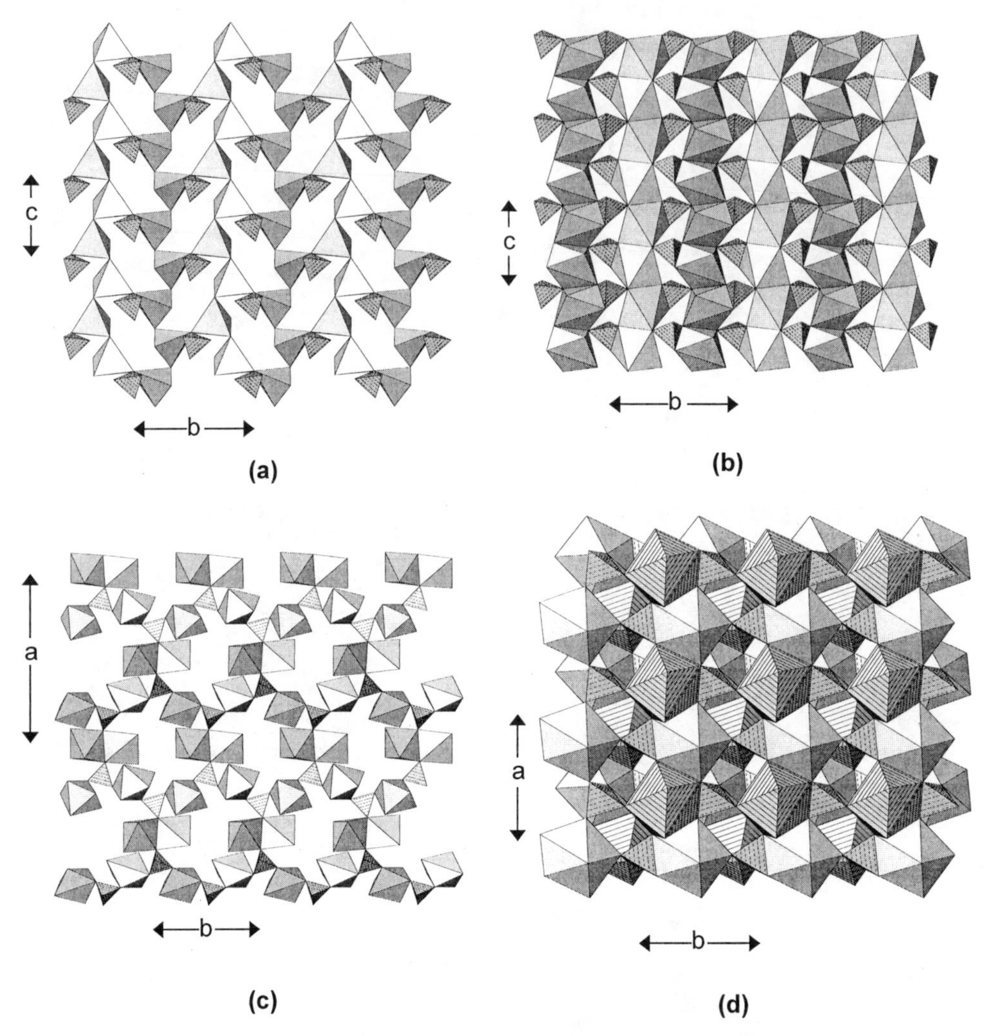

Figure 62. The crystal structures of dorfmanite and monetite: (a), (b) dorfmanite sheets projected onto (100); (c) dorfmanite projected onto (001); ($Na\phi_5$) and ($Na\phi_6$) polyhedra are shadow-shaded; (d) monetite projected onto (001); ($Ca\phi_7$) polyhedra are shadow-shaded, ($Ca\phi_8$) polyhedra are line-shaded.

arctite, quadruphite (Table 13, below) and several alkali-sulfate minerals (Sokolova and Hawthorne 2001).

Arctite, $(Na_5Ca)Ca_6Ba(PO_4)_6F$, contains one [12]-coordinated Ba, one [7]-coordinated Ca, and one [7]-coordinated site occupied by both Na and Ca. The (CaO_7) and ($\{Na,Ca\}O_7$) polyhedra link to form trimers (Fig. 63c), and these trimers link in the (001) plane to form a sheet. These sheets stack in the *c*-direction, the trimers linking to form truncated columns parallel to the *c*-direction; the result is a thick slab parallel to (001). The (BaO_{12}) icosahedra form a hexagonal array parallel to (001) and are linked by corner-sharing with (PO_4) groups in an arrangement that is also found in the glaserite (and related) structures. This layer is intercalated with the thick slabs to form the rather densely packed arctite arrangement (Fig. 63d).

Nabaphite, $NaBa(PO_4)(H_2O)_9$, and **nastrophite**, $NaSr(PO_4)(H_2O)_9$, are isostructural. Their cation positions have been located but the (PO_4) groups show extensive orientational disorder. There is one Na site and one Ba(Sr) site; the former is octahedrally coordinated and the latter is [9]-coordinated with the anions in a triaugmented triangular-prismatic arrangement. Baturin et al. (1981) note that nastrophite dehydrates easily 'in air,' and the (PO_4)-group disorder may be associated with incipient dehydration.

Lithiophosphate, Li_3PO_4, consists of a framework of (LiO_4) and (PO_4) tetrahedra.

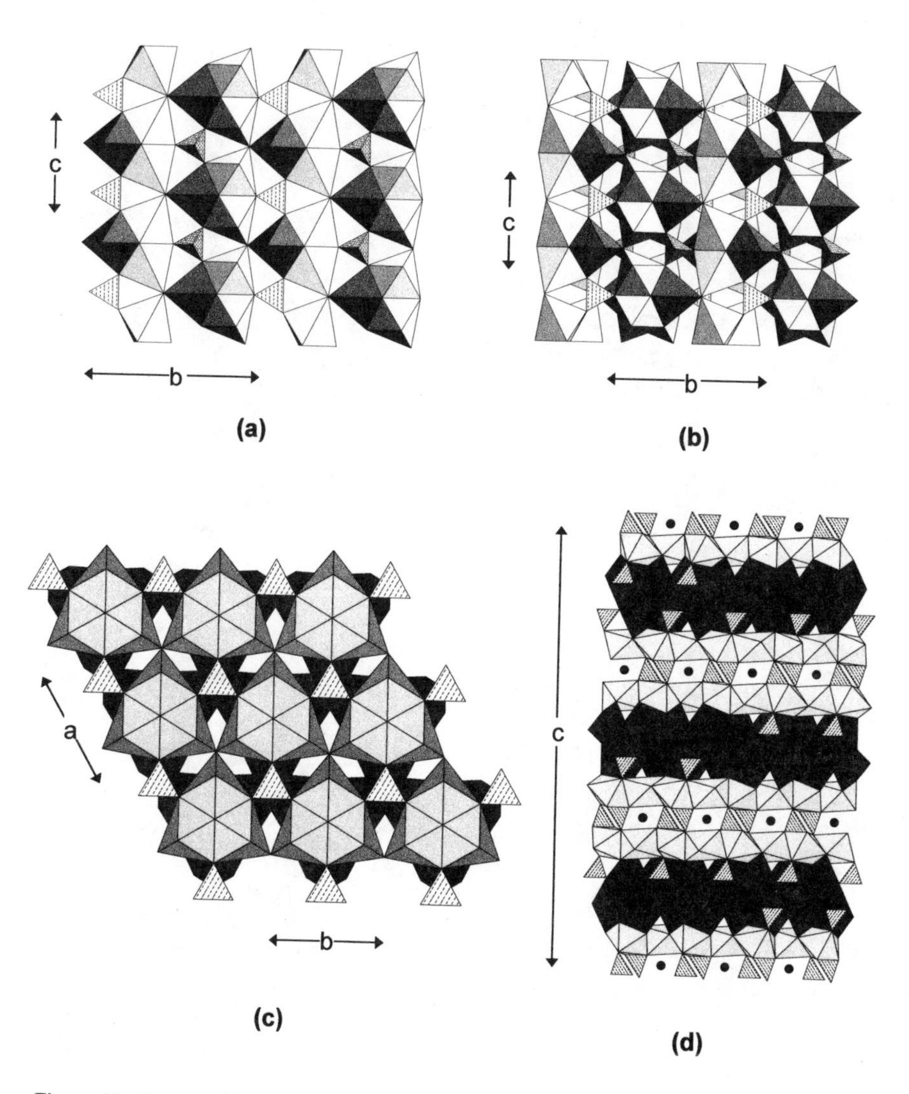

Figure 63. The crystal structures of nacaphite and arctite: (a) one layer of nacaphite projected onto (100); (b) two layers of nacaphite projected onto (100); (c) two layers of arctite projected onto (001); (d) the stacking of layers along [001] in arctite. ($\{Na,Ca\}\phi_n$) polyhedra are shadow-shaded, Ba are shown as dark circles.

From a geometrical perspective, it could also be classified as a member of the class of structures with polymerized (TO_4) groups (i.e., a framework structure of the types listed in Table 4). However, because of the low bond-valence (~0.25 vu) of the (LiO_4) groups, we have chosen to classify it as a 'large-cation' phosphate. (LiO_4) and (PO_4) tetrahedra are arranged at the vertices of a 3^6 plane net (Fig. 64a) and link by sharing corners; each (PO_4) group is surrounded by six (LiO_4) groups, and each (LiO_4) group is surrounded by two (PO_4) groups and four (LiO_4) groups. Both types of tetrahedra point both up and down the *c*-axis and link between adjacent sheets that stack in the *c*-direction (Fig. 64b).

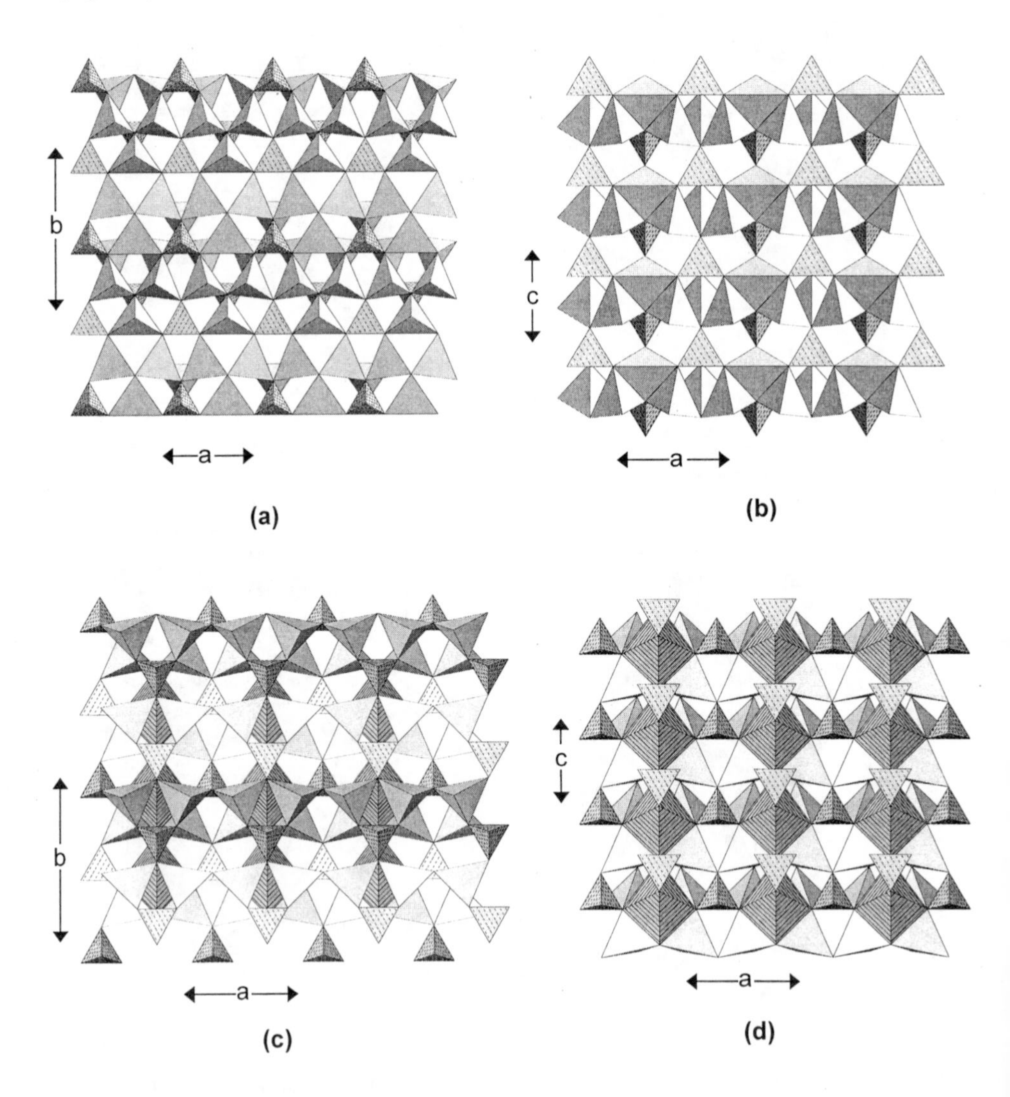

Figure 64. The crystal structures of lithiophosphate and nalipoite: (a) lithiophosphate projected onto (001); (b) lithiophosphate projected onto (010); (c) nalipoite projected onto (001); (d) nalipoite projected onto (010). (LiO_4) are shadow-shaded, (NaO_6) are line-shaded.

Nalipoite, $NaLi_2(PO_4)$, contains octahedrally coordinated Na and tetrahedrally coordinated Li. From a geometrical perspective, nalipoite could be classified as a structure involving polymerization of tetrahedra. Conversely, it can be classified as a large-cation structure from a bond-valence perspective, as the bond valence of $^{[4]}$Li-O is only 0.25 vu. We adopt the latter approach here. The (PO_4) and (LiO_4) tetrahedra occur at the vertices of a (6·3·6·3)(3·6·6·3) net; all 3-rings consist of two (LiO_4) tetrahedra and one (PO_4) tetrahedron, and the 6-rings show the sequence (Li-Li-P-Li-Li-P). The result is a layer of tetrahedra parallel to (100) (Fig. 64c,d) in which chains of corner-sharing (LiO_8) tetrahedra extend in the *a*-direction, and are linked in the *b*-direction by (PO_4) tetrahedra. Tetrahedra point along $\pm c$, and layers of tetrahedra meld in this direction to form a framework. Octahedrally coordinated Na occupies the interstices within this framework.

Nefedovite, $Na_5Ca_4(PO_4)_4F$, consists of [8]-coordinated Ca and both [7]- and [10]-coordinated Na. The $(Na\phi_{10})$ polyhedra share apical corners to form chains that extend in the *c*-direction (Fig. 65a) with (PO_4) tetrahedra linked to half of the meridional vertices. The (NaO_7) polyhedra share apical corners to form chains extending in the *c*-direction, and the individual polyhedra share corners with the $(Na\phi_{10})$ polyhedra, each bridging adjacent polyhedra along each chain (Fig. 65a). These chains are linked in the (001) plane by $(Ca\phi_8)$ polyhedra. The decorated $(Na\phi_{10})$-polyhedron chains have a square pinwheel appearance when viewed down [001] (Fig. 65b), and they are surrounded by a dense edge-sharing array of (NaO_7) and $(Ca\phi_8)$ polyhedra.

Olgite, $NaSr(PO_4)$, contains one [12]-coordinated site occupied by Sr (+ Ba), one [6]-coordinated site occupied by Na, and two [10]-coordinated sites occupied by both Sr and Na. The Sr site is situated at the vertices of a 3^6 plane net and is icosahedrally coordinated; adjacent icosahedra share edges to form a continuous sheet parallel to (001) (Fig. 65c) that is decorated by (PO_4) tetrahedra that share edges with the icosahedra. The [10]-coordinated Sr, Na sites share corners to form a sheet parallel to (001) (Fig. 65d). The [10]-coordinated polyhedra each have six peripheral anions, one apical anion along $+c$ and three apical anions along $-c$ (obscured in Fig. 65d). (NaO_6) octahedra are embedded on the underside of this sheet, sharing faces with the (Sr,NaO_{10}) polyhedra, and only one octahedron face is visible in Figure 65d (except at the edges of the sheet). This sheet is also decorated with (PO_4) tetrahedra. These two types of sheets stack alternately in the *c*-direction (Fig. 65e).

Phosphammite, $(NH_4)_2(PO_3\{OH\})$, consists of isolated $(P\phi_4)$ groups linked by hydrogen bonds involving (NH_4) groups. Khan et al. (1972) show that there are five oxygen atoms closer than 3.4 Å, but give a persuasive argument (based on stereochemistry) that there are only four hydrogen bonds from the (NH_4) group to the coordinating oxygen atoms. Consequently, we have drawn the 'large-cation' polyhedron as an $\{(NH_4)O_4\}$ tetrahedron (Figs. 66a,b). The $\{(NH_4)O_4\}$ tetrahedra occur at the vertices of a 6^3 net, linking to adjacent $\{(NH_4)O_4\}$ tetrahedra by sharing corners. The $(PO_3\{OH\})$ tetrahedra link to the $\{(NH_4)O_4\}$ tetrahedra, bridging across the six-membered rings of $\{(NH_4)O_4\}$ tetrahedra (Fig. 66a). The resultant layers link in the *c*-direction by corner-sharing between both $\{(NH_4)O_4\}$ tetrahedra, and between $\{(NH_4)O_4\}$ and $(PO_3\{OH\})$ tetrahedra (Fig. 66b).

Vitusite-(Ce), $Na_3Ce(PO_4)_2$, is a modulated structure, the substructure of which is related to the glaserite structure-type. The substructure contains two [8]-coordinated REE sites and six Na sites, two of which are [6]-coordinated and four of which are [7]-coordinated, together with four distinct (PO_4) groups. The typical unit of the glaserite arrangement is a large-cation polyhedron surrounded by a 'pinwheel' of six tetrahedra. In

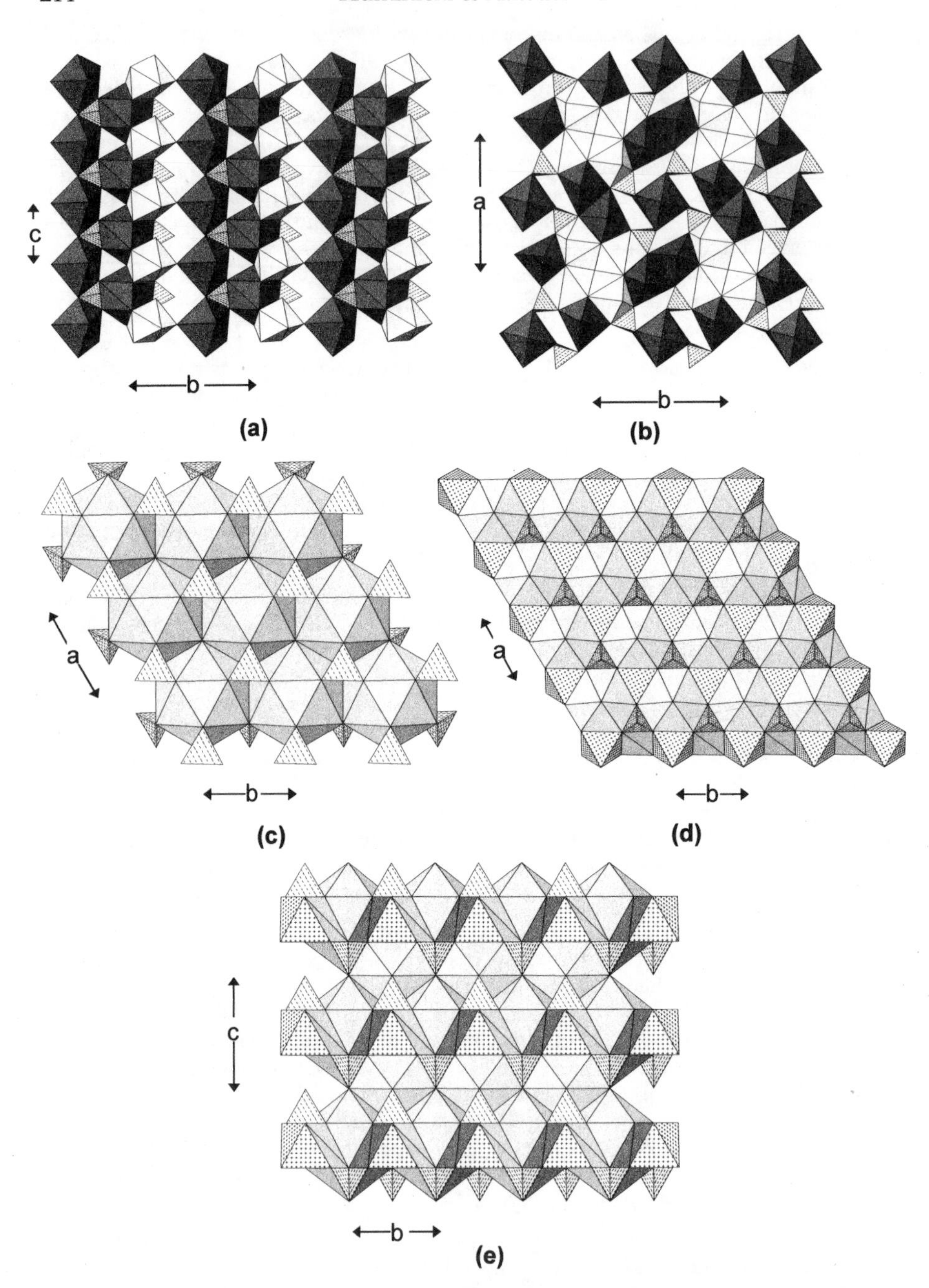

Figure 65. The crystal structures of nefedovite and olgite: (a) nefedovite projected onto (100); (b) nefedovite projected onto (001); ($Na\phi_n$) polyhedra are dark-shadow-shaded, ($Ca\phi_8$) polyhedra are light-shadow-shaded; (c) (SrO_{12})-icosahedron layer in olgite projected onto (001); (d) the (Sr,NaO_{10})-(NaO_7)-polyhedron layer in olgite projected onto (001); (e) the stacking of layers along [001] in olgite; (SrO_{12}) icosahedra are dark-shadow-shaded, (NaO_{10}) polyhedra are dark-shaded, (NaO_6) octahedra are dot-shaded.

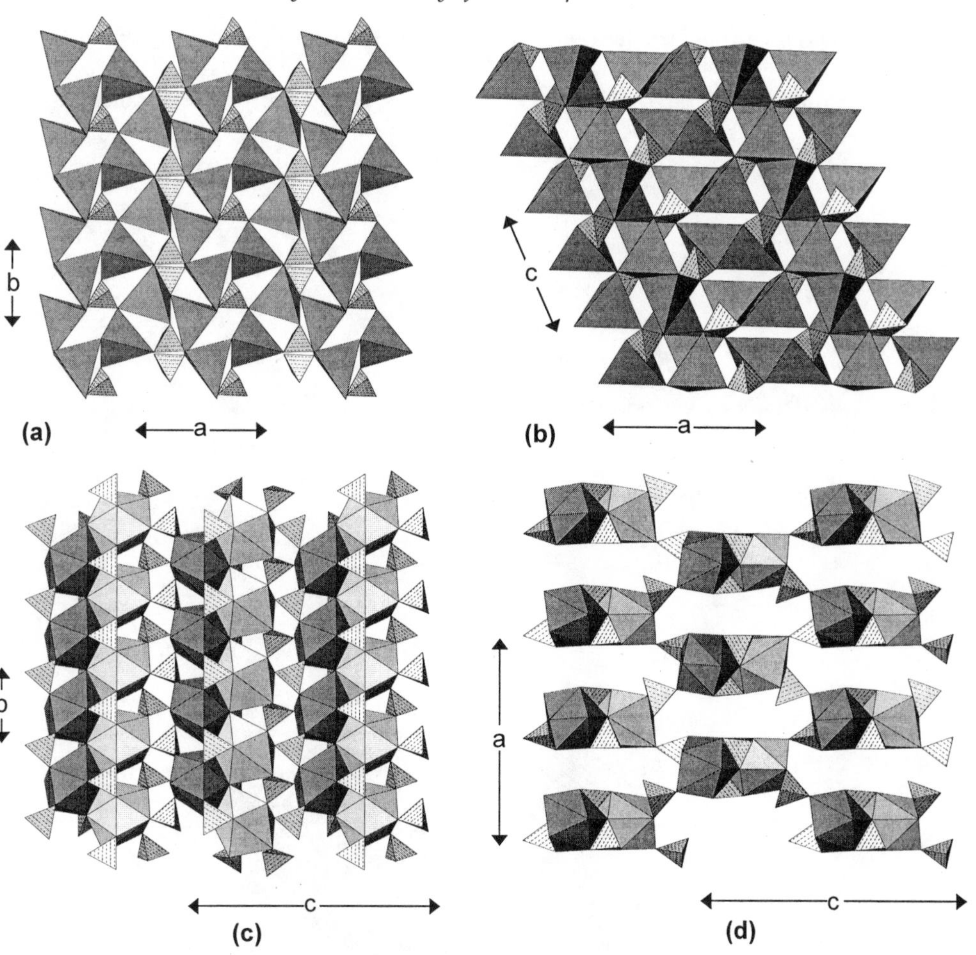

Figure 66. The crystal structures of phosphammite and vitusite: (a) phosphammite projected onto (001); (b) phosphammite projected onto (010); ($\{NH_4\}O_4$) tetrahedra are shadow-shaded; (c) vitusite-(Ce) projected onto (100); (d) vitusite-(Ce) projected onto (010). ($REEO_8$) polyhedra are shadow-shaded.

vitusite-(Ce), this unit is present (Fig. 66c), but is perturbed by edge-sharing between the large-cation polyhedra. The layers of Figure 65e stack in the *a*-direction by linkage between (PO_4) groups and the large-cation polyhedra. The structure is modulated in the *a*-direction, producing 5*a*, 8*a* and 11*a* modulations; the 8*a*-modulated structure seems to be the most common, and it was characterized by Mazzi and Ungaretti (1994). The modulations involve displacements of some O-atoms of the structure such that there are changes in the large-cation coordinations from those observed in the substructure.

Stercorite, $(NH_4)Na(H_2O)_3(PO_3\{OH\})(H_2O)$, consists of octahedrally coordinated Na that polymerizes by sharing *trans* edges to form an [$Na\phi_4$] chain that extends in the *b*-direction and is decorated by acid ($P\phi_4$) groups that share vertices with the octahedra and are arranged in a staggered fashion on each side of the chain (Figs. 67a,b). All ligands not involving ($P\phi_4$) groups are (H_2O) groups, i.e., ($NaO\{H_2O\}_5$). Chains are linked in the *a*- and *c*-directions by a complicated network of hydrogen bonds (Ferraris and Franchini-

Angela 1974) involving the interstitial (NH_4) group and an interstitial (H_2O) group that is held in the structure solely by hydrogen bonds.

Natrophosphate, $Na_7(PO_4)_2F(H_2O)_{19}$, contains octahedrally and tetrahedrally

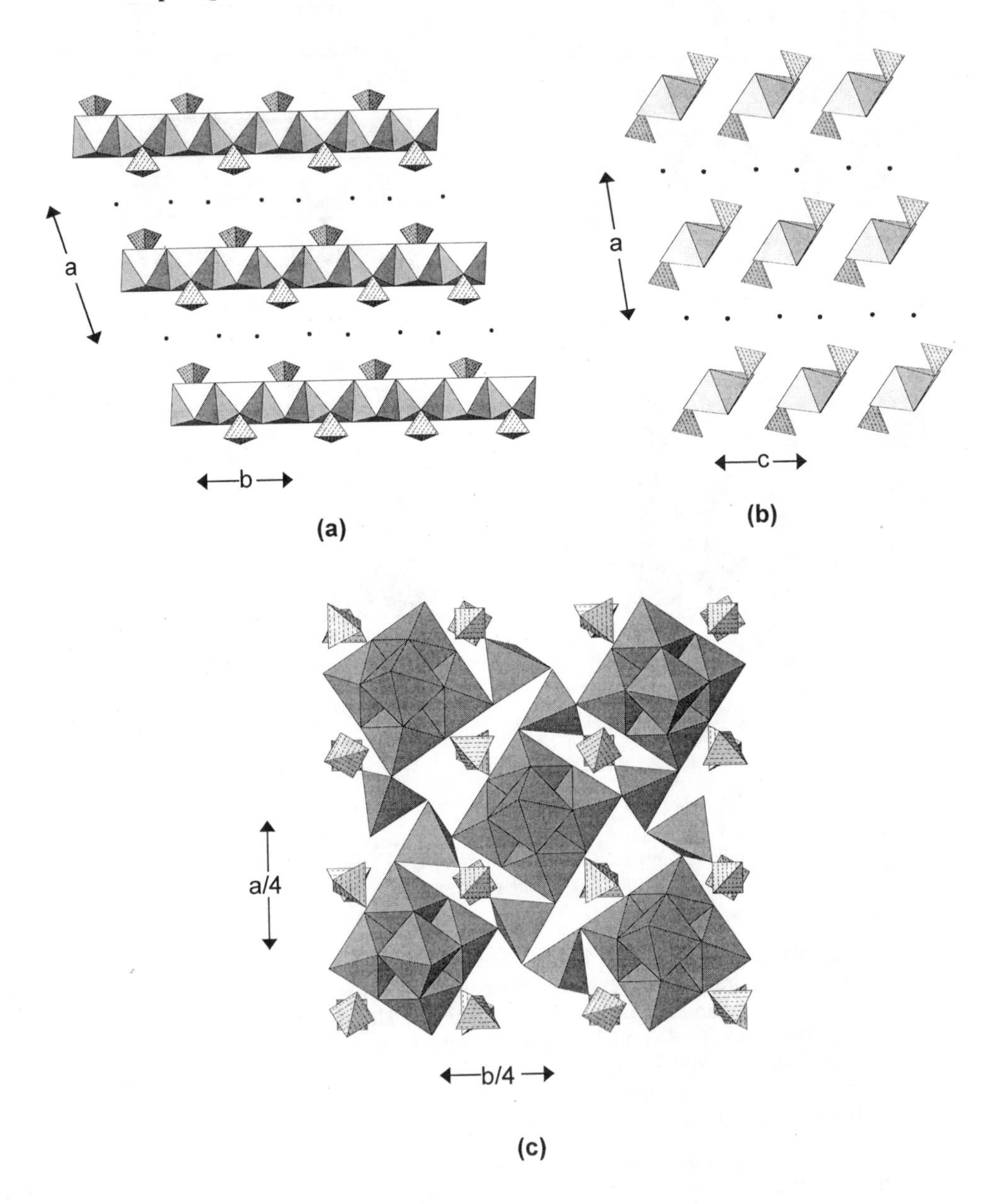

Figure 67. The crystal structures of stercorite and natrophosphate: (a) stercorite projected onto (001); (b) stercorite projected onto (010); (c) natrophosphate projected onto (001). $(Na\phi_6)$ octahedra are shadow-shaded, $(Na\phi_5)$ polyhedron is shadow-shaded, (NH_4) are shown as small black circles.

coordinated Na. Six ($Na\phi_6$) octahedra share edges to form a compact cluster of the form [$Na_6\phi_{18}$] (Fig. 67c). At the centre of this cluster is one F atom that is coordinated by six Na atoms; the remaining anions of the cluster are either [2]- or [1]-coordinated and hence are (H_2O) groups. The (PO_4) groups do not link directly to these clusters, but link to ($Na\phi_4$) tetrahedra that also bridge adjacent clusters (Fig. 67c). The [4]-coordinated site is only half-occupied (as required for electroneutrality) by Na, and is also half-occupied by (H_2O), giving rise to the rather unusual stoichiometry (for such a high-symmetry mineral).

Buchwaldite, $NaCa(PO_4)$, contains three unique Ca atoms, each with a coordination number of [8], and three distinct Na atoms with coordination numbers of [7], [6] and [9], respectively. The (CaO_8) polyhedra share edges to form two distinct chains that extend in the *a*-direction (Fig. 68a). These chains link in the *c*-direction through (NaO_n) polyhedra and (PO_4) groups. The (CaO_8) polyhedra share both edges and vertices with (PO_4) groups to link in the *b*-direction (Fig. 68b), further linkage being provided by (NaO_n) polyhedra to produce a densely packed structure.

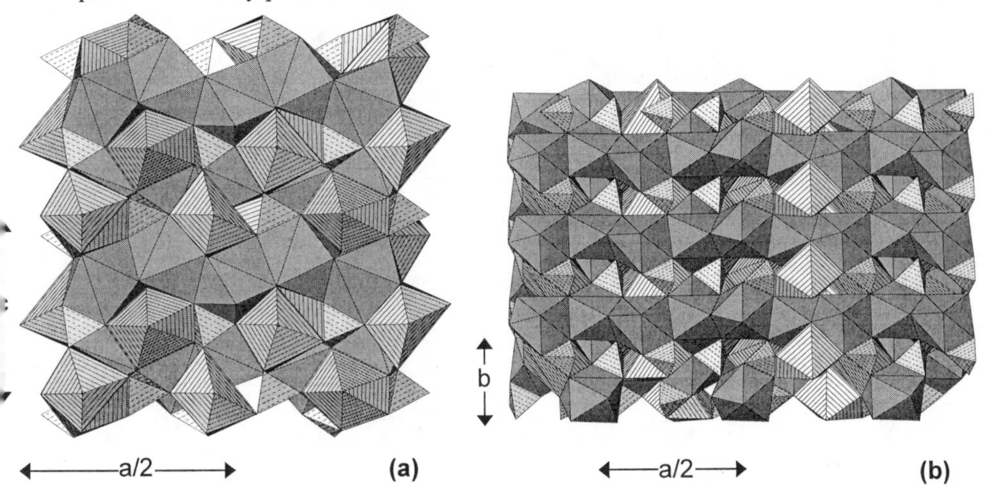

Figure 68. The crystal structure of buchwaldite: (a) projected onto (010); (b) projected onto (001). (CaO_8) polyhedra are shadow-shaded, (NaO_n) polyhedra are line-shaded.

APATITE-RELATED MINERALS

The minerals with apatite-like structures are dealt with in detail elsewhere in this volume and will not be considered here, except to list the relevant minerals (for completeness) in Table 12.

SILICOPHOSPHATE MINERALS

The silicophosphate (and related phosphate) minerals are a small group of extremely complicated structures that we will not describe here, because their complexity requires extensive illustration. For completeness, these minerals are listed in Table 13.

HEXAVALENT-URANIUM PHOSPHATE MINERALS

The hexavalent-uranium phosphate minerals (Table 14) are important and widespread uranyl-oxysalt minerals. Their structures and behavior are dominated by the crystal chemistry of the $(U^{6+}O_2)^{2+}$ uranyl group; they have been described in detail by Burns (1999) and will not be considered any further here.

Table 12. Minerals with apatite-like structures.

Mineral	*Formula*
Alforsite	$Ba_5(PO_4)_3Cl$
Belovite-(Ce)	$Sr_3NaCe(PO_4)_3(OH)$
Belovite-(La)	$Sr_3NaLa(PO_4)_3F$
Carbonate-fluorapatite	$Ca_5(PO_4,CO_3)_3F$
Carbonate-hydroxylapatite	$Ca_5(PO_4,CO_3)_3(OH)$
Chlorapatite	$Ca_5(PO_4)_3(Cl,F)$
Fluorapatite	$Ca_5(PO_4)_3F$
Hydroxylapatite	$Ca_5(PO_4)_3(OH)$
Pyromorphite	$Pb_5(PO_4)_3Cl$
Strontiumapatite	$Sr_6Ca_4(PO_4)_6F_2$

Table 13. Silicophosphate (and related phosphate) minerals.

Name	*Formula*	*Space group*
Attakolite	$CaMn^{2+}Al_4(SiO_3\{OH\})(PO_4)_3(OH)_4$	$C2/m$
Clinophosinaite	$Na_3Ca(SiO_3)(PO_4)$	$P2/c$
Harrisonite	$CaFe^{2+}(SiO_4)_2(PO_4)_2$	$R\bar{3}m$
Lomonosovite	$Na_5Ti^{4+}{}_2(Si_2O_7)(PO_4)O_2$	$P\bar{1}$
Polyphite	$Na_{17}Ca_3MgTi_4(Si_2O_7)_2(PO_4)_6O_3F_5$	$P\bar{1}$
Quadruphite	$Na_{14}CaMgTi^{4+}{}_4(Si_2O_7)_2(PO_4)_4O_4F_2$	$P\bar{1}$
Sobolevite	$Na_{11}(Na,Ca)_4(Mg,Mn^{2+})Ti^{4+}{}_4(Si_2O_7)_2(PO_4)_4O_3F_3$	$P\bar{1}$
Steenstrupine	$Na_{14}Ce_6Mn^{2+}Mn^{3+}Fe^{2+}{}_2Zr(Si_6O_{18})_2(PO_4)_7(H_2O)_3$	$R\bar{3}m$
Vuonnemite	$Na_{11}Ti^{4+}Nb_2(Si_2O_7)_2(PO_4)_2O_3(OH)$	$P\bar{1}$
Yoshimuraite	$Ba_2Mn_2TiO(Si_2O_7)(PO_4)(OH)$	$P\bar{1}$
Benyacarite	$(H_2O,K)_2TiMn^{2+}{}_2(Fe^{3+},Ti)_2(PO_4)_4(OH)_2(H_2O)_{14}$	$Pbca$

AN ADDENDUM ON CRYSTAL-CHEMICAL RELATIONS AMONG PHOSPHATE MINERALS

In providing a (fairly) complete hierarchical ordering of phosphate structures in this chapter, of necessity we have adopted a fairly broad-brush approach. There is a wealth of structural relations that we have not considered, and much of this is to be found in the original literature. Some papers have considered broader issues of structure and variations in bond topology in phosphates (and related minerals): Moore (1965b, 1970, 1973a,b; 1975b, 1976, 1984), Moore and Araki (1977c), Fanfani et al. (1978), Hawthorne (1979a, 1983a, 1985a,b; 1990, 1994, 1997, 1998). However, much work remains to be done in this area.

Table 14. Hexavalent-uranium phosphate minerals.

Meta-autunite group		***Phosphuranylite group***	
Meta-autunite*	$^{[6]}Ca(H_2O)_5[(UO_2)(PO_4)]_2(H_2O)$	Phuralumite	$^{[6]}Al_2(H_2O)_4(OH)_4[(UO_2)_3(PO_4)_2(OH)_2](H_2O)_6$
Meta-uranocircite	$^{[9]}Ba(H_2O)_4[(UO_2)(PO_4)]_2(H_2O)_2$		
Meta-ankoleite	$^{[8]}K(D_2O)_2[(UO_2)(PO_4)](LT)$	Upalite	$^{[6]}Al(H_2O)_6[(UO_2)_3(PO_4)_2(OH)O](H_2O)_1$
Meta-torbernite	$^{[6]}Cu^{2+}(H_2O)_4[(UO_2)(PO_4)]_2(H_2O)_4$	Mundite	$Al[(UO_2)_3(PO_4)_2(OH)O](H_2O)_{6.5}$
Saléeite	$^{[6]}Cu^{2+}(H_2O)_4[(UO_2)(PO_4)]_2(H_2O)_4$	Dewindite	$^{[8]}Pb^{2+}{}_2{}^{[11]}Pb^{2+}(H_2O)_{11}[(UO_2)_3(PO_4)_2(OH)O]_2(H_2O)_1$
Uramphite	$(NH_4)_2[(UO_2)(PO_4)]_2(H_2O)_6$	Francoisite(-Nd)	$^{[9]}Nd(H_2O)_5[(UO_2)_3(PO_4)_2(OH)O](H_2O)_1$
Na-meta-autunite	$Na_2[(UO_2)(PO_4)]_2(H_2O)_5$		
Bassetite	$Fe^{2+}[(UO_2)(PO_4)]_2(H_2O)_8$	Dumonite	$^{[8]}Pb^{2+}(H_2O)_4[(UO_2)_3(PO_4)_2O_2](H_2O)_1$
Lehnerite	$Mn^{2+}[(UO_2)(PO_4)]_2(H_2O)_8$	Phurcalite	$^{[8]}Ca^{[7]}Ca(H_2O)_7[(UO_2)_3(PO_4)_2O_2]$
Meta-autunite II	$Ca[(UO_2)(PO_4)]_2(H_2O)_2$	Bergenite	$(Ba,Ca)_2[(UO_2)_3O_2(PO_4)_2](H_2O)_{6.5}$
Przhevaskite	$Pb^{2+}[(UO_2)(PO_4)]_2(H_2O)_4$	Huegelite	$Pb^{2+}{}_2[(UO_2)_3(PO_4)_2O_2](H_2O)_5$
Chernikovite	$(H_3O)_2[(UO_2)(PO_4)]_2(H_2O)_6$		
Vochtenite	$(Fe^{2+},Mg)Fe^{3+}[(UO_2)(PO_4)]_4(OH)(H_2O)_{12-13}$	Althupite	$^{[6]}Al^{[9]}Th(H_2O)_8(OH)_3[(UO_2)\{(UO_2)_3(PO_4)_2(OH)O\}_2](H_2O)_7$
Ranunculite	$Al(OH)_3H[(UO_2)(PO_4)](H_2O)_4$	Yingjiangite	$K_2Ca[(UO_2)\{(UO_2)_3(PO_4)_2(OH)O_2\}_2](H_2O)_8$
		Renardite	$Pb^{2+}[(UO_2)\{(UO_2)_3(PO_4)_2O_2\}](H_2O)_9$
Walpurgite group		Phosphuranylite*	$^{[9]}K^{[8]}Ca(^{[3]}H_3O)_3(H_2O)_8[(UO_2)\{(UO_2)_3(PO_4)_2O_2\}_2]$
Walpurgite*	$^{[6]}Bi^{3+}{}_2{}^{[7]}Bi_2(H_2O)_2O_4[(UO_2)(AsO_4)_2]$	Meta-vanmeersscheite	$[U^{6+}(UO_2)_3(PO_4)(OH)_6](H_2O)_2$
Parsonite	$^{[9]}Pb^{2+}\,{}^{[6]}Pb[(UO_2)(PO_4)_2]$	Vanmeersscheite	$[U^{6+}(UO_2)_3(PO_4)(OH)_6](H_2O)_4$
Ulrichite	$CaCu^{2+}[(UO_2)(PO_4)_2](H_2O)_4$		
Pseudo-autunite	$Ca_2[(UO_2)_2(PO_4)_4](H_2O)_9$	***Miscellaneous***	
		Hallimondite	$Pb^{2+}[(UO_2)(AsO_4)_2]$

ACKNOWLEDGMENTS

We thank John Hughes for persuading us to write this chapter, for reviewing it, and for his indulgence regarding its length. We also thank Donna Danyluk for her assistance with compiling the references and the appendix. This work was supported by a Canada Research Chair in Crystallography and Mineralogy and by a Research Grant from the Natural Sciences and Engineering Research Council of Canada to FCH.

REFERENCES

Abbona F, Calleri M, Ivaldi G (1984) Synthetic struvite, $MgNH_4PO_4{\cdot}6H_2O$: correct polarity and surface features of some complementary forms. Acta Crystallogr B40:223-227

Abrahams SC (1966) Ferromagnetic and crystal structure of ludlamite, $Fe_3(PO_4)_2(H_2O)_4$ at 4.2 K. J Chem Phys 44:2230-2237

Adiwidjaja G, Friese K, Klaska K-H, Schlueter J (1999) The crystal structure of kastningite $(Fe_{0.5}Mn_{0.5})(H_2O)_4(Al_2(OH)_2(H_2O)_2(PO_4)_2)(H_2O)_2$, a new hydroxyl aquated orthophosphate hydrate mineral. Z Kristallogr 214:465-468

Alberti A (1976) Crystal structure of ferrisicklerite, $Li_{<1}(Fe^{(3+)},Mn^{(2+)})PO_4$. Acta Crystallogr B32: 2761-2764

Alkemper J, Fuess H (1998) The crystal structures of $NaMgPO_4$, $Na_2CaMg(PO_4)_2$ and $Na_{18}Ca_{13}Mg_5(PO_4)_{18}$: new examples for glaserite related structures. Z Kristallogr 213:282-287

Anderson JB, Shoemaker GL, Kostiner E, Ruszala FA (1977) The crystal structure of synthetic $Cu_5(PO_4)_2(OH)_4$, a polymorph of pseudomalachite. Am Mineral 62:115-121

Ankinovich EA, Bekanova GK, Shabanova TA, Zazubina IS, Sandomirsakaya SM (1997) Mitryaevaite, $Al_{10}[(PO_4)_{8.7}(SO_3OH)_{1.3}]_{10}AlF_3{\cdot}30H_2O$, a new mineral species from a Cambrian carbonaceous chert formation, Karatau Range and Zhabagly Mountains, southern Kazakhstan. Can Mineral 35:1415-1419

Antenucci D, Fontan F, Fransolet A-M (1989) X-ray powder diffraction data for wolfeite: $(Fe_{0.59}Mn_{0.40}Mg_{0.01})_2PO_4(OH)$. Powder Diffraction 4:34-35

Araki T, Moore PB (1981) Fillowite, $Na_2Ca(Mn, Fe)_7(PO_4)_6$: its crystal structure. Am Mineral 66:827-842

Araki T, Zoltai T (1968) The crystal structure of wavellite. Z Kristallogr 127:21-33

Araki T, Finney JJ, Zoltai T (1968) The crystal structure of augelite. Am Mineral 53:1096-1103

Arnold H (1986) Crystal structure of $FePO_4$ at 294 and 20 K. Z Kristallogr 177:139-142

Atencio D (1988) Chernikovite, a new mineral name for $(H_3O)_2(UO_2)_2(PO_4)_2{\cdot}6H_2O$ superseding "hydorgen autunite." Mineral Rec 19:249-252

Atencio D, Neumann R, Silva AJGC, Mascarenhas YP (1991) Phurcalite from Perus, Sao Paulo, Brazil, and redetermination of its crystal structure. Can Mineral 29:95-105

Bakakin VV, Rylov GM, Alekseev VI (1974) Refinement of the crystal structure of hurlbutite $CaBe_2P_2O_8$. Kristallografiya 19:1283-1285 (in Russian)

Bartl H (1989) Water of crystallization and its hydrogen-bonded cross linking in vivianite $Fe_3(PO_4)_2{\cdot}8(H_2O)$; a neutron diffraction investigation. Z Anal Chem 333:401-403

Baturin SV, Malinovskii YA, Belov NV (1981) The crystal structure of nastrophite $Na(Sr,Ba)(PO_4)(H_2O)_9$. Dokl Akad Nauk SSSR 261:619-623

Baturin SV, Malinovskii YA, Belov NV (1982) The crystal structure of nabaphite $NaBa(PO_4)(H_2O)_9$. Dokl Akad Nauk SSSR 266:624-627

Baur WH (1969a) The crystal structure of paravauxite, $FeAl_2(PO_4)_2(OH)_2(OH_2)_6(H_2O)_2$. N Jahrb Mineral Monatsh 430-433

Baur WH (1969b) A comparison of the crystal structures of pseudolaueite and laueite. Am Mineral 54: 1312-1322

Baur WH (1970) Bond length variation and distorted coordination polyhedra in inorganic crystals. Trans Am Crystallogr Assoc. 6:129-155

Baur WH (1974) The geometry of polyhedral distortions. Predictive relationships for the phosphate group. Acta Crystallogr B30:1195-1215

Baur WH (1977) Computer simulation of crystal structures. Phys Chem Minerals 2:3-20

Baur WH, Rama Rao B (1967) The crystal structure of metavauxite. Naturwiss 54:561

Baur WH, Rama Rao B (1968) The crystal structure and the chemical composition of vauxite. Am Mineral 53:1025-1028

Ben Amara M, Vlasse M, le Flem G, Hagenmueller P (1983) Structure of the low-temperature variety of calcium sodium orthophosphate, $NaCaPO_4$. Acta Crystallogr C39:1483-1485

Bernhard F, Walter F, Ettinger K, Taucher J, Mereiter K (1998) Pretulite, $ScPO_4$, a new scandium mineral from the Styrian and lower Austrian lazulite occurrences, Austria. Am Mineral 83:625-630

Birch WD, Mumme WG, Segnit ER (1988) Ulrichite: A new copper calcium uranium phosphate from Lake Boga, Victoria, Australia. Aust Mineral 3:125-131

Birch WD, Pring A, Bevan DJM, Kharisun (1994) Wycheproofite: a new hydrated sodium aluminum zirconium phosphate frmom Wycheproof, Victoria, Australia, and a new occurrence of kosnarite. Mineral Mag 58:635-639

Birch WD, Pring A, Foord EE (1995) Selwynite, $NaK(Be,Al)Zr_2(PO_4)_4 \cdot 2H_2O$, a new gainesite-like mineral from Wycheproof, Victoria, Australia. Can Mineral 33:55-58

Birch WD, Pring A, Self PG, Gibbs RB, Keck E, Jensen MC, Foord EE (1996) Meurigite, a new fibrous iron phosphate resembling kidwellite. Mineral Mag 60:787-793

Birch WD, Pring A, Kolitsch U (1999) Bleasdaleite $(Ca,Fe^{3+})_2$ Cu_5 (Bi,Cu) $(PO_4)_4$ $(H_2O,OH,Cl)_{13}$, a new mineral from Lake Boga, Victoria, Australia. Aust J Mineral 5:69-75

Bjoerling CO, Westgren A (1938) Minerals of the Varutraesk pegmatite. IX. X-ray studies on triphylite, varulite and their oxidation products. Geolog Foereningen Stockholm Foerhandlingar 412

Blanchard F (1981) ICDD Card # 33-0802 (sicklerite)

Blount AM (1974) The crystal structure of crandallite. Am Mineral 59:41-47

Borodin LS, Kazakova ME (1954) Belovite-(Ce): a new mineral from an alkaline pegmatite. Dokl Akad Nauk SSSR 96:613-616

Bowen NL (1928) The Evolution of Igneous Rocks. Princeton University Press, Princeton, New Jersey

Bragg WL (1930) The structure of silicates. Z Kristallogr 74:237-305

Bridge PJ, Clark RM (1983) Mundrabillaite, a new cave mineral from Western Australia. Mineral Mag 47: 80-81

Bridge PJ, Robinson BW (1983) Niahite—a new mineral from Malaysia. Mineral Mag 47:79-80

Britvin SN, Pakhomovskii YA, Bogdanova AN, Skiba VI (1991) Strontiowhitlockite, $Sr_9Mg(PO_3OH)$-$(PO_4)_6$, a new mineral species from the Kovdor deposit, Kola Peninsula, U.S.S.R. Can Mineral 29: 87-93

Britvin SN, Pakhomovskii YaA, Bogdanova AN, Khomyakov AP, Krasnova NI (1995) Rimkorolgite $(Mg,Mn)_5(Ba,Sr,Ca)(PO_4)_4 \cdot 8H_2O$—a new mineral from Kovdor iron deposit, Kola Peninsula. Zap Vser Mineral Obshch 124:90-95 (in Russian)

Britvin SN, Pakhomovskii YaA, Bogdanova AN (1996) Krasnovite $Ba(Al,Mg)(PO_4,CO_3)(OH)_2 \cdot H_2O$—a new mineral. Zap Vser Mineral Obshch 125:110-112

Brotherton PD, Maslen EN, Pryce MW, White AH (1974) Crystal structure of collinsite. Aust J Chem 27: 653-656

Brown ID (1981) The bond-valence method: an empirical approach to chemical structure and bonding. *In* Structure and Bonding in Crystals II. O'Keeffe M, Navrotsky A (eds) Academic Press, New York, p 1-30

Brown ID, Shannon RD (1973) Empirical bond strength-bond length curves for oxides. Acta Crystallogr A29:266-282

Brownfield ME, Foord EE, Sutley SJ, Bottinelly T (1993) Kosnarite, $KZr_2(PO_4)_3$, a new mineral from Mount Mica and Black Mountain, Oxford County, Maine. Am Mineral 78:653-656

Buchwald VF (1990) A new mineral, arupite, $Ni_3(PO_4)_2 \cdot 8H_2O$, the nickel analogue of vivianite. N Jahrb Mineral Monatsh 76-80

Buck HM, Cooper MA, _Cerny_ P, Grice JD, Hawthorne FC (1999) Xenotime-(Yb), $YbPO_4$, a new mineral species from the Shatford Lake pegmatite group, southeastern Manitoba. Can Mineral 37:1303-1306

Burdett JK, Hawthorne FC (1993) An orbital approach to the theory of bond valence. Am Mineral 78: 884-892

Burns PC (1999) The crystal chemistry of uranium. Rev. Mineral 38: 23-91

Burns PC (2000) A new uranyl phosphate chain in the structure of parsonite. Am Mineral 85:801-805

Burns PC, Hawthorne FC (1995) The crystal structure of sinkankasite, a complex heteropolyhedral sheet mineral. Am Mineral 80:620-627

Byrappa K (1983) The possible reasons for absence of condensed phosphates in Nature. Phys Chem Minerals 10:94-95

Cahill CL, Krivovichev SV, Burns PC, Bekenova GK, Shabanova TA (2001) The crystal structure of mitryaevaite, $Al_5(PO_4)_2[(P,S)O_3(OH,O)]_2F_2(OH)_2(H_2O)_8 \cdot 6.48H_2O$, determined from a microcrystal using synchrotron radiation. Can Mineral 39:179-186

Calvo C (1968) The crystal structure of graftonite. Am Mineral 53:742-750

Carling SG, Day P, Visser D (1995) Crystal and magnetic structures of layer transition metal phosphate hydrates. Inorg Chem 34:3917-3927

Catti M, Franchini-Angela M (1976) Hydrogen bonding in the crystalline state. Structure of $Mg_3(NH_4)_2(HPO_4)_4(H_2O)_8$ (hannayite), and crystal-chemical relationships with schertelite and struvite. Acta Crystallogr B32:2842-2848

Catti M, Ferraris G, Filhol A (1977a) Hydrogen bonding in the crystalline state. $CaHPO_4$ (monetite), $P\bar{1}$ or $P1$? A novel neutron diffraction study. Acta Crystallogr B33:1223-1229

Catti M, Ferraris G, Franchini-Angela M (1977b) The crystal structure of $Na_2HPO_4(H_2O)_2$. Competition between coordination and hydrogen bonds. Acta Crystallogr B33:3449-3452

Catti M, Ferraris G, Ivaldi G (1977c) Hydrogen bondng in the crystalline state. Structure of talmessite, $Ca_2(Mg,Co)(AsO_4)_2{\cdot}2H_2O$, and crystal chemistry of related minerals. Bull Minéral 100:230-236

Catti M, Ferraris G, Ivaldi G (1979) Refinement of the crystal structure of anapaite, $Ca_2Fe(PO_4)_2(H_2O)_4$. Hydrogen bonding and relationships with the bihydrated phase. Bull Minéral 102:314-318

Cavellec M, Riou D, Ferey G (1994) Synthetic spheniscidite. Acta Crystallogr C50:1379-1381

Cech F, Povondra P (1979) A re-examination of borickyite. Tschermaks Mineral Petrogr Mitt 26:79-86

Chao GY (1969) Refinement of the crystal structure of parahopeite. Z Kristallogr 130:261-266

Chen Z, Huang Y, Gu X (1990) A new uranium mineral-yingjiangite. Acta Mineral Sinica 10:102-105

Chernorukov N, Karyakin N, Suleimanov E, Belova Yu, Russ J (1997) ICDD Card #50-1561 (uranocircite)

Chopin C, Brunet F, Gebert W, Medenbach O, Tillmanns E (1993) Bearthite, $Ca_2Al(PO_4)_2(OH)$, a new mineral from high-pressure terranes of the western Alps. Schweiz Mineral Petrogr Mitt 73:1-9

Chopin C, Ferraris G, Prencipe M, Brunet F, Medenbach O (2001) Raadeite, $Mg_7(PO_4)_2(OH)_8$: A new dense-packed phosphate from Modum (Norway). Eur J Mineral 13:319-327

Cid-Dresdner H (1965) Determination and refinement of the crystal structure of turquois, $CuAl_6(PO_4)(OH)_8(H_2O)_4$. Z Kristallogr 121:87-113

Cipriani C, Mellini M, Pratesi G, Viti C (1997) Rodolicoite and grattarolaite, two new phosphate minerals from Santa Barbara mine, Italy. Eur J Mineral 9:1101-1106

Clark AM, Couper AG, Embrey PG, Fejer EE (1986) Waylandite: New data, from an occurrence in Cornwall, with a note on "agnesite." Mineral Mag 50:731-733

Cocco G, Fanfani L, Zanzzi PF (1966) The crystal structure of tarbuttite, $Zn_2(OH)PO_4$. Z Kristallogr 123: 321-329

Coda A, Guiseppetti G, Tadini C, Carobbi SG (1967) The crystal structure of wagnerite. Atti Accad Naz Lincei 43:212-224

Coleman LC, Robertson BT (1981) Nahpoite, Na_2HPO_4, a new mineral from the Big Fish River area, Yukon Territory. Can Mineral 19:373-376

Cooper M, Hawthorne FC (1994a) Refinement of the crystal structure of kulanite. Can Mineral 32:15-19

Cooper M, Hawthorne FC (1994b) The crystal structure of curetonite, a complex heteropolyhedral sheet mineral. Am Mineral 79:545-549

Cooper MA, Hawthorne FC (1999) The crystal structure of wooldridgeite, $Na_2CaCu^{(2+)}{}_2(P_2O_7)_2(H_2O)_{10}$, a novel copper pyrophosphate mineral. Can Mineral 37:73-81

Cooper MA, Hawthorne FC, Cerny P (2000) Refinement of the crystal structure of cyrilovite from Cyrilov, Western Moravia, Czech Republic. J Czech Geol Soc 45:95-100

Corbin DR, Abrams L, Jones GA, Harlow RL, Dunn PJ (1991) Flexibility of the zeolite RHO framework: effect of dehydration on the crystal structure of the beryllophosphate mineral, pahasapaite. Zeolites 11: 364-367

Corbridge DEC (1985) Phosphorous. An outline of its chemistry, biochemistry and technology (3rd edn). Elsevier, Amsterdam

Cozzupoli D, Grubessi O, Mottan A, Zanazzi PF (1987) Cyrilovite from Italy: Structure and crystal chemistry. Mineral Petrol 37:1-14

Curry NA, Jones DW (1971) Crystal structure of brushite, calcium hydrogen orthophosphate dihydrate: A neutron-diffraction investigation. J Chem Soc 3725-3729

Dai Y, Hughes JM (1989) Crystal-structure refinements of vanadinite and pyromorphite. Can Mineral 27: 189-192

de Bruiyn H, Beukes GJ, van der Westhuizen WA, Tordiffe EAW (1989) Unit cell dimensions of the hydrated aluminium phosphate-sulphate minerals sanjuanite, kribergite and hotsonite. Mineral Mag 53:385-386

Deliens M, Piret P (1981) Les phosphates d'uranyle et d'aluminum de Kobokobo. V. La mundite, nouveau mineral. Bull Minéral 104:669-671

Deliens M, Piret P (1982) Les phosphates d'uranyle et d'aluminum de Kobokobo. VI. La triangulite, $Al_3(OH)_5[(UO_2)(PO_4)]_4(H_2O)_5$, nouveau mineral. Bull Minéral 105:611-614

Deliens M, Piret P (1985) Les phosphates d'uranyle et d'aluminum de Kobokobo. V. La moreauite, $Al_3UO_2(PO_4)_3(OH)_2{\cdot}13H_2O$, nouveau mineral. Bull Minéral 108:9-13

Demartin F, Diella V, Donzelli S, Gramaccioli CM, Pilati T (1991) The importance of accurate crystal structure determination of uranium minerals. I. Phosphuranylite $KCa(H_3O)_3(UO_2)_7(PO_4)_4O_{4.8}H_2O$. Acta Crystallogr B47:439-446

Demartin F, Pilati T, Gay DD, Gramaccioli CM (1993) The crystal structure of a mineral related to paulkerrite. Z Kristallogr 208:57-71

Demartin F, Gramaccioli CM, Pilati T, Sciesa E (1996) Sigismundite, (Ba,K,Pb)Na_3(Ca,Sr)(Fe,Mg,Mn)$_{14}$Al$(OH)_2(PO_4)_{12}$, a new Ba-rich member of the arrojadite group from Spluga Valley, Italy. Can Mineral 34:827-834

Demartin F, Gay HD, Gramaccioli CM, Pilati T (1997) Benyacarite, a new titanium-bearing phosphate mineral species from Cerro Blanco, Argentina. Can Mineral 35:707-712

Dempsey MJ, Strens RGJ (1976) Modelling crystal structures. *In* Physics and Chemistry of Rocks and Minerals. Strens RGJ (ed) Wiley, New York, p 443-458

Di Cossato YMF, Orlandi P, Pasero M (1989a) Manganese-bearing beraunite from Mangualde, Portugal: Mineral data and structure refinement. Can Mineral 27:441-446

Di Cossato YMF, Orlandi P, Vezzalini G (1989b) Rittmanite, a new mineral species of the whiteite group from the Mangualde granitic pegmatite, Portugal. Can Mineral 27:447-449

Dick S (1999) Ueber die Struktur von synthetischem Tinsleyit $K(Al_2(PO_4)_2(OH)(H_2O))\cdot(H_2O)$. Z Naturforsch Anorg Chem Org Chem 54:1358-1390

Dick S, Zeiske T (1997) Leucophosphite $K(Fe_2(PO_4)_2(OH)(H_2O))\cdot(H_2O)$; hydrogen bonding and structural relationships. J Solid State Chem 133:508-515

Dick S, Zeiske T (1998) Francoanellit $K_3Al_5(HPO_4)_6(PO_4)_2\cdot 12H_2O$: Struktur und Synthese durch topochemische Entwaesserung von Taranakit. Z Naturforsch Anorg Chem Org Chem 53:711-719

Dick S, Gossner U, Weis A, Robl C, Grossmann G, Ohms G, Zeiske T (1998) Taranakite—the mineral with the longest crystallographic axis. Inorg Chim Acta 269:47-57

Dickens B, Brown WE (1971) The crystal structure of $Ca_7Mg_9(Ca, Mg)_2(PO_4)_{12}$. Tschermaks Mineral Petrogr Mitt 16:79-104

Dooley JR Jr, Hathaway JC (1961) Two occurrences of thorium-bearing minerals with rhabdophane-like structure. U S Geol Surv Prof Paper 424C:339

Dormann J, Gasperin M, Poullen JF (1982) Etude structurale de la sequence d'oxydation de la vivianite $Fe_3(PO_4)_2(H_2O)_8$. Bull Minéral 105:147-160

Dunn PJ, Rouse RC, Campbell TJ, Roberts WL (1984a) Tinsleyite, the aluminous analogue of leucophosphite, from the Tip Top pegmatite in South Dakota. Am Mineral 69:374-376

Dunn PJ, Rouse RC, Nelen JA (1984b) Englishite: New chemical data and a second occurrence, from the Tip Top pegmatite, Custer, South Dakota. Can Mineral 22:469-470

Dunn PJ, Peacor DR, Sturman DB, Ramik RA, Roberts WL, Nelen JA (1986) Johnwalkite, the Mn-analogue of olmsteadite, from South Dakota. N. Jahrb Mineral Monatsh 115-120

Eby RK, Hawthorne FC (1989) Cornetite: Modulated densely-packed Cu^{2+} oxysalt. Mineral Petrol 40: 127-136

Effenberger H, Mereiter K, Pimminger M, Zemann J (1982) Machatschkiite: Crystal structure and revision of the chemical formula. Tschermaks Mineral Petrogr Mitt 30:145-155

Effenberger H, Krause W, Belendorff K, Bernhardt HBJ, Medenbach O, Hybler J, Petricek V (1994) Revision of the crystal structure of mrazekite, $Bi_2Cu_3(OH)_2O_2(PO_4)_{2.2}(H_2O)$. Can Mineral 32:365-372

Egorov BL, Dara AD, Senderova VM (1969) Melkovite, a new phosphate-molybdate from the zone of oxidation. Zap Vses Mineral Obshch 98:207-212 (in Russian)

Ercit TS (1991) The crystal structure of nalipoite. Can Mineral 29:569-573

Ercit TS, Hawthorne FC, Cerny P (1986a) The crystal structure of bobfergusonite. Can Mineral 24:605-614

Ercit TS, Anderson AJ, Cerny P, Hawthorne FC (1986b) Bobfergusonite, a new primary phosphate mineral from Cross Lake, Manitoba. Can Mineral 24:599-604

Ercit TS, Cooper MA, Hawthorne FC (1998) The crystal structure of vuonnemite, $Na_{11}Ti^{(4+)}Nb_2(Si_2O_7)_2(PO_4)_2O_3(F, OH)$, a phosphate-bearing sorosilicate of the lomonosovite group. Can Mineral 36:1311-1320

Eventoff W, Martin R, Peacor DR (1972) The crystal structure of heterosite. Am Mineral 57:45-51

Eversheim VP, Kleber W (1953) Morphologie und Struktur des Reddingits, $P_2O_5\cdot 3FeO\cdot 3H_2O$. Acta Crystallogr 6:215-216

Fanfani L, Zanazzi PF (1966) La struttura cristallina della metastrengite. Atti Accad Naz Lincei Serie 8 40:889

Fanfani L, Zanazzi PF (1967a) The crystal structure of beraunite. Acta Crystallogr 22:173-181

Fanfani L, Zanazzi PF (1967b) Structural similarities of some secondary lead minerals. Mineral Mag 36: 522-529

Fanfani L, Zanazzi PF (1968) The crystal structure of vauquelinite and the relationships to fornacite. Z Kristallogr 126:433-443

Fanfani L, Zanazzi PF (1979) Switzerite: its chemical formula and crystal structure. Tschermaks Mineral Petrogr Mitt 26:255-269

Fanfani L, Nunzi A, Zanazzi PF (1970a) The crystal structure of wardite. Mineral Mag 37:598-605

Fanfani L, Nunzi A, Zanazzi PF (1970b) The crystal structure of fairfieldite. Acta Crystallogr B26:640-645

Fanfani L, Nunzi A, Zanazzi PF (1972) Structure and twinning in spencerite. Mineral Mag 38:687-692

Fanfani L, Nunzi A, Zanazzi PF, Zanzari AR (1975) The crystal structure of roscherite. Tschermaks Mineral Petrogr Mitt 22:266-277

Fanfani L, Nunzi A, Zanazzi PF, Zanzari AR (1976) Additional data on the crystal structure of montgomeryite. Am Mineral 61:12-14

Fanfani L, Zanazzi PF, Zanzari AR (1977) The crystal structure of a triclinic roscherite. Tschermaks Mineral Petrogr Mitt 24:169-178

Fanfani L, Tomassini M, Zanazzi PF, Zanzari AR (1978) The crystal structure of strunzite, a contribution to the crystal chemistry of basic ferric-manganous hydrated phosphates. Tschermaks Mineral Petrogr Mitt 25:77-87

Ferraris G, Franchini-Angela M (1973) Hydrogen bonding in the crystalline state. Crystal structure of $MgHAsO_4 \cdot 7H_2O$, rosslerite. Acta Crystallogr B29:286-292

Ferraris G, Franchini-Angela M (1974) Hydrogen bonding in the crystalline state. Crystal structure and twinning of $NaNH_4HPO_4(H_2O)_4$ (stercorite). Acta Crystallogr B30:504-510

Ferraris G, Fuess H, Joswig W (1986) Neutron diffraction study of $MgNH_4PO_4(H_2O)_6$ (struvite) and survey of water molecules donating short hydrogen bonds. Acta Crystallogr B42:253-258

Fisher DJ (1956) Hagendorfite unit cell. Geol Soc Am Bull 67:1694-1695

Fisher DJ (1965) Dickinsonites, fillowite and alluaudites. Am Mineral 50:1647-1669

Fisher FG, Meyrowitz R (1962) Brockite, a new calcium thorium phosphate from the Wet Mountains, Colorado. Am Mineral 47:1346-1355

Fitch AN, Cole M (1991) The structure of $KUO_2PO_{4.3}D_2O$ refined from neutron and synchrotron-radiation powder diffraction data. Material Res Bull 26:407-414

Fitch AN, Fender BEF (1983) The structure of deuterated ammonium uranyl phosphate trihydrate, $ND_4UO_2PO_4(D_2O)_3$ by powder neutron diffraction. Acta Crystallogr C39:162-166

Flachsbart I (1963) Zur Kristallstruktur von Phosphoferrit $(Fe, Mn)_3(PO_4)_2(H_2O)_3$. Z Kristallogr 118: 327-331

Fontan F, Pillard F, Permingeat F (1982) Natrodufrenite, $(Na,Fe^{3+},Fe^{2+})(Fe^{3+},Al)_5(PO_4)_4(OH)_6 \cdot 2(H_2O)$, a new mineral species of the dufrenite group. Bull Mineral 105:321 -326

Foord EE, Taggart JE Jr (1998) A reexamination of the turquoise group, the mineral aheylite, planerite (redefined), turquoise and coeruleolactite. Mineral Mag 62:93-111

Foord EE, Brownfield ME, Lichte FE, Davis AM, Sutley SJ (1994) McCrillisite, $NaCs(Be,Li)Zr_2(PO_4)_4 \cdot 1\text{-}2H_2O$, a new mineral species from Mount Mica, Oxford County, Maine, and new data for gainesite. Can Mineral 32:839-842

Fransolet A-M (1989) The problem of Na-Li substitution in primary Li-Al phosphates: New data on lacroixite, a relatively widespread mineral. Can Mineral 27:211-217

Fransolet A-M (1995) Wyllieite et rosemaryite dans la pegmatite de Buranga, Rwanda. Eur J Mineral 7: 567-575

Fransolet A-M, Cooper MA, Cerny_ P, Hawthorne FC, Chapman R, Grice JD (2000) The Tanco pegmatite at Bernic Lake, southeastern Manitoba. XV. Ercitite, $NaMn^{3+}PO_4(OH)(H_2O)_2$, a new phosphate mineral species. Can Mineral 38:893-898

Freeman K, Bayliss P (1991) ICDD card # 44-1429 (natromontebrasite)

Frondel C, Riska DD, Frondel JW (1956) X-ray powder data for uranium and thorium minerals. U S Geol Surv Bull B 1036-G:91-153

Galliski MA, Cooper MA, Hawthorne FC, Cerny P (1999) Bederite, a new pegmatite phosphate mineral from Nevados de Palermo, Argentina: description and crystal structure. Am Mineral 84:1674-1679

Galliski MA, Hawthorne FC, Cooper MA (2001) Refinement of the crystal structure of ushkovite from Nevados de Palermo, Republica Argentina. Can Mineral (accepted)

García-Guinea J, Chagoyen AM, Nickel EH (1995) A re-investigation of bolivarite and evansite. Can Mineral 33:59-65

Gatehouse BM, Miskin BK (1974) The crystal structure of brazilianite, $NaAl_3(PO_4)_2(OH)_4$. Acta Crystallogr B30:1311-1317

Geller S, Durand JL (1960) Refinement of the structure of $LiMnPO_4$. Acta Crystallogr 13:325-331

Genkina EA, Khomyakov AP (1992) Refinement of the structure of natural sodiumphosphate. Kristallografiya 37:1559-1560

Genkina EA, Kabalov YuK, Maksimov BA, Mel'nikov OK (1984) The crystal structure of synthetic tavorite $LiFe(PO_4)(OH,F)$. Kristallografiya 29:50-55

Genkina EA, Maksimov BA, Melnikov OK (1985) Crystal structure of synthetic tarbuttite $Zn_2(PO_4)(OH)$. Dokl Akad Nauk SSSR 282:314-317

Gheith MA (1953) Lipscombite: a new synthetic iron lazulite. Am Mineral 38:612-628

Ghose S, Leo SR, Wan C (1974) Structural chemistry of copper and zinc minerals. Part I. Veszelyite, $(Cu_{0.5}Zn_{0.5})_2ZnPO_4NOH)_3(H_2O)_3$: a novel type of sheet structure and crystal chemistry of copper-zinc substitution. Am Mineral 59:573-581

Giuseppetti G, Tadini C (1973) Refinement of the crystal structure of beryllonite, $NaBePO_4$. Tschermaks Mineral Petrogr Mitt 20:1-12

Giuseppetti G, Tadini C (1983) Lazulite, (Mg,Fe) $Al_2(OH)_2(PO_4)_2$: structure refinement and hydrogen bonding. N Jahrb Mineral Monatsh 410-416

Giuseppetti G, Tadini C (1984) The crystal structure of childrenite from Tavistock (SW England), $Ch_{89}Eo_{11}$ term of childrenite-eosphorite series. N Jahrb Mineral Monatsh 263-271

Giusepetti G, Tadini C (1987) Corkite, $PbFe_3(SO_4)(PO_4)(OH)_6$, its crystal structure and ordered arrangement of the tetrahedral cations. N Jahrb Mineral Monatsh 71-81

Giuseppetti G, Mazzi F, Tadini C (1989) The crystal structure of chalcosiderite $CuFe_6(PO_4)_4(OH)_8(H_2O)_4$. N Jahrb Mineral Monatsh 227-239

Gopal R, Calvo C, Ito J, Sabine WK (1974) Crystal structure of synthetic Mg-whitlockite, $Ca_{18}Mg_2H_2(PO_4)_{14}$. Can J Chem 52:1152-1164

Grice JD, Dunn PJ (1992) Attakolite: new data and crystal-structure determination. Am Mineral 77: 1285-1291

Grice JD, Groat LA (1988) Crystal structure of paulkellerite. Am Mineral 73:873-875

Grice JD, Roberts AC (1993) Harrisonite, a well-ordered silico-phosphate with a layered crystal structure. Can Mineral 31:781-785

Grice JD, Peacor DR, Robinson GW, van Velthuizen J, Roberts WL, Campbell TJ, Dunn PJ (1985) Tiptopite, $(Li,K,Na,Ca,G)_8Be_6(PO_4)_6(OH)_4$, a new mineral species from the Black Hills, South Dakota. Can Mineral 23:43-46

Grice JD, Dunn PJ, Ramik RA (1989) Whiteite-(CaMnMg), a new mineral species from the Tip Top pegmatite, Custer, South Dakota. Can Mineral 27:699-702

Grice JD, Dunn PJ, Ramik RA (1990) Jahnsite-(CaMnMn), a new member of the whiteite group from Mangualde, Beira, Portugal. Am Mineral 75:401-404

Griffen DT, Ribbe PH (1979) Distortions in the tetrahedral oxyanions of crystalline substances. N Jahrb Mineral Abh 137:54-73

Groat LA, Hawthorne FC (1990) The crystal structure of nissonite. Am Mineral 75:1170-1175

Groat LA, Raudsepp M, Hawthorne FC, Ercit TS, Sherriff BL, Hartman JS (1990) The amblygonite-montebrasite series: characterization by single-crystal structure refinement, infrared spectroscopy and multinuclear MAS-NMR spectroscopy. Am Mineral 75:992-1008

Guy BB, Jeffrey GA (1966) The crystal structure of fluellite, $Al_2PO_4F_2(OH)(H_2O)_7$. Am Mineral 51: 1579-1592

Hanson AW (1960) The crystal structure of eosphorite. Acta Crystallogr 13:384-387

Harrowfield IR, Segnit ER, Watts JA (1981) Aldermanite, a new magnesium aluminum phosphate. Mineral Mag 44:59-62

Hata M, Marumo F, Iwai SI (1979) Structure of barium chlorapatite. Acta Crystallogr B35:2382-2384

Hawthorne FC (1976) Refinement of the crystal structure of adamite. Can Mineral 14:143-148

Hawthorne FC (1979a) The crystal structure of morinite. Can Mineral 17:93-102

Hawthorne FC (1979b) Paradamite. Acta Crystallogr B35:720-722

Hawthorne FC (1982) The crystal structure of bøggildite. Can Mineral 20:263-270

Hawthorne FC (1983a) Enumeration of polyhedral clusters. Acta Crystallogr A39:724-736

Hawthorne FC (1983b) The crystal structure of tancoite. Tschermaks Mineral Petrogr Mitt 31:121-135

Hawthorne FC (1984) The crystal structure of stenonite and the classification of the aluminofluoride minerals. Can Mineral 22:245-251

Hawthorne FC (1985a) Towards a structural classification of minerals: the $^{vi}M^{iv}T_2O_n$ minerals. Am Mineral 70:455-473

Hawthorne FC (1985b) Refinement of the crystal structure of bloedite: structural similarities in the $[^{VI}M(^{IV}T\phi_4)_2\phi_n]$ finite-cluster minerals. Can Mineral 23:669-674

Hawthorne FC (1988) Sigloite: The oxidation mechanism in $[(M_2^{3+}(PO_4)_2(OH)_2(H_2O)_2)]^{2-}$ structures. Mineral Petrol 38:201-211

Hawthorne FC (1990) Structural hierarchy in $^{[6]}M^{[4]}TO_4$ minerals. Z Kristallogr 192:1-52

Hawthorne FC (1992) The role of OH and H_2O in oxide and oxysalt minerals. Z Kristallogr 201:183-206

Hawthorne FC (1994) Structural aspects of oxide and oxysalt crystals. Acta Crystallogr B50:481-510

Hawthorne FC (1997) Structural aspects of oxide and oxysalt minerals. *In* European Mineralogical Union Notes in Mineralogy, Vol.1, "Modular Aspects of Minerals". Merlino S (ed) Eötvös University Press, Budapest, p 373-429

Hawthorne FC (1998) Structure and chemistry of phosphate minerals. Mineral Mag 62:141-164

Hawthorne FC (2002) The use of end-member charge-arrangements in defining new minerals and heterovalent substitutions in complex minerals. Can Mineral (accepted)

Hawthorne FC, Ferguson RB (1975) Anhydrous sulphates. II Refinement of the crystal structure of anhydrite. Can Mineral 13:289-292

Hawthorne FC, Grice JD (1987) The crystal structure of ehrleite, a tetrahedral sheet structure. Can Mineral 25:767-774
Hawthorne FC, Schindler MS (2000) Topological enumeration of decorated $[Cu^{2+}\phi_2]_N$ sheets in hydroxy-hydrated copper-oxysalt minerals. Can Mineral 38:751-761
Hawthorne FC, Sokolova EV (2002) Simonkolleite, $Zn_5(OH)_8Cl_2(H_2O)$, a decorated interrupted-sheet structure of the form $[M\phi_2]_4$. Can Mineral (submitted)
Hawthorne FC, Groat LA, Raudsepp M, Ercit TS (1987) Kieserite, a titanite-group mineral. N Jahrb Mineral Abh 157:121-132
Hawthorne FC, Cooper MA, Green DI, Starkey RE, Roberts AC, Grice JD (1999) Wooldridgeite, $Na_2CaCu^{2+}_2(P_2O_7)_2(H_2O)_{10}$: a new mineral from Judkins quarry, Warwickshire, England. Mineral Mag 63:13-16
Hawthorne FC, Krivovichev SV, Burns PC (2000) The crystal chemistry of sulfate minerals. Rev Mineral 40:1-112
Hey MH, Milton C, Dwornik WJ (1982) Eggonite (kolbeckite, sterrettite), $ScPO_4{\cdot}2H_2O$. Mineral Mag 46: 493-497
Hill RG (1977) The crystal structure of phosphophyllite. Am Mineral 63:812-817
Hill RG, Jones JB (1976) The crystal structure of hopeite. Am Mineral 61:987-995
Hoyos MA, Calderon T, Vergara I, Garcia-Sole J (1993) New structural and spectroscopic data for eosphorite. Mineral Mag 57:329-336
Hughes JM, Drexler JW (1991) Cation substitution in the apatite tetrahedral site: crystal structures of type hydroxylellestadtite and type fermorite. N Jahrb Mineral Monatsh 327-336
Hughes JM, Cameron M, Crowley KD (1990) Crystal structures of natural ternary apatites: solid solution in the $Ca_5(PO_4)_3X$ (X= F, OH, Cl) system. Am Mineral 75:295-304
Hughes JM, Foord EE, Hubbard MA, Ni YX (1995) The crystal structure of cheralite-(Ce), (LREE, Ca, Th, U)(P, Si)O_4, a monazite-group mineral. N Jahrb Mineral Monatsh 344-350
Huminicki DMC, Hawthorne FC (2000) Refinement of the crystal structure of väyrynenite. Can Mineral 38:1425-1432
Huminicki DMC, Hawthorne FC (2002) Hydrogen bonding in the crystal structure of seamanite. Can Mineral (accepted)
Hurlbut CS (1942) Sampleite, a new mineral from Chuquicamata, Chile. Am Mineral 27:586-589
Ilyukhin AB, Katser SB, Levin AA (1995) Structure refinement of two crystals from the KDP family $(ND_4)D_2PO_4$ and KH_2AsO_4 in the paraphase. Z Neorg Khim 40:1599-1600
Isaaks AM, Peacor DR (1981) Panasquieratite, a new mineral: The OH-equivalent of isokite. Can Mineral 19:389-392
Isaaks AM, Peacor DR (1982) The crystal structure of thadeuite, $Mg(Ca,Mn)(Mg, Fe, Mn)_2(PO_4)_2(OH,F)_2$. Am Mineral 67:120-125
Johan Z, Slansky E, Povondra P (1983) Vashegyite, a sheet aluminum phosphate: new data. Can Mineral 21:489-498
Kabalov YuK, Sokolova EV, Pekov IV (1997) Crystal structure of belovite-(La). Phys Dokl 42:344-348
Kampf AR (1977a) Schoonerite: its atomic arrangement. Am Mineral 62:250-255
Kampf AR (1977b) Minyulite: its atomic arrangement. Am Mineral 62:256-262
Kampf AR (1977c) A new mineral: perloffite, the Fe^{3+} analogue of bjarebyite. Mineral Rec 8:112-114
Kampf AR (1992) Beryllophosphate chains in the structures of fransoletite, parafransoletite, and ehrleite and some general comments on beryllophosphate linkages. Am Mineral 77:848-856
Kampf AR, Moore PB (1976) The crystal structure of bermanite, a hydrated manganese phosphate. Am Mineral 61:1241-1248
Kampf AR, Moore PB (1977) Melonjosephite, calcium iron hydroxy phosphate: its crystal structure. Am Mineral 62:60-66
Kampf AR, Dunn PJ, Foord EE (1992) Parafransoletite, a new dimorph of fransoletite from the Tip Top Pegmatite, Custer, South Dakota. Am Mineral 77:843-847
Kato T (1970) Cell dimensions of the hydrated phosphate, kingite. Am Mineral 55:515-517
Kato T (1971) The crystal structures of goyazite and woodhouseite. N Jahrb Mineral Monatsh 241-247
Kato T (1987) Further refinement of the goyazite structure. Mineral J 13:390-396
Kato T (1990) The crystal structure of florencite. N Jahrb Mineral Monatsh 227-231
Kato T, Miura Y (1977) The crystal structures of jarosite and svanbergite. Mineral J 8:418-430
Katz L, Lipscomb WN (1951) The crystal structure of iron lazulite, a synthetic mineral related to lazulite. Acta Crystallogr 4:345-348
Keegan TD, Araki T, Moore PB (1979) Senegalite, $Al_2(OH)_3(H_2O)(PO_4)$, a novel structure type. Am Mineral 64:1243-1247

Keller P, Fontan F, Velasco Roldan F, Melgarejo I, Draper JC (1997) Stanékite, $Fe^{3+}(Mn,Fe^{2+},Mg)(PO_4)O$: A new phosphate mineral in pegmatites at Karibib (Namibia) and French Pyrénées (France). Eur J Mineral 9:475-482

Khan AA, Baur WH (1972) Salt hydrates. VIII. The crystal structures of sodium ammonium orthochromate dihydrate and magnesium diammonium bis(hydrogen ortho phosphate) tetrahydrate and a discussion of the ammonium ion. Acta Crystallogr B28:683-693

Khan AA, Roux JP, James WJ (1972) The crystal structure of diammonium hydrogen phosphate, $(NH_4)_2HPO_4$. Acta Crystallogr B28:2065-2069

Kharisun, Taylor MR, Bevan DJM, Pring A (1997) The crystal structure of kintoreite, $PbFe_3(PO_4)_2(OH,H_2O)_6$. Mineral Mag 61:123-129

Khomyakov AP, Aleksandrov VV, Krasnova NI, Ermilov VV, Smolyaninova (1982) Bonshtedtite, $Na_3Fe(PO_4)(CO_3)$, a new mineral. Zap Vses Mineral Obshch 111:486-490 (in Russian)

Khomyakov AP, Nechelyustov GN, Sokolova EV, Dorokhova GI (1992) Quadruphite, $Na_{14}CaMgTi_4[Si_2O_7]_2[PO_4]_4O_4F_2$ and polyphite $Na_{17}Ca_3Mg\ (Ti,Mn)_4[Si_2O_7]_2\ [PO_4]_6\ O_2F_6$—new minerals of the lomonovosite family. Zap Vser Mineral Obshch 121:105-112 (in Russian)

Khomyakov AP, Polezhaeva LI, Sokolova EV (1994) Crawfordite, $Na_3Sr(PO_4)(CO_3)$—a new mineral from the bradleyite. Zap Vser Mineral Obshch 123:107-111 (in Russian)

Khomyakov AP, Lisitsin DV, Kulikova IM, Rastsvetsaeva RK (1996) Deloneite-(Ce) $NaCa_2SrCe(PO_4)_3F$—a new mineral with a belovite-like structure. Zap Vser Mineral Obshch 125:83-94 (in Russian)

Khomyakov AP, Kulikova IM, Rastsvetaeva RK (1997) Fluorcaphite, $Ca(Sr,Na,Ca)(Ca,Sr,Ce)_3(PO_4)_3F$—a new mineral with the apatite structural motif. Zap Vser Mineral Obshch 126:87-97 (in Russian)

Khosrawan-Sazedj F (1982a) The crystal structure of meta-uranocircite II, $Ba(UO_2)_2(PO_4)_2(H_2O)_6$. Tschermaks Mineral Petrogr Mitt 29:193-204

Khosrawan-Sazedj F (1982b) On the space group of threadgoldite. Tschermaks Mineral Petrogr Mitt 30: 111-115

Kim Y, Kirkpatrick RJ (1996) Application of MAS NMR spectroscopy to poorly crystalline materials: Viséite. Mineral Mag 60:957-962

King GSD, Sengier Roberts L (1988) Drugmanite, $Pb_2(Fe_{0.78}Al_{0.22})H(PO_4)_2(OH)_2$: Its crystal structure and place in the datolite group. Bull Minéral 111:431-437

Klevtsova RF (1964) About the crystal structure of strontiumapatite. Z Strukt Khim 5:318-320

Kniep R, Mootz D (1973) Metavariscite—a redetermination of its crystal structure. Acta Crystallogr B29: 2292-2294

Kniep R, Mootz D, Vegas A (1977) Variscite. Acta Crystallogr B33:263-265

Kohlmann M, Sowa H, Reithmayer K, Schulz H, Krueger RBR, Abriel W (1994) Structure of a $[Y_{(1-x)}(Gd,Dy,Er)_x]PO_4{\cdot}2H_2O$ microcrystal using synchrotron radiation. Acta Crystallogr C50: 1651-1652

Kolitsch U, Giester G (2000) The crystal structure of faustite and its copper analogue turquoise. Mineral Mag 64:905-913

Kolitsch U, Taylor MR, Fallon GD, Pring A (1999a) Springcreekite, $BaV^{3+}{}_3(PO_4)_2(OH,H_2O)_6$, a new member of the crandalite group, from the Spring Creek mine, South Australia: the first natural V^{3+}-member of the alunite family and its crystal structure. N Jahrb Mineral Monatsh 529-544

Kolitsch U, Tiekink ERT, Slade PG, Taylor MR, Pring A (1999b) Hinsdalite and plumbogummite, their atomic arrangements and disordered lead sites. Eur J Mineral 11:513-520

Kolitsch U, Pring A, Tiekink ERT (2000) Johntomaite, a new member of the bjarebyite group of barium phosphates: description and structure refinement. Mineral Petrol 70:1-14

Kolkovski B (1971) ICDD card # 29-0756 (orpheite)

Krause W, Belendorff K, Bernhardt H-J (1993) Petitjeanite, $Bi_3O(OH)(PO_4)_2$, a new mineral, and additional data for the corresponding arsenate and vanadate, preisingerite and schumacherite. Neues Jahrb Mineral Monatsh 487-503

Krause W, Belendorff K, Bernhardt H-J, Petitjean K (1998a) Phosphogartrellite, $PbCuFe^{3+}$-$(PO_4)_2(OH){\cdot}H_2O$, a new member of the tsumcorite group. N Jahrb Mineral Monatsh 111-118

Krause W, Belendorff K, Bernhardt H-J, McCammon C, Effenberger H, Mikenda W (1998b) Crystal chemistry of the tsumcorite-group minerals. New data on ferrilotharmeyerite, tsumcorite, thometzekite, mounanaite, helmutwinklerite, and a redefinition of gartrellite. Eur J Mineral 10:179-206

Krause W, Bernhardt H-J, McCammon C, Effenberger H (1998c) Brendelite, $(Bi,Pb)_2Fe^{(3+,2+)}O_2(OH)(PO_4)$, a new mineral from Schneeberg, Germany: description and crystal structure. Mineral Petrol 63:263-277

Krutik VM, Pushcharovskii DYu, Khomyakov AP, Pobedimskaya EA, Belov NV (1980) Anion radical of mixed type (four (S_4O_{12}) rings and P orthotetrahedral) in the structure of monoclinic phosinaite. Kristallografiya 25:240-247

Kumbasar I, Finney JJ (1968) The crystal structure of parahopeite. Mineral Mag 36:621-624
Kurova TA, Shumyatskaya NG, Voronkov AA, Pyatenko YA (1980) Crystal structure of sidorenkite $Na_3Mn(PO_4)(CO_3)$. Dokl Akad Nauk SSSR 251:605-607
Lager GA, Gibbs GV (1974) A refinement of the crystal structure of herderite, $CaBePO_4OH$. Am Mineral 59:919-925
Lahti SI (1981) The granite pegmatites of the Eräjärevi area in Orivesi, southern Finland. Geol Surv Finland Bull 314:1-82
Lahti SI, Pajunen A (1985) New data on lacroixite, $NaAlFPO_4$. I. Occurrence, physical properties and chemical composition. Am Mineral 70:849-855
Leavens PB, Rheingold AL (1988) Crystal structures of gordonite, $MgAl_2(PO_4)_2(OH)_2(H_2O)_6\ (H_2O)_2$, and its Mn analog. N Jahrb Mineral Monatsh 265-270
Leavens PB, White JS, Nelen JA (1990) Zanazziite, a new mineral from Minas Gerais, Brazil. Mineral Rec 21:413-417
Lefebvre J-J, Gasparrini C (1980) Florencite, an occurrence in the Zairian copperbelt. Can Mineral 18: 301-311
Le Page Y, Donnay G (1977) The crystal structure of the new mineral maricite $NaFePO_4$. Can Mineral 15: 518-521
Liebau F (1985) Structural Chemistry of Silicates. Springer-Verlag, Berlin
Liferovich RP, Yakovenchuk VN, Pakhomovsky YaA, Bogdanova AN, Britvin SN (1997) Juonniite, a new mineral of scandium from dolomitic carbonatites of the Kovdor massif. Zap Vser Mineral Obshch 126: 80-88
Liferovich RP, Sokolova EV, Hawthorne FC, Laajoki K, Gehör S, Pakhomovsky YuA, Sorokhtina NV (2000a) Gladiusite, $Fe^{3+}{}_2(Fe^{2+},Mg_4)(PO_4)(OH)_{11}(H_2O)$, a new hydrothermal mineral from the phoscorite-carbonatite unit, Kovdor Complex, Kola Peninsula, Russia. Can Mineral 38:1477-1485
Liferovich RP, Pakhomovsky YaA, Yakubovich OV, Massa W, Laajoki K, Gehör S, Bogdanova AN, Sorokhtina NV (2000b) Bakhchisaraitsevite, $Na_2Mg_5[PO_4]_4{\cdot}7H_2O$, a new mineral from hydrothermal assemblages related to phoscorite-carbonatite complex of the Kovdor massif, Russia. N Jahrb Mineral Monatsh 402-418
Lightfoot P, Cheetham AK, Sleight AW (1987) Structure of $MnPO_4{\cdot}H_2O$ by synchrotron X-ray powder diffraction. Inorg Chem 26:3544-3547
Lindberg ML (1949) Frondelite and the frondelite-rockbridgeite series. Am Mineral 34:541-549
Lindberg ML (1958) The beryllium content of roscherite from the Sapucaia pegmatite mine, Minas Gerais, Brazil and from other localities. Am Mineral 43:824-838
Lindberg (1962) Manganoan lipscombite from the Sapucaia pegmatite mine, Minas Gerais, Brazil. First occurrence of lipscombite in nature. Am Mineral 47:353-359
Lindberg ML, Christ CL (1959) Crystal structures of the isostructural minerals lazulite, scorzalite and barbosalite. Acta Crystallogr 12:695-697
Livingstone A (1980) Johnsomervilleite, a new transition-metal phosphate mineral from the Loch Quoich area, Scotland. Mineral Mag 43:833-836
Makarov YS, Ivanov VI (1960) The crystal structure of meta-autunite, $Ca(UO_2)_2(PO_4)_2{\cdot}6H_2O$. Dokl Akad Nauk SSSR 132:601-603
Makarov YS, Tobelko KI (1960) The crystal structure of metatorbernite. Dokl Akad Nauk SSSR 131:87-89
Malinovskii YuA, Genkina EA (1992) Crystal structure of olympite $LiNa_5[PO_4]_2$. Sov Phys Crystallogr 37: 772-782
Mandarino JA, Sturman BD (1976) Kulanite, a new barium iron aluminum phosphate from the Yukon territory, Canada. Can Mineral 14:127-131
Mandarino JA, Sturman BD, Corlett MI (1977) Penikisite, the magnesium analogue of kulanite, from Yukon Territory. Can Mineral 15:393-395
Mandarino JA, Sturman BD, Corlett MI (1978) Satterlyite, a new hydroxyl-bearing ferrous phosphate from the Big Fish area, Yukon Territory. Can Mineral 16:411-413
Martini JEJ (1978) Sasaite, a new phosphate mineral from West Driefontein Cave, Transvaal, South Africa. Mineral Mag 42:401-404
Martini JEJ (1991) Swaknoite [$Ca(NH_4)_2(HPO_4)_2{\cdot}H_2O$, orthorhombic]: a new mineral from Arnheim Cave, Namibia. Bull S African Speleol Assoc 32:72-74
Martini JEJ (1993) ICDD Card # 45-1411 (swaknoite)
Matsubara S (2000) Vivianite nodules and secondary phosphates in Pliocene-Pleistocene clay deposits from Hime-Shima, Oita Prefecture and Kobe, Hyogo Prefecture, eastern Japan. Mem National Sci Mus, Tokyo 33:15-27
Mazzi F, Ungaretti L (1994) The crystal structure of vitusite from Illimaussaq (South Greenland): $Na_3REE(PO_4)_2$. N Jahrb Mineral Monatsh 49-66

McConnell D (1952) Viséite, a zeolite with the analcime structure and containing linked SiO_4, PO_4 and H_xO_4 groups. Am Mineral 37:609-617

McConnell D (1963) Thermocrystallization of richellite to produce a lazulite structure (calcium lipscombite). Am Mineral 48:300-307

McConnell (1990) Kehoeite and viséite reviewed: comments on dahllite and francolite. Mineral Mag 54: 657-658

McCoy TJ, Steele IM, Keil K, Leonard BF, Endres M (1994) Chladniite, $Na_2CaMg_7(PO_4)_6$: a new mineral from the Carlton (IIICD) iron meteorite. Am Mineral 79:375-380

McDonald AM, Chao GY, Grice JD (1994) Abenakiite-(Ce), a new silicophosphate carbonate mineral from Mont Saint-Hilaire, Quebec: Description and structure determination. Can Mineral 32:843-854

McDonald AM, Chao GY, Grice JD (1996) Phosinaite-(Ce) from Mont Saint-Hilaire, Quebec: New data and structure refinement. Can Mineral 34:107-114

McDonald AM, Grice JD, Chao GY (2000) The crystal structure of yoshimuraite, a layered Ba-Mn-Ti silicophosphate, with comments on five-coordinated Ti^{4+}. Can Mineral 38:649-656

Meagher EP, Gibbons CS, Trotter J (1974) The crystal structure of jagowerite. $BaAl_2P_2O_8(OH)_2$. Am Mineral 59:291-295

Medrano MD, Evans HTJr, Wenk H-R, Piper DZ (1998) Phosphovanadylite: a new vanadium phosphate mineral with a zeolite-type structure. Am Mineral 83:889-895

Mereiter K, Niedermayr G, Walter F (1994) Uralolite, $Ca_2Be_4(PO_4)_3(OH)\cdot3.5(H_2O)$: new data and crystal structure. Eur J Mineral 6:887-896

Merlino S, Pasero M (1992) Crystal chemistry of beryllophosphates: The crystal structure of moraesite, $Be_2(PO_4)(OH)\cdot4H_2O$. Z Kristallogr 201:253-262

Merlino S, Mellini M, Zanazzi PF (1981) Structure of arrojadite, $KNa_4CaMn_4Fe_{10}Al(PO_4)_{12}(OH)_2$. Acta Crystallogr B37:1733-1736

Miller SA, Taylor JC (1986) The crystal structure of saleeite, $Mg(UO_2PO_4)_2\cdot10H_2O$. Z Kristallogr 177: 247-253

Milton DJ, Bastron H (1971) Churchite and florencite (Nd) from Sausalito, California. Mineral Rec 2: 166-168

Milton C, McGee JJ, Evans HT Jr (1993) Mahlmoodite, $FeZr(PO_4)_2\cdot4H_2O$, a new iron zirconium phosphate mineral from Wilson Springs, Arkansas. Am Mineral 78:437-440

Moëlo Y, Lasnier B, Palvadeau P, Léone P, Fontan F (2000) Lulzacite, $Sr_2Fe^{2+}(Fe^{2+},Mg)_2Al_4(PO_4)_4(OH)_{10}$, a new strontium phosphate (Saint Aubin-des-Châteaux, Loire-Atlantique, France). CR Acad Sci Paris, Earth Planet Sci 330:317-324

Mooney RCL (1948) Crystal structure of a series of rare earth phosphates. J Chem Phys 16:1003

Mooney RCL (1950) X-ray diffraction study of cerous phosphate and related crystals. I. Hexagonal modification. Acta Crystallogr 3:337-340

Moore PB (1965a) The crystal structure of laueite, $MnFe_2(OH)_2(PO_4)_2(H_2O)_6(H_2O)_2$. Am Mineral 50: 1884-1892

Moore PB (1965b) A structural classification of Fe-Mn orthophosphate hydrates. Am Mineral 50: 2052-2062

Moore PB (1966) The crystal structure of metastrengite and its relationship to strengite and phosphophyllite. Am Mineral 51:168-176

Moore PB (1970) Crystal chemistry of the basic iron phosphates. Am Mineral 55:135-169

Moore PB (1971a) The $Fe_3^{2+}(H_2O)_n(PO_4)_2$ homologous series: Crystal-chemical relationships and oxidized equivalents. Am Mineral 56:1-16

Moore PB (1971b) Crystal chemistry of the alluaudite structure type: Contribution to the paragenesis of pegmatite phosphate giant crystals. Am Mineral 56:1955-1975

Moore PB (1972a) Natrophilite, $NaMn(PO_4)$, has ordered cations. Am Mineral 57:1333-1344

Moore PB (1972b) Octahedral tetramer in the crystal structure of leucophosphite, $K_2[Fe^{3+}_4(OH)_2(H_2O)_2(PO_4)_4]\cdot2H_2O$. Am Mineral 57:397-410

Moore PB (1973a) Pegmatite phosphates. Descriptive mineralogy and crystal chemistry. Mineral Rec 4: 103-130

Moore PB (1973b) Bracelets and pinwheels: A topological-geometrical approach to the calcium orthosilicate and alkali sulfate structures. Am Mineral 58:32-42

Moore PB (1974) I. Jahnsite, segelerite, and robertsite, three new transition metal phosphate species. II. Redefinition of overite, an isotype of segelerite. III. Isotypy of robertsite, mitridatite, and arseniosiderite. Am Mineral 59:48-59, 640

Moore PB (1975a) Brianite, $Na_2CaMg[PO_4]_2$: a phosphate analog of merwinite, $Ca_2CaMg[SiO_4]_2$. Am Mineral 60:717-718

Moore PB (1975b) Laueite, pseudolaueite, stewartite and metavauxite: a study in combinatorial polymorphism. N Jahrb Mineral Abh 123:148-59

Moore PB (1976) Derivative structures based on the alunite octahedral sheet: mitridatite and englishite. Mineral Mag 40:863-866

Moore PB (1984) Crystallochemical aspects of the phosphate minerals. *In* Phosphate Minerals. Nriagu JO, Moore PB (eds) Springer-Verlag, Berlin, p 155-170

Moore PB, Araki T (1973) Hureaulite, $(Mn^{2+})_5(H_2O)_4(PO_3(OH))_2(PO_4)_2$: Its atomic arrangement. Am Mineral 58:302-307

Moore PB, Araki T (1974a) Jahnsite, $CaMnMg_2(H_2O)_8Fe_2(OH)_2(PO_4)_4$. A novel stereoisomerism of ligands about octahedral corner-chains. Am Mineral 59:964-973

Moore PB, Araki T (1974b) Trolleite, $Al_4(OH)_3(PO_4)_3$. A very dense structure with octahedral face-sharing dimers. Am Mineral 59:974-984

Moore PB, Araki T (1974c) Bjarebyite, $Ba(Mn,Fe)_2Al_3(OH)_3(PO_4)_3$. Its atomic arrangement. Am Mineral 59:567-572

Moore PB, Araki T (1974d) Stewartite, $Mn^{2+}Fe_2^{3+}(OH)_2(H_2O)_6(PO_4)_2(H_2O)_2$. Its atomic arrangement. Am Mineral 59:1272-1276

Moore PB, Araki T (1974e) Montgomeryite, $Ca_4Mg(H_2O)_{12}[Al_4(OH)_4(PO_4)_6]$: Its crystal structure and relaton to vauxite, $Fe^{2+}{}_2(H_2O)_4[Al_4(OH)_4(H_2O)_4(PO_4)_4]{\cdot}4H_2O$. Am Mineral 59:843-850

Moore PB, Araki T (1975) Palermoite, $SrLi_2(Al_4(OH)_4(PO_4)$. Its atomic arrangement and relationship to carminite, $Pb_2(Fe_4(OH)_4(AsO_4)_4)$. Am Mineral 60:460-465

Moore PB, Araki T (1976) A mixed-valence solid solution series: Crystal structures of phosphoferrite and kryzhanovskite. Inorg Chem 15:316-321

Moore PB, Araki T (1977a) Samuelsonite: its crystal structure and relation to apatite and octacalcium phosphate. Am Mineral 62:229-245

Moore PB, Araki T (1977b) Mitridatite, $Ca_6(H_2O)_6(Fe_9O_6(PO_4)_9)(H_2O)_3$. A noteworthy octahedral sheet structure. Mineral Mag 41:527-528

Moore PB, Araki T (1977c) Overite, segelerite, and jahnsite: a study in combinatorial polymorphism. Am Mineral 62:692-702

Moore PB, Ghose S (1971) A novel face-sharing octahedral trimer in the crystal structure of seamanite. Am Mineral 56:1527-1538

Moore PB, Ito J (1978) I. Whiteite, a new species, and a proposed nomenclature for the jahnsite-whiteite complex series. II. New data on xanthoxenite. Mineral Mag 42:309-316

Moore PB, Ito J (1979) Alluaudites, wyllieites, arrojadites: crystal chemistry and nomenclature. Mineral Mag 43:227-35

Moore PB, Kampf AR (1977) Schoonerite, a new zinc-manganese-iron-phosphate mineral. Am Mineral 62:246-249

Moore PB, Molin-Case J (1974) Contribution to pegmatite phosphate giant crystal paragenesis. II. The crystal chemistry of wyllieite, $Na_2Fe(II)_2Al(PO_4)_3$, a primary phase. Am Mineral 59:280-290

Moore PB, Shen J (1983a) An X-ray structural study of cacoxenite, a mineral phosphate. Nature 306: 356-358

Moore Pb, Shen J (1983b) Crystal structure of steenstrupine: a rod structure of unusual complexity. Tschermaks Mineral Petrogr Mitt 31:47-67

Moore PB, Kampf AR, Irving AJ (1974) Whitmoreite, $Fe(II)Fe(III)_2(OH)_2(H_2O)_4(PO_4)_2$, a new species. Its description and atomic arrangement. Am Mineral 59:900-905

Moore PB, Irving AJ, Kampf AR (1975a) Foggite, $CaAl(OH)_2(H_2O)[PO_4]$; goedkenite, $(Sr,Ca)_2Al(OH)[PO_4]_2$; and samuelsonite, $(Ca,Ba)Fe^{2+}{}_2Mn^{2+}{}_2Ca_8Al_2(OH)_2[PO_4]_{10}$: Three new species from the Palermo No. 1 Pegmatite, North Groton, New Hampshire. Am Mineral 60:957-964

Moore PB, Kampf AR, Araki T (1975b) Foggite, $(CaH_2O)_2(CaAl_2(OH)_4(PO_4)_2)$. Its atomic arrangement and relationship to calcium Tschermak's pyroxene. Am Mineral 60:965-971

Moore PB, Araki T, Kampf AR, Steele IM (1976) Olmsteadite, $K_2(Fe^{2+})_2((Fe^{2+})_2((Nb^{6+})(Ta^{6+}))_2O_4(H_2O)_4(PO_4)_4)$, a new species, its crystal structure and relation to vauxite and montgomeryite. Am Mineral 61:5-11

Moore PB, Araki T, Kampf AR (1980) Nomenclature of the phosphoferrite structure type: refinements of landesite and kryzhanovskite. Mineral Mag 43:789-795

Moore PB, Araki T, Merlino S, Mellini M, Zanazzi PF (1981) The arrojadite-dickinsonite series, $KNa_4Ca(Fe,Mn)^{2+}{}_{14}Al(OH)_2(PO_4)_{12}$: crystal structure and crystal chemistry. Am Mineral 66:1034-1049

Moore PB, Araki T, Steele IM, Swihart GH, Kampf AR (1983) Gainesite, sodium zirkonium beryllophosphate: a new mineral and its crystal structure. Am Mineral 68:1022-1028

Moring J, Kostiner E (1986) The crystal structure of $NaMnPO_4$. J Sol State Chem 61:379-383

Mrose ME (1971) Dittmarite. US Geol Surv Prof Pap 750-A:A115

Mücke A (1979) Keckit, $(Ca,Mg)(Mn,Zn)_2Fe^{3+}{}_3[(OH)_3(PO_4)_4]{\cdot}2H_2O$, ein neues Mineral von Hagendorf/Opf, und seine genetische Stellung. N Jahrb Mineral Abh 134:183-192

Mücke A (1983) Wilhelmvierlingite, (Ca,Zn)$MnFe^{3+}[OH(PO_4)_2]\cdot2H_2O$, a new mineral from Hagendorf/Oberpfalz. Aufschluss 34:267-274

Mücke A (1988) Lehnerit, $Mn[UO_2/PO_4]_2\cdot8H_2O$, ein neues Mineral aus dem Pegmatit von Hagendorf/Oberpfalz. Aufschluss 39:209-217

Muto T, Meyrowitz R, Pommer AM, Murano T (1959) Ningyoite, a new uranous phosphate mineral from Japan. Am Mineral 44:633-650

Ng HN, Calvo C (1976) X-ray study of the alpha-beta transformation of berlinite, $AlPO_4$. Can J Phys 54:638-647

Ni YX, Hughes JM, Mariano AN (1995) Crystal chemistry of the monazite and xenotime structures. Am Mineral 80:21-26

Nord AG, Kierkegaard P (1968) The crystal structure of $Mg_3(PO_4)_2$. Acta Chem Scand 22:1466-1474

Nriagu JO (1984) Phosphate minerals: Their properties and general modes of occurrence. *In* Phosphate Minerals. Nriagu JO, Moore PB (eds) Springer-Verlag, Berlin, p 1-136

Olsen EJ, Steele IM (1997) Galileiite: a new meteoritic phosphate mineral. Meteor Planet Sci 32: A155-A156

Ono Y, Yamada N (1991) A structural study of the mixed crystal $K_{0.77}(NH_4)_{0.23}H_2PO_4$. J Phys Soc Japan 60:533-538

Ovchinnikov VE, Solov'eva LP, Pudovkina ZV, Kapustin YuL, Belov NV (1980) The crystal structure of kovdorskite $Mg_2(PO_4)(OH)(H_2O)_3$. Dokl Akad Nauk SSSR 255:351-354

Owens JP, Altschuler ZS, Berman R (1960) Millisite in phosphorite from Homeland, Florida. Am Mineral 45:547-561

Pajunen A, Lahti SI (1984) The crystal structure of viitaniemiite. Am Mineral 69:961-966

Pajunen A, Lahti SI (1985) New data on lacroixite, $NaAlFPO_4$. II. Crystal structure. Am Mineral 70: 849-855

Pauling LS (1929) The principles determining the structure of complex ionic crystals. J Am Chem Soc 51:1010-1026

Pauling L (1960) The nature of the chemical bond. 3rd ed. Cornell University Press, Ithaca, New York

Pavlov PV, Belov NV (1957) The crystal structures of herderite, datolite and gadolinite. Dokl Akad Nauk SSSR 114:884-887

Peacor DR, Dunn PJ (1982) Petersite, a REE and phospahte analogue of mixite. Am Mineral 67:1039-1042

Peacor DR, Dunn PJ, Simmons WB (1983) Ferrostrunzite, the ferrous iron analogue of strunzite from Mullica Hill, New Jersey N Jahrb Mineral Monatsh 524-528

Peacor DR, Dunn PJ, Simmons WB (1984) Earlshannonite, the Mn analogue of whitmoreite, from North Carolina. Can Mineral 22:471-474

Peacor DR, Rouse RC, Ahn T-H (1987) Crystal structure of tiptopite, a framework beryllophosphate isotypic with basic cancrinite. Am Mineral 72:816-820

Peacor DR, Rouse RC, Coskren TD, Essene EJ (1999) Destinezite ("diadochite"), $Fe_2(PO_4)(SO_4)$-$(OH)\cdot6(H_2O)$: Its crystal structure and role as a soil mineral at Alum Cave Bluff, Tennessee. Clays Clay Mineral 47:1-11

Pekov IV, Kulikova IM, Kabalov YuK, Eletskaya OV, Chukanov NV, Menshikov YuP, Khomyakov AP (1996) Belovite-(La), $Sr_3Na(La,Ce)[PO_4]_3(F,OH)$—a new rare earth mineral in the apatite group. Zap Vser Mineral Obshch 125:101-109 (in Russian)

Piret P, Declercq J-P (1983) Structure cristalline de l'upalite $Al((UO_2)_3O(OH)(PO_4)_2\cdot7(H_2O)$. Un exemple de macle mimetique. Bull Minéral 106:383-389

Piret P, Deliens M (1981) New data on holotype bergenite. Bull Mineral 104:16-18

Piret P, Deliens M (1982) La Vanmeersscheite, $U(UO_2)_3(PO_4)_2(OH)_6\cdot4H_2O$ et la meta-vanmeersscheite, $U(UO_2)_3(PO_4)_2(OH)_6\cdot2H_2O$, nouveaux mineraux. Bull Minéral 105:125-128

Piret P, Deliens M (1987) Les phosphates d'uranyle et d'aluminium de Kobokobo IX. L'althupite $AlTh(UO_2)((UO_2)_3O(OH)(PO_4)_2)_2(OH)_3(H_2O)_{15}$, nouveau mineral; proprietes et structure crystalline. Bull Minéral 110:65-72

Piret P, Deliens M (1988) Description of ludjibaite, a polymorph of pseudomalachite, $Cu_5(PO_4)_2(OH)_4$. Bull Minéral 111:167-171

Piret P, Piret-Meunier J (1988) Nouvelle determination de la structure cristalline de la dumontite $Pb_2((UO_2)_3O_2(PO_4)_3)(H_2O)5)$ Bull Minéral 111:439-442

Piret P, Piret-Meunier J, Declercq JP (1979) Structure of phuralumite. Acta Crystallogr B35:1880-1882

Piret P, Deliens M, Piret-Meunier J (1985) Occurrence and crystal structure of kipushite, a new copper-zinc phosphate from Kipushi, Zaire. Can Mineral 23:35-42

Piret P, Deliens M, Piret-Meunier J (1988) La francoisite-(Nd), nouveau phosphate d'uranyle et de terres rares; proprietes et structure crystalline. Bull Minéral 111:443-449

Piret P, Piret-Meunier J, Deliens M (1990) Composition chimique et structure cristalline de la dewindtite $Pb_3(H(UO_2)_3O_2(PO_4)_2)_2\cdot12H_2O$. Eur J Mineral 2:399-405

Popova VI, Popov VA, Sokolova EV, Ferraris G, Chukanov NV (2001) Kanonerovite $MnNa_3P_3O_{10}\cdot12H_2O$, first triphosphate mineral (Kazennitsa, Middle Urals, Russia). N Jahrb Mineral Monatsh 117-127

Potenza MF (1958) Autunite e metatorbernite nella Sienite di Biella. Rend Soc Mineral Ital 14:215-223

Pring A, Birch WD (1993) Gatehouseite, a new manganese hydroxy phosphate from Iron Monarch, South Australia. 57:309-313

Pring A, Birch WD, Dawe J, Taylor M, Deliens M, Walenta K (1995) Kintoreite, $PbFe_2(PO_4)_2(OH,H_2O)_6$, a new mineral of the jarosite-alunite family, and lusungite discredited. Mineral Mag 59:143-148

Pring A, Kolitsch U, Birch WD, Beyer BD, Elliott P, Ayyappan P, Ramanan A (1999) Bariosincosite, a new hydrated barium vanadium phosphate, from the Spring Creek mine, South Australia. Mineral Mag 63:735-741

Raade G, Roemming C, Medenbach O (1998) Carbonate-substituted phosphoellenbergerite from Modum, Norway: description and crystal structure. Mineral Petrol 62:89-101

Radoslovich EW, Slade PG (1980) Pseudo-trigonal symmetry and the structure of gorceixite. N Jahrb Mineral Monatsh 157-170

Rastsvetaeva RK (1971) Kristallicheskaya struktura lomonosovita $Na_5Ti_2(Si_2O_7)(PO_4)O_2$. Dokl Akad Nauk SSSR 197:81-84

Rastsvetsaeva RK, Khomyakov AP (1996) Crystal structure of deloneite-(Ce), the highly ordered Ca analogue of belovite. Dokl Ross Akad Nauk 349:354-357 (in Russian)

Richardson JM, Roberts AC, Grice JD, Ramik RA (1988) Mcauslanite, a supergene hydrated iron aluminum fluorophosphate from the East Kemptville tin mine, Yarmouth County, Nova Scotia. Can Mineral 26: 917-921

Rídkosil T, Sejkora J, _rein VL (1996) Smrkovecite, monoclinic $Bi_2O(OH)(PO_4)$, a new mineral of the atelesite group. N Jahrb Mineral Monatsh 97-102

Rinaldi R (1978) The crystal structure of griphite, a complex phosphate, not a garnetoid. Bull Minéral 101: 543-547

Rius J, Louër D, Mouër M, Galí S, Melgarejo JC (2000) Structure solution from powder data of the phosphate hydrate tinticite. Eur J Mineral 12:581-588

Roberts AC, Dunn PJ, Grice JD, Newbury DE, Roberts WL (1988) The X-ray crystallography of tavorite from the Tip Top pegmatite, Custer, South Dakota. Powder Diffraction 3:93-95

Roberts AC, Sturman BD, Dunn PJ, Roberts WL (1989) Pararobertsite, $Ca_2Mn^{3+}_3(PO_4)_3O_2\cdot3H_2O$, a new mineral species form the Tip Top pegmatite, Custer Country, South Dakota, and its relationship to robertsite. Can Mineral 27:451-455

Roca M, Marcos MD, Amorós P, Alamo J, Beltrán-Porter A, Beltrán-Porter D (1997) Synthesis and crystal structure of a novel lamellar barium derivative: $Ba(VOPO_4)_2\cdot4H_2O$. Synthetic pathways for layered oxovanadium phosphate hydrates $M(VOPO_4)_2\cdot nH_2O$. Inorg Chem 36:3414-3421

Romming C, Raade G (1980) The crystal structure of althausite, $Mg_4(PO_4)_2(OH,O)F$. Am Mineral 65: 488-498

Romming C, Raade G (1986) The crystal structure of heneuite, $CaMg_5(CO_3)(PO_4)_3(OH)$. N Jahrb Mineral Monatsh 351-359

Romming C, Raade G (1989) The crystal structure of natural and synthetic holtedahlite. Mineral Petrol 40: 91-100

Rose D (1980) Brabantite, $CaTh[PO_4]_2$, a new mineral of the monazite group. N Jahrb Mineral Monatsh 247-257

Ross V (1956) Studies of uranium minerals XXII: Synthetic calcium and lead uranyl phosphate minerals. Am Mineral 41:915-926

Rouse RC, Peacor DR, Freed RL (1988) Pyrophosphate groups in the structure of canaphite, $CaNa_2P_2O_7.4(H_2O)$: The first occurrence of a condensed phosphate as a mineral. Am Mineral 73: 168-171

Rouse RC, Peacor DR, Merlino S (1989) Crystal structure of pahasapaite, a beryllophosphate mineral with a distorted zeolite rho framework. Am Mineral 74:1195-1202

Sakae T, Nagata H, Sudo T (1978) The crystal structure of synthetic calcium phosphate sulfate hydrate, $Ca_2(HPO_4)(SO_4)(H_2O)_4$, and its relation to brushite and gypsum. Am Mineral 63:520-527

Sales BC, Chakoumakos BC, Boatner LA, Ramey JO (1993) Structural properties of the amorphous phases produced by heating crystalline $MgHPO_4\cdot3H_2O$. J Non-Crystalline Solids 159:121-139

Schindler M, Hawthorne FC (2001a) A bond-valence approach to the structure, chemistry and paragenesis of hydroxy-hydrated oxysalt minerals: I. Theory. Can Mineral 39:1225-1242

Schindler M, Hawthorne FC (2001b) A bond-valence approach to the structure, chemistry and paragenesis of hydroxy-hydrated oxysalt minerals: II. Crystal structure and chemical composition of borate minerals. Can Mineral 39:1243-1256

Schindler M, Hawthorne FC (2001c) A bond-valence approach to the structure, chemistry and paragenesis of hydroxy-hydrated oxysalt minerals: III. Paragenesis of borate minerals. Can Mineral 39:1257-1274

Schindler M, Huminicki DMC, Hawthorne FC (2002) A bond-valence approach to the chemical composition and occurrence of sulfate minerals. Chem Geol (submitted)

Schlüter J, Klaska K-H, Friese K, Adiwidjaja G (1999) Kastningite, $(Mn,Fe,Mg)Al_2(PO_4)_2(OH)_2 \cdot 8H_2O$, a new phosphate mineral from Waidhaus, Bavaria, Germany. N Jahrb Mineral Monatsh 40-48

Sebais M, Dorokhova GI, Pobedimskaya EA, Khomyakov AP (1984) The crystal structure of nefedovite and its typomorphism. Dokl Akad Nauk SSSR 278:353-357

Selway JB, Cooper MA, Hawthorne FC (1997) Refinement of the crystal structure of burangaite. Can Mineral 35:1515-1522

Sen Gupta PK, Swihart GH, Dimitrijevic R, Hossain MB (1991) The crystal structure of lueneburgite, $Mg_3(H_2O)_6(B_2(OH)_6(PO_4)_2)$. Am Mineral 76:1400-1407

Sergeev AS (1964) Pseudo-autunite, a new hydrous calcium phosphate. Mineral Geokhim 1:31-39 (in Russian)

Shannon RD (1976) Revised effective ionic radii and systematic studies of interatomic distances in halides and chalcogenides. Acta Crystallogr A32:751-767

Shen J, Peng Z (1981) The crystal structure of furongite. Acta Crystallogr A37, supp C-186

Shoemaker GL, Anderson JB, Kostiner E (1977) Refinement of the crystal structure of pseudomalachite. Am Mineral 62:1042-1048

Shoemaker GL, Anderson JB, Kostiner E (1981) The crystal structure of a third polymorph of $Cu_5(PO_4)_3(OH)_4$. Am Mineral 66:169-181

Sieber NHW, Tillmanns E, Medenbach O (1987) Hentschelite, $CuFe_2(PO_4)_2(OH)_2$, a new member of the lazulite group, and reichenbachite, $Cu_5(PO_4)_2(OH)_4$, a polymorph of pseudomalachite, two new copper phosphate minerals from Reichenbach, Germany. Am Mineral 72:404-408

Simonov MA, Egorov-Tismenko YK, Belov NV (1980) Use of modern X-ray equipment to solve fine problems of structural mineralogy by the example of the crystal RE of structure of babefphite $BaBe(PO_4)F$. Kristallografiya 25:55-59 (in Russian)

Slavik F (1914) Neue Phosphate vom Greifenstein bei Ehrenfriedersdorf. Ak Ceská, Bull intern ac sci Bohême 19:108-123

Sljukic M, Matkovic B, Prodic B, Anderson D (1969) The crystal structure of $KZr_2(PO_4)_3$. Z Kristallogr 130:148-161

Soboleva MV, Pudovkina IA (1957) Mineralogy of Uranium Handbook. Moscow: Dept Tech Lit USSR

Sokolova EV, Egorov-Tismenko YuK (1990) Crystal structure of girvasite. Dokl Akad Nauk SSSR 311: 1372-1376

Sokolova EV, Hawthorne FC (2001) The crystal chemistry of the $[M_3\phi_{11\text{-}14}]$ trimeric structures: from hyperagpaitic complexes to saline lakes. Can Mineral 39:1275-1294

Sokolova EV, Khomyakov AP (1992) Crystal structure of a new mineral, $Na_3Sr(PO_4)(CO_3)$, from the bredliit group. Dokl Akad Nauk SSSR 322:531-535

Sokolova EV, Yamnova NA, Egorov-Tismenko YK, Khomyakov AP (1984a) The crystal structure of a new phosphate of Na, Ca and $Ba(Na_5Ca)Ca_6Ba(PO_4)_6F_3$. Dokl Akad Nauk SSSR 274:78-83

Sokolova EV, Egorov-Tismenko YK, Yamnova NA, Simonov MA (1984b) The crystal structure of olgite $Na(Sr_{0.52}Ba_{0.48})(Sr_{0.58}Na_{0.42})(Na_{0.81}Sr_{0.19})(PO_{3.40})(P_{0.76}O_{3.88})$. Kristallografiya 29:1079-1083

Sokolova EV, Egorov-Tismenko YuK, Khomyakov AP (1987a) Crystal structure of lomonosovite and sulfohalite as a homolog of the structures of $Na_{14}CaMgTi_4(Si_2O_7)_2(PO_4)_4O_4F_2$. Mineral Z 9:28-35

Sokolova EV, Egorov-Tismenko YK, Khomyakov AP (1987b) Crystal structure of $Na_{17}Ca_3Mg(Ti,Mn)_4(Si_2O_7)_2(PO_4)_3O_2F_6$, a new representative of the family of layered titanium silicates. Dokl Akad Nauk SSSR 294:357-362

Sokolova EV, Egorov-Tismenko YuK, Khomyakov AP (1988) Crystal structure of sobolevite, Sov Phys Dokl 33:711-714

Sokolova EV, Hawthorne FC, McCammon C, Liferovich RP (2001) The crystal structure of gladiusite, $(Fe^{2+},Mg)_4Fe^{3+}{}_2(PO_4)(OH)_{11}(H_2O)$. Can Mineral 39:1121-1130

Steele IM, Olsen E, Pluth J, Davis AM (1991) Occurrence and crystal structure of Ca-free beusite in the El Sampal IIIA iron meteorite. Am Mineral 76:1985-1989

Stergiou AC, Rentzeperis PJ, Sklavounos S (1993) Refinement of the crystal structure of metatorbernite. Z Kristallogr 205:1-7

Street RLT, Whitaker A (1973) The isostructurality of rosslerite and phosphorosslerite. Z Kristallgr 137: 246-255

Sturman BD, Mandarino JA, Mrose ME, Dunn PJ (1981a) Gormanite, $Fe^{2+}{}_3Al_4(PO_4)_4(OH)_6 \cdot 2H_2O$, the ferrous analogue of souzalite, and new data for souzalite. Can Mineral 19:381-387

Sturman BD, Peacor DR, Dunn PJ (1981b) Wicksite, a new mineral from northeastern Yukon Territory. Can Mineral 19:377-380

Sutor DJ (1967) The crystal and molecular structure of newberyite, $MgHPO_4(H_2O)_3$. Acta Crystallogr 23: 418-422

Szymanski JT, Roberts AC (1990) The crystal structure of voggite, a new hydrated Na-Zr hydroxide-phosphate-carbonate mineral. Mineral Mag 54:495-500

Tadini C (1981) Magniotriplite: its crystal structure and relation to the triplite-triploidite group. Bull Minéral 104:677-680

Takagi S, Mathew M, Brown WE (1986) Crystal structures of bobierrite and synthetic $Mg_3(PO_4)_2(H_2O)_8$. Am Mineral 71:1229-1233

Taxer KJ (1975) Structural investigations on scholzite. Am Mineral 60:1019-1022

Taxer KJ, Bartl H (1997) Die "geordnete gemittelte" Kristallstruktur von Parascholzit. Zur Dimorphie von $CaZn_2(PO_4)_2 \cdot 2(H_2O)$, Parascholzit-Scholzit. Z Kristallogr 212:197-202

van der Westhuizen WA, deBruiyn H, Beukes GJ, Strydom D (1990) Dufrenite in iron-formation on the Kangnas farm, Aggeneys district, Bushmanland, South Africa. Mineral Mag 54:419-424

van Tassel R (1968) Données cristallographiques sur la koninckite. Bull Minéral 91:487-489

van Wambeke L (1972) Eylettersite, un nouveau phosphate de thorium appartenant à la série de la crandallite. Bull Mineral 95:98-105

van Wambeke L (1975) La zairite, un nouveau minéral appartenant à la serie de la crandallite. Bull Minéral 98:351-353

Vencato I, Mattievich E, Mascarenhas YP (1989) Crystal structure of synthetic lipscombite: A redetermination. Am Mineral 74:456-460

Vochten RD, De Grave E (1981) Crystallographic, Mössbauer, and electrokinetic study of synthetic lipscombite. Phys Chem Minerals 7:197-203

Vochten R, Pelsmaekers J (1983) Synthesis, solubility, electrokinetic properties and refined crystallographic data of sabugalite. Phys Chem Mineral 9:23-29

Vochten RF, van Acker P, De Grave E (1983) Mössbauer, electrokinetic and refine parameters study of synthetic manganoan lipscombite. Phys Chem Minerals 9:263-268

Vochten R, de Grave E, Pelsmaekers J (1984) Mineralogical study of bassetite in relation to its oxidation. Am Mineral 69:967-978

Voloshin AV, Pakhomovskiy YuA, Tyusheva FN (1983) Lun'okite; a new phosphate, the manganese analog of overite from granitic pegmatites of the Kola Peninsula. Int Geol Rev 25:1131-1136

Voloshin AV, Pakhomovskii YaA, Tyusheva EN (1992) Manganosegelerite (Mn,Ca)(Mn,Fe,Mg)-$Fe^{3+}(PO_4)_2(OH)$ _ $4H_2O$-a new phosphate of the overite group from granitic pegmatites of the Kola Peninsula. Zap Vser Mineral Obshch 121:95-103

von Knorring O, Fransolet A-M (1977) Gatumbaite, $CaAl_2(PO_4)_2(OH)_2 \cdot H_2O$: a new species from Buranga pegmatite, Rwanda. N Jahrb Mineral Monatsh 561-568

von Knorring O, Mrose ME (1966) Bertossaite, $(Li,Na)_2(Ca,Fe,Mn)Al_4(PO_4)_4(OH,F)_4$, a new mineral from Rwanda (Africa). Can Mineral 8:668

Waldrop L (1968a) Crystal structure of triplite. Naturwiss 55:178

Waldrop L (1968b) Crystal structure of triploidite. Naturwiss 55:296-297

Waldrop L (1970) The crystal structure of triploidite and its relations to the structures of other minerals of the triplite-triploidite group. Z Kristallogr 131:1-20

Walenta K (1978) Uranospathite and arsenuranospathite. Mineral Mag 42:117-28

Walenta K, Dunn PJ (1984) Phosphofibrite, ein neues Eisenphosphat aus der Grube Clara im mittleren Schwarzwald (BRD) Chem Erde 43:11-16

Walenta K, Theye T (1999) Haigerachit, ein neues Phosphatmineral von der Grube Silberbrünnle bei Gengenbach im mittleren Schwarzwald. Aufschluss 50:1-7

Walenta K, Birch WD, Dunn PJ (1996) Benauite, a new mineral of the crandallite group from the Clara mine in the central Black Forest, Germany. Chem Erde 56:171-176

Walter F (1992) Weinebeneite, $CaBe_3(PO_4)_2(OH)_2 \cdot 4(H_2O)$, a new mineral species: mineral data and crystal structure. Eur J Mineral 4:1275-1283

Warner JK, Cheetham AK, Nord AG, von Dreele RB, Yethiraj M (1992) Magnetic structure of iron(II) phosphate, sarcopside, $Fe_3(PO_4)_2$. J Mater Chem 2:191-196

White JS Jr, Henderson EP, Mason B (1967) Secondary minerals produced by weathering of the Wolf Creek meteorite. Am Mineral 52:1190-1197

White JS Jr, Leavens PB, Zanazzi PF (1986) Switzerite redefined as $Mn_3(PO_4)_2 \cdot 7H_2O$, and metaswitzerite, $Mn_3(PO_4)_2 \cdot 4H_2O$. Am Mineral 71:1221-1223

Wiench DM, Jansen M (1983) Kristallstruktur von wasserfreiem Na_2HPO_4. Z Anorg Allg Chem 501: 95-101

Wise MA, Hawthorne FC, _Cerny_ P (1990) Crystal structure of a Ca-rich beusite from the Yellowknife pegmatite field, Northwest Territories. Can Mineral 28:141-146

Witzke T, Wegner R, Doering T, Pöllmann H, Schuckmann W (2000) Serrabrancaite, $MnPO_4 \cdot H_2O$, a new mineral from the Alto Serra Branca pegmatite, Pedra Lavrada, Paraiba, Brazil. Am Mineral 85: 847-849

Yakubovich OV, Mel'nikov OK (1993) Libethenite $Cu_2(PO_4)(OH)$: Synthesis, crystal structure refinement, comparative crystal chemistry. Kristallografiya 38:63-70

Yakubovich OV, Urusov VS (1997) Electron density distribution in lithiophosphatite Li_3PO_4: Crystallochemical features of orthophosphates with hexagonal close packing. Kristallografiya 42: 301-308

Yakubovich OV, Simonov MA, Belov NV (1977) The crystal structure of a synthetic triphylite $LiFe(PO_4)$. Dokl Akad Nauk SSSR 235:93-95

Yakubovich OV, Simonov MA, Matvienko EN, Belov NV (1978) The crystal structure of the synthetic finite Fe-term of the series triplite B zwieselite $Fe_2(PO_4)$ F. Dokl Akad Nauk SSSR 238:576-579

Yakubovich OV, Massa W, Liferovich RP, Pakhomovsky YA (2000) The crystal structure of bakhchisaraitsevite, $[Na_2(H_2O)_2]$ $\{(Mg_{4.5}Fe_{0.5})(PO_4)_4(H_2O)_5\}$, a new mineral species of hydrothermal origin from the Kovdor phoscorite-carbonatite complex, Russia. Can Mineral 38:831-838

Yakubovich OV, Massa W, Liferovich RP, McCammon CA (2001) The crystal structure of bari_ite, $(Mg_{1.70}Fe_{1.30})(PO_4)_2{\cdot}8H_2O$, the magnesium-dominant member of the vivianite group. Can Mineral 39: 1317-1324

Young EJ, Weeks AD, Meyrowitz R (1966) Coconinoite, a new uranium mineral from Utah and Arizona. Am Mineral 51:651-663

Zanazzi PF, Leavens PB, White JS (1986) Crystal structure of switzerite, $Mn_3(PO_4)_2(H_2O)_7$ and its relationship to metaswitzerite, $Mn_3(PO_4)_2(H_2O)_4$. Am Mineral 71:1224-1228

Zhang J, Wan A, Gong W (1992) New data on yingjiangite. Acta Petrol Mineral 11:178-184

Zhesheng M, Nicheng S, Zhizhong P (1983) Crystal structure of a new phosphatic mineral—qingheiite. Scientia Sinica (Series B) 25:876-884

Zolensky ME (1985) New data on sincosite. Am Mineral 70:409-410

Zoltai T (1960) Classification of silicates and other minerals with tetrahedral structures. Am Mineral 45: 960-973

Zwaan PC, Arps CES, de Grave E (1989) Vochtenite, $(Fe^{2+},Mg)(Fe^{3+}[UO_2/PO_4]_4(OH){\cdot}12\text{-}13H_2O$, a new uranyl phosphate mineral from Wheal Basset, Redruth, Cornwall, England. Mineral Mag 53:473-478

APPENDIX

A tabulation of data and references for selected phosphate minerals

(on the following 17 pages)

APPENDIX. Data and references for selected phosphate minerals.

Mineral Name	Formula	*a* (Å)	*b* (Å)	*c* (Å)	α (°)	β (°)	γ (°)	Space Group	Ref.
Abenakiite-(Ce)	$Na_{26}(Ce_3Nd_2La)(SO_2)(SiO_3)_6(PO_4)_6(CO_3)_6$	16.018(2)	*a*	19.761(4)	90	90	120	$R\bar{3}$	(1)
Aheylite	$Fe^{2+}Al_6(PO_4)_4(OH)_8(H_2O)_4$	7.400(1)	9.896(1)	7.627(1)	110.87	115.00	69.96	$P\bar{1}$	(2)
Aldermanite	$Mg_5Al_{12}(PO_4)_8(OH)_{22}(H_2O)_{32}$	15.000(7)	8.330(6)	2.660(1)	90	90	90	*P* - - -	(3)
Alforsite	$Ba_5(PO_4)_3Cl$	10.284(2)	*a*	7.651(3)	90	90	120	$P6_3/m$	(4)
Alluaudite	$(Na,Ca)[Fe^{2+}(Mn^{2+},Fe^{2+},Fe^{3+},Mg)_2(PO_4)_3]$	12.004(2)	12.533(4)	6.404(1)	90	114.4(1)	90	$C2/c$	(5)
Althausite	$Mg_4(PO_4)_2(OH)F$	8.258(2)	6.054(2)	14.383(5)	90	90	90	$Pnma$	(6)
Althupite	$AlTh(UO_2)[(UO_2)_3(PO_4)_2O(OH)]_2(OH)_3(H_2O)_{15}$	10.953(3)	18.567(4)	13.504(3)	72.6(0)	68.2(0)	84.2(0)	$P\bar{1}$	(7)
Amblygonite	$Li[Al(PO_4)F]$	5.060	5.160	7.080	109.9	107.5	97.9	$P\bar{1}$	(8)
Anapaite	$Ca_2[Fe^{2+}(PO_4)_2(H_2O)_4]$	6.477(1)	6.816(1)	5.898(1)	101.64(3)	104.24(3)	70.76(4)	$P\bar{1}$	(9)
Archerite	$K[H_2(PO_4)]$	7.427(2)	*a*	7.046(2)	90	90	90	$I\bar{4}2d$	(10)
Arctite	$(Na_5Ca)Ca_6Ba(PO_4)_6F_3$	14.366(9)	*a*	14.366(9)	28.6(0)	28.6(0)	28.6(0)	$R\bar{3}m$	(11)
Ardealite	$Ca_2(PO_3\{OH\})(SO_4)(H_2O)_4$	5.721(5)	30.992(5)	6.250(4)	90	117.3(1)	90	Cc	(12)
Arrojadite	$KNa_4CaMn^{2+}{}_4Fe^{2+}{}_{10}Al(PO_4)_{12}(OH)_2$	16.526(4)	10.057(3)	24.730(5)	90	105.8	90	$C2/c$	(13)
Arupite	$Ni_3(PO_4)_2(H_2O)_8$	9.889	13.225	4.645	90	102.41	90	$I2/m$	(14)
Attakolite	$CaMn^{2+}Al_4(SiO_3\{OH\})(PO_4)_3(OH)_4$	17.188(4)	11.477(8)	7.322(5)	90	113.8(0)	90	$C2/m$	(15)
Augelite	$[Al_2PO_4(OH)_3]$	13.124(6)	7.988(5)	5.066(3)	90	112.3(0)	90	$C2/m$	(16)
Autunite	$Ca[(UO_2)(PO_4)]_2(H_2O)_{10\text{-}12}$	7.027	*a*	20.790	90	90	90	$I4/mmm$	(17)
Babefphite	$Ba[BePO_4F]$	6.889(3)	16.814(7)	6.902(3)	90.0(0)	90.0(0)	90.3(0)	$F1$	(18)
Bakhchisaraitsevite	$Na_2Mg_5(PO_4)_4(H_2O)_7$	8.3086(8)	12.906(1)	17.486(2)	90	102.01(1)	90	$P2_1/c$	(19)
Barbosalite	$Fe^{2+}[Fe^{3+}(PO_4)(OH)]_2$	7.250(20)	7.460(20)	7.490(20)	90	120.2(85)	90	$P2_1/c$	(20)
Baričite	$Mg_3(PO_4)_2(H_2O)_8$	10.085(2)	13.390(3)	4.6713(9)	90	104.96(3)	90	$C2/m$	(21)
Bariosincosite	$Ba(V^{4+}OPO_4)_2(H_2O)_4$	9.031(6)	*a*	12.755(8)	90	90	90	$P4/nmm$	(22)
Bassetite	$Fe^{2+}[(UO_2)(PO_4)]_2(H_2O)_8$	6.98(4)	17.07(4)	7.01(7)	90	90.53(1)	90	$P2_1/m$	(23)
Bearthite	$Ca_2Al(PO_4)_2(OH)$	7.231(3)	5.734(2)	8.263(4)	90	112.6(1)	90	$P2_1/m$	(24)
Bederite	$Ca_2(Mn^{2+}{}_2Fe^{3+}{}_2Mn^{3+}{}_2)(PO_4)_6(H_2O)_2$	12.559(2)	12.834(1)	11.714(2)	90	90	90	$Pcab$	(25)
Belovite-(Ce)	$Sr_3NaCe(PO_4)_3(OH)$	9.664(0)	*a*	7.182(0)	90	90	120	$P\bar{3}$	(26)
Belovite-(La)	$Sr_3NaLa(PO_4)_3F$	9.647	*a*	7.170	90	90	120	$P\bar{3}$	(27)

APPENDIX continued

Mineral Name	Formula	*a* (Å)	*b* (Å)	*c* (Å)	α (°)	β (°)	γ (°)	Space Group	Ref.
Benauite	$HSrFe^{3+}_3(PO_4)_2(OH)_6$	7.28	*a*	16.85	90	90	120	$R\bar{3}m$	(28)
Benyacarite	$(H_2O,K)_2TiMn^{2+}_2(Fe^{3+},Ti)_2(PO_4)_4(OH)_2(H_2O)_{14}$	10.561(5)	20.585(8)	12.516(2)	90	90	90	*Pbca*	(29)
Beraunite	$Fe^{2+}Fe^{3+}_5(PO_4)_4(OH)_5(H_2O)_6$	20.646(5)	5.129(7)	19.213(5)	90	93.62(7)	90	*C2/c*	(30)
Bergenite	$Ba[(UO_2)_3O_2(PO_4)_2](H_2O)_{6.5}$	22.32	17.19	20.63	90	93.0	90	$P2_1/c$	(31)
Berlinite	$[Al(PO_4)]$	4.943(0)	*a*	10.948(0)	90	90	120	$P3_12$	(32)
Bermanite	$Mn^{2+}[Mn^{3+}(PO_4)(OH)]_2(H_2O)_4$	5.446(3)	19.250(10)	5.428(3)	90	110.3(0)	90	$P2_1$	(33)
Bertossaite	$CaLi_2[Al(PO_4)(OH)]_4$	11.48(1)	15.73(2)	7.23(1)	90	90	90	*I*aa*	(34)
Beryllonite	$Na[BePO_4]$	8.178(3)	7.818(2)	14.114(6)	90	90	90	$P2_1/n$	(35)
Beusite	$(Mn^{2+},Fe^{2+})_3(PO_4)_2$	8.757(3)	11.381(4)	6.136(1)	90	99.1(0)	90	$P2_1/c$	(36)
Biphosphammite	$(NH_4)H_2PO_4$	7.514(0)	*a*	7.539(1)	90	90	90	$I\bar{4}2d$	(37)
Bjarebyite	$BaMn_2Al_2(OH)_3(PO_4)_3$	8.930(14)	12.073(24)	4.917(9)	90	100.2(1)	90	$P2_1/m$	(38)
Bleasdaleite	$(Ca,Fe^{3+})_2Cu^{2+}_5(Bi,Cu^{2+})(PO_4)_4(H_2O,OH,Cl)_{13}$	14.200(7)	13.832(7)	14.971(10)	90	102.08(8)	90	*C2/m*	(39)
Bobfergusonite	$Na_2Mn^{2+}_5Fe^{3+}Al(PO_4)_6$	12.776(2)	12.488(2)	11.035(2)	90	97.2(0)	90	$P2_1/n$	(40)
Bobierrite	$Mg_3(PO_4)_2(H_2O)_8$	4.667(1)	27.926(8)	10.067(3)	90	105.0(0)	90	*C2/c*	(41)
Böggildite	$Sr_2Na_2Al_2(PO_4)F_9$	5.251(3)	10.464(5)	18.577(9)	90	107.5(0)	90	$P2_1/c$	(42)
Bolivarite	$Al_2(PO_4)(OH)_3(H_2O)_4$	Amorphous	-	-	-	-	-	-	(43)
Bonshtedtite	$Na_3Fe^{2+}(PO_4)(CO_3)$	8.921	6.631	5.151	90	90.42	90	$P2_1/m$	(44)
Brabantite	$CaTh(PO_4)_2$	6.726(6)	6.933(5)	6.447(12)	90	103.89(3)	90	$P2_1$	(45)
Bradleyite	$Na_3Sr(PO_4)(CO_3)$	9.187(3)	5.279(1)	6.707(2)	90	90	90	$P2_1$	(46)
Brazilianite	$NaAl_3(PO_4)_2(OH)_4$	11.233(6)	10.142(5)	7.097(4)	90	97.4(0)	90	$P2_1/n$	(47)
Brendelite	$Bi^{3+}Fe^{3+}O_2(OH)(PO_4)$	12.278(2)	3.185(1)	6.899(1)	90	111.0(0)	90	*C2/m*	(48)
Brianite	$Na_2Ca[Mg(PO_4)_2]$	9.120(3)	5.198(2)	13.370(4)	90	90.8(0)	90	$P2_1/c$	(49)
Brockite	$(Ca,Th,REE)(PO_4)(H_2O)$	6.98(3)	6.98(3)	6.40(3)	90	90	90	*Aa*	(50)
Brushite	$Ca(PO_3\{OH\})(H_2O)_2$	5.812(2)	15.180(3)	6.239(2)	90	116.4(0)	90	*Ia*	(51)
Buchwaldite	$NaCa(PO_4)$	20.397(10)	5.412(4)	9.161(5)	90	90	90	$Pn2_1a$	(52)
Burangaite	$Na_2Fe^{2+}_2Al_{10}(PO_4)_8(OH)_{12}(H_2O)_4$	25.099(2)	5.049(1)	13.438(1)	90	110.9(0)	90	*C2/c*	(53)
Cacoxenite	$Fe^{3+}_{25}(PO_4)_{17}O_6(OH)_{12}(H_2O)_{75}$	27.559(1)	*a*	10.550(1)	90	90	120	$P6_3/m$	(54)

APPENDIX continued

Mineral Name	Formula	*a* (Å)	*b* (Å)	*c* (Å)	α (°)	β (°)	γ (°)	Space Group	Ref.
Canaphite	$CaNa_2P_2O_7(H_2O)_4$	5.673(4)	8.480(10)	10.529(5)	90	106.13(6)	90	*Pc*	(55)
Cassidyite	$Ca_2[Ni(PO_4)_2(H_2O)_2]$	5.71	6.73	5.41	96.83	107.36	104.58	$P\bar{1}$	(56)
Chalcosiderite	$Cu^{2+}Fe^{3+}{}_6(PO_4)_4(OH)_8(H_2O)_4$	7.653(4)	7.873(4)	10.190(4)	67.6(0)	69.2(0)	64.9(0)	$P\bar{1}$	(57)
Cheralite-(Ce)	$Ce(PO_4)$	6.747(2)	6.960(2)	6.453(1)	90	103.7(0)	90	$P2_1/n$	(58)
Chernikovite	$(H_3O)[(UO_2)(PO_4)]_2(H_2O)_8$	7.030(6)	*a*	9.034(8)	90	90	90	*P4/nmm*	(59)
Childrenite	$Mn^{2+}[Al(PO_4)(OH)_2(H_2O)]$	10.395(1)	13.394(1)	6.918(1)	90	90	90	*Bba2*	(60)
Chladniite	$Na_2CaMg_7(PO_4)_6$	14.967(2)	*a*	42.595(4)	90	90	120	$R\bar{3}$	(61)
Chlorapatite	$Ca_5(PO_4)_3Cl$	9.620(1)	*a*	6.776(1)	90	90	120	$P6_3/m$	(62)
Churchite-(Y)	$Y(PO_4)(H_2O)_2$	5.578(1)	15.006(3)	6.275(2)	90	117.8(0)	90	*I2/a*	(63)
Clinophosinaite	$Na_3Ca(SiO_3)(PO_4)$	7.303(2)	12.201(5)	14.715(4)	90	91.9	90	*P2/c*	(64)
Coconinoite	$Fe^{3+}{}_2Al_2(UO_2)_6(PO_4)_4(SO_4)(OH)_2(H_2O)_{20}$	12.50	12.97	23.00	90	106.6	90	*C2/c*	(65)
Coeruleolactite	$CaAl_6(PO_4)_4(OH)_8(H_2O)_{4\text{-}5}$	Existence dubious		-	-	-	-	-	(2)
Collinsite	$Ca_2[Mg(PO_4)_2(H_2O)_2]$	5.734(1)	6.780(1)	5.441(1)	97.3(0)	108.6(0)	107.3(0)	$P\bar{1}$	(66)
Corkite	$PbFe^{3+}{}_3(SO_4)(PO_4)(OH)_6$	7.280(1)	*a*	16.821(1)	90	90	120	$R\bar{3}m$	(67)
Cornetite	$Cu^{2+}{}_3PO_4(OH)_3$	10.854(1)	14.053(3)	7.086(2)	90	90	90	*Pbca*	(68)
Crandallite	$CaAl_3(PO_4)_2(OH)_5(H_2O)$	7.006(15)	*a*	16.192(32)	90	90	120	$R\bar{3}m$	(69)
Crawfordite	$Na_3Sr(PO_4)(CO_3)$	9.187	6.707	5.279	90	90	90	$P2_1$	(70)
Curetonite	$BaAl(PO_4)(OH)F$	6.977(2)	12.564(4)	5.223(1)	90	102.2(0)	90	$P2_1/n$	(71)
Cyrilovite	$NaFe^{3+}{}_3(OH)_4(PO_4)_2(H_2O)_2$	7.3255(4)	*a*	19.328(2)	90	90	90	$P4_12_12$	(72)
Deloneite-(Ce)	$NaCr_2SrCe(PO_4)_3F$	9.51	*a*	7.01	90	90	120	*P3*	(73)
Delvauxite	$CaFe^{3+}{}_4(PO_4)_2(OH)_8(H_2O)_{4\text{-}6}$	Amorphous	-	-	-	-	-	-	(74)
Destinezite	$Fe^{3+}{}_2(PO_4)(SO_4)(OH)(H_2O)_6$	9.570(1)	9.716(1)	7.313(1)	98.7(0)	107.9(0)	63.9(0)	$P\bar{1}$	(75)
Dewindtite	$Pb^{2+}{}_3[H(UO_2)_3O_2(PO_4)_2]_2(H_2O)_{12}$	16.031(6)	17.264(6)	13.605(2)	90	90	90	*Bmmb*	(76)
Dickinsonite	$(Na,Ca)_5(Mn,Fe,Mg)_{14}Al(PO_4)_{12}(OH)_2$	24.940(6)	10.131(4)	16.722(2)	90	105.6(0)	90	*A2/a*	(77)
Dittmarite	$(NH_4)Mg(PO_4)(H_2O)$	5.606	8.758	4.788	90	90	90	$Pmn2_1$	(78)
Dorfmanite	$Na_2(PO_3\{OH\})(H_2O)_2$	16.872(9)	10.359(4)	6.599(3)	90	90	90	*Pbca*	(79)
Drugmanite	$Pb^{2+}{}_2Fe^{3+}H(PO_4)_2(OH)_2$	11.111(5)	7.986(5)	4.643(3)	90	90.4(0)	90	$P2_1/a$	(80)

APPENDIX continued

Mineral Name	Formula	a (Å)	b (Å)	c (Å)	α (°)	β (°)	γ (°)	Space Group	Ref.
Dufrénite	$Fe^{2+}Fe^{3+}_5(PO_4)_3(OH)_5(H_2O)_2$	25.840(20)	5.126(3)	13.780(10)	90	111.2(1)	90	$C2/c$	(81)
Dumontite	$Pb^{2+}_2[(UO_2)_3(PO_4)_2O_2](H_2O)_5$	8.118(6)	16.819(8)	6.983(3)	90	109.0(0)	90	$P2_1/m$	(82)
Earlshannonite	$Mn^{2+}[Fe^{3+}(PO_4)(OH)]_2(H_2O)_4$	9.910(13)	9.669(8)	5.455(9)	90	93.95(9)	90	$P2_1/c$	(83)
Ehrleite	$Ca_2ZnBe(PO_4)_2(PO_3OH)(H_2O)_4$	7.130(4)	7.430(4)	12.479(9)	94.31(5)	102.07(4)	82.65(4)	$P\bar{1}$	(84)
Englishite	$Na_2K_3Ca_{10}Al_{15}(PO_4)_{21}(OH)_7(H_2O)_{26}$	38.43(2)	11.86	20.67	90	111.27	90	$A2/a$	(85)
Eosphorite	$Fe^{2+}[Al(PO_4)(OH)_2(H_2O)]$	10.445(1)	13.501(2)	6.970(30)	90	90	90	$Bba2$	(86)
Ercitite	$Na[Mn^{3+}(PO_4)(OH)](H_2O)_2$	5.362(5)	19.89(1)	5.362(5)	90	108.97(8)	90	$P2_1/n$	(87)
Evansite	$Al_3(PO_4)(OH)_6(H_2O)_8$	Amorphous	-	-	-	-	-	-	(43)
Eylettersite	$(Th,Pb)_{1Bx}Al_3(PO_4,SiO_4)_2(OH)_6$	6.99	a	16.70	90	90	90	$R\bar{3}m$	(88)
Fairfieldite	$Ca_2[Mn^{2+}(PO_4)_2(H_2O)_2]$	5.790(10)	6.570(10)	5.510(10)	102.3(2)	108.7(2)	90.3(2)	$P\bar{1}$	(89)
Farringtonite	$Mg_3(PO_4)_2$	7.596(1)	8.231(1)	5.077(1)	90	94.1(0)	90	$P2_1/n$	(90)
Faustite	$ZnAl_6(PO_4)_4(OH)_8(H_2O)_4$	7.419(2)	7.629(3)	9.905(3)	69.17(2)	69.88(2)	64.98(2)	$P\bar{1}$	(91)
Fermorite	$Ca_4Sr(PO_4)_3(OH)$	9.594(2)	9.597(2)	6.975(2)	90	90	120(0)	$P2_1/m$	(92)
Ferrisicklerite	$Li(Fe^{3+},Mn^{2+})(PO_4)$	5.918	10.037	4.798	90	90	90	$Pmnb$	(93)
Ferrostrunzite	$Fe^{2+}[Fe^{3+}(PO_4)(OH)(H_2O)]_2(H_2O)_4$	10.23(2)	9.77(3)	7.37(1)	89.65(16)	98.28(12)	117.26(16)	$P\bar{1}$	(94)
Fillowite	$Na_2CaMn^{2+}_7(PO_4)_6$	15.282(2)	a	43.507(3)	90	90	120	$R\bar{3}$	(95)
Florencite-(Ce)	$CeAl_3(PO_4)_2(OH)_6$	6.972(2)	a	16.261(6)	90	90	120	$R\bar{3}m$	(96)
Florencite-(La)	$LaAl_3(PO_4)_2(OH)_6$	6.987(2)	a	16.248(6)	90	90	120	$R\bar{3}m$	(97)
Florencite-(Nd)	$NdAl_3(PO_4)_2(OH)_6$	-	-	-	-	-	-	-	(98)
Fluellite	$Al_2(PO_4)F_2(OH)(H_2O)_7$	8.546(8)	11.222(5)	21.158(5)	90	90	90	$Fddd$	(99)
Fluorapatite	$Ca_5(PO_4)_3F$	9.367	a	6.884	90	90	90	$P6_3/m$	(62)
Fluorcaphite	$CaSrCa_3(PO_4)_3F$	9.485	a	7.000	90	90	120	$P6_3$	(100)
Foggite	$Ca[Al(PO_4)(OH)_2](H_2O)$	9.270(2)	21.324(7)	5.190(2)	90	90	90	$A2_122$	(101)
Francoanellite	$K_3Al_5(PO_3\{OH\})_6(PO_4)_2(H_2O)_{12}$	8.690(2)	a	82.271(13)	90	90	120	$R\bar{3}c$	(102)
Françoisite-(Nd)	$Nd[(UO_2)_3(PO_4)_2O(OH)](H_2O)_6$	9.298(2)	15.605(4)	13.668(2)	90	112.8(0)	90	$P2_1/c$	(103)
Fransoletite	$Ca_3[Be_2(PO_4)_2(PO_3\{OH\})_2](H_2O)_4$	7.348(1)	15.052(3)	7.068(1)	90	96.5(0)	90	$P2_1/a$	(104)
Frondellite	$Mn^{2+}Fe^{3+}_4(PO_4)_3(OH)_5$	13.89	17.01	5.21	90	90	90	$B22_12$	(105)

APPENDIX continued

Mineral Name	Formula	*a* (Å)	*b* (Å)	*c* (Å)	α (°)	β (°)	γ (°)	Space Group	Ref.
Furongite	$Al_2(OH)_2[(UO_2)(PO_4)_2](H_2O)_8$	17.87	14.18	12.18	67.8	77.5	79.9	$P\bar{1}$	(106)
Gainesite	$NaKZr_2[Be(P_4O_{16})]$	6.567(3)	*a*	17.119(5)	90	90	90	$I4_1/amd$	(107)
Galileiite	$NaFe^{2+}{}_4(PO_4)_3$	14.98	*a*	41.66	90	90	120	$R\bar{3}$	(108)
Gatehouseite	$Mn^{2+}{}_5(OH)_4(PO_4)_2$	9.097(2)	5.693(2)	18.002(10)	90	90	90	$P2_12_12_1$	(109)
Gatumbaite	$CaAl_2(PO_4)_2(OH)_2$	6.907(2)	5.095(2)	10.764(3)	90.68(8)	99.17(8)	90.17(8)	$P2/m$	(110)
Girvasite	$NaCa_2Mg_3(PO_4)_2[PO_2(OH)_2](CO_3)(OH)_2(H_2O)_4$	6.522(3)	12.250(30)	21.560(20)	90	89.5(0)	90	$P2_1/c$	(111)
Gladiusite	$Fe^{2+}{}_4Fe^{3+}{}_2(PO_4)(OH)_{11}(H_2O)$	16.950(2)	11.650(1)	6.2660(6)		90.000(4)		$P2_1/m$	(112)
Goedkenite	$Sr_2[Al(PO_4)_2(OH)]$	8.45(2)	5.74(2)	7.26(2)	90	113.7(1)	90	$P2_1/m$	(113)
Gorceixite	$BaAl_3(PO_4)(PO_3OH)(OH)_6$	7.036(0)	*a*	17.282(0)	90	90	120	$R3m$	(114)
Gordonite	$Mg[Al_2(PO_4)_2(OH)_2(H_2O)_2](H_2O)_4(H_2O)_2$	5.246(2)	10.532(5)	6.975(3)	107.5(0)	111.0(0)	72.2(0)	$P\bar{1}$	(115)
Gormanite	$Fe^{2+}{}_3Al_4(PO_4)_4(OH)_6(H_2O)_2$	11.76(1)	5.10(1)	13.57(1)	90.68(8)	99.17(8)	90.17(8)	$P\bar{1}$	(116)
Goyazite	$SrAl_3(PO_4)_2(OH)_5(H_2O)$	7.021(3)	*a*	16.505(15)	90	90	120	$R\bar{3}m$	(117)
Graftonite	$(Fe^{2+},Mn^{2+},Ca)_3(PO_4)_2$	8.910(10)	11.580(10)	6.239(8)	90	98.9(1)	90	$P2_1/c$	(118)
Grattarolaite	$Fe^{3+}{}_3O_3(PO_4)$	7.994(4)	*a*	6.855(4)	90	90	120	$R3m$	(119)
Grayite	$ThCa(PO_4)_2(H_2O)_2$	6.957	*a*	6.396	90	90	120	$P6_222$	(120)
Griphite	$Na_4Ca_6(Mn,Fe^{2+},Mg)_{19}Li_2Al_8(PO_4)_{24}F_8$	12.205(8)	*a*	*a*	90	90	90	$Pa\bar{3}$	(121)
Hagendorfite	$(Na,Ca)[Mn^{2+}(Fe^{2+},Mg,Fe^{3+})_2(PO_4)_3]$	11.92	12.59	6.52	90	114.7	90	$C2/c$	(122)
Haigerachite	$KFe^{3+}{}_3(PO_2\{OH\}_2)_6(PO_3\{OH\})_2(H_2O)_4$	16.95	9.59	17.57	90	90.85	90	$C2/c$	(123)
Hannayite	$Mg_3(NH_4)_2(PO_3\{OH\})_4(H_2O)_8$	10.728	7.670	6.702	97.87	96.97	104.74	$P\bar{1}$	(124)
Harrisonite	$CaFe^{2+}{}_6(SiO_4)_2(PO_4)_2$	6.248(1)	*a*	26.802(7)	90	90	120	$R\bar{3}m$	(125)
Heneuite	$CaMg_5(CO_3)(PO_4)_3(OH)$	6.311(1)	10.843(1)	8.676(1)	95.0(0)	93.4(0)	101.0(0)	$P\bar{1}$	(126)
Hentschelite	$Cu^{2+}Fe^{3+}{}_2(OH)_2(PO_4)_2$	6.984(3)	7.786(3)	7.266(3)	90	117.68(2)	90	$P2_1/n$	(127)
Herderite	$Ca[BePO_4F]$	9.800	7.680	4.800	90	90	90	$P2_1/a$	(128)
Heterosite	$Fe^{3+}(PO_4)$	5.830(10)	9.760(10)	4.769(5)	90	90	90	$Pmnb$	(129)
Hinsdalite	$Pb^{2+}[Al_3(OH)_6(PO_4)(SO_4)]$	7.029	*a*	16.789	90	90	120	$R\bar{3}m$	(130)
Holtedahlite	$Mg_{12}(PO_3\{OH\})(PO_4)_5(OH)_6$	11.203(3)	*a*	4.977(1)	90	90	90	$P31m$	(131)
Hopeite	$Zn_3(PO_4)_2(H_2O)_4$	10.597(3)	18.318(8)	5.031(1)	90	90	90	$Pnma$	(132)

APPENDIX continued

Mineral Name	Formula	*a* (Å)	*b* (Å)	*c* (Å)	α (°)	β (°)	γ (°)	Space Group	Ref.
Hotsonite	$Al_5(PO_4)(SO_4)(OH)_{10}$	11.29(6)	11.66(6)	10.55(7)	112.54(5)	107.52(5)	64.45(5)	$P\bar{1}$	(133)
Hureaulite	$Mn^{2+}_5(PO_3\{OH\})_2(PO_4)_2(H_2O)_4$	17.594(10)	9.086(5)	9.404(5)	90	96.67(8)	90	*C2/c*	(134)
Hurlbutite	$Ca[Be_2P_2O_8]$	8.306(1)	8.790(1)	7.804(1)	90	89.5(0)	90	$P2_1/a$	(135)
Hydroxylapatite	$Ca_5(PO_4)_3(OH)$	9.418	*a*	6.875	90	90	120	$P6_3/m$	(62)
Hydroxylherderite	$Ca[BePO_4(OH)]$	9.789(2)	7.661(1)	4.804(1)	90	90.02(1)	90	$P2_1/a$	(136)
Isokite	$Ca[Mg(PO_4)F]$	6.909	8.746	6.518	90	112.2	90	*A2/a*	(137)
Jagowerite	$Ba[Al(PO_4)(OH)]_2$	6.049(2)	6.964(3)	4.971(2)	116.51(4)	86.06(4)	112.59(3)	$P\bar{1}$	(138)
Jahnsite	$CaMnMg_2[Fe^{3+}(PO_4)_2(OH)]_2(H_2O)_8$	14.940(20)	7.140(10)	9.930(10)	90	110.16(8)	90	*P2/a*	(139)
Jahnsite-(CaMnMn)	$CaMn^{2+}Mn^{2+}_2Fe^{3+}_2(PO_4)_4(OH)_2(H_2O)_8$	14.887(8)	7.152(7)	9.966(6)	90	109.77(5)	90	*P2/a*	(140)
Johnsomervilleite	$Na_{10}Ca_6Mg_{18}(Fe^{2+},Mn^{2+})_{25}(PO_4)_{36}$	15.00	*a*	42.75	90	90	120	Hex	(141)
Johntomaite	$BaFe^{2+}_2Fe^{3+}_2(PO_4)_3(OH)_3$	9.199(9)	12.359(8)	5.004(2)	90	100.19(6)	90	$P2_1/m$	(142)
Johnwalkite	$KMn^{2+}_2[Nb(PO_4)_2O_2](H_2O)_2$	7.516(4)	10.023(8)	6.502(4)	90	90	90	$Pb2_1m$	(143)
Juonniite	$CaMgSc(PO_4)_2(OH)(H_2O)_4$	15.03	18.95	7.59	90	90	90	*Pbca*	(144)
Kanonerovite	$MnNa_3P_3O_{10}(H_2O)_{12}$	14.71(1)	9.33(1)	15.13(2)	90	89.8(1)	90	$P2_1/n$	(145)
Kastningite	$Mn(H_2O)_4[Al_2(OH)_2(H_2O)_2(PO_4)_2](H_2O)_2$	10.205(1)	10.504(1)	7.010(1)	90.38(1)	110.10(1)	71.82(1)	$P\bar{1}$	(146)
Keckite	$CaMn^{2+}_2Fe^{3+}_3(PO_4)_4(OH)_3(H_2O)_2$	15.02	7.19	19.74	90	110.5	90	$P2_1/a$	(147)
Kingite	$Al_3(PO_4)_2(OH)_3(H_2O)_9$	9.15(1)	10.00(1)	7.24(2)	98.6	93.6	93.2	$P\bar{1}$	(148)
Kintoreite	$Pb^{2+}Fe^{3+}_3(PO_4)_2(OH)_5(H_2O)$	7.331(1)	*a*	16.885(2)	90	90	120	$R\bar{3}m$	(149)
Kipushite	$[Cu^{2+}_5Zn(PO_4)_2](OH)_6(H_2O)$	12.197(2)	9.156(2)	10.667(2)	90	96.8(0)	90	$P2_1/c$	(150)
Kolbeckite	$[Sc(PO_4)(H_2O)_2]$	5.418	10.25	8.893	90	90.7	90	$P2_1/n$	(151)
Koninckite	$Fe^{3+}(PO_4)(H_2O)_3$	11.95	*a*	14.52	90	90	90	Tetragonal	(152)
Kosnarite	$KZr_2(PO_4)_3$	8.687(2)	*a*	23.877(7)	90	90	120	$R\bar{3}c$	(153)
Kovdorskite	$Mg_2(PO_4)(OH)(H_2O)_3$	10.350(40)	12.900(40)	4.730(20)	90	102.0(5)	90	$P2_1/a$	(154)
Krasnovite	$BaAl(PO_4)(OH)_2(H_2O)$	8.939	5.669	11.073	90	90	90	*Pnna/Pnnn*	(155)
Kribergite	$Al_5(PO_4)_3(SO_4)(OH)_4(H_2O)_4$	18.13(3)	13.5(2)	7.50(1)	70.50	117.87	136.58	$P\bar{1}$	(133)
Kryzhanovskite	$Mn^{2+}Fe^{3+}_2(PO_4)_2(OH)_2(H_2O)$	9.450(2)	10.013(2)	8.179(2)	90	90	90	*Pbna*	(156)
Kulanite	$BaFe^{2+}_2Al_2(PO_4)_3(OH)_3$	9.014(1)	12.074(1)	4.926(1)	90	100.48(1)	90	$P2_1/m$	(157)

APPENDIX continued

Mineral Name	Formula	a (Å)	b (Å)	c (Å)	α (°)	β (°)	γ (°)	Space Group	Ref.
Lacroixite	$Na[Al(PO_4)F]$	6.414(2)	8.207(2)	6.885(2)	90	115.5	90	*C2/c*	(158)
Landesite	$Fe^{3+}Mn^{2+}{}_2(PO_4)_2(OH)(H_2O)_2$	9.458(3)	10.185(2)	8.543(2)	90	90	90	*Pbna*	(159)
Laueite	$Mn^{2+}[Fe^{3+}{}_2(PO_4)_2(OH)_2(H_2O)_2](H_2O)_4(H_2O)_2$	5.280	10.660	7.140	107.9	111.0	71.1	$P\bar{1}$	(160)
Lazulite	$Mg[Al(PO_4)(OH)]_2$	7.144(1)	7.278(1)	7.228(1)	90	120.5(0)	90	$P2_1/c$	(161)
Lehnerite	$Mn^{2+}[(UO_2)(PO_4)]_2(H_2O)_8$	7.04(2)	17.16(4)	6.95(2)	90	90.18	90	$P2_1/n$	(162)
Leucophosphite	$K[Fe^{3+}{}_2(PO_4)_2(OH)(H_2O)](H_2O)_2$	9.782	9.658	9.751	90	102.24	90	$P2_1/n$	(163)
Libethenite	$Cu^{2+}{}_2(PO_4)(OH)$	8.071(2)	8.403(4)	5.898(3)	90	90	90	*Pnnm*	(164)
Lipscombite	$Fe^{2+}Fe^{3+}{}_2(PO_4)_2(OH)_2$	7.310	*a*	13.212	90	90	90	$P4_32_12_1$	(165)
Lithiophosphate	$Li_3(PO_4)$	10.490(3)	6.120(2)	4.9266(7)	90	90	90	*Pnma*	(166)
Lithiophyllite	$LiMn^{2+}(PO_4)$	6.100(20)	10.460(30)	4.744(10)	90	90	90	*Pmnb*	(167)
Lomonosovite	$Na_5Ti^{4+}{}_2(Si_2O_7)(PO_4)O_2$	5.440	7.163	14.830	99.0	106.0	90	$P\bar{1}$	(168)
Ludjibaite	$Cu^{2+}{}_5(PO_4)_2(OH)_4$	4.445(1)	5.873(1)	8.668(3)	103.6(0)	90.3(0)	93.0(0)	$P\bar{1}$	(169)
Ludlamite	$Fe^{2+}{}_3(PO_4)_2(H_2O)_4$	10.541(10)	4.638(8)	9.285(10)	90	100.7(1)	90	$P2_1/a$	(170)
Lulzacite	$Sr_2Fe^{2+}Fe^{2+}{}_2Al_4(PO_4)_4(OH)_{10}$	5.457(1)	9.131(2)	9.769(2)	108.47(3)	91.72(3)	97.44(3)	$P\bar{1}$	(171)
Lüneburgite	$Mg_3B_2(OH)_6(PO_4)_2(H_2O)_6$	6.347(1)	9.803(1)	6.298(1)	84.5(0)	106.4(0)	96.4(0)	$P\bar{1}$	(172)
Lun'okite	$Mn^{2+}{}_2Mg_2[Al(PO_4)_2(OH)]_2(H_2O)_8$	14.95	18.71	6.96	90	90	90	*Pbca*	(173)
Machatschkiite	$Ca_6(AsO_4)(AsO_3\{OH\})_3(PO_4)(H_2O)_{15}$	15.127(2)	*a*	22.471(3)	90	90	120	*R3c*	(174)
Maghagendorfite	$Na[Mn^{2+}MgFe^{2+}{}_2(PO_4)_3]$	-	-	-	-	-	-	-	(175)
Magniotriplite	$Mg_2(PO_4)F$	12.035(5)	6.432(4)	9.799(2)	90	108.1(0)	90	*I2/a*	(176)
Mahlmoodite	$FeZr(PO_4)_2(H_2O)_4$	9.12(2)	5.42(1)	19.17(2)	90	94.8(1)	90	$P2_1/c$	(177)
Mangangordonite	$Mn^{2+}[Al_2(PO_4)_2(OH)_2(H_2O)_6](H_2O)_2$	5.257(3)	10.363(4)	7.040(3)	105.4(0)	113.1(0)	78.7(0)	$P\bar{1}$	(115)
Manganosegelerite	$Mn^{2+}Mn^{2+}Fe^{3+}(PO_4)_2(OH)(H_2O)_4$	14.89	18.79	7.408	90	90	90	*Pbca*	(178)
Marićite	$NaFe^{2+}(PO_4)$	6.861(1)	8.987(1)	5.045(1)	90	90	90	*Pmnb*	(179)
Mcauslanite	$Fe^{2+}{}_3Al_2H(PO_4)_4F(H_2O)_{18}$	10.055(5)	11.568(5)	6.888(5)	105.84(6)	93.66(6)	106.47(5)	$P\bar{1}$	(180)
Mccrillisite	$NaCs[BeZr_2(PO_4)_4](H_2O)_{1\text{-}2}$	6.573(2)	*a*	17.28(2)	90	90	90	$I4_1/amd$	(181)
Melkovite	$CaFe^{3+}H_6(MoO_4)_4(PO_4)(H_2O)_6$	17.46	18.48	10.93	90	94.5	90	Mono	(182)
Mélonjosephite	$Ca[Fe^{2+}Fe^{3+}(PO_4)_2(OH)]$	9.542(1)	10.834(1)	6.374(2)	90	90	90	*Pbam*	(183)

APPENDIX continued

Mineral Name	Formula	*a* (Å)	*b* (Å)	*c* (Å)	α (°)	β (°)	γ (°)	Space Group	Ref.
Messelite	$Ca_2[Fe^{2+}(PO_4)_2(H_2O)_2]$	5.95(2)	6.52(2)	5.45(2)	102.3(4)	107.5(4)	90.8(2)	$P\bar{1}$	(184)
Meta-ankoleite	$K(UO_2)(PO_4)(H_2O)_3$	6.994(0)	*a*	17.784(0)	90	90	90	$P4/ncc$	(185)
Meta-autunite	$Ca[(UO_2)(PO_4)]_2(H_2O)_6$	6.960(10)	*a*	8.400(20)	90	90	90	$P4/nmm$	(186)
Metaswitzerite	$Mn^{2+}{}_3(PO_4)_2(H_2O)_4$	8.496(3)	13.173(3)	17.214(4)	90	96.7(0)	90	$P2_1/c$	(187)
Metatorbernite	$Cu^{2+}[(UO_2)(PO_4)]_2(H_2O)_8$	6.972(1)	*a*	17.277	90	90	90	$P4/n$	(188)
Meta-uranocircite	$Ba[(UO_2)(PO_4)]_2(H_2O)_6$	9.789(3)	9.882(3)	16.868(3)	90	90	89.9(0)	$P2_1/a$	(189)
Metavanmeerscheite	$U(OH)_4[(UO_2)_3(PO_4)(OH)_2](H_2O)_2$	34.18	33.88	14.074	90	90	90	$Fddd$	(190)
Metavariscite	$[Al(PO_4)(H_2O)_2]$	5.178(2)	9.514(2)	8.454(2)	90	90.35(2)	90	$P2_1/n$	(191)
Metavauxite	$Fe^{2+}(H_2O)_6[Al(PO_4)(OH)(H_2O)]_2(H_2O)_6$	10.220	9.560	6.940	90	97.9	90	$P2_1/c$	(192)
Metavivianite	$Fe^{2+}{}_3(PO_4)_2(H_2O)_8$	7.840(10)	9.110(10)	4.670(10)	95.0(0)	96.9(0)	107.7(0)	$P\bar{1}$	(193)
Meurigite	$KFe^{3+}{}_7(PO_4)_5(OH)_7(H_2O)_8$	29.52(4)	5.249(6)	18.26(1)	90	109.27(7)	90	$C2/m$	(194)
Millisite	$NaCaAl_6(PO_4)_4(OH)_9(H_2O)_3$	7.00	*a*	19.07	90	90	90	$P4_12_12$	(195)
Minyulite	$K[Al_2(PO_4)_2F(H_2O)_4]$	9.337(5)	9.740(5)	5.522(3)	90	90	90	$Pba2$	(196)
Mitridatite	$Ca_2[Fe^{3+}{}_3(PO_4)_3O_2](H_2O)_3$	17.553(2)	19.354(3)	11.248(2)	90	95.84(1)	90	Aa	(197)
Mitryaevaite	$Al_5(PO_4)_2(PO_3\{OH\})_2F_2(OH)_2(H_2O)_8(H_2O)_{6.5}$	6.918(1)	10.127(2)	10.296(2)	77.036(3)	73.989(4)	76.272(4)	$P\bar{1}$	(198)
Monazite-(Ce)	$Ce(PO_4)$	6.7902(10)	7.0203(6)	6.4674(7)	90	103.38(1)	90	$P2_1/n$	(199)
Monetite	$CaH[PO_4]$	6.910(1)	6.627(2)	6.998(2)	96.34(2)	103.82(2)	88.33(2)	$P\bar{1}$	(200)
Montebrasite	$Li[Al(PO_4)(OH)]$	6.713(1)	7.708(1)	7.019(1)	91.31(1)	117.93(1)	91.77(1)	$C\bar{1}$	(8)
Montgomeryite	$Ca_4MgAl_4(PO_4)_6(OH)_4(H_2O)_{12}$	10.023(1)	24.121(3)	6.243(1)	90	91.55(1)	90	$C2/c$	(201)
Moraesite	$[Be_2(PO_4)(OH)](H_2O)_4$	8.553(6)	12.319(6)	7.155(8)	90	97.9(1)	90	$C2/c$	(202)
Moreauite	$Al_3(UO_2)(PO_4)_3(OH)_2(H_2O)_{13}$	23.41	21.44	18.34	90	92.0	90	$P2_1/c$	(203)
Morinite	$Ca_2Na[Al_2(PO_4)_2F_4(OH)(H_2O)_2]$	9.454(3)	10.692(4)	5.444(2)	90	105.46(2)	90	$P2_1/m$	(204)
Mrázekite	$Bi^{3+}{}_2Cu^{2+}{}_3(OH)_2O_2(PO_4)_2(H_2O)_2$	9.065(1)	6.340(1)	21.239(3)	90	101.6(0)	90	$P2_1/n$	(205)
Mundite	$Al(OH)[(UO_2)_3(OH)_2(PO_4)_2](H_2O)_{5.5}$	17.08	30.98	13.76	90	90	90	$Pmcn$	(206)
Mundrabillaite	$(NH_4)_2Ca(PO_3\{OH\})_2(H_2O)$	8.643	8.184	6.411	90	98.0	90	$P2/m$	(207)
Nabaphite	$NaBa(PO_4)(H_2O)_9$	10.712(1)	*a*	*a*	90	90	90	$P2_13$	(208)
Nacaphite	$Na(Na\ Ca)(PO_4)F$	5.3232(2)	12.2103(4)	7.0961(2)	90.002(1)	89.998(1)	89.965(1)	$P\bar{1}$	(209)

APPENDIX continued

Mineral Name	Formula	*a* (Å)	*b* (Å)	*c* (Å)	α (°)	β (°)	γ (°)	Space Group	Ref.
Nahpoite	$Na_2H[PO_4]$	5.451(1)	6.847(2)	5.473(1)	90	116.33(8)	90	$P2_1/m$	(210)
Nalipoite	$NaLi_2(PO_4)$	6.884(2)	9.976(4)	4.927(2)	90	90	90	$Pmnb$	(211)
Nastrophite	$NaSr(PO_4)(H_2O)_9$	10.559(1)	*a*	*a*	90	90	90	$P2_13$	(212)
Natrodufrénite	$NaFe^{2+}Fe^{3+}{}_5(PO_4)_4(OH)_6(H_2O)_2$	25.83	5.150	13.772	90	111.53	90	$C2/c$	(213)
Natromontebrasite	$Na[Al(PO_4)(OH)]$	5.266	7.174	5.042	112.3	97.70	67.13	$P\bar{1}$	(214)
Natrophilite	$NaMn^{2+}(PO_4)$	10.523(5)	4.987(2)	6.312(3)	90	90	90	$Pnam$	(215)
Natrophosphate	$Na_7(PO_4)_2F(H_2O)_{19}$	27.712(2)	*a*	*a*	90	90	90	$Fd\bar{3}c$	(216)
Nefedovite	$Na_5Ca_4(PO_4)_4F$	11.644(2)	*a*	5.396(1)	90	90	90	$I\bar{4}$	(217)
Newberyite	$[Mg(PO_3\{OH\})(H_2O)_3]$	10.215(2)	10.681(2)	10.014(2)	90	90	90	$Pbca$	(218)
Niahite	$(NH_4)Mn^{2+}(PO_4)(H_2O)$	5.68	8.78	4.88	90	90	90	$Pmn2_1$	(219)
Ningyoite	$(U,Ca,Ce,Fe)_2(PO_4)_2(H_2O)_{1\text{-}2}$	6.78	12.10	6.38	90	90	90	$P222$	(220)
Nissonite	$Cu^{2+}{}_2Mg_2(PO_4)_2(OH)_2(H_2O)_5$	22.523(5)	5.015(2)	10.506(3)	90	99.62(2)	90	$C2/c$	(221)
Olgite	$NaSr(PO_4)$	5.565(2)	*a*	7.050(3)	90	90	120	$P3$	(222)
Olmsteadite	$KFe^{2+}{}_2(H_2O)_2[Nb(PO_4)_2O_2]$	7.512(1)	10.000(3)	6.492(2)	90	90	90	$Pb2_1m$	(223)
Olympite	$LiNa_5(PO_4)_2$	10.143(1)	14.819(3)	10.154(5)	90	90	90	$Pcmn$	(224)
Orpheite	$H_6Pb^{2+}{}_{10}Al_{20}(PO_4)_{12}(SO_4)_5(OH)_{40}(H_2O)_{11}$	7.016	*a*	16.730	90	90	120	$R\bar{3}m$	(225)
Overite	$Ca_2Mg_2[Al(PO_4)_2(OH)]_2(H_2O)_8$	14.723(14)	18.746(16)	7.107(4)	90	90	90	$Pbca$	(226)
Pahasapaite	$Ca_8Li_8[Be_{24}P_{24}O_{96}](H_2O)_{38}$	13.781(4)	*a*	13.783(1)	90	90	90	$I23$	(227)
Palermoite	$SrLi_2[Al(PO_4)(OH)]_4$	11.556(5)	15.847(7)	7.315(4)	90	90	90	$Imcb$	(228)
Panasqueiraite	$Ca[Mg(PO_4)(OH)]$	6.535(3)	8.753(4)	6.919(4)	90	112.33(4)	90	$C2/c$	(137)
Parafransoletite	$Ca_3[Be_2(PO_4)_2(PO_3\{OH\})_2](H_2O)_4$	7.327(1)	7.696(1)	7.061(1)	94.9(0)	96.8(0)	101.9(0)	$P\bar{1}$	(104)
Parahopeite	$Zn_3(PO_4)_2(H_2O)_4$	5.768(5)	7.550(5)	5.276(5)	93.42	91.18	91.37	$P\bar{1}$	(229)
Pararobertsite	$Ca_2Mn^{3+}{}_3(PO_4)_3O_2(H_2O)_3$	8.825(3)	13.258(4)	11.087(3)	90	101.19(4)	90	$P2_1/c$	(230)
Parascholzite	$CaZn_2(PO_4)_2(H_2O)_2$	17.186(6)	7.413(3)	6.663(2)	90	95.4(0)	90	$I2/c$	(231)
Paravauxite	$Fe^{2+}[Al_2(PO_4)_2(OH)_2(H_2O)_2](H_2O)_4(H_2O)_2$	5.233	10.541	6.962	106.9	110.8	72.1	$P\bar{1}$	(232)
Parsonsite	$Pb^{2+}{}_2[(UO_2)(PO_4)_2]$	6.842(4)	10.383(6)	6.670(4)	101.26(7)	98.17(7)	86.38(7)	$P\bar{1}$	(233)
Paulkellerite	$Bi^{3+}{}_2Fe^{3+}(PO_4)O_2(OH)_2$	11.380(3)	6.660(3)	9.653(3)	90	115.3(0)	90	$C2/c$	(234)

APPENDIX continued

Mineral Name	Formula	*a* (Å)	*b* (Å)	*c* (Å)	α (°)	β (°)	γ (°)	Space Group	Ref.
Penikisite	$BaMg_2Al_2(PO_4)_3(OH)_3$	8.999	12.069	4.921	90	100.52	90	$P2_1/m$	(235)
Perloffite	$BaMg_2Fe^{3+}{}_2(PO_4)_3(OH)_3$	9.223(5)	12.422(8)	4.995(7)	90	100.39(4)	90	$P2_1/m$	(236)
Petersite-(Y)	$YCu^{2+}{}_6(PO_4)_3(OH)_6(H_2O)_3$	13.288(5)	*a*	5.877(5)	90	90	120	$P6_3/m$	(237)
Petitjeanite	$Bi^{3+}{}_3(PO_4)_2O(OH)$	9.798	7.250	6.866	88.28	115.27	110.70	$P\bar{1}$	(238)
Phosinaite-(Ce)	$Na_{13}Ca_2Ce[Si_4O_{12}](PO_4)_4$	12.297(2)	14.660(3)	7.245(1)	90	90	90	$P22_12_1$	(239)
Phosphammite	$(NH_4)_2(PO_3\{OH\})$	11.043(6)	6.700(3)	8.031(4)	90	113.4(0)	90	$P2_1/c$	(240)
Phosphoellenbergite	$Mg_{14}(PO_4)_6(PO_3\{OH\})_2(OH)_6$	12.467(2)	*a*	5.044(0)	90	90	120	$P6_3mc$	(241)
Phosphoferrite	$Fe^{2+}{}_3(PO_4)_2(H_2O)_3$	8.660(30)	10.060(30)	9.410(30)	90	90	90	$Pcn2$	(242)
Phosphofibrite	$KCu^{2+}Fe^{3+}{}_{15}(PO_4)_{12}(OH)_{12}(H_2O)_{12}$	14.40	18.76	10.40	90	90	90	$Pbnm$	(243)
Phosphogartrellite	$Pb^{2+}Cu^{2+}Fe^{3+}(PO_4)_2(OH,H_2O)_2$	5.320	5.528	7.434	67.61	69.68	70.65	$P\bar{1}$	(244)
Phosphophyllite	$Zn_2Fe^{2+}(PO_4)_2(H_2O)_4$	10.378(3)	5.084(1)	10.553(3)	90	121.14(2)	90	$P2_1/c$	(245)
Phosphorrösslerite	$\{Mg(H_2O)_6\}(PO_3\{OH\})(H_2O)$	6.60	25.36	11.35	90	95	90	$C2/c$	(246)
Phosphosiderite	$[Fe^{3+}(PO_4)(H_2O)_2]$	5.330(3)	9.809(4)	8.714(5)	90	90.6(1)	90	$P2_1/n$	(247)
Phosphovanadylite	$V^{4+}{}_4P_2O_{10}(OH)_6(H_2O)_{12}$	15.470(4)	*a*	15.470(4)	90	90	90	$I\bar{4}3m$	(248)
Phosphuranylite	$KCa(H_3O)_3(UO_2)[(UO_2)_3(PO_4)_2O_2]_2(H_2O)_8$	15.899(2)	13.740(2)	17.300(3)	90	90	90	$Cmcm$	(249)
Phuralumite	$Al_2[(UO_2)_3(PO_4)_2(OH)_2](OH)_4(H_2O)_{10}$	13.836(6)	20.918(6)	9.428(3)	90	112.44	90	$P2_1/a$	(250)
Phurcalite	$Ca_2[(UO_2)_3O_2(PO_4)_2](H_2O)_7$	17.415(2)	16.035(3)	13.598(3)	90	90	90	$Pbca$	(251)
Planerite	$Al_6(PO_4)_2(PO_3\{OH\})_2(OH)_8(H_2O)_4$	7.505(2)	9.723(3)	7.814(2)	111.43	115.56	68.69	$P\bar{1}$	(2)
Plumbogummite	$Pb^{2+}Al_3(PO_4)_2(OH)_5(H_2O)$	7.039(5)	*a*	16.761(3)	90	90	120	$R\bar{3}m$	(130)
Polyphite	$Na_{17}Ca_3MgTi^{4+}{}_4(Si_2O_7)_2(PO_4)_6O_3F_5$	5.412(2)	7.079(3)	26.560(10)	95.2(0)	93.5(0)	90.1(0)	$P1$	(252)
Pretulite	$Sc(PO_4)$	6.589(1)	*a*	5.806(1)	90	90	90	$I4/amd$	(253)
Przhevalskite	$Pb[(UO_2)(PO_4)]_2(H_2O)_4$	7.24	*a*	18.22	90	90	90	Tetra	(254)
Pseudo-autunite	$Ca_2[(UO_2)_2(PO_4)_4](H_2O)_9$	6.964	*a*	12.85	90	90	120	Hexa	(255)
Pseudolaueite	$Mn^{2+}[Fe^{3+}(PO_4)(OH)(H_2O)]_2(H_2O)_4(H_2O)_2$	9.647	7.428	10.194	90	104.63	90	$P2_1/a$	(256)
Pseudomalachite	$Cu^{2+}{}_5(PO_4)_2(OH)_4$	4.4728(4)	5.7469(5)	17.032(3)	90	91.043(7)	90	$P2_1/c$	(257)
Purpurite	$Mn^{3+}(PO_4)$	4.760	9.680	5.819	90	90	90	$Pbnm$	(258)
Pyromorphite	$Pb_5(PO_4)_3Cl$	9.977(1)	9.976(1)	7.351(2)	90	90	120	$P6_3/m$	(259)

APPENDIX continued

Mineral Name	Formula	a (Å)	b (Å)	c (Å)	α (°)	β (°)	γ (°)	Space Group	Ref.
Qingheiite	$Na_2NaMn^{2+}_2Mg_2Al_2(PO_4)_6$	11.856(3)	12.411(3)	6.421(1)	90	114.45(2)	90	$P2_1/n$	(260)
Quadruphite	$Na_{14}CaMgTi^{4+}_4(Si_2O_7)_2(PO_4)_4O_4F_2$	5.4206(2)	7.0846(2)	20.364(1)	86.89(1)	94.42(1)	89.94(1)	$P1$	(209)
Raadeite	$Mg_7(PO_4)_2(OH)_8$	5.250(1)	11.647(2)	9.655(2)	90	95.94(1)	90	$P2_1/n$	(261)
Reddingite	$Mn^{2+}_3(PO_4)_2(H_2O)_3$	8.750(20)	10.173(8)	9.590(20)	90	90	90	$Pcmb$	(262)
Reichenbachite	$Cu^{2+}_5(PO_4)_2(OH)_4$	9.186(2)	10.684(2)	4.461(1)	90	92.31(1)	90	$P2_1/a$	(263)
Renardite	$Pb^{2+}(UO_2)[(UO_2)_3O_2(PO_4)_2](H_2O)_9$	15.9	17.6	13.8	90	90	90	$Bmmb$	(264)
Rhabdophane	$Ce(PO_4)(H_2O)$	7.055(3)	a	6.439(5)	90	90	120	$P6_222$	(265)
Richellite	$Ca_3Fe^{3+}_{10}(PO_4)_8(OH)_{12}(H_2O)_n$	Amorphous	-	-	-	-	-	-	(266)
Rimkorolgite	$Mg_5Ba(PO_4)_4(H_2O)_8$	12.829	8.335	18.312	90	90	90	$Pcmm$	(267)
Rittmanite	$Mn^{2+}Mn^{2+}Fe^{2+}_2[Al(PO_4)_2(OH)]_2(H_2O)_8$	15.01(4)	6.89(3)	10.16(3)	90	112.82(25)	90	$P2_1/a$	(268)
Robertsite	$Ca_2[Mn^{3+}_3(PO_4)_3O_2](H_2O)_3$	17.36(2)	19.53(5)	11.30(3)	90	96.0	90	Aa	(269)
Rockbridgeite	$Fe^{2+}Fe^{3+}_4(PO_4)_3(OH)_5$	13.873(12)	16.805(9)	5.172(4)	90	90	90	$Bbmm$	(81)
Rodolicoite	$Fe^{3+}(PO_4)$	5.048(3)	a	11.215(8)	90	90	120	$R3_121$	(270)
Roscherite	$Ca_2Mn^{2+}_5[Be_4P_6O_{24}(OH)_4](H_2O)_6$	15.874(4)	11.854(3)	6.605(1)	90	95.35(3)	90	$C2/c$	(271)
Rosemaryite	$NaMn^{2+}Fe^{3+}Al(PO_4)_3$	11.977(2)	12.388(2)	6.320(1)	90	114.45(2)	90	$P2_1/n$	(272)
Sabugalite	$Al[(UO_2)_4(PO_3\{OH\})(PO_4)_3](H_2O)_{16}$	19.426	9.843	9.850	90	96.16	90	$C2/m$	(273)
Saléeite	$Mg[(UO_2)(PO_4)]_2(H_2O)_{10}$	6.951(3)	19.947(8)	9.896(4)	90	135.17	90	$P2_1/c$	(274)
Sampleite	$NaCaCu^{2+}_5(PO_4)_4Cl(H_2O)_5$	9.70	38.40	9.65	90	90	90	Orth	(275)
Samuelsonite	$BaCa_8Fe^{2+}Al_2(OH)_2(PO_4)_{10}$	18.495(10)	6.804(4)	14.000(8)	90	112.8(1)	90	$C2/m$	(276)
Sanjuanite	$Al_2(PO_4)(SO_4)(OH)(H_2O)_9$	11.314(11)	9.018(9)	7.376(7)	93.07(1)	95.77(7)	105.66(7)	$P\bar{1}$	(133)
Sarcopside	$(Fe^{2+},Mn^{2+},Mg)_3(PO_4)_2$	6.019(0)	4.777(0)	10.419(1)	90	91.0(0)	90	$P2_1/c$	(277)
Sasaite	$Al_6(PO_4)_5(OH)_3(H_2O)_{35\text{-}36}$	10.75	15.02	46.03	90	90	90	P- - -	(278)
Satterlyite	$Fe^{2+}_2(PO_4)(OH)$	11.361	a	5.041	90	90	120	$P\bar{3}1m$	(279)
Schertelite	$(NH_4)_2[Mg(PO_3\{OH\})_2(H_2O)_4]$	11.49(2)	23.66(6)	8.62(1)	90	90	90	$Pbca$	(280)
Scholzite	$CaZn_2(PO_4)_2(H_2O)_2$	17.149(3)	22.236(2)	6.667(1)	90	90	90	$Pbc2_1$	(281)
Schoonerite	$Fe^{2+}_2ZnMn^{2+}(PO_4)_3(OH)_2(H_2O)_9$	11.119(4)	25.546(11)	6.437(3)	90	90	90	$Pmab$	(282)
Scorzalite	$Fe^{2+}[Al(PO_4)(OH)]_2$	7.15(2)	7.31(2)	7.25(2)	90	120.7(1)	90	$P2_1/c$	(20)

APPENDIX continued

Mineral Name	Formula	*a* (Å)	*b* (Å)	*c* (Å)	α (°)	β (°)	γ (°)	Space Group	Ref.
Seamanite	$[Mn^{2+}_3(B\{OH\}_4)(PO_4)(OH)_2]$	7.8231(9)	15.1405(14)	6.6999(7)	90	90	90	*Pbnm*	(283)
Segelerite	$Ca_2Mg_2[Fe^{3+}(PO_4)_2(OH)]_2(H_2O)_8$	14.826(5)	18.751(4)	7.307(1)	90	90	90	*Pbca*	(226)
Selwynite	$NaK[BeZr_2(PO_4)_4](H_2O)_2$	6.570(3)	*a*	17.142(6)	90	90	90	$I4_1/amd$	(284)
Senegalite	$Al_2(OH)_3(PO_4)(H_2O)$	7.675(4)	9.711(4)	7.635(4)	90	90	90	$P2_1nb$	(285)
Serrabrancaite	$Mn^{3+}(PO_4)(H_2O)$	6.914(2)	7.468(2)	7.364(2)	90	112.29(3)	90	*C2/c*	(286)
Sicklerite	$LiMn^{2+}(PO_4)$	4.794	10.063	5.947	90	90	90	*Pbnm*	(287)
Sidorenkite	$Na_3Mn^{2+}(PO_4)(CO_3)$	8.997(4)	6.741(2)	5.163(2)	90	90.16(4)	90	$P2_1/m$	(288)
Sigismundite	$BaNa_3CaFe^{2+}_{14}Al(PO_4)_{12}(OH)_2$	16.406(5)	9.945(3)	24.470(5)	90	105.73(2)	90	*C2/c*	(289)
Sigloite	$Fe^{3+}[Al_2(PO_4)_2(OH)_2(H_2O)_2](H_2O)_3(OH)(H_2O)_2$	5.190(2)	10.419(4)	7.033(3)	105.00(3)	111.31(3)	70.87(3)	$P\bar{1}$	(290)
Sincosite	$Ca(V^{4+}O)_2(PO_4)_2(OH)_4(H_2O)_3$	8.895(3)	*a*	12.747(2)	90	90	90	*P4/nmm*	(291)
Sinkankasite	$(Mn^{2+}(H_2O)_4)(Al(PO_3\{OH\})_2(OH))(H_2O)_2$	9.590(2)	9.818(2)	6.860(1)	108.0(0)	99.6(0)	98.9(0)	$P\bar{1}$	(292)
Smrkovecite	$Bi^{3+}_2O(PO_4)(OH)$	6.954	7.494	10.869	90	107.00	90	$P2_1/c$	(293)
Sobolevite	$Na_{11}Na_4MgTi^{4+}_4(Si_2O_7)_2(PO_4)_4O_3F_3$	7.078(1)	5.411(1)	40.618(10)	90	93.2(0)	90	*P*1	(294)
Souzalite	$Mg_3Al_4(PO_4)_4(OH)_6(H_2O)_2$	11.74(1)	5.11(1)	13.58(1)	90.83(8)	99.08(8)	90.33(8)	$P\bar{1}$	(116)
Spencerite	$Zn_4(PO_4)_2(OH)_2(H_2O)_3$	10.448(3)	5.282(1)	11.208(3)	90	116.73(3)	90	$P2_1/c$	(295)
Spheniscidite	$(NH_4)Fe^{3+}_2(OH)(PO_4)_2(H_2O)_2$	9.75	9.63	9.70	90	102.57	90	$P2_1/n$	(296)
Springcreekite	$BaV^{3+}_3(PO_4)_2(OH)_5(H_2O)$	7.258(1)	*a*	17.361(9)	90	90	120	$R\bar{3}m$	(297)
Stanekite	$Fe^{3+}Mn^{2+}(PO_4)O$	11.844(3)	12.662(3)	9.989(3)	90	105.93(2)	90	$P2_1/a$	(298)
Stanfieldite	$Mg_3Ca_3(PO_4)_4$	22.841(3)	9.994(1)	17.088(5)	90	99.6(0)	90	*C2/c*	(299)
Steenstrupine-(Ce)	$Na_{14}Ce_6Mn^{2+}Mn^{3+}Fe^{2+}_2Zr(Si_6O_{18})_2(PO_4)_7(H_2O)_3$	10.460(4)	*a*	45.479(15)	90	90	120	$R\bar{3}m$	(300)
Stercorite	$Na(NH_4)H[PO_4](H_2O)_4$	10.636(2)	6.919(1)	6.436(1)	90.46(3)	97.87(3)	109.20(3)	$P\bar{1}$	(301)
Stewartite	$Mn^{2+}[Fe^{3+}_2(PO_4)_2(OH)_2(H_2O)_2](H_2O)_4(H_2O)_2$	10.398(2)	10.672(3)	7.223(3)	90.10(3)	109.10(2)	71.83(2)	$P\bar{1}$	(302)
Strengite	$[Fe^{3+}(PO_4)(H_2O)_2]$	10.05	9.80	8.65	90	90	90	*Pcab*	(303)
Strontiowhitlockite	$Sr_9Mg(PO_4)_6(PO_3\{OH\})$	10.644(9)	*a*	39.54(6)	90	90	120	*R3c*	(304)
Strontium-apatite	$Sr_6Ca_4(PO_4)_6F_2$	9.630	*a*	7.220	90	90	120	$P6_3$	(305)
Strunzite	$Mn^{2+}[Fe^{3+}(PO_4)(OH)(H_2O)]_2(H_2O)_4$	10.228(5)	9.837(3)	7.284(5)	90.17(5)	98.44(5)	117.44(2)	$P\bar{1}$	(306)
Strüvite	$(NH_4)[Mg(H_2O)_6][PO_4]$	6.941(2)	6.941(2)	11.199(4)	90	90	90	$Pmn2_1$	(307)

APPENDIX continued

Mineral Name	Formula	*a* (Å)	*b* (Å)	*c* (Å)	α (°)	β (°)	γ (°)	Space Group	Ref.
Svanbergite	$Sr[Al_3(SO_4)(PO_4)(OH)_6]$	6.890	*a*	*a*	60.6	60.6	60.6	$R\bar{3}m$	(308)
Swaknoite	$Ca(NH_4)_2(PO_3\{OH\})_2(H_2O)$	20.959	7.403	6.478	90	90	90	*C- - -*	(309)
Switzerite	$Mn^{2+}_3(PO_4)_2(H_2O)_7$	8.528(4)	13.166(5)	11.812(4)	90	110.05(3)	90	$P2_1/a$	(310)
Tancoite	$Na_2Li[Al(PO_4)_2(OH)]H$	6.948(2)	14.089(4)	14.065(3)	90	90	90	*Pbcb*	(311)
Taranakite	$K_3Al_5(PO_3\{OH\})_6(PO_4)_2(H_2O)_{18}$	8.703(1)	*a*	95.050(10)	90	90	120	$R\bar{3}c$	(312)
Tarbuttite	$Zn_2(PO_4)(OH)$	5.499	5.654	6.465	102.85	102.77	86.83	$P\bar{1}$	(313)
Tavorite	$Li[Fe^{3+}(PO_4)(OH)]$	5.340(2)	7.283(2)	5.110(2)	109.29(2)	97.86(3)	106.32(3)	$P\bar{1}$	(314)
Thadeuite	$CaMg_3(PO_4)_2(OH)_2$	6.412(3)	13.563(8)	8.545(5)	90	90	90	$C222_1$	(315)
Threadgoldite	$Al[(UO_2)(PO_4)]_2(OH)(H_2O)_8$	20.168(8)	9.847(2)	19.719(4)	90	110.7(0)	90	$C2/c$	(316)
Tinsleyite	$K[Al_2(PO_4)_2(OH)(H_2O)](H_2O)$	9.499(2)	9.503(2)	9.535(2)	90	103.3(0)	90	$P2_1/n$	(317)
Tinticite	$Fe^{3+}_4(PO_4)_3(H_2O)_5$	7.965(2)	9.999(2)	7.644(2)	103.94(2)	115.91(2)	67.86(2)	$P\bar{1}$	(318)
Tiptopite	$K_2NaCaLi_3[Be_6P_6O_{24}(OH)_2](H_2O)_4$	11.655(5)	*a*	4.692(2)	90	90	120	$P6_3$	(319)
Torbernite	$Cu^{2+}[(UO_2)(PO_4)]_2(H_2O)_8$	7.06	*a*	20.5	90	90	90	$I4/mmm$	(320)
Triangulite	$Al_3(OH)_5[(UO_2(PO_4)]_4(H_2O)_5$	10.39	10.56	8.82	101.25	109.58	113.4	$P\bar{1}$	(321)
Triphylite	$LiFe^{2+}(PO_4)$	4.704	10.347	6.0189	90	90	90	*Pbnm*	(322)
Triplite	$Mn^{2+}_2(PO_4)F$	12.065	6.454	9.937	90	107.1	90	$I2/c$	(323)
Triploidite	$Mn^{2+}_2(PO_4)(OH)$	12.366(1)	13.276(2)	9.943(2)	90	108.2(0)	90	$P2_1/a$	(324)
Trolleite	$Al_4(OH)_3(PO_4)_3$	18.894(5)	7.161(1)	7.162(2)	90	99.99(2)	90	$I2/c$	(325)
Tsumebite	$Pb^{2+}_2[Cu^{2+}(PO_4)(SO_4)(OH)]$	7.85	5.80	8.70	90	111.5	90	$P2_1/m$	(326)
Turquoise	$Cu^{2+}Al_6(PO_4)_4(OH)_8(H_2O)_4$	7.410(1)	7.633(1)	9.904(1)	68.42(1)	69.65(1)	65.05(1)	$P\bar{1}$	(327)
Ulrichite	$Cu^{2+}[Ca(UO_2)(PO_4)_2](H_2O)_4$	12.790(30)	6.850(20)	13.020(30)	90	91.0(1)	90	$C2/m$	(328)
Upalite	$Al[(UO_2)_3O(OH)(PO_4)_2](H_2O)_7$	13.704	16.82	9.332	90	111.5	90	$P2_1/a$	(329)
Uralolite	$Ca_2[Be_4(PO_4)_3(OH)_3](H_2O)_5$	6.550(1)	16.005(3)	15.969(4)	90	101.64(2)	90	$P2_1/n$	(330)
Uramphite	$(NH_4)(UO_2)(PO_4)(H_2O)_3$	7.022(0)	*a*	18.091(0)	90	90	90	$P4/ncc$	(331)
Uranocircite	$Ba[(UO_2)(PO_4)]_2(H_2O)_{10}$	7.02	*a*	20.58	90	90	90	$P4/ncc$	(332)
Uranospathite	$AlH[(UO_2)(PO_4)]_4(H_2O)_{40}$	7.00	*a*	30.02	90	90	90	$P4_2/n$	(333)
Ushkovite	$Mg[Fe^{3+}_2(PO_4)_2(OH)_2(H_2O)_2](H_2O)_4(H_2O)_2$	5.3468(4)	10.592(1)	7.2251(7)	108.278(7)	111.739(7)	71.626(7)	$P\bar{1}$	(334)

APPENDIX continued

Mineral Name	Formula	a (Å)	b (Å)	c (Å)	α (°)	β (°)	γ (°)	Space Group	Ref.
Vanmeerscheite	$U(OH)_4[(UO_2)_3(PO_4)_2(OH)_2](H_2O)_4$	17.060(50)	16.760(30)	7.023(3)	90	90	90	$P2_1mn$	(190)
Variscite	$[Al(PO_4)(H_2O)_2]$	9.822(3)	8.561(3)	9.630(3)	90	90	90	*Pbca*	(335)
Varulite	$(Na,Ca)[Mn^{2+}(Mn,Fe^{2+},Fe^{3+})_2(PO_4)_3]$	11.99	12.64	6.51	90	114.64	90	*C2/c*	(336)
Vashegyite	$Pb^{2+}_2Cu^{2+}(CrO_4)(PO_4)(OH)$	10.773(3)	14.971(5)	20.626(6)	90	90	90	$Pna2_1$	(337)
Vauquelinite	$Pb^{2+}_2[Cu^{2+}(PO_4)(CrO_4)(OH)]$	13.754(5)	5.806(6)	9.563(3)	90	94.56(3)	90	$P2_1/n$	(338)
Vauxite	$Fe^{2+}Al_2(PO_4)_2(OH)_2(H_2O)_6$	9.13	11.59	6.14	98.3	92.0	108.4	$P\bar{1}$	(339)
Väyrynenite	$Mn^{2+}[Be(PO_4)(OH)]$	5.4044(6)	14.5145(12)	4.7052(6)	90	102.798(9)	90	$P2_1/a$	(340)
Veszelyite	$Cu^{2+}_3(PO_4)(OH)_3(H_2O)_2$	9.828(3)	10.224(3)	7.532(2)	90	103.18(2)	90	$P2_1/a$	(341)
Viitaniemiite	$NaCaAl(PO_4)(OH)F_2$	5.457(2)	7.151(2)	6.836(2)	90	109.36(3)	90	$P2_1/n$	(342)
Viséite	$Ca_{10}Al_{24}(SiO_4)_6(PO_4)_7O_{22}F_3(H_2O)_{72}$	6.89	*a*	18.065	90	90	120	$R\bar{3}m$	(343)
Vitusite-(Ce)	$Na_3Ce(PO_4)_2$	14.091(4)	5.357(1)	18.740(3)	90	90	90	$Pca2_1$	(344)
Vivianite	$Fe^{2+}_3(PO_4)_2(H_2O)_8$	10.021(5)	13.441(6)	4.721(3)	90	102.8(0)	90	*I2/m*	(345)
Vochtenite	$Fe^{2+}Fe^{3+}[(UO_2)(PO_4)]_4(OH)(H_2O)_{12\text{-}13}$	12.606	19.990	9.990	90	102.52	90	Mono	(346)
Voggite	$Na_2Zr(PO_4)(CO_3)(OH)(H_2O)_2$	12.261(2)	6.561(1)	11.757(2)	90	116.2(0)	90	*I2/m*	(347)
Vuonnemite	$Na_{11}Ti^{4+}Nb_2(Si_2O_7)_2(PO_4)_2O_3(OH)$	5.4984(6)	7.161(1)	14.450(2)	92.60(1)	95.30(1)	90.60(1)	$P\bar{1}$	(348)
Wagnerite	$Mg_2(PO_4)F$	11.957(8)	12.679(8)	9.644(7)	90	108.3(2)	90	$P2_1/a$	(349)
Wardite	$NaAl_3(OH)_4(PO_4)_2(H_2O)_2$	7.030(10)	*a*	19.040(10)	90	90	90	$P4_12_12$	(350)
Wavellite	$Al_3(OH)_3(PO_4)_2(H_2O)_5$	9.621(2)	17.363(4)	6.994(3)	90	90	90	*Pcmn*	(351)
Waylandite	$(Bi,Ca)Al_3(PO_4,SiO_4)_2(OH)_6$	6.983(3)	*a*	16.175(1)	90	90	120	$R\bar{3}m$	(352)
Weinbeneite	$Ca[Be_3(PO_4)_2(OH)_2](H_2O)_4$	11.897(2)	9.707(1)	9.633(1)	90	95.76(1)	90	*Cc*	(353)
Whiteite-(CaFeMg)	$CaFe^{2+}Mg_2[Fe^{3+}(PO_4)_2(OH)]_2(H_2O)_8$	14.90(4)	6.98(2)	10.13(2)	90	113.11(9)	90	*P2/a*	(354)
Whiteite-(CaMnMg)	$CaMn^{2+}Mg_2Al_2(PO_4)_4(OH)_2(H_2O)_8$	14.842(9)	6.976(1)	10.109(4)	90	112.59(5)	90	*P2/a*	(355)
Whitlockite	$Ca_9Mg(PO_4)_6(PO_3\{OH\})$	10.350(5)	*a*	37.085(12)	90	90	120	*R3c*	(356)
Whitmoreite	$Fe^{2+}[Fe^{3+}(PO_4)(OH)]_2(H_2O)_4$	10.00(2)	9.73(2)	5.471(8)	90	93.8(1)	90	$P2_1/c$	(357)
Wicksite	$NaCa_2Fe^{2+}_4MgFe^{3+}(PO_4)_6$	12.896(3)	12.511(3)	11.634(3)	90	90	90	*Pbca*	(358)
Wilhelmvierlingite	$Ca_2Mn_2[Fe^{3+}(PO_4)_2(OH)]_2(H_2O)_8$	14.80(5)	18.50(5)	7.31(2)	90	90	90	*Pbca*	(359)
Wolfeite	$Fe^{2+}_2(PO_4)(OH)$	12.319	13.230	9.840	90	108.40	90	$P2_1/a$	(360)

APPENDIX continued

Mineral Name	Formula	*a* (Å)	*b* (Å)	*c* (Å)	α (°)	β (°)	γ (°)	Space Group	Ref.
Woodhouseite	$CaAl_3(PO_4)(SO_4)(OH)_6$	6.976(2)	*a*	16.235(8)	90	90	120	$R\bar{3}m$	(361)
Wooldridgeite	$Na_2CaCu^{2+}_2(P_2O_7)_2(H_2O)_{10}$	11.938(1)	32.854(2)	11.017(1)	90	90	90	*Fdd*2	(362)
Wycheproofite	$NaAlZr(PO_4)_2(OH)_2(H_2O)$	10.926(5)	10.986(5)	12.479(9)	71.37(4)	77.39(4)	87.54(3)	$P1$ or $P\bar{1}$	(363)
Wyllieite	$Na(Mn^{2+},Fe^{2+})(Fe^{2+},Fe^{3+},Mg)Al(PO_4)_3$	11.868(15)	12.382(12)	6.354(9)	90	114.52(8)	90	$P2_1/n$	(364)
Xanthoxenite	$Ca_4Fe^{3+}_2(PO_4)_4(OH)_2(H_2O)_3$	6.70(4)	8.85(4)	6.54(3)	92.1(2)	110.2(2)	93.2(2)	$P\bar{1}$	(354)
Xenotime-(Y)	$Y(PO_4)$	6.895(1)	*a*	6.0276(6)	90	90	90	$I4_1/amd$	(199)
Xenotime-(Yb)	$YbPO_4$	6.866(2)	*a*	6.004(3)	90	90	90	$I4_1/amd$	(365)
Yingjiangite	$K_2Ca(UO_2)_7(PO_4)_4(OH)_6(H_2O)_6$	15.707	17.424	13.692	90	90	90	*Bmmb*	(366)
Yoshimuraite	$Ba_2Mn_2TiO(Si_2O_7)(PO_4)(OH)$	5.386(1)	6.999(1)	14.748(3)	89.98(1)	93.62(2)	95.50(2)	$P\bar{1}$	(367)
Zairite	$Bi(Fe^{3+},Al)_3(PO_4)_2(OH)_6$	7.015(5)	*a*	16.365(15)	90	90	120	$R\bar{3}m$	(368)
Zanazziite	$CaMg_5[Be_4P_6O_{24}(OH)_4](H_2O)_6$	15.874(4)	11.854(3)	6.605(1)	90	95.3(0)	90	*C*2/*c*	(369)
Zwieselite	$Fe_2(PO_4)F$	11.999(3)	9.890(3)	6.489(1)	90	90	107.7(0)	*I*2/*a*	(370)

References for mineral entries (1) through (370)

(1) McDonald et al. (1994), (2) Foord and Taggart (1998), (3) Harrowfield et al. (1981), (4) Hata et al. (1979), (5) Moore (1971b), (6) Romming and Raade (1980), (7) Piret and Deliens (1987), (8) Groat et al. (1990), (9) Catti et al. (1979), (10) Ono and Yamada (1991), (11) Sokolova et al. (1984a), (12) Sakae et al. (1978), (13) Moore et al. (1981), Merlino et al. (1981), (14) Buchwald (1990), (15) Grice and Dunn (1992), (16) Araki et al. (1968), (17) Potenza (1958), (18) Simonov et al. (1980), (19) Yakubovich et al. (2000), Liferovich et al. (2000b), (20) Lindberg and Christ (1959), (21) Yakubovich et al. (2001), (22) Roca et al. (1997), Pring et al. (1999), (23) Vochten et al. (1984), (24) Chopin et al. (1993), (25) Galliski et al. (1999), (26) Borodin and Kazakova (1954), (27) Pekov et al. (1996), Kabalov et al. (1997), (28) Walenta et al. (1996), (29) Demartin et al. (1993), Demartin et al. (1997), (30) di Cossato et al. (1989a), Fanfani and Zanazzi (1967a),

(31) Piret and Deliens (1981), (32) Ng and Calvo (1976), (33) Kampf and Moore (1976), (34) von Knorring and Mrose (1966), (35) Giuseppetti and Tadini (1973), (36) Steele et al. (1991), Wise et al. (1990), (37) Ilyukhin et al. (1995), (38) Moore and Araki (1974c), (39) Birch et al. (1999), (40) Ercit et al. (1986a), (41) Takagi et al. (1986), (42) Hawthorne (1982), (43) Garcia-Guinea et al. (1995), (44) Khomyakov et al. (1982), (45) Rose (1980), (46) Sokolova and Khomyakov (1992), (47) Gatehouse and Miskin (1974), (48) Krause et al. (1998c), (49) Moore (1975a), Alkemper and Fuess (1998), (50) Fisher and Meyrowitz (1962), (51) Curry and Jones (1971), (52) Ben Amara et al. (1983), (53) Selway et al. (1997), (54) Moore and Shen (1983a), (55) Rouse et al. (1988), (56) White et al. (1967), (57) Giuseppetti et al. (1989), (58) Hughes et al. (1995), (59) Atencio (1988), (60) Giuseppetti and Tadini (1984),

(61) McCoy et al. (1994), (62) Hughes et al. (1990), (63) Kohlmann et al. (1994), (64) Krutik et al. (1980), (65) Young et al. (1966), (66) Brotherton et al. (1974), (67) Giuseppetti and Tadini (1987), (68) Eby and Hawthorne (1989), (69) Blount (1974), (70) Khomyakov et al. (1994), (71) Cooper and Hawthorne (1994b), (72) Cozzupoli et al. (1987), Cooper et al. (2000), (73) Khomyakov et al. (1996), Rastsvetaeva and Khomyakov (1996), (74) Cech and Povondra (1979), (75) Peacor et al. (1999), (76) Piret et al. (1990), (77) Moore et al. (1981), (78) Mrose (1971), Carling et al. (1995), (79) Catti et al. (1977b), (80) King and Sengier Roberts (1988), (81) Moore (1970), (82) Piret and Piret-Meunier (1988), (83) Peacor et al. (1984), (84) Hawthorne and Grice (1987), (85) Dunn et al. (1984b), Moore (1976), (86) Hansen (1960), Hoyos et al. (1993), (87) Fransolet et al. (2000), (88) van Wambeke (1972), (89) Fanfani et al. (1970b), (90) Nord and Kierkegaard (1968),

(91) Kolitsch and Giester (2000), (92) Hughes and Drexler (1991), (93) Alberti (1976), (94) Peacor et al. (1983), (95) Araki and Moore (1981), (96) Kato (1990), (97) Lefebvre and Gasparrini (1980), (98) Milton and Bastron (1971), (99) Guy and Jeffrey (1966), (100) Khomyakov et al. (1997), (101) Moore et al. (1975b), (102) Dick and Zeiske (1998), (103) Piret et al. (1988), (104) Kampf (1992), Kampf et al. (1992), (105) Lindberg (1949), (106) Shen and Peng (1981), (107) Moore et al. (1983), (108) Olsen and Steele (1997), (109) Pring and Birch (1993), (110) von Knorring and Fransolet (1977), (111) Sokolova and Egorov-Tismenko (1990), (112) Liferovich et al. (2000a), Sokolova et al. (2001), (113) Moore et al. (1975a), (114) Radoslovich and Slade (1980), (115) Leavens and Rheingold (1988), (116) Sturman et al. (1981a), (117) Kato (1971), Kato (1987), (118) Calvo (1968), (119) Cipriani et al. (1997), (120) Dooley and Hathaway (1961),

(121) Rinaldi (1978), (122) Fisher (1956), (123) Walenta and Theye (1999), (124) Catti and Franchini-Angela (1976), (125) Grice and Roberts (1993), (126) Romming and Raade (1986), (127) Sieber et al. (1987), (128) Pavlov and Belov (1957), (129) Eventoff et al. (1972), (130) Kolitsch et al. (1999b), (131) Romming and Raade (1989), (132) Hill and Jones (1976), (133) de Bruiyn et al. (1989), (134) Moore and Araki (1973), (135) Bakakin et al. (1974), (136) Lager and Gibbs (1974), (137) Isaaks and Peacor (1981), (138) Meagher et al. (1974), (139) Moore (1974), Moore and Araki (1974a), (140) Grice et al. (1990), (141) Livingstone (1980), (142) Kolitsch et al. (2000), (143) Dunn et al. (1986), (144) Liferovich et al. (1997), (145) Popova et al. (2001), (146) Adiwidjaja et al. (1999), Schluter et al. (1999), (147) Mücke (1979), (148) Kato (1970), (149) Kharisun et al. (1997), Pring et al. (1995), (150) Piret et al. (1985),

(151) Hey et al. (1982), (152) van Tassel (1968), (153) Brownfield et al. (1993), Sljukic et al. (1969), (154) Ovchinnikov et al. (1980), (155) Britvin et al. (1996), (156) Moore (1971a), Moore and Araki (1976), Moore et al. (1980), (157) Cooper and Hawthorne (1994a), (158) Lahti and Pajunen (1985), Pajunen and Lahti (1985), Fransolet (1989), (159) Moore et al. (1980), (160) Moore (1965a), (161) Giuseppetti and Tadini (1983), (162) Mucke (1988), (163) Dick and Zeiske (1997), (164) Yakubovich and Mel=nikov (1993), (165) Vencato et al. (1989), (166) Yakubovich and Urusov (1997), (167) Geller and Durand (1960), (168) Rastvetaeva (1971), Sokolova et al. (1987a), (169) Piret and Deliens (1988), Shoemaker et al. (1981), (170) Abrahams (1966), (171) Moelo et al. (2000), (172) Sen Gupta et al. (1991), (173) Voloshin et al. (1983), (174) Effenberger et al. (1982), (175) Moore and Ito (1979), (176) Tadini (1981), (177) Milton et al. (1993), (178) Voloshin et al. (1992), (179) Le Page and Donnay (1977), (180) Richardson et al. (1988),

(181) Foord et al. (1994), (182) Egorov et al. (1969), (183) Kampf and Moore (1977), (184) Catti et al. (1977c), (185) Fitch and Cole (1991), (186) Makarov and Ivanov (1960), (187) Zanazzi et al. (1986), Fanfani and Zanazzi (1979), White et al. (1986), (188) Stergiou et al. (1993), Makarov and Tobelko (1960), (189) Khosrawan-Sazedj (1982a), (190) Piret and Deliens (1982), (191) Kniep and Mootz (1973), (192) Baur and Rama Rao (1967), (193) Dormann et al. (1982), (194) Birch et al. (1996), (195) Owens et al. (1960), (196) Kampf (1977b), (197) Moore and Araki (1977b), Moore (1976), (198) Cahill et al. (2001), Ankinovich et al. (1997), (199) Ni et al. (1995), (200) Catti et al. (1977a), (201) Fanfani et al. (1976), Moore and Araki (1974e), (202) Merlino and Pasero (1992), (203) Deliens and Piret (1985), (204) Hawthorne (1979a), (205) Effenberger et al. (1994), (206) Deliens and Piret (1981), (207) Bridge and Clark (1983), (208) Baturin et al. (1982), (209) Sokolova and Hawthorne (2001), (210) Coleman and Robertson (1981), Wiench and Jansen (1983),

(211) Ercit (1991), (212) Baturin et al. (1981), (213) Fontan et al. (1982), (214) Freeman and Bayliss (1991), (215) Moore (1972a), Moring and Kostiner (1986), (216) Genkina and Khomyakov (1992), (217) Sebais et al. (1984), (218) Sutor (1967), Sales et al. (1993), (219) Bridge and Robinson (1983), (220) Muto et al. (1959), (221) Groat and Hawthorne (1990), (222) Sokolova et al. (1984b), (223) Moore et al. (1976), (224) Malinovskii and Genkina (1992), (225) Kolkovski (1971), (226) Moore and Araki (1977c), (227) Corbin et al. (1991), Rouse et al. (1989), (228) Moore and Araki (1975), (229) Chao (1969), Kumbasar and Finney (1968), (230) Roberts et al. (1989), (231) Taxer and Bartl (1997), (232) Baur (1969a), (233) Burns (2000), (234) Grice and Groat (1988), (235) Mandarino et al. (1977 (see text)), (236) Kampf (1977c), (237) Peacor and Dunn (1982), (238) Krause et al. (1993), (239) McDonald et al. (1996), (240) Khan et al. (1972),

(241) Raade et al. (1998), (242) Flachsbart (1963), Moore and Araki (1976), (243) Walenta and Dunn (1984), (244) Krause et al. (1998a),

Krause et al. (1998b), (245) Hill (1977), (246) Ferraris and Franchini-Angela (1973), Street and Whitaker (1973), (247) Moore (1966), Fanfani and Zanazzi (1966), (248) Medrano et al. (1998), (249) Demartin et al. (1991), (250) Piret et al. (1979), (251) Atencio et al. (1991), (252) Sokolova et al. (1987b), Khomyakov et al. (1992), (253) Bernhard et al. (1998), (254) Soboleva and Pudovkina (1957), (255) Sergeev (1964), (256) Baur (1969b), (257) Shoemaker et al. (1977), (258) Bjoerling and Westgren (1938), (259) Dai and Hughes (1989), (260) Zhesheng et al. (1983), (261) Chopin et al. (2001), (262) Eversheim and Kleber (1953), (263) Anderson et al. (1977), Sieber et al. (1987), (264) Ross (1956), (265) Mooney (1948), (1950), (266) McConnell (1963), (267) Britvin et al. (1995), (268) di Cossato et al. (1989b), (269) Moore (1974), (270) Cipriani et al. (1997), Arnold (1986),

(271) Fanfani et al. (1975), (1977), (272) Fransolet (1995), (273) Vochten and Pelsmaekers (1983), (274) Miller and Taylor (1986), (275) Hurlbut (1942), (276) Moore and Araki (1977a), Moore et al. (1975a), (277) Warner et al. (1992), (278) Martini (1978), Johan et al. (1983), (279) Mandarino et al. (1978), (280) Khan and Baur (1972), (281) Taxer (1975), (282) Kampf (1977a), Moore and Kampf (1977), (283) Huminicki and Hawthorne (2002), (284) Birch et al. (1995), (285) Keegan et al. (1979), (286) Lightfoot et al. (1987), Witzke et al. (2000), (287) Blanchard (1981), (288) Kurova et al. (1980), (289) Demartin et al. (1996), (290) Hawthorne (1988), (291) Zolensky (1985), (292) Burns and Hawthorne (1995), (293) Rídko_il et al. (1996), (294) Sokolova et al. (1988), (295) Fanfani et al. (1972), (296) Cavellec et al. (1994), (297) Kolitsch et al. (1999a), (298) Keller et al. (1997), (299) Dickens and Brown (1971), (300) Moore and Shen (1983b),

(301) Ferraris and Franchini-Angela (1974), (302) Moore and Araki (1974d), (303) Moore (1966), (304) Britvin et al. (1991), (305) Klevtsova (1964), (306) Fanfani et al. (1978), (307) Abbona et al. (1984), Ferraris et al. (1986), (308) Kato and Miura (1977), (309) Martini (1991), (1993), (310) Fanfani and Zanazzi (1979), Zanazzi et al. (1986), White et al. (1986), (311) Hawthorne (1983b), (312) Dick et al. (1998), (313) Cocco et al. (1966), Genkina et al. (1985), (314) Roberts et al. (1988), Genkina et al. (1984), (315) Isaacs and Peacor (1982), (316) Khosrawan-Sazedj (1982b), (317) Dick (1999), Dunn et al. (1984a), (318) Rius et al. (2000), (319) Peacor et al. (1987), Grice et al. (1985), (320) Frondel et al. (1956), (321) Deliens and Piret (1982), (322) Yakubovich et al. (1977), (323) Waldrop (1968a), (324) Waldrop (1968b), Waldrop (1970), (325) Moore and Araki (1974b), (326) Fanfani and Zanazzi (1967b), (327) Cid-Dresner (1965), Kolitsch and Geister (2000), (328) Birch et al. (1988), (329) Piret and Declerq (1983), (330) Mereiter et al. (1994),

(331) Fitch and Fender (1983), (332) Chernorukov et al. (1997), (333) Walenta (1978), (334) Galliski et al. (2001), (335) Kniep et al. (1977), (336) Fisher (1965), (337) Johan et al. (1983), (338) Fanfani and Zanazzi (1968), (339) Baur and Rama Rao (1968), (340) Huminicki and Hawthorne (2000), (341) Ghose et al. (1974), (342) Pajunen and Lahti (1984), Lahti (1981), (343) McConnell (1952), Kim and Kirkpatrick (1996), (344) Mazzi and Ungaretti (1994), (345) Bartl (1989), (346) Zwaan et al. (1989), (347) Szymanski and Roberts (1990), (348) Ercit et al. (1998), (349) Coda et al. (1967), (350) Fanfani et al. (1970a), (351) Araki and Zoltai (1968), (352) Clark et al. (1986), (353) Walter (1992), (354) Moore and Ito (1978), (355) Grice et al. (1989), (356) Gopal et al. (1974), (357) Moore et al. (1974), (358) Sturman et al. (1981b), (359) Mücke (1983), (360) Antenucci et al. (1989), (361) Kato (1971), (362) Cooper and Hawthorne (1999), Hawthorne et al. (1999), (363) Birch et al. (1994), (364) Moore and Molin-Case (1974), Fransolet (1995), (365) Buck et al. (1999), (366) Zhang et al. (1992), (367) McDonald et al. (2000), (368) van Wambeke (1975), (369) Leavens et al. (1990), Fanfani et al. (1975), (370) Yakubovich et al. (1978).

6 Apatite in Igneous Systems

Philip M. Piccoli and Philip A. Candela

Department of Geology
University of Maryland
College Park, Maryland 20742

INTRODUCTION

Apatite is a minor but ubiquitous mineral in most igneous rocks. Although the modal proportion of apatite in common rocks is generally low, it can reach high concentrations in enclaves, cumulates, and other rocks of low abundance (i.e., rocks that constitute a small volume of the crust and mantle; e.g., nelsonites). The presence of apatite in most rocks is due not only to its low solubility in naturally occurring melts and aqueous solutions, but also to the limited ability of common rock-forming minerals to accept the amount of phosphorus that occurs in most rocks into their structure. In this paper, we will discuss some aspects of the occurrence, texture, composition, physical chemistry and petrogenetic significance of apatite in felsic rocks (i.e., andesite to rhyolite, and their plutonic analogs), in mafic (i.e., basalts and related rocks, and plutonic analogs), and ultramafic rocks of the Earth's crust and mantle.

Fluorapatite is by far the most common member of the apatite family found in igneous rocks. However, most natural fluorapatite contains some chlorine and hydroxyl as well, and these constituents can attain high concentrations in some cases. The other halogens, bromine and iodine, also occur in apatite, but their concentrations are much lower than chlorine and fluorine. Many cations commonly substitute for calcium and phosphorus in apatite; however, they rarely reach concentrations that warrant the definition of a separate mineral species.

Apatite can be described by the general formula $A_5(XO_4)_3Z$ (following Sommerauer and Katz-Lehnert 1985). The A-site accommodates large cations (e.g., Ca^{2+}, Sr^{2+}, Pb^{2+}, Ba^{2+}, Mg^{2+}, Mn^{2+}, Fe^{2+}, REE^{3+}, Eu^{2+}, Cd^{2+}, Na^{+}) (Pan and Fleet, Ch. 2 in this volume), and comprises two sites that exhibit VII-fold (Ca2) and IX-fold (Ca1) coordination (Hughes and Rakovan, Ch. 1 in this volume). The X-site, occupied primarily by P^{5+} (as PO_4^{3+}) exhibits IV-fold coordination and can accommodate other small highly charged cations (e.g., Si^{4+}, S^{6+}, As^{5+}, V^{5+}). The Z site is occupied by the halogens F^- and Cl^-, as well as OH^-.

Accessory minerals in igneous and other rocks are important beyond their modal proportion because they commonly contain elements that do not fit well into the major rock forming minerals. For example, zircon sequesters uranium and thorium, as well as zirconium, and therefore is of geochronological importance, and monazite can harbor high concentrations of the rare earth elements (REE). Apatite is an important constituent of most rocks not only because of its role as the main repository for whole rock phosphorus, but also because of its affinity for the halogens, sulfate, carbonate, strontium, the REE, and other elements (Pan and Fleet, Ch. 2 in this volume). Studies of apatite in igneous rocks are important in radiometric dating, thermochronology, trace element geochemistry, studies of weathering and hydrothermal activity, calculations of the composition of chloride and fluoride -bearing vapors and brines, fugacity ratios, fission track dating, the genesis of ore deposits, estimating sulfate concentrations of igneous rocks, and other problems.

1529-6466/00/0048-0006$05.00

As part of this work we have compiled chemical data from the literature on apatite chemistry. In that compilation we consider analyses of apatite from the literature to be of "unaltered" species, unless the authors specifically state otherwise. In addition, only a limited number of studies (Coulson et al. 2001, Fournelle et al. 1996, Loferski and Ayuso 1995, Meurer and Boudreau 1996, Parry et al. 1978, Piccoli and Candela 1994, Smith 1999) take into consideration the petrographic association of apatite, even though it has been suggested that apatite embedded in hydrous phases can experience retrograde exchange with their hosts, and may therefore have compositions different from apatite grains surrounded by nominally anhydrous phases. For example, apatite within biotite has been found to contain elevated F relative to apatite included within anhydrous phases (e.g., Loferski and Ayuso 1995).

This chapter is not meant to be exhaustive, for there are over 6000 published articles on apatite; over 2000 citations for igneous apatite alone exist in literature as of 2002. Bibliographies have been produced on apatite (Bridges et al. 1983), and several books have been written on the subject (e.g., Elliott 1994, McConnell 1973). We hope to transmit some salient features of igneous apatite petrology especially on topics not developed sufficiently in the literature, as well as some fundamentals and general background, so that interested readers can research topics further on their own.

HABIT AND TEXTURE OF IGNEOUS APATITE

Apatite is commonly observed in igneous rocks in one of two dominant habits as either equant to sub-equant, or acicular grains, that rarely exceed 1 mm in their longest dimension. In an experimental study, Wyllie et al. report that two habits crystallized from their compositionally simple melt (Wyllie et al. 1962). They found: (1) near-equilibrium growth generally produced small equant apatite in the presence of a liquid or vapor; (2) growth at large undercooling (i.e., quench) produced parallel and skeletal acicular forms (aspect ratios up to 20:1); (3) highly acicular quench apatites often have an elongate, slightly offset central cavity (similar cavities have also been observed in natural systems, where they may be occupied by glass); and (4) acicular apatite can precipitate from experiments *without* a rapid quench, but the proportion of acicular to stubby apatite increases with increasing cooling rate (a fact commonly overlooked in the literature). More recent experiments have affirmed some of these findings. According to Hoche et al. (2001), needles result from surface-controlled growth at high supersaturation (spiral-controlled growth parallel to the *c*-axis) whereas equant crystals grow at lower degrees of supersaturation under conditions where growth is diffusion controlled. Following the experimental study by Wyllie et al. (1962), many workers have inferred rapid cooling and/or the presence of a fluid, from the presence of acicular apatite in granitic rocks (e.g., Bargossi et al. 1999, Capdevila 1967, Girault 1966). In the granite massif of Neira, Spain, the interior of the granitic body contains stubby, equant apatite, whereas the margins contain highly acicular forms, a feature interpreted to indicate that the margins were much more rapidly cooled than the interior (Capdevila 1967). Similarly, in a study of the Oka Carbonatite, Canada, the variability of apatite habit, in conjunction with the experimental results of Wyllie et al. (1962), was used to infer that the carbonatite crystallized from a medium that was at least partially fluid (Girault 1966). More recently, the presence of acicular apatite in mafic enclaves in felsic rocks (or in the felsic rocks located near enclaves) was used to infer rapid cooling upon emplacement of the mafic melt into the magmatic system (e.g., Bargossi et al. 1999, Holden et al. 1987, Reid et al. 1983, Sha 1995). In some cases, the presence of acicular apatite in and around mafic enclaves included in felsic rocks has been interpreted to indicate a large temperature difference between mafic and felsic melt at the time of formation (Brown and Peckett 1977, Sha 1995). In other cases, the lack of an acicular habit has been used, in concert with other data, to suggest a cumulate genesis for enclaves (see, for example, Dorais et al. 1997) or even a non-magmatic origin (e.g., country rock digestion, Vernon 1983). Caution should be counseled here, for acicular apatites can also be found in otherwise normally textured granites or porphyries that otherwise do not have acicular crystals (for example, the Wangrah Suite, King et al. 2001). Clearly, apatite habit is dependent on environment and cooling history, but especially in light of point (4) above, the occurrence of acicular apatite can be over-interpreted.

Concentric zoning

Apatite in igneous rocks may be zoned or unzoned given the precision of the electron microprobe; chemically zoned apatites have been reported by Jolliff et al. (1989), Jacobs and Parry (1973), Roegge et al. (1974), Coulson and Chambers (1996), amongst others. Furthermore, sensitive analytical techniques such as XANES, cathodoluminescence, LA-ICPMS, have revealed that chemical zoning of minor and trace constituents is not uncommon (e.g., Jolliff et al. 1989, Rakovan et al. 1997, Rakovan and Reeder 1994). However, chemical zoning among Z-site elements (F, Cl, OH) in igneous and hydrothermal apatite is not common. Still, examples of limited zoning with respect to halogens in plutonic apatite grains do exist in the literature. For example, in their study of the La Gloria pluton, Chile, Cornejo and Mahood (1997) found that cores of apatite are slightly depleted in fluorine relative to rims, a trend crudely antithetical to chlorine. Although there is considerable scatter, they found a continuous compositional gradation in the well-exposed (over 2500 m) pluton which grades upward from granodioritc to quartz-monzonite, with apatite in the upper reaches of the pluton higher in fluorine relative to chlorine (Cornejo and Mahood 1997).

Indeed, considerations of halogen zoning or lack thereof indicate that diffusion needs to be examined more closely. Possibly the diffusion of halogens within apatite occurs during the protracted cooling history of igneous plutons, obliterating primary zonation. A detailed experimental study of the diffusion of halogens in apatite has been performed (Brenan 1993). The results of their experiments, performed at 1 atm and 1 GPa on natural fluorapatite reveal a directional variation in the diffusion coefficient, such that diffusion is faster parallel to the *c*-axis than perpendicular to it. They indicate that only under conditions of rapid cooling would apatites preserve zoning inherited at the time of crystallization. However, volcanic apatites are rarely zoned with respect to the halogens. We therefore suggest that the lack of halogen zoning in volcanic apatites indicates that subsequent diffusion is *not* important in altering halogen concentrations in plutonic apatite.

A considerable number of other studies exist in the literature on diffusion-related issues in apatite, including diffusion of Pb (e.g., Cherniak et al. 1991, Watson et al. 1985), Sr (e.g., Cherniak and Ryerson 1993, Farver and Giletti 1989, Watson et al. 1985), rare earth elements (e.g., Cherniak 2000, Watson et al. 1985), He (e.g., Farley 2000, Wolf et al. 1996), and oxygen (e.g., Farver and Giletti 1989). Clearly, interpretation of zoning patterns discerned in natural apatite (e.g., Jolliff et al. 1989, Tepper and Kuehner 1999) need to be made in light of these data.

SOLUBILITY OF APATITE IN MELTS

Many experiments have been performed on the crystallization of apatite in igneous systems (Ayers and Watson 1990, 1991; Baig et al. 1999, Harrison and Watson 1984, Jahnke 1984, Montel et al. 1988, Philpotts 1984, Watson 1979,1980; Wolf and London 1993,1994,1995). These experiments cover a wide variety of melt compositions ranging from peralkaline to peraluminous, ultramafic to rhyolitic. Generally, the solubility of apatite in melts increases with increasing temperature, with decreasing polymerization of the melt, and decreasing silica concentration of the melt (London et al. 1999).

Experiments have demonstrated that apatite crystallization is primarily a function of the SiO_2 and P_2O_5 concentrations and temperature in silicate melts for metaluminous melts with 45-75 wt % SiO_2 and 0-10 wt % water at crustal pressures (Harrison and Watson 1984). The effects of water concentration, pressure, and especially the concentration of calcium on apatite solubility were not significant *independent* controls on apatite solubility within the confines of the basalt-dacite-rhyolite temperature-composition space and the limits of precision of their experiments. The equilibrium concentration of P_2O_5 in peralkaline silicate melts saturated with apatite ranges between 0.04 and 0.28 wt % P_2O_5 at 1 kbar and temperatures between 750 and 900°C (Watson and Capobianco 1981). Furthermore, for high-silica peralkaline systems, the effects of temperature and SiO_2 concentration (between 72 and 79 wt %) on apatite solubility are small (Watson and Capobianco 1981). In general, the solubility of phosphate minerals, at a given temperature and activity of anorthite and silica in the magmatic system, can be expressed as a function of the aluminum saturation index (ASI), by virtue of equilibria such as:

$$\text{fluorapatite} + 13\ AlO_{1.5}^{melt} + 10\ SiO_2^{melt} = 5\ An^{plag} + 3\ AlPO_4^{melt} + 0.5\ F_2O_{-1}$$

which indicates the importance of ASI in apatite solubility for a given value of the ratio f_{HF}/f_{H2O}. The fugacity ratio can be related to the chemical potential of the F_2O_{-1} component by way of the relation: $F_2O_{-1} = 2\ HF - H_2O$. If instead, we were to cast the problem in terms of the fluorine concentration in the melt, we would need to consider the relationship between melt species, concentrations and activities, which may involve complex interactions as are implied by species such as Na_3AlF_6 (London et al. 1999).

Experiments have been performed on apatite solubilities in peraluminous melts which suggest higher solubilities than for metaluminous melts (Montel et al. 1988, Pichavant et al. 1992, Wolf and London 1994). Wolf and London (1994) performed solubility experiments over a wide range of aluminum saturation index (ASI = $Al_2O_3/Na_2O + K_2O + CaO$, on a molar basis). They found that in the range of ASI of 1 to 1.1, the solubility of apatite (as measured by the P_2O_5 concentration in the melt), was 0.1 wt %, similar to that determined by Harrison and Watson (1984). Furthermore, at higher ASI, they demonstrated a linear relationship between ASI and the concentration of phosphorus in the melt: P_2O_5 (wt %) = -3.4 + 3.1·ASI, consistent with saturation at considerably lower temperatures than for metaluminous systems. For example, the solubility of apatite in an anatectic melt with ASI = 1.3-1.4, and 750°C, is on the order of ~0.7 wt % P_2O_5, considerably higher than the case for metaluminous and peralkaline systems described above. For a rock with 0.2 wt % apatite, 50% melting would require 0.35 wt % P_2O_5 from the original rock to provide for apatite saturation; London (1997) concluded that most such anatectic melts will be initially undersaturated with respect to apatite (although he does not consider a given percent melting), and will therefore first fractionate plagioclase before apatite, yielding increasing P/Ca with crystallization. For Ca-poor anatectic melts, apatite may not saturate at all, and ferromagnesian phosphates may be stabilized, (as in the beryl-columbite-phosphate pegmatite sub-type, London 1999). More commonly, however, fractionation of Ca-bearing components yields reaction relationships such that melts become undersaturated with respect to apatite, and saturated with respect to xenotime and/or monazite.

Estimation of apatite saturation temperature (AST)

A strong temperature dependence on apatite/melt exchange equilibria has been recognized for some time, necessitating the determination of the temperature at which apatite crystallizes when making estimates of magmatic halogen concentrations or fugacities. Apatite crystallizes over a wide temperature range in most felsic metaluminous magmatic systems because it saturates relatively early relative to other minerals and does not undergo reaction to other phosphate-bearing minerals until very late, if at all. See Wolf and London (1995) for further discussion of xenotime crystallization at the expense of apatite. Estimations of apatite crystallization temperature can be made by comparing natural and experimental systems (e.g., Clemens et al. 1986, Kogarko 1990). Suitable experiments have been performed (Green and Watson 1982, Harrison and Watson 1984, Montel et al. 1988, Pichavant et al. 1992, Watson 1979,1980; Watson and Capobianco 1981, Wolf and London 1994,1995) on the solubility of apatite in melts of natural composition.

For metaluminous to slightly peraluminous systems, the Nernst partition coefficients presented by Harrison and Watson (1984) can be recast as an empirical solubility expression, and solved for temperature. Their results can be represented as

$$T = \frac{\left[26{,}400 \cdot C^{l}_{SiO_2} - 4{,}800\right]}{\left[12.4 \cdot C^{l}_{SiO_2} - \ln\left(C^{l}_{P_2O_5}\right) - 3.97\right]} \quad (1)$$

where T is the apatite saturation temperature (in Kelvin), and $C^{l}_{SiO_2}$ and $C^{l}_{P_2O_5}$ represent the concentration of phosphorus and silica in the melt at the temperature at which apatite begins to crystallize; these are expressed as weight fractions (wt % / 100). Generally, the whole rock concentrations will not be the same as melt concentrations, except in those cases where apatite is a liquidus or near-liquidus phase (see discussions in Coulson et al. 2001, Piccoli and Candela 1994). If apatite is the liquidus phase, the calculation of the temperature at which apatite begins to crystallize is straightforward given whole rock phosphorus and silica as estimators of melt composition. Whereas this is true for near-minimum felsic melts, it is not true universally. The solubility of apatite is higher in more mafic magmas (e.g., in a tonalite) and apatite may not appear until 1/3 or more of the melt has crystallized. Therefore, before a model AST can be calculated for most intermediate to mafic melt compositions, the instantaneous concentration of P_2O_5 and SiO_2 as a function of progressive crystallization upon cooling must be estimated, and the state of saturation with respect to apatite must be evaluated step-wise at each arbitrarily chosen increment. There is no standard method by which these parameters can be estimated for melts in plutonic systems, and therefore experimental data aid in these estimations (see discussion in Piccoli and Candela 1994). Another issue that complicates the calculation is that in some instances, especially in the case of high ASI systems, feldspars can incorporate high concentrations of phosphorus by way of the berlinite exchange component $AlPSi_{-1}$ (see London et al. 1990), and if protracted crystallization of feldspars occurs prior to apatite saturation, the calculated AST will be inherently too high. This effect can be accounted for (Sha and Chappell 1998). Although apatite crystallizes over a wide temperature interval in felsic magmas, most of the apatite crystallizes over a relatively restricted interval of within 60-100°C of the apatite saturation temperature (see Piccoli and Candela 1994 for further details). We suggest the thermodynamic calculations involving igneous apatite be performed at some temperature within this temperature interval, or at the apatite saturation temperature.

In most cases, workers who calculate apatite saturation temperatures fail to make these corrections (in part due to their difficulty); although in some systems the corrections are small (for example, volcanic systems with low degrees of crystallization), however, they may be significant for mafic systems where initiation of crystallization of apatite occurs significantly below the liquidus. For the calculations that follow, either the AST can be used, or calculations can be performed using *model* temperatures (e.g., see

Coulson et al. 2001).

COMPOSITION OF IGNEOUS APATITE

In an effort to better understand the chemistry of apatite in igneous and hydrothermal systems, apatite analyses from the literature were accumulated. In the dataset compiled as part of this effort, we have made an attempt to group apatite compositions based on the petrology of the associated rock unit and their relationship with ores, or the lack thereof. This was performed, in part, because of historic efforts in using apatite as an indicator of ore-potential (e.g., Roegge et al. 1974, Williams and Cesbron 1977), something that has continued to present (e.g., Belousova et al. 2002, Boudreau 1993). The categories and the associated abbreviations used are: felsic-intermediate rocks not associated with ore-deposits (F-NA OD), felsic pegmatites (F-PEGS), barren (i.e., not containing metals in sufficient quantity to be extracted and sold at a profit) felsic-intermediate igneous bodies that are related to, but distinct from, igneous rocks (ore-producing) rocks that are closely associated with ore (F-OD NOP), felsic-intermediate ore-producing igneous rocks closely associated with ore (F-OD OP), mafic rocks not associated with ore-deposits (M-NA OD), barren mafic rocks associated with ore-producing rocks (M-NOP), mafic ore-producing rocks (M-OD-OP), carbonatites (C), and low temperature, hydrothermally altered rocks (Low-T) (see Table 1). Each of the groups are presented in Table 2 and in the form of plots (on a mole basis) in ternary compositional spaces along which, in our opinion, large variations in apatite composition occur: Ca-Mn-Na (Fig. 1), Ca-Mn-Fe (Fig. 2), Ce-Mn-Fe (Fig. 3), P-Si-S (Fig. 4) and F-Cl-OH (Fig. 5) [discussed below]. Note that because the concentrations are normalized for plotting, absolute concentrations of the elements are not indicated directly, and more than one composition of apatite may plot at the same location within a figure if those components are present in the same ratio, but different absolute concentrations.

No attempt was made to group the data based on tectonic environment (e.g., apatite from mafic rocks associated with alpine ophiolites and apatite from flood basalts are in the same group). Ore-bearing and associated non-ore-bearing rocks have been discriminated in order to evaluate the role of chemical differences as a function of ore-forming potential. Felsic rocks are defined here as compositions between high-silica rhyolite (and their plutonic analogs) through andesite (and their plutonic analogs); although the equivalence of intrusive and extrusive rocks is not always straightforward, this approach has been adopted for simplicity. We have not considered the petrographic association of the apatites discussed in the literature (e.g., apatites included in other phases; as intercumulus phase, etc.), mainly because most papers do not report this information in any systematic manner; we strongly suggest that all apatite data collected in the future be done on apatite that has been well characterized petrographically, including notes on the identity and, if possible, the composition of all minerals that are observed to touch the analyzed apatite grains. Data for the Stillwater Complex in Montana have been subdivided such that apatite from just above the basal series through the *Pt-bearing* J-M reef have been classified as ore-bearing (following the rationale by Boudreau and McCallum 1989).

A summary of over 1000 analyses included in the database can be found in Table 3. This compilation includes arithmetic means for each category of analyses in the database, the corresponding standard deviation (1σ) and number of analyses. No rigorous attempt has been made to evaluate the quality of data included in the compilation. Clearly when using an electron microprobe to determine chemistry (by far the most common analytical technique used for apatite analyses presented in this overview), there can be problems with the accurate and precise analyses of apatite due to anisotropy effects possibly due to diffusion (e.g., F, Cl, P and Ca; Stormer et al. 1993), peak overlaps (e.g., F and P; Potts and Tindle1989, Raudsepp 1995), and heterogeneity in standards (e.g., REE-PO4; Donovan et al. 2003), amongst other analytical problems. For a thorough discussion of some of these issues, and with recommendations on how to properly analyze phosphates using an electron microprobe, the reader is referred to Pyle et al. (Ch. 8, p. 337ff.).

Apatite chemistry in Ca-Mn-Na space

Figure 1 demonstrates the distinction in composition among apatite from mafic rocks, carbonatites, low-T hydrothermal systems, and felsic rocks. The primary exchange that explains most of the variation in apatite from mafic rocks in Ca-Mn-Na space is the substitution of Na for Ca, which requires contribution of other elements to retain charge balance, e.g., $NaSP_{-1}Ca_{-1}$, which indicates $S^{6+} + Na^{+} = P^{5+} + Ca^{2+}$ (Liu and Comodi 1993), or $(LREE)SiP_{-1}Ca_{-1}$, and $(HREE)NaCa_{-2}$ (Sha and Chappell 1999). Apatite from carbonatites and low-T hydrothermal deposits (although the data are limited) exhibit the same dominant exchanges; however, apatites from carbonatites can exhibit about twice the extent of substitution as of those from mafic rocks, consistent with the higher average concentration of Na_2O in carbonatites (0.61 wt %) relative to mafic rocks not associated with ore deposits (0.33 wt %). Furthermore, apatites from mafic rocks associated with ore deposits (M OD-NOP/OP) have considerably lower Na_2O (0.05 wt %) and MgO (~0.06 wt %) than apatites from mafic rocks not associated with ores (0.33 wt % Na_2O, 0.34 wt % MgO; Table 2).

These trends are in sharp contrast to apatite in felsic rocks (Fig. 1a), which display considerable substitution of Mn for Ca, a substitution that does not require a charge balance. This is consistent with the large range found for MnO concentrations in apatite (F-NAOD, 0.01-5.4 wt % MnO: F-PEGS, 0.05-7.9 wt %). Note, however, that it is not clear from this figure which exchange operators are controlling sodium substitution. In felsic systems associated with ore systems, there is a clear distinction between Mn-rich and Na-rich apatite based on the type of associated ore (Fig. 1b). The apatites in the highly evolved, F-rich magmas associated with the Climax-type porphyry Mo deposits have Mn equal to or in excess of Na (on a molar basis), whereas the less evolved rocks associated with porphyry Cu deposits have apatite with Na that exceeds Mn. This suggests that fractionation more strongly affects Mn/Ca in apatite relative to Na/Ca, especially in highly fractionated systems, wherein the chemical potential of the Na-bearing component in limited by saturation with albite-rich feldspar, whereas Mn is not solubility-limited in most felsic magmatic systems. Further, it must be kept in mind that the substitution of sodium in apatite requires coupled substitution, and the changes in more than just sodium or Na/Ca in melt must be considered. Apatite from the ore-related plutons of iron oxide Cu-Au deposits of the Mt. Isa/Conclurry District (Queensland, Australia), are distinctly different from both porphyry Mo and Cu systems, and have apatite that is relatively depleted in both Mn and Na. This lends support to the hypothesis that iron oxide Cu-Au deposits are genetically distinct from porphyry-type ore deposits.

Apatite chemistry in Ca-Mn-Fe space

Discrimination between apatite from felsic rocks not associated with ore-deposits and felsic pegmatites is displayed on a plot of Ca/(100-Mn-Fe) (Fig. 2a). Apatites from felsic pegmatites have Mn concentrations much higher than Fe, and have a relatively restricted range in iron concentrations (0.01-0.75 wt % as FeO), compared to those from other felsic rocks (0.01-2.67 wt % FeO). In general, pegmatitic apatite is lower in iron. Experimental studies (Johannes and Holtz 1996) have shown that the iron concentration in aluminosilicate melts is limited by saturation with respect to iron-rich minerals such as biotite (annite) or magnetite, and that the equilibrium iron concentration in the melt decreases with decreasing temperature at the oxygen fugacities normally encountered in granitic systems. This effect is magnified for pegmatites, which probably crystallize at

Table 1. Composition of igneous and hydrothermal apatite.

Rock Composition/Association	Representative Examples	References
felsic-intermediate rocks not associated with ore-deposits (F-NA OD)	Trachyandesitic El Chichón pumices; dacites from Mt. Pinatubo; dacites and rhyolites of the Rabaul Caldera (PNG); augite syenites and sodalite foyalite, Ilimaussaq; granite gneisses and nephaline syenites of North Qôroq, Greenland; Bishop Tuff Rhyolite; Tuolumne Intrusive Suite; biotite-granodiorite of the Idaho batholith; Kirchberg Granite of the Erzgebirge district; rhyolitic Richtermont Dike (France); Achala batholith (Argentina); diorite, granoidiorite, adamellite and aplite of the Boggy Plain Suite, Australia; granitic rocks from South Korea; diorites to quartz monzon ites of the Odenwald Region, Germany; Turtle Pluton, CA; granites of the Emerald Lake pluton, Yukon	(Allen 1991; Barth and Dorais 2000; Belousova et al. 2001; Chang et al. 1996; Coulson and Chambers 1996; Coulson et al. 2001; Cruft 1966; Dorais et al. 1997; Fournelle et al. 1996; Gottesmann and Wirth 1997; Heming and Carmichael 1973; Hildreth 1977; Hoskin et al. 2000; Imai et al. 1996; Luhr et al. 1984; Parry and Jacobs 1975; Piccoli and Candela 1994; Piccoli 1992; Raimbault and Burnol 1998; Ronsbo 1989; Sallet 2000; Taborszky 1962; Tepper and Kuehner 1999; Tsusue et al. 1981)
felsic pegmatites (F-PEGS)	Tin Mountain Pegmatite, South Dakota; Bob Ingersoll Pegmatite, South Dakota; Crystal Lode pegmatite (CO); Gloserheia Granite Pegmatite, Norway.	(Amli 1975; Binder and Troll 1989; Chang et al. 1996; Cruft 1966; Jolliff et al. 1989; Liu and Comodi 1993; Shmakin and Shiryaye 1968; Walker 1984; Walker et al. 1986; Young and Munson 1966)
barren felsic-intermediate rocks associated with ore-producing rocks (F OD-NOP)	Rocks related to Climax-type porphyry Mo deposits (Silver Plume granite, Red Mountain Porphyry); granites of the Mt. Isa Inlier (Queen Elizabeth granite, Kalkadoon granodiorite)	(Belousova et al. 2001; Dilles 1987; Gunow 1983)
felsic-intermediate ore-producing rocks (F-OD-OP)	Rocks related to Climax-type porphyry Mo deposits (Henderson, Primose Intrusions); granites of the Mt. Isa Inlier (Wimberu, Mt. Angelay, Malakoff); rocks of the Providencia District, Mexico (Providencia, Concepcion del Oro and Noche Buena stocks)	(Belousova et al. 2001; Binder and Troll 1989; Dilles 1987; Gunow 1983; Parry and Jacobs 1975; Roegge et al. 1974; Streck and Dilles 1998)

Sources for information about apatite chemistry, including representatives of each of the locations used in the compilation, and the notation used in the figures.

Table 1 (continued)

Rock Composition/Association	Representative Examples	References
mafic rocks not associated with ore-deposits (M-NA OD)	Skaergaard Intrusion; gabbros of the Appenine Ophiolite; gabbros of the Odenwald Region; alkaline rocks (phonolites, basanites, foidites, tephrite, melilites) from Italy; pyroxenites, lhezolites and mantle peridotites; alkali basalts.	(Binder and Troll 1989; Brown and Peckett 1977; Chang et al. 1996; Coulson and Chambers 1996; Cruft 1966; Exley and Smith 1982; Huntington 1979; Liu and Comodi 1993; Nash 1976; O'Reilly and Griffin 2000; Pearson and Taylor 1996; Sassi et al. 2000; Stoppa and Liu 1995; Taborszky 1962; Tribuzio et al. 1999)
barren mafic rocks associated with ore-producing rocks (M-OD-NOP)	Basalts of the North Qôroq, Greenland; ultrabasic Lyons River sill, (Australia); upper mantle alkali basalts; rocks from the Skaergaard Intrusion, Greenland; basalts of the Kiglipait Intrusion, Labrador; mantle xenoliths from a variety of locations; gabbros from Ostenwald, Germany	(Boudreau and Kruger 1990; Boudreau et al. 1986; Boudreau and McCallum 1989,1990; Dilles 1987; Meurer and Boudreau 1996)
mafic ore-producing rocks (M-OD-OP)	A variety of rock types from the Bushveld (South Africa) and Stillwater, Montana	(Boudreau and Kruger 1990; Boudreau et al. 1986; Boudreau and McCallum 1989,1990)
carbonatites (C)	Carbonatites from the following localities: Quebec, Ontario, Germany, Brazil, Uganda, South Africa, Italy, Finland	(Binder and Troll 1989; Buhn et al. 2001; Chang et al. 1996; Cruft 1966; Hogarth et al. 1985; Liu and Comodi 1993; Riley et al. 1996; Seifert et al. 2000; Stoppa and Liu 1995)
Low temperature, hydrothermally altered rocks (Low-T)	Cerro de Mercado, Mexico; Kauhajarvi, Finland; locations within the Ural Mountains; El Laco, Chile; Bayan Obo, Inner Mongolia	(Andersen and Austrheim 1991; Binder and Troll 1989; Cruft 1966; Fominykh 1974; Karkkainen and Appelqvist 1999; Liu and Comodi 1993; Roegge et al. 1974; Smith 1999; Young et al. 1969)

Table 2. Compilation of major and trace elements for apatite from igneous and hydrothermal environments.

	F-NA OD N= 475		F-PEGS N= 69		F OD-Non OP N= 29		F OD-OP N= 97		M-NA OD N= 86		M OD- Non OP N= 71		M OD- OP N= 18		Carbonatites N= 88		Low-T, HT Dep N= 35	
	Avg	*Std*	Avg	*Std*	Avg	*Std*	Avg	*Std*	Avg	*Std*	Avg	*Std*	Avg	*Std*	Avg	*Std*	Avg	*Std*
Major Elements (wt%)																		
SiO_2	0.42	*0.72*	0.41	*0.25*	0.78	*0.05*	0.46	*0.29*	0.85	*0.67*	0.29	*0.24*	0.30	*0.22*	0.80	*1.09*	0.33	*0.41*
Al_2O_3	0.12	*0.24*	0.15	*0.13*	--	--	0.06	*0.07*	0.19	*0.26*	--	--	--	--	0.25	*0.29*	0.12	*0.12*
Fe_2O_3	0.10	*0.06*	0.21	*0.16*	--	--	--	--	0.27	--	--	--	--	--	0.22	--	--	--
FeO	0.30	*0.37*	0.21	*0.20*	0.05	*0.17*	0.15	*0.19*	0.38	*0.26*	0.15	*0.18*	0.17	*0.12*	0.13	*0.11*	0.22	*0.19*
MnO	0.24	*0.50*	2.01	*2.21*	0.16	*0.04*	0.40	*0.89*	0.05	*0.02*	0.05	*0.03*	0.07	*0.03*	0.09	*0.06*	0.07	*0.08*
MgO	0.07	*0.07*	0.20	*0.13*	--	--	0.05	*0.07*	0.34	*0.38*	0.06	*0.09*	0.09	*0.08*	0.14	*0.13*	0.12	*0.10*
CaO	53.81	*1.80*	52.65	*3.57*	54.83	*53.91*	54.59	*1.04*	53.75	*2.35*	54.36	*0.76*	54.04	*1.11*	51.95	*2.20*	53.60	*1.17*
SrO	0.07	*0.05*	0.57	*2.35*	0.01	*0.03*	0.03	*0.03*	0.97	*1.83*	0.07	*0.03*	0.05	*0.03*	1.85	*1.53*	0.77	*0.54*
Na_2O	0.16	*0.31*	0.15	*0.13*	0.07	*0.03*	0.13	*0.13*	0.33	*0.40*	0.05	*0.03*	0.05	*0.06*	0.61	*0.99*	0.30	*0.18*
K_2O	0.05	*0.04*	0.04	*0.03*	--	--	0.02	*0.02*	0.05	*0.04*	--	--	--	--	0.04	--	0.01	--
P_2O_5	40.44	*1.39*	41.18	*1.15*	40.21	*41.46*	40.85	*0.70*	40.43	*1.69*	40.70	*0.71*	41.11	*2.40*	38.97	*1.92*	41.24	*1.00*
SO_3	0.28	*0.24*	0.17	*0.18*	0.03	*0.01*	0.19	*0.21*	0.61	*0.53*	--	--	0.01	*0.00*	0.23	*0.27*	0.20	*0.24*
CO_2	1.72	--	0.47	*0.41*	0.00	*0.00*	0.65	*0.31*	0.52	*0.55*	--	--	--	--	0.72	*0.51*	0.32	*0.28*
F	2.74	*0.59*	3.15	*0.61*	2.40	*2.30*	2.49	*0.69*	2.52	*1.00*	1.16	*0.95*	1.16	*0.71*	3.19	*0.96*	2.84	*1.17*
Cl	0.27	*0.37*	0.14	*0.24*	0.01	*0.01*	0.38	*0.39*	0.52	*0.64*	2.72	*2.07*	3.48	*2.18*	0.06	*0.06*	0.46	*0.45*
OH^-	0.59	*0.26*	0.33	*0.17*	--	--	--	--	0.71	*0.44*	--	--	--	--	0.22	*0.11*	0.13	*0.09*
Total	99.08	*1.37*	99.05	*2.30*	100.33	*98.72*	100.81	*0.83*	100.02	*1.72*	99.61	*1.20*	100.90	*2.93*	99.39	*1.58*	101.19	*0.47*
-O=F,Cl	-1.20	*0.25*	-0.80	*0.67*	-1.01	*-0.97*	-1.17	*0.24*	-1.13	*0.36*	-1.10	*0.25*	-1.27	*0.24*	-1.05	*0.42*	-1.52	*0.31*
Total	97.88	*1.26*	98.25	*1.66*	99.32	*97.75*	99.64	*0.72*	98.89	*1.58*	98.51	*1.03*	99.63	*2.90*	98.34	*1.33*	99.67	*0.46*
X(Fap)	0.722	*0.158*	0.810	*0.187*	0.637	*0.610*	0.658	*0.179*	0.635	*0.273*	0.309	*0.253*	0.309	*0.187*	0.695	*0.268*	0.724	*0.284*
X(Cap	0.038	*0.054*	0.018	*0.034*	0.001	*0.001*	0.051	*0.057*	0.070	*0.092*	0.397	*0.299*	0.506	*0.314*	0.006	*0.008*	0.058	*0.065*
X(Hap)	0.230	*0.122*	0.137	*0.091*	0.362	*0.388*	0.290	*0.158*	0.295	*0.212*	0.294	*0.157*	0.185	*0.157*	0.299	*0.264*	0.218	*0.252*

The values are reported as averages; however, this in not meant to infer that apatite chemistry displays a normal distribution about a mean, but is displayed as such for simplicity sake. Sufficient analyses of mafic pegmatites were not present in the literature to be included. The compilation represents over 1000 analyses and is not meant to be all-inclusive. The following abbreviations are used: N is the number of analyses represented (however, not all elements were analyzed for in all analyses); Avg is the mean; Std is 1σ standard deviation about the mean, and -- indicates that sufficient data within a petrogenetic group are not available in the present dataset to appropriately define.

Table 2 (continued).

	F-NA OD N= 475		F-PEGS N= 69		F OD-Non OP N= 29		F OD-OP N= 97		M-NA OD N= 86		M OD- Non OP N= 71		M OD- OP N= 18		Carbonatites N= 88		Low-T, HT Dep N= 35	
	Avg	*Std*	Avg	*Std*	Avg	*Std*	Avg	*Std*	Avg	*Std*	Avg	*Std*	Avg	*Std*	Avg	*Std*	Avg	*Std*
Trace Elements (ppm)																		
Y	1858	*2005*	1883	*1826*	9514	*474*	2307	*1781*	891	*1781*	705	*537*	703	*404*	1210	*1245*	1425	*921*
La	3845	*7599*	492	*439*	110	*402*	2363	*1935*	1649	*3549*	629	*422*	1033	*908*	2636	*2714*	2189	*1729*
Ce	6535	*12890*	377	*694*	382	*1508*	4068	*3996*	3543	*6937*	1087	*644*	2388	*1765*	5861	*5291*	5325	*4020*
Pr	1336	*1748*	--	--	73	*202*	1118	*1175*	548	*960*	--	--	--	--	726	*645*	394	*387*
Nd	3490	*5714*	81	*159*	450	*723*	1847	*1153*	1584	*2913*	--	--	--	--	3220	*2417*	3842	*3178*
Sm	557	*1211*	81	*159*	234	*136*	228	*143*	454	*958*	--	--	--	--	604	*466*	983	*847*
Eu	19	*9*	8	*10*	53	*9*	25	*7*	29	*16*	--	--	--	--	151	*74*	24	*6*
Gd	155	*71*	315	*338*	504	*114*	228	*186*	251	*278*	--	--	--	--	497	*257*	147	*70*
Tb	--	--	608	--	--	--	--	--	16	*12*	--	--	--	--	58	*31*	35	*44*
Dy	229	*341*	25	*30*	1114	*80*	201	*215*	99	*70*	--	--	--	--	316	*208*	63	*16*
Ho	--	--	--	--	272	*16*	42	*40*	6	*3*	--	--	--	--	44	*28*	36	*51*
Er	110	*98*	26	*75*	815	*41*	117	*107*	34	*44*	--	--	--	--	108	*70*	26	*8*
Tm	6	*4*	--	--	119	*5*	17	*15*	--	--	--	--	--	--	18	*12*	4	*3*
Yb	75	*70*	146	*279*	763	*31*	112	*97*	21	*24*	--	--	--	--	65	*59*	25	*18*
Lu	6	*4*	--	--	103	*4*	18	*14*	1	*1*	--	--	--	--	11	*8*	2	*1*
V	--	--	--	--	--	--	--	--	--	--	--	--	--	--	--	--	--	--
Li	2.1	*0.9*	--	--	--	--	--	--	--	--	--	--	--	--	--	--	--	--
Rb	--	--	--	--	--	--	--	--	15.0	*10.0*	--	--	--	--	--	--	--	--
Ni	--	--	--	--	--	--	--	--	36.0	*33.1*	--	--	--	--	--	--	15.0	--
Sc	--	--	--	--	--	--	--	--	0.5	*0.4*	--	--	--	--	--	--	--	--
Cu	--	--	2.2	*0.9*	--	--	--	--	4.5	*4.0*	--	--	--	--	3.6	--	1.9	*0.2*
Cr	136.8	--	--	--	--	--	--	--	29.8	*14.0*	--	--	--	--	--	--	--	--
Hf	0.10	*0.10*	--	--	--	--	--	--	--	--	--	--	--	--	--	--	--	--
Ta	0.01	*0.01*	--	--	--	--	--	--	--	--	--	--	--	--	--	--	--	--
Nb	0.26	*0.14*	--	--	--	--	--	--	389	*401*	--	--	--	--	--	--	--	--
Ba	92	*432*	2	*1*	--	--	590	*195*	185	*185*	--	--	--	--	855	1606	--	--
Zr	203	*154*	15	*11*	--	--	--	--	21	*28*	--	--	--	--	93	--	--	--
Pb	--	--	--	--	--	--	13	*7*	38	*26*	--	--	--	--	--	--	--	--
Th	26	*11*	--	--	--	--	100	*60*	245	*237*	--	--	--	--	437	*299*	--	--
U	169	*1078*	--	--	--	--	32	*15*	176	*232*	--	--	--	--	151	*67*	--	--
Ti	313	*200*	229	*100*	--	--	69	*36*	--	--	--	--	--	--	--	--	--	--

Table 3. Major elements in apatite from felsic-intermediate ore-producing rocks.

	Climax-Type Porphyry Mo		Yerrington Porphyry Cu		Bingham Porphyry Cu		Basin and Range Porphyry Cu		Providencia District Skarn and Mantos		Mt. Isa Iron-Oxide Cu-Au	
	Avg	*Std*	*Avg*	*Std*	*Avg*	*Std*	*Avg*	*Std*	*Avg*	*Std*	*Avg*	*Std*
SiO_2	0.53	0.42	0.38	0.05	0.35	0.11	--	--	--	--	0.47	0.14
Al_2O_3	--	--	0.18	0.03	0.15	0.03	--	--	--	--	0.01	0.01
Fe_2O_3	--	--	--	--	--	--	--	--	--	--	--	--
FeO	0.18	0.19	0.35	0.27	0.39	0.30	--	--	--	--	0.04	0.02
MnO	0.83	1.23	0.17	0.04	0.10	0.04	--	--	--	--	0.04	0.04
MgO	--	--	0.21	0.02	0.12	0.05	--	--	--	--	0.01	0.00
CaO	53.86	1.20	54.87	0.93	54.77	0.59	--	--	--	--	55.18	0.38
SrO	0.06	0.04	--	--	--	--	--	--	--	--	0.02	0.01
Na_2O	0.10	0.06	0.15	0.10	0.33	0.14	--	--	--	--	0.05	0.04
K_2O	--	--	0.02	0.00	0.04	0.02	--	--	--	--	0.01	0.00
P_2O_5	40.70	0.79	41.42	0.43	41.46	0.37	--	--	--	--	40.67	0.37
SO_3	0.14	0.14	0.60	0.36	--	--	--	--	--	--	0.13	0.09
F	2.91	0.56	2.67	0.72	2.77	0.91	2.61	0.21	1.94	0.51	2.26	0.14
Cl	0.08	0.05	0.28	0.37	0.39	0.33	0.42	0.31	0.72	0.39	0.07	0.07
Total	101.30	0.60	100.72	1.17	100.76	0.87	--	--	--	--	100.05	0.47
X(FAp)	0.766	0.140	0.707	0.191	0.726	0.231	0.692	0.056	0.514	0.136	0.600	0.036
X(CAp)	0.007	0.009	0.041	0.054	0.058	0.049	0.062	0.045	0.106	0.057	0.010	0.010
X(HAp)	0.227	0.406	0.251	0.140	0.216	0.188	0.246	0.069	0.379	0.152	0.390	0.029

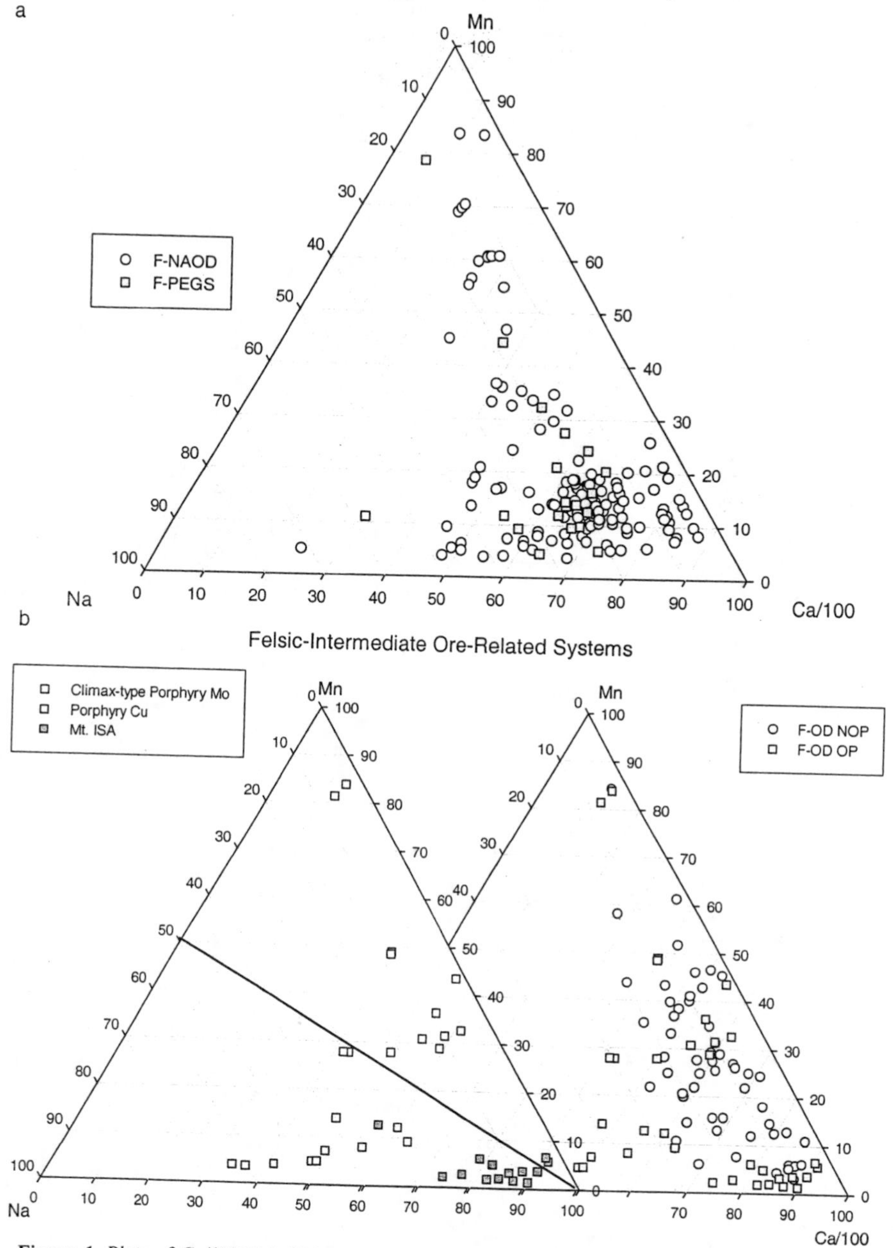

Figure 1. Plots of Ca/(100-Mn-Na) variability in the Ca-site in apatite as a function of host rock composition and association. This and subsequent plots use values from mineral formulas calculated on the basis of 26 (O, OH, F, Cl). (a) F-NAOD and F-PEGS; (b) F OD-NOP and F OD-OP (the plot on the left is a subset of the plot on the right, containing data only from porphyry Cu, Climax-type porphyry Mo deposits, and ore-related plutons of iron-oxide Cu-Au deposits of the Mt. Isa/Conclurry District, Australia)

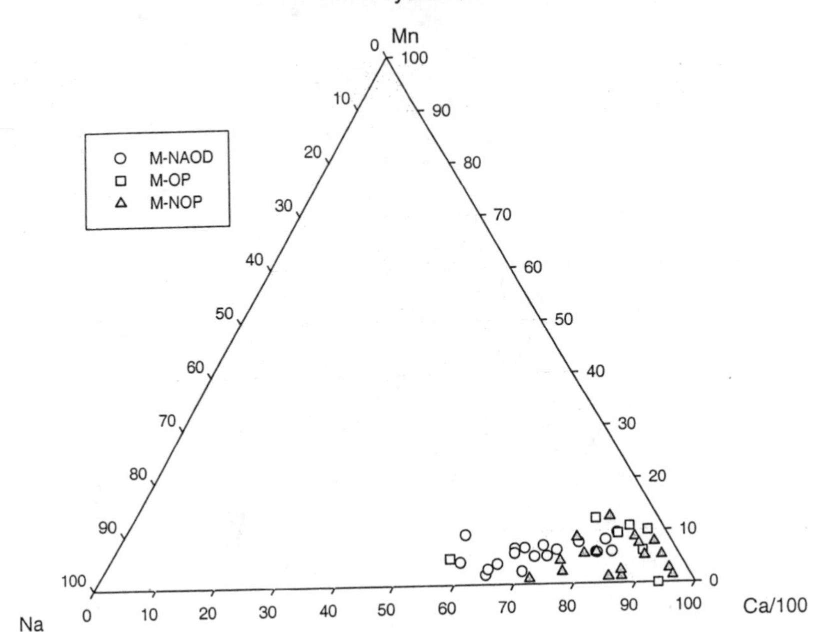

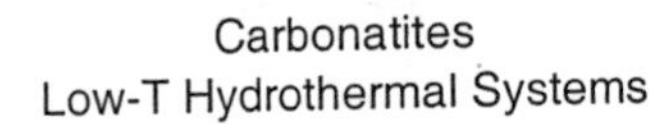

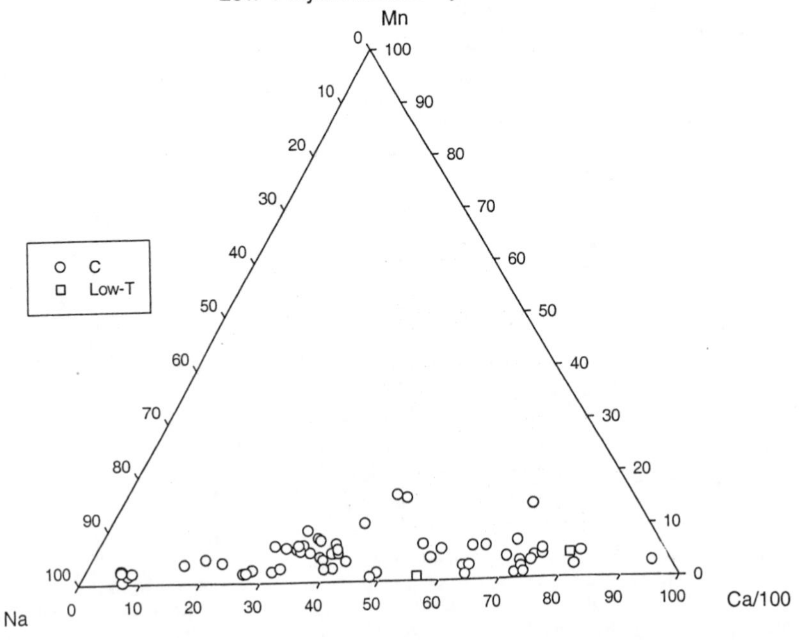

Figure 1, continued. (c) M-NAOD, M-OP and M-NOP; and (d) C and Low-T (abbreviations in Table 1).

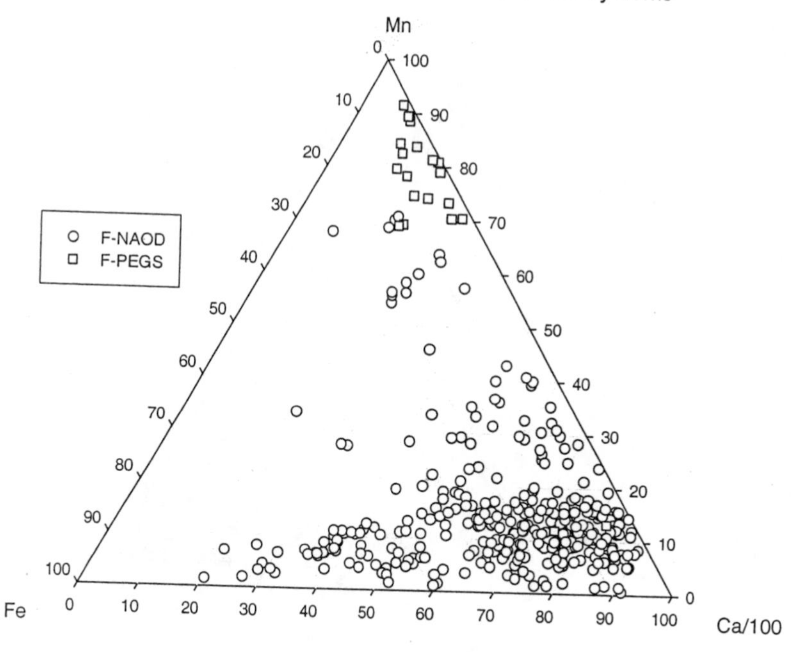

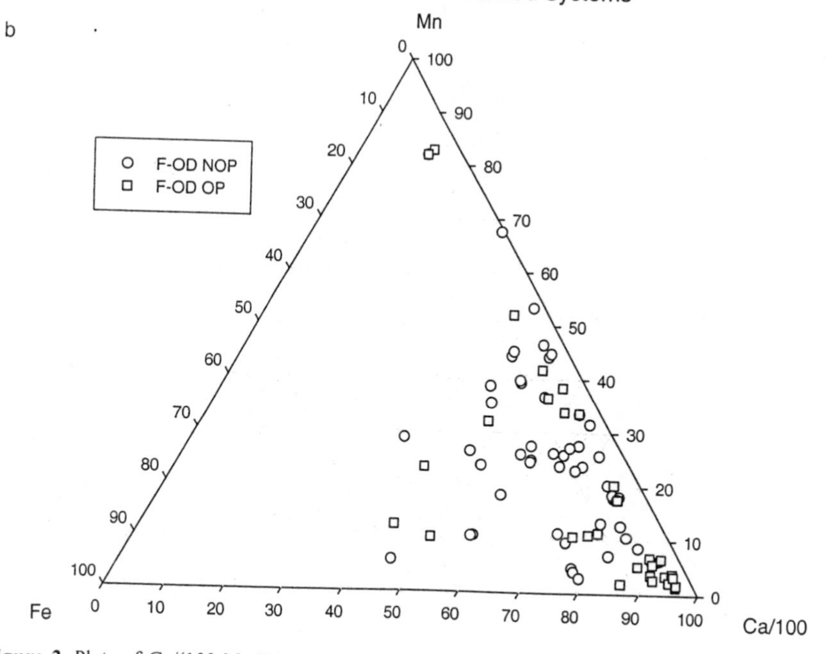

Figure 2. Plots of Ca/(100-Mn-Fe) variability in the Ca-site in apatite as a function of host rock composition and association. (a) F-NAOD and F-PEGS; (b) F OD-NOP and F OD-OP.

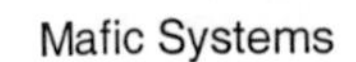

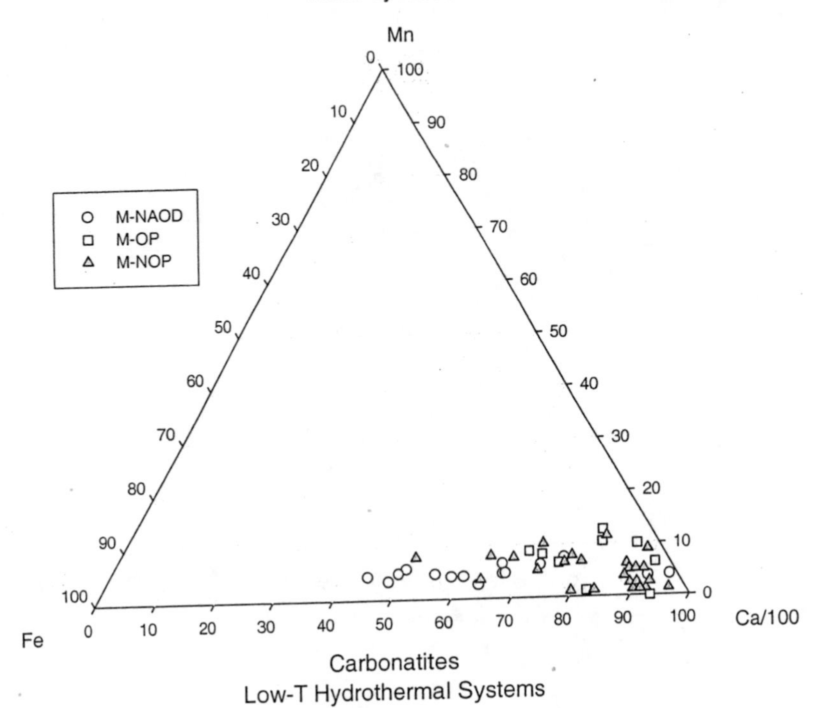

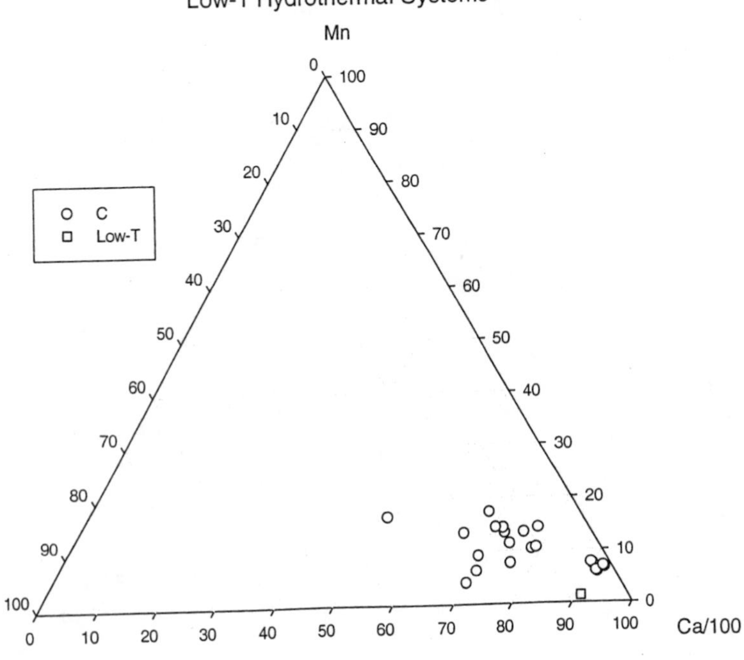

Figure 2, continued. (c) M-NAOD, M-OP and M-NOP; and (d) C and Low-T (abbreviations in Table 1).

temperatures significantly below normal granitic temperatures (London 1992). This is similar to the effect suggested for Na in the last section. Again, Mn is not limited in this regard, and we might expect Mn/Fe to increase in the melt, and in coexisting apatite as down temperature crystallization proceeds in many pegmatitic systems, even with some incorporation of Mn in iron bearing phases. Apatites from ore-related and non-ore producing systems plot in virtually identical fields in this composition space. However, within the ore producing systems, we find once again that apatite from the ore-related plutons of iron-oxide, Cu-Au deposits of the Conclurry District are depleted in Mn and Fe, relative to apatites from porphyry Cu and Mo systems (which are indistinguishable from one another).

In mafic systems, the trend in apatite chemistry in Ca/(100-Mn-Fe) is analogous to that in Ca/(100-Mn-Na) space (Fig. 2c). However, in carbonatites, the extent of substitution of Fe for Ca (Fig. 2d) is more restricted than that of Ca for Fe (Fig. 1d), consistent with the Fe-poor nature of the carbonatites.

Apatite chemistry in Ce-Mn-Fe space

In Figure 3, cerium is used as a proxy for the LREE. In apatites from felsic rocks, the bulk of the substitution in this space occurs along the Mn-Fe join. Based on a limited number of data, apatite from felsic pegmatites have lower Ce than from felsic-intermediate rocks not associated with ore deposits. This may appear surprising, but again, the details of coupled substitution, together with the lower magmatic temperatures that occur in pegmatitic systems, probably limits Ce substitution in pegmatitic apatite.

When considering felsic ore-related plutons (Fig. 3b), there is considerable overlap in the chemistry of apatite in the ore-related and non-ore producing systems. When considering only ore-producing systems, the apatites from the Conclurry district plot near the Ce end of the diagram; this is not because they have unusually high Ce (3,550 ppm), but because they have such low Mn (0.04 wt % MnO) and Fe (0.04 wt % FeO) when compared to apatite from Climax-type porphyry Mo systems (4,950 ppm Ce, 0.10 wt % MnO, 0.30 wt % FeO).

Apatite from mafic systems, irrespective of their relation to ore-systems, displays considerable variability and overlap on a plot of Ce-Mn-Fe (Fig. 3c), but apatite from systems not associated with ore-deposits may have lower concentrations of Mn. Carbonatites appear to have considerable exchange of Ce for Fe (Fig. 3d), a substitution that may require charge balance depending on the oxidation state of Fe substituting in the apatite.

Apatite chemistry in P-Si-S space

Apatites from felsic-intermediate rocks display considerable variability on a plot of P/100-Si-S (Fig. 4a). The ratio of Si to S is highly variable in the apatite, but the bulk of the analyses cluster crudely near a value of 1, consistent with the substitution $Si^{4+} + S^{6+} = 2P^{5+}$. Apatite associated with ore-systems (non- and ore-producing), have an excess of Si relative to S, which in all likelihood is due to additional substitutions in this site, for example $REE^{3+} + Si^{4+} = Ca^{2+} + P^{5+}$ (Roeder et al. 1987, Ronsbo 1989, Sha and Chappell 1999) (see also discussion above) or $SiO_4^{4-} + CO_3^{2-} = 2\ PO_4^{3+}$ (Sommerauer and Katz-Lehnert 1985). This excess of Si relative to S in apatite from ore-related systems may seem curious given the high S contents (bulk) associated with many porphyry-type Cu and Mo deposits; however, some studies (Core et al. 2001) suggest that some S in the melts is in a reduced form.

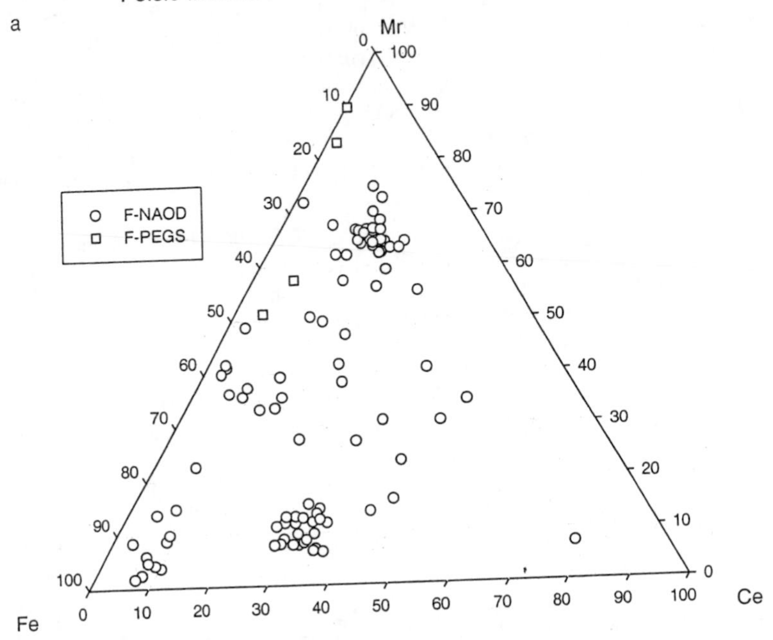

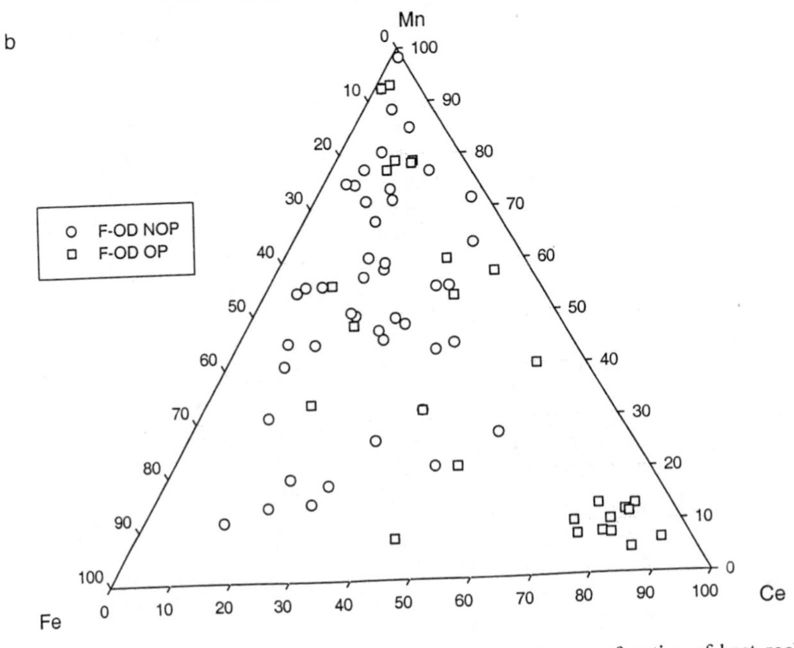

Figure 3. Plots of Ce-Mn-Fe variability in the Ca-site in apatite as a function of host rock composition and association. (a) F-NAOD and F-PEGS; (b) F OD-NOP and F OD-OP.

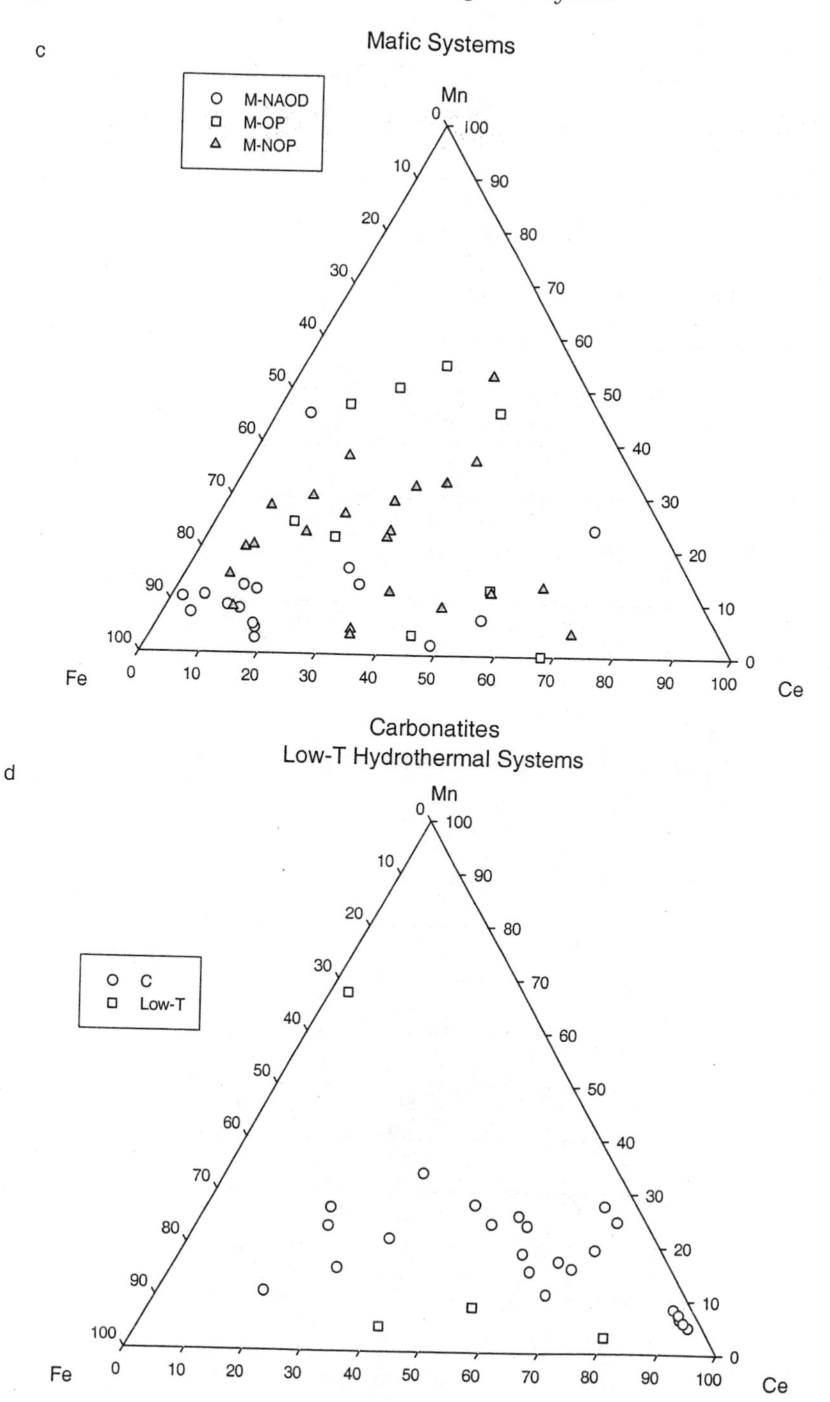

Figure 3, continued. (c) M-NAOD, M-OP and M-NOP; and (d) C and Low-T (abbreviations in Table 1).

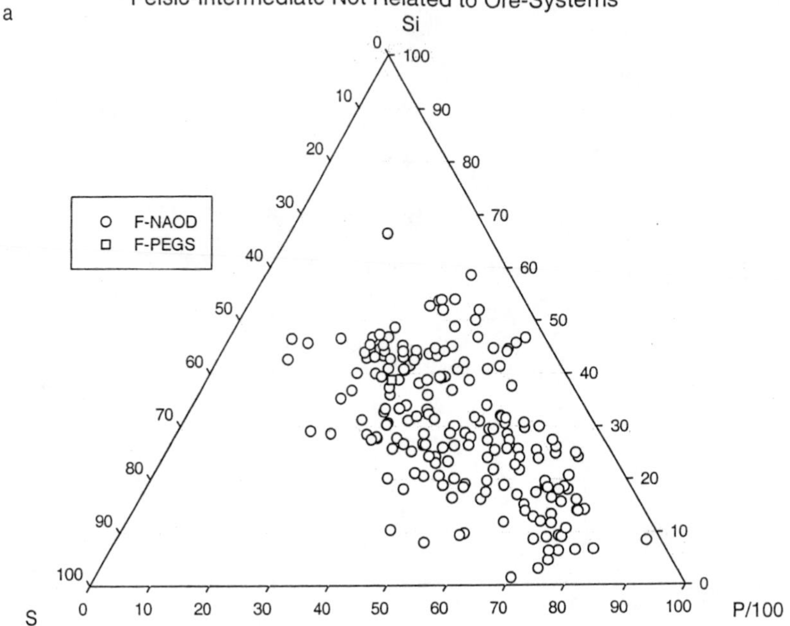

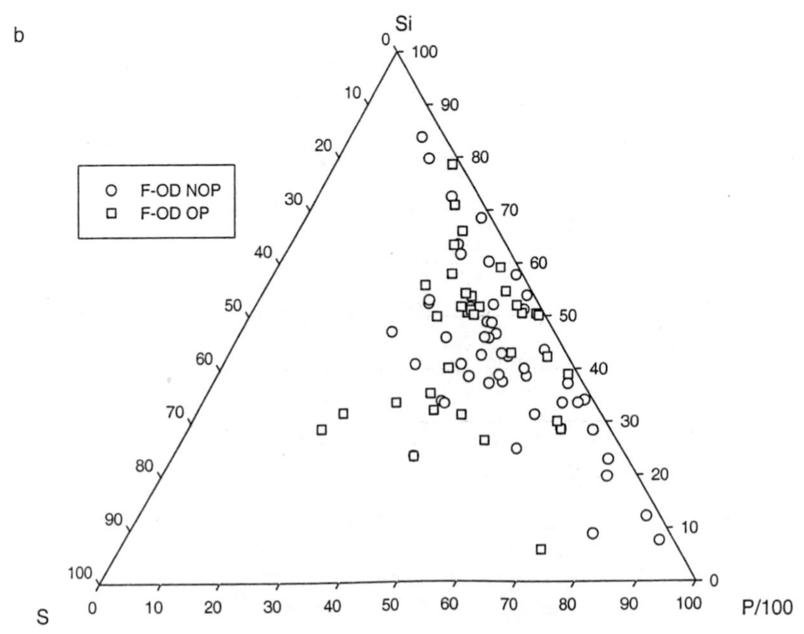

Figure 4. Plots of P/100-Si-S variability in the P-site in apatite as a function of host rock composition and association. (a) F-NAOD and F-PEGS; (b) F OD-NOP and F OD-OP.

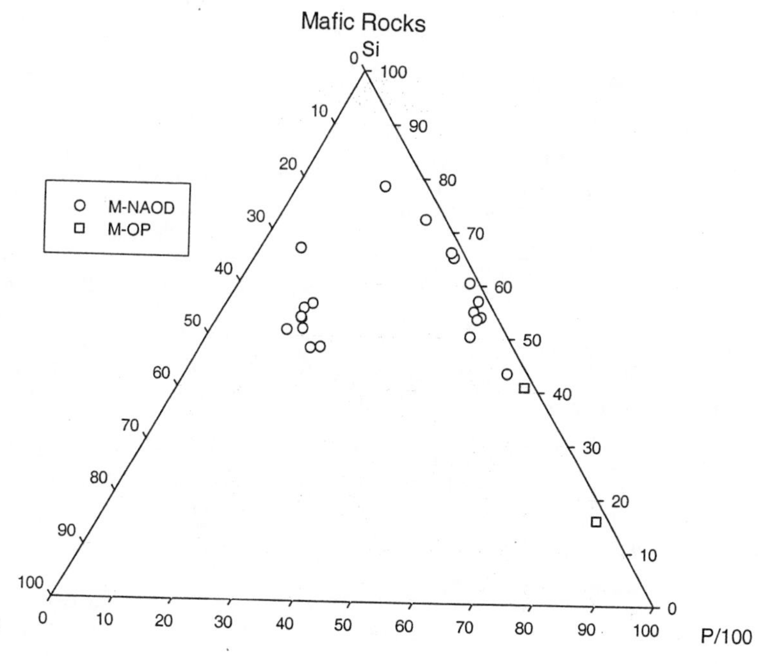

d

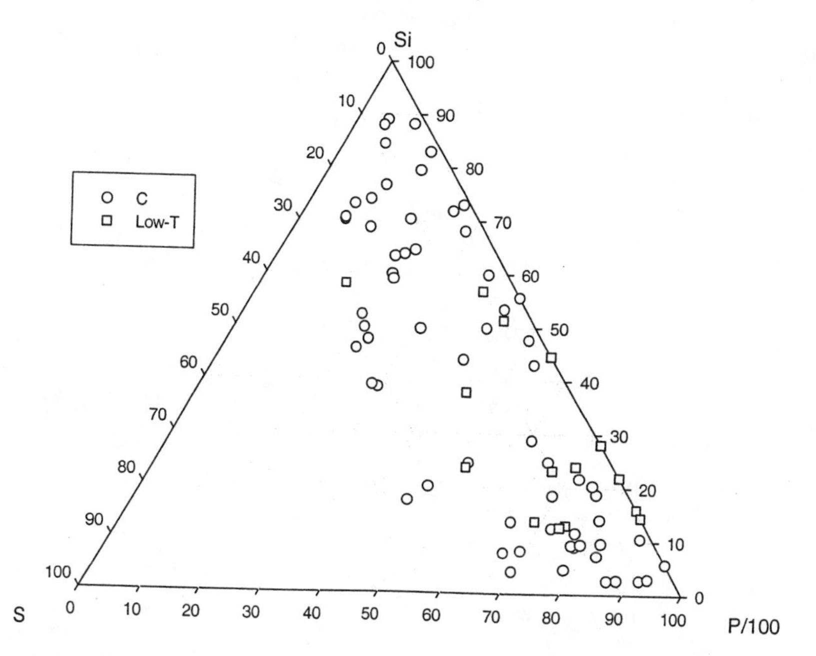

Figure 4, continued. (c) M-NAOD, M-OP and M-NOP; and (d) C and Low-T (abbreviations in Table 1).

Apatite chemistry in F-Cl-OH space

When considering igneous systems, *apatite* is nearly synonymous with *fluorapatite* (Fig. 5a-5d). There are reasons to believe that the halogen site (Z-site) in apatite is filled in apatite, i.e. vacancies are not common (see the Appendix). The data have been recast, so that F, Cl and OH fill the site. That is, we assume that O^{2-}, CO_3^{2-}, vacancies, and other anions in the site, are not significant. In some cases F appears to overfill the site (e.g., in some carbonatites), especially when significant CO_3^{2-} is present; some of this has been interpreted to be the result of analytical problems, a topic that will be discussed subsequently (a brief discussion of the methods used to calculate the mole fractions of fluor- (FAp), chlor- (CAp) and hydroxylapatite (HAp) can be found in the Appendix). There are two normalization methods: (1) the mole fractions of fluor- and chlorapatite can be calculated directly from the concentration of F and Cl, respectively, with the mole fraction of hydroxylapatite calculated by difference (given that very few apatites have been analyzed for H_2O); or (2) the mole fractions can be calculated along with the rest of the mineral formula, and mole fractions of FAp, CAp and HAp calculated as the final step. These techniques produce slightly different results, in part because in many cases incomplete analyses are reported in the literature. The former method was used in the generation of the plots in this chapter (Fig. 4). Although there are problems with this method of calculation, the biases induced are necessary in order for comparisons to be made. Analyses which do not report either F or Cl were excluded in the plots.

Apatites associated with felsic-intermediate intrusive and extrusive rocks exhibit a relatively restricted range in composition F-Cl-OH space (Fig. 5a). Apatites from these rocks contain between 50-100% F, 0-50% OH, and < 20% Cl, with few exceptions. The dominant exchange operating is $F(OH)_{-1}$, due in part to the similarity in radius for F^- and $(OH)^-$ as compared to Cl. Apatites from felsic-intermediate pegmatites display an even more restricted range of compositions, but possibly with less Cl; however, this may not be due to volatile exsolution (a topic discussed later). It is interesting to note that in some meteorites, for example the achondritic Nakhlite, the Z site is filled with F (1.6 wt %) and Cl (4.0-4.3 wt %) with no detectable OH (Bunch and Reid 1975) (not on figure).

Thermodynamics of F-Cl-OH exchange

The use of halogen concentrations in apatite to make estimates of aqueous fluid compositions is well developed (Korzhinskiy 1981, Zhu and Sverjensky 1991,1992). Given the temperature of apatite crystallization (calculated as an AST), or some other independent estimate of temperature, and an estimate of the pressure of crystallization (although the calculation is fairly insensitive to pressure), the fugacity ratio

$$\frac{f_{HCl}^{aq}}{f_{H_2O}^{aq}} = \frac{X_{CAp}^{Ap}}{X_{HAp}^{Ap}} \cdot \frac{1}{10^{\left[0.04661+\frac{2535.8}{T}-\frac{0.0303\cdot(P-1)}{T}\right]}} \tag{2}$$

can be calculated (following Piccoli and Candela 1994). Invoking the Lewis and Randall rule, and assuming that the fugacity coefficient ratio is 1, yields

$$m_{HCl}^{aq} \cong \frac{X_{CAp}^{Ap}}{X_{HAp}^{Ap}} \cdot \frac{1000}{18} \cdot \frac{1}{10^{\left[0.04661+\frac{2535.8}{T}-\frac{0.0303\cdot(P-1)}{T}\right]}} \tag{3}$$

where m_{HCl}^{aq} is the molality of HCl in the aqueous phase, X_{CAp}^{Ap} and X_{HAp}^{Ap} are the mole fractions of chlor- and hydroxylapatite, T is the temperature of apatite crystallization (Kelvin), and P is the pressure (bars) which can be used to calculate the concentration of HCl in an aqueous phase in equilibrium with the crystallizing apatite. In some instances where appropriate experimental data exist, estimates of the concentration of Cl (not HCl)

a

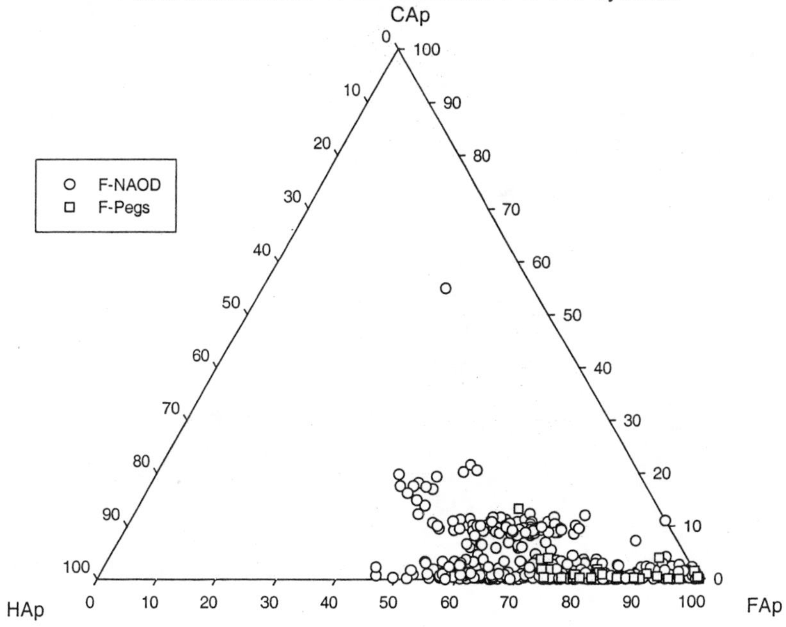

b

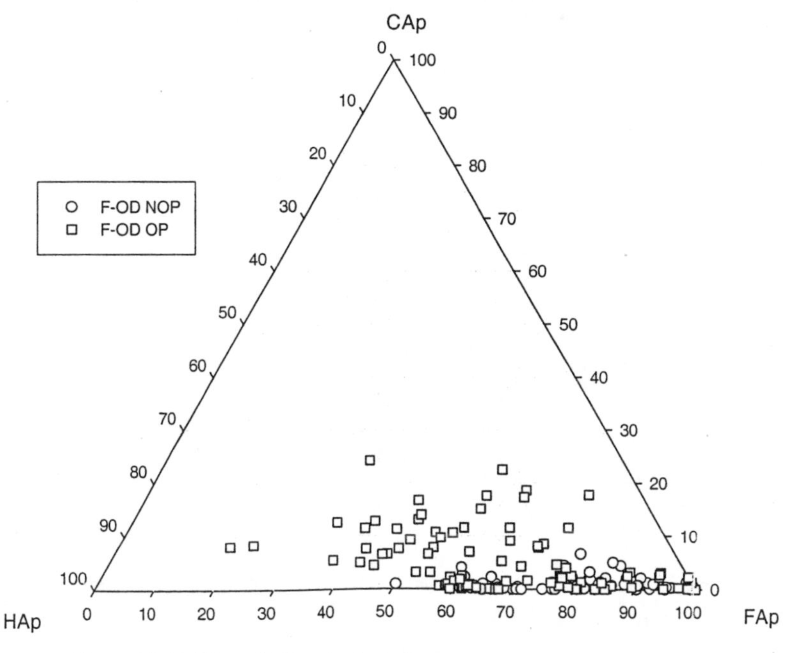

Figure 5. Plots of F-Cl-OH variability in the halogen-site in apatite as a function of host rock composition and association. (a) F-NAOD and F-PEGS; (b) F OD-NOP and F OD-OP.

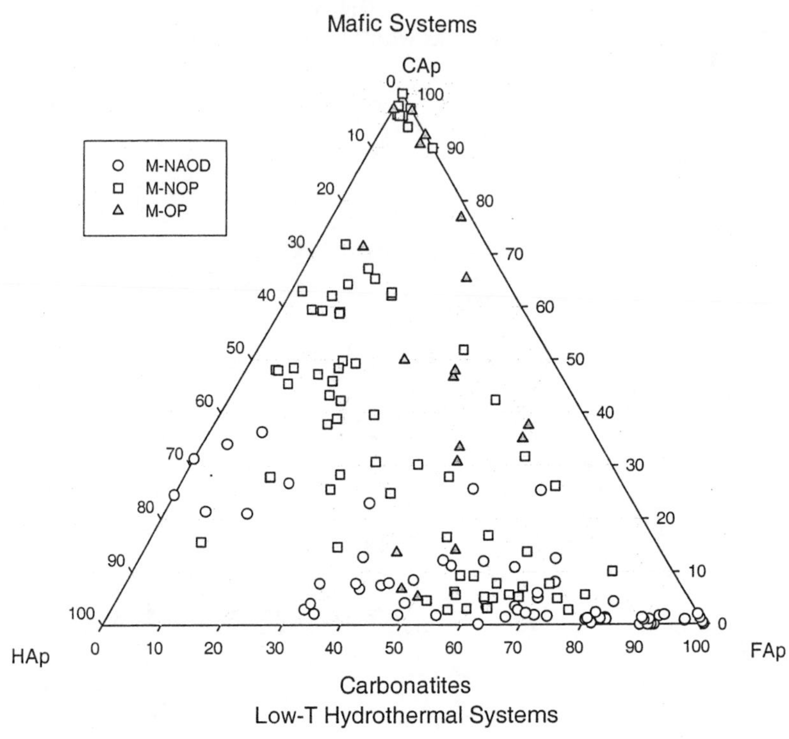

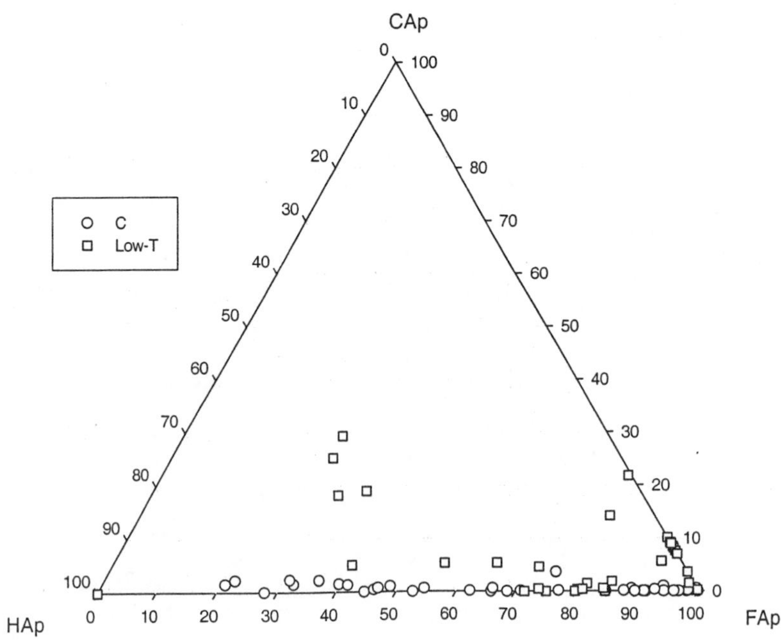

Figure 5, continued. (c) M-NAOD, M-OP and M-NOP; and (d) C and Low-T (abbreviations in Table 1).

in the aqueous vapor and silicate melt can be made; this technique was used to successfully predict the Cl concentration in melt inclusions in the plinian phase of the Bishop Tuff rhyolite, California (Piccoli and Candela 1994).

Equations (2) and (3) were used to generate Figure 6. The calculations were performed using constant Cl/OH in apatite of 4, 1.5, 0.67 and 0.25 (as mole fractions) at temperatures of 800 and 1000°C, at a pressure of 1000 bars. These calculations are independent of melt composition. The ratio f_{HCl}/f_{H_2O} is valid whether or not the melt is saturated with an aqueous phase. The m_{HCl} represents the HCl molality in an aqueous phase that would be in equilibrium with both the analyzed apatite as well as the melt.

The diagrams can be used to evaluate the relationship between temperature, apatite chemistry and the concentration of HCl in the associated magmatic aqueous phase. For example, assuming apatite with Cl/OH of 0.25 (molar) begins to crystallize at 1000°C, the m_{HCl} of the associated fluid is ~0.13 molal. At 800°C, the apatite would need to have a Cl/OH of 0.36 in order to remain in equilibrium with a 0.13 m_{HCl} aqueous fluid. That is, the higher the temperature of apatite crystallization for a given Cl/OH in apatite, the higher the concentration of HCl in the associated aqueous phase. It is important to note that these calculations are independent of the F concentration of the apatite (other than to calculate the molar proportion of the end-member apatites). Similar plots could be formulated containing information about HF and H_2O.

Figure 5b shows felsic-intermediate ore-related systems, and includes analyses from both felsic-intermediate non-ore-producing (NOP) and ore-producing (OP) as defined by the authors of the original papers. The apatites from OP systems are from the porphyry deposits of Yerrington, Nevada; Bingham, Utah; igneous rocks associated with porphyry and related deposits of the Basin and Range; Climax-type porphyry Mo deposits; apatite from the Primose intrusion associated with the Henderson porphyry Mo deposit, Colorado; the Concenpcion del Oro, Providencia and Noche Buena stocks (Mexico) associated with the skarn, mantos and related deposits; granites associated with iron-oxide Cu-Au deposits of the Conclurry District, Queensland, Australia; and from Erzgebirge and Waldstein, Germany. The NOP include apatite from non-ore producing granites in the ore-producing systems, and are represented by granitic rocks from Yerrington, Nevada, the porphyry Mo deposits of Colorado, and granites from Queensland. The characteristics of this admittedly limited data set are as follows. Only OP plutons have CAp > 5%, and CAp > 7% in systems with CAp + HAp > 30 mol %. Also, the ore-producing plutons have model ore-fluids with HCl molalities of 0.05 or greater. Some of the highest Cl concentrations in apatite in this data set are for the ore-related plutons of the Bingham district (2-14% CAp), and the Concepcion del Oro-Providencia District (4-24% CAp). Note that the apatites from Bingham are relatively low in FAp as compared to others in the OP-NOP data set. Ague and Brimhall (1987) relate inferred f_{HF}/f_{H_2O} from biotite to the proportion of crustal component in the magmatic system; extending their model to apatite chemistry indicates that systems like Bingham with lower FAp/HAp represent less evolved magmatic systems. The fact that the higher Cl/F apatites from some rocks associated with ores are also lower in F/OH may be indicative of the more primitive mantle component of these rocks; note however, that other interpretations are possible. Apatites from granites associated with Climax-type porphyry Mo deposits plot near the FAp apex, attesting to the fluorine-rich, highly evolved, nature of the magma.

Within a given granitic rock suite, the halogen concentrations in apatites commonly vary as a function of whole rock SiO_2. For example, in a study of the gold-mineralized Emerald Lake pluton, Tombstone Suite, Canada, Coulson et al. (2001) attempted to use fluid composition determined using apatite and biotite chemistry to infer the source of

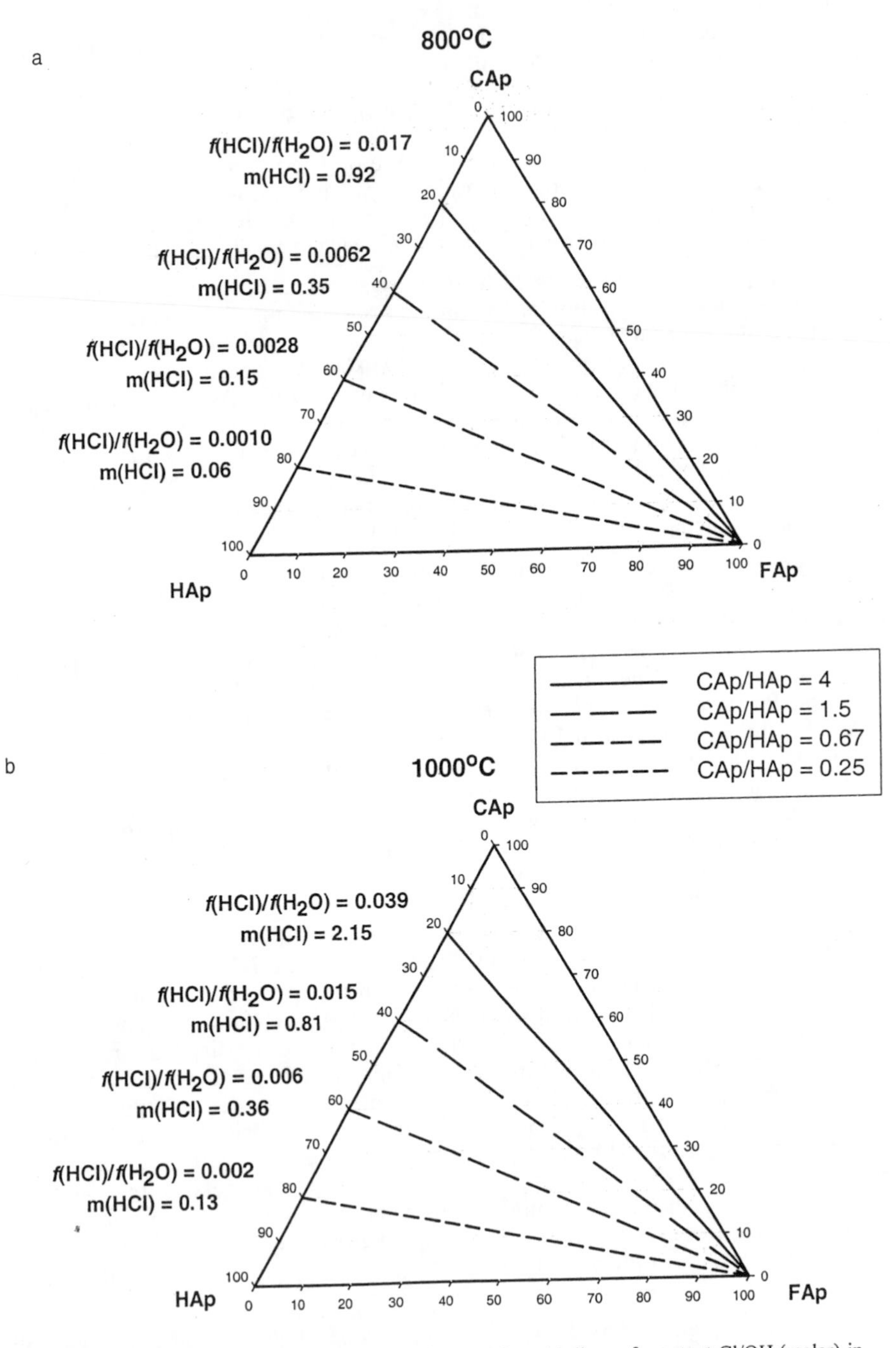

Figure 6. Plot of composition of the halogen site in apatite with lines of constant Cl/OH (molar) in apatite. These can be related to f_{HCl}/f_{H_2O} and m_{HCl} using Equations (2) and (3). In this figure, for display purposes, a pressure of 1000 bars and temperatures of 800°C (top) and 1000°C (bottom) have been used. Also plotted is the molality of HCl in the associated volatile phase.

gold-bearing fluids. Furthermore, they found that the halogen concentrations in apatite and biotite independently yield similar estimates of HCl/HF through the magmatic to hydrothermal transition. In the Tuolumne Intrusive Suite, there is a systematic variation in the concentration of halogens in apatite from the Kuna Crest granodiorite (58.7 wt % SiO_2 [WR (whole rock)]; 2.29 wt % F, 0.30 wt % Cl) to the Johnson Granite Porphyry [74.2 wt % SiO_2 (WR); 3.34 wt % F, 0.01 wt % Cl] (Piccoli and Candela 1994). This is consistent with the hypotheses that the magma giving rise to the particular pluton had lower cumulative Cl, or, progressive loss of volatiles occurred during the protracted magmatic evolution of the system, or both. A similar trend has been identified in the Debouille pluton, Maine, where Loferski and Ayuso (1995) found that apatites from the syenite (56.5 wt % SiO_2) had lower F (~2.5 wt %) and higher Cl (~1.1 wt %), than from the associated granodiorite (~67.7 wt % SiO_2) (~3.1 wt % F, ~0.15 wt % Cl).

The variations in the concentrations of the halogens in apatite from within suites of rocks associated with granite-associated ores is rather complex. The data from Dilles (1987) and Streck and Dilles (1998) show that within a given ore-related system, ore-related plutons can have lower CAp/HAp than the associated non-ore related (NOP) plutons, and this points out a major problem in the interpretation of apatite chemistry in magmatic-hydrothermal ore systems. Whereas Cl-rich systems in general may be more prospective for base and precious metal mineralization (e.g., Rogge et al. 1974), specific plutons that have devolatilized to a significant extent will be characterized by apatites with lower CAp/HAp ratios (Piccoli and Candela 1994). Therefore, apatite data need to be interpreted with caution, especially with regard to trends in apatite chemistry in plutonic systems. Apatite associated with the Yerrington ore-producing granites have lower Cl than the associated non-ore producing granites; however the data on the Conclurry granites associated with the iron oxide Cu-Au deposits show the opposite trend. Note that multiple effects are operative in controlling the Cl/F or CAp mole fraction in plots of this type (Fig. 5). First, magmas may indeed be enriched in Cl; second, these elements may be variably incorporated into crystallizing phases during fractional crystallization; thirdly, the effects of volatile exsolution may be superposed upon these trends. A reduction in Cl/F in crystallizing apatite is sometimes inferred to indicate devolatilization (due to the high partitioning of Cl into a volatile phase relative to F); however, the same trend can also be produced by a reduction in temperature. We therefore suggest that the reductions in Cl/OH in apatite be used as an indicator, but this also has limitations. For devolatilization at pressures of hundreds of bars (tens of MPa) or less, a significant reduction in Cl/OH may not occur (Piccoli et al. 1999) due to the lower partition coefficient for chlorine at low pressures, the potential effects of differential hydrous mineral/melt partitioning of chlorine and hydroxyl, and the effect of temperature, all of which may counteract the small to moderate reduction in melt Cl/H_2O produced by the low pressure exsolution of magmatic volatiles.

SURVEY OF APATITE PROVENANCE

Apatite in intermediate to felsic systems

Trace element concentrations in apatite in this broad range of magmatic products depend sensitively on many important parameters that include, not surprisingly, the initial magma composition, degree of fractionation, and oxygen fugacity. A detailed study of the distribution of trace elements in apatites from the granites of the Mt. Isa Inlier, performed by Belousova et al. (2002, 2001), is illuminating in this regard. They found that the Mn concentrations in apatite increase significantly from mafic to felsic rocks, and appears to be affected by f_{O2} (inferred from whole rock Fe^{3+}/Fe^{2+}), with higher Mn in apatite at lower whole rock Fe^{3+}/Fe^{2+} (i.e. more reduced conditions), consistent with the direct substitution of Mn^{2+} for Ca^{2+}. However, many highly fractionated granites have apatites with high Mn/Ca, regardless of f_{O_2}. They found that cerium concentrations in apatite from oxidized rocks may exceed 10,000 ppm, while Ce values over 2000 ppm are rare in apatites from reduced plutonic rocks. The data show only very weak correlations between the size of the Eu and Ce anomalies and the oxidation state of the whole rock as expressed by the Fe_2O_3/FeO ratio. The amplitude of these anomalies is probably more strongly controlled by other factors such as the crystallization of feldspars, which concentrate Eu^{2+} from the melt (Budzinski and Tischendorf 1989), and the presence of other accessory phases such as monazite and fluorite that are also able to concentrate Ce. Belousova et al. (2001) also found that the trends in the Sr content of apatite appears to mimic the trends of whole-rock Sr.

In a study of apatite from the I and S type granites of the Lachlan Fold Belt (LFB) of southeast Australia, Sha and Chappell (1999) found that apatite from the S-type granites have lower Cl/F ratios than apatites from I-type granites. They suggest this is due to lower Cl in S-type source rocks, with Cl lost due to weathering, and ultimately residing in the oceans, where Cl has a long residence time. However, they point out that this can also result for fractionation, noting that S-types are generally more silicic than I-types. They also found higher Mn in apatite from the more felsic S-types relative to the apatite from more mafic S-types. Sodium in apatite also increases in concentration with increasing fractionation (i.e., it correlates positively with Mn), consistent qualitatively with trends we reported and discussed earlier in this chapter. Strontium in apatite seems to be a function of whole rock Sr, and therefore is an indicator of fractionation, similar to the findings of Belousova et al. (2001). Iron is higher in S-types as compared to I-types, and is probably controlled by f_{O_2} (lower f_{O_2} leads to higher Fe^{2+}/Fe^{3+} in the melt, and for a given CaO concentrationin the melt, yields higher Fe^{2+} concentrations in apatite). We think that these data for S-types from a variety of suites within the Wagga Batholith (their VB samples), follow, roughly, a trend of increasing iron concentration in apatite with decreasing apatite Sr, suggesting that Fe/Ca ratios increase during fractionation, at least in low f_{O_2}, magnetite-undersaturated systems such the S-type granites of the LFB. However, this conclusion is rather tentative, and requires further evaluation. The presence of high concentrations of iron (>1.2 oxide wt %) in apatite usually occur concomitantly with an inversion of the apatite structure from hexagonal to monoclinic (Hughes and Cameron 1993). Ferrous iron (Fe^{2+}) has been thought to preferentially enter the Ca2 site, surprisingly similar to the substitution of Pb^{2+} into apatite. This is in contrast to Mn^{2+}, which prefers the Ca1 site, a site with larger coordination (Hughes et al. 1993, see also Hughes and Rakovan, Ch.1 in this volume).

The sulfate concentration in early-formed apatite in some volcanic rocks is elevated relative to later formed apatite. In the trachyandesitic pumices erupted during the 1982 eruptions of El Chichón, apatite poikilitically enclosed within phenocrysts has higher SO_3 (0.43 wt %) compared to those in the groundmass surrounded by glass (0.31 wt %) (Piccoli and Candela 1988). Similarly, in the intermediate plutonic rocks sampled by drilling as part of the Cajon Pass Deep Scientific Drillhole, California, apatite present as inclusions within hornblende, plagioclase and biotite have equivalent or considerably higher SO_3 than matrix apatite, possibly due to greater sulfate being present in the magma during earlier stages of its evolution (Barth and Dorais 2000). Recently, the S concentration in apatite has been used to estimate the sulfate/sulfide ratio (Core et al. 2001), a ratio that has important implications on the ore-producing potential of some magmas. Their approach was to evaluate the S (total) content of two associated plutons in the Wasatch and Oquirrh Ranges of Utah by using the S content of apatite. In the Last Chance Stock, apatite contains 1 wt % S in apatite, whereas in the Clayton Peak stock, the S concentration in apatite is below detection. Following their calculations, 60% and 10% of the S (maximum) in the stocks is present as sulfate (with the remainder present as sulfide species) in the Last Chance and Clayton Peak stocks, respectively. These estimates of S speciation do not agree with the speciation predicted by using oxygen fugacities (determined using Fe-oxide and Fe-silicate phases).

Experiments by Peng et al. (1997) show that, in trachyandesitic rocks, the SO_3 concentration in apatite increases with increasing oxygen fugacity from less than 0.04 wt % at reducing conditions (QFM: quartz-fayalite-magnetite) to 1-2.6 wt % in oxidizing conditions (MNH: manganosite-hausmanite or MH: magnetite-hematite). The measured partition coefficient for SO_3 between apatite and melt increases with decreasing pressure. Unfortunately, volcanic apatite does not always record high SO_3 activities. They suggest this is due to low f_{O_2} and low magmatic S concentration at the time of apatite growth (Peng et al. 1997), and the slow rate of re-equilibration under later high S and oxidizing conditions, consistent with the findings of Piccoli and Candela (1994), who suggested that apatite probably precipitates early in magmatic systems, and most of it grows over a restricted range of temperature. On the other hand, Tepper and Kuehner (1999) reported that the biotite granodiorite of the Idaho Batholith contains early high SO_3 in apatite cores.

Upon recognition of carbon in apatite, there was skepticism about whether carbon was a constituent of apatite, or occurred as fine inclusions within the apatite. It has been demonstrated, primarily through the use of XRD, that the C in apatite is an integral component of the apatite structure and produces distinct XRD patterns (see discussion in Elliott 1994). Attempts have been made to *calculate* the carbon content of apatite. This has been done by assuming an ideal stoichiometry in the phosphorus site ($P + Si + S + C = 6$) (Sommerauer and Katz-Lehnert 1985). Although this is difficult because of coupled substitutions among sites, it has been found to be consistent with measurements by coulometric titration in some cases (Sommerauer and Katz-Lehnert 1985).

Inclusions in felsic rocks may also be apatite bearing. Apatite constitutes up to 30 modal % in layered enclaves within the granitic Achala batholith, Argentina (Dorais et al. 1997). The enclaves, which contain up to 8 wt % P_2O_5 and 1.8 wt % F, contain millimetric, stubby apatite, within a biotite (50 modal %) matrix. The habit of the apatite, in conjunction with radiometric dating of zircons, mineralogy and bulk-rock characteristics in the enclaves and the host granite, led the authors to surmise that the enclaves represent magmatic segregations or cumulates. The apatites are F-rich (3.31-3.72 wt %), and contain little Cl (0.01-0.16 wt %). Interestingly, apatites in enclaves from two of three plutons contain similar compositions to their hosts in Ca-Mn-Fe space, but apatites in enclaves from the third are distinctly different than in their host (Mn-rich, Ca-poor).

Durango apatite

Whereas apatite is ubiquitous in igneous rocks, especially those of the more felsic end of the spectrum, it also occurs in many hydrothermal systems. The now famous Durango apatite is of hydrothermal origin, and occurs in the Tertiary iron deposits of the Cerro de Mercado deposit of Mexico, an ore mined intermittently for over 170 years. Cerro de Mercado is known for its large, gemmy fluorapatite crystals and octahedral martite specimens. Durango apatite has been extremely well characterized as a standard for electron probe microanalsis studies (discussed in a further section) (Jarosewich et al. 1980), as a standard in helium-chrononmetry studies (Farley 2000, House et al. 2000), for use in thermoluminesence studies (Vaz 1980), as fission-track annealing studies (Grivet et al. 1993, Jonckheere et al. 1993, Naeser and Fleischer 1975) and as a phosphate standard in spectrochemical laboratories (Young et al. 1969). The Cerro de Mercado deposits have been interpreted by Lyons (1975) as iron-oxide lava flows, with additional magnetite deposited hydrothermally. Barton and Johnson (1996), however, consider these deposits to be dominantly hydrothermal in origin, involving the circulation of non-magmatic brines by magmatic heat.

Apatite in mafic rocks

Apatite is often the only mineral in mafic rocks that can yield information about volatile constituents. Whereas biotite and hornblende can reach concentrations up to the order of 10 vol.% in felsic rocks, they are not common rock-forming minerals in most mafic systems. In mafic rocks, apatite occurs as inclusions in cumulus minerals, as cumulate apatite (up to ~10 vol %), and as interstitial (post-cumulus) apatite (usually less than 1 vol %) (see discussions in Barkov 1995, Boudreau and Kruger 1990, Boudreau et al. 1986). Apatite is a minor but ubiquitous accessory phase in the 2.06 Ga-old Bushveld Complex in the Republic of South Africa. Like the Bushveld, the Stillwater Complex in Montana also has chlorine-rich apatites. Similarly, the Imandrovsky and Lukkulaisvaara Layered Intrusions, Russia, contain apatite with highly variable composition but some discrete grains are nearly end-member chlorapatite (Barkov et al. 1995). However, these intrusions appear to be the exceptions. Most mafic complexes have fluorapatite with <20% Cl in the halogen site (e.g., the Skaergaard Intrusion, Greenland, and Great Dyke, Zimbabwe, Boudreau 2002). In the Bushveld, the Main Zone is a unit distinct from the Lower Zone (dunite, harzburgite, orthopyroxenite) and the feldspar-bearing Critical Zone. The Main Zone appears to involve some crustal input, as evidenced by a jump in $^{87}Sr/^{86}Sr$. This coincides with a sharp drop in the mole fraction of chlorapatite from an average of 0.8 to 0.4 across the Merensky Reef unit.

According to Meurer and Boudreau (1996) apatite in the middle banded series of the Stillwater shows distinct stratigraphic variations. Chlorapatite increases up section from less than 0.2 (mole fraction of chlorapatite in apatite) to greater than 0.6 in two sections. The authors state: "FAp varies inversely with CAp with average values ranging from 0.00 to 0.70, while HAp remains relatively constant near 0.40 or decreases slightly with height. These variations are remarkable given that no appreciable stratigraphic variations in either the major or trace element compositions of any of the cumulus minerals are found in the 800 m of section." Further, they state: "The data are best explained by a model involving the degassing of a Cl-rich volatile phase from the crystallizing interstitial liquid. The up-section migration of this fluid resulted in the crystallization of F-rich apatite in the lower portion and progressive Cl-enrichment in the apatite with height." (Meurer and Boudreau 1996). However, above the ultramafic zones of the Bushveld and the Stillwater, and in most other layered complexes, CAp decreases upsection (Warner et al. 1998). Halogens in mafic systems may be active in remobilizing platinum in the Stillwater and Bushveld, for apatites from those intrusive complexes show evidence of anomalous chlorine/water and chlorine/fluorine ratios (Boudreau et al. 1986). Chromatographic processes during degassing of intercumulus melt may play a role.

Mantle apatite

Apatite is a widespread accessory mineral in xenoliths of Phanerozoic lithospheric mantle brought to the surface by volcanic processes. It is important in understanding the mantle residence of volatiles such as Cl and F; further, it is a major host for REE and some LILE (such as U, Th, Sr) in the mantle (O'Reilly and Griffin 2000).

The apatite in mantle rocks studied by O'Reilly and Griffin (2000) include an unusual carbonate-bearing (0.7-1.7 wt %) hydroxyl-chlorapatite (1.5-2.5 wt % Cl, 6-40 ppm Br, 1 wt % F) associated with CO_2-rich fluid inclusion-bearing veins in mantle lherzolites (from both Australia and Alaska), and also a hydroxyl-fluorapatite variant that is igneous in origin. The former occurs in metasomatized mantle wall-rock peridotites, and were probably precipitated from CO_2- bearing, H_2O-rich metasomatizing fluids derived from a primitive mantle source region, whereas the latter may represent apatite crystallized from a fractionating alkaline magma that differentiated within the spinel lherzolite stability field in the lithospheric mantle. The parent magma was inferred by O'Reilly and Griffin (2000) to be a low-SiO_2, volatile-[CO_2-, F-, Cl-]rich magma of kimberlitic-carbonatitic affinity. The presence of Cl and OH-rich apatite in lithospheric mantle veins may suggest that deep metasomatizing fluids, when making a significant contribution to melting of fertile mantle, may be important in generating the high Cl systems similar to the layered mafic intrusions such as the Bushveld and the Stillwater. Further, O'Reilly et al. (2000) note that their mantle apatites are rich in U and Th and can contribute significantly to the radiogenic heat production of apatite-rich mantle. They also speculate that the abundance of apatite in Phanerozoic lithospheric mantle is greatly underestimated.

CONCLUDING REMARKS REGARDING THE HALOGENS IN APATITE

A number of workers have stressed the important role of Cl, F and H_2O in magmatic/hydrothermal processes, and the role of apatite in aiding in the determination of both the fugacities and concentrations of the halogens and their related compounds. We would like to emphasize some salient points about apatite chemistry in magmatic and hydrothermal systems:

1) For metaluminous systems, apatite crystallizes relatively early in the crystallization sequence of felsic magmas, and crystallizes comparatively later in more mafic melts. For example, it was estimated that apatite is a near liquidus phase in the Johnson Granite Porphyry, but began to crystallize after about 1/3 of the Kuna Crest (quartz diorite) magma crystallized (Piccoli and Candela 1994). Ultimately, apatite often first appears as an intercumulus phase in gabbros and related rocks (e.g., Boudreau et al. 1986).
2) If Cl, F and H_2O were to behave as perfectly incompatible substances *and therefore increase in concentration* in the melt, then the ratio of Cl, F and OH in successive aliquots of apatite will be *constant* as crystallization progresses (Candela 1986), if the effect of temperature is neglected.
3) Considering the effect of temperature, again with Cl, F and with H_2O behaving as perfectly incompatible substances in the melt, variations in Cl:F:OH in apatite reflect *nothing* other than how equilibrium constants change with temperature (Piccoli and Candela 1994).
4) Upon volatile exsolution, the ratios H_2O/F, Cl/F and (except at very shallow levels in the Earth's crust) Cl/H_2O decrease in the *melt*. Systematics for the composition of successive aliquots of apatite are not so simple because of the effect of temperature (and variable solid/melt partitioning). The use of the Cl/F ratio in apatite to deduce volatile saturation is particularly problematic because a reduction in temperature causes the Cl/F ratio in apatite to decrease independent of whether volatile saturation has been attained. A decrease in the Cl/OH ratio in apatite is probably the best indicator of volatile saturation, although certainly is not foolproof.
5) Information about the salinity of a magmatic volatile phase based on apatite chemistry can be attained *only* if the ratio of hydrogen to other cations in the volatile phase is known or assumed because ratios of Cl/H_2O and F/H_2O yield fugacity ratios of HCl/H_2O and HF/H_2O, respectively and not salinity.
6) Chemical data on apatite from igneous rocks need to be interpreted with caution, especially when the apatite analyses are not considered with regard to detailed petrography, whole rock composition and position within the rock unit. Only when

detailed petrographic information is available do workers have a chance to unravel the integrated effects of initial melt composition, protracted magmatic evolution, and volatile exsolution, ultimate plutonic consolidation, and subsequent subsolidus cooling and hydrothermal alteration.

SUMMARY

Apatite is a nearly ubiquitous phase in igneous rocks, due in part to the low solubility of P_2O_5 in silicate (and other) melts, and the limited amount of phosphorus accepted by the major rock-forming minerals. The solubility characteristics have been well characterized and can be used to make estimates of the temperature at which apatite crystallizes in magmatic systems. Apatite is a major repository of the halogens in rocks of the Earth's crust and mantle, and is one of the few phases that can be used to infer information about the composition of crustal and mantle fluids.

The analyses compiled for this work, and from detailed site-specific studies, suggest a number of petrogenetically and metallogenetically significant trends. The concentration of Mn in apatite appears to be related to the degree of fractionation of the host granite; Na behaves similarly, but appears to be limited by saturation of the melt with respect to Na-bearing phases. As a broad generalization, our plots show that Climax-type Mo-related porphyries tend to have apatites with Mn/Na ≥ 1, whereas Cu-related porphyries have Mn/Na ≤ 1. Strontium in apatite also seems to reflect fractionation of the host, and correlates fairly well with whole rock Sr. With fractionation, Fe/Ca in apatite may increase, but this trend may be inhibited at higher oxygen fugacities. Regarding sulfur, our compilation is broadly consistent with S being incorporated into apatite as a coupled substitution with Si ($SiSP_{-2}$). Considering the halogens, the mole fraction of chlorapatite in most igneous systems, is generally restricted to 0.2 or less; noticeable exceptions are the Stillwater, Bushveld, and related complexes. Fluids with a mantle provenance may be relatively enriched in Cl relative to F and water, consistent with apatite compositions from lithospheric mantle veins, mafic complexes, and the more primitive members of intermediate-felsic fractionation suites. Data in our compilation show that some (but not all) ore-related felsic plutons have Cl-rich apatites (CAp > 5%), and further, the magmas were capable of releasing volatiles with greater than 0.05 molal HCl, a constituent essential for hydrothermal alteration in magmatic-hydrothermal systems.

APPENDIX:

Comments on the Calculation of Mineral Formulas for End-Member F-, Cl- and OH-Apatites

In this appendix the following notation will be used: FAp is the mole fraction of fluorapatite in apatite, calculated using $X^{Ap}_{FAp} = C^{Ap}_{F} / 3.767$, where C^{Ap}_{F} is the concentration of F in apatite in wt %; CAp is the mole fraction of chlorapatite ($X^{Ap}_{ClAp} = C^{Ap}_{Cl} / 6.809$); and HAp is the mole fraction of hydroxylapatite ($X^{Ap}_{HAp} = 1 - X^{Ap}_{FAp} - X^{Ap}_{ClAp}$). Inherent in these calculations is the assumption that the halogen site is filled with Cl, F and OH, only, and that the major constituents in apatite don't vary appreciably from calcium and phosphorus only (a simplification that may not be appropriate where considerable substitution in apatite occurs; e.g., apatite from Mineville, New York, that contains ~10 wt % REE_2O_3) (McKeown and Klemic 1956, Roeder et al. 1987).

Alternatively, there are several ways in which one can calculate the OH concentration in apatite (given that concentrations of H_2O/OH are not easily obtained by using many common microanalytical techniques). One method is to calculate the OH concentration by assuming charge balance in apatite (e.g., see study by Stoppa and Liu 1995). This method has limitations if other anion species in apatite are not measured (e.g., CO_3^{2-} which is rarely measured, and SO_4^{2-} which is sometimes measured), or if there is uncertainty in the oxidation state of Fe (which can lead to an overestimation of OH). Alternatively, the OH concentration in apatite can be calculated assuming that the halogen site is filled (by Cl, F and OH). Although there is evidence that this is not the case in some apatites (e.g., Young and Munson 1966), this is thought to be an appropriate approximation (Piccoli and Candela 1994, Sommerauer and Katz-Lehnert 1985, see discussions in Tacker and Stormer 1989). This calculation, however, is also not without problems; it has been clearly demonstrated that in carbonate-rich apatites, the F concentration in apatite is greater than that in stoichiometric FAp. Although the amount of F bound to CO_3 can be calculated (using the appropriate operative exchange), carbonate is rarely measured, and therefore excess F can not be accurately determined. In the section on geochemistry, we have used the second method to determine the OH concentration, and made no attempt to determine the concentration of F (bound to CO_3) due to the paucity of carbonate data in the literature.

Apatite mole fractions of the end-members have been calculated, but with the following stipulation. If the F content of the apatite is >3.767, as is often the case with apatite from carbonatites, then the mole-fraction of FAp is assumed to be 1, with the remainder of the F thought to be present bound to the carbonate radical. From that, the appropriate amount of CAp is calculated, and an equivalent amount is removed from the calculated FAp, and an equivalent amount of the (absolute value) of CAp is added to the F in the carbonate site (if Cl is measured). Given that the HAp is calculated from the FAp and CAp, this has the possibility of ignoring any OH in the apatite. However, we feel that this is merited because Cl in carbonate apatite is small, and because in rare cases where OH has been measured in carbonate apatite, along with Cl and F, our calculations produce similar results.

REFERENCES

Ague JJ, Brimhall GH (1987) Granites of the batholiths of California: products of local assimilation and regional-scale crustal contamination. Geology 15:63-66

Allen CM (1991) Local equilibrium of mafic enclaves and granitoids of the Turtle Pluton, southeast California: mineral, chemical, and isotopic evidence. Am Mineral 76:574-588

Amli R (1975) Mineralogy and rare-earth geochemistry of apatite and xenotime from Gloserheia Granite Pegmatite, Froland, southern Norway. Am Mineral 60:607-620

Andersen T, Austrheim H (1991) Temperature HF fugacity trends during crystallization of calcite carbonatite magma in the Fen Complex, Norway. Mineral Mag 55:81-94

Ayers JC, Watson EB (1990) Solubility of accessory minerals in aqueous fluids at 1.0-2.8 GPa and 1000-1200°C. Geol Soc Am Ann Meet, Progr Abstr 22:158-159

Ayers JC, Watson EB (1991) Solubility of apatite, monazite, zircon, and rutile in supercritical aqueous fluids with implications for subduction zone geochemistry. *In* The behaviour and influence of fluids in subduction zones. Tarney J, Pickering KT, Knipe RJ, Dewey JF (eds) Phil Trans Royal Soc London, London, p. 365-375

Baig AA, Fox JL, Young RA, Wang Z, Hsu J, Higuchi WI, Chhettry A, Zhuang H, Otsuka M (1999) Relationships among carbonated apatite solubility, crystallite size, and microstrain parameters. Calcified Tissue Intl 64:437-449

Bargossi GM, Del Moro A, Ferrari M, Gasparotto G, Mordenti A, Rottura A, Tateo F (1999) Caratterizzazione petrografico-geochimica e significato dell'associazione monzogranito-inclusi femici microgranulari della Vetta di Cima d'Asta (Alpi Meridionali). Mineral Petrog Acta 42:155-179

Barkov AY, Savchenko YE, Zhangurov AA (1995) Fluid migration and its role in the formation of platinum-group minerals—evidence from the Imandrovsky and Lukkulaisvaara layered intrusions, Russia. Mineral Petrol 54:249-260

Barth AP, Dorais MJ (2000) Magmatic anhydrite in granitic rocks: first occurrence and potential petrologic consequences. Am Mineral 85:430-435

Barton MD, Johnson DA (1996) Evaporitic-source model for igneous-related Fe oxide-(REE-Cu-Au-U) mineralization. Geology 24:259-262

Belousova EA, Griffin WL, O'Reilly SY, Fisher NI (2002) Apatite as an indicator mineral for mineral exploration: trace-element compositions and their relationship to host rock type. J Geochem Explor 76:45-69

Belousova EA, Walters S, Griffin WL, O'Reilly SY (2001) Trace-element signatures of apatites in granitoids from the Mt. Isa Inlier, northwestern Queensland. Austral J Earth Sci 48:603-619

Binder G, Troll G (1989) Coupled anion substitution in natural carbon-bearing apatites. Contrib Mineral Petrol 101:394-401

Boudreau A (2002) The Stillwater and Bushveld magmas were wet! 9th Intl Platinum Symp, Duke University, Durham, North Carolina, p 57-60

Boudreau AE (1993) Chlorine as an exploration guide for the platinum-group elements in layered intrusions. J Geochem Explor 48:21-37

Boudreau AE, Kruger FJ (1990) Variation in the composition of apatite through the Merensky cyclic unit in the western Bushveld Complex. Econ Geol 85:737-745

Boudreau AE, Mathez EA, McCallum IS (1986) Halogen geochemistry of the Stillwater and Bushveld Complexes—evidence for transport of the platinum-group elements by Cl-rich fluids. J Petrol 27: 967-986

Boudreau AE, McCallum IS (1989) Investigations of the Stillwater Complex, Part 5: Apatites as indicators of evolving fluid composition. Contrib Mineral Petrol 102:138-153

Boudreau AE, McCallum IS (1990) Low-temperature alteration of REE-rich chlorapatite from the Stillwater Complex, Montana. Am Mineral 75:687-693

Brenan JM (1993) Partitioning of fluorine and chlorine between apatite and aqueous fluids at high pressure and temperature: implications for the F and Cl content of high P-T fluids. Earth Planet Sci Lett 117:251-263

Bridges NJ, Jones MY, Lee AIN, Bowen RW, Sheldon RP (1983) Bibliography of the geology of sedimentary phosphorite and igneous apatite. U S Geol Surv Open-File Rpt 83-841:925

Brown GM, Peckett A (1977) Fluorapatites from Skaergaard Intrusion, East Greenland. Mineral Mag 41:227-232

Budzinski H, Tischendorf G (1989) Distribution of REE among minerals in the Hercynian postkinematic granites of Westerzgebirge-Vogtland, GDR. Z Geol Wissen 17:1019-1031

Buhn B, Wall F, Le Bas MJ (2001) Rare-earth element systematics of carbonatitic fluorapatites, and their significance for carbonatite magma evolution. Contrib Mineral Petrol 141:572-591

Bunch TE, Reid AM (1975) The Nakhlites Part I: Petrography and mineral chemistry. Meteoritics 10: 303-315

Candela PA (1986) Toward a thermodynamic model for the halogens in magmatic systems: an application to melt-vapor-apatite equilibria. Chem Geol 57:289-301

Capdevila R (1967) Repartition et habitus de l'apatite dans le granite de Neira (Espagne): comparaisons avec des donnees experimentales et applications petrogenetiques. C R Acad Sci Paris 264:1694-1697

Chang LLY, Howie RA, Zussman J (1996) Non-silicates: Sulphates, Carbonates, Phosphates, Halides. Longman Group Limited, Essex, UK

Cherniak DJ (2000) Rare earth element diffusion in apatite. Geochim Cosmochim Acta 64:3871-3885

Cherniak DJ, Lanford WA, Ryerson FJ (1991) Lead diffusion in apatite and zircon using ion implantation and Rutherford backscattering techniques. Geochim Cosmochim Acta 55:1663-1673

Cherniak DJ, Ryerson FJ (1993) A study of strontium diffusion in apatite using Rutherford backscattering spectroscopy and ion implantation. Geochim Cosmochim Acta 57:4653-4662

Clemens JD, Holloway JR, White AJR (1986) Origin of an A-type granite: experimental constraints. Am Mineral 71:317-324

Core DP, Kesler SE, Essene EJ (2001) Oxygen fugacity and sulfur speciation in felsic intrusive rocks from the Wasatch and Oquirrh Ranges, Utah. 11th Annual V.M. Goldschmidt Conference, Lunar and Planetary Institute, Houston, Abstr 1088:3455

Cornejo PC, Mahood GA (1997) Seeing past the effects of re-equilibration to reconstruct magmatic gradients in plutons: La Gloria Pluton, central Chilean Andes. Contrib Mineral Petrol 127:159-175

Coulson IM, Chambers AD (1996) Patterns of zonation in rare-earth-bearing minerals in nepheline syenites of the North Qoroq Center, South Greenland. Can Mineral 34:1163-1178

Coulson IM, Dipple GM, Raudsepp M (2001) Evolution of HF and HCl activity in magmatic volatiles of the gold-mineralized Emerald Lake pluton, Yukon Territory, Canada. Mineral Dep 36:594-606

Cruft EF (1966) Minor elements in igneous and metamorphic apatite. Geochim Cosmochim Acta 30: 375-398

Dilles JH (1987) Petrology of the Yerington Batholith, Nevada—evidence for evolution of porphyry copper ore fluids. Econ Geol 82:1750-1789

Donovan J, Hanchar J, Piccoli P, Schrier MD, Boatner L, Jarosewich E (2003) A re-examination of the rare-earth element orthophosphate reference samples for electron microprobe analysis. Can Mineral (accepted):

Dorais MJ, Lira R, Chen Y, Tingey D (1997) Origin of biotite-apatite-rich enclaves, Achala Batholith, Argentina. Contrib Mineral Petrol 130:31-46

Elliott JC (1994) Structure and chemistry of the apatites and other calcium orthophosphates. Elsevier, Amsterdam–New York

Exley RA, Smith JV (1982) The role of apatite in mantle enrichment processes and in the petrogenesis of some alkali basalt suites. Geochim Cosmochim Acta 46:1375-1384

Farley KA (2000) Helium diffusion from apatite: general behavior as illustrated by Durango fluorapatite. J Geophys Res 105:2903-2914

Farver JR, Giletti BJ (1989) Oxygen and strontium diffusion kinetics in apatite and potential applications to thermal history determinations. Geochim Cosmochim Acta 53:1621-1631

Fominykh VG (1974) Fluorine and chlorine in the coexisting apatites and amphiboles in the rocks and ores of the titanomagnetite deposits of the Urals. Geochem Intl 11:354-356

Fournelle J, Carmody RW, Daag AS (1996) Anhydrite-bearing pumices from the June 15, 1991, eruption of Mount Pinatubo: geochemistry, mineralogy and petrology. *In* Fire and mud: eruptions and lahars of Mount Pinatubo, Philippines. Fournelle J, Carmody RW, Daag AS (eds) Philippine Institute of Volcanology and Seismology, Quezon City, p 845-863

Girault J (1966) Sur la genese des cristaux d'apatite des carbonatites d'Oka (Canada). C R Acad Sci Paris 263:97-100

Gottesmann B, Wirth R (1997) Pyrrhotite inclusions in dark pigmented apatite from granitic rocks. Eur J Mineral 9:491-500

Green TH, Watson EB (1982) Crystallization of apatite in natural magmas under high-pressure hydrous conditions, with particular reference to orogenic rock series. Contrib Mineral Petrol 79:96-105

Grivet M, Rebetez M, Ben Ghouma N, Chambaudet A, Jonckheere R, Mars M (1993) Apatite fission-track age correction and thermal history analysis from projected track length distributions. Chem Geol 103:157-169

Gunow AJ (1983) Trace element mineralogy in the porphyry molybdenum environment. PhD dissertation, University of Colorado, Boulder, Colorado

Harrison TM, Watson EB (1984) The behavior of apatite during crustal anatexis: equilibrium and kinetic considerations. Geochim Cosmochim Acta 48:1467-1477

Heming RF, Carmichael IS (1973) High-temperature pumice flows from Rabaul Caldera-Papua, New Guinea. Contrib Mineral Petrol 38:1-20

Hildreth EW (1977) The magma chamber of the Bishop Tuff: gradients in temperature, pressure, and composition. PhD dissertation, University of California, Berkeley, California

Hoche T, Moisescu C, Avramov I, Russel C, Heerdegen WD (2001) Microstructure of SiO_2-Al_2O_3-CaO-P_2O_5-K_2O-F- glass ceramics. 1. Needlelike versus isometric morphology of apatite crystals. Chem Mater 13:1312-1319

Hogarth DD, Hartree R, Loop J, Solberg TN (1985) Rare-earth element minerals in 4 carbonatites near Gatineau, Quebec. Am Mineral 70:1135-1142

Holden P, Halliday AN, Stephens WE (1987) Neodymium and strontium isotope content of microdiorite enclaves points to mantle input to granitoid production. Nature 330:53-56

Hoskin PWO, Kinny PD, Wyborn D, Chappell BW (2000) Identifying accessory mineral saturation during differentiation in granitoid magmas: an integrated approach. J Petrol 41:1365-1396

House MA, Farley KA, Stockli D (2000) Helium chronometry of apatite and titanite using Nd-YAG laser heating. Earth Planet Sci Lett 183:365-368

Hughes JM, Cameron M (1993) Order/disorder in the apatite anion columns: crystal chemical constraints on F, Cl, OH. Geol Soc Am Ann Meet, Progr Abstr 25:371

Hughes JM, Fransolet AM, Schreyer W (1993) The atomic arrangement of iron-bearing apatite. N Jahrb Mineral Monat 1993:504-510

Huntington HD (1979) Kiglapait mineralogy: I, apatite, biotite, and volatiles. J Petrol 20:625-652

Imai A, Listanco EL, Fujii T (1996) Highly oxidized and sulfur-rich dacitic magma of Mount Pinatubo: implications for metallognesis of porphyry copper mineralization in the western Luzon Arc. *In* Fire and mud: eruptions and lahars of Mount Pinatubo, Philippines. Fournelle J, Carmody RW, Daag AS (eds) Philippine Institute of Volcanology and Seismology, Quezon City, p 865-974

Jacobs DC, Parry WT (1973) Geochemistry of biotites from porphyry copper deposits. Geol Soc Am Ann Meet, Progr Abstr 5:681-682

Jahnke RA (1984) The synthesis and solubility of carbonate fluorapatite. Am J Sci 284:58-78

Jarosewich E, Nelen JA, Norberg JA (1980) Reference samples for electron microprobe analysis. Geostandards Newsletter 4:43-47

Johannes W, Holtz F (1996) Petrogenesis and experimental petrology of granitic rocks. Springer Verlag, Berlin, New York
Jolliff BL, Papike JJ, Shearer CK, Shimizu N (1989) Inter- and intra-crystal REE variations in apatite from the Bob Ingersoll Pegmatite, Black Hills, South Dakota. Geochim Cosmochim Acta 53:429-441
Jonckheere R, Marș M, Van den Haute P, Rebetez M, Chambaudet A (1993) L'apatite de Durango (Mexique): analyse d'un mineral standard pour la datation par traces de fission. Chem Geol 103: 141-154
Karkkainen N, Appelqvist H (1999) Genesis of a low-grade apatite-ilmenite-magnetite deposit in the Kauhajarvi gabbro, western Finland. Mineral Dep 34:754-769
King PL, Chappell BW, Allen CM, White AJR (2001) Are A-type granites the high-temperature felsic granites? Evidence from fractionated granites of the Wangrah Suite. Austral J Earth Sci 48:501-514
Kogarko LN (1990) Ore-forming potential of alkaline magmas. *In:* Alkaline igneous rocks and carbonatites. Woolley AR, Ross M (eds) Elsevier, p. 167-175
Korzhinskiy MA (1981) Apatite solid solutions as indicators of the fugacity of HCl° and HF° in hydrothermal fluids. Geochem Int 18:44-60
Liu Y, Comodi P (1993) Some aspects of the crystal-chemistry of apatites. Mineral Mag 57:709-719
Loferski PJ, Ayuso RA (1995) Petrography and mineral chemistry of the composite Deboullie Pluton, northern Maine, U.S.A.: implications for the genesis of Cu-Mo mineralization. Chem Geol 123:89-105
London D (1992) The Application of Experimental Petrology to the Genesis and Crystallization of Granitic Pegmatites. Can Mineral 30:499-540
London D, Cerny P, Loomis J, Pan JJ (1990) Phosphorus in alkali feldspars of rare-element granitic pegmatites. Can Mineral 28:771-786
London D, Wolf MB, Morgan GBVI, Garrido MG (1999) Experimental silicate-phosphate equilibria in peraluminous granitic magmas, with a case study of Albuquerque Batholith at Tres Arroyos, Badajoz, Spain. J Petrol 40:215-240
Luhr JF, Carmichael ISE, Varekamp JC (1984) The 1982 eruptions of El Chichon Volcano, Chiapas, Mexico: mineralogy and petrology of the anhydrite-bearing pumices. J Volc Geotherm Res 23:69-108
Lyons JI, Jr. (1975) Volcanogenic iron ore of Cerro de Mercado and its setting within the Chupaderos Caldera, Durango, Mexico. MS thesis, University of Texas, Austin, Texas
McConnell D (1973) Apatite: Its Crystal Chemistry, Mineralogy, Utilization, and Geologic and Biologic Occurrences. Springer Verlag, Vienna–New York
McKeown FA, Klemic H (1956) Rare-earth-bearing apatite at Mineville, Essex County, New York. U S Geol Surv Bull 9-23
Meurer WP, Boudreau AE (1996) An evaluation of models of apatite compositional variability using apatite from the Middle Banded Series of the Stillwater Complex, Montana. Contrib Mineral Petrol 125:225-236
Montel JM, Mouchel R, Pichavant M (1988) High apatite solubilities in peraluminous melts. Terra Cognita 8:71
Naeser CW, Fleischer RL (1975) Age of the apatite at Cerro de Mercado, Mexico: a problem for fission-track annealing corrections. Geophys Res Lett 2:67-70
Nash WP (1976) Fluorine, chlorine, and OH-bearing minerals in the Skaergaard Intrusion. Am J Sci 276:546-556
O'Reilly SY, Griffin WL (2000) Apatite in the mantle: implications for metasomatic processes and high heat production in Phanerozoic mantle. Lithos 53:217-232
Parry WT, Ballantyne GH, Wilson JC (1978) Chemistry of biotite and apatite from a vesicular quartz latite porphyry plug at Bingham, Utah. Econ Geol 73:1308-1314
Parry WT, Jacobs DC (1975) Fluorine and chlorine in biotite from Basin and Range plutons. Econ Geol 70:554-558
Pearson JM, Taylor WR (1996) Mineralogy and geochemistry of fenitized alkaline ultrabasic sills of the Gifford Creek Complex, Gascoyne Province, Western Australia. Can Mineral 34:201-219
Peng G, Luhr JF, McGee JJ (1997) Factors controlling sulfur concentrations in volcanic apatite. Am Mineral 82:1210-1224
Philpotts AR (1984) Solubility of apatite in iron-silicate liquids. Geol Soc Am Ann Meet, Progr Abstr 16:55
Piccoli P, Candela P (1994) Apatite in felsic rocks: a model for the estimation of initial halogen concentrations in the Bishop Tuff (Long Valley) and Tuolumne Intrusive Suite (Sierra Nevada Batholith) magmas. Am J Sci 294:92-135
Piccoli PM (1992) Apatite chemistry in felsic magmatic systems. PhD dissertation, University of Maryland, College Park, Maryland
Piccoli PM, Candela PA (1988) Trends in apatite chemistry from El Chichon: implications for vapor evolution. Geol Soc Am Ann Meet, Progr Abstr 20:194-195

Piccoli PM, Candela PA, Williams TJ (1999) Estimation of aqueous HCl and Cl concentrations in felsic systems. Lithos 46 3:591-604
Pichavant M, Montel J-M, Richard LR (1992) Apatite solubility in peraluminous liquids: experimental data and an extension of the Harrison-Watson model. Geochim Cosmochim Acta 56:3588-3861
Potts PJ, Tindle AG (1989) Analytical characteristics of a multilayer dispersion element ($2d$ = 60 Å) in the determination of fluorine in minerals by electron microprobe. Mineral Mag 53:357-362
Raimbault L, Burnol L (1998) The Richemont rhyolite dyke, Massif Central, France: a subvolcanic equivalent of rare-metal granites. Can Mineral 36:265-282
Rakovan J, McDaniel DK, Reeder RJ (1997) Use of surface-controlled REE sectoral zoning in apatite from Llallagua, Bolivia, to determine a single-crystal Sm-Nd age. Earth Planet Sci Lett 146:329-336
Rakovan J, Reeder RJ (1994) Differential incorporation of trace-elements and dissymmetrization in apatite—the role of surface-structure during growth. Am Mineral 79:892-903
Raudsepp M (1995) Recent advances in the electron-probe microanalysis of minerals for the light-elements. Can Mineral 33:203-218
Reid JB, Jr., Evans OC, Fates DG (1983) Magma mixing in granitic rocks of the central Sierra Nevada, California. Earth Planet Sci Lett 243-261
Riley TR, Bailey DK, Lloyd FE (1996) Extrusive carbonatite from the quaternary Rockeskyll complex, West Eifel, Germany. Can Mineral 34:389-401
Roeder PL, MacArthur D, Ma X-P, Palmer GR, Mariano AN (1987) Cathodoluminescence and microprobe study of rare-earth elements in apatite. Am Mineral 72:801-811
Roegge JS, Logsdon MJ, Young HS, Barr HB, Borcsik M, Holland HD (1974) Halogens in apatites from the Providencia Area, Mexico. Econ Geol 69:229-240
Ronsbo JG (1989) Coupled substitutions involving REEs and Na and Si in apatites in alkaline rocks from the Ilimaussaq Intrusion, South Greenland, and the petrological implications. Am Mineral 74:896-901
Sallet R (2000) Fluorine as a tool in the petrogenesis of quartz-bearing magmatic associations: applications of an improved F-OH biotite-apatite thermometer grid. Lithos 50:241-253
Sassi R, Harte B, Carswell DA, Yujing H (2000) Trace element distribution in Central Dabie eclogites. Contrib Mineral Petrol 139:298-315
Seifert W, Kaempf H, Wasternack J (2000) Compositional variation in apatite, phlogopite and other accessory minerals of the ultramafic Delitzsch Complex, Germany: implication for cooling history of carbonatites. Lithos 53:81-100
Sha L-K (1995) Genesis of zoned hydrous ultramafic/mafic-silicic intrusive complexes: an MHFC hypothesis. Earth Sci Rev 39:59-90
Sha L-K, Chappell BW (1998) Contribution of feldspars to the whole-rock phosphorus budget of I- and S-type granites: a quantitative estimation. Proc Intl Conf Genetic Significance of Phosphorus Fractionated Granites 42:125-128
Sha L-K, Chappell BW (1999) Apatite chemical composition, determined by electron microprobe and laser-ablation inductively coupled plasma mass spectrometry, as a probe into granite petrogenesis. Geochim Cosmochim Acta 63:3861-3881
Shmakin BM, Shiryaye V (1968) Distribution of rare earths and some other elements in apatites of muscovite pegmatites, eastern Siberia. Geochem Intl 5:796
Smith MP (1999) Reaction relationships in the Bayan Obo Fe-REE-Nb deposit Inner Mongolia, China: implications for the relative stability of rare-earth element phosphates and fluorcarbonates. Contrib Mineral Petrol 134:294-310.
Sommerauer J, Katz-Lehnert K (1985) A new partial substitution mechanism of $CO_3^{(2-)}$ / $CO_3OH^{(3-)}$ and $SiO4^{(4-)}$ for the $PO_4^{(3-)}$ group in hydroxyapatite from the Kaiserstuhl alkaline complex (SW-Germany). Contrib Mineral Petrol 91:360-368
Stoppa F, Liu Y (1995) Chemical-composition and petrogenetic implications of apatites from some ultra-alkaline Italian rocks. Eur J Mineral 7:391-402
Stormer JC, Jr., Pierson ML, Tacker RC (1993) Variation of F and Cl x-ray intensity due to anisotropic diffusion in apatite during electron microprobe analysis. Am Mineral 78:641-648
Streck MJ, Dilles JH (1998) Sulfur evolution of oxidized arc magmas as recorded in apatite from a porphyry copper batholith. Geology 26:523-526
Taborszky FK (1962) Geochemie des Apatits in Tiefengesteinen am Beispiel des Odenwaldes. Beiträge Mineral Petrogr 8:354-392
Tacker RC, Stormer JC, Jr. (1989) A thermodynamic model for apatite solid solutions, applicable to high-temperature geologic problems. Am Mineral 74:877-888
Tepper JH, Kuehner SM (1999) Complex zoning in apatite from the Idaho Batholith: a record of magma mixing and intracrystalline trace element diffusion. Am Mineral 84:581-595

Tribuzio R, Tiepolo M, Vannucci R, Bottazzi P (1999) Trace element distribution within olivine-bearing gabbros from the Northern Apennine ophiolites (Italy): Evidence for post-cumulus crystallization in MOR-type gabbroic rocks. Contrib Mineral Petrol 134:123-133
Tsusue A, Nedachi M, Hashimoto K (1981) Geochemistry of apatites in the granitic-rocks of the molybdenum, tungsten, and barren provinces of southwest Japan. J Geochem Explor 15:285-294
Vaz JE (1980) Effects of natural radioactivity on the thermoluminescence of apatite crystals at Cerro del Mercado, Mexico. Modern Geology 7:171-175
Vernon RH (1983) Restite, xenoliths and microgranitoid enclaves in granites. Journal and Proc Royal Soc New South Wales 77-103
Walker RJ (1984) The origin of the Tin Mountain Pegmatite, Black Hills, South Dakota. PhD dissertation, State University of New York, Stony Brook, New York
Walker RJ, Hanson GN, Papike JJ, O'Neil JR, Laul JC (1986) Internal evolution of the Tin Mountain Pegmatite, Black Hills, South Dakota. Am Mineral 71:440-459
Warner S, Martin RF, Abdel-Rahman A-FM, Doig R (1998) Apatite as a monitor of fractionation, degassing, and metamorphism in the Sudbury igneous complex, Ontario. Can Mineral 36:981-999
Watson EB (1979) Apatite saturation in basic to intermediate magmas. Geophys Res Lett 6:937-940
Watson EB (1980) Apatite and phosphorus in mantle source regions: an experimental study of apatite/melt equilibria at pressures to 25 kbar. Earth Planet Sci Lett 51:322-335
Watson EB, Capobianco CJ (1981) Phosphorus and the rare earth elements in felsic magmas: an assessment of the role of apatite. Geochim Cosmochim Acta 45:2349-2358
Watson EB, Harrison TM, Ryerson FJ (1985) Diffusion of Sm, Sr, and Pb in fluorapatite. Geochim Cosmochim Acta 49:1813-1823
Williams SA, Cesbron FP (1977) Rutile and apatite: useful prospecting guides for porphyry copper deposits. Mineral Mag 41:288-292
Wolf MB, London D (1993) Apatite solubility in the peraluminous haplogranite system: not deja vu all over again. EOS Trans, Am Geophys Union 74 16:341
Wolf MB, London D (1994) Apatite dissolution into peraluminous haplogranitic melts: an experimental study of solubilities and mechanisms. Geochim Cosmochim Acta 58:4127-4145
Wolf MB, London D (1995) Incongruent dissolution of REE- and Sr-rich apatite in peraluminous granitic liquids: differential apatite, monazite, and xenotime solubilities during anatexis. Am Mineral 80: 765-775
Wolf RA, Farley KA, Silver LT (1996) Helium diffusion and low-temperature thermochronometry of apatite. Geochim Cosmochim Acta 60:4231-4240
Wyllie PJ, Cox KG, Biggar GM (1962) The habit of apatite in synthetic systems and igneous rocks. J Petrol 3:238-242
Young EJ, Munson EL (1966) Fluor-chlor-oxy-apatite and sphene from Crystal Lode pegmatite near Eagle, Colorado. Am Mineral 51:1476-1493
Young EJ, Myers AT, Munson EL, Conklin NM (1969) Mineralogy and geochemistry of fluorapatite from Cerro de Mercado, Durango, Mexico. U S Geol Surv Prof Paper 650:84-93
Zhu C, Sverjensky DA (1991) Partitioning of F-Cl-OH between minerals and hydrothermal fluids. Geochim Cosmochim Acta 55:1837-1858
Zhu C, Sverjensky DA (1992) F-Cl-OH partitioning between biotite and apatite. Geochim Cosmochim Acta 56:3435-3467

7 Apatite, Monazite, and Xenotime in Metamorphic Rocks

Frank S. Spear and Joseph M. Pyle

Department of Earth and Environmental Sciences
Rensselaer Polytechnic Institute
Troy, New York 12180

INTRODUCTION

This chapter focuses on phosphates that are significant in metamorphic rocks. A quick survey of phosphate mineral descriptions at a commercial mineral web-site revealed over 500 phosphate mineral names. Remarkably, only three are common in metamorphic rocks: apatite, monazite, and xenotime, and this chapter is restricted to discussion of these minerals.

Apatite, monazite, and, to a lesser extent, xenotime, have enjoyed intensive study during the previous half-century. Following World War II, considerable study was made of minerals that contain fissionable materials, and the sometimes large concentrations of U and Th in monazite culminated in a number of seminal papers on the occurrence of that mineral (e.g., Overstreet 1967). During the 1970s and early 1980s, the study of apatite and monazite turned towards their role as a sink for REEs and other trace elements in rocks (e.g. Watson 1980, Watson and Capobianco 1981). Throughout this period, monazite (and to some extent xenotime) enjoyed attention as a geochronometer (e.g., Parrish 1990), and through today this attention has continued to grow.

Accessory minerals such as monazite, xenotime, and apatite have come to the forefront of research in metamorphic petrology in recent years for two reasons. First, it has recently been recognized that trace elements in metamorphic rocks, and especially trace element zoning in garnet and other major phases, contain considerable detailed information about the reaction history a rock has experienced. Coupled with the fact that diffusion of many trace elements in metamorphic minerals is, in general, considerably slower than diffusion of major elements in those same minerals, the possibility has recently emerged of recovering details of metamorphic petrogenesis that were previously unattainable. Inasmuch as accessory minerals play a dominant role in the mass budget of many trace elements, attention has naturally turned to their paragenesis. It has also recently been discovered that accessory minerals themselves contain a paragenetic record in their chemical zoning profiles, adding further impetus to their study.

Second, the past decade has seen the development of *in situ* dating techniques (e.g., Secondary Ion Mass Spectrometry (SIMS), Laser Ablation-Inductively Coupled Plasma-Mass Spectrometry (LA-ICP-MS), and Electron Microprobe (EMP) analysis that make it possible to obtain Th/Pb, U/Pb, or Pb/Pb ages on individual spots in a grain of monazite, xenotime, apatite or zircon. This is significant for two reasons. First, *in situ* dating permits correlation of the age of a mineral and its textural setting. For example, it is possible to measure the age difference between monazites that are included within garnet and those in a rock's matrix (e.g., Foster et al. 2000, Catlos et al. 2001). Second, the complex chemical zoning observed in monazite and zircon requires a technique that permits age determination on individual chemical zones within a crystal, otherwise mixed ages are inevitable.

Research currently focuses on melding these two approaches. A major goal of accessory mineral studies in metamorphic rocks is the determination of age (t), pressure (P), and temperature (T) on an individual spot in a mineral such as monazite. In principle, this should be possible. Monazite records chemical evidence of involvement in major-

1529-6466/00/0048-0007$05.00

phase reactions, for which P and T information may be gleaned by conventional methods. Monazite itself contains temperature-sensitive information in its Y and HREE content if it coexists with xenotime (i.e., the monazite-xenotime miscibility gap; Heinrich et al. 1997, Seydoux-Guillaume et al. 2002) or garnet (garnet-monazite net transfer thermometry; Pyle et al. 2001). Measurement of an age in a monazite growth zone for which P and T have been determined results in a specific P-T-t point in a rock's history. Two or more such points results in such fundamental information as heating and burial rates and would provide quantitative constraints on the rates of tectonic processes.

This chapter provides a summary of the current state of knowledge on metamorphic monazite, xenotime, and apatite. Details of geochronological approaches are addressed elsewhere (Harrison et al. this volume), and this chapter will focus on petrology. The chemistries of these minerals will be summarized, and examples of zoning and paragenesis presented. As will be seen, much is as yet unknown about the behavior of these accessories during metamorphism, and directions for future research to address the outstanding questions will be discussed.

OCCURRENCE AND CHEMISTRY OF METAMORPHIC APATITE, MONAZITE AND XENOTIME

Electron microprobe analysis of apatite, monazite, and xenotime

The first issue with respect to analyzing accessory phosphates is to obtain accurate chemical analyses. This is not a simple task. Rare-earth and other elements in these phosphates have a large number of X-ray peaks with numerous overlaps, and accurate analysis using EMP takes considerable care in selecting X-ray lines and locating background measurements. Apatite is additionally problematic because of the rapid diffusion of F under the electron beam (e.g., Stormer et al. 1993), a problem that affects accurate analysis of all elements. These issues must be kept in mind when evaluating published analyses of metamorphic phosphates. Details of analytical methods for the analysis of metamorphic phosphates using the EMP are discussed elsewhere in this volume (Pyle et al. this volume).

Apatite

Although apatite has been intensely studied for such diverse applications as fission track dating and igneous petrogenesis (Gleadow et al. this volume, Piccoli and Candela this volume), studies concerned exclusively with compositional variation of metamorphic apatite are relatively few in number in comparison to the other types of apatite-focused studies in the literature (e.g., Kapustin 1987). Apatite papers applicable to metamorphic rocks include studies of the fugacity of components in metamorphic fluids in apatite-bearing samples (Korzhinskiy 1981, Yardley 1985, Smith and Yardley 1999), apatite-biotite OH-F exchange thermometry (Stormer and Carmichael 1971, Ludington 1978, Sallet and Sabatier 1996, Sallet 2000), and apatite-fluid partitioning (Zhu and Sverjensky 1991, 1992; Brennan 1993). Studies of metamorphic apatite composition generally focus on either halogen content (e.g., Yardley 1985, Smith and Yardley 1999) or REE content (e.g., Cruft 1966, Puchelt and Emmerman 1976, Bingen et al. 1996), although some studies present a more or less complete apatite analysis including Ca, P, halogens, and REEs (e.g., Bea and Montero 1999).

The literature data presented in this chapter represent a wide variety of metamorphic rock types including metabasites, metapelites, marbles, calc-silicate rocks, ultramafic rocks, and metagranites. Hydrothermal or low-temperature apatite occurrences were not considered for this chapter.

Occurrence and stability. Apatite, $Ca_5(PO_4)_3(OH,F,Cl)$, is common in rocks of diverse bulk composition, and is reported in metamorphic rocks of pelitic, carbonate, basaltic, and ultramafic composition. The essential constituents of apatite are present in amounts greater than trace quantities in these types of rocks, and apatite forms easily as a result.

Apatite is found at all metamorphic grades from transitional diagenetic environments and low-temperature alteration to migmatites, and in ultra high-pressure (diamond-bearing; e.g., Liou et al. 1998) samples. A list of all papers that report metamorphic apatite would be prohibitively long. The occurrence of apatite is not apparently dictated by its stability relative to other phosphates, but rather by the availability of essential constituents (P, Ca, and F).

Apatite composition. The compositions of metamorphic apatite typically fall along the F-OH join, although apatite with small amounts of Cl has been reported from metamorphic rocks (e.g., Kapustin 1987). Most, if not all, metamorphic apatite is dominated by the F end-member (i.e., fluor-apatite). A survey of analyses is shown in Figures 1 and 2. The F/(F+OH) typically ranges from 0.4 to 1.0 with a median value of 0.85 (Fig. 1a). Thus, it appears that the widespread presence of apatite in metamorphic rocks may be as much a function of the availability of F as it is the availability of P. (It should be noted that many published apatite analyses have reported F values in excess of the maximum permitted by stoichiometry, presumably due to analytical difficulties measuring F in apatite on the EMP as discussed by Stormer et al. 1993).

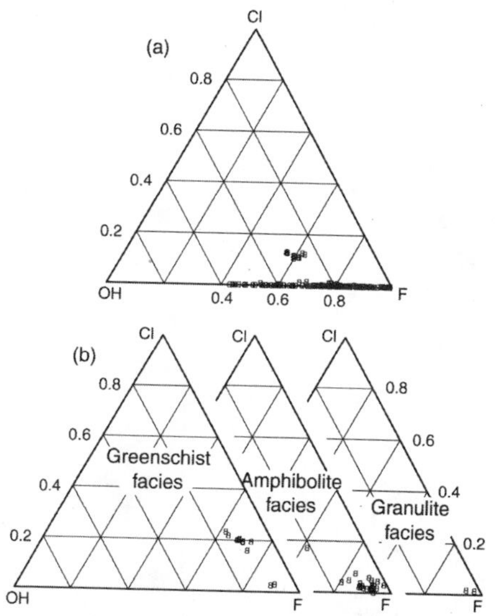

Figure 1. (a) Ternary F-OH-Cl plot of metamorphic apatite compositions from pelites from central New England, USA (from Pyle 2001). (b) Compositions of metamorphic apatite separated by metamorphic grade. Note that maximum Cl content decreases and F/(F+OH) increases with metamorphic grade. Data from Kapustin (1984).

Figure 1b shows the compositions of metamorphic apatite separated by metamorphic grade from Kapustin (1987). The analyses were obtained by wet chemistry, circumventing the problems with F analysis that are exhibited by EMP analyses. As can be seen from the plots, the composition of metamorphic apatite appears to be a function of grade. The greenschist facies apatites have modest Cl contents and F/(F+OH) of 0.80-0.94. By the granulite facies, apatites are virtually Cl-free, and the F/(F+OH) is 0.94-0.98. The strong preference of metamorphic apatite for F over Cl can be understood in the context of F-Cl-OH partitioning between apatite and fluid (see below) as well as a crystallographic context, as the F^- ion has a better fit within the central "column" anion site than either Cl^- or $(OH)^-$ (Hughes and Rakovan, this volume), and the decrease in Cl with grade could be due to Rayleigh distillation of Cl accompanying dewatering of the metamorphic rocks.

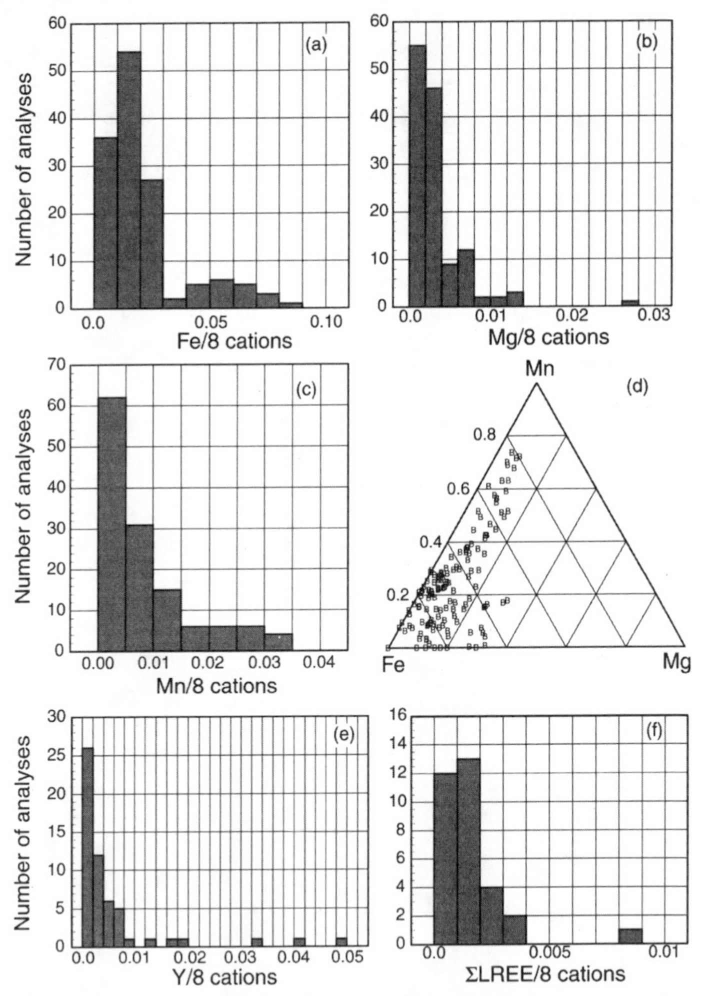

Figure 2. Plots showing compositions of metamorphic apatite from pelites from central New England (data from Pyle 2001). Units are cations / 8 oxygens. (a) Fe; (b) Mg; (c) Mn; (d) Fe-Mg-Mn ternary plot; (e) Y; and (f) ∑LREEs (La-Sm).

The minor elements Fe, Mg and Mn are in concentrations less than 0.1 cations/8 oxygens (Figs. 2a-d). Additionally, there appears to be a systematic relationship between Fe/(Fe+Mg) of apatite and coexisting phases, which has potential for petrogenetic studies. Y and LREE concentrations are also typically low (Figs. 2e,f), but sometimes show zoning (see below). Apatite that coexists with monazite in pelitic samples typically contains much lower concentrations of REEs (Pyle 2001, Yang and Rivers 2002) than does apatite in marbles and metaperidotites (Cruft 1966) or metabasalts (Bingen et al. 1996) with or without allanite. However, spuriously high REE concentrations in apatite may arise from analysis of minute inclusions of REE-rich phases (monazite, xenotime; Amli 1975, Bea et al. 1994). Therefore, apatite REE compositional data derived from non *in situ* analytical methods are suspect (Cruft 1966), and even *in situ* derived analytical data must be viewed with caution (Bea et al. 1994). REE patterns of metamorphic apatites (Fig. 3a,d) are relatively flat or display minor HREE depletion.

Compositional zoning. Zoning has been described in some metamorphic apatite crystals (Fig. 4), although most crystals examined by the authors display little or no zoning in BSE images taken at up to 100 nA current (unpublished results). Smith and Yardley (1999) report zoned apatites that were interpreted as having detrital cores (Fig. 4a). Yang and Rivers (2002) report Y zoned apatite crystals with core Y higher than rims (Fig. 4b). Apatite from a leucosome in a migmatite from central New Hampshire (Figs. 4c,d; unpublished data of the authors') displays zoning in F and Fe/(Fe+Mg). No zoning of P, Y, or Ca was observed in this crystal. The Fe/(Fe+Mg) zoning is believed to reflect growth zoning during melt crystallization.

Monazite and xenotime

Occurrence and stability. Monazite [(LREE)PO_4], and xenotime [(Y,HREE)PO_4] are common in metapelitic rocks and less abundant in mafic and calcic bulk compositions. Monazite and xenotime occur as detrital grains (beach sands comprised largely of monazite are reported) but are apparently not stable during diagenesis (e.g., Overstreet 1967), except as detrital grains (Suzuki and Adachi 1991, Suzuki et al. 1994, Hawkins and Bowring 1999, Ferry 2000, Catlos et al. 2001, Rubatto et al. 2001, Wing et al. 2002). Monazite appears to be stable in rocks of appropriate bulk composition at all metamorphic grades at and above the greenschist facies. Monazite has been reported from low-pressure contact aureoles (e.g., Ferry 2000, Wing et al. 2002), cordierite + K-feldspar rocks (e.g., Franz et al. 1996), granulite facies migmatites (e.g., Watt and Harley 1993, Bea et al. 1994, Hawkins and Bowring 1999, Pyle et al. 2001), and ultra high-pressure (coesite- and diamond-bearing) pelitic rocks (Terry et al. 2000). Liou et al. (1998) report monazite exsolution from apatite from a sample of ultra high-pressure clinopyroxenite from the Dabie Shan, China, which suggests that the solubility of REEs in apatite (i.e., monazite components) increases with increasing pressure.

Xenotime is reported from rocks of the chlorite zone to the cordierite + K-feldspar zone (e.g., Franz et al. 1996). Pyle and Spear (1999) report on xenotime textural occurrences in a suite of metapelitic samples from central New England that range in grade from garnet through cordierite + garnet zone. In the garnet zone, xenotime is common in the matrix of the rock and as inclusions within garnet. In garnet-bearing rocks, xenotime reacts out as garnet grows and is often consumed by the mid-garnet zone (Fig. 5a). Because of the incorporation of Y and HREEs into garnet, xenotime is often not present in the matrix of garnet-rich rocks above the mid garnet zone, although it may still be present as inclusions within garnet. Matrix xenotime appears again in some rocks of the migmatite zone as a product of melt crystallization (Fig. 5d). Finally, alteration of Y-rich garnet may release sufficient Y to stabilize xenotime during retrogression (Figs. 5b,c). In garnet-absent samples, xenotime may persist throughout all metamorphic grades (e. g. Bea and Montero 1999, Pyle and Spear 1999).

The crystal size of monazite and xenotime tends to increase with increasing metamorphic grade (e.g., Franz et al. 1996, Rubatto et al. 2001). Ferry (2000) and Wing et al. (2002) report xenoblastic monazite at lower grades in a contact aureole, a low-P

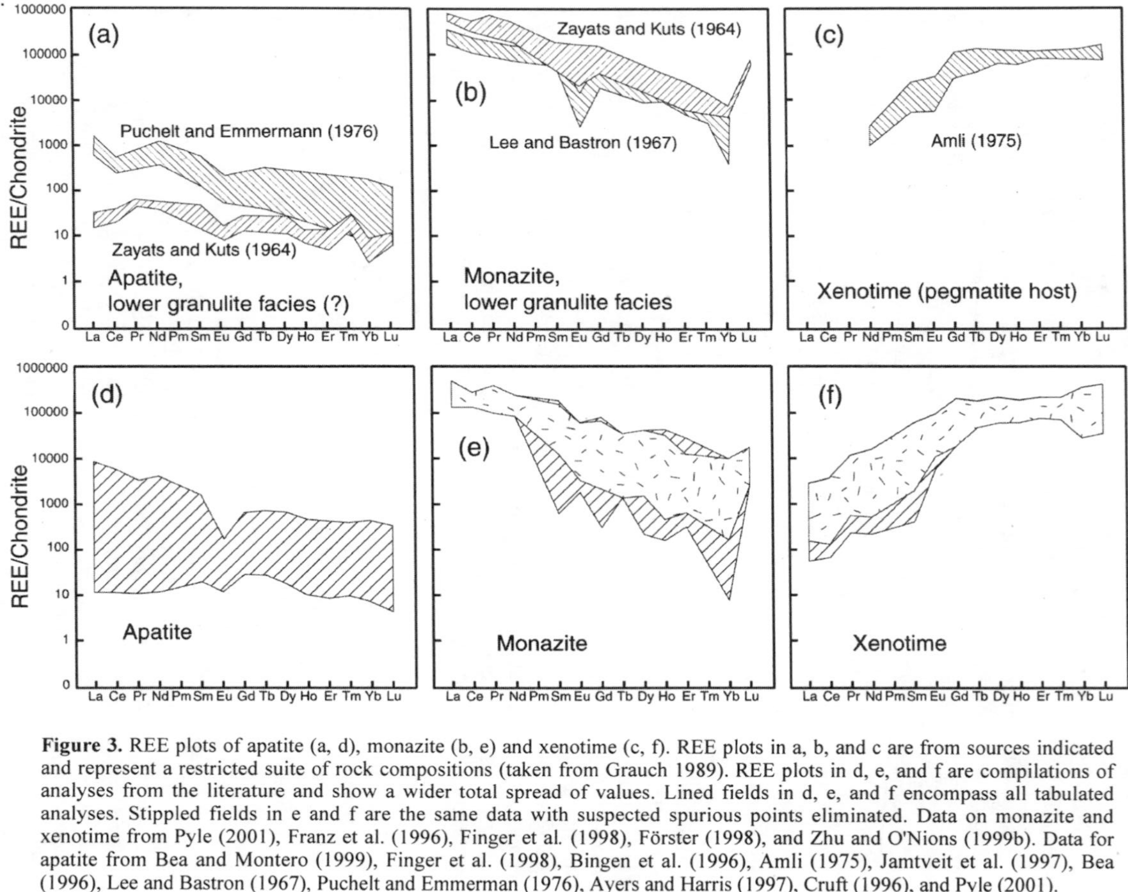

Figure 3. REE plots of apatite (a, d), monazite (b, e) and xenotime (c, f). REE plots in a, b, and c are from sources indicated and represent a restricted suite of rock compositions (taken from Grauch 1989). REE plots in d, e, and f are compilations of analyses from the literature and show a wider total spread of values. Lined fields in d, e, and f encompass all tabulated analyses. Stippled fields in e and f are the same data with suspected spurious points eliminated. Data on monazite and xenotime from Pyle (2001), Franz et al. (1996), Finger et al. (1998), Förster (1998), and Zhu and O'Nions (1999b). Data for apatite from Bea and Montero (1999), Finger et al. (1998), Bingen et al. (1996), Amli (1975), Jamtveit et al. (1997), Bea (1996), Lee and Bastron (1967), Puchelt and Emmerman (1976), Ayers and Harris (1997), Cruft (1996), and Pyle (2001).

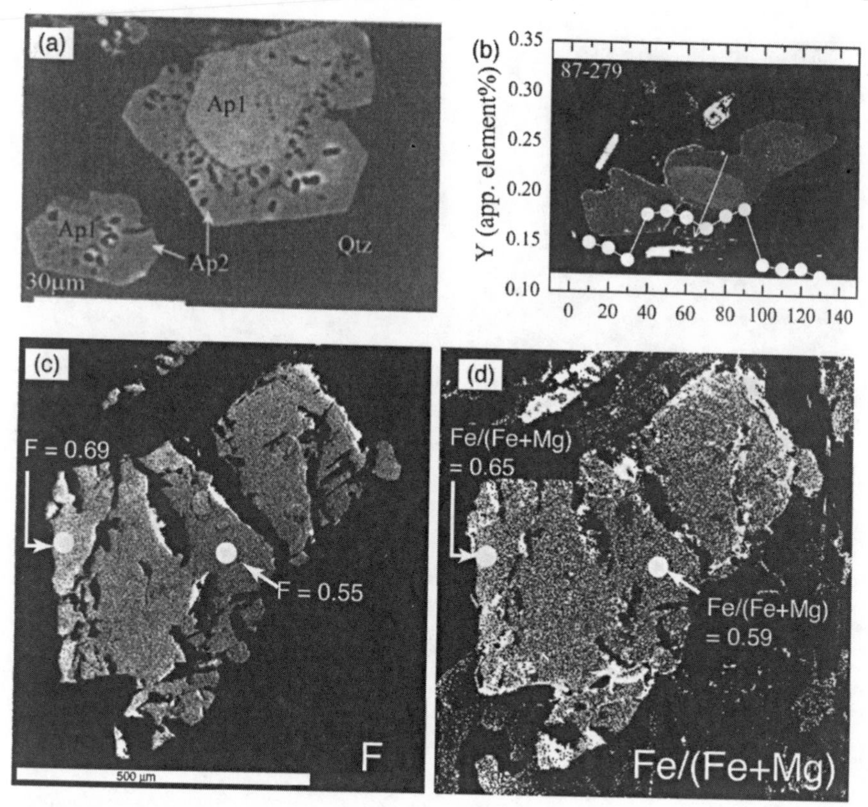

Figure 4. Zoning in metamorphic apatite. (a) Detrital Cl-rich apatite core and F-rich low-grade (pumpellyite-actinolite facies) metamorphic overgrowth. From Smith and Yardley (1999). (b) Zoned metamorphic apatite from Yang and Rivers (2002). Bright areas correspond to higher Y concentrations. Overlay of dots connected by lines show traverse analyzed for Y concentration. (c) F and (d) Fe/(Fe+Mg) zoning in apatite from leucosome, west-central New Hampshire (Pyle, unpublished data).

Buchan terrane, and a Barrovian terrane, which is replaced by allanite with increasing grade before an idioblastic variety appears at the Al_2SiO_5 isograd.

A detailed discussion of the paragenesis of monazite and xenotime in metamorphic rocks will be taken up below.

Composition. Monazite has the nominal composition $(LREE)PO_4$ and the LREEs (primarily La + Ce + Nd) generally comprise approximately 75% of the total cation proportions (exclusive of P) of most metamorphic monazites (Fig. 6). Indeed, the range of observed LREE concentrations is quite restricted (Figs. 6, 8a, 10a) with La, Ce and Nd averaging around 0.2, 0.43, and 0.17 cations/4 oxygens, respectively. Most monazites also contain additional Th, U, Ca, Si, HREEs, Y, and Pb. Y concentrations up to 0.1 cations/4 oxygens have been reported (e.g., Heinrich et al. 1997), although HREE concentrations are generally less than 0.02 cations/4 oxygens (Fig. 6e, 10a,b).

Xenotime has the nominal composition YPO_4, and Y comprises 75-85% of the total (Fig. 7a) with the HREEs comprising the bulk of the remainder (Figs. 7b-e; 8b,c; 10c,d). The range of HREE contents is actually rather restricted with Gd, Dy, Er, and Yb all

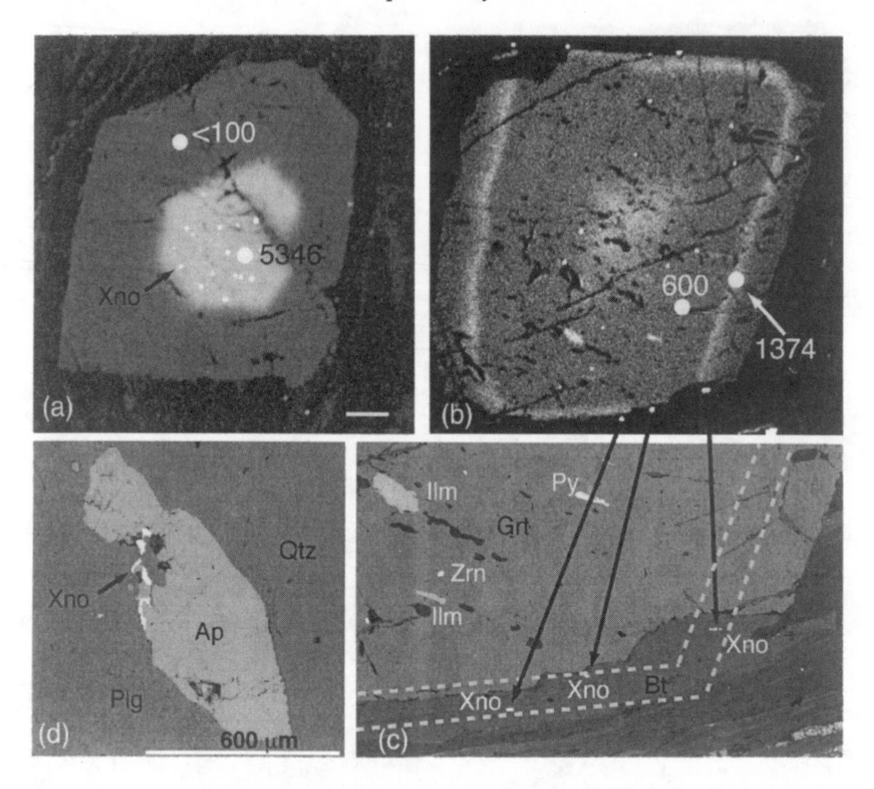

Figure 5. Backscatter electron (BSE) images and X-ray composition maps of metamorphic xenotime textures. (a) X-ray map of Y concentration in garnet from eastern Vermont showing high Y in the core where xenotime inclusions are located (white spots in garnet core) and low Y elsewhere. (b) Yttrium X-ray composition map of garnet in from staurolite grade pelite from New Hampshire. Note truncation of high-Y annulus and three xenotime crystals (three white dots) where annulus is truncated on lower garnet rim. (c) BSE image of lower corner of (b) showing xenotime grains located in the relict annulus (dotted lines). (d) Xenotime formed along grain-boundaries of late generation apatite grains occurring in leucosomes, migmatite zone, west-central New Hampshire. Numbers on images are Y concentration in ppm. Grt = garnet; Bt = biotite; Mnz = monazite; Xno = xenotime; Ap = apatite; Plg = plagioclase; Qtz = quartz; Zrn = zircon; Ilm = ilmenite; Py = pyrite. From Pyle and Spear (1999) and Pyle (2001).

showing a limited range of concentrations (Figs. 7, 8, 10c,d). As discussed below, Y and HREE concentrations in monazite coexisting with xenotime are temperature sensitive and may be used for geothermometry.

REE patterns of metamorphic monazite (Figs. 3b,e) have negative slopes, sometimes with a negative Eu anomaly. Xenotime shows a "dog leg" pattern with a positive slope through the MREEs and flat in the HREEs (Figs. 3c,f).

Th and U are found in metamorphic monazites with concentrations that range up to 0.25 atoms per four oxygen formula unit, although most metamorphic monazites have less than 0.05 atoms (Figs. 9a, 10b). U concentrations range up to approximately 0.01 atoms per formula unit, but most are less than 0.005 atoms (Fig. 9b). Xenotime Th and U concentrations are much more restricted with the maximum being approximately 0.02 and 0.01 atoms per formula unit, respectively (Figs. 9c,d; 10d).

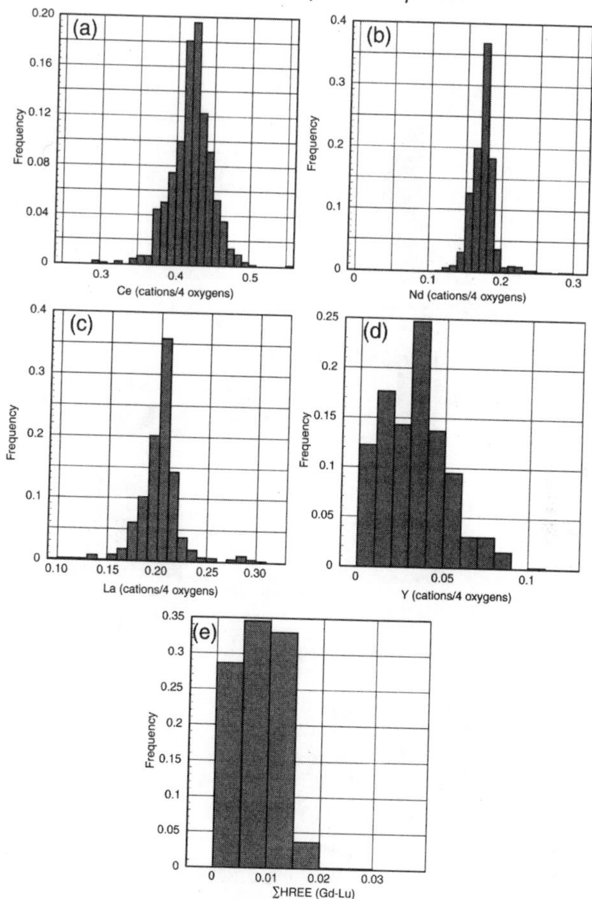

Figure 6. Histograms of monazite compositions (cations/4 oxygens). (a) Ce, (b) Nd, (c) La, (d) Y, (e) ∑HREEs (Gd-Lu). Note the restricted range of composition displayed for LREEs. Data sources as in Figure 3.

Pb is present in metamorphic monazite and xenotime, but it is believed that nearly all of the Pb is radiogenic (e.g., Parrish 1990). The amount of Pb mainly depends on the initial concentrations of Th and U and the age of the sample, and values greater than 1 wt % have been measured (see Harrison et al. this volume).

Exchange vectors in monazite and xenotime. Trivalent Y substitutes for REEs simply: Y = REE. Th and U both have charges of +4 and require coupling to achieve charge balance. Two substitutions have been proposed:

(1) Th or U + Si = REE + P (huttonite in monazite; thorite in xenotime)
with Si replacing P in the tetrahedral site and

(2) Th or U + Ca = 2 REE (brabantite)
with Ca replacing an additional REE in the 8-fold site.

Figure 11 shows plots of Th + U, Ca, and Si for monazite to illustrate the exchange vectors. Figure 11a shows an excellent correlation between Th + U versus Si + Ca, indicating that nearly all of the Th + U can be accommodated by the two substitution

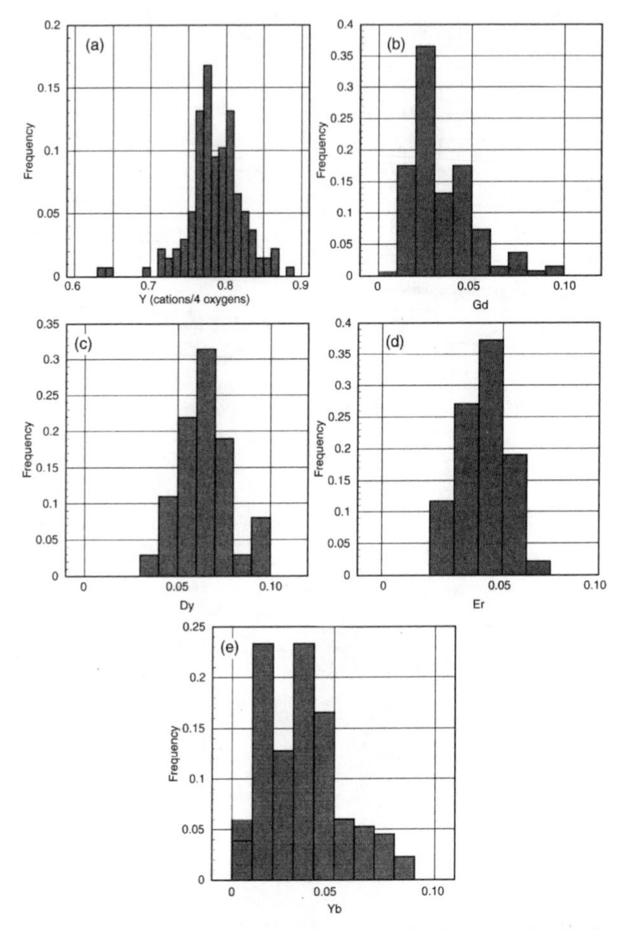

Figure 7. Histograms of xenotime compositions (cations/4 oxygens). (a) Y, (b) Gd, (c) Dy, (d) Er, (e) Yb. Most metamorphic xenotimes are 75-80 mole % YPO_4 with limited ranges of HREEs. Data sources as in Figure 3.

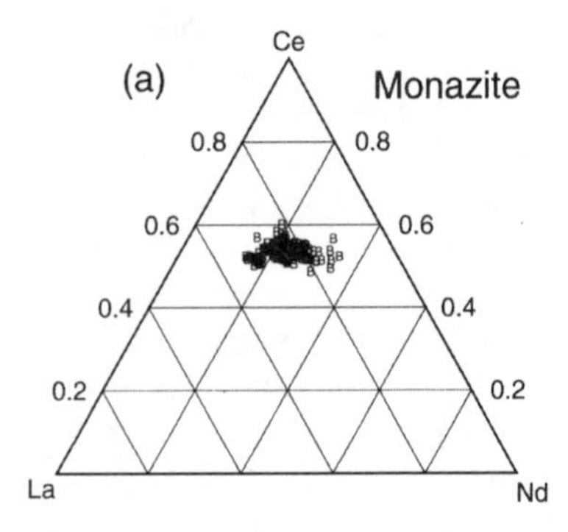

Figure 8. Ternary plots of (a) monazite and (b,c – next page) xenotime compositions. Note restricted range of monazite compositions, plus the nearly constant Ce values. Note the strong correlation in xenotime (b) between Dy and Yb. Data sources as in Figure 3.

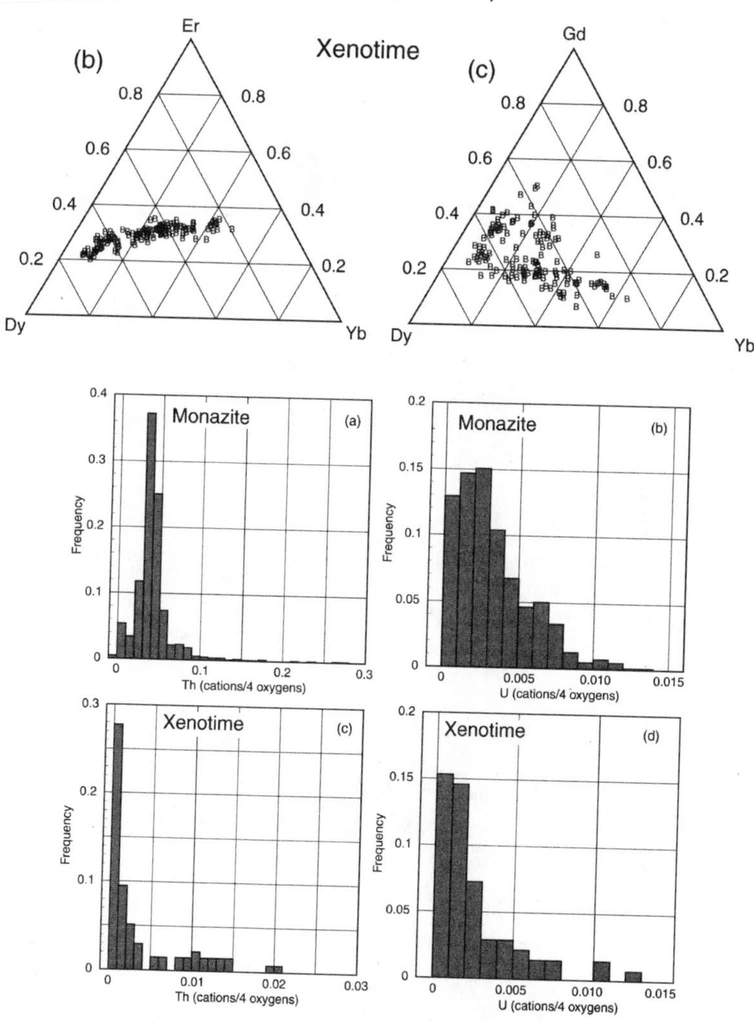

Figure 9. Histograms of Th and U concentrations in metamorphic (a,b) monazite and (c,d) xenotime. (a) Metamorphic monazites have average Th concentrations of 0.03-0.05 cations/4 oxygens, although maximum values are over 0.25 atoms/4 oxygens. (b) U values in monazite are typically less than 0.005 cations/4 oxygens. Xenotime Th and U values are generally smaller than those in monazite. Note the difference in Th scales between monazite (a) and xenotime (c). Data sources as in Figure 3.

mechanisms. Figures 11b and c show that the brabantite substitution dominates in monazite. This can be seen even more clearly in Figures 11d and 11e. Figure 11d shows the residual Th + U after the huttonite component is subtracted, and the strong correlation with Ca is evident. In Figure 11e the contribution of the brabantite component has been subtracted from Th + U and it can be seen that there is very little residual Th + U that needs to be accommodated. A trend in a subset of the data, however, indicates that in some samples the huttonite substitution may be important.

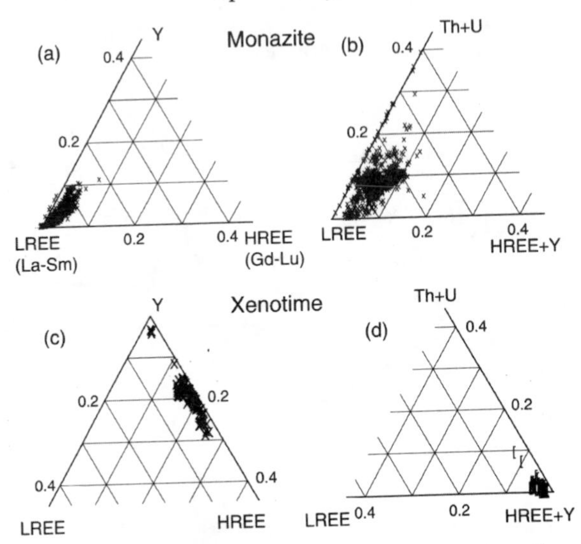

Figure 10. Ternary plots of metamorphic monazite and xenotime analyses from the literature. (a) LREE-HREE-Y in monazite; (b) LREE-Th+U-HREE+Y in monazite; (c) LREE-HREE-Y in xenotime; (d) LREE-Th+U-HREE+Y in xenotime. Data sources as in Figure 3.

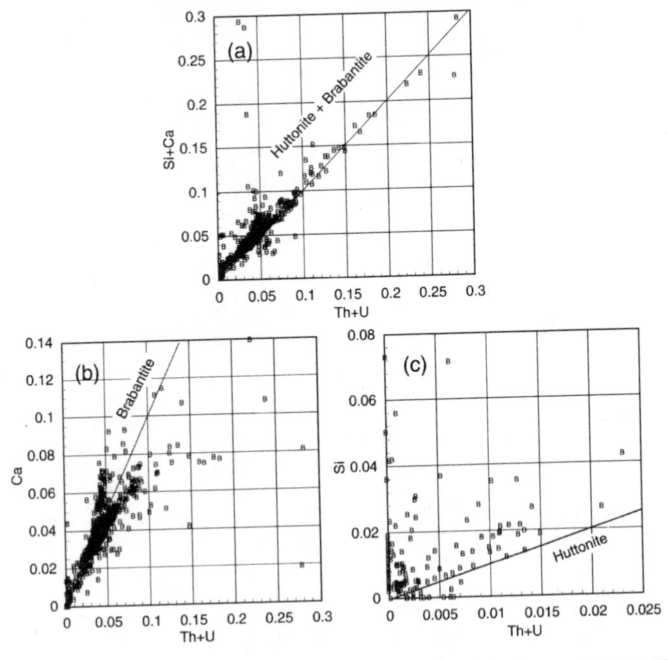

Figure 11. Plots of monazite composition illustrating chemical exchanges. (a) Th + U versus Si + Ca. The strong linear trend suggests that most Th + U is accommodated by a combination of the brabantite ($(Th,U)Ca(PO_4)_2$) and huttonite ($(Th,U)SiO_4$) exchanges. (b) Th + U versus Ca. (c) Th + U versus Si. The strong correlation of Th + U with Ca in (b) suggests the brabantite exchange dominates in metamorphic monazite. (d) Th + U – Si. (e) Th + U – Ca. These "residual" plots show the amount of Th + U that is charge compensated by either Ca (d) or Si (e). Data sources as in Figure 3.

Figure 11, continued

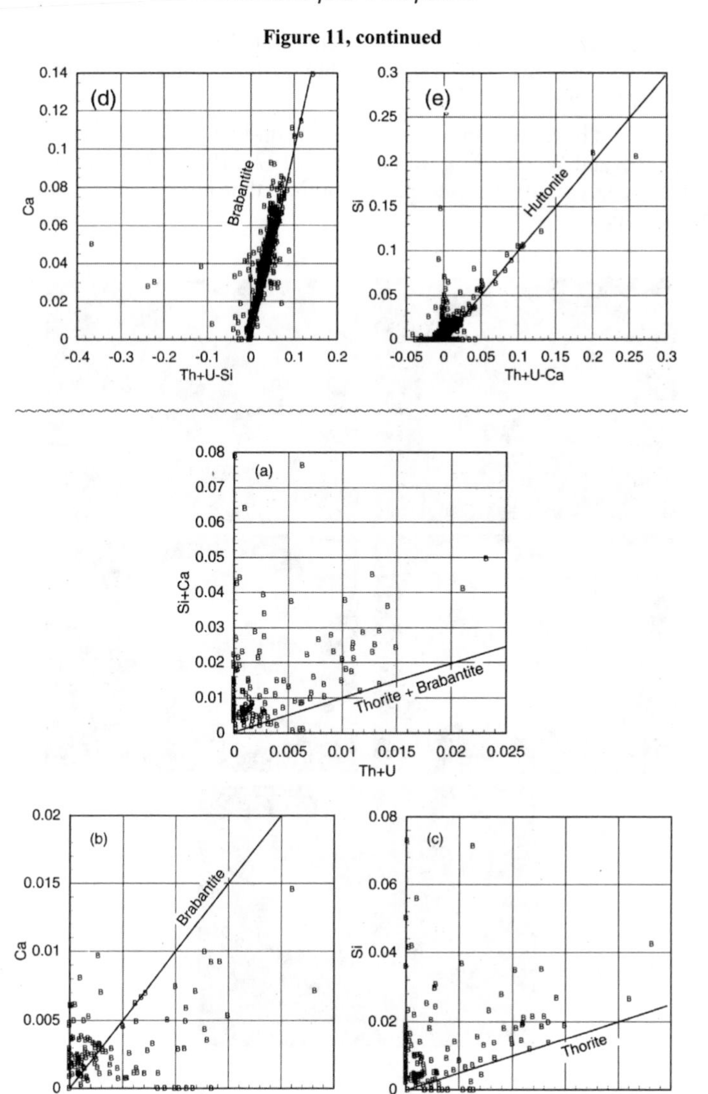

Figure 12. Plots of xenotime composition illustrating chemical exchanges. (a) Th + U versus Si + Ca. (b) Th + U versus Ca. (c) Th + U versus Si. The lack of a strong correlation of Th + U with Si, Ca or Si + Ca suggests that additional elements must be considered in the charge compensation of metamorphic xenotime. The most likely candidate is Zr component in xenotime, which would be compensated by Si in a "zircon" molecule ($ZrSiO_4$). Unfortunately, few analyses of metamorphic xenotime include analyses for Zr. Data sources as in Figure 3.

Plots of xenotime compositions show much less obvious correlations (Fig. 12). Figure 12a reveals that there is an excess of Si + Ca over what is required for incorporation of Th + U. The plot of Ca against Th + U (Fig. 12b) reveals no obvious correlation, but the plot of Si against Th + U reveals an excess of Si over what is required for charge balance. This suggests that Si is incorporated into metamorphic xenotime by an additional substitution mechanism suspected to be the zircon substitution:

Zr + Si = REE + P (zircon).

Zircon and xenotime have the same structure, and solid solution between these phases is evident (e.g., Görz and White 1970, Romans et al. 1975, Bea 1996). Unfortunately, there are no complete analyses of metamorphic xenotime that include Zr, so this suggestion must remain speculative.

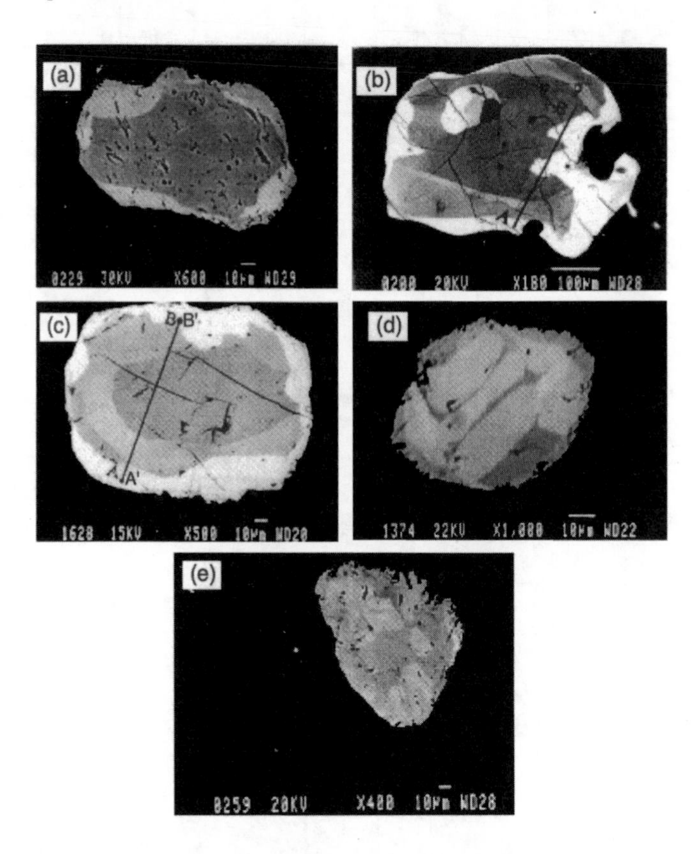

Figure 13. BSE images and chemical zoning profiles of monazite from Zhu and O'Nions (1999a) showing three major types of zoning observed. (a) Simple concentric zoning. (b,c) Complex concentric zoning. (d) intergrowth zoning. (e) Patchy zoning.

Chemical zoning in monazite. A large number of papers have reported chemical zoning in metamorphic monazite (e.g., Parrish 1990, Watt and Harley 1993, Crowley and Ghent 1999, Williams et al. 1999, Zhu and O'Nions 1999a, Hawkins and Bowring 1999, Bea and Montero 1999, Ayers et al. 1999, Townsend et al. 2000, Pyle et al. 2001). Zoning in monazite is often readily apparent in back-scattered electron (BSE) images. Based on a survey of monazites from two samples of high grade gneisses from the Lewisian terrain, Scotland, Zhu and O'Nions (1999a) identified three distinct zoning types: concentric, patchy, and "intergrowth-like". Representative examples of each are shown in Figures 13a-e. Concentric zoning was the most common and was observed to be either simple (Fig. 13a) or complex (Figs. 13b,c). Rims were typically higher in Th but the opposite (low-Th rims) was also observed. Line traverses (Figs. 13f,g) across two grains showing concentric zoning (Figs. 13b,c) reveal distinct chemical domains with sharp boundaries. Furthermore, Th and U concentrations are positively correlated with Ca and Si and are negatively correlated with the LREEs (La, Ce, Sm, Nd), supporting the brabantite and huttonite substitution mechanisms discussed above. Pb is also positively correlated with Th and U. Most important are the sharp boundaries between the chemical domains, which suggests that post-crystallization diffusional modification of the zoning is below the spatial resolution of the electron microprobe (~2 μm).

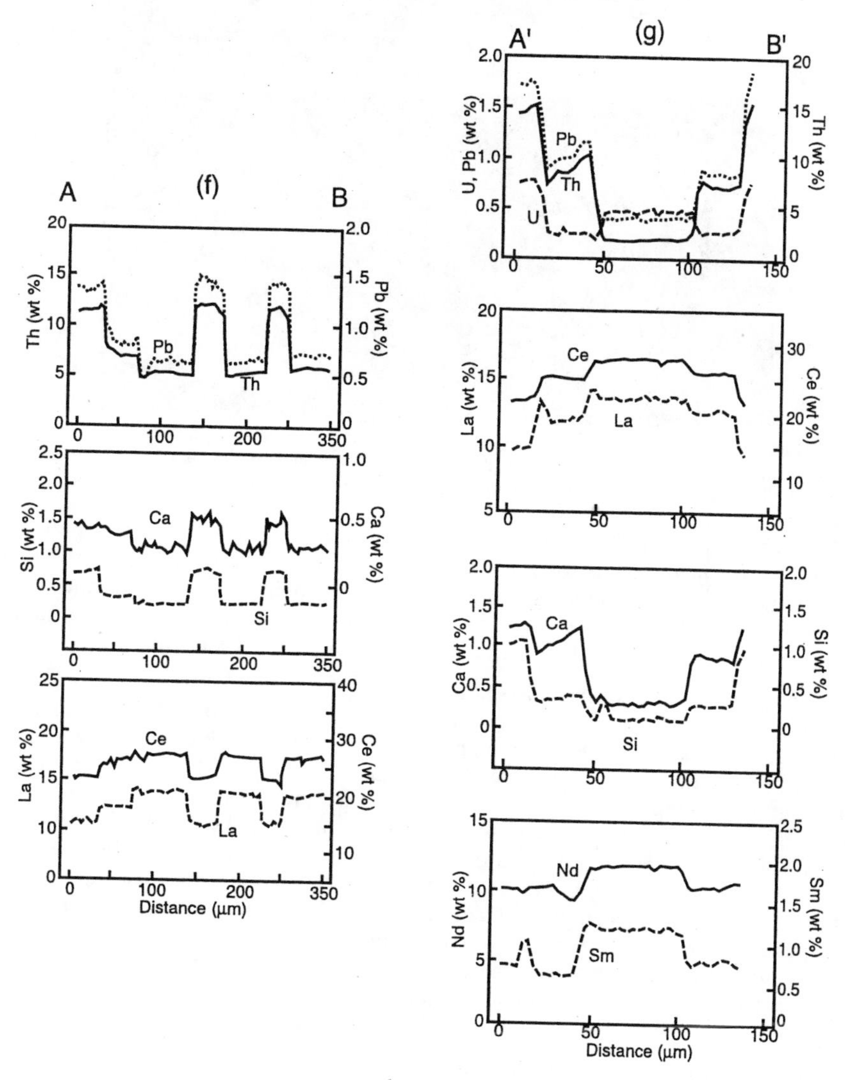

Figure 13, continued (f) Composition profiles of the monazite in (b) along the line A-B. (g) Composition profiles of the monazite in (c) along the line A′-B′. Note the sympathetic zoning of Pb with Th and U and Si + Ca with Th and U. Also note the sharp gradients in all elements indicating very little diffusional relaxation.

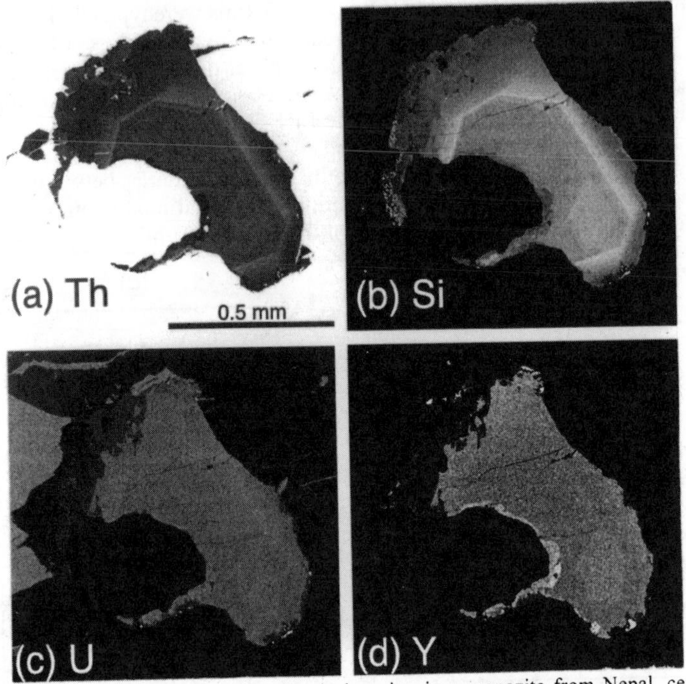

Figure 14. X-ray maps showing chemical zoning in a monazite from Nepal, central Himalaya. (a) Th; (b) Si; (c) U; and (d) Y. Note the sympathetic zoning of Th and Si and, to a lesser degree, U. Y zoning is similar, but not as sharp. (M. Kohn, unpublished data).

The correlation of Th and U with Si is shown dramatically in the X-ray maps of Figure 14. An intermediate zone in this monazite crystal has a euhedral outline with initially high Th, which decreases steadily outward. Si zoning (Fig. 14b) matches the Th zoning perfectly. U is only weakly zoned, but shows a similar pattern. Y zoning is indistinct, and does not show the sharp chemical zones seen in Th and Si.

Zoning in monazite from schists ranging from garnet to migmatite grade from west-central New Hampshire, USA, is shown in Figure 15 (Pyle 2000, Pyle et al. 2001). Typical garnet zone monazite has a high-Th core with little zoning of Y (Fig. 15a,b,c). Sillimanite zone monazites often show low-Th, high-Y overgrowths (Figs. 15d,e,f). The most complex zoning was observed in monazite from the migmatite zone (Figs. 15g-l). These high-grade monazites are characterized by a number of chemically distinct growth zones, which are especially obvious on Y maps (Figs. 15h,k). In one suite of samples from cordierite + garnet migmatites from central New Hampshire, as many as four distinct growth zones in a single monazite grain have been documented (Fig 15k). These high-grade monazites are interpreted to have undergone a series of growth and resorption reactions throughout their history, which can be correlated with reactions involving major phases in the rock. Similarly complex zoning in high-grade monazite was reported by Bea and Montero (1999) and Hawkins and Bowring (1999).

Compositionally distinct material observed in the cores of some migmatite zone monazite crystals (e.g., Figs. 15g,h,i) could readily be interpreted as representing the initially-formed monazite core. However, Pyle et al. (2001) have suggested that this texture, in some instances, may arise from an extremely irregular crystal shape and the orientation of the thin section cut through the crystal (Fig. 16). Monazite at low grades is

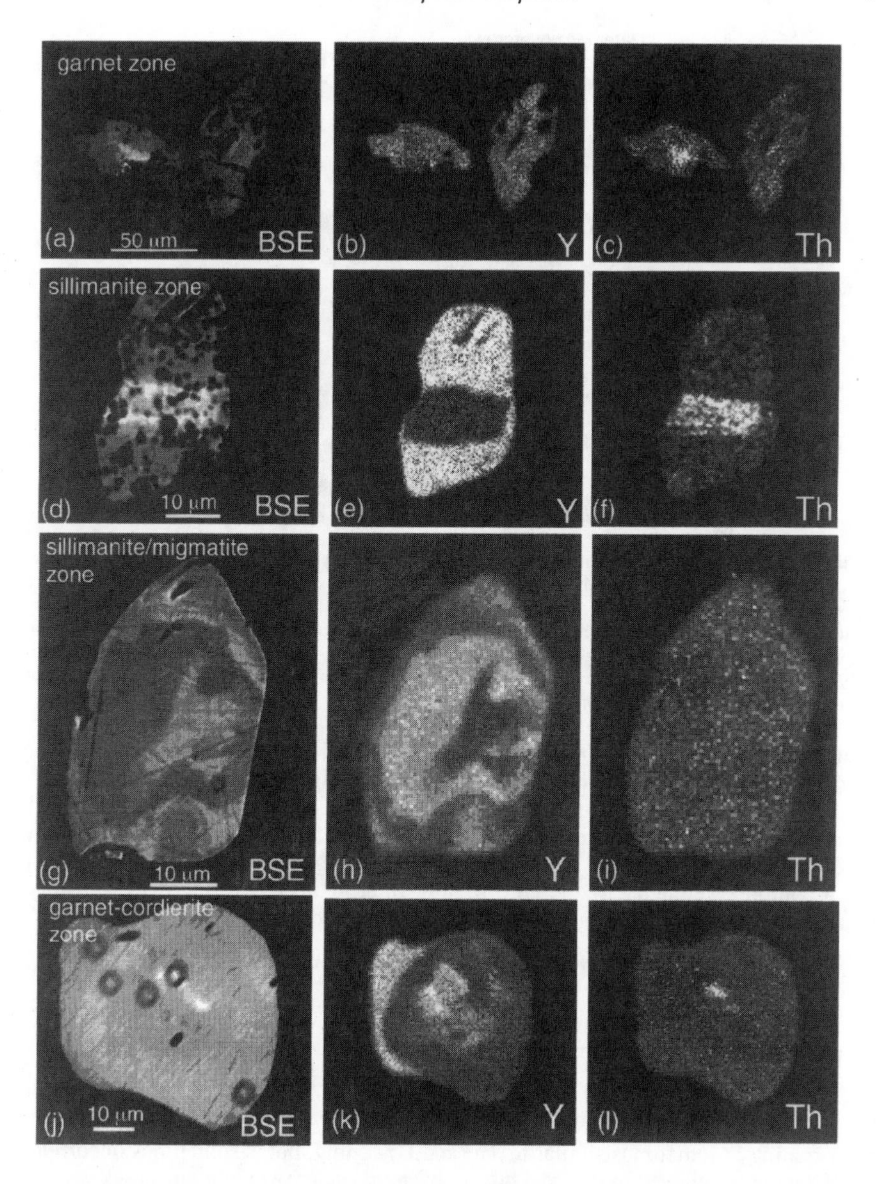

Figure 15. Images showing compositional zoning in metamorphic monazite. (a, d, g, j) BSE images; (b, e, h, k) Y-distribution maps; (c, f, i, l) Th-distribution maps. Brighter areas indicates higher concentration of element. (a-c) Monazite from the garnet zone of eastern Vermont. Y is roughly homogeneous, and Th decreases towards the monazite rim, corresponding with back-scatter zoning. (d-f) Monazite from the sillimanite zone, west-central New Hampshire. Back-scatter brightness corresponds to Th enrichment, and Y zoning is antithetic to Th zoning. (g-i) Monazite from the transitional sillimanite/migmatite zone, west-central New Hampshire. Th is nearly homogeneous, but Y is complexly zoned (corresponding to back-scatter variation), with at least four distinct compositional zones visible in the map. (j-l) Monazite from cordierite + garnet migmatites, central New Hampshire. High Y rim on left of grain is monazite crystallized from melt. Three zones of intermediate Y concentration were produced during prograde metamorphism. The high Th core may be a detrital relict. From Pyle et al. (2001) and Pyle (2001).

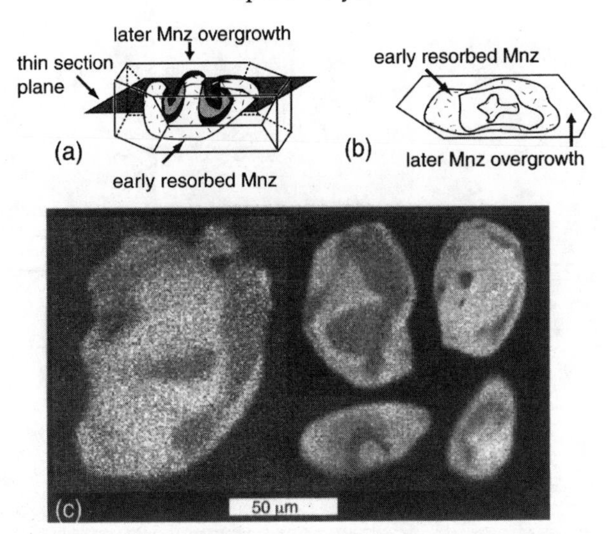

Figure 16. (a,b) Cartoon depicting complex monazite growth geometry (a) and resulting 2-D image from arbitrary thin section cut through the grain (b). Only two monazite compositional zones are present in (a), but the zoning observed in the cut through the complex geometry could be misinterpreted as four growth zones (b). (c) Yttrium X-ray maps of monazites with complex internal zoning (same sample as Fig. 15g-i) showing two to four growth zones, supporting the interpretation that crystals displaying four growth zones are the result of the orientation of the thin section cut through a complex geometry. From Pyle (2001).

often embayed due to either irregularities in growth in different directions, or to resorption due to monazite-consuming reactions. A highly embayed monazite crystal (Fig. 16a) that experiences later growth that fills in the embayments, would, after cutting in certain orientations, have the appearance of an internal core with a composition similar to the rim. Thus, what appears at first to be four distinct growth zones is, in fact, only two with a complex geometry of the core region. Interpretations of the growth history of monazite in complexly zoned grains must take into consideration this possibility.

Several generalizations can be made based on zoning observed in BSE and X-ray images of monazite:

- Th and U zoning are correlated with Ca, Si, and LREE zoning, consistent with the proposed exchange mechanisms;
- Th and U are often zoned, but their zoning is not always correlated;
- Pb is correlated with Th and U concentrations and with age;
- Y zoning is sometimes similar to Th (and U) zoning, but is sometimes uncorrelated. Y zoning is believed to record a history of monazite growth in many cases.
- Compositional heterogeneity observed in BSE images is sometimes due to zoning in Th and/or U and sometimes due to zoning in Y. Grains must be tested individually by spot analyses or X-ray mapping to determine the cause of BSE zoning.
- Very sharp boundaries exist between chemically distinct zones. This is especially true of Th, U, Si, Ca and Pb zoning. Y zoning is also typically sharp, but sometimes not as distinct as the others. That is, there is no evidence for diffusional homogenization (except possibly for Y), despite the high metamorphic grade of some samples.

This last observation has important implications with respect to monazite geochronology (e.g., Harrison et al., this volume). The available evidence from natural monazites, including those from the granulite facies, suggests that diffusion has not modified the chemical zoning, including that of Pb, at a scale of >~2 μm. This observation implies that the closure temperature of monazite for Th-U-Pb geochronology is above 800°C, consistent with the recent diffusion study of Cherniak et al. (2000).

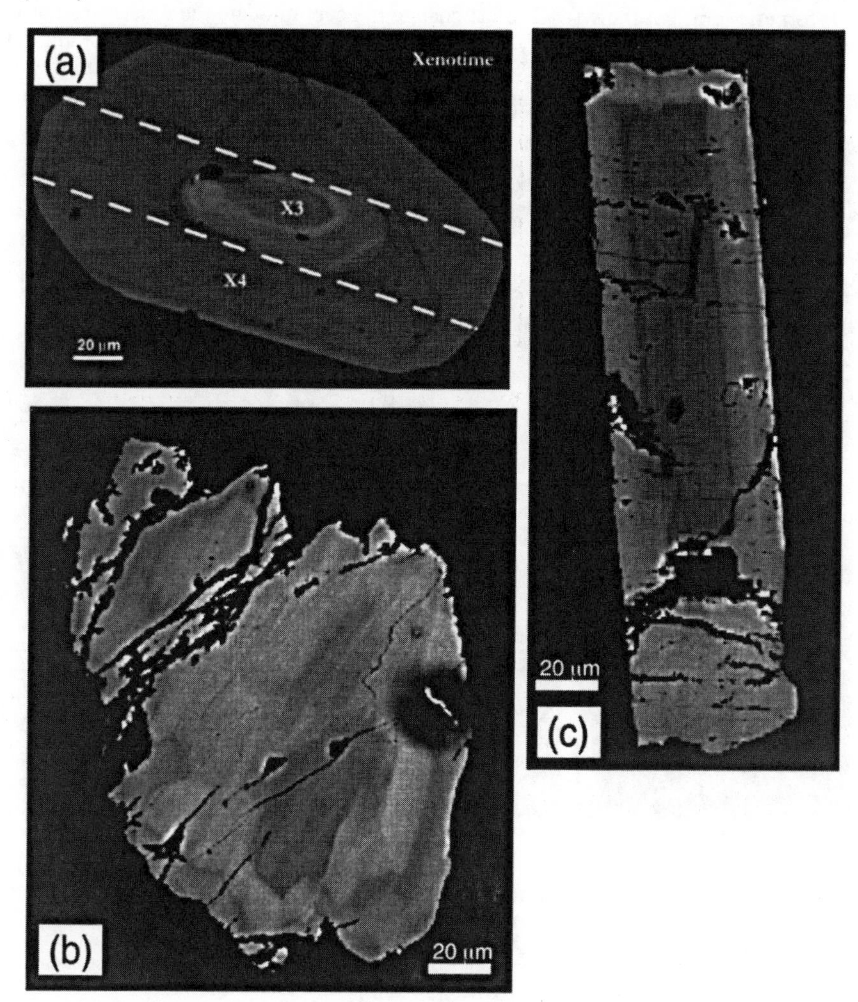

Figure 17. BSE images of xenotime showing complex internal geometries. (a) Xenotime from Nepal, Himalaya (from Viskupic and Hodges 2001). (b,c) Xenotimes from kyanite metaquartzite, New Mexico (from C. Daniel and J. Pyle unpublished data).

Chemical zoning in xenotime. Chemical zoning in xenotime is not easy to observe because most metamorphic crystals are smaller than 20 μm. An exception is a 300 μm xenotime described by Viskupic and Hodges (2001), which is zoned in BSE imaging (Fig. 17a). Although no X-ray maps are available for this xenotime, spot analyses reveal that the BSE contrast is due to variations in Y (and presumably HREE) content. Figures

17b and 17c are BSE images of two xenotime crystals in kyanite-bearing metaquartzite from New Mexico (C. Daniel and J. Pyle unpublished data). X-ray maps of one of these (Fig. 17b) reveal chemical zoning similar to the observed BSE contrast (Fig. 18). Zoning in P, Y, Yb and Dy mimic the BSE contrast. The HREEs Yb and Dy are antithetic to Y, and the zoning in P reflects the change in average atomic weight of the crystal due to the substitution of HREEs for Y. That is, there is no evidence that this crystal has less than 1.0 P/formula unit, but the wt % P must change to accommodate the other substitutions. Si and U are zoned sympathetically, but do not correlate as a whole with variation in the other elements. The correlation of Si and U supports the substitution mechanism Si + U = P + REE; the elements Th, Zr, Ca, and Yb were mapped but show no visible variation.

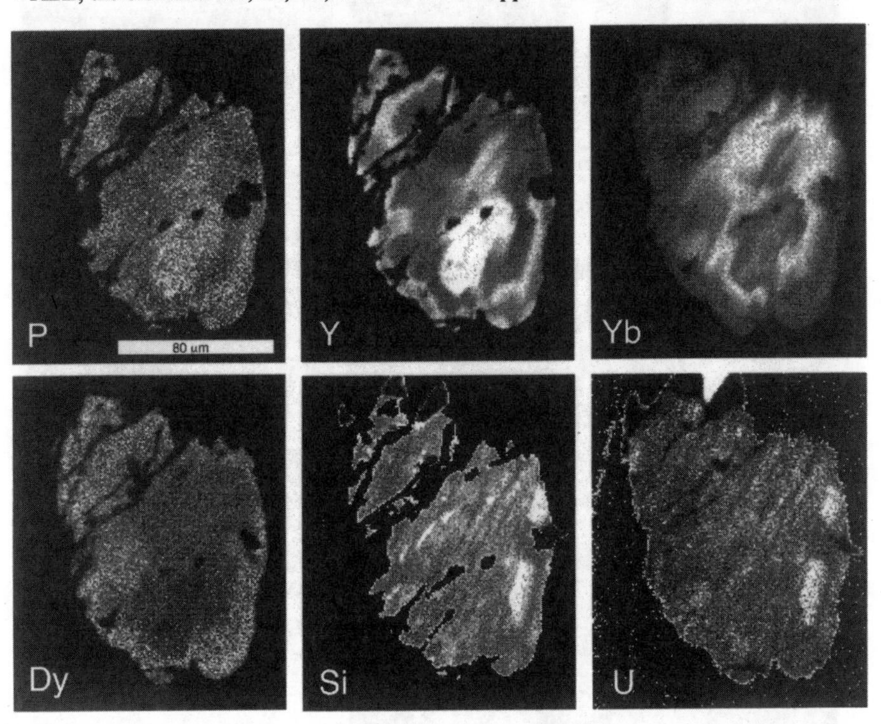

Figure 18. X-ray composition maps of xenotime (Fig. 17b) showing complex internal zoning. Y and Yb are antithetic, and P zoning reflects change in average atomic weight sympathetic to Y, Yb and Dy substitutions. Zoning of Si and U are sympathetic, indicative of U + Si = REE + P substitution and are apparently decoupled from Y, Yb and Dy zoning. Th, Zr, Pb and Ca do not display observable zoning in this grain. Comparison with Figure 17b reveals that BSE contrast is due to variations in Y+HREEs. (From C. Daniel and J. Pyle, unpublished data).

Age zoning in monazite. Numerous studies have reported age zoning in monazite, based on SIMS analysis, Thermal Ionization Mass Spectrometry (TIMS) analysis on parts of individual monazite grains, and X-ray mapping of Th, U, and Pb, coupled with EMP spot age determinations (e.g., Parrish 1990, Williams et al. 1999, Crowley and Ghent 1999, Terry et al. 2000, Viskupic and Hodges 2001, Zhu and O'Nions 1999a). Geochronology of phosphate minerals is addressed in detail elsewhere in this volume, but examples that show the texture of the age zoning based on EMP analysis and X-ray mapping are discussed here to illustrate the relationship between compositional zoning and age zoning in monazite (e.g., Williams et al. 1999).

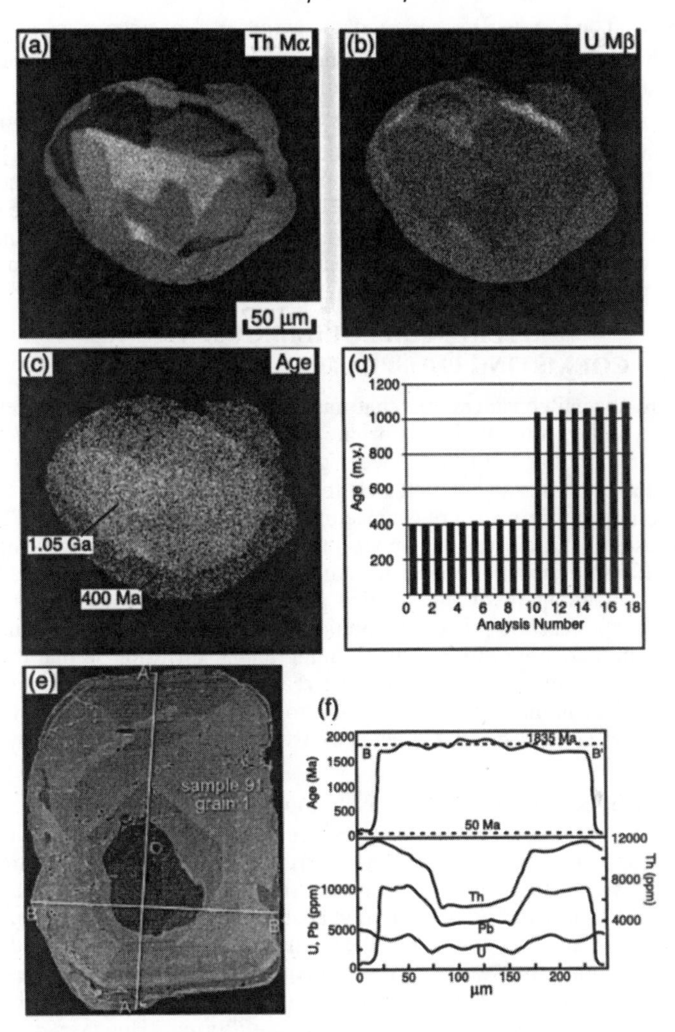

Figure 19. Age zoning in monazite. (a-d) Monazite from Western Gneiss Region, Norway, shows complex internal zoning in Th (a) and U (b), but only two age domains (c and d). Age traverse in (d) taken from core (right) to rim (left). From Williams et al. (1999). (e) BSE image and (f) chemical and age zoning in monazite from the Monashee complex, British Columbia, Canada. Concentrically zoned core (e) of monazite shows variation in Th and Pb, but little age variation. Rim, which displays a composition similar to the near-rim, is considerably younger than the bulk of the grain. From Crowley and Ghent (1999).

Figure 19 illustrates two examples of age zoning in monazite. The grain illustrated in Figures 19a-c is from a sample of ultra high-pressure gneiss from the Western Gneiss Region, Norway (Williams et al. 1999, Terry et al. 2000). The core of the grain displays irregular, patchy zoning in Th and U with three distinct concentrations of Th separated by sharp boundaries. The rim is a fourth chemical zone. The age map shows two distinct age zones, the core at approximately 1.05 Ga and the rim at 400 Ma (Figs. 19c,d). Particularly revealing is the observation that, despite the complex Th zoning in the core of the monazite, the calculated ages are all relatively similar (i.e., 1.05 Ga). That is, the zoning present in the core must have originated during a Grenville metamorphic event and survived ultra high-pressure metamorphism during the Caledonian.

The second example (Figs. 19e,f), from Crowley and Ghent (1999), shows a monazite from the Monashee complex, southern Canadian Cordillera, with a low-Th core surrounded by a high-Th rim. Pb is low in the core, increases in the near-rim and plummets at the rim. The age plot (Fig. 19f) reveals a relatively constant age of 1935 Ma from core to rim with the age plummeting to ~50 Ma at the rim. This is impossible to predict from the BSE image (Fig. 19e), and it is worth noting that there is no evidence in these samples for modification of the ages due to diffusion of Th, U, or Pb.

CRITERIA FOR EQUILIBRIUM AMONG COEXISTING PHOSPHATES AND OTHER PHASES

There are special considerations that must be borne in mind when interpreting the compositions and zoning in metamorphic phosphates and certain other minerals. Petrologists are familiar with the concept of a refractory phase in rocks undergoing melting: refractory phases do not melt and are left in the residue or "restite." In solid state reactions, "refractory" implies a reluctance to react, which carries a kinetic connotation. For the purposes of this discussion, we wish to present another definition that focuses on the ability of a solid solution phase to maintain a homogeneous composition during reaction. Most reactions involving solution phases result in changes in composition of those phases as the reaction proceeds. Solution phases that possess rapid diffusivity can change their internal composition through intragranular diffusion so that the interior of the phase has the same composition as the reacting rim. Solution phases with slow diffusivities do not homogenize as reaction proceeds, leading to zoned crystals if the phase is a product in the reaction. Garnet is an excellent example of this phenomenon in a metamorphic rock.

Accessory phosphates possess relatively low diffusivities for many elements under metamorphic conditions, based on the presence of nearly ubiquitous chemical zoning (Suzuki et al. 1994) and corroborated by experimental determinations of trace element (REEs, U, Pb, Th) diffusivities in phosphates (e.g., Smith and Giletti 1997, Cherniak 2001). Chemical zoning has been reported in apatite (Fig. 4), monazite (Figs. 13-16), and xenotime (Figs. 17 and 18), and unzoned crystals are the exception rather than the rule. Special considerations apply when evaluating equilibrium between two phases with low intracrystalline diffusivities. If both phases are growing during reaction, then the outer shell of both phases can be considered to be in equilibrium. However, if one phase is growing at the expense of the other, then it is likely that no measurable part of either phase is in equilibrium with the other. Even if intragranular transport is rapid such that local equilibrium between the outer, infinitesimal rim of each phase is in equilibrium, the interiors of the phases will not be. Considering that the electron microprobe is only capable of measuring to within approximately 1-2 μm of the rim of a mineral, the equilibrated parts may not be measurable.

Figure 20 shows four possibilities for reaction between two phases with slow diffusivities. Of the four reaction histories shown, the only combination that yields coexisting equilibrium compositions is the one in which both phases grow (Fig. 20a).

The point of this discussion is to emphasize that the mere presence of a mineral such as monazite, xenotime or apatite does not ensure that it is part of an equilibrium assemblage. For example, the presence of an included phase such as monazite in a garnet porphyroblast does not assure two-phase equilibrium between these phases. Furthermore, the complexity of zoning observed in monazite, xenotime and apatite suggest complex reaction histories with several episodes of growth and resorption. It is not a simple task to demonstrate that a specific growth zone within one phase (e.g., monazite) could have formed in equilibrium with any other phase. This point is especially significant when one attempts to apply monazite-xenotime, xenotime-garnet, or monazite-garnet thermometry, which requires equilibration between the phases to be valid. Indeed, achieving the goal of obtaining an age and a metamorphic temperature from a single spot on a monazite (e.g., Viskupic and Hodges 2001) requires careful evaluation of equilibrium.

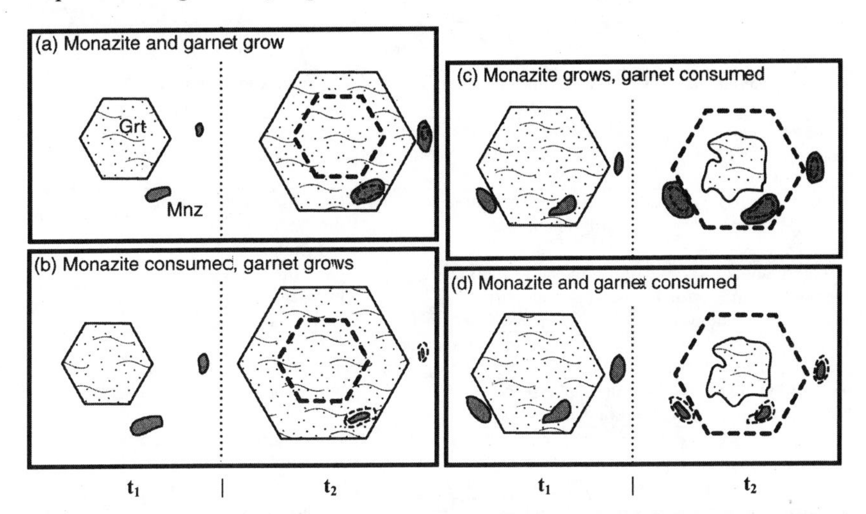

Figure 20. Schematic representation of the four possible combinations of garnet and monazite reaction relationships and the implications of these for finding equilibrated garnet-monazite pairs. At the left in each panel: Solid outlines show grain boundary positions at time 1 (t_1). At the right in each panel: Solid lines show grain positions at time 2 (t_2), and dashed lines show former (t_1) grain boundary positions. Garnet and monazite are assumed to be in equilibrium at t_1, and both are assumed to be refractory (zoned) phases. (a) Garnet and monazite both grow between t_1 and t_2. The grain boundary of included monazite is in equilibrium with some portion of the occluding garnet between the monazite-garnet grain boundary and the garnet-matrix grain boundary. Rim of matrix monazite is in equilibrium with rim of garnet. (b) Garnet grows and monazite is consumed between t_1 and t_2. (c) Garnet is consumed and monazite grows between t_1 and t_2. (d) Garnet and monazite both consumed between t_1 and t_2. In (b), (c), and (d), no existing portion of garnet is in equilibrium with monazite. From Pyle et al. (2001).

Several approaches may be utilized to assess equilibrium, including textural analysis, examination of element partitioning, evaluation of chemical zoning, and thermodynamic modeling. Pyle et al. (2001) addressed the issue of textural equilibrium between monazite and xenotime and between monazite and garnet and proposed the criteria listed in Tables 1 and 2, ranked in order of confidence level. Because of the limited diffusivities in these three minerals, however, textural criteria alone are not particularly robust indicators of chemical equilibrium, and need to be combined with other approaches, as discussed below.

ELEMENT PARTITIONING AMONG COEXISTING PHOSPHATES AND OTHER PHASES

The study of partitioning of elements between and among metamorphic phases has two useful objectives. First, it ascertains whether the compositions are consistent with the phases having achieved chemical equilibrium. Second, the partitioning may be sensitive to temperature and/or pressure, and thus be useful for thermobarometry.

Several studies have reported on the partitioning of various elements between coexisting metamorphic phosphates and between phosphates and other metamorphic phases, and are summarized here.

Table 1. Ranked textural and compositional criteria for assumption of monazite-xenotime compositional equilibrium. From Pyle et al. (2001).

Observed textural and/or compositional criteria	Assumption	Relative rank
Physical contact of monazite and xenotime grains	Both grains in compositional equilibrium	1
Y-homogeneous monazite and xenotime coexist in sample matrix (not necessarily in contact)	All matrix monazite and xenotime in equilibrium	2
Y-homogenous monazite and xenotime included in same garnet, with no garnet yttrium discontinuities between the two inclusions	Included monazite and xenotime in equilibrium	3
Discontinuously Y-zoned monazite and xenotime coexist in sample matrix (not necessarily in contact)	Only high-Y portion of monazite in equilibrium with xenotime	4
Discontinuously Y-zoned monazite and xenotime included in same garnet, with no garnet yttrium discontinuities between the two inclusions	Only high-Y portion of monazite in equilibrium with xenotime	5
Discontinuously Y-zoned monazite included in specific garnet, and xenotime included in different grain of garnet, and both inclusions present in same type of "compositional domain" within garnets	Only high-Y portion of monazite in equilibrium with xenotime	6
Discontinuously Y-zoned monazite included in garnet, and xenotime present in matrix	Only high-Y portion of monazite in equilibrium with xenotime	7
Monazite and some portion of occluding garnet are approx. homogeneous in Y, and monazite Y content is comparable to that of monazite in xenotime-bearing samples of the same metamorphic grade	Monazite grew with xenotime present in the mineral assemblage (now absent?)	8

Table 2. Ranked textural and compositional criteria for assumption of monazite-garnet compositional equilibrium. From Pyle et al. (2001)

Observed textural and/or compositional criteria	Assumption	Relative rank
Inclusion of homogeneous monazite in homogeneous or continuously Y-zoned garnet	Both grains in (compositional) equilibrium	1
Discontinuously Y-zoned monazite included in homogeneous or continuously Y-zoned garnet	Garnet in equilibrium with outer portion of monazite	2
Matrix monazite and garnet both homogeneous or continuously Y-zoned	Both grains in equilibrium	3
Garnet homogeneous or continuously Y-zoned; matrix monazite discontinuously Y-zoned	Garnet in equilibrium with outer portion of monazite	4
Garnet discontinuously Y-zoned, included monazite discontinuously Y-zoned	None: knowledge of reaction relationship needed	--
Garnet discontinuously Y-zoned; matrix monazite discontinuously Y-zoned	None: knowledge of reaction relationship needed	--
Textural evidence for monazite-garnet reaction relationship	Monazite and portion of garnet involved in reaction in equilibrium if reaction produces both phases	Used in conjunction with all of above

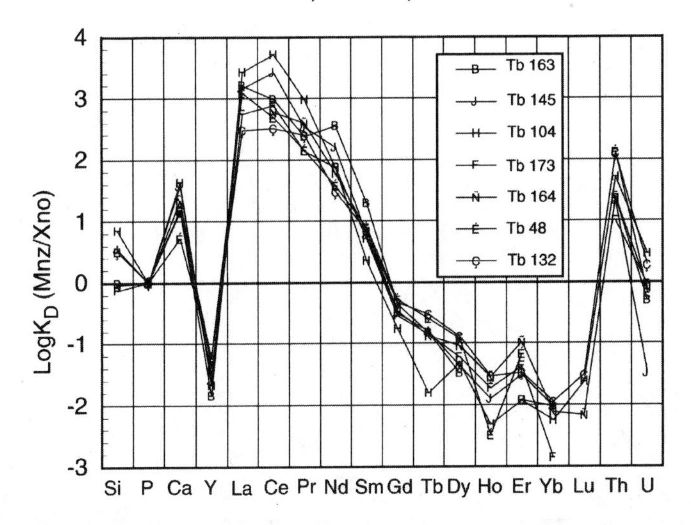

Figure 21. Plot of log(K_D)(monazite/xenotime) for coexisting monazite-xenotime pairs from NE Bavaria, Germany. Modified from Franz et al. (1996).

Monazite–xenotime partitioning

Franz et al. (1996) present a plot of element distributions between monazite and xenotime formulated as $K_D = (El_{Mnz}/El_{Xno})$, which is reproduced here with modification in Figure 21. Monazite strongly favors the LREEs with log(K_D) values between 2.5 and 3.5, whereas xenotime favors the HREEs with log(K_D) between -0.5 and -2.5. Si is favored slightly in monazite, which also strongly partitions Th with log(K_D) between 1 and 2. U partitioning is variable. In some samples, U is favored in monazite, whereas in others it is favored in xenotime.

Figure 22 shows compositions of coexisting monazite and xenotime from a suite of metapelitic rocks from central New England (Pyle and Spear 1999, Pyle et al. 2001). Criteria used to select pairs of coexisting phosphates are listed in Table 1. In general, there is a regular array of tie lines between monazite and xenotime, and monazite and xenotime with the highest HREE concentrations coexist. The range of Y concentrations displayed by monazite coexisting with xenotime is a function of metamorphic grade, as will be discussed below.

Partitioning of Y, Gd, and Dy between monazite and xenotime as a function of metamorphic grade has been discussed by Pyle et al. (2001) (Fig. 23). In this study, K_{eq} is defined as the equilibrium constant for the two exchange reactions,

$(Gd, Dy)PO_4(Mnz) + YPO_4(Xno) = YPO_4(Mnz) + (Gd,Dy)PO_4(Xno)$, as:

$$K_{eq} = \frac{\left(\dfrac{X_{YPO_4}}{X_{(Gd,Dy)PO_4}}\right)_{Mnz}}{\left(\dfrac{X_{YPO_4}}{X_{(Gd,Dy)PO_4}}\right)_{Xno}}.$$

The variation in K_{eq} across metamorphic grade (Fig. 23) is small. Although there is a small increase in K_{eq} with grade for the Y-Dy equilibrium constant, the variation is within analytical uncertainty.

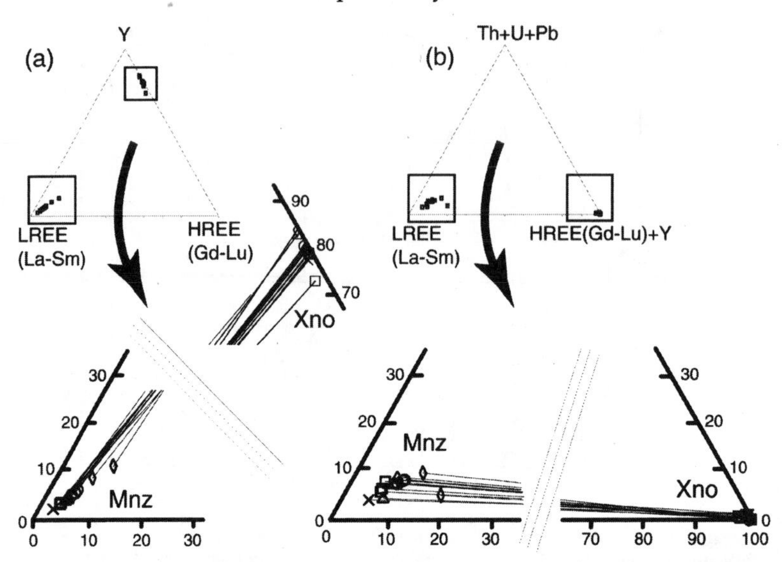

Figure 22. Composition plots of coexisting monazite and xenotime. (a) LREE–HREE–Y ternary showing tie lines between coexisting monazite and xenotime. (b) LREE–HREE+Y–(Th+U+Pb) ternary. × = biotite + chlorite zone, □ = garnet zone, Δ = staurolite zone, ○ = sillimanite zone, diamonds = migmatite zone. From Pyle et al. (2001).

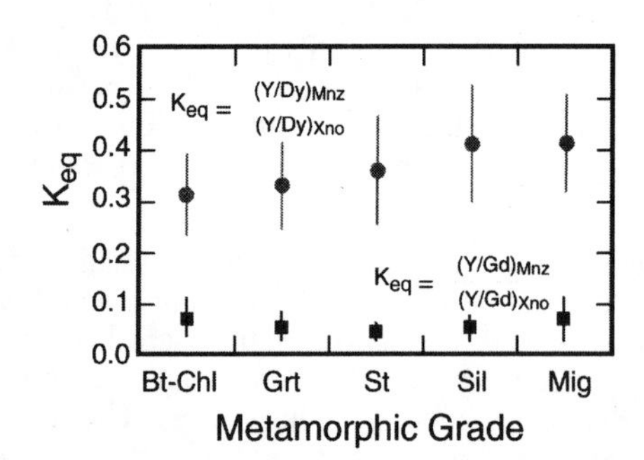

Figure 23. Plots of average $K_{eq} = (Y/(Gd,Dy))_{Mnz}/(Y/(Gd,Dy))_{Xno}$ vs. metamorphic grade. Error bars represent ±1 standard deviation on the average value of K_{eq} for each metamorphic zone. Number of monazite analyses averaged per zone is as follows: Bt-Chl (28); Grt (97), St (120), Sil (123), Mig (88). From Pyle et al. (2001).

Monazite-xenotime miscibility gap. A number of studies have documented the temperature dependence of Y or Y + HREEs in monazite coexisting with xenotime, and suggested its application as a geothermometer (Gratz and Heinrich 1997, 1998; Andrehs and Heinrich 1998, Heinrich et al. 1997, Pyle et al. 2001, Seydoux-Guillaume et al. 2002). An accurate thermometer based on monazite composition would be enormously valuable because it would open the possibility of obtaining a temperature-time point on a single mineral grain, with obvious application to petrogenesis, thermochronology, and tectonics.

Figure 24 reproduces the experimental and empirical data on the monazite-xenotime miscibility gap. The diagrams have been redrafted from the originals to the same scale to

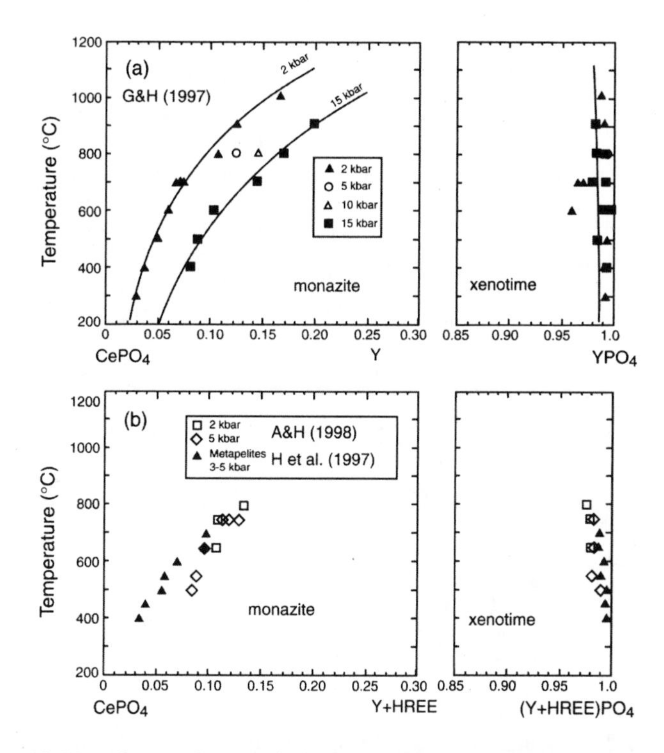

Figure 24. Monazite-xenotime miscibility gap. (a) Experimental data from Gratz and Heinrich (1997). (b) Experimental and natural monazite-xenotime pairs from Heinrich et al. (1997) and Andrehs and Heinrich (1998). Note the x axis here is Y + HREEs.

facilitate comparisons. Gratz and Heinrich (1997; Fig. 24a) reported experiments in the simple $CePO_4$-YPO_4 system. The miscibility gap is modestly pressure sensitive and displays a temperature sensitivity that permits temperature estimates to approximately ±25°C, provided the effects of additional components are properly accounted for. Heinrich et al. (1997; Fig. 24b, solid symbols) and Pyle et al. (2001; Fig. 24c) report empirical miscibility gaps where the temperatures were estimated from phase relations and thermometry in the samples. Both studies show gaps with similar trends, but the Pyle et al. (2001) data show slightly higher Y + HREEs at a given temperature than do the data of Heinrich et al. (1997). Andrehs and Heinrich (1998) conducted experiments in which the partitioning of Y + HREEs was measured between monazite and xenotime of more or less natural compositions (Fig. 24b, open symbols). These data show a narrower gap than do the natural data at temperatures below 600°C, the cause of which, although unknown, is likely to be the presence of additional components such as Ca, U, Si, and Th in the natural monazites. Seydoux-Guillaume et al. (2002; Fig. 24d) examined experimentally the effect of the $ThSiO_4$ substitution on the monazite-xenotime gap. Incorporation of Th into monazite narrows the gap somewhat, and is clearly an important consideration in the application of monazite-xenotime thermometry. Seydoux-Guillaume et al. (2002) also present phase diagrams for the system $CePO_4$–YPO_4–$ThSiO_4$ (Fig. 25) in which they illustrate the effect of the $ThSiO_4$ component in the ternary system and show the limit of $ThSiO_4$ substitution at thorite saturation.

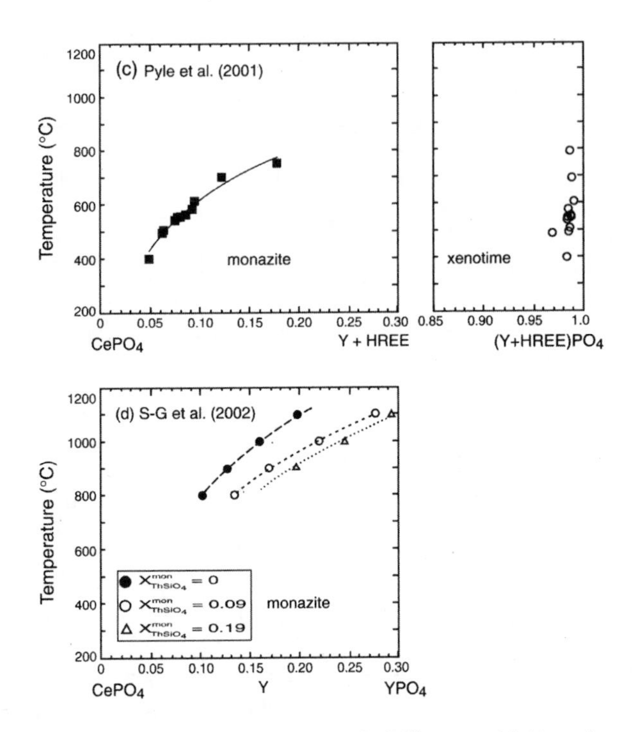

Figure 24, continued. Monazite-xenotime miscibility gap. (c) Natural coexisting monazite and xenotime from Pyle et al. (2001). (d) The effect of Th+Si substitution in monazite on the miscibility gap (Seydoux-Guillaume et al. 2002).

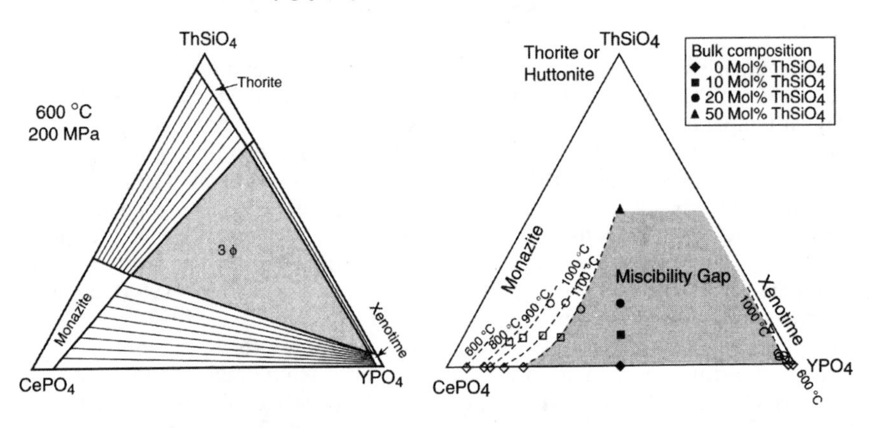

Figure 25. Ternary (CePO4-YPO4-ThSiO4) plots showing the stability fields for monazite, xenotime and thorite as a function of temperature (Seydoux-Guillaume et al. 2002).

Accuracy of the monazite-xenotime thermometer is difficult to assess at this time. It is clear that HREEs and $ThSiO_4$ affect the position of the gap relative to the pure $CePO_4$–YPO_4 system. The brabantite component [$CaTh(PO_4)_2$], which is primarily responsible for Th and U substitution into monazite, may also be significant, but its effect has not been examined. A solution model for monazite that incorporates the major compositional effects is an obvious goal for future research. Equally important, it must be emphasized that unless it can be documented that xenotime was, in fact, in equilibrium with monazite (not always easy to do), the thermometer can only give minimum temperature estimates. Nevertheless, the potential of this thermometer is large, and merits considerable further study.

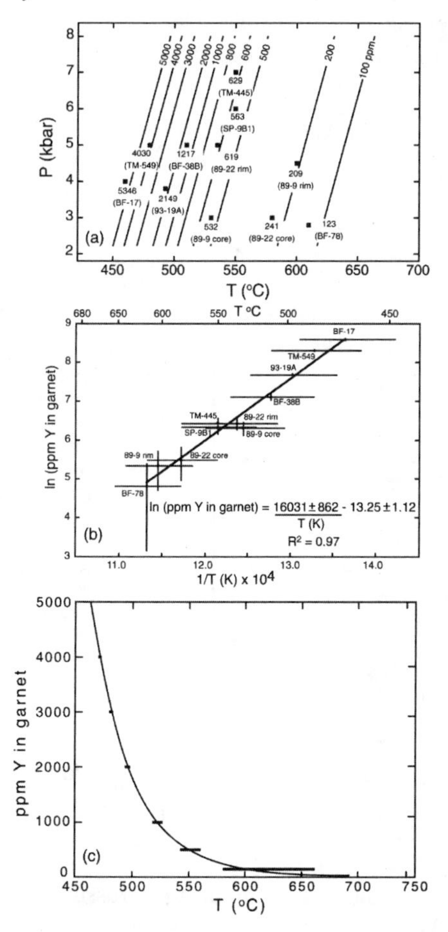

Figure 26. YAG – Xenotime thermometer. (a) P-T plot of eleven garnet yttrium analyses from nine xenotime-bearing pelites showing empirical P-T dependence of garnet YAG composition coexisting with xenotime. Labels indicate concentration of Y in garnet (ppm). Lines are contours of Y concentration in garnet (ppm) drawn by inspection. (b) Plot of ln(ppm Y in garnet) vs. reciprocal absolute temperature for xenotime-bearing pelites. Error bars represent assumed accuracy measurements on temperature estimates from thermobarometry (±30°C) and electron microprobe measurements of garnet composition (± 100 ppm Y). (c) Graphical representation of the garnet-xenotime thermometer as a plot of garnet composition (ppm Y in garnet) vs. calculated temperature, using the regression from (b). Bars show precision of temperature estimate at different Y concentrations assuming $\sigma_{[Y]_{Grt}}$ = ±100 ppm. From Pyle and Spear (2000).

Xenotime/monazite – garnet Y distribution

Yttrium partitions strongly into garnet relative to other silicates. Pyle and Spear (2000) report a strong correlation between metamorphic grade and the Y concentration in garnet coexisting with xenotime (Fig. 26a), and Pyle et al. (2001) describe a similar relationship for garnet coexisting with monazite (Fig. 27). Two semi-empirical thermometers were derived (Figs. 26b and 27) that are based on compositions measured in a well-characterized suite of samples from central New England.A net transfer reaction can be written among garnet, apatite, quartz, plagioclase and fluid that describes the dependence of Y concentration in garnet on grade:

$$\underset{\text{YAG (in garnet)}}{Y_3Al_2Al_3O_{12}} + \underset{\text{OH-apatite}}{Ca_5(PO_4)_3(OH)} + \underset{\text{Qtz}}{25/4\ SiO_2} =$$

$$\underset{\text{Xno or Mnz}}{3\ YPO_4} + \underset{\text{Grs (in garnet)}}{5/4\ Ca_3Al_2Si_3O_{12}} + \underset{\text{An (in plagioclase)}}{5/4\ CaAl_2Si_2O_8} + \underset{\text{fluid}}{1/2\ H_2O}$$

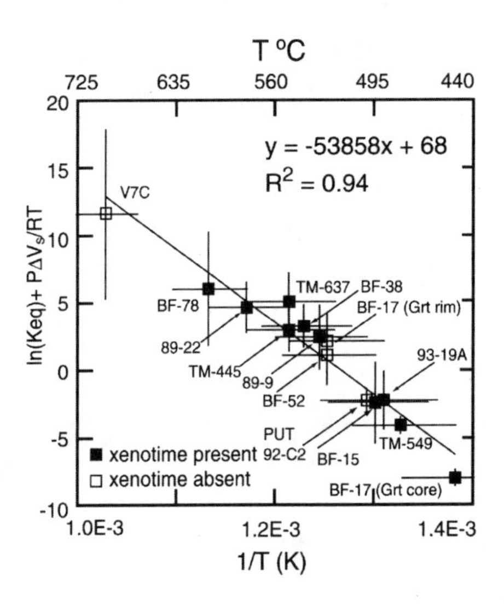

Figure 27. Plot of $\ln(K_{Eq}) + P\Delta V_S/RT$ vs. reciprocal temperature for the reaction YAG + OH-apatite + 25/4 quartz = 5/4 grossular + 5/4 anorthite + 3 YPO_4 monazite + 1/2 H_2O. Solid squares = xenotime-bearing assemblages, open squares = xenotime-absent assemblages. Least squares regression line is fit to all data points. Horizontal error bars represent temperature uncertainty of ±30°C. Vertical error bars are ±1σ ($\ln K_{Eq} + P\Delta V/RT$), derived from propagation of uncertainties in P (± 1000 bars), T (±30°C), ΔV_{rxn} (1%), compositional parameters (0.001 mole fraction YAG, 0.01 mole fraction all others), and $f(H_2O)$ (±7.5; 1000 trial Monte Carlo simulation). Labels on graph indicate sample numbers. From Pyle et al. (2001).

The YAG-xenotime thermometer is based solely on the Y concentration (in ppm) in garnet (Fig. 26b), but the YAG-monazite thermometer incorporates activity models for the components of the above reaction (Fig. 27). The systematics of the YAG-xenotime thermometer imply that the activities of YPO_4 in xenotime, OH-apatite, and H_2O are relatively constant and that the activity product of grossular and anorthite is relatively constant. For the YAG-monazite thermometer, measured compositions of YPO_4 in monazite, grossular in garnet, anorthite in plagioclase, and OH in apatite (via subtraction of F^- and Cl^- components) were used, but a constant activity of H_2O was assumed.

These thermometers are quite sensitive to changes in temperature. In the garnet zone, the YAG-xenotime thermometer has a sensitivity of only a few degrees, based on analytical precision of ±100 ppm (Fig. 26c). At higher grades, the temperature sensitivity is considerably worse, increasing to several tens of degrees by the sillimanite zone, owing to the much smaller Y concentrations of garnet at higher grades (Fig. 26c). The YAG-monazite thermometer has larger uncertainty (±20-30 degrees in the garnet zone) because of the smaller concentration of YPO_4 in monazite as compared with xenotime and the similarly smaller concentrations of YAG in garnet. More sensitive analytical methods to measure YAG in garnet such as LA-ICP-MS would reduce these errors, especially in higher-grade samples where the YAG concentrations in garnet are lowest.

APATITE-FLUID AND APATITE-BIOTITE PARTITIONING

F and Cl are partitioned among apatite, metamorphic fluids, and micas, and a number of studies have examined this partitioning as either a monitor of fluid composition (e.g., Yardley 1985, Sisson 1987, Nijland et al. 1993, Smith and Yardley 1999) or as a geothermometer (e.g., Stormer and Carmichael 1971, Ludington 1978, Zhu and Sverjensky 1992, Sallet 2000).

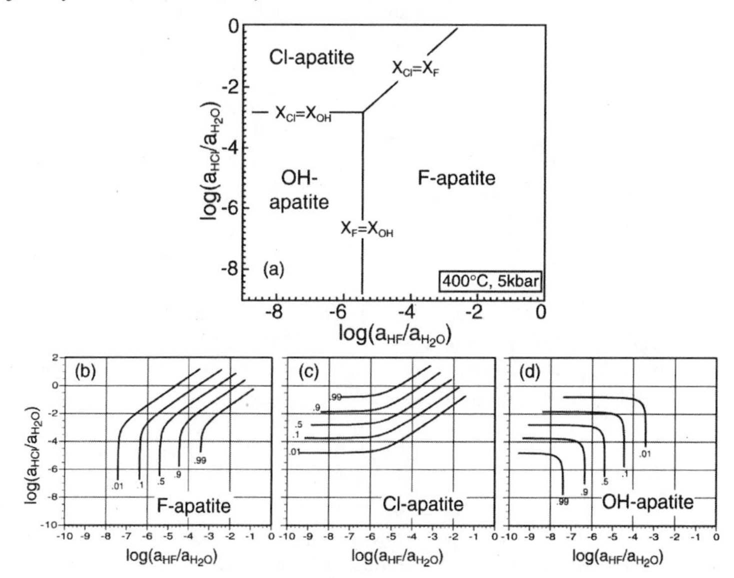

Figure 28. Apatite-fluid partitioning. (a) Plot of logs of activity ratios (HF/H_2O versus HCl/H_2O) in fluid with composition fields of apatite indicated. From Zhu and Sverjensky (1991). (b-d) Isopleths showing mole fractions of apatite components calculated from data in (a) and assuming ideal mixing. Note that high concentrations of F-apatite are possible in fluids considerably more dilute in HF than is the case for Cl-apatite.

F is partitioned into apatite considerably more strongly than is Cl (Fig. 28). Smith and Yardley (1999) calculated a species predominance diagram based on data in Zhu and Sverjensky (1992) that illustrates the magnitude of this partitioning. For example, at the intersection of the three equal molar lines in Figure 28a, the composition of the apatite is $X_F = X_{Cl} = X_{OH} = 1/3$ (assuming ideal mixing). The fluid composition at this point has activity ratios of $a_{HF}/a_{H2O} = 10^{-5.2}$ and $a_{Cl}/a_{H2O} = 10^{-2.8}$. That is, the fluid has over two orders of magnitude more Cl than F, despite their equal concentrations in apatite. This can also be seen in the isopleth diagrams calculated assuming ideal mixing (Figs. 28b,c,d). F-rich apatites (e.g., with F-apatite compositions greater than 0.9 X_F) are in equilibrium with relatively dilute metamorphic fluids ($a_{HF}/a_{H2O} < 10^{-4}$) (Fig. 28b), whereas fluids with $a_{HCl}/a_{H2O} < 10^{-4}$ are in equilibrium with apatite with $X_{Cl} < 0.1$ (Fig. 28c). This strong partitioning accounts for the observation that metamorphic apatite is typically F-rich (i.e., Fig. 1; Smith and Yardley 1999), in contrast to igneous apatites, which are commonly Cl-rich (e.g., Candela 1986). It should be noted, however, that the apatite-fluid partitioning is expected to be a strong function of T and P (Zhu and Sverjensky 1991).

Whether apatite represents the tail or the dog in halogen mass balance during metamorphism most likely depends on the environment. In high fluid/rock environments where fluid composition controls rock composition (e.g., veins, skarns and ore deposits), apatite composition likely reflects the composition of the infiltrating fluid. In typical regional metamorphic environments where dewatering of fluids and progressive drying out of rocks predominates, apatite likely controls the halogen content of the fluid. For example, consider a rock with ≤1% porosity and 1% modal apatite with a composition $X_F = X_{Cl} = X_{OH} = 1/3$. The fluid composition will be approximately $10^{-5.2}$ m HF and $10^{-2.8}$ m HCl (assuming unit activity of H_2O and ideal mixing). Removal of this fluid and generation of additional H_2O by dewatering of hydrous silicates will progressively deplete the rock in Cl relative to F. Resultant apatite and coexisting biotite should become increasingly F-rich with progressive dewatering, as indicated by the set of apatite analyses from Kapustin (1987; Fig. 1b).

Smith and Yardley (1999) have applied this type of analysis to the situation in which a Cl-rich detrital apatite is present (Fig. 4a). Their calculations indicate that exchange of Cl with a limited volume of metamorphic fluid can change the concentration of Cl in the fluid by several orders of magnitude.

The partitioning of OH, F, and Cl between apatite and hydrous silicates has also been shown to be temperature sensitive, and has been used as a geothermometer (e.g., Stormer and Carmichael 1971, Ludington 1978, Zhu and Sverjensky 1992, Sallet and Sabatier 1996, Sallet 2000). The incorporation of F and Cl into biotite is well known to be a function of Fe/Mg (e.g., Munoz 1984), and presumably is also a function of Fe/Mg in apatite (although the total quantities of Fe and Mg in apatite are minor and probably have a negligible affect). Therefore, any thermometer that utilizes the F and Cl content of Fe-Mg silicates must consider this effect. Figure 29 shows a plot from Zhu and Sverjensky (1992) that illustrates the influence of Fe/Mg in biotite on the partitioning of F and Cl between apatite and biotite. Temperature estimates from Figure 29 are quoted as having an uncertainty of 20-40°C (Zhu and Sverjensky 1992). Conversely, it should be noted that F and Cl in biotite will influence Fe/Mg exchange thermometers involving biotite, and must be accommodated in a fashion such as discussed by Zhu and Sverjensky (1992).

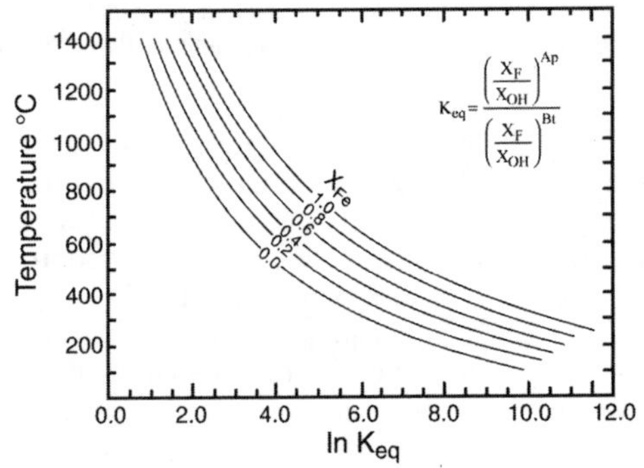

Figure 29. Plot of T versus $\ln K_{eq}$ for coexisting biotite and apatite. Note the strong dependence of K_{eq} on the Fe/(Fe+Mg) of biotite. From Zhu and Sverjensky (1992).

PARAGENESIS OF METAMORPHIC APATITE, MONAZITE AND XENOTIME

There are several mechanisms by which metamorphic phosphates can grow. In the case of monazite this is especially important because recrystallization will reset the Th-U-Pb systematics, and the age measured will be the age of the recrystallization event. Recrystallization can occur by heterogeneous reactions in response to changes in pressure and temperature (prograde or retrograde metamorphism). Recrystallization is also possible in shear zones as a result of grain size reduction, influx of fluids and high strain. Minerals can coarsen by a process such as Ostwald's ripening, possibly enhanced by the presence of a fluid or melt. Growth can occur by intersection with a solvus limb during cooling or by reaching the limits of solubility in a more complex system (e.g., heterogeneous unmixing). Minerals can also be precipitated from a fluid or vapor phase. This section reviews some of the literature pertaining to apatite, monazite, and xenotime recrystallization in metamorphic rocks.

PROGRADE METAMORPHISM OF APATITE, MONAZITE AND XENOTIME

Apatite, monazite, and xenotime are found in rocks of nearly every metamorphic grade and, apparently, have wide P-T stability ranges. Despite their widespread occurrence, surprisingly few studies on the prograde metamorphic evolution of phosphates have been published. Kapustin (1987) reported on apatite compositions as a function of metamorphic grade (e.g., Fig. 1b).

Monazite is known to be a detrital mineral, and some low-to-medium grade occurrences of monazite are detrital relics (e.g., Overstreet 1967, Suzuki and Adachi 1991, Suzuki et al. 1994, Hawkins and Bowring 1999, Ferry 2000, Rubatto et al. 2001, Wing et al. 2002). Authigenic monazite has been reported from several low-grade rocks of various bulk compositions. Cabella et al. (2001) reported coexisting monazite and xenotime from pelitic layers in pumpellyite-actinolite facies metacherts (3 kbar $< P \leq 4$ kbar, 300°C $< T \leq$ 350°C) of the Sestri-Voltaggio zone, Italy. Rasmussen et al. (2001) identified metamorphic monazite in weakly metamorphosed quartz-sericite-chlorite phyllites from the contact aureole of the 1830 Ma Mt. Bundy granite, N. Australia. Overstreet (1967) concluded that, although detrital monazite was not uncommon, monazite is not thermodynamically stable during diagenesis and occurrences of monazite in very low-grade samples were probably detrital relics.

Kingsbury et al. (1993) examined the paragenesis of monazite in a prograde sequence of pelitic schists (unmetamorphosed shale to sillimanite zone) from the eastern Mojave Desert (California, USA). A single detrital grain of monazite was found in the unmetamorphosed shale, along with abundant detrital zircon and apatite and lesser amounts of xenotime and rutile. These authors addressed the issue of stable "sinks" for REEs, Y, Th and U at conditions below monazite appearance. They reported the absence of LREE or Th-bearing phases in unmetamorphosed shales of their study, and suggested that these constituents are located along grain boundaries (adsorbed?) or in clay minerals. In the greenschist facies samples, Kingsbury et al. (1993) reported thorite, thorianite, a possible Ce oxide, and LREE phosphates (not apatite). Monazite was absent from the greenschist facies samples (biotite + chlorite zone) and did not appear until the staurolite zone ($T \approx$ 550°C). Although allanite was not positively identified as a precursor, its former presence was suspected based on pseudomorphic shapes. At higher grade ($T \approx$ 650°C) monazite grains were found to increase in number and to coarsen to their maximum size of ~30 μm diameter.

Hawkins and Bowring (1999) reported on the textural development of monazite, xenotime, and titanite from migmatitic gneisses of the Grand Canyon, Arizona. They describe complexly zoned grains with internal truncations of zoning and suggested that dissolution and redeposition in the presence of a silicate melt was responsible for much of the monazite growth. Additionally, the vestiges of monazite cores that record isotopic inheritance were used to infer that the closure temperature of Th-U-Pb is above the peak temperature experienced by these rocks (725°C).

Rubatto et al. (2001) reported on the prograde metamorphism of monazite and zircon from pelitic schists from the Reynolds Range, central Australia. They identified monazite from greenschist-facies samples, which was interpreted to be detrital. Significant growth of new monazite did not occur until the amphibolite facies (sillimanite zone), although no specific reactions were postulated. At the highest grades (granulite facies: Grt + Crd + Bt + Sil + melt), abundant, large monazite grains were found that contain complex internal zoning, suggesting that solution and reprecipitation in the presence of a melt phase was an important mechanism in monazite growth.

Ferry (2000) and Wing et al. (2002) mapped the monazite isograd in three locations: the contact aureole of the Onawa pluton, Maine, the low-P regional Buchan terrane of south-central Maine, and the medium-P Barrovian terrane of central Vermont. Monazite that was characterized as having patchy chemical zoning and irregular habit was found to occur in the lowest grades (chlorite zone) in each terrane, and was interpreted as being detrital. With increasing grade, this monazite was replaced by allanite, and, very nearly coincident with the Al_2SiO_5 isograd (andalusite in the low-P terranes and kyanite in Vermont), allanite was replaced by monazite. These authors also report that samples with lower than average Al and/or Ca contents contain the low-grade (detrital) monazite rather than allanite to grades as high as the garnet zone, and samples with higher Al and/or Ca contents contain allanite to higher grades than the Al_2SiO_5 isograd.

Pyle and Spear (1999), Pyle et al. (2001), and Pyle (2001) report on the relationship between accessory mineral association (monazite, xenotime, and apatite) and trace element zoning (especially Y and P) in garnet from a suite of pelitic schists of different metamorphic grade from central New England. In these samples, monazite and xenotime are both present in chlorite-biotite zone (garnet-absent) samples, and in the matrix and as inclusions in garnet in samples from the garnet zone. Yttrium contents of garnet from the garnet zone coexisting with xenotime are high (up to several thousand ppm Y), and decrease rapidly with increasing temperature (Pyle and Spear 2000). Xenotime reacts out continuously as garnet grows, and is typically gone from the matrix assemblage in samples from the middle garnet zone. Monazite grows and displays increasing Y content with increasing T in xenotime-bearing samples, consistent with the monazite-xenotime miscibility gap (Fig. 24). When xenotime reacts out of the assemblage (usually by the middle of the garnet zone), monazite and garnet show a dramatic drop in Y concentration with monazite displaying evidence of continued growth along with garnet. When garnet-consuming reactions are encountered, such as garnet + chlorite + muscovite = staurolite + biotite + H_2O or garnet + chlorite + muscovite = Al_2SiO_5 + biotite + H_2O, there is clear textural evidence that monazite is produced at the expense of garnet, such as the presence of monazite crystals or overgrowths on old crystals in the garnet reaction halo. Monazite was also interpreted to have precipitated in and around leucosomes as melt crystallized.

Phosphates in the melting region

Solubility of phosphates in anatectic melts. The solubility of phosphate is greatest in ultramafic and mafic melts and smallest in granitic melts at identical temperatures and pressures (e.g., Watson 1979, Ryerson and Hess 1980). It is also well established that the solubility of apatite in silicate liquids is a strong function of the alumina saturation index (ASI = molar Al/[Na + K + 2Ca]) of the melt (e.g., Bea et al. 1992, Gan and Hess 1992, Pichivant et al. 1992, Wolf and London 1994) because of the formation of the aluminum phosphate complex $AlPO_4$. In metaluminous compositions (ASI = 1), the solubility of apatite is small and dissolved P_2O_5 is less than 0.1 wt % but increases to 0.7 wt % P_2O_5 at ASI = 1.3 (Wolf and London 1994). Therefore, in metapelitic rocks that contain an Al_2SiO_5 polymorph in the anatectic region in which the ASI is high, the solubility of apatite should likewise be high. For example, using the data of Patiño Douce and Johnston (1991), the ASI of sillimanite-saturated pelitic melts ranges from 1.15 to 1.6, which implies a solubility of P_2O_5 of 0.1 to 1.7 wt % based on the data of Wolf and

London (1994). This amount of P_2O_5 is equivalent to the dissolution of 0.3 to over 3.0 modal per cent apatite. If the melt does not leave the rock, the phosphates will precipitate as the melt crystallizes. Apatite solubility is also enhanced in low ASI melts (peralkaline), because of the presence of alkali phosphate complexes (Ellison and Hess 1988).

In contrast to apatite, the solubilities of monazite and xenotime are much more limited in peraluminous melts compared with apatite solubility, with maximum solubilities of <0.05 wt % P_2O_5 (Wolf and London 1995). Indeed, Wolf and London (1995) report precipitation of monazite at sites of apatite dissolution in melting experiments. REE solubilities, in contrast, are higher in mafic compositions relative to granitic compositions (Ryerson and Hess 1978, 1980), and monazite solubility is greater in peralkaline melts than in meta or peraluminous melts (Montel 1986, Ellison and Hess 1988). Rapp and Watson (1986) also report a strong temperature dependence of monazite solubility.

Solubility of phosphates in anatectic melts can be fruitfully analyzed by use of the concept of saturation index (SI) (see also Watt and Harley 1993). The saturation index is defined as the log of the ratio of the ion activity product (IAP) to the solubility constant (K_{sp}), SI = log (IAP/K_{sp}), and is a measure of the degree of oversaturation (positive SI) or undersaturation (negative SI) of a mineral. Dissolution of the phosphates apatite, monazite and xenotime into a silicate liquid may be described by the reactions

$$\underset{\textit{apatite}}{Ca_5(PO_4)_3(F)} = 5\,Ca + 3\,PO_4 + F \qquad IAP_{A_p} = (a_{Ca})^5\,(a_{PO_4})^3\,(a_F)$$

$$\underset{\textit{monazite}}{(LREE)PO_4} = LREE + PO_4 \qquad IAP_{Mnz} = (a_{LREE})(a_{PO_4})$$

$$\underset{\textit{xenotime}}{(Y{+}HREE)PO_4} = Y + HREE + PO_4 \qquad IAP_{Xno} = (a_{Y+HREE})(a_{PO_4})$$

Although simplistic, this approach emphasizes a very important point: phosphate solubility is a function not only of phosphate activity, but the activity of Ca, LREEs, or Y + HREEs as well. Therefore, factors that change the activities of any of these species in the melt will affect the total solubility of the solid.

Complexing in the silicate melt imparts a first-order effect on the activities of melt species. For example, aluminum phosphate complexes, produced by reactions such as $PO_4 + Al = AlPO_4$, have the effect of decreasing the activity of PO_4, thus lowering the IAP and SI of apatite resulting in an increase in apatite solubility. Based on this analysis, it is easy to understand why apatite solubility is such a strong function of ASI: the more free Al that is available for production of aluminum phosphate complexes, the more the PO_4 activity will be lowered, thus enhancing apatite solubility. In peralkaline melts, REE phosphate complexes will lower the activity of REEs in the liquid, increasing monazite solubility.

Kinetic considerations. Studies of phosphate solubility reveal kinetic limitations to dissolution rates. Harrison and Watson (1984) and Rapp and Watson (1986) measured the dissolution rates of apatite and monazite, respectively, and found that the dissolution rate is limited by diffusion of P or LREEs away from the dissolving apatite or monazite. Furthermore, the diffusivity, and hence dissolution rate, is strongly dependent on the H_2O content of the melt. In dry melts, dissolution is so slow that complete dissolution of even small crystals of apatite or monazite is unlikely. In melts produced by dehydration melting of muscovite or biotite, where the H_2O content is in the range of 4-8 wt % H_2O, apatite crystals on the order of 500 μm diameter will dissolve in 100-1000 years.

However, monazite dissolution is slower and in a similar melt, a monazite of 50 μm diameter will persist for several millions of years at 700°C (Rapp and Watson 1986), thus making it likely that monazite would survive an anatectic event.

Inferences from studies of natural samples. Considerable evidence exists from natural samples to shed light on the behavior of natural phosphates in the anatectic region. For example, Watt and Harley (1993) examined the geochemistry of two different lower crustal leucogneisses from East Antarctica that were derived from anatexis of pelites, and evaluated the geochemical signatures of each with respect to monazite and apatite reaction with the melt. The geochemistry of one leucogneiss was consistent with little dissolution of accessory phases and removal of melt before complete equilibration. The geochemistry of the other leucogneiss was consistent with accessory phase entrainment resulting from disequilibrium melting. Bea and Montero (1999) examined the geochemistry of host rocks and leucosomes and their major and accessory mineral associations from the Ivrea-Verbano zone, NW Italy. Monazite, apatite, and xenotime were present in samples of kinzigite (incipient migmatites) but their average grain size decreased markedly in the stronolites (high-grade migmatites). Furthermore, the fraction of phosphate crystals contained as inclusions within other minerals (notably garnet) increased dramatically in the stronolites relative to the kinzigites. Xenotime, in particular, was only found as inclusions in garnet in the stronolites, indicating that xenotime consumption occurred as garnet grew. Modal apatite also decreased significantly in stronolites relative to kinzigites. Composition changes were also noted in monazite, notably an increase in the Th/U by a factor of two. Based on textures, modal changes, and melt composition, Bea and Montero (1999) inferred the melts to have been saturated with respect to apatite and, except at the earliest stages, monazite.

A number of other papers address important aspects of accessory mineral behavior in the melting region, and the reader is encouraged to examine Montel (1993), Dirks and Hand (1991), Miller (1985), Miller and Mittlefehldt, (1982), Watson et al. (1989), Watson and Harrison (1984), Pichavant et al. (1992), Bea 1996, Bea et al. 1992, and Pyle et al. (2001).

Hydrothermal origin of apatite, monazite and xenotime

The hydrothermal alteration of monazite has received increased attention in recent years in part because of concerns over how alteration will disturb Th-U-Pb systematics and in part because of the proposed application of monazite-like materials as repositories of nuclear waste. Towards this end, Poitrasson et al. (1996) reported on the hydrothermal alteration of monazites from two granites in England, one of which underwent chloritization at T ≈ 284°C and the other which underwent greisenization at T ≈ 200°C. Chloritization resulted in a leaching of the REEs, leaving the phosphate framework and the Th-U-Pb systematics intact. Greisenization partially destroyed the phosphate framework, but left the Th concentration intact or even enriched. It was also suggested that the loss of Pb during greisenization might open the possibility of dating the fluid-rock event by $^{207}Pb/^{206}Pb$ *in situ* methods.

Finger et al. (1998) describe hydrothermal alteration of monazite from the eastern Alps. Monazite was replaced by a corona consisting of apatite, allanite, and epidote (outward from monazite). Elements released from monazite (REEs, Y, P, Th and U) were conserved in the corona, suggesting limited transport by the fluid. Additionally, very little radiogenic Pb was lost from the unaltered part of the monazite, and reasonable crystallization ages were obtained using chemical dating with the electron probe. In contrast, Lanzirotti and Hanson (1996) report on monazite altered to apatite during chloritization that yielded strongly discordant U-Pb ages, suggesting Pb-loss during alteration. Hawkins and Bowring (1997, 1999) and Fitzsimmons et al. (1997) have called upon magmatic fluids as responsible for overgrowths on monazite. Ayers et al. (1999) and Townsend et al. (2000) have suggested that recrystallization during hydrothermal alteration of pre-existing monazite can, in many cases, result in distinctive patchy zoning.

Experimental studies of accessory mineral solubility in aqueous fluids suggests that apatite and monazite solubility in pure H_2O is low, and increases with decreasing pH (Ayers and Watson 1991). Ayers et al. (1999), in an experimental study of the coarsening kinetics of monazite in the presence of aqueous fluids, found that monazite either recrystallizes and moves as host grain boundaries migrate, or is trapped as inclusions within the host phase. Inclusion monazites, once isolated from the rock matrix, should record the time of entrapment whereas matrix monazite should record the time of final recrystallization, or, if growth zoned, a complex age spectrum.

Implications for monazite and xenotime geochronology

It is clear from the above studies that the accessory minerals monazite, xenotime, and apatite do not behave as an isolated subsystem in metamorphic rocks (e.g., Pyle and Spear 1999, Ferry 2000, Pyle et al. 2001, Wing et al. 2002). Rather, reactions that involve major phases also affect the amounts and compositions of phosphate minerals, because phosphorus and REEs exist in phases other than the phosphates, a notable example of which is garnet. Garnet contains up to several hundred ppm P, which can be a considerable fraction of the total P in the rock. For example, consider a rock with 10 modal % garnet with 200 ppm P and 1000 ppm Y, 1 modal % apatite, 0.1 modal % monazite, and 0.01 modal % xenotime (Table 3). Garnet contains only 6% of the total phosphorus budget in this example, similar to xenotime. However, garnet contains 30% of the total Y budget and dominates the mass balance of this element. For this reason, the growth, consumption, and chemical zoning in phosphates clearly is influenced by reactions involving other phases.

Table 3. Mass balance calculations for P and Y.

	density	*mode*	*grams phase*	*wt% P*	*total gm P*	*% mass of P*	*wt % Y*	*total gm Y*	*% mass of Y*
Garnet	4.2	10	42	0.02	0.840	6.0	0.1	4.2	29.9
Apatite	3.2	0.1	0.32	18.452	5.904	42.1	0.2	0.064	0.5
Xenotime	4.5	0.01	0.045	16.939	0.762	5.4	48	2.16	15.4
Monazite	5	0.1	0.5	13.044	6.522	46.5	1	0.5	3.6

The significant consequence of this coupling between the phosphate and silicate minerals in a rock is that the phosphates will grow or be consumed as reactions among the major phases proceed. The simplest observational evidence of this is the nearly ubiquitous zoning observed in the phosphate minerals, especially well represented by monazite. If monazite grew only once during a metamorphic episode it might be systematically zoned from core to rim, but it would never display the complex style of zoning that is commonly observed.

To realize the potential of monazite (and xenotime) as a geochronometer, it is necessary to identify the monazite-producing reactions, and, ideally, to locate them in P-T space. It is clear from the summary above that there are a number of reactions by which monazite may grow during metamorphism, and no simple generalizations are possible.

The first appearance of monazite. The uncertainty surrounding monazite-producing reactions is perhaps best illustrated by the question of what reaction produces the first appearance of metamorphic monazite. The fact that monazite is sometimes detrital in origin is not in dispute, but the conditions at which the first metamorphic monazite appears are controversial.

In some rocks, metamorphic monazite is clearly preceded by allanite, and the transition from allanite to monazite appears to occur in the middle amphibolite facies (staurolite or Al_2SiO_5 zone) (Overstreet 1967, Smith and Barreiro 1990, Kingsbury et al. 1993, Bingen et al. 1996, Harrison et al. 1997, Ferry 2000, Rubatto et al. 2001, Pyle 2001, Wing et al. 2002). For the purpose of geochronology, the replacement of allanite by monazite should represent a well-defined metamorphic condition, lending significance to ages determined on such grains (provided inherited cores are not present; Hawkins and Bowring 1999, Rubatto et al. 2001).

In contrast, several studies have reported what has been interpreted to be metamorphic monazite at grades as low as the chlorite zone, without allanite as a precursor. For example, Franz et al. (1996) report monazite and xenotime from rocks of the chlorite zone in NE Bavaria, and Pyle et al. (2001) report monazite from the chlorite-biotite zone and as inclusions inside garnet from garnet-zone samples in central New England, implying monazite growth in the sub-garnet zone. The precursors to these low-grade monazites are not at all clear. Rhabdophane has been suggested as a precursor to monazite (Sawka et al. 1986), but its instability in the greenschist facies rules out this possibility (Akers et al. 1993). Another possibility is REE enriched clays (e.g., Copeland et al. 1971), oxides, or hydrous phosphates (e.g., montmorillonite, thorite, thorianite etc.) or Th and REE oxides (Kingsbury et al. 1993).

The range of conditions suggested for the first appearance of monazite during metamorphism (~350-550°C) probably reflects differences in sample bulk compositions and especially the bulk rock Ca (and possibly Al) concentration. In rocks with little or no Ca but some phosphate and REEs, monazite and xenotime appear in the lower greenschist facies because the potential sinks for REEs and Th (clays, thorite, thorianite, LREE oxides, or hydrous phosphates) become unstable at those grades. In rocks with higher Ca (and Al and Ti) content, allanite and possibly titanite apparently harbor the REEs up to the staurolite or Al_2SiO_5 isograds. For example, the rocks examined by Ferry (2000) were calc-pelites and the monazite-in isograd was suppressed to the amphibolite facies. In contrast, the rocks studied by Franz et al. (1996) contained CaO of only 0.10 to 0.37 wt %, and both monazite and xenotime appeared in the greenschist facies. Significantly, the rocks studied by Kingsbury et al. (1993) were quite low in Ca (0.15-0.39 wt %) and the monazite precursors were Th- and Ce-oxides rather than allanite. Monazite did not appear until the staurolite zone in these rocks, but the high Al (19.6-24.0 wt %) content of the samples may have stabilized staurolite at a relatively low temperature.

Monazite growth beyond the monazite-in isograd. Once stable, monazite continues to grow, dissolve, and reprecipitate with increasing grade, as evidenced by the complex growth histories recorded in the chemical zoning in many grains. In the sub-solidus, reactions that consume garnet will, in some instances, produce monazite (e.g., Pyle et al. 2001). In the melting region, monazite may dissolve into early-formed silicate melt and recrystallize on melt solidification, but the relationships are not well understood (e.g., Hawkins and Bowring 1999, Rubatto et al. 2001, Pyle et al. 2001). Hydrothermal fluids also have the potential of causing monazite to precipitate, whether by dissolving and reprecipitating material from existing grains, or by transporting necessary constituents to the deposition site (e.g., Ayers et al. 1999). Formation of shear zones may also induce new monazite growth (e.g., Lanzirotti and Hansen 1996).

CONCLUDING REMARKS

The past decade has seen an enormous growth of information about metamorphic phosphates. Particularly encouraging is the wealth of information contained in these minerals regarding metamorphic histories. These studies point towards a number of fruitful areas for further research:

(1) New studies examining the textures, chemistry, zoning, and petrogenesis of phosphates in well-constrained metamorphic terranes to supplement the current, sparse data sets. Complete, accurate chemical analyses of natural phosphates from these studies would be especially revealing about the geochemical controls on monazite stability.

(2) Additional experimental and empirical studies of the monazite-xenotime miscibility gap including the incorporation of the effects of brabantite substitution [$CaTh(PO_4)_2$] into monazite will further refine the accuracy of this thermometer.

(3) Refinement of the YAG-xenotime and YAG-monazite thermometers to improve accuracy of calibrations. Currently, these are limited to empirical calibrations from a single metamorphic terrane.

(4) Studies of the incorporation of trace quantities of REEs, P, Th and U into phases other than phosphates will be necessary to better model the interactions between major and accessory phases during metamorphism.

(5) Additional studies on the melting of phosphates will help quantify phosphate solubility and will aid greatly in the interpretation of these minerals from migmatites.

The goal of these studies is to link the paragenesis of these (and other) phases to events in a rock's history which, when combined with texture-sensitive geochronology (e.g., SIMS, LA-ICP-MS, and EMP), will enable the determination of the time of these events. These data will allow direct inferences of fundamental tectonic parameters such as the duration of metamorphic events, the duration of melting in the crust, the rates of heating/cooling and burial/exhumation of orogenic belts, and the time and duration of hydrothermal activity.

ACKNOWLEDGMENTS

The authors thank Dave Wark for assistance with electron microprobe analysis of complex REE phosphates and E. Bruce Watson for many informative discussions about accessory mineral paragenesis. Careful reviews by L. Storm, G. Franz and J. M. Ferry are also gratefully acknowledged, as well as both the review and editorial handling by M. Kohn. This work was supported in part by NSF grant EAR-0106738.

REFERENCES

Akers WT, Grove M, Harrison T, Ryerson FJ (1993) The instability of rhabdophane and its unimportance in monazite paragenesis. Chem Geol 110:169-176

Amli R (1975) Mineralogy and rare earth geochemistry of apatite and xenotime from the Gloserheia granite pegmatite, Froland, southern Norway. Am Mineral 60:607-620

Andrehs G, Heinrich W (1998) Experimental determination of REE distributions between monazite and xenotime: potential for temperature-calibrated geochronology. Chem Geol 149:83-96

Ayres M, Harris N (1997) REE fractionation and Nd-isotope disequilibrium during crustal anatexis from Himalayan leucogranites. Chem Geol 139:249-269

Ayers JC, Miller CF, Gorisch B, Milleman J (1999) Textural development of monazite during high-grade metamorphism: hydrothermal growth kinetics, with implications for U, Th-Pb geochronology. Am Mineral 84:1766-1780

Ayers JC, Watson EB (1991) Solubility of apatite, monazite, zircon, and rutile in supercritical aqueous fluids with implications for subduction zone geochemistry. Phil Trans R Soc London 335:365-375

Bea F (1996) Residence of REE, Y, Th, and U in granites and crustal protoliths: implications for the chemistry of crustal melts. J Petrol 37:521-552

Bea F, Fershtater G, Corretge´ LG (1992) The geochemistry of phosphorus in granite rocks and the effect of aluminum. Lithos 29:43-56

Bea F, Montero P (1999) Behavior of accessory phase and redistribution of Zr, REE, Y, Th, and U during metamorphism and partial melting of metapelites in the lower crust: an example from the Kinzigite Formation of Ivrea-Verbano, NW Italy. Geochim Cosmochim Acta 63:1133-1153

Bea F, Pereira MD, Stroh A (1994) Mineral/leucosome trace-element partitioning in a peraluminous migmatite (a laser ablation-ICP-MS study). Chem Geol 117:291-312

Bingen B, Demaiffe D, Hertogen J (1996) Redistribution of rare earth elements, thorium, and uranium over accessory minerals in the course of amphibolite to granulite facies metamorphism: the role of apatite and monazite in orthogneisses from southwestern Norway. Geochim Cosmochim Acta 60:1341-1354

Brenan JM (1993) Partitioning of fluorine and chlorine between apatite and aqueous fluids at high pressure and temperature: implications for the F and Cl content of high P-T fluids. Earth Planet Sci Lett 117:251-263

Cabella R, Lucchetti G, Marescotti P (2001) Authigenic Monazite and xenotime from pelitic metacherts in pumpellyite-actinolite-facies conditions, Sestri-Voltaggio Zone, central Liguria, Italy. Can Mineral 39:717-727

Candela PA (1986) Towards a thermodynamic model for the halogens in magmatic systems—an application to melt vapour apatite equilibria. Chem Geol 57:289-301

Catlos EJ, Harrison TM, Kohn MJ, Grove M, Ryerson FJ, Manning CE, Upreti BN (2001) Geochronologic and thermobarometric constraints on the evolution of the Main Central Thrust, central Nepal Himalaya. J Geophys Res 106:16177-16204

Cherniak DJ (2001) Rare earth element diffusion in apatite. Geochim Cosmochim Acta 64:3871-3885

Cherniak DJ, Watson EB, Harrison TM, Grove M (2000) Pb diffusion in monazite: A progress report on a combined RBS/SIMS study. EOS Trans Am Geophys Union 81:S25

Copeland RA, Frey FA, Wones DR (1971) Origin of clay minerals in a mid-Atlantic ridge sediment. Earth Planet Sci Lett 10:186-192

Crowley JL, Ghent ED (1999) An electron microprobe study of the U-Pb-Th systematics of metamorphosed monazite: the role of Pb diffusion versus overgrowth and recrystallization. Chem Geol 157:285-302

Cruft EF (1966) Minor elements in igneous and metamorphic apatite. Geochim Cosmochim Acta 30: 375-398

Dirks P, Hand M (1991) Structural and metamorphic controls on the distribution of zircon in an evolving quartzofeldspathic migmatite: an example from the Reynolds Range, central Australia. J Metamorph Geol 9:191-201

Ellison AJG, Hess PC (1988) Peraluminous and peralkaline effects upon "monazite" solubility in high-silica liquids. EOS Trans Am Geophys Union 69:498

Ferry JM (2000) Patterns of mineral occurrence in metamorphic rocks. Am Mineral 85:1573-1588

Finger F, Broska I, Roberts MP, Schermaier A (1998) Replacement of primary monazite by apatite-allanite-epidote coronas in an amphibolite facies granite gneiss from the eastern Alps. Am Mineral 83:248-258

Fitzsimmons ICW, Kinny PD, Harley SL (1997) Two stages of zircon and monazite growth in anatectic leucogneiss: SHRIMP constraints on the duration and intensity of Pan-African metamorphism in Prydz Bay, east Antarctica. Terra Nova 9:47-51

Förster H-J (1998) The chemical composition of REE-Y-Th-U-rich accessory minerals in peraluminous granites of the Erzgebirge-Fichtelgebirge region, Germany, Part I: The monazite-(Ce)-brabantite solid solution series. Am Mineral 83:259-272

Foster G, Kinny P, Vance D, Prince C, Harris N (2000) The significance of monazite U-Pb-Th age data in metamorphic assemblages: A combined study of monazite and garnet chronometry. Earth Planet Sci Lett 181:327-340

Franz G, Andrehs G, Rhede D (1996) Crystal chemistry of monazite and xenotime from Saxothuringian-Moldanubian metapelites, NE Bavaria, Germany. Eur J Mineral 8:1097-1108

Gan H, Hess PC (1992) Phosphate speciation in potassium aluminosilicate glass. Am Mineral 77:495-506

Görz H, White EH (1970) Minor and trace elements in HF-soluble zircons. Contrib Mineral Petrol 29: 180-182

Gratz R, Heinrich W (1997) Monazite-xenotime thermobarometry: experimental calibration of the miscibility gap in the system $CePO_4$-YPO_4. Am Mineral 82:772-780

Gratz R, Heinrich W (1998) Monazite-xenotime thermometry, III: experimental calibration of the partitioning of gadolinium between monazite and xenotime. Eur J Mineral 10:579-588

Grauch RI (1989) Rare earth elements in metamorphic rocks. Rev Mineral 21:147-167

Harrison TM, Ryerson FJ, Le Fort P, Yin A, Lovera OM, Catlos EJ (1997) A Late Miocene-Pliocene origin for the Central Himalayan inverted metamorphism. Earth Planet Sci Lett 146:1-7

Harrison TM, Watson EB (1984) The behavior of apatite during crustal anatexis: equilibrium and kinetic considerations. Geochim Cosmochim Acta 48:1467-1477

Hawkins DP, Bowring SA (1997) U-Pb systematics of monazite and xenotime: case studies from the Paleoproterozoic of the Grand Canyon, Arizona. Contrib Mineral Petrol 127:87-103

Hawkins DP, Bowring SA (1999) U-Pb monazite, xenotime, and titanite geochronological constraints on the prograde to post-peak metamorphic thermal history of Paleoproterozoic migmatites from the Grand Canyon, Arizona. Contrib Mineral Petrol 134:150-169

Heinrich W, Andrehs G, Franz G (1997) Monazite-xenotime miscibility gap thermometry. I. An empirical calibration. J Metamorph Geol 15:3-16

Jamtveit B, Dahlgren S, Austrheim H (1997) High-grade contact metamorphism of calcareous rocks from the Oslo Rift, southern Norway. Am Mineral 82:1241-1254

Kapustin YL (1987) The composition of apatite from metamorphic rocks. Geochem Int'l 24:45-51

Kingsbury JA, Miller CF, Wooden JL, Harrison TM (1993) Monazite paragenesis and U-Pb systematics in rocks of the eastern Mojave Desert, California, USA: implications for thermochronometry. Chem Geol 110: 147-167

Korzhinskiy MA (1981) Apatite solid solutions as indicators of the fugacity of HCl and HF in hydrothermal fluids. Geochem Intl 3:45-60

Lanzirotti A, Hanson GN (1996) Geochronology and geochemistry of multiple generations of monazite from the Wepawaug Schist, Connecticut, USA: implications for monazite stability in metamorphic rocks. Contrib Mineral Petrol 125:332-340

Lee DE, Bastron H (1967) Fractionation of rare-earth elements in allanite and monazite as related to geology of the Mt. Wheeler mine area, Nevada. Geochim Cosmochim Acta 31:339-356

Liou JG, Zhang RV, Ernst WG, Rumble DI, Maruyama S (1998) High-pressure minerals from deeply subducted metamorphic rocks. Rev Mineral 37: 33-96

Ludington S (1978) The biotite-apatite geothermometer revisited. Am Mineral 63:551-553

Miller CF (1985) Are strongly peraluminous magmas derived from pelitic sedimentary sources? J Geol 93:673-689

Miller CF, Mittlefehldt DW (1982) Depletion of LREE in felsic magmas. Geology 10:129-133

Montel J-M (1986) Experimental determination of the solubility of Ce-monazite in SiO_2-Al_2O_3-K_2O-Na_2O melts at 800°C, 2 kbar, under H_2O-saturated conditions. Geol 14:659-662

Montel J-M (1993) A model for monazite/melt equilibrium and application to the generation of granitic magmas. Chem Geol 110:127-146

Munoz JL (1984) F-OH and Cl-OH exchange in micas with applications to hydrothermal ore deposits. Rev Mineral 13:469-493

Nijland TG, Jansen JBH, Maijer C (1993) Halogen geochemistry of fluid during amphibolite-granulite metamorphism as indicated by apatite and hydrous silicates in basic rocks from the Bamble sector, South Norway. Lithos 30:167-189

Overstreet WC (1967) The geologic occurrence of monazite. U S Geol Surv Prof Paper 530

Parrish R (1990) U-Pb dating of monazite and its application to geological problems. Can J Earth Sci 27:1431-1450

Patiño Douce AE, Johnston AD (1991) Phase equilibria and melt productivity in the pelitic system: implications for the origin of peraluminous granitoids and aluminous granulites. Contrib Mineral Petrol 107:202-218

Pichavant M, Montel J-M, Richard LR (1992) Apatite solubility in peraluminous liquids: experimental data and an extension of the Harrison-Watson model. Geochim Cosmochim Acta 56:3855-3861

Poitrasson F, Chenery SR, Bland DJ (1996) Contrasted monazite hydrothermal alteration mechanisms and their geochemical implications. Earth Planet Sci Lett 145:79-96

Poitrasson F, Chenery SR, Shepherd TJ (2000) Electron microprobe and LA-ICP-MS study of monazite hydrothermal alteration: implications for U-Th-Pb geochronology and nuclear ceramics. Geochim Cosmochim Acta 64:3283-3297

Puchelt H, Emmermann R (1976) Bearing of rare earth patterns of apatites from igneous and metamorphic rocks. Earth Planet Sci Lett 31:279-286

Pyle JM (2001) Distribution of select trace elements in pelitic metamorphic rocks: pressure, temperature, mineral assemblage, and reaction-history controls. PhD dissertation, Rensselaer Polytechnic Institute, Troy, New York

Pyle JM, Spear FS (1999) Yttrium zoning in garnet: coupling of major and accessory phases during metamorphic reactions. Geol Mater Res 1:1-49

Pyle JM, Spear FS (2000) An empirical garnet (YAG)–xenotime thermometer. Contrib Mineral Petrol 138:51-58

Pyle JM, Spear FS, Rudnick RL, McDonough WF (2001) Monazite–xenotime–garnet equilibrium in metapelites and a new monazite–garnet thermometer. J Petrol 42:2083-2107
Rapp RP, Watson EB (1986) Monazite solubility and dissolution kinetics: implications for the thorium and light rare earth chemistry of felsic magmas. Contrib Mineral Petrol 94:304-316
Rasmussen B, Fletcher IR, McNaughton NJ (2001) Dating low-grade metamorphic events by SHRIMP U-Pb analysis of monazite in shales. Geol 29:963-966
Romans PA, Brown LL, White JC (1975) An electron microprobe study of yttrium, rare earth, and phosphorus distribution in zoned and ordinary zircon. Am Mineral 60:475-480
Rubatto D, Williams IS, Buick IS (2001) Zircon and monazite response to prograde metamorphism in the Reynolds Range, central Australia. Contrib Mineral Petrol 140:458-468
Ryerson FJ, Hess PC (1978) Implications of liquid-liquid distribution coefficients to mineral-liquid partitioning. Geochim Cosmochim Acta 42:921-932
Ryerson FJ, Hess PC (1980) The role of P_2O_5 in silicate melts. Geochim Cosmochim Acta 44:611-624
Sallet R (2000) Fluorine as a tool in the petrogenesis of quartz-bearing magmatic associations: applications of an improved F-OH biotite-apatite thermometer grid. Lithos 50:241-253
Sallet R, Sabatier H (1996) A new formulation of the biotite-apatite geothermometer. Applications to magmatic and sub-solidus conditions: Santa Catarina, Brazil, and Bingham, USA mining districts: Bishop Tuff, USA and Ploumanac'h granite, France. Sociedade Brasileira de Geologia, 39° Congresso Brasileiro de Geologia 2:115-119
Sawka WN, Banfield JF, Chappell BW (1986) A weathering-related origin of widespread monazite in S-type granites. Geochim Cosmochim Acta 50:171-175
Seydoux-Guillaume A-M, Wirth R, Heinrich W, Montel J-M (in press) Experimental determination of thorium partitioning between monazite and xenotime using analytical electron microscopy. Eur J Mineral
Sisson VB (1987) Halogen chemistry as an indicator of metamorphic fluid interaction with the Ponder Pluton, Coast Plutonic Complex, British Columbia, Canada. Contrib Mineral Petrol 95:123-131
Smith HA, Barreiro B (1990) Monazite U-Pb dating of staurolite grade metamorphism in pelitic schists. Contrib Mineral Petrol 105:602-615
Smith HA, Giletti BJ (1997) Lead diffusion in monazite. Geochim Cosmochim Acta 61:1047-1055
Smith MP, Yardley BWD (1999) Fluid evolution during metamorphism of the Otago Schist, New Zealand: (II), Influence of detrital apatite on fluid salinity. J Metamorph Geol 17:187-193
Stormer JC, Carmichael ISE (1971) Fluorine-hydroxyl exchange in apatite and biotite: A potential igneous geothermometer. Contrib Mineral Petrol 31:121-131
Stormer JCJ, Pierson MJ, Tacker RC (1993) Variation of F and Cl X-ray intensity due to anisotropic diffusion of apatite during electron microprobe analysis. Am Mineral 78:641-648
Suzuki K, Adachi M (1991) Precambrian provenance and Silurian metamorphism of the Tsubonosawa paragneiss in the south Kitakami terrane, northeast Japan, revealed by the chemical Th-U-total Pb isochron ages of monazite, zircon and xenotime. Geochem J 25:357-376
Suzuki K, Adachi M, Kajizuka I (1994) Electron microprobe observations of Pb diffusion in metamorphosed detrital monazites. Earth Planet Sci Lett 128:391-405
Terry MP, Robinson P, Hamilton MA, Jercinovic MJ (2000) Monazite geochronology of UHP and HP metamorphism, deformation, and exhumation, Nordøyane, Western Gneiss Region, Norway. Am Mineral 85:1651-1664
Townsend KJ, Miller CF, D'Andrea JL, Ayers JC, Harrison TM, Coath CD (2000) Low temperature replacement of monazite in the Ireteba granite, southern Nevada: geochronological implications. Chem Geol 172:95-112
Viskupic K, Hodges KV (2001) Monazite-xenotime thermochronometry: Methodology and an example from the Nepalese Himalaya. Contrib Mineral Petrol 141:233-247
Watson EB (1979) Apatite saturation in basic to intermediate magmas. Geophys Res Lett 6:937-940
Watson EB (1980) Apatite and phosphorus in mantle source regions: an experimental study of apatite/melt equilibria at pressures to 25 kbar. Earth Planet Sci Lett 51:322-335
Watson EB, Capobianco CJ (1981) Phosphorus and the rare earth elements in felsic magmas: An assessment of the role of apatite. Geochim Cosmochim Acta 45:2349-2358
Watson EB, Harrison TM (1984) What can accessory minerals tell us about felsic magma evolution? A framework for experimental study. Proc 27th Intl Cong 11:503-520
Watson EB, Vicenzi EP, Rapp RP (1989) Inclusion/host relations involving accessory minerals in high-grade metamorphic and anatectic rocks. Contrib Mineral Petrol 101:220-231
Watt GL, Harley SL (1993) Accessory phase controls on the geochemistry of crustal melts and restites produced during water-undersaturated partial melting. Contrib Mineral Petrol 114:550-566
Williams ML, Jercinovic MJ, Terry MP (1999) Age mapping and dating of monazite on the electron microprobe: deconvoluting multistage tectonic histories. Geol 27:1023-1026

Wing BA, Ferry JM, Harrison TM (in press) Prograde destruction and formation of monazite and allanite during contact and regional metamorphism of pelites: petrology and geochronology. Contrib Mineral Petrol

Wolf MB, London D (1994) Apatite dissolution into peraluminous haplogranitic melts: An experimental study of solubilities and mechanisms. Geochim Cosmochim Acta 58:4127-4145

Wolf MB, London D (1995) Incongruent dissolution of REE- and Sr- rich apatite in peraluminous granitic liquids: Differential apatite, monazite, and xenotime solubilities during anatexis. Am Mineral 80: 765-775

Yang P, Rivers T (2002) Trace element zoning in pelitic garnet, apatite and epidote group minerals: The origin of Y annuli and P zoning in garnet. Geol Mat Res 4:1-35

Yardley BWD (1985) Apatite composition and fugacities of HF and HCl in metamorphic fluids. Mineral Mag 49:77-79

Zayats AP, Kuts VP (1964) Rare earth elements in the accessory minerals of gneisses in the Ukrainian crystalline shield. Geochem Intl 1:1126-1128

Zhu C, Sverjensky DA (1991) Partitioning of F-Cl-OH between minerals and hydrothermal fluids. Geochim Cosmochim Acta 55:1837-1858

Zhu C, Sverjensky DA (1992) F-Cl-OH partitioning between biotite and apatite. Geochim Cosmochim Acta 56:3435-3467

Zhu XK, O'Nions RK (1999a) Zonation of monazite in metamorphic rocks and its implications for high temperature thermochronology: a case study from the Lewisian terrain. Earth Planet Sci Lett 171: 209-220

Zhu XK, O'Nions RK (1999b) Monazite chemical composition: some implications for monazite geochronology. Contrib Mineral Petrol 137:351-363

8 Electron Microprobe Analysis of REE in Apatite, Monazite and Xenotime: Protocols and Pitfalls

Joseph M. Pyle, Frank S. Spear and David A. Wark

Department of Earth and Environmental Sciences
Rensselaer Polytechnic Institute
Troy, New York 12180

INTRODUCTION

Phosphate accessory phases contain a wealth of petrologic and chronological information, but each possesses particular properties that make accurate quantitative microanalysis difficult. This chapter provides an overview of the literature concerned with the main problems of electron microprobe (EMP) phosphate analysis. These include the volatility of fluorine in F-bearing apatite, and the mutual interference of L- and M-line X-rays from the major and trace elements in monazite and xenotime, along with consideration of standards, detection limits, absorption edges, and ZAF corrections in REE phosphates.

FLUORINE EXCITATION DURING APATITE ANALYSIS

The typical apatite in metamorphic parageneses is enriched in F^- (e.g., Smith and Yardley 1999), though enrichments in OH^- and Cl^- have been noted, either as inherited grains (Smith and Yardley 1999), or original growth (Harlov et al. 2002). Diffusive volatility of light anions such as F and Cl, and light cations such as K and Na, is a well known problem in electron microprobe analysis (Goldstein et al. 1984), but the relatively high F content of typical metamorphic apatite combined with its particular crystalline structure generate a unique analytical problem, namely, time-dependent variation in apatite F X-ray intensity driven by surface-ward diffusion of F ions in response to the electric field produced by implantation of primary beam electrons at a depth below the analyzed region (Stormer et al. 1993).

Stormer et al. (1993) examined variations in apatite F and Cl X-ray intensity as a function of a number of factors including analysis time, analysis current, accelerating voltage, and crystallographic orientation. Under typical EMP operating conditions (15 keV, 15 nA), Stormer et al. (1993) found that for sections oriented perpendicular to the *c* crystallographic axis, F X-ray intensity doubles in the first 60-120 s of cumulative beam exposure (Fig. 1). There is a strong crystallographic control over this F X-ray intensity increase. F Kα X-ray intensity nearly doubles in the first 120 s on sections of apatite oriented parallel to (0001) (perpendicular to *c*) (Fig.1). Apatite grains oriented parallel to (100) (i.e., with *c* in the plane of the section) display only a 20% increase in F X-ray intensity over the first 300 s of analysis time (Fig. 1). After long continuous exposure times, the F-X ray intensities decrease below the original values, with ~240 s required for F X-ray intensity to decrease below original values in (0001)-parallel sections (Fig. 1), and 20 times as long required in (100)-parallel sections (Stormer et al. 1993). Moreover, the increase in F X-ray excitation does not appear to decay with time after the beam is turned off. Apatites analyzed by Stormer et al. (1993) for a cumulative total of 150 s, then re-analyzed after a 16-hour hiatus, display no decay of F X-ray intensity, suggesting that the damage is permanent.

1529-6466/00/0048-0008$05.00

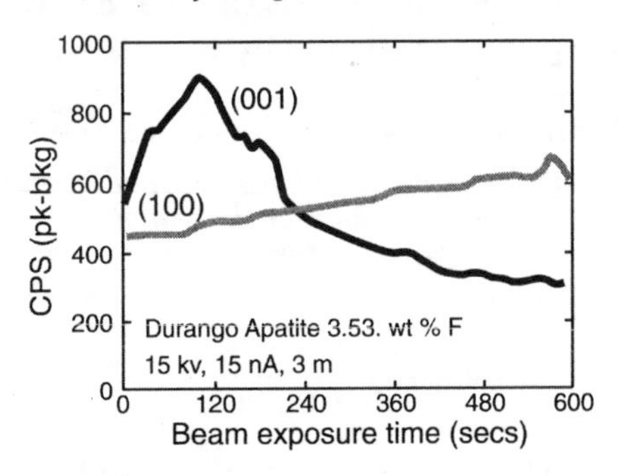

Figure 1. F Kα intensity plotted against total cumulative beam exposure time for (001) (black) and (100) (grey) sections of Durango fluorapatite. Data were obtained in 6 s counting intervals separated by 3 s of beam blanking. Background was determined by separate measurement at wavelengths above and below the peak on immediately adjacent spots. Figure from Stormer et al. (1993).

The magnitude of the intensity increase of apatite F X-rays is also affected by beam-diameter and accelerating voltage. Increasing the beam diameter *decreases* the rate of intensity rise and falloff, but results in a greater absolute increase in F X-ray intensity (Stormer et al. 1993). Varying accelerating voltage appears to have little effect on maximum F X-ray intensity (Stormer et al. 1993), but higher accelerating voltages may result in quicker decay of F X-ray intensity after the intensity maximum. Stormer et al. (1993) suggest that this decay may be a result of beam contamination or destruction of the carbon coat, rather than an intrinsic response of the apatite matrix to increased accelerating voltages.

Stormer et al. (1993) also conclude that apatite Cl X-rays respond similarly. The initial Cl X-ray intensity increase takes place on a similar time scale (Stormer et al. 1993). However, Cl X-rays are not absorbed as strongly as F X-rays in the apatite matrix, so as the Cl X-rays are brought surfaceward, their emergent intensity is not significantly increased.

A survey of several analyses of apatite from the geological literature demonstrates that the increase in apatite F X-ray intensity was unrecognized, or ignored. Stoichiometric Ca-fluorapatite contains 3.76 wt % F, but several analyses from the literature (Amli 1975, Bea and Montero 1999, Kifer and Wolf 2000) report apatite F contents well in excess of 4 wt %, with extreme values approaching 6.5 wt % (Bea and Montero 1999). When normalized to a +25 total cationic charge, apatite analyses such as these yield F atom totals far in excess of 1.0, indicating major problems with the apatite analytical routines used in these studies.

The apparent, non-reversible excitation of F X-rays in fluorapatite poses serious obstacles to EMP analysis of apatite, especially if accompanied by high current and/or long beam exposure time. Thus, apatites that have been imaged and mapped, may already have their "true" F content compromised. To minimize this effect, Stormer et al. (1993) suggest: (1) that apatite should not be used as a fluorine standard for microprobe analysis, as measured F concentration of the standard will vary with increased beam exposure, and; (2) that unknown grains should be analyzed in different spots with an increased (10-20 μm) beam diameter. A time series of F X-ray intensity using 10-20 s counting time should be collected and used to extrapolate back to an initial F X-ray intensity, and "true" F concentration. However, small grain size may preclude this type of analysis. If several apatite grains exist in a given sample, one possible analytical strategy is to "reserve" several texturally similar apatite grains for BSE imaging and/or element mapping, and others for quantitative analysis. Initial, short-time (10-15 s on-peak) analyses can be used to determine major components (Ca, P, F, Cl), and once these numbers are fixed, a second, long-time (1-2 min on-peak) analysis can be performed to determine minor- and trace-element content of apatite.

ANALYSIS OF REE PHOSPHATES

Introduction

Analysis of the most common REE phosphates, monazite and xenotime, is made difficult by a number of factors both mineralogical and EMP-specific, including choice of standard materials, inter-element interferences arising from high concentrations of elements with numerous L-lines and M-lines, detector artifacts (pulse energy shifts and gas absorption edges), X-ray collimator settings, element detection limits, and compositional variation arising from application of different ZAF (or $\phi(\rho z)$) correction schemes.

The discussion in this section is based on general EMP analysis (Goldstein et al. 1984, Scott et al. 1995) and REE-phosphate-specific analysis (Amli 1975, Amli and Griffin 1975, Exley 1980, Roeder 1985, Scherrer et al. 2000, Pyle 2001). Table 1 lists the preferred analytical peaks for monazite and xenotime along with a brief statement indicating the reasoning for selection (or avoidance) of a particular peak or peaks. Further detail is presented below.

STANDARDS

Complete analysis of monazite and xenotime requires detection of between 15-20 elements, including Si, Ca, Y, REE, Th, U, and Pb. For REE analysis, typical standards are either REE glasses with or without Ca, Si, and Al (Drake and Weill 1972), or apatite and stoichiometric Y and REE phosphates (Jarosewich and Boatner 1991). Th standards used for calibration include synthetic Th metal (Förster 1998), Th oxide (Broska et al. 2000), $ThSiO_4$ (Pyle et al. 2001) and ThP_2O_7 (Scherrer et al. 2000). UO_2 is the typical uranium standard (Scherrer et al. 2000, Pyle et al. 2001). Standards used for lead analysis include Pb-doped glasses (Förster 1998), well-characterized natural samples of crocoite and/or vanadinite (Scherrer et al. 2000), pyromorphite (Williams et al. 1999), and synthetic $PbSiO_3$ (Spear and Pyle unpublished). Galena should be avoided as a lead standard because of Pb-S interference (Scherrer et al. 2000). In order to minimize the magnitude of ZAF corrections, as many phosphates as possible should be used as standards; the use of different standards for a single element and their overall effect on ZAF corrections for that element will be discussed in a later section. Synthesis of REE+Y phosphate is relatively straightforward (Jarosewich and Boatner 1991), and, with the exception of Th, constitutes the majority of typical monazite or xenotime components.

X-RAY COUNT RATE AND DETECTOR ENERGY SHIFT

Typically, analysis protocols for monazite and/or xenotime call for the same current to be used both for calibration of standards and analysis of unknowns (e.g. Scherrer et al. 2000), though adopted current values vary widely (e.g., Finger and Broska 1999, Gratz and Heinrich 1997). However, this approach may cause problems if a standard and unknown generate very different count rates.

Energy distributions in gas proportional counters—whether gas flow or sealed counters—undergo a noticeable shift to lower energies when X-ray counting rates (and, concomitantly, detector gas ionization rates) increase. The exact origin of this phenome-

Table 1. Analytical peaks for monazite (Mnz) and xenotime (Xno).

	Mnz	*Comments*	*Xno*	*Comments*
Si	$K\alpha_1$ (TAP)	Separation from Y $L\alpha$, Th M4-O2	$K\alpha_1$(TAP)	Separation from Y $L\alpha$, Th M4-O2
P	$K\alpha_1$(TAP)	Low Y content—no Y $L\beta$ interference	$K\alpha_1$ (PET)	Avoids Y $L\beta$ interference on TAP
Ca	$K\alpha_1$(PET)	No interfering peaks in this region of spectrum	$K\alpha_1$(PET)	No interfering peaks in this region of spectrum
Y	$L\alpha_1$(PET)	Lower counts, but separated from P $K\alpha$	$L\alpha_1$(PET)	Y abundant, and separation from P $K\alpha$
La	$L\alpha_1$(LIF)	No interfering peaks in typical monazite	$L\alpha 1$(LIF)	Near or below EMP detection limit in xenotime
Ce	$L\alpha_1$(LIF)	No interfering peaks in typical monazite	$L\alpha_1$(LIF)	Near or below EMP detection limit in xenotime
Pr	$L\beta_1$(LIF)	Avoids La $L\beta$ interference on Pr $L\alpha$	$L\beta_1$(LIF)	Near or below EMP detection limit in xenotime
Nd	$L\beta_1$(LIF)	Avoids Ce $L\beta 1$ interference on Nd $L\alpha$	$L\alpha_1$(LIF)	Low Ce in xenotime minimizes Ce $L\beta_1$ interference
Sm	$L\beta_1$(LIF)	Avoids Ce $L\beta_{2,15}$ interference on Sm La	$L\alpha_1$(LIF)	Low Ce in xenotime minimizes Ce $L\beta_{2,15}$ interference
Eu	$L\beta_1$(LiF)	Avoids Nd $L\beta_3$, Pr $L\beta_{2,15}$. Correct for Dy $L\alpha_{1,2}$ interference	$L\alpha_1$(LiF)	Low Nd, Pr minimizes Nd $L\beta_3$, Pr $L\beta_{2,15}$ Interferences, but should be corrected
Gd	$L\beta_1$(LiF)	Avoids La $L\gamma_{2,3}$ and Ce $L\gamma_1$ interference. Ho $L\alpha$ interference minimal	$L\alpha_1$(LiF)	Avoids Ho $L\alpha$ interference on Gd $L\beta$. Low La, Ce minimizes La $L\gamma_{2,3}$, Ce $L\gamma_1$ interferences.
Tb	$L\beta_1$(LiF)	Corr'n necessary for Er $L\alpha$, but Er Concentration in monazite is low	$L\alpha_1$(LiF)	Low Sm content reduces Sm $L\beta_{1,4}$ interference
Dy	$L\alpha_1$(LIF)	Minor corr'n for Eu $L\beta$ interference	$L\alpha_1$(LIF)	Minor correction for Eu L_β interference
Ho	$L\beta_1$(LIF)	Near or below EMP detection limit in monazite	$L\beta_1$(LIF)	Avoids Gd La interference. Minor corr'n for Eu $L\gamma_1$
Er	$L\alpha_1$(LIF)	Correct for Tb $L\beta$ interference. Avoids Gd $L\gamma_1$, Eu $L\gamma_3$	$L\alpha_1$(LIF)	Correct for Tb L_β interference. Avoids Gd $L\gamma_1$, Eu$L\gamma_3$
Tm	$L\alpha_1$(LIF)	Near or below EMP detection limit in monazite	$L\alpha_1$(LIF)	Minor corr'ns for Sm $L\gamma_1$, Dy $L\beta_4$ interference
Yb	$L\alpha_1$(LIF)	Near or below EMP detection limit in monazite	$L\alpha_1$(LIF)	Minor corr'ns for Tb $L\beta_2$, Dy $L\beta_3$ interference
Lu	$L\beta_1$(LIF)	Near or below EMP detection limit in monazite	$L\beta_1$(LIF)	Corr'n for Dy $L\gamma_2$, Dy $L\gamma_3$, Yb $L\beta_2$, Ho $L\gamma_1$ interference
Pb	$M\beta_1$(PET)	Avoids Th $M\zeta_1$, Th $M\zeta_2$, Y $L\gamma_{2,3}$. Minor U $M\zeta_2$, Ce $L\alpha_1$ (2nd-order)	$M\beta_1$(PET)	Avoids Th $M\zeta_1$, Th $M\zeta_2$, Y $L\gamma_{2,3}$. Minor U $M\zeta_2$, Ce $L\alpha_1$ (2nd-order)
Th	$M\alpha_1$(PET)	No interfering peaks in this region of spectrum	$M\alpha_1$(PET)	No interfering peaks in this region of spectrum
U	$M\beta_1$(PET)	Avoids Th $M\beta_1$. Correct for Th $M\gamma_1$ interference	$M\beta_1$(PET)	Avoids Th $M\beta_1$. Correct for Th $M\gamma_1$ interference

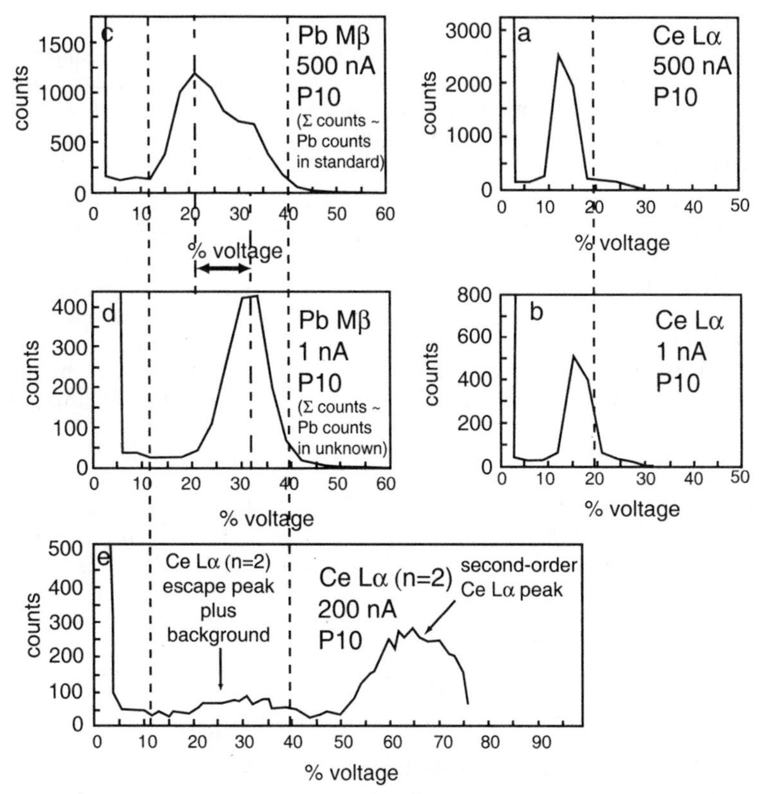

Figure 2. Single-Channel Analyzer (SCA) scans of Trebilcock monazite (Tomascak et al., 1996) on a JEOL 733 Superprobe. (a) Ce Lα, LiF, 500 nA. (b) Ce Lα, LiF, 1 nA. (c) Pb Mβ, PET, 500 nA. (d) Pb Mβ, PET, 1 nA, (e) Ce Lα (n = 2), PET, 200 nA Scan times are 10 sec/channel for the 1 nA scans, 0.1 sec/channel for the 500 nA scans and 5 sec/channel for the 200 nA scan. The scans reveal the slight shift in peak energy distribution with changes in counting rate due to increased sample current. High current scan of Pb is analogous to scan of Pb-rich standard material at moderate current, where as low-current scan of Pb is analogous to scan of Pb-poor unknown at the same current, and is illustrative of peak shift that can occur between calibration and analysis at identical current. Dotted lines represent energy window width; short dash-long dash lines in (c) and (d) show magnitude of peak shift.

non is not understood, but it is probably related to either main amplifier base line stability, or a decrease in effective potential across the counter tube caused by ion buildup around the counter tube walls (Goldstein et al. 1984). For elemental analysis with wide-open voltage windows, this energy shift does not pose serious problems, but if narrow voltage windows are required for sample-specific analysis, problems can arise, as discussed below.

Figures 2a-d show two pairs of single-channel analyzer (SCA) scans—one for Ce Lα, and one for Pb Mβ—at two different sample currents (1 nA and 500 nA). In the case of Ce (Fig. 2a-b), there is a slight, but noticeable, shift in the peak energy distribution towards lower energies with increased current and count rate. With voltage windows set as shown, an analysis at 1 nA current returns a nearly 25 percent lower intensity (41.72 cts/nA·sec vs. 55.42 cts/nA·sec) than the 500 nA analysis, due to cutoff of the right shoulder of the energy peak at lower current (Fig. 2b). If narrow SCA windows are required in analysis of a particular element, energy distribution shifts are likely to filter out detector gas voltage pulses generated by characteristic X-rays from the element of interest if: (1) the element of interest has a high concentration in the standard and is a trace element in the unknown (such as lead in monazite), or (2) the element of interest has approximately the same concentration in the unknown as the standard (e.g., cerium in monazite), and the unknown is analyzed at a significantly higher current than the calibration current.

Even if voltage windows are sufficiently narrow to filter out signals from unwanted higher-order peaks, yet wide enough to account for energy distribution shifts due to count rate changes, problems may still arise as a result of the type of detector gas used. Figures 2c and 2d show SCA scans of the Pb Mβ peak for $PbSiO_3$. Analysis at 500 nA and 1 nA is analogous to calibration of a Pb-rich standard and analysis of an unknown with trace amounts of Pb at the same current, with the lower count rate of the unknown causing the energy distribution to shift (indicated by the double arrow) to higher energies. The voltage window as shown is the narrowest window which includes all Pb Mβ (5.076 Å)-generated voltage pulses, and it also filters out the adjacent (2 λ = 5.123 Å) second-order Ce Lα peak (Fig. 2e). However, the use of P10 ($Ar_{90}CH4_{10}$) as a detector gas induces generation of a second-order Ce Lα escape peak in the same region occupied by Pb Mβ. As long as Ar (or P10) is used as a detector gas and the Pb Mβ peak is analyzed, X-ray counts from the second-order Ce Lα escape peak cannot be filtered out for a voltage window of any width without also excluding signal from Pb Mβ. This is one of the phenomena related to choice of X-ray detector gas; the Ce escape peak generated by xenon counter gas occurs at a much lower energy (i.e., 4.84 keV minus 4.11 keV) and does not interfere with Pb Mβ. Another phenomenon related to choice of detector gas (absorption edges) and its influence over REE phosphate analysis is described below.

DETECTOR GAS ABSORPTION EDGES

Another phenomenon related to the choice of detector gas arises from detector gas absorption edges. The two gases commonly used in X-ray detectors are xenon and argon (or P10). The use of either one of these gases influences the shape of the continuum radiation, as the ionization efficiency of each gas varies with of X-ray wavelength. P10 is commonly used to detect "soft" X-rays (λ > 3 Å). Xe gas has a higher quantum counter efficiency for shorter-wavelength X-rays (λ < 2 Å) (Goldstein et al. 1984). Ar may be used effectively over the entire X-ray energy range of interest if the gas pressure is increased to 2-3 bars, so as to increase its absorbing power for short-wavelength X-rays (Scott et al. 1995). However, the ionization efficiency of each of these gases undergoes a discontinuous decrease at its particular absorption edge(s) (Fig. 3). The absorption is especially notable at the Ar K absorption edge (Fig. 3), where ionization efficiency loss is roughly 40 % (Goldstein et al. 1984).

These absorption edges are present in critical portions of the spectrum for REE phosphate analysis. The three Xe absorption edges span the range of the $L\alpha_1$ and lower intensity peaks for La, Ce, and Nd. The Ar K absorption edge is adjacent to U Mα, U Mβ, Th Mβ, and Th Mγ (Fig. 4a). This factor makes placement of backgrounds for uranium analysis especially critical, as linear interpolation of background counts under the peak position that does not account for the absorption edge will result in higher than actual (if U Mα is used) or lower than actual (if U Mβ is used) peak minus background counts. For example, Scherrer et al. (2000) suggested U Mβ for monazite analysis with a high background position of +3390 sin units × 10000 (+9.5 mm) relative to the U Mβ peak and a low background position of –1505 sin units ×10000 (-4.2 mm), which puts the high background above the Th Mβ peak, but also on the high side of the Ar K absorption edge.

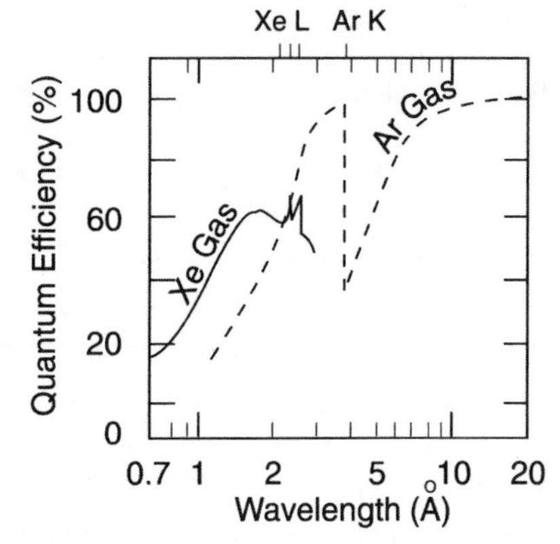

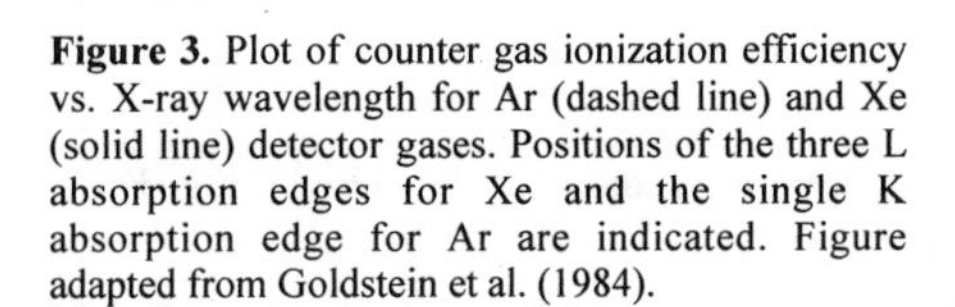

Figure 3. Plot of counter gas ionization efficiency vs. X-ray wavelength for Ar (dashed line) and Xe (solid line) detector gases. Positions of the three L absorption edges for Xe and the single K absorption edge for Ar are indicated. Figure adapted from Goldstein et al. (1984).

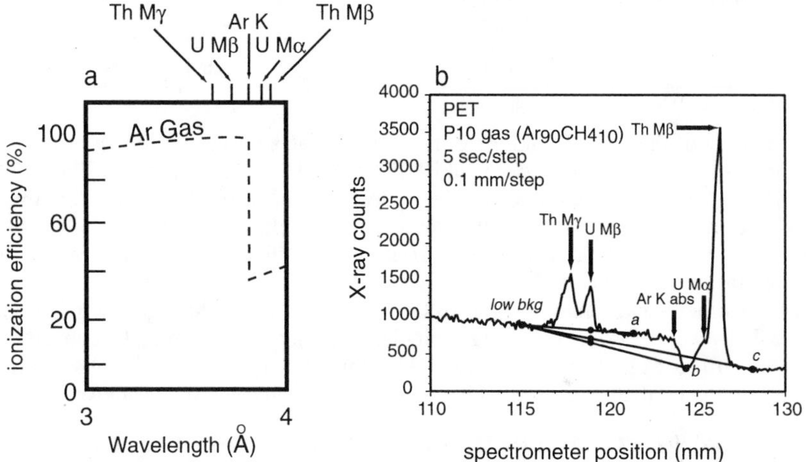

Figure 4. (a) A portion of Figure 3 shown at expanded scale, highlighting U and Th peak positions in the vicinity of the Ar K absorption edge. (b) WDS scan of a portion of the X-ray spectrum for Trebilcock monazite (Tomascak et al. 1996) using a PET crystal, P10 detector gas, a step size of 0.1 mm and a dwell time of 5 s per step. The scan shows the presence of the Ar K absorption edge between the U Mβ and U Mα peaks, as well as the greater resolution of the U Mβ peak from its neighboring Th Mγ peak. Peak-to-background count ratios for the different high-background positions for U Mβ (a,b,c) are discussed in the text.

Figure 4b shows a WDS scan of Trebilcock monazite (Tomascak et al. 1996), using a P10 gas-flow counter, for the portion of the X-ray spectrum discussed above. The figure shows the greater resolution between U Mβ and Th Mγ vs. U Mα and Th Mβ, which necessitates a smaller correction for Th Mγ interference (Scherrer et al. 2000, this study Table 2). Clearly, for elements present at minor or trace levels, background positions on opposite sides of the Ar K absorption edge have a pronounced effect on calculated peak-to-background count ratios.

Peak minus background counts/background counts (P-B/B) ratios are calculated for 3 different background positions on one side of the analysis peak (a, b, and c for U Mβ), generating three different P-B/B ratios (Table 2). One background position is on the same side of the absorption edge, and two background positions are on the opposite side of the absorption edge. As a reference, the high and low U Mβ background positions suggested by Scherrer et al. (2000) are approximated by points "c" and "low bkg", respectively. Background counts at the peak position were determined by linear interpolation using the counts at each background position. The measured P-B/B ratios range from 0.73 to 1.10 (50 % variation), with the best value (point a) being the lowest.

Table 2. Peak counts with respect to Ar K absorption edge (123.98 on PET) (Fig. 3). U Mβ ages on Trebilcock monazite, with U corrections to bk #2 and bk #3

	Position (mm)	*Counts*	*Bkgrd cts under peak*	*P-B*	*(P-B)/B*	*Δ (%)*	*ppm U*	*Age (Ma)*
pk	119.0	1428						
low bk	115.0	952						
hi bk a	121.5	748	826 (1)	826	0.73		5465	280
hi bk b	124.4	316	681 (2)	681	1.10	50.6	8230	262
hi bk c	128.1	270	744 (3)	744	0.92	26.4	6908	270

The effect of the background measurement translates into a significant effect (Table 2) on the calculated chemical age (Suzuki and Adachi 1991, Montel et al. 1996, Williams et al. 1999, Scherrer et al. 2000). The oldest age (280 Ma) is given by the true background measurement (a), and the youngest (262 Ma) by the background position suggested by Scherrer et al. (2000) This variation is less than age variations arising from analytical uncertainties, but with decreasing concentration of U, the increase of the P-B/B ratio for the other background positions relative to the "correct" background position will increase. However the effect on the age calculation is also a function of Th concentration; monazites with lower Th concentrations will contain larger age variations due to the above phenomenon than will a high-Th monazite, e.g., Table 4, part B (below).

ANALYTICAL PRECISION AND DETECTION LIMITS

The concentration of elements typically analyzed in monazite and/or xenotime spans the range from major components (~42 wt. Y_2O_3 in xenotime, ~28-30 wt % Ce_2O_3 in monazite) to trace components (e.g., ≤0.1 wt % Er_2O_3 in monazite, ≤0.1 wt % La_2O_3 in xenotime). Accurate detection of elements at low concentrations requires a combination of long analysis time and/or high analysis current. However, high-current analysis runs the risk of sample damage and/or carbon coating (Stormer et al. 1993).

Two measurable quantities relate to precision in EMP analysis of phosphates. The first quantity is the uncertainty of the analysis itself, i.e., standard deviation, which is the main measure of the uncertainty of either thermometry estimates or chemical age. The second quantity is the minimum detection limit, i.e., the smallest concentration of an element of interest generating characteristic X-ray counts that are statistically distinguishable from background X-rays.

Characteristic X-rays are emitted in time, and, consequently, obey Poisson statistics, with the standard deviation σ for N X-ray counts equal to $(N)^{1/2}$, and a standard deviation on the count rate (counts per s) of $(N)^{1/2}/t$. The concentration of an element is generally assumed to be proportional to the total number of peak minus background counts (Scott et al. 1995), and for a count rates of N_P/t_P and N_B/t_B (peak and background, respectively), the standard deviation (σ) and standard error (ε) are given by, respectively:

$$\sigma_{P-B} = \sqrt{\left(\frac{N_P}{t_P^2}\right)+\left(\frac{N_B}{t_B^2}\right)} \quad (1a)$$

$$\varepsilon_{P-B} = \sqrt{\left(\frac{N_P}{t_P^2}\right)+\left(\frac{N_B}{t_B^2}\right)} \Big/ \left(\frac{N_P}{t_P}\right)-\left(\frac{N_B}{t_B}\right) \quad (1b)$$

The standard error is identical to the relative (%) standard deviation, since $\sigma_{P\text{-}B}$ relative $= \sigma/N = \varepsilon_{P\text{-}B}$. The standard deviation and standard error of the k-ratio are calculated by summing the uncertainties in both calibration of the standard (std) and analysis of the unknown (unk), i.e.,

$$\sigma_{P-B,k-ratio} = \sqrt{\sigma^2_{P-B,std} + \sigma^2_{P-B,unk}} \quad (2a)$$

$$\varepsilon_{P-B,k-ratio} = \sqrt{\varepsilon^2_{P-B,std} + \varepsilon^2_{P-B,unk}} \quad (2b)$$

For a relative standard deviation on X-ray counts, and by extension, composition, of 1%, 10,000 peak minus background counts of the desired element must be accumulated. Assuming that X-ray production scales linearly with analysis current, count accumulation to the desired precision can be viewed as a function of the nA·sec counting product.

For high concentration elements (e.g., Ce, Y, Th), the standard deviation is the primary uncertainty variable, and as such, should be considered in propagation of uncertainties in temperature or age calculations. However, as elemental concentration decreases it becomes more difficult to distinguish characteristic X-rays from continuum radiation. For large numbers of generated X-rays, the distribution of X-ray count frequency about the mean is approximately Gaussian (Goldstein et al. 1984). Therefore, to be more than 99 % certain that the peak of a particular element is present in the unknown, the peak count on the element standard must exceed 3 standard deviations of the background count, or $N_P \geq 3(N_B)^{1/2}$. Using this relationship together with the assumption that X-ray intensity is proportional to concentration, the minimum detection limit (MDL) for a given element in the unknown may be defined as (Scott et al. 1984):

$$C_{MDL,unk} = C_{std} * \frac{3\sqrt{N_{B,std}}}{N_{P,std}} \quad (3)$$

with the same concentration unit C used for both the standard and the unknown.

Figure 5 shows detection limits for Y, Th, U, Pb, and several REE. Detection limits will vary depending on standard composition, analytical line, spectrometer resolution, diffraction crystal, X-ray collimator setting, and detector efficiency. Figure 5 is based on standards given in Pyle (2001) on the JEOL 733 Superprobe at RPI.

Detection limits of less than 0.1 wt % (1000) ppm for LREE require a roughly 1000-2000 nA·sec analysis, and the same analysis time-current product generates detection limits ranging from approximately 620 ppm for U, 1200 ppm for Th, 820 ppm for Pb, 240 ppm for Y, and 800-3500 for the HREE. A detection limit of less than 500 ppm for all elements in REE phosphate requires roughly 150000 nA·sec analyses, and a detection limit of 200 ppm Pb (suitable for chemical dating of low-Th Paleozoic monazite) requires approximately 35000 nA·sec analyses, or 140 s on each peak and background at an analysis current of 250 nA.

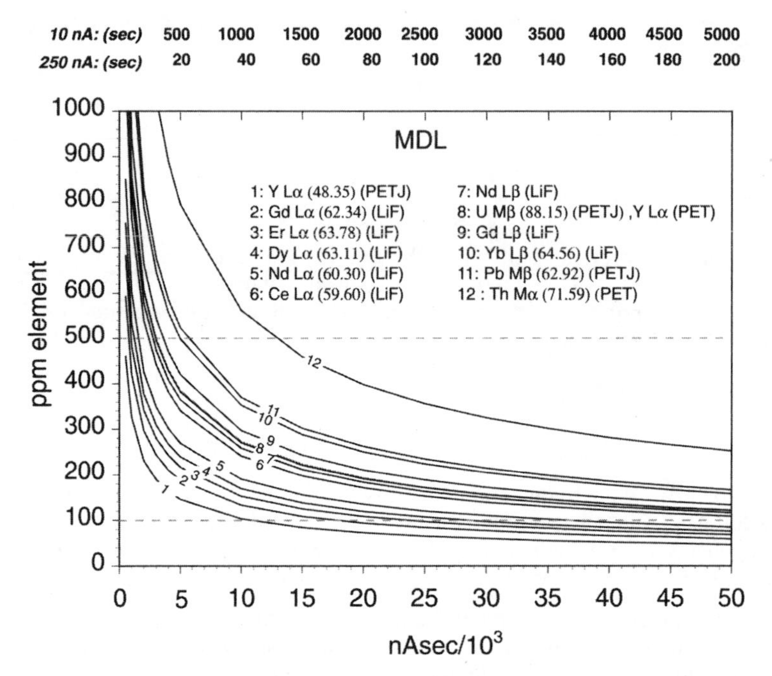

Figure 5. Minimum detection limits for a variety of REE, plus Y, U, Th, and Pb. Detection limits were calculated for by standardization of synthetic REE and Y phosphates, Th and Pb silicates and U oxide, using an equation from Scott et al. (1995). Figure illustrates effect of (1) larger crystal (PETJ vs. PET) and (2) more intense X-ray line (α vs. β) on minimum detection limit. Detection limits shown here are specific to the JEOL 733 Superprobe. EMP and standards of the given composition (wt % element in parentheses) for the elements listed.

The elements (and associated peaks) shown in Figure 5 are the optimal peaks for monazite and/or xenotime analysis, based on avoidance of interferences with other peaks. Any factor influencing the peak count rate will ultimately affect the MDL. Several REE Lα peaks are subject to inter-element interference (see below) and therefore neglected in favor of Lβ peak analysis, despite the higher MDL associated with Lβ peaks.

Mutual interference of L- and M-line elements in monazite and xenotime

Analysis of REE requires avoiding interference on both peak and background for several families of L lines. An abundant REE, such as Ce in monazite, generates over a dozen resolvable characteristic lines. The interference of many families of L lines (and M lines for U, Th, and Pb) (Fig. 6) is the single largest factor impeding accurate quantitative analysis of REE phosphates. Selection of peaks is a trade-off between maximizing X-ray counts and minimizing interferences. Analytical protocols and element interference correction procedures for REE phosphates are discussed at length in Amli and Griffin (1975), Roeder (1985), Scherrer et al. (2000), and Pyle et al. (2001) and summarized here.

Recent software developments have made it possible to examine "virtual" minerals of any composition for theoretical inter-element interferences (e.g., Virtual WDS; Reed and Buckley 1996). User input of analytical conditions (composition, accelerating voltage, diffraction crystal, and detector gas) results in a background-free virtual WDS spectrum as an aid in creating analytical routines. Figure 7 shows a pair of such diagrams. Gd Lα_1 in monazite has a Ce Lγ_1 interference (Fig. 7a), but the Gd Lβ_1 peak (Fig. 7b) is free of major interferences, and thus preferred. In contrast, Gd Lβ_1 in xenotime is subject to a very large Ho Lα_1 overlap, but Gd Lα_1 is virtually free from interference due to negligible Ce concentration in xenotime.

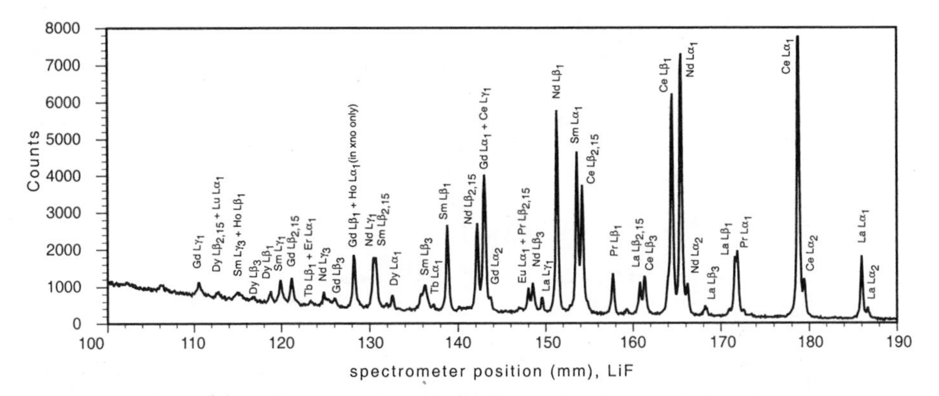

Figure 6. WDS scans of natural monazite (Rapp, 1985) with LiF crystal over the Rowland circle range 100-190 mm. Scan performed with sealed counter using Xe detector gas, an accelerating voltage of 25 keV, a 0.1 mm step size with a dwell time of 5 s/step, and SCA settings including a bias of 1780 V, a gain of 8 V, an SCA window baseline of 1 keV, and window width of 9 keV.

The use of β lines in monazite avoids the need for several element interference corrections. Exley (1980) suggests using Lβ lines for analysis of Pr, Nd, Sm, Gd, and Dy, and Lα lines for Y, La, Ce, Er, and Yb. For MREE in xenotime—specifically Nd, Sm, Eu, Gd, and Tb, the Lα peaks may be used, because of the low Ce and La concentrations. In addition, the use of Gd Lα in xenotime avoids the Ho Lα interference with Gd Lβ.

The elements present in REE phosphates for which M lines are usually measured are Th, U, and Pb. Th Mα is free of any major interferences for typical monazite and xenotime compositions (Scherrer et al. 2000, Pyle et al. 2001). U Mα and U Mβ are both subject to Th interference (Th Mβ_1 and Th Mγ_1, respectively), but the correction is smaller for the case of U Mβ (Scherrer et al. 2000; Table 2), and the Ar K absorption edge compromises U Mα analysis.

The choice of the analytical line for Pb is extremely problematic, and should depend on whether the analytical protocol is for a conventional 15-20 element compositional analysis (Pyle 2001), or optimized for Th-U-total Pb chemical dating of monazite (Williams et al. 1999). Both major Pb M peaks are subject to interferences; Th Mζ_1 and Mζ_2 plus Y L$\gamma_{2,3}$ interfere with Pb Mα, and U Mζ_1 interferes with Pb Mβ. Additionally, second-order La Lα and Ce Lα interfere with Pb Mα and Pb Mβ, respectively, if P10 detector gas is used. The overall interferences on Pb Mβ are of smaller magnitude, especially if Xe detector gas is used, and the use of Pb Mβ is suggested for conventional monazite analysis. However, for chemical monazite dating, unpublished work of Spear et al. suggests that Pb Mα is the superior analytical line if analysis is performed with Xe detector gas. Exploratory work comparing Xe and P10 detector gases shows: (1) less background curvature in the Pb M region of monazite spectra for Xe gas, and: (2) elimination of second order Ce and La escape peaks in the Pb M region if Xe gas is used.

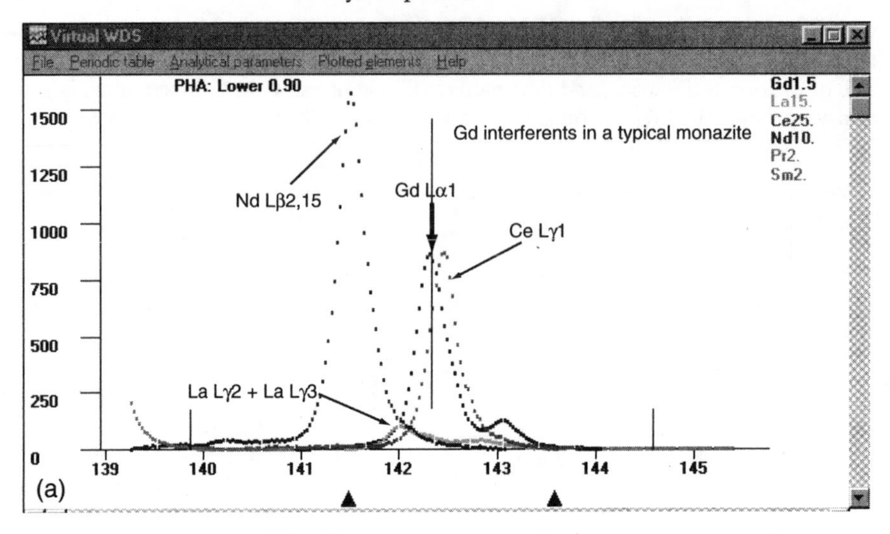

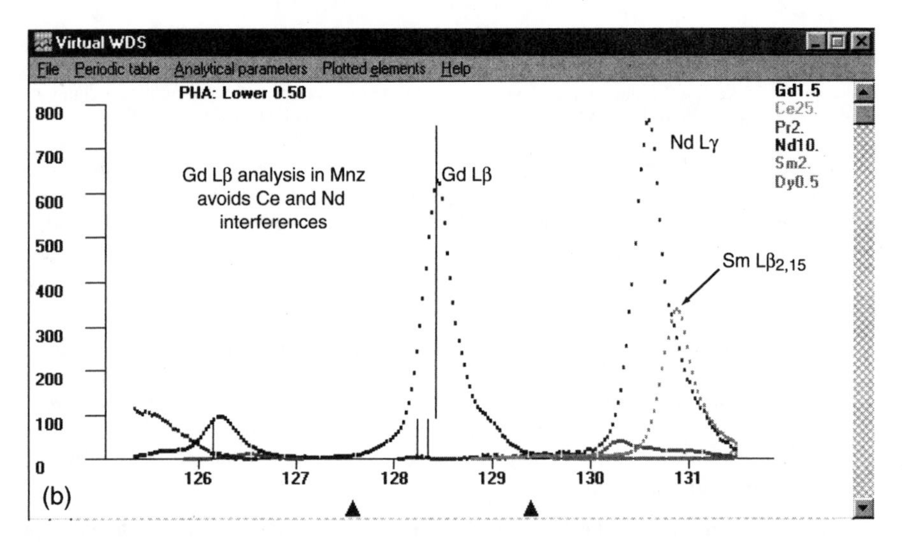

Figure 7. (a) Virtual WDS (Reed and Buckley, 1996) plot for monazite in the region of Gd Lα, LiF crystal. Virtual counter is sealed, with Xe detector gas. Model monazite has typical LREE (La-Gd) content. Minor interference of Nd $L\beta_{2,15}$ and La $L\gamma_2$ + La $L\gamma_3$ peaks, and major interference of the Ce $L\gamma_1$ peak render the Gd Lα peak unusable in monazite analysis. (b) Virtual WDS plot for monazite in the region of Gd Lβ, LiF crystal. Virtual counter is sealed, with Xe detector gas. Model monazite has typical LREE (La-Gd) content. Gd Lβ avoids the interferences of the Gd Lα peak, and is sufficiently resolved from Nd Lγ to use for quantitative analysis of monazite.

Where element interference is unavoidable, several possible corrections exist. The interference correction scheme of Roeder (1985) is outlined below, followed by a listing of theoretical and empirical correction factors (Table 3) for inter-element interferences from multiple sources (Amli and Griffin 1975, Roeder 1985, Reed and Buckley 1996, Scherrer et al. 2000, Pyle 2001).

Table 3. Correction factors for inter-element interference.

Line	*interferent*	*C.F. #1*	*C.F. #2*	*C.F. #3*	*C.F. #4*	*C.F.#5*
P Kα_1	Y Lβ_1				0.050	0.1528
La Lβ_1	Pr Lα_1	0.220				
Nd Lα_1	Ce Lβ_1					0.0027
Pr Lα_1	La Lβ_1	0.124	0.136	0.127	0.156	
Sm Lα_1	Ce Lβ_2	0.002		0.019	0.012	
Eu Lα_1	Pr Lβ_2	0.217	0.233	0.203	0.268	
	Nd Lβ_3	0.012			0.021	
Eu Lβ_1	Dy Lα_2	0.099				
	Dy Lα_1	0.003				
Gd Lα_1	La Lγ_2	0.010		0.028	0.014*	
	La Lγ_3	0.002			*	
	Ce Lγ_1	0.095	0.109	0.118	0.0840	
Gd Lβ_1	Ho Lα_1	0.875			0.780	1.2817
	Ho Lα_2	0.003				
Tb Lβ_1	Er Lα_1	0.117				
Dy Lα_1	Eu Lβ_1	0.001			0.012	0.0062
Ho Lα_1	Gd Lβ_1	0.408		0.421	0.435*	
	Gd Lβ_4	0.002			*	
Ho Lβ_1	Eu Lγ_1	0.003			0.004	
Er Lα_1	Tb Lβ_1	0.052		0.03	0.105*	0.0333
	Tb Lβ_4	0.042		0.03	*	
Er Lβ_1	Gd Lγ_1	0.036			0.025	
	Eu Lγ_3	0.007			0.018	
Tm Lα_1	Sm Lγ_1	0.093	0.091	0.087	0.139	
	Dy Lβ_4	0.015			0.018	
Tm Lβ_1	Gd Lγ_1	0.010			0.035*	
	Gd Lγ_2	0.008			*	
	Tb Lγ_1	0.079			0.132	
Yb Lα_1	Tb Lβ_2	0.003			0.009	
	Dy Lβ_3	0.002			0.008	

Correction for inter-element interference requires knowledge of the wavelength of both the desired peak (λ_d) and interfering peak (λ_i), the width of the peak at 1 standard deviation (width at half height, W), and the intensity of each peak. The amount of peak overlap, X, is given by

$$X = \left(\frac{\lambda_d - \lambda_i}{W - 2} \right) \tag{4}$$

(Bolz and Tuve 1976). The width of the spectral line is a function of spectrometer resolution, and as both peaks are essentially the same wavelength, W will be approximately equal for both peaks. The fraction of counts contributed by the interfering peak to the total count, F, is given by:

$$\boldsymbol{F} = e^{-X2/2} \tag{5}$$

Table 3 (cont)

Line	*interferent*	*C.F. #1*	*C.F. #2*	*C.F. #3*	*C.F. #4*	*C.F.#5*
Yb $L\beta_1$	Tb $L\gamma_2$	0.010			0.028*	
	Tb $L\gamma_3$	0.007			*	
	Dy $L\gamma_1$	0.057			0.087	
	Tm $L\beta_3$	0.003			0.007	
Lu $L\alpha_1$	Dy $L\beta_2$	0.111	0.122	0.090	0.078	
	Ho $L\beta_2$	0.065			0.107	
Lu $L\beta_1$	Dy $L\gamma_2$	0.010			0.024*	
	Dy $L\gamma_3$	0.002			*	
	Yb $L\beta_2$	0.028			0.020	
	Ho $L\gamma_1$	0.024			0.022	
Pb $M\alpha_1$	Y $L\gamma_1$				0.003	0.0031
	Th $M\zeta_{1,2}$				0.001	
Pb $M\beta_1$	U $M\zeta_2$				0.001	
	Ce $L\alpha_1$ (2)				0.000§	
U $M\alpha_1$	Th $M\beta_1$				0.013	0.0101
U $M\beta_1$	Th $M\gamma_1$				0.006	

C.F.: correction factor. Corrected concentration is calculated using the relation $\text{corrected}_{x,unk} = \text{measured}_{x,unk} - \text{CF}^* \text{measured}_{y,unk}$, where x is the desired, y is the interfering element, and concentrations are in wt % oxide, except for CF #5, where correction applies to k-ratios, to which ZAF corrections are subsequently applied

C.F. #1: Roeder (1985) calculated

C.F. #2 Roeder (1985) measured

C.F. #3 Amli and Griffin (1975) measured

C.F. #4: calculated with Virtual WDS (Reed and Buckley, 1996) for 25 kV accelerating voltage, Ar detector gas, and indicated diffraction crystal

C.F.#5: Pyle (2001): measured by analysis of synthetic phosphate, silicate, sulfide, and oxide standards.

*Virtual WDS unable to deconvolute adjacent peaks: C. F. represents sum of counts due to convolution of $\alpha_1 + \alpha_2$, $\beta_1 + \beta_4$, etc.

§ filtered out with proper PHA settings

This assumes an approximately Gaussian peak shape. ***F*** is then multiplied by the ratio of the intensity of the interfering peak to the intensity of the measured peak for that element. Typically, the interfering peak cannot be measured directly since it sits on top of the desired peak, so an alternate peak must be measured and the theoretical intensity of the interfering peak calculated from the known ratio. For example, La $L\beta_1$ interferes with Pr $L\alpha_1$ in monazite. To calculate the intensity of La $L\beta_1$, the La $L\alpha_1$ peak is measured and multiplied by 57 % , the intensity ratio of La $L\beta_1$ to La $L\alpha_1$ as reported by Roeder (1985). Figure 8 (adapted from Roeder 1985) shows intensity ratios of various L lines relative to $L\alpha_1$ for different REE with increasing atomic number. Spectrometer response and efficiency vary among EMPs; thus, the relative intensities shown in Figure 7 will vary slightly from machine to machine.

For corrections based on theoretical X-ray intensity ratios, an additional problem for several REE arises due to $L\beta_4$ overlap of $L\beta_1$. The crosses on the $L\beta_1$ trend-line show the measured intensity ratio before subtraction of the counts due to $L\beta_4$ overlap of $L\beta_1$. This interference is most severe for Nd Lβ and Sm Lβ, where overlap of $L\beta_4$ and $L\beta_1$ is nearly total. White and Johnson (1979) give relative intensities of $L\beta_1$ and $L\beta_4$ REE peaks, and the contribution of $L\beta_4$ to $L\beta_1$ can be determined using Equations (4) and (5) above.

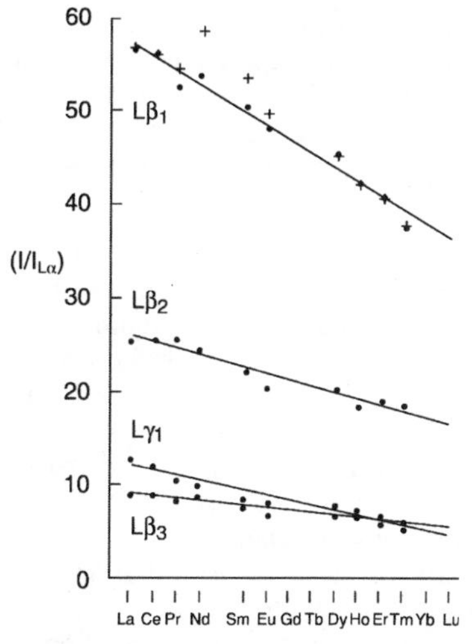

Figure 8. Plot of X-ray line intensity ratios for REE Lα and lower intensity lines ($I/I_{L\alpha}$), redrafted from Roeder (1985). For the $L\beta_1$ trend line, + = measured intensity of $L\beta_1$ lines, and • = $L\beta_1$ intensity corrected by subtraction of $L\beta_4$ interference.

The calculation of peak-overlap corrections assumes that relative peak intensities remain the same regardless of matrix (Roeder 1985). In addition, samples containing other elements with peaks in the Lα to Lγ region for any particular REE may affect the relative intensities due to secondary fluorescence and absorption. Roeder (1985) measured the relative intensities of Dy $L\beta_1$ and Dy $L\alpha_1$ in pure Dy and an alloy of 50 wt % Fe and 50 wt % Dy to investigate the effect of Fe absorption of Dy X-rays. The intensity ratio decreased from 0.44 (pure Dy) to 0.425 (50-50 Dy-Fe), showing that the absorption effect is detectable, but minor. In addition, detector gas absorption may cause changes in intensity ratios.

As an example of REE interference, Roeder (1985) calculated the contribution of La $L\beta_1$ interference to total Pr, as measured on Pr $L\alpha_1$. The values of λ_d for Pr$L\alpha_1$ (2.4630 Å), λ_i for La $L\beta_1$ (2.4589 Å), and W (0.0047 Å) yield an F value of 0.22. This F value is multiplied by the ratio of intensities of La $L\beta_1$ to La $L\alpha_1$, the peak measured to obtain La concentration, (0.57, from Fig. 7) to give a final correction factor (CF) of 0.125. The corrected Pr concentration is then calculated as:

$$\text{corrected}_{\text{Pr,unk}} = \text{measured}_{\text{Pr,unk}} - \text{CF} \times \text{measured}_{\text{La,unk}} \qquad (6)$$

Alternately, interferences between two elements, x and y, may be measured empirically. EMP-specific correction factors, which incorporate machine specific effects such as spectrometer resolution, diffraction crystal quality, X-ray collimator settings, detector gas, and detector electronics, may be calculated by measuring the apparent intensity of element x in the x-free standard for element y. The apparent intensity of x is due to the interference of some peak λ_i of element y with the desired peak λ_d of element x. For example, Pyle (2001) measured Y $L\beta_1$ interference on P $K\alpha_1$. Analysis of pure Y_2O_3 (Geller Analytical) yields an apparent P k-ratio of 0.220 which arises from measuring some number of Y $L\beta_1$ counts at the P $K\alpha_1$ peak position, and a Y k-ratio of 1.44 (in this instance, the normal Y standard YPO_4 (k-ratio ≈ 1) can not be used because it contains phosphorus). The resulting correction factor is the apparent k-ratio of P in Y_2O_3 divided by the k-ratio of Y in Y_2O_3 = (0.22/1.44) = 0.1528. The correction factors measured in the above manner should be applied to the raw k-ratios themselves, rather than the post-ZAF wt % oxide values, i.e. (for the above example),

$$\text{corrected k-ratio}_{\text{P,unk}} = \text{measured k-ratio}_{\text{P,unk}} - \text{CF} \times \text{measured k-ratio}_{\text{Y,unk}} \qquad (7)$$

The k-ratios corrected for elemental interference should then have ZAF corrections applied.

Empirical correction factors measured for a number of interferences are listed in Table 3, column 7 (C.F. #5). A survey of the theoretical and empirical correction factors (Table 3) shows varying degrees of agreement among the different sets. In some cases, the agreement is quite good (less than 10% relative difference), as in the case of correction factors 1-3 (Roeder 1985, Amli and Griffin 1975) for La Lβ interference on Pr Lα. In others, there are large disparities between theoretical and empirical corrections for the same interference (e.g. Ho Lα interference on Gd Lβ, and Tb Lβ interference on Er Lα). These differences likely arise due to machine specific factors such as quality of diffraction crystal, detector gas and electronic settings, operating conditions, and, perhaps most importantly, X-ray collimator settings.

The entire surface of an X-ray diffraction crystal will lie on a Rowland circle of radius R only if the crystal is bent to a radius of 2 R and the surface of the crystal is ground to a radius of R (the so-called Johansson (1933) focusing). To avoid crystal degradation (especially in the case of PET and TAP crystals), the surface of the diffraction crystal is not typically ground. This results in a geometry known as Johann (1931) focusing, where much of the crystal surface does not lie on the Rowland circle. As the X-ray beam strikes farther from the center of the diffraction crystal, the departure from the Bragg angle increases, and X-ray wavelength resolution decreases. Placement of an X-ray collimator in front of the X-ray detector improves wavelength resolution (at the expense of intensity) and also decreases the intensity of background radiation (thus improving the peak-to-background ratio, as well). As a result, empirical element interference correction factors are highly sensitive to collimator setting and positioning. Thus: (1) machine-specific element interference corrections are applicable only if the identical collimator settings are used in every set of analyses, and (2) empirical corrections measured on one EMP are likely not to be portable to other EMPs.

One way to assess the effect of various correction factors is to apply them to a single analysis and examine the effect that the corrections have on composition-dependent parameters such as thermometry estimates or chemical ages. Table 4 shows an analysis of monazite from a sample of sillimanite-zone schist from west-central New Hampshire (Spear et al. 1995). This monazite shows the scale of interferences on particular lines (Pr Lα, Gd Lα, Pb Mα, U Mα) and the range of compositions generated by different correction factors. Two sets of corrections are applied to the original analysis: a correction for La Lβ interference on Pr Lα plus Ce Lγ interference on Gd Lα (Table 4, part A), and corrections for Y Lγ interference on Pb Mα plus Th Mβ interference on U Mα (Table 4, part B). The high analysis total of the uncorrected analysis is an indication of the existence of these interferences.

The REE corrections show a wide range of values (0.124-0.156 for La interference, 0.084-0.118 for Ce interference), and the effect on final Gd concentration is notable. For monazite thermometry, Gd is partitioned more strongly into xenotime than coexisting monazite. Given the coincidence of the position of the monazite solvus limb in the system $CePO_4$-YPO_4 (Gratz and Heinrich 1997) with the empirical monazite limb of Heinrich et al. (1997), Gratz and Heinrich (1997) suggest that application of their experimental monazite solvus thermometer to natural samples include Y + total HREE (Gd-Lu) in calculation of the monazite mole fraction to be used in the thermometer expression. The correction for Ce interference on Gd results in substantially lower values for X_{Y+HREE} of monazite (by 0.025 to 0.035), and reducing temperature estimates by 110° to 175°C. This is significant, but may be avoided by measuring the Gd Lβ peak in monazite, where Ce Lγ interference is negligible, and low Ho concentrations minimize Ho Lα interference. Other mutual HREE interferences exist (Er-Tb, Yb-Tb, Dy-Lu), but the concentrations of both interferent and desired analyte in monazite are generally too

Table 4. Effect of interference corrections on thermometry and age estimates, central New England (NH) monazite 89-22.

A. La Lβ on Pr Lα *and* Ce Lγ on Gd Lα

element	*original*	*C.F.#1*	*C.F.#2*	*C.F. #3*	*C.F. #4*
C.F. (Pr)		0.124	0.136	0.127	0.156
C.F. (Gd)		0.095	0.109	0.118	0.0840
SiO_2	0.10				
P_2O_5	30.29				
CaO	1.05				
Y_2O_3	2.22				
La_2O_3	13.94				
Ce_2O_3	28.50				
Pr_2O_3	4.42	2.69	2.52	2.65	2.24
Nd_2O_3	12.76				
Sm_2O_3	2.12				
Gd_2O_3	4.45	1.74	1.34	1.09	2.06
Tb_2O_3	n.a.				
Dy_2O_3	0.76				
Er_2O_3	0.10				
ppm Pb	857				
ppm Th	29523				
ppm U	5025				
Total	104.73	100.29	99.72	99.82	100.16
X_{Y+H}	0.1088	0.0806	0.0761	0.0729	0.0848
T (°C)	812	681	655	636	703

B. Y Lγ on Pb Mα *and* Th Mβ on U Mα

element	*original*	*C.F.#4*	*C.F.#5*
C.F. (Pb)		0.003	0.0031
C.F. (U)		0.013	0.0101
SiO_2	0.10		
P_2O_5	30.29		
CaO	1.05		
Y_2O_3	2.22		
La_2O_3	13.94		
Ce_2O_3	28.50		
Pr_2O_3	2.24		
Nd_2O_3	12.76		
Sm_2O_3	2.12		
Gd_2O_3	2.06		
Tb_2O_3	n.a.		
Dy_2O_3	0.76		
Er_2O_3	0.10		
ppm Pb	857	778	778
ppm Th	29523		
ppm U	5025	4563	4666
Total	100.16	100.10	100.11
Age (Ma)	417	392	389

small (<<1 wt %) to have much effect on monazite HREE content. Gratz and Heinrich (1997) quote an uncertainty on X_Y of ±0.01, so the uncertainty in composition and temperature due to inter-element interference corrections is roughly 2 to 3 times the uncertainty associated with analytical imprecision.

The second set of corrections (Table 4, Part B) is for Y interference on Pb and Th interference on U, as applied to chemical U-Th-Pb monazite ages. The uncorrected U, Pb and Th concentrations yield a U-Th-Pb chemical age of 417 Ma, and the corrected compositions yield chemical U-Th-Pb ages of 392 Ma (Virtual WDS correction) and 389 Ma (empirical corrections from Pyle 2001). The empirical and Virtual WDS correction factors are virtually identical for Y $L\gamma_{2,3}$ interference on Pb $M\alpha_1$ (0.0031 and 0.0030, respectively), so most of the corrected age estimate difference arises from the difference in Virtual WDS (0.0130) and empirically measured (0.0101) correction factors for Th $M\beta_1$ interference on U $M\alpha_1$.

Scherrer et al. (2000) found overall older ages and larger age variations associated with analysis of Pb $M\alpha_1$ vs. Pb $M\beta_1$. Use of the Pb $M\beta_1$ line (corrected and uncorrected) combined with either U $M\alpha_1$ or U $M\beta_1$ resulted in lower overall ages (411-426 Ma) and a smaller spread in ages than did the use of Pb Ma in the same manner (456-473 Ma). The uncertainty on counting statistics for chemical age of mid-Paleozoic monazite with ~40000 ppm Th, ~5500 ppm U, and ~1000 ppm Pb, determined on the JEOL 733 Superprobe at RPI (Pyle unpublished) is on the order of 390±30 Ma (8 % relative), a factor of 2 greater than corrections due to mutual element interference; for older and/or more thorogenic monazite, the age uncertainties approach those generated by elemental interference.

ZAF corrections

Because raw X-ray count intensities in an unknown material are ratioed to the count intensity for the same element in a standard, the ratio of the ZAF factor in the unknown relative to the standard is critical. Minimization of the difference in matrix composition between standard and unknown will minimize errors due to inaccurate ZAF factors. ZAF corrections account for: (1) differences in average atomic number between standard and unknown (Z factor); (2) absorption of excited X-rays by matrix components (A factor), and; (3) secondary fluorescence of X-rays by other characteristic X-rays, rather than primary beam electrons (F correction). For REE phosphates, these corrections are significant for some elements (Table 5), e.g., Y and Pb. This raises the question as to the effect that the accuracy of the ZAF factors and/or application of different ZAF correction models have on the ultimate composition estimate.

For standards, corrections for fluorescence are minor, but absorption corrections for the light element standards (Si, P) are very significant (2.61 and 2.32, respectively), and atomic number corrections are on the order of ±10-20%. In monazite, fluorescence corrections are again minor, but absorption corrections are extreme (>100 %) for Si and P, and very large (>20 %) for Ca, Y, and Pb. The final ZAF correction factors (Table 5, last column), show significant (>10 %) corrections for Si, Y, the MREE, U, Pb, Th, and U, with very high calculated corrections for Y (1.75) and Pb (1.36), which are two of the more critical elements measured in monazite, due to their use in thermometry (Gratz and Heinrich 1997, Pyle et al. 2001) and geochronology (Parrish 1990, Suzuki and Adachi 1991, Montel et al. 1996).

As a variety of potential standard materials exist for each element, the choice of standard material for "critical" elements should be, in part, based on the ZAF factor for a potential standard material relative to that of a typical unknown. Potential standards for Pb include pure Pb as well as lead sulfide, silicate, phosphate, and chromate.

Table 5. *ZAF factors from analysis of monazite standards and Trebilcock monazite.

el	*Standard*	$[Z]_{std}$	$[A]_{std}$	$[F]_{std}$	$[ZAF]_{std}$	$[Z]_{unk}$	$[A]_{unk}$	$[F]_{unk}$	$[ZAF]_{unk}$	$[ZAF]_{unk\ std}$
Si	$ThSiO_4$	0.8080	2.6135	1.0000	2.1116	0.8673	2.9265	0.9938	2.5223	1.1945
P	$CePO_4$	0.9078	2.3171	0.9078	2.0959	0.9008	2.2504	0.9978	2.0226	0.9651
Ca	apatite	1.0258	1.0561	1.0000	1.0834	0.8930	1.3546	0.9752	1.1797	1.0889
Y	YPO_4	1.1066	0.9284	1.0000	1.0274	1.035	1.7387	0.9982	1.7962	1.7483
La	$LaPO_4$	1.1131	0.9397	1.0000	1.0460	1.1075	0.9987	0.9945	1.0999	1.0515
Ce	$CePO_4$	1.1087	0.9433	1.0000	1.0459	1.0998	1.006	0.9976	1.1038	1.0554
Pr	$PrPO_4$	1.1030	0.9461	1.0000	1.0435	1.0894	0.9993	0.9973	1.0856	1.0403
Nd	$NdPO_4$	1.1071	0.9459	1.0000	1.0472	1.0990	0.9966	0.9991	1.0943	1.0450
Sm	$SmPO_4$	1.1131	0.9461	1.0000	1.0531	1.1141	0.9908	0.9999	1.1037	1.0481
Gd	$GdPO_4$	1.1218	0.9523	1.0000	1.0683	1.1351	1.0373	1.0000	1.1774	1.1021
Tb	$TbPO_4$	1.1205	0.9541	1.0000	1.0691	1.1330	1.0731	1.0000	1.2158	1.1372
Dy	$DyPO_4$	1.1252	0.9550	1.0000	1.0746	1.1446	1.0601	0.9999	1.2134	1.1292
Er	$ErPO_4$	1.1275	0.9580	1.0000	1.0801	1.1514	1.0623	0.9999	1.2230	1.1323
Pb	$PbSiO_3$	1.0782	0.9583	1.0000	1.0333	1.1445	1.2298	0.9998	1.4073	1.3619
Th	$ThSiO_4$	1.0800	0.9047	1.0000	0.9771	1.1496	0.9835	0.9995	1.1301	1.1566
U	UO_2	1.0298	0.9344	1.0000	0.9622	1.1492	0.9474	0.9995	1.0881	1.1309

*Equations for [Z], [A], and [F] from Armstrong (1984)

Table 6. *ZAF factors for various lead standards and Trebilcock monazite

Material	$[Z]_{std}$	$[A]_{std}$	$[F]_{std}$	$[ZAF]_{std}$	$[ZAF]_{unk/std}$
Trebilcock (unknown)	1.1445	1.2298	0.9998	1.4073	1.0000
Pb	1.0000	1.0000	1.0000	1.0000	1.4073
PbS	1.0336	0.9212	1.0000	0.9521	1.4781
$PbSiO_3$	1.0782	0.9583	1.0000	1.0333	1.3619
$Pb_5(PO_4)_3Cl$	1.0623	0.9569	0.9999	1.0164	1.3846
$PbCrO_4$	1.0861	0.8876	0.9998	0.9638	1.4602
NIST 610 (426 ppm Pb)	1.3336	0.9201	0.9995	1.2264	1.1475
†CaAlSiO glass with 10 wt % PbO	1.2964	0.9006	0.9992	1.1666	1.2063

*Equations for [Z], [A], and [F] from Armstrong (1984)

† Composition 14.87 wt % Ca, 7.18 wt % Al, 26.00 wt % Si, 9.28 wt % Pb, 42.67 wt % O

Table 6 shows calculated ZAF factors for these materials as well as a hypothetical calcium aluminosilicate glass with a CaO-Al_2O_3-SiO_2 ternary eutectic ratio plus 10 wt % PbO, and NIST 610 glass, a sodium-calcium aluminosilicate glass with a certified Pb concentration of 426±1 ppm (see also Pearce et al. 1997). ZAF factor ratios for the standard materials and Trebilcock monazite (Tomascak et al. 1996) are also given in Table 6.

Although the unknown-to-standard ZAF product ratio is important, and should be minimized if at all possible, the ratios of the individual factors are critical for two reasons. First, errors in ZAF factors are propagated through individual terms, rather than the product of the terms. Thus, even if the [ZAF] ratios are close to 1.0, individual terms may be quite different in the standard and the unknown, and larger differences will propagate accordingly.

Second, in empirically correcting for peak interferences (e.g. Y $L\gamma_{2,3}$ on Pb $M\alpha_1$), it is implicitly assumed that the ZAF correction for the measured line of an element (e.g. Y $L\alpha$) also applies to the interfering line (e.g. Y $L\gamma$). Thus, changes in the intensities of Y $L\alpha$ and Y $L\gamma$ are assumed to be proportional. However, for a pure phosphorus absorber, the mass attenuation coefficient for Y $L\alpha_1$ is 356 cm^2/gm, as opposed to 2154 cm^2/gm for Y $L\gamma_{2,3}$ absorption by phosphorus (Seltzer 1993), because the K absorption edge for phosphorus (2.1455 keV) lies between the energies of Y $L\alpha_1$ (1.92256 KeV) and Y $L\gamma_{2,3}$ (2.6638 keV). Application of a correction factor involving a mass absorption coefficient for Y $L\alpha_1$ in phosphorus actually underestimates (to nearly a factor of 10) the degree to which the Y $L\gamma_{2,3}$ X-rays that produce the apparent lead are absorbed. Ergo, absorption and fluorescence can be very different for different lines, which can potentially yield different ZAF factors.

Table 6 shows ZAF factors calculated for Pb in Trebilcock monazite and a series of potential Pb standards. Overall, the crystalline lead standards are better match to the [Z] of the unknown than are the lead-bearing glasses, but the larger values of [Z] for the Pb-bearing glasses offset the low [A] values of all standards relative to [A] for Trebilcock monazite. In terms of matching individual [Z], [A], and [F] terms between standard and unknown, Pb silicate is the best standard choice, but the selection of either glass as a Pb standard results in the lower [ZAF] unknown: [ZAF] standard (1.15-1.20). The low Pb content of NIST 610 makes it unsuitable for a lead standard, unless calibration time is long enough to generate sufficient lead counts at the desired statistical precision. The Ca-Al-Si glass with 10 wt % PbO has sufficient lead to be used as a standard, whereas NIST 610 may be used to determine the accuracy of microprobe lead analyses at low (< 500 ppm) levels of concentration. To correct for differences in [ZAF] between standard and unknown, element interference correction factors should be multiplied by the unknown ZAF: standard ZAF ratio for the interfering element, subject to the caveat discussed above.

The theory and implementation of various ZAF corrections has been described by Goldstein et al. (1984) and Scott et al. (1995). Although a full discussion is beyond the scope of this paper, it is worth noting that small values of $f(\chi)$ (absorption coefficient) result in large [A] values. $f(\chi)$ is generally minimized by maximizing take-off angle (Ψ) and minimizing overvoltage (beam energy minus critical shell excitation energy). Most EMPs have take-off angles between 40° and 52.5°, and many monazite analytical protocols require accelerating voltages of 20 keV or greater (Scherrer et al. 2000, Pyle 2001), resulting in large overvoltages for most analyzed elements. Yakowitz and Heinrich (1968) suggest maintaining $f(\chi)$ at values at or above 0.8 in order to avoid serious analytical errors arising from uncertainties in input parameters, and to minimize discrepancies between different equations for $f(\chi)$. For our standards and analytical conditions at RPI, $f(\chi)$ for Y, Th , U, and Pb are 0.58, 0.58, 0.60, and 0.50, suggesting that these elements are subject to both large absorption corrections and large errors due to propagation of ZAF input data uncertainty.

Different ZAF expressions will obviously lead to different apparent compositions, and Table 7 shows an analysis of Th-rich Trebilcock monazite (Tomascak et al. 1996) pegmatite with uncorrected and corrected k-ratios converted to compositions according to six different ZAF correction routines. Each ZAF formulation uses different forms of the equations constituting the [Z] and [A] portions of the correction; $f(\chi)$ (or $\phi(\rho z)$) and $\phi(0)$ expressions for [A], backscatter correction (R), stopping power (S), backscatter coefficient (η), mean ionization potential (J), and ionization cross-section (Q) for [Z], along with corrections for both characteristic and continuum fluorescence. Variations in the final oxide concentrations among the 6 correction schemes are negligible for P and the REE (<2 % relative), but higher for Th (4% relative), Ca (4 % relative), Pb (6% relative), and Si (6% relative), and highest for Y (10% relative) and U (13% relative), reflecting the large $f(\chi)$ values for these elements, and the resultant divergence arising from application of different ZAF correction schemes (Yakowitz and Heinrich 1968).

The bottom of Table 7 shows two sets of calculations using both the raw data and the various concentrations of (Y+HREE) and (U+Th+Pb) generated by each ZAF correction scheme. The first set is a temperature calculation employing the Gratz and Heinrich (1997) monazite-xenotime thermometer for the various values of $X_{(YPO4+HREEPO4)}$, and the second shows the U-Th-Pb chemical age of the analyzed spot (e.g. Montel et al. 1996), assuming all lead is derived by decay of U and Th. The absolute spread on the temperatures for the 6 different ZAF corrections is ~20°C (856-875°C), despite the fact that the ZAF corrections themselves are large for individual elements such as Y and Pb. The Gratz and Heinrich (1997) calibration uncertainty is ~±60°C at this temperature. Thus, ZAF corrections are a minor contributor to thermometric error. The Armstrong (1984), Heinrich (1985), and Bastin et al. (1984) ZAF corrections are in good agreement with each other, whereas both Love-Scott (Sewell et al. 1985) corrections and the Duncumb and Reed (1968) correction yield lower temperatures than the former group of ZAF corrections.

The total range in chemical U-Th-Pb ages is approximately 20 million years (259-278), with the Armstrong (1984) ZAF correction yielding the highest age (278 Ma). The

Table 7. Effect of ZAF correction on Trebilcock monazite composition.

Oxide	*Uncorrected*	*Armstrong*	*Heinrich*	*Bastin*	*Love/Scott I*	*Love/Scott II*	*Duncumb/Reed*
SiO_2	1.27	1.52	1.53	1.54	1.60	1.63	1.56
P_2O_5	29.26	28.21	28.28	28.22	28.05	28.00	28.36
CaO	0.79	0.83	0.83	0.83	0.80	0.81	0.79
Y_2O_3	1.64	2.86	2.89	2.84	2.67	2.67	2.60
La_2O_3	7.98	8.39	8.46	8.34	8.33	8.37	8.43
Ce_2O_3	21.55	22.74	22.89	22.72	22.59	22.72	22.84
Pr_2O_3	2.84	2.96	2.97	2.98	2.94	2.96	2.97
Nd_2O_3	10.96	11.45	11.50	11.50	11.50	11.39	11.46
Sm_2O_3	4.00	4.19	4.21	4.20	4.18	4.21	4.21
Gd_2O_3	2.57	2.83	2.84	2.82	2.81	2.85	2.85
Tb_2O_3	0.36	0.41	0.41	0.41	0.40	0.41	0.41
Dy_2O_3	0.91	1.03	1.03	1.02	1.02	1.04	1.04
Er_2O_3	0.13	0.14	0.14	0.14	0.14	0.14	0.14
PbO	0.10	0.17	0.17	0.17	0.17	0.17	0.17
ThO_2	10.68	12.35	12.61	12.49	12.61	12.70	12.81
UO_2	0.55	0.62	0.67	0.67	0.65	0.66	0.71
Total	95.59	100.72	101.43	100.89	100.46	100.73	101.35
Temp, °C	778	875	875	873	862	862	856
Age, Ma	191	278	266	273	272	271	259

Compositions in weight percent.

Bastin et al. (1984) and Love-Scott I and II (Sewell et al. 1984) corrections yield nearly identical ages (273, 272, 271 Ma), and the Heinrich (1985) and Duncumb and Reed (1968) ZAF corrections yield the lowest ages (266, 259 Ma). Thus, the uncertainty associated with the various ZAF corrections is of the same order as the statistical uncertainty.

SUMMARY

Accurate EMP analysis of phosphates is an exacting procedure, but problems may be alleviated by employing the analytical procedures outlined in this chapter. Short analysis times are suggested for major elements (Ca, P, F, Cl) in apatite in order to minimize F excitation, followed by longer analyses for trace constituents of apatite.

The analysis of monazite requires detailed WDS scans of both standards and "typical" unknowns for placement of critical background positions, knowledge of desired detection limits and the analytical time and current needed to achieve these limits, and careful selection of analytical peaks to avoid mutual element interference. Changes in count rate between standard and unknown may cause PHA peak shifts and loss of counts when narrow energy windows are used. Placement of background positions on either side of Ar absorption peaks should be avoided.

The lines used in analyzing monazite should not always be used for xenotime as the change in composition between monazite (LREE-rich) and xenotime (HREE-rich) lessens the severity of some interferences and increases the severity of others. Where interferences are unavoidable, correction errors may be as large as counting statistics errors. ZAF factors are a function of differences in composition between standard and unknown, but also machine operating conditions. ZAF corrections can be reduced by selecting calibration standards compositionally similar to the unknowns, but the compositional complexity of most metamorphic (REE) phosphates makes standard selection non-trivial. Furthermore, incorporation of many elements with a wide range in characteristic X-ray energies necessitates high accelerating voltage to excite high-energy X-rays, at the cost of large overvoltages for elements with lower energy X-rays. This results in larger sample excitation volumes and an overall increase in ZAF correction factors. However, the range of compositions generated by using different ZAF models is roughly on the order of compositional uncertainty due to counting statistics.

The following procedures for undertaking EMP analysis of REE phosphates are suggested, although many of the procedures and/or corrections summarized here are *machine-specific*:

Standards. REE phosphates should be used as standards for the REE. Silicates or phosphates of U, Th, and Pb are also suggested as standards for these elements. For EMP dating of monazite, use of lead silicate or phosphate results in smaller ZAF corrections for typical monazite composition than does the use of lead chromate or sulfide. Lead-doped (500-5000 ppm) alkaline earth aluminosilicate glass produces the smallest ZAF corrections for typical monazite compositions, but may generate insufficient X-ray counts due to low lead concentration, compromising its use as a standard.

Accelerating voltage. An accelerating voltage between 20-30 keV should be adopted. Higher accelerating voltages ensure sufficient intensity of HREE X-rays, but lower accelerating voltages minimize overvoltage for elements with lower energy critical excitation energies, and result in lower ZAF correction factors for these elements.

Analysis current (Fig. 5). A Faraday cup current of 50 nA (at 25 keV) combined with 1-2 min counting on both peak and high and low background yields MDLs on the order of 200-1000 ppm for REE, Y, U, Pb, and Th for the standards given here. For EMP dating of monazite, a 250 nA cup current, 3 min of analysis on both peak and background, and calibration with a lead silicate standard yields an MDL for Pb of approximately 200 ppm. Lα and Mα peaks yield lower MDL than associated β peaks, but should not be used if subject to inter-element interference.

X-ray detector electronic settings (Fig. 2). For elements where narrow voltage windows are required to filter out interferences from higher-order peaks, calibration and analysis current should be varied so as to: (a) minimize changes in X-ray count rate, and (b) minimize voltage pulse distribution shift. Detector settings are EMP-specific.

Detector gas absorption edges (Figs. 3,4; Table 2). If Ar (or P10) is used as a detector gas, *high* background placement for analysis of U Mβ should be on the *low* side of the Ar K absorption edge, else linear interpolation will result in an underestimation of background counts under the U Mβ peak.

Element interferences and corrections (Tables 1 and 3). For conventional monazite analysis, the Lα peaks of the MREE Pr-Tb should be avoided due to La, Ce and Nd interferences, but low concentrations of La, Ce, and Nd permit analysis of the MREE Lα peaks in xenotime; Gd Lα in xenotime avoids the complete overlap of Gd Lβ with Ho Lα. Th Mα may be analyzed in monazite and xenotime, but U Mβ and P Mβ should be analyzed in both REE phosphates so as to avoid significant Y, Th and U interferences. Element interference corrections are required for U Mβ and Pb Mβ analysis, but these are minor corrections. Empirical correction factors should be calculated on a machine-specific basis, require that session-to-session X-ray collimator settings be identical, and corrections should be performed on raw k-ratio values rather than ZAF-corrected oxide or element concentrations. Use of Xe detector gas in monazite chemical age analysis greatly reduces interference from second-order La Lα and Ce Lα escape peaks.

ZAF corrections (Tables 5-7). ZAF corrections for certain elements (Si, Ca, Y, U, Pb) are inherently large due to overvoltage requirements and typical monazite compositions resulting in large [A] factors for unknowns. However, variation in compositions resulting from application of different ZAF models is equal to or less than the compositional variation associated with analytical uncertainty.

ACKNOWLEDGMENTS

We thank K. Becker for assistance with analyses and maintenance of the JEOL 733 Superprobe at Rensselaer Polytechnic Institute, and P. Roeder and J. Armstrong for helpful discussions. This work was partially funded by NSF grant EAR-0106738. The review and editorial handling of M. Kohn is gratefully acknowledged.

REFERENCES

Amli R (1975) Mineralogy and rare earth geochemistry of apatite and xenotime from the Gloserheia granite pegmatite, Froland, southern Norway. Am Mineral 60:607-620

Amli R, Griffin WL (1975) Microprobe analysis of REE minerals using empirical correction factors. Am Mineral 60:599-606

Armstrong JT (1984) Quantitative analysis of silicate and oxide minerals: a reevaluation of ZAF corrections and proposal for new Bence-Albee coefficients. Microbeam Anal-1984 19:208-212

Bastin GF, vanLoo FJJ, Heijligers HJM (1984) An evaluation of the use of Gaussian prz curves in quantitative electron probe microanalysis. X-Ray Spectrom 13:91-97

Bea F, Montero P (1999) Behavior of accessory phase and redistribution of Zr, REE, Y, Th, and U during metamorphism and partial melting of metapelites in the lower crust: an example from the Kinzigite Formation of Ivrea-Verbano, NW Italy. Geochim Cosmochim Acta 63:1133-1153

Bolz RE, Tuve GL (1976) CRC Handbook of Tables of Applied Engineering Science. CRC Press, Cleveland, Ohio

Broska I, Petrík I, Williams CT (2000) Coexisting monazite and allanite in peraluminous granitoids of the Tribec Mountains, Western Carpathians. Am Mineral 85:22-32

Drake MJ, Weill DF (1972) New rare earth element standards for electron microprobe analysis. Chem Geol 10:179-181

Duncumb P, Reed SJB (1968) The calculation of stopping power and backscattering effects in electron microprobe analysis. In Heinrich KFJ (Ed) Quantitative Electron Microprobe Analysis, NBS Spec Pub 298. U S Dept Commerce Washington, DC, 133-154

Exley RA (1980) Microprobe studies on REE-rich accessory minerals; implications for Skye granite petrogenesis and REE mobility in hydrothermal systems. Earth Planet Sci Lett 48:97-110

Finger F, Broska I (1999) The Gemeric S-type granites in southeastern Slovakia: late Paleozoic or Alpine intrusions? Evidence from electron microprobe dating of monazite. Schweiz mineral petrogr Mitt 79:439-443

Förster HJ (1998) The chemical composition of REE-Y-Th-U-rich accessory minerals in peraluminous granites of the Erzgebirge-Fichtelgebirge region, Germany, Part I: The monazite-(Ce)-brabantite solid solution series. Am Mineral 83:259-272

Goldstein J, Newbury D, Echlin P, Joy D, Fiori D, Lifshin E (1984) Scanning Electron Microscopy and X-Ray Microanalysis. Plenum, New York

Gratz R, Heinrich W (1997) Monazite-xenotime thermobarometry: experimental calibration of the miscibility gap in the system $CePO_4$-YPO_4. Am Mineral 82:772-780

Harlov DE, Förster HJ, Nijland TG (2002) Fluid-induced nucleation of (Y+REE)-phosphate minerals within apatite: nature and experiment. Am Mineral 87:245-261

Heinrich KFJ (1985) A simple, accurate absorbtion model.Microbeam Analysis–1985 20:79-81

Heinrich W, Andrehs G, Franz G (1997) Monazite-xenotime miscibility gap thermometry. I. An empirical calibration. J Metamor Geol 15:3-16

Jarosewich E, Boatner LA (1991) Rare-earth element reference samples for electron microprobe analysis. Geostd Newslett 15:397-399

Johann HH (1931) Die Erzeugung lichstarker Röntgenspektren mit Hilfe von Konkavkristallen. Z Phys 69:185-206

Johannson T (1933) Über ein neuartiges genau fokussierendes Röntgenspektrometer. Z Phys 82:507-528

Kifer J, Wolf D (2000) Fluorine partitioning between apatite and granitic melts. Geol Soc Am North-Central Sect Mtg Abstr Prog 32:21

Montel JM, Foret S, Veschambre M, Nicollet C, Provost A (1996) Electron microprobe dating of monazite. Chem Geol 131:37-53

Parrish R (1990) U-Pb dating of monazite and its application to geological problems. Can J Earth Sci 27:1431-1450

Pearce NJG, Perkins W, Westgate JA, Gorton MP, Jackson SE, Neal CR, Chenery SP (1997) A compilation of new and published major and trace element data for NIST SRM 610 and NIST SRM 612 glass reference materials. Geostd Newslett 21:115-144

Podor R, Cuney M (1997) Experimental study of Th-bearing $LaPO_4$ (780°C, 200 MPa): Implications for monazite and actinide orthophosphate stability. Am Mineral 82:765-771

Pyle JM (2001) Distribution of select trace elements in pelitic metamorphic rocks: pressure, temperature, mineral assemblage, and reaction-history controls. PhD dissertation, Rensselaer Polytechnic Inst, Troy, New York

Pyle JM, Spear FS, Rudnick RL, McDonough WF (2001) Monazite–xenotime–garnet equilibrium in metapelites and a new monazite–garnet thermometer. J Petrol 42:2083-2107

Rapp R (1985) An experimental investigation of the solubility and dissolution kinetics of monazite and its implications for the thorium and rare earth element chemistry of felsic magmas. MS thesis, Rensselaer Polytechnic Inst, Troy, New York

Reed SJB, Buckley A (1996) Virtual WDS. Mikrochim Acta 13:479-483

Roeder PL (1985) Electron-microprobe analysis of minerals for rare-earth elements: use of calculated peak-overlap corrections. Can Mineral 23:263-271

Scherrer NC, Engi M, Gnos E, Jakob V, Leichti A (2000) Monazite analysis; from sample preparation to microprobe age dating and REE quantification. Schweiz mineral petrog Mitt 80:93-105

Scott VD, Love G, Reed SJB (1995) Quantitative Electron-Probe Microanalysis (2nd edn). Ellis Horwood Ltd, Chichester, UK

Seltzer SM (1993) Calculation of photon mass energy-transfer and mass energy-absorption coefficients. Radiation Res 136:147-170

Sewell DA, Love G, Scott VD (1985) Universal correction procedure for electron-probe microanalysis. I. Measurement of X-ray depth-distributions in solids. J Phys D 18:1233-1243

Smith MP, Yardley BWD (1999) Fluid evolution during metamorphism of the Otago schist, New Zealand; (II). Influence of detrital apatite on fluid salinity. J Metamor Geol 17:187-193

Spear FS, Kohn MJ, Paetzold S (1995) Petrology of the regional sillimanite zone, west-central New Hampshire, U.S.A., with implications for the development of inverted isograds. Am Mineral 80: 361-376

Stormer JCJr, Pierson MJ, Tacker RC (1993) Variation of F and Cl X-ray intensity due to anisotropic diffusion of apatite during electron microprobe analysis. Am Mineral 78:641-648

Suzuki K, Adachi M (1991) Precambrian provenance and Silurian metamorphism of the Tsubonosawa paragneiss in the south Kitakami terrane, northeast Japan, revealed by the chemical Th-U-total Pb isochron ages of monazite, zircon, and xenotime. Geochem J 25:357-376

Tomascak PB, Krogstad EJ, Walker RJ (1996) U-Pb monazite geochronology of granitic rocks from Maine: implications for late Paleozoic tectonics in the northern Appalachians. J Geol 104:185-195

White EW, Johnson GG Jr (1979) X-ray emission and absorption edge wavelengths and interchange settings for LiF geared curved crystal spectrometer. Earth Mineral Sci Spec Publ, The Pennsylvania State University, University Park, Pennsylvania

Williams ML, Jercinovic MJ, Terry MP (1999) Age mapping and dating of monazite on the electron microprobe: deconvoluting multistage tectonic histories. Geology 27:1023-1026

Yakowitz H, Heinrich KFJ (1968) Quantitative electron probe microanalysis: absorption correction uncertainty. Mikrochim Acta 1968:182-200

9 Sedimentary Phosphorites—An Example: Phosphoria Formation, Southeastern Idaho, U.S.A.

Andrew C. Knudsen and Mickey E. Gunter

Department of Geological Sciences
University of Idaho
Moscow, Idaho 83844

INTRODUCTION

The phosphorous cycle filters through both the biological and geological worlds, and sedimentary phosphates exist at the interface between these two portions of the global phosphorous cycle. Sedimentary phosphates form only when the proper physical conditions and biological activity coincide. Once deposited, the phosphate (PO_4^{3-}) crystallizes into the mineral carbonate-fluorapatite (CFA). This mineral, with substitution of CO_3^{2-} for PO_4^{3-}, compositionally falls between the familiar forms of fluorapatite found in igneous and metamorphic rocks and the apatites found in biological material, including our own bones and teeth. Finally, through mining and processing, it is these sedimentary phosphates that are used as fertilizers and additives for food production and other activities and are subsequently returned to the biosphere.

The unique settings in which phosphorites form, those areas where phosphate is concentrated millions of times above normal and sedimentation rates are drastically slowed, are not fully understood, and have been the subject of sometimes intense scientific debate over the last century and a half. Furthermore, the mineralogy of these deposits is unique, difficult to study, and contentious. Because of our dependence on phosphate for use in fertilizers and other agricultural and chemical uses, these deposits are interesting not only from a scientific viewpoint but from an economic standpoint as well. However, like many mining operations, phosphate mining has environmental problems that must be addressed.

This chapter provides an overview of the salient points in phosphorite formation, mineralogy, and mining and puts into context the current work being undertaken on the Phosphoria Formation in southeastern Idaho. This chapter does not provide a full description of all the sedimentalogical, mineralogical, and industrial work that has been done on sedimentary phosphates; indeed numerous full volumes have been published on phosphorites and more will undoubtedly follow.

PHOSPHOGENESIS

For large-scale phosphorites to form, two major obstacles must be over come. First, phosphorous must be concentrated at levels from 1 to 2 million times that of average seawater. Secondly, once phosphate (PO_4^{3-}) is accumulated, the stable CFA phase must crystallize. For much of the history of phosphorite study, the main objective has been to discover the unique method for the formation of these deposits. However, through analysis and reanalysis of the many deposits being formed today and those formed throughout the geologic record, it has come to be accepted that phosphorites have formed in a variety of ways throughout Earth's history. However, there are important processes that connect all, or at least most, of the known deposits.

Phosphorous accumulation and sedimentation

Sedimentary phosphorites are predominantly marine sediments formed by an upwelling of phosphate-rich waters into relatively shallow marine settings, where

1529-6466/00/0048-0009$05.00

biological accumulation further concentrates the phosphate and eventually deposits it on the sea floor. There are also cases of marine phosphorites formed from terrigenous supplies of phosphate and even fresh water phosphorites in lacustrine settings. A brief discussion of the phosphate cycle as it applies to phosphorites is provided below, although a more complete discussion can be found in Filippelli (this volume).

Bushinskiy (1964) noted the supply of a total of 1.8×10^7 kg of phosphorous per year deposited by the Volga River into the Caspian Sea as evidence of the importance of terrigenous supply to phosphorite formation. The importance of an accompanying low sediment load in the river and a low sediment deposition rate in the zone of phosphorite has also been noted. The ideal situation for a fluvial-supplied phosphate deposit would be that of a river with its source in a humid, biologically active environment, an environment that loads the river with phosphate from organic sources. The river would then flow through an arid region downstream, preventing an influx of terrigenous material in the sediment load. The importance of rivers has been shown in some deposits (Baturin 1988, Ruttenberg and Berner 1993, Lucotte et al. 1994) where CFA forms in deltaic and estuarine environments not fed by upwelling. However, because rivers carry only about 20 ppb dissolved P, only one fourth the load in upwelling deep marine water (Bentor 1980), most workers have focused on marine upwelling zones.

The beginning of modern phosphorite research was ushered in with the work of Kazakov (1937). He theorized that phosphorites form as upwelling cold phosphorous-rich waters are brought to the shallow warm water zone in the oceans. When these phosphorous-rich waters reach the warm shallow layer in the ocean, CO_2 diffuses into the air, increasing the pH and leaving the waters super-saturated with CFA that precipitates directly on the sea floor. Bentor (1980) however, points out that deep marine water, when brought to the surface, is still some 10^4 to 10^5 times below the levels of super-saturation necessary for direct precipitation. Whereas upwelling does not lead to the direct precipitation of CFA, it does replenish the phosphate supply to biologically active shallow waters. Although not entirely accurate in his theory, Kazakov did provide a foundation for modern phosphorite research.

The association between organic productivity and phosphorite deposition was noted in the pioneering expedition of the H.M.S. Challenger of 1873-1876. On that voyage, Murray and Renard (1891) noticed the association of phosphorites and organic material and theorized that these large deposits of phosphate-rich organic sediments were the result of massive die-offs of fish and other marine creatures. Like Kazakov's theory, the connection to a more general principle was correct but the theory itself was wrong. Since that earlier time, one of the fundamental tenants of phosphorite formation is that the deposits are quite slow in developing, deduced from the observation that at no time could the concentration of P ever be so high in the waters (including the concentrations contained in the biota) that phosphorites could be quickly deposited (Bentor 1980).

It is generally accepted that most phosphorites form as the result of a series of events that include biological activity and upwelling. High biological activity (including but not limited to bacteria, plankton, and algae) consumes, processes, and concentrates dissolved phosphate in bottom sediments, depleting the supply in the water column. This phosphate supply is then replenished by marine upwelling, or in some rare cases fluvial supply, which does not on its own saturate the waters with phosphate, but rather supplies the active biota with more phosphate, allowing them to further concentrate the phosphate. These processes are accompanied by very low sedimentation rates of other materials, allowing the deposited sediment to become enriched in organics, including phosphorous. This organic-rich sediment is enriched relative to standard sea water by a factor of about 1.5×10^5, requiring further enrichment of 10-20 times to precipitate CFA.

CFA crystallization

Crystallization processes. The saturation index of fluorapatite [$Ca_5(PO_4)_3F$] is estimated to lie within the range 5×10^{-8} to 2.5×10^{-7} g/l, between pH = 7.1 to 7.8, and T between 5° and 20°C (Atlas and Pytkowicz 1977, Van Cappellen and Berner 1988). However, because the super-saturation levels that are widely used have been established in laboratory work on pure fluorapatite, it is unknown what the exact effect biological activity and other inorganic phases have on the crystallization rates. In addition, the slow kinetics of CFA formation may require concentrations well above equilibrium. Increased concentrations can be attained in several ways, including biogenic concentration, bioturbation, mechanical mixing, and chemical concentration. On the basis of geochemical models and observations in active phosphorite forming zones, it appears that most of the phosphate crystallization occurs just above the oxygen minimum zone, just below the sediment water interface, in the interstitial waters (Slansky 1986).

Theories of direct precipitation can be divided into two categories: (1) CFA as a direct precipitant and (2) CFA that evolved from an initial phosphatic gel (Slansky 1986). On the basis of the work of Martens and Harris (1970) regarding the Mg-inhibition discussed below, either phosphatic gel or CFA will precipitate from the interstitial waters, depending on pH and the Ca/Mg ratio. At pH levels of 7.5-8, the Ca/Mg ratio must be above 5.2; at neutral pH, the value decreases to 1.2 whereas the ratio in natural seawater is 0.2. Whereas these Ca/Mg ratios are too high to appear in open seawater, they could exist in the interstitial waters of the organic-rich sediments where the organic nature of the sediment may decrease the Ca/Mg ratio. Also, the formation of Mg-containing minerals such as montmorillonite (Wollast and Debroeu 1971, Van Bennekom and Van der Gaast 1976) may in some cases, such as in the Gafsa Bay in Tunisia, sufficiently increase the Ca/Mg ratio locally in the interstitial waters to allow for CFA growth. Likewise, in the Santa Barbara Basin off the California coast absorption of Mg^{2+} ions by clay minerals may significantly increase the Ca/Mg ratio (Sholkovitz 1973). These models for increasing the Ca/Mg ratio rely on the inability of interstitial waters to exchange with the seawater and equilibrate their Ca/Mg ratio, as Schwennicke et al. (2000) noted in the deposits near La Paz, Baja California.

Replacement. CFA may also crystallize through replacement of other minerals, particularly calcite or gypsum. This was first shown in a series of experiments in which powdered calcite was exposed to a phosphatic solution and converted to CFA (Ames 1959). This process has been widely reported in nature with phosphatization of foraminiferal and other shell material. Initially, this process was thought to involve PO_4^{3-} ions replacing CO_3^{2-} ions within the structure, a realignment of the Ca^{2+}, and addition of F^-. This method is unlikely because of the size differences of the phosphate and carbonate, the coordination difference of Ca^{2+} between the two minerals, and completely different crystal structures. It has since been shown instead that the calcite or gypsum dissolves first before apatite crystallization (Nathan and Lucas 1972). Dissolution of either of these minerals would increase the Ca/Mg ratio and encourage crystallization of apatite.

Phosphatic pump. The phosphatic pump trap theory is based on the tendency of iron and manganese oxyhydroxides to adsorb phosphate. These complexes, once buried in sediments, migrate upward with mechanical reworking and bioturbation. As they reach the oxygen minimum zone, they are reduced and release their phosphate, which accumulates at the sea floor (Berner 1973, Krom and Berner 1980, Krom and Berner 1981, O'Brien and Heggie 1988, Heggie et al. 1990, O'Brien et al. 1990, Lucotte et al. 1994). The dense reduced iron and manganese oxyhydroxides migrate downward, where more phosphate is adsorbed onto their surface, and the cycle continues accumulating sufficient phosphate to form CFA or a phosphatic gel (Baturin 1999).

Microbial activity. One of the most important ways in which phosphate is concentrated in sediments is by microbial activity (Föllmi 1996). Microbes break down and release phosphate into the pore waters of the oxygen minimum zone (Berner et al. 1993, Compton et al. 1993, Van Cappellen et al. 1993, Krajewski et al. 1994). Formation of microbial mats and eventually stromatolites has been associated with formation of phosphorite beds (Williams and Reimers 1983), as seen in modern phosphorite formation off the Peru-Chile coast (Froelich et al. 1988), in French Polynesia (Schwennicke et al. 2000), and off the coast of India (Purnachandra Rao et al. 2000). These modern formations have been used as models for some ancient phosphorites such as those from the early Cambrian in western China's Zhongyicum Formation (Schwenicke et al. 2000), and in the Betic Cordillera of Spain (Martin-Algarra and Sanchez-Navas 2000). In the case of the microbial mat off of the Peru-Chile continental margin, the mat is composed of the sulfide-oxidizing microbe Thioploca, which alters the water chemistry at the water sediment interface, thus allowing for phosphate precipitation (Froelich et al. 1988).

The methods and theories presented above are the most common for apatite crystallization. However, other workers have examined the role of fluoridation (Baturin and Shishkina 1973, Froelich et al. 1983, Ong and Davidson 1992) and still others have correlated phosphate formation at the oxygen minimum zone and humic acids (Slansky 1986). In addition, there are less popular theories not mentioned here, and certainly others still to be proposed.

Mg-inhibition. Even if sediments or interstitial waters are at or near saturation with phosphate, there are still impediments to CFA crystallization. One of the most significant is the role of Mg-inhibition. Martens and Harris (1970) demonstrated in the laboratory that when phosphate and fluorine are added to a solution with a composition similar to that of ocean water in Mg^{2+}/Ca^{2+} ratio, apatite does not crystallize. Instead, a Ca-phosphate gel precipitates with a Ca/P ratio of 1.35 compared to the ratio of 1.67 found in apatite. In their experiments, the gel was kept in the proxy seawater solution for 8 months with no appearance of apatite. However, when the gel was placed in a Mg-free solution, apatite quickly formed. Mg-inhibition results from the smaller Mg^{2+} ion replacing Ca^{2+} in the apatite structure, the resultant lattice distortion prevents effective crystal growth (Martens and Harris 1970). Later work by Gulbrandsen et al. (1984) showed that given enough time, in their experimentation seven years, apatite would indeed crystallize from the gel despite the Mg^{2+} in solution.

Geological settings. Sedimentary phosphates are also characterized by a wide array of sedimentary structures, from finely layered to massive beds, to a variety of sizes of pelletal phosphates. These structures are linked to phosphogenesis and subsequent diagenetic alterations. Although these structures and what causes them have not been discussed here, a thorough review can be found in Slansky (1986) or Föllmi (1996). They are a necessary component to a full understanding of any individual deposit's sedimentological and geochemical history, and for a complete view of the sedimentology of phosphorites around the world.

Phosphorite deposits are sensitive to temporal and spatial conditions. Sedimentary phosphate deposits are found throughout geologic time from the Precambrian to the present, however they are generally restricted to the warm waters of low latitudes. Of the major sedimentary phosphate deposits of the world, including those forming today, nearly all are below 40°, and all are below 60° paleolatitude (Fig. 1). This relationship to low latitudes is not surprising given the dependence on warm nutrient-rich waters to create the organic-rich sediments from which phosphates crystallize. Zones of phosphorite formation are also seen to be preferentially located in areas of upwelling as discussed above. This preference tends to concentrate phosphorite formation along west coasts at low latitude, where the trade winds encourage oceanic upwelling, as is seen today in areas such as the Peru-Chile margin. Upwelling has also aided phosphogenesis in areas such as off the coast of the southeastern United States, where currents from the Gulf of Mexico provide sufficient upwelling and nutrient-rich water. In some less common settings upwelling has been found not to be active in phosphorite formation (Baturin 1988, Ruttenberg and Berner 1993, Lucotte et al. 1994), instead these areas depend on sufficient supplies from continental sources.

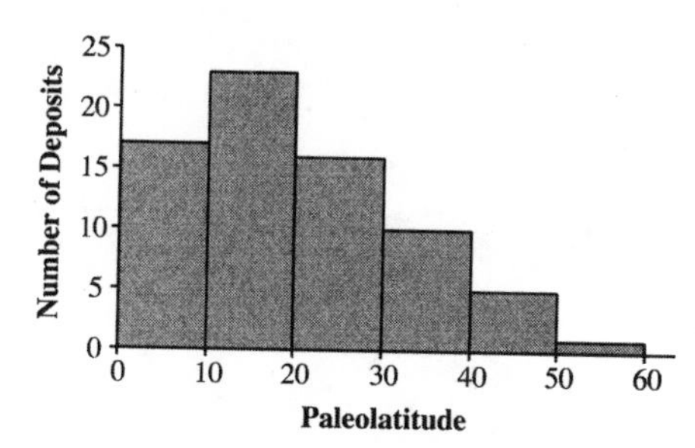

Figure 1. Paleolatitudes of 72 major phosphate deposits (modified from Slansky 1986).

MINERALOGY AND CRYSTAL CHEMISTRY OF PHOSPHORITES

Carbonate-fluorapatite nomenclature

Sedimentary apatites can be generalized as fluorapatites with extensive substitutions and vacancies throughout the structure. The most significant deviation from pure fluorapatite is the addition of CO_3^{2-} into the structure, but also of significance are Ca and F variations from their stoichiometric fluorapatite concentrations. Whereas the ideal formula for fluorapatite is $Ca_{10}(PO_4)_6F_2$, the formula for substitution-rich sedimentary apatites is $Ca_{10\text{-}a\text{-}b\text{-}c}\,Na_a\,Mg_b\,(PO_4)_{6\text{-}x}\,(CO_3)_{x\text{-}y\text{-}z}\,(CO_3,F)_y\,(SO_4)_z\,F_2$, where $x = y + a + 2c$ (c represents vacancies in the Ca site) demonstrating the extensive substitutions in the fluorapatite structure. The name francolite was introduced (Sandell et al. 1939) and has since been widely used to identify sedimentary apatites that contain at least 1% fluorine and appreciable CO_3^{2-}. However, this name is not rigorously defined and is not formally recognized as a mineral name, instead the more general term carbonate-fluorapatite (CFA) is preferred, and used herein.

The carbonate presence

The presence of CO_3^{2-} in the apatite structure has been one of the most studied areas in phosphorite mineralogy. The small crystallite size, on the order of tens to hundreds of Ångstroms, and the great complexity of substitutions, has made definitive determinations of the crystal structure and chemistry quite difficult. The presence of carbonate ions has long been associated with sedimentary apatite; determining how and where it resides in the structure has proved difficult. Originally, three schools of thought dominated the debate on this issue: (1) those believing that the carbonate exists as a distinct and separate phase (Thewlis et al. 1939, Brausseur et al. 1946), (2) those believing the carbonate is adsorbed onto the apatite surface (Neuman and Neuman 1953), and (3) those claiming that the carbonate is incorporated into the structure of the apatite (Gruner and McConnell 1937). Altschuler et al.'s (1952) X-ray work showing a contraction of the *a*-cell parameter presented the first proof that carbonate was present within and altering the apatite structure.

The presence of carbonate in fluorapatite is accepted as the product of a substitution of CO_3^{2-} for PO_4^{3-}, as first suggested by McConnell (1938). McClellan and Lehr (1969) showed a statistical correlation between increased carbonate and decreased phosphate with 110 samples showing the appearance and disappearance of the respective ions occurring on approximately a 1:1 basis (Fig. 2). Later work by McClellan and Van Kauwenbergh (1990) further supported these findings, with an additional 150 samples, where the data show a variance of $r^2 = 0.938$. They found the slope of the regression line to be 0.998 (where 1 represents a pure 1:1 substitution) and an intercept of 5.996 (where 6 represents the ideal case).

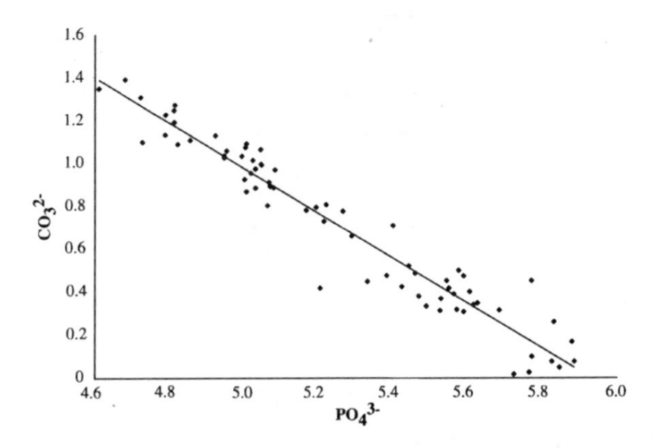

Figure 2. The relation of CO_3 to PO_4 showing approximately a 1:1 increase of CO_3 with decreased PO_4 (modified from McLellan and Van Kauwenbergh 1990).

The substitution of CO_3^{2-} for PO_4^{3-} poses two problems: first, how is charge balance maintained, and second, what is the effect of substituting a trigonal-planar carbonate for a tetrahedral phosphate? Borneman-Starynkevitch and Belov (1940) presented the theory that would answer both of these questions, suggesting that the trigonal carbonate is coupled with a F^- ion to form a pseudo-tetrahedron. This proposal was based on excess F found in chemical analyses, which would maintain electroneutrality and satisfy the bonds vacated by the phosphate. Gulbrandsen (1966), Smith and Lehr (1966), Treuman (1966), and Elliot (1969) were among the first to support this theory. The association of the CO_3F^{3-} tetrahedron in apatite was first determined analytically using ESR (Nathan 1984) and supported statistically by Binder and Troll (1989).

Although this theory seems to be a perfect fit, it is not without problems. McClellan and Lehr (1969) calculated that despite the fact that most CFAs do indeed have excess F^-, there is only (on average) 0.4 F^- in excess of standard fluorapatite stoichiometry for every substituted CO_3^{2-} group. In addition, some CFAs are even deficient in F^- compared to standard fluorapatite. Thus even if F^- and CO_3^{2-} do indeed couple to form the pseudo-tetrahedron, the additional charge imbalance must be accounted for. Three common explanations exist to supplement the role of F^- in charge-balancing the CFA structure. First, McClellan (1980) among others suggests that the difficult-to-measure OH^- anion follows the F^- anion and creates another distorted tetrahedron. Although this seems plausible, little to no OH^- has been confirmed in most CFAs. Second, the minor, though ubiquitous, substitution of monovalent Na^+ for divalent Ca^{2+} in CFA helps achieve charge balance. Finally, Gulbrandsen (1966) suggested that CFAs might have voids in the Ca^{2+} site. Although little has been done to confirm this in geologic samples, it has been noted in biological carbonate-hydroxyapatites (Posner and Perloff 1957, Berry 1967). This would prove quite efficient, as a single vacancy would account for two substituting carbonate anions.

Further complicating the chemistry and structure of CFA is the prevalence of other substitutions. Sulfur has been found in many sedimentary apatites as SO_4^{2-}. Like carbonate, sulfate is divalent, and would leave a charge imbalance when substituting for phosphate. However, the similar size and shape of phosphate and sulfate anionic complex would not cause the strain on the structure produced by carbonate substitution. Whereas sulfate is commonly found in the structure of sedimentary apatites, it is rarely as abundant as carbonate (McClellan and Lehr 1969). It has also been suggested that SiO_4^{4-} substitutes for phosphate as seen in the mineral wilkeite, $Ca_{10}(PO_4, SiO_4, SO_4)_6(F, OH)_2$. However this substitution has never been reported in sedimentary apatite. Also of note is the substitution of Mg^{2+} for Ca^{2+}. Although this substitution is important in its pertinence to the role of Mg inhibiting growth of marine apatites as discussed above, it has no effect on charge balance.

Even given the excess of F^- in the structure the theory of the distorted tetrahedron is not universally accepted. Regnier et al. (1994) used FTIR spectroscopy, ^{13}C NMR, and quantum mechanical calculations on synthetic and natural samples to demonstrate that the repulsion between the F^- anion and the carbonate anionic complex is too great to form the distorted tetrahedron. They do not dismiss the possibility that F^- couples with CO_3^{2-} to balance the charge, but that if they do both enter the structure, the F^- enters independently and fills a vacancy in the structure. Nathan (1996) however points out that all of the samples used in these experiments are actually fluorine-deficient apatites and thus unlikely to have the CO_3F^{3-} pseudo-tetrahedron, so clearly more research is needed before final judgments can be made.

Measurement of CO_3^{2-} substitution

Measurement of the carbonate content in sedimentary apatites is complicated by the difficulty in separating the small single crystals from the phosphorite matrix. The crystals must be completely removed from any contaminating material, particularly because of the common relationship with calcite and dolomite in the rock matrix, which would skew any direct chemical measurements. The substitution of carbonate into the apatite structure causes systematic changes in the mineral properties, which can be observed to estimate the extent of carbonate substitution. The birefringence of apatite is directly related to the degree of carbonate substitution (Nathan 1984). However, because of the fine-grained nature of phosphorite samples, measurement of retardation is not a practical method for the proxy measurement of carbonate content. Thus developments in X-ray methods have become the most widely used. McClellan and Lehr (1969) observed a variation in the *a*-cell parameter of CFAs, ranging from 9.392 Å for a sample from Sweden with 0.11% CO_2 by weight to 9.322 Å for a sample from South Carolina with 5.91% CO_2. The *c*-cell parameter, however, shifted only from 6.896 Å to 6.900 Å for the two samples, respectively. In that same paper, McClellan and Lehr (1969) introduced equations for determining the carbonate content from the cell parameters. McClellan (1980) later refined this work where the carbonate content was determined to match the equation:

$$(CO_3/PO_4)\,(Z/6\text{-}Z) = (9.369 - a_{obs})\,/\,0.185. \qquad (1)$$

McClellan (1980) suggested that the degree of substitution of Na and Mg is linked to carbonate substitution (Fig. 3) and thus is also linked to the *a*-cell parameter shift by the following equations:

$$(\mathrm{Na})\ X = 7.173\,(9.369 - a_{obs}) \qquad (2)$$

$$(\mathrm{Mg})\ Y = 2.784\,(9.369 - a_{obs}). \qquad (3)$$

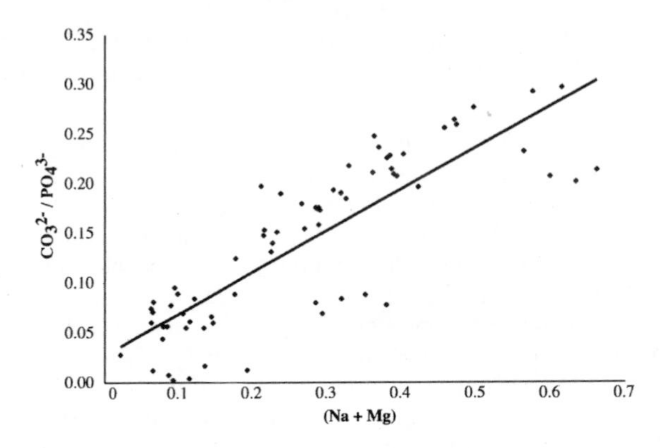

Figure 3. The molar ratio of CO_3/PO_4 versus the sum (Na + Mg) in CFA (modified from Lehr et al. 1968).

On this basis McClellan (1980) gave the general formula for carbonate fluorapatite as: $Ca_{10-X-Y}\,Na_X\,Mg_Y\,(PO_4)_{6-Z}\,(CO_3)_Z\,F_{0.4Z}\,F_2$. Note that this generalized formula is similar but not identical to the one given previously in that it does not account for Ca site vacancy.

Gulbrandsen (1970) also proposed a method for determination of carbonate content using 74 of the samples of McClellan and Lehr (1969), in which he used the shift in the *a*-cell parameter and relative stability of the *c*-cell parameter to create a method that does not require an internal standard. He compared the (410) and (004) diffraction peaks, and found that the distance between them (in °2θ) could be related to carbonate substitution by:

$$CO_2\,(wt) = 23.641 - 14.7369\,\Delta 2\theta°_{(004\text{-}410)} \tag{4}$$

using Cu $K\alpha$ with a standard error estimate of 0.56.

Schuffert et al. (1990) introduce two more equations for measurement of the carbonate content based on cell parameters of synthetic CFAs. The first uses the same (004)-(410) separation used by Gulbrandsen (1970), arriving at the relationship of:

$$CO_2\,(wt) = 10.643\,\Delta 2\theta°_{(004\text{-}410)}{}^2 - 52.512\,\Delta 2\theta°_{(004\text{-}410)} + 56.982 \tag{5}$$

with an $r^2 = 0.97$ for this equation and an estimated standard error of CO_2 (wt) = 0.49. In the same paper, the (300) and (002) peaks were compared, and found that they could be related by the equation:

$$CO_2\,(wt) = 38.381\,\Delta 2\theta°_{(300\text{-}002)} - 276.343 \tag{6}$$

The variation was found to have an $r^2 = 0.97$ and an estimated standard error of CO_2 (wt) = 0.57. Equation (6) is not as reliable as those based on the (004)-(410) split because the greater distance between the peaks allows for more error. In addition, the movement of the peak locations away from ideal is smaller for the (300),(002) pair as carbonate substitutes.

These methods based on peak separation have proven popular and easy to use. Unfortunately, however, they are hampered by the overlap of CFA peaks with those associated with dolomite. Although this method is effective in samples without dolomite, many phosphorites contain both dolomite and CFA, making the method unreliable and ineffective at determining variability of carbonate substitution into apatite in a more carbonate-rich environment. Our current work on samples from the Phosphoria Formation in southeastern Idaho (Knudsen and Gunter 2002), have made use of Rietveld analysis to determine cell parameters, and thus directly compare the *a* and *c* unit-cell parameters to determine the carbonate content. On the basis of the measured cell-parameters, the locations of the peaks are calculated and not measured. This eliminates the problems of peak overlap, irregular peak shape, and errors in measurement. McClellan and Van Kauwenbergh (1990) note that peak comparison is only effective on samples with relatively low carbonate substitution levels and those enriched in F^-, making it ideal for the Phosphoria Formation but not for some other samples.

Infrared spectra have also been useful in determination of carbonate substitution in apatites. Gulbrandsen (1966) used the splitting of the ν_3 vibration between 1,470 and 1,410 cm^{-1} as well as a less distinguishable split between 890 and 850 cm^{-1} peaks. This double splitting has been used as evidence to suggest that the carbonate enters two different sites (Emerson and Fisher 1962, White 1974). By comparing the C–O and P–O absorption intensities, the degree of substitution can be estimated with the CO_2 index (Lehr et al. 1968). This CO_2 index is defined as the ratio of the C–O and P–O absorption intensities. The C–O intensity is measured from the ν_3 doublet at 1,453 and 1,420 cm^{-1} and the P–O intensity is measured by the intensity of the ν_4 vibration band at 602 cm^{-1} (Lehr et al. 1968). This method also has drawbacks, however. In natural samples, sulfates, silicates, carbonates, and water all interfere with the spectra, and all of these species are commonly found in phosphorites. Nathan (1984) describes the use of Fourier transform infrared spectroscopy, and a stripping of interfering spectra to index the carbonate content. Integration of the spectra between 1,550 and 1,375 cm^{-1} for C–O and 690–535 cm^{-1} for P–O, show the following equation reflects carbonate substitution:

$$CO_2 \text{ index} = 0.0342 + (0.226 \times \%CO_2) \tag{7}$$

Extent of CO_3^{2-} substitution

Generally, CFAs contain between 4 and 7% CO_2 by weight, although these values vary between localities. Moroccan phosphates have been reported to contain as much as 16.1% CO_2 (Elderfield et al. 1972), which may be too high, but McArthur (1978) reported CO_2 levels between 8.1 and 8.9% for the same area. Table 1 summarizes some of the more significant deposits and their reported CO_2 content.

Table 1. Carbonate content in CFA of global phosphorites.

Location	*CO_2 content*
Morocco	8.1%-8.9% (McArthur 1978)
Spain	6% (Lucas et al. 1978)
North and South Carolina	5%-6% (McClellan & Lehr 1969)
Israel	4% (Nathan 1984)
Turkey	4% (Lucas et al. 1980)
Western U.S. (Idaho, Wyoming, Utah)	2% (Gulbrandsen 1970)

Many have pointed to differences in weathering, diagenesis, and metamorphism as the primary causes of the chemical variability (McArthur 1978, 1985; McClellan 1980). Further evidence for the post-depositional alteration of carbonate content was offered by Axelrod et al. (1980) through work on diagenetically altered bone fragments and Matthews and Nathan's (1977) work on thermally-lost CO_2 in the "Mottled Zone." These arguments suggest that CFAs are subject to alteration, which is not surprising given the well known destabilizing effect that carbonate has on the apatite structure. The best evidence for limited variability of CO_2 substitution is that when in equilibrium with sea water CFA will contain between 5% and 8% CO_2 (Nathan 1984). However, this range could expand with variability in the geochemistry and biological activity of the depositional environment.

Gulbrandsen (1970) examined 368 samples from the Phosphoria Formation and found a strong correlation between carbonate substitution and lithology. Samples from southeastern Idaho, where the lithology is dominated by mudstones, chert, and phosphorite, showed relatively low levels of substitution of around 1.5%, yet samples from further east in Wyoming that are dominated by dolostones, limestones, and phosphorites have higher substitution rates of 3.2-3.4%. Gulbrandsen used this to suggest that variability in carbonate content is a result of depositional environment.

One concern surrounding Gulbrandsen's work was that the samples he analyzed came from different regions within the Phosphoria Formation, which leaves the possibility that different weathering patterns have affected the eastern and western portions of the Phosphoria Formation to produce the results Gulbrandsen reported. Current work on the Meade Peak Member of the Phosphoria Formation (Knudsen and Gunter 2002) supports Gulbrandsen's (1970) theory.

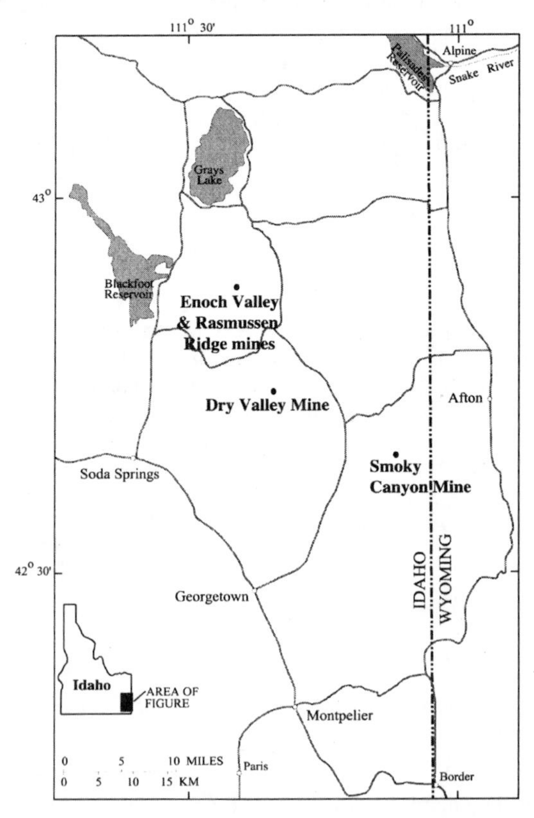

Figure 4. Location map showing the four active phosphate mines in southeastern Idaho. "Less-weathered" and "more-weathered" sections were collected at the Enoch Valley, Dry Valley, and Smoky Canyon mines. A deep, "least-weathered" section was collected at the Enoch Valley mine (modified form Herring et al. 1999).

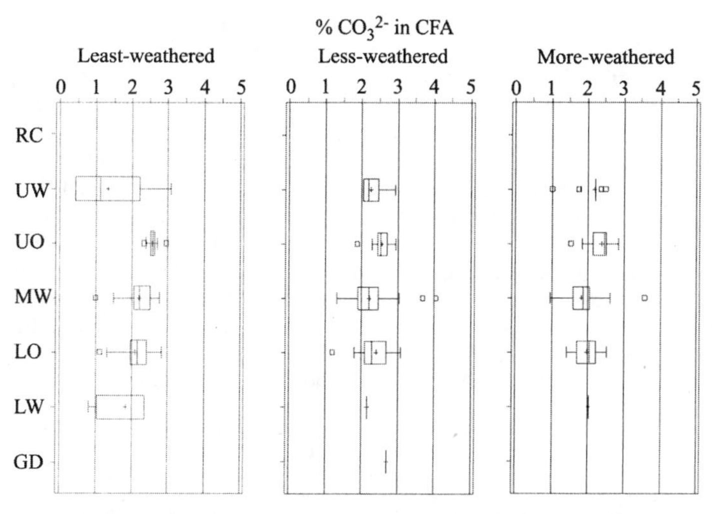

Figure 5. Schematic box and whisker plot showing the amount of CO_3^{2-} in the CFA structure through the "least-," "less-," and "more-weathered" sections from three of the four active phosphate mines in southeastern Idaho. Boxes extend from the lower quartile to the upper quartile (i.e., the middle 50 % of the data), separated by the median. The whiskers extend to the most outlying value within the range of 1.5× the inner quartile range. Outliers are designated with boxes and a cross represents the mean. The vertical axis is divided into units bordering or within the Meade Peak Member of the Phosphoria Formation. The units include the Rex Chert (RC), upper waste (UW), upper ore (UO), middle waste (MW), lower ore (LO), lower waste (LW), and the Grandeur Dolostone (GD) member of the Park City Formation, the lithology and stratigraphic relationships of these units are shown in Figure 9.

Four active phosphate mines in southeastern Idaho (Fig. 4, opposite page) produced pairs of measured stratigraphic sections, one "more-weathered" and one "less-weathered." In addition, a deep core was collected at the Enoch valley mines as a "least-weathered" section. The effect of weathering on these sections is discussed in Knudsen and Gunter (2002) as well as the final section of this chapter. The CO_3^{2-} content in CFA is plotted on box and whisker plots (Fig. 5, above), with the vertical axis separated into continuous stratigraphic units that are recognized at each of the three sampling localities. These units include the Grandeur Dolostone (GD) Member of the Permian Park City Formation, which underlies the Meade Peak Member of the Phosphoria Formation and the Rex Chert (RC) Member of the Phosphoria Formation, which overlies the Meade Peak Member. Within the Meade Peak Member are the upper waste (UW), upper ore (UO), middle waste (MW), lower ore (LO), and the lower waste (LW). Figure 5 does not show a significant pattern of change in the CO_3^{2-} content in CFA that can be attributed to weathering. The box and whisker plots do, however, show a pattern in that the ore zones tend to have higher levels of CO_3^{2-} substitution in CFA than is seen in the waste zones.

Generally, the recognized ore zones are primarily phosphorites whereas strata in the waste zones include more dolostones and silicic layers. Factor analysis to determine the relationships of the lithology and the CO_3^{2-} content in CFA considered four variables: (1) the degree of CO_3^{2-} substitution for PO_4^{3-} in CFA, (2) CFA content, (3) silicate mineral content (quartz + sheet silicates + feldspars), and (4) carbonate mineral content (dolomite + calcite). Because the "more-," and "less-weathered" strata have lost much of their original carbonate mineral content, only the "least-weathered" samples are used for factor analysis, as they are the best proxy for the unweathered Phosphoria Formation.

Factor analysis shows that two factors describe most of the variability among these four variables (Table 2a). The first factor has a strong positive weighting by both the degree of substitution in CFA and the amount of CFA in a sample, a strong negative weighting on the silicate mineral content, and very little weighting on the carbonate minerals (Table 2b). The second factor also has a positive weighting on CO_3^{2-} in CFA, a positive weighting for the carbonate minerals, and again a negative weighting against the silicate minerals (Table 2b). This shows a strong positive link between the amount of CO_3^{2-} in CFA and the concentrations of CFA and to a lesser extent the carbonate minerals, and there is a negative link between CO_3^{2-} in CFA and higher silicate mineral content in samples. Although neither study (Gulbrandsen 1970, Knudsen and Gunter 2002) explain why the amount of CO_3^{2-} in CFA is so much lower in samples from the Phosphoria Formation than is seen in other phosphorites, there is clearly a link between the composition of the strata, or the original depositional environment, and the amount of CO_3^{2-} in CFA; similar studies to these at other phosphorites could prove very interesting.

Table 2a. The eigenvalues of a correlation matrix for the variables: (1) CO_3^{2-} in CFA, (2) CFA concentration, (3) total carbonate minerals concentration, and (4) total silicate minerals concentrations. The eigenvalues for each factor are listed, along with the difference between a factors eigenvalue and the eigenvalue for the next factor, the proportion of an eigenvalue to the total variance of 4, and finally the cumulative variance described by for each factor plus the preceding factors. The first two factors are retained as they account for over 80% of the total variance (from Knudsen and Gunter 2002).

Eigenvalues of the correlation matrix: Total = 4, Average = 1				
	Eigenvalue	***Difference***	***Proportion***	***Cumulative***
1	1.98	0.73	0.49	0.49
2	1.24	0.47	0.31	0.80
3	0.78	0.77	0.19	1.00
4	0.01		0.00	1

Table 2b. Factor pattern for each variable from within each factor, showing the weighting of each variable within the two factors that are retained.

	Factor 1	*Factor 2*
CO_3^{2-}	0.57	0.27
CFA	0.90	-0.39
Carbonates	-0.02	0.98
Silicates	-0.91	-0.25

Other phases

In general, CFA is the only significant phosphate mineral found in phosphorite deposits. Upon extensive weathering and leaching, the original CFA can be altered into alumino-calcic phosphate minerals such as millisite and crandallite, and eventually into aluminous minerals such as augelite and wavellite. However, these minerals only occur in highly leached environments found in tropical settings associated with aluminum ores.

Silica is an abundant component of phosphorite deposits, found as both authigenic silica (later crystallized to quartz) and detrital quartz. Interbeds of chert and phosphorite are common in these sedimentary deposits. Most of the world's phosphorites are old enough that the original opal has all been converted to quartz, although some of the most recent deposits still contain opal C-T (Nathan 1984). The presence of other silicate minerals varies, although feldspars and illite/muscovite are common in many deposits. Other clay minerals are also common, and as discussed previously, clays such as montmorillonite may play a role in the original concentration and formation of CFA as a transporter of adsorbed phosphate. Carbonate minerals are also commonly associated with phosphorites, particularly calcite and dolomite. Sulfides such as pyrite and sphalerite are also common in the black shales of some phosphorite deposits. Other unique minerals are found from location to location, such as the ammonium feldspar buddingtonite in the Phosphoria Formation, but these minerals are not generally important in the sense of global phosphorite mineralogy.

Finally, organic matter is a significant component of phosphorites. These organic phases have been of interest because phosphorites have been shown to be effective sources for hydrocarbon development (Maughan 1980). Organics have also been of interest to those studying the residency of trace elements in phosphorites, such as the interest in Se in the Phosphoria Formation, which will be discussed later.

PHOSPHORITE AS AN ORE

Phosphorous is one of the essential nutrients for both plant and animal life. As the human population grows, humankind has become increasingly dependent on the use of manufactured fertilizers, and thus on phosphorous. Sedimentary phosphates account for over 90% of the total mined phosphorous. Like any ore body, economic phosphorites must be highly concentrated and relatively easy to extract. These extra demands make ore-producing phosphorites an important subgroup of phosphorites.

Very slow deposition rates are necessary to create economic-grade phosphorites. Slansky (1986) estimates sediment accumulation rates of economic-grade phosphorites between 2×10^5 and 1×10^6 years per meter of deposit. This slow deposition rate is seen in Permian sediments of the Rocky Mountains where the Phosphoria Formation is about 30 m thick in the most phosphatic zone, whereas the same time period is represented by over 1000 m in non-phosphatic sections of the Permian in the Rocky Mountains.

Economic-grade phosphorites contain 20% or more P_2O_5 (Harben and Bates 1990). Weathering and leaching can play an important role in increasing the grade of phosphorite ores beyond the initial concentrations of P_2O_5. Weathering and leaching of gangue components such as organics and carbonates elevate the concentration of P_2O_5, increasing ore grade. In the ores of the Tennessee Valley, Ordovician phosphatic limestone, initially containing 5-6% P_2O_5 has been naturally beneficiated through weathering, resulting in an economic grade ore with 18-24% P_2O_5 (Slansky 1986). Weathering and leaching alter CFA as well, decreasing the amount of carbonate in the CFA, yielding a more pure fluorapatite. Such a decrease is seen in Moroccan phosphorites, where unaltered phosphorites have a CO_3^{2-} / PO_4^{3-} ratio of 0.27, which is decreased to 0.06 in the most weathered zones (Lucas et al. 1979).

The majority of phosphorite mining is undertaken in large open-pit mines. The mining methods used at a deposit depend on the thickness of the ore and the structural attitude of the sediments. In North Carolina, mining is undertaken using large draglines because of thick beds with little to no dip. In contrast, the Phosphoria Formation of the western United States with moderate to steep dips requires the use of 85-ton trucks and large shovels which can follow the dipping strata (Gillerman and Bennett 1997). Whereas underground mines once dominated phosphate mining, these have all but disappeared, as open-pit mining has become more economical.

Once removed, phosphate ore is processed to make phosphoric acid or elemental phosphorous by two separate methods. When necessary, ore is crushed, sized, and floated in order to increase ore grade, and calcined to remove water and organic material. To create phosphoric acid, which is the source of phosphate fertilizers, the ore is mixed with sulfuric acid, creating phosphoric acid (H_3PO_4) and gypsum as a by product (Gillerman and Bennett 1997). About 5-10% of phosphate is reduced to elemental phosphorous for use in animal feed, food additives, fire extinguishers, and numerous other products. For elemental phosphorous, beneficiated ore is mixed with coke and silica and melted in a furnace at over 2500°C. The phosphorous and other elements are volatilized, and the gas is cleaned and condensed into liquid elemental P_4 (Gillerman and Bennett 1997).

PHOSPHORIA FORMATION

Background

Phosphogenesis. The Permian Phosphoria Formation was deposited off the western shores of the Pangean super continent in what is today southeastern Idaho and surrounding states (Fig. 6). The Phosphoria Formation is considered a "super giant" among phosphorites, covering about 340,000 km^2 and containing five to six times more phosphorous than the total phosphorous budget in today's oceans. As an economic resource, the Phosphoria Formation contains approximately 25 billion tons (Gt) of ore shallower than 300 m, and an additional 500 Gt below 300 m (Cathcart 1991).

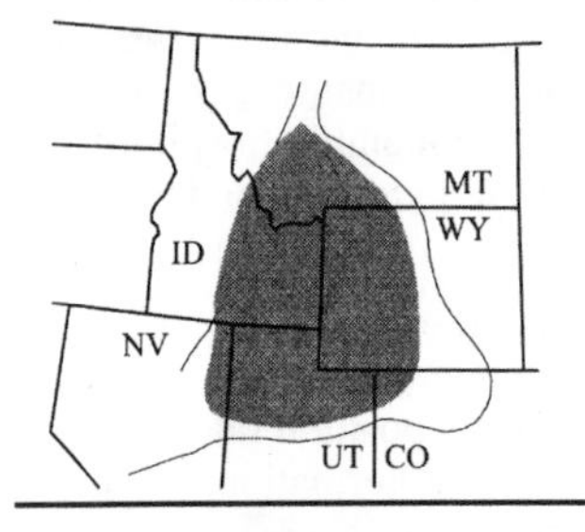

Figure 6. Geographic extent of western phosphate fields, and the Permian Phosphoria Formation (modified from Herring 1995).

Coastal upwelling has long been cited as the source of the P-rich waters that helped form the Phosphoria Formation. This theory has been supported by models for atmospheric circulation around an oceanic subtropical high resulting in shore-parallel winds which would induce upwelling currents up the continental margin via Ekman transport (Carroll et al. 1998 and references therein). Direct evidence of these atmospheric currents and the resulting oceanic upwelling has been difficult to find, although recent work has identified clay-poor siltstones in the Phosphoria Formation as eolian derived silts which support this model of atmospheric and oceanic currents (Carroll et al. 1998).

The Phosphoria Formation was originally thought to have formed as cold P-rich waters were delivered via upwelling to the continental shelf margin analogous to modern systems such as the one of the Peru-Chile coast. However, more recent research has shown that the Phosphoria Sea was more likely a relatively shallow (less then 200 m deep), semi-restricted epicontinental embayment (Hiatt and Budd 2001 and references therein). Within this restricted sea, phosphorite formation occurred primarily in the deeper portions of the basin, although phosphorites also formed in the shallower waters (Hiatt and Budd 2001).

Modern models of the Phosphoria Basin generally depict a warm, shallow sea. Based on the presence of contemporaneous evaporites and organic carbon levels as high as 32.9%, Stephens and Carroll (1999) developed a model for a system with both salinity stratified waters and upwelling (Fig. 7). Analysis of $\delta^{18}O_P$ values have shown water temperatures during phosphogenesis to have been temperate (14-26°C) in the deeper zones of the Phosphoria Sea and between 30-37°C in the shallower waters, which matches previous estimates of the air temperatures (Hiatt and Budd 2001). These models have shown important differences between the believed mechanisms of Phosphoria Formation phosphogenesis and modern systems, allowing a better understanding of the productivity in ancient oceans.

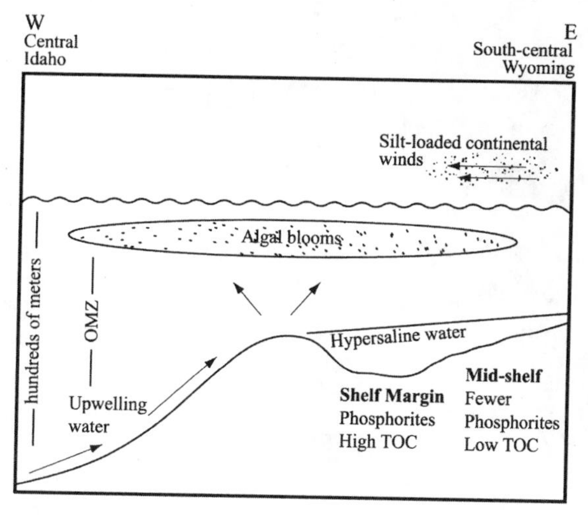

Figure 7. Proposed model of the Phosphoria sea showing the interplay of continental wind currents, upwelling ocean currents, algal blooms, and salinity stratification (modified from Stephens and Carroll 1999).

Many questions remain regarding the relationship of the Phosphoria Formation in size and age to other phosphorites. Estimates of the rate of burial of P in Phosphoria sediments relative to the total P-flux into the contemporary oceans vary widely (Trappe 1994) with estimates of 0.8 % (Filippelli and Delaney 1992), 4% (Herring 1995), and 10% (Arthur and Jenkyns 1981). The largest values are comparable to the burial rates of P relative to P-flux rates at the modern analogs off Peru and Namibia, suggesting that the burial rate of the Phosphoria Formation was less important than the extent and duration of phosphogenesis (Trappe 1994).

General geology. The Phosphoria Formation lies above the Permian Park City Formation, formed largely of limestone, and is overlain by the Triassic Dinwoody Formation (Fig. 8). The Phosphoria Formation contains two primarily phosphatic shale members, the Retort and the Meade Peak. The Meade Peak Member is the larger of the two main phosphorite bodies, and is the source of samples considered in our current work discussed here. The Meade Peak Phosphatic Shale Member is comprised of phosphorites, phosphatic shales, dolostones, siltstones, and mudstones (Fig. 9).

Extensive work has been undertaken to understand the general geology and economic prospects of the Phosphoria Formation. Early works that described the geology, including the stratigraphy, lateral extent, and economic viability of the Phosphoria were part of a large study by the U. S. Geological Survey (USGS) and other scientists, among them McKelvey et al. (1953, 1959), Cressman and Swanson (1964), and Service (1966). More detailed work on the mineralogy and bulk chemistry of the various lithologies throughout the phosphatic intervals, and broadened for pertinence to phosphorites as a whole, was performed by Gulbrandsen (1966, 1970). These and other works established a geologic understanding of the area critical to the progress of mining operations in the area, and current scientific work on the Phosphoria Formation.

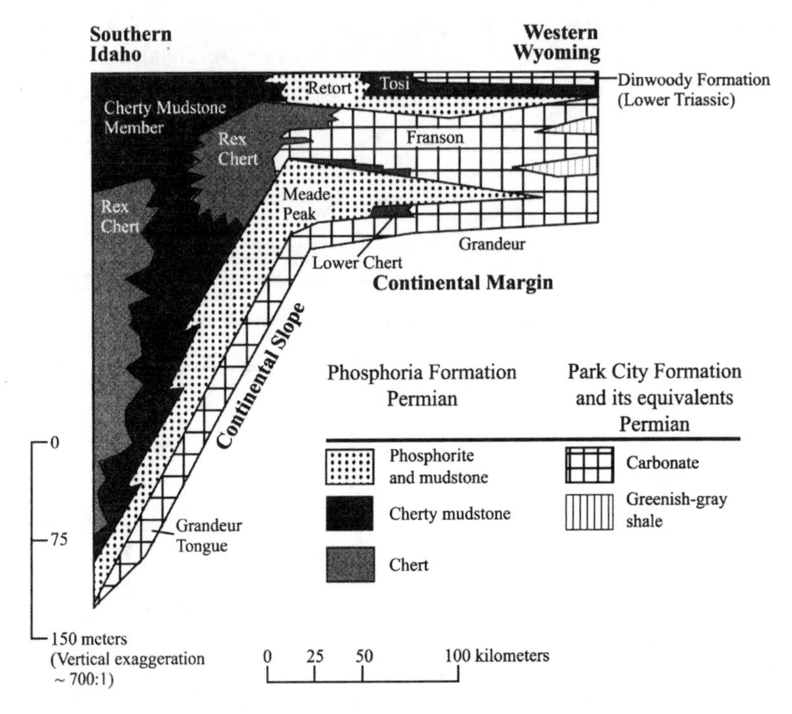

Figure 8. Cross-sectional view of the Phosphoria Formation and surrounding units from southern Idaho to western Wyoming (modified from Herring 1995).

The mineralogy of the Phosphoria Formation is much like that of most phosphorites, but it is also unique. As in all phosphorites, the primary phosphate-bearing mineral is CFA. Phosphoria Formation CFA is enriched in F^-, however it has relatively low levels of carbonate substitution for phosphate (Gulbrandsen 1966, 1970; Knudsen and Gunter 2002). Among the common suite of gangue minerals including silicates, carbonates, and sulfides is the ammonium feldspar buddingtonite. The presence of buddingtonite is of interest for its unique composition and rarity as well as its ties to the diagenesis of organic materials and hydrocarbon formation (Loughnan and Roberts 1983).

Current work

Mining of the Phosphoria Formation accounts for approximately 5% of the world's total phosphate production (Herring 1995) and 15% of U.S. production. This production accounts for by far the largest mineral resource in the state of Idaho, bringing in an estimated $600 million per year (Gillerman and Bennett 1997). The phosphate ore is primarily within the Meade Peak Member of the Phosphoria Formation, with the highest grade ores containing around 30% P_2O_5. Ore grade is greatly influenced by weathering, as increased alteration eliminates carbonate minerals and some organic material, thus increasing the relative amount of phosphate. There are four active mines in the area, each run by a different mining company. The companies and their mines are Agrium U.S. Inc. (Rasmussen Ridge mine), Astaris LLC (Dry Valley mine), J.R. Simplot Company (Smoky Canyon mine), and Monsanto Corporation (Enoch Valley mine) (Fig. 4).

Like many mining and refining operations there have been environmental concerns surrounding phosphate mining in Idaho. Until recently the most significant problems had been the air quality of calciner emissions, water quality surrounding waste piles, slight

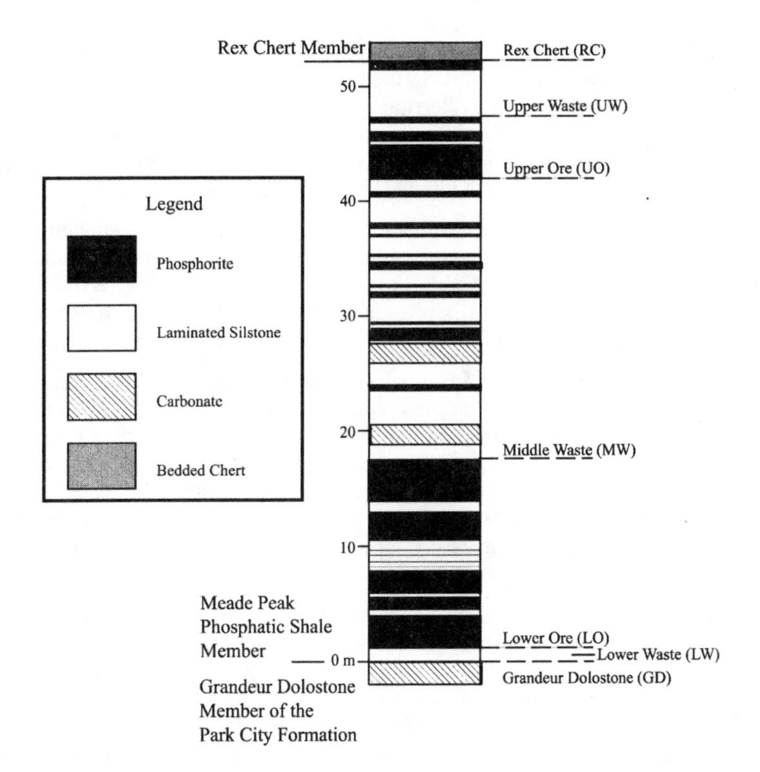

Figure 9. Idealized stratigraphic section of the Meade Peak Phosphatic Shale Member of the Permian Phosphoria Formation at the Enoch Valley Mine (Fig. 4). The major ore and waste units within the Meade Peak, which are recognized throughout the southeastern Idaho phosphate-mining district, are shown. This stratigraphic section (modified from Carroll et al. 1998) is most comparable to the "least-weathered" or "less-weathered" sections collected for our current work; the "more-weathered" section generally does not contain the carbonate strata.

radioactivity in refinement wastes as a result of uranium inclusion in apatite, and mine pit reclamation (Gillerman and Bennett 1997). However, a new and more threatening problem has arisen: selenium and other trace elements are naturally incorporated in these deposits. When exposed and transported during mining and milling processes, the selenium can be mobilized and enter the environment where it becomes bioavailable. Selenium, along with other trace elements (As, Cd, Cu, Mo, U, V, and Zn), is a major environmental concern surrounding the Idaho Phosphate fields. Concerns about the environmental impacts of Se related to mining in the Phosphoria Formation are tied to the deaths of seven horses and possibly more than sixty sheep that had been grazing in fields adjacent to mine waste piles; the deaths occurred from selenium toxicosis, which results from an over-consumption of Se. It is presumed that Se-enriched waters ran off the waste piles, and Se was concentrated in certain plants, which were consumed by the animals.

Selenium, like many trace elements, at low levels is essential for life, but elevated levels can be fatal. Humans and other animals, such as sheep, horses, cattle, and deer, need 0.1-0.3 ppm Se in their diet, yet levels of 3-15 ppm can prove deadly (Mayland 1994). In fact, not far from the area in which these livestock died of Se toxicosis, ranchers use Se-enriched salt blocks for their cattle to avoid Se deficiencies. These livestock deaths have raised concerns about the effects on fish, birds, and other wildlife, as well as humans. This new environmental concern has initiated a series of intense studies by a coalition headed by the USGS. and the phosphate industry. Workers have begun studying the concentrations of Se in soils (Amacher et al. 2001), plants (Herring and Amacher 2001), aquatic plants (Herring et al. 2001a), and air samples (Lamonthe and Herring 2000), as well as wildlife studies on elk and deer populations, fish, and birds. There have also been studies on the mechanisms in which these trace elements are released into the hydrosphere and the biota (Piper et al. 2000).

There are also efforts to further characterize the geology and geochemistry of the Phosphoria Formation in order to better understand Se residence and this unique deposit as a whole. Selenium is spread throughout the stratigraphic section (Herring et al. 1999, 2000a, 2000b, 2000c, 2001b), and is present in many phases (Desborough et al. 1999, Grauch 2002) including: sulfides, sulfates, native selenium, and other minor phases.

We analyzed channel samples from pairs of "more-weathered" and "less-weathered" measured stratigraphic sections from three active phosphate mines in southeastern Idaho (Fig. 4), and a deep core from one of the mines (Enoch Valley mine) was sampled as a "least-weathered" section. The sections were each measured and sampled by the USGS. for petrological, mineralogical, and geochemical studies. Samples were analyzed using powder X-ray diffraction (XRD) and Rietveld refinement (Knudsen and Gunter 2002) and geochemically using inductively coupled plasma-atomic emission spectrometry (Herring et al. 1999, 2000a, 2000b, 2001b).

The Meade Peak Phosphatic Shale Member of the Permian Phosphoria Formation is comprised generally of phosphorites, dolostones, shales, and siltstones, overlain by the Rex Chert Member of the Phosphoria Formation, and it overlies the Grandeur Dolostone Member of the Permian Park City Formation (Figs. 8 and 9). Because individual strata are difficult to trace between the measured sections, ore zones dominated by phosphorites, and waste zones dominated by dolostones and silicic-clastics, are used to compare the measured sections. Schematic box and whisker plots of the concentrations of CFA, total carbonates, and total silicates plotted across the recognized ore and waste zones (Fig. 10) summarize the dominant mineralogy and lithology of the "more-," "less-," and "least-weathered" sections.

The role weathering plays on the nature of the Meade Peak Phosphatic Shale is evident with the decreasing concentrations of the carbonate minerals with increased weathering. Weathering has important implications in that it increases the ore grade of the phosphatic rocks by concentrating P_2O_5 as gangue components are removed by weathering (Fig. 11a). However, the concentration of CFA does not increase with increased weathering and removal of other phases (Fig. 11b). As weathering increases the P_2O_5 / CFA ratio tends to increase (Fig. 11c). The increase in the P_2O_5 / CFA ratio (Fig. 11c) is most likely the result of a breakdown of a portion of the original CFA and the subsequent release of P_2O_5 into the new nondiffracting phases.

An estimate of the total nondiffracting, or amorphous, component was calculated by comparing the quantified mineralogical data (Knudsen and Gunter 2002) and chemical data (Herring et al. 1999, 2000a,b; 2001b) for these samples (Fig. 11d). To compare these two data sets, we determined, on the basis of chemical data, what the maximum concentration of a given phase in the whole sample would be based on a given element. For example, if there is only enough silica to account for 40% quartz and our quantified mineralogical data reports 50% quartz, then we assume that the sample is, at maximum, 80% crystalline (Knudsen and Gunter 2002, Knudsen 2002). Because of the many assumptions made, including perfect stoichiometry, and the compounding of errors by combining two separate data sets, these values can only be considered semi-quantitative. Figure 11d shows a trend in the increasing amount of the nondiffracting component,

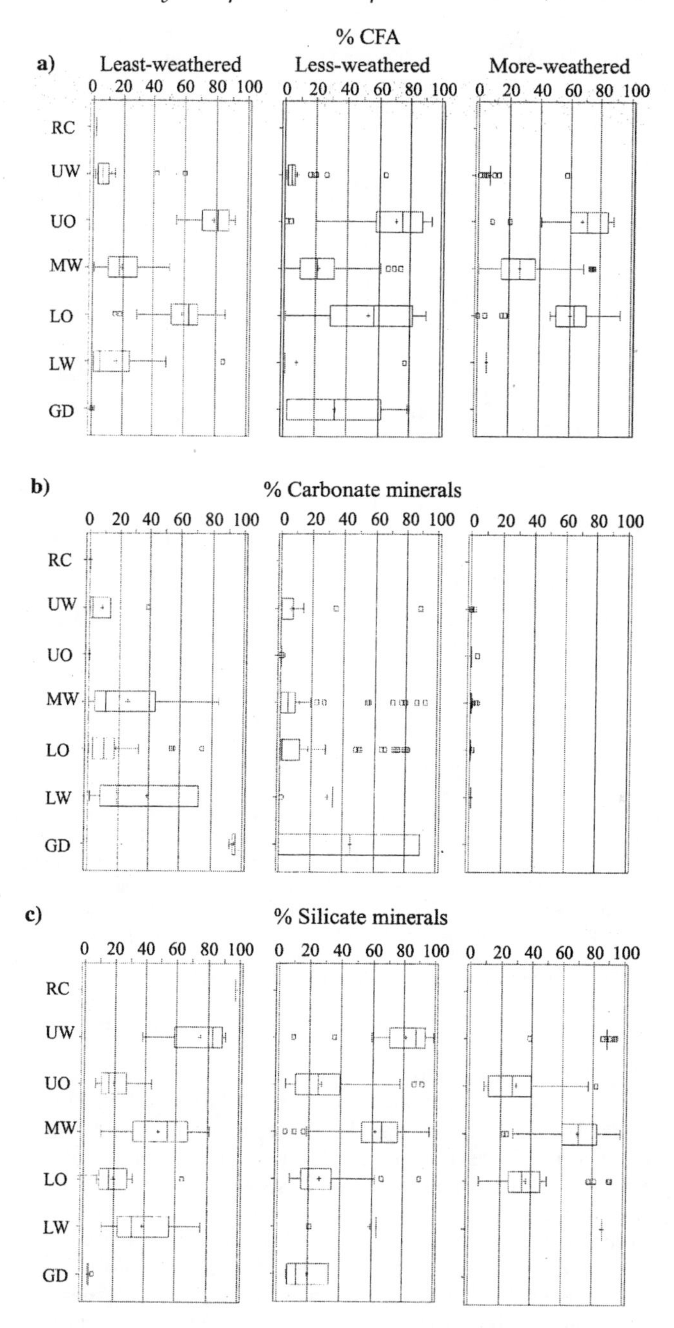

Figure 10. Schematic box and whisker plots of: (a) CFA, (b) carbonate minerals (i.e., calcite and dolomite), and (c) total silicates (i.e., quartz, muscovite, illite, albite, orthoclase, buddingtonite, kaolinite, and montmorillonite) in the "least-," "less-," and "more-weathered" sections from three active phosphate mines in southeastern Idaho (Fig. 4). [Refer to Figure 5 for a description of the schematic box and whisker plot.] Data from Knudsen (2002).

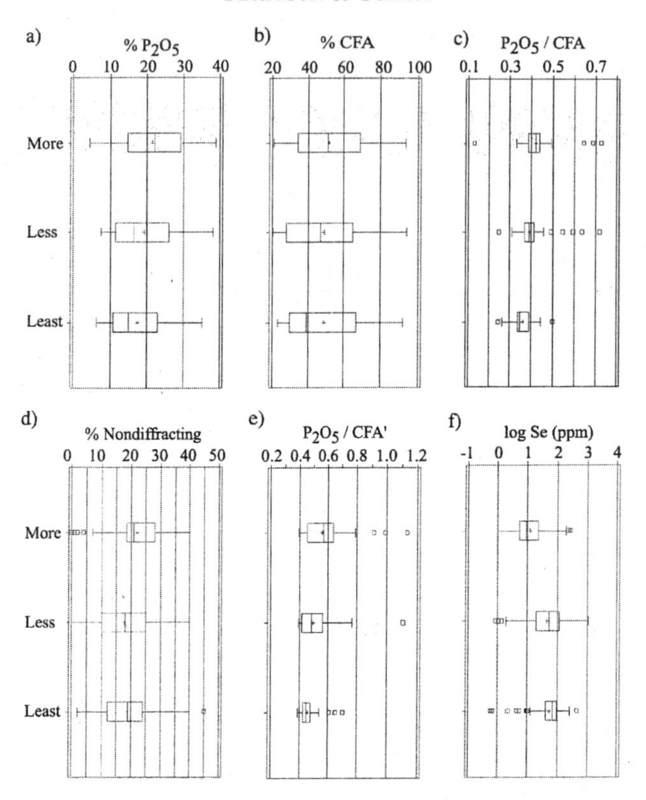

Figure 11. Schematic box and whisker plots of: (a) P_2O_5, (b) CFA, (c) P_2O_5 / CFA ratio, (d) estimated nondiffracting component, (e) P_2O_5 / CFA' ratio, where CFA' is obtained by multiplying the CFA concentration by the maximum crystallinity (i.e., 1 – nondiffracting component), and (f) log of the selenium concentrations in strata with greater than 20% CFA in the "least-," "less-," and "more-weathered" sections from three active phosphate mines in southeastern Idaho (Fig. 4). [Refer to Figure 5 for a description of the schematic box and whisker plot.] Chemical data from Herring et al. (1999, 2000a,b; 2001b); mineralogical data from Knudsen (2002).

likely the result of crystalline phases breaking down and remaining in the rock.

Figure 11c can be adjusted to account for the nondiffracting component by multiplying the CFA concentration by the maximum crystallinity (Fig. 11e). This adjusted P_2O_5 / CFA ratio has two notable features: (1) the trend seen in Figure 11c becomes more apparent, and (2) the ratio increases for each of the three sections. This suggests that in each of the three groups of data, and increasingly so with more weathering, the reported CFA concentrations, and all other minerals, are over-stated as a result of a significant nondiffracting component present in the samples. The increase in the ratio for each of the three groups of samples suggests that a significant amount of P_2O_5 is present in the nondiffracting component. In Figure 11e, the values tend to center near the 0.5 value, whereas they are centered around the 0.4 value in Figure 11c. The ratio of 0.4 is the approximate ratio that would be present in stoichiometric CFA, ranging from ~0.39 to 0.42 depending upon substitutions. A value of 0.5 suggests that there is more P_2O_5 present than can be accounted for on the basis of the CFA concentration. It is important to note that for each of the Figures 11a-c and 11e, only the samples with greater than 20%

reported CFA are used; this technique focuses on phosphorite samples, and eliminates the very large errors which are inevitable when comparing small numbers.

Considering Figure 11c, the unadjusted P_2O_5 / CFA ratio, the relationship of the P_2O_5 and the CFA is in accord with what would be predicted on the basis of stoichiometry. However, this relationship breaks down when the P_2O_5 / CFA ratio is adjusted (Fig. 11e). This observation is explained if P_2O_5 exists in approximately the same proportions in the crystalline and non-crystalline portions of sample, suggesting that nondiffracting phosphate phases may be considerably more abundant than has been previously considered. One possible explanation for this phenomenon is the removal of CO_3^{2-} from the CFA structure. As weathering increases, the degree of CO_3^{2-} in the CFA structure does not show the systematic reduction as expected (Lucas et al. 1979). If, in the Phosphoria Formation, the CFA structure breaks down into an apatite-like nondiffracting phase when the CO_3^{2-} is removed, the high concentration of nondiffracting phosphatic material, as well as the lack of a systematic reduction in the measured CO_3^{2-} in the CFA, can be explained.

Weathering also impacts the environmental concerns surrounding the Phosphoria Formation. Three general observations can be made relating to concentration of Se within and between the "least-," "less-," and "more-weathered" strata. First, average Se concentration decreases with increased weathering (Fig. 11f). Secondly, the variability of concentrations does not decrease, as seen particularly in the "less-weathered" strata with values greater than 1000 ppm Se. Finally, with increased weathering Se concentrations change from a near even distribution across the "least-weathered" strata to the "more-weathered" and "less-weathered" strata where the Middle Waste zone has notably more Se, and concentrations tend to decrease gradually both above and below the Middle Waste zone (Fig. 12a).

Selenium has been observed in a number of phases, particularly the sulfide minerals, pyrite and sphalerite, organic material, and native selenium, which, based upon its fragile morphology, is considered to be strictly a secondary phase (Grauch 2002, Desborough et al. 1999). Both the gradually decreasing average Se concentrations and the increased variability with increased weathering can be explained by the decreasing concentrations of the sulfides and organic C (Fig. 12b-c). Both of these important Se-bearing phases decrease in their concentrations with increased weathering. For the sulfides (Fig. 12b), only the sections from the Enoch Valley mine are shown because samples from the Dry Valley mine are considerably higher in pyrite, and thus both the "more-weathered" and "less-weathered" strata show high sulfide levels. Still, all three sites show a decrease from the "more-weathered" to the "less-weathered" strata (Knudsen and Gunter 2002). As these Se-bearing phases are weathered, the Se is mobilized, and either leaches out of the system or forms secondary phases such as native selenium or becomes associated with the increasing nondiffracting component (Fig. 11d).

Also, with increased weathering Se becomes more concentrated in the Middle Waste (Fig. 12a), which is characterized primarily by organic-rich shales and siltstones with lesser amounts of phosphate- and carbonate-bearing strata, than the other ore and other waste zones. The Middle Waste is also higher in both sulfides and organic C (Fig. 12b-c), although, unlike Se concentrations, these preferential concentrations are seen before weathering as well as after. The cause for this preferential concentration is not known. They may be related to lithological features, such as lower permeability in the Middle Waste, protecting the Se-bearing phases from weathering. It may also be a geochemical phenomenon in which these anoxic black shales create an environment that helps remove selenium from solution, including selenium that originated within and outside the Middle Waste shales. These questions will be considered in ongoing work on the Phosphoria Formation and the environmental problems surrounding phosphate extraction.

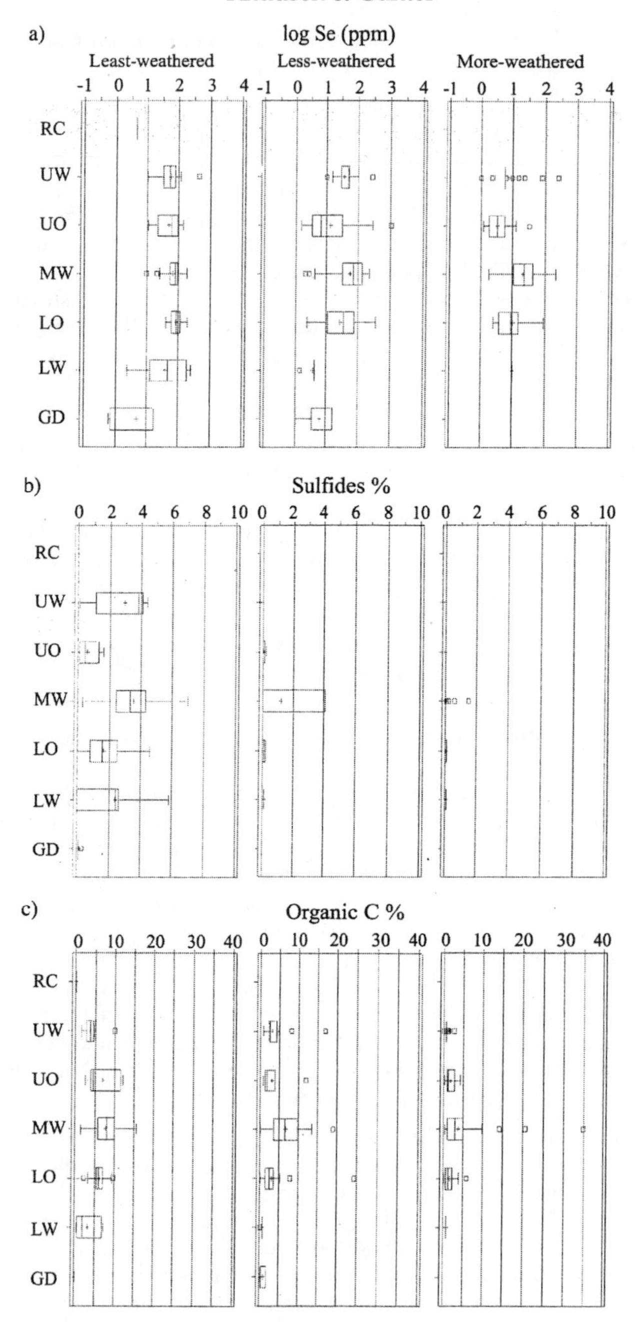

Figure 12. Schematic box and whisker plots of: (a) the log of selenium concentration, (b) sulfide minerals (i.e., pyrite and sphalerite), and (c) total concentration of organic carbon in the "least-," "less-," and "more-weathered" sections from three active phosphate mines in southeastern Idaho (Fig. 4). [Refer to Figure 5 for a description of the schematic box and whisker plot.] Data sources as in Figure 11.

ACKNOWLEDGMENTS

We thank the U. S. Geological Survey for their financial and scientific support in our work on the Phosphoria Formation, in particular, Phil Moyle, Dick Grauch, and Jim Herring; and we thank Tom Williams, University of Idaho, for his help with our XRD work. Also, we thank the four operating phosphate mines in southeastern Idaho for samples. Thorough reviews of early manuscripts by Karl Föllmi, Earl Bennet, and an anonymous reviewer helped to greatly improve this work. Finally, we thank the "Johns" (both Hughes and Rakovan) for helping to make this chapter a reality.

REFERENCES

Altschuler ZS, Cusney EA, Barlow IH (1952) X-ray evidence the nature of carbonate-apatite. Geol Soc Am Bull 63:1230-1231

Amacher MC, Herring JR, Stillings LL (2001) Total recoverable selenium and other elements by HNO_3 and $HClO_4$ digestion and other soil characterization data from Wooley Valley units 3 and 4 waste rock dumps and Dairy Syncline lease area soils, southeast Idaho. U S Geol Surv Open-File Report 01-69

Ames LL (1959) The genesis of carbonate apatites. Econ Geol 54:829-841

Arthur MA, Jenkyns HC (1981) Phosphorites and paleoceanography. Oceanol Acta, Proc 26th Intl Geol Congr, Paris, p 83-96

Atlas E, Pytkowicz RM (1977) Solubility behavior of apatites in seawater. Limnol Oceanogr 22:290-300

Axelrod S, Metzer A, Rohrlich V (1980) The petrography of Israeli phosphorites as related to their beneficiation. *In* Marine Phosphorites. Bentor YK (ed) SEPM Spec Pub 29:153-165

Baturin GN (1988) Disseminated phosphorous in ocean sediments—A review. Mar Geol 84:95-104

Baturin GN (1999) Hypotheses of phosphogenesis and oceanic environment. Lith Min Res 34:411-430

Baturin GN, Shishkina OV (1973) Behaviour of fluorine during phosphorite formation in the ocean. Oceanol 13:523-527

Bentor YK (1980) Phosphorites—the unsolved problems. *In* Marine phosphorites. Bentor YK (ed) Soc Econ Paleontol Mineral Spec Pub 29:3-18

Berner RA (1973) Phosphate removal from seawater by adsorption on volcanogenic ferric oxides. Earth Planet Sci Lett 18:84-86

Berner RA, Ruttenberg KC, Ingal ED, Rao JL (1993) The nature of phosphorous burial in modern marine sediments. *In* Interaction of C, N, P and S biochemical cycles and global change. Wollast R, Mackenzie FT, Chou L (eds) NATO ASI Series I4. Springer, Berlin, p 365-378

Berry EE (1967) The structure and composition of some calcium-deficient apatites. J Inorg Nucl Chem 29:3-18

Binder G, Troll G (1989) Coupled anion substitution in natural carbon-bearing apatites. Contrib Mineral Petrol 101:394-401

Borneman-Starynkevitch ID, Belov NV (1940) Isomorphic substitutions in carbonate-apatite. Dokl Acad Sci URSS XXVI 18:804-806

Brausseur H, Dallemagne MJ, Melon J (1946) Chemical nature of salts from bones and teeth and of tricalcium phosphate precipitates. Nature 157:453

Bushinskiy GI (1964) On shallow water origin of phosphorite sediments, deltaic and shallow marine deposits. Develop Sediment 1:62-70

Carroll AR, Stephens NP, Hendrix MS, Glenn CR (1998) Eolian-derived siltstone in the Upper Permian Phosphoria Formation: implications for marine upwelling. Geol 26:1023-1026

Cathcart JB (1991) Phosphate deposits of the United States—discovery, development, economic geology and outlook for the future. *In* The Geology of North America. Gluskoter HJ, Rice DD, Taylor RB (eds) Geol Soc Am, Boulder, Colorado, P-2:153-164

Compton JS, Hodell DA, Garrido JR, Mallinson DJ (1993) Origin and age of phosphorite from the south-central Florida platform: relation of phosphogenesis to sea-level fluctuations and $\delta^{13}C$ excursions. Geochim Cosmochim Acta 57:131-146

Cressman ER, Swanson RW (1964) Stratigraphy and petrology of the Permian rocks of southwestern Montana. U S Geol Surv Prof Paper 313-C:275-C569

Desborough GA, DeWitt E, Jones J, Meier A, Meeker G (1999) Preliminary mineralogical and chemical studies related to the potential mobility of selenium and associated elements in Phosphoria Formation strata, southeastern Idaho. U S Geol Surv Open-File Report 99-129

Elderfield H, Holmefjord T, Summerhayes CP (1972) Enhanced CO_2 substitution in carbonate-apatite from the Moroccan continental margin. Univ Leeds, Inst African Geol Res 16th Ann Report, p 51-52

Elliot JC (1969) Recent progress in the chemistry, crystal chemistry and structure of the apatites. Calcif Tissue Res 3:297-307

Emerson WH, Fisher EE (1962) The infra-red absorption spectra of carbonate in calcified tissues. Arch Oral Biol 7:671-683

Filippelli GM, Delaney ML (1992) Similar phosphorous flux in ancient phosphorite deposits and a modern phosphogenetic environment. Geol 20:709-712

Föllmi KB (1996) The phosphorous cycle, phosphogenesis and marine phosphate-rich deposits. Earth Sci Rev 40:55-124

Froelich PN, Kim KH, Jahnke R, Burnett WC, Soutar A, Deakin M (1983) Pore water fluoride in Peru continental margin sediments: uptake from seawater. Geochim Cosmochim Acta 47:1605-1612

Froelich PN, Arthur MA, Burnett WC, Deakin M, Hensley V, Jahnke R, Kaul L, Kim KH, Roe K, Soutar A, Vathakanon C (1988) Early diagenesis of organic matter in Peru continental margin sediments: phosphorite precipitation. Mar Geol 80:309-343

Gillerman VS, Bennett EH (1997) Industrial minerals of Idaho. *In* Proc 32nd Annual Forum on the Geology of Industrial Minerals. Jones RW, Harris RE (eds) Wyoming State Geol Surv, Public Info Circ 38: 207-218

Grauch RI (2002) Trace-element mineral residence and paragenetic framework of the Meade Peak Phosphatic Shale Member of the Phosphoria Formation, southeastern Idaho. *In* Life cycle of the Phosphoria Formation: from deposition to the post-mining environment. Hein JR (ed) Handbook of Exploration Geochemistry series (in press)

Gruner JW, McConnell D (1937) The problem of the carbonate apatites: The structure of francolite. Z Kristallogr Z27:208-215

Gulbrandsen RA (1966) Chemical composition of phosphorites of the Phosphoria Formation. Geochim Cosmochim Acta 30:769-778

Gulbrandsen RA (1970) Relation of carbon dioxide content of apatite of the Phosphoria Formation to regional facies. U S Geol Surv Prof Paper 700-B:B9-B13

Gulbrandsen RA, Roberson CE, Neil ST (1984) Time and crystallization of apatite in seawater. Geochim Cosmochim Acta 48:213-218

Harben PW, Bates RL (1990) Industrial Minerals Geology and World Deposits. Industrial Minerals Division Metal Bulletin, London

Heggie DT, Shyring GW, O'Brien GW, Reimers C, Herczeg A, Moriarty DJW, Burnett WC, Milnes AR (1990) Organic carbon cycling and modern phosphorite formation on the east Australian continental margin: an overview. *In* Phosphorite Research and Development. Notholt AJG, Jarvis I (eds) Geol Soc (London) Spec Pub 52:87-117

Herring JR (1995) Permian phosphorites: a paradox of phosphogenesis. *In* The Permian of northern Pangea: vol 2, sedimentary basins and economic resources. Peryt TM, Ulmer-Scholle DS (eds) Springer-Verlag, Berlin, p 293-312

Herring JR, Amacher MC (2001) Chemical composition of plants growing on the Wooley Valley phosphate mine waste pile and on similar rocks in nearby Dairy Syncline, Caribou County, southeast Idaho. U S Geol Surv Open-File Report 01-25

Herring JR, Desborough GA, Wilson SA, Tysdal RG, Grauch RI, and Gunter ME (1999) Chemical composition of weathered and unweathered strata of the Meade Peak Phosphatic Shale Member of the Permian Phosphoria Formation. A. Measured sections A and B, central part of Rasmussen Ridge, Caribou County, Idaho. U S Geol Surv Open-File Report 99-147-A

Herring JR, Wilson SA, Stillings LA, Knudsen AC, Gunter ME, Tysdal RG, Grauch RI, Desborough GA, Zielinski RA (2000a) Chemical composition of weathered and less weathered strata of the Meade Peak Phosphatic Shale Member of the Permian Phosphoria Formation—B: measured sections C and D, Dry Valley, Caribou County, Idaho. U S Geol Surv Open-File Report 99-147-B

Herring JR, Grauch RI, Tysdal RG, Wilson SA, Desborough GA (2000b) Chemical composition of weathered and less weathered strata of the Meade Peak Phosphatic Shale Member of the Permian Phosphoria Formation—D: measured sections G and H, Sage Creek area of the Webster Range, Caribou County, Idaho. U S Geol Surv Open-File Report 99-147-D

Herring JR, Grauch RI, Desborough GA, Wilson SA, Tysdal RG (2000c) Chemical composition of weathered and less weathered strata of the Meade Peak Phosphatic Shale Member of the Permian Phosphoria Formation—C: measured sections E and F, Rasmussen Ridge, Caribou County, Idaho. U S Geol Surv Open-File Report 99-147-C

Herring JR, Castle CJ, Brown ZA, Briggs PH (2001a) Chemical composition of deployed and indigenous aquatic bryophytes in a seep flowing from a phosphate mine waste pile and in the associated Angus Creek drainage, Caribou County, southeast Idaho. U S Geol Surv Open-File Report 01-26

Herring JR, Grauch RI, Seims DF, Tysdal RG, Johnson EA, Zielinski RA, Desborough GA, Knudsen AC, Gunter ME (2001b) Chemical composition of strata of the Meade Peak phosphatic shale member of the Permian Phosphoria Formation channel-composited and individual rock samples of measured section J and their relationship to measured sections A and B, central part of Rasmussen Ridge, Caribou County, Idaho. U S Geol Surv Open-File Report 01-195

Hiatt EE, Budd DA (2001) Sedimentary phosphate formation in warm shallow waters: new insights into the paleoceanography of the Permian Phosphoria Sea from analysis of phosphate oxygen isotopes. Sed Geol 145:119-133

Kazakov AV (1937) The phosphorite facies and the genesis of phosphorites. Trans Sci Inst Fertil Insecto-Fung 142:95-113

Knudsen AC (2002) A mineralogical investigation of the Permian Phosphoria Formation, southeastern Idaho: characterization, environmental concerns, and weathering. PhD dissertation, University of Idaho, Moscow

Knudsen AC, Gunter ME (2002) Mineralogy of the Phosphoria Formation: effects of weathering on bulk mineralogy and carbonate-fluorapatite substitution. *In* Life cycle of the Phosphoria Formation: from deposition to the post-mining environment. Hein JR (ed) Handbook of exploration geochemistry series (in press)

Krajewski KP, Van Cappellen P, Trichet J, Kuhn O, Lucas J, Martín-Algarra A, Prévôt L, Tewari VC, Gaspar L, Knight RI, Lamboy M (1994) Biological processes and apatite formation in sedimentary environments. *In* Concepts and controversies in phosphogenesis. Föllmi KB (ed) Eclogae Geol Helv 87:701-745

Krom MD, Berner RA (1980) Adsorption of phosphate in anoxic marine sediments. Limnol Oceanogr 25:797-806

Krom MD, Berner RA (1981) The diagenesis of phosphorous in a near-shore marine sediment. Geochim Cosmochim Acta 45:207-216

Lamonthe PJ, Herring JR (2000) Selenium and other trace elements in air samples collected near the Wooley Valley phosphate mine waste pile. U S Geol Surv Open-File Report 00-514

Lehr JR, McClellan GH, Smith JP, Frasier AW (1968) Characterization of apatites in commercial phosphate rocks. *In* Coll Int Phosphates Mineraux Solides, Toulouse 1967 Vol. 2, Paris, p 29-44

Loughan F, Roberts I (1983) Buddingtonite (NH_4-feldspar) in the Condor oilshale deposit, Queensland, Australia. Mineral Mag 47:327-334

Lucas J, Prèvôt L, Lamboy M (1978) Les Phosphorites de la marge nord de l'Espagne. Chimie, Minéralogie, Génèse. Oceanol Acta 1:55-72

Lucas J, Prèvôt L, El. Mountassir M (1979) Les phosphorites rubéfiées de Sidi Daoui. Transformation météorique locale du gisement de phosphate des Ouled Abdoun (Maroc). Sci Geol Bull 32:21-37

Lucas J, Prèvôt L, Ataman G, Gundogdu N (1980) Mineralogical and geochemical studies of phosphatic formations in southeastern Turkey (Mazidagl-Mardin). *In* Marine Phosphorites. Bentor YK (ed) SEPM Spec Pub 29, Tulsa, p 142-152

Lucotte M, Mucci A, Hillaire-Marcel C, Tran S (1994) Early diagenetic processes in deep Labrador Sea sediments: reactive and nonreactive iron and phosphorus. Can J Earth Sci 31:14-27

Martens CS, Harris R (1970) Inhibition of apatite precipitation in the marine environment by magnesium ions. Geochim Cosmochim Acta 34:621-625

Martin-Algarra A, Sanchez-Navas A (2000) Bacterially mediated authigenesis in Mesozoic stromatolites from condensed "pelagic" sediments. *In* Marine Authigenesis: From Global to Microbial. Glenn CR, Prévôt-Lucas L, Lucas J (eds) SEPM Spec Pub 66, Tulsa, p 499-526

Matthews A, Nathan Y (1977) The decarbonation of carbonate-fluorapatite (francolite). Am Mineral 62:565-573

Maughan EK (1980) Relation of phosphorite, organic carbon and hydrocarbons in the Permian Phosphoria Formation, western USA. *In* Geologie comparèe des gisements de phosphates et de pètrole. Doc B.R.G.M 24:63-91

Mayland HF (1994) Selenium in plant and animal nutrition. *In* Selenium in the environment. Frankenberger Jr. WT, Benson S (eds) Marcel Dekker, New York, p 29-45

McArthur JM (1978) Systematic variations in the contents of Na, Sr, CO_2, and SO_4 in marine carbonate-fluorapatite and their relation to weathering. Chem Geol 21:41-52

McArthur JM (1985) Francolite geochemistry-compositional controls during formation, diagenesis, metamorphism and weathering. Geochim Cosmochim Acta 49:23-35

McClellan GH (1980) Mineralogy of carbonate fluorapatites. J Geol Soc London 137:675-681

McClellan GH, Lehr JR (1969) Crystal chemical investigation of natural apatites. Am Mineral 54: 1374-1391

McClellan GH, Van Kauwenbergh SJ (1990) Mineralogy of sedimentary apatites. *In* Phosphorite research and development. Notholt AJG, Jarvis I (eds) Geol Soc Spec Pubs 52:23-31

McConnell D (1938) A structural investigation of the isomorphism of the apatite group. Am Mineral 23: 1-19

McKelvey VE, Davidson DF, O'Malley FW, Smith LE (1953) Stratigraphic sections of the Phosphoria Formation in Idaho, 1947-48, Part 1. U S Geol Surv Circular 208

McKelvey VE, Williams JS, Sheldon RP, Cressman ER, Gheney TM, Swanson RW (1959) The Phosphoria Formation, and Shedhorn Formations in the Western Phosphate Field. US Geol Surv Prof Paper 313-A

Murray J, Renard AF (1891) Scientific results, HMS Challenger. Deep Sea Deposits p 391-400

Nathan Y (1984) The mineralogy and geochemistry of phosphorites. *In* Phosphate minerals. Nriagu JO, Moore PB (eds) Springer-Verlag, Berlin, Heidelberg, New York, Tokyo, p 275-291

Nathan Y (1996) Mechanism of CO_3^{2-} substitution in carbonate fluorapatite: evidence from FTIR spectroscopy, ^{13}C NMR, and quantum mechanical calculations—discussion. Am Mineral 81:513-514

Nathan Y, Lucas J (1972) Synthèse de l'apatite à partir du gypse: application au problème de la formation des apatites carbonates par precipitation directe. Chem Geol 9:99-112

Neuman WF, Neuman MW (1953) Nature of the mineral phase of bone. Chem Rev 53:1-38

O'Brien GW, Heggie D (1988) East Australian continental margins phosphorites. EOS Trans, Am Geophys Union 69:2

O'Brien GW, Milnes AR, Veeh HH, Heggie DT, Riggs SR, Cullen DJ, Marshall JF, Cook PJ (1990) Sedimentation dynamics and redox iron-cycling: controlling factors for the apatite-glauconite association on the east Australian margin. *In* Phosphorite research and development. Notholt AJG, Jarvis I (eds) Geol Soc Spec Publ 52:61-86

Ong RG, Davidson DM Jr. (1992) Fluoridation of North Carolina pelletal phosphorites implications for phosphogenesis. Econ Geol 87:1166-1173

Piper DZ, Skorupa JP, Presser TS, Hardy MA, Hamilton SJ, Huebner M, Gulbrandsen RA (2000) The Phosphoria Formation at Hot Springs mine in southeast Idaho: a source of selenium and other trace elements to surface water, ground water, vegetation, and biota. U S Geol Surv Open-File Report 00-50

Posner AS, Perloff A (1957) Apatites deficient in divalent cations. Nat Bur Stand J Res 58:279-286

Purnachandra Rao P, Mohan Rao K, Raju DSN (2000) Quaternary phosphorites from the continental margin off Chennai, southeast India: analogs of ancient phosphate stromatolites. J Sed Res 70: 1197-1209

Regnier P, Lasaga AC, Berner RA, Han OH, Zilm KW (1994) Mechanism of CO_3^{2-} substitution in carbonate fluorapatite: evidence from FTIR spectroscopy, ^{13}C NMR, and quantum mechanical calculations. Am Mineral 79:809-818

Ruttenberg KC, Berner RA (1993) Authigenic apatite formation and burial in sediments from non-upwelling, continental margin environments. Geochim Cosmochim Acta 57:991-1007

Sandell EB, Hey MH, McConnell D (1939) The composition of francolite. Mineral Mag 25:395-401

Schuffert JD, Kastner M, Emanuele G, Jahnke RA (1990) Carbonate-ion substitution in francolite: A new equation. Geochim Cosmochim Acta 54:2323-2328

Schwennicke T, Siegmund H, Jehl C (2000) Marine phosphogenesis in shallow-water environments: Cambrian, Tertiary, and Recent examples. *In* Marine authigenesis: from global to microbial. Glenn CR, Prévôt-Lucas L, Lucas J (eds) SEPM Spec Pub 66, Tulsa, Oklahoma, p 481-498

Service AL (1966) An evaluation of the western phosphate industry and its resources (in five parts), 3. Idaho. U S Bur Mines Report Invest 6801

Sholkovitz E (1973) Interstitial water chemistry of the Santa Barbara Basin sediments. Geochim Cosmochim Acta 51:1861-1866

Slansky M (1986) Geology of Sedimentary Phosphates. North Oxford Academic, London

Smith JP, Lehr JR (1966) An X-ray investigations of carbonate apatite. J Agric Food Chem 14:342-349

Stephens NP, Carroll AR (1999) Salinity stratification in the Permian Phosphoria sea: a proposed paleoceanographic model. Geology 27:899-902

Thewlis J, Glock GE, Murray MM (1939) Chemical and X-ray analysis of dental, mineral and synthetic apatites. Trans Faraday Soc 35:358-363

Trappe J (1994) Pangean phosphorites—ordinary phosphorite genesis in an extraordinary world? *In* Pangea: global environments and resources. Embry AF, Glass DJ (eds) Memoir, Can Soc Petrol Geol 17:469-478

Treuman NA (1966) Substitutions for phosphate ions in apatite. Nature 210:937-938

Van Bennekom AJ, Van der Gaast SJ (1976) Possible clay structures in frustules of living diatoms. Geochim Cosmochim Acta 40:1149-1152

Van Cappellen P, Berner RA (1988) A mathematical model for the early diagenesis of phosphorous and fluorine in marine sediments: apatite precipitation. Am J Sci 288:289-333

Van Cappellen P, Gaillard JF, Rabouille C (1993) Biogeochemical transformation in sediments: kinetic models of early diagenesis. *In* Interactions of C, N, P and S biochemical cycles and global change. Wollast R, Mackenzie FT (eds) NATO ASI Series I4, Springer, Berlin, p 401-455

White WB (1974) The carbonate minerals. Mineral Soc 4:227-284

Williams LA, Reimers C (1983) Role of bacterial mats in oxygen-deficient marine basins and coastal upwelling regimes: Preliminary report. Geol 11:267-269

Wollast R, Debroeu F (1971) Study of the behavior of dissolved silica in the estuary of the Scheldt. Geochim Cosmochim Acta 35:613-620

10 The Global Phosphorus Cycle

Gabriel M. Filippelli

Department of Geology
Indiana University - Purdue University Indianapolis
723 West Michigan Street
Indianapolis, Indiana 46202

INTRODUCTION

Phosphorus (P) is a limiting nutrient for terrestrial biological productivity that commonly plays a key role in net carbon uptake in terrestrial ecosystems (Tiessen et al. 1984, Roberts et al. 1985, Lajtha and Schlesinger 1988). Unlike nitrogen (another limiting nutrient but one with an abundant atmospheric pool), the availability of "new" P in ecosystems is restricted by the rate of release of this element during soil weathering. Because of the limitations of P availability, P is generally recycled to various extents in ecosystems depending on climate, soil type, and ecosystem level. The release of P from apatite dissolution is a key control on ecosystem productivity (Cole et al. 1977, Tiessen et al. 1984, Roberts et al. 1985, Crews et al. 1995, Vitousek et al. 1997, Schlesinger et al. 1998), which in turn is critical to terrestrial carbon balances (e.g., Kump and Alley 1994, Adams 1995). Furthermore, the weathering of P from the terrestrial system and transport by rivers is the only appreciable source of P to the oceans. On longer time scales, this supply of P also limits the total amount of primary production in the ocean (Holland 1978, Broecker 1982, Smith 1984, Filippelli and Delaney 1994). Thus, understanding the controls on P weathering from land and transport to the ocean is important for models of global change. In this paper, I will present an overview of the natural (pre-human) and modern (syn-human) global P mass balances, followed by in-depth examinations of several current areas of research in P cycling, including climatic controls on ecosystem dynamics and soil development, the control of oxygen on coupled P and Carbon (C) cycling in continental margins, and the role that P plays in controlling ocean productivity on Cenozoic timescales.

GLOBAL PHOSPHORUS CYCLING

Natural (pre-human) phosphorus cycle

The human impact on the global P cycle has been substantial over the last 150 years and will continue to dominate the natural cycle of P on the globe for the foreseeable future. Because this anthropogenic modification began well before scientific efforts to quantify the cycle of P, we can only guess at the "pre-anthropogenic" mass balance of P. Several aspects of the pre-anthropogenic sources and sinks of P are relatively well constrained (Fig. 1). The initial source of P to the global system is via the weathering of P during soil development, whereby P is released mainly from apatite minerals and made soluble and bioavailable (this process will be discussed at considerable length later). In contrast to this process of chemical weathering, the physical weathering and erosion of material from continents results in P that is typically unavailable to biota. An exception to this, however, is the role that physical weathering plays in producing fine materials with extremely high surface area/mass ratios. Phosphorus and other components may be rapidly weathered if this fine material is deposited in continental environments (i.e., floodplains and delta systems) where it undergoes subsequent chemical weathering and/or soil development. Thus, the total amount of P weathered from continents may be very different from the amount of potentially bioavailable P.

Apatite minerals, the dominate weathering source of P, vary widely in chemistry and

1529-6466/00/0048-0010$05.00

structure and can form in igneous, metamorphic, sedimentary, and biogenic environments. All apatite minerals contain phosphate oxyanions linked by Ca^{2+} cations to form a hexagonal framework, but they differ in elemental composition at the corners of the hexagonal cell (Table 1; McClellan and Lehr 1969, Hughes and Rakovan, this volume). By far the most abundant of the igneous apatite minerals, fluorapatite, is an early-formed accessory mineral that appears as tiny euhedral crystals associated with ferromagnesian minerals (McConnell 1973).

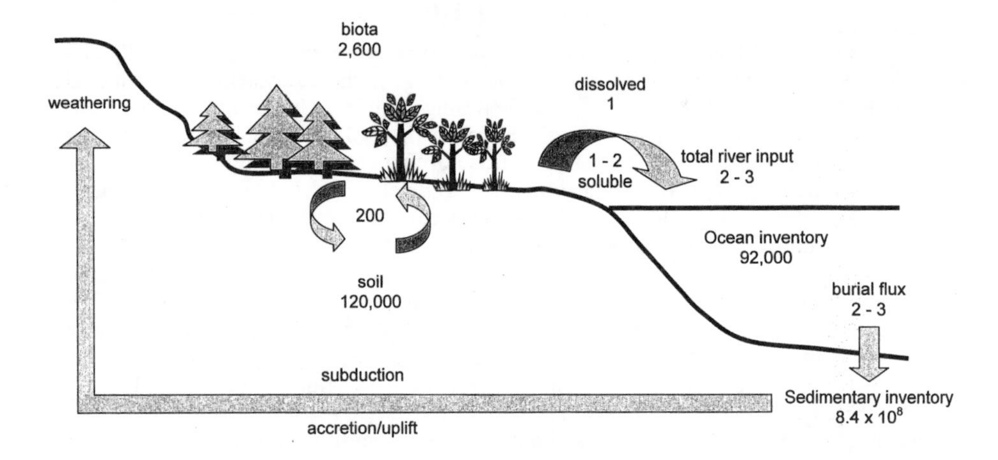

Figure 1. The natural (pre-human) dissolved phosphorus cycle, showing reservoirs (in Tg P) and fluxes (denoted by arrows, in Tg P/yr) in the P mass balance.

Table 1. Terminology and crystal chemistry of apatites.

Mineral name	*Chemical composition*
Fluorapatite (FAP)	$Ca_5\,(PO_4)_3F$
Hydroxylapatite (HAP)	$Ca_5\,(PO_4)_3OH$
Chlorapatite	$Ca_5\,(PO_4)_3Cl$
Dahllite	$Ca_5\,(PO_4, CO_3OH)_3(OH)$
Francolite (CFAP)	$Ca_5\,(PO_4, CO_3OH)_3(F)$

After Phillips and Griffen (1981).

Apatite dissolution: phosphate rock and synthetic hydroxyapatite (HAP). The dissolution rate of apatite minerals has been the focus of a variety of relatively disparate studies. For example, the importance of P as a fertilizer and as an integral component of teeth and bone material has led to a wide range of P dissolution and retention studies. Most of the agricultural studies have focused on the dissolution of fertilizers, in which P has already been leached from phosphate rock ore (usually a mixed lithology of marine sediments containing high percentages of carbonate fluorapatite (CFAP), $Ca_5\,(PO_4, CO_3OH)_3(F)$, as described in Filippelli and Delaney 1992a) and re-precipitated as a highly soluble phosphoric salts. Some studies, however, have investigated the dissolution of phosphate rock in laboratory experiments (Smith et al. 1974, 1977; Olsen 1975, Chien et al. 1980, Onken and Matheson 1982). The medical studies have investigated the dissolution behavior of synthetic HAP (similar in composition to dahllite) in acidic solutions, especially to understand the processes of bone resorption and formation, as well as the formation of dental caries (cavities) (Christoffersen et al. 1978, Christoffersen 1981, Christoffersen and Christoffersen 1985, Fox et al. 1978, Nancollas et al. 1987, Constantz et al. 1995).

Several workers have attempted to develop dissolution rate equations to model apatite dissolution (Olsen 1975, Smith et al. 1977, Christoffersen et al. 1978, Fox et al. 1978, Chien et al. 1980, Onken and Metheson 1982, Hull and Lerman 1985, Hull and Hull 1987, Chin and Nancollas 1991). Rate equations from these models include zero order, first order, parabolic diffusion, mixed order, and other forms. The most current model (Hull and Hull 1987) focuses on surface dissolution geometry, which the authors argue fit the experimental results better than previous dissolution models. These experiments and the dissolution rate equations derived from them are missing the experimental conditions that replicate the natural dissolution processes and agents in soils, as they do not include the range of apatite mineralogies likely to be naturally weathering in soils.

Apatite dissolution: marine sediments and CFAP. Perhaps the most comprehensive and geologically appropriate laboratory examination of apatite dissolution has been performed on CFAP (francolite). Researchers focused on the marine P cycle have performed several studies on CFAP dissolution (Lane and Mackenzie 1990, 1991; Tribble et al. 1995), as well as CFAP precipitation in laboratory settings (Jahnke 1984, Van Cappellen and Berner 1988, Van Cappellen 1991, Filippelli and Delaney 1992b, 1993). These studies have focused on CFAP because of its importance as an authigenic marine mineral phase in terms of the oceanic P cycle (Ruttenberg 1993, Filippelli and Delaney 1996). Several general observations about apatite dissolution/ precipitation have been made. For example, the presence of Mg has been determined to retard CFAP formation (Atlas and Pytkowicz 1977), but the presence of other trace elements (e.g., Fe and Mn) appears to have little effect (Filippelli and Delaney 1992b, 1993).

In a study of the pH-dependence of CFAP dissolution, Lane and Mackenzie (1990, 1991) used a fluidized bed reactor to determine dissolution chemistry. As presented in Tribble et al. (1995), a Ca and fluorine (F)-depleted surface layer appears to form during early dissolution, followed by a later stage when stoichiometric dissolution is achieved. The incongruent initial dissolution is probably due to the removal of most or all of the F and some Ca in the depleted surface layer, and the formation of a hydrogen-calcium phosphate phase, which has also been observed by other workers (Atlas and Pytkowicz 1977, Smith et al. 1974, Driessens and Verbeeck 1981, Thirioux et al. 1990). This surface controlled reaction eventually achieves a steady state—the depleted surface layer does not change in depth and the solid effectively dissolves congruently (Tribble et al. 1995). Ruttenberg (1990, 1992) performed extensive dissolution experiments on fluorapatite (FAP), HAP, and CFAP, with a goal of developing an extraction scheme for the characterization of these mineral phases in marine sediments. Although geared toward extraction technique development, these studies showed increased dissolution rate with decreasing grain sizes, and higher dissolution rates of HAP and CFAP than FAP with moderately acidic pH values.

Phosphorus cycling in soils. The cycling of P in soils (see Fig. 1) has received much attention, both in terms of fertilization and the natural development of ecosystems. Of the approximately 122,600 Tg P within the soil/biota system on the continents, nearly 98% is held in soils in a variety of forms. The exchange of P between biota and soils is relatively rapid, with an average residence time of 13 years, whereas the average residence time of

P in soils is 600 years (Fig. 1). As noted above, the only significant weathering source for phosphorus in soils is apatite minerals. These minerals can be congruently weathered as a result of reaction with dissolved carbon dioxide:

$$Ca_5(PO_4)_3OH + 4\ CO_2 + 3H_2O \rightarrow 5\ Ca^{2+} + 3\ HPO_4^{2-} + 4\ HCO_3^{1-}$$

In soils, P is released from mineral grains by several processes. First, the reduced pH produced from respiration-related CO_2 in the vicinity of both degrading organic matter and root hairs dissolves P-bearing minerals (mainly apatites) and releases P to root pore spaces (e.g., Schlesinger 1997). Second, organic acids released by plant roots also can dissolve apatite minerals and release P to soil pore spaces (Jurinak et al. 1986). Phosphorus is very immobile in most soils, and its slow rate of diffusion from dissolved form in pore spaces strongly limits its supply to rootlet surfaces (Robinson 1986). Furthermore, much of the available P in soils is in organic matter, which is not directly accessible for plant nutrition. Plants have developed two specific tactics to increase the supply of P to roots. First, plants and soil microbes secrete phosphatase, an enzyme that can release bio-available inorganic P from organic matter (McGill and Cole 1981, Malcolm 1983, Kroehler and Linkins 1988, Tarafdar and Claasen 1988). Second, the symbiotic fungi *mycorrhizae* can coat plant rootlets, excreting phosphatase and organic acids to release P and providing an active uptake site for the rapid diffusion of P from soil pore spaces to the root surface (Antibus et al. 1981, Bolan et al. 1984, Dodd et al. 1987). In exchange, the plant provides carbohydrates to the *mycorrhizal* fungi (Schlesinger 1997).

Phosphorus in soils is present in a variety of forms, and the distribution of P between these forms changes dramatically with time and soil development. The forms of soil P can be grouped into refractory (not readily bio-available) and labile (readily bio-available). The refractory forms include P in apatite minerals and P co-precipitated with and/or adsorbed onto iron and manganese oxyhydroxides (termed "occluded" P). The reducible oxyhydroxides have large binding capacities for phosphate, due to their immense surface area and numerous delocalized positively charged sites (e.g., Froelich 1988, Smeck 1985). The labile forms include P in soil pore spaces (as dissolved phosphate ion) and adsorbed onto soil particle surfaces (these forms are termed "non-occluded" P), as well as P incorporated in soil organic matter. On a newly exposed lithic surface, nearly all of the P is present as P in apatite. With time and soil development, however, P is increasingly released from this form and incorporated in the others (Fig. 2). Over time, the total amount of P available in the soil profile decreases, as soil P is lost through surface and subsurface runoff. Eventually, the soil reaches a terminal steady state, when soil P is heavily recycled and any P lost through runoff is slowly replaced by new P weathered from apatites at the base of the soil column.

Riverine transport of particulate and dissolved phosphorus. The eventual erosion of soil material and transport by rivers delivers P to the oceans. Riverine P occurs in two main forms: particulate and dissolved. Most of the P contained in the particulate load of rivers is held within mineral lattices and never participates in the active biogenic cycle of P. This will also be its fate once it is delivered to the oceans, because dissolution rates in the high pH and heavily buffered waters of the sea are exceedingly low. Thus, much of the net P physically eroded from continents is delivered relatively unaltered to the oceans, where it is sedimented on continental margins and in the deep sea, until subduction or accretion eventually returns P to be exposed on land again. Some of the particulate P is adsorbed onto soil surfaces, held within soil oxide, and incorporated into particulate organic matter. This P likely interacts with the biotic P cycle on land, and its fate upon transfer to the ocean is poorly understood. For example, P adsorbed onto soil surfaces may be effectively displaced by the high ionic strength of ocean water, providing an additional source of P into the ocean. Furthermore, a small amount of the P incorporated into terrestrial organic matter may be released in certain environments during bacterial oxidation after sediment burial. Finally, some sedimentary environments along continental margins are suboxic or even anoxic, conditions favorable for oxide dissolution and release of the incorporated P.

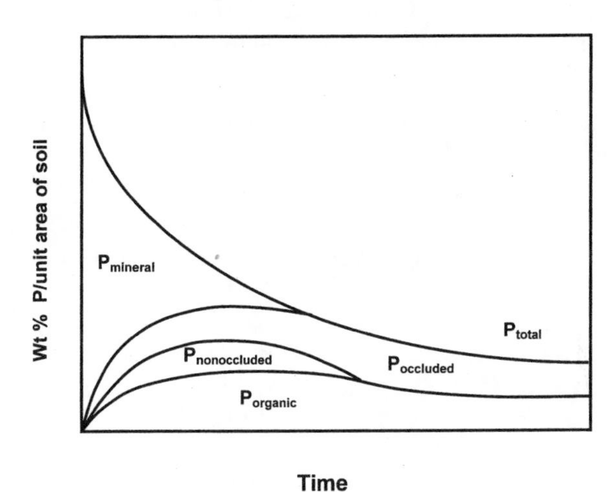

Figure 2. Modeled changes in soil phosphorus geochemistry over time (based on Walker and Syers 1976), showing transformation of mineral phosphorus into non-occluded and organic forms before eventual dominance of occluded (oxide-bound) and organic forms. The relative bio-reactivity of phosphorus increases from mineral to occluded to organic forms of phosphorus. Note the continual loss of total phosphorus from the system. From Filippelli and Souch (1999).

Several important studies have examined the transfer of P between terrestrial and marine environments (e.g., Froelich 1988, Berner and Rao 1994, Ruttenberg and Goñi 1997), but more work clearly needs to be done to quantify the interactions between dissolved and particulate P forms and the aquatic/marine interface. The net pre-human flux of dissolved P to the oceans is ~ 1 Tg P/year, with an additional 1-2 Tg P/year of potentially soluble P, bringing the total to about 2-3 Tg P/year. Thus, the residence time of biologically available P on land is about 40-60 kyr with respect to export to the oceans. It may be no coincidence that this residence time is of a glacial timescale—later I discuss the control of climate on the terrestrial P cycle. But clearly, the interaction between biologically available P on land and loss of this P to the oceans is relatively dynamic and speaks to the relatively rapid cycling of P on land.

Marine sedimentation. Once in the marine system, dissolved P acts as the critical long-term nutrient limiting biological productivity. Phosphate concentrations are near zero in most surface waters, as this element is taken up by phytoplankton as a vital component of their photosystems (phosphate forms the base for ATP and ADP, required for photosynthetic energy transfer) and their cells (cell walls are comprised of phospholipids). Dissolved P has a nutrient profile in the ocean, with a surface depletion and a deep enrichment. Furthermore, deep phosphate concentrations increase with the age of deep water, and thus values in young deep waters of the Atlantic are typically ~1.5 μM whereas those in the older Pacific are ~2.5 μM (Broecker and Peng 1982). Once incorporated into plant material, P roughly follows the organic matter loop, undergoing active recycling in the water column and at the sediment/water interface.

Phosphorus input and output are driven to steady state mass balance in the ocean by biological productivity. As mentioned previously, one of the difficulties with accurately determining the pre-anthropogenic residence time of P in the ocean is that input has been nearly doubled due to anthropogenic activities, and thus we must resort to estimating the burial output of P, a technique plagued by site-to-site variability in deposition rates and poor age control. Estimates of P burial rates have been performed by a variety of methods, including determination via P sedimentary sinks (e.g., Froelich et al. 1982, 1983) and riverine suspended matter fluxes calibrated to the P geochemistry of those fluxes (Ruttenberg 1993). We have used an areal approach to quantifying the modern P mass balance (Table 2; Filippelli and Delaney 1996). Using this areal approach, we find that the output term for the P mass balance might be higher than previously estimated, and therefore the P residence time might be shorter than previously thought (Table 2).

Table 2. Phosphorus accumulation, burial, and the modern oceanic phosphorus mass balance (from Filippelli and Delaney 1996).

	Equat. Pacific[a]	*All mod/ high prod. pelagic*[b]	*Low prod. pelagic regions*[c]	*Cont. margins*[d]	*Whole Ocean*[e]	*Previous estimates Ruttenberg (1993)*	*Froelich et al. (1982)*[f]
P acc. rate *(μmol cm^{-2} kyr^{-1})*	19	28	10	500	--	--	--
Area (10^6 km^2)	30	276	60	25	361	--	--
P burial rate[g] *(10^{10} mol P/yr)*	0.6	7.7	0.6	12.5	**21.0**	8.0–18.5	5.0
P residence time *(kyr)*[h]					**15**	16–38	60

[a] Mean of all samples from 0 to 1 Ma in Filippelli and Delaney (1996).

[b] Defined as a region with a primary productivity greater than 35 g C m^{-2} yr^{-1} (Berger and Wefer 1991). Phosphorus accumulation rate is a mean of various moderate to high productivity pelagic regions (e.g., Southern Ocean, equatorial Atlantic, equatorial Pacific; compiled by Filippelli 1997a).

[c] Defined as a region with a primary productivity less than 35 g C m^{-2} yr^{-1} (Berger and Wefer 1991). Phosphorus accumulation rate is a mean of various low productivity pelagic regions (e.g., pelagic red clay provinces; compiled by Filippelli 1997a).

[d] Defined as area above 200-m water depth (Menard and Smith 1966). Phosphorus accumulation rate is the low range value of various continental margin regions (e.g., upwelling and non-upwelling environments; compiled by Filippelli 1997a). The low range was used because many of the published phosphorus concentration values included non-reactive phosphorus components, which do not interact with the dissolved marine phosphorus cycle.

[e] Filippelli and Delaney (1996). Whole ocean is sum of all moderate to high productivity pelagic regions, low productivity pelagic regions, and continental margins.

[f] As revised in Froelich (1984).

[g] Phosphorus accumulation rate multiplied by area.

[h] Calculated as whole ocean phosphorus inventory (3.1×10^{15} mol P) divided by whole ocean phosphorus burial rate (21×10^{10} mol P/yr).

Averaging across regions with varying P accumulation rates undoubtedly imparts a relative error of at least 50% to the burial rate value derived here. Also, some restricted environments exhibit quite high but spatially variable P burial rates (e.g., phosphatic continental margin environments, Froelich et al. 1982; hydrothermal iron sediments, Froelich et al. 1977, Wheat et al. 1996). The areal approach is a more direct route to quantifying the burial terms in the P mass balance, and indicates that reactive P burial in continental margin sediments accounts for about 60% of the oceanic output, with deep sea sediments nearly equivalent as a sink.

Although the areal approach is a promising and direct method for quantifying the modern P mass balance and potentially determining past variations in the balance, we are still greatly limited by the sparse data that presently exist for P geochemistry and accumulation rates, especially along continental margins. Interestingly, the concentration of reactive P in sediments displays a relatively narrow range in both deep-ocean and continental margin environments (Filippelli 1997a). The major force driving variations in accumulation rates from site to site is the sedimentation rate, termed the 'master variable' in P accumulation by Krajewski et al. (1995) (Fig. 3). Thus, the P accumulation rate at any given site is most responsive to sedimentation rate—this also results in reactive P burial rates being approximately equally distributed between continental margin and deep ocean regions, even though continental margins represent only a fraction of the total ocean area (Filippelli and Delaney 1996, Compton et al. 2000).

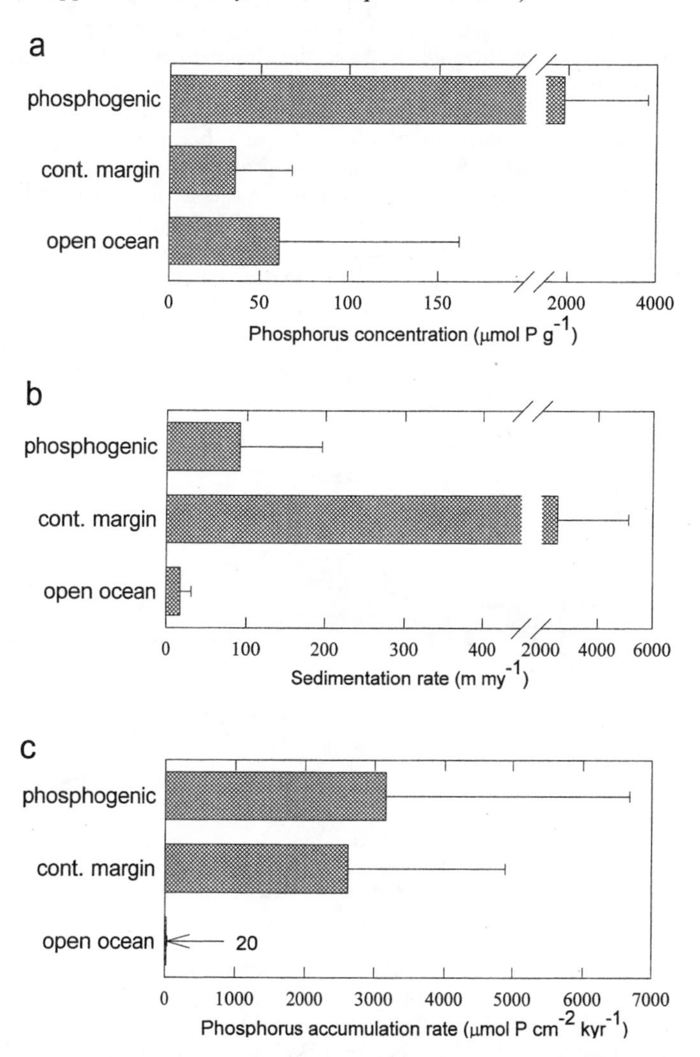

Figure 3. Plots of (a) phosphorus concentration, (b) sedimentation rate, and (c) phosphorus accumulation rate for open ocean, continental margin, and phosphogenic environments.

Modern phosphorus cycle

Human impacts have been manifested for decades at least in a variety of settings.Since the awareness that P is an important plant nutrient, the application of P directly to fields to increase crop yields has increasingly dominated the P cycle. The early source of this fertilizer P was from rock phosphate, generally phosphate-rich sedimentary rocks (phosphorites) mined in the 19th century from a variety of settings in Europe and the US. This P-rich rock was ground and applied to fields, where it significantly enhanced crop yield after initial application. Farmers soon found, however, that crop production significantly decreased after several years of application of this rock fertilizer. Unbeknownst to the farmers, the P-rich rock they were applying to their fields was also rich in cadmium and uranium, both naturally found in the organic-rich marine sediments that are precursors to phosphorite rocks but also toxic to plants (Schlesinger 1997). Many previously productive fields became barren due to this practice, and the awareness of the heavy-metal enriched P fertilizers led to the development of a variety of leaching techniques to separate the beneficial P from the toxic heavy metals. The perfection of this technique and the co-application of P with nitrogen (N), potassium (K), and other micronutrients in commercially available fertilizers boomed during the "Green Revolution," the period after WW II that saw prosperity in some countries and exponential growth in global populations. The irony of the moniker "Green Revolution" has become apparent, as the production of enormous amounts of food fueled by fertilizers to feed a growing global population has caused a variety of detrimental environmental conditions, including eutrophication of surface water supplies, significant soil loss, and expanding coastal "dead zones." These regions of hypoxia and fish mortality, exemplified by the Gulf of Mexico near the outflow of the Mississippi River, are likely caused by fertilizer runoff from agricultural practices in the Great Plains and the Midwest of the United States (see Wetzel et al. 2001 for an excellent review).

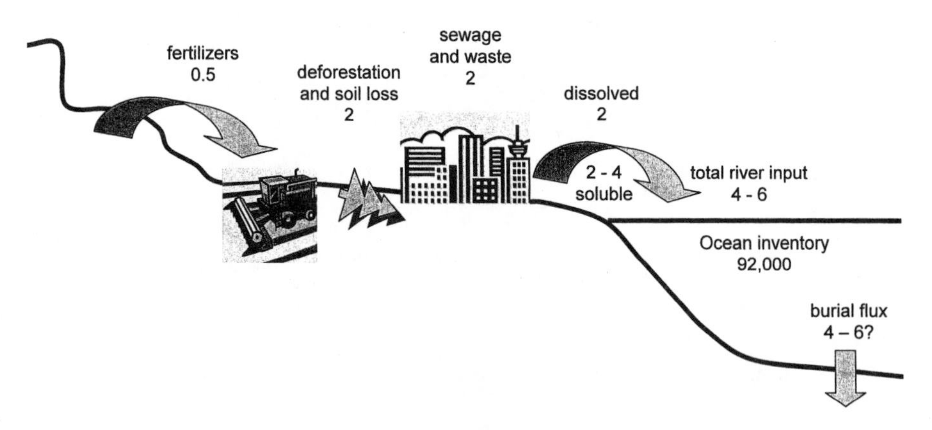

Figure 4. The modern (syn-human) phosphorus cycle, showing reservoirs (in Tg P) and fluxes (denoted by arrows, in Tg P/yr) associated with human activities, which have effectively doubled the natural dissolved P fluxes (cf. Fig. 1).

The net increase in dissolved-P release from land due to human activities also includes deforestation (plus concurrent soil loss), sewage, and waste sources (Fig. 4). Deforestation, typically by burning after selective tree harvesting, converts the standing stock of P in plant matter to ash. This P is rapidly leached from the ash and transported as dissolved loads in rivers; this transfer can happen on timescales of a year or two (Schlesinger 1997). Furthermore, the lack of rooted stability on the landscape results in the loss of the relatively organic P-rich O and A soil horizons from many of these areas, from which some of the P is solubilized during transport. Sewage and waste are additional anthropogenic contributors to the terrestrial P cycle. Human waste, waste from processing of foodstuffs, and waste from industrial uses of detergents (which now are generally low-P but in the past contained up to 7 wt % P) contribute roughly equally to deforestation and soil loss.

The syn-human terrestrial P cycle (Fig. 4), therefore, is substantially different from the pre-human cycle (Fig.1). This is evidenced by high loads of dissolved P in rivers (about 2× estimated natural values) and higher loads of P-bearing particulates. Together, the net input of dissolved P from land to the oceans is 4-6 Tg P/yr, representing a doubling of pre-human input fluxes. As noted previously, this increased flux of P does influence coastal regions directly via eutrophication, but presumably an impact will be seen on the whole ocean. One likely scenario is that the P output flux and the oceanic P reservoir will both increase in response to higher inputs. This effect will be minimal, however, given the limited sources of P available for exploitation (one projection indicates the depletion of phosphate rocks as fertilizer sources by 3400 AD; Filippelli 1995) and the relatively long residence time of P in the ocean (about 15-30 kyr: Ruttenberg 1993, Filippelli and Delaney 1996, Colman and Holland 2000).

ECOSYSTEM DYNAMICS AND SOIL DEVELOPMENT

The effect of climate and soil development on P availability has been a focus of several excellent papers (Walker and Syers 1976, Gardner 1990, Crews et al. 1995, Cross and Schlesinger 1995, Vitousek et al. 1997, Schlesinger et al. 1998, Chadwick et al. 1999). For the most part, these studies have used P extraction techniques to determine the biogeochemical forms of P within soils. These techniques differentiate P in similar fractions, as displayed in Figure 2 (e.g., Tiessen and Moir 1993). The extraction techniques have been applied to depth and age profiles in soils to assess the rate of soil P transformations, the role of climate on these processes, the bioavailability of P in these systems, and the limiting controls on plant productivity. As the current geochemical state of a given soil is an integration of all conditions acting since soil development, most efforts have focused on settings in which climate is likely to have been constant (i.e., tropical settings), and the beginning state of the system and its age are very well known (i.e., soil developing on lava flows). These studies have thus made the classic substitution of space for time, with all the inherent assumptions of constancy in climate and landscape history.

Another approach to assessing terrestrial P cycling is to examine P geochemistry in lake sediments, using the same extraction techniques as the soil studies. This technique adds several dimensions to the soil work outlined above. First, lake sediment records allow us to examine an integrated record of watershed-scale processes associated with P cycling on the landscape. Second, it allows discrete temporal resolution at a given site, providing an actual record of local processes including landscape stability, soil development, and ecosystem development. Third, it extends our understanding of terrestrial P cycling to alpine and glaciated systems. The soil chronosequence approach is not likely to be successful here because of the climatic and slope variability between various sites (i.e., no substitution of space for time is possible). This third dimension is perhaps the most critical in terms of the P mass balance, as the greatest degree of variations in climate have occurred in these settings, and thus they hold the key to understanding the terrestrial P cycle on glacial/interglacial timescales.

Several limitations also exist in this approach. First, the lake sediment records analyzed must be dominated by terrestrial input, with very little lake productivity (i.e., very oligotrophic) and/or diagenetic processes occurring in the lake sediments. This is critical because the clearest signal of the state of soils in the surrounding landscape will come from a lake receiving this sediment as the primary source. This situation holds for numerous small lakes in headwater catchments, where local surface sedimentation dominates, dissolved nutrient inputs are so low that *in situ* organic production is at a minimum, and low amounts of labile organic matter in the sediments limits the degree of diagenetic overprinting on the original sediment record. Lowland lakes that integrate several watersheds and streams have a complex sedimentation history, and often have high rates of *in situ* productivity because of high nutrient inputs. Thus, these lakes are poor candidates for this approach. However, carefully chosen headwater lakes that characterize a relatively large range in climatic conditions, bedrock lithologies, and local landscape relief will provide important information on the weathering processes of P on land.

Example of the lake history approach to terrestrial P cycling

We have applied the lake sediment approach described above to several settings, and have found exciting results with important implications for the effects of climate and landscape development on the terrestrial P cycle. In a recent study (Filippelli and Souch 1999), we analyzed sediments from several small upland lakes in headwater catchments, selected to represent contrasting climates and glacial/postglacial histories. The longest records (about 20,000 years) came from two different lakes in the western Appalachian Plateau of the mid-western US: Jackson Pond, in central Kentucky (289 m above sea level), and Anderson Pond, in north-central Tennessee (300 m above sea level). Climate has strongly affected the ecological development of both of these systems (Delcourt 1979, Mills and Delcourt 1991, Wilkins et al. 1991), although neither was directly glaciated. In contrast, a 11,500 year record from Kokwaskey Lake in British Columbia (1050 m asl), clearly reveals the deglacial history of the region and retains the effect of alpine glaciers throughout the record (Souch 1994).

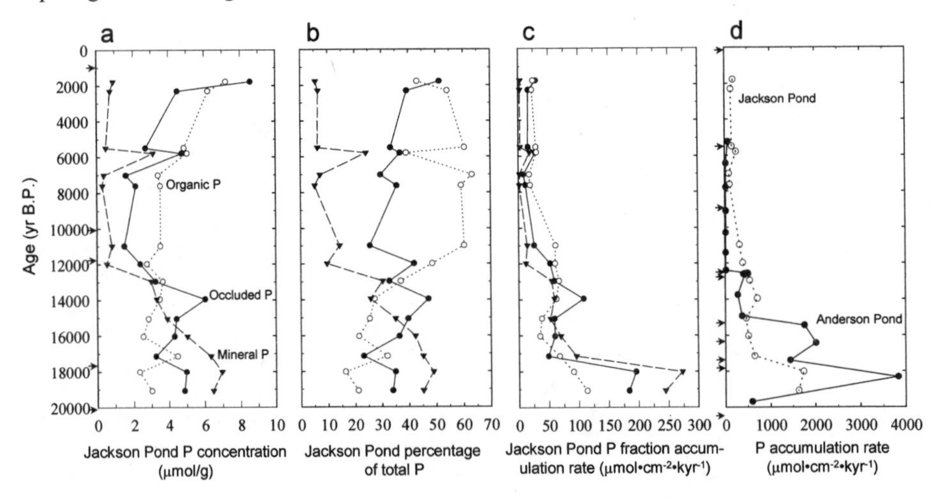

Figure 5. (a) P concentration and (b) percent of total P for each P-bearing fraction in sediments from Jackson Pond. (c) Accumulation rate of P fractions at Jackson Pond. (d) Total P accumulation rate at Anderson Pond and Jackson Pond. ^{14}C age-control points for Jackson Pond (Wilkins et al. 1991) are shown as arrows on age axis in (a). Lowermost point is 20,330 yr. ^{14}C age-control points are also shown for Anderson Pond (Delcourt 1979, Mills and Delcourt 1991) as arrows in (d). Lowermost point is 25,000 yr. The nonoccluded fraction (Fig. 2) is small and is combined with the occluded fraction in this analysis.

The geochemical profile for Jackson Pond reveals extreme changes through time (Fig. 5). In the early part of the record, marked by full glacial conditions (Wilkins et al. 1991), mineralized P was the dominant form entering the lake, with occluded and organic forms of lesser importance. During this interval, the landscape was marked by thin soils, high surface runoff, and closed boreal forests (Wilkins et al. 1991). With landscape stabilization and the onset of soil development (17 ka to 10 ka), the dominant forms of P entering the lake changed significantly, marked by a decline in the proportion of mineralized P and an increase in organic P (occluded P varied but exhibited no clear trend with age during this interval). In this interval, the closed boreal forests gave way to more open boreal woodland and a rapidly thickening soil cover (Wilkins et al. 1991). From the early mid-Holocene to the present, the concentration of mineral P shows little variation, while that of organic P and occluded P increased. Meanwhile, the ecosystem became dominated by deciduous hardwood forests and grasses, and a thick and stable soil existed (Wilkins et al. 1991). In terms of the percent of total P reflected by each fraction, the early Holocene marks a stabilization of the system to one dominated by organic and occluded P. By comparison, the concentration of mineral P is higher throughout the Kokwaskey Lake record, with occluded P of secondary importance and organic P in relatively low concentrations (Fig. 6). Mineral P exhibits a decrease during the interval of landscape stabilization and soil development (between about 9 and 6 ka), while the proportion of occluded P, and to a lesser extent organic P, increase during this interval. From the mid- to late- Holocene (6 ka to 1 ka), characterized by cooler/wetter conditions (Pellatt and Mathewes 1997), each fraction varies slightly. The last 1,000 years of this record, marked by the most extensive Holocene neoglacial activity (Ryder and Thompson 1986), is characterized by a rapid return of P geochemistry to glacial/ deglacial conditions. Throughout the record, the ecosystem in this site is dominated by boreal forest, where slopes are steep, relatively unstable, and have much thinner soils than in the western Appalachian example.

These two examples suggest several controlling factors on P geochemical cycles.

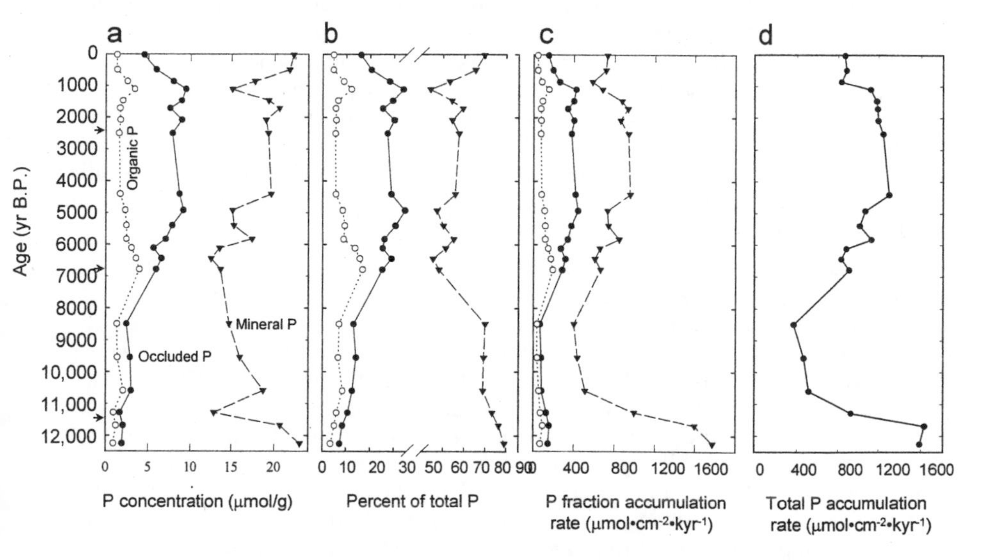

Figure 6. (a) P concentration, (b) percent of total P for each P-bearing fraction, (c) accumulation rate of P fractions, and (d) total P accumulation rate in Kokwaskey Lake sediments. Age-control points (Souch, 1994) are shown as arrows on age axis in A; the lowermost point represents a ^{14}C datum, and the other two points indicate tephra dates (Mazama and Bridge River).

First, the transformation of mineral P to other (more bio-available) forms is more complete in a relatively wet, warm, and low relief system like western Appalachian than in a wet but cold and high relief system like the Coast Mountains. Thus, mineralized P decreases from 50% to 5% of total P at Jackson Pond compared to 90% to 50% at Kokwaskey Lake. This is consistent with the primary importance of physical weathering and erosion in the highly unstable environment in the Coast Mountains of British Columbia. Second, the relative starting points and stabilization points of these two systems are drastically different, driven by their contrasting landscape history and stability. Kokwaskey Lake was completely glacier covered at 12 ka, and thus the initial starting point for this system is nearly completely mineral P from rock flour. Throughout the Holocene, the high relief in the watershed led to constant loss of surface sediments with poorly developed soils and relatively little organic P. In contrast, low relief and relatively rapid development of a stabilized landscape and soil profiles in the Jackson Pond watershed supports the development of a mature soil in terms of P geochemistry. Additionally, because the Jackson Pond watershed started with a climate and ecosystem similar to that of Kokwaskey Lake during the interval from 6 ka to 1 ka, their comparative P geochemistries are similar in these intervals. The model presented earlier (Fig. 2) is therefore useful for tracking soil P transformations from the lake cores presented here. For the western Appalachian setting, the soil system begins in a slightly pre-weathered stage, and progresses to an end-point dominated by the organic and occluded fractions (Fig. 2). For the alpine setting, the soil system begins as rock flour, and shifts slightly toward more developed P forms as warming and more landscape stabilization occurs in the mid-Holocene followed by extensive neoglaciation over the last 1,000 years (Fig. 6). Finally, the P geochemistry of these systems can be reset quickly. During the last 1000 years, readvance of alpine glaciers during the Little Ice Age has been sufficient to return the P geochemical cycle in this region back to near glacial conditions (Fig. 6).

Patterns in P weathering and release are similar to those observed in several tropical settings. For example, a chronosequence of Hawaiian soils reveals significant transitions in mineral P after about 1,000 years of soil development (Crews et al. 1995, Vitousek et al. 1997), while significant transformations occur in P biogeochemistry during soil development on a Krakatau lava flow in just over 100 years (Schlesinger et al. 1998). In the temperate settings presented here, this decrease occurs over timescales of 3,000 to 5,000 years. One of the key differences is that the temperate settings are periodically reset due to glacial climate conditions. This resetting keeps these soil systems in a 'building phase.' But like the tropical environments, the progressive loss of P during soil development leads to P limitation to terrestrial ecosystems.

Large changes in the accumulation rate of P over time are reflected in all the lake records. These accumulation rate changes are driven partly by changes in bulk sedimentation rate, but several of the rapid shifts in P accumulation occur between age control points and thus are not driven just by rates of sedimentation. The two western Appalachian sites reveal a rapid pulse of P input during the glacial and initial deglacial interval (Fig. 5), a period of enhanced colluvial activity when soil development was just commencing. Upon landscape and soil stability (by 10 ka), P inputs had decreased 10-40 fold, and remained low and constant to the top of the record (2 ka). The Kokwaskey alpine site also reveals significant changes in P input (Fig. 6), beginning with a rapid early deglacial pulse consisting mainly of mineralized P originating from glacial rock flour. From approximately 9 to 6 ka, landscape stabilization begins, although the landscape is still marked by high relief and rapid rates of erosion. Phosphorus input rates are high but decrease slightly from 4 to about 1 ka. In this high relief, alpine setting, soil development is retarded by rapid denudation, and thus the terrestrial P cycle is stuck in an 'initial development stage' with high mineral P and high P release rates from the landscape.

Although the temporal records differ between the sites, some generalizations may be made about terrestrial P cycling, climate, and landscape development as a function of glacial/interglacial cycles. First, the lake records reflect input of solid-phase P released from watersheds. Some portion of this released P is bioreactive in terrestrial systems and upon transport, also bioreactive in the ocean. The P geochemical results indicate that the more bioreactive forms dominate in later stages of soil development. However, this increase in the bioreactive nature of the P is offset by very low total P release at these times, as reflected by the total P accumulation rate records from the lakes. In the case of dissolved phosphate, we relate the initial stage of soil development to high solid-phase loss that also likely leads to a relatively poor recycling of the dissolved phase (from a lack of oxide-bound occluded pools), a so-called 'leaky ecosystem.' Second, the western Appalachian sites have reached a degree of landscape stabilization due to relatively stable climates. Although the elevation, local relief, mean temperature, rainfall, and ecosystem are quite different for these sites, the net result of soil stabilization is to reduce the rate of P loss from these systems. The alpine British Columbia site, on the other hand, has never achieved the relative stabilization of the other sites due to high relief and neoglacial activity. Thus, the trends in P release from this system lags, and P release remains relatively high and constant. We believe that the results presented here characterize mid- and high- latitude watersheds that have been directly glaciated or strongly affected by proximal glacial conditions. Tropical environments may experience less dramatic climate change through glacial/interglacial intervals, but recent terrestrial evidence reveals that even these systems experience notably cooler and drier conditions during glacial intervals (Thompson et al. 1997), which may lead to changes in the extent of soil development and P release in these systems as well.

The lake sediment records presented here indicate a terrestrial P mass balance that is not near steady state on glacial/interglacial timescales, with important implications for the functioning of terrestrial and oceanic systems. Coupling these records of solid-phase P changes over time with oceanic records of dissolved inputs from rivers will eventually provide important constraints for the influence of climate on chemical weathering and the global P cycle.

CONTROL OF OXYGEN ON COUPLED P AND C CYCLING IN CONTINENTAL MARGINS: A CASE STUDY FROM THE ANOXIC SAANICH INLET

Uncertainty currently exists about the removal of C and P from the oceanic reservoir. The ratio of organic C (Corg) to organic P (Porg) in sediments is a critical parameter in many paleoceanographic and paleochemical models (Holland 1984, Sarmiento and Toggweiler 1984, Delaney and Boyle 1988, Boyle 1990, Delaney and Filippelli 1994, Van Cappellen and Ingall 1994, 1996), with estimates of the global average Corg/Porg in marine sediments ranging from 367 (Williams et al. 1976) to 496 (Ramirez and Rose 1992), with at least a 70% range on each of these averages. Also, these values contrast with the value of about 106 in marine organic matter (e.g., Redfield et al. 1963), which is the dominant source of organic C and P to the marine sediments. Several recent studies suggest that P may exhibit a significant preferential regeneration compared to C during diagenesis of organic matter in low oxygen continental-margin sediments (Ingall et al. 1993, Compton et al. 1993, Ingall and Jahnke 1994, Schenau et al. 2000). If true, decoupling of these elements may occur in their long-term biogeochemical cycles (Van Cappellen and Ingall 1994, 1996). As noted by Colman et al. (1997), accurate measurements of this ratio in a variety of marine settings are required to fully understand coupled C and P cycling and burial, and much controversy still exists as to the viability of the preferential regeneration model for P (e.g., Anderson et al. 2001).

Recent research on the P geochemistry of both deep-sea (Filippelli and Delaney 1996, Delaney 1998) and continental-margin sediments (Ruttenberg and Berner 1993, Filippelli 2001) suggests that a better understanding of the diagenesis of P in sediments may help to resolve many questions concerning what happens to P and C after burial. Several diagenetic transformations of P have only recently been detected (Fig. 7). First, organic P concentrations in sediments decrease dramatically with both depth and age; values at depth typically are less than 10% of near-surface values (Filippelli and Delaney 1996). These decreases are probably caused by microbial regeneration of organic matter and release of P from its organic form. Based only on these results, a net loss of P from the sediments may be assumed. However, concurrent with the decrease in reactive forms of organic P is an increase in authigenic P in the sediments. Authigenic P most likely occurs as carbonate fluorapatite (CFA), an authigenic mineral that forms predominantly in continental margin sediments under regions of high primary productivity (Kazakov 1937, Veeh et al. 1973, Burnett et al. 1982, Froelich et al. 1982, Baturin 1988, Garrison and Kastner 1990, Filippelli and Delaney 1992, Filippelli et al. 1994, Schuffert et al. 1994, Reimers et al. 1996, Kim et al. 1999), but has also recently been identified in non-upwelling continental-margin sediments (O'Brien and Veeh 1980, Ruttenberg and Berner 1993, Berner et al. 1993, Lucotte et al. 1994, Kim et al. 1994, Krajewski et al. 1994) and even in deep-sea sediments (Filippelli and Delaney 1996).

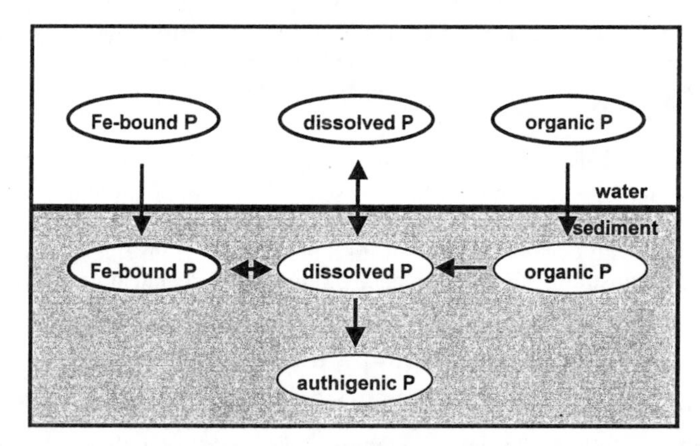

Figure 7. A conceptual diagram of P geochemistry in oceanic sediments (from Filippelli and Delaney 1996). Phosphorus associated with organic matter is the primary source of P to the sediments, although the depositional flux of Fe-P is commonly similar to that of organic P in continental margin settings. Organic matter degradation leads to the release of P to interstitial waters. Dissolved P in interstitial waters appears to be involved in several processes, including adsorption to grain surfaces, diffusion back to bottom waters, binding to iron oxyhydroxide minerals (which also scavenge some P from deep waters during particle formation in the water column), and incorporation in an authigenic mineral phase. In Saanich Inlet, the most important burial sink for P is the authigenic component, where P is probably mineralized authigenically as carbonate fluorapatite (CFA).

The increase in authigenic P with depth largely offsets the decrease in organic P, resulting in limited loss of P compared to C in sediments after burial below the sediment-water interface (Fig. 8). Important implications of these results include:

(1) Organic P is altered in marine sediments during diagenesis, but P is redistributed within the sediments to other burial pools rather than lost.

(2) Because Porg decreases, post-diagenetic Corg to Porg ratios may be higher than pre-diagenetic ratios. This increase, however, may not represent real changes in total organic C to reactive P ratios in sediments because of the transfer of P into other burial pools.

(3) Previous studies have assumed no fractionation of Corg compared to Porg during diagenesis (Mach et al. 1987, Ramirez and Rose 1992); the diagenesis of Porg discussed above indicates that this might not be the case.

(4) The variability in published Corg:Porg ratios in marine sediments may be due to sampling across zones of organic P diagenesis, the depth trends of which may vary from region to region depending on sediment type, original organic matter content, and sedimentation rate (e.g., Ruttenberg and Berner 1993, Filippelli and Delaney 1996).

Saanich Inlet, British Columbia

In this discussion, I will examine the cycling of C and P and C:P ratios in sediments from the anoxic Saanich Inlet (located near Victoria on Vancouver Island, British Columbia, Canada). This basin was cored by Ocean Drilling Program (ODP) Leg 169S in 1996 at two sites (Bornhold et al. 1998). The blue-gray silty sediments in the lower part

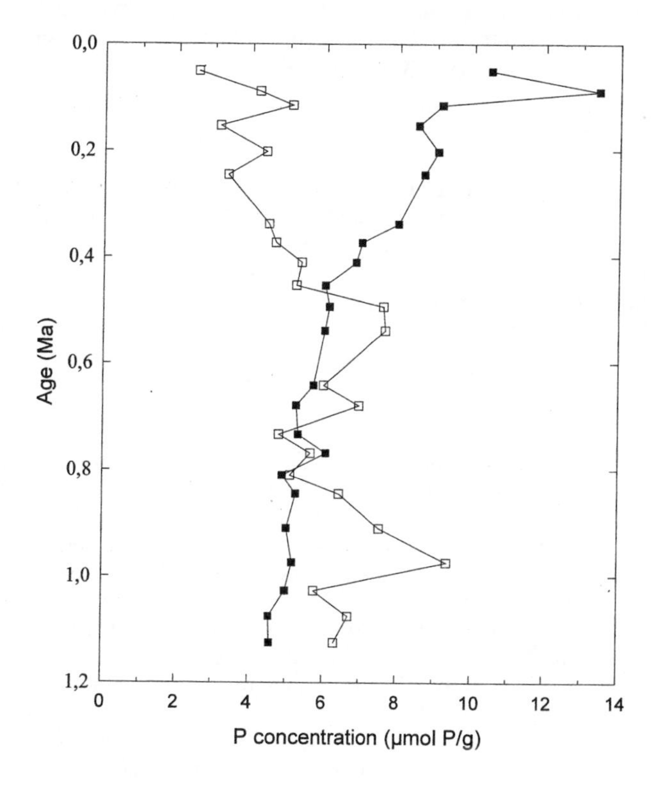

Figure 8. Detailed view of phosphorus geochemistry at Leg 138 Site 846B, eastern equatorial Pacific, for the past 1.2 Myr. (From Filippelli and Delaney 1996). Authigenic component (□) and the sum of the other reactive components (■, including adsorbed, iron-bound, and organic phosphorus) versus age, showing decreases in other reactive forms offset by increases in the authigenic component.

of the cores reflect the input of glacially derived material from regional and local deglaciation. Clear laminations begin about 9,000 years ago and persist in the basin from about 7,000 years ago to present. This transition marks the onset of mainly marine sedimentation and extreme sill-controlled restriction of circulation in the Saanich Inlet, which resulted in anoxia. The laminations are varves, exhibiting seasonal stratification related to winter runoff and spring and summer biological blooms (Dean et al. 2001, Johnson and Grimm 2001), from which annual-scale sedimentation rates can be determined. The average rates of sedimentation are quite high (up to 2.5 cm/year), and thus accumulation rates of organic matter are high. These factors make the Saanich Inlet cores recovered during ODP Leg 169S excellent end-member examples of coupled C and P cycling. Results presented here reveal that although Corg:Porg ratios are high and increase with depth, this effect is due largely to a remobilization of P from an organic matter sink to an authigenic sink. Reducible sedimentary components act as temporary shuttles in this process even in this anoxic setting, with the ultimate burial sink for the remobilized P being carbonate fluorapatite. The effective ratio of organic C to reactive P (the sum of P-bearing sedimentary components considered potentially chemically reactive in marine settings) appears to be about 150-200, indicating some preferential loss of P compared to carbon during organic matter degradation, but not approaching previously reported values of over 3,000 in black shales (e.g., Ingall et al. 1993). Reactive P burial rates in this basin range from 10,000-60,000 $\mu mol/cm^2/kyr$, greatly exceeding the range of 500-8000 $\mu mol/cm^2/kyr$ found in most continental margin settings, including regions of modern phosphogenesis (Filippelli 1997a).

Age determinations were made based on the ^{14}C stratigraphy developed for this site (Blaise-Stevens et al. 2001). A variety of high-resolution carbon and isotopic records were also produced for this site (Calvert et al. 2001, McQuoid and Whiticar 2001). In order to develop geochemical accumulation rate estimates that take into account real sedimentation variability in this setting, I used a different technique to calculate sedimentation rates. The laminations preserved in this core are annual, an interpretation based on years of box and piston core work in this setting. Using core photos, I measured the lamination width within the 10 cm span of each sample location used in this study and reported this average lamination width as the sedimentation rate. Thus, instantaneous sedimentation rates are used for calculations of P accumulation rates instead of sedimentation rates averaged over large age control ranges.

Organic carbon and phosphorus geochemistry

Organic carbon concentrations are high in the upper marine-dominated sediments and decrease with increasing depth and age, ranging from 2.0-2.4 wt% at the surface to 1.2-1.8 wt% at 7,000 years ago (Fig. 9). This decrease continues to the base of the cores, and very low organic carbon values are found in the terrigenously derived clays near the lower portions of the cores, with values as low as 0.2 wt%. Values are consistently higher in the shallower Site 1033 by about 0.3 wt%. The uppermost sample at Site 1033 is relatively low (0.9 wt%) and deviates from the general trend of the other organic-rich samples near the surface. Organic P values decrease with increasing age and depth, ranging from values of 4-6 µmol/g at the surface to 1-2 µmol/g at depth (Fig. 9). This decrease virtually parallels that of Corg, although the decrease in Porg with depth and age is more rapid than that of Corg (Fig. 9). It should be noted that for these two sites, with different average rates of sediment accumulation, geochemical trends with age match, whereas trends with depth do not. This indicates that organic matter degradation, the dominant factor driving these geochemical parameters, is related to sediment age and not depth. This further implies that the process of organic matter degradation is kinetically controlled rather than diffusionally controlled at this site.

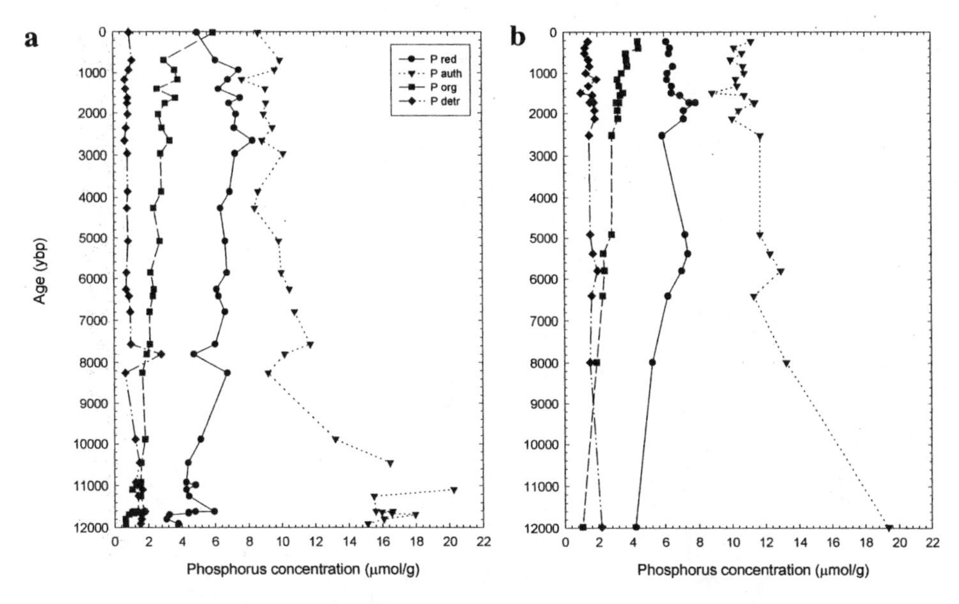

Figure 9. Geochemistry of P with age at Sites 1033 (a) and 1034 (b). The geochemical fractionation was performed via a sequential extraction technique (Ruttenberg 1992, Filippelli and Delaney 1996, Anderson and Delaney 2000), which separates reducible and adsorbed phases (P red), authigenic phases (likely CFA--P auth), organically-bound P (P org) and detrital P (P detr). Note an overall decrease in the reducible and organic fractions and an increase in the authigenic fraction in the older sediments.

The molar ratio of Corg:Porg is elevated compared to the values of Redfield et al. (1963) throughout the Saanich Inlet record (Fig. 10). Values as high as 800 exist in the lower marine-dominated sediments. Corg:Porg ratios decrease in the younger sediments, approaching projected values of 150-250 in surface sediments (note that a true sediment/water interface was not successfully recovered in these cores due to the soupy nature of the sediments, and the youngest sample shown for Site 1033 is likely a composite of sediments deposited over the last century). Relatively high values of 400 are seen in the glacially-derived clays in the older sections, likely reflecting typical Corg:Porg ratios in terrestrial soils and sediments (Schlesinger 1997).

The increase in Corg:Porg ratios with depth and increasing age could be interpreted to represent a net loss of P in comparison to carbon from the sedimentary sink in this environment. Certainly, the values greater than Redfield et al. (1963) projected even for the surficial sediments indicate a small preferential loss of P from this system during organic matter degradation, supported by P flux measurements in this and other margin settings (e.g., Murray et al. 1978, Ingall and Jahnke 1994, 1997; McManus et al. 1997, Colman and Holland 1999). The Corg:Porg ratios for surficial sediments found here are much lower than those determined for several other anoxic settings, indicating that preferential release of P may not occur to the extent previously suggested in anoxic settings (e.g., Ingall et al. 1993; note, however, that this study was performed on black shale in the rock record, not modern marine sediments). It should be noted, however, that the Saanich Inlet is an end-member example of marginal sedimentary environments due to its extremely high sedimentation rates. Furthermore, organic matter degradation is very rapid in this setting, and ODP-type coring will not be able to capture a true snapshot of

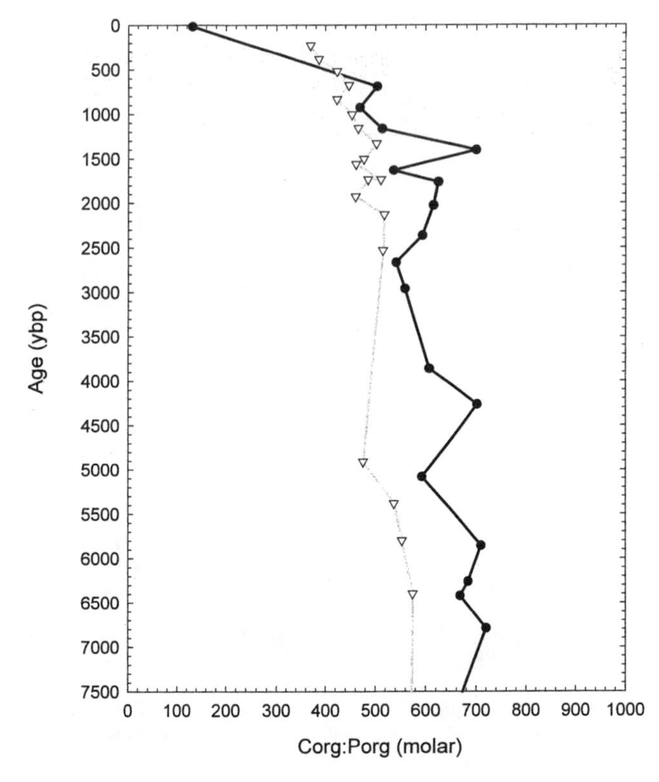

Figure 10. Corg:Porg ratio (molar) with age at Sites 1033 (●) and 1034 (Δ) in the Saanich Inlet. This ratio increases sharply over the first ~2,000 years of the record, driven by the redistribution of P to authigenic phases with no net loss from the sediment column.

the geochemical cycles of C and P on timescales of years to decades in surface sediments, because of the soupy nature of these sediments and their disturbance due to blow-off during coring. Present efforts at box-coring coupled with benthic chambers may be the most fruitful approach for this purpose.

The increase in Corg:Porg ratios with depth and age (Fig. 10) is the result of preferential loss of P compared to C from organic matter. However, the P released from organic matter is not simply lost as a benthic dissolved flux. Instead, this P is incorporated into several other sedimentary phases with a net retention of total reactive P (Fig. 11). At shallower depths and younger ages, the released P is adsorbed onto reducible iron and manganese-bearing phases. This is supported by the initial increase in reducible-related P as well as by peaks in iron and manganese released from a reductant treatment. Iron oxyhydroxides can coprecipitate a large amount of phosphate due to their large surface areas (~500 m^2/g) and numerous delocalized charges (Froelich 1988, Feely et al. 1990). Iron redox cycling in sediments appears to be a strong control on benthic P retention and release, with disseminated iron oxyhydroxides acting as a temporary shuttle in the interface between oxic and iron-reducing zones within continental margin and deep sea sediments (Sundby et al. 1992, Lucotte et al. 1994, Filippelli and Delaney 1996). The presence of reducible iron phases with depth in anoxic sediments has been verified by Canfield (1989). The exact nature of this reducible phase is debatable. The ferric phases

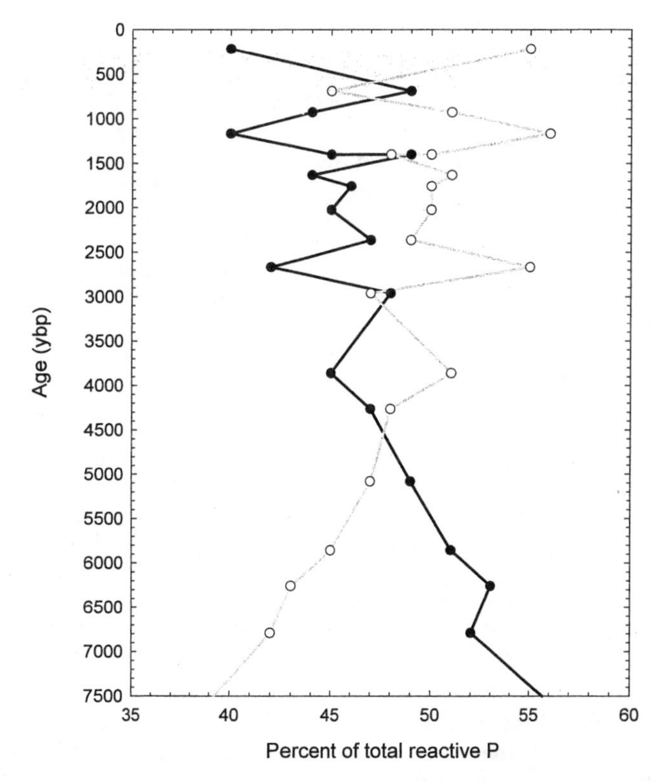

Figure 11. Percent of total reactive P in this authigenic phases (filled circles) and in the reducible and organic phases (open circles) for Site 1033. This comparison reveals the sink-switching that occurs during sedimentary P cycling, with the ingrowth of an authigenic P phase and a loss of P from other reactive phases with age and depth.

documented by Canfield (1989) were well crystallized and likely relatively low in P. It is not likely that amorphous oxyhydroxides, which are important for adsorbing/co-precipitating P in oxic settings, are contributing at depth in this anoxic setting: Reimers et al. (1996) found little or no ferric oxyhydroxides below 6 cm in sediments from the anoxic Santa Barbara Basin. It could be argued that the aggressive reducing agent used in this extraction (buffered citrate dithionite bicarbonate solution—CDB) can partially dissolve iron silicates, although this would lead only to elevated iron and not to the observed P enrichment (due to the low concentration of P associated with iron silicates). Regardless of the exact nature of the reducible pool, it appears only to be a relatively temporary sink for P. At greater depths and ages, P is lost from the reducible phase and incorporated into an authigenic P-bearing mineral phase. This authigenic phase is likely carbonate fluorapatite (CFA), a suggestion supported by depletion in pore water fluoride at depth in the Saanich Inlet cores (Mahn and Gieskes 2001) and, as previously mentioned, documented in several other marine settings. Interestingly, CFA appears to form slowly and at relatively great depth in the Saanich Inlet compared to the Santa Barbara Basin, where porewater chemistry indicates that CFA growth is restricted to the upper few centimeters (Reimers et al. 1996).

In a marginal setting like the Saanich Inlet, changes in depositional characteristics of P have to be considered along with diagenetic remobilization of P. For example, how much do the down-core records of organic P reflect changes in the chemistry of the continental input? It appears that for this setting, all of the changes in P chemistry observed are driven by internal (diagenetic) processes rather than external (depositional) processes. This bold statement can be made because we have also examined the dynamics of terrestrial P weathering in this region. As noted previously, we analyzed down-core P geochemistry in a sediment core from Kokwaskey Lake, a headwater lake of the Kwoiek watershed in the Coast Mountains of British Columbia, just east of the Saanich Inlet (Filippelli and Souch 1999). We found an increase in the fraction of organic P weathered from the landscape from 12,000 - 8,000 years ago in response to soil development, nearly constant values until about 1,000 years ago, then a sharp decrease in organic P over the last 1,000 years in response to the Little Ice Age and neoglacial advances. In comparison, the organic P record from the Saanich Inlet shows an increase over the last 7,000 years, and not the constant values followed by a sharp decrease we would expect if this record was driven primarily by depositional processes.

Phosphorus burial

Short-term versus long-term burial. Because of active diagenesis of organic matter in the Saanich Inlet, several aspects of C and P burial must be considered. The longer records presented above speak to issues of C and P burial on long (~1,000 to 10,000 yr) timescales. This is a critical parameter for understanding the role of continental margins for storing these components in terms of glacial/interglacial mass balances and climate, and will be addressed in the next section. The very active processes occurring at the sediment-water interface, however, are not addressed by the long-term burial trends. Thus, it might be instructive to compare long-term burial with instantaneous measurements of the reflux of dissolved phosphate from sediments back to the water column.

Pore water determinations and benthic flux chamber experiments can both provide insight into how P cycling relates to organic matter degradation occurring in the surface tens of centimeters to several meters of sediment. For pore water determinations, the concentration of dissolved phosphate is determined from pore water squeezed from numerous depth intervals. Trends in the pore water phosphate concentration with depth can be modeled using diffusion models, typically related to organic matter degradation, and these modeled trends can be used to estimate the flux of dissolved phosphate out of the sediments. This type of analysis has been performed for P in a variety of settings, and is summarized comprehensively in Colman and Holland (2000). They find that P reflux from sediments is quite large in terms of oceanic mass balances, with high sedimentation rate, high C regeneration rate, and the redox state of the sediments enhancing P reflux from continental margin sediments. Given sedimentation rates for surface sediments in the Saanich Inlet, the P reflux would be about 0.2 mol P/m^2/yr, based on the global relationship of Colman and Holland (2000). Murray et al. (1978) performed pore water chemistry on sediments from the Saanich Inlet sampled in April 1976. Using the sediment parameters from this study, which are quite similar to those for the ODP cores analyzed here, Colman and Holland (2000) calculated a seasonally averaged annual mean P reflux rate of 0.18 mol P/m^2/yr. The long-term burial rates of P found in this study range from ~0.1 to 0.6 mol P/m^2/yr. Thus, in a simplistic comparison, approximately 35-75% of the total reactive P reaching the seafloor is buried (i.e., 25-65% is released back to the water column). This assumes, of course, a predictable relationship between short-term and long-term burial processes. But as noted earlier, P remobilization and incorporation into authigenic phases appears to be the largest control on long-term P cycling, whereas P remobilization and reflux back to the overlying water column is the largest control on short-term P cycling. Although C and P cycling pathways during diagenesis are different, taken together, a benthic loss of 25-65% of the reactive P with little net change in the Corg:Preactive ratio indicates a substantial benthic loss of C as well. This indicates that organic matter regeneration and C and P cycling are very active in anoxic settings, supporting suggestions by Pederson and Calvert (1990) and Calvert et al. (1992) to that effect.

Benthic flux chamber experiments measure the loss or gain of various dissolved species from a fixed container on the seafloor, thus yielding instantaneous measurements of P cycling. A benthic flux chamber deployment in the Saanich Inlet in summer, 1997

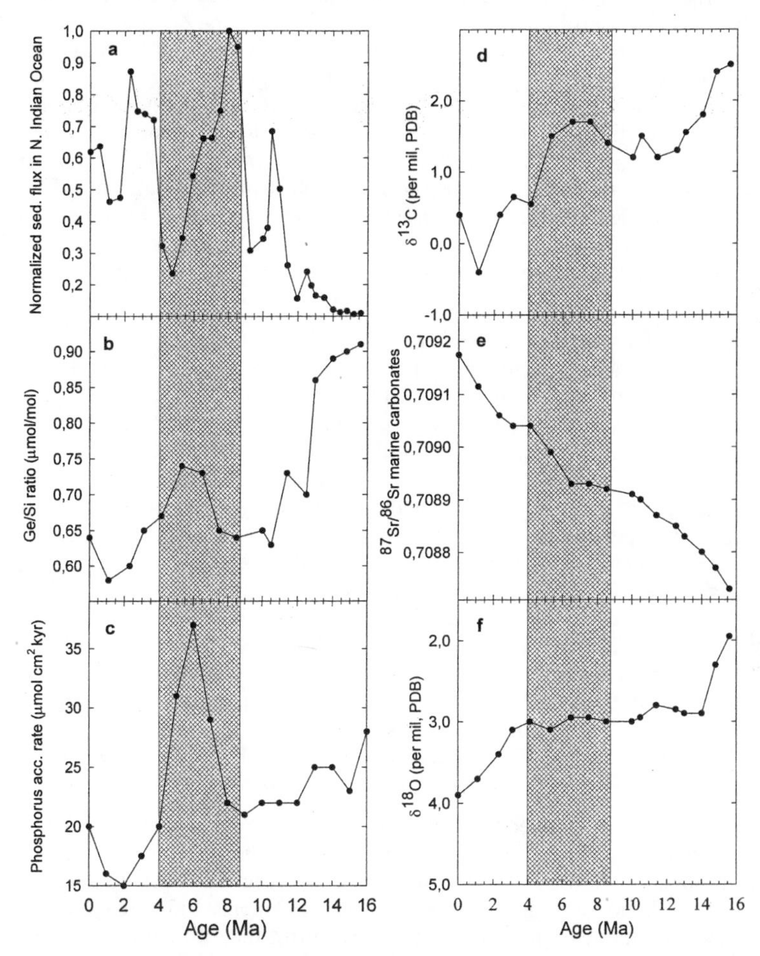

Figure 12. Sedimentary and geochemical records from oceans, showing dramatic transient shifts in most records in an interval from just before 8 Ma to 4 Ma (shaded), from Filippelli (1997b). Symbols in all records represent averages of 1 Myr intervals, except for normalized sediment flux curve, which represents 0.5 Myr averages. After interval averaging, all records were adjusted to time scale of Cande and Kent (1992) for consistency. (a) Normalized sediment flux in northern Indian Ocean (Rea 1992). (b) Ge/Si ratio in opaline silica from diatoms (Shemesh et al. 1989). (c) $\delta^{13}C$ of bulk marine carbonates (Shackleton 1987). Although details of different carbon isotope records differ, general trends revealed in this low-resolution record are robust. PDB is Pee Dee belemnite. (d) Phosphorus accumulation rates in equatorial Pacific (Filippelli and Delaney 1994). Peak in accumulation rates is also observed in other parts of Pacific (Moody et al. 1988) and western Atlantic (Delaney and Anderson 1997). These peaks are linked with increased phosphorus input rates from continental weathering (e.g., Filippelli and Delaney 1994). (e) $^{87}Sr/^{86}Sr$ record from marine carbonates (Hodell et al. 1990, 1991). (f) $\delta^{18}O$ of benthic foraminifera (Miller et al. 1987).

yielded a P reflux rate of ~2 mol P/m^2/yr (E. Ingall and T. Lyons, pers. comm.). This is extremely high, compared to short-term rates of ~0.2 mol P/m^2/yr, and compared to long-term burial rates would indicate that only about 5% of the reactive P reaching the seafloor is buried. More likely, this instantaneous summer flux is a pulsed release of P released from biogenic bloom detritus and represents a non-steady state cycling event. The degradation of the bloom detritus would contribute a significant amount of P related to organic matter. Rapid degradation could also result in a lowering of the redox boundaries in the upper several millimeters of the sediments, releasing some of the iron-bound P (e.g., Colman and Holland 2000). Note also that the bottom waters of the Saanich Inlet are seasonally oxic. This could lead to seasonal storage of P in a thin surficial layer of oxic sediment that would have fresh iron oxyhydroxides—formed by oxidation of pore water ferrous iron diffusing from the anoxic portion of the sediments. Coprecipitation and adsorption of phosphate could lead to a substantial amount of P in this thin layer. Rapid deposition of organic matter to the sediment following a bloom would not only be followed by rapid degradation and release of associated organic P, but could contribute to the release of any surficial oxyhydroxide-associated P. This kind of seasonal behavior for P diagenesis is well documented for Cape Lookout Bight sediments (Klump and Martens 1981).

Long-term P burial and the global P cycle. Continental margins are important burial sites for oceanic P. Margin sediments likely constitute between 50% and 80% of the total P removal from the ocean (e.g., Baturin 1988, Ruttenberg 1993, Filippelli 1997a, Schuffert et al. 1998, Compton et al. 2000), even given their small surface area compared to the deep ocean. As such, sedimentologic, tectonic, and eustatic influences on continental margins can greatly affect the oceanic P cycle. The Saanich Inlet is an end-member example of a dynamic continental margin setting, experiencing a sharp transition from glacial-terrigenous to marine sedimentation, silling of the outlet and restricted circulation, and local glacial rebound in the face of the eustatic sea level rise associated with deglaciation. The Saanich Inlet is also an end-member example of P burial, with accumulation rates of 10,000 to 60,000 μmol/cm^2/kyr, about 2 to 50 times that typically observed in continental margin settings (Filippelli 1997a). Although the Saanich Inlet is not a typical margin setting, it can serve to illustrate the extreme example of the effects of continental margin dynamics on marine P cycling during a glacial to interglacial (G/IG) transition. This may serve to help us link several G/IG records developed in the deep ocean (e.g., Murray et al. 1993, Latimer and Filippelli, 2001) with a new awareness of the dynamic nature of P cycling on land (Filippelli and Souch 1999).

The Saanich Inlet evolved rapidly from a basin that accumulated no reactive P from the marine reservoir during full glacial and deglacial time to one with extremely high reactive P accumulation by 7,000 years ago. High reactive P accumulation rates have persisted from 7,000 years ago to present, driven by intense biological blooms and very high sedimentation rates. This progression may mirror that of many continental margins, especially those at or above sea level at the last glacial maximum. In these settings, the locus of upwelling-induced productivity would be centered over deeper waters well off the continental shelf. Given that the major initial input of P to sediments is in the form of marine organic matter (with the exception of regions proximal to river outflows, where terrestrial organic matter and oxides are important additional sources; e.g., Berner and Rao 1994, Compton et al. 2000, Goñi et al. 1998), the additional sinking time and slower sediment burial at the sediment/water interface would increase the regeneration of P from this organic matter back to the water column. Thus, the net internal effect of lowered sea level would be to enhance P regeneration, reduce the continental margin sink for P, and increase P burial in the deep ocean sink (assuming steady state). Because of enhanced internal recycling, this process could also lead to a greater inventory (higher concentration) of P in the ocean. The few studies of P accumulation rates on G/IG timescales in deep sea sediments do indicate higher P accumulation rates during glacial intervals (e.g., Murray et al. 1993, Murray et al. 2000, Latimer and Filippelli, 2001), but these studies were performed in high productivity regions (the equatorial Pacific, the Southern Ocean) which respond strongly to changes in upwelling intensity. The only reliable proxy for P concentration in the ocean is the Cd/Ca ratio of foraminifera, which showed a slightly higher P concentration in the deep Atlantic waters during glacial intervals, but changed little in other basins (e.g., Boyle 1990). The whole scenario is further complicated, however, by the assumption of steady state input flux of P to the oceans and constant P cycling on continental margins on G/IG timescales. Neither assumption is good, and in fact we see quite large variations in terrestrial P cycling on G/IG timescales as the result of glaciation, soil, and ecosystem development (Filippelli and Souch 1999), and much has been said about margin-basin fractionation of nutrients on these timescales (e.g., Broecker 1982). More studies of C and P cycling and burial will need to be performed on a greater variety of continental margins to constrain the internal dynamics of the marine P cycle on G/IG timescales, with these results tied to deep sea and terrestrial records.

PHOSPHORUS AND MARINE PRODUCTIVITY ON CENOZOIC TIMESCALES

Ocean geochemical and sedimentological records are among the best evidence available for reconstructing past climates and weathering events on land. An intriguing hypothesis developed from studies of Cenozoic marine records is the Himalayan uplift hypothesis (e.g., Raymo et al. 1988, Raymo and Ruddiman 1992), which argues that the uplift of the Himalayan-Tibetan Plateau during the Cenozoic increased net continental-chemical weathering rates, increased drawdown of atmospheric CO_2 by weathering reactions, resulting in global cooling and increased continental ice buildup through the Cenozoic. This hypothesis is largely based on the rapid rate of increase in the strontium isotopic ratio of marine carbonates since 40 Ma, cited as evidence for an increase in the rate of radiogenic (continental) strontium input to the ocean. Although the uplift and weathering of the Himalayan-Tibetan Plateau must have had profound effects on global ocean biogeochemical cycles, especially considering the large drainage area and unique geochemical characteristics of this region (Raymo and Ruddiman 1992, Edmond 1992), some studies have called into question the direct link between Sr isotopic values and net chemical and sedimentological fluxes to the ocean (Rea 1992, Filippelli and Delaney 1994, Delaney and Filippelli 1994, Raymo 1994a, Kump and Arthur 1997). A short-coming of many of these studies is the attempt to link one or several of these oceanic geochemical indicators to records of climate change and tectonics throughout the Cenozoic or a large portion of it.

The goal of this discussion is to trace the effect that the proposed intensification of the Asian monsoon at about 8 Ma had on continental weathering, geochemical and sedimentological fluxes to the ocean, biogeochemical cycles in the ocean, and the possible feedbacks to global climate (Filippelli 1997b). The Asian monsoon is a major climate system that currently affects global weather patterns. Massive warm rainfall events in late summer (from the Asian monsoon) combined with high elevations result in a disproportionate amount of chemical weathering occurring in the Himalayan-Tibetan Plateau; dissolved chemical fluxes in rivers draining this region account for about 25% of the global total, even though this region constitutes only slightly more than 4% of the global drainage area (Raymo and Ruddiman 1992). Oceanic and continental records suggest that the Asian monsoon may have intensified ~8 Myr ago (Kroon et al. 1991, Prell et al. 1992, Molnar et al. 1993), possibly as a result of an uplift event in the Himalayan-Tibetan Plateau (Prell and Kutzbach 1992). Several researchers have highlighted the potential importance of this major climatic event on global cycles (Janecek and Rea 1983, France-Lanord and Derry 1994, Derry and France-Lanord 1996), and Pisias et al. (1995) called directly on increased chemical weathering rates in the Himalayan-Tibetan Plateau, in response to the Asian monsoon, as a cause for the "biogenic bloom" event documented recently in many ocean basins at this time.

Weathering changes in the Himalayan-Tibetan plateau

Several records of weathering changes in the Himalayan-Tibetan Plateau centering around 8 Ma have been extracted from the Indus and Bengal fans of the northern Indian Ocean. In a study of temporal trends in sediment flux rates to both the Indus and Bengal fans, Rea (1992) determined two main peaks in sediment input derived from the Himalayan-Tibetan Plateau. The largest of these peaks begins between 9 and 8 Ma and persists to about 5 Ma (Fig. 12). Sediments associated with this peak are clearly Himalayan in origin, and the sediment input spike is probably related to uplift and erosion of the Himalayan-Tibetan Plateau (Rea 1992). France-Lanord and Derry (1994) and Derry and France-Lanord (1996) presented clay mineralogy and grain-size data from the Bengal fan in the Indian Ocean indicating that chemical weathering intensity in the Ganges-Brahmaputra watersheds increased to a peak between 8 and 7 Ma. The increased chemical weathering intensity in material weathering from the Himalayan-Tibetan Plateau is likely to have resulted in large increases in dissolved material, due to the effective leaching of interlayer cations and oxyanions during clay formation and transformation processes (e.g., Birkeland 1984). Thus, it is plausible that an uplift event produced more physical weathering products and intensified the Asian monsoon, causing these weathering products to undergo intense chemical weathering. Recent evidence from ocean drilling in the western Atlantic Ocean indicates a similar uplift and weathering history for the Andes, with uplift causing orographic precipitation from trade winds and resulting in greatly increased sediment flux to the Atlantic at 8 Ma (Curry et al. 1995). Increased rainfall in the Amazon basin may have combined with the higher amount of weatherable material in increasing chemical weathering rates from this region. In order to explore the global effect of these regional phenomena, I will examine several key oceanic geochemical records.

Continental weathering and increased chemical fluxes to the ocean

Several records reveal increased rates of elemental input to the ocean beginning at about 8 Ma. The global decrease in the oceanic Ge/Si ratio of opaline diatoms over the past 35 Myr is interpreted to reflect increased rates of total weathering, particularly an increase in the rate of physical weathering versus chemical weathering (Shemesh et al. 1989). Superimposed on this overall decrease is a temporary reversal of this trend beginning at 8 Ma and reaching a peak at about 6 Ma (Fig. 12). This short-term increase, not discussed by Shemesh et al. (1989), whose focus was on the Cenozoic record, indicates a rapid increase in the proportion of chemical versus physical weathering. Using the mass-balance assumptions of Delaney and Filippelli (1994), the relative increase of 0.1 μmol/mol in the oceanic Ge/Si ratio from about 9 to 6 Ma was caused either by up to a 27% increase in Ge river input (assuming constant Si river flux), a 14% lower Si river flux (assuming constant Ge/Si ratio of river input), or some combination of the two. Several lines of evidence suggest that, if anything, river fluxes of Si from the Himalayan-Tibetan Plateau increased during this interval. First, the peak in sediment input argues for a greater net surface area of eroded material from the Himalayan-Tibetan Plateau, and thus higher chemical leaching rates of Si. Second, the "biogenic bloom" event noted by Pisias et al. (1995) and Farrell et al. (1995a) occurs during this interval. Carbonate accumulation rates peak in many ocean basins during this interval (Pacific: Rea et al. 1995, Berger et al. 1993, Pisias et al. 1995, Farrell et al. 1995b, Indian: Peterson et al.

1992), and opal accumulation rates are also high (Leinen 1979, Kemp and Baldauf 1993, Farrell et al. 1995b, Rea et al. 1995, Rea and Snoeckx 1995). For all of these accumulation rate records, a net increase in global oceanic burial rate on these time scales is likely driven by an increase in the dissolved input of these elements from continental weathering. Thus, a marked increase in the Ge/Si ratio of river input beginning about 8 Ma is the most likely cause of the increase in oceanic Ge/Si ratios, and, using the arguments presented in Murnane and Stallard (1990), this increase was probably caused by a relative intensification of chemical weathering at this time.

Nutrient fluxes to the ocean

Given the role of P as a limiting nutrient in the ocean (Smith 1984, Codispoti 1989), geologic records of the accumulation rate of P in the ocean can be a window not only on P input rates from weathering but also on paleoproductivity (Filippelli and Delaney 1994, Delaney and Filippelli 1994, Föllmi 1995, Delaney 1998). The Neogene record of phosphorus accumulation rates in the equatorial Pacific reveals a large increase beginning at about 8 Ma and peaking by 6 Ma (Fig. 12). Through a series of mass-balance calculations, Filippelli and Delaney (1994) argued that this increase must reflect a net increase in the rate of phosphorus input to the ocean, and cannot be accounted for simply by changing the distribution of phosphorus within the ocean. Furthermore, other records from the Pacific (Moody et al. 1988), the western Atlantic (Delaney and Anderson 1997), and the Indian Oceans (Hermoyian and Owen 2000) reveal this increase in phosphorus accumulation rates. This phosphorus peak coincides with high rates of physical and chemical weathering likely deriving from the Himalayan-Tibetan Plateau (and possibly the Andes and Amazon Basin), as well as high biogenic productivity recorded in ocean sediments during this interval.

The mechanism for high phosphorus weathering rates during intense chemical weathering is revealed by a modern analog in the Amazon basin. In the lower Amazon, the transport-limited regime yields soil material that is nearly completely leached of phosphorus (Stallard and Edmond 1983), a nutrient limitation that requires ecosystems in this region to extensively recycle this element (e.g., Schlesinger 1997). Thus, an episode of high physical weathering rates in the Himalayan-Tibetan Plateau and other uplifting regions, coupled with high water fluxes and extensive chemical weathering in associated lowlands (Derry and France-Lanord 1996), would likely result in high phosphorus weathering rates and higher inputs of this nutrient to the ocean as reflected in the phosphorus accumulation rate trend.

The nutrient pulse to the ocean would be expected to increase net primary productivity and the export of organic carbon to the seafloor on these time scales. This hypothesis is supported by two observations. First, Dickens and Owen (1994) suggested that a broad expansion of the oxygen minima zone during this interval was caused by an increase in carbon export and oxidation at the sea floor, possibly driven by an increase in the net flux of nutrients to the ocean. Second, a transient positive excursion in the ocean-wide $\delta^{13}C$ of bulk carbonates coincides with the nutrient pulse (Fig. 12). The carbon isotopic composition of bulk carbonates is a function of the proportion of organic to inorganic carbon burial in the ocean, as well as the carbon isotopic values and fluxes of carbon to the ocean from rivers. The Neogene trend of more negative values toward the present has been interpreted as a decrease in the proportion of organic to inorganic carbon burial (Shackleton 1987). By analogy, the relatively short-term positive excursion beginning in the late Miocene may reflect a temporary increase in organic to inorganic carbon burial. Raymo (1994a), using the carbon isotope curve from Shackleton (1987, Fig. 12) and a strontium isotope mass balance, calculated a short-term maxima in organic carbon burial rates from 8 to 5 Ma, with rates 16% higher than preceding values. Another approach, using the increase in reactive phosphorus input rates to the ocean suggested for this interval (Filippelli and Delaney 1994) and assuming that this increase directly drives organic matter burial, yields an increase in organic carbon burial of 20%. Thus, two different approaches result in similar magnitudes of organic carbon burial increases during the late Miocene to early Pliocene. At this point, however, changes in the isotopic composition of carbon input to the ocean cannot be ruled out (especially given evidence for major changes in C3/C4 vegetation assemblages in the late Miocene), which may themselves have been changed due to different rates of carbon weathering and/or intensification of chemical weathering at this time (e.g., Derry and France-Lanord 1996, Cerling et al. 1997, Pagani et al. 1999).

Late Miocene – early Pliocene weathering event and climate

A significant increase in the rate of chemical weathering during the late Miocene - early Pliocene may have had an impact on the long-term carbon cycle. First, the direct effect of the weathering of silicate minerals (combined with calcium carbonate sedimentation in the ocean) is an important long-term sink for atmospheric carbon (e.g., Berner et al. 1983). Second, an increase in organic matter production and burial in the ocean due to an increase in the nutrient flux is an additional sink for atmospheric carbon. Most models of Cenozoic climate invoke just such long-term carbon sinks for the observed deterioration of global climate (e.g., Raymo and Ruddiman 1992). Did the increased drawdown of atmospheric CO_2 inferred from the geochemical records presented here (Fig. 12) have any effect on climate trends in the Neogene? Hodell et al. (1989) linked higher chemical weathering rates in the late Miocene - early Pliocene to increased glaciation in Antarctica. The cause of the decrease in the rate of increase in the marine $^{87}Sr/^{86}Sr$ ratio during the interval between 6 and 4 Ma (Fig. 12) is unclear—in fact, Taylor et al. (2000) suggested that this flattening was due to a basalt-weathering event from the Columbia Plateau. Miller et al. (1987) suggested a possible ice growth event near the middle to late Miocene boundary. The oxygen isotope record from benthic foraminifera shows little change during the late Miocene to early Pliocene (Fig. 12). This record suggests a relatively stable climate (in terms of temperature and ice volume) during this interval until about 2.8 Ma, at which time extensive Northern Hemisphere glaciation commenced (e.g., Raymo 1994b and references therein), and the $\delta^{18}O$ of the ocean increased. The continental weathering event appears to have had little immediate effect on climate, although a long-term drawdown of atmospheric CO_2 associated with this weathering event may have tipped the climate balance from its previous stable state (14 to 3 Ma) to the Pliocene - Pleistocene situation of lower temperatures and greater ice volume. It should be noted, however, that several important factors need to be considered related to this scenario. Recent work by Pagani et al. (1999) indicates little change of atmospheric CO_2 through the Miocene. Furthermore, Wells et al. (1999) suggested that the biogenic bloom observed in the Pacific was a result of increased loads of upwelled iron, derived from submarine volcanism related to the tectonic evolution of Papua New Guinea (in a process similar to upwelled iron in the Southern Ocean; Latimer and Filippelli 2001). Thus, the crystal ball of science is murkier than usual during this interval, and will only be cleared up by more work on productivity and weathering records.

CONCLUSIONS

Significant advances have been made in our understanding of the dynamics of the global P cycle. Interestingly, many of these advancements are derived from the widespread application of P geochemical techniques developed for soils and sediments over the last several decades, but only recently adopted by oceanographers. These techniques have allowed us to further elucidate the dynamics of P transformations during weathering, transport, and deposition; the role that P bioavailability plays on terrestrial and marine ecosystems; and the impacts of P availability on biotic systems and climate change in the past. Despite these recent advances, several gaps exist in our knowledge of the global P cycle. I believe that several specific questions will drive research on the P cycle over the next decade.

What happens to P during transport from river mouths to sediments? Continental margins and marginal sediments play a significant role in the cycling of P, with about 50% of total P input to the ocean deposited in marginal sediments. Given this significance, we have only a few spotty records of P cycling on margins, and most focus on low oxygen settings. A more comprehensive examination of P sedimentation, sedimentary cycling, and eventual burial on continental margins is called for to address this lack of data. This effort should examine margins with large river inputs like the Mississippi, Amazon, and Ganges, together with smaller, more well constrained systems with a variety of sedimentation rates. Sampling should include coring and benthic flux chambers, which together would constrain sedimentation, cycling, and burial, placed along the margin and across the margin. Water column sampling and experimentation (e.g., Benitez-Nelson 2000) should be tied in with the core locations, involving sediment traps deployed over several years. Analytical methods could include P geochemistry, pore fluid analysis for the cores, isotopic analysis of organic matter (a critical component of the sedimentary P cycle) and sedimentology. This "source-to-sink" approach for P could be coupled to C and N to provide a clearer picture of nutrient and carbon cycling on continental margins. This approach would basically capture the anthropogenically modified P cycle, but could provide critical information about cycling across a wide range of systems on monthly, seasonal, and annual timescales. With this information, we could also hindcast the effects of changes in river input and more importantly the effects of changing sea level on continental margin P cycling.

How is the record of P accumulation in deep-sea sediments related to P input from land? As outlined here, a number of efforts have used records of P accumulation in deep-sea sediments to determine changes in P input to the ocean, and hence potential changes in marine productivity. But this "picture" of terrestrial P weathering is filtered through the dynamic continental margin system. Unfortunately, it is difficult to separate the terrestrial and marginal influences—for example, terrestrial P cycling clearly changed due to continental glaciation through the Pleistocene G/IG cycles, but so did sea level and thus marginal sedimentation. Perhaps the best way to answer this is to focus on several intervals with significant changes in sea level, but perhaps similar net climatic regimes. A comprehensive examination, for example, of deep sea P accumulation rates from the late Oligocene to the middle Miocene would allow one to examine the effects of significant second order changes in eustatic sea level (and thus changing deposition on the continental margins) on the oceanic P cycle in the face of more stable climatic conditions on land (compared to the rest of the Cenozoic, at least). This analysis would also be able to elucidate the true role of the deposition of the P-rich Miocene Monterey Formation on the oceanic P cycle. Adequate coverage of both high productivity and low productivity, above CCD and below CCD, and opal versus carbonate versus terrigenous sedimentation dominance would be required to truly close the deep sea P mass balance in such a study.

How did terrestrial P cycling change during the evolution of rootedness in land plants? Much has been said about, and debated on, the development of rootedness and its effects on soil development, chemical weathering, and climate during the Phanerozoic (e.g., Algeo et al 1995, Martin 1996, Berner 1997). A large gap exists, however, in determining the characteristics of soil development, weathering, and terrestrial nutrient cycling during this time. One exciting but undeveloped aspect of this is an examination of P geochemistry of Phanerozoic paleosols (old soil horizons) versus nearby lacustrine or marginal basin sediments. Although diagenesis and rock-forming processes have likely altered the original composition, it might still be possible to extract a record of P cycling in paleosols in a similar manner as presented using lake sediment records earlier in this paper and outlined by Filippelli and Souch (1999). But to expand on this, one could apply geochemical proxies of weathering extent (e.g., Sr and Pb isotopes, Ti and Zr records) in conjunction with P geochemical analyses to examine whether the development of rootedness drastically altered the net chemical weathering and the release of nutrients from landscapes, thus permanently speeding up the soil development process and initiating a new and enhanced steady state for nutrients on land.

Although a few open and important research directions are noted above, it is likely that research on P and P-cycling will expand in a number of different and important ways over the next decade. These may include closing the biological production, recycling, sedimentation, and burial mass balances for P in a number of settings; improving the use of the oxygen isotopic composition of the phosphate molecule as a paleoproductivity proxy; improving our knowledge of the role of microbes in apatite weathering and determining soil-appropriate apatite weathering rate constants; clarifying the roles of P, N, and silica (Si) as limiting nutrients in different oceanic environments and on different timescales; and comparing the reliability of a variety of paleoproductivity proxies (e.g., P accumulation rates, P/Ti ratios, Ba accumulation rates, micropaleontological and biochemical proxies, and carbon and nitrogen isotopic records) in the same place and on the same temporal scales. As with all endeavors, however, it will be with a driving curiosity and sense of wonder that we will slowly unlock the mysteries wrapped around this deceptively simple but critically important element.

ACKNOWLEDGMENTS

Funding for much of the research and thoughts that have gone into my quest for an answer to the question "Why P?" have come from the National Science Foundation, the Donors of the Petroleum Research Fund, administered by the American Chemical Society, the Ocean Drilling Program through JOI/USSSP, and the Department of Defense, and is gratefully acknowledged here. My initial forays into P were spurred by Jeffrey Mount, shepherded by Robert Garrison, and supported and expanded significantly by Peggy Delaney; numerous helpful, sometimes critical but always sage comments by reviewers, friends, and students through the years have helped me to steer a research path that at first glance may appear scattered, but that I hope reflects some of the methods to my madness. I thank Lee Kump, Rick Murray, and John Rakovan, whose reviews helped to improve this manuscript.

REFERENCES

Adams J (1995) Weathering and glacial cycles. Nature 373:100

Algeo TJ, Berner RA, Maynard JB, Scheckler SE (1995) Late Devonian oceanic anoxic events and biotic crises: "Rooted" in the evolution of vascular land plants? GSA Today 5:1-5

Anderson LD, Delaney ML (2000) Sequential extraction and analysis of phosphorus in marine sediments: Streamlining of the SEDEX procedure. Limnol Oceanogr 45:509-515

Anderson LD, Delaney ML, Faul KL (2001) Carbon to phosphorus ratios in sediments: Implications for nutrient cycling. Global Biogeochem Cycles 15:65-79

Antibus RK, Croxdale JG, Miller OK, Linkins AE (1981) Ecotomycorrhizal fungi of *Salix rotundifolia*. III. Resynthesized mycorrhizal complexes and their surface phosphatase activities. Can J Botany 59: 2458-2465

Atlas EL, Pytkowicz RM (1977) Solubility behavior of apatites. Limnol Oceanogr 22:290-300

Baturin GN (1988) Disseminated phosphorus in oceanic sediments—A review. Mar Geol 84:95-104

Benitez-Nelson CR (2000) The biogeochemical cycling of phosphorus in marine systems. Earth-Science

Rev 51:109-135
Berger WH, Wefer G (1991) Productivity of the glacial ocean: Discussion of the iron hypothesis. Limnol Oceanogr 36:1899-1918
Berger WH, Leckie RM, Janecek TR, Stax R, Takayama T (1993) Neogene carbonate sedimentation on Ontong Java Plateau: Highlights and open questions. *In* Proceedings of the Ocean Drilling Program: Scientific Results. College Station, Texas, Ocean Drilling Program 130:711-744
Berner RA Lasaga AC, Garrels RM (1983) The carbonate-silicate geochemical cycle and its effect on atmospheric carbon dioxide over the past 100 million years. Am J Sci 283:641-683
Berner RA, Rao J-L (1994) Phosphorus in sediments of the Amazon River and estuary: Implications for the global flux of phosphorus to the sea. Geochim Cosmochim Acta 58:2333-2340
Berner RA, Ruttenberg KC, Ingall ED, Rao J-L (1993) The nature of phosphorus burial in modern marine sediments. *In* Wollast R., Mackenzie FT, Chou L (eds) Interactions of C, N, P, and S, Biogeochemical Cycles and Global Change. NATO ASI Series 14:365-378
Berner RA (1997) The rise of plants and their effect on weathering and atmospheric CO_2. Science 276: 544-546
Birkeland PW (1984) Soils and Geomorphology. Oxford University Press, New York
Blaise-Stevens A, Bornhold BD, Kemp AES, Dean JM, Vaan AA (2001) Overview of Later Quaternary stratigraphy in Saanich Inlet, British Columbia: Results of the Ocean Drilling Program Leg 169S. Mar Geol 174:27-41
Bolan NS, Robson AD, Barrow NJ, Aylmore LAG (1984) Specific activity of phosphorus in mycorrhizal and non-mycorrhizal plants in relation to the availability of phosphorus to plants. Soil Biol Biochem 16:299-304
Bornhold B, et al. (1998) Proceedings ODP, Initial Reports, 169S. College Station, Texas (Ocean Drilling Program)
Boyle EA (1990) Quaternary deepwater paleoceanography. Science 249:863-870
Broecker WS (1982) Ocean chemistry during glacial time. Geochim Cosmochim Acta 46:1689-1705
Broecker WS, Peng T-H (1982) Tracers in the Sea. Eldigio Press, Palisades, New York
Burnett WC, Beers MJ, Roe KK (1982) Growth rates of phosphate nodules from the continental margin of Peru. Science 215:1616-1618
Calvert SE, Bustin RM, Pedersen TF (1992) Lack of evidence for enhanced preservation of sedimentary organic matter in the oxygen minimum of the Gulf of California. Geology 20:757-760
Calvert SE, Pederson TF, Karlin RE (2001) Geochemical and isotopic evidence for post-glacial paleoceanographic changes in Saanich Inlet, British Columbia. Mar Geol 174:287-306
Cande SC, Kent DV (1992) A new geomagnetic polarity time scale for the Late Cretaceous and Cenozoic. J Geophys Res 97:13917-13951
Canfield DE (1989) Reactive iron in marine sediments. Geochim Cosmochim Acta 53:619-632
Cerling TE, Harris JM, MacFadden BJ, Leakey MG, Quade J, Eisenmann V, Ehleringer JR (1997) Global vegetation change through the Miocene/Pliocene boundary. Nature 389:153-158
Chadwick OA, Derry LA, Vitousek PM, Huebert BJ, Hedin LO (1999) Changing sources of nutrients during four millions years of ecosystem development. Nature 397:491-497
Chien SH, Clayton WR, McClellan GH (1980) Kinetics of dissolution of phosphate rocks in soils. Soil Sci Am J 44:260-264
Chin KOA, Nancollas GH (1991) Dissolution of fluorapatite. A constant-composition kinetics study. Langmuir-ACS J Surfaces Colloids 7:2175-2179
Christoffersen J (1981) Dissolution of calcium hydroxyapatite. Calcified Tissue Intl 33:557-560
Christoffersen MR, Christoffersen J (1985) The effect of aluminum on the rate of dissolution of calcium hydroxyapatite--a contribution to the understanding of aluminum-induced bone diseases. Calcified Tissue Intl 37:673-676
Christoffersen J, Christoffersen MR, Kjaergaard N (1978) The kinetics of dissolution of calcium hydroxyapatite in water at constant pH. J Crystal Growth 43:501-511
Codispoti LA (1989) Phosphorus vs. nitrogen limitation of new and export production. *In* Berger WH, et al. (eds) Productivity of the Ocean: Present and Past. New York, Wiley & Sons, p 377-394
Cole CV, Innis GS, Stewart JWB (1977) Simulation of phosphorus cycling in semiarid grasslands. Ecology 58:3-15
Colman AS, Holland HD (2000) The global diagenetic flux of phosphorus from marine sediments to the oceans: Redox sensitivity and the control of atmospheric oxygen levels. *In* Glenn C, Prévôt-Lucas L, and Lucas J (eds) SEPM Spec Publ 66, Marine Authigenesis: From Microbial to Global, p 53-76
Colman AS, Mackenzie FT, Holland HD (1997) Reply to: Redox stabilization of the atmosphere and oceans and marine productivity (Van Cappellen and Ingall, 1996). Science 275:406-408
Compton JS, Hodell DA, Garrido JR, Mallinson DJ (1993) Origin and age of phosphorite from the south-central Florida Platform: Relation of phosphogenesis to sea-level fluctuations and $\delta^{13}C$ excursions.

Geochim Cosmochim Acta 57:131-146
Compton J, Mallinson D, Glenn C, Filippelli G, Föllmi K, Shields G, Zanin Y (2000) Variations in the global phosphorus cycle. *In* Glenn C, Prévôt-Lucas L, and Lucas J (eds) SEPM Spec Publ 66, Marine Authigenesis: From Microbial to Global, p 21-34
Constantz BR, Ison IC, Rosenthal DI (1995) Skeletal repair by *in situ* formation of the mineral phase of bone. Science 267:1796-1799
Crews TE, Kitayama K, Fownes JH, Riley RH, Herbert DA, Mueller-Dombois D, Vitousek PM (1995) Changes in soil phosphorus fractions and ecosystem dynamics across a long chronosequence in Hawaii. Ecology 76:1407-1424
Cross AF, Schlesinger WH (1995) A literature review and evaluation of the Hedley fractionation: Applications to the biogeochemical cycle of soil phosphorus in natural ecosystems. Geoderma 64: 197-214
Curry WB, Shackleton NJ, Richter C, et al. (1995) Leg 154 synthesis. *In* Proceedings of the Ocean Drilling Program: Initial results, College Station, Texas, Ocean Drilling Program 154:421-442
Dean JM, Kemp AES, Pearce RB (2001) Palaeo-flux records from electron microscope studies of Holocene laminated sediments, Saanich Inlet, British Columbia. Mar Geol 174:139-158
Delaney ML (1998) Phosphorus accumulation in marine sediments and the oceanic phosphorus cycle. Global Biogeochem Cycles 12:563-572
Delaney ML, Anderson LD (1997). Phosphorus geochemistry in Ceara Rise sediments. *In* Proceedings of the Ocean Drilling Program: Scientific results, College Station, Texas, Ocean Drilling Program 154:475-482
Delaney ML, Boyle EA (1988) Tertiary paleoceanic chemical variability: Unintended consequences of simple geochemical models. Paleoceanogr 3:137-156
Delaney ML, Filippelli GM (1994) An apparent contradiction in the role of phosphorus in Cenozoic chemical mass balances for the world ocean. Paleoceanogr 9:513-527
Delcourt HH (1979) Late-Quaternary vegetation history of the eastern Highland Rim and adjacent Cumberland Plateau of Tennessee. Ecol Monogr 49:255-280
Derry LA, France-Lanord C (1996) Neogene Himalayan weathering history and river $^{87}Sr/^{86}Sr$: Impact on the marine Sr record. Earth Planet Sci Lett 142:59-76
Dickens GR, Owens RM (1994) Late Miocene-early Pliocene manganese redirection in the central Indian Ocean: Expansion of the intermediate water oxygen minimum zone. Paleoceanogr 9:169-181
Dodd JC, Burton CC, Burns RG, Jeffries P (1987) Phosphatase activity associated with the roots and the rhizosphere of plants infected with vascular-arbuscular mycorrhizal fungi. New Phytologist 107: 163-172
Driessens FCM, Verbeeck RMH (1981) Metastable states in calcium phosphate-aqueous phase equilibria. J Crystal Growth 53:55-62
Edmond JM (1992) Himalayan tectonics, weathering processes, and the strontium isotope record of marine limestones. Science 258:1594-1597
Farrell JW, Clemens SC, Gromet LP (1995a) Improved chronostratigraphic reference curve of the late Neogene seawater $^{87}Sr/^{86}Sr$. Geology 23:403-407
Farrell J, and nine others (1995b) Late Neogene sedimentation patterns in the eastern equatorial Pacific Ocean (Leg 138). *In* Proceedings of the Ocean Drilling Program: Scientific results, College Station, Texas, Ocean Drilling Program 138:717-756
Feely RA, Massoth GJ, Baker ET, Cowen JP, Lamb MF, Krogsland KA (1990) The effect of hydrothermal processes on midwater phosphorus distributions in the northeast Pacific. Earth Planet Sci Lett 96:305-318
Filippelli GM (1995) Assessing the future impacts of nutrient release from agricultural runoff on ocean productivity. SEPM Congress on Sedimentary Geology, St. Petersburg, Florida, p 53
Filippelli GM (1997a) Controls on phosphorus concentration and accumulation in oceanic sediments. Mar Geol 139:231-240
Filippelli GM (1997b) Intensification of the Asian monsoon and a chemical weathering event in the late Miocene-early Pliocene: Implications for late Neogene climate change. Geology 25:27-30
Filippelli GM (2001) Carbon and phosphorus cycling in anoxic sediments of the Saanich Inlet, British Columbia. Mar Geol 174:307-321
Filippelli GM, Delaney ML (1992a) Similar phosphorus fluxes in ancient phosphorite deposits and a modern phosphogenic environment. Geology 20:709-712
Filippelli GM, Delaney ML (1992b) Quantifying cathodoluminescent intensity with an on-line camera and exposure meter. J Sediment Petrol 62:724-725
Filippelli GM, Delaney ML (1993) The effects of manganese (II) and iron (II) on the cathodoluminescence signal in synthetic apatite. J Sediment Petrol 63:167-173
Filippelli GM, Delaney ML (1994) The oceanic phosphorus cycle and continental weathering during the

Neogene. Paleoceanogr 9:643-652

Filippelli GM, Delaney ML (1995) Phosphorus geochemistry and accumulation rates in the eastern equatorial Pacific Ocean: Results from ODP Leg 138. *In* Mayer et al. (eds) Proc ODP Scientific Results 138: 757-768

Filippelli GM, Delaney ML (1996) Phosphorus geochemistry of equatorial Pacific sediments. Geochim Cosmochim Acta 60:1479-1495

Filippelli GM, Souch C. (1999) Effects of climate and landscape development on the terrestrial phosphorus cycle. Geology 27:171-174

Filippelli GM, Delaney ML, Garrison RE, Omarzai SK, Behl RJ (1994) Phosphorus accumulation rates in a Miocene low oxygen basin: The Monterey Formation (Pismo Basin), California. Mar Geol 116: 419-430

Föllmi KB (1995) 160 m.y. record of marine sedimentary phosphorus burial: Coupling of climate and continental weathering under greenhouse and icehouse conditions. Geology 23:859-862

Fox JL, Higuchi WI, Fawz MB, Wu MS (1978) A new two-site model for hydroxyapatite dissolution in acidic media. J Colloid Interface Sci 67:312-330

France-Lanord C, Derry LA (1994) $\delta^{13}C$ of organic carbon in the Bengal fan: Source evolution and transport of C3 and C4 plant carbon to marine sediments. Geochim Cosmochim Acta 58:4809-4814

Froelich PN (1988) Kinetic controls of dissolved phosphate in natural rivers and estuaries: A primer on the phosphate buffer mechanism. Limnol Oceanogr 33:649-668

Froelich PN, Bender ML, Heath GR (1977) Phosphorus accumulation rates in metalliferous sediments on the East Pacific Rise. Earth Planet Sci Lett 34:351-359

Froelich PN, Bender ML, Luedtke NA, Heath GR, DeVries T (1982) The marine phosphorus cycle. Am J Sci 282:474-511

Froelich PN, Kim K-H, Jahnke R, Burnett WC, Soutar A, Deakin M (1983) Pore water flouride uptake in Peru continental margin sediments: Uptake from seawater. Geochim Cosmochim Acta 47:1605-1612

Gardner LR (1990) The role of rock weathering in the phosphorus budget of terrestrial watersheds. Biogeochem 11:97-110

Garrison RE, Kastner M (1990) Phosphatic sediments and rocks recovered from the Peru margin during ODP Leg 112. *In* Proceedings of the Ocean Drilling Program: College Station, Texas, Ocean Drilling Program 112:111-134

Gibbs MT, Kump LR (1994) Global chemical denudation during the last glacial maximum and the present: Sensitivity to changes in lithology and hydrology. Paleoceanogr 9:529-544

Goñi MA, Ruttenberg KC, Eglinton TI (1998) A reassessment of the sources and importance of land-derived organic matter in surface sediments from the Gulf of Mexico. Geochim Cosmochim Acta 62:3055-3075

Hermoyian CS, Owen RM (2001) Late Miocene-early Pliocene biogenic bloom: Evidence from low-productivity regions of the Indian and Atlantic Oceans. Paleoceanogr 16:95-100

Hodell DA, Mueller PA, McKenzie JA, Mead GA (1989) Strontium isotope stratigraphy and geochemistry of the late Neogene ocean. Earth Planet Sci Lett 92:165-178

Hodell DA, Mead GA, Mueller PA (1990) Variation in the strontium isotopic composition of seawater (8 Ma to present): Implications for chemical weathering rates and dissolved fluxes to the oceans. Chem Geol 80:291-307

Hodell DA, Mueller PA, Garrido JR (1991) Variations in the strontium isotopic composition of seawater during the Neogene. Geology 19:24-27

Holland HD (1978) The Chemistry of the Atmosphere and Oceans. Wiley Interscience, New York

Holland HD (1984) The Chemical Evolution of the Atmosphere and Oceans. Princeton University Press, Princeton, New Jersey

Hull AB, Lerman A (1985) The kinetics of apatite dissolution: Application to natural aqueous systems. Natl Mtg Am Chem Soc Div Environ Chem 25:421-424

Hull AB, Hull JR (1987) Geometric modeling of dissolution kinetics: Application to apatite. Water Resources Res 23:707-714

Ingall ED, Jahnke R (1994) Evidence for enhanced phosphorus regeneration from sediments overlain by oxygen depleted waters. Geochim Cosmochim Acta 54:2617-2620

Ingall ED, Jahnke R (1997) Influence of water-column anoxia on the elemental fractionation of carbon and phosphorus during sediment diagenesis. Mar Geol 139:219-229

Ingall ED, Bustin RM, Van Cappellen P (1993) Influence of water column anoxia on the burial and preservation of carbon and phosphorus in marine shales. Geochim Cosmochim Acta 57:303-316

Jahnke RA (1984) The synthesis and solubility of carbonate fluorapatite. Am J Sci 284:58-78

Janecek TR, Rea DK (1983) Eolian deposition in the northeast Pacific Ocean: Cenozoic history of atmospheric circulation. Geol Soc Am Bull 94:730-738

Johnson KM, Grimm KA (2001) Opal and organic carbon in laminated diatomaceous sediments: Saanich

Inlet, Santa Barbara Basin, and the Miocene Monterey Formation. Mar Geol 174:159-176

Jurinak JJ, Dudley LM, Allen MF, Knight WG (1986) The role of calcium oxalate in the availability of phosphorus in soils of semiarid regions: A thermodynamic study. Soil Sci 142:255-261

Kazakov AV (1937) The phosphorite facies and the genesis of phosphorites. Soviet Geol 8:33-47

Kemp AES, Baldauf JG (1993) Vast Neogene laminated diatom mat deposits from the eastern equatorial Pacific Ocean. Nature 362:141-144

Kim D, Shuffert JD, Kastner M (1999) Francolite authigenesis in California continental slope sediments and its implications for the marine P cycle. Geochim Cosmochim Acta 63:3477-3485

Klump JV, Martens CS (1981) Biogeochemical cycling in an organic rich coastal marine basin—II: Nutrient sediment-water exchange processes. Geochim Cosmochim Acta 45:101-121

Krajewski KP, Van Cappellen P, Trichet J, Kuhn O, Lucas J, Martín-Algarra A, Prévôt L, Tewari VC, Gaspar L, Knight RI, Lamboy M (1994) Biological processes and apatite formation in sedimentary environments. Eclogae Geol Helv 87:701-745

Kroehler CJ, Linkins AE (1988) The root surface phosphatases of *Eriophorum vaginatum*: Effects of temperature, pH, substrate concentration and inorganic phosphorus. Plant Soil 105:3-10

Kroon, D., Steens, T., and Troelstra, S. R., 1991, Onset of monsoonal related upwelling in the western Arabian Sea as revealed by planktonic foraminifera. *In* Proceedings of the Ocean Drilling Program, Scientific results: College Station, Texas, Ocean Drilling Program 117:257-263

Kump LR, Alley RB (1994) Global chemical weathering on glacial time scales. *In* Board on Earth Sciences and Resources, National Research Council, Material fluxes on the surface of the Earth. Washington, DC, National Academy of Sciences, p 46-60

Kump LR, Arthur MA (1997) Global chemical erosion during the Cenozoic: Weatherability balances the budget. *In* Ruddiman W (ed) Tectonics, uplift, and climate change, Plenum Publishing, p 399-426

Lajtha K, Schlesinger WH (1988) The biogeochemistry of phosphorus cycling and phosphorus availability in a desert shrubland ecosystem. Biogeochem 2:29-37

Lane M, Mackenzie FT (1990) Mechanisms and rates of natural carbonate fluorapatite dissolution. Bull Geol Soc Am Abstr, A208

Lane M, Mackenzie FT (1991) Kinetics of carbonate fluorapatite dissolution: Application to natural systems. Bull Geol Soc Am Abstr.:A151

Latimer JC, Filippelli GM (2001) Terrigenous input and productivity in the Southern Ocean. Paleoceanogr 16:627-643

Leinen M (1979) Biogenic silica accumulation in the central equatorial Pacific and its implications for Cenozoic paleoceanography. Geol Soc Am Bull 90:1310-1376

Lucotte M, Mucci A, Hillaire-Marcel C, Tran S (1994) Early diagenetic processes in deep Labrador Sea sediments: Reactive and non reactive iron and phosphorus. Can J Earth Sci 31:14-27

McClellan GH, Lehr JR (1969) Crystal chemical investigation of natural apatites. Am Mineral 54: 1372-1389

McConnell D (1973) Apatite: Its Crystal Chemistry, Mineralogy, Utilization, and Geologic and Biologic Occurrences. Springer-Verlag, New York

McGill WB, Cole CV (1981) Comparative cycling of organic C, N, S, and P through soil organic matter. Geoderma 26:267-286

McManus J, Berelson WM, Coale KH, Johnson KS, Kilgore TE (1997) Phosphorus regeneration in continental margin sediments. Geochim Cosmochim Acta 61:2891-2907

McQuoid MR, Whiticar MJ, Calvert SE, Pederson TF (2001) A post-glacial isotope record of primary productivity and accumulation in the organic sediments of Saanich Inlet, ODP Leg 169S. Mar Geol 174:273-286

Mach DL, Ramirez AJ, Holland HD (1987) Organic phosphorus and carbon in marine sediments. Am J Sci 287:429-441

Mahn CL, Gieskes JM (2001) Halide systematics in comparison with nutrient distributions in Sites 1033B and 1034B, Saanich Inlet: ODP Leg 169S. Mar Geol 174:323-340

Malcolm RE (1983) Assessment of phosphatase activity in soils. Soil Biol Biochem 15:403-408

Martin RE (1996) Secular increase in nutrient levels through the Phanerozoic: Implications for productivity, biomass, and diversity of the marine biosphere. Palaios 11:209-219

Menard HW, Smith SM (1966) Hypsometry of ocean basin provinces. J Geophys Res 71:4305-4326

Miller KG, Fairbanks RG, Mountain GS (1987) Tertiary oxygen isotope synthesis, sea level history, and continental margin erosion. Paleoceanogr 2:1-19

Mills HH, Delcourt PA (1991) Quaternary geology of the Appalachian Highlands and interior low plateaus. *In* Morrison RB (ed) Quaternary non-glacial geology: Conterminous U.S. Boulder, Colorado, Geological Society of America, The Geology of North America, K-2:611-628

Molnar P, England P, Martinod J (1993) Mantle dynamics, uplift of the Tibetan Plateau, and the Indian Monsoon. Rev Geophys 31:357-396

Moody JB, Chaboudy LR, Worsley TR (1988) Pacific pelagic phosphorus accumulation during the last 10 m.y. Paleoceanogr 3:113-136

Murnane RJ, Stallard RF (1990) Germanium and silicon in rivers of the Orinoco drainage basin. Nature 344:749-752

Murray JW, Grundmanis V, Smethie WM (1978) Interstitial water chemistry in the sediments of Saanich Inlet. Geochim Cosmochim Acta 42:1011-1026

Murray RW, Leinen M, Isern AR (1993) Biogenic flux of Al in the central equatorial Pacific Ocean: Evidence for increased productivity during glacial periods. Paleoceanogr 8:651-670

Murray RW, Knowlton C, Leinen M, Mix AC, Polski CH (2000) Export production and carbonate dissolution in the central equatorial Pacific Ocean over the past 1 Ma. Paleoceanogr 15:570-592

Nancollas GH, Amjad Z, Koutsoukas P (1979) Calcium phosphates-speciation, solubility, and kinetic considerations. Am Chem Soc Symp Ser 93:475-497

O'Brien GW, Veeh HH (1980) Holocene phosphorite on the East Australian continental margin. Nature 288:690-692

Olsen RA (1975) Rate of dissolution of phosphate from minerals and soils. Soil Sci Am J 39:634-639

Onken AB, Matheson RL (1982) Dissolution rate of EDTA-extractable phosphate from soils. Soil Sci Am J 46:276-279

Pagani M, Freeman KH, Arthur MA (1999) Late Miocene Atmospheric CO_2 Concentrations and the Expansion of C4 Grasses. Science 285:876-879

Pederson TF, Calvert SE (1990) Anoxia vs. productivity: What controls the formation of organic-carbon-rich sediments and sedimentary rocks. AAPG Bull 74:454-466

Pellatt MG, Mathewes RW (1997) Holocene tree line and climate change on the Queen Charlotte Islands, Canada. Quat Res 48:88-99

Peterson LC, Murray DW, Ehrmann WU, Hempel P (1992) Cenozoic carbonate accumulation and compensation depth changes in the Indian Ocean. *In* Duncan R, Rea D, Kidd R, von Rad U, Weissel J (eds) Synthesis of Results from Scientific Drilling in the Indian Ocean. Am Geophys Union, Geophys Monogr 70:311-333

Phillips WR, Griffen DT (1981) Optical Mineralogy: The Nonopaque Minerals. W.H. Freeman, San Francisco

Pisias NG, Layer LA, Mix AC (1995) Paleoceanography of the eastern equatorial Pacific during the Neogene: Synthesis of Leg 138 drilling results. *In* Proceedings of the Ocean Drilling Program, Scientific Results, College Station, Texas, Ocean Drilling Program 138:5-21

Prell WL, Kutzbach JE (1992) Sensitivity of the Indian monsoon to forcing parameters and implications for its evolution. Nature 360:646-650

Prell WL, Murray DW, Clemens SC (1992) Evolution and variability of the Indian Ocean Summer Monsoon: Evidence from the Western Arabian Sea Drilling Program. *In* Duncan R, Rea D, Kidd R, von Rad U, Weissel J (eds), Synthesis of results from scientific drilling in the Indian Ocean: Am Geophys Union Geophys Monogr 70:447-469

Ramirez AJ, Rose AW (1992) Analytical geochemistry of organic phosphorus and its correlation with organic carbon in marine and fluvial sediments and soils. Am J Sci 292:421-454

Raymo ME (1994a) The Himalayas, organic carbon burial, and climate in the Miocene. Paleoceanogr 9:399-404

Raymo ME (1994b) The initiation of northern hemisphere glaciation. Ann Rev Earth Planet Sci 22:353-383

Raymo ME, Ruddiman WF (1992) Tectonic forcing of late Cenozoic climate. Nature 359:117-122

Raymo ME, Ruddiman WF, Froelich PN (1988) Influence of late Cenozoic mountain building on ocean geochemical cycles. Geology 16:649-653

Rea DK (1992) Delivery of Himalayan sediment to the northern Indian Ocean and its relation to global climate, sea level, uplift, and seawater strontium. *In* Duncan R, Rea D, Kidd R, von Rad U, Weissel J, eds., Synthesis of results from scientific drilling in the Indian Ocean: Am Geophys Union Geophys Monogr 70, p. 387-402

Rea DK, Snoeckx H (1995) Sediment fluxes in the Gulf of Alaska: Paleoceanographic record from Site 887 on the Patton - Murray Seamount platform. *In* Proceedings of the Ocean Drilling Program, Scientific results: College Station, Texas, Ocean Drilling Program 145:247-256

Rea DK, Basov V, Krissek LA, and Shipboard Party (1995) Scientific results of drilling the north Pacific transect. *In* Proceedings of the Ocean Drilling Program, Scientific results: College Station, Texas, Ocean Drilling Program 145:577-596

Redfield AC, Ketchum BH, Richards FA (1963) The influence of organisms in the composition of seawater. *In* Hill MN (ed) The Sea: Composition of Sea-water Comparative and Descriptive Oceanography 2:26-77. Wiley Interscience, New York

Reimers CE, Ruttenberg KC, Canfield DE, Christiansen MB, Martin JB (1996) Porewater pH and authigenic phases formed in the uppermost sediments of the Santa Barbara Basin. Geochim

Cosmochim Acta 60:4037-4057
Roberts TL, Stewart JWB, Bettany JR (1985) The influence of topography on the distribution of organic and inorganic soil phosphorus across a narrow environmental gradient. Can J Soil Sci 65:651-665
Robinson D (1986) Limits to nutrient inflow rates in roots and root systems. Physiolog Plantar 68:551-559
Ruttenberg KC (1990) Diagenesis and burial of phosphorus in marine sediments: Implications for the marine phosphorus budget. PhD dissertation, Yale University, New Haven, Connecticut
Ruttenberg KC (1992) Development of a sequential extraction method for different forms of phosphorus in marine sediments. Limnol Oceanogr 37:1460-1482
Ruttenberg KC (1993) Reassessment of the oceanic residence time of phosphorus. Chem. Geol. 107: 405-409
Ruttenberg KC, Berner RA (1993) Authigenic apatite formation and burial in sediments from non-upwelling continental margin environments. Geochim Cosmochim Acta 57:991-1007
Ruttenberg KC, Goni MA (1997) Phosphorus distribution, C:N:P ratios, and $\delta^{13}C_{oc}$ in arctic, temperate, and tropical coastal sediments: Tools for characterizing bulk sedimentary organic matter. Mar Geol 139:123-146
Ryder JM, Thompson B (1986) Neoglaciation in the southern Coast Mountains of British Columbia: Chronology prior to the late-Neoglacial maximum. Can J Earth Sci 23:273-287
Sarmiento JL, Toggweiler JR (1984) A new model for the role of the oceans in determining atmospheric p_{CO_2}. Nature 308:620-624
Schenau SJ, Slomp CP, De Lange GJ (2000) Phosphogenesis and active phosphorite formation in sediments form the Arabian Sea oxygen minimum zone. Mar Geol 169:1-20
Schlesinger WH (1997) Biogeochemistry: An Analysis of Global Change. Academic Press, San Diego
Schlesinger WH, Bruijnzeel LA, Bush MB, Klein EM, Mace KA, Raikes JA, Whittaker RJ (1998) The biogeochemistry of phosphorus after the first century of soil development on Rakata Island, Krakatau, Indonesia. Biogeochem 40:37-55
Schuffert JD, Jahnke RA, Kastner M, Leather J, Sturz A, Wing MR (1994) Rates of formation of modern phosphorite off western Mexico. Geochim Cosmochim Acta 58:5001-5010
Schuffert JD, Kastner M, Jahnke RA (1998) Carbon and phosphorus burial associated with modern phosphorite formation. Mar Geol 146:21-31
Shackleton NJ (1987) The carbon isotope record of the Cenozoic: History of organic carbon burial and of oxygen in the ocean and atmosphere. *In* Fleet AJ, Brooks J (eds), Marine petroleum source rocks: Geol Soc London Spec Publ 26:423-434
Shemesh A, Mortlock RA, Froelich PN (1989) Late Cenozoic Ge/Si record of marine biogenic opal: Implications for variations of riverine fluxes to the ocean. Paleoceanogr 4:221-234
Smeck NE (1985) Phosphorus dynamics in soils and landscapes. Geoderma 36:185-199
Smith AN, Posner AM, Quirk JP (1974) Incongruent dissolution of surface complexes if hydroxyapatite. J Colloid Inter Sci 48:442-449
Smith AN, Posner AM, Quirk JP (1977) A model describing the kinetics of dissolution of hydroxyapatite. J Colloid Inter Sci 62:475-494
Smith SV (1984) Phosphorus versus nitrogen limitation in the marine environment. Limnol Oceanogr 29:1149-1160
Souch C (1994) A methodology to interpret downvalley lake sediments as records of neoglacial activity: Coast Mountains, British Columbia, Canada. Geografiska Annaler 76A:169-186
Stallard RF, Edmond JM (1983) Geochemistry of the Amazon. 2. Influence of geology and weathering environment on the dissolved load. J Geophys Res 88:9671-9688
Sundby B, Gobeil C, Silverberg N, Mucci A (1992) The phosphorus cycle in coastal marine sediments. Limnol Oceanogr 37:1129-1145
Tarafdar JC, Claasen N (1988) Organic phosphorus compounds as a phosphorus source for higher plants through the activity of phosphatase produced by plant roots and microorganisms. Biol Fertil Soils 5:308-312
Taylor A, Blum JD, Lasaga A, MacInnis IN (2000) Kinetics of dissolution and Sr release during biotite and phlogopite weathering. Geochim Cosmochim Acta 64:1191-1208
Thirioux L, Baillif P, Touray JC, Ildefouse JP (1990) Surface reactions during fluorapatite dissolution-recrystallization in acid media (hydrochloric and citric acids). Geochim Cosmochim Acta 54: 1969-1977
Thompson LG, Yao T, Davis ME, Henderson KA, Mosley-Thompson E, Lin P-N, Synal H-A, Cole-Dai J, Bolzan JF (1997) Tropical climate instability: The last glacial cycle from a Qinghai-Tibetan ice core. Science 276:1821-1825
Tiessen H, Moir JO (1993) Characterization of available P by sequential extraction. *In* Carter M (ed) Soil Sampling and Methods of Analysis. Lewis, Boca Raton, Florida, p 75-86
Tiessen H, Stewart JWB, Cole CV (1984) Pathways of phosphorus transformations in soils of differing

pedogenesis. Soil Sci Soc Am J 48:853-858
Tribble JS, Arvidson RS, Lane M, Mackenzie FT (1995) Crystal chemistry, and thermodynamic and kinetic properties of calcite, dolomite, apatite, and biogenic silica: applications to petrologic problems. Sediment Geol 95:11-37
Van Cappellen PV (1991) The formation of marine apatite: A kinetic study. PhD dissertation, Yale University, New Haven, Connecticut
Van Cappellen PV, Berner RA (1988) A mathematical model for the early diagenesis of phosphorus and fluorine in marine sediments: apatite precipitation. Am J Sci 288:289-333
Van Cappellen PV, Ingall ED (1994) Benthic phosphorus regeneration, net primary production, and ocean anoxia: A model of the coupled marine biogeochemical cycles of carbon and phosphorus. Paleoceanogr 9:677-692
Van Cappellen PV, Ingall ED (1996) Redox stabilization of the atmosphere and oceans by phosphorus-limited marine productivity. Science 271:493-496
Veeh HH, Burnett WC, Soutar A (1973) Contemporary phosphorites on the continental margin of Peru. Science 181:844-845
Vitousek PM, Chadwick OA, Crews TE, Fownes JH, Hendricks DM, Herbert D (1997) Soil and ecosystem development across the Hawaiian Islands. GSA Today 7:1-8
Walker TW, Syers JK (1976) The fate of phosphorus during pedogenesis. Geoderma 15:1-19
Wells ML, Vallis GK, Silver EA (1999) Tectonic processes in Papua New Guinea and past productivity in the eastern equatorial Pacific Ocean. Nature 398:601-604
Wetzel MA, Fleeger JW, Powers SP (2001) Effects of hypoxia and anoxia on meiofauna: A review with new data from the Gulf of Mexico. *In* Turner RE (ed) Coastal hypoxia: Consequences for living resources and ecosystems. Coastal Estuarine Studies 58:165-184
Wheat CG, Feely RA, Mottl MJ (1996) Phosphate removal by oceanic hydrothermal processes: An update of the phosphate budget in the oceans. Geochim Cosmochim Acta 60:3593-3608
Wilkins GR, Delcourt PA, Delcourt HR, Harrison FW, Turner MR (1991) Paleoecology of central Kentucky since the last glacial maximum. Quat Res 36:224-239
Williams JDH, Murphy TP, Mayer T (1976) Rates of accumulation of phosphorus forms in Lake Erie sediments. J Fisheries Res Board Canada 33:430-439

11 Calcium Phosphate Biominerals

James C. Elliott

Dental Biophysics
Medical Sciences Building
Queen Mary, University of London
London E1 4NS, United Kingdom

BIOMINERALS—OVERVIEW

Tissues and minerals

On a quantitative basis, the most important of the calcium phosphates (Table 1) is an apatite closely related to hydroxylapatite (HAP). This is better described as an impure carbonate-containing apatite (CO_3Ap) and forms the inorganic component of bones and teeth. Table 1 also includes calcium phosphates (including two pyrophosphates) that occur in pathological mineralizations and those that are used for the repair of mineralized tissues. Unlike the well-controlled process of normal mineralization in bones and teeth (see later), pathological mineralizations are usually poorly controlled with the result that several calcium phosphates may occur together. In addition, their crystallographic orientations are often random.

Table 1. Occurrence of calcium phosphates.

Name	*Formula*	*Examples of occurrence*
Hydroxylapatite	$Ca_{10}(PO_4)_6(OH)_2$	In an impure form (main impurities: CO_3^{2-}, Mg^{2+} and HPO_4^{2-}) as the mineral in bones and teeth. In most pathological calcifications. Pure HAP as a synthetic biomaterial
Amorphous calcium phosphate, ACP		One time thought to be a major component of bone mineral (see text). Occurs in plasma-sprayed HAP coatings on prosthetic implants[a]
Octacalcium phosphate	$Ca_8H_2(PO_4)_6{\cdot}5H_2O$	Dental calculus[b] and possibly a transient phase in precipitation of biological HAPs[c]
Brushite	$CaHPO_4{\cdot}2H_2O$	Dental and urinary calculi
Magnesian whitlockite	Approximately $Ca_{18}Mg_2H_2(PO_4)_{14}$	Dental calculus, urinary and salivary calculi, carious lesions in teeth and other pathological calcifications
Calcium pyrophosphate dihydrate	$Ca_2P_2O_7{\cdot}2H_2O$ triclinic and monoclinic	Joints, particularly knee or other large peripheral joints[d]
Tetracalcium phosphate	$Ca_4(PO_4)_2O$	Used in some bone cements
Monetite	$CaHPO_4$	Used in some bone cements
α- and β-tricalcium phosphate	α- and β-$Ca_3(PO_4)_2$	Used in some bone cements

[a]Gross and Phillips (1998) [b]Schroeder (1969) [c]Chow and Eanes (2001)
[d]Numerous anhydrous, dihydrate and tetrahydrate polymorphs are known, only those listed occur in joints (Pritzker 1998)

The solubility isotherms of the calcium phosphates in the system $Ca(OH)_2$-H_3PO_4-H_2O at 37°C are shown in Figure 1. Monetite occurs in the phase diagram, although its occurrence in normal or pathological calcifications has never been confirmed, one contributing reason being that its nucleation and growth is more difficult than brushite

1529-6466/00/0048-0011$05.00

under biological conditions, so brushite forms in preference, even though brushite is less stable. β-$Ca_3(PO_4)_2$ is a high-temperature phase that does not precipitate directly in aqueous systems; however, it is sufficiently stable in water for a solubility product to be determined so that an isotherm can be calculated, which is the origin of the isotherm in Figure 1. However, if the aqueous system contains Mg^{2+} (1 or 2 mmol/l), the solubility product of "$Ca_3(PO_4)_2$" is dramatically reduced (more than a thousand-fold) with the result that the structurally related whitlockite can easily precipitate (Hamad and Heughebaert 1986, LeGeros et al. 1989). Fe^{2+} ions have a similar effect. Whitlockite has superficially the same X-ray diffraction (XRD) powder pattern as β-$Ca_3(PO_4)_2$, but has some of the Ca^{2+} replaced by Mg^{2+} (or Fe^{2+}) and—importantly—also contains HPO_4^{2-} ions [see idealized formula in Table 1 and Elliott (1994) for a discussion of structural relations between whitlockite and β-$Ca_3(PO_4)_2$]. α-$Ca_3(PO_4)_2$ and tetracalcium phosphate are high-temperature phases and thus do not form directly in aqueous systems, although solubility products have been determined experimentally and used to calculate isotherms for inclusion in a $Ca(OH)_2$-H_3PO_4-H_2O phase diagram (Chow and Eanes 2001).

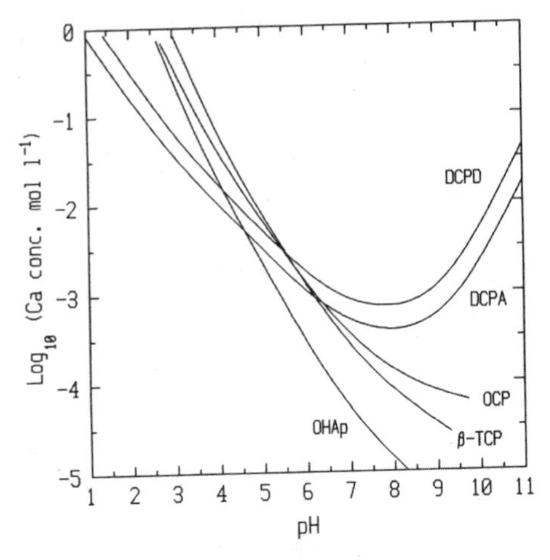

Figure 1. Solubility isotherms of calcium phosphate phases in the system $Ca(OH)_2$-H_3PO_4-H_2O at 37°C. DCPD is dicalcium phosphate dihydrate, brushite; DCPA is dicalcium phosphate anhydrous, monetite; OCP is octacalcium phosphate; β-TCP is β-tricalcium phosphate; and OHAp is hydroxylapatite. [Figure 1.1 from Elliott (1994), reprinted with permission from Elsevier Science.]

As mentioned above, the mineral in bones and teeth contains CO_3^{2-} ions so its solubility properties will not be the same as depicted for HAP in Figure 1. In fact, the presence of CO_3^{2-} ions increases the solubility product and the rate of dissolution in acids significantly. Taking dental enamel because it is better crystallized than bone mineral, an activity product of 5.5×10^{-55} mol^9 l^{-9} (presumably at 25°C) has been reported (Patel and Brown 1975). This compares with 3.04×10^{-59} mol^9 l^{-9} at 25°C for stoichiometric HAP (McDowell et al. 1977). Both activity products are based on $(Ca^{2+})^5(PO_4^{3-})^3(OH^-)$ where the parentheses indicate activities. In more recent studies, the activity of the CO_3^{2-} ion has been included in the calculation of activity product of enamel (Aoba and Moreno 1992). This necessitated making solubility measurements under controlled CO_2 partial pressures.

Bone and dentin are living tissues that are metabolically active during life; bone is capable of regeneration and dentin can respond in a defensive manner. Most bone undergoes continuous remodelling through the process of resorption by osteoclasts and formation of new bone by osteoblasts. On the other hand, enamel is acellular, so is not a living tissue. Nevertheless, it is sufficiently porous for diffusion and chemical reactions to occur within its structure, particularly acidic dissolution, as in dental caries, and remineralization from saliva, as in the "healing" of caries lesions. Bone is a complex composite of mineral and collagen, the main function of the mineral being to provide it with stiffness, but bone mineral also provides a reservoir of ions (Ca^{2+}, PO_4^{3-}, HCO_3^-, etc.) to maintain the concentrations of these ions in the extracelluar fluid (see below). Bone, surprisingly contains 0.8-0.9 wt % citrate, probably adsorbed on the apatite crystals (Dallemagne and Richelle 1973). Enamel is the most highly mineralized tissue and forms a hard, wear-resistant surface for mastication which overlies the more resilient, less mineralized, dentin.

Table 2 lists the major features of the apatite in bones and teeth. Note particularly the very small size of the crystals: this means that their structure is not easy to determine by diffraction methods. Furthermore, the very small size of the crystals in bone gives a surface area of at least 100 m^2/g (Neuman and Neuman 1958, Posner et al. 1979). The result is that a substantial fraction of the ions occupy surface positions which makes it difficult to correlate the chemical composition of the bone mineral with its possible crystallographic structure. The large surface area however is advantageous for one important function of the mineral, namely long-term acid-base regulation, particularly in acidosis, through the release of CO_3^{2-} and HCO_3^- ions. Another noteworthy feature seen from Table 2 is that the apatite crystals are highly oriented. The explanation of how this orientation arises clearly has to be an important aspect of any understanding of the mineralization process.

Table 2. Major components and features of enamel, dentin and bone.

	Developing enamel	*Mature enamel*	*Dentin and bone*
Inorganic wt %	37	96	72
Organic wt %	19	0.1	20
Major organic component(s)	Amelogenins and enamelins	Enamelin-like protein	Type I collagen
Size of crystals	Flattened hexagons about 70 × 25 nm and 0.1-5 μm or more long		Plates 50 × 30 nm, 4 nm thick
Orientation of crystals	Approx. parallel with enamel rods which are very approx. normal to enamel surface		Most with long dimension parallel with collagen fiber direction

Biological apatite crystals are too small to allow direct measurement of their density. However, the calculated density from estimated unit cell contents (see later section on atomic structure of biological apatites) is about 5% less than HAP (Elliott 1997).

The embryological origin of bone and dentin is different from enamel. The former tissues are of mesodermal origin whilst enamel is of ectodermal origin. Thus bone and dentin have collagen as their main organic matrix, as expected. On the other hand, enamel, though at one time thought to contain keratin as is found in other ectodermal tissues, in fact contains amelogenins and enamelins, proteins not found elsewhere in the body (Simmer and Fincham 1995, Fincham et al. 2000). The larger size of the crystals in enamel compared with dentin and bone (Table 2) is undoubtedly associated with differences in its organic matrix compared with collagen. There is also a marked difference in the mineralization process in enamel compared with dentin and bone (see later references on biomineralization).

The typical chemical composition of the mineral in calcified tissues is given in Table 3. These values are only approximate; the main constituents can vary by a percent or more. Driessens and Verbeeck (1990) give other tabulations and literature references. Whilst most of the phosphorus is present as PO_4^{3-} ions, several percent can be present as HPO_4^{2-} ions. The evidence for this is the formation of pyrophosphate on heating (typically at 300-600°C) through the condensation of HPO_4^{2-} ions with the loss of water (Herman and Dallemagne 1961). Also, it may well be that some of the carbonate in bone mineral is present as HCO_3^- ions, as its carbonate content is rather labile (Poyart et al. 1975).

Table 3. Typical compositions of inorganic fraction of enamel, dentin and bone (wt %).[a]

Component	*Enamel*[b]	*Dentin*[c]	*Bone*[d]
Ca	37.6	40.3	36.6
P	18.3	18.6	17.1
CO_2[e]	3.0	4.8	4.8
Na	0.7	0.1	1.0
K	0.05	0.07	0.07
Mg	0.2	1.1	0.6
Sr	0.03	0.04	0.05
Cl	0.4	0.27	0.1
F	0.01	0.07	0.1
Ca/P molar	1.59	1.67	1.65

[a]see Driessens and Verbeeck (1990) for a detailed discussion of changes in chemical composition with age, type of tissue and diet.
[b]based on Tables 1 and 2 in Weatherell and Robinson (1973), but expressed as a percentage of the ash content of whole enamel (96 wt %).
[c]as for footnote (b), with an ash content of whole dentin of 73 wt %.
[d]bovine cortical bone, based on Table 3 in Dallemagne and Richelle (1973), but expressed as a percentage of the mineral fraction (taken as 73 wt % of dry fat-free bone).
[e]present as carbonate

The listing of mean chemical compositions in Table 3 hides anatomical variations within the tissues and changes that might take place with diet and age. As regards enamel (in a study on pigs), the HPO_4^{2-} and CO_3^{2-} contents fall from the level found in the initially secreted enamel as it matures (Aoba and Moreno 1992). For human dental enamel, the mineral and fluoride (F) contents are higher at the surface, but the CO_3 and Mg contents are lower (Miles 1967, Williams and Elliott 1989). Once formed, little change generally takes place in the interior of enamel, though the composition may reflect aspects of the ionic composition of the extracelluar fluid at the time of mineral deposition (e.g., ^{90}Sr from radioactive fall-out and Pb in children from living near motorways). The exceptions to this are the preferential loss of CO_3 and Mg during the acid dissolution of mineral in dental caries, and the marked uptake of F in the very surface of enamel that can take place before a tooth erupts and also in carious lesions after eruption (see Williams and Elliott 1989).

The mineral composition of bone is much more susceptible to change as a result of metabolic activity, particularly because of the large surface area of the apatite crystals in contact with extracelluar fluid. For example, Pb and Cd (both toxic elements) can accumulate in bone (Driessens and Verbeeck 1990), as well as a range of bone-seeking radioactive isotopes (Neuman and Neuman 1958). F^- ions also accumulate in bone mineral (Driessens and Verbeeck 1990) and even a modest F content in the drinking water supply can lead to a substantial amount of F in the skeleton during a life-time (Wix and Mohamedally 1980). In patients with uraemia and acidosis, there is a significant reduction in the Ca and CO_3 in cortical bone that is proportional to the severity and duration of the condition (Pellegrino and Biltz 1965). The Ca/P ratio in very young bone mineral has a Ca/P mole ratio significantly less than 1.67, the value for stoichiometric HAP, whilst the acid phosphate content is higher (Glimcher 1998). A similar high acid phosphate content was noted above for immature enamel.

In studying changes in bone mineral with age, it is the age of the mineral that counts (Glimcher 1998), rather than the chronological age of the bone, the reason being the remodelling of bone noted earlier that usually (but not always) takes place. The difference between the age of the crystals and age of the bone is reflected in a microscopic heterogeneity of bone. Thus in order to produce samples in which the mineral crystals are of reasonably similar ages, finely ground samples of bone can be density fractionated (Richelle 1964, see Bonar et al. 1983 for other references).

The reduction in dental caries as a result of fluoridation of water supplies and use of fluoride containing toothpastes is well known (Shellis and Duckworth 1994). This is clearly linked in part to the fact that the solubility product of fluorapatite is less than that of HAP (Moreno et al. 1977). For a partial replacement of OH^- by F^- ions, the solubility product for $Ca_5(PO_4)_3(OH)_{1-x}F_x$ is a minimum for $x = 0.56$ (Moreno et al. 1977). Chow and Eanes (2001) and LeGeros (1991) discuss further the effects of F^- ions on the solubility, rate of dissolution and formation of apatites. Fluoride has also been used in attempts to rebuild bone lost as a result of osteoporosis (see Grynpas and Cheng 1988 and Baud et al. 1988 for references). Fluoride has effects on both bone mineral and cellular activity (Baylink et al. 1970, Eanes and Reddi 1979). For example, F^- ions reduce the rate of dissolution of the mineral in acidic buffers (Grynpas and Cheng 1988). Other effects on the mineral will be mentioned later.

A number of features of normal biomineralization have already been mentioned, namely that it is very well controlled, the crystals are usually highly oriented, and the process in enamel is different from that in bone and dentin. The orientation of biominerals is usually determined with respect to an organic protein matrix, so the way in which macromolecules interact with biominerals is of great interest (see for example the ^{13}C Nuclear Magnetic Resonance (NMR) study (Shaw et al. 2000) of dynamics in a hydrated salivary peptide adsorbed to HAP). There must also be specific mechanisms that initiate the nucleation of biominerals and terminate their growth as necessary, processes that are likely to involve interaction of the mineral with organic macromolecules and the control of local ion concentrations. Although apatites are the main subject of this review, many other inorganic crystals occur as biominerals. Study of biominerals and biomineralization is currently a very active area of research. Several excellent reviews (Mann et al. 1989, Lowenstam and Weiner 1989, Simmer and Fincham 1995, Mann 1996, 2001; Fincham et al. 2000) and a conference proceeding (Li et al. 2000) have been published.

CRYSTAL SIZE, SHAPE, ORIENTATION, AND LATTICE PERFECTION

In this section, we consider the various characteristics of the mineral crystals above atomic dimensions, particularly their sizes and shapes (habits). The relationship between the crystallographic axes and other microscopic structures (preferred orientation), particularly proteins that might be involved in biomineralization are also considered. The perfection of the apatite lattice is discussed as this affects chemical reactivity. Many of the above characteristics will also have significant effects on the mechanical properties of the mineralized tissues.

There are a number of experimental methods, mainly based on XRD and electron microscopy (EM) that can be used to explore the character of crystals. However, different techniques, although nominally measuring the same parameter, measure different things. Further, differences between techniques can be accentuated by characteristics of the crystals, for example, very anisotropic dimensions. For these reasons, the main experimental methods are reviewed prior to discussing results. This section concludes with a summary.

Wide-angle diffraction: Theory

The diffraction maxima in the wide-angle diffraction pattern derive from the periodicity of the crystal lattice (Cullity and Stock 2001). Their angular pattern is characteristic of the crystal lattice, so is ideal for identification and detection purposes. For monochromatic radiation (X-rays, neutrons or electrons) of wavelength λ, Bragg's Law gives

$$n\lambda = 2d \sin\theta_B$$

where n is the order of diffraction, d the interplanar spacing, and θ_B the Bragg diffraction angle. Accurate measurement of the θ_B from the diffraction pattern allows the unit cell dimensions to be determined using a least-squares fit of preferably 10 or more diffraction maxima (lines). However, more accurate cell parameters can often be derived by Rietveld whole-pattern fitting methods (Young 1993) which are usually directed at the determination of unit cell structure (see later). The intensity distributions of the maxima about the incident beam allow the preferred orientation of the crystals to be determined (Cullity and Stock 2001).

The widths of the diffraction maxima give information about the size and perfection (e.g., microstrain) of the crystals (Cullity and Stock 2001, Snyder et al. 1999). Assuming perfect crystals apart from a small size, the Scherrer formula is,

$$t = 0.9\, \lambda / (B \cos\theta_B)$$

where t is the size of the crystal in the direction normal to the diffracting planes and B the full-width at half maximum (FWHM). B is expressed in terms of 2θ in units of radians. t is sometimes referred to as the coherence length of the crystal, that is, the length over which the planes diffract essentially in phase. The multiplier is usually taken as 0.9 as in the above equation, though the exact value is dependent on the shape. As an example, taking the smallest dimension of enamel crystals (200 Å), λ = 1.5417 Å (Cu K_α) and θ = 16° (corresponding to the strongest peak), the calculated value of B is 0.41° in 2θ, which is easily measured.

If a crystal is bent or otherwise mechanically distorted, the lattice spacing d of the diffracting planes will vary over the crystal. If the range in d is Δd, and the resultant spread in the positions of the diffraction maxima is $\Delta 2\theta$, then by differentiating Bragg's Law

$$\Delta 2\theta = -2 \tan\theta\ \Delta d/d$$

This equation allows the spread in microstrain, $\Delta d/d$, to be calculated from the observed broadening assuming that size broadening is absent.

In practice in biological apatites, it is most likely that both size- and microstrain-broadening are present. Inhomogeneities in the chemical composition of the crystals may also cause broadening of maxima, particularly those maxima whose positions are mainly dependent on the *a* parameter, as this parameters usually changes more than *c* as a result of lattice substitutions in apatites. Little is known about the magnitude of such effects in biological apatites and they will not be discussed further. Size and microstrain broadening can be separated in principle if the broadening is measured over a wide angular range in 2θ as they have different angular dependencies (see equations above). Various methods have been developed for this using line widths measured directly from the diffractometer output (Snyder et al. 1999). However, Rietveld whole pattern fitting methods have been recently developed in which the size and microstrain are determined by modelling them in the calculated diffraction pattern that is fitted to the observed pattern (Snyder et al. 1999). Given the difficulties of separating size and microstrain effects, this is rarely attempted for the poorly crystallized biological apatites, though examples are given below. If the contribution of microstrain to line broadening is neglected, and the whole broadening attributed to size effects, the size calculated from the Scherrer formula will be smaller than the physical dimension of the mineral crystals observed in the EM.

Differences in the broadening (assumed to be entirely from size effects) of the diffraction lines originating from different lattice planes give information about differences in the dimensions of the crystal in directions normal to the planes, and hence about the habit of the crystals. However, if crystals have long and short dimensions for planes with the same or nearly the same *d* spacing, the result can be deceptive. The reason is that the diffraction maxima from these planes will coincide, but the experimentally measured line width for such maxima with multiple components will be dominated by the contribution from the smallest crystal dimension provided this dimension is very short. The result is that it appears erroneously that all the crystal dimensions contributing to the maximum are small. Such a coincidence of diffraction maxima with different widths is generally unlikely, but seems to apply to the apatite crystals in bone, which resulted in confusion about whether they are needles or plates (see line-broadening section under *Wide-angle scattering studies of bone*, below).

Wide-angle scattering studies of enamel

It has been known for many years (see Miles 1967) that the lattice parameters of dental enamel, particularly *a*, differ significantly from those of stoichiometric HAP. For example, for human dental enamel, *a* = 9.441(2) Å and *c* = 6.878(1) Å for a fraction with density > 2.95 g cm^{-3} (Young and Mackie 1980), and *a* = 9.4555(76) Å and *c* = 6.8809(47) Å (Wilson et al. 1999) for pooled enamel from permanent teeth have been reported. Both sets of measurements used Rietveld methods applied to XRD data. These values compare with *a* = 9.4243(55) Å and *c* = 6.8856(35) Å (Morgan et al. 2000) for stoichiometric HAP, also determined from Rietveld analysis. Thus the *a*-axis parameter for dental enamel is about 0.3% larger than for HAP.

X-ray diffraction patterns from enamel sections cut in the longitudinal direction of a tooth show that the *c*-axes (the hexagonal axes) of the apatite crystals (also their long direction) are highly oriented in the direction of the enamel rods (sometimes called prisms) (Miles 1967). The rods run from the enamel surface (to which they are nearly normal) to the enamel-dentin junction and are about 4-7 μm in diameter.

Glas and Omnell (1960) made use of the high degree of preferred orientation of the apatite crystals in hippopotamus enamel to measure crystal dimensions over several orders of diffraction. They found that the crystal dimension in the *a*-axis direction (the width of the crystals) from the Scherrer formula (410 Å, approximate precision limits 370 and 450 Å) was independent of the order of diffraction. However, dimensions in the *c*-axis direction (the length of the crystals) calculated from the 002, 004, 006, and 008 peaks were 741±100, 466±40, 361±35 and 295±25 Å respectively. After correction for microstrain effects, the size in the *c*-axis direction was 1600 Å (approximate precision limits 1040 and 2700 Å). This example gives some idea of the effect of neglect of microstrain effects on the calculation of mineral crystal dimensions. Note that, even if corrections are made for microstrain, the dimension is the coherence length for diffraction which may be much less than the dimension observed in the EM. For example, the coherence length in the direction of the length of a long enamel crystal will be much less that the length seen in the EM, particularly if the crystal is bent. Moriwaki et al. (1976) have also attempted to separate size and microstrain effects for the 00*l* and *hk*0 XRD peaks from powder samples of human enamel. Again the crystals in the *c*-axis direction were larger than in the *a*-axis direction (920 and 600 Å respectively in one case) with evidence of microstrain in both directions. Aoba et al. (1981) showed that permanent human enamel is slightly better crystallized than deciduous enamel.

Lattice parameters calculated from wide-angle XRD patterns of human dental enamel showed no measurable shift when the F content increased from 70 to 670 parts per million (Frazier 1967). However, the 002, 211, 300 and 202 peaks became slightly sharper. The Scherrer formula showed increases in size from 1420±100 to 2740±240, 780±25 to 1030±25 and from 780±20 to 1000±20 Å from the 002, 300 and 211 peaks respectively. In the absence of further study, it is not possible to say whether these changes were due to a real increase in size or to a reduction in microstrain.

Wide-angle scattering studies of bone

The broad diffraction lines of the mineral in bone make accurate measurement of lattice parameters problematical, hence such measurements are rarely attempted. Nevertheless, Handschin and Stern (1992) measured the lattice parameters of human iliac crest samples of 87 individuals aged 0-90 years. Determinations were based on a weighted least-squares analysis of the positions of 6 lines (002, 102, 310, 222, 213, 004) and gave average statistical errors of ±0.002 Å in *a* and ±0.003 Å in *c*. The parameters showed a slight reduction with age (*a* by 0.00015 Å and *c* by 0.00005 Å, both per year). Chemical analyses of such samples have been reported (Handschin and Stern 1994 and 1995). It may be that the reduction in *a*-axis is in part caused by the increase in F content with age noted earlier (Wix and Mohamedally 1980).

It is well known (Elliott 1994) from studies on model systems that HAP is not usually precipitated directly, but is more likely to be formed via an unstable intermediate of octacalcium phosphate (OCP) or amorphous calcium phosphate (ACP). There is therefore considerable interest in whether such phases occur in bone.

Brown (1966) originally made the suggestion that OCP was a precursor of biological apatites. The principal support (Brown 1966, Brown et al. 1987; see also Chow and Eanes 2001) for this concept derives from the following: (1) the close structural similarity of OCP and HAP; (2) formation of interlayered "single crystals" of OCP and HAP (pseudomorphs of OCP); (3) the easier precipitation of OCP compared with HAP; (4) the apparent plate- or lath-like habit of biological apatites (see later) that does not conform to hexagonal symmetry, but looks like a pseudomorph of triclinic OCP; and (5) the presence of HPO_4^{2-} in bone mineral, particularly newly formed mineral. However, despite careful XRD and spectroscopic studies of the initially formed mineral, no direct evidence for the presence of OCP has been found (Rey et al. 1995a, Glimcher 1998). Likewise, crystalline brushite could not be detected in embryonic chick or bovine bone by XRD at a level greater than 1% (Bonar et al. 1984), although earlier work had suggested its presence (Roufosse et al. 1979). Further studies of the HPO_4^{2-} ion environment in freshly deposited bone mineral have been made with ^{31}P NMR spectroscopy (see *Spectroscopy Studies* section).

Studies on the possible presence of ACP in bone have been reviewed (Glimcher et al. 1981, Grynpas et al. 1984, Glimcher 1998). The original postulate (Termine and Posner 1967; see also Posner and Betts 1975 and Glimcher et al. 1981 for other references) was that the first mineral in bone was ACP and that the observed improvement in crystallinity with the age of the bone mineral (see later discussion on line broadening) was a result of a progressive reduction in the ACP content. Young bone was thought to contain 50% or more ACP and even mature bone 25-30% based on the observation that the observed XRD intensity seemed to be less than expected for crystalline HAP. However, it subsequently became clear that the apparent amount of ACP in bone was considerably less if the observed intensity was compared with an apatite that more closely matched the characteristics of bone mineral (crystal size/microstrain and CO_3 content) (Termine et al. 1973, Aoba et al. 1980). Posner and Betts (1975) and Grynpas et al. (1984) used XRD radial distribution function techniques to show that, if there were ACP in bone, it constituted less than 10% of the mineral.

It has been known for a long time that the *c*-axes (hexagonal axes) of the apatite crystals in bone are strongly oriented in the direction of the collagen fibers (see Fernández-Morán and Engström 1957 for references). For long bones, this will also be the morphological axis of the bone. More recently, Sasaki and Sudoh (1997) have studied the detailed relation between the apatite crystals and the helical axis of collagen fibers in bovine femurs by comparing pole figures for the (002) plane (*c*-axis) and {210} plane of the apatite crystals with pole figures for the collagen helical axis direction (the 0.3 nm periodicity). The collagen pole figure was measured on the original samples after demineralization in ethylene diamine tetraacetic acid (pH 8). As expected, the apatite *c*-axis and the collagen helical axis had a strong preferred orientation in the direction of the anatomical long axis of the bone, though the distribution was wider for the collagen. However, the apatite *c*-axis also had an appreciable preferred orientation in the radial and tangential directions to the long axis of the bone, which was hardly seen for the collagen. This suggested that there were two types of apatite crystal (with different morphologies, see later), the predominant one with the *c*-axis parallel to the long axis of the bone (the collagen axis) and another with the *c*-axis perpendicular to this direction. The {210} pole figures showed appreciable preferred orientation in the radial and tangential directions to the long axis of the bone. Sasaki and Sudoh commented that in order for {210} planes to have this preferred orientation, the set of mineral particles with their *c*-axes perpendicular to the long axis of the bone would have to have an anisotropic cross section when viewed down their *c*-axes, say rectangular rather than a regular hexagon. This observation was consistent with a plate-, rather than a needle-like morphology (Sasaki and Sudoh 1997).

A synchrotron X-ray source has also been used to quantitate the preferred orientation of the crystals in calcified turkey tendon, and in trabecular and cortical regions of osteonal bone in the bovine metacarpus (Wenk and Heidelbach 1999). Although the observed preferred orientation was as expected, the synchrotron source allowed much improved quantitation and use of an X-ray beam of diameter 2 or 10 μm. As regards diffraction experiments, neutrons have a reduced absorption compared with X-rays, so can be used to study the preferred orientation of apatite crystals in much larger samples of bone, several millimeters thick (Bacon et al. 1979).

Line broadening. Ziv and Weiner (1994) and Glimcher (1998) have reviewed studies of the size and shape of the apatite crystals in bone. One of the controversial questions is whether the crystals are needles or plates. For the needles, the *c*-axis (hexagonal) direction would be parallel to the long direction of the needles, but for the plates, the hexagonal *c*-axis would lie in the plane of the plates but oriented to coincide with the plate's largest dimension. Measurements by direct observation in the EM will be reviewed below, but, as described previously, line broadening of diffraction peaks can also give this information and will now be considered. In one of the earliest such studies, Fernández-Morán and Engström (1957) commented: "…since in X-ray diagrams the 00*l*-reflections are less broadened than the *h*00-lines, it is clear that the crystallites are elongated with their *c*-axis parallel with the long axis of the crystallites," and later in the same paper: "Thus the evidence derived from electron microscopy conforms with the results of polarized light and X-ray diffraction studies and lends support to the view that we are dealing with rod- or needle-shaped crystallites, 30-60 Å in diameter and with an average length of about 200 Å." As discussed in the theoretical section above, the fact that the 00*l* lines are much sharper than *h*00 lines is most likely to mean the crystals are needles elongated in the *c*-axis direction, but this is not absolutely necessary. Such an observation would also be consistent with the plates described above in which the *c*-axes have hexagonal symmetry even though they are in the plane of (very thin) plates. Such plates could originate as pseudomorphs of triclinic OCP, which has a platy habit. These OCP plates can hydrolyse to hexagonal HAP with the apatite hexagonal *c*-axis in the plane of the plate (Chow and Eanes 2001).

Lundy and Eanes (1973) reported lengths in the *c*-axis direction of 200-400 Å and microstrains of 3 to 4×10^{-3} determined from line broadening of the 002 and 004 apatite reflections from mineralized turkey leg tendon. They also tabulated earlier measurements of the dimensions of biological apatites in the *c*-axis direction. Wide-angle XRD also shows that older bone mineral is better crystallized than newly formed mineral. For example, Bonar et al. (1983) measured the improvement in crystallinity for bone from embryonic to 2-year-old chickens. They reported that the coherence length (neglecting microstrain) calculated from the Scherrer formula from the 002 peak increased from 107 to 199 Å. In a similar study of bone from 3-week to 2-year-old rats (Ziv and Weiner 1994), the change in the same quantity was from approximately 200 to 280 Å (means of 3 measurements).

Fluoride. As with dental enamel (above), an increase in the F content of bone mineral is associated with an improvement in crystal size/reduction in microstrain. The effect however seems to be restricted to directions perpendicular to the *c*-axis (Posner et al. 1963). In a study of human bone with increasing F contents, the mean crystal size remained at 96±9.6 Å in the *c*-axis direction, but the crystallinity improved perpendicular to this direction (Posner et al. 1963). The line profiles were analyzed by Fourier methods, so the dimension quoted above is not directly comparable to the value obtained from the Scherrer formula (this gave 192Å). Samples with higher F contents showed a reduction in the *a*-axis parameter, indicating that OH^- ions were partially replaced by F^- ions in the lattice (Posner et al. 1963). Similar results on crystal size and lattice parameters have been reported in humans treated with fluoride for osteoporosis (Baud et al. 1988). Turner et al. (1997) also demonstrated an increase in crystal size (uncorrected for microstrain) from 61.2±0.4 to 66.2±0.9 Å in a direction perpendicular to the *c*-axis when the F content in bone increased from 0.1 to 0.7 wt % in rabbits given a dietary fluoride supplement, whereas again the crystal dimension in the *c*-axis direction was little affected. In laboratory studies, Eanes and Hailer (1998) showed that the effect of F^- ions on HAP crystal growth was primarily restricted to the width (perpendicular to the *c*-axis), with little effect on the length (parallel to the *c*-axis). They concluded that the effect of F^- ions seen in vivo is of physicochemical, rather than cellular, origin.

Small-angle scattering: Theory

Scattering close to the X-ray beam, typically at angles less than 5° 2θ, is described as small-angle X-ray scattering (SAXS). As with wide-angle scattering, monochromatic radiation is used. The equivalent phenomenon with neutrons is SANS (small-angle neutron scattering). Periodic structures with long spacings, say greater than 50 Å, are one source of low-angle scattering. This is directly equivalent to the wide-angle scattering described above. A good example is given by the many meridional maxima with ~640 Å spacing (in the dry state) given by collagen (see below). However, it is not necessary to have periodicities to get low-angle scattering. In fact, many SAXS studies are of particles without any periodic structure. Low-angle scattering will occur if the material in the beam has discontinuities in electron density in the dimensional range 1-100 nm. These discontinuities can either be due to the external boundary of particles, i.e., the particle size is in the range 1-100 nm, or originate from internal inhomogeneities in this dimensional range.

The curve of scattering intensity, I, against scattering angle, 2θ, is conveniently divided into the Guinier region, near the direct beam, and the Porod region further out (Cullity and Stock 2001). Regardless of the presence or absence of periodicities, these two regions give information about the size (e.g., the radius of gyration, R_{gs}) and total surface area of the scatterers in the beam, respectively. The radius of gyration can be estimated from a Guinier plot of $\log_e I$ against $16\pi^2\theta^2/\lambda^2$ as the slope is $-(R_{gs})^2$. If there is a range of particle sizes, an average radius of gyration is obtained which is heavily weighted towards particles with larger sizes (Cullity and Stock 2001). Guinier plots generally do not have a linear region very close to $2\theta = 0$, so the closest linear region is chosen for measurement of the slope. The derivation of the equation for R_{gs} assumes that interparticle interference effects can be neglected, a condition not likely to be fully achieved for mineralized tissue.

Small-angle scattering studies

Small-angle scattering, like wide-angle scattering, also gives detailed information about the relation between the mineral and collagen in mineralized tissues, but in this case, it can be advantageous to use both X-rays and neutrons (White et al. 1977, Berthet-Columinas et al. 1979). The meridional XRD pattern of collagen shows many orders of the 670 Å (D) spacing, which originate from the axial quarter stagger of the tropo-collagen molecules, which forms the overlap, and gap regions. In XRD patterns, the D periodicity is substantially enhanced if the collagen is mineralized. Consideration of the diffracted intensities shows that the mineral is arranged with a periodicity of D in the form of a step function along the collagen with a step length of about 0.5 D (about the length of the gap region). Observation of the neutron diffraction pattern in different D_2O/H_2O concentrations allows the neutron diffraction from the mineral and collagen to be measured separately. The reason for this is that, unlike X-rays, the scattering of the mineral and collagen are of comparable magnitude with neutrons (White et al. 1977). Further, the scattering length of deuterium and hydrogen are of opposite sign, so that the scattering of the water background can be varied by changing the D_2O/H_2O ratio to match either the mineral or collagen. If it matches the mineral, the collagen pattern is seen and vice versa. The X-ray and neutron results together showed that the mineral steps coincide with the collagen gaps and therefore that the mineral occupies the gap regions (White et al. 1977, Berthet-Columinas et al. 1979).

The equatorial neutron diffraction pattern also gives information about the site of the mineral in bone (Lees et al. 1984, Bonar et al. 1985). An equatorial spacing of 1.24 nm in wet mineralized bovine bone and of 1.53 nm after demineralization was observed. The corresponding figures for dry bone and demineralized bone (i.e., collagen) were 1.16 and 1.12 nm respectively. These spacings are determined by the radial packing of the collagen molecules and showed that the collagen in fully mineralized bone was considerably more closely packed than had been thought. From these figures, it was possible to calculate the collagen intermolecular spacing and hence estimate the volume available between the collagen fibers and the volume of the gap regions within the fibers. The conclusion was that the major fraction of the mineral must be outside the collagen fibers, a result that appears to be at variance with studies of the meridional pattern outlined above. However, Lees and Prostak (1988) pointed out that there is probably no inconsistency as the crystals, even when outside the fibrils, seem to be in phase with the gap regions in the fiber axis direction and would therefore enhance the *D* periodicity.

Early SAXS studies of bone were interpreted as showing that the crystals in bone were elongated in the *c*-axis (hexagonal) direction, which was aligned parallel to the collagen fiber direction (Engström and Finean 1953). The particles were estimated to be 60-80 Å wide and the long dimension about three times the width. It was also shown from SAXS studies of the Porod region that a rise in the F content of human bone was accompanied by a decrease in the area of the interface between the mineral apatite phase and the remaining bone tissue (Eanes et al. 1965). Detailed analysis suggested this was attributable to an increase in the mean crystal volume and concomitant decrease in the number of apatite crystals (Eanes et al. 1965). More recent SAXS studies of the size, shape and predominant orientation of the bone crystals have been published (Fratzl et al. 1991, 1996; Rinnerthaler et al. 1999). The scattering followed the Porod law, so it was possible to calculate the ratio of surface to volume of the crystals without making any assumption about their shape. This allowed a parameter characteristic of the smallest crystal dimension to be computed. As expected, this parameter increased in going from very young to older bone (Fratzl et al. 1991). Theoretical scattering curves gave a good fit to experimental data if the bone crystals were assumed to be needles (Fratzl et al. 1991). However, Fratzl et al. (1996) later reported that, at least for some bone samples, the fit was not as good as expected, possibly because they were not isotropic about their long axis or because of inter-particle interference effects. Subsequently, Rinnerthaler et al. (1999) reported that SAXS patterns from human bone were consistent with a plate-like morphology of the crystals. The work reported in 1999 used a 2-D scanning system that allowed maps of SAXS patterns to be obtained from bone with a probe diameter of 200 μm. A scanning system was also used in a very recent SAXS study of sections of dentin using a synchrotron X-ray source (Kinney et al. 2001). The beam dimensions were 0.3×0.35 mm^2 and the section was scanned from the pulp to dentin-enamel junction (DEJ) in 0.5 mm steps. Analysis of the data indicated that the crystals were approximately needle-like and became more plate-like near the DEJ, but with an invariant thickness of ~5 nm. However, Kinney et al. (2001) cautioned that, in this context, "needles" and "plates" indicated the particles were roughly one-dimensional or two-dimensional because of the lack of a unique solution for the inverse of the scattering profiles. The above discussion of SAXS experiments to determine crystal shapes clearly shows the ambiguity that can occur in the detailed interpretation of SAXS patterns.

Electron microscopy, including atomic force microscopy

Enamel. There have been many EM studies of the apatite crystals in dental enamel (Nylen 1964, Daculsi and Kerebel 1978, Daculsi et al. 1984, Cuisinier et al. 1992). In the early stages of enamel formation (secretory stage of amelogenesis), long tape- or ribbon-like crystals can be seen. For example, lengths of at least 100 μm for crystals from embryonic human enamel have been reported (Daculsi et al. 1984). In cross-section, they are flattened hexagons. Electron diffraction shows that the Miller indices of the largest face of the hexagon is (100) so the *b*-axis (crystallographically equivalent to the *a*-axis) lies in the plane of the largest face, whilst the *c*-axis (hexagonal axis) is parallel to the length of the crystal (Elliott and Fearnhead 1961). Typical dimensions for human fetal enamel are: width (*b*-axis direction) 2.3-160 nm (mean 70±39 nm), thickness 2.4-12 (mean 7±3 nm) and length (*c*-axis direction) 40-100 nm (mean 74±34 nm) with nearly half of the crystals being bent (Cuisinier et al. 1992). The crystals grow rapidly in width and slowly in thickness after their initial formation (Daculsi and Kerebel 1978). After this, there is a much slower process in which they grow only in thickness up to their mature dimensions. The mature crystals still have a flattened hexagon cross-section. Daculsi and Kerebel (1978) reported an average thickness of 263±22 Å and width 683±134 Å.

Lattice periodicities have frequently been seen in high-resolution EM images of enamel crystals and synthetic analogues (Nelson et al. 1986, 1989; Daculsi et al. 1989, Iijima et al. 1992, Cuisinier et al. 1992, Brès et al. 1993b, Aoba 1996, Reyes-Gasga et al. 1997 and references therein). Good correlation has been reported between the atomic detail in calculated images based on the HAP structure and the images actually seen (Nelson et al. 1986, Brès et al. 1993b). Various lattice defects can be seen in images of fetal human enamel crystals, including screw dislocations and high- and low-angle boundaries (Cuisinier et al. 1992). The preferential acid dissolution of the centers of enamel crystals seen in dental caries probably originates from a screw dislocation running down the length (*c*-axis direction) of the crystal (Daculsi et al. 1989 and references therein). In images of the cross-section of early enamel crystals (i.e., cut perpendicular to the *c*-axis direction), a dark line of width 0.8 to 1.5 nm that bisects the crystal in the long direction of the flattened hexagon cross-section can often be seen. This line (often called the central dark line) is also sometimes seen running down the length of the crystals. Although originally reported in enamel crystals (Nylen 1964), it is sometimes seen in synthetic precipitated carbonate-containing apatites (CO_3Aps) (Nelson et al. 1986, 1989). Nelson et al. (1986) calculated high-resolution EM images for four different defect models for the central dark line for comparison with the images actually observed. Typical images were consistent with a two-dimensional OCP inclusion, one unit cell thick, embedded in an apatite matrix. In studies by Iijima et al. (1996) in which the F concentration in the precipitating solution was incremented from 0.1 to 1.0 parts per million, the F content and the amount of apatite in the crystalline product increased, whereas the amount of OCP and total amount of product decreased. Models based on differential uptake and effects on growth rate between HAP and the sandwiched OCP were proposed to explain these and other results (Iijima et al. 1996).

Ordering of the OH^- ions that leads to the monoclinic space group $P2_1/b$ in pure HAP (Elliott et al. 1973) might also occur in dental enamel. This could lead to a symmetry lower than $P6_3/m$, the space group found in slightly nonstoichiometric HAP, such as that from Holly Springs (Sudarsanan and Young 1969). High-resolution TEM studies (Brès et al. 1993b) suggested a loss of hexagonal symmetry in human enamel, but this was not confirmed using convergent-beam electron diffraction (Brès et al. 1993a). The matter is clearly not settled.

Bone. As mentioned in the *Line broadening* section under *Wide-angle scattering studies of bone* (above), Ziv and Weiner (1994) and Glimcher (1998) commented that there was controversy in the literature as to whether the crystals in bone were needles or plates. Early EM studies of bone crystals indicated that they had a plate-like habit (Robinson and Watson 1952), but Fernández-Morán and Engström (1957) subsequently reported that EM studies supported the view based primarily on wide and small-angle XRD, that bone crystals had a needle habit with elongation in the *c*-axis direction (see quotation in *Line broadening* section). However, studies shortly afterwards in which the specimen was rotated showed that the supposed needles were in fact plates seen edge on (Johansen and Parks 1960). This interpretation was subsequently confirmed by others (e.g., Landis et al. 1977). Nevertheless, details of the exact shapes and particularly the dimensions of bone crystals are still missing (Ziv and Weiner 1994). However, Landis et al. (1993) have provided further significant clarity in the understanding the 3-dimensional relationship between the crystals and collagen, and the shape of the crystals. They used tomographic reconstructions of high voltage (1 MV) EM images of 0.5 μm sections of mineralizing turkey leg tendon taken over ±60° at intervals of 2°. Reconstructions showed irregularly shaped plates generally parallel to each other directed with their long dimension (the *c*-axis direction) along the collagen fiber axis. The faces of the plates were (100) planes with the *b*-axis in the plane of the plate. The length was variable (~40 nm, but some crystals in the range 40 to 170 nm), width 30-45 nm and thickness uniform (~4-6 nm, at the limit of resolution). More crystals (those with length ~40 nm) were located in the collagen hole regions than overlap regions at the earliest stages of mineralization. Ziv and Weiner (1994) measured the length and widths of the crystals in rat bone crystals of different ages by transmission electron microscopy. Measurements on well-separated crystals from rats aged 3 weeks, 5 months and 2 years were (length × width, in Å): 298±123 × 169±54 (70), 311±128 × 178±64 (129) and 366±168 × 217±82 (119), respectively, where the error is the standard deviation and the number of crystals observed is given in parentheses. Lengths from the Scherrer formula for two of the same samples have been given in the earlier XRD line broadening section on bone crystals.

Very recently, atomic force microscopy (AFM) has been used to investigate the size and shape of isolated bone crystals from mature bovine bone (Eppell et al. 2001). These authors reported that their results were largely consistent with earlier EM studies. Approximately 98% of the crystals had a plate-like habit with thickness less than 2 nm. Both thickness and width distributions had single peaks, with means of 0.61±0.19 nm and 10±3 nm respectively. The length distribution was wider with a mean of 15±5 nm taken over all lengths of the 98% subset and appeared to be multi-modal with peaks separated by ~6 nm. Thus the conclusion was that the crystals had mean dimensions of 0.61 × 10 × 12 nm (thickness × width × length), with the length and width parallel to the *c*- and *a*-axes respectively, and a thickness of about one unit cell. It was suggested that the extreme thinness of the crystals might be associated with their location within the collagen fibrils. The remaining 2% of the crystals were generally a few tens of a nanometer thick (range 12-115 nm, mean 37 nm). The width and length ranges were 27-172 nm (mean 64 nm) and 43-226 nm (mean 90 nm), respectively. These dimensions precluded location of the larger crystals within the collagen fibers without complete disruption of the fiber.

Summary

As outlined above, there has been controversy about whether bone crystals are needles (elongated in the hexagonal *c*-axis direction) or plates (*c*-axis in the plane of plate and lying in the direction of the plate's longest dimension). The obvious interpretation of wide and small-angle X-ray scattering experiments is that they are needles, but there is in fact no inconsistency between the results of these scattering experiments and the thin plates clearly seen in the EM, particularly when the specimen is tilted either for direct viewing or computed tomography.

Enamel crystals have a flattened hexagonal cross-section, and are very long in the *c*-axis direction. On the other hand, crystals associated with collagen in bone and tendon have platy habits with the *c*-axis in the plane of the plate, which is again extended in the *c*-axis direction. Both habits deviate from the hexagonal cross-section expected for a crystal belonging to the hexagonal system of apatite, though the deviation is much more extreme for bone crystals. As discussed above, the explanation may be that the first formed mineral is OCP (or OCP-like) which has a platy or tape-like habit which then converts to pseudomorphs of OCP, or alternatively, that the OCP nucleates the crystal growth of HAP in such a way that, in further growth, it maintains the platy or tape-like habit of OCP. The presence of the central dark line in enamel crystals is indicative of a role for OCP, but definitive proof for this role in either enamel or bone crystals is frustratingly absent.

Electron microscopy and AFM studies give detailed, and generally consistent, information about the shape and dimensions (also orientation and lattice imperfections for EM) of individual crystals, though care must always be taken to ensure that specimen preparation does not change the mineral. X-ray (and neutron) scattering methods look at much larger volumes than EM and atomic force microscopy methods. This confers two advantages, the statistical sampling is much better and sample preparation is very much less likely to degrade the sample. The disadvantage is that the shape of crystals cannot be determined unambiguously, nor can the distributions of shapes be determined. However, a number of parameters can be accurately and reproducibly measured, these parameters being dependent on the dimensions, lattice imperfections, surface area and orientation of the crystals. This makes scattering methods ideal for investigating changes in the apatite crystals with age, drug treatment and anatomical location.

Table 4 gives representative results from several experimental methods for the dimensions and shapes of the apatite crystals in enamel and bone.

Table 4. Summary of representative results for dimensions and habit of crystals in enamel and bone.

Tissue	*Method*	*Results*
Enamel (mature hippopotamus)	XRD[a]	1600 Å in *c*-axis direction × 410 Å wide (Glas and Omnell 1960)
Enamel (mature human)	XRD[a]	920 Å in *c*-axis direction × 600Å wide (Moriwaki et al. 1976)
Enamel (mature human)	EM	Flattened hexagons 263±22 Å thick, 683±134 Å wide (Daculsi & Kerebel 1978)
Enamel (various fetal)	EM	At least 100 μm long in *c*-axis direction (Daculsi et al. 1984)
Enamel (fetal human)	EM	7×70×74 nm for mean thickness, width and length respectively (Cuisinier et al. 1992)
Bone (various)	XRD[b]	Needles 280 Å long in *c*-axis direction aligned with collagen fiber axis × 30-60 Å wide (Fernández-Morán & Engström 1957)
Bone (rat, 3 weeks and 2 years)	XRD[b]	Coherence lengths 200 and 280 Å (means of 3 determinations), respectively, in *c*-axis direction. Lengths perpendicular to *c*-axis not measurable (Ziv & Weiner 1994)
Mineralized leg tendon (turkey)	EM[c]	Irregularly edged plates, generally parallel in sheets. Lengths ~4-170 nm (in *c*-axis direction which is generally parallel to collagen axis). Width 30-45 nm (in *b*-axis direction which is perpendicular to collagen axis) and thickness ~4-6 nm. Crystals (~40 nm long) preferentially in hole zones, but may grow beyond (Landis et al. 1993)
Bone (mature cortical bovine)	AFM	Plates 15±5 nm long in *c*-axis direction with multi-modal distribution with peaks separated by ~6 nm. Width (in *a*-axis direction, crystallographically equivalent to the *b*-axis) 10±3 nm and thickness 0.61 nm (Eppell et al. 2001)

[a] wide angle, corrected for microstrain [b] wide angle, uncorrected for microstrain
[c] 3-D tomographic reconstructions

SPECTROSCOPIC STUDIES

Single-crystal XRD is the usual, and preferable, method of determining structure at the atomic level in solids. Unfortunately biological apatite crystals are too small for such studies, though Rietveld methods applied to powder diffraction data are being increasingly used to give structural information (see later). However, spectroscopic methods, such as infrared (IR), Raman (R) and NMR spectroscopy, which do not require a well-ordered crystal structure, can give detailed information about specific aspects of the local atomic structure. As a result, spectroscopic methods have been the main source of detailed structural information for biological apatites. Infrared and—to a lesser extent—Raman spectroscopy have been particularly informative, particularly about the OH^-, CO_3^{2-} and PO_4^{3-} ions (LeGeros 1991, Elliott 1994, 1998). Much of this work has relied on comparisons between synthetic model systems and biological apatites. This will be summarized here (see Elliott 1994 for more details and references), followed by an account of more recent developments. Mainly dental enamel is considered for the following reasons: (1) there are no complications from overlapping collagen bands; (2) enamel spectra are generally better defined by comparison with dentin and bone, presumably because the structure is better ordered; and (3) the good orientation of the crystals in enamel (see above) allows the orientation of the ions with respect to the lattice to be obtained from polarized IR spectra (Fig. 2). Presumably bone and dentin apatite crystals are similar to enamel, except for effects related to their much greater surface areas.

The sharp band at 3570 cm^{-1} (Fig. 2) is assigned to the OH stretch, based on its move to 2610 cm^{-1} on deuteration and the presence of a similar band in synthetic HAP. The

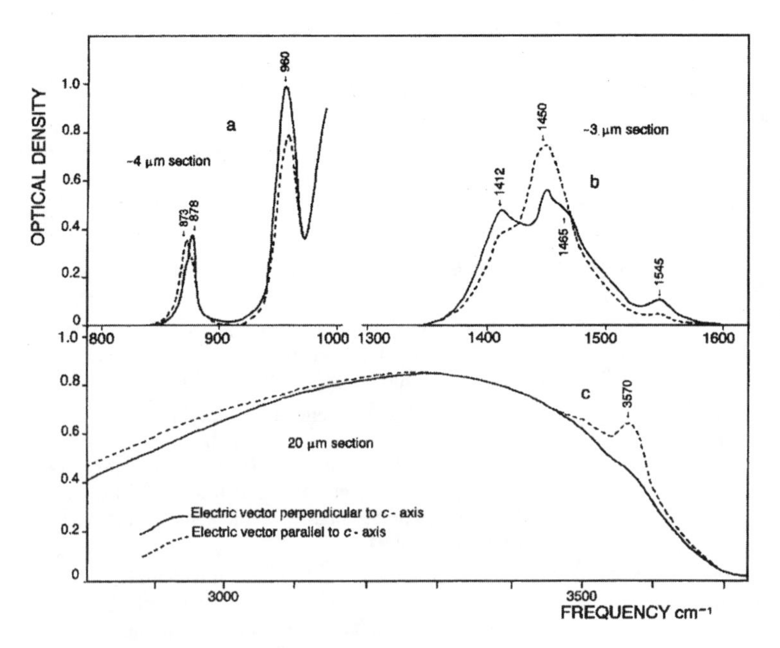

Figure 2. Polarized IR spectra of sections of human dental enamel. Data collected with ×10 magnification Schwarzschild-type reflecting microscope, NaCl prism spectrometer and selenium polarizer (Ford et al. 1958). [Figure 4.7 from Elliott (1994), reprinted with permission from Elsevier Science.]

parallel dichroism (Fig. 2) shows that the proton is located on the *c*-axis, in agreement with single-crystal neutron diffraction studies of Holly Springs HAP (Sudarsanan and Young 1969). The OH stretch is not seen in Fourier transform infrared (FTIR) spectra of the mineral in bone and cartilage at any stage of maturation, nor in the very youngest crystals from enamel (Rey et al. 1995a). This absence of OH^- ions in bone has been confirmed by NMR and neutron inelastic scattering studies (see later). The OH librational mode at 630 cm^{-1} in synthetic HAP is much weaker in enamel (not shown), and is similarly weak in Ca-deficient apatites (Ca-def Aps, see later) and CO_3Aps. The satellite at 3500 cm^{-1} (Fig. 2) is assigned to the OH stretch of an OH^- ion perturbed by a neighboring Cl^- ion on the *c*-axis (note the 0.4 wt % Cl content of enamel, Table 3) as the same perturbation is seen in model systems. This band also shifts on deuteration. The broad band from 3700 to 2700 cm^{-1} and a weaker broad band at 1640 cm^{-1} (not shown) are assigned to the H_2O stretching and bending modes, respectively.

The undistorted CO_3^{2-} ion is planar with three equal symmetrically placed C-O bonds. There are four normal modes, ν_1 at 1063 cm^{-1} (only R active), ν_2 at 879 cm^{-1} (only IR active, transition moment normal to plane of ion), ν_3 at 1415 cm^{-1} (doubly degenerate, R and IR active, transition moment in plane of ion) and ν_4 at 680 cm^{-1} (doubly degenerate, R and IR active, transition moment in plane of ion). ν_4 is the weakest IR band and has not been seen in biological apatites. ν_3 and ν_2 are the strongest and next strongest IR bands respectively (Fig. 2). The fact that two ν_2 peaks were observed (873 and 880 cm^{-1}) for this nondegenerate mode indicated that the CO_3^{2-} ion is in two different environments. The complete assignments (cm^{-1}) that have been made (with dichroism in parentheses) are: 1465 (perpendicular), 1412 (perpendicular) and 873 (parallel) to ν_{3b}, ν_{3a} and ν_2 with the plane approximately perpendicular to the *c*-axis; and 1545 (perpendicular), 1450 (parallel) and 880 (perpendicular) to ν_{3b}, ν_{3a} and ν_2 with the plane approximately parallel to the *c*-axis respectively. Here ν_{3a} is the symmetric mode with transition moment parallel to a C-O bond and ν_{3b} is the antisymmetric mode with its transition mode perpendicular to the same C-O bond. Comparison of the dichroism and band positions with francolite (B-type substitution, CO_3 for PO_4) and A-type CO_3Ap (CO_3 for OH substitution) indicated that, respectively, the first environment was B-type with the plane of the ion approximately occupying the sloping face of a vacated PO_4 tetrahedron, whilst the second was A-type, with its plane approximately parallel to the *c*-axis, as space requirements dictated. The match of frequencies with francolite was not exact, but the match was better with synthetic B-type CO_3Aps that were F-free. The CO_3 band intensities, particularly the weakness of the 1545 cm^{-1} band, indicate that only a small fraction of the CO_3^{2-} ions are in A-sites. Two estimates of the extent of A-type substitution based on measuring or matching band intensities using samples of A-type and B-type CO_3Aps with known CO_3 contents have been made. The two estimates were that $11\pm1\%$ and 15% of the total CO_3 were in A-sites in dental enamel. In bone, the ratio of A- to B-type sites appears not to change with the age of the bone as judged by the intensity ratios of bands at 878 and 871 cm^{-1} assigned to these site respectively (Rey et al. 1989, 1991a).

It was early recognized that the concept of CO_3^{2-} ions in two well-defined environments in dental enamel and synthetic analogues must be an oversimplification because multiple bands were sometimes seen in the ν_3 CO_3 region of synthetic CO_3Aps on the high frequency flank of the peak at 1450 cm^{-1}. There were also inconsistencies in the relative change in intensities of the ν_3 bands in precipitated CO_3Aps with an increasing fraction of A-type substitution, and also in the relative intensities of the bands in the ν_3 region of the polarized IR spectrum of enamel; both observations suggested that a third environment was contributing to the 1450 cm^{-1} peak. There were also difficulties in determining the quantity of A-type substitution in enamel in that the A-type bands in enamel could not be completely matched with those in an external standard.

A weak band at 866 cm^{-1} seen in deconvoluted FTIR spectra of enamel, bone and synthetic analogues has been assigned to CO_3^{2-} ions in a labile environment (Rey et al. 1989). This environment is probably additional to those that might be responsible for the effects noted above. The intensity of this band with respect to other ν_2 bands fell with increasing age of enamel (Rey et al. 1991b) and bone (Rey et al. 1991a).

ν_1 cannot be seen in the IR in biological apatites because of overlap from the strong ν_3 PO_4, but a band at 1105 cm^{-1} and part of a band at ~1070 cm^{-1} in the Raman spectrum of enamel have been assigned to ν_1 from A- and B-type substitutions respectively (Fowler 1977).

The undistorted PO_4^{3-} ion is a regular tetrahedron. There are four normal modes, ν_1 at 938 cm^{-1} (only R active), ν_2 at 420 cm^{-1} (doubly degenerate, only R active), ν_3 at 1017 cm^{-1} (triply degenerate, R and IR active) and ν_4 at 567 cm^{-1} (triply degenerate, R and IR active). ν_3 is seen in biological apatites as a very intense band (see the shoulder of this band in Fig. 2) with a very much weaker overtone at ~2000 cm^{-1}. ν_1 appears as a weak band on the low frequency side of ν_3 (at 960 cm^{-1} in enamel, Fig. 2). This IR-inactive band is seen because the PO_4^{3-} ion is distorted by the apatite structure. ν_4, split into two bands within the range 650 to 500 cm^{-1}, is seen as a strong band in biological and synthetic apatites. The degree of splitting of these two bands has been related to the degree of crystallinity of the apatite (originally to the ACP content at the time when this was thought to be present as a major constituent), and has in the past been used as a measure of crystallinity (Posner et al. 1979).

Many synthetic precipitated Ca-def Aps, as well as biological apatites, contain HPO_4^{2-} ions that contribute several additional bands to the IR spectrum (Berry 1967). The P-(OH) stretch is at 870 cm^{-1}, very close to ν_2 CO_3, which can cause difficulties with their individual study. As regards ν_3, weak bands at 1020 cm^{-1}, 1100 cm^{-1} and 1145 cm^{-1} have been assigned to HPO_4^{2-} ions (Rey et al. 1991c). Two further bands at 1110 and 1125 cm^{-1} disappeared progressively during maturation of synthetic and biological apatites and were assigned to HPO_4^{2-} ions in poorly crystallized regions that were easily solubilized. In studies of ν_4 PO_4, two bands at 520-530 and 540-550 on the low frequency side of ν_4 were assigned to HPO_4^{2-} (Rey et al. 1990). In biological apatites, their intensity fell with maturation (Rey et al. 1990). More detailed information about the environment of the HPO_4^{2-} ion in freshly deposited bone mineral can be obtained from ^{31}P MAS NMR spectroscopy (Wu et al. 1994). These authors used differential cross polarization to suppress the dominant PO_4^{3-} signal to reveal the spectra of the minor phosphate constituents. They were able to measure directly both the proportion of HPO_4^{2-} and the parameters that characterize its isotropic and anisotropic chemical shift. They found that its ^{31}P isotropic chemical shift corresponded precisely to that of the HPO_4^{2-} in OCP and differed from that in brushite. In contrast, the anisotropic shift was close to that seen in brushite and differed significantly from that in OCP. They concluded that the HPO_4^{2-} environment in bone was unique, and did not correspond to that in any of the common synthetic calcium phosphates.

More recent spectroscopic work. Most of the recent advances in IR and Raman studies of mineralized tissue have been possible as a result of improvements in data analysis and instrumentation, particularly microscopic methods that allow spectra to be obtained from micron sized regions of tissue. For example, Paschalis et al. (1996) used FTIR microspectroscopy to study the mineral/matrix, (total carbonate)/phosphate, and CO_3 Type A/Type B ratios across human osteons in 10 μm steps. The resolution was 4 cm^{-1} for spectra from 20×20 μm^2 areas. Similar instrumentation and data analysis have been used to study changes in mineral content and composition in biopsies of non-osteoporotic (Paschalis et al. 1997a) and osteoporotic (Paschalis et al. 1997b) human cortical and trabecular bone from iliac crest biopsies. Amongst the results obtained, the osteoporotic bone mineral appeared to be more crystalline/mature than normal bone mineral. In this context, a parameter describing "crystallinity/maturity/perfection" was determined from the spectra from different bone areas at a spatial resolution of 20 μm. This parameter was related, amongst other things, to a greater HAP-like stoichiometry and reduced substitution by ions such as CO_3^{2-}.

Infrared microscopes have also been used with array detectors to give 2-D images of the spectral absorption in the fingerprint region of the IR spectrum. The data can be displayed as either a series of spectroscopic images collected at individual wavelengths, or as a collection of IR spectra obtained at each image pixel position in the image. For example, a 64 × 64 mercury-cadmium-telluride detector array has been used with a Fourier transform (FT) IR microscope coupled to a FT-IR spectrometer to determine the spatial distributions of the mineral-to-matrix ratio and mineral maturity as a function of distance from an osteon (Marcott et al. 1998). Infrared microscopes have also been used with synchrotrons, which are powerful sources of continuous radiation. One such microscope is at the National Synchrotron Light Source at Brookhaven (Carr 2001) and allows spectra to be obtained from specific regions approaching the diffraction limit (3-20 μm). Commercial IR microspectrometers do not normally allow observation at sufficiently low wavenumbers to permit study of the ν_4 PO_4 bands at 650-500 cm^{-1}. However, the Brookhaven microscope is equipped with a detector to go down to these frequencies and enabled changes in the mineral to be mapped out across a single osteon (Miller et al. 2001). Study of a series of bone apatites of different ages and synthetic apatites allowed a number of spectra-structure correlations to be developed, including the assignment of a band near 540 cm^{-1} to HPO_4^{2-}. These, and previously known assignments, showed that when the crystallinity increased and the HPO_4^{2-} content decreased in the bone apatite, the CO_3^{2-} occupation of A-sites fell and occupation of B-sites increased. Spectra were collected at 10 μm intervals in line scans across the center of an osteon and in area scans (20 × 25 points). The crystallinity increase and HPO_4^{2-} content decrease occurred in the direction from the center of the osteon towards the older bone at the periphery (Miller et al. 2001).

Raman spectra can also be measured from microscopic areas of bones and teeth, typically from a 1-μm diameter spot. Although laser Raman spectrometers were developed before high-resolution IR microscopes, less use has been made of them for the study of mineralized tissue. Nevertheless, good high-resolution spectra can be obtained. For example, Penel et al. (1998) studied the small shifts in the ν_1 PO_4 band from 960 cm^{-1} due to CO_3^{2-} substitution in A and B sites. The ν_3 PO_4 band also showed measurable shifts, whereas ν_2 and ν_4 were unaffected. The authors reported that the phosphate band shifts seen in biological apatites were consistent with CO_3^{2-} ions in both A and B sites. Spectral evidence could be seen for HPO_4^{2-} ions, with quantities increasing in the sequence dental enamel, dentin and bone. Freeman et al. (2001) have recently used Laser Raman Spectroscopy to study the mineral in small regions of mouse femora as a function of age and in vitro fluoride treatment. They reported that observed shifts in the ν_1 PO_4 band on fluoride treatment were consistent with a transformation from carbonated HAP to carbonated fluorapatite.

ATOMIC STRUCTURE OF BIOLOGICAL APATITES

It is clear from the earlier discussion, particularly spectroscopic studies, that the mineral in bones and teeth differs from HAP in many respects, though it is clearly apatitic. Quantitatively, the main differences are the variable Ca/P ratio, and the HPO_4^{2-} and CO_3 contents found in biological apatites.

Synthetic CO_3-free apatites are easily precipitated with Ca/P mole ratios in the range 1.5 to 1.667 (and even outside this range). These preparations are conveniently referred to as Ca-deficient apatites (Ca-def Aps) and probably provide a good model for the variable Ca/P ratio found in biological apatites. Ca-def Aps form pyrophosphate on heating and have a P-(OH) band at 870 cm^{-1}, showing that they contain HPO_4^{2-} ions (Berry 1967). Well-crystallized Ca-def Aps (micron sized crystals) can be prepared, so an explanation of their variable Ca/P ratio and HPO_4^{2-} content must be found in the unit cell, rather than ions adsorbed on a large crystal surface. These well-crystallized Ca-def Aps typically have a slightly larger *a* parameter (+0.16%) and smaller *c* parameter (-0.09%) compared with high-temperature stoichiometric HAPs.

Models of various degrees of complexity have been proposed for Ca-def Aps (see Elliott 1994). One of the earliest and simplest,

$$Ca_{10-x}(HPO_4)_x(PO_4)_{6-x}(OH)_{2-x}, \text{ where } 0 \le x \le 1,$$

probably represents a good approximation to the unit cell contents. Notice that the loss of one Ca^{2+} ion is compensated by a gain of an HPO_4^{2-} ion plus the loss of an OH^- ion. It is likely that water molecules occupy some of the vacated OH sites. The model is consistent with compositional changes seen during the maturation of biological apatites, namely a fall in HPO_4^{2-} content and concomitant rise in Ca/P ratio.

In apatite, there are two crystallographically independent Ca sites, Ca1 and Ca2 (see Pan and Fleet, this volume, p. 13). Rietveld analysis of XRD powder data shows that it is predominantly Ca2 that is lost in Ca-def Aps (Jeanjean et al. 1996). This preferential loss has recently been confirmed in a study of well-crystallized Ca-def Aps prepared at 80-95°C precipitated from solution by a slow increase in pH caused by the hydrolysis of urea (Wilson, Elliott, Dowker and Rodriguez—unpublished results) or formamide (Wilson, Elliott and Dowker—unpublished results).

Historically, the best evidence that CO_3^{2-} ions can replace PO_4^{3-} ions in the apatite lattice (B-type substitution) has been the decrease in the *a*-axis cell parameter that results. However, as regards the mineral in dental enamel, the *a* parameter is actually somewhat larger than for pure HAP. The *a* parameter in CO_3-free Ca-def Aps is also expanded (see above), and it may be that the factors that cause this expansion (probably mostly HPO_4^{2-} and/or H_2O in the lattice) also cause at least part of the expansion seen in enamel. Dental enamel also contains some Cl^- ions (Table 3) and CO_3^{2-} replacing OH^- ions that will increase the *a* parameter. Detailed consideration of these factors (Elliott 1994) resulted in the conclusion that the difference in *a* parameter between enamel and HAP seems mainly due to HPO_4^{2-} and/or H_2O, whilst the effects of the lattice substitutions of Cl^- and CO_3^{2-} almost cancel each other out. In any case, a B-type substitution is not ruled out by the expansion of the *a* parameter. As reported above, the polarized IR spectrum is consistent with CO_3^{2-} replacing PO_4^{3-} ions. Direct confirmation of B-type substitution in enamel has recently been obtained through Rietveld whole pattern fitting methods for powder XRD data (Wilson et al. 1999). Though it was not possible to include the CO_3^{2-} ion explicitly in the refinement, there was clear evidence for the replacement of PO_4. The most direct evidence was that the apparent PO_4 tetrahedral volume was reduced from 1.87 Å^3 for stoichiometric HAP to 1.787 Å^3 in dental enamel. The P occupancy was reduced by 8% and the Ca2 occupancy by 5% (the Ca1 occupancy was held at the stoichiometric value).

Rietveld structure refinements have also been undertaken on synthetic precipitated CO_3Aps as model systems for biological apatites. The reduction in apparent PO_4 volume, and P and Ca2 occupancies seen in dental enamel were also seen in Na-free precipitated CO_3Aps with CO_3 contents of ~4 wt % produced by the reaction between $CaCO_3$ and $CaHPO_4$ at 100°C (Morgan et al. 2000). CO_3Aps precipitated in the presence of Na^+ ions can have CO_3 contents of up to 22 wt %, so more extreme effects might be expected in these compounds. Rietveld refinements of such a preparation with 17.3 wt % CO_3 again showed a reduction in the apparent PO_4 volume and P occupancy, but the reduction in P occupancy was much less than expected had all the CO_3^{2-} ions replaced PO_4^{3-} ions in the lattice (Elliott et al., in press). A suggested explanation for this was that a substantial amount of CO_3 was not in PO_4 sites, but was possibly present as amorphous $CaCO_3$, though this was not supported by independent evidence (Elliott et al., in press). This preparation also had a deficiency in Ca1 sites, rather than Ca2 as observed in the Na-free CO_3Aps and dental enamel (above).

Although the variable Ca/P ratio and CO_3 content form the most conspicuous difference between biological apatites and HAP, there are other differences, for example in the OH^- ion. It has already been commented on that OH^- ions cannot be detected in the IR spectrum of bone (Rey et al. 1995a). This was confirmed by Rey et al. (1995b) who could not detect structural OH^- ions by ^1H MAS NMR (magic angle spinning nuclear magnetic resonance spectroscopy) and by Loong et al. (2000) who found no evidence from inelastic neutron scattering experiments for the sharp excitations characteristic of the vibrational and stretch modes of OH^- ions in HAP (~80 and 450 MeV, 645 and 3630 cm^{-1}, respectively). Even in dental enamel, which clearly contains OH^- ions, Rietveld analysis shows that the ion is displaced 0.079 Å further away from the mirror plane at $z = 1/4$ compared with synthetic HAP (Wilson et al. 1999). A similar shift is also seen in the precipitated CO_3Aps referred to above (Morgan et al. 2000). There is no evidence from Rietveld analysis for a reduction in the OH^- ion content of dental enamel, though the distribution of electron density along the *c*-axis is wider than in HAP. This has been ascribed to Cl^- ions and H_2O molecules (Young and Mackie 1980, Wilson et al. 1999).

Mineralized tissues contain a significant amount of Mg (Table 3). Although Mg seems to be important in mineralized tissues, there has always been doubt about whether it was all adsorbed on the apatites crystal surfaces, or whether a significant amount replaced Ca^{2+} ions in the apatite lattice (Glimcher 1998). Lattice substitution is doubted because the radius of the Mg^{2+} ion (0.72 Å) is much less than that of the Ca^{2+} ion (1.00 Å) and because definitive proof of the possibility of such a substitution was lacking. However, Bigi et al. (1996) have used Rietveld analysis of XRD data to show that a limited replacement (at most ~10 atom %) appeared to be possible. It seemed that Mg^{2+} had replaced some Ca^{2+} ions in Ca2 sites.

As mentioned at the beginning of this review, the very large surface area of the crystals in bone makes it very problematical to try to determine the unit cell contents from the chemical composition of the mineral. With the much smaller surface area of the crystals in enamel, it is much more feasible and several formulae have been published (Elliott 1997). As the density of enamel apatite is not known, the absolute contents cannot be determined, so it is usually assumed that the number of ions in the phosphate sites (PO_4^{3-} and CO_3^{2-}) sum to 6, whilst the OH^- ion content is derived by the requirement of charge balance. An example for late mature porcine enamel that included acid phosphate is (Aoba and Moreno 1992):

$$Ca_{9.26}(HPO_4)_{0.22}(CO_3)_{0.5}(PO_4)_{5.63}(OH)_{1.26}.$$

An equivalent formula based on published analyses for mature human dental enamel is (after Table 3, Elliott 1997)

$$Ca_{8.86}Mg_{0.09}Na_{0.29}K_{0.01}(HPO_4)_{0.28}(CO_3)_{0.41}(PO_4)_{5.31}(OH)_{0.70}Cl_{0.08}(CO_3)_{0.05}.$$

Notice that account has been taken of the small fraction of CO_3^{2-} ions that are believed to replace OH^- ions. The density of the apatite in dental enamel has been estimated at ~3.0

g cm^{-3} from the above unit cell contents and published lattice parameters (Elliott 1997). The density is significantly less than 3.16 g cm^{-3} calculated for stoichiometric HAP (Elliott 1994). This difference can be mainly attributed to the presence of vacancies in Ca2 sites mentioned earlier as determined by Rietveld analysis (Wilson et al. 1999).

CONCLUSIONS

The basic apatite structure is well known. Though there has been a general understanding for many years of how the structure changes with the incorporation of CO_3^{2-} ions and with a variable Ca/P ratio, details were lacking. However, the changes in atomic positions and occupancies that take place in the unit cell are beginning to emerge from Rietveld whole diffraction pattern fitting methods, particularly for the well-crystallized dental enamel apatite. How applicable these ideas are to bone mineral, with its much larger surface area, is not yet clear.

There seems now to be better agreement about the shapes and sizes of the biological apatites. The reason why these minerals do not exhibit the regular hexagonal cross-section expected of the hexagonal apatite structure is not definitely established, though it is difficult to escape the conclusion that this is associated with OCP. However, the possibility of a non-hexagonal space group, at least for dental enamel, still needs to be settled.

The development during the last few years of methods to obtain wide- and low-angle XRD patterns and IR and Raman spectra from micron-sized regions of tissue is quite notable. These techniques have already provided information on the changes in mineral in different regions of tissue and various stages of mineralization, but much more should be achieved in the future.

Perhaps one of the greatest challenges, at least from a biological standpoint, will be to understand the factors that control the nucleation, growth, and growth inhibition of the biological apatites at the molecular level.

ACKNOWLEDGMENT

The author's research is supported by the Medical Research Council Grant No. G9824467.

REFERENCES

Aoba T (1996) Recent observations on enamel crystal formation during mammalian amelogenesis. Anatomical Record 245:208-18

Aoba T, Moreno EC, (1992) Changes in the solubility of enamel mineral at various stages of porcine amelogenesis. Calcif Tissue Intl 50:266-272

Aoba T, Moriwaki Y, Doi Y, Okazaki M, Takahashi J, Yagi T (1980) Diffuse X-ray scattering from apatite crystals and its relation to amorphous bone mineral. J Osaka Univ Dental School 20:81-90

Aoba T, Yagi T, Okazaki M, Takahashi J, Doi Y, Moriwaki Y (1981) Crystallinity of enamel apatite: an X-ray diffraction study of human and bovine-fetus teeth. J Osaka Univ Dental School 21:87-98

Bacon GE, Bacon PJ, Griffiths RK (1979) Stress distribution in the scapula studied by neutron diffraction. Proc Roy Soc London B204:355-362

Baud CA, Very JM, Courvoisier B (1988) Biophysical study of bone mineral in biopsies of osteoporotic patients before and after long-term treatment with fluoride. Bone 9:361-365

Baylink D, Wergedal J, Stauffer M, Rich C (1970) Effects of fluoride on bone formation, mineralization, and resorption in the rat. *In* Fluorine in Medicine. Vischer T (ed) Hans Huber Publisher, Bern, p 37-69

Berthet-Columinas C, Miller A, White SW (1979) Structural study of the calcifying collagen in turkey leg tendons. J Mol Biol 134:431-445

Berry EE (1967) The structure and composition of some calcium-deficient apatites. Arch Oral Biol 29: 317-327

Bigi A, Falini G, Foresti E, Gazzano M, Ripamonti A, Roveri N (1996) Rietveld structure refinements of calcium hydroxyapatite containing magnesium. Acta Crystallogr B52:87-92

Bonar LC, Grynpas MD, Glimcher MJ (1984) Failure to detect crystalline brushite in embryonic chick and bovine bone by X-ray diffraction. J Ultrastructure Res 86:93-99

Bonar LC, Lees S, Mook HA (1985) Neutron diffraction studies of collagen in fully mineralized bone. J Mol Biol 181:265-270

Bonar LC, Roufosse AH, Sabine WK, Grynpas MD, Glimcher MJ (1983) X-ray diffraction studies of the crystallinity of bone mineral in newly synthesized and density fractionated bone. Calcif Tissue Intl 35:202-209

Brès EF, Cherns D, Vincent R, Morniroli J-P (1993a) Space group determination of human tooth-enamel crystals. Acta Crystallogr B49:56-62

Brès EF, Steuer P, Voegel J-C, Frank RM, Cuisinier FJG (1993b) Observation of the loss of the hydroxyapatite six-fold symmetry in a human fetal tooth enamel crystal. J Micros 170:147-154

Brown WE (1966) Crystal growth of bone mineral. Clin Orthopaedics 44:205-220

Brown WE, Eidelman N, Tomazic B (1987) Octacalcium phosphate as a precursor in biomineral formation. Adv Dent Res 1:306-313

Carr GL (2001) Resolution limits for infrared microspectroscopy explored with synchrotron radiation. Rev Sci Instru 72:1613-1619

Chow LC, Eanes ED (eds) (2001) Octacalcium Phosphate. Monographs in Oral Science, Vol 18. Karger, Basel

Cuisinier FJG, Steuer P, Senger B, Voegel JC, Frank RM (1992) Human amelogenesis I: High-resolution electron microscopy study of ribbon-like crystals. Calcif Tissue Intl 51:259-268

Cullity BD, Stock SR (2001) Elements of X-ray Diffraction (3rd edn). Prentice Hall, Upper Saddle River, New Jersey

Daculsi G, Kerebel B (1978) High-resolution electron microscope study of human enamel crystallites: Size, shape and growth. J Ultrastructure Res 65:163-172

Daculsi G, LeGeros RZ, Mitre D (1989) Crystal dissolution of biological and ceramic apatites. Calcif Tissue Intl 45:95-103

Daculsi G, Mentanteau J, Kerebel LM, Mitre D (1984) Length and shape of enamel crystals. Calcif Tissue Intl 36:550-555

Dallemagne MJ, Richelle LJ (1973) Inorganic chemistry of bone. *In* Biological Mineralization. Zipkin I (ed) John Wiley, New York, p 23-42

Driessens FCM, Verbeeck RMH (1990) Biominerals. CRC Press, Boca Raton, Florida

Eanes ED, Hailer AW (1998) The effect of fluoride on the size and morphology of apatite crystals grown from physiologic solutions. Calcif Tissue Intl 63:250-257

Eanes ED, Reddi AH (1979) The effect of fluoride on bone mineral apatite. Metabolic Bone Disease Related Res 2:3-10

Eanes ED, Zipkin I, Harper RA, Posner AS (1965) Small-angle X-ray diffraction analysis of the effect of fluoride on human bone apatite. Arch Oral Biol 10:161-173

Elliott JC (1994) Structure and Chemistry of the Apatites and Other Calcium Orthophosphates. Elsevier, Amsterdam

Elliott JC (1997) Structure, crystal chemistry and density of enamel apatites. *In* Dental enamel (Ciba Foundation Symp 205). Chadwick D, Cardew G (eds) John Wiley & Sons, Chichester, UK, p 54-67

Elliott JC (1998) Recent studies of apatites and other calcium orthophosphates. *In* Les matériaux en phosphate de calcium. Aspect fondamentaux. Brès E, Hardouin P (eds) Sauramps Médical, Montpelier, p 25-66

Elliott JC, Fearnhead RW (1962) Electron microscope and electron diffraction study of developing enamel crystals. J Dent Res 41:1266

Elliott JC, Mackie PE, Young RA (1973) Monoclinic hydroxyapatite. Science 180:1055-1057

Elliott JC, Wilson RM, Dowker SEP (in press) Apatite structures. *In* Advances in X-ray analysis. Vol 45. Proc 50th Annual Denver X-ray Conference, Huang TC (ed) International Centre for Diffraction Data, Newtown Square, Pennsylvania

Eppell SJ, Tong W, Katz JL, Kuhn L, Glimcher ML (2001) Shape and size of isolated bone mineralites measured using atomic force microscopy. J Orthopaedic Res 19:1027-1034

Fernández-Morán H, Engström A (1957) Electron microscopy and X-ray diffraction of bone. Biochem Biophys Acta 23:260-264

Engström A, Finean JB (1953) Low-angle X-ray diffraction of bone. Nature 171:564

Fincham AG, Luo W, Moradian-Oldak J, Paine ML, Snead ML, Zeichner-David M (2000) Enamel biomineralization: the assembly and disassembly of the protein extracellular matrix. *In* Development, Function and Evolution of Teeth. Teaford MF, Smith MM, Ferguson MWJ (eds) Cambridge University Press, Cambridge, p 37-61

Ford MA, Price WC, Seeds WE, Wilkinson GR (1958) Double-beam infrared microspectrometer. J Optical Soc Am 48:249-254

Fowler BO (1977) I. Polarized Raman spectra of apatites. II. Raman bands of carbonate ions in human tooth enamel. Mineralized Tissue Research Communications Vol 3, no. 68

Fratzl P, Fratzl-Zelman N, Klaushofer K, Vogl G, Koller K (1991) Nucleation and growth of mineral crystals in bone studied by small-angle X-ray scattering. Calcif Tissue Intl 48:407-413

Fratzl P, Schreiber S, Boyde A (1996) Characterization of bone mineral crystals in horse radius by small-angle X-ray scattering. Calcif Tissue Intl 58:341-346

Frazier PD (1967) X-ray diffraction analysis of human enamel containing different amounts of fluoride. Arch Oral Biol 12:35-42

Freeman JJ, Wopenka B, Silva MJ, Pasteris JD (2001) Raman spectroscopic detection of changes in bioapatite in mouse femora as a function of age and *in vitro* fluoride treatment. Calcif Tissue Intl 68:156-162

Glas J-E, Omnell K-Å (1960) Studies on the ultrastructure of dental enamel. I. Size and shape of the apatite crystallites as deduced from X-ray diffraction data. J Ultrastructure Res 3:334-344.

Glimcher MJ (1998) The nature of the mineral phase in bone: biological and clinical implications. *In* Metabolic Bone Disease. Avioli LV, Krane SM (eds) Academic Press, New York, p 23-50

Glimcher MJ, Bonar LC, Grynpas MD, Landis WJ, Roufosse AH (1981) Recent studies of bone mineral: is the amorphous calcium phosphate theory valid? J Cryst Growth 53:100-119

Gross KA, Phillips MR (1998) Identification and mapping of the amorphous phase in plasma-sprayed hydroxyapatite coatings using scanning cathodoluminescence microscopy. J Mater Sci Mater Medicine 9:797-802

Grynpas MD, Bonar LC, Glimcher MJ (1984) X-ray diffraction radial distribution function studies on bone mineral and synthetic calcium phosphates. J Mater Sci 19:723-736

Grynpas MD, Cheng P-T (1988) Fluoride reduces the rate of dissolution of bone. Bone and Mineral 5:1-9

Hamad M, Heughebaert J-C (1986) The growth of whitlockite. J Cryst Growth 79:192-197

Handschin RG, Stern WB (1992) Crystallographic lattice refinement of human bone. Calcif Tissue Intl 51:111-120

Handschin RG, Stern WB (1994) Crystallographic and chemical analysis of human bone apatite (Crista Iliaca). Clin Rheumatology 13 (Suppl 1), p 75-90

Handschin RG, Stern WB (1995) X-ray diffraction studies on the lattice perfection of human bone apatite (Crista iliaca). Bone 16 (Suppl):355S-363S

Herman H, Dallemagne MJ (1961) The main mineral constituent of bone and teeth. Arch Oral Biol 5: 137-144

Iijima M, Nelson DGA, Pan Y, Kreinbrink AT, Adachi M, Goto T, Moriwaki Y (1996) Fluoride analysis of apatite crystals with a central planar OCP inclusion: concerning the role of F^- ions on apatite/OCP/apatite structure formation. Calcif Tissue Intl 59:377-384

Iijima M, Tohda H, Moriwaki Y (1992) Growth and structure of lamella mixed crystals of octacalcium phosphate and apatite in a model system of enamel formation. J Cryst Growth 116:319-326

Jeanjean J, McGrellis S, Rouchaud JC, Fedoroff M, Rondeau A, Perocheau S, Dubis A (1996) A crystallographic study of the sorption of cadmium on calcium hydroxyapatites: incidence of cationic vacancies. J Solid State Chem 126:195-201

Johansen E, Parks HF (1960) Electron microscopic observations on the three-dimensional morphology of apatite crystallites of human dentine and bone. J Biophysic Biochem Cytol 7:743-753

Kinney JH, Pople JA, Marshall GW, Marshall SJ (2001) Collagen orientation and crystallite size in human dentin: A small-angle X-ray scattering study. Calcif Tissue Intl 69:31-37

Landis WJ, Paine MC, Glimcher MJ (1977) Electron microscopic observations of bone tissue prepared anhydrously in organic solvents. J Ultrastructure Res 59:1-30

Landis WJ, Song MJ, Leith A, McEwen L, McEwen BF (1993) Mineral and organic matrix interaction in normally calcifying tendon visualized in three dimensions by high-voltage electron microscopic tomography and graphic image reconstruction. J Struct Biol 110:39-54

LeGeros RZ (1991) Calcium Phosphates in Oral Biology and Medicine. Monographs in Oral Science, Vol 15. Karger, Basel

LeGeros RZ, Daculsi G, Kijkowska R, Kerebel B (1989) The effect of magnesium on the formation of apatites and whitlockites. *In* Magnesium in health and disease. Itokawa Y, Durlach J (eds) John Libbey & Co Ltd, London, p 11-19

Lees S, Bonar LC, Mook HA (1984) A study of dense mineralized tissue by neutron diffraction. Intl J Biol Macromol 6:321-326

Lees S, Prostak K (1988) The locus of mineral crystallites in bone. Connective Tissue Res 18:41-54

Li P, Calvert P, Kokubo T, Levy R, Scheid C (eds) (2000) Mineralization in Natural and Synthetic Biomaterials. Mater Res Soc Symp Proc Vol 599. Materials Research Society, Warrendale, Pennsylvania

Loong C-K, Rey C, Kuhn LT, Combes C, Wu Y, Chen S-H, Glimcher MJ (2000) Evidence of hydroxyl-ion deficiency in bone apatites: an inelastic neutron-scattering study. Bone 26:599-602

Lowenstam HA, Weiner S (1989) On Biomineralization. Oxford University Press, New York

Lundy DR, Eanes ED (1973) An X-ray line-broadening study of turkey leg tendon. Arch Oral Biol 18: 813-826

Mann S (ed) (1996) Biomimetic Materials Chemistry. VCH, New York

Mann S (2001) Biomineralization, Principles and Concepts in Bioinorganic Materials Chemistry. Oxford University Press, Oxford

Mann S, Webb J, Williams RJP (eds) (1989) Biomineralization, Chemical and Biochemical Perspectives. VCH Verlagsgesellschaft, Weinheim, Germany

Marcott C, Reeder RC, Paschalis EP, Tatakis DN, Boskey AL, Mendelsohn R (1998) Infrared microspectroscopic imaging of biomineralized tissues using a mercury-cadmium-telluride focal-plane array detector. Cellular Mol Biol 44:109-115

McDowell H, Gregory TM, Brown WE (1977) Solubility of $Ca_5(PO_4)_3OH$ in the system $Ca(OH)_2$-H_3PO_4-H_2O at 5, 15, 25, and 37°C. J Res Natl Bur Std A Phys Sci 81A:273-281

Miles AEW (ed) (1967) Structural and Chemical Organisation of Teeth. Vols I and II. Academic Press, New York

Miller LM, Vairavamurthy V, Chance MR, Mendelsohn R, Paschalis EP, Betts F, Boskey AL (2001) *In situ* analysis of mineral content and crystallinity in bone using infrared micro-spectroscopy of the ν_4 PO_4^{3-} vibration. Biochem Biophys Acta 1527:11-19

Moreno EC, Kresak M, Zahradnik RT (1977) Physicochemical aspects of fluoride-apatite systems relevant to the study of dental caries. Caries Res (Suppl 1) 11:142-160

Morgan H, Wilson RM, Elliott JC, Dowker SEP, Anderson P (2000) Preparation and characterisation of monoclinic hydroxyapatite and its precipitated carbonate apatite intermediate. Biomater 21:617-627

Moriwaki Y, Aoba T, Tsutsumi S, Yamaga R (1976) X-ray diffraction studies on the lattice imperfection of biological apatites. J Osaka Univ Dental School 16:33-45

Nelson DGA, Barry JC, Shields CP, Glena R, Featherstone JDB (1989) Crystal morphology, composition and dissolution behaviour of carbonated apatites prepared at controlled pH and temperature. J Coll Interface Sci 130:467-479

Nelson DGA, Wood GJ, Barry JC (1986) The structure of (100) defects in carbonated apatite crystallites: a high-resolution electron microscope study. Ultramicroscopy 19:253-266

Neuman WF, Neuman MW (1958) The Chemical Dynamics of Bone Mineral. University of Chicago Press, Chicago

Nylen MU (1964) Electron microscope and allied biophysical approaches to the study of enamel mineralization. J Roy Micros Soc 83:135-141

Paschalis EP, Betts F, DiCarlo E, Mendelsohn R, Boskey AL (1997a) FTIR microspectroscopic analysis of human cortical and trabecular bone. Calcif Tissue Intl 61:480-486

Paschalis EP, Betts F, DiCarlo E, Mendelsohn R, Boskey AL (1997b) FTIR microspectroscopic analysis of human iliac crest biopsies from untreated osteoporotic bone. Calcif Tissue Intl 61:487-492

Paschalis EP, DiCarlo E, Betts F, Sherman P, Mendelsohn R, Boskey AL (1996). FTIR microspectroscopic analysis of human osteonal bone. Calcif Tissue Intl 59:480-487

Patel PR, Brown WE (1975) Thermodynamic solubility product of human tooth enamel powder: powdered sample. J Dent Res 54:728-736

Pellegrino ED, Biltz RM (1965) The composition of human bone in uremia. Medicine 44:397-418

Penel G, Leroy G, Rey C, Brès, E (1998) MicroRaman spectral study of the PO_4 and CO_3 vibrational modes in synthetic and biological apatites. Calcif Tissue Intl 63:475-481

Posner AS, Betts F (1975) Synthetic amorphous calcium phosphate and its relation to bone mineral structure. Accounts Chem Res 8:273-281

Posner AS, Betts F, Blumenthal NC (1979) Bone mineral composition and structure. *In* Skeletal Research, Ch 9. Simmonds DJ, Kunin AS (eds) Academic Press, New York, p 167-192

Posner AS, Eanes ED, Harper RA, Zipkin I (1963) X-ray diffraction analysis of the effect of fluoride on human bone apatite. Arch Oral Biol 8:549-570

Poyart CF, Bursaux E, Fréminet A (1975) The bone CO_2 compartment: evidence for a bicarbonate pool. Respiration Physiol 25:89-99

Pritzker, KPH (1998) Calcium pyrophosphate crystal formation and dissolution. *In* Calcium Phosphates in Biological and Industrial Systems, Ch 12. Amjad Z (ed) Kluwer Academic Publishers, Boston, p 277-301

Rey C, Collins B, Goehl T, Dickens IR, Glimcher MJ (1989) The carbonate environment in bone mineral: a resolution-enhanced Fourier transform infrared spectroscopy study. Calcif Tissue Intl 45:157-164

Rey C, Hina A, Togfighi A, Glimcher ML (1995a) Maturation of poorly crystalline apatites: chemical and structural aspects in vivo and in vitro. Cells Mater 5:345-356

Rey C, Miquel JL, Facchini L, Legrand AP, Glimcher MJ (1995b) Hydroxyl groups in bone mineral. Bone 16:583-586

Rey C, Renugopalakrishnan V, Collins B, Glimcher MJ (1991a) Fourier transform infrared spectroscopic study of the carbonate ions in bone mineral during aging. Calcif Tissue Intl 49:251-258

Rey C, Renugopalakrishnan V, Shimizu M, Collins B, Glimcher MJ (1991b) A resolution-enhanced Fourier transform infrared spectroscopic study of the environment of the CO_3^{2-} ion in the mineral phase of enamel during its formation and maturation. Calcif Tissue Intl 49:259-268

Rey C, Shimizu M, Collins B, Glimcher MJ (1990). Resolution-enhanced Fourier-transform infrared spectroscopy study of the environment of phosphate ions in the early deposits of a solid phase of calcium-phosphate in bone and enamel, and their evolution with age. I: Investigations in the ν_4 PO_4 domain. Calcif Tissue Intl 46:384-394

Rey C, Shimizu M, Collins B, Glimcher MJ (1991c). Resolution-enhanced Fourier transform infrared spectroscopy study of the environment of phosphate ion in the early deposits of a solid phase of calcium phosphate in bone and enamel and their evolution with age: 2. Investigations in the ν_3 PO_4 domain. Calcif Tissue Intl 49:383-388

Reyes-Gasga J, Alcantara-Rodriguez CM, Gonzalez-Trejo AM, Madrigal-Colin A (1997) Child, adult and aged human tooth enamel characterized by electron microscopy. Acta Microscopica 6:24-38

Richelle LJ (1964) One possible solution to the problem of the biochemistry of bone mineral. Clin Orthopaedics 33:211-219

Rinnerthaler S, Roschger P, Jakob HF, Nader A, Klaushofer K, Fratzl P (1999) Scanning small-angle X-ray scattering analysis of human bone sections. Calcif Tissue Intl 64:422-429

Robinson RA, Watson ML (1952) Collagen-crystal relationships in bone as seen in the electron microscope. Anat Record 114:385-410

Roufosse AH, Landis WJ, Sabine WK, Glimcher MJ (1979) Identification of brushite in newly deposited mineral from embryonic chicks. J Ultrastruct Res 68:235-255

Sasaki N, Sudoh Y (1997) X-ray pole figure analysis of apatite crystals and collagen molecules in bone. Calcif Tissue Intl 60:361-367

Schroeder HE (1969) Formation and Inhibition of Dental Calculus. Hans Huber, Berne

Shellis RP, Duckworth RM (1994) Studies on the cariostatic mechanisms of fluoride. Intl Dent J. 44 (Suppl 1), p 263-73

Shaw WJ, Long JR, Campbell AA, Stayton PS, Drobny GP (2000) A solid state NMR study of dynamics in a hydrated salivary peptide adsorbed to hydroxyapatite. J Am Chem Soc 122:7118-7119

Simmer JP and Fincham AG (1995) Molecular mechanisms of dental enamel formation. Crit Rev Oral Biol Med 6:84-108

Snyder RL, Fiala J, Bunge HJ (eds) (1999) Defect and Microstructure Analysis by Diffraction. Oxford University Press, Oxford

Sudarsanan K and Young RA (1969) Significant precision in crystal structural details: Holly Springs hydroxyapatite. Acta Crystallogr B25:1534-1543

Termine JD, Eanes ED, Greenfield DJ, Nylen MU, Harper RA (1973) Hydrazine-deproteinated bone mineral: Physical and chemical properties. Calcif Tissue Res 12:73-90

Termine JD, Posner AS (1967) Amorphous/crystalline interrelationships in bone mineral. Calcif Tissue Res 1:8-23

Turner CH, Garetto LP, Dunipace AJ, Zhang W, Wilson ME, Grynpas MD, Chachra D, McClintock R, Peacock M, Stookey GK (1997) Fluoride treatment increased serum IGF-1, bone turnover, and bone mass, but not bone strength, in rabbits. Calcif Tissue Intl 61:77-83

Weatherell JA, Robinson C (1973) The inorganic composition of teeth. *In* Biological Mineralization. Zipkin I (ed) John Wiley, New York, p 43-74

Wenk H-R, Heidelbach F (1999) Crystal alignment of carbonated apatite in bone and calcified tendon: results from quantitative texture analysis. Bone 24:361-369

White SW, Hulmes DJS, Miller A (1977) Collagen-mineral axial relationship in calcified turkey leg tendon by X-ray and neutron diffraction. Nature 266:421-425

Williams RAD, Elliott JC (1989) Basic and Applied Dental Biochemistry, 2nd edn. Churchill Livingstone, Edinburgh

Wilson RM, Elliott JC, Dowker SEP (1999) Rietveld refinement of the crystallographic structure of human dental enamel apatites. Am Mineral 84:1406-1414

Wix P, Mohamedally SM (1980) The significance of age-dependent fluoride accumulation in bone in relation to daily intake of fluoride. Fluoride 13:100-104

Wu Y, Glimcher MJ, Rey C, Ackerman JL (1994) A unique protonated phosphate group in bone mineral not present in synthetic calcium phosphates. Identification by phosphorus-31 solid-state NMR spectroscopy. J Mol Biol 244:423-435

Young RA (1993) The Rietveld Method. Oxford University Press, Oxford

Young RA, Mackie PE (1980) Crystallography of human tooth enamel: initial structure refinement. Mater Res Bull 15:17-29

Ziv V, Weiner S (1994) Bone crystal sizes: a comparison of transmission electron microscopic and X-ray diffraction line width broadening techniques. Conn Tissue Res 30:165-175

12 Stable Isotope Compositions of Biological Apatite

Matthew J. Kohn

Department of Geological Sciences
University of South Carolina
Columbia, South Carolina 29208

Thure E. Cerling

Department of Geology and Geophysics
University of Utah
Salt Lake City, Utah 84112

INTRODUCTION

The stable isotope compositions of biogenic materials record a combination of environmental parameters and biological processes. In general, the environment provides a range of isotopic compositional inputs, and an animal processes those signals through dietary preference, physiology, behavior, etc. Geochemists then use isotope signals preserved in biogenic materials to infer either what the biologic filter was (i.e., a species-specific biologic process), or the environment in which the animal lived (e.g., see review of Koch 1998). Although stable isotopes of hydrogen, carbon, nitrogen, oxygen, and strontium in modern-day animals have provided fruitful information on environment or biology, preservation of hydrogen and nitrogen is poor for most fossil materials, especially those older than a few million years. Consequently, nearly all stable isotope studies focus on the best preserved tissues, which are biological apatites (or bioapatites)— bone, dentin, enamel, scales, etc.—and on the most diagenetically resistant isotopes–oxygen, carbon, and occasionally strontium, which occur as principal or substitutional components in bioapatite. Because of the inherent synergy between biology and environment, the scientific scope of stable isotope research on bioapatites is quite broad. In addition to studies of terrestrial paleoclimate and dietary preference, stable isotopes in bioapatite have helped elucidate such diverse issues as dinosaur thermoregulation (Barrick and Showers 1994, 1995; Barrick et al. 1996, 1998; Fricke and Rogers 2000), the global rise of C_4 plants (Cerling et al. 1993, 1997), pinniped migration (Burton and Koch 1999), cetacean osmoregulation (Thewissen et al. 1996, Roe et al. 1998), herding practices (Bocherens et al. 2001), topographic uplift (Dettman et al. 2001, Kohn et al. 2003), the demise of Norse colonies in Greenland (Fricke et al. 1995, Arneborg et al. 1999), Miocene "rhinoceros" ecology (MacFadden 1998), and mastodon migration (Hoppe et al., 1999)! In this chapter, we first describe some analytical basics, and discuss the different materials that can be analyzed, focusing on special concerns when dealing with fossils. We then outline the principles underlying C, O, and Sr isotopic variations in bioapatites, and conclude with a few examples from the literature.

ANALYSIS

CO_3 component

The CO_3 component of bioapatites is analyzed for C and O isotopes by dissolution in H_3PO_4 (McArthur et al. 1980, Land et al. 1980), similar to the analysis of carbonates (McCrea 1950). Phosphates contain several CO_3 components, including structural CO_3 that substitutes for PO_4 and OH, as well as "labile" components whose structural identity is ambiguous (e.g., Rey et al. 1991). Only the PO_4- and OH-substituting CO_3 is believed to have any diagenetic resistance, so the labile CO_3 is removed by acid pretreatment,

1529-6466/00/0048-0012$05.00

usually in a (buffered) acetic acid solution (e.g., see Koch et al. 1997), which also removes any calcite or dolomite contamination. Dissolution of the purified bioapatite ordinarily produces ~0.5 μmole of CO_2/mg, so commonly ~1 mg of bioapatite is analyzed in dual-inlet machines. Sulfur compounds may be liberated by reaction of contaminants in fossils, and their masses can interfere with those of CO_2, biasing analyses. Sulfur contamination can be avoided if Ag metal is present during dissolution or (in off-line systems) by reacting the mixed CO_2 and sulfur gases with Ag foil (Cerling and Sharp 1996; MacFadden et al. 1996).

PO_4 component

The PO_4 component is more difficult to analyze, and several decades of research have been required to develop techniques for its routine O isotope analysis. PO_4 is separated from other O-bearing components by dissolution in acid (usually HNO_3 or HF), and further purified on ion exchange columns and/or directly chemically processed to produce either $BiPO_4$ (Tudge 1960; Longinelli 1965; Longinelli and Nuti 1973a,b; Longinelli et al. 1976; Kolodny et al. 1983; Karhu and Epstein 1986) or Ag_3PO_4 (Firsching 1961, Wright and Hoering 1989, Crowson et al. 1991, Lécuyer et al. 1993, O'Neil et al. 1994, Dettman et al. 2001). Ag_3PO_4 is now the material of choice because it is much less hydroscopic (Wright and Hoering 1989) and conveniently decomposes at > 1000 °C to produce O_2 gas (Anbar et al. 1960, O'Neil et al. 1994). The $BiPO_4$ or Ag_3PO_4 is then fluorinated (Longinelli 1965, Longinelli and Sordi 1966, Luz et al. 1984a, Wright and Hoering 1989, Crowson et al. 1991), brominated (Stuart-Williams and Schwarcz 1995), or thermally decomposed in the presence of C using furnaces or lasers, to produce either CO_2 or CO (O'Neil et al. 1994, Farquhar et al. 1997, Wenzel et al. 2000, Vennemann et al. 2002), which is then analyzed in a mass spectrometer. The amount of Ag_3PO_4 needed for analysis ranges from <<1 to ~20 mg. Although small quantities of Ag_3PO_4 appear to be precisely measurable by continuous-flow mass spectrometry, it is unclear whether small amounts of PO_4 can be reproducibly precipitated as Ag_3PO_4.

Combined/bulk components. Bioapatites may also be analyzed in bulk for O and/or C isotopes via laser fluorination (CO_2 laser: Kohn et al. 1996; UV laser: Jones et al. 1999), direct thermal decomposition (Cerling and Sharp 1996, Sharp and Cerling 1996, Sharp et al. 2000), or ion probe (Eiler et al. 1997). These techniques all offer reduced sample size, and the ion probe and UV laser accommodate *in situ* analysis. Bulk analysis has disadvantages that include accuracy corrections (Kohn et al. 1996, Eiler et al. 1997), pretreatment requirements that can potentially bias compositions (Lindars et al. 2001), and, most importantly, a difficulty in assessing diagenetic alteration and contamination of fossils. Although ~85% of oxygen in bioapatite is P-bound, bulk analysis includes all other oxygen components (OH, CO_3, diagenetic oxides and silicates, etc.). If different components react differently to diagenesis, then samples will be isotopically biased, possibly by several permil, depending on the diagenetic environment (Kohn et al. 1999).

MATERIALS AND DIAGENESIS

Basically, five bioapatite materials may be analyzed: bone, dentin, enamel(oid), fish scales, and invertebrate shells. Cementum is very rarely analyzed. Each material has its proponents and critics, in part depending on whether PO_4 or CO_3 is analyzed. The mineralogy of bone, dentin, and enamel is reviewed in Elliott (this volume), and crystallite size is briefly discussed here because it influences diagenetic susceptibility. Bones and shells are extremely common, but small crystallite size (a few nm wide, a few tens of nm long, and possibly less than 1 nm thick) and high porosity and organic content (for bone) make them extremely susceptible to recrystallization and isotopic alteration. Different types of bone do have different porosities, however, and cancellous ("spongy") bone is likely far more susceptible to alteration than cortical ("compact") bone (e.g., Sealy et al. 1991). Dentin has a similar crystallite size to bone, but much lower porosity, potentially reducing susceptibility to alteration. In contrast to dentin and bone, enamel is extremely compact and has larger crystallites (tens of nm wide and thick, and possibly hundreds of nm long), so it is least likely to be affected diagenetically. Fish scales are rarely analyzed (but see Fricke et al. 1998a), so diagenetic concerns are less clear.

The different chemical components in bioapatite also exhibit different chemical susceptibilities to alteration. F-, U- and REE-substitution for OH and Ca is thousands of times greater in fossils than in original bioapatites (e.g., see Tuross and Trueman, this volume) attesting to extreme alteration potential for some components, but at issue is whether other chemical constituents of bioapatite are affected, specifically heavy isotope ratios, and the CO_3 and PO_4 components. Because of the strength of P-O bonds, the PO_4 component is widely believed to have great resistance to diagenetic alteration, whereas the CO_3 component is less resistant (Shemesh et al. 1983; Kolodny et al. 1983; Luz et al. 1984b; Nelson et al. 1986; Kolodny and Luz 1991; Barrick and Showers 1994, 1995; Barrick et al. 1996; Fricke et al. 1998a; Kohn et al. 1999). Bulk strontium is known to be altered in virtually all fossil materials, and retrieval of a biologic signal rests solely on a chemical separation of components, based on differential solubilities.

Identifying physical or chemical alteration of bioapatite is extremely contentious and scientifically polarized, as some research is based on either the retrieval of pristine biogenic signals, or the assumption that diagenetic alteration overwhelms original compositions. Table 1 summarizes 46 studies of diagenesis. Generally six diagenetic tests have been applied:

(1) Do different samples from a single locality exhibit unexpected compositional heterogeneity or homogeneity? Depending on the environment and species, different animals may be expected to have similar or different compositions, and so unexpected compositional heterogeneity or homogeneity has been explained by diagenetic alteration. Such an approach is, of course, predicated on a previous knowledge of "normal" variability. The growth of different tissues at different times in an animal may compromise definitive identification of diagenesis. In the case of Bryant et al. (1994), a compositional difference between enamel and bone from the same quarry was ascribed to diagenetic alteration of enamel, whereas later Bryant et al. (1996) interpreted compositional variation in enamel from different teeth as reflecting excellent preservation of biological signals. The reason for this interpretational disparity was Bryant et al.'s later realization that enamel *should* preserve compositional variation and could exhibit compositional differences compared to other tissues. Generally, interpretations based on homogeneity and/or heterogeneity have suggested that bone, dentin, and cementum can be readily altered even for Pleistocene samples (Ayliffe et al. 1992), whereas there is no evidence for alteration of enamel CO_3 or PO_4 (Ayliffe et al. 1992, Bryant et al. 1996). Extremely young samples can sometimes preserve original CO_3 and PO_4 isotopic compositions (Iacumin et al. 1996a). Work by Barrick and coworkers suggests that cortical bone PO_4 may preserve isotopic integrity, even for Cretaceous samples (Barrick 1998; Barrick and Showers 1994, 1995, 1999; Barrick et al. 1996).

(2) Do different sympatric animals preserve expected isotopic differences? Some animals have known isotopic compositional differences, because of diet (for C) or habitat (terrestrial vs. marine, for Sr), and a change in the magnitude of original compositional offset is ascribed to alteration. Bone CO_3 $\delta^{13}C$ can be affected for Pleistocene and younger samples (Nelson et al. 1986, Lee-Thorpe and van der Merwe 1991), although some Pleistocene samples apparently retain original com-

Table 1. Summary of diagenetic tests and results.

Test	*Summary of results*	*References*
Multi-sample heterogeneity or homogeneity	Pleistocene bone, dentin and cementum PO_4 altered, but Cretaceous bone PO_4 may be unaltered. No evidence for enamel PO_4 alteration	1, 2, 4-8 11, 12, 20, 22
Different sympatric animals retain expected isotopic differences	Bone and dentin CO_3 affected for Pliocene samples, no evidence for enamel CO_3 alteration. Sr in Miocene teeth is not preserved.	3, 10, 13, 15, 16, 26, 30, 32, 33, 36, 43, 46
Different tissues from a single animal retain biological fractionations	Bone CO_3 altered for archeological and Pleistocene bulk samples, and for some pretreated Pliocene samples; Pleistocene bone PO_4 and dentin PO_4 and CO_3 may be altered; no evidence for enamel CO_3 or PO_4 alteration	1, 2, 11, 14, 23, 25, 26, 28, 29, 30, 37, 39, 40, 43, 45
Crystallinity	Bone altered within years; Pleistocene dentin may be altered; enamel unaltered	2, 9, 11, 14, 19, 27, 30, 31, 34, 35, 38, 41, 42, 44
Comparison to surrounding sediment or expected value	Some Pleistocene bone CO_3 altered; some Plio-Pleistocene bone PO_4 may be altered, but Cretaceous bone PO_4 may be unaltered; bone Sr altered immediately; late Pliocene bulk tooth Sr altered	3, 4-8, 24, 28, 29, 30, 33, 36, 37, 42, 44, 45
Isotopic correlation between different chemical components	Most bone CO_3 altered, but some Recent bone CO_3 unaltered; Eocene enamel CO_3 may be altered	4-8, 17, 18, 20, 21

References: 1. Ayliffe et al. (1992); 2. Ayliffe et al. (1994); 3. Barrat et al. (2000); 4. Barrick (1998); 5. Barrick and Showers (1994); 6. Barrick and Showers (1995); 7. Barrick and Showers (1999); 8. Barrick et al. (1996); 9. Bartsiokas and Middleton (1992); 10. Bocherens et al. (1996); 11. Bryant et al. (1994); 12. Bryant et al. (1996); 13. Cerling et al. (1997); 14. Delgado-Huertas et al. (1997); 15. Ericson et al. (1981); 16. Feranec and MacFadden (2000); 17. Fox and Fisher (2001); 18. Fricke et al. (1998a); 19. Hedges et al. (1995); 20. Iacumin et al. (1996a); 21. Iacumin et al. (1996b); 22. Iacumin et al. (1996c); 23. Iacumin et al. (1997); 24. Koch et al. (1992); 25. Koch et al. (1997); 26. Koch et al. (1998); 27. Kyle (1986); 28. Land et al. (1980); 29. Lee-Thorpe and van der Merwe (1987); 30. Lee-Thorpe and van der Merwe (1991); 31. Michel et al. (1996); 32. Morgan et al. (1994); 33. Nelson et al. (1986); 34. Person et al. (1995); 35. Person et al. (1996); 36. Quade et al. (1992); 37. Sánchez-Chillón et al. (1994); 38. Schoeninger (1982); 39. Schoeninger and DeNiro (1982); 40. Sharp et al. (2000); 41. Sillen (1986); 42. Stuart-Williams et al. (1996); 43. Sullivan and Krueger (1981); 44. Tuross et al. (1989); 45. Wang and Cerling (1994); 46. Zazzo et al. (2000).

positions (Ericson et al. 1981, Sullivan and Krueger 1981, Bocherens et al. 1996). There is no evidence for isotopic alteration of enamel CO_3 for samples up to Miocene in age (Quade et al. 1992, Morgan et al. 1994, Bocherens et al. 1996, Cerling et al. 1997, Koch et al. 1998, Feranec and MacFadden 2000, Zazzo et al. 2000). Strontium isotopes are affected for Miocene teeth (Barrat et al. 2000).

(3) Do different tissues from the same individual retain biological fractionations? Different tissues have known isotopic offsets in modern samples. Intercomparison of different tissues (e.g., $\delta^{13}C$ of collagen vs. bone CO_3) indicates that bone and possibly dentin CO_3 is already altered for Pleistocene samples (Land et al. 1980, Lee-Thorpe and van der Merwe 1991), although some Plio-Pleistocene bone CO_3

may preserve isotopic integrity (Ericson et al. 1981; Sullivan and Krueger 1981; Lee-Thorpe and van der Merwe 1987). Pleistocene bone, dentin, and cementum PO_4 can also be altered (Ayliffe et al. 1992, 1994; Sánchez-Chillón et al. 1994). There is no evidence for CO_3 or PO_4 alteration in enamel as old as the Oligocene (Wang and Cerling 1994).

(4) Do samples retain biological crystal size distributions? Because of their small size, bioapatite crystallites have a strong tendency to coarsen post-mortem. Because coarser crystals have sharper X-ray diffraction peaks, Shemesh (1990) proposed using peak sharpness as a crystallinity index, or semiquantitative measure of recrystallization. Crystallinity index does depend on sample preparation (Roe et al. 1998), and furthermore a change in crystallinity will not affect oxygen isotopes in PO_4 if it occurs abiotically at $T < 75\text{-}100$ °C and does not involve incorporation of exogenous material (Tudge 1960; Kolodny et al. 1983; Blake et al. 1997, 1998; Lécuyer et al 1999). Nonetheless bone undergoes recrystallization within a few years post-mortem (Tuross et al. 1989), and dentin and cementum are affected in Pleistocene samples (Ayliffe et al. 1994; Fig. 1). Fresh enamel is essentially fully crystalline, and so no change in crystallinity is apparent (Ayliffe et al. 1994; Fig. 1).

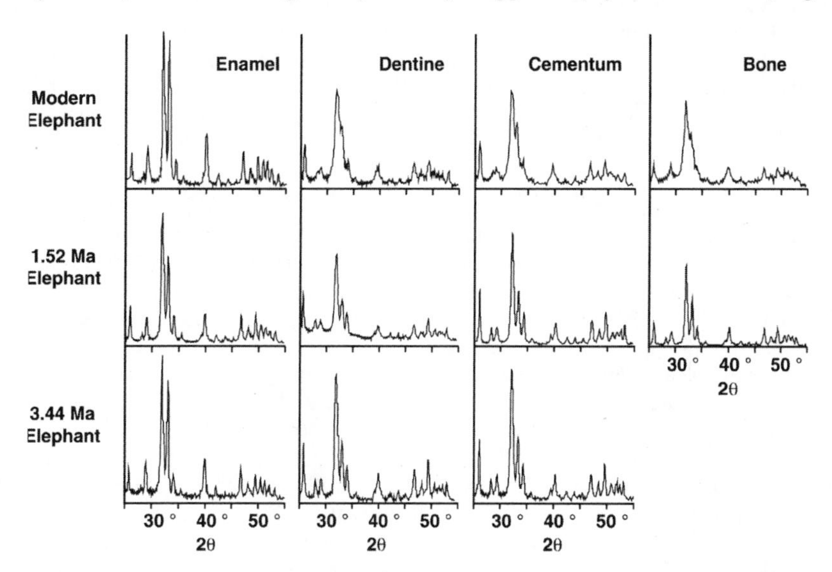

Figure 1. XRD traces of different bioapatite materials from modern and fossil elephants. Fossil dentin, cementum and bone all show a decrease in peak width compared to modern materials, indicating diagenetic recrystallization and coarsening of original apatite. Enamel peaks are essentially unaffected. From Ayliffe et al. (1994).

(5) Do compositions of fossils remain distinct from surrounding sediments or retain a known original composition? Diagenetic alteration is expected to shift the compositions of a fossil towards equilibrium with the surrounding sediment. Such a shift is quite evident for Sr in archeological bone (e.g., Nelson et al. 1986). Marine teeth of known age have a known original Sr isotope composition, and whereas modern fish teeth have isotopic compositions indistinguishable from modern-day seawater (e.g., Schmitz et al. 1991, Vennemann and Hegner 1998, Vennemann et al. 2001), fish teeth as young as Plio-Pleistocene deviate significantly from expected values (Staudigel et al. 1985). Pleistocene or even archeological bone CO_3 can be altered

(Land et al. 1980, Schoeninger and DeNiro 1982, Nelson et al. 1986), but Cretaceous cortical bone PO_4 may be unaltered (Barrick 1998; Barrick and Showers 1994, 1995, 1999; Barrick et al. 1996).

(6) Are compositions of different chemical components of the same tissue correlated and do they retain a biological offset? As discussed below, the $\delta^{18}O$ of CO_3 and PO_4 components have a systematic offset that is not commonly retained in bone (Iacumin et al. 1996a) except for Recent materials in extraordinary settings (Iacumin et al. 1996b). Tooth enamel preserves the expected offset for Miocene samples (Fox and Fisher 2001), but not for Eocene samples (Fricke et al. 1998a).

In summary, bone hardly ever preserves original isotopic compositions, except in young and/or exceptionally well-preserved samples, although the work of Barrick and coworkers implies that some bone PO_4 may be diagenetically resistant. The CO_3 component of dentin is not likely to be reliable in samples older than the Plio-Pleistocene. There is no evidence for isotopic change to the PO_4 component of enamel for any sample, and enamel CO_3 is apparently resistant for Oligocene samples, albeit not for some Eocene samples. Strontium isotopes are highly suspect in bone of any age, and in dentin that is Pliocene or perhaps younger in age. Enamel $^{87}Sr/^{86}Sr$ has not been investigated thoroughly, but is likely to be much better preserved than in other tissues. Because enamel and PO_4 are physically and chemically most resistant to alteration, many researchers now restrict analysis and research scope to fossil enamel and/or the PO_4 component. Other researchers persevere with other phosphatic tissues and components, burdened as they are with diagenetic concerns.

CARBON ISOTOPES IN BIOAPATITES

Carbon isotope analysis of bioapatites was first applied to terrestrial mammals in the early 1980s (Land et al. 1980, Ericson et al. 1981, Sullivan and Krueger 1981). The work on fossils by Ericson, Sullivan, Krueger and coworkers was immediately criticized as having ignored diagenesis because results from collagen did not agree with results from bone from archeological sites (Schoeninger and DeNiro 1982). While it is now known that some bone does undergo C-isotope exchange extremely readily, collagen, bone, and enamel record different periods of time in the life of a single individual, and diet may change. That is, there is a fundamental ambiguity (preservation vs. normal intra-individual differences) in interpreting isotopic differences among different tissues. Unfortunately, early results were taken to imply that all bioapatites are unreliable, and it was not until the 1990s that it became accepted that tooth enamel, at least, is a robust recorder of diet. Thus, the early work of Lee-Thorp and van der Merwe (Lee-Thorp and van der Merwe 1987, 1991; Lee-Thorp et al. 1989) was a struggle against the tide of misplaced opinion.

Ecological trends

One interesting story involving carbon isotopes in bioapatites concerns the different photosynthetic pathways used by terrestrial plants (Fig. 2) and the animals that eat them. Carbon-isotope discrimination among plants has been described well in a pair of comprehensive reviews by O'Leary (1988) and Farquhar et al. (1989). The main controls on fractionation are the action of a particular enzyme and the "leakiness" of cells.

Atmospheric CO_2 first moves through the stomata, dissolves into leaf water, and so enters the mesophyll cell. Most dicotyledonous plants (dicots) as well as monocotyledenous plants (monocots) in regions with cool growing seasons use the C_3-photosynthetic pathway. Mesophyll CO_2 is directly combined with ribulose bisphosphate (RuBP—a 5-carbon molecule), in a reaction catalyzed by the enzyme ribulose bisphos-

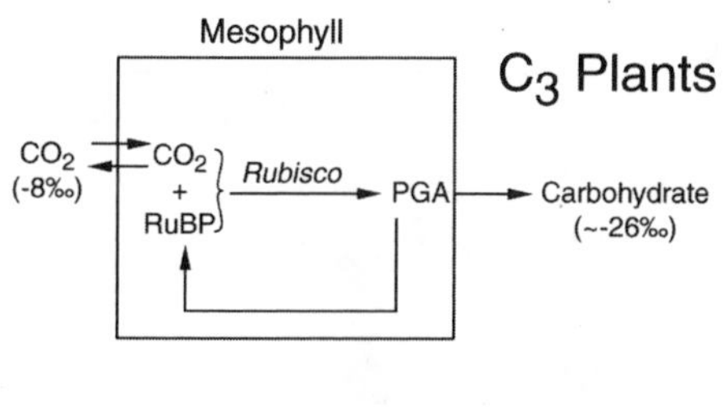

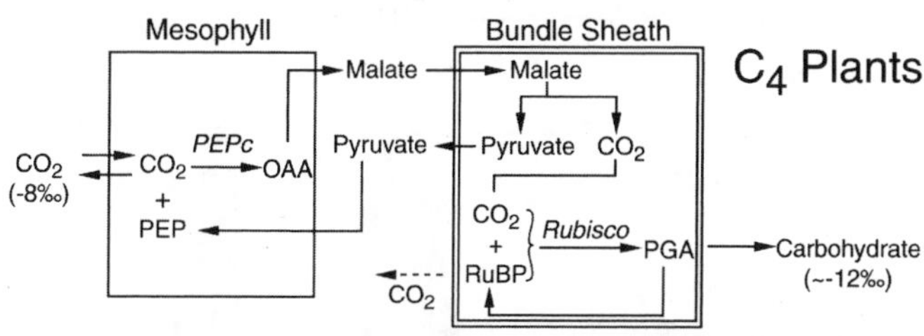

Figure 2. Schematic diagram showing differences in photosynthetic pathways for C_3 vs. C_4 terrestrial plants, which yield different C-isotope fractionations in plant tissue. Phosphoglycerate (PGA), oxaloacetate (OAA), malate, and pyruvate are organic compounds that participate in photosynthesis, and Rubisco and PEPc are enzymes that catalyze reactions among these organic compounds and CO_2. Solid arrows show flow of organic molecules within and across cell; dashed arrow shows that some CO_2 can leak out of bundle sheath cells. See text for more detail. Permil values are all relative to Vienna Peedee Belemnite (V-PDB).

phate carboxylase/oxygenase ("Rubisco"). The resulting 6-carbon molecule is then cleaved into 2 molecules of phosphoglycerate (PGA), each with 3 carbon atoms (hence "C_3"). Most PGA is recycled to make RuBP, but some is used to make carbohydrates. There is fairly free exchange between external and mesophyll CO_2 (i.e., the mesophyll cell is very "leaky") so the large isotope fractionation associated with the carboxylation of RuBP (~-25‰) contributes directly to the isotope composition of end-product phosphoglycerate (PGA) and carbohydrates.

Many monocots use the C_4 photosynthetic pathway. CO_2 in mesophyll cells is first combined with phosphoenolpyruvate (PEP) via the enzyme PEP carboxylase (PEPc) to make the molecule oxaloacetate (OAA), which has 4 carbon atoms (hence "C_4"). The OAA is usually transformed into malate, which is transported into bundle sheath cells and cleaved to pyruvate and CO_2 again. The pyruvate is recycled back into the mesophyll cells to reform PEP. Unlike mesophyll cells, the bundle sheath cells in C_4 plants are able to concentrate CO_2 (i.e., they are not very leaky), so that most of the CO_2 is fixed, and less fractionation occurs in forming PGA. If bundle sheath cells were perfectly gas tight, there would be zero fractionation from Rubisco, whereas if the cells were completely permeable to CO_2, the isotope fractionation from Rubisco would be ~-25‰. In reality, bundle sheath cells can exhibit some "leakiness," so there can be some Rubisco discrimination, but far less than in mesophyl cells.

Other photosynthetic steps also have isotope fractionations, particularly diffusion of CO_2 across stomata (~4‰). However, the isotopic difference between C_3 vs. C_4 plants is largely influenced by the degree to which the Rubisco isotope discrimination is dominant, which depends on the degree to which cells are "CO_2-leaky." Because mesophyll cells are very "leaky" but bundle sheath cells are not, C_3 vs. C_4 plants have ^{13}C depletions of ~18‰ vs. ~4‰ relative to atmospheric CO_2. Therefore, in the modern world ($\delta^{13}C$ of atmospheric CO_2 is -8‰), most C_3 dicots have $\delta^{13}C$ values from about –25 to –30‰

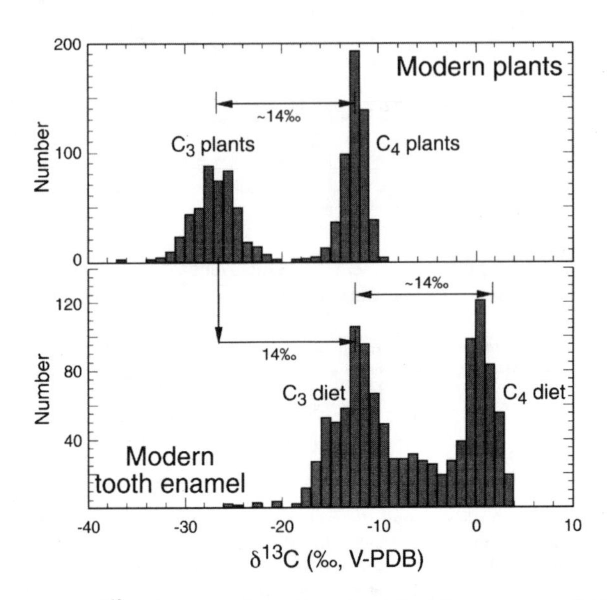

Figure 3. Histograms of $\delta^{13}C$ of modern plants (above) and the CO_3 component of African mammal tooth enamel (below). C_3 plants have an average isotope composition of ~−26‰, whereas C_4 plants cluster ~14‰ higher at ~−12‰. This isotope difference is passed onto herbivores: C_4 consumers have $\delta^{13}C$ values that cluster ~14‰ higher than C_3 consumers (0 to +2‰ vs. ~−12‰). The 14‰ offset between plant and enamel $\delta^{13}C$ (e.g., C_3 plants vs. enamel of C_3 plant consumers) reflects two components: the bulk $\delta^{13}C$ enrichment of an animal over its food source (2-4‰, DeNiro and Epstein 1978) and a large isotopic fractionation between the bulk $\delta^{13}C$ of an animal and the $\delta^{13}C$ of the carbonate component of bioapatite. Modern tooth compositions that fall between C_3 and C_4 diet end-members probably reflect mixed feeding.

(depending on local conditions) whereas most C_4 monocots have $\delta^{13}C$ values between −11 and −13‰ (Fig. 3).

In general, an animal's isotope composition depends on what it consumes (i.e., "You are what you eat" isotopically; DeNiro and Epstein 1976, 1978), and herbivores have dietary preferences—given a choice, some eat monocots, others eat dicots, and still others eat a mix. The resulting dietary C-isotope composition is transmitted to and recorded in all body tissues, including bioapatites. Thus monocot (C_4) consumers in warm regions will have a higher $\delta^{13}C$ than dicot (C_3) consumers, with mixed ($C_3 + C_4$) feeders in between (Fig. 3). Bone, tooth dentin, and tooth enamel are possible recorders of diet preferences in mammals, but the $\delta^{13}C$ of bioapatite is offset from the $\delta^{13}C$ of animal tissues. Studies of the isotope enrichment from diet to bioapatite are still underway. Field studies of large herbivores suggest an isotopic enrichment of ~14‰ (see summary of Koch 1998; Cerling and Harris 1999), whereas controlled feeding studies of mice, rats, and pigs suggest a smaller enrichment of ~9‰ (see summary of Koch 1998). No single herbivore has been studied both in the laboratory and field to reconcile these differences. The few data for carnivores suggest an isotopic enrichment of ~9‰ (Lee-Thorp et al. 1989). Because most C_4 plants are tropical grasses, at lower latitudes (< ~40° N or S latitude), stable carbon isotopes can readily distinguish between grazing mammals, which consume grass, from browsing mammals, which do not consume grass. At increasingly higher latitudes, temperate grasses use the C_3 photosynthetic mechanism, which ultimately limits identification of browsing vs. grazing.

Other factors can influence the $\delta^{13}C$ of plant tissues and hence bioapatite. CO_2 in closed canopy forests has a lower $\delta^{13}C$ because of degrading low-$\delta^{13}C$ plant material at the forest floor, and this ultimately stratifies the $\delta^{13}C$ of leaves within the forest (e.g., Medina and Minchin 1980, Medina et al. 1986). There are also physiological and isotopic responses to drought, salinity and light levels (e.g., see Farquhar et al. 1989), and ocean organic matter exhibits ^{13}C variations, both latitudinal and onshore vs. offshore, which ultimately impacts the isotopic compositions of marine animals farther up the food chain (Burton and Koch 1999). Because of these links among environment, diet, and isotopic compositions, $\delta^{13}C$ in bioapatite has enjoyed extensive application in a variety of paleoecological and paleodietary studies, including investigation of (a) feeding ecology, specifically C_3 vs. C_4 consumption (Ericson et al. 1981, Lee-Thorp et al. 1994, MacFadden and Cerling 1996, Koch et al. 1998, MacFadden 1998, MacFadden et al. 1999a, Sponheimer and Lee-Thorp 1999, Feranec and MacFadden 2000, Bocherens et al. 2001), (b) other aspects of animal feeding ecology, including marine vs. terrestrial components, canopy effects, etc. (Iacumin et al. 1997, Koch et al. 1998, Roe et al. 1998, Clementz and Koch 2001, Gadbury et al., 2000), and (c) terrestrial ecosystems or environmental conditions, using $\delta^{13}C$ of fossil bioapatites as a monitor (Thackeray et al. 1990; Koch et al. 1992, 1995b; Quade et al. 1992, 1994; Wang et al. 1993, 1994; Bocherens et al. 1996; MacFadden et al. 1996, 1998b; Cerling et al. 1997; Iacumin et al. 1997; Latorre et al. 1997; Sponheimer and Lee-Thorp 1999; Gadbury et al. 2000; MacFadden 2000; Luyt et al. 2000; Zazzo et al. 2000; Passey et al. 2002)

Tooth growth and isotopic zoning

Some phosphatic tissues such as bone can remodel during an animal's lifetime and hence provide an average composition, albeit possibly biased towards seasons of preferential tissue growth and/or bioapatite mineralization (Luz et al. 1990, Stuart-Williams and Schwarcz 1997). However other tissues such as teeth and shells precipitate progressively and are chemically and physically invariant after formation (aside from normal wear). Teeth are particularly well studied, and the following description of tooth growth is based on Hillson (1996). Enamel and dentin first nucleate at the dentin-enamel junction at the crown of a tooth, and grow away from each other (inward for dentin, outward for enamel), towards the tooth base through time (Fig. 4). In both materials an organic matrix seeded with crystallites is laid down, with an average organic content of about 30%. In enamel, the organic matrix is progressively replaced, ultimately creating a dense, compact structure, whereas dentin retains a high organic content. Tubular holes (tubules), ~1 μm in diameter, are moderately abundant in dentin but virtually absent in enamel. Once formed, enamel and dentin are nearly invariant chemically and structurally except for wear, and (unfortunately) dissolution of enamel to form caries. Later dentin can precipitate interior to tubules in some mammal orders (e.g., primates, but not rodents) and on the inner boundary with the pulp chamber.

A single tooth from a small animal may take a few weeks to months to form; a single tooth from a large herbivore such as an elk or horse may require over a year to form fully; and elephant tusks can grow for decades. Thus, by virtue of excellent preservation, teeth have the potential to record seasonal changes in diet long into the geological past. Blood chemistry changes very quickly, so that some tissues (e.g. hair) can record significant changes in diet over a few days time (Cerling and Cook 1998). Although it is attractive to suppose that diets could be recorded with such high fidelity in teeth, each volume of enamel or dentin forms over a significant time period, so that the diet signal is attenuated over weeks or months (Fisher and Fox 1998; Hoppe et al. 2001, Passey and Cerling, 2002). Despite these limitations, useful seasonal isotope changes have been identified using tooth enamel and dentin (Koch et al. 1989, 1998; Cerling and Sharp 1996; Fricke and O'Neil 1996; Sharp and Cerling 1996; Fricke et al. 1998a,b; Kohn et al. 1998, 2003;

Sharp and Cerling 1998; Feranec and MacFadden 2000, Gadbury et al. 2000, MacFadden 2000, Dettman et al. 2001; Bocherens et al. 2001; Fox and Fisher 2001).

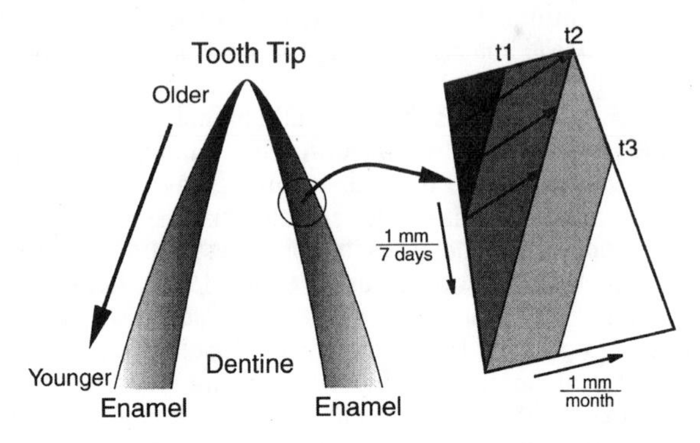

Figure 4. Sketch of how teeth from large herbivores grow. Enamel nucleates at dentin-enamel interface (DEI) at crown of tooth, and grows downward and outward through time. Crown-root growth rate is ~1 mm/week (Fricke et al. 1998b, Kohn et al. 1998). Angle between DEI and incremental growth lines in enamel is ~15° (e.g., Kierdorf and Kierdorf 1997), implying an outward growth rate of ~1 mm/25 days. From Kohn et al. (2003).

OXYGEN ISOTOPES IN BIOAPATITES

The oxygen isotopic composition of bioapatite depends on temperature and the isotopic composition of the fluid from which it precipitates. For mammals, there is a constant offset between the $\delta^{18}O$ of body water and PO_4 (~18‰ for mammals), and between PO_4 and CO_3 components of bioapatite (~8‰; Bryant et al. 1996, Iacumin et al. 1996a). Therefore, for modern samples or isotopically unaltered fossils, the PO_4 and CO_3 components are both equivalent measures of the $\delta^{18}O$ of the contributing organism. Both have been used to delineate modern isotopic trends.

Ecological trends

Oxygen isotope systematics in bioapatite were first investigated with carbonate shells (Longinelli 1965), which contain a small percentage of crystalline bioapatite. These data showed a clear dependence of isotope composition on temperature (Fig. 5a). It had been hoped that the isotope fractionations of PO_4 and CO_3 would have different temperature dependencies, so that a single shell could be used to determine both local water composition and temperature. Unfortunately, temperature dependencies of PO_4 and CO_3 fractionations are too similar (~0.23‰/°C). However, later work (Longinelli and Nuti 1973a, Kolodny et al. 1983, Luz et al. 1984, Luz and Kolodny 1985, Lécuyer et al. 1996a) also showed that the temperature-dependence of PO_4 fractionations is indistinguishable among invertebrates, fish, and mammals. In general it is believed that rapid isotopic equilibration of PO_4 with body water at body temperature is achieved intracellularly via mechanisms involving ATP, which is common to all organisms. Certainly, studies of microbial use of PO_4 indicate very rapid changes to dissolved PO_4 $\delta^{18}O$ (Blake et al. 1997, 1998). Precipitation of Ca-phosphates then preserves the equilibrated PO_4 composition. Although there has been some investigation of ancient conodonts and brachiopods (Luz et al. 1984b, Picard et al. 1998, Wenzel et al. 2000), most work has focused on systematics in thermoregulating terrestrial vertebrates, because the known temperature of precipitation simplifies isotope interpretation.

The first bioapatite work on thermoregulators indicated, as expected, that isotope composition varies systematically with local water composition (Longinelli 1984, Luz et

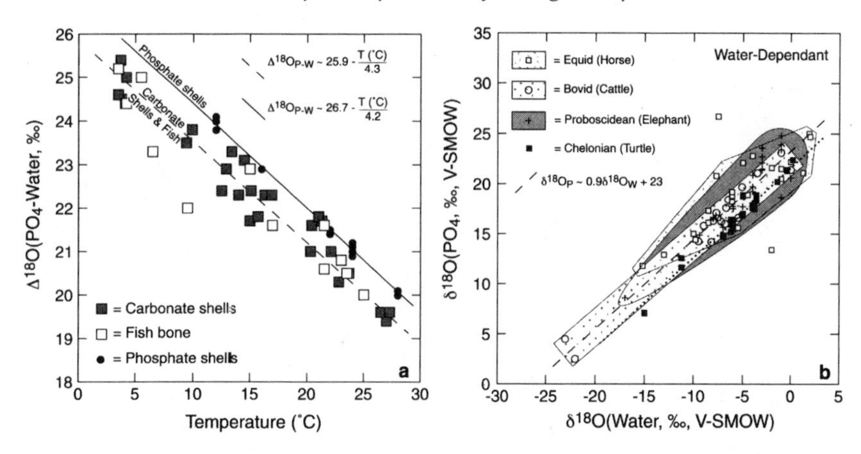

Figure 5. Plots of $\delta^{18}O$ of bioapatite vs. various environmental parameters. Almost all compositions for vertebrates are from modern bone PO_4. (a) Dependence of the $\delta^{18}O$ of the PO_4 component on temperature for invertebrates and fish. Isotopic offset between phosphate shells and other invertebrate bioapatite could be a species-specific difference, or could be an analytical artifact, as Lécuyer et al. (1996a) used the Ag_3PO_4 method, whereas Longinelli and Nuti (1973) and Kolodny et al. (1983) used the $BiPO_4$ method. Solid line is regression for phosphatic shells; dashed line is regression for carbonate shells and fish bone. Data from Longinelli and Nuti (1973a), Kolodny et al. (1983), and Lécuyer et al. (1996). (b-d) Plot of the $\delta^{18}O$ of the PO_4 component of bioapatite vs. local water for selected animals. Dashed line is regression of global data set of vertebrates, omitting turtles, deer, rabbits, and macropodids. (b) Obligate drinkers and turtles with greatest water dependence show a strong positive correlation between $\delta^{18}O$ of bioapatite and local water. The similar slope for turtles (dotted line) compared to mammals (dashed line) probably reflects behavioral thermoregulation and/or shell precipitation over a restricted temperature range. Data from Luz and Kolodny (1985), D'Angela and Longinelli (1990), Koch et al. (1991), Ayliffe et al. (1992), Bryant et al. (1994), Sanchez-Chillon et al. (1994), Delgado Huertas et al. (1995), Bocherens et al. (1996), Iacumin et al. (1996a), Barrick et al. (1999).

al. 1984a, Luz and Kolodny 1985). Humans were the first sources analyzed, but soon other animals were investigated, including cattle, pigs, sheep, mice, or whatever else was convenient and available (Longinelli 1984, Luz et al. 1984a, D'Angela and Longinelli 1990). These data all showed a reasonably simple relationship between bioapatite PO_4 and local water $\delta^{18}O$ (Fig. 5b-c). Bioapatite PO_4 calibration was solved and paleowater research could begin! However, three papers published nearly simultaneously and independently (Luz et al. 1990, Ayliffe and Chivas 1990, Koch et al. 1991) indicated that bioapatite isotopic systematics were not so simple after all. Bioapatite compositions can also depend critically on humidity (*h*, Fig. 5d-e), and different animals living in the same environment (i.e., same water and *h*) can have different isotope compositions (Fig. 5f). The variation seemed to be linked to the response of food source (plant) or surface water compositions to *h*, and possibly isotopic differences attending C_3 vs. C_4 diets.

More recent work has corroborated previous results and expanded the list of organisms. Rabbits show a strong *h*-dependence (Delgado Huertas et al. 1995), and sympatric animals with different diets and physiologies do have different isotope compositions, especially in arid settings (Kohn et al. 1996; Fig. 5f). Although an *h*-dependence is not obvious for some species, this may simply reflect small data sets. Humidity-dependencies have been found wherever they have been sought out, especially where data sets are large, although emphasis has been placed on drought-tolerant animals where *h*-dependencies should be highest. In the case of C_3 vs. C_4 consumption, the data of Bocherens et al. (1996) support a general trend towards a lower $\delta^{18}O$ with increased C_3

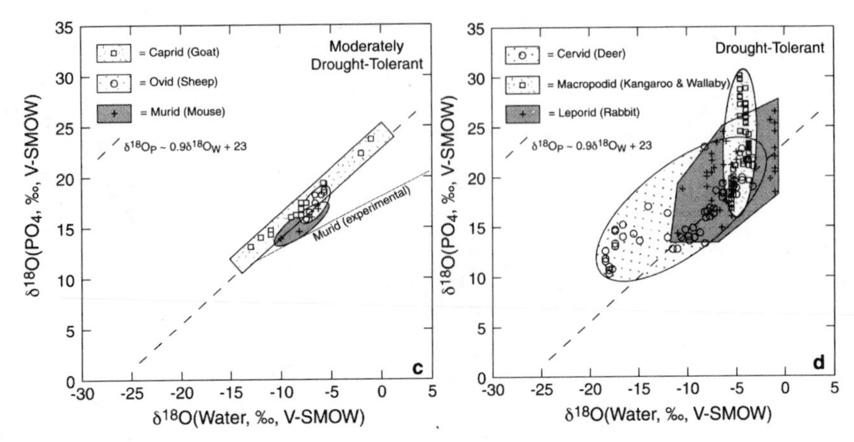

Figure 5, continued. Plots of $\delta^{18}O$ of bioapatite vs. various environmental parameters. Almost all compositions for vertebrates are from modern bone PO_4. (c and d) are plots of the $\delta^{18}O$ of the PO_4 component of bioapatite vs. local water for selected animals. Dashed line is regression of global data set of vertebrates, omitting turtles, deer, rabbits, and macropodids. (c) Moderately drought-tolerant animals show a strong positive correlation between $\delta^{18}O$ of bioapatite and local water. The thin solid line represents experimental results from laboratory mice; deviation from natural data may reflect how human-controlled settings cause isotopic deviations compared to natural settings. Data from Delgado Huertas et al. (1995) and D'Angela and Longinelli (1990). (d) Drought-tolerant animals show a poor correlation between $\delta^{18}O$ of bioapatite and local water. Data from Luz et al. (1990), Ayliffe and Chivas (1990), D'Angela and Longinelli (1990), and Delgado Huertas et al. (1995). (e) The $\delta^{18}O$ of the PO_4 component of bone for deer and macropodids re-plotted vs. humidity (*h*) after factoring out local water composition. The strong, negative correlation between $\delta^{18}O$ and *h* reflects influence of leaf water and leaf tissue composition on herbivore $\delta^{18}O$, as well as evaporative enrichment of local water at low *h*. The large variability in bioapatite composition at low humidity probably reflects uncertainties in assumed values of local water $\delta^{18}O$. Data from Luz et al. (1990) and Ayliffe and Chivas (1990). Lines show model predictions of humidity-dependencies (Kohn 1996). (f) $\delta^{18}O$ vs. $\delta^{13}C$ for the CO_3 component of bone and enamel for different sympatric species from East Africa. Boxes (multiple analyses) and circled star (single analysis) are for animals from Amboseli National Park (Koch et al. 1991, Bocherens et al. 1996), and illustrate a weak trend towards decreasing $\delta^{18}O$ of bioapatite CO_3 with decreasing $\delta^{13}C$, assuming that anomalously low $\delta^{18}O$ for hippos reflects aquatic adaptation or feeding ecology (Bocherens et al. 1996). More extensive data from tooth enamel for the Athi plains (individual points; T. Cerling, unpublished data) show greater scatter, and a tendency for some animals, such as gazelle, to plot far from the Amboseli line.

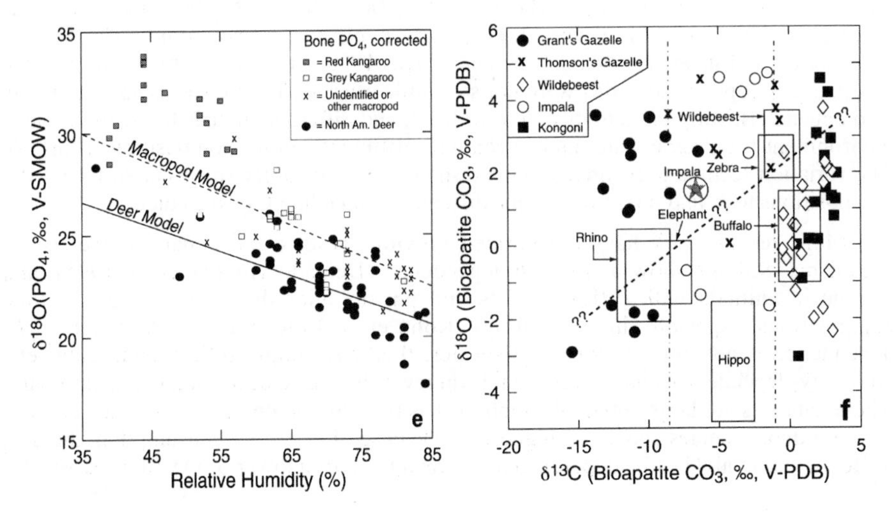

consumption. Because C_3 plants have a lower $\delta^{18}O$ than coexisting C_4 plants (Sternberg et al. 1994), the isotope trend was assumed to result from diet and incorporated in theoretical models (Kohn 1996). However, deviations from a simple correlation between $\delta^{18}O$ and $\delta^{13}C$ (Kohn et al. 1996) and the extreme variability exhibited by other animals (Fig. 5f) suggest that other factors play a major role. Data are now available for horses, goats, elephants, and turtles (Fig. 5b-c), as well as cetaceans (Yoshida and Miyazaki 1991, Barrick et al. 1992) and sharks (Vennemann and Hegner 1998, Vennemann et al. 2001). Interestingly, if bioapatite PO_4 compositions of turtles reflected mean annual temperature, then the turtle slope should be much shallower than that of mammals (Kohn 1996), because $\Delta^{18}O(PO_4\text{-water})$ increases with decreasing temperature (Fig. 5a), and lower temperatures correlate with lower meteoric water $\delta^{18}O$ (Dansgaard 1964). The steep turtle slope probably reflects behavioral maintenance of nearly constant body temperature, and/or preferential shell precipitation over a restricted temperature range. Thus it should not be assumed that heterothermic ("cold-blooded") animals can record different temperatures in different environments (i.e., contra models of Kohn 1996).

In sum, modern studies show that bioapatite $\delta^{18}O$ depends on: (1) water composition and temperature for non-thermoregulating aquatic organisms (as is well described for carbonate-secreters), or (2) water composition, humidity and—in some terrestrial vertebrates—diet.

Mass balance

Mass balance models (Luz and Kolodny 1985, Luz et al. 1990, Ayliffe and Chivas 1990, Cormie et al. 1994, Bryant and Froehlich 1995, Kohn 1996) provide a simple way of understanding oxygen isotope systematics in animal body water. Body water is the dominant control on bioapatite $\delta^{18}O$, because the isotopic compositions of the PO_4 and CO_3 components have a constant offset relative to the fluid from which the bioapatite precipitates. The key to these models is the identification of input and output oxygen components, and fractionations of these components with respect to two variables: local water [$\delta^{18}O(LW)$] and body water [$\delta^{18}O(BW)$]. The main input and output components are listed in Table 2, and are based on water turnover and energy expenditure studies of wild terrestrial herbivores. There are many other components, including water in plant stems (which is not very fractionated), oxygen in plant tissues (which is extremely fractionated), water vapor in inspired air, oxygen in urea, etc., but the seven factors listed in Table 2, normalized to 100%, are sufficient for illustration.

To maintain isotope equilibrium, $M(in)*\delta^{18}O(in) = M(out)*\delta^{18}O(out)$, where M is moles of oxygen. Summing terms in Table 2:

$$0.25 * [\delta^{18}O(BW) + 38.65] + 0.4 * [\delta^{18}O(BW) + 0] + 0.25 * [\delta^{18}O(BW) - 8.5] =$$
$$0.25 * 15 + 0.45 * [\delta^{18}O(LW) + 26.2 * (1 - h)] + 0.3 * \delta^{18}O(LW) \quad (1)$$

or, propagating physiological variability and uncertainties,

$$\delta^{18}O(BW) = 9.85 \pm 4.6 - 11.8 \pm 3.9 * h + 0.75 \pm 0.3 * \delta^{18}O(LW) \quad (2)$$

Several points are worth noting: (1) Predicting an isotope composition a priori is extremely uncertain for an animal whose physiology is poorly known, such as extinct taxa. Theoretically, $\delta^{18}O$(Bioapatite) for a species could be distinct, due to different offsets, h dependencies, or slopes on a $\delta^{18}O$(Bioapatite) vs. $\delta^{18}O(LW)$ plot. However, the fact that many animals have similar compositions (Fig. 5b-c) indicates that physiological factors commonly offset each other. (2) There is a linear relationship among local water composition, h, and body water composition, with a direct h-dependence reflecting a leaf water source. (3) The slope of $\delta^{18}O(BW)$ or $\delta^{18}O$(Bioapatite) vs. $\delta^{18}O(LW)$ will be essentially 1 – (% air O_2), and so should be ~0.75 (Kohn 1996). The occurrence of many isotope slopes ~0.9 (Fig. 5b-c) could indicate that animals take in more water than expected. In fact, domestic animals do consume more water per mole of air O_2 than their wild counterparts (e.g., Nagy and Petersen 1988), which would steepen slopes. Alternatively, not accounting for h could bias slopes if most of the high $\delta^{18}O$(Bioapatite) values occur in drier environments. (4) Measured h-dependencies are more extreme than predicted for some species (Fig. 5e). This probably reflects a h-dependency to the $\delta^{18}O$ of local water compositions, as local water will be enriched by evaporation in dry environments, a feature which is not considered in the predictive models.

Table 2. Most important oxygen input and output components affecting animal isotope compositions

Inputs	*Air oxygen*	*Leaf water*	*Drinking water*	
% of total[1]	25 ± 5	45 ± 15	30 ± 30	
$\Delta^{18}O$[2]	(fixed ~15‰)[3]	~26.2[4] * (1 – h)	0	
Outputs	*CO_2*	*Urine + sweat + fecal water*	*Exhaled water vapor*	*Transcutaneous water vapor*
% of total[1]	25 ± 5	40 ± 15	25 ± 10	10 ± 5
$\Delta^{18}O$[5]	38.65[6]	0	-8.5[7]	-18 ± 6[8]

[1]The percentage variability corresponds to actual different physiologies of animals (e.g., some are drought-tolerant, others require drinking water).

[2]Isotope fractionation of input component with respect to local water.

[3]Air O_2 has a composition of ~22.5‰, but there is a kinetic isotope fractionation of ~-7.5‰ associated with O_2 uptake in the lungs (Epstein and Zeiri 1988, Zanconato et al. 1992).

[4]Recent research is working to understand humidity dependencies better.

[5]Isotope fractionation of output component with respect to body water.

[6]Equilibrium fractionation between CO_2 and liquid water at 37°C.

[7]Assumes most loss is oral, but some is nasal.

[8]Virtually unmeasured; range indicates uncertainty in the correct value(s).

Isotopic zoning

Any perturbation to oxygen input or output amounts or $\delta^{18}O$ compositions will change the isotopic composition of bioapatite (Table 2, Eqn. 1, Eqn. 2) and will induce isotope zoning in teeth or other progressively precipitated phosphatic tissues. The most important perturbations are seasonal variations in local water composition and h. Other factors, such as temperature, do affect physiology (e.g., water turnover), but their isotopic effects are not well characterized. The isotopic expression of seasonal variations provides an isotopic proxy for seasonal climate, both in modern and fossil materials. This isotope zoning is extremely important, both because it must be accounted for when analyzing fossils (i.e., a single analysis is inadequate for average compositions unless it samples the entire length of the tooth), and because isotope seasonality can be a useful paleoclimatic indicator. Zoning studies have been conducted using both modern and fossil teeth (Koch et al. 1989, 1998; Fricke and O'Neil 1996; Fricke et al. 1998b; Kohn et al. 1998; Sharp and Cerling 1998; Feranec and MacFadden 2000, Gadbury et al. 2000, MacFadden 2000, Dettman et al. 2001; Bocherens et al. 2001; Fox and Fisher 2001). In general, large teeth can require an entire year to grow, whereas smaller teeth may form in only a single season. However, because different teeth may start growing at different times, analysis of several smaller teeth can sometimes retrieve the yearly seasonal signal (Kohn et al. 1998, Bocherens et al. 2001). Tusks are especially useful for seasonal isotope studies (Koch et al. 1989, Koch et al. 1995b, Fox 2000, Fox and Fisher 2001) because the dentin (e.g., for elephants) and enamel (e.g., for gomphotheres) can potentially record decades of isotopic information.

Several factors damp climatically induced isotope seasonality. The most important are reservoir effects within the local environment and within the organism itself. Large water bodies such as rivers and lakes have less isotopic variability than seasonal precipitation. Drinking water from rivers and lakes or plants growing nearby will show less isotopic variability than would be predicted solely on the basis of seasonal variations in h or precipitation $\delta^{18}O$. Similarly, an organism cannot respond instantaneously to a change in the environment. The oxygen flux per day through an organism is ordinarily a small percentage (5-10%) of its total oxygen content. Therefore, compositional extremes within the environment will be smoothed, and a time-lag will develop between the predicted (instantaneous) and actual isotope composition of the species. For a sinusoidal variation in environmental composition with compositional amplitude ΔC, and a residence time τ, the damping factor D and time lag Δt will be (Albarède 1995),

$$D = \frac{\Delta C_{out}}{\Delta C_{in}} = \frac{1}{\left[\frac{2\pi\tau}{365} + 1\right]^{1/2}} \quad (1)$$

$$\Delta t = \frac{365}{2\pi}\tan^{-1}\left[\frac{2\pi\tau}{365}\right] \quad (2)$$

where τ (in days) is the total oxygen pool divided by the daily flux. If 5-10% of total oxygen is turned over daily, then τ is 10-20 days. Applying Equation (2), this implies that the isotope composition of an animal varies by only 85-90% of the total seasonal signal, and has a time lag of 2-3 weeks. Drought-tolerant animals (e.g., camels with τ~50 days) will have greater damping and a longer time lag, whereas extremely water-dependent animals (e.g., humans with $\tau \approx 7$ days) can potentially record climate in their phosphatic tissues with higher fidelity. However, a last damping factor is the precipitation of the enamel itself. The enamel maturation process can lead to significant time averaging, and thick enamel crowns can take months or even a year to form (Hillson 1996). Thus the highest fidelity can be obtained only for relatively thin enamel, or by using sampling techniques that remove a shallow band of enamel, where maturation is relatively fast.

Modern gazelle teeth from near Nairobi illustrate how teeth record yearly seasonality (Kohn et al. 1998; Fig. 6). In Kenya, the gazelle-birthing season and timing of tooth eruptions are well known, which allows the zoning patterns to be placed in a rough chronology. International Atomic Energy Agency (IAEA) stations provide seasonal water compositions for East Africa. Most locations in the world have either a single rainy or cold season, leading to a single isotope low, and a quasi-sinusoidal yearly isotope variation. However, this part of East Africa is a little unusual because it experiences a double rainy season, so that each year there are two isotope lows. Combining IAEA data with observations on gazelle diet, physiology, etc., permits construction of a theoretical isotope model. Such a model has large absolute uncertainties (e.g., predicted average composition for the year could vary by a couple permil), but relative differences among seasons are well resolved (uncertainties of much less than 1 permil). There are also uncertainties in the temporal placement of compositions (within the dashed line plus solid bar), because the same tooth can grow at different times in different individuals. Nonetheless, the data can be reconciled extremely well with the seasonal model, implying that seasonal climate change likely drives compositional changes recorded in this tooth enamel. Fricke et al. (1998b) also analyzed teeth from areas close to IAEA stations, and although they found that tooth enamel is ubiquitously zoned, the compositional zoning could not always be modeled based solely on seasonal climate factors. Many teeth recorded smaller variability than predicted, especially in settings where water compositions changed very rapidly. The damping likely reflects reservoir effects, within the environment, the animal, or both. Apparently terrestrial vertebrates do not always record extremely rapid environmental changes well.

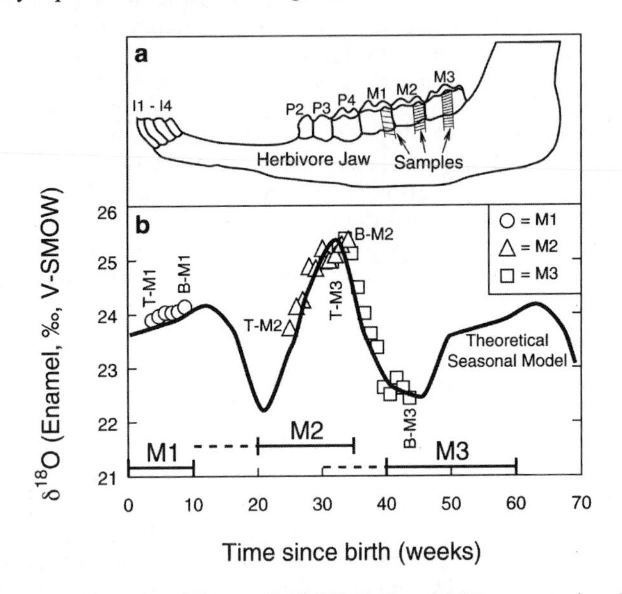

Figure 6. (a) Identity of teeth in a typical herbivore jaw. P2-P4 are premolars, M1-M3 are molars, and I1-I4 are incisors. Relative order of premolar and molar formation is: M1 < M2 < M3 < P2 ≤ P3 ≤ P4. Samples were collected by cutting a strip of enamel from molars, and sub-sampling at ~1–2 mm resolution. (b) Oxygen isotope zoning for bulk enamel from molars of a modern Nairobi gazelle vs. time since birth. By knowing the birth season of gazelles and when different teeth first erupt (i.e., appear above the gum line; solid bars), and by accounting for earlier growth prior to eruption (dashed line), data can be reconciled with a theoretical model of seasonal isotope compositions (thick black line). "T" and "B" refer to compositions from the top and base of each tooth crown. From Kohn et al. (1998), but with modifications to include reservoir effects within gazelles (Eqn. 1 and 2).

STRONTIUM ISOTOPES IN BIOAPATITES

Ordinarily, strontium isotope ratios ($^{87}Sr/^{86}Sr$) are considered radiogenic, because of the decay of ^{87}Rb to produce ^{87}Sr. However, because apatite accepts virtually no Rb, there is no radiogenic source able to change $^{87}Sr/^{86}Sr$ *in situ*. Thus, strontium can be considered a stable isotope, useful as an ecological or environmental tracer. Strontium is taken up significantly when bioapatite forms, but concentrations and isotopic compositions can also change after burial, with dentin and bone more strongly affected than enamel (see Tuross and Trueman, this volume). Therefore, applications must either (a) separate the original unaltered bioapatite crystallites, using chemical leaching (e.g., Sillen 1986, Sealy et al. 1991), or (b) assume that burial and diagenesis have no isotope effect or completely reset isotope compositions to pore-fluid compositions (e.g., Staudigel et al. 1985, Schmitz et al. 1991).

One application of Sr isotopes makes use of the change in seawater Sr isotope composition through time for age determinations (e.g., DePaolo and Ingram 1985). The isotope composition of modern marine bioapatite is indistinguishable from modern seawater (e.g., Staudigel et al. 1985, Schmitz et al. 1991, Koch et al. 1992, Vennemann et al. 2001), so if unaltered material is analyzed, it can in theory be matched to the seawater Sr curve to define an age (Staudigel et al. 1985, Koch et al. 1992, Ingram 1995, Holmden et al. 1996, Vennemann and Hegner 1998, Martin and Haley 2000; Fig. 7). However, Sr isotope compositions of some bioapatites give erratic or inconsistent results compared to expected values for samples of known age, or in comparison with coexisting carbonate, possibly due to PO_4 recycling and Sr overprinting (Staudigel et al. 1985; Schmitz et al. 1991, 1997; Denison et al. 1993). Chemical leaching does not always remove extraneous Sr, even for enamel (Tuross et al. 1989, Koch et al. 1992, Elliot et al. 1998, Barrat et al. 2000). The solubility product of fluorapatite is less than that of hydroxyapatite (Moreno et al. 1977), and so the last material to be dissolved may well be diagenetic, rather than biologic.

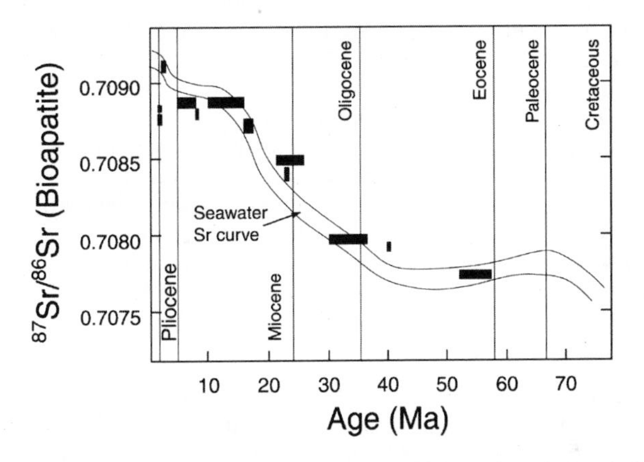

Figure 7. Strontium isotope compositions of fossil teeth and phosphatic debris vs. time, with the carbonate-based seawater Sr curve for reference. General correspondence between bioapatite and seawater $^{87}Sr/^{86}Sr$, and large variation in seawater $^{87}Sr/^{86}Sr$ suggests that isotope compositions of fossil marine bioapatites could be used geochronologically. However, several bioapatite compositions depart significantly from the seawater curve, even for Plio-Pleistocene samples, suggesting diagenetic alteration in many cases. From Staudigel et al. (1985).

Terrestrial $^{87}Sr/^{86}Sr$ is commonly elevated compared to marine $^{87}Sr/^{86}Sr$, because of radiogenic ^{87}Sr from old continental crust, and it is possible to use this $^{87}Sr/^{86}Sr$ difference as a measure of fish origin (Schmitz et al. 1991, 1997). Ocean water has a Sr concentration that is ~100 times higher than fresh water, so even a small marine component would overwhelm any freshwater isotope composition. However, freshwater fish consistently show higher $^{87}Sr/^{86}Sr$ compared to marine fish, permitting a freshwater vs. brackish/marine distinction (Schmitz et al. 1991, 1997). Similarly, fish living in different drainages exhibit isotopic differences correlated with bedrock $^{87}Sr/^{86}Sr$ (Kennedy et al. 1997).

Several studies have used isotope zoning, or differences in isotope compositions of samples of different individuals found together, to identify the occurrence and geographic patterns of migration (Ericson 1985; Sealy et al. 1991; Price et al. 1994a,b, 2000; Koch et al. 1995a; Ezzo et al. 1997; Grupe et al. 1997; Sillen et al. 1998; Hoppe et al. 1999). The main problem in validating such studies for fossil materials is whether the diagenetic overprint can be convincingly eliminated, and the general consensus is that it cannot for bone (e.g., see Tuross and Trueman, this volume), except perhaps in some extraordinary settings. Enamel has greater potential, as it is less susceptible to diagenetic alteration. However, it is not immune to changes in trace element concentrations either (e.g., Kohn et al. 1999).

EXAMPLES AND APPLICATIONS

The Rapid Increase in C_4 Ecosystems (RICE)

The C_4 photosynthetic mechanism is a geologically rather recent development. Grasses had already evolved possibly as early as the Late Cretaceous and certainly by the early Eocene (Linder 1986, Crepet and Feldman 1991, Crepet and Herendeen 1992; see review of Jacobs et al. 1999). However, the earliest definitive C_4 macrofossils are from ~12 Ma (Dove Spring Formation, middle of Chron C5A; Nambudiri et al. 1978, Whistler and Burbank 1992, Jacobs et al. 1999; D. Whistler, pers. comm., 2001). All pre-mid-Miocene grasses apparently were C_3, yet today C_4 grasses dominate many low latitude ecosystems. Thus the origins and rate(s) of increased C_4 abundance worldwide constitute a major paleoecological issue. Because of the $\delta^{13}C$ difference attending C_3 vs. C_4 photosynthesis and because isotopic compositions of animals reflect the available plant ecosystem (i.e., their diet), $\delta^{13}C$ compositions of fossil bioapatites provide a critical line of evidence regarding when C_4 plants arose and how rapidly they spread. Reconstructing plant biomass from herbivore diet is not completely straightforward, however, because dietary selectivity filters the ecological signal (e.g., MacFadden et al. 1999). An absence of high $\delta^{13}C$ values does not imply an absence of C_4 plants, if animals simply chose not to eat them, or if they were not abundant. Nonetheless, if high $\delta^{13}C$ values are found, especially for a large animal with large daily food requirements, then C_4 plants must have been locally abundant.

Identifying clear C_4 consumption from stable isotopes in teeth first requires delineating the range of $\delta^{13}C$ values expected for pure C_3 consumers: only if $\delta^{13}C$ values exceed the range for a pure C_3 diet can a clear C_4 dietary component be inferred. The highest $\delta^{13}C$ values for enamel of modern-day large C_3 consumers is ~-8‰ [$\delta^{13}C$ (C_3 plant) $\leq$ -22‰ plus a 14‰ offset; Fig. 3]. However fossil fuel burning since the industrial revolution has decreased the $\delta^{13}C$ of atmospheric CO_2 and modern plant tissues by at least 1‰ (Friedli et al. 1986, Marino and McElroy 1991). Thus a conservative isotopic cutoff for identifying ancient C4 consumption from the $\delta^{13}C$ of herbivore tooth enamel is ~-7‰. This cutoff is extremely conservative because modern C_3 plants with $\delta^{13}C \geq$ -22‰ are rare. A pre-industrial bioapatite $\delta^{13}C$ of ~-7‰ for a pure C_3 feeder would imply either that an entire ecosystem was anomalous, or that an animal selectively fed on the anomalous plant(s) within an ecosystem.

Figure 8 summarizes the work of Cerling et al. (1997) on the $\delta^{13}C$ of fossil herbivore teeth from ~20 Ma to the present, emphasizing equids and proboscideans, which consume C_4 plants if available. Clearly C_4 plants must have been rare on Earth before 8 Ma, as herbivore $\delta^{13}C$ values are all lower than -7‰ and all but one are lower than -8‰. Of course, if C_3 plants had $\delta^{13}C$ values of -24 to -26‰, then a small dietary component of C_4 may have been present since 14-19 Ma (Fig. 8; Morgan et al. 1994). However, by 6-8 Ma, herbivores in Pakistan, Africa, South America, and southern North America show a significant C_4 dietary component, indicating a Rapid Increase in C_4 Ecosystems (RICE[1])

---

[1] Ironically, rice (*Oryza sativa*) is C_3, but efforts are underway to genetically modify it to be more C_4-like in order to improve grain yields.

worldwide. It is especially important that RICE was geologically rapid, but apparently not completely synchronous. C_4 plants were first evident at low latitudes and appeared later at higher latitudes (Cerling et al. 1997, Fig. 8).

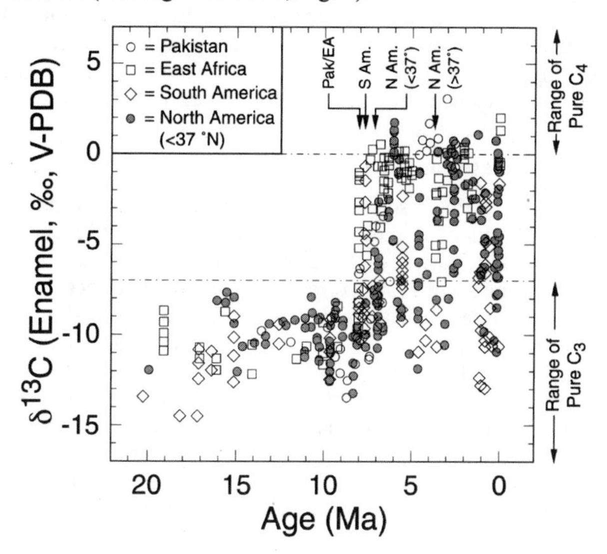

Figure 8. $\delta^{13}C$ of the CO_3 component of tooth enamel vs. age for global dataset of equids and elephantids at low latitudes, showing significant dietary change at 7-8 Ma. Arrows show times of shift to C_4-rich diets. High latitude data from western Europe show no isotopic shift, and data from North America north of 37°N show a smaller shift at 3-4 Ma (data not shown). From Cerling et al. (1997).

Despite the occurrence of RICE during the latest Miocene, the mid-Miocene C_4 grass fossils (Nambudiri et al. 1978) and sporadic high $\delta^{13}C$ values up to -4‰ reported recently for 10-12 Ma teeth (Clementz and Koch 2001) imply that C_4 plants were already present on Earth before ~8 Ma. Thus the time between 8 and 6 million years ago evidently was a period of rapid C_4 biomass expansion, and not initial evolution. Although the shift observed in the first data from Pakistan was originally ascribed to tectonism (Quade et al. 1989, 1992), the later recognition of RICE's global character shifted focus to a global rather than regional chemical and/or climatic causes. It is extremely important that one of the main differences in C_3 vs C_4 plant physiology is the ability of C_4 plants to concentrate CO_2 in bundle sheath cells (Fig. 2). Because of this unique capability, decreasing CO_2 levels are a particularly attractive explanation for RICE (e.g., Ehleringer et al. 1991, Ehleringer and Cerling 1995, Cerling et al. 1997). If CO_2 levels dropped below some critical threshold, C_4 plants would have had a photosynthetic advantage, and could have spread rapidly. Furthermore, because C_4 plants compete better with C_3 plants at higher light levels, the CO_2 hypothesis also explains why RICE occurred first at low latitudes and spread to higher latitudes (Cerling et al. 1997).

The decreasing CO_2 hypothesis for RICE remains controversial. Recent work supports low CO_2 levels during the Miocene (Pagani et al. 1999a,b; Pearson and Palmer 2000) so that CO_2-limitation may well have promoted Miocene C_4 diversification and increased abundance. However, some studies suggest that CO_2 levels have been uniformly low since at least 20-25 Ma, prior to RICE, with no pronounced dip or trend to lower values at 6-8 Ma, (Pagani et al. 1999a,b; Pearson and Palmer 2000). So whereas a critical threshold may have been crossed at 7-8 Ma, other climatic triggers besides CO_2

levels continue to be considered. Two possibilities (Freeman and Colarusso 2001) are increased aridity, as C_4 plants are more drought-tolerant than C_3 plants, or increased seasonality with longer or warmer growing seasons. For the western US and Indian subcontinent, regional tectonics strongly influences aridity and climate seasonality, and mountain ranges and plateaus in these areas were evolving in the late Miocene. Thus, the rise of C_4 plants may perhaps in part reflect tectonic factors after all (Quade et al. 1989, 1992), although such a mechanism still has difficulty explaining the nearly simultaneous occurrence of RICE on several continents.

Terrestrial-marine climate coupling

An important goal in Earth science is to characterize past climates and to identify chemical or physical causes of their change (e.g., atmospheric chemistry, mountain building, ocean circulation patterns, orbital forcing, etc.). Most global climate research has focused on the marine record for two reasons: (1) climate directly impacts sea surface temperatures, ice volume, and carbon balance, and (2) marine carbonates are extremely common and well-preserved, and so provide a continuous record of Cenozoic climate via species abundances and stable isotopes of carbon and oxygen. However it is difficult to extrapolate from the oceans to the continents, and biogenic carbonates simply are not well represented or preserved in terrestrial sequences. In contrast, terrestrial bioapatites have been sampled on the continents for centuries in the form of fossil bones and teeth, and are well curated in paleontological collections. Thus, bioapatites provide an unparalleled resource for investigating continental paleoclimate, and allow the terrestrial and marine records of global climate to be linked.

One of the best-characterized links between marine and terrestrial paleoclimate concerns the Paleocene-Eocene thermal maximum (PETM). High-resolution marine cores delineated a rapid drop in benthic foraminifera $\delta^{13}C$ and $\delta^{18}O$ at ~55 Ma. The $\delta^{18}O$ drop corresponds to a deep-sea temperature rise of 5-6°C in ten thousand years or less (Kennett and Stott 1991, Bralower et al. 1995, Thomas and Shackleton 1996). The cause is now inferred to be catastrophic methane release from clathrates on the continental margins (Dickens et al. 1995, 1997; Dickens 1999). Because methane is a greenhouse gas and has an extremely low $\delta^{13}C$ value (~-60‰), a large release would affect both temperature (oxygen isotopes) and carbon isotopes.

To examine the terrestrial record during the latest Paleocene, fossil teeth from the Paleocene-Eocene sequence of the Bighorn Basin, Wyoming were analyzed for $\delta^{13}C$ and $\delta^{18}O$ (Koch et al. 1992, 1995b; Fricke et al. 1998a). These data (Fig. 9) show that there was a simultaneous dip in carbon isotopes and bump in oxygen isotopes at the end of the Paleocene. Paleosol carbonates show similar trends. These isotope dips and bumps correspond temporally with the $\delta^{13}C$ and $\delta^{18}O$ marine spikes, as expected if terrestrial and marine climates responded in concert with a global PETM event. Specifically, if the global bioreservoir $\delta^{13}C$ dipped at 55 Ma, then all organic materials should record this event, including plant communities, herbivores, and marine carbonates. The increase in temperature should have caused a decrease in the $\delta^{18}O$ of marine carbonate, assuming ocean water composition did not change significantly, because Δ(carbonate-water) decreases with increasing temperature. However, because meteoric water $\delta^{18}O$ increases with increasing temperature, a global rise in temperature should have caused an increase in terrestrial bioapatite $\delta^{18}O$, as shown by the teeth. The PETM event also corresponds to an enormous faunal radiation at the boundary between the Clarkforkian and Wasatchian land mammal ages (Gingerich 1989). In North America, this boundary marks the first appearance of artiodactyls (antelopes, pigs, cattle, etc.), perissodactyls (horses, tapirs and rhinos), and true primates. This radiation implies that global climate change either helped drive evolution directly, or promoted major immigration and intercontinental faunal exchange (Koch et al. 1995b). Terrestrial-marine paleoclimate links have also been investigated at the Permo-Triassic boundary (Thackeray et al. 1990, MacLeod et al. 2000).

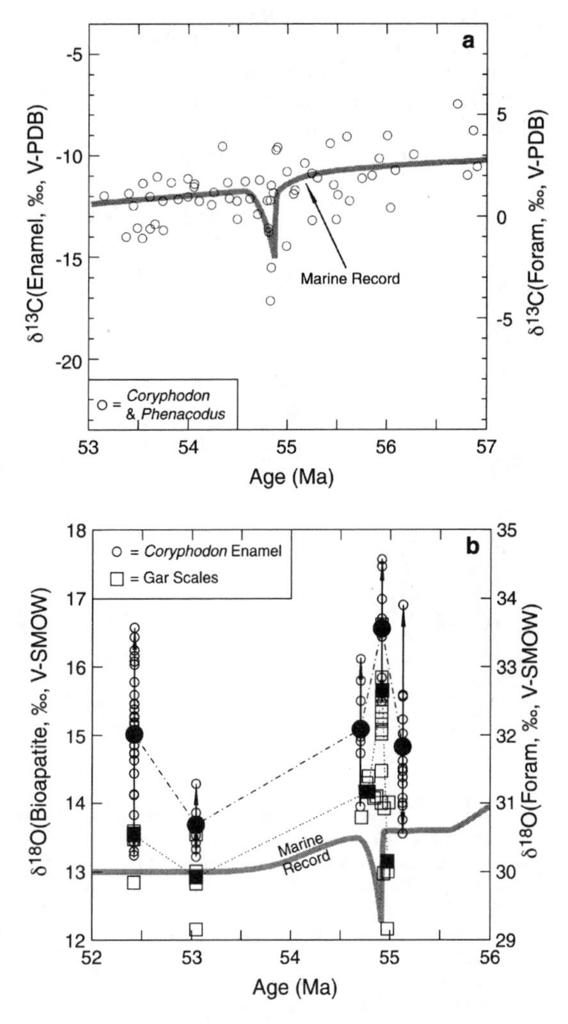

Figure 9. Marine vs. terrestrial isotope variations across the Paleocene-Eocene thermal maximum for (a) $\delta^{13}C$ of the CO_3 component of tooth enamel, and (b) $\delta^{18}O$ of the PO_4 component of tooth enamel and fish scales. Solid symbols in (b) are average $\delta^{18}O$. Note correlated dip in $\delta^{13}C$ for marine carbonates and for the large herbivores *Phenacodus* and *Coryphodon*, and correlated drop in marine carbonate $\delta^{18}O$ and rise in *Phenacodus* and gar $\delta^{18}O$. Terrestrial data from Koch et al. (1992, 1995b) and Fricke et al. (1998a). Marine data averaged from Shackleton et al. (1985), Kennett and Stott (1991), Bralower et al. (1995), and Thomas and Shackleton (1996).

Tectonics

Stable isotopes of bioapatites can be used to investigate tectonics via a paleoclimate link. One direct climate-tectonics link is the generation of rain shadows by mountain ranges (Kohn et al. 2003). Mountain ranges in western North America profoundly affect isotope compositions of meteoric water. The N-S trending Sierra Nevada, Cascades, and

Coast Ranges force east-traveling weather systems upward, cooling clouds and causing intense precipitation on western slopes, which produces a rain shadow on the eastern slopes and inland areas. Rainfall can be five times higher and humidity 30% greater on the coast compared to the eastern interior. Rayleigh distillation during rainout depletes clouds and subsequent rainfall in ^{18}O (Dansgaard 1964), so that precipitation in the eastern interior has a much lower $\delta^{18}O$ than along the coast (e.g., Sheppard et al. 1969, Coplen and Kendall 2000; Fig. 10). The degree of distillation (isotope depletion) is principally dependent on range height and lapse rate, but for modern heights, the difference in meteoric water compositions across these ranges is ~7-8‰.

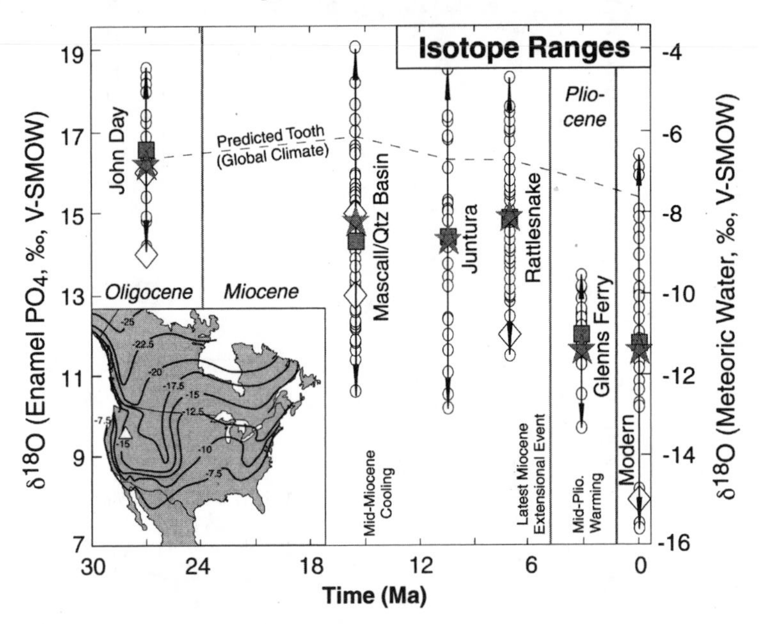

Figure 10. $\delta^{18}O$ of enamel PO_4 for fossil equids from central Oregon and western Idaho, illustrating a systematic shift through time. Open circles are individual measurements, shaded squares are mean and median compositions (indistinguishable), and stars are midpoint compositions. Compositional ranges reflect isotopic zoning within teeth as well as isotopic differences among teeth of same age. Water compositions (open diamonds) are inferred based on bioapatite compositions and humidity estimates from paleosols and paleoflora. Dashed line shows predicted shift to tooth $\delta^{18}O$ resulting solely from secular changes to global climate. Data from Kohn et al. (2003). Inset shows $\delta^{18}O$ isopleths for meteoric water across North America (modified from Sheppard et al. 1969, Coplen and Kendall 2000). Triangle on the map shows the approximate study area.

Kohn et al. (2003) investigated the development of the Cascade rain shadow by analyzing fossil equid teeth in interior Oregon and western Idaho. These teeth show a ~5‰ decrease in midpoint and mean $\delta^{18}O$ compositions since the late Oligocene, essentially in 2 periods: between 27 and 15.4, and between 7.2 and 0 Ma (Fig. 10). In the intervening mid- to late-Miocene, tooth $\delta^{18}O$ essentially stabilized. Large changes in isotope ranges are also apparent, with ≥7‰ seasonal and yearly variability in the mid-Miocene, decreasing to only 3-4‰ in the mid-Pliocene, and expanding to ~8‰ in modern samples.

In contrast to the PETM, there is little evidence for a global climate signal in these data. Marine records of warming from the Oligocene to Miocene are not reflected in any increase in bioapatite $\delta^{18}O$, and cooling through the Miocene caused no obvious change to mean or median compositions. Semiquantitative modeling of predicted meteoric water and tooth isotope compositions, accounting for both temperature and ice volume effects suggests that global climate change since 27 Ma should have decreased tooth $\delta^{18}O$ by only ~-1‰. That is, of the 5‰ decrease in tooth $\delta^{18}O$ that occurred since 27 Ma, only ~1‰ can be ascribed to changes in source (coastal) water compositions resulting from global climate change. The remaining ~4‰ must instead result from monotonic uplift in topography, first between 27 and 15.4 Ma, and then further between 7.2 and 0 Ma. One important feature is the constancy of mean isotope compositions between 15.4 and 7.2 Ma, which corresponds with a change in volcanic style and/or abundance (e.g., Priest 1990). Arc volcanism was abundant from ~35 to ~18 Ma, and ~8 to 0 Ma. During 18 and 8 Ma, the Columbia River flood basalts erupted (mainly 18-14 Ma). Thus, the Cascades appear to have gained height and impacted isotope compositions when the arc was volcanically active, but changed height very little when the Columbia River Basalts erupted.

The observed isotope shift in tooth compositions (~4‰) is much smaller than the modern-day isotope difference for rainwater across the Cascades (~7‰). This could be because the range had already begun forming prior to the oldest samples at 27 Ma, but a more likely explanation involves the dependence of bioapatite $\delta^{18}O$ on water composition and relative humidity. Global correlations between bioapatite and local water $\delta^{18}O$ suggest a ~4‰ shift in bioapatite composition corresponds to a larger ~4.5‰ shift in water composition (Fig. 5b). Development of the rain-shadow over central Oregon also caused a decrease in relative humidity of ≥~15%, as indicated by paleoflora and paleosols (e.g., Ashwill 1983, Retallack et al. 2000). Theoretical models (Kohn 1996) and observations for deer (Luz et al. 1990; Fig. 5e) indicate a ~0.1-0.2‰ increase in bioapatite $\delta^{18}O$ for each 1% decrease in relative humidity. The ≥~15% decrease in humidity documented by flora and paleosols translates into a ≥1.5 to 3‰ increase in bioapatite $\delta^{18}O$ since 27 Ma. Subtracting off the humidity effect and accounting for the fact that bioapatite and local water do not have a 1:1 correlation implies that the 4‰ decrease in bioapatite $\delta^{18}O$ likely reflects a decrease in meteoric water $\delta^{18}O$ of ≥6-8‰. The similarity of the inferred shift to the modern isotope effect across the range implies that most topographic uplift occurred subsequent to 27 Ma.

Another link between climate and tectonics is the generation of monsoons by plateaus. The three major plateaus on Earth (Tibet, Puna/Altiplano, and Colorado) are each associated with a strong monsoon (in southeast Asia, central South America, and southwestern North America, respectively). The two other monsoons on Earth are either very weak (West Africa), or arguably driven by the Tibetan plateau (northern Australia). So investigations of isotope seasonality as a proxy for climatic seasonality can potentially elucidate the rates of development of a seasonal monsoon, and its attendant plateau. In general this climate-tectonics relationship has been rarely exploited, but Dettman et al. (2001) did show that isotope seasonality in fossil equid teeth and freshwater clam shells of the Indian subcontinent has remained essentially unchanged for over 10 million years, implying that the Tibetan plateau was already sufficiently broad and high by then to cause a monsoon. Thus, Dettman et al.'s data contradict popular models of geologically rapid rise at 7-8 Ma or later (although the plateau may have gained additional height or extent after 10 Ma), and limit the range of tectonic processes that led to its formation.

Dinosaur thermoregulation

A perennial physiological question concerning dinosaurs is whether they were capable of maintaining relatively uniform temperatures throughout their bodies (homeothermy; nominally taken to be ±2°C) via intrinsic mechanisms (endothermy). Thermoregulation in fact can be achieved by a variety of mechanisms. Even ectotherms (animals that are incapable of internally maintaining temperature) can sometimes roughly regulate temperature via behavior (e.g., turtles, crocodiles, etc.), or high thermal mass (gigantothermy; Spotila et al. 1991). Therefore, assessing endothermy is not simple. Two isotopic approaches have been used.

Barrick and Showers (1994) noted that endotherms maintain a nearly constant temperature *throughout* their bodies, whereas homeothermic ectotherms maintain uniform temperatures only in their body cores. Temperatures in ectotherms could be significantly different in their core vs. extremities, in contrast to endotherms. Because oxygen isotopic compositions of bioapatites precipitated from a uniform fluid reservoir are temperature-dependent (Longinelli 1965), ectotherms are expected to have more variable compositions, whereas endotherms should have uniform compositions. Thus, Barrick and Showers (1994) suggested that isotopic differences within and among bones of an individual could be used to identify endothermy. Application to several different dinosaur genera shows relatively uniform isotopic compositions, at least within their cores (core homeothermy), and similar but somewhat cooler or more variable temperatures in their limbs (regional heterothermy), in contrast to lizards that showed variable temperatures throughout their bodies (Barrick and Showers 1994, 1995, 1999; Barrick et al. 1996, 1998; Fig. 11). From these data Barrick and coworkers conclude that dinosaurs were basically endothermic, and that endothermy was accomplished by high metabolisms, but that metabolisms were perhaps not as high as in modern day birds or mammals. Others have criticized their results, partly because of issues regarding the potential for alteration of PO_4 oxygen in bones (Kolodny et al. 1996), and partly because a > ± 2 °C temperature variability for the "dinosaurian endotherms" cannot be rejected statistically (Ruxton 2000). Nonetheless, the apparent preservation of isotopic differences among bones of an ecotherm, but not in dinosaurs, does suggest that dinosaurs had a different metabolism, and were capable of at least crude endothermy.

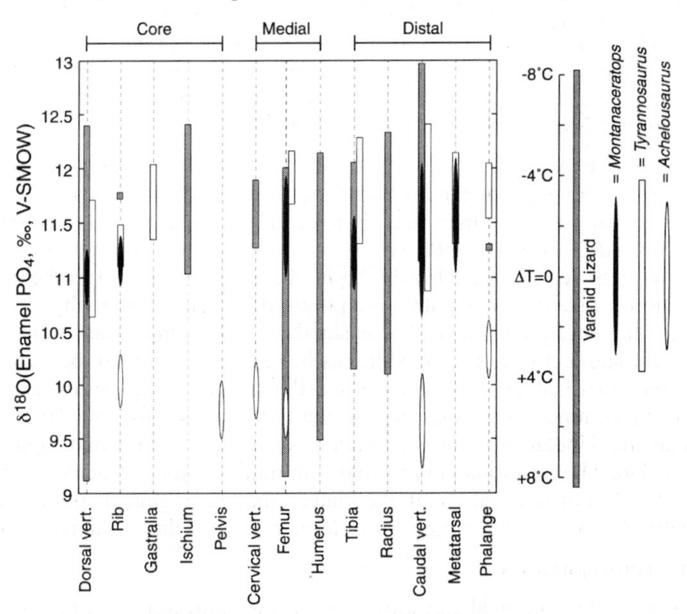

Figure 11. $\delta^{18}O$ of the PO_4 component of bone vs. bone identity, from core to extremities for a ~1 m long heterothermic lizard (Varanid) and three dinosaurs: a 1-2 m *Montanaceratops*, a <6 m juvenile *Achelousaurus*, and a ~12 m *Tyrannosaurus*. Large variation in bioapatite composition for lizard corresponds to ~±4°C (1σ) variability (right side) and is consistent with heterothermy. Smaller →→

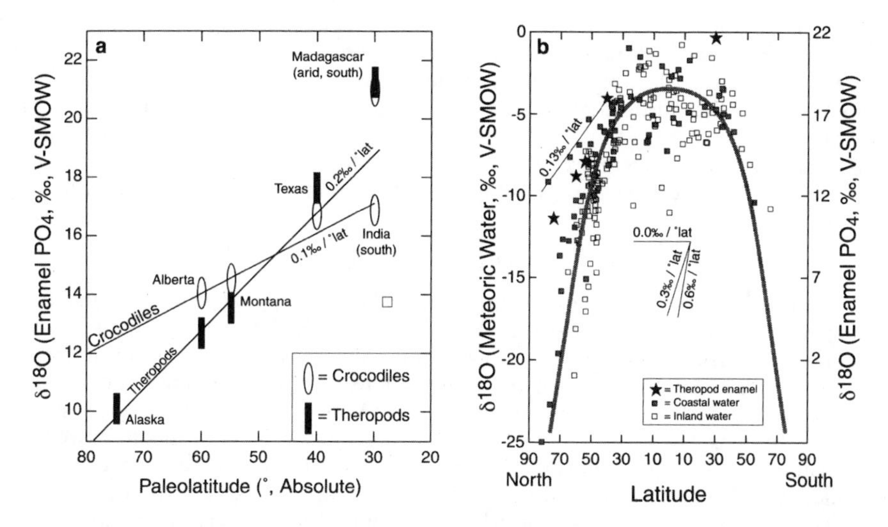

Figure 12. (a) $\delta^{18}O$ of the PO_4 component of theropod and crocodile teeth vs. latitude. Fossil theropod dinosaur enamel shows a steeper isotope gradient than do coexisting crocodile teeth. $\delta^{18}O$ should be shifted to higher values at colder temperatures (Fig. 5a). Assuming that heterothermic crocodile body temperature parallels local temperature, the different slopes imply theropods were better able to regulate body temperature than crocodiles, i.e., dinosaurs may have been homeotherms. (b) $\delta^{18}O$ of meteoric water and the PO_4 component of theropod enamel vs. latitude. Assuming homeothermy, the shallower slope for theropods (~0.2‰/° lat.) compared to average meteoric water at similar latitudes today (~0.3-0.6‰/° lat.) implies greater poleward heat transport in Cretaceous oceans and atmosphere compared to today. The shallowest modern slope for modern meteoric water at intermediate to high latitudes (0.13‰/° lat.; thin line) reflects unusually high poleward heat transport in the north Atlantic. Fossil tooth data from Fricke and Rogers (2000). Squares are meteoric water data from Rozanski et al. (1993).

Fricke and Rogers (2000) proposed that tooth compositions of dinosaurs at different latitudes (i.e., at differing mean annual temperatures) could be compared to those of sympatric crocodiles, which are known ectotherms. They reasoned that because bioapatite $\delta^{18}O$ fractionations increase with decreasing temperature (Longinelli 1965; Fig. 5a), a variable-temperature ectotherm should have a higher $\delta^{18}O$ relative to a constant-temperature endotherm at progressively colder temperatures. A similar isotopic offset, independent of latitude would indicate that dinosaurs had similar average temperatures as crocodiles, so dinosaurs probably were ectotherms. If instead dinosaurs had progressively lower $\delta^{18}O$ compositions at higher latitudes in comparison to crocodiles, then dinosaurs were probably endotherms. Fricke and Rogers indeed found that Cretaceous theropod dinosaurs show less isotopic change with latitude than crocodiles (Fig. 12), implying a greater degree of homeothermy. Criticism (Barrick and Kohn 2001) has focused on some of the details of interpreting isotope compositions for different species: do crocodile compositions accurately reflect mean annual temperatures, could behavioral or adaptive strategies produce different isotope slopes for different species, etc.? Nonetheless, data from Barrick, Fricke, and coworkers have generally supported dinosaurian endothermy. Certainly compositional comparisons of dinosaurs vs. known ectotherms show clear differences (Fig. 11, 12).

variation in bioapatite composition for dinosaurs corresponds to $<\pm 2$°C (1σ) temperature variability (right side) and is consistent with homeothermy. Bioapatite compositional variability in dinosaurs is independent of both body size and ontogeny, strongly suggesting that dinosaurs were endotherms. Data from Barrick and Showers (1994, 1995) and Barrick et al. (1996).

Fricke and Rogers' work also has implications for global heat distributions in the past. The Cretaceous latitudinal gradient in theropod isotope composition (~0.2‰ per degree latitude) is shallower than observed for meteoric water on Earth today, except in settings such as the North Atlantic, where ocean currents transport an unusual amount of heat pole-ward (Fig. 12). If Fricke and Rogers' results are typical for the Cretaceous, then bioapatite isotope compositions can be used to identify increased or reduced pole-ward heat transport in the past and possibly latitudinal temperature gradients within the continents.

CONCLUSIONS

The isotope compositions of bioapatite depend on ecology, physiology, climate, and geology, and consequently bioapatite $\delta^{13}C$, $\delta^{18}O$, and $^{87}Sr/^{86}Sr$ compositions have the potential to inform processes within and at the interfaces of all these disciplines. In general, the basic controls on isotopic compositions of bioapatites are known. Carbon isotopes closely reflect diet, and can be used in a variety of (paleo)ecological studies. Oxygen isotopes are more complicated, but are known to depend on local water compositions, climate (humidity), and diet. In some instances, these dependencies are sufficiently well known to investigate ecology and climate. Original strontium isotopes reflect the soil/bedrock substrate on which an animal lives, but can be rapidly altered diagenetically, at least for bone and dentin. The original biogenic signal can potentially help delineate sample age (for marine settings), migratory patterns, and/or habitat use. In terms of materials, diagenetic bias is common for bone and dentin, but by all indications enamel is extremely resistant to isotopic exchange. Consequently, most workers are focusing on enamel for retrieving original compositions, although this does require assessing seasonal isotope variations. Note, however, that the diagenetic imprint has its own usefulness (Tuross and Trueman, this volume).

Lingering unresolved issues remain regarding isotope studies of bioapatite. The biological fractionation of carbon isotopes remains ambiguous, and whereas a relatively pure C_3 diet can be readily distinguished from a relatively pure C_4 diet, definitive identification of differing ecosystems (e.g., closed canopy vs. savannah) is not always possible. A reconciliation of laboratory and field observations is necessary. The dependence of oxygen isotope compositions on water composition, climate, and diet is rather rudimentary, which essentially restricts work to comparisons of one taxon through time (to eliminate physiological effects), or of sympatric taxa at selected times. There are ongoing studies of large animals, both wild and domestic (e.g., Hoppe and Amundsen, 2001), but complete results are not yet available because of the long times necessary for the animals to mature. Last, diagenesis is obvious both physically and chemically in virtually all fossils, yet the mechanisms by which bioapatite crystallites change size and composition post-mortem is only vaguely understood. A detailed understanding of the crystal-chemical mechanisms of alteration might permit improved methods of retrieving original biologic signals, and help fully realize the potential of bioapatites in studying (paleo)ecology, (paleo)climate, and (paleo)diet.

ACKNOWLEDGMENTS

We thank B. Jacobs and D. Whistler for clarifying the age and setting of the earliest C_4 plant fossils. P. Koch, L. Roe, and J. Rakovan provided excellent, detailed reviews that greatly improved the manuscript. MJK acknowledges funding from National Science Foundation grant EAR 9909568.

REFERENCES

Albarède F (1995) Introduction to geochemical modeling. Cambridge University Press, Cambridge–New York–Melbourne

Anbar M, Halmann M,Silver B (1960) Determination of oxygen-18 in phosphate ion. Analyt Chem 32: 841-842

Arneborg J, Heinemeier J, Lynnerup N, Nielsen HL, Rud N,Sveinbjornsdottir AE (1999) Change of diet of the Greenland Vikings determined from stable carbon isotope analysis and ^{14}C dating of their bones. Radiocarbon 41:157-168

Ashwill M (1983) Seven fossil floras in the rain shadow of the Cascade Mountains, Oregon. Oregon Geology 45:107-111

Ayliffe LK, Chivas AR (1990) Oxygen isotope composition of the bone phosphate of Australian kangaroos: potential as a palaeoenvironmental recorder. Geochim Cosmochim Acta 54:2603-2609

Ayliffe LK, Lister AM, Chivas AR (1992) The preservation of glacial-interglacial climatic signatures in the oxygen isotopes of elephant skeletal phosphate. Palaeogeog Palaeoclim Palaeoecol 99:179-191

Ayliffe LK, Chivas AR, Leakey MG (1994) The retention of primary oxygen isotope compositions of fossil elephant skeletal phosphate. Geochim Cosmochim Acta 58:5291-5298

Barrat JA, Taylor RN, Andre JP, Nesbitt RW, Lécuyer C (2000) Strontium isotopes in biogenic phosphates from a Neogene marine formation: Implications for palaeoseawater studies. Chem Geol 168:325-332

Barrick R (1998) Isotope paleobiology of the vertebrates: ecology physiology, and diagenesis. Paleontol Soc Papers 4:101-137

Barrick RE, Kohn MJ (2001) Multiple taxon-multiple locality approach to providing oxygen isotope evidence for warm-blooded theropod dinosaurs: discussion. Geology 29:565-566

Barrick RE, Showers WJ (1994) Thermophysiology of *Tyrannosaurus rex*: Evidence from oxygen isotopes. Science 265:222-224

Barrick RE, Showers WJ (1995) Oxygen isotope variability in juvenile dinosaurs (*Hypacrosaurus*): evidence for thermoregulation. Paleobiology 21:552-560

Barrick RE, Showers WJ (1999) Thermophysiology and biology of *Giganotosaurus*: comparison with *Tyrannosaurus*. Palaeontologia Electronica: http://www.palaeo-electronica.org/1999_2/gigan/issue2_99.htm

Barrick RE, Fischer AG, Kolodny Y, Luz B, Bohaska D (1992) Cetacean bone oxygen isotopes as proxies for Miocene ocean composition and glaciation. Palaios 7:521-531

Barrick RE, Showers WJ, Fischer AG (1996) Comparison of thermoregulation of four ornithischian dinosaurs and a varanid lizard from the Cretaceous Two Medicine Formation: evidence from oxygen isotopes. Palaios 11:295-305

Barrick RE, Stoskopf MK, Marcot JD, Russell DA, Showers WJ (1998) The thermoregulatory functions of the Triceratops frill and horns: heat flow measured with oxygen isotopes. J Vert Paleo 18:746-750

Barrick RE, Fischer AG, Showers WJ (1999) Oxygen isotopes from turtle bone: applications for terrestrial paleoclimates? Palaios 14:186-191

Bartsiokas A, Middleton AP (1992) Characterization and dating of Recent and fossil bone by X-ray diffraction. J Archaeol Sci 19:63-72

Berggren WA, Kent DV, Flynn JJ (1985) Jurassic to Paleogene, Part 2: Paleogene geochronology and chronostratigraphy. *In* The chronology of the geological record. Snelling NJ (ed) Geological Society of London Memoir, London, United Kingdom, p 141-195

Blake RE, O'Neil JR, Garcia GA (1997) Oxygen isotope systematics of microbially mediated reactions of phosphate I.: Degradation of organophosphorus compounds. Geochim Cosmochim Acta 61:4411-4422

Blake RE, O'Neil JR, Garcia GA (1998) Effects of microbial activity on the δ^{18}O of dissolved inorganic phosphate and textural features of synthetic apatites. Am Mineral 83:1516-1531

Blondel C, Bocherens H, Mariotti A (1997) Stable carbon and oxygen isotope ratios in ungulate teeth from French Eocene and Oligocene localities. Bull Soc Geol France 168:775-781

Bocherens K, Koch PL, Mariotti A, Geraads D, Jaeger JJ (1996) Isotopic biogeochemistry (^{13}C, ^{18}O) of mammalian enamel from African Pleistocene hominid sites. Palaios 11:306-318

Bocherens H, Machkour M, Billiou D, Pelle E, Mariotti A (2001) A new approach for studying prehistoric herd management in arid areas: intra-tooth isotopic analyses of archaeological caprine from Iran. Earth Planet Sci 332:67-74

Bralower TJ, Zachos JC, Thomas E, Parrow M, Paull CK, Kelly DC, Premoli Silva I, Sliter WV, Lohmann KC (1995) Late Paleocene to Eocene paleoceanography of the Equatorial Pacific Ocean: stable isotopes recorded at Ocean Drilling Program Site 865, Allison Guyot. Paleoceanog 10:841-865

Bryant JD, Froelich PN (1995) A model of oxygen isotope fractionation in body water of large mammals. Geochim Cosmochim Acta 59:4523-4537

Bryant JD, Luz B, Froelich PN (1994) Oxygen isotopic composition of fossil horse tooth phosphate as a record of continental paleoclimate. Palaeogeog Palaeoclim Palaeoecol 107:303-316

Bryant JD, Koch PL, Froelich PN, Showers WJ, Genna BJ (1996) Oxygen isotope partitioning between phosphate and carbonate in mammalian apatite. Geochim Cosmochim Acta 60:5145-5148

Burton RK, Koch PL (1999) Isotopic tracking of foraging and long-distance migration in northeastern Pacific pinnipeds. Oecologia 119:578-585

Cerling and Cook (1998) You are what you eat: A traveler's diet in Mongolia. Finnigan Analyt News 2:4-5

Cerling TE, Harris JM (1999) Carbon isotope fractionation between diet and bioapatite in ungulate mammals and implications for ecological and paleoecological studies. Oecologia 120:347-363

Cerling TE, Sharp ZD (1996) Stable carbon and oxygen isotope analysis of fossil tooth enamel using laser ablation. Palaeogeog Palaeoclim Palaeoecol 126:173-186

Cerling TE, Wang Y, Quade J (1993) Expansion of C_4 ecosystems as an indicator of global ecological change in the late Miocene. Nature 361:344-345

Cerling TE, Harris JM, MacFadden BJ, Leakey MG, Quade J, Eisenmann V, Ehleringer JR (1997) Global vegetation change through the Miocene/Pliocene boundary. Nature 389:153-158

Clementz MT, Koch PL (2001) Early occurrence of C_4 grasses in middle Miocene North America based on stable isotopes in tooth enamel. North Am Paleontol Conf Abstr: http://www.ucmp.berkeley.edu/napc/abs6.html

Clyde WC, Sheldon ND, Koch PL, Gunnell GF, Bartels WS (2001) Linking the Wasatchian/Bridgerian boundary to the Cenozoic global climate optimum: new magnetostratigraphic and isotopic results from South Pass, Wyoming. Palaeogeog Palaeoclim Palaeoecol 167:175-199

Coplen TB, Kendall C (2000) Stable Hydrogen and Oxygen Isotope Ratios for Selected Sites of the U.S. Geological Survey's NASQAN and Benchmark Surface-water Networks. U S Geol Surv Open-File Report

Cormie AB, Luz B, Schwarcz HP (1994) Relationship between the hydrogen and oxygen isotopes of deer bone and their use in the estimation of relative humidity. Geochim Cosmochim Acta 58:3439-3449

Crepet WL, Feldman GD (1991) The earliest remains of grasses in the fossil record. Am J Botany 78: 1010-1014

Crepet WL, Herendeen PS (1992) Papilionoid flowers from the early Eocene of southeastern North America. *In* Advances in Legume Systematics, Part 4: The Fossil Record. Herendeen PS Dilcher DL (eds) The Royal Botanic Gardens, Kew, UK

Crowson RA, Showers WJ, Wright EK, Hoering TC (1991) Preparation of phosphate samples for oxygen isotope analysis. Anal Chem 63:2397-2400

D'Angela D, Longinelli A (1990) Oxygen isotopes in living mammal's bone phosphate: Further results. Chem Geol 86:75-82

Dansgaard W (1964) Stable isotopes in precipitation. Tellus 16:436-468

Delgado Huertas A, Iacumin P, Stenni B, Sanchez Chillon B, Longinelli A (1995) Oxygen isotope variations of phosphate in mammalian bone and tooth enamel. Geochim Cosmochim Acta 59: 4299-4305

Delgado Huertas A, Iacumin P, Longinelli A (1997) A stable isotope study of fossil mammal remains from the Paglicci Cave, southern Italy, 13 to 33ka BP: palaeoclimatological considerations. Chem Geol 141:211-223

Dettman DL, Kohn MJ, Quade J, Ryerson FJ, Ojha TP, Hamidullah S (2001) Seasonal stable isotope evidence for a strong Asian monsoon throughout the past 10.7 Myr. Geology 29:31-34

DeNiro MJ, Epstein S (1976) You are what you eat (plus a few ‰): the carbon isotope cycle in food chains. Geol Soc Am Abstr Progr 8:834-835

DeNiro MJ, Epstein S (1978) Influence of diet on the distribution of carbon isotopes in animals. Geochim Cosmochim Acta 42:495-506

Denison RE, Hetherington EA, Bishop BA, Dahl DA, Koepnick RB (1993) The use of strontium isotopes in stratigraphic studies: an example from North Carolina. Southeastern Geol 33:53-69

DePaolo DJ, Ingram BL (1985) High-resolution stratigraphy with strontium isotopes. Science 227:938-941

Dickens GR (1999) The blast in the past. Nature 401:752-753, 755

Dickens GR, O'Neil JR, Rea DK, Owen RM (1995) Dissociation of oceanic methane hydrate as a cause of the carbon isotope excursion at the end of the Paleocene. Paleoceanog 10:965-971

Dickens GR, Castillo MM, Walker JCG (1997) A blast of gas in the latest Paleocene: simulating first-order effects of massive dissociation of oceanic methane hydrate. Geology 25:259-262

Ehleringer JR, Sage RF, Flanagan LB, Pearcy RW (1991) Climate change and the evolution of C_4 photosynthesis. Trends Ecol Evol 6:95-99

Ehleringer JR, Cerling TE (1995) Atmospheric CO_2 and the ratio of intercellular to ambient CO_2 concentrations in plants. Tree Physiol 15:105-111

Eiler JM, Graham CM, Valley JW (1997) SIMS analysis of oxygen isotopes: matrix effects in complex minerals and glasses. Chem Geol 138:221-244

Elliott TA, Forey PL, Williams CT, Werdelin L (1998) Application of the solubility profiling technique to Recent and fossil fish teeth. Bull Soc Geol France 169:443-451

Epstein S, Zeiri L (1988) Oxygen and carbon isotopic compositions of gases respired by humans. Proc Nat Acad Sci 85:1727-1731

Ericson JE, Sullivan CH, Boaz NT (1981) Diets of Pliocene mammals from Omo, Ethiopia, deduced from carbon isotopic ratios in tooth apatite. Palaeogeog Palaeoclim Palaeoecol 36:69-73

Ericson JE (1985) Strontium isotope characterization in the study of prehistoric human ecology. J Human Evol 14:503-514

Ezzo JA, Johnson CJ, Price TD (1997) Analytical perspectives on prehistoric migration: a case study from East-Central Arizona. J Archaeol Sci 24:447-466

Farquhar GD, Ehleringer JR, Hubick KT (1989) Carbon isotope discrimination and photosynthesis. Ann Rev Plant Physiol Plant Mol Biol 40:503-537

Farquhar GD, Henry BK, Styles JM (1997) A rapid on-line technique for determination of oxygen isotope composition of nitrogen-containing organic matter and water. Rapid Comm Mass Spec 11:1554-1560

Feranec RS, MacFadden BJ (2000) Evolution of the grazing niche in Pleistocene mammals from Florida: evidence from stable isotopes. Palaeogeog Palaeoclim Palaeoecol 162:155-169

Firsching FH (1961) Precipitation of silver phosphate from homogeneous solution. Analyt Chem 33: 873-874

Fisher DC, Fox DL (1998) Oxygen isotopes in mammoth teeth: sample design, mineralization patterns, and enamel-dentin comparisons. J Vert Paleontol 18 (suppl):41A-42A

Fox DL (2000) Growth increments in Gomphotherium tusks and implications for late Miocene climate change in North America. Palaeogeog Palaeoclim Palaeoecol 156:327-348

Fox DL, Fisher DC (2001) Stable isotope ecology of a late Miocene population of *Gomphotherium productus* (Mammalia, Proboscidea)from Port of Entry Pit, Oklahoma. Palaios 16:279-293

Freeman KH, Colarusso LA (2001) Molecular and isotopic records of C_4 grassland expansion in the late Miocene. Geochim Cosmochim Acta 65:1439-1454

Fricke HC, O'Neil JR (1996) Inter-and intra-tooth variation in the oxygen isotope composition of mammalian tooth-enamel phosphate: Implications for palaeoclimatological and palaeobiological research. Palaeogeog Palaeoclim Palaeoecol 126:91-99

Fricke HC, O'Neil JR, Lynnerup N (1995) Oxygen isotope composition of human tooth enamel from medieval Greenland: linking climate and society. Geology 23:869-872

Fricke HC, Clyde WC, O'Neil JR, Gingerich PD (1998a) Evidence for rapid climate change in North America during the latest Paleocene thermal maximum: oxygen isotope compositions of biogenic phosphate from the Bighorn Basin (Wyoming). Earth Planet Sci Lett 160:193-208

Fricke HC, Clyde WC, O'Neil JR (1998b) Intra-tooth variations in $\delta^{18}O$ (PO_4) of mammalian tooth enamel as a record of seasonal variations in continental climate variables. Geochim Cosmochim Acta 62: 1839-1850

Fricke DC, Rogers RR (2000) Multiple taxon-multiple locality approach to providing oxygen isotope evidence for warm-blooded theropod dinosaurs. Geology 28:799-802

Friedli H, Loetscher H, Oeschger H, Siegenthaler U, Stauffer B (1986) Ice core record of the $^{13}C/^{12}C$ ratio of atmospheric CO_2 in the past two centuries. Nature 324:237-238

Gadbury C, Todd L, Jahren AH, Amundson R (2000) Spatial and temporal variations in the isotopic composition of Bison tooth enamel from the early Holocene Hudson-Meng bone bed, Nebraska. Palaeogeog Palaeoclim Palaeoecol 157:79-93

Gingerich PD (1989) New earliest Wasatchian mammalian fauna from the Eocene of northwestern Wyoming: composition and diversity in a rarely sampled high-floodplain assemblage. Univ Mich Papers Paleontol 28:1-97

Grupe G, Price TD, Schroeter P, Soellner F, Johnson CM, Beard BL (1997) Mobility of Bell Beaker people revealed by strontium isotope ratios of tooth and bone: a study of southern Bavarian skeletal remains. Appl Geochem 12:517-525

Hedges REM, Millard AR, Pike AWG (1995) Measurements and relationships of diagenetic alteration of bone from three archaeological sites. J Archaeol Sci 22:201-209

Hillson S (1996) Teeth. Cambridge University Press, Cambridge, UK

Holmden C, Creaser RA, Muehlenbachs K, Bergstrom SM, Leslie SA (1996) Isotopic and elemental systematics of Sr and Nd in 454 Ma biogenic apatites: implications for paleoseawater studies. Earth Planet Sci Lett 142:425-437

Hoppe KA, Amundsen R (2001) Interpreting the significance of stable isotopic variations within mammalian teeth: evaluating the influence of biological vs. environmental factors. Geol Soc Am Abstr Prog 33:A113-A114

Hoppe KA, Koch PL, Carlson RW, Webb SD (1999) Tracking mammoths and mastodons: reconstruction of migratory behavior using strontium isotope ratios. Geology 27:439-442

Hoppe KA, Amundson R, Todd LC (2001) Patterns of tooth enamel formation in hypsodont teeth: implications for isotopic microsampling. J Vert Paleo 21:63

Iacumin P, Bocherens H, Mariotti A, Longinelli A (1996a) Oxygen isotope analyses of co-existing carbonate and phosphate in biogenic apatite: a way to monitor diagenetic alteration of bone phosphate? Earth Planet Sci Lett 142:1-6

Iacumin P, Bocherens H, Mariotti A, Longinelli A (1996b) An isotopic palaeoenvironmental study of human skeletal remains from the Nile Valley. Palaeogeog Palaeoclim Palaeoecol 126:15-30

Iacumin P, Cominotto D, Longinelli A (1996c) A stable isotope study of mammal skeletal remains of mid-Pleistocene age, Arago Cave, eastern Pyrenees, France: evidence of taphonomic and diagenenetic effects. Palaeogeog Palaeoclim Palaeoecol 126:151-160

Iacumin P, Bocherens H, Delgado Huertas A, Mariotti A, Longinelli A (1997) A stable isotope study of fossil mammal remains from the Paglicci Cave, southern Italy: N and C as palaeoenvironmental indicators. Earth Planet Sci Lett:349-357

Ingram BL (1995) High-resolution dating of deep-sea clays using Sr isotopes in fossil fish teeth. Earth Planet Sci Lett 134:545-555

Jacobs BF, Kingston JD, Jacobs LL (1999) The origin of grass-dominated ecosystems. Ann Missouri Bot Garden 86:590-643

Jones AM, Iacumin P, Young ED (1999) High-resolution $\delta^{18}O$ analysis of tooth enamel phosphate by isotope ratio monitoring gas chromatography mass spectrometry and ultraviolet laser fluorination. Chem Geol 153:241-248

Karhu J, Epstein S (1986) The implication of the oxygen isotope records in coexisting cherts and phosphates. Geochim Cosmochim Acta 50:1745-1756

Kennedy BP, Folt CL, Blum JD, Chamberlain CP (1997) Natural isotope markers in salmon. Nature 387:766-767

Kennett JP, Stott LD (1991) Abrupt deep-sea warming, palaeoceanographic changes and benthic extinctions at the end of the Palaeocene. Nature 353:225-229

Kierdorf H, Kierdorf U (1997) Disturbances of the secretory stage of amelogenesis in fluorosed deer teeth: a scanning electron-microscopic study. Cell Tiss Res 289:125-135

Koch PL (1998) Isotopic reconstruction of past continental environments. Ann Rev Earth Planet Sci 26:573-613

Koch PL, Fisher DC, Dettman D (1989) Oxygen isotope variation in the tusks of extinct proboscideans: A measure of season of death and seasonality. Geology 17:515-519

Koch PL, Behrensmeyer AK, Tuross N, Fogel ML (1991) Isotopic fidelity during bone weathering and burial. Carnegie Inst Wash Yrbk:105-110

Koch PL, Halliday AN, Walter LM, Stearley RF, Huston TJ, Smith GR (1992a) Sr isotopic composition of hydroxyapatite from Recent and fossil salmon: the record of lifetime migration and diagenesis. Earth Planet Sci Lett 108:277-287

Koch PL, Zachos JC, Gingerich PD (1992b) Correlation between isotope records in marine and continental carbon reservoirs near the Paleocene/Eocene boundary. Nature 358:319-322

Koch PL, Heisinger J, Moss C, Carlson RW, Fogel ML, Behrensmeyer AK (1995) Isotopic tracking of change in diet and habitat use in African elephants. Science 267:1340-1343

Koch PL, Zachos JC, Dettman DL (1995b) Stable isotope stratigraphy and paleoclimatology of the Paleogene Bighorn Basin (Wyoming, USA). Palaeogeog Palaeoclim Palaeoecol 115:61-89

Koch PL, Tuross N, Fogel ML (1997) The effects of sample treatment and diagenesis on the isotopic integrity of carbonate in biogenic hydroxylapatite. J Arch Sci 24:417-429

Koch PL, Hoppe KA, Webb SD (1998) The isotopic ecology of late Pleistocene mammals in North America: Part 1, Florida. Chem Geol 152:119-138

Kohn MJ (1996) Predicting animal $\delta^{18}O$: Accounting for diet and physiological adaptation. Geochim Cosmochim Acta 60:4811-4829

Kohn MJ, Schoeninger MJ, Valley JW (1996) Herbivore tooth oxygen isotope compositions: Effects of diet and physiology. Geochim Cosmochim Acta 60:3889-3896

Kohn MJ, Schoeninger MJ, Valley JW (1998) Variability in herbivore tooth oxygen isotope compositions: reflections of seasonality or developmental physiology? Chem Geol 152:97-112

Kohn MJ, Schoeninger MJ, Barker WW (1999) Altered states: effects of diagenesis on fossil tooth chemistry. Geochim Cosmochim Acta 18:2737-2747

Kohn MJ, Miselis JL, Fremd TJ (2003) Oxygen isotope evidence for progressive uplift of the Cascade Range, Oregon. Earth Planet Sci Lett (in press)

Kolodny Y, Luz B (1991) Oxygen isotopes in phosphates of fossil fish: Devonian to Recent. *In* Stable Isotope Geochemistry: a Tribute to Samuel Epstein. Taylor HP Jr, O'Neil JR, Kaplan IR (eds) Geochemical Society, University Park, Pennsylvania, p 105-119

Kolodny Y, Luz B, Navon O (1983) Oxygen isotope variations in phosphate of biogenic apatites, I. Fish bone apatite-rechecking the rules of the game. Earth Planet Sci Lett 64:398-404

Kolodny Y, Luz B, Sander M, Clemens WA (1996) Dinosaur bones: fossils of pseudomorphs? The pitfalls of physiology reconstruction from apatitic fossils. Palaeogeog Palaeoclim Palaeoecol 126:161-171

Kyle JH (1986) Effect of post-burial contamination on the concentrations of major and minor elements in human bones and teeth: the implications for palaeodietary research. J Archaeol Sci 13:403-416

Land LS, Lundelius EL Jr., Valastro S (1980) Isotopic ecology of deer bones. Palaeogeog Palaeoclim Palaeoecol 32:143-151

Latorre C, Quade J, McIntosh WC (1997) The expansion of C_4 grasses and global change in the late Miocene: stable isotope evidence from the Americas. Earth Planet Sci Lett 146:83-96

Lécuyer C, Grandjean P, O'Neil JR, Cappetta H, Martineau F (1993) Thermal excursions in the ocean at the Cretaceous-Tertiary boundary (northern Morocco): $\delta^{18}O$ record of phosphatic fish debris. Palaeogeog Palaeoclim Palaeoecol 105:235-243

Lécuyer C, Grandjean R,Emig CC (1996a) Determination of oxygen isotope fractionation between water and phosphate from living lingulids: potential application to palaeoenvironmental studies. Palaeogeog Palaeoclim Palaeoecol 126:101-108

Lécuyer C, Grandjean P, Paris F, Robardet M,Robineau D (1996b) Deciphering "temperature" and "salinity" from biogenic phosphates:the $\delta^{18}O$ of coexisting fishes and mammals of the middle Miocene sea of western France. Palaeogeog Palaeoclim Palaeoecol 126:61-74

Lécuyer C, Grandjean P, Sheppard SMF (1999) Oxygen isotope exchange between dissolved phosphate and water at temperatures $\leq$135 degrees C: inorganic versus biological fractionations. Geochim Cosmochim Acta 63:855-862

Lee-Thorp JA, van der Merwe NJ (1987) Carbon isotope analysis of fossil bone apatite. South Afr J Sci 83:712-715

Lee-Thorp JA, van der Merwe NJ (1991) Aspects of the chemistry of modern and fossil biogenic apatites. J Archaeol Sci 18:343-354

Lee-Thorp JA, Sealy JC, van der Merwe NJ (1989) Stable carbon isotope ratio differences between bone collagen and bone apatite, and their relationship to diet. J Archaeol Sci 16:585-599

Lee-Thorp JA, van der Merwe NJ, Brain CK (1994) Diet of *Australopithecus robustus* at Swartkrans from stable carbon isotopic analysis. J Human Evol 27:361-372

Lindars ES, Grimes ST, Mattey DP, Collinson ME, Hooker JJ, Jones TP (2001) Phosphate $\delta^{18}O$ determination of modern rodent teeth by direct laser fluorination: An appraisal of methodology and potential application to palaeoclimate reconstruction. Geochim Cosmochim Acta 65:2535-2548

Linder HP (1986) The evolutionary history of the Poales/Restionales—A hypothesis. Kew Bull 42: 297-318

Longinelli A (1965) Oxygen isotopic composition of orthophosphate from shells of living marine organisms. Nature 207:716-719

Longinelli A (1984) Oxygen isotopes in mammal bone phosphate: a new tool for paleohydrological and paleoclimatological research? Geochim Cosmochim Acta 48:385-390

Longinelli A, Nuti S (1973a) Revised phosphate-water isotopic temperature scale. Earth Planet Sci Lett 19:373-376

Longinelli A, Nuti S (1973b) Oxygen isotope measurements of phosphate from fish teeth and bones. Earth Planet Sci Lett 20:337-340

Longinelli A, Sordi M (1966) Oxygen isotopic composition of phosphate from shells of some living crustaceans. Nature 211:727-728

Longinelli A, Bartelloni M, Cortecci G (1976) The isotopic cycle of oceanic phosphate: I. Earth Planet Sci Lett 32:389-392

Luyt J, Lee-Thorp JA, Avery G (2000) New light on middle Pleistocene west coast environments from Elandsfontein, Western Cape Province, South Africa. South Afr J Sci 96:399-403

Luz B, Kolodny Y (1985) Oxygen isotope variations in phosphate of biogenic apatites, IV. Mammal teeth and bones. Earth Planet Sci Lett 75:29-36

Luz B, Kolodny Y, Horowitz M (1984a) Fractionation of oxygen isotopes between mammalian bone-phosphate and environmental drinking water. Geochim Cosmochim Acta 48:1689-1693

Luz B, Kolodny Y, Kovach J (1984b) Oxygen isotope variations in phosphate of biogenic apatites, III: Conodonts. Earth Planet Sci Lett 69:255-262

Luz B, Cormie AB, Schwarcz HP (1990) Oxygen isotope variations in phosphate of deer bones. Geochim Cosmochim Acta 54:1723-1728

MacFadden BJ (1998) Tale of two rhinos: isotopic ecology, paleodiet, and niche differentiation of Aphelops and Teleoceras from the Florida Neogene. Paleobiol 24:274-286

MacFadden BJ (2000) Middle Pleistocene climate changes recorded in fossil mammal teeth from Tarija, Bolivia, and upper limit of the Ensenadan land-mammal age. Quat Res 54:121-131

MacFadden BJ, Cerling TE (1996) Mammalian herbivore communities, ancient feeding ecology, and carbon isotopes: a 10 million-year sequence from the Neogene of Florida. J Vert Paleo 16:103-115
MacFadden BJ, Cerling TE, Prado J (1996) Cenozoic terrestrial ecosystem evolution in Argentina: evidence from carbon isotopes of fossil mammal teeth. Palaios 11:319-327
MacFadden BJ, Solounias N, Cerling TE (1999a) Ancient diets, ecology, and extinction of 5-million-year-old horses from Florida. Science 283:824-827
MacFadden BJ (1999b) Ancient latitudinal gradients of C_3/C_4 grasses interpreted from stable isotopes of New World Pleistocene horse (*Equus*) teeth. Glob Ecol Biogeog 8:137-149
MacLeod KG, Smith RMH, Koch PL, Ward PD (2000) Timing of mammal-like reptile extinctions across the Permian-Triassic boundary in South Africa. Geology 28:227-230
Marino BD, McElroy MB (1991) Isotopic composition of atmospheric CO_2 inferred from carbon in C_4 plant cellulose. Nature 349:127-131
Martin EE, Haley BA (2000) Fossil fish teeth as proxies for seawater Sr and Nd isotopes. Geochim Cosmochim Acta 64:835-847
McArthur JM, Coleman ML, Bremner JM (1980) Carbon and oxygen isotopic composition of structural carbonate in sedimentary francolite. J Geol Soc London 137:669-673
McCrea JM (1950) On the isotope chemistry of carbonates and a paleotemperature scale. J Chem Phys 18:849-857
Medina E, Minchin P (1980) Stratification of $\delta^{13}C$ values of leaves in Amazonian rain forests. Oecologia 45:377-378
Medina E, Montes G, Cuevas E, Rokczandic Z (1986) Profiles of CO_2 concentration and $\delta^{13}C$ values in tropical rain forests of the Upper Rio Negro Basin, Venezuela. J Tropical Ecol 2:207-217
Michel V, Ildefonse P, Morin G (1996) Assessment of archaeological bone and dentine preservation from Lazaret Cave (middle Pleistocene) in France. Palaeogeog Palaeoclim Palaeoecol 126:109-119
Moreno EC, Kresak M, Zahradnik RT (1977) Physicochemical aspects of fluoride-apatite systems relevant to the study of dental caries. Caries Res (suppl 1) 11:142-160
Morgan ME, Kingston JD, Marino BD (1994) Carbon isotopic evidence for the emergence of C_4 plants in the Neogene from Pakistan and Kenya. Nature 367:162-165
Nagy KA, Peterson CC (1988) Scaling of water flux rate in animals. Univ Calif Publ Zool 120:1-172
Nambudiri EMV, Tidwell WD, Smith BN, Hebbert NP (1978) A C_4 plant from the Pliocene. Nature 276:816-817
Nelson BK, DeNiro MJ, Schoeninger MJ (1986) Effects of diagenesis on strontium, carbon, nitrogen, and oxygen concentration and isotopic composition of bone. Geochim Cosmochim Acta 50:1941-1949
O'Leary MH (1988) Carbon isotopes in photosynthesis. Biosci 38:328-336
O'Neil JR, Roe LJ, Reinhard E, Blake RE (1994) A rapid and precise method of oxygen isotope analysis of biogenic phosphate. Israel J Earth-Sci 43:203-212
Pagani M, Arthur MA, Freeman KH (1999a) Miocene evolution of atmospheric carbon dioxide. Paleoceanog 14:273-292
Pagani M, Freeman KH, Arthur MA (1999b) Late Miocene atmospheric CO_2 concentrations and the expansion of C_4 grasses. Science 285:876-879
Passey BH, Cerling TE (2002) Tooth enamel mineralization in ungulates: implications for recovering a primary isotopic time-series. Geochim Cosmochim Acta 66 (in press)
Passey BH, Cerling TE, Perkins ME, Voorhies MR, Harris JM, Tucker ST (2002) Environmental change in the Great Plains: an isotopic record from fossil horses. J Geol 110:123-140
Pearson PN, Palmer MR (2000) Atmospheric carbon dioxide concentrations over the past 60 million years. Nature 406:695-699
Person A, Bocherens H, Saliège J-F, Paris F, Zeitoun V, Gérard M (1995) Early diagenetic evolution of bone phosphate: an X-ray diffractometry analysis. J Archaeol Sci 22:211-221
Person A, Bocherens H, Mariotti A, Renard M (1996) Diagenetic evolution and experimental heating of bone phosphate. Palaeogeog Palaeoclim Palaeoecol 126:135-149
Picard S, Garcia J-P, Lecuyer C, Sheppard SMF, Cappetta H, Emig CC (1998) $\delta^{18}O$ values of coexisting brachiopods and fish: temperature differences and estimates of paleo-water depths. Geology 26: 975-978
Price TD, Johnson CM, Ezzo JA, Ericson J, Burton JH (1994a) Residential mobility in the prehistoric southwest United States: a preliminary study using strontium isotope analysis. J Archaeol Sci 21: 315-330
Price TD, Grupe G, Schroeter P (1994b) Reconstruction of migration patterns in the Bell Beaker Period by stable strontium isotope analysis. Appl Geochem 9:413-417
Price TD, Manzanilla L, Middleton WD (2000) Immigration and the ancient city of Teotihuacan in Mexico: A study using strontium isotope ratios in human bone and teeth. J Archaeol Sci 27:903-913

Priest GR (1990) Volcanic and tectonic evolution of the Cascade volcanic arc, central Oregon. J Geophys Res 95:19583-19599

Quade J, Cerling TE, Bowman JR (1989) Development of Asian monsoon revealed by marked ecological shift during the latest Miocene in northern Pakistan. Nature 342:163-166

Quade J, Cerling TE, Barry JC, Morgan ME, Pilbeam DR, Chivas AR, Lee-Thorp JA, van der Merwe NJ (1992) A 16-Ma record of paleodiet using carbon and oxygen isotopes in fossil teeth from Pakistan. Chem Geol 94:183-192

Quade J, Solounias N, Cerling TE (1994) Stable isotopic evidence from Paleosol carbonates and fossil teeth in Greece for forest or woodlands over the past 11 Ma. Palaeogeog Palaeoclim Palaeoecol 108:41-53

Retallack GJ, Bestland EA, Fremd TJ (2000) Eocene and Oligocene paleosols of central Oregon. Geol Soc Am Spec Paper 344

Rey C, Renugopalakrishnan V, Shimizu M, Collins B, Glimcher MJ (1991) A resolution-enhanced fourier transform infrared spectroscopic study of the environment of the CO_3^{2-} ion in the mineral phase of enamel during its formation and maturation. Calcif Tiss Int 49:259-268

Roe LJ, Thewissen JGM, Quade J, O'Neil J, Bajpai S, Sahni A, Hussain ST (1998) Isotopic approaches to understanding the terrestrial-to-marine transition of the earliest cetaceans. *In* The Emergence of Whales. Thewissen JGM (ed) Plenum Press, p 399-422

Rozanski K, Araguas-Araguas L, Gonfiantini R (1993) Isotopic patterns in modern global precipitation. *In* Climate change in continental isotopic records. Swart PK, Lohmann KC, McKenzie JA, Savin S (eds) American Geophysical Union, Washington, DC, p 1-36

Ruxton GD (2000) Statistical power analysis: application to an investigation of dinosaur thermal physiology. J Zool 252:239-241

Sánchez-Chillón BS, Alberdi MT, Leone G, Bonadonna FP, Stenni B, Longinelli A (1994) Oxygen isotopic composition of fossil equid tooth and bone phosphate: an archive of difficult interpretation. Palaeogeog Palaeoclim Palaeoecol 107:317-328

Schmitz B, Aberg G, Werdelin L, Forey P, Bendix-Almgreen SE (1991) $^{87}Sr/^{86}Sr$, Na, F, Sr, and La in skeletal fish debris as a measure of the paleosalinity of fossil-fish habitats. Geol Soc Am Bull 103: 786-794

Schmitz B, Ingram SL, Dockery DT III, Aberg G (1997) Testing $^{87}Sr/^{86}Sr$ as a paleosalinity indicator on mixed marine, brackish-water and terrestrial vertebrate skeletal apatite in late Paleocene-early Eocene near-coastal sediments, Mississippi. Chem Geol 140:275-287

Schoeninger MJ (1982) Diet and the evolution of modern human form in the Middle East. Am J Phys Anthropol 58:37-52

Schoeninger MJ, Deniro MJ (1982) Carbon isotope ratios of apatite from fossil bone cannot be used to reconstruct diets of animals. Nature 297:577-578

Schoeninger MJ, Moore KM, Murray ML, Kingston JD (1989) Detection of bone preservation in archaeological and fossil samples. Appl Geochem 4:281-292

Sealy JC, van der Merwe NJ, Sillen A, Kruger FJ, Krueger HW (1991) $^{87}Sr/^{86}Sr$ as a dietary indicator in modern and archaeological bone. J Archaeol Sci 18:399-416

Shackleton NJ, Corfield RM, Hall MA (1985) Stable isotope data and the ontogeny of Paleocene planktonic foraminifera. J Foram Res 15:321-336

Sharp ZD, Cerling TE (1996) A laser GC-IRMS technique for in situ stable isotope analyses of carbonates and phosphates. Geochim Cosmochim Acta 60:2909-2916

Sharp ZD, Cerling TE (1998) Fossil isotope records of seasonal climate and ecology, straight from the horse's mouth. Geology 26:219-222

Sharp ZD, Atudorei V, Furrer H (2000) The effect of diagenesis on oxygen isotope ratios of biogenic phosphates. Am J Sci 300:222-237

Shemesh A (1990) Crystallinity and diagenesis of sedimentary apatites. Geochim Cosmochim Acta 54:2433-2438

Shemesh A, Kolodny Y, Luz B (1983) Oxygen isotope variations in phosphate of biogenic apatites, II: Phosphorite rocks. Earth Planet Sci Lett 64:405-416

Sheppard SMF, Nielsen RL, Taylor HP, Jr. (1969) Oxygen and hydrogen isotope ratios of clay minerals from porphyry copper deposits. Econ Geol 64:755-777

Sillen A (1986) Biogenic and diagenetic Sr/Ca in Plio-Pleistocene fossils of the Omo Shungura Formation. Paleobiol 12:311-323

Sillen A, Hall G, Richardson S, Armstrong R (1998) $^{87}Sr/^{86}Sr$ ratios in modern and fossil food-webs of the Sterkfontein Valley: implications for early hominid habitat preference. Geochim Cosmochim Acta 62:2463-2473

Sponheimer M, Lee-Thorp JA (1999) Isotopic evidence for the diet of an early hominid, *Australopithecus africanus*. Science 283:368-370

Spotila JR, O'Connor MP, Dodson P, Paladino FV (1991) Hot and cold running dinosaurs: body size, metabolism and migration. Mod Geol 16:203-227

Staudigel H, Doyle P, Zindler A (1985) Sr and Nd isotope systematics in fish teeth. Earth Planet Sci Lett 76:45-56

Sternberg LO, DeNiro MJ, Johnson HB (1984) Isotope ratios of cellulose from plants having different photosynthetic pathways. Plant Physiol 74:557-561

Stuart-Williams HLQ, Schwarcz HP (1995) Oxygen isotopic analysis of silver orthophosphate using a reaction with bromine. Geochim Cosmochim Acta 59:3837-3841

Stuart-Williams HLQ, Schwarcz HP (1997) Oxygen isotopic determination of climatic variation using phosphate from beaver bone, tooth enamel, and dentine. Geochim Cosmochim Acta 61:2539-2550

Stuart-Williams HLQ, Schwarcz HP, White CD, Spence MW (1996) The isotopic composition and diagenesis of human bone from Teotihuacan and Oaxaca, Mexico. Palaeogeog Palaeoclim Palaeoecol 126:1-14

Sullivan CH, Krueger HW (1981) Carbon isotope analysis of separate chemical phases in modern and fossil bone. Nature 292:333-335

Thackeray JF, van der Merwe NJ, Lee-Thorp JA, Sillen A, Lanham JL, Smith R, Keyser A, Monteiro PMS (1990) Changes in carbon isotope ratios in the Late Permian recorded in therapsid tooth apatite. Nature 347:751-753

Thewissen JGM, Roe LJ, O'Neil J, Hussain ST, Sahni A, Bajpai S (1996) Evolution of cetacean osmoregulation. Nature 381:379-380

Thomas E, Shackleton NJ (1996) The Paleocene-Eocene benthic foraminiferal extinction and stable isotope anomalies. *In* Correlation of the early Paleogene in Northwest Europe. Knox RWO'B, Corfield RM, Dunay RE (eds) Geological Society, London, p 401-441

Tieszen LL, Fagre T (1993) Effect of diet quality and composition on the isotopic composition of respiratory CO_2, bone collagen, bioapatite and soft tissue. *In* Prehistoric Human Bone, Archaeology at the Molecular Level. Lambert J, Grupe G (eds) Springer-Verlag, Berlin, p 121-155

Tudge AP (1960) A method of analysis of oxygen isotopes in orthophosphates—its use in the measurement of paleotemperatures. Geochim Cosmochim Acta 18:81-83

Tuross N, Behrensmeyer AK, Eanes ED (1989) Strontium increases and crystallinity changes in taphonomic and archaeological bone. J Archaeol Sci 16:661-672

Vennemann TW, Hegner E (1998) Oxygen, strontium, and neodymium isotope composition of fossil shark teeth as a proxy for the palaeoceanography and palaeoclimatology of the Miocene northern Alpine Paratethys. Palaeogeog Palaeoclim Palaeoecol 142:107-121

Vennemann TW, Hegner E, Cliff G, Benz GW (2001) Isotopic composition of Recent shark teeth as a proxy for environmental conditions. Geochim Cosmochim Acta 65:1583-1599

Vennemann TW, Fricke HC, Blake RE, O'Neil JR, Colman A (2002) Oxygen isotope analysis of phosphates: a comparison of techniques for analysis of Ag_3PO_4. Chem Geol (in press)

Wang Y, Cerling TE (1994) A model for fossil tooth and bone diagenesis: implications for paleodiet reconstruction from stable isotopes. Palaeogeog Palaeoclim Palaeoecol 107:281-289

Wang Y, Cerling TE, Quade J, Bowman JR, Smith GA, Lindsay EH (1993) Stable isotopes of Paleosols and fossil teeth as paleoecology and paleoclimate indicators: an example from the St. David Formation, Arizona. *In* Climate change in continental isotopic records. Swart PK, Lohmann KC, McKenzie JA, Savin S (eds) Am Geophys Union, Washington, DC, p 241-248

Wang Y, Cerling TE, MacFadden BJ (1994) Fossil horses and carbon isotopes: new evidence for Cenozoic dietary, habitat, and ecosystem changes in North America. Palaeogeog Palaeoclim Palaeoecol 107: 269-279

Wenzel B, Lécuyer C, Joachimski MM (2000) Comparing oxygen isotope resords of Silurian calcite and phosphate-$\delta^{18}O$ compositions of brachiopods and conodonts. Geochim Cosmochim Acta 64: 1859-1872

Whistler DP, Burbank DW (1992) Miocene biostratigraphy and biochronology of the Dove Spring Formation, Mojave Desert, California, and characterization of the Clarendonian mammal age (late Miocene) in California. Geol Soc Am Bull 104:644-658

Wright EK, Hoering TC (1989) Separation and purification of phosphates for oxygen isotope analysis. Carnegie Inst Wash Yrbk:137-141

Yoshida N, Miyazaki N (1991) Oxygen isotope correlation of cetacean bone phosphate with environmental water. J Geophys Res 96:815-820

Zanconato S, Cooper DM, Armon Y, Epstein S (1992) Effect of increased metabolic rate on oxygen isotopic fractionation. Resp Physiol 89:319-327

Zazzo A, Bocherens H, Brunet M, Beauvilain A, Billiou D, Mackaye HT, Vignaud P, Mariotti A (2000) Herbivore paleodiet and paleoenvironmental changes in Chad during the Pliocene using stable isotope ratios of tooth enamel carbonate. Paleobiol 26:294-309

13 Trace Elements in Recent and Fossil Bone Apatite

Clive N. Trueman and Noreen Tuross

Smithsonian Center for Materials Research and Education
Smithsonian Institution
Silver Hill Road, Suitland, Maryland 20746

and

National Museum of Natural History
10th and Constitution Ave, NW
Washington DC 20560

INTRODUCTION

The biomineral component of bones and teeth is synthesized within the body, under direct physiological control, and therefore the chemistry of bone (including the trace element chemistry) reflects aspects of the animal's biology, particularly the trace metal load, but also the state of metabolism of specific trace metals. Bone mineral is relatively reactive due principally to its small crystal size, and consequently bone is unstable once removed from the body. Bone undergoes a complex set of diagenetic processes post mortem, which generally lead to dissolution, however bones and teeth are relatively common components in (post-Ordovician) rocks. The trace element content of bone is susceptible to alteration immediately upon exposure, and while these changes reduce the usefulness of ancient bone as a monitor of the physiology or diet of ancient animals, the trace element composition of ancient bone can yield useful paleo-environmental information. This chapter is concerned principally with archaeological and geological applications of trace element chemistry of bone, rather than physiological or medical implications.

BONE MINERAL

The inorganic (mineral) component of bone is carbonated calcium phosphate ($Ca_{10}(CO_3,PO_4)_6(OH)_2$), similar to the mineral dahllite. The calcium phosphates form the vast majority of all vertebrate hard tissues (Young and Brown 1982), and exhibit a very wide range of physical and chemical properties. Stoichiometric hydroxyapatite is not known biologically, and is only produced artificially or geologically at high temperature and pressure. The mineralogy of bone, dentine, and enamel is reviewed in Elliott (this volume). Kohn and Cerling (this volume) highlight the fact that apatite crystal size and shape varies between bone and enamel, and point out that variations in crystal size are in part responsible for the greater diagenetic susceptibility of bone compared to enamel.

Bone crystallites are essentially plate-shaped, with average dimensions of 350-400 Å × 350-400 Å × 25-50 Å (Weiner and Price 1986). Variation in crystallite size occurs with age, within different regions in the same bone, between different bones in the same animal, and between species (Tannembaum and Termine 1965; Robinson and Cameron 1964; Posner et al. 1965). Nonetheless, the surface area of bone crystallites has been estimated at 200 m^2/g (Weiner and Price 1986). This large surface area accounts for the relatively reactive nature of bone apatite, and its correspondingly high rates of dissolution in natural waters (Nriagu 1983). Enamel crystals are two orders of magnitude larger than bone crystallites (Brudevold and Söremark 1967), and have a much lower surface area, and therefore lower solubility.

1529-6466/00/0048-0013$05.00

TRACE ELEMENTS IN LIVING BONE

Trace element composition of bone

The cations Sr and Ba concentrate in the vertebrate skeleton, and the amounts of these elements vary as a function of mineral structure. *In vivo,* strontium has been found to accumulate in bone by exchange onto crystal surfaces, and is rapidly washed out after exogenous strontium is withdrawn (Dahl et al. 2001). Incorporation of strontium into the crystal lattice as a substitute of calcium occurs at a low level *in vivo*, in contrast to the extensive lattice substitution of strontium for calcium in fossil bone. Strontium is not easily washed out of subfossil bone (Tuross et al. 1989), and the uptake of strontium into biological apatite was once proposed as a potentially useful chronometer analogous to fluorine uptake (Turekian and Kulp 1956). The combined uptake of strontium and fluorine into vertebrate calcified tissue may in no small part account for the existence of a fossil record. Both of these elements stabilize biological apatite, and add substantially to the crystal stability of apatite under acidic conditions (Curzon 1988).

Table 1. Strontium and barium contents of living bone.

Strontium levels/Tissue	*Barium levels/Tissue*	*Reference*
79-114 ppm/bone	2-6 ppm/bone	Manea-Kritchen et al. 1991
70-550 ppm/bone		Bryant et al. 1960
	1.3 ppm/bone	Yoshinaga et al. 1995
210-390 ppm/bone		Hodges et al. 1950
~10-200 ppm/bone (wet weight)		D'Haese et al. 1999
53-146 ppm/bone	4-11.4 ppm/bone	Sowden & Stitch 1956
25-600 ppm/enamel		Steadman et al. 1958

The amount of strontium and barium in bones and teeth differs by two orders of magnitude (Table 1). Human strontium and barium levels in bones and teeth exist in a bimodal literature that derives from an interest in the incorporation of ^{90}Sr in the 1950s, and a more recent group of studies that focus on strontium's effect on mineralization. A wide range in strontium content in human bone is reported in groups of studies that are fifty years apart, but, in general, modern human bone and enamel tends to have strontium content <200 ppm, while barium content is commonly reported at less than 10 ppm. The amount of strontium in a normal human diet ranges from 0.023 to 0.046 mmol/Sr/day (Marie et al. 2001), and when administered with calcium, strontium is adsorbed to a lesser extent (Milsom et al. 1987).

This impact of calcium content in the diet on strontium uptake has led to a reevaluation of paleodietary interpretations that relied on a biopurification model (Burton and Wright 1995). It was originally proposed that the selective transport of calcium (over barium and strontium) in the vertebrate gut would result in a decline in Ba/Ca and Sr/Ca up a terrestrial food chain, and furthermore that carnivory, omnivory and herbivory could been distinguished on this basis (Elias et al. 1982). The need to account for the relative amount of calcium containing foods in multicomponent diets is not the only confounding factor in a biopurification model of dietary interpretation. Bone itself exhibits a preferential use of calcium in favor of either strontium or barium (Lengemann 1960) through an active uptake system that is heat labile. Age and osteonal remodeling cause an increase in barium with age (Manea-Kritchen et al. 1991). In rats, ethanol consumption and protein deficiency alter bone strontium and barium levels: ethanol consumption decreased both barium and strontium in bone, while a low protein diet increased bone strontium levels (Gonzalez-Reimers et al. 1999). Strontium levels in bone vary according to anatomical unit after strontium administration (Dahl et al. 2001) with higher amounts of strontium found in cancellous compared to cortical bone. The duration and/or evenness of the strontium exposure, as well as the sex of the animal can play a role in the final concen-tration of this cation in bones and teeth.

Lead is another element of interest in a number of applications. Most of the ingested lead is stored in bone, and is difficult to remove once it is incorporated into the mineral phase. The half-life of lead in bone can be up to 20 years (Anderson and Danylchuck 1977, Drasch 1982). Lead is not distributed equally among the bones of the skeleton, although there seems to be a relationship among anatomical units within one skeleton that would allow an estimation of total skeletal lead burden (Wittmers et al. 1988). In a modern population of humans (n = 240) that had not been exposed to lead occupationally, the mean lead content in the femur was 3.86 mg/kg bone wet weight as compared to the temporal bone (5.59 mg/kg) and the pelvic bone (1.65 mg/kg) (Drasch et al. 1987). An occupationally exposed population had approximately ten to twenty times the amount of lead in bone compared to the unexposed population (Brito et al. 2000).

Cautionary reports regarding the postmortem accumulation of lead into bones have not always been headed. Correlations between soil and bone lead content from a depositional environment, and failures to remove diagenetic lead from buried bone have been reported (Waldron 1982, Patterson et al. 1987). Lead does not only accumulate in bone diagenetically, but can also be lost through migration out of bone (Jaworowski et al. 1985). The possibility of both uptake and loss of any element, without knowledge of uptake or leaching profiles within a bone, makes dietary or environmental interpretations suspect.

Paleodietary and paleoenvironmental interpretations of strontium and barium content from vertebrate fossils have two major analytical and interpretive hurdles. First, a number of biological variables must be accounted for that are known to impact the concentrations of these cations in biological apatites. Second, the diagenetic alterations that occur due to uptake and substitution of barium and strontium into the apatite lattice must be mitigated. In practice, both the complexities in the biological use of strontium and barium, as well as the post-mortem changes in trace element content, can be extremely difficult to deconvolute (see later section). Some progress has been made in assessing diagenetic alterations through the use of strontium isotopes (Koch et al. 1992) although the problems of validating postmortem alterations in these isotopes often recapitulate the history of strontium content studies (see Kohn and Cerling, this volume).

Trace elements as provenancing tools. As trace elements are incorporated into bone *in vivo* from both diet and environmental water, variations in environmental trace elements are also passed into bone, and can serve as natural tracers. The stable isotope (e.g., O, C, Sr, Nd) composition of bone has been used in this way (e.g., to source ivory products, Van der Merwe et al. 1990, Vogel et al. 1990), and the trace element composition of bone in living animals can also potentially be used to identify an animal's geographical habitat, or to monitor trace metal exposure. In contrast to mammalian tooth enamel, bone is continually resorbed and re-deposited throughout life, so that the trace element composition of bone is an integrated record of trace metal exposure. For this reason, provenance studies based on bone are limited to animals that have a limited geographical range (and therefore experience a narrow range of trace metal compositions). Despite these methodological restrictions, bone has some advantages as an environmental tracer over other target tissues such as otoliths or internal organs—first, as we have seen, many trace metals are readily incorporated into the apatite lattice, including metals that have relatively minor physiological uses. Secondly and crucially, many organisms have external apatite hard tissues (e.g., fish scales, crocodile dermal scutes, turtle shells, teeth) that can be removed non-lethally (with a little courage in some cases). The concentrations of heavy metals in the bodies of marine turtles, for instance, can be monitored using carapace tissues (Sakai et al. 2000). Similarly, fish scales can be used to distinguish breeding stocks of commercially valuable fish, where removing alternative tracer tissues such as otoliths requires both the death of the fish, and also results in loss in commercial value of the carcass (Farrell et al. 2000, Wells et al. 2000, Gillanders 2001, Thorrold et al. 2001). Distinguishing between breeding stocks requires sustained differences in environmental concentrations of trace metals. Such differences are maximized in rivers running over distinct local geologies. Exploiting this situation, Markich et al. (2001) were able to distinguish salt-water crocodiles from three catchment areas on the basis of trace metal concentrations in osteoderms (boney plates in the skin) sampled non-lethally. Biomonitoring of trace elements in phosphate tissues can therefore provide wildlife managers with valuable information on migration and movement of animals, help to identify wild stocks threatened from over-exploitation, and be used to assess metal pollution, bioavailability and exposure. Trace metal metabolism is complex, however, and rather poorly understood. Incorporation of metals into biominerals is not necessary a passive process, and any physiological control complicates the interpretation of *in vivo* metal signatures in biogenic phosphates.

TRACE ELEMENTS IN FOSSIL BONE

Bone mineral is unstable in most burial environments, and alters during diagenesis. Alteration either leads to dissolution of the bone, or to eventual preservation as a fossil bone (fossilization). During fossilization trace elements may be added or removed from bone, depending on the concentration of the element in question in fresh bone, the chemistry of the surrounding pore waters (and speciation of the element in question), the partition or adsorption coefficient for the element (or elemental complex) between apatite and water, and its diffusion through bone. Concentrations of trace elements in fossil bones vary over several orders of magnitude, although relatively few studies have measured concentrations of trace metals other than Sr, Ba and the rare earth elements. Typical values are shown in Table 2 (compare values of Sr and Ba with Table 1).

As noted elsewhere in this volume, apatite is capable of admitting a wide range of metal cations into the lattice, largely as substitutions for Ca^{2+}(see following discussion), and consequently the concentration of many trace metals in bone apatite increases during diagenesis. The concentrations of some essential trace elements that are relatively abundant in bone may decrease during diagenesis (most likely as a consequence of growth or recrystallization of apatite with low trace metal concentrations rather than leaching from bone). Most recent work with fossil bone, however, has focused on elements that are rare in fresh bone, and are added during diagenesis so that the trace element composition of the fossil bone can be assumed to be a diagenetic or post mortem feature. Typically, most workers have focused on U, Th and the REE, and this section concentrates on the uptake of these elements into bone, and their applications to some geological problems. To understand the uses and limitations of studies of the trace element composition of fossil bone apatite, it is useful to first outline the current understanding of the processes that lead to the fossilization of bone.

Fossilization of bone

Fossil bones contain at best trace amounts of collagen and evidently collagen has been lost during diagenesis, probably through hydrolysis to gelatin (Collins et al. 1995, Nielsen-Marsh et al. 2001, Trueman and Martill 2002). Hydrolysis of collagen would theoretically expose apatite crystallite surfaces and should lead to rapid dissolution of the bone, but in fossil bones the space left by collagen does not remain as pore space (Fig. 1).

Table 2. Concentrations (ppm) of selected trace metals in fossil bones from terrestrial localities (values in plain type: 1 standard deviation).

Locality	*Sr*	*Ba*	*Pb*	*La*	*Sm*	*Yb*	*U*	*Reference*
Olduvai FLKN N=17	**9700**, 2100	**4000**, 575		**70**, 260			**1400**, 5800	Denys et al. 1996
Olduvai FLKNN N=17	**7300**, 2100	**4000**, 1324		**5000**, 3900	**500**, 640		**3210**, 7200	Denys et al. 1996
Naran Bulak N=22	**2500**, 390	**900**, 170	**130**, 20	**1400**, 200	**480**, 120	**300**, 230		Samoilov et al. 2001
Dinosaur Provincial Park N=34				**280**, 300	**50**, 70	**50**, 60		Trueman 1997
Two Medicine Formation N=45				**300**, 500	**40**, 80	**15**, 30		Trueman 1997
Dinosaur National Monument N=11	**2300**, 460	**350**, 82	**180**, 170	**440**, 300			**280**, 200	Hubert et al. 1996
Laño, Iberian Peninsula N=3	**1194**, 95	**268**, 45	**225**, 18	**160**, 46	**21**, 9	**14**, 5	**473**, 219	Eloza et al. 1999
Gobi Desert N=105	**1820**, NA	**735**, NA	**77**, NA	**1000**, NA	**301**, NA	**46**, Na	**287**, NA	Samoilov & Benjamini 1996

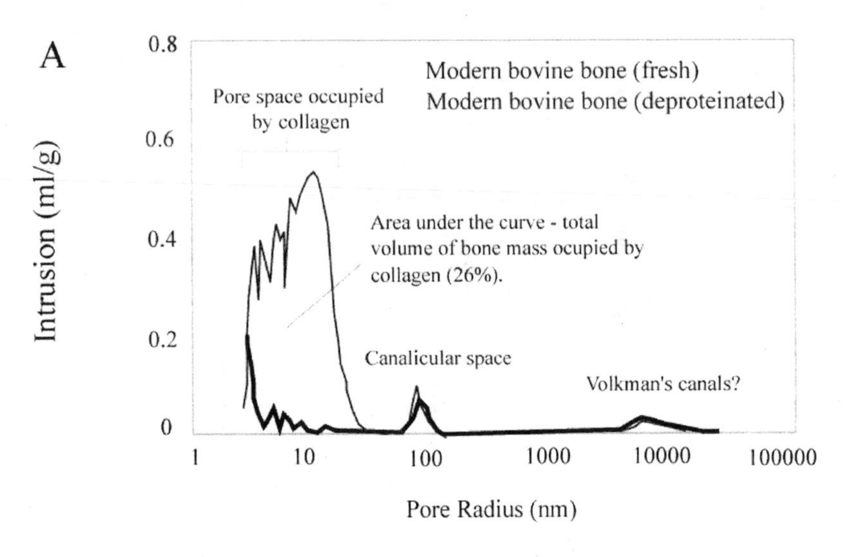

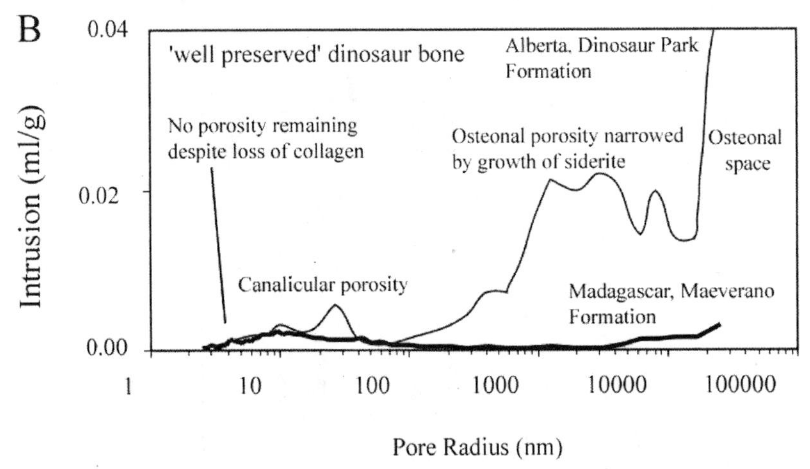

Figure 1. Porosimetry measurements for fresh and deproteinated modern bovine bone (A) and dinosaur bone from the Cretaceous Dinosaur Park Formation of Alberta, Canada, and Maverano Formation, Madagascar (B). The porosity of modern cow bone increases by ~26% after collagen is removed. However, even well preserved or unpermineralized dinosaur bones (bones that contain essentially no collagen and no minerals other than apatite) do not preserve the porosity originally occupied by collagen. Therefore around 26% of the volume of these fossil bones must be composed of authigenic apatite added post mortem.

This suggests that fossilization (or stabilization) of bone occurs when the size of exposed apatite crystals increases, reducing solubility and porosity. This 'recrystallization' or growth of new apatite must occur relatively rapidly, as exposed crystallites would soon dissolve or be exploited biologically in most pore-water environments. It is likely that recrystallization proceeds together with hydrolysis of collagen (Collins et al. 1995, Bocherens et al. 1997), however the relative rates of collagen loss and mineral growth are undoubtedly environment-specific and are currently unknown. Changes in crystal size and/or perfection can be monitored optically (Pfretzschner 2000), by X-ray diffraction

(XRD; e.g., Tuross et al. 1989), or by FTIR spectroscopy (e.g., Weiner and Bar-Yosef 1990), and analyses of large numbers of archaeological bones suggest that it may take anything from a thousand to hundreds of thousands of years for fresh bone to achieve the levels of recrystallization seen in fossil bones (Person et al. 1996). Certainly, changes in the crystallinity of bone apatite occur very rapidly upon exposure, and can be detected within 10 years *post mortem* (Tuross et al. 1989, Shinomiya et al. 1998). Archaeological bone samples do not, however, necessarily provide a good analogue for the processes that led to preservation of most fossil bones, as the majority of archaeological samples are derived from human burials, and the taphonomy and biochemistry of deliberate burial is very different from that of 'natural' accumulation and burial (Trueman and Martill 2002).

Several mechanisms have been proposed to account for the observed increase in mineral density that accompanies (and to some extent defines) fossilization of bone – all of these mechanisms come under the broad term recrystallization. One mechanism involves dissolution-reprecipitation of apatite, whereby small, reactive bone crystallites dissolve, and subsequently new authigenic francolite (carbonate fluor-apatite) grows in place of bone crystals. In this case fossil bones are essentially pseudomorphs of the original bone (Kolodny et al. 1996). However, most fossil bones retain exquisite details of their histology—notably retaining birefringence patterns characteristic of fresh bone. These patterns relate to the biogenic orientation of apatite crystallite *c*-axes along collagen fibrils, and their preservation in fossil bone argues against widespread dissolution-reprecipitation of apatite during fossilization (e.g., Hubert et al. 1996). Preferential (biogenic) alignment of apatite crystallites in fossil bone has also been shown using TEM (Zocco and Schwartz 1995, Schweitzer et al. 1999). An alternative mechanism for bone recrystallization was proposed by Hubert et al. (1996), who argued

(1) Fresh bone – apatite crystallites arranged parallel to collagen fibers, with intercrystalline spaces.

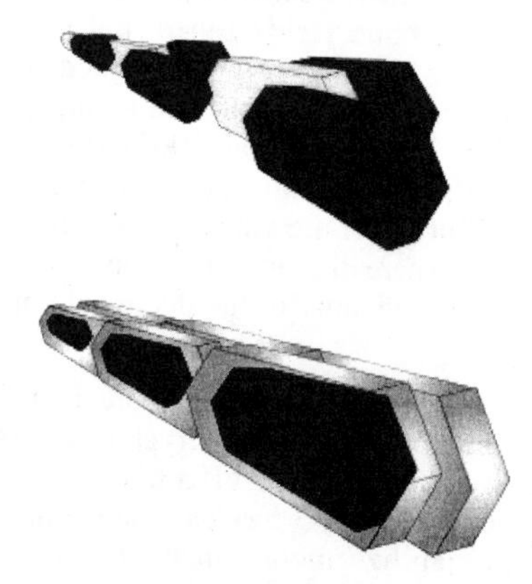

(2) During recrystallization, trace elements are concentrated in new authigenic apatite.

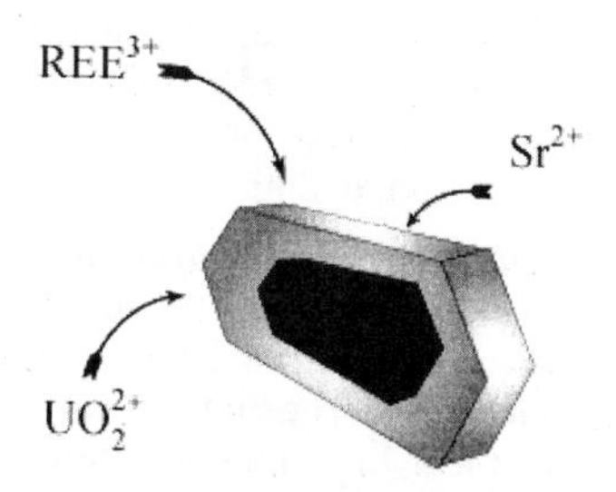

(3) Fossil bone – seeded growth of authigenic apatite around original bone crystallites preserves original alignment and closes intercrystalline porosity.

New apatite may be treated as an authigenic mineral.

Figure 2. Schematic model of bone recrystallization by addition of authigenic apatite (note: original bone crystals are on the order of 40×40×10nm, and are mostly poorly formed, plate-shaped crystals of dahllite).

that apatite was essentially added to biogenic 'seed' crystals, preserving apatite crystallite alignment. This seeded apatite therefore requires addition of new authigenic phosphate, and where conditions allow, growth would continue until all inter-crystallite porosity has been infilled with authigenic apatite (Fig. 2).

Permineralization (infilling of vascular spaces within the bone by authigenic minerals) is a common but not essential feature of bone fossilization. Common permineralizing minerals are calcite, apatite (francolite), pyrite, barite, clay minerals, and quartz. Permineralization will further reduce porosity of the bone, and therefore reduce the potential for later bone-water interaction. This may complicate preparation of bone samples for trace element analysis.

Post-mortem uptake of trace metals into bone

Three main mechanisms have been proposed for introduction of metal cations into bone, and each may play a part in incorporation of certain trace metals into bone under differing environmental conditions:

1. Adsorbtion of metal cations onto crystal surfaces (presumably displacing surface calcium).
2. Direct ion substitution (or coupled substitution e.g., $M^{3+} + Na^+ - 2\ Ca^{2+}$) in the apatite structure.
3. Growth of discrete trace metal-phosphate phases disseminated within the bone.

The last of these mechanisms has been observed experimentally (Valsami-Jones et al. 1997); however, these experiments were performed with synthetic hydroxyapatite, with very high dissolved concentrations of trace elements (Pb, Cd), where dissolution of apatite was driven by high concentrations of trace metals. Discrete trace metal-phosphate phases are rarely observed in fossil bone, despite very high total trace element concentrations (for instance REE concentrations may exceed 1 wt %). However Molleson et al. (1998) show a unique example of extensive replacement of dahllite with the lead phosphate mineral pyromorphite, apparently through diffusive-substitution of apatite lattice-bound Ca with Pb. Elemental mapping of fossil bone yields zoned, but finely disseminated distributions of trace metals, rather than the highly concentrated and localized distribution that would be found were trace metals concentrated in authigenic metal-phosphates (Williams 1988, Downing and Park 1998, Rogers et al. 2001). It seems unlikely, therefore, that trace metals are held as discrete metal-phosphate phases disseminated within an apatite matrix in fossil bones, but rather are contained within the apatite lattice as adsorbed or substituted ions. However, where dissolution of bone apatite occurs, insoluble metal-phosphate phases may form in the vicinity of the degraded bone (e.g., Karkanas et al. 2000).

Substitution and adsorption of metals. As noted above, and elsewhere in this volume, bone crystallites are small, and poorly crystalline, with many crystal defects. Bone crystals therefore have a high surface energy, proving an effective site for adsorption of ions. The high adsorption capacity of bone crystallites has led to attempts to use bone meal or synthetic apatite crystallites to stabilize metals in contaminated groundwater (Valsami-Jones et al. 1996; Chen et al. 1997a, b). Cations are likely adsorbed onto bone crystallites via cation exchange with Ca^{2+} at crystal surfaces. Cations adsorbed onto crystal surfaces are susceptible to later exchange, as long as the crystal surface remains exposed. However, if inter-crystalline porosity is closed during diagenesis, then individual crystallite surfaces will be protected from further exchange, and adsorbed cations will be stabilized within the bone.

Direct substitution of cations in pore waters for Ca^{2+} in the apatite lattice is a third possible mechanism to introduce cations into bone post mortem. In this mechanism, cations diffuse into the lattice from external pore waters, substituting for Ca^{2+} within lattice sites, and potentially altering the crystal chemistry. The replacement of dahllite by pyromorphite discussed by Molleson et al. (1998) appears to be an extreme example of this process. Finally, cations may be introduced into bone via growth of authigenic apatite phases with cations such as Sr^{2+} substituting for Ca^{2+} in the lattice.

Substitution and adsorption mechanisms arguably constitute the most important mechanisms for incorporation of most trace metals into bone during diagenesis. However, as both mechanisms involve displacement of Ca^{2+} ions, it is very difficult to distinguish between these two mechanisms *a posteriori*. This distinction is important, however, as the method of incorporation of trace metals has implications for the retrieval of biogenic signals through sequential washes (e.g., Sillen and LeGeros 1991).

An appreciation of the mechanisms of trace element-bone interaction can be gained by comparing uptake of REE in bone with experimental and analytical studies of high temperature apatite-melt REE partitioning behavior. Apatite-melt partition coefficients derived from experimental studies, when plotted against ion radius, show parabolic traces with maxima around Nd-Sm (Fig. 3). These REE ions have radii closest to Ca, and indicate that REE are held in Ca sites in the lattice. Blundy and Wood (1994) developed a predictive model for calculation of equilibrium partition coefficients of trace elements between crystals and melt in igneous systems. In this model, the partitioning behavior of isovalent cations between a mineral and liquid at a particular pressure, temperature, and composition of interest can be explained by a version of the Brice (1975) equation, in which the size and elasticity of the crystal lattice sites play a critical role:

$$D_i(P,T,X) = D_o(P,T,X)\exp\left[\frac{-4\pi E N_A\left[\frac{r_o}{2}\left(r_i - r_o\right)^2 + \frac{1}{3}\left(r_i - r_o\right)^3\right]}{RT}\right] \quad (1)$$

where D_i is the partition coefficient, defined as $[i]$ mineral / $[i]$ fluid, and D_o is a "strain-compensated partition coefficient" that describes strain-free cation substitution (i.e., the radius of the substituent cation r_i is equal to that of the optimum radius of the cation site r_o) at the pressure, temperature and composition of interest. E is the Young's modulus of the host cation site, N_A is Avagadro's number, R is the gas constant, and T is in Kelvin.

Fitting Equation (1) to plots of measured partition coefficients against ion radius therefore yields estimates of the Young's modulus of the cation site (the 'tightness' of the parabola), and the optimum ion radius (Figs. 3a-c). Values of the average bulk modulus ($K = E(apa)/1.5$, Wood and Blundy 1997) of apatite Ca sites as derived by applying equation (1) to equilibrium partition coefficients (Nagasawa 1970, Watson and Green 1981, Fujimaki 1986), are shown in Figure 3. Also shown in Figure 3 are the values of the site bulk modulus K determined by direct physical measurement of apatite crystals (Bass 1995). The fitted values of K are consistent between the different partition experiments, but they are significantly lower than values of the crystal bulk modulus determined by direct physical measurements (after correction for substitution of a trivalent cation into a divalent site). The fitted r_o values also yield metal-oxygen distances that are consistently higher than the weighted mean Ca-O distances for FAP and HAP (calculated for 4 Ca (I) and 6 Ca (II) sites). This occurs because the curves are fitted against the REE ionic radii in IX-fold coordination rather than a weighted ionic radius reflecting the relative site occupancy in apatite. The simpler case was chosen because no site occupancy information was available for the biogenic apatites studied.

Non-equilibrium partitioning of trace metals between seawater and bone apatite can be studied by normalizing the REE concentrations of the fish debris to those of overlying

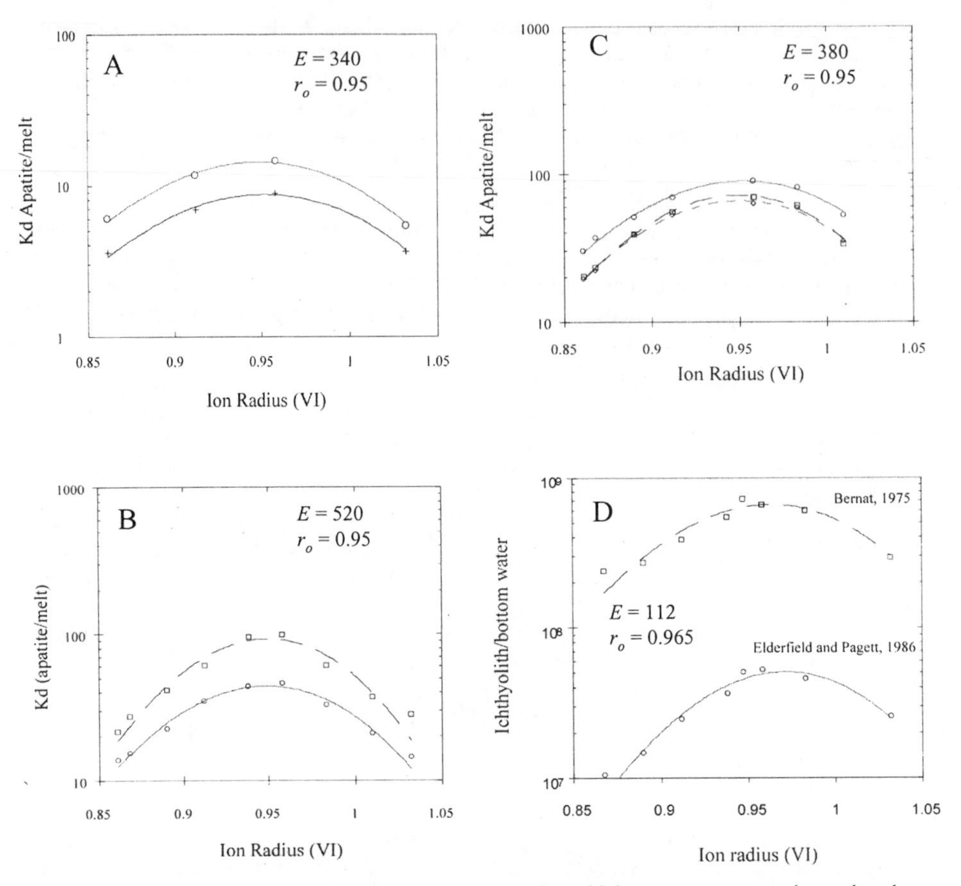

Figure 3. Apatite-melt partition coefficients derived from high temperature experimental and natural systems (A-C) and estimates of partitioning between ichthyoliths and bottom water (D), together with the best fit of Equation (1) (Blundy and Wood 1994) to this data. Values taken from; (A) Watson and Green (1981), (B) Nagasawa (1970), (C) Fujimaki (1986), and (D) Bernat (1975), Elderfield and Pagett (1985), and Peipgras and Jacobsen (1992).

ocean water. The best fit curves of Equation (1) to the estimated partition coefficients are shown in Figure 3d. In both cases the data are best described by a single smooth curve that represents the average REE content of both of the two Ca sites in apatite, rather than by two separate curves for the two different Ca sites.

The fitted values of r_o are similar, but not identical for both low- and high-temperature data. Mean r_o values are 0.95 Å (S.D. = 0.01 Å) and 0.965 Å (S.D. = 0.01 Å) for high and low temperature datasets respectively. These values indicate that maximum partition coefficients for the REE in apatite occur around Nd-Sm in both low and high temperature environments, but suggest that the Ca(2)-O distances are slightly larger on average in carbonate-hydroxylapatite and carbonate fluorapatite than in HAP or FAP. Hughes et al. (1991) suggested that maximum partition coefficients should be centered around Nd, however substitution in the F-OH sites significantly alters the Ca(2)-O distances. Substitution of CO_3^{2-} into both the phosphate and hydroxyl sites and charge-balanced coupled substation of REE+Na for Ca also likely alter the Ca-O distances, and may thus explain the observed shift in r_o values towards Nd (and thus larger site sizes) in the bone data. These results indicate clearly that the REE in fossil bone are associated directly with Ca^{2+} sites in the apatite lattice rather than forming discrete REE phosphate phases.

The ichthyolith studies show that the REE are partitioned very strongly into bone from seawater. Measured "partition coefficients" between ichthyoliths and seawater are many orders of magnitude higher than equilibrium partition coefficients between apatite and fluids predicted for low temperatures and measured at high temperatures. Concentrations of REE in ichthyoliths are approximately six orders of magnitude higher than typical concentrations in seawater (Keto and Jacobsen 1987), comparable to adsorption coefficients measured experimentally between seawater and apatite particles (Koeppenkastrop and Decarlo 1992). Experimental studies of REE adsorption onto apatite particles in seawater show that REE-apatite adsorption coefficients vary smoothly according to ion radius, with maxima around Nd. This strongly suggests that adsorptive capture of REE onto apatite occurs via REE-Ca cation exchange at crystal surfaces. Therefore it seems likely that in bone-pore water systems, adsorption is the dominant mechanism of REE capture, rather than lattice substitution, and the REE are incorporated into bone via adsorptive cation exchange at apatite crystal surfaces.

This argument was explored by Reynard et al. (1999), using values of E and r_o obtained from the experimental partitioning data of Fujimaki (1986). Reynard et al. (1999) used Equation (1) to predict equilibrium REE-apatite partition coefficients at surface temperature and pressure, assuming that the crystal chemistry of bone apatite is broadly similar to that of HAP, and that crystal-melt partition coefficients can be used to estimate crystal-water partitioning. Reynard et al. (1999) then compared the predicted partition coefficients with measured adsorption coefficients for the REE between seawater and HAP derived by Koeppenkastrop and DeCarlo (1992), and concluded that incorporation of REE into bone via a substitution mechanism produces 'bell shaped' REE patterns with relatively little fractionation between La and Lu. Incorporation of REE into bone via an adsorption mechanism, on the other hand, produces significant fractionation between La and Lu (La/Lu = 5). Based on REE patterns found in fossil fish teeth, they concluded that REE uptake in fossil bone was dominated by adsorption mechanisms, but that subsequent recrystallization may superimpose a degree of substitution-related fractionation over the initial, adsorption related REE pattern. It is important to note, however, that these predictions are based on crystal chemistry of hydroxyapatite and fluorapatite, and not dahllite and francolite. Variations in E and r_o will affect relative adsorption and/or partition coefficients, and may alter the predicted partition coefficient ratios (e.g., La/Lu and La/Sm).

Given the very high adsorption coefficients, and relatively low concentrations in surface fluids, the REE (and by analogy other trace metals with similarly high M/apatite adsorption coefficients) will be effectively scavenged from circulating fluids. Incorporation of trace elements into bone will be limited by the relative rate of supply of trace metals, rate of diffusion/flow of trace metals through the bone, and rate of recrystallization of bone. This final point is well illustrated by profiles of trace element concentrations through bone cortices.

Concentration profiles of trace metals in ancient bone. In many cases the concentrations of elements such as U, Th and the REE show marked heterogeneities with depth perpendicular to the external cortical (periosteal) surface. The shape of these concentration profiles has received considerable study, as calculations of the age of bone based on ESR and U-series dating techniques depend upon knowledge of the style of uptake of U into bone (e.g., Millard and Hedges 1999).

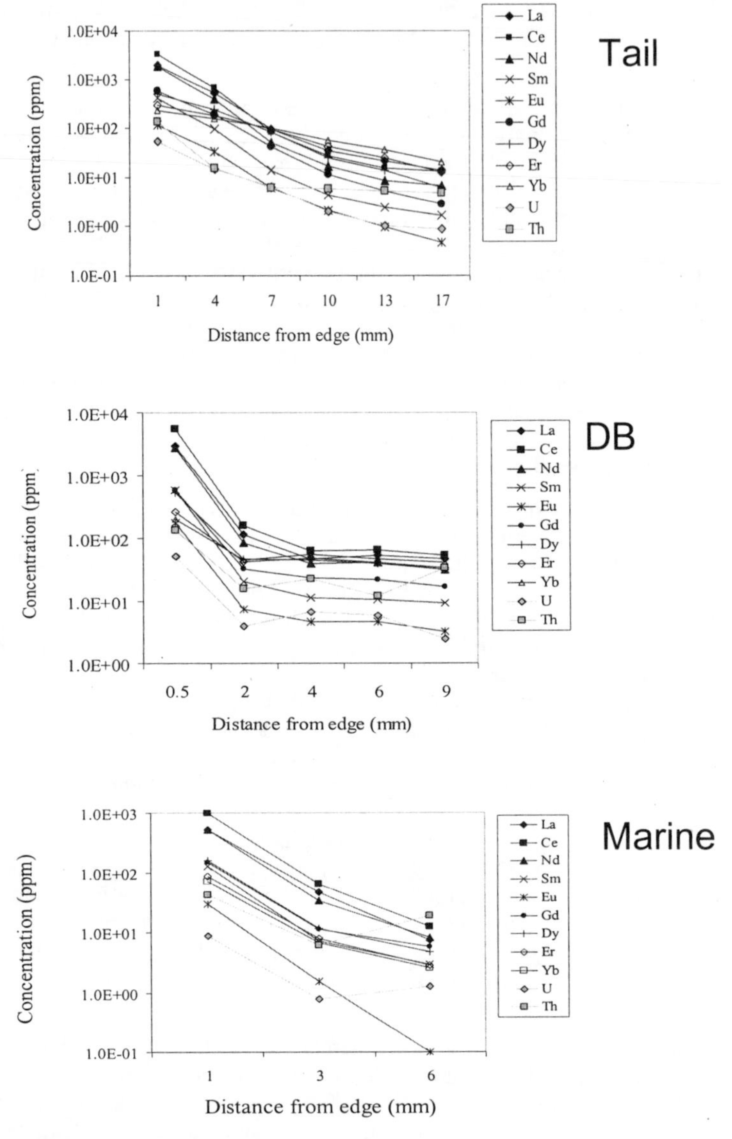

Figure 4. Concentration of rare earth elements, U and Th with depth in three bones from the Createceous Dinosaur Park Formation of Alberta, Canada (details of samples given in Trueman et al., submitted).

Under equilibrium conditions, the concentration profile of any given element will be a flat line, the concentration determined by the partition coefficient and concentration of that element (or elemental complex) in groundwaters, and its rate of diffusion through bone (Millard and Hedges 1999). Flat profiles are fairly commonly found for uranium in older archaeological bone (Hedges et al. 1995), but the few analyses made of fossil bone suggest that many elements (including U) often show steep concentration gradients (e.g., Fig. 4), with the highest concentrations found at bone margins, and lowest concentrations found at depth within the bone cortex (Williams 1988, Williams et al. 1989, Elliott and Grime 1993, Janssens et al. 1999). These concentration profiles suggest that despite the porous nature of bone, most trace elements are distributed within a bone via diffusion directed from the outer surfaces perpendicularly into the thickness of the cortex. The vascular network of bone (the osteonal canals for instance) appear to have a minor effect on passage of trace metals, with less well developed concentration gradients extending out orthogonally from the osteonal margins, superimposed on the main surface to center concentration gradient. This pattern is most easily explained if the bone is effectively saturated, with very little active flow of water within the bone, and thus little chance to refresh the bone-pore waters (Millard and Hedges 1999).

The steep concentration profiles seen in many fossil bones, particularly those from terrestrial environments (e.g., Fig. 4), indicate that fossil bones often do not reach equilibrium in terms of metal partitioning, despite millions of years of potential exchange with groundwaters. Evidently exchange with groundwaters is halted, presumably by growth of authigenic mineral phases and reduction of porosity.

Millard and Hedges (1999) produced a mathematical model based on the rate of diffusion of uranyl ions through water saturated bone and adsorption coefficients between uranyl and bone. This model was designed to calculate the amount of time taken to develop an observed elemental concentration profile within a single bone, and can be used to estimate the amount of time needed for a particular bone to reach equilibrium. In the case of fossil bones, where equilibrium is clearly not reached despite millions of years of potential exchange, such a model predicts the amount of time that the bone remained as an open system, able to incorporate trace elements via diffusion. Further incorporation of metals is halted by the closure of bone porosity, presumably after growth of authigenic apatite.

Applying the Millard and Hedges model to measured U profiles in fossil bones yields rates of recrystallization on the order of 1 kyr to 10 Myr depending on the thickness of the bone, and estimates of the burial environment and diagenetic condition of the bone (Trueman et al. in prep). These may serve as a first order approximation of the amount of time that a single bone may be exposed to trace element uptake.

From the argument above it follows that those bones that eventually become fossils do not reach equilibrium with the surrounding pore fluids, and that the concentration of a trace element at a single point within a single fossil bone is determined largely by the relative rates of recrystallization of bone and diffusion of metals through the bone cortex. As discussed above, many trace elements are effectively scavenged by bone via an adsorption mechanism at the bone/pore water interface, and trace element patterns in the deep cortex may be very different from those at the periosteal or endosteal margins (Williams and Marlow 1987, Williams and Potts 1988, Williams 1988, Janssens et al. 1999). Elements with very high M-bone adsorption coefficients will be readily adsorbed onto the external bone cortex, and removed from solution so that deeper parts of the cortex receive waters progressively depleted in those elements most compatible in the apatite lattice (Fig. 5). Therefore, the external portions of bone cortices will inherit a largely quantitative record of the external pore water chemistry as competition for cation-binding sites is limited, whereas deeper parts of the cortex will inherit a more derived trace element pattern, reflecting selective removal of compatible trace elements (at least in environments where transport of metal ions into bone is dominated by diffusion).

Trace element composition of ancient bone: Archaeological applications

Paleodiet analyses. The majority of work on the trace-element chemistry of bone apatite has come from attempts to trace the diet of ancient humans. The intake of trace

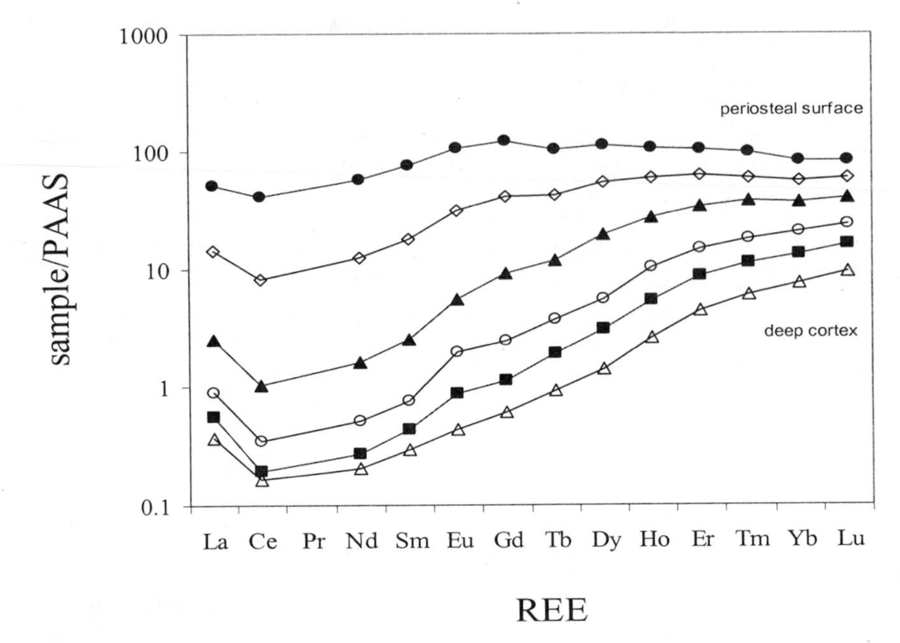

Figure 5. Fractionation of REE with depth in the cortex of a single dinosaur vertebra from the Cretaceous Dinosaur Park Formation of Alberta, Canada. Sample points are ~5 mm apart (from Trueman et al., in review).

elements occurs from two sources, food and water. In marine animals the most important source of trace elements is seawater (which tends to be rather buffered and stable), whereas terrestrial animals obtain a very small amount of their trace elements from water, compared to the intake from food sources. Different food sources contain very different concentrations and types of trace elements, and so the diet of an animal will determine the trace elements available for substitution into bone (Parker and Toots 1980).

Initially, strontium was seen as a potentially useful element in paleodietary analysis, as it was seemingly unaffected by diagenesis. Sr content is low in animal flesh, so carnivores should have low bone Sr contents compared to herbivores, and animals with a significant animal hard-part component in their diet (e.g., shellfish eaters) should have higher Sr than flesh eaters. Omnivores should be able to be split into groups that have high and low plant content diets. Toots and Voorhies (1965) pioneered the use of bone Sr levels to deduce paleodiet. Later, Brown (1973) emphasized the stability of Sr in archaeological bones, and used this as evidence that the Sr content had not changed appreciably during diagenesis. This has subsequently been very controversial in paleodiet studies, with many conflicting results and interpretations (e.g., Nelson et al. 1986, Tuross et al. 1989, Buikstra et al. 1990, Price et al. 1992, Radosevich 1993, Sandford and Weaver 2000). If the strontium values in bones can remain unaltered throughout diagenesis, then the Ca sites (the sites of Sr/Ca substitution) must not be implicated during bone recrystallization. If, however, the Sr content or Sr/Ca ratio has changed, then recrystallization and mobility of the apatite must have occurred. The early optimism surrounding the stability of Sr in apatite and its potential to record biogenic signals has been replaced by the realization that biogenic Sr signals are not generally recorded in fossil or archaeological bones (Nelson et al. 1986, Tuross et al. 1989, Locock et al. 1992, Elliott and Grime 1993). This is because of extensive overprinting, contamination or substitution of diagenetic Sr, which suggests that the Ca site is not stable during diagenesis, but undergoes exchange with pore-water-held ions (e.g., Pate and Hutton 1988, Price 1989). Attempts have been made to retrieve biogenic Sr isotope ratios from diagenetically altered bone through sequential acid washes (e.g., Sillen 1986, Sillen and Le Geros 1991). These techniques rely on the assumption that all diagenetic addition of strontium into bone is associated with authigenic mineral phases that are more soluble than fresh bone apatite, and not exchanged with Ca (or Sr) in bone crystals. However, as discussed above, diagenetic addition of trace metals such as Sr, Ba and the REE is associated with adsorptive substitution on apatite crystal surfaces, and growth of authigenic apatite.

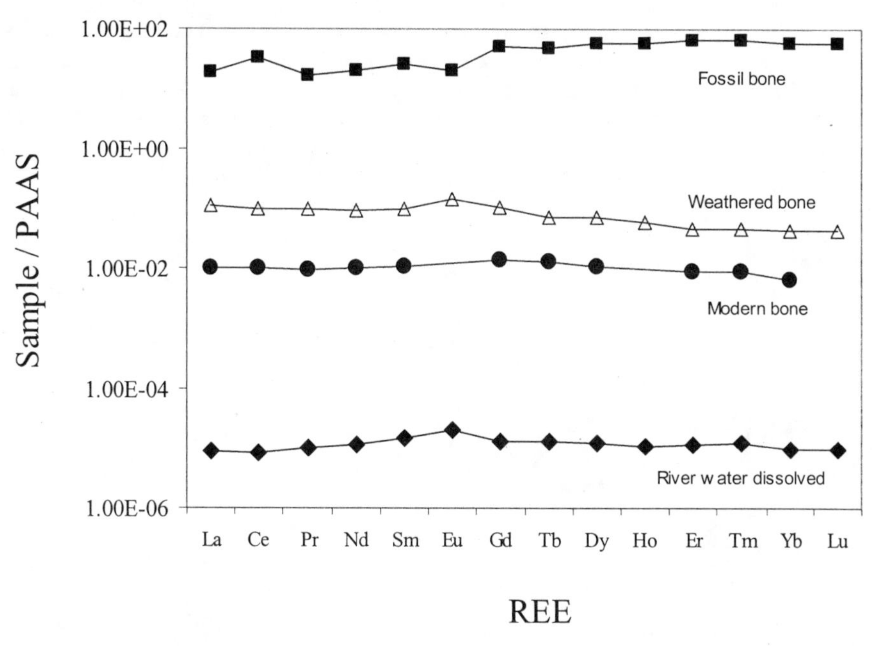

Figure 6. Concentration of REE in wetland waters (dissolved phase, Dupré et al. 1999), modern bone (NIST SRM 1400 Hinners et al. 1998), modern weathered bone (bone exposed for 25 years on Amboseli plain, Kenya; Trueman—unpublished data), and fossil bone (Olorgesailie Formation, Pleistocene, Kenya,; Trueman et al. 2001).

The susceptibility of the Ca^{2+} sites in bone apatite to cation substitution post mortem is dramatically demonstrated by the incorporation of ions not present in living bone. The REE perhaps show the clearest evidence for cation substitution during growth of secondary diagenetic apatite (Fig. 6). All fossil bone samples so far analyzed show large (several orders of magnitude) enrichments in REE over *in vivo* levels and usually over sediment and pore-water levels (e.g., Arrhenius et al. 1957, Henderson et al. 1983, Shaw and Wasserburg 1985, Staudigel et al. 1985, Williams 1988, Kolodny et al. 1996, Hubert et al. 1996, Trueman and Benton 1997, Kohn et al. 1999, Trueman 1999, Staron et al. 2001, Samoilov and Benjamini 2001). The REE in fossil bones must be derived from pore waters, and so the high levels of REE found in all fossil bones prove that a significant portion of the Ca-site associated trace element content of a fossil bone has been derived from ground waters (Kolodny et al. 1996). Furthermore, studies of recent exposed bone clearly show that this incorporation of trace metals begins almost immediately upon exposure (Tuross et al. 1989, Shinomiya et al. 1998). Increases in REE concentrations in bone appear to be linked with changes in mineralogy towards a bulk francolite composition (Martin and Haley 2000), which most likely reflects seeded growth of authigenic apatite within intercrystalline porosity (Hubert et al. 1996). However, trace metals are adsorbed onto bone crystallites almost immediately upon exposure, so bulk mineralogy is no guide to the extent of alteration of trace metal signals in bone. As preservation of bone requires growth of authigenic apatite in contact with groundwaters, all fossil bone will contain a significant and probably inseparable diagenetic trace metal content (Tuross et al. 1989; Kolodny et al. 1996).

It is not clear that authigenic apatite added post mortem is more soluble than biogenic bone apatite crystals (in fact this seems rather unlikely as addition of authigenic francolite stabilizes bone, and increases its survival potential, although see Sandford and Weaver 2000 for a recent and more optimistic review). More soluble phases (such as calcite) that bear diagenetic trace element patterns are certainly removed during the initial stages of the wash technique, but there is little convincing evidence that an unambiguously biogenic apatite phase is ever retrieved, and some convincing evidence that all wash leachates are derived at least partially from diagenetically altered apatite phases (Tuross et al. 1989, Koch et al. 1992, Elliot et al. 1998, Budd et al. 2000, Fabig and Herrmann 2002).

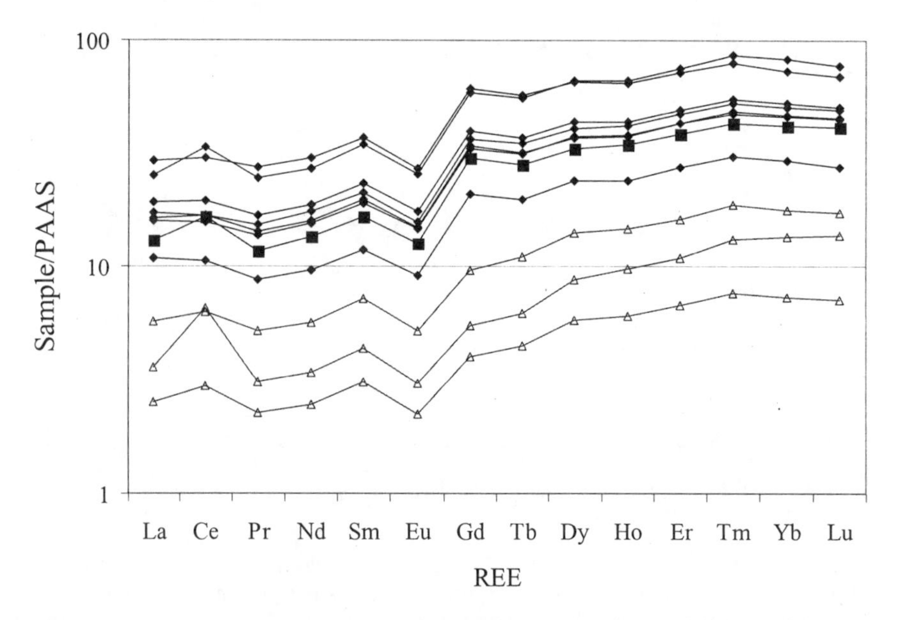

Figure 7. Concentration of REE in contemporaneous bone and tooth enamel from the Pleistocene Olorgesailie Formation. Concentrations of REE are approximately 5-10× lower in fossil enamel than fossil bone from the same sites, but still 2-3 orders of magnitude greater than typical concentrations in modern bone.

Enamel is a potentially more robust source for biogenic trace metal signals than bone, by virtue of its large crystal size, and reduced solubility and porosity. However, trace elements such as the REE, Pb and U are also added to enamel during diagenesis (Fig. 7). Consequently it is difficult to argue that any particular sample of ancient enamel or bone retains an entirely unaltered biogenic trace metal signal. Growth of authigenic minerals in vascular spaces can also severely complicate trace metal analyses. Bones typically contain a fairly diverse suite of authigenic minerals that begin to form immediately after death (e.g., Barker et al. 1997). Preliminary work has identified calcite, barite, sepiolite, crandallite, and trona growing in bones exposed (unburied) for up to 30 years post mortem. These minerals have very different solubilities, and in some cases (e.g., crandallite) are extremely difficult to separate chemically from bone dahllite during preparation. Obviously, even minor amounts of barite present in bone will severely compromise trace element analyses.

Despite these difficulties, the trace element composition of ancient bones continues to be used as an indication of body water chemistry (Price et al. 1992, Grupe et al. 1997, Arnay-De-La-Rosa et al. 1998, Safont et al. 1998, Schutkowski et al. 1999, Baraybar 1999). Many studies attempt to control for diagenetic alterations in (particularly) Sr and Ba, by comparison with 'controls' either from modern populations or sites with contrasting diagenetic parameters (e.g., Price et al. 1989). These comparisons are not true controls, however, as the investigator cannot control the full complexity of natural diagenetic conditions in specific bones, and in many cases the diagenetic parameters being compared operate independently, so that an 'unaltered' signal in one parameter does not imply pristine values in other parameters. For instance, Ca/P ratios are often cited as a test for 'pristine' bone (e.g., Sillen 1986, 1989; Price 1989; Price et al. 1992). The Ca/P ratio in modern bone is around 2.15 (LeGeros and LeGeros 1984), however dissolution of ancient bone often yields very high Ca/P ratios. This is almost certainly caused by growth of calcite in most cases (Sillen 1986, 1989). This excess calcite is easily removed by weak acid wash, and later washes contain a Ca/P ratio close to 2. However, this does not imply a return to pristine bone compositions, as diagenetic francolite has similar Ca/P ratios. This point has been demonstrated elegantly by Fabig and Herrmann (2002), who compiled a dataset of some 580 individual bone analyses, previously treated using accepted methods to 'remove' diagenetic Sr. All bones yielded acceptable Ca/P ratios between 2 and 2.3, and therefore cleaning was assumed to be successful, and Sr/Ca and Ba/Ca ratios were previously interpreted as representing biogenic values. However, Fabig and Herrmann (2002) show that Sr/Ca and Ba/Ca ratios in fact co-vary significantly with the degree of post mortem microbial damage to the bone. Bones that suffer more histological destruction yield higher Sr and Ba concentrations. Evidently, the cleaning procedure did not remove diagenetic Sr and Ba bearing phases, and the Ca/P ratio is not a suitable criterion for identifying unaltered biogenic bone apatite (Radosevich 1993, Fabig and Herrmann 2002). It is likely that microbial decomposition of bone promotes dissolution of bone apatite, and subsequent re-precipitation of apatite in newly created pore spaces. Soil-derived Sr and Ba will be incorporated into these new carbonate-apatite phases that have Ca/P ratios similar to bone, and are not removed during standard acid wash procedures. Consequently, removal of more soluble phases in buried bone is not sufficient to retrieve an unambiguous biogenic trace element signal.

Several studies show that bones recovered from herbivorous animals yield statistically significant differences in some trace metal levels than bones from known carnivorous animals (e.g., Safont et al. 1998). Even if the reported differences really do represent differences in the biogenic trace metal concentration (and this is debatable), there is usually significant overlap between the pre-determined 'herbivore' and 'carnivore' categories. Single bones can rarely be confidently assigned to one or other dietary group (as expected if the biogenic trace metal concentrations are overprinted by diagenetic environmental signals to varying and essentially random degrees within the total assemblage) and thus the value of such low-resolution analyses is debatable.

A more convincing (independent) test for preservation of pristine trace metal concentrations could be found in the levels of metal ions such as the REE and U. The rare earth element lanthanum is commonly found in living bone at the 10-1000 ppb level, but is rapidly incorporated into bone *post-mortem* associated with the Ca sites. Increased La concentrations (e.g., >1 ppm) in bone (or acid wash solutions) are suggestive of post mortem alteration in Ca sites. It follows that bones that have been treated to remove diagenetic strontium should also yield La concentrations significantly <1 ppm as well as Ca/P ratios equivalent to carbonated apatite. This La proxy was suggested by Kohn et al. (1999) and Kolodny et al. (1996) with reference to the stable isotope composition of oxygen in ancient bone, but to our knowledge their suggestion has not been widely adopted. Unfortunately, low REE concentrations do not prove unambiguously that a bone preserves a pristine Sr or Ba signal. Concentration profiles of REE, U, Ba, and Sr may be very different in single bones (Janssens et al. 1999), indicating that the rates of transport of trace metals within bones are different. The REE are effectively scavenged at the bone, pore water interface, whereas Sr is able to penetrate deeper into the cortex (U may behave either as the REE or as Sr depending on the redox state and species [Millard and Hedges 1999, Janssens et al. 1999]). However, low concentrations of REE and U near the exterior margins of bones certainly argue strongly for minimal bone-pore water interaction.

Unfortunately, often the only criterion for judging that a signal is 'unaltered' is the value of the signal itself, either in isolation or in comparison with other bones. Small changes in fluid flow, pore water pH, ionic strength, or any number of diagenetic and soil parameters will affect the bone-pore water interaction, and therefore can produce changes in (or homogenize) trace metal ratios in bone. We argue therefore, that at present it is extremely dangerous to assume that the trace element composition of any bone collected several years post mortem truly reflects biogenic trace element concentrations. Furthermore, there is no independent *a priori* criterion that can be used to test whether a biogenic signal has been recovered. We are forced to conclude that the trace element content of ancient bone cannot yield reliable information regarding paleodiet.

Uranium -series dating of bone. Uranium concentrations in living bone are typically ~0.1 ppm (Aitken 1990), whereas concentrations in fossil bone may exceed 1000 ppm. Theoretically, therefore, Quaternary age bones could be dated directly by the U-series method.

U-series dating is based on the decay series of U. Natural ^{238}U decays to ^{206}Pb via a number of intermediate radioactive daughter isotopes. For the purposes of dating, the most important isotopes are ^{238}U, ^{234}U and ^{230}Th.

^{238}U half-life: 4500 Myr

^{234}U half-life: 245 kyr

^{230}Th half-life: 75.4 kyr

U-series dating hinges on the ratio of $^{230}Th/^{234}U$ in a sample. ^{230}Th is very insoluble, and is generally assumed to be absent in pore waters, and all ^{230}Th in an ancient sample is assumed to derive from decay of ^{234}U. The $^{230}Th/^{234}U$ ratio increases with age at a rate determined by the relatively short half -life of ^{230}Th. Eventually an equilibrium will be reached between the rate of decay of ^{230}Th and the rate of production of ^{230}Th from decay of ^{234}U. This stage marks the older limit of ages suitable for U-series dating (generally around 350-500 Ka). U-series dating is performed by analyzing ^{234}U and ^{230}Th concentrations in the sample to be dated.

Successful U-series dating demands that two basic conditions are met: the sample must be a closed system with uranium present at the time of deposition. Bone contravenes both of these assumptions (Millard and Hedges 1995). However, the benefits of direct dating of bone are obvious, and many attempts have been made to overcome the problems posed by U series of bone. In the simplest models U is added to the bone rapidly with respect to the total age of the bone, roughly equivalent to the time of deposition. This is the early uptake (EU) model. Alternatively, U may be added gradually, often modeled as constant or linear uptake (LU). Bischoff et al. (1995) compared the effects of different uptake models on calculated U-series ages of bones, and showed that LU ages are about twice EU ages for EU ages younger than 50 Ka, rising to 3 times EU ages as isotopic equilibrium is approached. Evidently, the method and style of U uptake profoundly affects the reliability of the final derived date, and this partly explains why only 60% of published U-series dates on fossil bone agree with comparative dates at the 2σ range (Millard 1993). Several attempts to refine the basic uptake models have been made (Szabo-Rosholt 1969, Chen and Yuan 1988). The success of these models could be tested against bones of known age from a range of depositional environments, but this has not been done, partly due to the difficulty of obtaining suitable test samples (Millard and Hedges 1999).

Recently, U concentration profiles have been measured to pre-select bones for U-series dating that conform to simple uptake models, and mathematical models have been developed that predict the spatial distribution of U within a single bone cortex as a function of time (Millard and Hedges 1999). The use of concentration profiles and quantitative models of U-bone interaction provide a better rationale for choosing a particular uptake model when calculating a U-series age. Diffusion-adsorption models also predict that the U-series age will vary within a single bone with a U-shaped concentration profile (central portions of the bone yielding a younger age than the external margin). Differences in U-series age between internal and external portions of a bone cortex could therefore be used to further constrain the chosen uptake model. However, U-series dating of fossil bone remains rather complex.

Paleoenvironmental and geological applications

Fish debris (ichthyoliths) as a proxy for paleo-seawater chemistry. Following the pioneering work of Wright et al. (1984), interest in the trace element chemistry of ancient bone switched to elements that are sparingly present in bone in life, but readily enriched during diagenesis. These elements can be used to reconstruct aspects of the environment of burial, and, given the apparently rapid recrystallization rate of bone (~10,000 yr) and subsequent resistance to later overprinting, bone is a valuable proxy for water chemistry.

The possibility of using fossil bone as a proxy for water chemistry arguably stems from Arrhenius (1957), who determined the concentrations of several trace metals (REE, U) in fish debris from the ocean floor, and compared the concentrations to seawater. He concluded that ichthyoliths readily incorporate trace metals from seawater and thus might serve as a record of ancient seawater composition. Subsequently Goldberg (1962) confirmed that the REE composition of fish teeth from marine basins mirrored that of the overlying water column. Bernat (1975) and Shaw and Wasserburg (1985) reported high REE concentrations in ichthyoliths (relative to biogenic concentrations) from the uppermost 4cm of sediment of ocean cores, which had a bulk REE pattern similar to the overlying waters. Furthermore, there is no record of systematic increase in REE content with burial (Bernat 1975, Staudigel et al. 1985, Elderfield and Pagett 1986, Wright et al. 1987). These results suggest that ichthyoliths inherit a REE composition directly from seawater, during early diagenesis, with little or no fractionation.

Fossil bone chemistry as a paleoredox indicator. Uniquely among the REE, cerium has the capability to adopt a tetravalent ion under the appropriate oxidizing conditions. The oxidized form of Ce (Ce^{4+}) is relatively insoluble compared to Ce^{3+}(de Baar et al. 1985), so that cerium takes part in active redox cycling. Under oxic conditions, Ce exists in the tetravalent state, and is readily removed from solution either onto particle surface coatings, or into authigenic minerals (Sholkovitz et al. 1993, Koeppenkastrop and

DeCarlo 1992). Under reducing conditions, Ce^{3+} may be released back into the water column or into pore waters. The anomalous behavior of Ce compared to its neighboring REEs (the cerium anomaly) is quantified by the ratio of the measured abundance of Ce to an expected value (Ce*) interpolated from the neighboring trivalent REEs (after de Baar et al. 1985):

$$Ce/Ce^* = (Ce/Ce_{shale}) / [2/3\ (La/La_{shale}) + 1/3\ (Nd/Nd_{shale})] \quad (2)$$

Ce/Ce* values that are significantly greater or less than 1 imply the presence of Ce^{4+} and therefore oxic conditions. Similarly, the lack of a cerium anomaly in concentrations of dissolved REE implies suboxic or anoxic conditions.

Wright et al. (1984) determined REE composition of hundreds of individual conodonts and ichthyoliths, and found consistent REE shapes in pre-Carboniferous conodonts, with enrichment of MREE, lesser enrichment of LREE, and no cerium anomaly. This pattern is fundamentally different to the REE pattern found in modern ocean bottom waters, and led Wright et al. (1984) to suggest that the REE chemistry and particularly Ce anomaly in ancient marine apatites could be used to infer widespread anoxia in pre-Carboniferous ocean basins.

However, interpreting Ce anomalies in ichthyoliths is not simple, Elderfield and Pagett (1986) pointed out that the relative rates of REE uptake and diagenesis in bone are critical to establishing the meaning of any REE pattern in fossil bone. The challenge facing those wishing to use bone as a proxy for ocean REE composition is to show that the bulk of all REEs are incorporated (and stabilized) early within this time-scale, and that the contribution of REE from later sources (presumably from diagenetic pore-fluids) is relatively minor (Kemp and Trueman in press).

German and Elderfield (1990) noted that pore waters below the sediment/water interface rapidly develop high REE concentrations, with REE abundances that are not related to the overlying bottom waters, and that authigenic minerals forming in contact with such pore waters would not record a 'seawater' signal. German and Elderfield (1990) support the suggestion of Elderfield and Pagett (1986) that in the modern (oxic) ocean, only ichthyoliths deposited in sedimentary environments undergoing oxic diagenesis have a chance of recording a negative cerium anomaly. Under reducing conditions (common in shelf pore-water environments), further incorporation of REE would tend to obscure any negative cerium anomaly inherited at the sediment water interface.

In summary, preservation of a seawater redox signal is favored by rapid mineralization and stabilization at the sediment/water interface; and uptake of REE from pore waters (rather than from seawater) is likely to reduce or eliminate any inherited cerium anomaly. Negative cerium anomalies in ancient marine biogenic apatite therefore suggest oxic conditions in the water column and possibly in the upper pore waters, but the lack of a negative cerium anomaly in biogenic apatite does not necessarily indicate sub-oxic or anoxic conditions in the water column (Kemp and Trueman in press).

REE patterns in ichthyoliths. Many studies have confirmed that modern 'seawater'-type REE patterns are rarely found in ichthyoliths recovered from pre-Triassic seawaters (Fig. 8). Instead, these tend to show MREE enriched traces, peaked at Eu-Gd, similar to those recorded by Wright et al. (1984, e.g., Grandjean and Albarède 1989, Bertram et al. 1992, Girard and Albarède 1996, Armstrong et al. 2001).

Estuaries are major sinks for the REE, preferentially removing light and middle REE (preferential removal of LREE and MREE onto particle surfaces is responsible for the HREE enrichment characteristic of open ocean waters). LREE are preferentially adsorbed

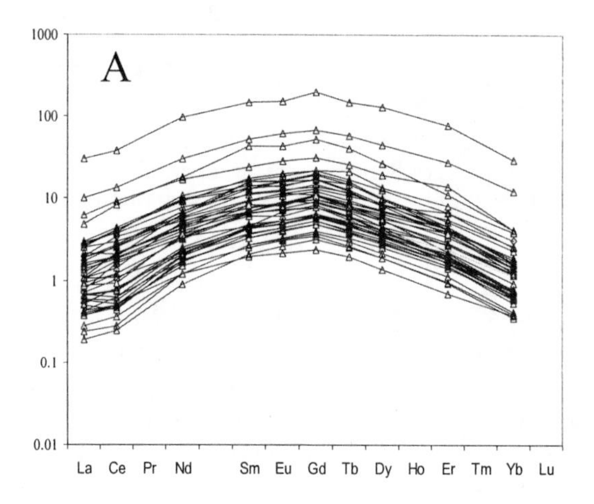

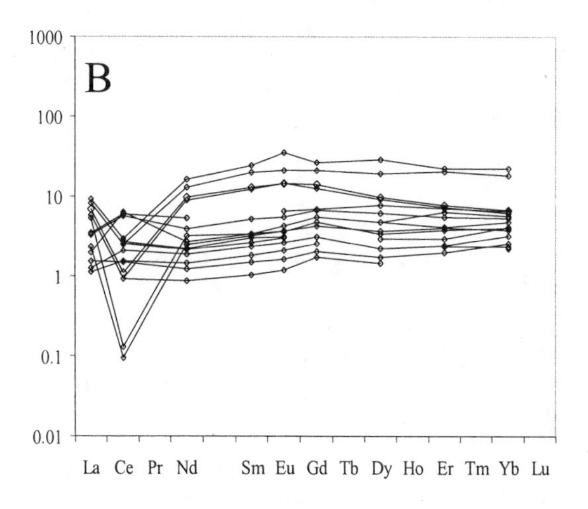

Figure 8. Shale (PAAS) normalized REE concentrations in ichthyoliths from pre (A) and post (B) Cretaceous sediments. (A) REE concentrations in conodonts from the Coumiac Limestone (Devonian, France). (B) Icthyoliths recovered from deep marine sediments from the Pacific and Atlantic oceans. Data from Girard and Albarede (1996) and Elderfield and Pagett (1986).

onto particle surfaces (e.g., Sholkovitz et al. 1994), and the MREE appear to be preferentially associated with colloidal Fe-organic particles (Hoyle et al. 1984). Major changes in the speciation and transport mechanism of the REE could account for the apparent differences seen in the REE patterns in pre- and post-Cretaceous ichthyoliths.

Grandjean-Lécuyer et al. (1992) suggested that the absence of HREE-enriched patterns in Palaeozoic ichthyoliths represented a fundamental difference in pre-100 Ma ocean chemistry, caused by the lack of modern phytoplankton. They argued that the primary transport mechanism of REE from the surface waters to sediment in modern oceans is through the action of phytoplankton and zooplankton (Fowler et al. 1987), and that since these groups radiated late in the Jurassic, Palaeozoic ocean REE dynamics would be different. Specifically they suggested that the primary supply of REE would be through inorganic co-precipitation of REE and Fe/Mn oxyhydroxides. However, in the modern ocean the principal primary transfer mechanism of REE also appears to be coprecipitation of REE with Mn oxides. Moffett (1990) suggested that the apparent correlation between Mn and Ce redox cycles is in fact caused not by inorganic co-precipitation, but by similar microbially mediated oxidation. Formation of Mn-oxide particles (and hence REE co-precipitates in the modern ocean) is thus biologically controlled, but not necessarily controlled by calcareous planktonic organisms.

Nonetheless, Grandjean-Lécuyer et al. (1992) attempted to model the development of MREE enriched seawater through adsorption of REE onto mineral particles (e.g., Fe or Mn oxyhydroxides) where the concentration of element C at the i'th step of entrainment by oxyhydroxide precipitation is given as follows:

$$C^i_l = (1-X_{i+1})C^{i+1}_l + X^{i+1}\, DC^{i+1}_l \qquad (3)$$

where the subscript l refers to seawater, X is the fraction of Fe/Mn removed at the i'th step, and D is the precipitate seawater fractionation coefficient for the element.

Grandjean-Lécuyer et al. (1992) assumed that log D varied linearly with atomic number, and D_{La} = 1000 and D_{Yb} = 10. Applying these equations, they were able to model seawater REE compositions evolving towards MREE-enriched patterns, peaked around Gd. Furthermore, Grandjean-Lécuyer et al. (1992) suggested that 'minute extents' of mineral precipitation are capable of producing MREE-enriched residual seawaters. However, the difference between D_{La} and D_{Yb} used by Grandjean-Lécuyer et al. (1992) is rather severe, as experimental (Koeppenkastrop and De Carlo 1992), and observed (Sholkovitz et al. 1994) fractionation coefficients for REE between seawater and manganese oxide surfaces vary much less (producing La/Lu ratios of 4.67 and 10, respectively). A more profound problem with the model outlined above is that it assumes closed system behavior, with no further additions of REE. It is difficult to justify this assumption in the open ocean, and indeed adsorption of REE onto mineral surfaces in natural systems does not produce the residual MREE-enriched waters predicted by Grandjean-Lécuyer et al. (1992). In closed systems, or REE-limiting environments, however, removal of REE by coprecipitation with Fe or Mn oxyhydroxide particles could produce MREE-enriched residual waters, and hence could be responsible for the heavy MREE enrichments seen in pre-Cretaceous and some more recent ichthyoliths.

Finally, Reynard et al. (1999) suggest that late diagenetic recrystallization of ichthyoliths might lead to fractionation of REE between water and apatite via a substitution mechanism. In this case, small increments of fractionation might drive REE patterns towards MREE-enriched 'bell-shaped' patterns. However, Keto and Jacobsen (1987) suggested that seawater signals were retained in conodonts with an alteration index (CAI) less than 5 (equivalent to heating to temperatures of ~350°C). Armstrong et al. (2001) investigated the effect of thermal metamorphism on REE patterns in single conodont elements, and showed that REE patterns only appeared to be altered in conodonts with an alteration index (CAI) greater than 6. Above ~500°C, REE patterns became relatively enriched in MREE, consistent with changes in the crystal lattice (and especially changes in Ca-O bond lengths).

In summary, the absence of HREE enriched 'seawater' patterns in pre-Cretaceous ichthyoliths is difficult to explain, and probably does point either to a major difference in the ocean chemistry of the REE, or widespread late diagenetic recrystallization and mobilization of REE in Palaeozoic conodonts. However, as some modern ichthyoliths also do not record 'seawater' patterns, and as pore waters frequently record REE patterns significantly fractionated with respect to 'seawater,' it is difficult to come up with a single, unifying explanation of REE patterns in ichthyoliths. Moreover pre-Cretaceous sedimentary phosphorites often record 'modern' seawater patterns, while ichthyoliths record MREE-enriched patterns. Apparently the trace element composition of pre-Cretaceous ichthyoliths is trying to tell us something, although it is rather difficult to discern precisely what that might be.

Despite the uncertainty regarding pre-Cretaceous ichthyolith data, several authors have noted that the La/Yb ratio recorded in fossil biogenic apatites (and cherts) decreases progressively from coastal settings to open ocean settings (e.g., Grandjean et al. 1987, Laenen et al. 1987, Murray et al. 1992). This suggests that the early diagenetic source of REE changes from a shale-like or LREE-enriched coastal component, dominated by terrigenous bound REEs, towards a more HREE-enriched pattern. This increasingly HREE-enriched pattern reflects progressive removal of more reactive LREE onto particle surfaces and evolution of seawater towards more HREE enriched patterns—thus the trace element chemistry of fossil bone has proven useful in tracing large scale transport mechanisms of the REE—presumably this approach could be extended to other elements that are readily adsorbed onto bone apatite post-mortem, and could prove to be a valuable tracer for metal cycling and transport within estuarine-coastal-basin systems.

Nd isotopes in ichthyoliths. Attention has also focused on the isotopic composition of the rare earth element Nd in ichthyoliths as a proxy for paleo-ocean circulation (Shaw and Wasserburg 1985, Keto and Jecobsen 1987, Martin and Haley 1999).

River water is the major source for Nd into the oceans, and consequently the isotopic composition of Nd in the oceans is controlled by the geology of the surrounding land. As the residence time of Nd in seawater is low with respect to ocean mixing, individual water masses develop distinct Nd isotopic compositions that reflect the average surrounding crustal composition. Changes in the isotopic composition of Nd in one location through time therefore reflect changes in ocean circulation and/or changes in the river input. Ocean circulation plays a very important role regulating global climate, and consequently there is a great deal of interest in reconstructing the circulation history of oceans (Martin and Haley 2000).

Nd is not easily accommodated in the calcite lattice, and therefore biogenic carbonates tend to have very low concentrations of Nd (e.g., Shaw and Wasserburg 1985). In contrast, ichthyoliths incorporate REE rapidly from seawater, and frequently contain very high Nd concentrations (typically between 100 and 1000 ppm). Once again, ichthyoliths only serve as a useful proxy for the isotopic composition of Nd in ancient seawater if it can be demonstrated that Nd is incorporated rapidly at the sediment/seawater interface, and that subsequent incorporation of Nd does not significantly alter the 'seawater' ratio. Several studies have attempted to demonstrate that this in fact is the case. Keto and Jacobsen (1987) showed that conodonts originally deposited on either side of the Iapetus Ocean yielded distinct Nd isotope ratios. This suggests that the Iapetus Ocean was composed of two separate basins or water masses until the onset of the Caledonian/Taconic Orogeny, after which Nd isotope ratios rapidly homogenized, suggesting mixing of the whole water mass. Martin and Haley (2000) attempted to test directly the influence of pore water chemistry on the isotopic composition of Nd in ichthyoliths. They collected ichthyoliths from two sites in the Pacific, which experienced similar seawater Nd compositions, but had very different sediment and presumably pore water Nd compositions. Ichthyoliths were recovered from a pure carbonate sediment in one case, and a volcaniclastic-rich clay in the other. The isotopic composition of Nd in ichthyoliths from both sites corresponded well, and also was in excellent agreement with the Nd isotopic composition of ferromanganese crusts (Martin and Haley 2000). However, on the basis of four samples, Bertram et al. (1992) are less confident, and suggest that the isotopic composition of Nd in ichthyoliths may be prone to contamination.

It seems that ichthyoliths from deep ocean basins do record seawater Nd isotopic ratios, and can be used to reconstruct ocean circulation. This is consistent with current understanding of the rates of recrystallization. Rates of sedimentation in deep ocean basins are frequently on the order of 1 cm/1000 yr (e.g., Bernat 1975). If major recrystallization of bone (and co-incident uptake of trace metals) occurs over ~10,000 years, then we would expect ichthyoliths to incorporate the majority of their REE at or very near the sediment water interface. The chemistry of Ce in the upper few centimeters of the sediment may differ significantly from bottom waters due to microbial redox cycling, but it is less likely that the isotopic composition of Nd will vary dramatically in the upper few centimeters (unless the sediment is very radiogenic). However, it may be dangerous to extend these analyses to environments with more rapid burial rates, as deeper pore-waters will bear less and less relation to the overlying seawater.

Bone trace metal compositions as a record of fractionation and/or speciation of metals during weathering. The trace element composition of fossil bone can only be used as an environmental proxy if the trace elements under study are fractionated significantly during earth surface processes. Otherwise bone will only record the broad source lithology, and will be of marginal use. Luckily elements such as the REE and U have relatively complex behavior during weathering. Notably, these elements may be transported as free ions, adsorbed onto particle surfaces, or as 'dissolved' colloidal complexes. Differences in the speciation of the REE, and the binding nature of any complexing agents may fractionate trace metals from one another in which case the REE and U content of fossil will not record bulk source rock compositions, but will reflect surface processes.

The REE are rapidly removed from solution in estuarine and coastal environments by adsorptive scavenging onto the surfaces of settling particles (Hoyle et al. 1984), and as LREE are preferentially adsorbed onto particle surfaces (Sholkovitz et al. 1994), estuarine and coastal sediments and their pore waters receive relatively more LREE than HREE, leaving the oceans relatively depleted in the LREE. This pattern is reflected in ichthyoliths recovered from estuarine, coastal and deep ocean environments (Fig. 9). Bones from terrestrial environments consistently record shale-normalized REE patterns that are relatively HREE enriched. This pattern is repeated in many different depositional settings, and therefore appears to be independent of local and regional geology (Fig. 9). This suggests that terrestrial pore waters are either enriched in HREE, or that the HREE component of terrestrial pore waters is more easily adsorbed onto bone. One exception to this general observation is that bones from environments influenced by aeolian sediment transport tend to display relatively LREE-enriched patterns. These patterns are similar to bones from estuarine and coastal marine settings, and probably reflect a greater contribution of LREE from REE adsorbed onto wind-blown particle surfaces. Bones from estuarine settings apparently plot intermediate between soil and coastal environments, although this is based on only one sample locality. The shale-normalized pattern of REE in fossil bones therefore appears to reflect the chemistry of the REE in soil or sediment pore waters, and may yield information concerning REE transport chemistry, weathering reactions, soil conditions, or sedimentology.

The REE composition of fossil bones also varies within single depositional environments. The REE composition of several hundred bones distributed across a single Pleistocene soil horizon at Olorgesailie, Kenya, was determined by Trueman et al. (2001). Bones were recovered from 11 discrete excavation sites, distributed across an ~1 km linear transect. Bones from some localities yielded REE signals that were unique to that locality and distinct from contemporaneous localities less than 50 m away. The REE composition of the associated soils were uniform across the sample area, and all REE were derived primarily through weathering of widespread ash deposits (i.e., REE source was uniform within the scale of the sample area). The spatial variation seen in bone REE compositions most likely reflects differences in REE speciation between sites across the soil, and therefore differences in the adsorption and diffusion coefficients between pore waters and bone. This level of geochemical separation suggests that the aqueous chemistry of the REE is relatively complex, and that REE are fractionated from one another during weathering.

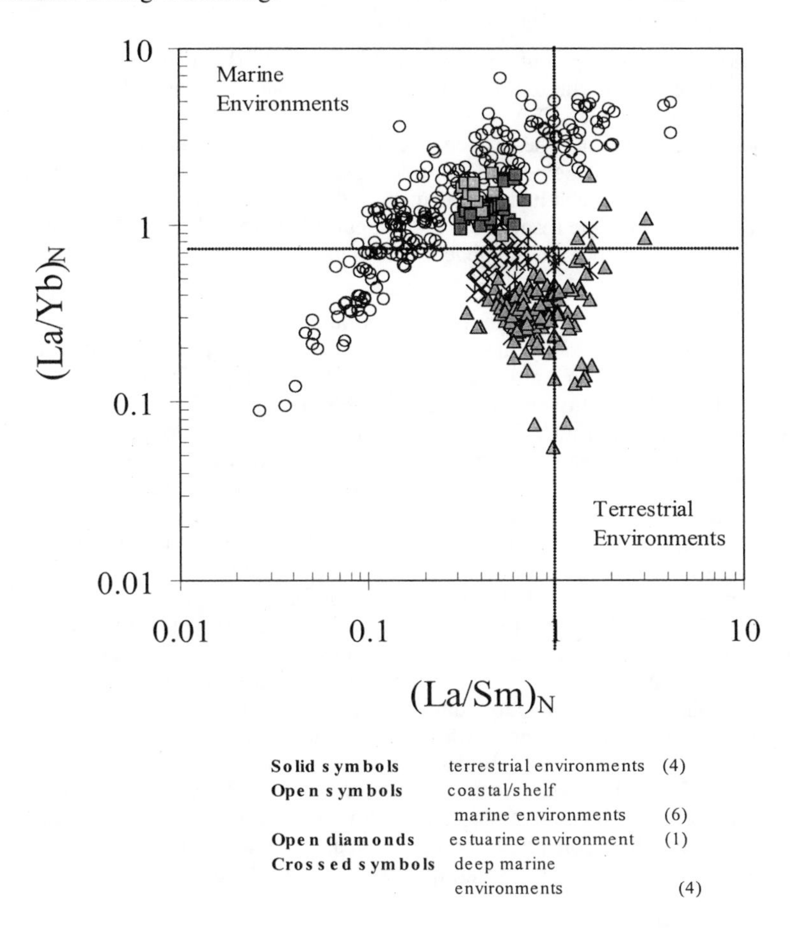

Figure 9. Relative abundances of REE in bones and teeth from a range of depositional settings. The REE composition of fossil bones and teeth can be used to group them according to their depositional environment. Thus coastal marine bones, teeth, and conodont elements all group towards the top left side in this REE ratio plot. Bones from aeolian, estuarine, deep marine and soil settings all form discrete groups (although there is some overlap between these environmental groups). Numbers in brackets refer to the number of localities sampled. Figure modified from Trueman et al. (submitted).

Other uses of REE data in fossil bone. As we have seen, there is a growing body of evidence suggesting that the timescale of active incorporation of trace metals into bone post mortem is relatively short compared to the geological age of fossil bones (likely on the order of 1-50,000 years). Furthermore, once a bone has ‘recrystallized’ it remains a largely closed system with respect to further addition of trace metals. In other words, once fossilized, bone retains a memory of its early diagenetic environment. As the incorporation of trace metals into bone is controlled by the chemistry and trace element composition of the pore water, bone can potentially record changes in pore water chemistry, changes in weathering regime, or changes in metal transport over 10-100 kyr timescales. This approach was adopted by Samoilov et al. (2001), who used changes in the trace element geochemistry of dinosaur and turtle bones recovered from a succession of lacustrine sediments from the Gobi desert to investigate changes in aridity across the K/T boundary linked to the aridization of Mongolia.

Bone records a trace element signal related to its early depositional environment. Bones fossilized together in the sample depositional environment should therefore inherit the same trace element signal. This argument is the foundation of a new and potentially powerful tool for establishing the provenance of a fossil bone. Trace element patterns have long been used in archaeological studies to establish the provenance of pottery, but until recently such an approach had not been applied to fossil bones. However, there are many problems that could be resolved if the temporal and spatial context of individual fossil bones (their provenance), or diagenetic associations between groups of bones could be established.

The provenance of a fossil bone, or a population of fossil bones within an assemblage, can be obscured by many processes natural and human. Bones may be reworked from one horizon to another or recovered from geologically complex or mixed sediments. A large proportion of the entire vertebrate fossil record is preserved in attritional depositional environments, which may mix fresh bones with previously buried bones. Consequently the remains of animals that never co-existed in life may become co-mingled in the fossil record (e.g., Behrensmeyer and Hook 1992, Lyman 1994). Occasionally the stratigraphic range of a species may be extended based on reworked fossils (e.g., Argast et al. 1987). In addition to these problems, the increasing value attached to fossils encourages commercial exploitation of limited and in many cases protected resources, and such exploitation can cause problems such as incomplete or false stratigraphic and locality information, and even deliberate alteration of fossil skeletons, often by addition of fossil bones from other, unrelated specimens. Finally, many skeletons in paleontological collections and displays are in fact made up from disarticulated isolated bones from any number of individual animals, and it is not always easy to identify which bones belong to which individuals. The trace element chemistry of fossil bones provides a quantitative comparative test of provenance that can be applied to establish whether bones in a test assemblage were derived from a single early diagenetic environment (and by implication, a single locality), or from a mixture of localities. The test assemblage might be a collection of bones from a single attritional deposit, or a set of bones from a single skeleton (very small quantities of sample are required so such tests are minimally destructive).

The basic premise of provenance studies requires that a population of potential sources be measured and analyzed statistically. If the trace element signals form a continuous population, then there is no evidence that the assemblage is derived from several sources. If, however, the trace element signals fall into discrete populations, then there is good evidence that the population is mixed, and furthermore the individual bones within the sample can be discriminated into separate groups that represent their original burial associations. Such techniques assume that any mixed assemblage was derived from sources with contrasting geochemistry. However, bones may have been introduced into that assemblage from an area with very similar geochemistry, or may not have spent enough time in separate environments to inherit a discrete geochemical signal. Consequently, while a positive result (trace element signals form several discrete populations) is good evidence for a composite or mixed assemblage of bones, a negative result (trace element signals form a single population) does not prove that the assemblage is not mixed. Such studies can be used to assign surface collected bones to particular stratigraphic levels (e.g., Plummer et al. 1994).

Recognizing reworking. Traditional methods for recognizing reworked or allocthonous bones are based on the condition of the surface of the bone. As bones are transported (usually by water) they are abraded, and the surface of the bone is altered. It follows that more heavily abraded bones are likely to have traveled further or suffered several cycles of erosion and transport (Fiorillo 1988). However, many factors contribute to the level of abrasion seen in any single bone, especially the nature of the sediment bed load, and the hydrology of the transport medium (Lyman 1994). While the degree of abrasion of a *population* of bones in an assemblage may suggest that the assemblage is attritional, and may indicate qualitatively the probability that any bone in the assemblage is allocthonous, surface morphology is an unreliable guide to the provenance of individual bones. Geochemical taphonomic techniques are ideally suited to identifying allochthonous bones or teeth as by definition these remains are derived from different depositional settings. Geochemical tests for reworking are of course another type of provenance test, whereby the trace metal composition of the disputed fossil elements is determined and compared to an autochthonous sample population from the host unit. If the disputed remains are statistically separate from the autochthonous unit, then a strong argument can be constructed in favor of reworking – indeed further sampling may even locate likely source areas for the allochthonous bones (e.g., Plummer et al. 1994, Trueman and Benton 1997, Staron et al. 2001).

SUMMARY

The mineral component of bone is composed of very small crystallites of non-stoichiometric carbonate-hydroxyapatite. These crystallites have a very large surface area and consequently a high cation exchange capacity. After death, the organic matrix of bone is lost, and the mineral component either breaks down or recrystallizes. Recrystallization involves addition of new apatite, most likely seeded onto biogenic apatite crystals, and this crystal growth continues until the porosity originally occupied by collagen is filled. Many trace metals are adsorbed onto bone crystallite surfaces post mortem, and are introduced through the bone principally via processes of diffusion and adsorption. The original biogenic trace element composition of bone is therefore rapidly (within 10 years post mortem in sites with high evaporation rates) complicated by uptake of diagenetic trace metals. However, while the trace metal composition of ancient bone may be unsuitable for paleodietary investigation, many metals (such as the REE) are concentrated in bone post mortem, so that the fossil bone reflects an exclusively diagenetic signal, controlled by the environment of burial. As uptake and recrystallization (chemical closure) of bone is relative rapid on geological scales (~1-50 kyr), bone can retain a trace metal signal that reflects the environment of burial. Most recent work on trace metal compositions of fossil bones and teeth has focused on the rare earth elements, and has shown that bone likely preserves a relatively unfractionated record of local pore water compositions. Fossil bones and teeth are a common component in a wide range of sedimentary environments, and consequently fossil bones (and to a lesser extent, teeth) could be used to study transport mechanisms and cycling of trace metals in a wide variety of depositional settings. Initial studies of the REE in fossil bones distributed across single terrestrial environments suggests a far more complex local weathering and transport chemistry for the REE than is traditionally assumed. This is a promising avenue for future work.

REFERENCES

Aitken MJ (1990) Science-based Dating in Archaeology. Longman Archaeology Series, Longman, London

Anderson C, Danylchuk KD (1977) The effect of chronic low level lead intoxication on the Haversian remodeling system in dogs. Lab Invest 37:466-469

Argast S, Farlow JD, Gabet RM, Brinkman DL (1987) Transport-induced abrasion of fossil reptilian teeth: implications for the existence of Tertiary dinosaurs in the Hell Creek Formation, Montana. Geology 15:927-930

Armstrong HA, Pearson DG, Griselin M (2001) Thermal effects on rare earth element and strontium isotope chemistry in single conodont elements. Geochim Cosmochim Acta 65:435-441

Arnay-De-La-Rosa M, Gonzalez-Reimers E, Velasco-Vazquez J, Barros-Lopez N, Galindo-Martin L (1998) Bone trace-element pattern in an 18th century population of Tenerife (Canary Islands)—Comparison with a prehistoric one. Biol Trace Elem Res 65:45-51

Arrhenius G, Bramlette MN, Piciotto E (1957) Localization of radioactive and stable heavy nuclides in ocean sediments. Nature 180:85-86

Baraybar JP (1999) Diet and death in a fog oasis site in central coastal Peru: A trace element study of Tomb 1 Malanche 22. J Arch Sci 26:471-482.

Barker MJ, Clarke JB, Martill DM (1997) Mesozoic reptile bones as diagenetic windows. Bull Soc Geol Fr 168:535-545

Bass JD (1995) Mineral Physics and Crystallography: A Handbook of Physical Constants. American Geophysical Union, Washington, DC

Behrensmeyer AK, Hook RW (1992) Paleoenvironmental contexts and taphonomic modes. *In* Terrestrial Ecosystems Through Time: Evolutionary Paleoecology of Terrestrial Plants and Animals. Behrensmeyer AK, Damuth JD, DiMichele WA, Potts R, Sues H-D, Wing SL (eds) University of Chicago Press, Chicago, p 15-136

Bernat M (1975) Les isotopes de l'uranium et du thorium et les terres rares dans l'environment marin. Cahiers OSTROM Series Geologie 7:65-83

Bertram CJ, Elderfield H, Aldridge RJ, Conway Morris S (1992) $^{87}Sr/^{86}Sr$, $^{143}Nd/^{144}Nd$ and REEs in Silurian phosphatic fossils. Earth Planet Sci Lett 113:239-249

Bischoff JL, Rosenbauer RJ, Moench AF, Ku TL (1995) U-series age equations for uranium assimilation by fossil bones. Radiochim Acta 69:127-135

Blundy JB, and Wood BJ (1994) Prediction of crystal-melt partition coefficients from elastic moduli. Nature 372:452-454

Bocherens H, Tresset A, Wiedemann F, Giligny F, Lafage F, Lanchon Y, Mariotti A (1997) Diagenetic evolution of mammal bone in two French Neolithic sites. Bull Soc Geol France 168:555-564

Brice JC (1975) Some thermodynamic aspects of the growth of strained crystals. J Crystal Growth 28: 249-253

Brito JA, McNeill FE, Chettle DR, Webber CE, Vaillancourt C (2000) Study of the relationships between bone lead levels and its variation with time and the cumulative bone lead index, in repeated bone lead survey. J Environ Monit 2:1-6

Brown AB (1973) Bone strontium content as a dietary indicator in human skeletal populations. PhD dissertation, University of Michigan, Ann Arbor

Brudevold, F, Söremark R (1967) Chemistry of the mineral phase of enamel *In* Structural and Chemical Organization of Teeth. Miles EW (ed.) Academic Press, New York

Bryant FJ, Henderson EH, Lee I, Lloyd GD, Webb MSW (1960) Radioactive and natural strontium in human bone: A.E.R.E Results for U.K. *In* Assay of Strontium-90 in Human Bone in the United Kingdom. Medical Research Council Monitoring Report, Ser 1. London, p 2-12

Budd P, Montgomery J, Barreiro B, Thomas RG (2000) Differential diagenesis of strontium in archaeological human dental tissues. Appl Geochem 15:687-694

Buikstra JE, Frankenberg S, Lambert JB, Xue L. (1990) Multiple elements: multiple expectations. *In* The Chemistry of Prehistoric Human Bone. Price TD (ed) Cambridge University Press, Cambridge, UK, p 155- 211

Burton JH, Wright LE (1995) Nonlinearity in the relationship between Sr/Ca and diet: paleodietary implications. Am J Phys Anthropol 96:273-282

Chen T–M, Yuan S (1988) Uranium series dating of bones from Chinese palaeolithic sites. Archaeometry 30:59-76

Chen X-B, Wright, JV, Conca JL, Peurrung LM (1997a) Effects of pH on Heavy Metal Sorption on Mineral Apatite. Environ Sci Technol 31:624-631

Chen X-B, Wright, JV, Conca JL, Peurrung LM (1997b) Evaluation of Heavy Metal Remediation Using Mineral Apatite. Water, Air Soil Pollut 98:57-78

Collins MJ, Riley MS, Child AM, Turner-Walker G (1995) A basic mathematical simulation of the chemical degredation of ancient collagen. J Archaeol Sci 22:175-183

Dahl SG, Allain R, Marie PJ, Mauras Y, Boiviin G, Ammann P, Tsouderos Y, Delmas PD, Christiansen C (2001) Incorporation and distribution of strontium in bone. Bone 28:446-453

D'Haese PC, Couttenye M, Lamberts LV, Elsiviers MM, Goodman WG, Schrooten I, Cabrera WE, DeBroe ME (1999) Aluminum, Iron, Lead, Cadmium, Copper, Zinc, Chromium, Magnesium, Strontium and Calcium content in bone of end-stage renal failure patients. Clinical Chem 45:1548-1556

DeBarr HJW, Bacon MP, Brewer PG, Bruland KW (1985) Rare earth elements in the Pacific and Atlantic Oceans. Geochim Cosmochim Acta 49:1943-1959

Denys C, William CT, Dauphin Y, Andrews P, Fernandez-Jalvo Y (1996) Diagenetical changes in Pleistocene small mammal bones from Olduvai Bed-1. Palaeogeogr Palaeoclim Palaeoecol 126: 121-134

Downing KF, Park L. (1998) Geochemistry and early diagenesis of mammal-bearing concretions from the Sucker Creek Formation (Miocene) of southeastern Oregon. Palaios 13:14-27

Drasch GA (1982) Lead burden in prehistorical, historical and modern human bones. Sci Total Environ 24:199-231

Drasch GA, Bohm J, Baur C (1987) Lead in human bones. Investigations on an occupationally non-exposed population in southern Bavaria (F.R.G). I. Adults. Sci Total Environ 64:303-315

Dupré B, Viers J, Dandurand J-L, Polve M, Bénézeth P, Vervier P, Braun J-J (1999) Major and trace elements associated with colloids in organic-rich river waters: Ultrafiltration of natural and spiked solutions. Chem Geol 160:63-80

Elderfield H, Greaves MJ (1982) The rare earth elements in seawater. Nature 296:214-219

Elderfield H, Pagett R (1986) Rare earth elements in ichthyoliths: Variations with redox conditions and depositional environment. Sci Total Environ 49:175-197

Elderfield H, Sholkovitz ER (1987) Rare earth elements in the pore waters of reducing near-shore sediments. Earth Planet Sci Lett 82:280-288

Elias RW, HiraoY, Patterson CC (1982) The circumvention of the natural biopurification of calcium along nutrient pathways by atmospheric inputs of industrial lead. Geochim Cosmochim Acta 46:2561-2580

Elliott TA, Grime GW (1993). Examining the diagenetic alteration of human bone material from a range of archaeological burial sites using nuclear microscopy. Nuclear Inst Phys Res B77:537-547

Elliott TA, Forey PL, Williams CT, Werdelin L (1998) Application of the solubility profiling technique to recent and fossil fish teeth. Bull Soc Geol France 169:443-451

Elorza J, Astibia H, Murelaga X, Pereda-Suberbiola X (1999) Francolite as a diagenetic mineral in dinosaur and other Upper Cretaceous reptile bones (Laño, Iberian Peninsula): microstructural, petrological and geochemical features. Cretaceous Res 20:169-187

Fabig A, Herrmann B (2002) Trace elements in buried human bones: intra-population variability of Sr/Ca and Ba/Ca ratios – diet or diagenesis? Naturwissenschaften 89:115-119

Farrell AP, Hodaly AH, Wang, S (2000) Metal analysis of scales taken from Arctic Grayling. Arch Environ Contam Toxicol 39:515-522

Fiorillo AR (1988) Taphonomy of Hazard Homestead Quarry (Ongalla Group), Hitchcock County, Nebraska. Contrib Geol Univ Wyoming 26:57-97

Fowler SW, Buat-Menard P, Yokoyama Y, Ballestra S, Holm E, Van Nguyen H (1987) Rapid removal of Chernobyl fallout from Mediterranean surface waters by biological activity. Nature 335:622-625

Fujimaki H (1986) Partition coefficients of Hf, Zr, and REE between zircon, apatite, and liquid. Contrib Mineral Petrol 94:42-45

German C R, Elderfield H (1990) Application of the Ce anomaly as a redox indicator: The ground rules. Paleoceanography 5:823-833

Gillanders BM (2001) Trace metals in four structures of fish and their use for estimates of stock structure. Fish Bull 99:410-419

Girard C, Albarède F. (1996) Trace elements in conodont phosphates from the Frasnian/Famennian boundary. Palaeogeogr Palaeoclim Palaeoecol 126:195-209

Glimcher MJ (1984) Recent studies of the mineral phase in bone and its possible linkage to the organic matrix by protein-bound phosphate bands. Phil Trans Roy Soc London B304:509-518

Goldberg ED Koide M Schmitt RA Smith J (1963) Rare earth distributions in the marine environment. J Geophys Res 68:4204-4217

Gonzalez-Reimers E, Rogriguez-Moreno F, Marinez-Riera A, Mas-Pascual A, Delgado-Ureta E, Galindo-Martin L, Arney-de la Rosa M, Samtolaria-Fernanadez, F (1999) Relative and combined effects of ethanol and protein deficiency on strontium and barium bone content and fecal urinary excretion. Biol Trace Element Res 68:41-49

Grandjean P, Cappetta H, Michard A, Albarède F (1987) The assessment of REE patterns and $^{143}Nd/^{144}Nd$ ratios in fish remains. Earth Planet Sci Lett 84:181-196

Grandjean P, Albarède F (1989) Ion probe measurement of rare earth elements in biogenic phosphates. Geochim Cosmochim Acta 53:3179-3183

Grandjean-Lécuyer P, Feist R, Albarède F (1993) Rare earth elements in old biogenic apatites. Geochim Cosmochim Acta 57:2507-2514

Grupe G, Price TD, Schroter P, Sollner F (1999) Mobility of Bell Beaker people revealed by Sr isotope ratios of tooth and bone: a study of southern Bavarian skeletal materials. Appl Geochem 12:517-525

Hassan AA, Termine JD, Vance-Haynes C Jr (1977) Mineralogical studies on bone apatite and their implications for radiocarbon dating. Radiocarbon 19:364-374

Hedges REM, Millard AR, Pike AWG (1995) Measurements and relationships of diagenetic alteration of bone from three archaeological sites. J Archaeol Sci 22:201-209

Henderson P, Marlow CA, Molleson TI, Williams, CT (1983) Patterns of chemical change during bone fossilization. Nature 306:358-360

Hinners TA, Hughes R, Outridge PM, Davis W, Simon K, Woolard DR (1998) Interlaboratory comparison of mass spectrometric methods for lead isotopes and trace elements in NIST SRM 1400 bone ash. J Analyt Atomic Spectros 13, 963-970

Hodges RM (1950) Strontium content of human bones. J Biol Chem 185:519

Hoyle J, Elderfield H, Gledhill A, Greaves M (1984) The behaviour of the rare earth elements during mixing of river and sea waters. Geochim Cosmochim Acta 48:143-149

Hubert JF, Panish PT, Chure DJ, Prostak KS (1996) Chemistry, microstructure, petrology, and diagenetic model of Jurassic dinosaur bones, Dinosaur National Monument, Utah. J Sed Res 66:531-547

Hughes JM, Cameron M, Mariano A N (1991) Rare-earth-element ordering and structural variations in natural rare-earth-bearing apatites. Am Mineral 76:1165-1173

Janssens K, Vincze L, Vekemans B, Williams CT, Radtke M, Haller M, Knöchel A (1999) The non-destructive determination of REE in fossilized bone using synchrotron radiation induced K-line X-ray microfluorescence analysis. Fres J Analytical Chem 363:413-420

Jaworowski Z, Barbalat F, Blain, C, Peyre E (1985) Heavy metals in human and animal bones from ancient and contemporary France. Sci Total Environ 43:103-126

Karkanas P, Bar-Josef O, Goldberg P, Weiner S (2000) Diagenesis in prehistoric caves: the use of minerals that form *in situ* to assess the completeness of the archaeological record. J Archaeol Sci 27:915-929

Keto LS Jacobsen SB (1987) Nd and Sr isotopic variations of Early Paleozoic oceans. Earth Planet Sci Lett 84:27-41

Kemp RA, Trueman CN (2002) Rare earth elements in Solnhofen biogenic apatite: Geochemical clues to the palaeoenvironment Sed Geol (in press)

Koeppenkastrop D, DeCarlo EH (1992) Sorption of rare-earth elements from seawater onto synthetic mineral particles - an experimental approach. Chem Geol 95:251-263

Koch PL, Halliday AN, Walter LM, Stearly RF, Huston TJ, Smith GR (1992) Sr isotopic composition of hydroxyapatite from recent and fossil salmon: the record of lifetime migration and diagenesis. Earth Planet Sci Lett 108:277-287

Kohn MJ, Schoeninger MJ, Barker WW (1999) Altered states: Effects of diagenesis on fossil tooth chemistry. Geochim Cosmochim Acta 63:2737-2747

Kolodny Y, Luz B, Sander M, Clemens WA (1996) Dinosaur bones: fossils or pseudomorphs? The pitfalls of physiology reconstruction from apatitic fossils. Palaeogeogr Palaeoclim Palaeoecol 126:161-171

Laenen B, Hertogen J, Vandenberghe N (1997) The variation of the trace-element content of fossil biogenic apatite through eustatic sea-level cycles. Palaeogeogr Palaeoclim Palaeoecol 32:325-342

LeGeros RZ (1965) Effect of carbonate in the lattice parameters of apatite. Nature 206:403-404

Lengemann FW (1960) Studies on the discrimination against strontium by bone grown *in vitro*. J Biol Chem 235:1859-1862

Locock M, Currie CK, Gray S (1992) Chemical changes in buried animal bone: Data from a post-medieval assemblage. Intl J Osteoarchaeol 2:297-304

Lyman RL (1994) Vertebrate Taphonomy. Cambridge Manuals in Archaeology: Cambridge University Press, Cambridge, p 523

McConnell D (1952) The Crystal chemistry of carbonate apatites and their relationship to the composition of calcified tissues. J Dental Res 31:53-63

Manea-Kritchen M, Patterson C, Miller G, Settle D, Erel Y (1991) Comparative increases of lead and barium with age in human tooth enamel, rib and ulna. Sci Total Environ 107:179-203

Markich SJ, Jeffree RA, Harch BD (2001) Catchment-specific element signatures in estuarine crocodiles (*Crocodylus porosus*) from the Alligator Rivers Region, northern Australia. Sci Total Environ 287: 83-95

Marie PJ, Ammann P, Boivin G, Rey C. (2001) Mechanisms of action and therapeutic potential of strontium in bone. Calcif Tiss Intl 69:121-129

Martin EE, Haley BA (2000) Fossil fish teeth as proxies for seawater Sr and Nd isotopes. Geochim Cosmochim Acta 64:835-847

Middleton J (1844) On the flourine in bones, its source and its application to the determination of the geological age of fossil bones. Proc Geol Soc London 4:148-157

Millard AR (1993) Diagenesis of archaeological bone: The case of uranium uptake. PhD dissertation, Univ Oxford, Oxford, UK

Millard AR, Hedges REM (1999) A diffusion-adsorption model of uranium uptake by archaeological bone. Geochim Cosmochim Acta 60:2139-2152

Moffett JW (1990) Microbially mediated cerium oxidation in seawater. Nature 345:421-423

Molleson TI, Williams CT, Cressey G, Din VK (1998) Radiographically opaque bones from lead-lined coffins at Christ Church, Spitalfields, London—An extreme example of diagenesis. Bull Soc Geol France 169:425-432

Murray RW, Tenbrink MRB, Gerlach DC Russ GP Jones DL (1992) Interoceanic variation in the rare-earth, major, and trace-element depositional chemistry of chert—perspectives gained from the DSDP and ODP record. Geochim Cosmochim Acta 56:1897-1913

Nagasawa H (1970) Rare earth concentrations in zircons and apatites and their host dacites. Earth Planet Sci Lett 9:359-364

Nelson BK, DeNiro, MJ, Schoeninger, M, DePaolo, DJ, Hare, PE (1986) Effects of diagenesis on strontium, carbon, nitrogen and oxygen concentration and isotopic composition of bone. Geochim Cosmochim Acta 50:1941-1949

Nielsen-Marsh CN, Hedges, REM (2000) Patterns of diagenesis in bone I: The effects of site environments. J Archaeol Sci 27:1139-1150

Nriagu J O (1983) Rapid decomposition of fish bones in Lake Erie sediments. Hydrobiologica 106:217-222

Parker RB, Toots H. (1980) Trace elements in bones as paleobiological indicators. *In* Fossils in the Making. Behrensmeyer AK, Hill AP (eds) University of Chicago Press, Chicago, p 197-207

Pate FD, Hutton JT, Norrish K (1989) Ionic exchange between soil solution and bone: Toward a predictive model. Appl Geochem 4:303-316

Pate FD, Hutton JT (1988) The use of soil chemistry data to address post-mortem diagenesis in bone mineral. J Archaeol Sci 15:729-739

Patterson CC, Shirahata H, Ericson JE (1987) Lead in ancient human bones and its relevance to historical developments of social problems with lead. Sci Total Environ 61:167-2000

Person A, Bocherens H, Mariotti A, Reynard M (1996) Diagenetic evolution and experimental heating of bone phosphate. Palaeogeogr Palaeoclim Palaeoecol 126:135-150

Piepgras DJ, Jacobsen SB (1992) The behavior of rare-earth elements in seawater—precise determination of variations in the North Pacific water column. Geochim Cosmochim Acta 56:1851-1862

Pfretszscnher H-U (2000) Microcracks and fossilization of haversian bone. N Jb Geol Palaeontol Abh 216:413-432

Plummer TW, Kinuyua AM, Potts R (1994) Provenancing of hominid and mammalian fossils from Kanjera, Kenya, using EDXRF. J Archaeol Sci 21:553-563

Posner AS, Harper RA, Muller SA (1965) Age changes in the crystal chemistry of bone apatite. Annals New York Acad Sci 131:737-742

Price TD (1989) Multi-element studies of diagenesis in prehistoric bone. *In* The chemistry of prehistoric human bone. Price TD (ed) Cambridge University Press, Cambridge, UK, p 126-154

Price TD, Blitz J, Burton J, Ezzo JA. (1992) Diagenesis in prehistoric bone: Problems and solutions. J Archaeol Sci 19:513-529

Radosevich SC (1993) The six deadly sins of trace element analysis: a case of wishful thinking in science. *In* Investigations of ancient human tissue: chemical analyses in anthropology. (Food and nutrition in history and anthropology, vol 10) Sanford MK (ed) Gordon and Breach, Langhorne, UK, p 269-332

Reynard B, Lécuyer C, Grandjean P (1999) Crystal-chemical controls on rare-earth element concentrations in fossil biogenic apatites and implications for paleoenvironmental reconstructions. Chem Geol 155:233-241

Robinson RA, Cameron DA (1964) Bone. *In* Electron Microscope Anatomy. Kurtz SM (ed) Academic Press, New York

Rogers RR, Arcucci AB, Abdafa F, Sereno PC, Forster CA, May CL (2001) Paleoenvironments and taphonomy of the Chañares tetropod assemblage (Middle Triassic) northwestern Argentina: Spectacular preservation in volcanogenic concretions. Palaios 16:461-481

Sakai H, Saeki K, Ichihashi H, Suganuma H, Tanabe S, Tatsukawa R (2000) Species-specific distribution of heavy metals in tissues and organs of loggerhead turtle (*Caretta caretta*) and green turtle (*Chelonia mydas*) from Japanese coastal waters. Marine Pollution Bull 40:701-709

Safont S, Malgosa A, Subira ME, Gilbert J (1998) Can trace elements in fossils provide information about palaeodiet? Intl J Osteoarchaeol 8:23-37

Samoilov VS, Benjamini C (1996) Geochemical features of dinosaur remains from the Gobi Desert, South Mongolia. Palaios 11:519-531

Samoilov VS, Benjamini C, Smirnova EV (2001) Early diagenetic stabilization of trace elements in reptile bone remains as an indicator of Maastrichtian-Late Paleocene climatic changes: evidence from the Naran Bulak locality, the Gobi Desert (South Mongolia). Sed Geol 143:15-39

Sandford MK, Weaver DS (2000) Trace element research in anthropology: new perspectives and challenges. *In* Biological Anthropology of the Human Skeleton. Katzenberg MA, Sanders SR (eds). Wiley-Liss, New York, p 329-350

Schutkowski H, Herrmann B, Wiedemann F, Bocherens H, Grupe G (1999) Diet, status and decomposition at Weingarten: Trace element and isotope analyses on early medieval skeletal material. J Archaeol Sci 26:675-685

Schweitzer MH, Johnson C, Zocco TG, Horner JR, Starkey JR (1997) Preservation of biomolecules in cancellous bone of *Tyrannosaurus rex*. J Vert Paleo 17:349-359

Shaw HF, Wasserburg GJ (1985) Sm-Nd in marine carbonates and phosphates: Implications for Nd isotopes in seawater and crustal ages. Geochim Cosmochim Acta 49:503-518

Shemesh A (1990) Crystallinity and diagenesis of sedimentary apatites. Geochim Cosmochim Acta 54:2433-2438

Shinomiya T, Shinomiya K, Orimoto C, Minami T, Tohno Y, Yamada M (1998) In- and out-flows of elements in bones embedded in reference soils. Forensic Sci Intl 98:109-118

Sholkovitz ER, Landing WM, Lewis BL (1994) Ocean particle chemistry—the fractionation of rare-earth elements between suspended particles and seawater. Geochim Cosmochim Acta 58:1567-1579

Sillen A (1986) Biogenic and diagenetic Sr/Ca in Plio-Pleistocene fossils of the Omo Shungura Formation. Paleobiol 12:311-323

Sillen A, Le Geros R (1991) Solubility profiles of synthetic apatites and of modern and fossil bones. J Archaeol Sci 18:385-397

Sillen A, Sealey JC (1995) Diagenesis of Sr in bone: a reconsideration of Nelson et al. (1986). J Archaeol Sci 22:313-320

Slansky M. (1986) Geology of Sedimentary Phosphates. Kogan Page Ltd., London

Smyth JR, Bish DL(1988) Crystal Structures and Cation Sites of the Rock-Forming Minerals. Allen and Unwin, Boston

Sowden EM, Stitch SR (1956) Trace elements in human tissue. Biochem J 67:104-109

Staron R, Granstaff B, Gallagher W, Grandstaff DE (2001) REE signals in vertebrate fossils from Sewel, NJ: Implications for location of the K-T boundary. Palaios 16:255-265

Staudigel H, Doyle P, Zindler A (1985) Sr and Nd isotope systematics in fish teeth. Earth Planet Sci Lett 76:45-56

Steadman LT, Brudevold F, Smith FA (1958) Distribution of strontium in teeth from different geographic areas. J Am Dental Assoc 57:340-344

Szabo BJ, Rosholt JN (1969) Uranium series dating of Pleistocene molluscan shells from southern California: an open system model. J Geophys Res 74:3253-3260

Tannenbaum PJ, Termine JD (1965) Statistical analysis of the effect of fluorine on bone apatite. Annals New York Acad Sci 154:1660-1661

Thorrold SR, Latkoczy C, Swart PK, Jones CM (2001) Natal homing in a marine fish population. Science 291:297-299

Toots H, Voorhies MR (1965) Strontium in fossil bones and the reconstruction of food chains. Science 149:854-855

Trueman CN (1999) Rare earth element geochemistry and taphonomy of terrestrial vertebrate assemblages Palaios 14:555-568

Trueman CN, Benton MJ (1997) A geochemical method to trace the taphonomic history of reworked bones in sedimentary settings. Geology 25:263-266

Trueman, CN, Martill DM (2002) The long term preservation of bone: The role of bioerosion. Archaeometry 44:371-382

Trueman CN, Behrensmeyer, AK, Potts R, Tuross N (2001) Soil heterogeneity in a Pleistocene wetland. J Vert Paleo (supplement SVP abstracts)

Turekian KK, Kulp JL (1956) Strontium content in human bones. Science 124:405-406

Tuross N, Behrensmeyer AK, Eanes ED (1989) Strontium increases and crystallinity changes in taphonomic and archaeological bone. J Archaeol Sci 16:661-672

Van der Merwe NJ, Lee-Thorp JA, Thackeray JF, Hall-Martin A, Kruger FJ, Coetzee H, Bell RHV, Lindeque M (1990) Source-area determination of elephant ivory by isotopic analysis. Nature 346: 744-746

Valsami-Jones E, Ragnarsdottir KV, Crewe-Read NO, Mann T, Kemp AJ, Allen GC (1996) An experimental investigation of the potential of apatite as radioactive and industrial waste scavenger. *In* Fourth Intl Symp Geochemistry of the Earth's Surface. Yorkshire, UK. Bottrells SH (ed) University of Leeds, Leeds, UK, p 686-689

Vogel JC, Eglington B, Auret JM (1990) Isotope fingerprints in elephant bone and ivory. Nature 346: 747-749

Waldron HA (1982) Lead in bones: A cautionary tale. Ecol Dis 1:191-196

Watson EB, Green TH (1981) Apatite/ liquid partition coefficients for the rare earth elements and strontium. Earth Planet Sci Lett 56:405-421

Weiner S, Goldberg P, Bar-Josef O. (1993) Bone preservation in Kebara Cave, Israel, using on-site Fourier Transform Infrared spectrometry. J Archaeol Sci 20:613-627

Weiner S, Price PA (1986) Disaggregation of bone into crystals. Calcif Tiss Intl 39:365-375

Weiner S, Bar-Josef O (1990) States of preservation of bones from prehistoric sites in the Near East: A survey. J Archaeol Sci 17:187-196

Wells BK, Thorrold SR, Jones CM (2000) Geographic variation in trace element composition of juvenile Weakfish scales. Trans Am Fish Soc 129:889-900

Williams CT (1988) Alteration of chemical composition of fossil bones by soil processes and groundwater. *In* Trace Elements in Environmental History. Grupe G, Herrmann B (eds) Springer-Verlag, p 27-40

Williams CT, Marlow C A (1987) Uranium and Thorium distributions in fossil bones from Olduvai Gorge, Tanzania and Kanam, Kenya. J Archaeol Sci 14:297-309

Williams CT, Henderson P, Marlow CA, Molleson TI (1997) The environment of deposition indicated by the distribution of rare earth elements in fossil bones from Olduvai Gorge, Tanzania. Appl Geochem 12:537-547

Williams CT, Potts PJ (1988) Element distribution maps in fossil bones. Archaeometry 30:237-247

Wittmers LE, Aufderheide AC, Wallgren J, Rapp G, Alich A (1988) Lead in Bone IV Distribution of lead in the human skeleton. Arch Environ Health 43:381-391

Wood BJ, Blundy JD (1997) A predictive model for rare earth element partitioning between clinopyroxene and anhydrous silicate melt. Contrib Mineral Petrol 129:166-181

Wright J, Seymour R S, Shaw HF (1984) REE and Nd isotopes in conodont apatite: variations with geological age and depositional environment. *In* Conodont Biofacies and Provincialism. Clark DL (ed) Geol Soc Am Spec Paper, p 325-340

Wright J, Schrader H, Holser W T (1987) Paleoredox variations in ancient oceans recorded by rare earth elements in fossil apatite. Geochim Cosmochim Acta 51:631-644

Yoshinaga J, Suzuki T, Morita M, Hayakawa M (1995) Trace elements in ribs of elderly people and elemental variation in the presence of chronic diseases. Sci Total Environ 162:239-252

Young RA, Brown WE (1982) Structures of biological minerals. *In* Biological Mineralisation and Deminerlisation. Nancollas GH (ed) Springer-Verlag, Berlin

Zocco TG, Schwartz HL (1994) Microstructural analysis of bone of the sauropod dinosaur *Seismosaurus* by transmission electron microscopy. Palaeontology 37:493-503

14 U-Th-Pb Dating of Phosphate Minerals

T. Mark Harrison

Research School of Earth Sciences
The Australian National University
Canberra, A.C.T. 0200 Australia

Elizabeth J. Catlos

School of Geology
Oklahoma State University
Stillwater, Oklahoma 74078

Jean-Marc Montel

LMTG-Minéralogie UMR
CNRS 5563, 39 allées J. Guesde
31000 Toulouse, France

BACKGROUND

Introduction

The dominant occurrence of phosphate minerals in crystalline rocks is as accessory phases, most notably apatite, monazite, and xenotime. Because these minerals tend, to varying degrees, to partition U and Th into their structures they can often contain the majority of those elements in a rock. These three phases, again to varying degrees, tend not to incorporate significant amounts of Pb during crystallization and thus were early candidates for utilization as U-Th-Pb geochronometers.

The ideal U-Th-Pb geochronometer would be a phase that is stable over all possible environmental conditions and is quantitatively retentive of parent and daughter isotopes. In fact, the silicate zircon comes reasonably close to meeting these criteria. Zircon has a broad stability field, is refractory under a wide variety of geologic environments (e.g., weathering, sedimentary transport, anatexis, and metamorphism), and can be highly retentive of daughter products in the U-Th-Pb decay system. However, the limit of zircon as an ideal chronometer lies only in its limited resistance to auto-irradiation damage that can render it metamict. The phosphate minerals apatite, monazite and xenotime have a more restricted range of stability and, to varying degrees, are incompletely retentive of Pb under crustal conditions, but they are resilient to radiation damage. Thus interpretation of results from these geochronometers requires a greater understanding of their structure, stability, and kinetic properties than does zircon. This chapter is aimed at providing the reader with an introduction to those characteristics that will facilitate interpretation of the occasionally problematic nature of phosphate U-Th-Pb dating results.

The U-Th-Pb dating system

U and Th decay. Nuclei approaching one hundred protons are unstable because the strong nuclear force, which acts to hold neutrons and protons together, is about 100 times greater than electromagnetic repulsion. Both uranium (92 protons) and thorium (90 protons) decay by emission of a ^{4}He nucleus (or α particle), which lowers the coulomb energy but changes the nuclear binding energy little. In certain cases, the intermediate daughter product is unstable with respect to β^- decay and transmutes to an isobar closer to the valley of beta stability.

Both long-lived isotopes of uranium, ^{238}U and ^{235}U, decay to isotopes of Pb with

1529-6466/00/0048-0014$05.00

half-lives of 4.4 Ga and 700 Ma (see Table 1), respectively, through a complex chain of intermediate daughter isotopes (Faure 1986). The overall decay relationships are:

$$^{238}U \rightarrow {}^{206}Pb + 8\,\alpha + 6\,\beta^- + 47.4\ \text{MeV}$$

$$^{235}U \rightarrow {}^{207}Pb + 7\,\alpha + 4\,\beta^- + 45.2\ \text{MeV}$$

Table 1. Solutions to U-Th-Pb age equations.

System	*Decay constant (Ma^{-1})*	*Solution*
$^{238}U \rightarrow {}^{206}Pb$	1.55125×10^{-4}	$^{206}Pb = {}^{238}U \cdot (e^{\lambda_{238}t} - 1)$
$^{235}U \rightarrow {}^{207}Pb$	9.8585×10^{-4}	$^{207}Pb = {}^{235}U \cdot (e^{\lambda_{235}t} - 1)$
$(^{207}Pb/^{206}Pb)$*		$\left(\frac{^{207}Pb}{^{206}Pb}\right)^* = \frac{1}{137.88}\left(\frac{e^{\lambda_{235}t} - 1}{e^{\lambda_{238}t} - 1}\right)$
$^{232}Th \rightarrow {}^{208}Pb$	4.9475×10^{-5}	$^{208}Pb = {}^{232}Th \cdot (e^{\lambda_{232}t} - 1)$

In order that the U-Pb date corresponds to the age of mineral formation, the series of intermediate daughter activities must be in secular equilibrium, which occurs after five half-lives of the intermediate isotope have elapsed. Most of the intermediate daughter isotopes are short-lived (i.e., half-lives of <10,000 years), but ^{234}U (in the ^{235}U decay chain) and ^{230}Th (in the ^{238}U decay chain) have half-lives of 247,000 and 76,000 years, respectively, thus requiring up to ~1 Myr for attainment of secular equilibrium. Therefore, secular equilibrium cannot be considered instantaneous over geologic time, and element partitioning during growth leads to preferential exclusion or uptake of intermediate daughters resulting in, respectively, daughter product deficiencies or excesses.

The single long-lived isotope of thorium, ^{232}Th, decays to ^{208}Pb (Table 1) with a half-life of 14 Ga through a series of intermediate daughters via the relationship:

$$^{232}Th \rightarrow {}^{208}Pb + 6\,\alpha + 4\,\beta^- + 39.8\ \text{MeV}$$

Unlike uranium, intermediate daughter isotopes from ^{232}Th decay all have short half-lives such that secular equilibrium is achieved across the chain within ~30 years.

The concordia plot. Because uranium isotopes are unlikely to be fractionated in nature, the paired decay of ^{238}U and ^{235}U permits the two decays systems to be plotted together (Fig. 1) yielding the so-called concordia plot (Wetherill 1956). The potential advantage of such a plot is that a single stage of open system behavior can in principle be recognized and corrected. For ancient samples, this approach appears to work well. However, more youthful samples are problematic and phosphates in particular can be pathological in this regard. Because of this, we focus particularly on the behavior of young (i.e., <100 Ma) phosphates.

Pb loss (or U gain) has the effect of moving data off concordia along a line whose lower intercept reflects the age of the disturbance (Fig. 1). For recent Pb loss (or U gain), that intercept is the origin. A sample that has experienced minor Pb loss will plot close to its origin on the concordia curve, while one that has lost more Pb will move down a discordia line towards the lower intercept. U-Pb ages of young samples can be difficult to interpret because the trajectory of Pb loss may not be resolvable from concordia. In the case of multiple Pb loss events, the data will fan out from the upper intercept on the

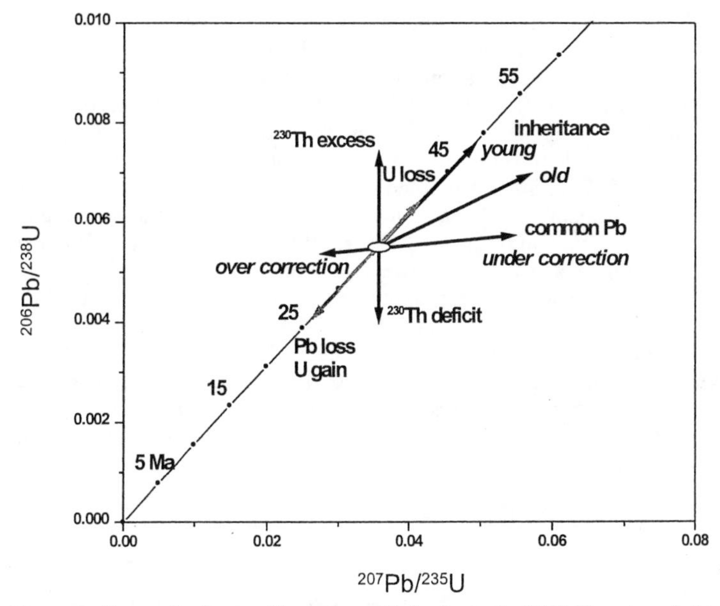

Figure 1. Concordia diagram illustrating misbehavior in the U-Th-Pb system that is relevant to U-Pb dating. Arrows show how concordant data can be affected by inheritance, Pb loss, disequilibrium, and common Pb (Pb°) corrections.

concordia curve. Continuous diffusive Pb loss results in a broadly linear array which resembles that of episodic Pb loss, cautioning against ascribing significance to a lower intercept unless independent evidence exists for a geologic event at that time (Faure 1986). Analysis of minerals that experienced and recorded complicated metamorphic histories can result in discordia lines with lower intercepts that have no geological meaning due to mixed ages.

Inheritance presents another complication that arises when protolith is incompletely consumed during a rock forming process and the restitic phase retains daughter Pb. Such occurrences are likely in crustal melts as solubilities of phosphate minerals such as apatite and monazite are generally low (Harrison and Watson 1984, Rapp and Watson 1986). Because the solidus temperature for many crustal magmas is higher than early estimates of closure temperature for Pb in monazite, inheritance of monazite in igneous rocks was once believed to be quite rare. However, numerous studies have documented the survival of restitic monazite in granitoids, dispelling this notion (Copeland et al. 1988, Harrison et al. 1995, Edwards and Harrison 1997). The preservation of inheritance in monazite saturated crustal magmas appears to be due to short timescales of crustal melting and the high degree of retentivity of Pb in monazite (Copeland et al. 1988, Montel 1993, Cherniak et al. 2000).

Incorporation of a restitic mineral containing radiogenic Pb (Pb*) during mineral growth in a melt will result in anomalously old ages that plot to the right of the concordia curve (Fig. 1). A more complicated case arises when a mineral has inherited Pb sometime in the past and then lost Pb more recently, potentially resulting in concordant data, which yield meaningless ages (e.g., Deniel et al. 1987). In the case of new mineral growing around an inherited core, bulk analysis methods will yield a range of ages, reflecting mixing between two or more age components. On a concordia diagram this would manifest itself as a mixing line (2 components) or mixing array (multiple components).

As we note later, such a mixture is potentially resolvable using (or combining isotope dilution with) a technique that provides high spatial resolution, such as ion or electron microprobe analysis, or laser ablation-inductively coupled plasma-mass spectrometry. Disequilibrium between different nuclides in the ^{238}U or ^{235}U decay schemes can seriously complicate young U-Pb age data interpretation (Schärer 1984). For example, monazite contains abundant ThO_2 (1 to >30 wt %) (Overstreet 1967, Deer et al. 1992, Van Emden et al. 1997) and has a marked preference for Th compared to U, thus often incorporating a component of ^{230}Th, a relatively short-lived intermediate daughter isotope in the ^{238}U decay chain. The decay of ^{230}Th results in unsupported ^{206}Pb, anomalously high $^{206}Pb/^{238}U$, and thus data plot above the concordia curve (Schärer 1984). On a concordia diagram, excesses in intermediate decay products in the ^{238}U chain will move data up parallel to the $^{206}Pb/^{238}U$ axis (Fig. 1). Schärer (1984) first documented and corrected for this so-called reverse discordance in Himalayan monazite (see Fig. 2), which has subsequently been widely recognized (e.g., Schärer et al. 1986, Parrish and Armstrong 1987). Xenotime also presents a problem due to its lack of incorporation of ^{230}Th and resulting deficiency of ^{206}Pb, whereas monazite can incorporate ^{231}Pa, resulting in unsupported ^{207}Pb (Schärer 1984).

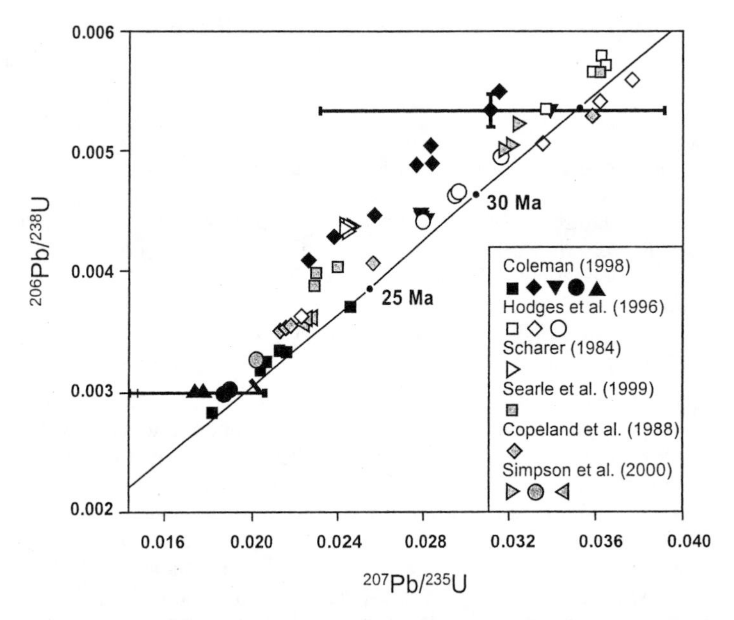

Figure 2. Concordia diagram of Himalayan monazite grains. References listed in the key. Note that a remarkable number these monazite grains show reverse discordance, consistent with incorporation of excess ^{206}Pb.

Complications due to inheritance, Pb loss, and disequilibrium highlighted in the above discussion can be particularly problematic for bulk or chemical measurement methods, but other techniques have their own drawbacks. For example, U-Pb ion microprobe data can be subject to calibration shifts that move a concordant data point along a line with a slope proportional to its $^{207}Pb/^{206}Pb$ age. Data can also be pulled to the left or right of concordia by over or under-correcting for common Pb (Pb^0) (Gilley 2001).

Note that ^{232}Th-^{208}Pb dating of monazite is insensitive to problems arising from disequilibrium of Pb because secular equilibrium among the intermediate daughter isotopes is reached within ~30 years, making it unlikely that monazite could contain unsupported $^{208}Pb^*$ (Harrison et al. 1995). In addition, the high ThO_2 concentrations in monazite result typically in high levels of $^{208}Pb^*$.

Early dating attempts. Because certain phosphate minerals strongly concentrate radioactive elements, early twentieth century geochronologists recognized their value. Monazite was an early candidate for U-Th-Pb dating (Fenner 1928, Nier et al. 1941) whereas the dating of apatite (Aldrich et al. 1955, Tilton et al. 1955) and xenotime (Lyakhovich 1961, Köppel and Grünenfelder 1975) were not attempted until later. Numerous methods were developed to obtain ages of phosphate minerals. Radiographic photos were used to document the distribution of radioactive elements within large monazite crystals and to ascertain the extent of alteration (see Fenner 1932, Marble 1935). Monazite grains so selected were washed with nitric acid to eliminate contamination or alteration products and then reacted with acids to isolate Th, U, and Pb (e.g., Fenner 1928, 1932). Prior to isotopic determinations (e.g., Nier et al. 1941), the fundamental equation for calculating ages was

$$[\log(U + 0.38\ Th + 1.156\ Pb) - \log(U + 0.38\ Th)]/(0.434\ \lambda)$$

where U, Th, and Pb are measured concentrations and λ is the total U decay constant, then assumed to be 1.5×10^{-11}/yr (Fenner 1932, Marble 1935). Constants arise in this equation via assumptions made about the equivalence of Th to U in radioactive decay. Although this age equation incorporated U decay, monazite was considered as a primarily Th-bearing phase and U was used as a proxy for alteration (Fenner 1928, Marble 1935). Uranium in monazite posed a problem for early geochronologists, who speculated that the element could be "brought in by circulating waters from adjacent uranium minerals" (Fenner 1932). Today, monazite is commonly dated using the U-Pb method (e.g., Schärer 1984, Copeland et al. 1988, Parrish and Tirrul 1989, Hodges et al. 1996, Parrish and Hodges 1996, Simpson et al. 2000) and more recently by Th-Pb (e.g., Wang et al. 1994, Harrison et al. 1995, Grove and Harrison 1999, Catlos et al. 2001, 2002a).

Monazite was also analyzed using "the Larsen method," in which the total Pb content was compared to measured alpha activity (Larsen et al. 1949, Gottfried et al. 1959, Sivaramakrishnan and Venkatasubramaniam 1959). Although the Larsen method was viewed as "considerably less precise than any of the isotopic methods currently employed in age work" (Gottfried et al. 1959), a large number of Pb-α age determinations could be made rapidly and inexpensively. Apatite was also analyzed by the Larsen method (Larsen et al. 1949) and later by U-Th-Pb dating (Aldrich et al. 1955, Tilton et al. 1955, Iskanderova and Legiyerskiy 1966, Vinogradov et al. 1966).

The existence of monazite and uraninite in the same geologic setting led to age comparisons and the recognition that uraninite was incompletely retentive of daughter Pb (Fenner 1932, Holmes et al. 1949, Ahrens 1955). The geochronologic significance of xenotime ages were evaluated in relation to monazite and Rb-Sr mica ages from the same or neighboring rocks (e.g., Köppel and Grünenfelder 1975).

When the capability arose to measure Th, U, and Pb isotopes (Nier 1939), it was recognized that many monazites yield different U-Pb and Th-Pb ages. Table 2 is a compilation of monazite age data obtained between 1941-1957. Tilton and Nicolaysen (1957) report "gross discrepancies" between individual isotopic ages that "greatly exceed experimental errors" and speculate that although "Pb loss is a more plausible explanation for low U-Pb and Th-Pb ages than U or Th gain…it is very difficult to give a definite proof for this hypothesis." Discordant monazite ages were attributed to Pb diffusion via extended time at elevated temperatures or prolonged chemical reactions occurring at varying rates and involving different agents (see Shestakov 1969). In natural samples,

Michot and Deutsch (1970) invoked Pb loss via recrystallization and episodic metamorphism to explain age discrepancies.

To ascertain the source of the monazite Th-Pb and U-Pb age difference, analytical techniques were evaluated for isotopic fractionation (e.g., Burger et al. 1967) and diffusion experiments were performed (e.g., Tilton 1960, Shestakov 1969).

Table 2. Mid-twentieth-century monazite age data (Ma)[a]

Sample name (locality)	*^{238}U-^{206}Pb*	*^{235}U-^{207}Pb*	*^{207}Pb-^{206}Pb*	*Th-^{208}Pb*	*% diff.*[b]
Ebonite Claims (S. Rhodesia)	2640	2670	2700	2640	1
Ebonite Claims (S. Rhodesia)	2620	2620	2620	2640	-1
Yadiur (India)	1410	1820	2330	1800	1
Goodhouse (S. Africa)	930	915	880(60)[b]	900	2
Steenkampskraal (S. Africa)	1080	nr[c]	nr	990	9
Soniana (India)	635	697	913	611	14
Mount Isa (Australia)	nr	nr	1160	1000	16
Jack Tin Claims (S. Rhodesia)	2210	2460	2660	1940	27
Houtenbek (S. Africa)	1400	1230	930(50)	940	31
Brown Derby (Colorado)	1590(30)	1420(10)	1170(50)	995(25)	43
Huron Claim (Canada)	3180	2840	2600	1830	55
Irumi Hills (N. Rhodesia)	1990	2330	2640	1380	69
Las Vegas (New Mexico)	1730	1560	1340	770	103
Antsirabe (Madagascar)	1370	1850	2450	610	203

a. Note most results have errors of 3-5% unless otherwise indicated (Neir et al. 1941, Holmes et al. 1949, Holmes, 1954, 1955; Holmes and Cahen 1955, Aldrich et al. 1956, Tilton and Nicolaysen 1957).

b. To quantify the difference between the ^{235}U-^{207}Pb and Th-Pb ages, we used {[(^{235}U-^{207}Pb age)-(Th-^{208}Pb age)]/(Th-^{208}Pb age)}*100. For the Steenkampskraal monazite, ^{238}U-^{206}Pb age and Mount Isa ^{207}Pb-^{206}Pb age were used instead of ^{235}U-^{207}Pb age. Note in many cases Th-Pb ages are younger than the U-Pb ages.

c. "nr" = not reported.

Despite potential problems with method, mineral ages of phosphates were used to address geologic problems. Apatite could be analyzed to understand crystallization history during metasomatism (e.g., Iskanderova and Legiyerskiy 1966). Xenotime and monazite U-Th-Pb ages were applied to problems regarding the timing of Alpine metamorphism (e.g., Köppel and Grünenfelder 1975). Monazite ages from the African continent were used to constrain the origin of life on Earth to ~2.6 Ga (Holmes 1954).

MINERAL STRUCTURE AND RADIATION DAMAGE CONSIDERATIONS

Monazite

Mineral structure. Monazite has an ABO_4 stoichiometry, where A-site is occupied by large cation, such as rare earth elements (REE^{3+}), Ca^{2+}, Th^{4+}, and B-site contains small, tetrahedrally coordinated cations, such as P^{5+}. The structure of monazite is comprised of chains of alternating PO_4 tetrahedra, and nine-fold coordinated REE in the A site (e.g., Ni et al. 1995, and Boatner, this volume). All REE are accepted in the A site, although, as far as pure end-members are concerned, only $LaPO_4$ to $GdPO_4$ have the monazite structure at room temperature.

The incorporation of U and Th is accommodated in the REE-site via two substitutions (Burt 1989, van Emden et al. 1997, Förster 1998a): the huttonitic substitution $REE^{3+}P^{5+}=Th^{4+}Si^{4+}$ and the brabantitic (or cheralitic) substitution $2REE^{3+}=Ca^{2+}Th^{4}$. Corresponding end-members are huttonite ($ThSiO_4$) and brabantite ($Ca_{0.5}Th_{0.5}PO_4$). The same substitution accounts for U but the corresponding end-members have not been found. Experimental studies have shown that the $LaPO_4$-$Ca_{0.5}Th_{0.5}PO_4$ and $LaPO_4$-$Ca_{0.5}U0_{.5}PO_4$ solid solutions are continuous (Podor et al. 1995, Podor and Cuney 1997). Some data suggest also that the $LaPO_4$-$ThSiO_4$ solid solution is continuous (Peiffert and Cuney 1999). Experiments show that the A site can accept various divalent cations instead of Ca including Cd, Sr, Pb, and Ba (Montel et al. 2002).

Because of its ability to accept U and Th, monazite is one of the most radioactive minerals after uraninite, thorianite or thorite. It is the most *common* radioactive mineral (Overstreet 1967), and in many rocks the main host of U and Th. The possibility for monazite to accept Pb in the same site as U and Th is obviously important for geochronology. Pb* produced by U and Th has a place in the structure (Quarton et al. 1984). Therefore there is a not a natural tendency for Pb to be released from monazite, as might be anticipated by considering the structure of zircon. Another consequence of the ability of monazite to incorporate various ions is that, in the three U and Th decay chains, most elements can be incorporated in the A site. If all actinides and a small amount of Ra (because Ba is favourably partitioned) can be accepted in monazite, all elements with half-lives greater than about one year are incorporated in the mineral structure. This suggests that at any moment in the radioactive decay chains of U and Th, there is little tendency for any intermediate decay products to escape from the mineral structure.

Radiation effects. Alpha decay is the primary cause of radiation damage in minerals. For example, ^{232}Th produces an alpha particle and a recoil nucleus of ^{228}Ra. The total energy of this decay is 4 MeV, and both the alpha particle and recoil nucleus are ejected from their initial lattice position in opposite directions (see Ewing 1994). The light alpha particle loses most of its energy by ionization along a 10,000-nm-long path, and its effect on the structure is small. In contrast, the recoil nucleus is big and heavy and displaces many atoms along a ~10 nm path, severely damaging the crystal lattice. Although spontaneous fission of ^{238}U is a rare event as the half-life is 10^7 longer than for alpha decay, it is much more destructive than alpha decay because it creates two heavy ions with high energy (~200 MeV).

Although strongly radioactive (average ThO_2: 8.79 wt %, average UO_2: 0.08 wt %, Van Emden et al. 1997), for poorly understood reasons, monazite does not become metamict (see transmission electron microscopy images in Black et al. 1984 and Seydoux-Guillaume 2002).

The X-ray patterns shown in Figure 3 was obtained on a 500 Ma monazite containing 13% Th and 0.25% U. It can be calculated that this monazite experienced about 7 x 10^{19} α/g. Meldrum et al. (1998) estimated that each α-decay in monazite produces 860 displacements. So this monazite should have suffered 3.7 displacements per atom, or one order of magnitude greater that the minimum dose necessary to make a mineral amorphous (Meldrum et al. 1997a). The crystalline integrity of this grain indicates that a mechanism must exist that is able to naturally restore monazite structure.

Recent work carried out to evaluate the role of monazite as a ceramic for nuclear waste storage suggest that the natural "radiation resistance" may be due to strong bonds between P and O and mineral structure, which has a lower symmetry compared to zircon, thereby mitigating α-particle damage (Krogstad and Walker 1994, Meldrum et al. 1996, 1997b). Thermal annealing appears to be an important mechanism to restore the damaged structure. For example, heavy ion irradiation experiments carried out at various temperatures demonstrate that it is not possible to amorphize a natural monazite grain above a critical temperature of 175°C (Meldrum et al. 1997b). For end-member $LaPO_4$, the critical temperature is as low as 60°C. Karioris et al. (1981) determined that the recrystallization of a totally amorphous monazite was completed at 300°C. Electron irradiation is an efficient way to anneal metamict monazite, as total recrystallization is observed in a few minutes for 0.3 A/cm^2 using 200 kV electrons (Meldrum et al. 1997c). Therefore, ionization created by the alpha particles, or electrons from α decays in the U and Th chains, may partially anneal damage created by recoil nuclei.

Whatever the mechanism, monazite easily heals radiation damage and, thus behaves as a perfect structure in which all elements are bound together by chemical forces in a continuum. This is true for the parent (U, Th) and the final daughter product (Pb) as well as most intermediate decay products. Fluid phases cannot generally permeate through the structure and there are no amorphous domains from which leachable elements (Pb or U^{+6}) can easily escape (Davis and Krogh 2000).

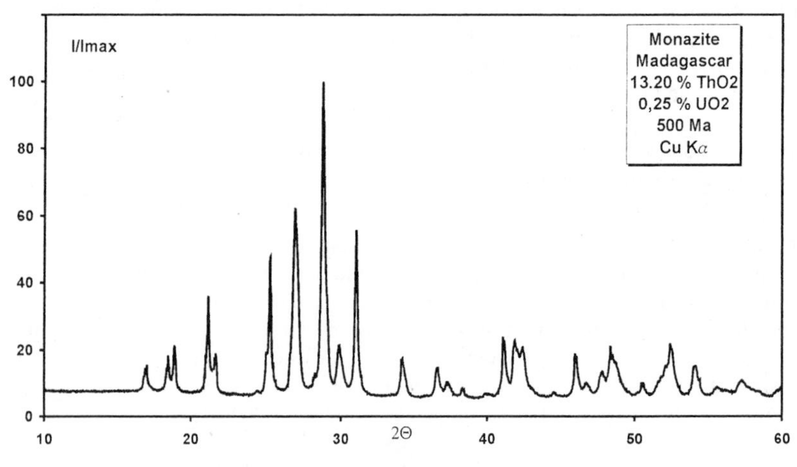

Figure 3. X-ray diffraction pattern of monazite.

Apatite

Mineral structure. The apatite structure is a complex association of a large Ca-site and PO_4 tetrahedra around a channel containing a big anion such as F, Cl, OH. In detail, the unit-cell has two types of Ca bearing sites: a seven-fold site, and a nine-fold site (see Boatner, this volume). The two sites are not equivalent, so large ions such as U and Pb are preferentially incorporated in the nine-fold site. Natural apatite contains mainly uranium as actinide element and almost no Th. The exact substitution mechanism is not fully understood. The replacement of Ca^{2+} by U^{4+} would require two charge compensation mechanisms. This can be achieved by complex substitutions such as 2 $Ca^{2+} + P^{5+} = U^{4+} + Na^{+} + Si^{4+}$, but numerous other possibilities exist as many ions are acceptable in the apatite structure. The U content of natural apatite thus remains limited and vacancies or defects could also play a role in this substitution. Pb is accepted in the apatite structure producing the end member phase pyromorphite. Unfortunately, this isostructural substitution may lead to high Pb^0 concentrations (Krogstad and Walker 1994). Thus, apatite, as with monazite, has little crystallographic tendency to behave as an open geochronometric system for U-Th-Pb.

Radiation effects. Radiation damage in apatite has been studied mainly for purposes of understanding the thermal behavior of fission tracks. Apatite is never found metamict, but this is surely in part due to its limited U-content. In rare cases, apatite can contain as much U as zircon. Fission-track studies as shown than thermal annealing in apatite occurs at ~90°C over geologic time scales (Gleadow et al. 1983).

It has been recently demonstrated in apatite that α particle activity also heals recoil damage (Ouchani et al. 1997). While not demonstrated for monazite, both minerals are fundamentally constructed of PO_4 tetrahedra and thus this mechanism could also operate in monazite as well.

Xenotime

Mineral structure. Xenotime has a zircon-type structure combining eight-coordinated site containing Y and PO_4 tetrahedra. The Y site accepts ions with a smaller ionic radius than monazite. As a consequence, it is the structure preferred by $REEPO_4$ end-members from Tb to Lu, and YPO_4. A second consequence is that, contrary to monazite, xenotime favourably partitions U relative to Th, with mechanisms of substitution similar to monazite (Van Emden et al. 1997, Förster 1998b). Little is known about the incorporation of Pb in xenotime, but the large size of Pb and the preference of xenotime for smaller ions, suggests that it is unlikely that xenotime accepts Pb as easily as monazite.

Radiation effects. The radioactive element content of xenotime is lower than in monazite (e.g., Bea and Montero 1999) but still high enough to produce significant radiation damage. However, as with monazite, no metamict xenotime has ever been described. External irradiation experiments (Meldrum et al. 1997b) have shown that radiation damage in synthetic phosphates with xenotime structure are also healed easily at a temperature about 100°C higher than monazite. Thus the conclusions drawn above for monazite also likely apply to xenotime.

KINETIC PROPERTIES

Diffusion

Background. The phosphate minerals apatite, monazite and xenotime have strongly variable retention properties for Pb under crustal conditions. The mechanisms by which the daughter product can be lost include dissolution/reprecipitation reactions, recrystallization, and diffusive loss. The latter mechanism is likely a common source of discrepancy between a mineral date and the age of the rock from which it formed.

At sufficiently high temperature, daughter Pb will tend to migrate out of a crystal rapidly by diffusion. Since diffusion is strongly temperature dependent, transport rates rapidly decrease as temperature drops, eventually becoming negligible. In certain cases, the apparent age recorded by a mineral corresponds to the temperature at which the daughter product ceased to be lost from the crystal.

The Arrhenius relationship. Atoms diffusing through a crystalline solid are transferred via diffusion jumps between point defects in the crystal lattice. The formation of point defects result from one of two processes: intrinsic defects are thermally controlled defects which maintain electroneutrality, whereas extrinsic defects are caused by chemical impurities which create vacancies in order to conserve charge. Above absolute zero temperature, there is a finite probability of an atom having sufficient local thermal energy to jump from its current position to an adjacent defect. As temperature is raised, the probability of an atom in the Boltzmann distribution acquiring the threshold energy to make this jump increases exponentially. Thus the number of atoms jumping into adjacent vacancies is a function of both the fraction of vacant sites and the fraction of atoms having the necessary thermal energy to overcome the activation barrier. Because both the rate of defect formation and migration involve an exponential dependence, the diffusivity is given by the Arrhenius relationship,

$$D = D_o \exp(-E/RT) \tag{1}$$

where E is the activation energy, D_o is the frequency factor, R is the gas constant, and T is absolute temperature. Note that on a plot of $\log D$ vs. $1/T$ (Fig. 4), the slope of the line is proportional to E and the y axis intercept is $\log(D_o)$.

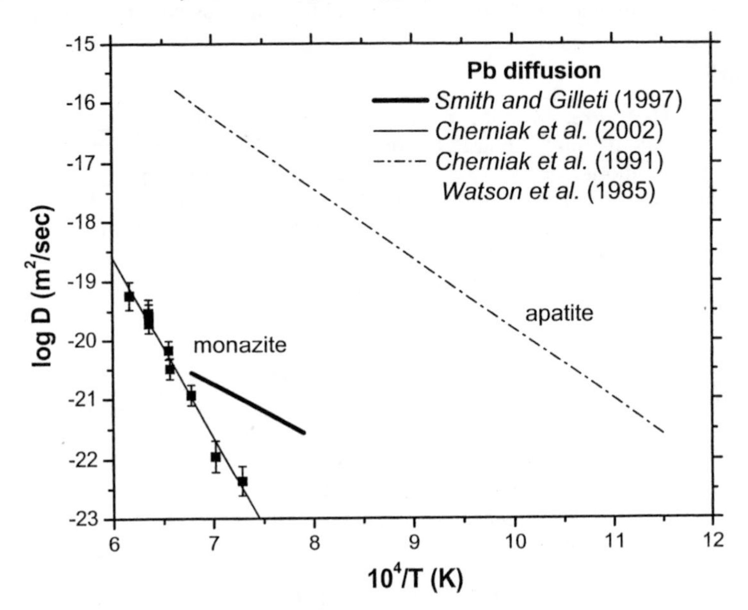

Figure 4. Arrhenius plot showing results of Pb diffusion experiments in monazite and apatite. The apatite data are combined from the studies of Watson et al. (1985) and Cherniak et al. (1991). There is a substantial discrepancy between the two Pb diffusion results for monazite that remains unexplained.

The diffusion equation. The general problem of unsteady-state diffusion within a solid involves the prediction of the concentration distribution $C(x,y,z)$ within a solid as a function of the space coordinates and time, t. To derive an equation that can be solved for $C(x,y,z,t)$, conservation of mass and Fick's first law (i.e., the rate of transfer of mass per unit area is proportional to the concentration gradient, see Fick 1855) are applied to a differential control volume. The resulting expression is the diffusion equation

$$\frac{\partial C}{\partial t} = D\left(\frac{\partial^2 C}{\partial x^2} + \frac{\partial^2 C}{\partial y^2} + \frac{\partial^2 C}{\partial z^2}\right) \tag{2}$$

where D is the diffusion coefficient. Solutions of the diffusion equation can be obtained for a number of initial and boundary conditions covering simple concentration distributions and diffusion geometries (Crank 1975). These solutions have two distinct types: either an infinite trigonometric series or a solution involving the error function. In one dimension, x, the diffusion equation has the form

$$\frac{\partial C}{\partial t} = D\frac{\partial^2 C}{\partial x^2} \tag{3}$$

Using a coordinate transformation, Equation (3) can be modified to describe radial flow in a sphere (Crank 1975). The solution of this equation for the case of a sphere of radius r with an initially uniform concentration, C_o, held in an infinite reservoir of zero concentration can then be translated into terms of fractional loss by integrating the mass loss occurring over the diffusion interval. This fractional loss, f, is given by

$$f = 1 - \frac{6}{\pi^2} \sum_{n=1}^{\infty} \frac{1}{n^2} \exp(-Dn^2\pi^2 t / r^2) \tag{4}$$

In the case of short diffusion times (i.e., only near surface penetration), it can be useful to approximate a mineral with a planar boundary as a semi-infinite medium. For the case of diffusion from a well-stirred semi-infinite reservoir at concentration C_o into a half space initially at zero concentration, the concentration distribution is given by

$$C = C_0 \operatorname{erfc} \frac{x}{\sqrt{4Dt}} \tag{5}$$

where x is the spatial position relative to the boundary at $x = 0$, t is the diffusion time, and erfc is the complimentary error function (erfc = 1 – erf z). Values of $x / \sqrt{4Dt}$ close to $^1/_2$ are essentially unmodified by the error function. Thus it is convenient to define the characteristic diffusion time necessary to change concentration at a specified position by half (i.e., $C/C_o = 0.5$) as $t = x^2/D$.

The closure equation. Minerals originating at deep crustal levels undergo a transition during slow cooling from temperatures that are sufficiently high that the daughter product escapes as fast as it is formed, to temperatures sufficiently low that diffusion is negligible and the retention of radiogenic isotopes by the mineral can be thought of as complete. Between these two states there is a continuous transition over which accumulation eventually balances loss, then exceeds it. The calculated age in this situation relates to the transition interval – the apparent age is the extrapolation of the total accumulation part of the curve to the time axis, which implicitly corresponds to an apparent temperature at which the bulk system became closed (Dodson 1973).

The exponential decay of daughter product loss with time, a consequence of the form of the Arrhenius equation, allows for a closed mathematical solution to the diffusion problem. Dodson (1973) derived a general solution of the accumulation–diffusion–cooling equation for a single diffusion length scale, r, using appropriate substitutions and variable boundary conditions that reduced the problem to an equation identical to Fick's second law. This equation is then be solved using general infinite series solutions (Carslaw and Jaeger 1959) for various geometries yielding expressions for the concentration distributions. Dodson (1973) then evaluated the coefficients, which reduce these expressions to a single characteristic value he called the closure temperature (T_c) which is given by

$$\frac{E}{\mathrm{R}T_c} = \ln\left(\frac{A\mathrm{R}T_c^2 D_0 / r^2}{E\, dT/dt}\right) \tag{6}$$

where A is a geometric constant (sphere = 55, cylinder = 27, plane sheet = 8.7), and dT/dt is cooling rate.

Experimental determination of diffusion parameters. Although isotope exchange can be enhanced by fast-path diffusion along crystalline defects, mechanical grain size reduction, or structural decomposition, volume diffusion is the rate limiting transport mechanism in crystalline solids. In those cases where non-diffusional effects are inactive, thermal history information can be extracted from geological materials if the volume diffusion parameters are known. The approach is to perform diffusion experiments at temperature well above those of geological relevance and then extrapolate the results to lower temperatures using the systematic temperature dependence of the Arrhenius relationship (Eqn. 1). Note, however, that extension of these data may have no meaning if, for example, a diffusion mechanism with lower temperature dependence is encountered over the extrapolated portion. We first briefly review the requirements for the

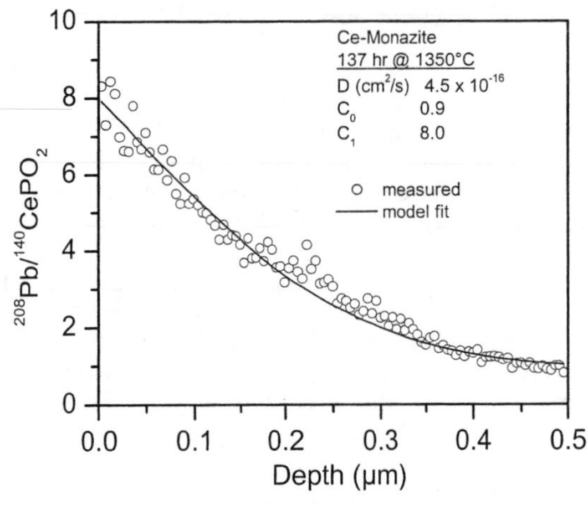

Figure 5. Characteristic form of an error function profile.

measurement of intercrystalline diffusion and summarize current knowledge of the diffusion of Pb in apatite and monazite.

The requirements of a diffusion study of a crystalline solid vary depending on the nature of the experiment, but two criteria must be met in all cases: the mineral must remain stable throughout the duration of the experiment and the initial distribution of the diffusing substance must be known. For purposes of interpretation, diffusion experiments should be designed to address questions regarding chemical vs. tracer diffusion or wet vs. dry conditions. If an isothermal experiment can be designed such that an effectively constant reservoir of diffusant is maintained on a planar surface of a mineral of known or zero concentration of the diffusing species, then knowledge of the concentration distribution (i.e., C as a function of x) can be translated into a value of D using Equation (5). The characteristic form of an error function profile (Fig. 5) can itself be used as a test of whether diffusion was the rate limiting transport mechanism during the experiment. Typically, analysis of these types of experiments is carried out with the ion microprobe using depth profiling mode, or with Rutherford backscattering (see Ryerson 1987).

Diffusion of Pb in monazite

Early geological estimates for the closure temperature of Pb in monazite range from 530°C (Black et al. 1984) to 600°C (Köppel et al. 1980). However, recent evidence suggests the U-Th-Pb system must remain closed to Pb loss through upper sillimanite zone conditions (Copeland et al. 1988, Smith and Barreiro 1990, Spear and Parrish 1996). A closure temperature of Pb in monazite was estimated by Copeland et al. (1988) at 720-750°C for 10-100 μm crystals cooled at 20°C/Ma, based on the presence of inherited monazite in a Himalayan leucogranite. Parrish (1990) suggested a closure temperature of 725±25°C on the basis of a preserved Pb loss profile in monazite from a paragneiss formed under upper amphibolite conditions. Suzuki et al. (1994) calculated closure temperatures for Pb diffusion in monazite as a function of grain size and cooling rate. For a 100 μm monazite, their results indicate closure temperatures of 650°C and 720°C, assuming respective cooling rates of 10 and 100°C/Myr.

In an early experimental study of sintered monazite, Shestakov (1969) estimated an activation energy of 60 kcal/mol by volatilizing Pb in a stream of nitrogen at temperatures between 800 and 1100°C. Smith and Giletti (1997) measured the tracer diffusion of Pb in natural monazites using ion microprobe depth-profiling and observed Arrhenius parameters of E = 43±11 kcal/mol and D_o = 6.6 × 10^{-15} m^2/sec in the temperature range of 1200 to 1000°C (Fig. 4). They found that transport parallel to the *c*-

axis is two to five times slower than perpendicular to c.

In a combined Rutherford backscattering and ion microprobe study, Cherniak et al. (2002) reported E = 149±9 kcal/mol and D_o = 0.94 m^2/sec for Pb diffusion in synthetic monazites in the temperature range 1150 to 1350°C (Fig. 4). This activation energy is over three times higher than the value reported by Smith and Giletti (1997), but diffusivities in the overlapping temperate range (i.e., between 1200 and 1150°C) were approximately the same. Subsequent studies utilizing natural monazites have confirmed this higher activation energy (Cherniak et al. 2002). The large discrepancy between the results of Smith and Giletti (1997) and Cherniak et al. (2000, 2002) is not fully understood, but analytical artifacts may play a role. In any case, the activation energy of Smith and Giletti (1997) is anomalously low.

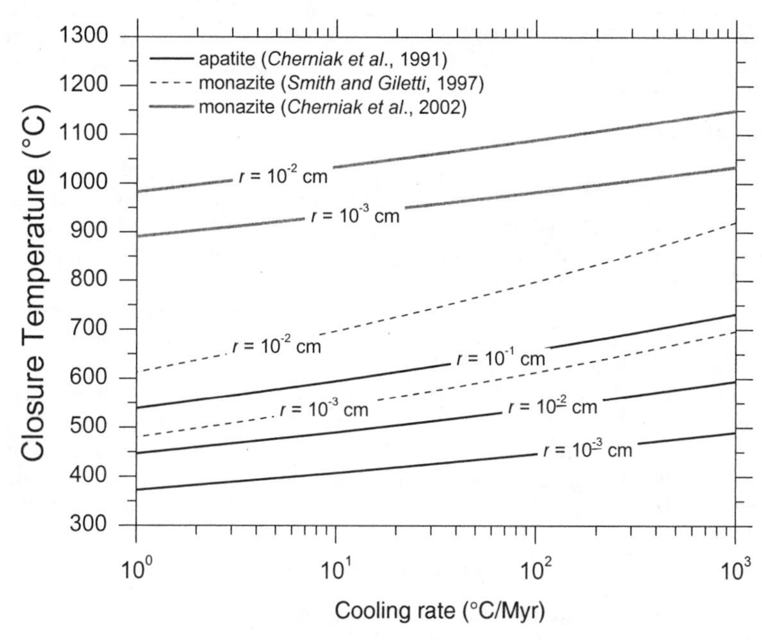

Figure 6. Closure temperature vs. cooling rate for apatite and monazite, assuming a spherical geometry with radius (r) indicated.

Using the diffusion data of Cherniak et al. (2000, 2002) and assuming spherical geometry, calculated T_c's for a range of cooling rates and diffusion domain sizes of 100 and 10 μm are shown in Figure 6. Assuming that the experimental study of Cherniak et al. (2000, 2002) is correct, the concept of closure is largely irrelevant to the U-Th-Pb monazite system under crustal conditions as Pb is predicted to be essentially immobile. Because disturbances to the U-Th-Pb system in monazite are widely reported (indeed, it was once believed that the T_c was <600°C, see Köppel et al. 1980, Black et al. 1984), mechanisms other than diffusive exchange (e.g., recrystallization, dissolution/reprecipitation) must be relatively common. Even prolonged thermal excursions at high crustal temperatures would have little effect on the U-Th-Pb monazite system. For example, a heating episode of 10-Myr duration at a temperature of 700°C would result in a diffusion profile with characteristic length-scale of only 0.25 μm (Eqn. 5).

Diffusion of Pb in apatite

Watson et al. (1985) used the electron microprobe to reveal gradients in Pb produced by annealing apatite in a dry environment in the temperature range 900 to 1250°C. The calculated Arrhenius parameters they obtained are: $E = 70$ kcal/mol and $D_o = 350$ m^2/sec. Subsequently, Cherniak et al. (1991) measured Pb diffusion in apatite in the broader temperature range 600 to 900°C using the significantly more sensitive Rutherford backscattering method and obtained results in good agreement with the data of Watson et al. (1985). The combined dataset yields an overall Arrhenius relationship of $E = 54.6\pm1.7$ kcal/mol and $D_o = 1.2$ m^2/sec.

Using the data of Cherniak et al. (1991) and assuming spherical geometry, calculated T_c's for a variety of cooling rates and diffusion domain sizes from 1000 to 10 μm are shown in Figure 6. For the case of a 100 μm diameter apatite, closure temperatures are in the range of 450 to 550°C for most geological relevant cooling rates. However, note that a heating episode of 1-Myr duration at a temperature of 600°C would cause a fractional Pb loss (Eqn. 4) of only 2.5% from an apatite of that size.

Diffusion of Pb in xenotime

In a Pb-isotopic transport and crystal chemical study of U-bearing minerals, Dahl (1997), suggests the closure temperature of Pb in xenotime is ≥ 750°C, similar to the results for monazite and epidote. Studies that date monazite and xenotime found in the same rock typically yield similar ages (e.g., Hawkins and Bowring 1997), suggesting that the minerals record similar closure temperatures or crystallization events. We speculate that the U-Pb system in xenotime is less retentive than monazite, as few examples of magmatic xenotime containing inherited U-Pb ages have been documented and the Y content in most magmatic rocks (which controls the solubility of xenotime) is about an order of magnitude less than LREE (which control monazite solubility), thus restitic xenotime from anatectic events may be rare. This idea may be invalid as Viskupic and Hodges (2001) report older ages preserved within both xenotime and monazite grains separated from a Himalayan orthogneiss. Clearly, obtaining more geochronologic data and experimental studies of the diffusion of Pb in xenotime are required to ascertain the significance of the mineral's age.

DATING APPROACHES

A variety of methods are currently being utilized to obtain U-Th-Pb ages of phosphates, including isotope dilution mass spectrometry, secondary ion mass spectrometry, chemical Pb dating, and laser ablation-inductively coupled plasma-mass spectrometry. We discuss these methodologies in turn and describe their relative merits in specific applications. Note that many recent studies consider combining the techniques described below as a means to maximize the benefits and circumvent drawbacks (e.g., Poitrasson et al. 1996, Cocherie et al. 1998, Crowley and Ghent 1999, Paquette et al. 1999, Poitrasson et al. 2000, White et al. 2001). Other recently developed techniques not presented here include proton microprobe analysis (e.g., Bruhn et al. 1999) and energy-dispersive, miniprobe, multielement analysis (Cheburkin et al. 1997).

Isotope dilution mass spectrometry

Mass spectrometers are fundamentally designed to determine isotope ratios. For example, the Pb isotopic composition of a U- and Th-bearing phosphate can be measured directly using a thermal ionization mass spectrometer provided the instrumental mass fractionation is known or can be assessed in the course of the analysis. Determining absolute abundances of isotopes requires a technique called isotope dilution in which a pre-determined quantity of a 'spike' of known but exotic isotope composition is added to an aliquot of sample containing an unknown amount of the same element. Thus the isotopic composition of the mixture can be used to determine the amount of the element present in the unknown. In the case of elements containing more than one primordial isotope, instrumental mass fractionation can be addressed by correcting the measured isotopic composition of isotopes with a constant ratio in nature to the known value. Because Pb contains only one isotope that is neither radioactive or the product of radioactive decay (i.e., ^{204}Pb), isotope ratio precision is limited by the level of control on instrumental discrimination (typically ~1 ‰) unless a double spike (e.g., a mixture of ^{210}Pb and ^{205}Pb) is utilized. In the case of determining U and Th abundances, mixed spikes of ^{234}U and ^{230Th}, for example, can be used. Precision of inter-element ratios (i.e., $^{208}Pb/^{232}Th$ or $^{206}Pb/^{238}U$) of ~0.2% are routinely achieved (e.g., Mattinson 1994).

High precision monazite, xenotime and apatite ages of single and multiple grains and fragments have been obtained using this approach (e.g., Krogstad and Walker 1994, Searle et al. 1997, Viskupic and Hodges 2001) and characterization of standard materials for other U-Th-Pb methods are generally based on isotope dilution measurements (e.g., Zhu et al. 1998, Sano et al. 1999a, Harrison et al. 1999a, Willigers et al. 2002). The high precision attainable by this approach has allowed the identification of important chronological characteristics of phosphate minerals such as the presence of an inherited component (e.g., Copeland et al. 1988) and unsupported daughter products from disequilibrium isotopes (e.g., Schärer 1984). These issues have the potential to limit the accuracy of data interpretations. If isotopic measurements are carried out using Faraday cup detectors, whole single grains, or more commonly aggregates of grains, are dissolved for analysis. This reliance on mineral separation can limit age interpretation textural relationships are destroyed, and unsupported Pb* from inclusions may be incorporated (Hawkins and Bowring 1997). In response to these concerns, less precise methods to date phosphate minerals have been developed that preserve textural relationships, and in some cases, the dated grain itself.

Secondary ion mass spectrometry

Spot analysis. Ion microprobe dating of monazite takes advantage of the kinetic energy distribution of U^+, Th^+, and Pb^+ ions sputtered from the grain using a primary oxygen (O^- or O_2^-) beam focused to a 10-20 μm diameter spot (Harrison et al. 1995, 1999a; Zhu et al. 1998, Stern and Sanborn 1998, Stern and Berman 2000, Petersson et al. 2001). At a mass resolving power of ~4500, all Pb and Th isotopes are resolved from any significant molecular interferences (Harrison et al. 1995). Zhu et al. (1998) report U and Pb peaks resolved in monazite at 6000, but operate with a mass resolving power of 2000 assuming interferences are negligible. However, an unidentified isobar at 203.960 amu complicates determination of $^{204}Pb^+$ in monazite with higher Th contents (Stern and Sanborn 1998, Stern and Berman 2000). In the case of Th-Pb dating, a linear relationship between $^{208}Pb^+/Th^+$ versus ThO_2^+/Th^+ is observed for monazite grains with a known uniform Pb/Th ratio (Harrison et al. 1995, 1999a; Catlos et al. 2002b). Isotopic data collected from several ion microprobe spots on the age standard define this linear relationship and allow a correction factor to be derived by dividing the measured $^{208}Pb^+/Th^+$ of a standard grain at a reference ThO_2^+/Th^+ value by its known daughter-to-parent ratio. This correction factor permits the determination of Pb/Th ratios of unknown grains measured under the same instrumental conditions. Precision is generally limited by the reproducibility of the calibration curve, which is typically ±2%. Stern and Berman (2000) report similar uncertainties for U-Th-Pb ion microprobe measurements. Rather than assessing the relative inter-element sensitivity, DeWolf et al. (1993) used an ion microprobe obtain $^{207}Pb/^{206}Pb$ spot ages on Precambrian monazites.

Ion microprobe analysis of apatite is outlined by Sano et al. (1999a), whereas xenotime ion microprobe dating is described by Petersson et al. (2001). A mass resolving power of 4500 is used to resolve U and Pb peaks in xenotime (Petersson et al. 2001) and 5800 in apatite (Sano et al. 1999b).

For single-grain spot analysis, mineral grains are typically mounted in epoxy, polished, and photographed in reflected and transmitted light (e.g., Stern and Berman 2000). Backscattered electron (BSE) images are taken of individual grains to record zoning patterns and the probe spot can be focused on selected sites within the mineral grain (e.g., Zhu et al. 1997, Ayers et al. 1999, Sano et al. 1999a, Zhu and O'Nions 1999a). The spatial resolution of the ion microprobe can be <10 μm and the crater depth is generally <2 μm during a typical 15-minute analysis. However, misleading results are possible if the beam overlaps areas of substantially different age. Detailed images of the grain may help to resolve this issue if a correlation exists between chemical composition and age (see Zhu and O'Nions 1999b).

Petersson et al. (2001) note that matrix effects limit the accuracy of the xenotime ion microprobe ages and they were unable to precisely reproduce the U-Pb ages of their standards. For example, xenotime standard 88102-5 has a concordant ^{238}U-^{206}Pb age of 919±5 Ma, whereas ion microprobe analyses yielded 934±4 Ma. Zhu et al. (1998) explores potential matrix effects during monazite ion microprobe analysis, but analytical methods outlined in their paper preclude conclusive interpretations. Stern and Sanborn (1998) report that the use of high-Th monazite standards may lead to errors in measuring ages of low-Th grains, and suggest the use of compositionally matched standards and unknowns.

Depth profiling. In the 'classical' approach to U-Pb dating (e.g., Compston et al. 1984), the spatial resolution of a measurement is defined by the diameter of the primary ion beam spot. Given the trace concentration levels of Pb expected, even in monazite, and reasonable analysis times, typical probe spot diameters of ca. 10 μm are often required. However, because virtually all secondary ions originate from the first or second atomic layer of the instantaneous sample surface, atomic mixing due to the impacting primary ions and geometric effects on crater production are the essential limiting factors on depth resolution. The resulting spatial resolution offered by depth profiling is potentially two orders of magnitude higher than spot analyses. Care must be taken in this analysis mode to ensure material sputtered from the crater walls is not analyzed. This can be accomplished by placing an aperture in front of the emergent ion beam that restricts entry into the mass spectrometer to only those ions originating on the crater floor.

Grove and Harrison (1999) investigated the feasibility of obtaining Th-Pb age profiles in the surface regions of natural monazites and found that ion intensities were adequate to resolve age differences of <1 Myr with better than 500 Å depth resolution in late Tertiary monazites. These age gradients were then used to extract continuous thermal history information from which they constrained the displacement history of a Himalayan thrust. The sputtering of natural surfaces was found to yield inter-element calibration plots of similar reproducibility to that of polished surfaces.

In situ ***ion microprobe dating.*** Numerous studies report U-Pb or Th-Pb ages of phosphate minerals grains in thin section (= *in situ*) using the ion microprobe (DeWolf et al. 1993, Harrison et al. 1997a, Zhu et al. 1997, Catlos 2000, Foster et al. 2000, Catlos et al. 2001, 2002a; England et al. 2001, Gilley 2001, Petersson et al. 2001, Robinson et al. 2001, Gilley et al. 2002). The spatial resolution allows preservation of textural relationships, and thus potentially less ambiguous age interpretation.

Monazite is an ideal mineral for *in situ* geochronology as it appears in pelitic lithologies at conditions ranging from the garnet to the aluminosilicate isograds (e.g.,

Smith and Barreiro 1990, Harrison et al. 1997a, Bingen and Van Bremen 1998, Ferry 2000, Wing et al. 2002) and inclusions in garnet appear armored against daughter product loss (e.g., Montel et al. 2000). Garnet-bearing assemblages allow the determination of pressure-temperature (P-T) conditions (e.g., Spear 1993), and when combined with monazite age data, suggest a powerful combination for ascertaining the evolution of metamorphic terranes (DeWolf et al. 1993, Harrison et al. 1997a, 1998; Foster et al. 2000, Terry et al. 2000, Catlos et al. 2001, Gilley 2001). Individual grains are ordinarily identified and characterized texturally and chemically by using the electron microprobe, then either cut out with a high-precision saw, or drilled out with a diamond drill corer. The fragments are cleaned and mounted in epoxy with a pre-polished block of age standards. Zhu et al. (1997) and Catlos et al. (2002b) briefly outline methods of *in situ* monazite chronometry, whereas Petersson et al. (2001) describe methods for obtaining *in situ* ion microprobe xenotime ages.

Chemical Pb dating

Background. The concept of chemical U-Th-Pb dating emerged shortly after the discovery of radioactivity (e.g., Holmes 1911). As U and Th disintegrate ultimately to Pb, it is possible to estimate an age solely from the elemental proportion of Pb, U and Th. This age has geological meaning only if two conditions are met: (1) all Pb is radiogenic, or at least, that the Pb° can be neglected, and (2) the system remained closed. The first condition is only approximated in minerals rich in U and Th and poor in Pb, such as zircon, monazite, thorite, uraninite, thorianite, and xenotime, whereas the second condition cannot be tested independently without isotopic data. Because isotopic dating is now so efficient, chemical dating is now generally restricted to *in situ* dating techniques.

If a mineral is sufficiently old and rich in Th and U, Pb* concentrations can be high enough to be detectable using an electron microprobe. Analysis of all three elements then provides an estimate of age. This method has been widely used in uranium-ore studies, as pitchblende produces high Pb* contents in relatively short times. Recently this technique has been applied with success to monazite (Suzuki and Adachi 1991, 1994; Montel et al. 1994, 1996; Williams and Jercinovic 2002), and, with more difficulty, to zircon, and xenotime (Suzuki and Adachi 1991, Geisler and Schleicher 2000)

The electron microprobe's detection limit for Pb can be a limitation in the precision of the chemical age results (e.g., Olsen and Livi 1998) as can potentially the assumption that monazite is "generally concordant" (e.g., Cocherie et al. 1998) and does not incorporate "appreciable amounts" of Pb° during growth (e.g., Scherrer et al. 2000). Finger and Helmy (1998) suggest that monazite is "resistant to post-crystallization disturbance and mostly free from inheritance." These assumptions are not always met as numerous studies indicate that the mineral can accumulate excess ^{206}Pb (Schärer 1984, Parrish 1990, Schärer et al. 1990, see also Fig. 2), Pb° (e.g., Coleman 1998, Catlos et al. 2001, 2002a; see also Table 3), contain inherited Pb* (Copeland et al. 1988) and sustain Pb loss (Catlos et al. 2001).

The attraction of electron microprobe monazite dating is that this apparatus is widely distributed and accessible to most geologists. Modern microprobes are highly automated, are easy to handle, have expanded mapping and profiling capacities, and have excellent electronic and optical visualizing devices. For example, electron microprobe maps showing the distribution of ages within single grains have been reported by several workers (Braun et al. 1998, Cocherie et al. 1998, Crowley and Ghent 1999, Williams et al. 1999, Williams and Jercinovic 2002). Cocherie et al. (1998) suggests electron microprobe dating has "no competitor for dating complex polygenetic monazite, especially when two events are recorded." Cocherie et al. (1998) warrant that the spatial resolution of the electron microprobe (1-5 μm) is ideal for measuring ages of regions

Table 3. Summary of U-Pb analytical data from two mineral separation/isotope dilution case studies.[a]

Sample[b]	*Description*	N[c]	$^{206}Pb/^{238}U$ *age (Ma),* $\pm 1\sigma$	$^{207}Pb/^{235}U$ *age (Ma),* $\pm 1\sigma$	$\%^{206}Pb^{*}$[dd]
			AS-31 Leucogranite orthogneiss		
MA		3	*36.7 (0.2)*	36.4 (0.2)	99.1
MB		9	34.4 (0.1)	33.7 (0.1)	99.1
MC	*visible core*	1	36.4 (0.1)	35.9 (0.1)	98.6
MD		1	37.2 (0.1)	36.2 (0.1)	99.1
			36.2 (0.1)144[e]	35.6 (0.1)138	
			AS-38 Deformed migmatitic leucosome, unit 1		
MA		>50	32.5 (0.1)	33.5 (0.1)	92.8
MB		4	23.3 (0.1)	23.3 (0.1)	98.1
MC		15	35.8 (0.1)	35.8 (0.1)	98.4
MD		7	34.8 (0.1)	34.8 (0.1)	98.4
			31.6 (0.1)3253	31.9 (0.1)3338	
			AS-23 Deformed migmatitic leucosome, unit 1		
M1		1	31.8 (0.1)	31.6 (0.1)	98.6
M3		1	29.9 (0.1)	29.7 (0.2)	98.0
M4	round	2	28.3 (0.3)	28.0 (0.8)	96.0
M5		2	29.5 (0.1)	29.2 (0.4)	97.5
M6		5	29.8 (0.1)	29.6 (0.2)	96.7
			29.9 (0.2)104	29.6 (0.4)99	
			MCT-1 migmatitic pelitic schist		
Ma	clear, incls	1	20.9 (0.9)	20.7 (0.9)	99.1
Mb	clear, incls	1	21.5 (2.0)	*21.4 (2.0)*	98.6
Mc	clear, incls	1	18.3 (1.8)	18.2 (1.8)	98.1
Md	clear, incls	3	21.5 (1.2)	21.7 (1.2)	99.0
Me	clear, incls	5	20.6 (0.8)	20.5 (0.8)	99.2
Mf	clear, incls	15	28.0 (0.9)	30.5 (0.9)	99.2
Mg	clear, incls	20	23.8 (1.0)	24.7 (1.1)	98.8
			22.1 (1.3)9	22.5 (1.3)17	
			MCT-93-88 foliated leucogranite		
M2a	clear	9	28.8 (1.8)	28.0 (1.7)	99.0
M2b	clear, incls	4	28.6 (9.4)	28.3 (9.3)	96.5
M2c	clear, incls	5	28.8 (3.1)	28.0 (3.1)	97.3
M2d	clear, incls	1	34.3 (1.4)	33.9 (1.4)	99.6
			30.1 (5.1)3	29.6 (5.0) 4	
			DK-1 undeformed leucogranite		
Mf	clear	5	*19.3 (2.0)*	18.7 (1.9)	97.5
Mg	clear	4	19.4 (2.4)	19.0 (2.4)	97.1
Mh	clear	5	19.3 (2.5)	18.9 (2.4)	97.0
			19.3 (2.3)<1	18.9 (2.3)<1	
			DK-2 undeformed leucogranite		
Mi	incls	>50	19 (73)	17 (67)	61.7
Mj	incls	>50	19 (70)	17 (64)	64.7
			19 (72)<1	17 (66)<1	

a. "AS" data from Hodges et al. (1996), whereas "MCT" and "DK" data from Coleman (1998). See Figure 2 for a concordia diagram of this and other Himalayan monazite data.
b. Sample name is followed by a description of the mineral grains if provided. "incls", inclusions.
c. Number of grains analyzed.
d. % radiogenic ^{206}Pb, calculated using $^{206}Pb/^{204}Pb$ of 18.7 (Stacey and Kramers 1975). Note that a majority of monazite grains have high (>95%) Pb* contents.
e. Average age, Ma ($\pm 1\sigma$) and Mean Square Weighted Deviation calculated using 1σ errors. The majority of data reported from single samples are inconsistent with a single population.

within monazite grains inaccessible to conventional U-Pb dating techniques and too small for the ion microprobe analysis. This view appears to be based on a misunderstanding of the inherent spatial resolution of the ion microprobe which can be as low as a few μm's in spot analysis mode or as much as two orders of magnitude lower in depth profiling mode (e.g., Harrison et al. 1997b, Grove and Harrison 1999).

The inability of electron microprobe to provide isotopic ratios is a potentially serious limitation of this technique. However the open/closed system problem can be, in part, assessed by comparing ages to chemical compositions (e.g., Suzuki et al. 1994, Braun et al. 1998). Finally we should note that the electron microprobe is a non-destructive method and thus can be combined with other techniques. For example, electron microprobe analyses in a thin section analyzed afterward by ion microprobe or grains mounted in epoxy, extracted and dated by thermal ionization mass spectrometry.

Technical and analytical basis for electron microprobe dating. Electron microprobe dating requires precise analyses of U, Th, and Pb. For these heavy elements, M-lines are used which are intrinsically weak and broad. Special attention must be paid to the Pb analyses because the precision and accuracy on this element directly controls the precision and accuracy of the calculated age. Refer to the original papers to find details of the analytical conditions (Suzuki and Adachi 1991, 1994; Montel et al. 1994, 1996; Rhede et al. 1996, Cocherie et al. 1998, Finger et al. 1998, Olsen and Livi 1998, Crowley and Ghent 1999, Williams et al. 1999, Scherrer et al. 2000).

As for any high-precision analysis with the electron microprobe, long counting times and high sample currents are necessary. Typical counting times are several minutes, especially on the Pb peaks and backgrounds. Typical probe currents range from 50 to 150 nA. The accelerating voltage used varies from 15 kV to 40 kV. When M lines are utilized, low accelerating voltage (15 kV) can be used. With such a low voltage, the excited domain is less than 2 μm in diameter, and matrix corrections are limited.

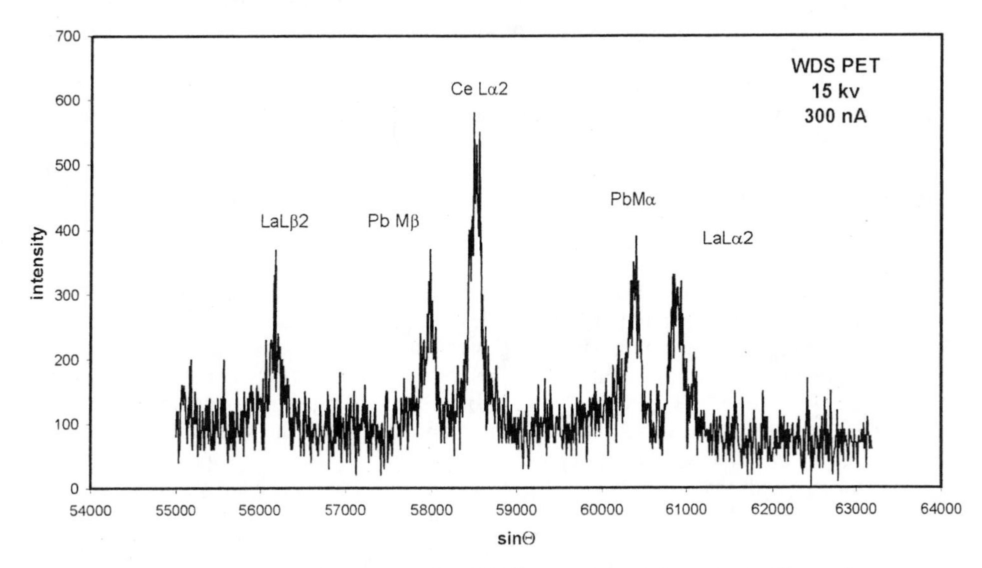

Figure 7. Wavelength dispersive spectrometry (WDS) scan of the region corresponding to the two M-lines for Pb. Additional interference that are not readily visible are the two Th-Mζ_1 and Th-Mζ_2 line around the Pb-Mα line and the Y-Lα line at the Pb-Mα position. The subscript 2 is for second-order lines.

The choice of lines and background positions are key to the procedure (Fig. 7). There is no general agreement on the best lines and the corresponding background positions, because, for each microprobe, as a function of analytical condition, the optimal positions are different and must be empirically determined by the user. The main difficulties arise from the presence of other X-rays lines interfering with peaks or background positions (Th M_β on U M_α, K K_α on U M_β, Y L_α and Th M_ζ on Pb M_α second order Ce L_α on Pb M_β). This require careful investigations of the Wavelength Dispersive Spectrometry (WDS) signal of monazite around the M-lines for Pb, U, Th. The theoretical WDS spectrum presented in Scherrer et al. (2000) should be used only as a general guide because the actual spectrum may differ significantly from those calculated ones. As an example, we report in Figure 7 the WDS diagram around the Pb lines for a monazite containing about 1% Pb, at 300 nA and 15 kV, on a Cameca SX-50 microprobe. [See Chapter 8 by Pyle et al., this volume, for additional information on microprobe analysis of phosphates.]

Data processing. The raw peaks and background data are processed through ZAF, PAP, or equivalent matrix correction programs. There are no particular difficulties at this stage but care must be taken to analyse the light elements of monazites (Ca, Y, Si, P), because they significantly influence matrix correction coefficients. A variety of standards are used for UO_2, ThO_2, and multiple compounds for Pb, including synthetic glass, PbS, and vanadinite. Errors in U, Th, and Pb contents, as well as detection limits are obtained by classical methods, the best being the mathematical analysis of Ancey et al. (1978). Detection limits can be as low as ±80 ppm on Pb, with use of long counting times, high probe currents, and a modern instrument.

The amount of Pb in a monazite of age τ is calculated from:

$$Pb/M_{Pb} = Pb^0 + Th/232 \cdot (e^{(\lambda_{232} \cdot \tau)} - 1) + U/235 \cdot (1 - R) \cdot (e^{(\lambda_{235} \cdot \tau)} - 1) + U/238 \cdot R \cdot e^{(\lambda_{238} \cdot \tau)} - 1 \quad (7)$$

where M_{Pb} is the molar weight of Pb, and 232, 235, 238 are the molar weights of ^{232}Th and ^{235}U and ^{238}U, respectively. Decay constants are represented by λ_{232}, λ_{235}, λ_{238} and R is the $^{238}U/^{235}U$ atomic ratio (137.88). Because Pb in monazite is primarily ^{208}Pb, we can assume 208 to be the molar weight of Pb. Equation (7) must be solved iteratively for τ. Age uncertainty depends on the uncertainty on Pb, Th, and U contents and is calculated by classical error propagation techniques, or by Monte-Carlo simulation. A property of the error in electron microprobe analysis is that the absolute error is almost constant for given analytical conditions (Ancey et al. 1978), thus relative errors are lower for higher Pb contents. As a consequence, absolute age errors are lower for higher U and Th contents.

A population of 10 to 50 ages is quickly and easily obtained from monazites using *in situ* electron microprobe dating techniques. Several statistical procedures for extracting as much information from this population as possible have been suggested. For example, the 0-dimension method (Montel et al. 1996) directly uses Equation (7), assuming Pb° can be neglected. Then each U, Th, Pb triad is transformed into an age with associated error. The statistical procedure is carried out on the age population with the goal of assessing two questions: (1) does the population represent a single geological event? and (2) what is the best estimate (with associated errors) for the age of this event? The answer to the second question is given by classical statistical formulae and the answer to the first by a statistical test. If the answer to the first question is no, another hypothesis must be developed. Montel et al. (1996) describe a procedure when two ages (or more) are suspected. The best representation of this method of calculation is the weighted-histogram representation that plots the probability as a function of age for the whole population (Fig. 8).

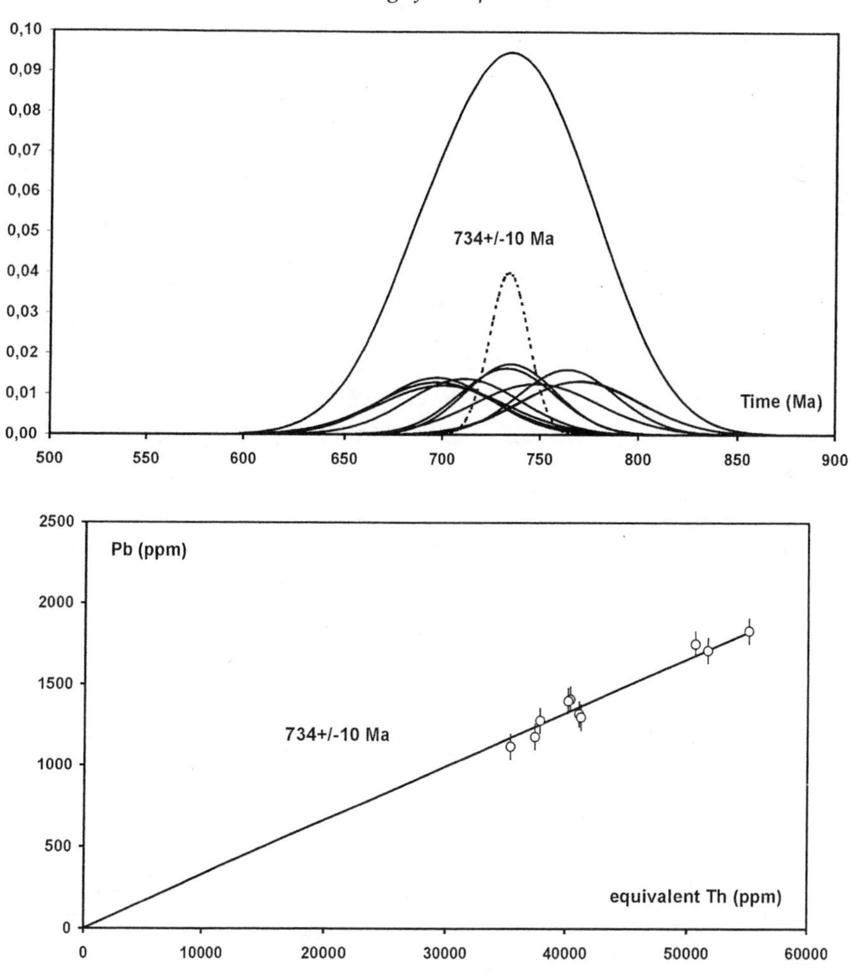

Figure 8. (Upper) weighted-histogram and (lower) isochron representation of a sample in which a single population of ages can be distinguished. The sample is a migmatite from Madagascar.

Suzuki and Adachi (1991, 1994) use a 1-dimension procedure in which Equation (7) is transformed into the equation of a straight line. They assume that Pb^0 is constant, and transform U content into a "fictive Th" content (i.e., the amount of Th that would give the same amount Pb than U for a given age). The sum of the real Th and fictive Th is an "equivalent Th" value, Th*. Plotting Pb versus Th* should give a straight line, the slope of which yields the age and the intercept at zero Th giving Pb^0. The advantage of this procedure is that it visually shows the entire population. However this procedure assumes that all grains have a unique Pb^0 content. A variant of this technique is to use the same representation, but to force the line through the origin. This should give the same age than the 0-dimension model, but provides an isochron-like representation. Two examples are reported in Figures 8 and 9. In both cases the weighted-histogram and the isochron-like representations are given.

Rhede et al. (1996) proposes a 2-dimensional procedure. However, Equation (7) can be considered a plane in Pb-Th-U space. If the ages given by Th-Pb system and U-Pb

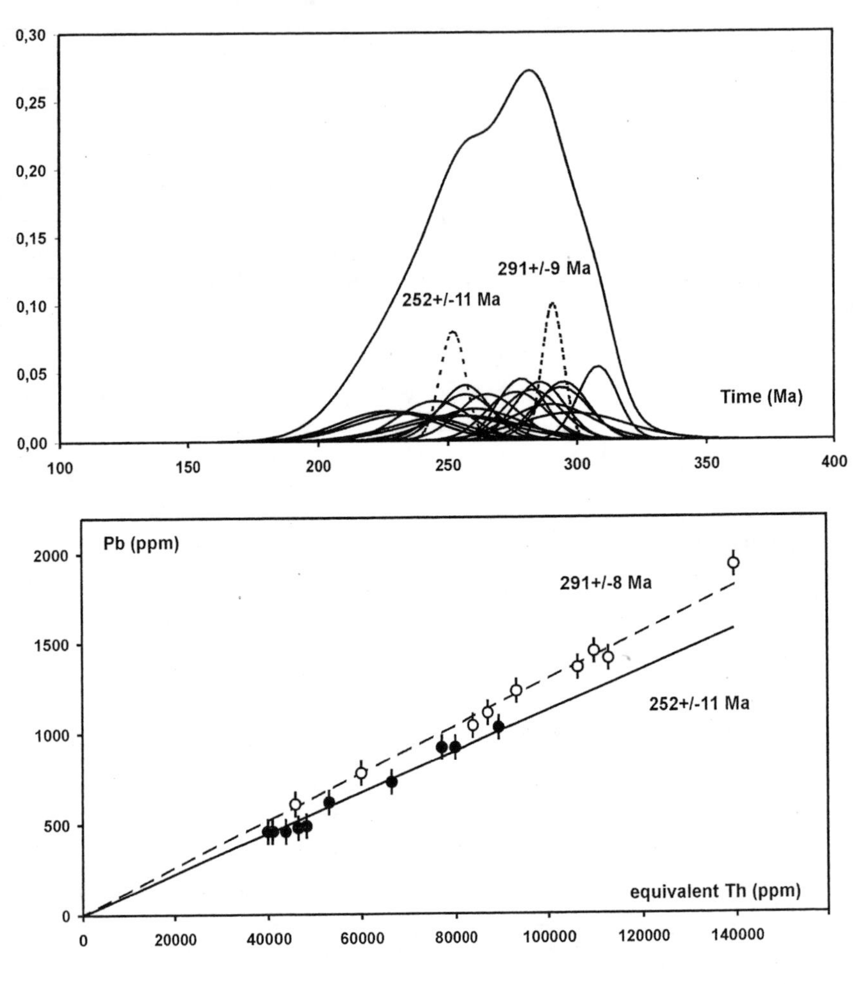

Figure 9. Upper—weighted-histogram. Lower—isochron representation of a sample in which two populations can be distinguished. The sample is a microgranite affected by a hydrothermal event. The older age is thought to be the magmatic age; the younger age may be the age of the hydrothermal event.

system are different, the best-fit plane in the Pb-Th-U space gives the two ages and Pb^0. The advantage of this method is to provide an estimate of system concordance by comparing the ages from derived from two different systems. However, this technique requires complex graphical representation and is more difficult to compute than the 0- or 1-dimensional methods. A modification proposed by Cocherie and Albarède (2001) considers the ratios Pb/U and Pb/Th instead of a Th-U-Pb diagram.

Whatever procedure used, a significant reduction of the error on ages is obtained when the whole data population is considered. With modern microprobes, statistically-reduced errors of ±10 Ma are commonly obtained; a value close to the maximum precision attainable as uncertainties related to matrix effect corrections, line interferences, Pb atomic weight, or standard composition, cannot be eliminated without changing analytical procedures.

Sample preparation. Two main types of samples have been used for monazite electron microprobe dating, both having advantages. The most direct method is to use uncovered rock thin sections prepared for electron microscopy. The advantages here are obvious: no specific preparation is required, and more importantly, the position of each dated grain in the thin section is known, and it is then possible to correlate the ages with the petrology of the sample. This is the preferred method when a complex history is anticipated in a sample. The best method to find monazite in thin section is electron imaging in BSE mode as monazite grains appear as bright grains (Scherrer et al. 2000). The exact nature of the grains must be confirmed by qualitative analyses using an energy dispersive system, or, on microprobe, by looking for high Th, Ce and P contents.

Another technique is to use separated grains mounted in epoxy resin. With this technique a great number of crystals can be analysed. However, small grains can be lost and the position of the grain in the rock is unknown, making the interpretation of the results difficult if the age population is mixed. This method however can be easily combined with a thermal ionization mass spectrometry dating study, for which monazite separation is always necessary.

Laser ablation-inductively coupled plasma-mass spectrometry

The lower precision results produced by *in situ* ion or electron microprobe dating of phosphate minerals compared to isotope dilution methods continue to impel development of alternative *in situ* approaches with the potential for high precision. Recently, laser ablation-inductively coupled plasma-mass spectrometry has been applied to monazite and apatite dating (Machado and Gauthier 1996, Poitrasson et al. 1996, 2000; Engi et al. 2001, Kosler et al. 2001, Willigers et al. 2002). With this method, a laser beam directly samples a selected portion of a mineral and ionized material is transferred into a mass spectrometer (see Willigers et al. 2002). The method excavates the mineral of interest with a laser spot size of 25-100 μm, making the technique applicable for *in situ* studies of phosphate minerals with larger grain sizes (Machado and Gauthier 1996, Willigers et al. 2002). Kosler et al. (2001) report U-Th-Pb monazite ages that are as young as several tens of million years with a precision better than 2%, whereas Willigers et al. (2002) reports apatite and monazite ages with uncertainties comparable to those determined by isotope dilution. These recent results contrast to the low precision reported by Machado and Gauthier (1996) testifying to the rapid development of this technique. Areas of continued development (see Willigers et al. 2002) include improvements to sensitivity, ion transmission, and laser methodology, which may lead to laser-ablation inductively-coupled plasma mass spectrometry becoming competitive for *in situ* U-Th-Pb dating with the spatial selectivity of the ion and electron microprobes.

U-TH-PB DATING OF MONAZITE

Prograde thermochronometry

In contrast to thermochronologic systems involving an isotopic closure transition, the use of prograde thermochronometers, such as U-Th-Pb monazite-in-garnet dating, permits mineral growth ages to be directly determined. Detrital monazite is unstable in most pelitic rocks at low grades (chlorite to biotite grades) as allanite or REE oxides appear to host REE and radiogenic isotopes (Smith and Barriero 1990, Kingsbury et al. 1993). Note that Rasmussen et al. (2001) reports ion microprobe U-Pb ages of monazite found in northern Australian Pine Creek shales, but the scarcity of the mineral in most low-grade metamorphic terranes preclude widespread application. In some pelitic rocks, neoformed monazite reappears at about the aluminosilicate isograd (Ferry 2000, Wing et al. 2002). Thus, U-Th-Pb dating of monazite potentially times aluminosilicate growth and formation of a metamorphic isograd. Transects along the Himalayan inverted metamorphic sequences show allanite as the dominant mineral at lower metamorphic grades, whereas monazite appears coincident with garnet, suggesting that in this case, monazite ages can time garnet growth (Harrison et al. 1997a, Foster et al. 2000, Catlos et al. 2001, Kohn et al. 2001). In each case, major phase compositions can yield the thermobarometric conditions of formation (Spear 1993), and suggest the utility of monazite as a geochronometer in metamorphic terranes.

While the diffusive transport of Pb in monazite is slow under most crustal conditions (Smith and Giletti 1997, Cherniak et al. 2000, 2002), in certain cases it appears that monazite inclusions in garnet are armored against Pb loss (Montel et al. 2000, Catlos et al. 2001). Exploration of monazite growth prior to garnet growth can be addressed through phase equilibria (e.g., Pyle and Spear 1999) and trace element studies (e.g., Foster et al. 2000). For example, because Y and heavy REE are preferentially concentrated in garnet, their concentrations in monazite have been used to assess paragenesis, although Catlos et al. (2002b) found no correlation between composition and age.

A major complication facing *in situ* monazite analysis is the interpretation age data inconsistent with a single population (see Fig. 10). To address this issue, monazite chemical variability has been speculated to reflect timing information (e.g., Ayers et al. 1999, Zhu and O'Nions 1999a,b). For example, Foster et al. (2000) surmises that lower Y

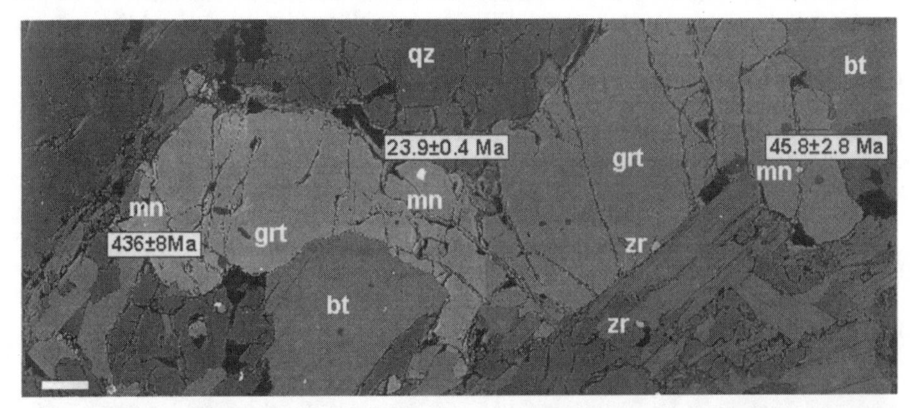

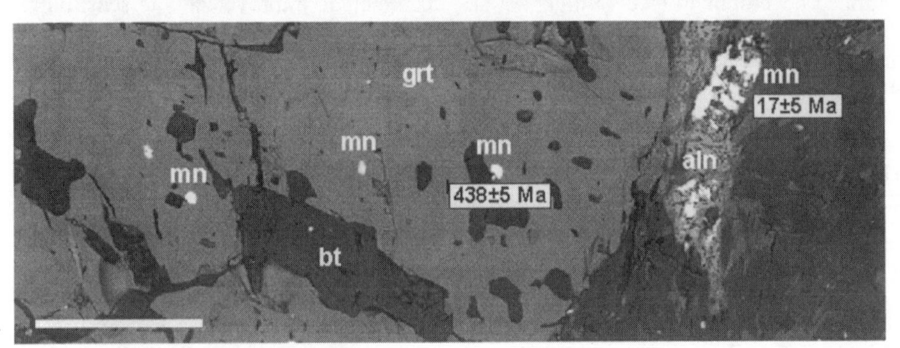

Figure 10. Backscattered electron images of (upper) Nepal Himalaya (Catlos et al. 2002a) and (lower) Turkey Menderes Massif samples. Monazite grains (mn) are seen as bright grains with ion microprobe Th-Pb ages are in boxes with 1σ errors. Other abbreviations include: grt, garnet; zr, zircon; bt, biotite; qz, quartz; aln, allanite. These samples were collected from regions that experienced complicated polymetamorphic histories, and they illustrate the benefit of an *in situ* technique for monazite geochronology.

concentrations in matrix monazites are clear evidence that the matrix grains crystallized at a later time than those included in the garnet (see also Bea and Montero 1999). Finger and Helmy (1998) suggest that monazite grains with higher Y contents at the rims indicate crystal growth under prograde temperature conditions. Zhu and O'Nions (1999b) found significantly lower amounts of Y and heavy REE in monazite grains in rocks containing garnet as a major mineral compared to grains found in samples without garnet and those speculated to form after garnet breakdown. However, Catlos et al. (2002b) report that monazite grains from several localities in Nepal and Vietnam, dated using the ion microprobe and chemically analyzed with the electron microprobe, show no correlation between LREE and Y composition and age. For example, an eastern Nepal sample contains ~500 Ma, ~45 Ma, and ~20 Ma monazite grains, consistent the timing of Pan-African and polymetamorphic Himalayan events (Fig. 10). The ~20 Ma grains contain 0.71-1.33 wt % Y_2O_3, ~45 Ma grains have 0.97-3.36 wt %, and a ~500 Ma grain has ~1.2 wt %. Instead of relying on monazite chemical composition, possibly influenced by bulk rock composition, reactions with other accessory minerals, crystal orientation, and the monazite-forming reactant (e.g., allanite), Catlos et al. (2002b) suggest garnet zoning patterns and peak P-T conditions facilitate age interpretation. Five sources of uncertainty explain complicated age distributions from monazite grains dated *in situ*: (1) Pb loss due to prolonged experienced above the closure temperature, (2) dissolution/reprecipitation along the retrograde path, (3) analytical uncertainties, (4) analyses of overlapping age domains, and (5) episodic monazite growth. Thermobarometric data and textural observations can assess potential polymetamorphism, retrogression, and in conjunction with diffusion studies, the predicted extent of monazite Pb loss.

Case studies

Isotope dilution. The Himalayan range has largely provided the backdrop for the development and understanding of the U-Th-Pb system in monazite. A large number of studies employ isotope dilution techniques to the analysis of separates of Himalayan monazites (e.g., Schärer 1984, Copeland et al. 1988, Parrish and Tirrul 1989, Nazarchuk 1993, Hodges et al. 1996, Parrish and Hodges 1996, Searle et al. 1997, Coleman 1998, Simpson et al. 2000). Many of these studies have addressed the temporal evolution of the Greater Himalayan Crystallines and High Himalayan leucogranites, rocks found in the hanging wall of the Main Central Thrust, a large-scale structure that largely created the Himalayan range (e.g., Harrison et al. 1999b). The Greater Himalayan Crystallines are speculated to have experienced a complicated metamorphic evolution (see Pêcher and Le Fort 1986). A first stage (Eocene-Oligocene) of Barrovian-type metamorphism, termed the Eohimalayan event, corresponds to burial of the nappe beneath the southern edge of Tibet (Le Fort 1996). During this stage the base reached 650-700°C and ~0.8 GPa. During a second stage (Miocene), termed the Neohimalayan event, the base experienced 550-600°C conditions, whereas the top records lower pressures and/or temperatures. The Neohimalaya has been associated with slip along the Main Central Thrust and the development of the High Himalayan leucogranites exposed at upper structural levels of the Greater Himalayan Crystallines (e.g., Harrison et al. 1997b).

High precision can be obtained from the isotope dilution approach, but the relative youth and low U contents of the Himalayan monazites often result in discordant ages with large uncertainties (see Nazarchuk 1993, Hodges et al. 1996, Coleman 1998). For example Table 3 summarizes several isotope dilution ages of aliquots of Himalayan monazite grains. Many ages are highly precise but inconsistent with a single population (e.g., samples AS-31, AS-38, AS-23). The number of grains analyzed range from 1 to >50, and most data sets incorporate a qualitative description of the grains prior to analysis (e.g., "inclusions," "visible core," or "clear grains"). Recently, BSE images of the grains are typically incorporated as a means to speculate on the textural relationships of the dated minerals (e.g., Simpson et al. 2000). Some ages have high uncertainties (e.g. 19±72 Ma 17±66 Ma, sample DK2) due to instrumental limitations in measuring small amounts of U. Despite their problematic behavior (Fig. 2), high precision results from Himalayan monazites were primarily responsible for the identification of inheritance (e.g., Copeland et al. 1988) and unsupported daughter products from U disequilibrium (e.g., Schärer 1984, Schärer et al. 1990) and spurred the development and use of other methods to obtain more accurate information from the Himalayas.

Electron microprobe. Numerous studies date monazite using an electron microprobe (e.g., Montel et al. 1996, Bindu et al. 1998, Braun et al. 1998, Cocherie et al. 1998, Finger et al. 1998, Crowley and Ghent 1999, Williams et al. 1999, Martelat et al. 2000, Grew et al. 2002). This *in situ* (= in thin section) methodology allows the preservation of the textural relationships of the grain being dated. Montel et al. (1996) suggest the technique is feasible for >100 Ma monazite, and typical precision is ±30-50 Ma for a total counting time of ~10 min.

Applications of electron-microprobe dating of monazite include exploratory studies as a means (1) to rapidly obtain a large number of ages (Finger and Helmy 1998, Martelat et al. 2000), (2) to identify rocks that experienced a complex geological history, in which several high temperature events are superimposed (Braun et al. 1998, Cocherie et al. 1998, Williams et al. 1999), and/or (3) to obtain age constraints from samples in which textural information is crucial (Finger et al. 1998, Montel et al. 2000).

An example that illustrates the application of electron microprobe dating of monazites is the study of the Beni Bousera granulite. Montel et al. (2000) reported electron-microprobe ages for monazites in metapelites. Monazite grains included in garnet are small (5-20 μm), rare, and difficult to localize in the thin section. Although analytical uncertainties associated with electron microprobe analysis are large, matrix monazites suggest significantly younger ages (Cenozoic) than monazites included in garnet (284±27 Ma). Based on these results, Montel et al. (2000) proposed that monazites entirely included in garnet have the potential to survive granulite facies conditions without being reset.

Ion microprobe. Ion microprobe analysis preserves textural relationships and is thereby ideal for dating monazite grains found in garnet-bearing assemblages (e.g., Harrison et al. 1997a). Using results from the Smith and Giletti (1997) or Cherniak et al. (2002) monazite diffusion studies, a garnet rim P-T data and observations of grain size allow an evaluation of the amount of Pb diffusion possibly experienced by the monazite, thus combining temporal and thermobarometric constraints is a powerful method to obtain the tectonic history. Monazite inclusions in garnet may be further armored against daughter product loss because of the low solubility and permeability of Pb in the host (Montel 1999, Montel et al. 2000).

In situ Th-Pb ion microprobe ages of monazite grains found as inclusions in garnet porphyroblasts and within the deformed rock matrix from samples collected in central Nepal (Catlos et al. 2001), eastern Nepal, and northwest India (Catlos et al. 2002a) indicate the shear zone largely responsible for the creation of the Himalayas was active during the Late Miocene-Pliocene. The age data was combined with P-T estimates from their garnet-bearing assemblages to constrain and develop models for the evolution of the range (e.g., Harrison et al. 1998, Johnson et al. 2001).

Many Himalayan Th-Pb monazite age data are difficult to interpret when results from a single rock are inconsistent with a single population (see also Foster et al. 2000). Monazite growth in metapelites may be due to dissolution of existing detrital grains or allanite breakdown. The dissolution/reprecipitation process is a possible age resetting and

Pb loss mechanism (e.g., Ayers et al. 1999, Townsend et al. 2001) that can be assessed by knowing the textural relationship of the monazite, the distribution of ages within the sample, and the rock's P-T history. Some Himalayan rocks contain evidence of a polyphase history, and dissolution/reprecipitation is speculated to have a strong influence in these samples.

To interpret ion microprobe ages, the P-T history of the rock, textural relationships of the monazite with other phases, uncertainty in the calibration curve, and fraction of Pb* need to be understood. The significance of the age, including those obtained from garnet-hosted and matrix grains, can be judged using models that evaluate potential Pb loss based on temperature, duration conditions, and grain size. Uncertainty in grain size is introduced because the thin section only provides a two-dimensional view of the monazite.

Zhu and O'Nions (1999b) present an example from granulites in Scotland that illustrates the armoring effect of garnet. They report monazite inclusions in garnet that yield U-Pb ages over 1000 Ma older than matrix monazites from the same sample (Zhu and O'Nions 1999b). Catlos (2000) documents an example from the Nepal Himalaya where monazite inclusions in garnet-bearing gneisses yield older ages (late Eocene) than monazites in the matrix (late Oligocene to middle Miocene). Foster et al. (2000) dated matrix monazites (34 to 25 Ma) coexisting with monazite inclusions in garnet (44 to 36 Ma) from a Himalayan metapelite.

The above examples illustrate the importance of textural position for the behavior of the U-Th-Pb system in monazite. Geochronologic studies that employ mineral separation fail to preserve textural relationships between accessory and host phases, and therefore destroy information critical to the interpretation of monazite ages. The common occurrence of monazite as inclusions in garnet provides a unique opportunity to directly date prograde garnet growth. The age patterns revealed between monazites in the matrix of a rock and monazites included in garnet can be used to constrain the timing of garnet growth. If monazite inclusions show an age progression from core to rim of their garnet host, monazite likely grew synchronously with garnet.

U-TH-PB DATING OF APATITE

Apatite has been dated by U-Pb ion microprobe (Sano et al. 1999a,b; Sano and Terada 1999), isotope dilution techniques (e.g., Romer 1996, Corfu and Stone 1998, Chamberlain and Bowring 2000), and laser ablation-inductively coupled plasma-mass spectrometry (Willigers et al. 2002). Low U content (~<50 ppm, see Fig. 11) and high Pb° (e.g., Corfu and Stone 1998) hinders apatite U-Pb dating. Pb° issues are addressed in isotope dilution studies by using the isotopic composition of coexisting feldspar as an approximation for the initial Pb in the apatite grain (Corfu and Stone 1998, Chamberlain and Bowring 2000, Corfu and Easton 2000). Corfu and Stone (1998) suggest that there is "no insurance" that Pb in related minerals is the same as that incorporated into the apatite, but Chamberlain and Bowring (2000) advocate the method as a "good approximation…in many geologic settings". Although Corfu and Easton (2000) report 15-19% precision for U-Pb ages of Precambrian(?) apatites, Chamberlain and Bowring (2000) obtained precisions for apatite U-Pb ages varying from ±1.2% to ±0.3% for Proterozoic or Archean grains. Mougeot et al. (1997) reports a precise age of French Massif Velay anatectic dome apatite grains dated using isotope dilution of 289±5 Ma, and suggest the age times the end of the thermal event. Rakovan et al. (1997) were able to date a bulk reference apatite (American Museum of Natural History #69739) as well as its crystal faces using Sm/Nd isotope dilution techniques, and outline a methodology that could be potentially applicable to U-Th-Pb dating.

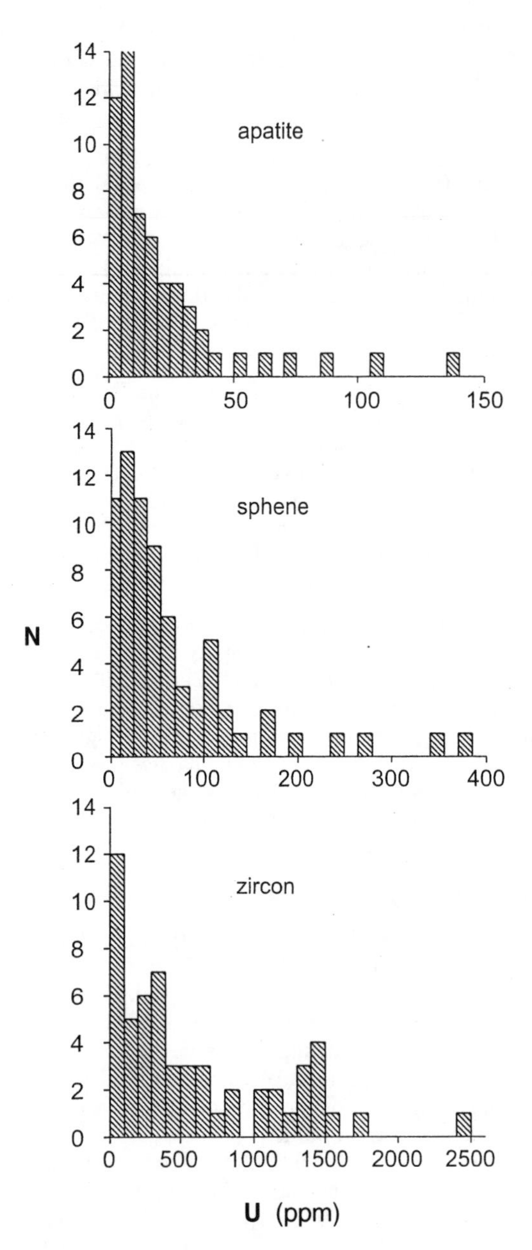

Figure 11. U concentrations (ppm) of apatite, sphene, and zircon from Romer (1996), Corfu and Stone (1998), Chamberlain and Bowring (2000), and Corfu and Easton (2000).

Sano et al. (1999a) report 3-5% precisions for apatite ages obtained using an ion microprobe. Sano and Terada (1999) used the method to date a fossil shark tooth and obtained a ^{238}U-^{206}Pb isochron age of 266±18 Ma, consistent with the highly imprecise ^{235}U-^{207}Pb age of 453±170 Ma and Th-Pb age of 235±310 Ma. Sano et al. (1999b) used the ion microprobe to date zircon and apatite grains from the Acasta Gneiss, Canada. The apatite ages are ~3 Ga younger than the zircon crystallization ages or correspond closely to the lower concordia intercept defined by some zircon analyses, suggesting that apatite records a later thermal event.

Willigers et al. (2002) report apatite and monazite ages from Paleoproterozoic metamorphic rocks from West Greenland obtained using laser-ablation inductively-coupled plasma mass spectrometry. Their ages were identical to those determined by isotope dilution, but also measured more Pb* · the solution-based isotope dilution. Willigers et al. (2002) suggest *in situ* analysis eliminates the need for mineral separation, thus limiting the potential of Pb° contamination during separation, chemical digestion and/or analysis.

U-Th-Pb DATING OF XENOTIME

Xenotime is commonly dated using isotope dilution methods (e.g., Schärer et al. 1990, Hawkins and Bowring 1997, 1999; Graessner et al. 2000), and, more recently, by ion microprobe techniques (Kamber et al. 1998, McNaughton et al. 1999, Petersson et al. 2001) and electron microprobe methods (Suzuki and Adachi 1991, Griffen et al. 2000, Asami et al. 2002, Grew et al. 2002). To aid in age interpretation, monazite is commonly analysed from the same rock or locality. Gehrels and Smith (1991) obtained a concordant age of 51±2 Ma for isotope dilution analyses of monazite and xenotime fractions collected from a Southern Arizona leucogranite, suggesting that these minerals record temporally similar closure temperature or crystallization events. Table 4 lists xenotime and monazite ages obtained by several workers. Himalayan monazites and xenotimes often yield different ages (Hodges et al. 1992, Simpson et al. 2000, Viskupic and Hodges 2001), probably because these grains were collected from lithologies that experienced complicated metamorphic histories (e.g., Le Fort 1996), thus recording multiple magmatic or metamorphic events, inheritance or Pb loss. Kamber et al. (1998) report older ages of xenotime inclusions in cordierite (2612±8 Ma) compared to those in the matrix (2551±5 Ma), and suggest that matrix grains were exposed to subsequent Pb loss via diffusion. Preservation of textural relationships of dated grains is key for deciphering the significance of the xenotime age.

Table 4. Summary of U-Pb analytical data from monazite and xenotime.[a]

Locality	*Xenotime age (Ma)*	*Monazite age (Ma)*
	$^{207}Pb/^{208}Pb$ (±2σ)	$^{207}Pb/^{208}Pb$ (±2σ)
	Hawkins & Bowring (1997)	
Grand Canyon	1676 (1)	1673 (1)
Grand Canyon	1681 (1)	1678 (2)
	Kamber et al. (1998)	
Limpopo	2612 (8)	2600 (7)
	Hawkins & Bowring (1999)	
Grand Canyon	1698 (1)	1698 (1)
	$^{207}Pb/^{235}U$ (±2σ)	$^{207}Pb/^{235}U$ (±2σ)
	Schärer et al. (1990)	
Ailao Shan	23.1 (0.3)	22.7 (0.3)
	Hodges et al. (1992)	
D33 Himalaya	21.1 (0.1)	21[b]
	Schärer et al. (1994)	
Diangcan Sahn	22.4 (0.1)	24.2 (0.2)
Jianchuan	23.0 (0.2)	22.7 (0.3)
	Hodges et al. (1996)	
Himalaya	23.4 (0.1)	35.6 (0.1)
	Schärer et al. (1996)	
Isorno-Orselina	24.7 (0.3)	26.5 (0.6)
	Simpson et al. (2000)	
21.0 (0.4)	44.2 (0.5)	
Himalaya[c]	21.0 (0.4)	20.3 (0.5)
	Viskupic & Hodges (2001)	
Himalaya	99.8 (0.4)	42.5 (0.5)
Himalaya[c]	22.3 (0.6)	24.1 (0.1)
	$^{206}Pb/^{238}U$ (±2σ)	$^{206}Pb/^{238}U$ (±2σ)
	Eliasson & Schoberg (1991)	
Bohus Granite	919 (10)	922 (10)

a. All analyses were obtained using isotope dilution methods, except Kamber et al. (1998) who analyzed the minerals using an ion microprobe.

b. Analyses by Copeland et al. (1988).

c. Average ages of only the Miocene grains.

SUMMARY

The widespread occurrence in crustal rocks of the accessory phosphate minerals apatite, monazite, and xenotime, their tendency to concentrate U and Th and exclude Pb, their generally high degree of retentivity of daughter Pb, and their resilience to radiation damage, make them excellent candidates as geochronometers. Furthermore, the restricted range of monazite stability provides a uniquely capable thermochronometer for understanding dynamic crustal processes. The various methodologies used to obtain U-Th-Pb ages each have certain attractions and limitations. Isotope dilution potentially provides high precision but with an accompanying loss of textural information. The electron microprobe is widely available and has high spatial resolution, but has poor precision and requires assumptions that cannot be rigorously tested without isotopic data. The ion microprobe has both high spatial resolution and *in situ* analysis capability but is restricted to a minimum inter-element precision of ~1%. Laser ablation inductively coupled plasma mass spectrometry promises high spatial resolution with precision potentially comparable to conventional isotope dilution studies. Depending on the application, the preferred approach is one that combines the best balance between spatial resolution, textural and sample preservation, precision, and accuracy.

REFERENCES

Ahrens LH (1955) The convergent lead ages of the oldest monazites and uraninites (Rhodesia, Manitoba, Madagascar, and Transvaal). Geochim Cosmochim Acta 7:294-300

Aldrich LT, Tilton GR, Davis GL, Nicolaysen LO, Patterson CC (1955) Comparison of U-Pb and Pb-Pb and Rb-Sr ages of Precambrian minerals. *In* Symposium on Precambrian Correlation and Dating. Vol. 7. Derry DR (ed) Canadian Geol Assoc, p 7-13

Aldrich LT, Davis GL, Tilton GR, Wetherill GW (1956) Radioactive ages of minerals from the Brown Derby Mine and the Quartz Creek Granite near Gunnison, Colorado. J Geophys Res 61:215-232

Ancey M, Bastenaire F, Tixier R (1978) Application des méthodes statistiques en microanalyse. *In* Microanalyse, microscopie electronique à balayage. Maurice F, Meny L, Tixier R (eds) Les Editions du Physicien Orsay, France, p 323-347

Asami M, Suzuki K, Grew ES (2002) Chemical Th-U-total Pb dating by electron microprobe analysis of monazite, xenotime, and zircon from the Archean Napier Complex, East Antarctica: Evidence for ultra-high-temperature metamorphism at 2400 Ma. Precambrian Res 114:249-275

Ayers JC, Miller C, Gorisch B, Milleman J (1999) Textural development of monazite during high-grade metamorphism: Hydrothermal growth kinetics, with implications for U, Th-Pb geochronology. Am Mineral 84:1766-1780

Bea F, Montero P (1999) Behavior of accessory phases and redistribution of Zr, REE, Y, Th, and U during metamorphism and partial melting of metapelites in the lower crust: An example from the Kinzigite Formation of Ivrea-Verbano, NW Italy. Geochim Cosmochim Acta 63:1133-1153

Bindu RS, Yoshida M, Santosh M (1998) Electron microprobe dating of monazite from the Chittikara Granulite, South India: Evidence for polymetamorphic events. J Geosci 41:77-83

Bingen B, Van Bremen O (1998) U-Pb monazite ages in amphibolite- to granulite-facies orthogneiss reflect hydrous mineral breakdown reactions: Sveconorwegian Province of SW Norway. Contrib Mineral Petrol 132:336-353

Black LP, Fitzgerald JD, Harley SL (1984) Pb isotopic composition, colour, and microstructure of monazites from a polymetamorphic rock in Antarctica. Contrib Mineral Petrol 85:141-148

Braun I, Montel J-M, Nicollet C (1998) Electron microprobe dating of monazites from high-grade gneisses and pegmatites of the Kerala Khondalite Belt, southern India. Chem Geol 146:65-85

Bruhn F, Moller A, Sie SH, Hensen BJ (1999) U-Th-Pb chemical dating of monazites using the proton microprobe. Nucl Instr Meth Phys Res B 158:616-620

Burger AJ, Nicolaysen LO, Ahrens LH (1967) Controlled leaching of monazites. J Geophys Res 72: 3585-3594

Burt DM (1989) Compositional and phase relations among rare earth element minerals. Rev Mineral 21:259-307

Carslaw HS, Jaeger JC (1959) Conduction of Heat in Solids, 2nd edn. Clarendon, Oxford, UK

Catlos EJ (2000) Thermobarometric and geochronologic constraints on the evolution of the Main Central Thrust, Himalayan orogen. PhD dissertation, University of California, Los Angeles, California

Catlos EJ, Gilley LD, Harrison TM (in press, 2002b) Interpretation of monazite ages obtained via *in situ* analysis. Chem Geol

Catlos EJ, Harrison TM, Kohn, MJ, Grove M, Ryerson FJ, Manning CE, Upreti BN (2001) Geochronologic and thermobarometric constraints on the evolution of the Main Central Thrust, central Nepal Himalaya. J Geophys Res 106:16177-16204

Catlos EJ, Harrison TM, Manning CE, Grove M, Rai SM, Hubbard MS, Upreti BN (2002a) Records of the evolution of the Himalayan orogen from in situ Th-Pb ion microprobe dating of monazite: Eastern Nepal and Garhwal. J Asian Earth Sci 20:459-479

Chamberlain KR, Bowring SA (2000) Apatite-feldspar U-Pb thermochronometer: A reliable, mid-range (~450°C), diffusion-controlled system. Chem Geol 172:173-200

Cheburkin AK, Frei R, Shotyk W (1997) An energy-dispersive miniprobe multielement analyzer (EMMA) for direct analysis of trace elements and chemical age dating of single mineral grains. Chem Geol 135:75-87

Cherniak DJ, Lanford WA, Ryerson FJ (1991) Pb diffusion in apatite and zircon using ion implantation and Rutherford backscattering techniques. Geochim Cosmochim Acta 55:1663-1673

Cherniak DJ, Watson EB, Grove M, Harrison TM (2002) Pb diffusion in monazite (manuscript)

Cherniak DJ, Watson EB, Harrison TM, Grove M (2000) Pb diffusion in monazite: A progress report on a combined RBS/SIMS study. EOS Trans, Am Geophys Union 81:S25

Cocherie A, Albarède F (2001) An improved U-Th-Pb age calculation for electron microprobe dating of monazite. Geochim Cosmochim Acta 65:4509-4522

Cocherie A, Legendre O, Peucat JJ, Kouamelan AN (1998) Geochronology of polygenetic monazites constrained by in situ electron microprobe Th-U-total lead determination: Implications for lead behaviour in monazite. Geochim Cosmochim Acta 62:2475-2497

Coleman ME (1998) U-Pb constraints on Oligocene-Miocene deformation and anatexis within the central Himalaya, Marsyandi Valley, Nepal. Am J Sci 298:553-571

Compston W, Williams IS, Meyer C (1984) U-Pb geochronology of zircons from Lunar Breccia 73217 using a sensitive, high mass resolution ion microprobe: Proceedings of the 14th Lunar and Planetary Science Conference, Part 2. J Geophys Res 89:8525-8534

Copeland P, Parrish RR, Harrison TM (1988) Identification of inherited radiogenic Pb in monazite and its implications for U-Pb systematics. Nature 333:760-763

Corfu F, Easton RM (2000) U-Pb evidence for polymetamorphic history of Huronian rocks within the Grenville front tectonic zone east of Sudbury, Ontario, Canada. Chem Geol 172:149-171

Corfu F, Stone D (1998) The significance of titanite and apatite U-Pb ages: Constraints for the post-magmatic thermal-hydrothermal evolution of a batholitic complex, Berens River area, northwestern Superior Province, Canada. Geochim Cosmochim Acta 62:2979-2995

Crank J (1975) The Mathematics of Diffusion. Oxford University Press, New York

Crowley JL, Ghent ED (1999) An electron microprobe study of the U-Th-Pb systematics of metamorphosed monazite: The role of Pb diffusion versus overgrowth and recrystallization. Chem Geol 157:285-302

Dahl PS (1997) A crystal-chemical basis for Pb retention and fission-track annealing systematics in U-bearing minerals, with implications for geochronology. Earth Planet Sci Lett 150:277-290

Davis DW, Krogh TE (2000). Preferential dissolution of ^{234}U and radiogenic Pb from a-recoil damages lattice sites in zircon: Implications for thermal histories and Pb-isotopic fractionation in the near-surface environment. Chem Geol 172:41-58

Deer WA, Howie RA, Zussman J (1992) An Introduction to the Rock-Forming Minerals, 2nd edn. Longman Scientific and Technical Press, Essex, UK

Deniel C, Vidal P, Fernandez A, Le Fort P, Peucat JJ (1987) Isotopic study of the Manaslu granite (Himalaya, Nepal): Inferences on the age and source of the Himalayan leucogranites. Contrib Mineral Petrol 96:78-92

DeWolf CP, Belshaw N, O'Nions RK (1993) A metamorphic history from micron-scale ^{207}Pb/^{206}Pb chronometry of Archaean monazite. Earth Planet Sci Lett 120:207-220

Dodson MH (1973) Closure temperature in cooling geochronological and petrological systems. Contrib Mineral Petrol 40:259-274

Edwards MA, Harrison TM (1997) When did the roof collapse? Late Miocene north-south extension in the High Himalaya revealed by Th-Pb monazite dating of the Khula Kangri granite. Geology 25:543-546

Eliasson T, Schoberg H (1991) U-Pb dating of the post-kinematic Sveconorwegian (Grenvillian) Bohus Granite, SW Sweden: Evidence for restitic zircon. Precambrian Res 51:337-350

Engi M, Scherrer NC, Burri T (2001) Metamorphic evolution of pelitic rocks of the Monte Rosa nappe: Constraints from petrology and single grain monazite age data. Schweiz mineral petrogr Mitt 81: 305-328

England GL, Rasmussen B, McNaughton NJ, Fletcher IR, Groves DI, Krapez B (2001) SHRIMP U-Pb ages of diagenetic and hydrothermal xenotime from the Archaean Witwatersrand Supergroup of South Africa. Terra Nova 13:360-367

Ewing RC (1994) The metamict state: 1993 the centennial. Nucl Inst Meth Phys Res B 91:22-29

Faure G (1986) Principles of Isotope Geology, 2nd ed. John Wiley and Sons, New York

Fenner CN (1928) The analytical determination of uranium, thorium, and lead as a basis for age-calculations. Am J Sci 26:369-381

Fenner CN (1932) Radioactive minerals from Divino de Uba, Brazil. Am J Sci 23:382-391

Ferry JM (2000) Patterns of mineral occurrence in metamorphic rocks. Am Mineral 85:1573-1588

Fick A (1855) Ueber diffusion. Ann Phys Chem 94:59-86

Finger F, Broska I, Roberts, MP, Schermaier A (1998) Replacement of primary monazite by apatite-allanite-epidote coronas in an amphibolite facies granite gneiss from the eastern Alps. Am Mineral 83:248-258

Finger F, Helmy HM (1998) Composition and total-Pb model ages of monazite from high-grade paragneisses in the Abu Swayel area, southern Eastern Desert, Egypt. Mineral Petrol 62:269-289

Förster HJ (1998a) The chemical composition of REE-Y-Th-U-rich accessory minerals in peraluminous granites of the Erzgebirge-Fichtelgebirge region, Germany, Part I: The monazite-(Ce)-brabantite solid solution series. Am Mineral 83:259–272

Förster HJ (1998b) The chemical composition of REE-Y-Th-U-rich accessory minerals in peraluminous granites of the Erzgebirge-Fichtelgebirge region, Germany. Part II: Xenotime. Am Mineral 83:1302–1315

Foster G, Kinny P, Vance D, Prince C, Harris N (2000) The significance of monazite U-Th-Pb age data in metamorphic assemblages: A combined study of monazite and garnet chronometry. Earth Planet Sci Lett 181:327-340

Gehrels GE, Smith CH (1991) U-Pb geochronological constraints on the age of thrusting, crustal extension, and peraluminous plutonism in the Little Rincon Mountains, southern Arizona. Geology 19:238-241

Geisler T, Schleicher H (2000) Improved U-Th-total-Pb dating of zircons by electron microprobe using a simple new background modeling procedure and Ca as a chemical criterion of fluid-induced U-Th-Pb discordance in zircon. Chem Geol 163:269-285

Gilley LD (2001) Timing of left-lateral shearing and prograde metamorphism along the Red River Shear Zone, China and Vietnam. PhD Dissertation, University of California, Los Angeles, California

Gilley LD, Harrison TM, Leloup PH, Ryerson FJ, Lovera OM, Wang J-JH (submitted, 2002) Direct dating of left-lateral deformation along the Red River shear zone, China and Vietnam. J Geophys Res

Gleadow AJW, Duddy IR, Lovering JF (1983) Fission track analysis: A new tool for the evaluation of thermal histories and hydrocarbon potential. Aust Petrol Explor Assoc J 23:93-102

Gottfried D, Jaffe HW, Senftle FE (1959) Evaluation of the lead-alpha (Larsen) method for determining ages of igneous rocks. Geol Surv Bull B 1097-A:1-63

Graessner T, Schenk V, Brocker M, Mezger K (2000) Geochronological constraints on the timing of granitoid magmatism, metamorphism, and post-metamorphic cooling in the Hercynian crustal cross-section of Calabria. J Metamorph Geol 18:409-421

Grew ES, Kazuhiro S, Masao A (2002) CHIME ages of xenotime, monazite, and zircon from beryllium pegmatites in the Napier Complex, Khmara Bay, Enderby Land, East Antarctica. Polar Geosci 14: 99-118

Griffen BJ, Forbes D, McNaughton NJ (2000) Evaluation of dating of diagenetic xenotime by electron microprobe. Microsc Microanal 6 Suppl 2:408-409

Grove M, Harrison TM (1999) Monazite Th-Pb age depth profiling. Geology 27:487–490

Harrison TM, Grove M, Lovera OM (1997b) New insights into the origin of two contrasting Himalayan granite belts. Geology 25:899-902

Harrison TM, Grove M, Lovera OM, Catlos EJ (1998) A model for the origin on Himalayan anatexis and inverted metamorphism. J Geophys Res 103:27017-27032

Harrison TM, Grove M, Lovera OM, Catlos EJ, D'Andrea J (1999b) The origin of Himalayan anatexis and inverted metamorphism: Models and constraints. J Asian Earth Sci 17:755-772

Harrison TM, Grove M, McKeegan KD, Coath CD, Lovera OM, Le Fort P (1999a) Origin and emplacement of the Manaslu intrusive complex, Central Himalaya. J Petrol 40:3-19

Harrison TM, McKeegan KD, Le Fort P (1995) Detection of inherited monazite in the Manaslu leucogranite by $^{208}Pb/^{232}Th$ ion microprobe dating: Crystallization age and tectonic implications. Earth Planet Sci Lett 133:271-282

Harrison TM, Ryerson FJ, Le Fort P, Yin A, Lovera OM, Catlos EJ (1997a) A Late Miocene-Pliocene origin for central Himalayan inverted metamorphism. Earth Planet Sci Lett 146:E1-E8

Harrison TM, Watson EB (1984) The behavior of apatite during crustal anatexis: Equilibrium and kinetic considerations. Geochim Cosmochim Acta 48:1467-1477

Hawkings DP, Bowring SA (1997) U-Pb systematics of monazite and xenotime: Case studies from the Paleoproterozoic of the Grand Canyon, Arizona. Contrib Mineral Petrol 127:87-103
Hawkings DP, Bowring SA (1999) U-Pb monazite, xenotime, and titanite geochronological constraints on the prograde to post-peak metamorphic thermal history of Paleoproterozoic migmatites from the Grand Canyon, Arizona. Contrib Mineral Petrol 134:150-169
Hodges KV, Parrish RR, Housh TB, Lux DR, Burchfiel BC, Royden LH, Chen Z (1992) Simultaneous Miocene extension and shortening in the Himalayan orogen. Science 258:1466-1470
Hodges KV, Parrish RR, Searle MP (1996) Tectonic evolution of the central Annapurna Range, Nepalese Himalayas. Tectonics 15:1264-1291
Holmes A (1911) The association of lead with uranium in rock-minerals and its application to the measurement of geological time. Proc Roy Soc London A 85:248-256
Holmes A (1954) The oldest dated minerals of the Rhodesian Shield. Nature 173:612
Holmes A (1955) Dating the Precambrian of peninsular India and Ceylon. Proc Geol Soc Canada 7:81-106
Holmes A, Cahen L (1955) African geochronology. Col Geol Min Res 5:3-38
Holmes A, Smales AA, Leland WT, Nier AO (1949) The age of uraninite and monazite from the post-Delhi pegmatites of Rajputana. Geol Mag 86:288-302
Iskanderova AD, Legiyerskiy Y (1966) Use of apatite for determination of the absolute age of geological formations by the lead-isotope method. Akad Nauk SSSR Kom Opred Absol Vozrasta Geol Form Tr 13:444-448
Johnson MRW, Oliver GJH, Parrish RR, Johnson SP (2001) Synthrusting metamorphism, cooling, and erosion of the Himalayan Kathmandu complex, Nepal. Tectonics 20:394-415
Kamber BS, Frei R, Gibb AJ (1998) Pitfalls and new approaches in granulite chronometry: An example from Limpopo Belt, Zimbabwe. Precambrian Res 91:269-285
Karioris FG, Gowda K, Cartz L (1981) Heavy ion bombardment on monoclinic $ThSiO_4$, ThO_2, and monazite. Radiat Eff Lett 58:1-3
Kingsbury JA, Miller CF, Wooden JL Harrison TM (1993) Monazite paragenesis and U-Pb systematics in rocks of the eastern Mojave Desert, California, USA: Implications for thermochronometry. Chem Geol 110:147-167
Kohn MJ, Catlos EJ, Ryerson FJ, Harrison TM (2001) P-T-t path discontinuity in the MCT Zone, Central Nepal. Geology 29:571-574
Köppel V, Grünenfelder M (1975) Concordant U-Pb ages of monazite and xenotime from the central Alps and the timing of the high temperature Alpine metamorphism, a preliminary report. Schweiz mineral petrogr Mitt 55:129-132
Köppel V, Gunthert A, Grünenfelder M (1980) Patterns of U-Pb zircon and monazite ages in polymetamorphic units of the Swiss Central Alps. Schweiz mineral petrogr Mitt 61:97-119
Kosler J, Tubrett MN, Sylvester PJ (2001) Application of laser ablation ICP-MS to U-Th-Pb dating of monazite. Geostand Newslett 25:375-386
Krogstad EJ, Walker RJ (1994) High closure temperatures of the U-Pb system in large apatites from the Tin Mountain pegmatite, Black Hills, South Dakota, USA. Geochim Cosmochim Acta 58:3845-3853
Larsen ES, Keevil NB, Harrison HC (1949) Preliminary report on determining the age of rocks by the lead-uranium ration of zircon, apatite, and sphene from the rocks using alpha counting and spectrographic methods. *In* Report of the Committee on Measurement of Geologic Time. National Research Council, Div Geol Geogr Ann Rep E, 1947-1948. Marble JP (ed) American Geological Institute, p 27-28
Le Fort P (1996) Evolution of the Himalaya. *In* The Tectonic Evolution of Asia. Yin A, Harrison TM (eds) Cambridge University Press, New York, p 95-109
Lyakhovich VV (1961) Accessory minerals and the absolute age of igneous rocks. Trudy Inst Mineral Geokhim Krisallokhim Redkikh Elementov 7:212-225
Machado N, Gauthier G (1996) Determination of $^{207}Pb/^{206}Pb$ ages on zircon and monazite by laser-ablation ICPMS and application to a study of sedimentary provenance and metamorphism in southeastern Brazil. Geochim Cosmochim Acta 60:5063-5073
Marble JP (1935) Possible age of monazite from Mars Hill, North Carolina. Am Mineral 21:724-732
Martelat J-E, Lardeaux J-M, Nicollet C, Rakotondrazafy R (2000) Strain pattern and late Precambrian deformation history in southern Madagascar. Precambrian Res 102:1-20
Mattinson JM (1994) A study of complex discordance in zircons using step-wise dissolution techniques. Contrib Mineral Petrol 116:117-129
McNaughton NJ, Rasmussen B, Fletcher IR (1999) SHRIMP uranium-lead dating of diagenetic xenotime in siliclastic sedimentary rocks. Science 285:78-80
Meldrum A, Boatner LA, Ewing RC (1997b) Displacive radiation effects in the monazite- and zircon-structure orthophosphates. Phys Rev B 56:13805-13814
Meldrum A, Boatner LA, Ewing RC (1997c) Electron-irradiation-induced nucleation and growth in amorphous $LaPO_4$, $ScPO_4$, and zircon. J Mater Sci 12:1816-1827

Meldrum A, Boatner LA, Wang LM, Ewing RC (1997a) Ion-beam-induced amorphisation of $LaPO_4$ and $ScPO_4$. Nucl Instr Meth B 127/128:160-165

Meldrum A, Boatner LA, Weber WJ, Ewing RC (1998) Radiation damage in zircon and monazite. Geochim Cosmochim Acta 62:2509-2520

Meldrum A, Wang LM, Ewing RC (1996) Ion beam induced amorphization of monazite. Nucl Instr Meth Phys Res B 116:220-224

Michot J, Deutsch S (1970) U-Pb zircon ages and polycylism of the Gneiss de Brest and the adjacent formations (Brittany). Eclog Geol Helv 63:215-227

Montel J-M (1993) A model for monazite/melt equilibrium and application to the generation of granitic magmas. Chem Geol 110:127-146

Montel J-M (1999) Some good reasons for monazite to be concordant. J Conf Abstr EUG 10 4:800

Montel JM, Devidal JL, Avignant, DC (2002, in press) X-ray diffraction study of brabantite-monazite solid solution. Chem Geol

Montel J-M, Foret S, Veschambrem M, Nicollet C, Provost A (1996) Electron microprobe dating of monazite. Chem Geol 131:37-53

Montel J-M, Kornprobst J, Vielzeuf D (2000). Preservation of old U-Th-Pb ages in shielded monazite: Example from Beni Bousera Hercynian kinzigites (Morocco). J Metamorph Geol 18:335-342

Montel J-M, Veschambrem M, Nicollet C (1994) Datation de la monazite à la microsonde électronique. C R Acad Sci Paris 318:1489

Mougeot R, Respaut JP, Ledru P, Marignac C (1997) U-Pb geochronology on accessory minerals of the Velay anatectic Dome (French Massif Central). Eur J Mineral 9:141-156

Nazarchuk JH (1993) Structure and geochronology of the Greater Himalaya, Kali Gandaki region, west-central Nepal. Masters Thesis, Carleton University, Ottawa, Ontario

Ni Y, Hughes JM, Mariano A (1995) Crystal chemistry of the monazite and xenotime structures. Am Mineral 80:21-26

Nier AO (1939) The isotopic constitution of radiogenic lead and the measurement of geologic time, II. Phys Rev 55:153-163

Nier AO, Thompson RW, Murphy BF (1941) The isotopic composition of lead and the measurement of geologic time, III. Phys Rev 60:789-793

Olsen SN, Livi K (1998) Dating of monazite from migmatites in the Aar Massif, Swiss Alps, by electron microprobe analyses. Abstr Geol Soc Am 30:231

Ouchani S, Dran JC, Chaumont J (1997) Evidence of ionization annealing upon helium-irradiation of pre-damaged fluorapatite. Nucl Instr Meth Phys Res B 132:447-451

Overstreet WC (1967) The geologic occurrence of monazite. U S Geol Surv Prof Pap 530:1-327

Paquette JL, Montel J-M, Chopin C (1999) U-Th-Pb dating of the Brossasco ultrahigh-pressure metagranite, Dora-Maira massif, western Alps. Eur J Mineral 11:69-77

Parrish RR (1990) U-Pb dating of monazite and its application to geological problems. Can J Earth Sci 27:1431-1450

Parrish RR, Armstrong RL (1987) The ca. 162 Ma Galena Bay Stock and its relationship to the Columbia River fault zone, southeast British Columbia. Radiogenic age and isotopic studies, Geol Surv Canada Rep 1 87-2:25-32

Parrish RR, Hodges KV (1996) Isotopic constraints on the age and provenance of the Lesser and Greater Himalaya sequences, Nepalese Himalaya. Geol Soc Am Bull 108:904-911

Pêcher A, Le Fort P (1986) The metamorphism in central Himalaya, its relations with the thrust tectonic. *In* Évolution des domains orogéniques d'Asie méridionale (de la Turquie à l'Indoneasie) Le Fort P, Colchen M, Montenat C (eds) Science de la Terre 47, p 285-309

Parrish RR, Tirrul R (1989) U-Pb age of the Baltoro granite, northwest Himalaya, and implications for monazite U-Pb systematics. Geology 17:1076-1079

Peiffert C, Cuney M (1999) hydrothermal synthesis of the complete solid solution between monazite ($LaPO_4$) and huttonite ($ThSiO_4$) at 780°C and 200 MPa. J Conf Abstr EUG 10 4:522

Petersson J, Whitehouse MJ, Eliasson T (2001) Ion microprobe U-Pb dating of hydrothermal xenotime from an episyenite: Evidence for rift-related reactivation. Chem Geol 175:703-712

Podor R, Cuney M (1997) Experimental study of Th-bearing $LaPO_4$ (780°C, 200 MPa): Implication for monazite and actinide orthophosphate stability. Am Mineral 82:765-771

Podor R, Cuney M, Nguyen TC (1995) Experimental study of the solid solution between monazite-(La) and $Ca_{0.5}U_{0.5}PO_4$ at 780°C and 200 MPa. Am Mineral 80:1261-1268

Poitrasson F, Chenery S, Bland DJ (1996) Contrasted monazite hydrothermal alteration mechanisms and their geochemical implications. Earth Planet Sci Lett 145:79-96

Poitrasson F, Chenery S, Shepherd T (2000) Electron microprobe and LA-ICP-MS study of monazite hydrothermal alteration: Implications for U-Th-Pb geochronology and nuclear ceramics. Geochim Cosmochim Acta 64:3283-3297

Pyle J, Spear FS (1999) Yttrium zoning in garnet: Coupling of major and accessory phases during metamorphic reactions. Geol Mater Res 1:1-49
Quarton M, Zouiri M, Freundlich W (1994) Cristallochimie des orthophosphates doubles de thorium et de plomb. C R Acad Sci Paris 229:785-788
Rakovan J, McDaniel DK, Reeder R (1997) Use of surface-controlled REE sectoral zoning in apatite from Llallagua, Bolivia, to determine a single-crystal Sm-Nd age. Earth Planet Sci Lett 146:329-336
Rapp RP, Watson EB (1986) Monazite solubility and dissolution kinetics: Implications for the thorium and light rare-earth chemistry of felsic magmas. Contrib Mineral Petrol 94:304-316
Rasmussen B, Fletcher IR, McNaughton NJ (2001) Dating low-grade metamorphic events by SHRIMP U-Pb analysis of monazite in shales. Geology 29:963–966
Rhede D, Wendt I, Forster H-J (1996) A three-dimensional method for calculating independent chemical U/Pb- and Th/Pb-ages of accessory minerals. Chem Geol 130:247-253
Robinson A, Yin A, Manning CE, Harrison TM, Hei W, Xiong MY, Feng, WX (2001) Geochronologic, thermochronologic, and thermobarometric constraints on the tectonic evolution of the northeastern Pamir. EOS Trans, Am Geophys Union 82:T11E-0888
Romer RL (1996) U-Pb systematics of stilbite-bearing low-temperature mineral assemblages from the Malmberget iron ore, northern Sweden. Geochim Cosmochim Acta 60:1951-1961
Ryerson FJ (1987) Diffusion Measurements: Experimental Methods. *In* Methods of Experimental Geophysics. Vol. 24. Sammis CG, Henyey T (eds) Academic Press, New York, p 89-129
Sano Y, Oyama T, Terada K, Hidaka H (1999a) Ion microprobe dating of apatite. Chem Geol 153:249-258
Sano Y, Terada K (1999) Direct ion microprobe U-Pb dating of fossil tooth of a Permian shark. Earth Planet Sci Lett 174:75-80
Sano Y, Terada K, Hidaka H, Yokoyama K, Nutman AP (1999b) Palaeoproterozoic thermal events recorded in the ~4.0-Ga Acasta gneiss, Canada: Evidence from SHRIMP U-Pb dating of apatite and zircon. Geochim Cosmochim Acta 63:899-905
Schärer U (1984) The effect of initial ^{230}Th disequilibrium on young U-Pb ages: The Makalu case, Himalaya. Earth Planet Sci Lett 67:191-204
Schärer U, Cosca M, Steck A, Hunziker J (1996) Termination of major ductile strike-slip shear and differential cooling along the Insubric line (central Alps): U-Pb, Rb-Sr, and ^{40}Ar/^{39}Ar ages of cross-cutting pegmatites. Earth Planet Sci Lett 142:331-351
Schärer U, Tapponier P, Lacassin R, Leloup PH, Dalai Z, Shaocheng J (1990) Intraplate tectonics in Asia: A precise age for large-scale Miocene movement along the Ailao Shan-Red River shear zone, China. Earth Planet Sci Lett 97:65-77
Schärer U, Xu RH, Allegre CJ (1986) U-(Th)-Pb systematics and ages of Himalayan leucogranites, South Tibet. Earth Planet Sci Lett 77:35-48
Schärer U, Zhang LS, Tapponier P (1994) Duration of strike-slip movements in large shear zones: The Red River belt, China. Earth Planet Sci Lett 126:379-397
Scherrer NC, Engi M, Gnos E, Jakob V, Liechti A (2000) Monazite analysis: From sample preparation to microprobe age dating and REE quantification. Schweiz mineral petrogr Mitt 80:93-105
Searle MP, Parrish RR, Hodges KV, Hurford A, Ayers MW, Whitehouse MJ (1997) Shisha Pangma leucogranite, South Tibetan Himalaya: Field relations, geochemistry, age, origin, and emplacement. J Geol 105:295-317
Seydoux-Guillaume AM, Wirth R, Nasdala L, Gottschalk M, Montel J-M, Heinrich W (2002) An XRD, TEM and Raman study of experimentally annealed natural monazite. Phys Chem Min 29:240-253
Shestakov GI (1969) On diffusional loss of lead from a radioactive mineral. Trans Geok 9:1103-1111
Simpson RL, Parrish RR, Searle MP, Waters DJ (2000) Two episodes of monazite crystallization during metamorphism and crustal melting in the Everest region of the Nepalese Himalaya. Geology 28: 403-406
Sivaramakrishnan V, Venkatasubramaniam VS (1959) Ages of some detrital monazites by the lead-alpha method of geochronology. Proc Nat Inst Sci India A Phys Sci 25:278-280
Smith HA, Barreiro B (1990) Monazite U-Pb dating of staurolite grade metamorphism in pelitic schists. Contrib Mineral Petrol 105:602-615
Smith HA, Giletti BJ (1997) Lead diffusion in monazite. Geochim Cosmochim Acta 61:1047-1055
Spear FS (1993) Metamorphic Phase Equilibria and Pressure-Temperature-Time Paths. Mineralogical Society of America Monograph, Washington, DC
Spear FS, Parrish RR (1996) Petrology and cooling rates of the Valhalla complex, British Columbia, Canada. J Petrol 37:733-765
Stacey JS, Kramers JD (1975) Approximate of terrestrial lead isotope evolution by a two-stage model. Earth Planet Sci Lett 26:207-221
Stern RA, Berman RG (2000) Monazite U-Pb and Th-Pb geochronology by ion microprobe, with an application to in situ dating of an Archean metasedimentary rock. Chem Geol 172:113-130

Stern RA, Sanborn N (1998) Monazite U-Pb and Th-Ph geochronology by high-resolution secondary ion mass spectrometry. *In* Radiogenic age and isotopic studies, Report 11, Curr Res Geol Surv Canada, Ottawa, 1998-F, p 1-18

Suzuki K, Adachi M (1991) Precambrian provenance and Silurian metamorphism of the Tsunosawa paragneiss in the South Kitakami terrane, northeast Japan, revealed by the chemical Th-U-total Pb isochron ages of monazite, zircon and xenotime. J Geochem 25:357

Suzuki K, Adachi M (1994) Middle Precambrian detrital monazite and zircon from the Hida gneiss on Oki-Dogo Island, Japan: Their implications for the correlation of basement gneiss of Southwest Japan and Korea. Tectonophysics 235:277-292

Suzuki K, Adachi M, Kajizuka I (1994) Electron microprobe observations of Pb diffusion in metamorphosed detrital monazites. Earth Planet Sci Lett 128:391-405

Terry MP, Robinson P, Hamilton MA, Jercinovic, MJ (2000) Monazite geochronology of UHP and HP metamorphism, deformation, and exhumation, Nordoyane, Western Gneiss Region, Norway. Am Mineral 85:1651-1664

Tilton GR (1960) Volume diffusion as a mechanism for discordant lead ages. J Geophys Res 65:2933-2945

Tilton GR, Nicolaysen LO (1957) The use of monazites for age determination. Geochim Cosmochim Acta 11:28-40

Tilton GR, Patterson CC, Brown H, Inghram M, Hayden R, Hess D, Larsen E (1955) Isotopic composition and distribution of lead, uranium and thorium in Precambrian granite (Ontario). Geol Soc Am Bull 66:1131-1148

Townsend KJ, Miller CF, D'Andrea JL, Ayers JC, Harrison TM, Coath CD (2001) Low temperature replacement of monazite in the Ireteba granite, Southern Nevada: geochronological implications. Chem Geol 172:95-112

Van Emden B, Graham J, Lincoln FJ (1997). The incorporation of actinides in monazite and xenotime from placer deposits in western Australia. Can Mineral 35:95-104

Vinogradov AP, Tugarinov AI, Zykov SI, Stupnikova NI, Bibikova YV, Knorre KG, Mel'nikova GL (1966) Geochronology of the Precambrian of India. Akad Nauk SSSR Kom Opred Absol Vozrasta Geol Form Tr 13:394-408

Viskupic K, Hodges KV (2001) Monazite-xenotime thermochronometry: Methodology and an example from the Nepalese Himalaya. Contrib Mineral Petrol 141:233-247

Wang JW, Tatsumoto M, Li X, Permo W, Chao ECT (1994) A precise Th-232-Pb-208 chronology of fine-grained monazite age of the Bayan Obo REE-Fe-Nb ore deposit, China. Geochim Cosmochim Acta 58:3155-3169

Watson EB, Harrison TM, Ryerson FJ (1985) Diffusion of Sm, Sr and Pb in fluorapatite. Geochim Cosmochim Acta 49:1813-1823

Wetherill GW (1956) Discordant Uranium-Lead ages, I. EOS Trans, Am Geophys Union 37:320-326

White NM, Parrish RR, Bickle MJ, Najman YMR, Burbank D, Maithani A (2001) Metamorphism and exhumation of the NW Himalaya constrained by U-Th-Pb analyses of detrital monazite grains from early foreland basin sediments. J Geol Soc 158:625-635

Williams ML, Jercinovic MJ, Terry MP (1999) Age mapping and dating of monazite on the electron microprobe: Deconvoluting multistage tectonic histories. Geology 27:1023-1026

Williams ML, Jercinovic MJ (2002) Microprobe monazite geochronology: Putting absolute time into microstructural analysis. J Struct Geol 24:1013-1028

Willigers BJA, Baker JA, Krogstad EJ, Peate DW (2002) Precise and accurate in situ Pb-Pb dating of apatite, monazite, and sphene by laser ablation multiple-collector ICP-MS. Geochim Cosmochim Acta 66:1051-1066

Wing BA, Ferry JM, Harrison TM (submitted, 2002) Prograde destruction and formation of monazite and allanite during contact and regional metamorphism of pelites: Petrology and geochronology. Contrib Mineral Petrol

Zhu XK, O'Nions RK (1999a) Zonation of monazite in metamorphic rocks and its implications for high temperature thermochronology: A case study from the Lewisian terrain. Earth Planet Sci Lett 171: 209-220

Zhu XK, O'Nions RK (1999b) Monazite chemical composition: Some implications for monazite geochronology. Contrib Mineral Petrol 137:351-363

Zhu XK, O'Nions RK, Belshaw NS, Gibb AJ (1997) Lewisian crustal history from in situ SIMS mineral chronometry and related metamorphic textures. Chem Geol 136:205-218

Zhu XK, O'Nions RK, Gibb AJ (1998) SIMS analysis of U-Pb isotopes in monazite: Matrix effects. Chem Geol 144:305-312

15 (U-Th)/He Dating of Phosphates: Apatite, Monazite, and Xenotime

Kenneth A. Farley

Division of Geological and Planetary Sciences
California Institute of Technology
Pasadena, California 91125

Daniel F. Stockli

Department of Geology
University of Kansas
Lawrence, Kansas 66045

INTRODUCTION

The common phosphate minerals, apatite $Ca_5(PO_4)_3(F,OH,Cl)$, monazite $(Ce,La,Th)PO_4$, and xenotime YPO_4, have found widespread use in geochronology because they incorporate U and Th into their structures. For example, apatite usually has a few tens of ppm of both U and Th, while monazite and xenotime usually have hundreds of ppm to weight percent levels of these elements. As a consequence, these phosphates can be dated using several fundamentally different isotopic techniques. Elsewhere in this volume Harrison et al. describe phosphate dating using ingrowth of radiogenic Pb, the final daughter of U and Th series decay, and Gleadow et al. describe dating based on damage tracks from the spontaneous fission of ^{238}U. The most recently developed dating technique applied to phosphates, described in this chapter, uses the accumulation of α particles from U and Th series decay, (U-Th)/He dating. While phosphate U-Th-Pb dating is usually used to date high temperature events such as crystallization of igneous rocks and the timing of prograde metamorphism, fission track and (U-Th)/He dating are more commonly used to establish cooling histories through low temperatures, for example, in the range ~110-40°C in the case of apatite.

Dating of minerals using radiogenic He was first explored shortly after the discovery of radioactivity (Strutt 1908) and was investigated extensively in the 1950s and 1960s, mostly on very U- and Th-rich minerals such as zircon and titanite (Hurley 1952, 1954; Damon and Kulp 1957, Damon and Green 1963). Apatite He dating was first investigated by Zeitler et al. (1987), who studied the diffusion rate of He from apatite and proposed that apatite He dating might provide a useful thermochronometer, recording cooling through about 100°C. Further studies both in the laboratory (Lippolt et al. 1994, Wolf et al. 1996b, Warnock et al. 1997, Farley 2000) and in the natural setting (House et al. 1999, Stockli et al. 2000) have confirmed this idea, and the technique has now been applied to a range of tectonic, geologic and geomorphologic problems (e.g., House et al. 1997, 1998; Spotila et al. 1997, Farley et al. 2001, Stockli et al. 2000). Monazite and xenotime have only recently come under scrutiny for He geochronology (e.g., Pik and Marty 1999), and little is yet known of their potential for routine geo- or thermochronometry.

DETAILS OF THE METHOD

The age equation

^{238}U, ^{235}U, and ^{232}Th decay through a series of short-lived radionuclides ultimately to Pb isotopes, along with 8, 7, and 6 α particles, respectively. Additional ^4He is contributed by the α decay of ^{147}Sm. Under conditions of secular equilibrium among the short-lived actinide daughters, the rate of ^4He ingrowth is:

1529-6466/00/0048-0015$05.00

$$d^4He/dt = 8\,\lambda_{238}\,^{238}U + 7\lambda_{235}\,^{235}U + 6\,\lambda_{232}\,^{232}Th + \lambda_{147}\,^{147}Sm \quad (1)$$

where λs are the decay constants. In most minerals the contribution of ^4He from ^{147}Sm is small compared to that from U and Th. Based on published tabulations of the chemistry of apatite, monazite, and xenotime (e.g., see Van Emden et al. 1997, Sabourdy et al. 1997, Sha and Chappell 1999), ^{147}Sm typically contributes <1% of the total He indicated by Equation (1). This contribution is small compared to the typical precision of He ages of a few percent (Farley et al. 2001), so it is common to ignore the Sm-derived component. However in some rare-earth-rich, U and Th poor phosphates it may be necessary to account for this α source. Because secular equilibrium among the U-series daughters is assured within ~350,000 years of crystallization, Equation (1) is generally valid except in the case of rocks with young crystallization ages. In young rocks, such as volcanics, the effects of secular disequilibrium on He ages can be large (Farley et al. 2002).

By measuring U, Th, and He contents, the He apparent age t can be calculated iteratively (ignoring Sm) from:

$$^4He = 8\,^{238}U\,(e^{\lambda_{238}t} - 1) + 7\,^{238}U/137.88\,(e^{\lambda_{235}t} - 1) + 6\,^{232}Th\,(e^{\lambda_{232}t} - 1) \quad (2)$$

where the factor of 137.88 is the $^{238}U/^{235}U$ ratio of natural uranium. This expression assumes an initial He content of zero in the dated grain. Unlike argon, helium is found in such low concentration in the atmosphere (5 ppm vs. 1% for ^{40}Ar) that air-derived He is unlikely to be present in most minerals. Of greater concern is excess He in fluid or mineral inclusions. Excess He in fluid inclusions has been reported in several cases, especially in hydrothermal phosphates (Lippolt et al. 1994, Stockli et al. 2000). Mineral inclusions are a more common problem, and are discussed in detail below. Visual screening for inclusions is important to eliminate these sources of excess He.

Effects of long α-stopping distances

α decay of actinides releases a large amount of energy which is taken up mostly in the form of motion: α-recoil of the parent nucleus, and energetic emission of the α particle. This presents a complication for the He dating method in that α particles may be ejected out of the crystal being dated, or, alternatively, injected into a crystal from decay occurring in surrounding grains. The distance required for α particles to come to rest depends on decay energy and mineral density and composition (Table 1, see also Farley et al. 1996). In apatite, the mean stopping distance in the ^{238}U-series is ~19.3 μm, while it is ~16.6 and ~16.0 μm in monazite and xenotime, respectively. The shorter stopping distance in xenotime compared with monazite reflects the greater stopping power of lighter relative to heavier elements (i.e., of Y vs. Ce). Because the mean decay energy is higher, the stopping distances are longer in the Th series: 22.8 19.0, and 18.3 μm in apatite, monazite and xenotime respectively. Stopping distances vary inversely with mineral density. The distances in Table 1 assume nominal densities, and so represent best estimates of the stopping distances in common samples. The reported range of densities in apatite (~±3%) and monazite (~±4%) (Chang et al. 1996) indicate that these nominal stopping distances are good to about ±4%. While it might be desirable to use actual densities of dated grains to estimate stopping distances, in practice measuring the density of individual grains might be quite challenging. As a result variations in density will introduce errors in the α-ejection correction discussed below. In most cases the resulting error is likely to be <1% (see Eqns. 3 and 4).

Phosphate specimens for dating are usually only ~3 to 30 times larger than these stopping distances, and as a result the "skin effect" of ejection or implantation can have a significant influence on He ages computed from Equation (2). U and Th tend to be concentrated in phosphates compared to surrounding minerals, so the net consequence of this α redistribution (relative to the parent nuclides) will usually be ejection. Put differently, U and Th located near grain boundaries cannot contribute the expected number of α particles to the crystal being dated (Fig. 1). Were this effect not taken into consideration, He ages of minerals that concentrate U and Th relative to the surrounding matrix would invariably underestimate the true age.

Table 1. Alpha stopping distances.

	Fluorapatite $Ca_5(PO_4)_3F$	***Monazite*** $CePO_4$	***Xenotime*** YPO_4
Assumed density (g/cm^3)	3.2	5.1	4.5
	Stopping distance (μm)*		
Mean ^{238}U Chain	19.3	16.6	16.0
Mean ^{235}U Chain	22.4	19.3	18.6
Mean ^{232}Th Chain	22.1	19.0	18.3

** Stopping distances revised from Farley et al. (1996) based on range data in ICRU (1993).*

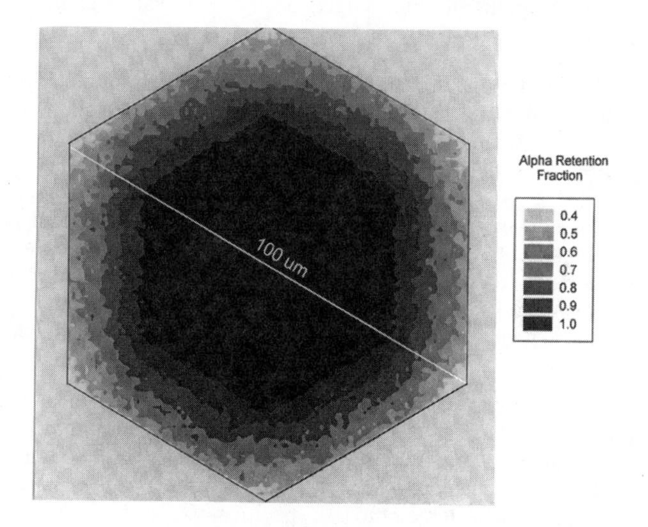

Figure 1. Results of a Monte Carlo simulation showing the spatial distribution of the fraction of ^{238}U decays leading to α retention in a hexagonal prism of apatite (see Farley et al. 1996 for model details). In the interior every α particle is retained, but near the edges the probability that the α particle will be launched on a trajectory leading to the outside of the crystal increases. Near the apices almost 60% of the α particles are ejected. The zone of ejection is one stopping distance wide, or about 20 μm. This model assumes an infinitely long prism.

α ejection can be handled in several different ways. Most obviously, a ~20 μm thick outer shell could be removed from the grain either physically or chemically. This approach would be most appropriate if the local He/(U+Th) ratio (i.e., the "age") of the residual core differs from that of the entire crystal only by the ejection effect. However, in the cases so far investigated (apatite, titanite) the He diffusion domain corresponds to the physical grain (Farley 2000, Reiners and Farley 1999), and as a result diffusional loss causes the grain edges to have lower He/(U+Th) ratios than the core, independent of ejection. Thus the He age of a grain from which the outer surfaces were removed would be older than the He age of the entire crystal. In most cases it is the age of the entire crystal which is sought, for example when interpreting cooling ages in terms of Dodson's (1973) closure temperature concept. In the case of rapidly cooled samples this approach might be feasible because the discrepancy between the age of the core and the age of the entire crystal would be small, but in slowly cooled samples the discrepancy can be large (Reiners and Farley 1999).

Quantitative modeling of α ejection is an alternative approach (Farley et al. 1996). The object of the modeling is to calculate what fraction (F_T) of the U and Th in a crystal is capable of yielding α particles that stop within the crystal. The importance of this quantity is that the measured U and Th must be reduced by this fraction when calculating a He age. Equivalently, an α ejection corrected age (t_{corr}) can be computed by increasing the age from Equation (2) by a factor of $1/F_T$:

$$t_{corr} = t\,(1/F_T) \quad (3)$$

F_T depends on the stopping distances of the α particles (hence on both the mineral of interest and its relative Th and U abundance), the size and geometry of the crystal, and how the parent nuclides are distributed. For homogeneously distributed parent nuclides in a sphere of radius r with stopping distance s it has been shown (Farley et al. 1996) that:

$$F_T = 1 - 0.75\,(s/r) + 0.0625\,(s/r)^3 \quad (4)$$

F_T is plotted as a function of radius for the three phosphates of interest in Figure 2a. The important feature here is the large magnitude of the α-ejection correction for common phosphate grain sizes. For example, an apatite sphere of radius 75 µm has $F_T \approx 0.80$ and the age calculated from Equation (2) must be corrected upward by 25% (i.e., 1/0.8) according to Equation (3). A sphere is not a particularly relevant geometry for real crystals, so Figure 2b shows results of numerical modeling for the hexagonal prism geometry common to apatite. In our experience apatite crystals typically have a length to radius ratio of ~6 and range from ~30 to ~150 µm in prism half width. Across this size range F_T increases from about 0.5 to about 0.9. Ordinary monazite and xenotime crystals are similarly small; thus in general phosphate He ages will require large corrections for the ejection of α particles.

Correction equations similar to (4) but valid for cylinders, hexagonal prisms, and cubes have been published (Farley et al. 1996). These equations are appropriate for some phosphates (especially apatite), but may not apply to others, e.g., to commonly low-symmetry crystals of monazite. An approximation that may be useful for low symmetry minerals relates F_T to surface to volume ratio (β), specifically:

$$F_T \approx 1 - (s/4)\beta \quad \text{valid for } \beta > 0.07 \qquad \text{(Farley et al. 1996)} \quad (5)$$

For $\beta < 0.07$, this approximation generally underestimates the true F_T value (Fig. 2a). Meesters and Dunai (2002b) present alternative means for assessing the effects of α ejection for various geometries.

At present, the practical solution to the α-ejection problem is to measure the physical dimensions of the dated grains prior to analysis, and correct resulting ages based on the computed F_T values (Farley et al. 1996). While this correction is based on well-understood physical phenomena, two critical assumptions are required. The first is that the grain being dated includes all of the original grain surfaces. For this reason it is critical to analyze samples that retain their original size and geometry. Fortunately many phosphate samples separate wholly and cleanly from the surrounding matrix, making the correction tractable. In our experience many samples yield apatite grains that are entire hexagonal prisms or prisms broken across the long axis of the crystal. In the latter case it is impossible to know the original grain length, but as shown in Figure 2b the F_T correction is not very sensitive to this length. The α-ejection correction also presents a challenge for dating of physically rounded detrital grains. If rounding occurs prior to substantial He accumulation, then the F_T modeling remains valid provided an appropriate geometry can be established. However, when significant He accumulation occurs prior to or during rounding, these models will yield inaccurate F_T values.

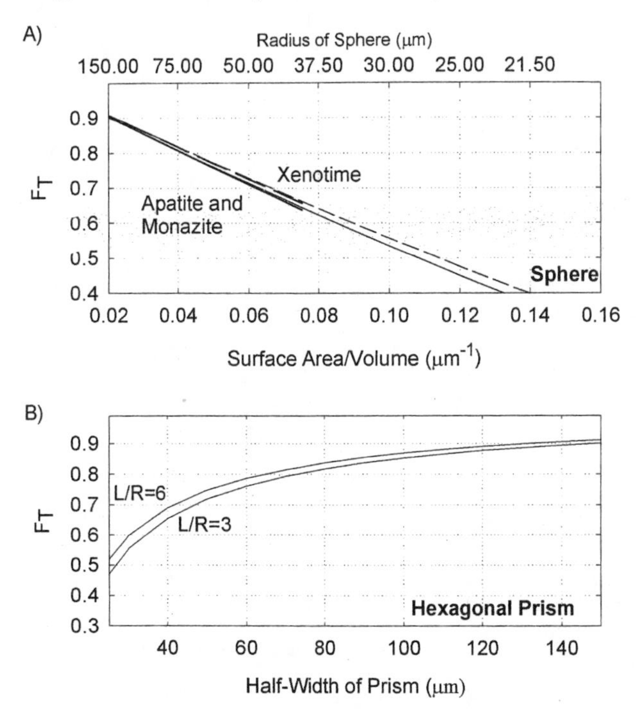

Figure 2. Results of α-ejection modeling, after Farley et al. (1996). F_T is fraction of α particles retained in an entire crystal. (A) shows results calculated for a sphere (narrow lines, radius indicated at top) of apatite, monazite, or xenotime composition. Bold lines give F_T approximation for arbitrary geometries of these minerals based on surface to volume ratio. (B) F_T for an apatite hexagonal prism as a function of half the distance between opposed apices, for two length/radius (L/R) ratios. Note the relative insensitivity to this quantity. For apatite the mean ^{238}U stopping distance was assumed, reflecting the dominant source of α particles in apatite. For monazite and xenotime the mean ^{232}Th stopping distance was used since the Th/U ratio of these minerals is high.

A second assumption of the F_T models is that the parent nuclide distribution is homogeneous. If U and Th concentrate on the rim of a grain, the F_T value computed, e.g., from Equation (4), will be too high. The opposite applies when U and Th concentrate deep in the grain interior. Equations for calculating the fraction of retained α's for inhomogeneous parent distributions have been presented elsewhere (Farley et al. 1996). Figure 3 illustrates several examples, computed assuming a 75-μm radius apatite sphere and ^{238}U series decay. In the absence of zonation, this sphere would have $F_T = 0.81$ based on Equation (4). In the worst cases of zonation, the fraction of retained α particles could be far lower (0.45, all U and Th on rim) or far higher (1, all U and Th in core). All other degrees and styles of zonation will yield fractional retention values between these two extremes.

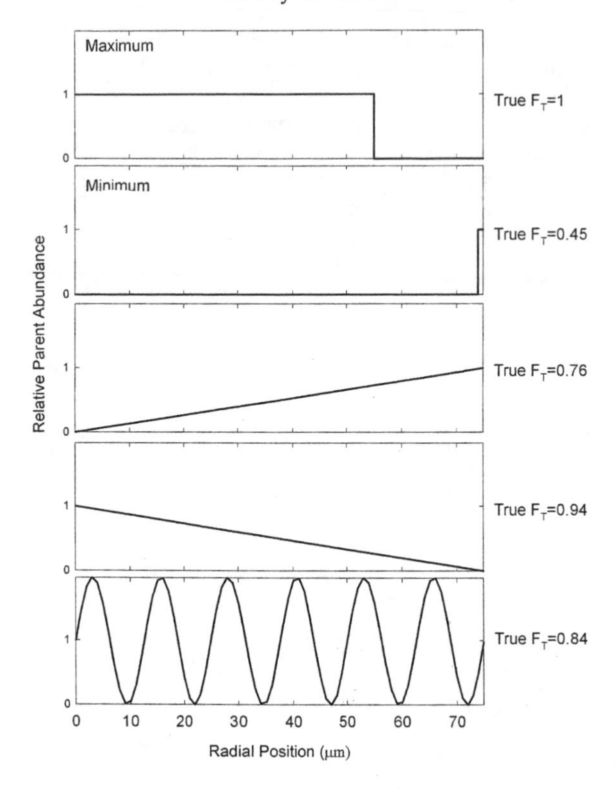

Figure 3. The effects of parent distribution on the fraction of α particles retained (F_T), following equations given by Farley et al. (1996). Calculations are for stopping of ^{238}U α particles in a 75 μm radius sphere of apatite composition. For a homogeneous parent distribution, F_T for this sphere would be 0.81. The top two panels show the maximum possible departure from this value: all the parent located more than 1 stopping distance from the rim (maximum), and all parent on the outermost 1 μm (minimum). All other distributions, such as the three shown, must fall between these bounds.

Unfortunately models that accommodate zonation are of limited practical utility in the absence of a non-destructive technique for measuring the U and Th distribution at μm resolution in the crystal to be dated. At best, aliquots analyzed by secondary ion mass spectrometry, electron microprobe, and fission track radiography may provide useful indications of the existence, degree, and style of zonation within an analyzed population. This remains a problematic area for the (U-Th)/He method.

The importance of the α-ejection correction should not be underestimated:

1. Grains to be analyzed must retain the original size and geometry found in the rock. In practice this usually means that whole (or nearly whole) euhedral grains must be selected. Figure 4 shows examples of acceptable and unacceptable grains from this perspective.
2. Grains to be analyzed must be large; the smaller the grain to be analyzed the bigger the correction. The correction increases very rapidly for grains with minimum dimension smaller than about 75 μm (Fig. 2).
3. Uncertainties arise from the possible existence of parent nuclide zonation. Errors arising from possible zonation may be large, up to about ±40% of F_T.

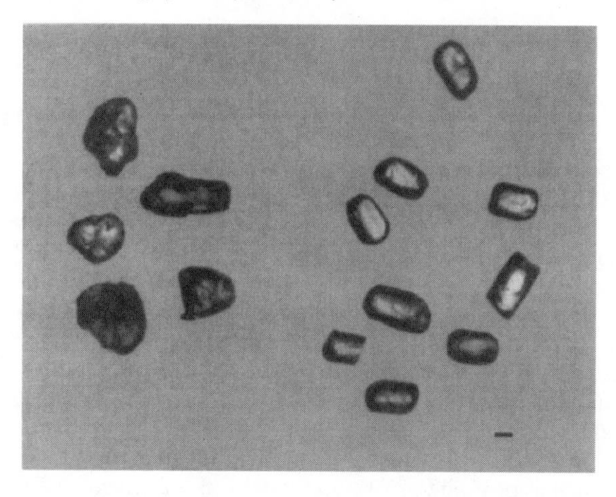

Figure 4. Fluorapatite grains showing morphologies that are acceptable (right) and unacceptable (left) for making an F_T correction. Note that "good grains" may have two complete terminations, or may be broken perpendicular to the c-axis. "Bad" grains include fragments, anhedral grains, and euhedral grains of complex morphology. Scale bar in lower right is 100 μm. These grains are from sample 96MR-47, a tonalite from the Coast Mountains of British Columbia, Canada (Farley et al. 2001).

Overall these effects substantially restrict what samples can be analyzed, and more importantly limit the precision of the (U-Th)/He method applied to small crystals. For samples with $F_T \approx 0.8$ (typical of most apatites we have analyzed) the ultimate precision of the (U-Th)/He method is unlikely to be better than a few percent.

Analytical procedures

Determination of a (U-Th)/He age involves three steps: measurement of grain dimensions for determination of the α-ejection correction, measurement of He content, and measurement of U and Th content. We routinely measure all three quantities on a single aliquot, which is frequently a single crystal. Use of a single aliquot eliminates uncertainties that arise from grain-to-grain heterogeneity, e.g., in U and Th content. The current technique we use at Caltech is described below. Broadly similar techniques are in use in several other laboratories.

F_T measurement. After hand-selection of sufficiently large and usually euhedral crystals free of fluid and mineral inclusions (see below), grain dimensions are measured under a 120× binocular microscope. In the case of euhedral hexagonal prisms of apatite, it is sufficient to measure the length and width of the prism; for minerals of lower symmetry additional measurements are required. Based on replicate observations we believe that grain dimensions of ~100 μm are precise and accurate to about 5 μm. This allows us to obtain F_T values with a precision of a few percent for F_T > ~0.70. As grains become smaller, precision rapidly degrades both because of increasing error in the grain measurement and the steepening slope of the F_T function (Fig. 2b). We have observed no significant observer bias in this measurement. It must be stressed that the *accuracy* of the F_T value obtained is sensitive to the assumption that U and Th are homogeneously distributed in the grain of interest. When multi-grain aliquots are analyzed, the F_T value assigned to the aliquot is the mass-weighted mean of the individual grain F_T values. This assumption adds another source of uncertainty: if the parent nuclide abundances vary strongly from grain to grain, then this weighting will not be appropriate. This source of error can be reduced by analyzing a single grain or by analyzing aliquots that contain a very restricted range of grain sizes and hence F_T values.

He, U, and Th. Helium is most commonly extracted from phosphates by in-vacuum heating, either in a resistance furnace (Wolf et al. 1996a, Warnock et al. 1997) or using a "laser microfurnace" in which a tiny Pt tube holding the grain is heated with a laser (House et al. 2000). The laser technique has the advantage of lower and more reproducible blanks. In addition, because only the microcrucible is heated, sample throughput is far higher. Samples are heated sufficiently to outgas He, but at a low enough temperature to prevent fusion. This permits removal of the sample from the furnace for U and Th analysis. Experience shows that 5 min at 950-1000°C is sufficient to extract all of the He from apatite; requisite temperatures for complete degassing of monazite and xenotime are higher (perhaps 20 min at ~1250-1300°C), but have not yet been well-documented. U and Th contents are apparently unaffected be heating to sub-melting temperatures (Wolf et al. 1996b). However, unambiguous evidence indicates that direct laser heating and/or fusion of apatite (Stuart and Persano 1999) and titanite (Reiners and Farley 1999) grains causes volatilization loss of U and/or Th; this is why the Pt microcrucible technique was developed.

Measurement of the evolved He is made by peak height comparison with standard gases on sector-type mass spectrometers such as the MAP 215-50 and VG-3600 (e.g., Wolf et al. 1996a, Warnock et al. 1997), or by ^{3}He isotope dilution (ID) on a quadrupole mass spectrometer (QMS). We find that the precision and sensitivity of the ID-QMS technique are superior to those of the sector MS-peak height method. Reproducibility of gas standards suggests that for typical amounts of He evolved from a sample (e.g., of order 1×10^{-9} cc STP), the ID-QMS technique has a precision of ~0.5% (1σ). The accuracy of this measurement depends on the accuracy of the standard used for calibration, which is probably better than 1% when capacitance manometry is used.

Upon removal from the vacuum chamber, samples are dissolved in acid (HNO_3 for apatite, HCl for monazite and xenotime) and spiked for U and Th analysis. At Caltech we spike with ^{235}U and ^{230}Th and analyze U and Th isotope ratios on a Finnigan Element double-focusing inductively-coupled plasma mass spectrometer. The accuracy and precision of these measurements is typically better than 0.5%.

Accuracy and precision of ages. The combined precision of parent and daughter abundance measurements yields an analytical precision of ~1.5% (2σ) for He ages uncorrected for α ejection. Replicate He age determinations on fragments of Durango apatite (large, gem-quality crystals described by Young et al. 1969) from which α-ejection-affected surfaces have been removed, are about 2× more variable. This observation indicates that other sources of error exist. One candidate is zonation of U and Th, which when coupled with the long stopping distance of α particles, could yield heterogeneity in the local daughter/parent ratio of the crystal. Such an effect has been proposed for the Durango standard by Boyce and Hodges (2001). Under the best of circumstances the F_T correction for typical-size phosphate grains adds perhaps another 1 to 2% to this uncertainty figure. For smaller grains or those that are morphologically complex the uncertainty will be larger. Replicate analyses of relatively large high quality apatites support a reproducibility of about 6% (2σ) (Farley et al. 2001).

In some cases we have observed substantially larger inter-aliquot variability. We suspect that in some cases this age spread results from a violation of the assumption of parent nuclide homogeneity on which the α-ejection correction is based (see above). In other cases it probably reflects differences in He closure temperature from grain to grain coupled with a thermal history that magnifies such differences. In still other cases, mineral inclusions are likely the problem. The latter two possibilities are described below.

As described here the (U-Th)/He method is an absolute dating technique based on fundamental measured quantities rather than comparisons to independently dated materials. Our best estimate of the accuracy of He age determinations is ~2% (2σ). Although He "age standards" have not yet been established, support for approximately this degree of accuracy comes from analyses of Durango apatite, and Fish Canyon titanite and zircon (House et al. 2000, Reiners and Farley 2002). Similarly, monazite analyses of a rapidly cooled Oligocene ash flow tuff from SE Peru (Mac-83) yield an average age of 23.98 ± 0.6 Ma (unpublished) that is in agreement with the monazite Th-Pb age of 24.21 ± 0.10 Ma (Villeneuve et al. 2000).

SIGNIFICANCE OF PHOSPHATE (U-TH)/HE AGES

The He concentration in a crystal reflects the balance between production by radioactive decay (modified by α ejection) and loss by diffusion, integrated along the entire time-temperature history of the crystal. Thus establishing the significance of a (U-Th)/He "apparent age" requires a thorough knowledge of He diffusion as a function of temperature. At present the only direct technique for establishing the He diffusion rate within a mineral is through incremental outgassing, in which the amount of He evolved from a sample held for a fixed time and temperature is measured. This amount can be used to compute a diffusion coefficient under certain assumptions (Fechtig and Kalbitzer 1966). Without further information on the diffusion process, the quantity obtained from such an experiment is D/a^2 where D is the diffusivity (in cm^2/sec) and a is the characteristic length scale of the diffusion domain. In order to obtain sufficient He in a reasonable amount of time, the temperatures at which diffusion coefficients are measured are usually far higher than those of interest in nature. To the extent that the measured diffusivity scales predictably with temperature, the diffusivity at lower temperatures can be estimated. For example, Figure 5 shows that He diffusion coefficients measured on apatite, monazite, and xenotime closely obey an Arrhenius relationship:

$$D/a^2 = D_o/a^2 e^{-Ea/RT} \quad \text{or} \quad \ln(D/a^2) = \ln(Do/a^2) - (E_a/R)(1/T) \tag{9}$$

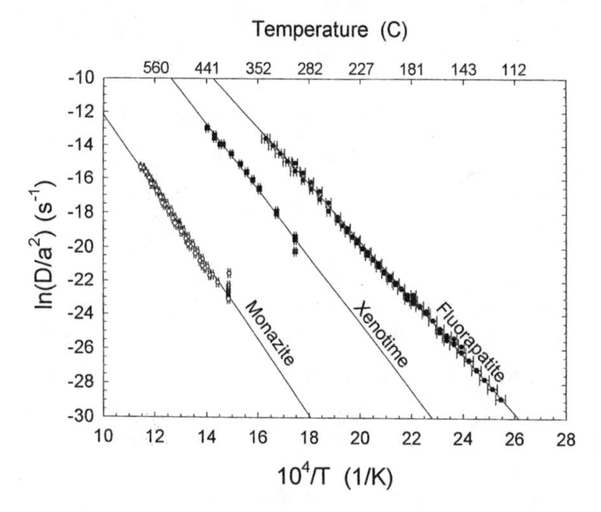

Figure 5. Arrhenius plot of He diffusion from Durango apatite (Farley 2000) and from QC-A xenotime (Miocene leucogranite, Tibet) and 554 monazite, commonly used SIMS standard (Tertiary granodiorite from Catalina Mtns, Arizona). Monazite and xenotime measurements are from Stockli and Farley (unpublished).

where D_o is the diffusivity at infinite temperature, E_a is the activation energy of the process, R is the gas constant, and T is the Kelvin temperature. From such data the unknown diffusion parameters D_o/a^2 and E_a can be obtained, allowing prediction of D/a^2 at other temperatures. These parameters are also used to calculate closure temperature (Dodson 1973).

By controlling the geometry and size of grains analyzed during the incremental outgassing experiment, it is possible to assess the significance of the quantity a in D/a^2. In the case of Ar diffusion from K-feldspar, a is the characteristic length of diffusion domains that are smaller than the grain itself (Zeitler 1987, Lovera et al. 1989). In contrast, in the few minerals that are well studied for He diffusion, the diffusion length scale a corresponds to the size of the grain itself (Farley 2000, Reiners and Farley 1999). The importance of this observation is that diffusion rate (and thus closure temperature) scales with the inverse square of the grain size. This is discussed more fully below.

There are several practical difficulties in translating incremental outgassing data into diffusion coefficients (Farley 2000). The most commonly used computational models require that the distribution of diffusant be uniform within the diffusion domain. This assumption is violated in many samples by the α-ejection effect and by He diffusion in nature, both of which act to round the concentration profile at the grain surface. As a consequence, the initial rate of He release from a sample is anomalously retarded relative to later release. Fortunately this effect can be identified and greatly reduced by incremental outgassing schedules that involve cycling from low to high temperatures and back (Farley 2000).

Similarly, diffusion models require specification of a geometry for the diffusion domain, commonly one of the analytically tractable geometries of sphere, infinite cylinder, and infinite slab (Crank 1975, McDougall and Harrison 1988). Neither whole, euhedral natural crystals nor their fragments are well described by these geometries. In practice, the effect of the assumed geometry on computed closure temperature is small, provided that a consistent geometry is assumed when the diffusion coefficients and the closure temperature are computed. For example, when spherical geometry is assumed the incremental release data for apatite yield the diffusion data in Figure 5, $\ln(D_o/a^2)$ = 14.0 s^{-1}, E_a = 33.3 kcal/mol, and from these, a closure temperature of 71°C according to the spherical formulation of Dodson (here and elsewhere assuming a 10°C/Myr cooling rate). The same incremental release data modeled as an infinite slab yields a linear Arrhenius plot with $\ln(D_o/a^2)$ = 15.0 s^{-1} and E_a =32.0 kcal/mol, implying a closure temperature for an infinite slab of 64°C. For most applications this distinction is unlikely to be significant and is probably within the uncertainty of the diffusivity data. At present it is standard practice to assume spherical geometry unless specific effort has been made to obtain some other geometry; all diffusion data described below were computed assuming a spherical diffusion domain. Recent numerical models presented by Meesters and Dunai (2002a) should provide a new method for handling more realistic geometries.

He diffusivity from phosphates

Apatite. The most detailed He diffusion measurements on phosphates are from apatite, especially on fragments of the widely available Durango apatite. The data in Figure 5 indicate an activation energy of about 33 kcal/mol (138 kj/mol) for He diffusion from this material. Additional experiments indicate that a corresponds to the physical grain size, which when coupled with the data in Figure 5 indicates that $D_o \approx 30$ cm^2/sec. Experiments on crystallographically controlled slabs of Durango apatite suggest that He diffusivity is very nearly isotropic (Farley 2000). Experiments on other apatites yield broadly similar results (Lippolt et al. 1994, Wolf et al. 1996b, Warnock et al. 1997). Attempts to identify what physical or chemical factors control the He diffusion rate have been inconclusive. For example, Warnock et al. (1997) reported variations in E_a between high F/Cl and low F/Cl apatites, but the variations did not exceed analytical uncertainty. Similarly, while radiation damage may affect some aspects of He diffusion from apatite (Farley 2000) there is at present no compelling evidence that it is a strong control. This is an active area of research.

Monazite and xenotime. Monazite and xenotime have received far less attention for He dating than has apatite, despite the fact that these minerals are important hosts of Th in Earth's crust and monazite is commonly analyzed in U-Th-Pb geochronology (Harrison et al., this volume). This situation probably reflects the general rarity of these minerals compared to apatite. Figure 5 shows preliminary He diffusion data on these phases obtained at Caltech (unpublished). He diffusion from both $(Ce,La,Th)PO_4$ and YPO_4 define linear arrays on the Arrhenius plot; at any given temperature He diffusivity increases in the order monazite, xenotime, apatite. Activation energies of 44 kcal/mol (184 kj/mol) for monazite and 38 kcal/mol (159 kj/mol) for xenotime are higher than the 33 kcal/mol found for apatite. No data are available to determine whether the diffusion domains in monazite and xenotime are the grain itself or some sub-grain structure.

Our diffusion experiments on monazite show well-behaved Arrhenius relationships and show no evidence of erratic diffusion behavior attributable to radiation damage, such as in zircon (Reiners and Farley 2002). Despite the high concentrations of U and Th, monazite rarely becomes metamict, unlike zircon, and it has been suggested that it self-anneals at low temperatures (see Gleadow et al., this volume). Diffusion experiments on monazite samples (<100 Ma) with extremely variable Th concentrations (1-13 wt %) are all characterized by well-behaved Arrhenius relationships yielding similar diffusivities and closure temperatures (unpublished data), suggesting that compositional variations and radiation damage in monazite are generally negligible.

INTERPRETATION

The diffusion data presented above indicate that all three phosphates quantitatively retain He at Earth surface temperatures (see Wolf et al. 1998 for a discussion of the minimal affects of diurnal heating and forest fires on He diffusion from apatite). As a result, all three minerals may be used to accurately date the formation of quickly cooled rocks, such as volcanics. Apatite has already been used for this purpose (Stockli, in preparation; Farley et al. 2002). Because these phases are rare in volcanic rocks, the more likely application of phosphate He chronometry is to assessment of cooling histories. In slowly cooled rocks, the quantity t computed from the age equation (Eqn. 4) is an apparent age, which may or may not correspond to a specific geologic event. The simplest way to interpret such an apparent age is to associate it with cooling through a particular closure temperature, computed from the diffusion parameters (Dodson 1973). The diffusion arrays in Figure 5 indicate closure temperatures of 115°C and 220°C for xenotime, and monazite, respectively. For apatite, in which diffusivity is known to scale with inverse square of grain radius, the diffusivity data indicate a closure temperature of ~73°C for a prism half-width of 75 μm. Half-widths of 50 and 100 μm yield closure temperatures of ~68 and ~77°C respectively.

The closure temperature computation assumes that He is uniformly distributed within the grain. However, this assumption is violated when the α-ejection-influenced grain edge and diffusion domain boundary correspond, as they do in apatite (there are no data for monazite or xenotime). This correspondence causes the diffusive loss rate to be lower than if the He concentration profile were uniform. Numerical modeling suggests that the effect is relatively small, equivalent to about a 2 or 3°C increase in the closure temperature (Farley 2000).

While simple and intuitive, the closure temperature approach provides limited insight because its computation assumes a constant cooling rate. In nature such cooling must be the exception rather than the rule. While closure temperatures provide a quick comparison among different thermochronometric systems and give some idea of the range of temperatures to which a particular system is sensitive, more detailed interpretation can only be made by actually integrating the production-diffusion equation along potential time-temperature paths (Wolf et al. 1998).

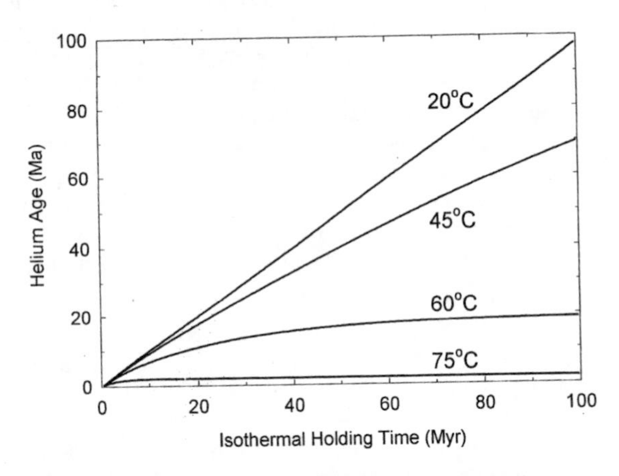

Figure 6. (U-Th)/He ages that result from holding an apatite at a fixed temperature (labeled) for 100 Myr, using computational technique of Wolf et al. (1998). Note the initial rise in He age along the 1:1 line, followed by an increasingly large deviation to low He age. This reflects the competing effects of radiogenic production and diffusion; the rate of loss increases as the concentration, and thus the concentration gradient, increases. Ultimately, regardless of temperature, a steady state between production and loss is achieved. These curves were calculated with diffusivities obtained on Durango apatite, corrected for α-ejection rounding (Farley 2000): $D_o = 32$ cm^2/sec, $E_a = 33$ kcal/mol, r = 75 μm.

Mathematical models for the computation of He ages in a system cooling on an arbitrary time-temperature path have been described by Wolf et al. (1998) and more recently by Meesters and Dunai (2002a). Here we use such a model to illustrate key aspects of low temperature thermochronometry in general and the (U-Th)/He system in particular, using apatite as an example. Figure 6 shows how He ages evolve as a function of time at a fixed temperature. At every temperature, the He ages initially increase in 1:1 proportion with calendar age, but as time progress the He ages drop progressively further below the 1:1 line. Ultimately a steady state is achieved between radiogenic production and diffusive loss, and the He age becomes invariant with time. The higher the temperature the more rapidly this steady state is achieved and the lower is the steady state age. For example, an apatite will achieve a steady state He age of 280 ka in 2.1 Myr at 90°C, but at 50°C the steady state age of 80 Ma will be achieved only after ~550 Myr. This trend toward steady state develops because the diffusive loss rate depends not only on the diffusion coefficient, but also on the concentration gradient. For a sample with no initial helium, the He concentration, and hence concentration gradient, is low, and so too is diffusive loss. As production outpaces diffusion, the concentration rises, causing diffusive loss to increase. Ultimately each atom added by radioactive decay is balanced by diffusive loss, and a steady state age is achieved. Similar logic dictates that a sample with a He age in excess of the steady state age will decrease until production and loss balance; this might apply to a detrital apatite subjected to elevated temperatures in a sedimentary sequence.

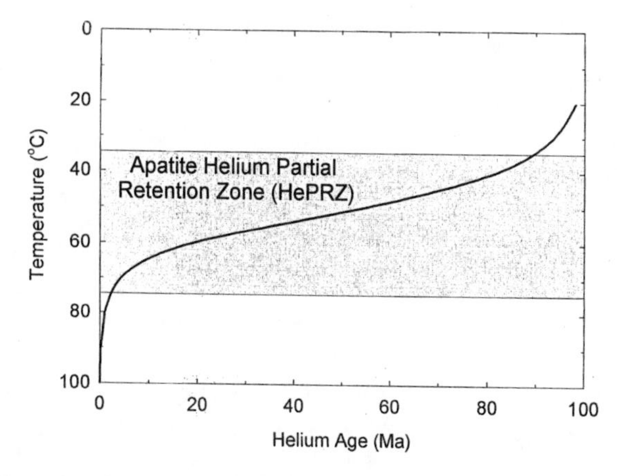

Figure 7. The apatite-helium partial retention zone (HePRZ). In an isothermal setting held for 100 Myr, apatite (U-Th)/He ages change most rapidly between ~75°C and 35°C. He ages are essentially zero at temperatures above 85°C. Note the extreme sensitivity of He ages to temperature in the middle of the partial retention zone. Computed using same diffusion parameters as Figure 6.

Figure 7 presents a different view of the same data: He ages are now plotted as a function of temperature for a specific holding time, in this case 100 Myr. The peculiar choice of x- and y-axes facilitates comparison to the age pattern expected in a borehole. He ages define a sigmoidal shape, with high retention and ages at low temperature and no retention and zero ages at high temperature. The zone that separates these two extremes, in which He ages decrease very rapidly with temperature, is termed the He partial retention zone (HePRZ). The exact temperature range of this zone depends on the holding time, but can rather arbitrarily be defined to lie between 35 and 75°C for periods of 10^7 to 10^8 years. In a typical geothermal gradient of 20°C/km, this region would reside between about 1 and 3 km depth in the crust. As shown in Figure 8, the calculated partial retention zone for xenotime lies between 75°C and 110°C, while the partial retention zone for monazite is between 150°C and 210°C.

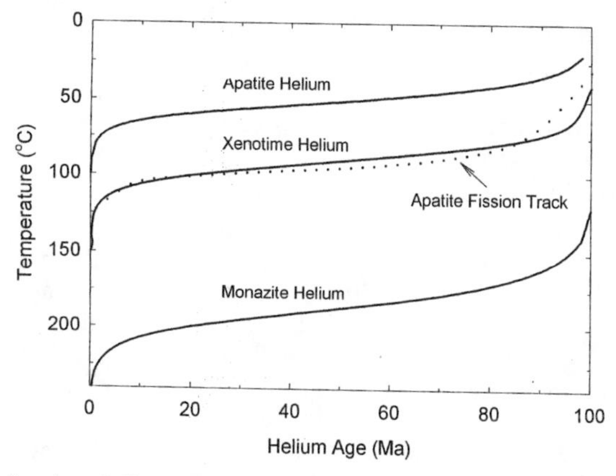

Figure 8. The phosphate helium partial retention zones, computed using the diffusion parameters described in the text, compensating for the effects of α ejection. The analogous apatite fission track partial annealing zone is shown for comparison. This figures provides an indication of the range of temperatures over which He ages from a given mineral will be sensitive.

An important characteristic of He ages in the HePRZ is their extreme sensitivity to small changes in temperature or diffusivity. For example an apatite held at 48°C for 100 Myr will have a He age ~15 Myr (~40%) older than one held at 52°C. Similarly, tiny variations in diffusivity will be greatly amplified in this region. At 50°C, the expected age difference between an apatite with a radius of 80 μm and one of 60 μm is 12 Myr, yet the difference in closure temperature between these two grain sizes is just 4°C. Thus we expect extreme scatter in He ages between and possibly within samples that have spent substantial amounts of time in the HePRZ. The amplification of small differences in temperature or diffusivity into large age differences within the HePRZ has been used to deduce important aspects of thermal histories of rocks (Reiners and Farley 2001), and also provides perhaps the most sensitive method available for assessing what controls He diffusion.

In addition to these isothermal cases, the numerical solution to the production-diffusion equation is useful for forward modeling of He ages for comparison with measured He ages. As with all thermochronometers, a general conclusion of such modeling is that an individual He age can be produced by essentially an infinite number of thermal histories (Wolf et al. 1998). There are several ways to reduce the number of acceptable paths. Most commonly, He ages are obtained from samples along a specific sampling transect, commonly a vertical profile down a mountain. The use of vertical profiles to constrain cooling histories and causative tectonic and geomorphic processes has been discussed in the fission track literature (see Gleadow et al., this volume), and these established techniques and principles are directly applicable to He-derived cooling ages. Indeed vertical profiles are by far the most common way that apatite He age patterns have been investigated (e.g., Wolf et al. 1996a, House et al. 1997, Stockli et al. 2000, Farley et al. 2001).

An example illustrates the approach. Let us imagine that a rigid package of rock sits isothermally from 40 Ma to 10 Ma, after which it experiences steady cooling at a rate of 5°C/Myr. How is this cooling recording in the block of rock? Using the forward model of Wolf et al. (1998) and assuming a geothermal gradient of 20°C/km, the predicted variation in He age with elevation is shown in Figure 9. The major features are an "exhumed HePRZ" with the same general shape as the HePRZ in Figure 7 except shifted to the right by "aging" occurring after cooling ensues, a "break in slope" (see Gleadow et al., this volume) indicating the onset of cooling at 10 Ma, and a linear age-elevation segment indicating fairly rapid and constant cooling. Thus the age elevation profile captures in reasonable detail the input cooling history and provides a context within which to interpret any individual age. An age elevation profile from the Coast Mountains of British Columbia yields a pattern almost identical to this prediction (inset and Farley et al. 2001). Additional examples of forward-modeled He ages are presented by Wolf et al. (1998) and Meesters and Dunai (2002a).

An alternative approach to limiting the number of acceptable cooling histories using just a single rock is to combine multiple thermochronometers. In practice, the logical pairing is apatite (U-Th)/He and apatite fission track methods. The apatite fission track method has an annealing temperature (see Gleadow et al., this volume) of about 100°C, and a partial annealing zone about 30°C hotter than the apatite HePRZ (Fig. 8). By analyzing apatites from a single rock it is thus possible to establish the time of cooling through ~100°C and 70°C, and thus the cooling rate. Combination of fission track length modeling (Gleadow et al., this volume), which is sensitive to temperatures between about 110 and 60°C, with He ages sensitive to temperatures from ~75°C to 35°C should be an especially powerful approach to establishing cooling histories. However, although some consistent results have been obtained by (U-Th)/He-fission track pairing (Stockli et al.

2000), in other cases it is hard to reconcile the two techniques (House et al. 1997). The most likely explanation is that the thermal sensitivity of one or the other method (or both) is not yet fully and accurately understood. This is an area of active research in both the (U-Th)/He and fission track communities. It may also be fruitful to combine phosphate (U-Th)/He ages with feldspar Ar-Ar dating to constrain cooling from >250 to ~35°C. Apparently no such studies have yet been published.

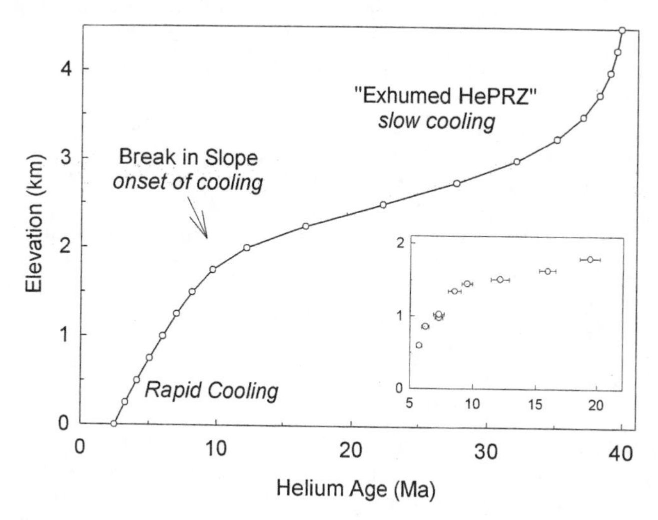

Figure 9. A vertical profile of modeled He ages in a system that was isothermal from 40 to 10 Ma, then cooled at a rate of 5°C/Myr to 0 Ma. The vertical profile carries a record of this cooling: an exhumed HePRZ developed during isothermal conditions, a break in slope indicating the onset of cooling at 10 Ma, and a linear age-elevation segment indicating rapid and steady cooling to the present. The inset shows a vertical profile of apatite He ages obtained in the Coast Mountains, British Columbia, Canada, which shows precisely this pattern (Farley et al. 2001). Ages were modeled assuming the same diffusion parameters in Figure 6 using the method described by Wolf et al. (1998). A geothermal gradient of 20°C/km was assumed to compute positions on the y-axis.

CHALLENGES

There is now substantial experience in applying the apatite (U-Th)/He method. At Caltech alone over a thousand samples have been dated from many different parts of the world. A variety of problems have been identified:

1. Some rocks yield only fractured and/or anhedral crystals. This seems especially common in gabbros, where apatites are often seen to be space-filling rather than euhedral. Such grains are difficult to use both because they make the α-ejection correction very difficult and because it is not obvious how to determine the original grain size. As a result the diffusion characteristics, which scale with grain size, are not well known.
2. Some rocks yield deeply pitted or very small crystals. These grains are problematic for the same reasons as described above. Sedimentary rocks as well as altered or weathered igneous rocks are the most common carriers of pitted grains. Small grains occur in all rock types we have investigated, but not surprisingly are most common in fine-grained rocks or those poor in phosphorous (e.g., true granites).

While apatites with these characteristics are problematic, it is readily apparent from simple visual inspection of the apatite separate. Of greater concern are problems that are difficult to detect and which may unexpectedly compromise the age analysis:

3. Strong zonation in parent nuclides invalidates the α-ejection correction, and will also affect the diffusive loss rate of helium, in effect changing the He closure temperature (see Meesters and Dunai 2002b for some model examples). At present we do not know how to detect zonation in the specific grains to be dated, though the presence of a zoned population can be established for example from fission track uranium mapping. We believe parent zonation is the cause of some cases of irreproducible He ages, an idea supported by both SEM and ion probe studies currently in progress.
4. Perhaps the greatest problem of all originates with mineral inclusions (House et al. 1997, McInnes et al. 1999, Axen et al. 2001). Unlike K/Ar dating which uses minerals with stoichiometric K, the (U-Th)/He technique utilizes radionuclides which are either minor or trace constituents in the dated minerals. As a result, inclusions of other phases with higher U and Th may dramatically affect the He budget of an apatite. Because α stopping distances are long, the He from small inclusions will mostly be found in the apatite, so the difference in closure temperature between the apatite and the inclusion is not an important factor. However, the commonly adopted dissolution techniques using HNO_3 are not effective at dissolving several important inclusion minerals, most notably zircon. (Use of HF bombing techniques to ensure dissolution is possible, but enormously more time consuming). In addition, a localized concentration of U and Th in an inclusion must to some extent degrade the accuracy of the α-ejection correction.

Through SEM and electron microprobe investigations we have found that inclusions in apatite consist of monazite ≈ zircon > quartz > feldspar > xenotime > sulfides. Of these, monazite, zircon and xenotime are of the greatest concern because they have high U and Th concentrations. Figure 10 shows several backscatter SEM images of these inclusions. In many cases inclusion-bearing grains can be detected and eliminated prior to dating, most readily by taking the apatite to extinction under cross-polars. (This must be done using unmounted, unpolished grains so they can still be dated after inspection). High birefringence causes even tiny inclusions to stand out against the extinct apatite background. However some types of inclusions, especially of monazite, remain problematic. Monazite commonly appears to be exsolved from the apatites, as indicated by the presence of sub-μm diameter rods oriented parallel to the apatite c-axis (Fig. 10c). These inclusions are essentially invisible under cross-polars. Ultimately the strongest assurance that no inclusion problems exist with a measured He age is the reproducibility of that age. Good reproducibility is unlikely if He ages are strongly affected by rare inclusions heterogeneously distributed among dated aliquots (House et al. 1997, Axen et al. 2001).

CONCLUSIONS

(U-Th)/He dating of phosphates is in an early stage of development, but appears likely to become a practical technique. Several phosphates concentrate U and Th, so readily measurable amounts of He accumulate within a short period. For example, a typical 150-μm diameter apatite crystal will have accumulated enough He to be dated after just a few Myr. Apatites as young as 7 kyr have been dated with good precision (Stockli, unpublished). Monazite and xenotime tend to have considerably higher U and Th contents, so He detection limits are not generally an issue. Current analytical techniques yield He ages reproducible to no better than a few percent (1σ; Farley et al. 2001, House et al. 2000); given that most phosphate crystals in nature are small, corrections of a few tens of percent for α ejection are required and these large corrections are likely to prevent the method from ever achieving much higher precision.

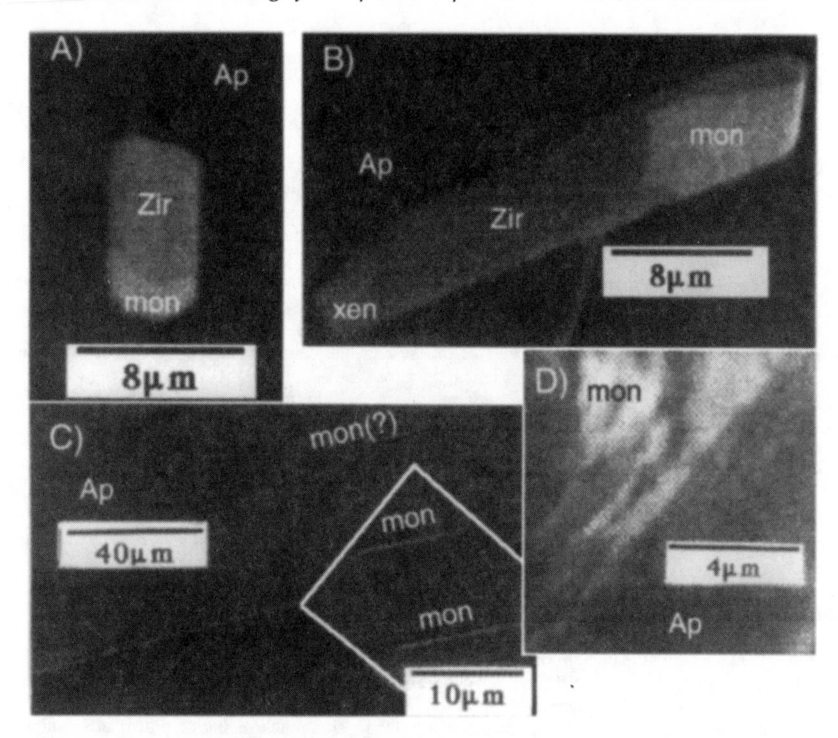

Figure 10. SEM backscatter images of inclusions in apatite. (A) Composite zircon-monazite inclusion; (B) composite zircon-monazite-xenotime inclusion; (C) monazite rods, mostly plucked in main image but intact in inset - rods are parallel to the apatite c-axis; (D) gradational transition from apatite to monazite inclusion.

Detailed work on He diffusion from apatite indicates a closure temperature of ~70°C. This system has the lowest closure temperature of currently known thermochronometric techniques; were the closure temperature much lower there would be significant loss at Earth surface temperatures. Information on cooling of rocks through such low temperatures is of great interest in tectonic and geomorphologic studies, and many applications of the apatite (U-Th)/He method have now been published. These studies amply demonstrate the value of the technique. From a methodological perspective, areas of current research include (1) assessment of what factors, such as chemistry and radiation damage, control He diffusion rate, and (2) development of robust methods for the identification and elimination of inclusion-bearing apatite grains. Because the technique provides insights to previously inaccessible portions of rock time-temperature paths, there is also interest in thermal modeling of very low temperature processes in the Earth's crust, for example arising from tectonics, geomorphology and the interaction of the two.

Other than the diffusion measurements reported here, we are aware of no published work on (U-Th)/He systematics of other phosphates such as monazite and xenotime. Our work suggests closure temperatures of ~220°C for monazite and ~115°C for xenotime. Because they tend to be relatively fine-grained, these phases will require relatively large α-ejection corrections, and so will necessarily have fairly large age uncertainties. Other phosphates which occur in more unusual environments, such as ore deposits (e.g., pyromorphite), are also interesting targets for further study.

ACKNOWLEDGMENTS

We thank Barry Kohn and Peter Zeitler for helpful comments.

REFERENCES

Axen GJ, Lam PS, Grove M, Stockli DF, Hassanzadeh J (2001) Exhumation of the west-central Alborz Mountains, Iran, Caspian subsidence, and collision-related tectonics. Geology 129:559-562

Boyce JW, Hodges KV (2001) Chemical variations in the Cerro de Mercado (Durango, Mexico) fluorapatite: Assessing the effect of heterogeneity on a geochronologic standard. EOS Trans Am Geophys Union 82:V22C-1061

Chang LLY, Howie RA, Zussman J (1996) Rock-Forming Minerals, Vol 5B, Non-silicates: Sulphates, Carbonates, Phosphates, Halides. Longman, Essex, UK

Crank J (1975) The mathematics of diffusion. Oxford University Press, New York

Damon PE, Green WD (1963). Investigations of the helium age dating method by stable isotope dilution technique. Radioactive Dating. IAEA, Vienna, p 55-69.

Damon PE, Kulp JL (1957) Determination of radiogenic helium in zircon by stable isotope dilution technique. Trans Roy Soc Edinburgh 38:945-953

Dodson MH (1973) Closure temperatures in cooling geological and petrological systems. Contrib Mineral Petrol 40:259-274

Farley KA, Wolf RA, Silver LT (1996) The effects of long alpha-stopping distances on (U-Th)/He dates. Geochim Cosmochim Acta 60:1-7

Farley KA (2000) Helium diffusion from apatite: general behavior as illustrated by Durango fluorapatite. J Geophys Res 105:2903-2914

Farley KA, Rusmore ME, Bogue SW (2001) Post-10 Ma uplift and exhumation of the Northern Coast Mountains, British Columbia. Geology 29:99-102

Farley KA, Kohn BP, Pillans B (2002) The effects of secular disequilibrium on (U-Th)/He systematics and dating of Quaternary volcanic zircon and apatite. Earth Planet Sci Lett (in press)

Fechtig H, Kalbitzer S (1966) The diffusion of argon in potassium bearing solids. *In* Potassium-Argon Dating. OA Schaeffer, J Zahringer (eds) Springer, Heidelberg

Gleadow AJW, Belton DX, Kohn BP, Brown RW (2002) Fission track dating of phosphate minerals and the thermochronology of apatite. Rev Mineral Geochem (this volume)

Harrison TM, Catlos EJ, Montel J-M (2002) U-Th-Pb dating of phosphate minerals. Rev Mineral Geochem 48:523-558 (this volume)

House MA, Farley KA, Stockli D (2000) Helium chronometry of apatite and titanite using Nd-YAG laser heating. Earth Planet Sci Lett 183:365-368

House MA, Farley KA, Kohn BP (1999) An empirical test of helium diffusion in apatite: borehole data from the Otway basin, Australia. Earth Planet Sci Lett 170:463-474

House MA, Wernicke BP, Farley KA (1998) Dating topography of the Sierra Nevada, California, using apatite (U-Th)/He ages. Nature 396:66-69

House MA, Wernicke BP, Farley KA, Dumitru TA (1997) Cenozoic thermal evolution of the central Sierra Nevada from (U-Th)/He thermochronometry. Earth Planet Sci Lett 151:167-179

Hurley PM (1952) Alpha ionization damage as a cause of low He ratios. EOS Trans, Am Geophys Union 33:174-183

Hurley PM (1954) The helium age method and the distribution and migration of helium in rocks. *In* Nuclear Geology. John Wiley & Sons, New York, Faul H (ed) p 301-329

ICRU (1993). International Commission on Radiation Units and Measurements. *In* ICRU Report 49, Stopping Powers and Ranges for Protons and Alpha Particles.

Lippolt HJ, Leitz M, Wernicke RS, Hagedorn B (1994) (U+Th)/He dating of apatite: experience with samples from different geochemical environments. Chem Geol 112:179-191

Lovera O, Richter F, Harrison T (1989) The $^{40}Ar/^{39}Ar$ thermochronometry of slowly cooled samples having a distribution of diffusion domain sizes. J Geophys Res 94:17917-17935

McDougall I, Harrison TM (1988) Geochronology and thermochronology by the $^{40}Ar/^{39}Ar$ method. Oxford, New York

McInnes BIA, Farley KA, Sillitoe RH, Kohn BP (1999) Application of apatite (U-Th)/He thermochronometry to the determination of the sense and amount of vertical fault displacement at the Chuqicamata porphyry copper deposit, Chile Econ Geol 94:937-947

Meesters AG, Dunai TJ (2002a) Solving the production-diffusion equation for finite diffusion domains of various shapes (part I): implications for low-temperature (U-Th)/He thermochronology. Chem Geol 186:333-344

Meesters AG, Dunai TJ (2002b) Solving the production-diffusion equation for finite diffusion domains of various shapes (part II): application to cases with α-ejection and non-homogeneous distribution of the source. Chem Geol 186:57-73

Pik R, Marty B (1999) (U-Th)/He thermochronometry: extension of the method to more U-bearing minerals. EOS Trans Am Geophys Union 80:F1169

Reiners PW, Farley, KA (1999) Helium diffusion and (U-Th)/He thermochronometry of titanite. Geochim Cosmochim Acta 63:3845-3859

Reiners PW, Farley KA (2002) (U-Th)/He thermochronometry of zircon: Initial results from Fish Canyon Tuff and Gold Butte, Nevada. Tectonophysics 349:297-308

Reiners PW, Farley KA (2001) Influence of crystal size on apatite (U-Th)/He thermochronology: An example from the Bighorn mountains, Wyoming. Earth Planet Sci Lett 188:413-420

Sabourdy G, Sagon JP, Patier P (1997) La composition chimique du xenotime en Limousin, Massif Central, France. Can Mineral 35:937-946

Sha LK, Chappell BW (1999) Apatite chemical composition, determined by electron microprobe and laser ablation inductively coupled plasma mass spectrometry, as a probe into granite petrogenesis. Geochim Cosmochim Acta 63:3861-3881

Spotila JA, Farley KA, Sieh K (1997) The exhumation and uplift history of the San Bernardino Mountains along the San Andreas fault, California, constrained by radiogenic helium thermochronometry. Tectonics 17:360-368

Stockli DF, Farley KA, Dumitru TA (2000) Calibration of the (U-Th)/He thermochronometer on an exhumed fault block, White Mountains, California. Geology 28:983-986

Strutt R (1908) On the accumulation of helium in geologic time. Proc Roy Soc London 81A:272-277

Stuart FM, Persano C, (1999) Laser melting of apatite for (U-Th)/He chronology: Progress to date. EOS Trans, Am Geophys Union 80:F1169

Van Emden B, Thornber MR, Graham J, Lincoln FJ (1997) The incorporation of actinides in monazite and xenotime from placer deposits in Western Australia. Can Mineral 35:97-104

Villeneuve M, Sandeman HA, Davis WJ (2000) A method for intercalibration of U-Th-Pb and ^{40}Ar-^{39}Ar ages in the Phanerozoic. Geochim Cosmochim Acta 64:4017-4030

Warnock AC, Zeitler PK, Wolf RA, Bergman SC (1997) An evaluation of low-temperature apatite U-Th/He thermochronometry. Geochim Cosmochim Acta 61:5371-5377

Wolf RA, Farley KA, Kass DM (1998) A sensitivity analysis of the apatite (U-Th)/He thermochronometer. Chem Geol 148:105-114

Wolf RA, Farley KA, Silver LT (1996a) Assessment of (U-Th)/He thermochronometry: the low-temperature history of the San Jacinto Mountains, California. Geology 25:65-68

Wolf RA, Farley KA, Silver LT (1996b) Helium diffusion and low temperature thermochronometry of apatite. Geochim Cosmochim Acta 60:4231-4240

Young EJ, Myers AT, Munson EL, Conklin NM (1969) Mineralogy and geochemistry of fluorapatite from Cerro de Mercado, Durango, Mexico. U S Geol Surv Prof Paper 650-D:D84-D93

Zeitler PK (1987) Argon diffusion in partially outgassed alkali feldspars: insights from $^{40}Ar/^{39}Ar$ analysis. Chem Geol 65:167-181

Zeitler PK, Herczig AL, McDougall I, Honda M (1987) U-Th-He dating of apatite: A potential thermochronometer. Geochim Cosmochim Acta 51:2865-2868

16

Fission Track Dating of Phosphate Minerals and the Thermochronology of Apatite

Andrew J.W. Gleadow, David X. Belton, Barry P. Kohn and Roderick W. Brown

School of Earth Sciences
The University of Melbourne
Melbourne, Australia 3010

INTRODUCTION

Several phosphate minerals have been investigated for their usefulness in geochronology and thermochronology by the fission track method. Of these, apatite [$Ca_5(PO_4)_3(F,Cl,OH)$], has proved pre-eminently suitable for this purpose for reasons discussed below. Apatite was one of the first minerals, amongst many others, to be investigated for fission track dating by Fleischer and Price (1964) and has subsequently become by far the most important of all the minerals used for dating by this method. The usefulness of apatites for fission track dating arises from their near-universal tendency to concentrate uranium within their structure at the time of crystallization and their widespread occurrence in all of the major rock groups.

Price and Walker (1963) first recognized the possibility that the accumulation of radiation damage tracks in natural minerals from the spontaneous nuclear fission of ^{238}U within their lattices could be used for geological dating. They also demonstrated that a simple chemical etching procedure served to enlarge these fission tracks to optical dimensions so that they could be observed and measured under an ordinary optical microscope (Price and Walker 1962b, 1963). This simple procedure quickly opened the way for a wide range of nuclear particle track studies in natural minerals and glasses (e.g., Fleischer et al 1975), the most important of which was fission track dating. Fleischer et al. (1964) also showed that spontaneous, or fossil, fission tracks provided an explanation for various 'anomalous' etch pits in apatite which had puzzled crystallographers over many years. The essence of a fission track age determination involves measuring the number of tracks that have accumulated over the lifetime of the mineral along with an estimate of the amount of uranium that is present. Knowing the rate of spontaneous fission decay, a geological age can be calculated.

Detailed studies on apatite for routine geological dating applications began with the pioneering work of Naeser (1967) and Wagner (1968, 1969). These studies established basic procedures that enabled the rapid and widespread adoption of apatite for a variety of fission track dating applications. It was realized very early that fission tracks in minerals displayed only limited stability under exposure to elevated temperatures and that apatite was one of the most sensitive minerals to thermal annealing of fission tracks (Fleischer and Price 1964). Later studies have greatly refined our understanding of track annealing behavior in apatite providing a basis for many of the applications of the method which are discussed below. A comprehensive overview of the general field of fission track dating and its applications has been provided by Wagner and Van den Haute (1992), and a very useful review of methods and interpretive strategies by Gallagher et al. (1998). Here we will focus specifically on the fundamental mineralogical aspects fission of track dating, specifically as applied to apatite and other phosphate minerals, and emphasize the more recent applications and future directions for this field.

1529-6466/00/0048-0016$05.00

FISSION TRACK DATING OF PHOSPHATE MINERALS

Relatively little work has been carried out on fission track dating of phosphate minerals other than apatite, and merrillite, β-$Ca_3(PO_4)_2$, is the only other example to have received any significant attention. Two early studies examined the potential of monazite, (Ce,La,Y,Th)PO_4, and pyromorphite $Pb_5(PO_4)_3Cl$, but these minerals have not been studied further. Pyromorphite was shown by Haack (1973) to contain numerous spontaneous fission tracks, but these were very unevenly distributed, making its use in dating very difficult.

Monazite

Monazite is known to contain high concentrations of uranium, so it is perhaps surprising that it has not been investigated further. The only reported fission track dating study of monazite was carried out by Shukoljukov and Komarov (1970) who reported ages from two specimens from Kazakhstan which turned out to be much younger than that expected for the host rock. This observation was attributed to a very low thermal stability of fission tracks in this mineral and consequent fading of fission tracks at relatively low temperatures. Such an observation of reduced fission track age is actually quite typical of the pattern frequently observed in other minerals, particularly apatite, although this was not well understood at the time this study on monazite was carried out.

The apparently low thermal stability of fission tracks in monazite would clearly repay further investigation, as the common occurrence of monazite would otherwise make it an attractive target for fission track analysis. Supporting this view is the general observation that monazite does not become metamict (i.e., have its lattice disordered by the effects of accumulated radiation damage), which is again similar to apatite. This indicates that radiation damage does not accumulate substantially in this mineral, implying that it has a low thermal stability. Monazite fission track dating may therefore have potential as a new low temperature thermochronometer.

Analytical considerations, however, are likely to be the major impediment to monazite dating by conventional fission track methods. These problems arise from the generally very high concentrations of heavy elements in this mineral, and include possible interference from ^{232}Th fission due to the very high concentrations of Th present, and the potential for incipient metamictization of the crystal lattice. High concentrations of rare earth elements, typical of this species, could also cause problems of neutron self-absorption by these elements during neutron irradiation (Wagner and Van den Haute1992), leading to an underestimate of the amount of uranium. However if alternative methods were applied to measure U and Th abundances directly then there would seem to be no reason that routine fission track dating could not be applied to this mineral. Such measurements could be achieved by electron microprobe (Montel et al. 1996, Williams et al. 1999) or by laser-ablation inductively coupled plasma mass spectrometry (Cox et al. 2000), making this approach a very fruitful avenue for future research.

Merrillite (“whitlockite”)

Merrillite, an anhydrous calcium phosphate mineral often coexisting with apatite in lunar and meteorite samples, has been used for a number of fission track dating studies of extraterrestrial materials. Following a paper by Fuchs (1962) this mineral was most commonly identified in meteorites as whitlockite, but Dowty (1977) has shown that it exhibits significant differences to terrestrial whitlockite, $(Ca,Mg)_3(PO_4)_2$, and should be distinguished from it. As a result, earlier publications use ‘whitlockite’ while later ones apply the name ‘merrillite’, for the same mineral. Merrillite is now the appropriate species name for the high-temperature phosphate mineral found predominantly in meteorites.

Fission track dating of merrillite has some unusual characteristics compared to the dating of terrestrial apatites due to the great age of these extraterrestrial samples. The fission track densities are extremely high, requiring modified track etching and counting procedures (e.g., Crozaz and Tasker 1981). Typically the tracks are very lightly etched compared to what would be used for normal observation in terrestrial minerals. As a result the tracks are very small and must be counted using scanning electron microscopy, either on the etched material itself or on a plastic replica of that surface, rather than by optical microscopy. An example of an etched merrillite grain from Apollo 12 lunar rock 12040 is shown in Figure 1 (Burnett et al. 1971).

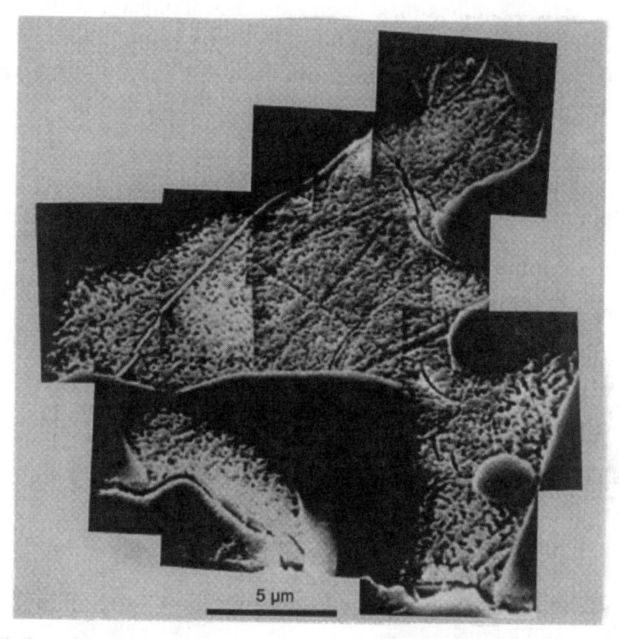

Figure 1. Scanning electron microscope image of the etched surface of a merrillite grain in lunar rock 12040. The sample has been very lightly etched in 0.12% HNO_3 for 20 s revealing an extremely high density of spontaneous fission track etch pits (numerous small dark dots). The grain has a considerable excess of ^{244}Pu tracks as well as ^{238}U tracks giving it an apparent fission track 'age' of 6.4 Ga. [Used by permission of the publisher, from Burnett et al. (1971) *Geochimica et Cosmochimica Acta*, Supplement 2, p. 1507.]

Two other factors unique to these extraterrestrial samples must also be accounted for. The first is that tracks from heavy cosmic ray particles (mainly Fe-group nuclei) and other nuclear interactions with cosmic rays may also be present, in addition to fission tracks. The simplest method for correcting for the cosmic ray background is to measure the track density in adjacent silicate mineral grains, such as feldspar and pyroxene, which do not contain uranium and are therefore free of fission tracks. The second factor is that the samples are so old as to contain tracks from the spontaneous fission of now-extinct transuranic elements, particularly ^{244}Pu with a half-life of 82 Myr. Such Pu tracks will only be present in samples older than about 3.9 Gyr (Crozaz and Tasker 1981).

The occurrence of extinct ^{244}Pu tracks first became apparent when a large excess of fission tracks over that which would be expected from ^{238}U spontaneous fission alone was discovered (e.g., Burnett et al. 1971). If calculated as a fission track age in the usual way, assuming all tracks were produced by ^{238}U spontaneous fission, these data led to apparent ages which were considerably greater than the age of the Solar System. Significantly an excess of tracks is not always observed in chlorapatites from similar lunar materials. In the case of some Apollo 12 lunar samples, apatites gave much younger fission track ages of only about 1,300 Myr (Burnett et al. 1971). It may be that the apatite ages represent a genuinely much younger age for the rock or the time of heating by a later thermal event. Another possible explanation would be that tracks in merrillite may be stable to higher temperatures than in apatite and so have survived later heating sufficient to substantially reduce the number of fission tracks in apatite. Mold et al (1984) have provided some experimental evidence that tracks in merrillite may indeed be stable to somewhat higher temperatures than in co-existing chlorapatites, but the difference is probably not significant over geological timescales.

A different explanation was provided by Pellas and Storzer (1975) who showed that uranium is more enriched in apatite than in merrillite, at least in meteorites, whereas plutonium is preferentially concentrated in merrillite. This would explain why a much greater ^{244}Pu track excess is typically observed in merrillites without the need for complex thermal histories. Burnett et al. (1971) found that in some Apollo 12 lunar samples, apatites and merrillites had similar uranium concentrations and generally smaller plutonium track excesses than are typically observed in meteorite phosphates. This difference may reflect the generally younger ages of the lunar rocks examined compared to most meteoritic materials. These observations have important implications for the early history of the lunar surface implying the formation and survival without later heating of crustal materials from the first few hundred Myr of its existence as a planetary body. Similarly, the retention of plutonium tracks and their relatively low thermal stability in merrillite have provided useful information about the cooling rates of meteorites in the early history of the solar system (e.g., Pellas et al. 1983, Storzer and Pellas 1977).

Apatite

Apatite initially accounted for a relatively minor fraction of the number of fission track age determinations, representing only about 20% of all measurements by the early 1970s, at which time natural glasses and micas were the most widely used materials. This fraction grew to around 40% by the mid 1980s, over 50% by 1990 and is probably in excess of 70% today. This increasing proportion is even more significant in that the total number of fission track analyses published has increased markedly throughout this period. The striking pre-eminence of apatite amongst the wide variety of minerals that have been utilised for fission track dating is because of the degree to which it consistently meets all of the following criteria. To be widely useful for routine fission track dating a mineral should be:

- well crystallized
- consistently enriched in uranium
- significantly larger in grainsize (>~60μm) than fission track dimensions
- uniform in its uranium distribution
- highly transparent and free of obscuring inclusions, and
- of common occurrence in all major rock groups.

Crystalline apatite meets all of these criteria extremely well, particularly the last which is a vitally important, but often under-appreciated, factor for any mineral to achieve widespread application. Many other minerals have been the subject of overly optimistic reports on their suitability for fission track dating but fail meet one or more of these requirements and so are limited in their application. For example, cryptocrystalline apatite has generally proved unsuitable because the grain size is too small and etching along grain boundaries obscures any tracks present. Any mineral that is not sufficiently common will always be severely limited in its application to real geological dating problems. Apatite on the other hand is a ubiquitous primary accessory phase in a wide variety of igneous and metamorphic rocks and is also a common detrital mineral in most clastic sedimentary rocks. For example, experience has shown that useful amounts and grain-sizes of apatite can be extracted from approximately 80% of sandstones (Dumitru et al. 1991).

Several additional characteristics of apatite have contributed to its dominance in fission track dating. These include its simplicity of handling in terms of mineral separation, mounting, polishing and etching procedures; its reproducible etching characteristics; and its well-characterized and reproducible response to elevated temperatures. It also has excellent optical characteristics and is mostly free of interfering inclusions and dislocation etch pits providing excellent conditions for observing fission tracks. Importantly apatite is also resistant to the effects of chemical weathering (Gleadow and Lovering 1974) which makes it a remarkably persistent detrital phase in sandstones, except where these have been affected by low pH groundwaters (Morton 1986, Weissbrod et al. 1987). Tracks in apatite are also resistant to the effects of pressure and brittle deformation (Fleischer et al. 1965a). Finally, apatite does not accumulate radiation damage from alpha decay and so does not become metamict. Incipient metamictization progressively modifies the properties of the next two most commonly dated minerals by the fission track method, zircon and titanite, adding complexity to their processing and limiting their applicability, especially in older rocks.

For all these reasons apatite can be considered an ideal mineral for fission track dating. It can be extracted in useful quantities from the great majority of crystalline and clastic rocks, and almost always contains suitable quantities of uranium (typically 1-50 μg/g). This means that it can be applied to a wide variety of geological studies, especially those requiring a significant regional coverage of sampling, and has thus become the standard mineral for the great majority of fission track dating studies. Apatite will be the only mineral considered further in this article.

FISSION TRACK DATING OF APATITE

The formation of fission tracks

In addition to the extinct isotope ^{244}Pu discussed above, the very heavy isotopes ^{232}Th, ^{235}U and ^{238}U all undergo spontaneous nuclear fission and remain in existence. In practice, only ^{238}U is a significant source of natural fission tracks in terrestrial minerals as it has a fission half-life of 8.2×10^{15} years (see Wagner and Van den haute 1992). While extraordinarily long, this is still 4 to 6 orders of magnitude shorter than the spontaneous fission half-lives of ^{232}Th and ^{235}U. In terrestrial apatites, therefore, it can safely be assumed that solely ^{238}U has produced all the fission tracks observed. These three heavy isotopes also decay through chains of alpha (^{4}He) and beta emissions to produce three different isotopes of lead. The alpha decay half-life is very much shorter for each of these isotopes than for fission so that for ^{238}U there are more than 10^{6} alpha decays for every fission event. However the alpha decay events do not themselves cause sufficient defects to produce etchable tracks and appear to produce no other permanent accumulation of lattice damage in apatite. It is worth emphasizing here that each spontaneous fission track represents a single ^{238}U nuclear fission event. It is this extreme sensitivity for the detection of individual nuclear events that makes the fission track dating method practical, despite the extraordinarily long half-life for the fission decay, which is around six order of magnitude longer than for the commonly used isotopic dating methods.

The mechanism of track formation is a combination of two simultaneous and very transient processes. Upon disintegration of the parent uranium nucleus, the fission fragments, with excess kinetic energy, are propelled from the reaction site in opposite directions. Each high-energy fragment interacts with the host lattice, fundamentally changing the properties of the atomic structure. The nature of this interaction changes as the particle loses energy and slows down Paretzke (1986). Total fragment energy imparted to the solid is the sum of both electronic and nuclear interactions by the fragments, in addition to some emitted radiation. Electronic stopping occurs when the fast moving ion undergoes inelastic collisions with lattice electrons. The number of these collisions is large and with each interaction the particle surrenders energy by stripping electrons from target atoms, by having its own electrons stripped away, and by raising the excitation levels of lattice electrons. As the projectile slows, nuclear stopping becomes more important. This regime is dominated by elastic collisions between two free particles (projectile ion and target atom) at low velocities, hence the "billiard ball" analogy (Kelly 1966). The ionization level of the projectile ion diminishes with continued loss of energy and by the time nuclear stopping becomes dominant, the ion is in a neutral state.

As Wagner and Van den haute (1992) pointed out, there is general agreement on the nature of the processes affecting the fission fragment as it slows and loses energy in a solid. What is less clear and has long been the subject of some debate is the nature of the processes suffered by atoms of the crystalline solid. Some authors (Vineyard 1976, Chadderton 1988, Spohr 1990) have supported a model widely accepted in ion implantation research: the "thermal spike". Here the term "spike" evokes the sense of a rapid, intense event and the term has found favor with many descriptions of the track formation process (Vineyard 1976, Fleischer et al. 1975, Chadderton 1988, Spohr 1990). The thermal spike concept treats the rapid deposition of energy into the lattice as a near-instantaneous heating event with a time span on the order of 10^{-12} seconds. According to the model, intense heat and multiple collision cascades generate a plasma in the track core (Wein 1989) followed by transport of the energy to the surrounding lattice by heat conduction processes (Chadderton 1988). Lattice defects created during the thermal activation process are quenched within the solid as the thermal perturbation rapidly decays (Spohr 1990) and the track core is left in a massively disordered state (Fleischer et al. 1975).

An alternative track formation model has long been accepted amongst fission track researchers (Wagner and Van den haute 1992). The "Coulomb explosion" model or "ion explosion spike", (Young 1958, Fleischer et al. 1965a,b; 1975), as its name suggests, is largely electrostatic in nature and explains the primary observation that nuclear tracks are only observed in dielectric solids. The massive positive charge on a rapidly moving fission fragment strips and excites lattice electrons along its path through the lattice (in this case apatite) leaving a core of positively-ionized lattice atoms. The resulting clusters of positive ions then explosively recoil away from each other as a result of coulomb repulsion as illustrated in Figure 2. Secondary cascades of atoms result from the barrage of ionized atoms that have been discharged from the track core. This is consistent with the results of small angle neutron experiments in other minerals, in which a significant density deficit within the track core is observed (Albrecht et al. 1982, 1985). The displacement cascades generate profuse interstitials and vacancies, which in turn cause distortions and elastic strain gradients within the lattice structure. This, it is argued, is responsible for the amorphous nature of lattice damage seen in latent/unetched tracks that have been examined by electron microscopy in apatite (Paul and Fitzgerald 1992).

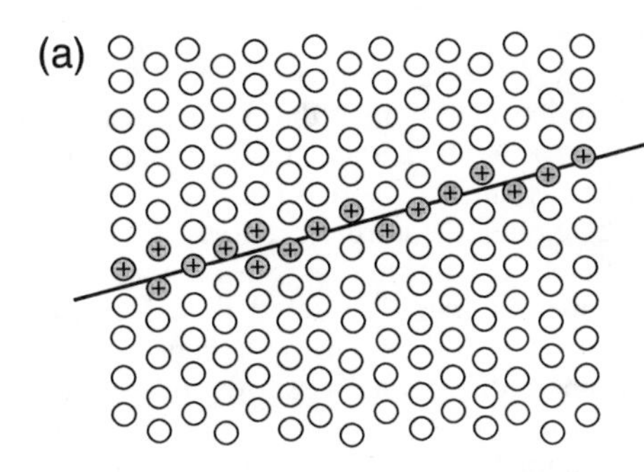

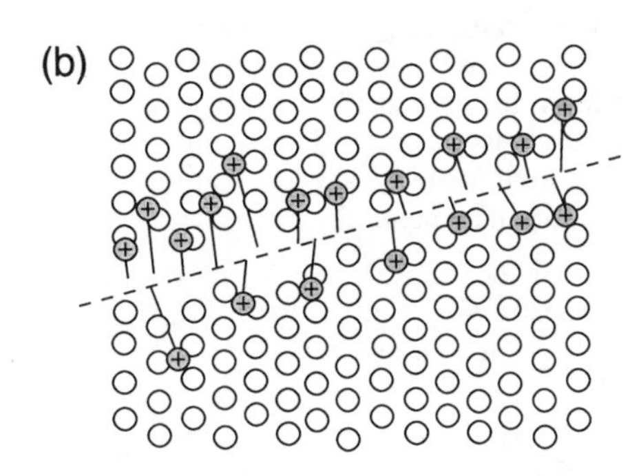

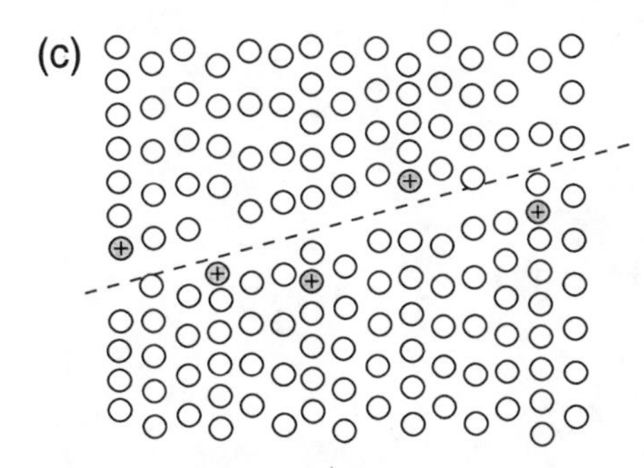

Figure 2. Three main stages of the ion explosion spike according to Fleischer et al. (1975). (a) The highly charged fission fragment ionizes lattice atoms along its trajectory. (b) Electrostatic repulsion causes displacement of lattice atoms along fragment path. (c) The matrix is strained elastically in proximity to defects and defect clusters. Some relaxation occurs in the lattice.

Track etching and observation

Apatites are usually prepared for fission track analysis by mineral separation from the host rock using conventional heavy liquid and magnetic techniques on a 60-200 μm size fraction of crushed rock. A sample of apatite crystals, usually several hundred grains, are then mounted in an epoxy film on a glass slide and mechanically ground and polished to reveal an internal surface cut through the interior of the grains. The tracks are then simply revealed by etching the polished mount in dilute HNO_3 (most commonly 5N) at room temperature for 20 seconds, although minor variations of etchant concentration and time are used in different laboratories. The form of the track etch pits varies according to the orientation of the polished surface relative to various crystallographic directions (Wagner 1968, Wagner and Van den Haute1992).

A preference was shown by some early workers for observation of fission tracks on basal surfaces of apatite under oil immersion. Tracks on these surfaces have an unusual and distinctive form, consisting of a broad hexagonal pit with a narrow track channel continuing into the crystal from the apex (Wagner 1968, 1969). Subsequently most workers have adopted observation using dry microscope objectives (100×) and using surfaces roughly parallel to prism faces (Gleadow et al. 1986) at a total magnification of 1000× or greater. Such observation conditions have several advantages, and are particularly favored for the external detector method of analysis that will be discussed below. Although oil immersion objectives are capable of higher optical resolution, the refractive index of immersion oil (n = 1.54) is very close to that of the apatites themselves, drastically reducing the optical contrast of the tracks relative to the host material. Dry objectives provide a substantially higher contrast in the image of fission tracks in apatite making them much easier to observe (Gleadow et al. 1986). In addition, dry objectives are much more convenient, particularly with motor drive microscope stage systems.

The typical prismatic habit of most apatite crystals means that they tend to align on prism faces during mounting in the epoxy mounting medium, so that after subsequent polishing many of the new surfaces exposed are roughly parallel to the *c*-axis. This means that a substantial fraction of the grains have surfaces which are in a consistent crystallographic orientation and reveal tracks as elongated narrow channels. Etching is usually continued until the tracks are approximately 1-2 μm in diameter as progressive etching experiments have shown that the number of tracks revealed is stable when counting at these dimensions (Gleadow and Lovering 1977). The appearance of well etched fission tracks on a polished surface of an apatite crystal is shown in Figure 3. Further etching reveals that the tracks have the form of faceted etch pits related to the underlying symmetry of the crystal as shown diagrammatically in Figure 4. Tracks on surfaces parallel to the crystallographic *c*-axis are needle-like when first revealed, and take on the form of faceted knife-blades with continued etching (Wagner 1969), contrasting with the broad hexagonal pits observed on basal surfaces (Fig. 4). These forms reflect the anisotropy of etching which follows different crystallographic directions. Even though the tracks are manifestly anisotropic in their etching behavior they appear to have a uniform distribution with angle to the *c*-axis so that tracks at all orientations are revealed satisfactorily for identification and counting.

Some spurious, non-track features, such as dislocations, may also be revealed by etching in apatite. Although these may be superficially similar to fission tracks in appearance, and very occasionally may may cause serious interference to track counting

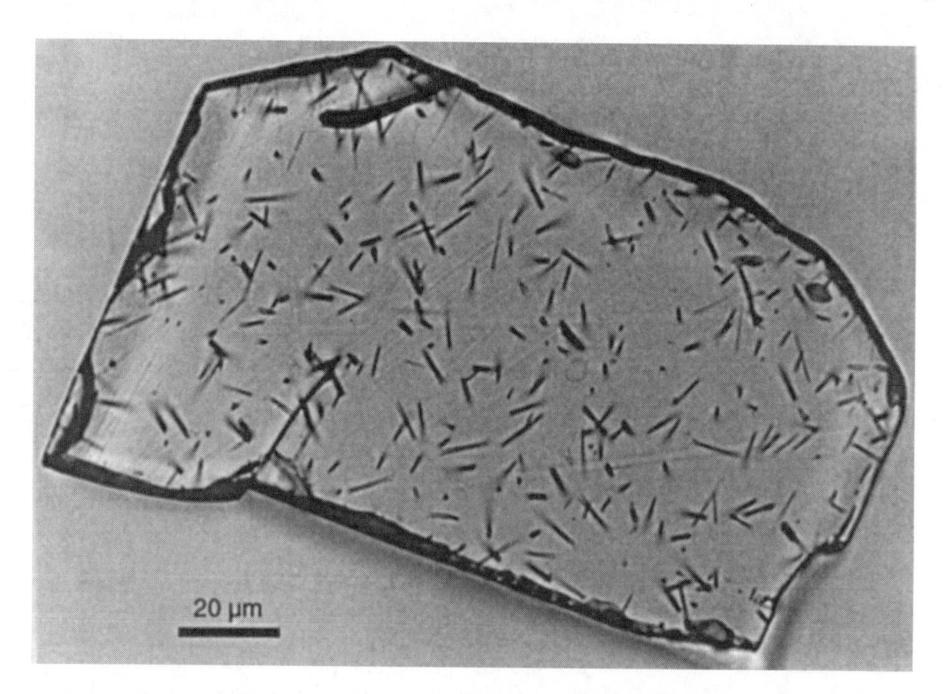

Figure 3. Etched fission tracks in an apatite grain from the Grassy Granodiorite of King Island, southeastern Australia. The tracks show their characteristic appearance as randomly oriented, straight-line etch channels up to a maximum length of around 16μm. The tracks appear as dark, high-contrast features in this image observed under a high magnification dry objective. Faint continuous lines across the surface are polishing scratches.

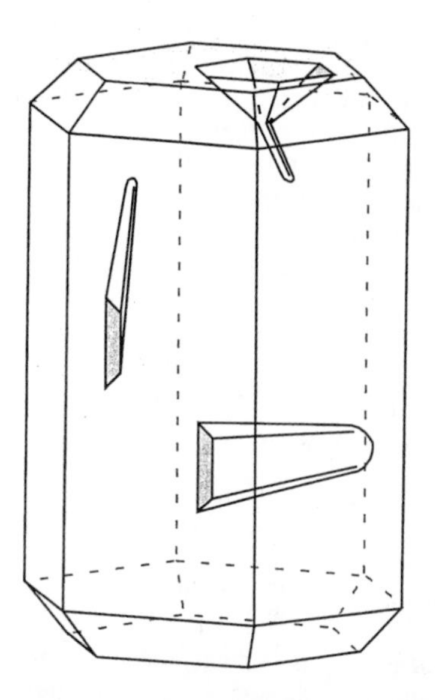

Figure 4. Schematic diagram showing the form of strongly-etched fission tracks etched on different crystal surfaces in apatite. The anisotropic etching which is characteristic of apatite produces knife-blade shapes for tracks on the preferred prismatic faces. Such features are only obvious after much longer etching times than usually applied for track counting.

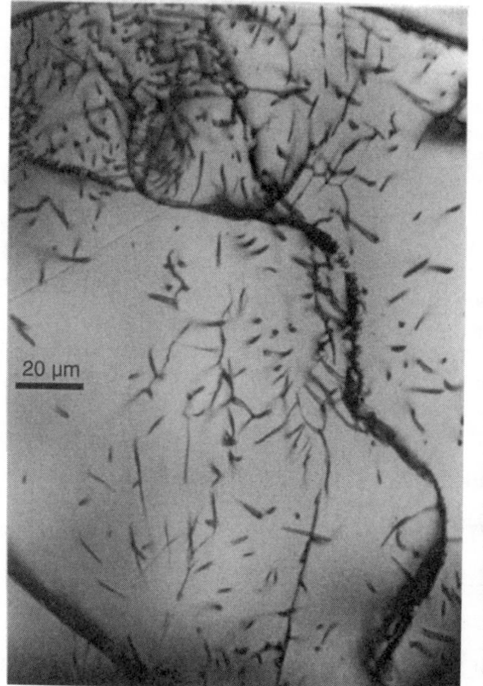

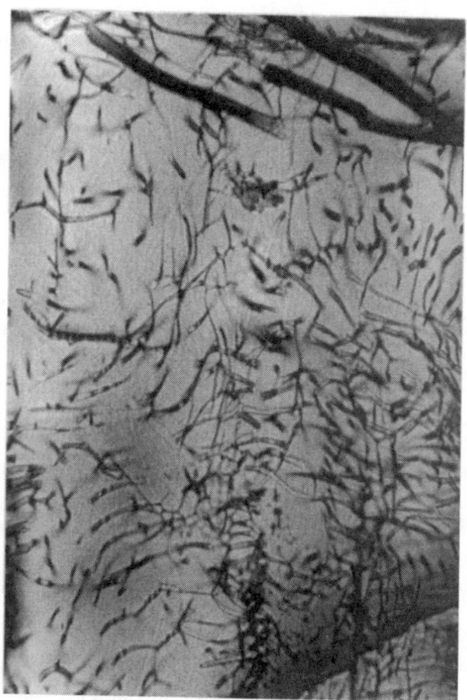

Figure 5. Dense swarms of dislocation etch pits in two apatite grains from the Early Cretaceous sandstones of the Otway Basin of southeastern Australia. Grains containing such high dislocation densities are most often observed in volcanic apatites and are best completely avoided for fission track counting. Typical dislocation features such as sub-parallel arrays, branching and extremely long etch channels can be seen. Fortunately, even in such samples, many grains are free of such features enabling reliable track counting.

(Fig. 5, above), tracks can be reliably discriminated from spurious etch features in that tracks are straight-line defects which are of limited length and randomly oriented. Dislocation etch pits, on the other hand, tend to be wavy or curved, are often of much greater length than tracks, frequently show branching, and often occur in swarms of parallel or sub-parallel features (Fig. 5). Such grains are best avoided altogether. Dense dislocation swarms are found most commonly in volcanic apatites but are extremely rare in apatites of plutonic origin, an observation that has not yet been fully explained. The distinction may relate to the different thermal histories of these two groups of apatite. Fortunately, however, the great majority of apatite grains, are almost entirely free of dislocation etch pits, even in the rock that produced the extreme examples in Figure 5. Therefore the selection of grains in which fission tracks can be identified is relatively straightforward and reliable in most apatite separates.

Fission track dating methods

The basic principles and practical methods of fission track age determination have been described elsewhere (e.g., Fleischer at al. 1975, Wagner and Van den Haute1992, Gallagher et al. 1998) and will be summarized only briefly here. Once fission tracks have been revealed by etching, the main parameter to be measured, which is representative of geological age, is the track density or the number of tracks per unit area on the etched surface. This is measured by counting the number of track intersections with the surface using a calibrated grid in the microscope eyepiece. For a given uranium concentration, the spontaneous fission track density will steadily increase through time, provided the tracks remain stable and are therefore quantitatively retained.

Determination of a fission track age requires several further experimental steps to measure the uranium concentration. The uranium concentration is not measured directly, but a second set of fission tracks is created artificially in the sample by a thermal neutron irradiation. This irradiation induces fission in a tiny fraction of the ^{235}U atoms, which are present in a constant ratio to ^{238}U in natural uranium. Knowing the total neutron fluence received during irradiation, the number of induced tracks provides a measure of the uranium concentration of the grain. Because the induced tracks are derived from a different isotope of uranium than the spontaneous tracks an important consideration in fission track dating is the assumption that the isotopic ratio of the two major isotopes of uranium, ^{235}U and ^{238}U, is constant in nature. With the notable exception of the unique “natural” nuclear reactors of Oklo in Gabon (Bros et al. 1998), where this isotopic ratio is disturbed, this is a very safe assumption. Numerous measurements have shown that ^{235}U and ^{238}U are always present in their natural abundances of 0.73% and 99.27%, respectively.

The fission track age, t, is then calculated from the ratio of spontaneous (ρ_s) to induced (ρ_i) track densities according to the standard fission track age equation (Fleischer and Price 1964, Naeser 1967):

$$t = \frac{1}{\lambda_D} \ln\left(1 + \frac{\lambda_D \phi \sigma I}{\lambda_f} \frac{\rho_s}{\rho_i}\right) \qquad (1)$$

where λ_D is the total decay constant for ^{238}U from all decay modes (effectively just the α-decay constant), λ_f is the fission decay constant, ϕ is the total neutron fluence received, σ is the thermal neutron cross section for ^{235}U, and I is the $^{235}U/^{238}U$ isotopic ratio for natural uranium (effectively a constant in nature). The neutron fluence is most conveniently measured by determining the induced track density, ρ_d, produced in a calibrated uranium-bearing glass irradiated along with the dating samples. The fluence is related to ρ_d, usually measured in an mica external track detector held adjacent to the standard glass, by the following:

$$\phi = B\rho_d \tag{2}$$

where B is a constant of proportionality. Substituting (2) into Equation (1) gives:

$$t = \frac{1}{\lambda_D}\ln\left(1 + \lambda_D \zeta \frac{\rho_s}{\rho_i}\rho_d\right) \tag{3}$$

where

$$\zeta = \sigma IB / \lambda_f \tag{4}$$

The aggregate constant ζ (zeta) is determined empirically by measurements on age standard materials following the initial suggestion of Fleischer and Hart (1972), subsequently elaborated by Hurford and Green (1982, 1983) and Green (1985). In principle it is possible to determine the component constants individually and calculate the ages absolutely (Wagner and Van den Haute1992) although various experimental difficulties with this approach have led to the emergence of the zeta calibration. The empirical zeta approach has been recommended by the IUGS Subcommission on Geochronology (Hurford 1990) and since adopted almost universally. The various issues involved in thermal neutron irradiation were discussed in detail by Green and Hurford (1984).

Experimental procedures

Measurement of a fission track age following Equation (3) requires the determination of three different track densities, ρ_s, ρ_i and ρ_d. The various different experimental strategies involved have been elaborated by Naeser (1979a), Gleadow (1981), Hurford and Green (1982) and Wagner and Van den Haute(1992) and are summarized in Figure 6. Of the five main alternatives, only the Population (PM) and

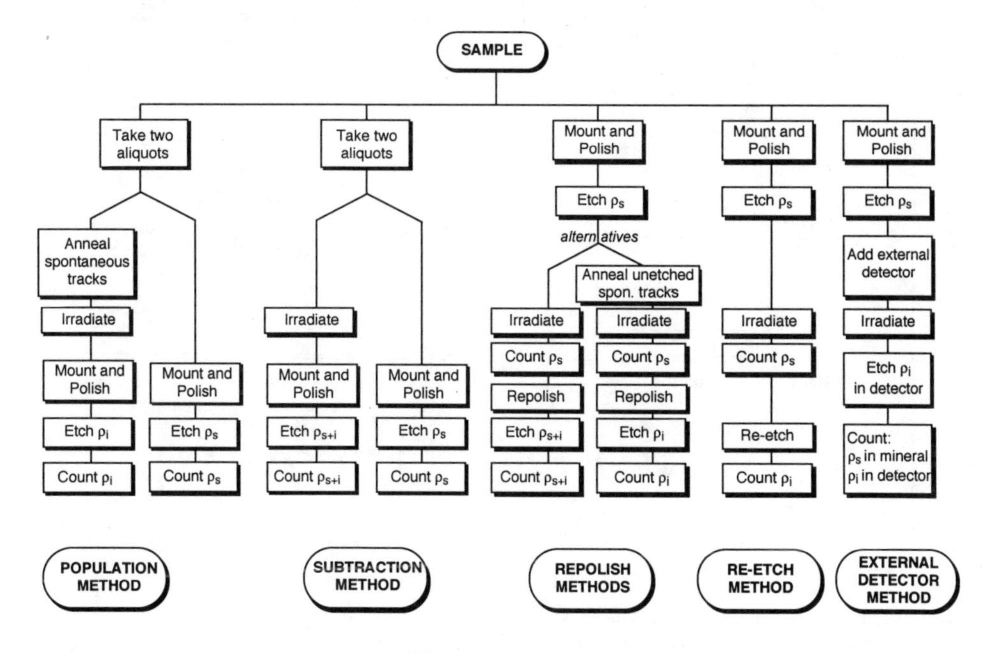

Figure 6. Schematic representation of different fission track dating procedures (after Hurford and Green 1982). Of these, only population and external detector methods have gained wide currency.

External Detector (EDM) methods have been extensively used for apatite and the EDM has now become the standard procedure in most laboratories. Galbraith (1984) describes the statistical treatment of analytical data derived by both these methods, and inter-laboratory comparisons have mostly demonstrated excellent reproducibility using both procedures (Miller et al. 1985, 1990).

The Population Method measures both ρ_s and ρ_i on internal surfaces within the apatite grains themselves, but on two separate aliquots, assuming that the uranium concentrations of the two aliquots are statistically equivalent. Whereas the PM was initially the preferred method for dating apatites (e.g., Naeser 1967, Wagner 1968) the EDM is now usually preferred because it provides age information on a grain-by-grain basis. The variability between single grain ages has turned out to be important (Galbraith and Green 1990, Galbraith and Laslett 1993), as expected in the case of dating detrital apatites from sedimentary rocks, where a spread of single grain ages could be anticipated, but surprisingly, also in application to igneous rocks (e.g., O'Sullivan and Parrish 1995).

The steps involved in the External Detector Method are illustrated in Figure 7. The spontaneous tracks are etched on an internal polished surface on the apatite grains and the induced tracks on a mica external detector attached to the grain surface during neutron irradiation. After irradiation the external detector is etched to reveal an induced fission track image corresponding to grains in the apatite mount. Even though this results in a second set of tracks being produced within the apatite grains themselves, these tracks will not be detected because they are not etched after the irradiation.

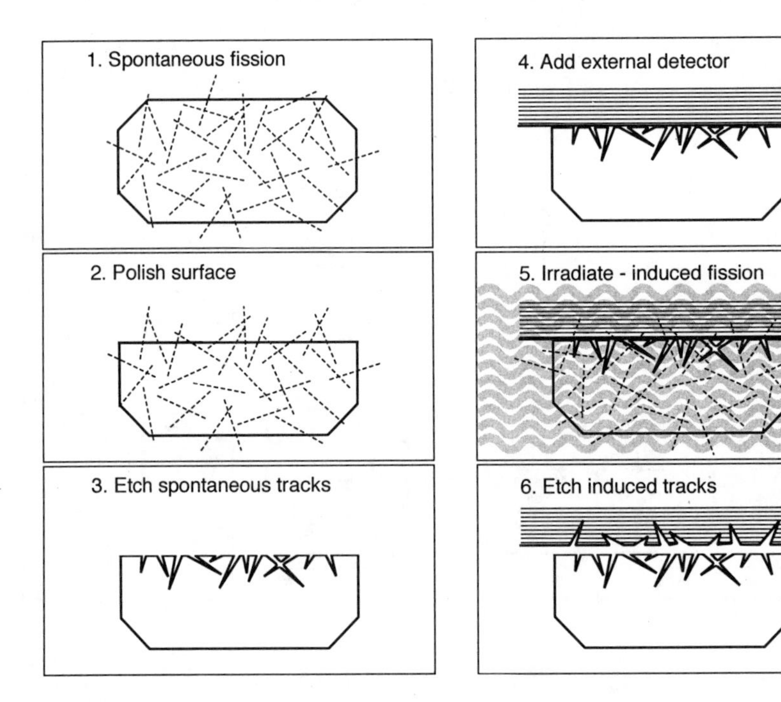

Figure 7. The sequence of steps involved in the external detector method of fission track dating. This method is now the dominant procedure used in most fission track dating laboratories for apatite because of its ease of handling, suitability for automation and its provision of single grain age information.

To measure the fission track age by the EDM involves determining ρ_s in a selected apatite grain and finding the mirror image area on the mica external detector where ρ_i is counted over exactly the same area. Because the geometry of track registration is not the same for the internal surface (4π) on which the spontaneous tracks are measured and the external detector surface (2π) used for induced tracks, a geometry factor must be introduced to correct for this difference. The geometry factor is ~0.5, but not exactly because of small differences in etching efficiency on the two surfaces and differences in the range of fission tracks in the two different materials (e.g., Gleadow and Lovering 1977, Green and Durrani 1978, Iwano et al. 1993). An individual age can be calculated for each grain counted using Equation (3). A combined age for the sample may also be calculated, usually using the 'central age' of Galbraith and Laslett (1993), which is a weighted mean of the log normal distribution of single grain ages. Errors on ages are usually calculated using the 'conventional method' described by Green (1981),

The EDM is broadly applicable to most minerals used in fission track dating and also allows for a high degree of automation to be applied. Computer-controlled microscope stage systems are now usually used for the purpose of matching areas of induced tracks on the mica external detector to the corresponding grains in the apatite mount (Gleadow et al. 1982, Smith and Leigh-Jones 1985, Crowley and Young 1988). Such systems provide a greatly enhanced productivity compared to older manual methods of grain location and additionally present an opportunity for systematic organization and management of data collection.

Track length measurements

Implicit in Equation (1) above is an assumption that the track lengths of spontaneous and induced tracks are the same. This is because the 3-D distribution of fission events within the apatite crystal is related to the observed track density on a 2-D surface via the average track length. In practice the lengths of spontaneous and induced tracks are never exactly the same for apatite (Gleadow et al. 1986) due to some shortening of the spontaneous tracks over their lifetime. As a result, some knowledge of the distribution of track lengths is essential in order to interpret properly the apparent fission track ages obtained.

From each fission event, the two fission fragments travel in exactly opposite directions to produce a single linear damage trail with an overall length equal to the combined range of both particles. The *etchable* length of each track is actually somewhat shorter than the combined range as there is a small 'range deficit' of up to several μm at each end of the track where the damage intensity is not sufficient to produce a continuous etchable track (e.g., Fleischer et al. 1975, Iwano et al. 1993). On an internal surface of an apatite crystal tracks at all lengths up to the maximum etchable length are observed, with the shortest tracks produced from a fission event almost one fission fragment range above the surface (e.g., Gleadow and Lovering 1977). Clearly such surface-intersecting tracks are randomly truncated at some arbitrary distance along their lengths, making their use in estimating the full etchable length difficult.

Several track length parameters have been utilized in fission track dating studies since the first use of projected track lengths in apatite by Wagner and Storzer (1972). They were the first to recognize that spontaneous tracks had different length distributions relative to fresh induced tracks in apatite. Dakowski (1978) established a clear understanding of the geometric properties of the various length parameters which was further extended by the work of Laslett et al. (1982). Following the work of Bhandari et al (1974) it has been shown that the greatest information about the true distribution of fission track lengths can be obtained from the measurement of horizontal "confined" tracks (Laslett et al. 1982, Gleadow et al.1986). Confined tracks do not intersect the polished surface, but are etched wholly within the body of the mineral where the etchant has gained access below the surface along other tracks or fractures. These have been identified as Track-in-Track (TINT) or Track-in-Cleavage (TINCLE) events by Lal et al. (1969). Several examples of such confined fission tracks, are shown in Figure 8 and procedures for measuring their lengths are described in detail by Gleadow et al. (1986).

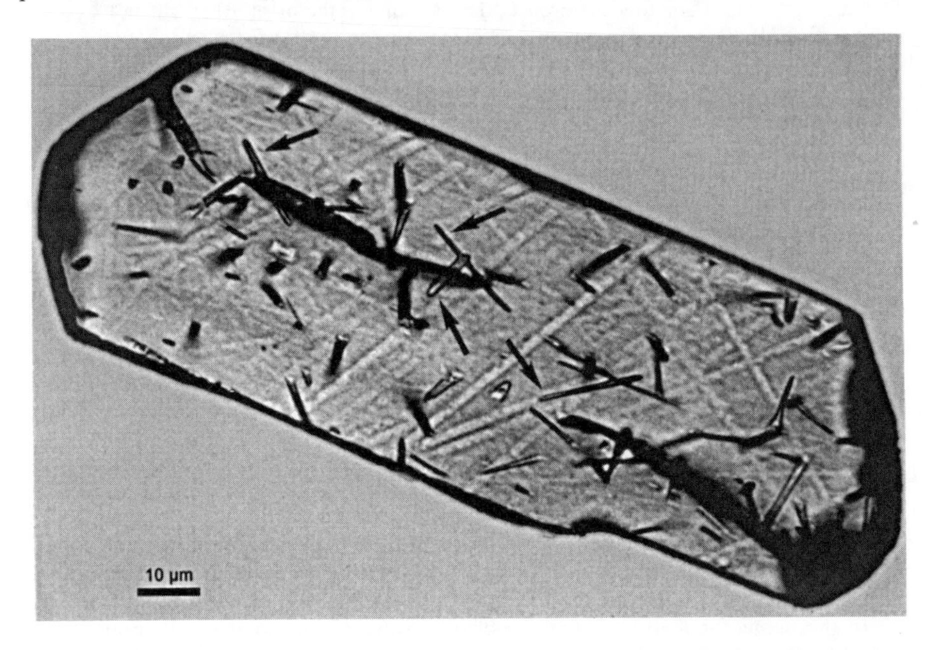

Figure 8. Etched spontaneous fission tracks on a polished internal surface cut in an apatite crystal, observed at high magnification using a dry microscope objective. Most of the visible tracks are surface intersecting spontaneous tracks which are used for age determination. Arrows point to four individual *confined tracks* which do not intersect the surface but are fully contained within the body of the apatite crystal and etched from fractures which allow passage of etchant from the surface to the tracks. Such confined tracks are used for length measurement and provide the closest approximation to the true distribution of etchable lengths of latent, unetched fission tracks. The center pair of confined tracks also illustrate the effect of anisotropic etching with the track across the grain being much wider than the narrow track which lies closer to the elongated c-axis of the crystal. After Gleadow et al. (1986).

The application of fission track length studies to the interpretation of fission track ages depends on three properties of spontaneous fission tracks.

1. All tracks in apatite have a very similar initial length (Gleadow et al. 1986), which is controlled by the energetics of the fission decay and the nature of the track recording material (e.g., apatite).
2. Tracks become progressively shorter during exposure to elevated temperatures so that the final length is controlled principally by the maximum temperature that each track has experienced.
3. New tracks are continually added to the sample through time so that each one has experienced a different fraction of the total thermal history.

These factors combine to give a final distribution of track lengths which contains a complete record of the temperatures experienced, below about 120°C. Different length distributions result from different styles of thermal history, as illustrated in Figure 9.

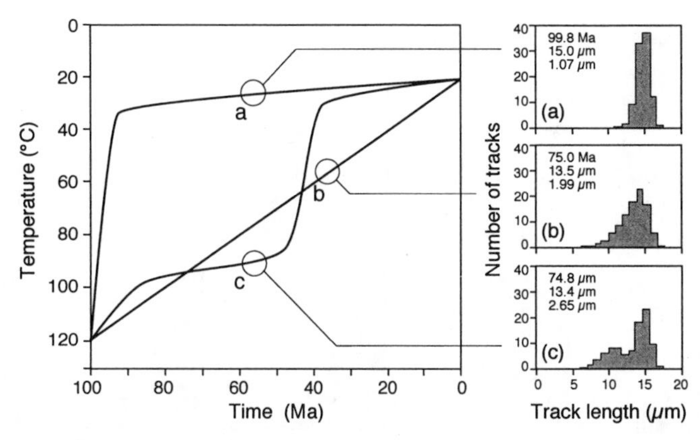

Figure 9. Predicted fission track age and length distributions resulting from three different hypothetical cooling histories, all cooling within the same temperature interval between 120° and 20°C. The resulting length distributions are shown as the three histograms on the right. The three numbers on each histogram represent the predicted fission track age (top), the mean track length, and the standard deviation of the length distribution (bottom). Case (a) shows an example of rapid cooling which results in an event age closely approximating the time of rapid cooling, (b) a steady cooling path leading to a broader, skewed distribution and a younger cooling age, and (c) a two-stage cooling history resulting in a mixed age and a bimodal track length distribution. Modified after Gleadow and Brown (2000).

When combined with the apparent fission track age, length distributions can be used to reconstruct the variation of temperature through time.

Observations of track lengths from a wide variety of surface rocks (Gleadow et al. 1986) show that distinctive patterns of track length characterize particular geological environments. Figure 10 illustrates the major length distribution categories identified by Gleadow et al. (1986) in rocks for which the thermal history is known, or can be inferred with reasonable accuracy. The undisturbed volcanic type is, as its name implies, characteristic of volcanic rocks that have remained undisturbed and at relatively low surface temperatures since their formation. A similar pattern will result in any rock which has cooled rapidly and not been re-heated thereafter. This type of distribution is similar to that shown by fresh induced tracks, although the mean length is slightly lower, indicating that some shortening occurs in spontaneous tracks even at ambient surface temperatures.

Apatite grains that have spent a significant period of time within the fission track annealing zone will show various patterns of broader length distribution, such as the undisturbed basement type, representing monotonic cooling from temperatures above about 120°C. More complex, multi-stage thermal histories will produce the even broader 'mixed' distributions. When the peaks in such a distribution are clearly resolved, as in the bimodal case, the distribution is indicative of a two-stage history with an older generation of tracks shortened during a later thermal event, and a new generation of long tracks produced subsequently. Such a bimodal distribution is particularly useful, giving information on the timing as well as the severity of the thermal event.

TRACK STABILITY

The phenomenon of fission track fading, or annealing, was first recognized by Silk and Barnes (1959) who showed that materials that had undergone heating contained shorter tracks than untreated specimens. The fading mechanism was initially interpreted

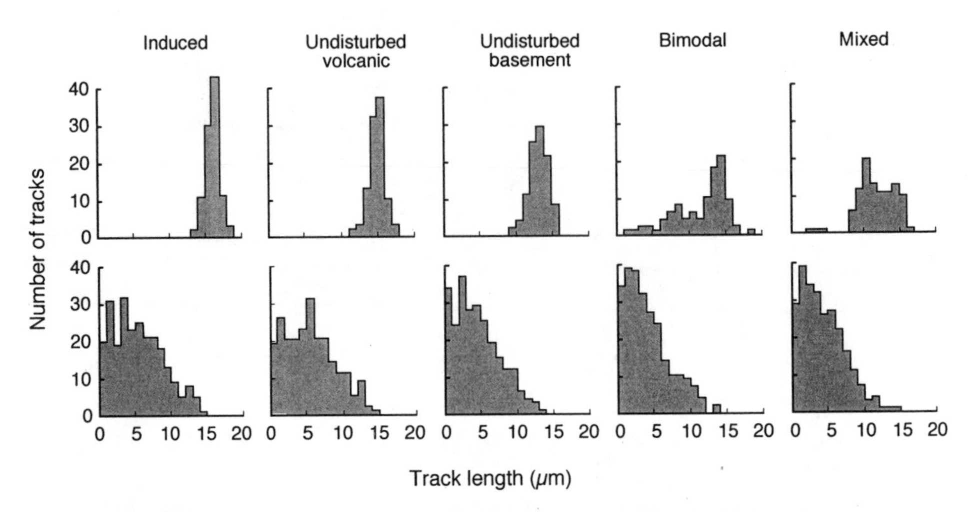

Figure 10. Representative track length distributions for spontaneous tracks in the various apatite length groups recognized by Gleadow et al. (1986). The top row represents measurements on horizontal confined tracks for which the differences between the different types are more distinctive than for the corresponding projected length distributions (bottom row). 100 confined tracks and 500 projected track lengths were measured in each case. After Gleadow et al. (1986).

by Fleischer et al. (1964) as a pinching out of segments along the track which prevented further access of the track etchant. Naeser (1979a) suggested that the diffusion rate within a heavily annealed track gradually approached that of the undamaged solid, so that it became increasingly difficult to etch out the track. It remains unclear precisely how the structure of the track, at the atomic level, may be altered in response to changes in environmental conditions, such as pressure, ionization, increased temperature, and duration of exposure (Fleischer et al. 1965b), but it is now generally accepted that temperature and time are the primary controls on track stability. Experimental studies by Green et al. (1986) confirmed that track fading was probably a two-part process, in which tracks at first begin to shrink from each end, while continuing to be etchable for the remainder of their length. Eventually track fading enters another phase, one of segmentation (Green et al. 1986), where the latent track cannot be fully etched and is apparently broken by small unetchable gaps (see also Hejl 1995). Transmission Electron Microscope observations of highly annealed tracks have provided additional evidence for unetchable gaps in highly annealed tracks (Paul and Fitzgerald 1992).

A fission track has its inception when two fission fragments pass violently through the crystal lattice at high velocities with an initial energy of around 1 MeV/nucleon (Fleischer et al. 1975). The metastable damage zone formed (Fig. 2) immediately begins to heal (Green 1980, Donelick et al. 1990) at a rate largely determined by the temperature of the sample and, to a lesser extent, the duration of the elevated temperature (Laslett et al. 1987). The tracks will continue to shorten until they cool to lower temperatures. The final length of each track therefore represents the integrated result of its passage through time-temperature so that each, in effect, behaves as maximum-recording thermometer.

Progressive shortening of the confined track length is accompanied by a reduction in the measured track density. This is due to the reduced probability of shortened tracks intersecting the polished surface of a grain and thus being exposed to the etchant. As the observed or apparent age of samples is determined on the basis of track density (Naeser and Faul 1969, Wagner and Reimer 1972, Nagpaul et al. 1974) considerable research aimed at determining the reduction of track density during annealing has been reported (Wagner and Storzer 1972, Bertel and Mark 1983, Laslett et al. 1984, Green 1988). The term "apparent" is emphasized since the observed age may be modified by the amount of track annealing that has occurred.

Variations in the observed fission track age with track fading initially seemed to be something of an impediment to mineral dating by this method. However, the systematic and progressive nature of the track length reduction presented an unparalleled opportunity to extract thermal history information in addition to the chronology (Wagner and Storzer 1972). Track length distributions observed in geological samples, were seen to be the net result of both track production and track fading processes over a span of geological time. Indeed, it was clear that there was a wealth of information available if the natural track distributions could be understood.

Annealing over geological time-scales

Typical crustal geothermal gradients are around 20-30°C/km so that the temperature at 4-5 km depth is in the range 100-120°C, allowing for surface temperatures of around 10-20°C (Pollack et al. 1993). The analysis of samples from deep bore holes (e.g., Naeser and Forbes 1976, Naeser 1981, Gleadow and Duddy 1981, Hammerschmidt et al. 1984), has provided direct evidence of natural thermal annealing of fission tracks in apatite over geological time-scales. Data from hydrocarbon exploration wells drilled within the Otway basin in southeastern Australia (Gleadow et al. 1983, Green et al. 1989a) clearly demonstrate a systematic reduction in the mean confined track length and apparent fission track age with increasing temperature (Fig. 11). At temperatures greater than about 120°C (depths > ~3km) no fission tracks are preserved within apatite and so the apparent fission track age and mean length are effectively zero. Both the apatite fission track age and mean track length decrease systematically with depth from the initial provenance age of ~125 Myr at the surface to zero at a depth of ~3.5 km forming a characteristic concave-up profile of apparent apatite age (Fig. 11).

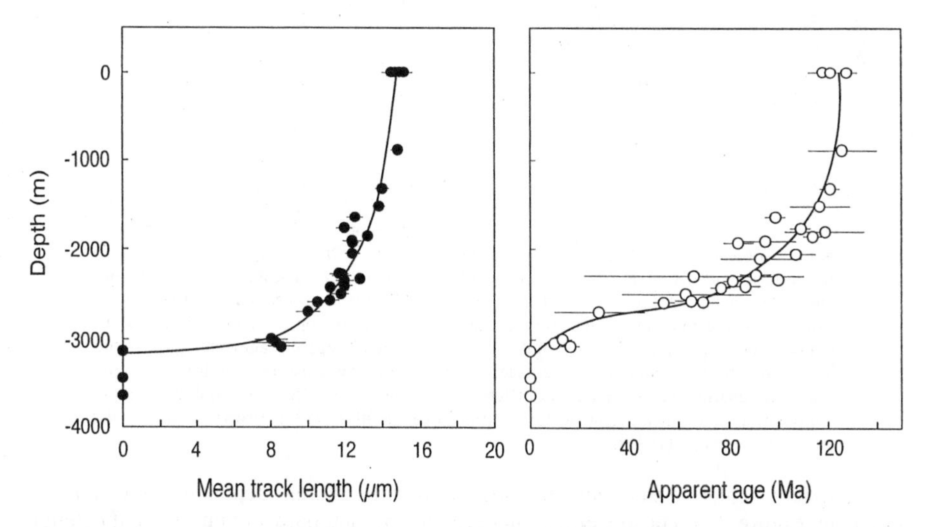

Figure 11. Composite apatite fission track crustal profiles of mean fission track length (●) and apparent apatite fission track age (○) plotted against depth for samples from several wells from the central Otway Basin in southeastern Australia. These clearly illustrate the progressive decrease in mean track length and apparent apatite fission track age with depth, and the characteristic concave-up form of both profiles. After Gleadow and Duddy (1981b) and Green et al. (1989a).

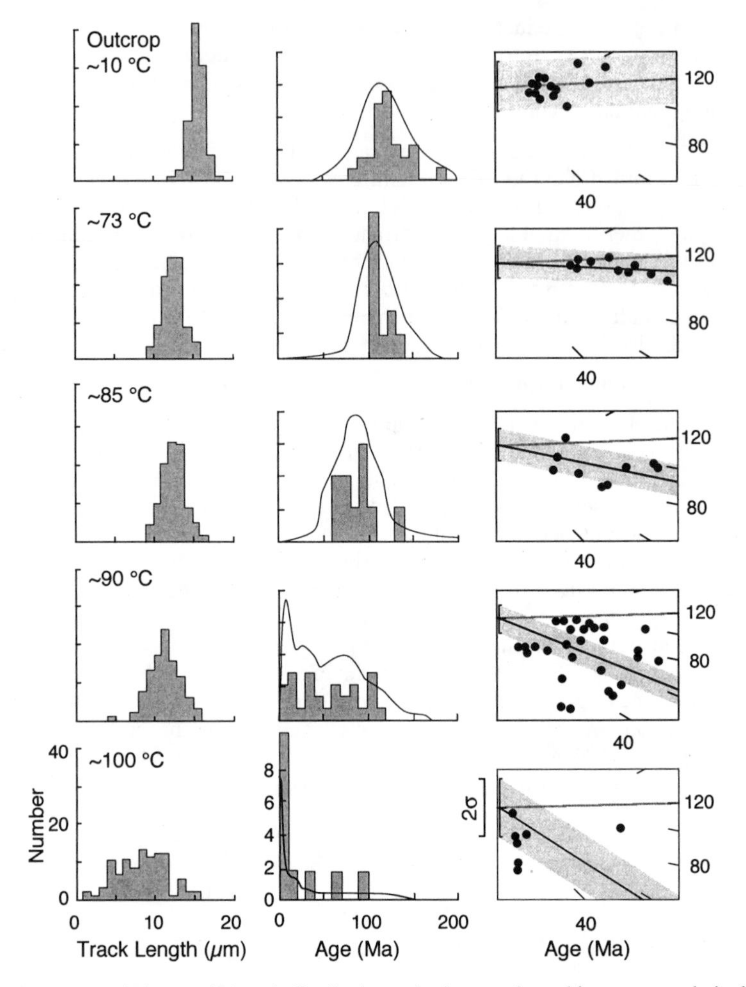

Figure 12. Fission track length distributions, single crystal age histograms and single crystal age radial plots (Galbraith 1990) for four individual samples from the central Otway Basin wells (data in Fig. 11). These are representative of successive degrees of thermal annealing and illustrate the progressive change in the shape of the track length distribution and dispersion in apparent single crystal fission track ages with increasing depth. The mean apparent age and the mean track length of the sample decreases with progressive thermal annealing from its original value of approximately 120 Ma. In addition the dispersion of track lengths and single crystal apparent ages increases as the degree of thermal annealing increases. This occurs because fission tracks in the individual apatite crystals anneal at different rates due to the effect of variable chemical composition and annealing anisotropy. After Green et al. (1989a), Brown et al. (1994), and Gallagher et al. (1998).

Four representative track length distributions for different depths in these wells are shown in Figure 12. The observed increase in the standard deviation of the length distribution with decreasing mean track length (higher T) is a consequence of the anisotropy of track shortening (tracks perpendicular to the *c*-crystallographic axis anneal faster than tracks parallel to the *c*-axis) as well as the variation in apatite composition between grains (e.g., Green et al. 1985, Green 1988). The thermal history for Otway basin samples can be reconstructed from the relatively simple burial history and indicates that this pattern was produced by heating times in the range of 10-100 Myr. Also shown in Figure 12 are a series of histograms of single grain ages with smoothed probability distributions and radial plots (Galbraith 1990) from the same four samples (Dumitru et al. 1991, Gallagher et al. 1998).

Other examples of natural thermal annealing can be seen in the vicinity of shallow level igneous intrusions. Calk and Naeser (1972), for example, demonstrated a systematic reduction in the apparent apatite fission track age of an 80 Myr old granitic pluton with increasing proximity to the contact with a small (~100 m) basaltic intrusion emplaced ~10 Myr ago. This pattern of age reduction within the granite was influenced by the thermal effect of the basalt intrusion and is consistent with the pattern of annealing observed in laboratory annealing experiments and deep drill holes.

In order to improve our understanding of the relationships between time, temperature and the observed track parameters in the natural environment, numerous experiments have been conducted over the last twenty-five years. Perhaps the most useful outcome of these annealing experiments has been the development of robust mathematical modeling.

Ever-more sophisticated models continue to be developed, from which geological interpretations of greater precision can be extracted. We will now examine some of the laboratory annealing experiments, as well as some of the current ideas on the processes involved, highlighting some the strengths and the weaknesses of the approach, and describing several problems that have been overcome to make the fission track technique a versatile and powerful tool in tectonic and landscape analysis. This review of some of the key developments in apatite annealing as it pertains to fission track thermochronology includes a selective and by no means exhaustive bibliography.

The process of annealing

When a solid contains defects that exceed the equilibrium concentration, as in the case of radiation or fission induced damage, the defects will react to reduce the free energy of the solid (Kelly 1966). These reactions include diffusion of defects to sinks, immobilization at traps, annihilation, and complexing and clustering with other defects such as vacancies, interstitials or impurities that may occur within the sample (Fig. 13), processes that are diffusion-limited rather than reaction-limited. The speeds of the reactions are controlled by the rate at which a defect species can move through the solid crystal lattice (Borg and Dienes 1988). The defect reactions proceed as a function of time, but the effect of temperature has long been known to play the most significant role (e.g., Haack 1977). Increases in temperature accelerate the process and, defects will be eliminated until some (quasi-) equilibrium state is established. The overall process of defect elimination is termed annealing (Kelly 1966).

Because of the observed relationship between time and temperature, fission track annealing data have traditionally been displayed on a variation of an Arrhenius plot (e.g., Haack 1972, Laslett et al. 1987) using Equation (5).

$$t = A\exp\left(\frac{E}{\mathrm{k}T}\right) \tag{5}$$

Where t is time, A is a mineral specific constant, E is activation energy and T is absolute temperature, with k being Boltzmann's constant. Early researchers (Fleischer et al. 1965a, Wagner 1968, Naeser and Faul 1969, Wagner and Reimer 1972, Nagpaul et al. 1974) noted that annealing data could be represented as iso-density contours (for track density data) or alternatively iso-length contours (for track length data) on such a plot (Fig. 14).

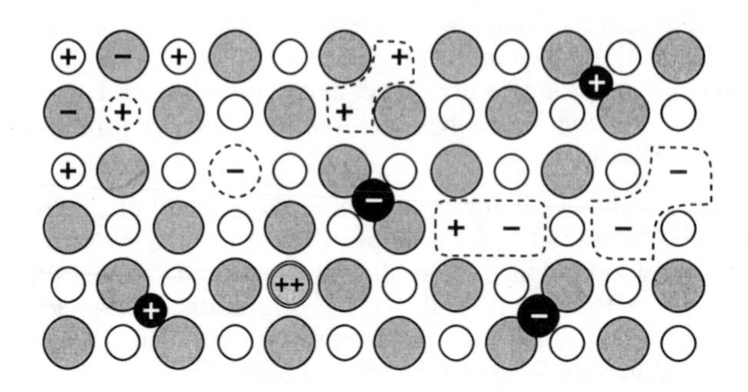

Figure 13. Schematic representation of some important diffusion species in ionic crystals, including cation and anion vacancies, cation and anion interstitials, cation and anion divacancies, cation-anion divacancy, allovalent cations (e.g., REE). These are by no means all the likely defect species but illustrate the complex kinetics involved (modified after Borg and Dienes 1988).

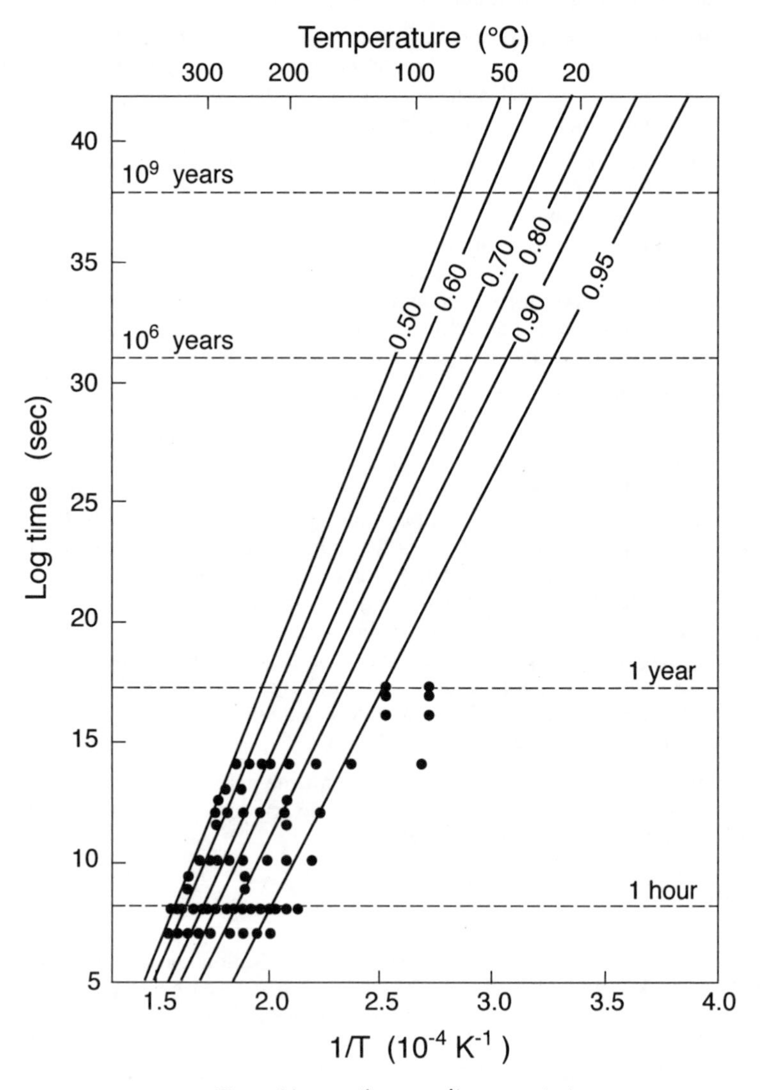

Figure 14 — caption opposite page. →→

The Arrhenius equation describes the kinetics of a reaction process but there has been some debate about how to establish the relevant activation energies. Gold et al. (1981) argued that track density measurements were unrelated to the rate constants for the underlying reaction. While Gold et al. (1981) accepted the empirical nature and usefulness of the derived values, they concluded that these parameters were in fact "physically meaningless". The use of the Arrhenius equation, however, gained some justification when Goswami et al. (1984) established a series of analytical equations for the quantitative assessment of variable-temperature track annealing. Goswami et al. (1984) explicitly assumed the validity of the Arrhenius equation in dealing with track annealing data, while Green et al. (1988) presented detailed arguments to refute the interpretation of first-order reaction kinetics for either the shortening of tracks or the reduction of track density. Green et al. argued that while the fundamental process of defect transport in the solid may be described as first-order, the diffusion process becomes extraordinarily complicated in an anisotropic, polyatomic crystal with a spectrum of defect species.

Problems in track measurement

Measurement of the individual diffusion parameters that affect fission track annealing is not routinely undertaken. Rather, we are forced to use a proxy, the tubular hole that remains in the crystal after chemical etching of the fission damage trail. As pointed out by Crowley et al. (1991), etching involves not only removal of the damage trail itself, but also an unknown amount of the host crystal. The diameter of the etched track (~1 μm) is orders of magnitude larger than the unetched latent track (~10 nm) with a consequent loss of chemical and structural detail. Early studies of track annealing were generally based on track densities, since a clear Boltzmann-law relationship between the track density, time and temperature (Eqn. 5) had been recognized (Fleischer and Price 1964, Naeser and Faul 1969, Haack 1972).

However, track density measurements (e.g., Bertel and Märk 1983) were insensitive to the subtle variations in track length that occur, particularly in samples with complex thermal histories, so that valuable information that may assist with interpretation was lost. And, although a direct relationship between length and density distributions was observed, there was considerable interest in the form of that relationship (Green 1988). Difficulties of inter-laboratory comparison of track density data also may have inhibited developments in fission track modeling (Green et al. 1988). As the method evolved, however, it became clear that confined track lengths provided greater precision in constraining annealing processes and were more amenable to modeling than track densities (Gleadow et al. 1983, 1986; Crowley 1985, Green et al. 1986). As a consequence, there has been widespread acceptance of the utility of combined track length and density measurements for fission track studies. Nevertheless, the track density method continues to find application in some specialized research fields (Carpéna 1998).

The nature of size distributions in etched track lengths is further complicated by the difficulty in determining the dimensions of unetched, or latent tracks. While there were some assumptions about the shape of a latent fission track (see Carlson 1990), their geometry has proven to be notoriously difficult to determine (Kobetich and Katz 1968). In a classic case of the act of observation modifying the observed property—radiolytic annealing was observed in apatite when samples were exposed to the electron beams

Figure 14. Illustration of the extrapolation involved from laboratory scale data (Green et al. 1985) to the geological scale (i.e., months to millions of years). In this case the iso-length contours are fitted using the fanning model of Laslett et al. (1987). Extrapolations of this order magnify the differences between each of the annealing models of Laslett et al. (1987), Carlson (1990) and Crowley (1991). Adapted from Laslett et al. (1987).

applied in TEM (Silk and Barnes 1959, Paul and Fitzgerald 1992) and microprobe analysis (Stormer et al. 1993). There has also been debate regarding the relationship of defect distribution to latent track geometry (Dartyge et al. 1981, Albrecht et al. 1982, 1985; Villa et al. 2000). Even the absolute, initial length of the latent tracks in apatite was (and remains) problematic, partly due to the very process of annealing. Fission fragment ranges have been calculated using range-energy codes (see Henk and Benton 1967, Green 1980, Crowley 1985), which produce an approximately Gaussian distribution. Ranges generated using the Ziegler et al. (1985) "SRIM" package, result in a negatively skewed distribution of fragment ranges.

In an unusual annealing experiment, Donelick et al. (1990) addressed the widespread observation of earlier researchers (e.g., Wagner and Storzer 1970, Bertel et al. 1977, Green 1980, Gleadow et al. 1986) that natural (spontaneous) fission tracks are always shorter than laboratory-induced tracks. Donelick et al. (1990) conducted a series of rapid irradiation and etching experiments and showed that an initial phase of track fading occurs on a remarkably short timescale, even at room temperature. This rapid shortening impacts on the common perception that tracks are completely stable at low temperatures (Naeser 1979a). A shortening of ~0.5 μm took place at 23°C over three weeks following irradiation, but beyond this time, additional track length shortening was undetectable. Issues of precisely what is a fission track and what are its fundamental, measurable properties are not the only ones facing researchers. The host mineral, apatite, presents us with a number of complexities that have challenged fission track researchers.

Compositional effects

Gleadow and Duddy (1981b) observed variability in the annealing properties of individual apatites from deep wells in the Otway Basin that they attributed to compositional differences. Subsequently, Green et al. (1985) clearly demonstrated a preferential retention of tracks in chlorine-rich apatites relative to fluorine-rich (Fig. 15). These effects have now been widely observed in sedimentary rocks, where the variable provenance of detrital grains contributes to inter-sample variation (Burtner et al. 1994, Corrigan 1993, Stockli et al. 2001), as well as igneous rocks, (e.g., O'Sullivan and Parrish 1995). While fluorine-rich apatites (such as Durango) typically show complete annealing of natural geological samples at temperatures of 90-100°C, chlorine-rich samples are characterized by an increase of the total annealing temperature to around 110-150°C (Burtner et al. 1994).

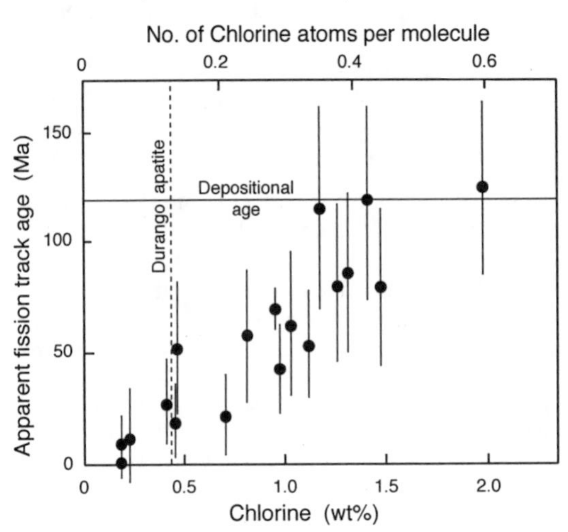

Figure 15. Variation of individual grain age with chlorine concentration in a single rock sample of volcanogenic sandstone from the Otway Basin in southern Victoria, Australia. The sample was recovered from a depth of 2585 m in the Flaxmans-1 well, where the current temperature is 92°C. The depositional age of the sandstone is indicated by a horizontal line, and the chlorine concentration of the Durango apatite is shown as a vertical dotted line, for comparison. High-chlorine grains resist annealing and thus record an older age than fluorine-rich grains. Cl concentrations are expressed as wt % and as number of atoms per $Ca_{10}(PO_4)_6(F,OH,Cl)_2$ molecule. Modified after Green et al. (1985).

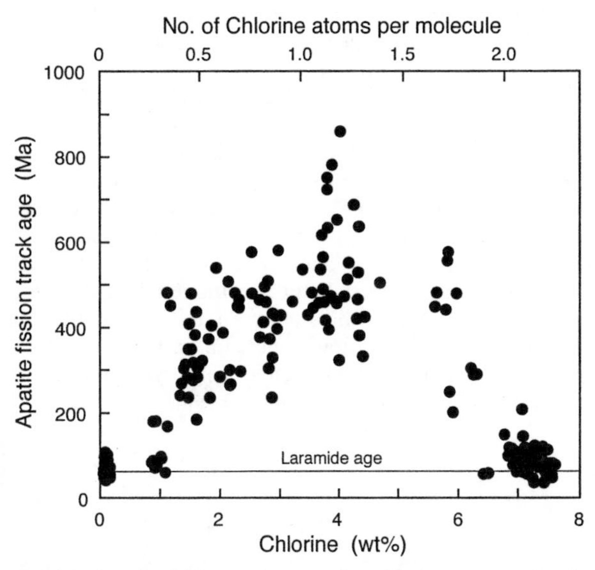

Figure 16. Plot of apatite fission track age against chlorine concentration for apatite grains in a suite of samples from the Stillwater Complex, Montana. Apatites from these rocks exhibit an unusually wide range of chlorine concentrations. A positive correlation is observed between fission track age and chlorine concentration up to about 4% Cl, reflecting an increasing resistance of the apatite grains to annealing. The most chlorine-rich apatites (~6-7.6 wt % Cl), however reverse this trend and generally yield late Cretaceous-early Tertiary apatite fission track ages similar to those for fluorine-rich apatites. The youngest ages are considered to date relatively rapid regional cooling related to the Laramide Orogeny, the approximate age of which is indicated. The reversal in trend is attributed to a change in apatite fission track annealing behavior accompanying the transition from hexagonal to monoclinic symmetry in the most chlorine-rich apatites. Scales as for Figure 15. Modified after Kohn et al. (2002a).

At temperatures below that of total annealing, chlorine-rich grains may retain more short track lengths (and thus have a shorter mean track length) than fluorine-rich grains experiencing the same thermal history. Thus chlorine-rich grains typically record older apparent ages despite experiencing the same thermal history. An example of this effect in apatites with an exceptional range of halogen compositional variation (sometimes at the hand-specimen level) is reported by Kohn et al. (2002a) from the mafic Archaean Stillwater Complex, Montana (Fig. 16). This compositional dependence response has two important implications. First, samples, particularly sedimentary ones rich in detrital apatite, should be analyzed for halide content to avoid misinterpretation of the ages and thermal history. Second, once the relative annealing kinetics are understood, the added complexity of multi-compositional grains provides the analyst with the power of additional thermochronometers contained in a single sample. Measurements of apatite chlorine content have become largely routine, and a variety of analytical approaches are used (Stormer et al. 1993, Sidall and Hurford 1998). Although the rare earth elements are a significant trace constituent of apatite (Roeder et al. 1987, Hughes et al. 1991b) their influence on annealing processes has been little studied (but see Hurford et al. 2000).

Crystallographic effects

The structural anisotropy of apatite (Sudarsanan and Young 1978, Hughes et al. 1989, 1990) affects the annealing properties of tracks lying in different crystallographic orientations (Green and Durrani 1977). It appears that the clearly defined channels (notably the anion columns—see Hughes et al. 1990) parallel to the *c*-axis in the crystal structure favor transport of diffusing species (Fig. 17). This results in more rapid annealing of tracks orthogonal to this axis (Green and Durrani 1977). In contrast, tracks oriented parallel to the *c*-axis, appear to be influenced by a less favorable diffusive transport across the crystal lattice, and hence shrink more slowly. The crystal retains a track length distribution spanning these two extremes. As annealing proceeds, the track length anisotropy becomes more obvious (see Wagner and Van den haute 1992). Green et al. (1986) pointed out that observation of horizontal confined tracks on a polished section parallel to the *c*-axis will expose the full angular spectrum of track lengths whereas basal sections will only sample the shortest lengths orthogonal to this axis. Green (1988) subsequently investigated the relationships between anisotropy and length-bias in track shortening and their effects on fission track ages. Later, Donelick and co-workers have significantly expanded the experimental database on crystallographic orientation and anisotropic annealing (Donelick et al. 1990, 1999; Donelick 1991).

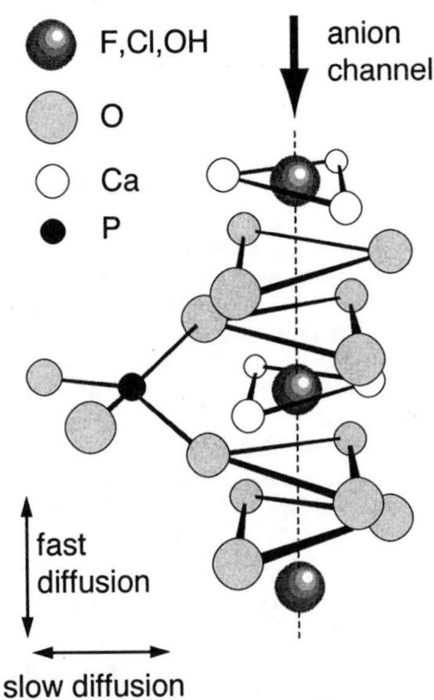

Figure 17. Approximate location of F, Cl and OH ions in the anion column parallel to the *c*-axis of apatite(Hughes et al.1989). The ion position shown is for F^-, whereas Cl^- and OH^- are accommodated by disorder above and below the mirror plane (modified after Sudarsanan and Young 1978).

Numerical annealing models

The development of numerical annealing algorithms has proven to be an enormous analytical breakthrough and is the key to interpretation of results in the majority of recent geological applications. What follows is intended as an overview of the most widely used algorithms, rather than a rigorous assessment of each individual approach. Several other authors provide greater detail and informative discussion on all of the models mentioned here (Gallagher 1995, Willet 1997, Ketcham et al. 1999). By the early 1980s, some classic papers (Wagner and Reimer 1972, Wagner et al. 1977, Naeser 1979b, Gleadow and Duddy 1981b) had already demonstrated the power of fission track analysis in addressing important problems in geology. Interpretation, however, remained a somewhat imprecise art, although several mathematical descriptions of the annealing process had been proposed.

Two contributions (Bertagnolli et al. 1983, Crowley 1985) in particular established formal mathematical treatments to describe the production and shortening of fission tracks in response to thermal history. The approach by Bertagnolli et al. (1983) did not gain wide acceptance, but remained central to much of the work by the French Besançon research group which applied a "convection-type" equation, (Chambaudet et al. 1993, Meillou et al. 1997, Igli et al. 1998). Crowley (1985) used a semi-analytical solution to the problem of track length shortening. He examined several characteristic track-length "signatures" for samples with simple thermal histories and, perhaps more importantly, recognized the applicability of inverse modeling for geological samples. The following year, Green et al. (1986) reported on an extensive laboratory data-set (Fig. 18) that derived from confined track lengths, signaling a major advance in the apatite fission track technique. They also observed that annealing was not accomplished by simple shortening but included a rapid segmentation at high degrees of annealing. The data covered samples heated for times ranging from 20 min to 500 days at temperatures between 95° and 400°C and provided the basis for an empirical, mathematical model published the following year —the widely cited "Laslett et al. (1987) model".

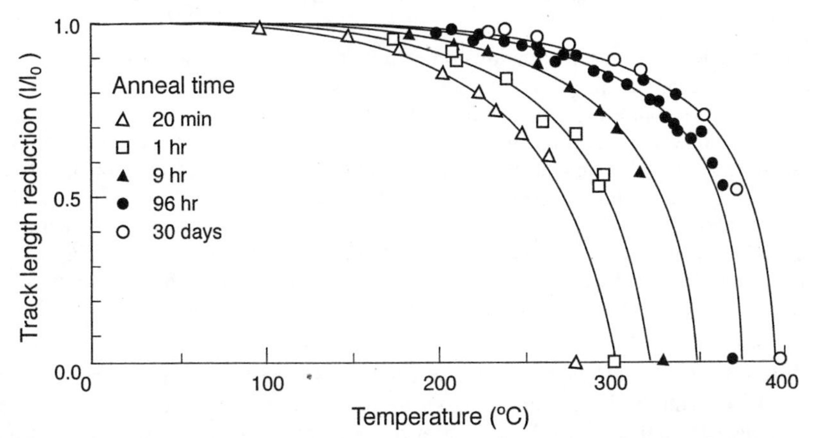

Figure 18. Summary of isochronal laboratory annealing data for confined track lengths in Durango Apatite (Green et al. 1985) for annealing experiments conducted over intervals ranging from 20 min to 30 d. Curves were fitted using the fanning Arrhenius model based on Equation (6a) in the text (adapted from Laslett et al. 1987).

Laslett et al. (1987) based their model on a "fanning Arrhenius relationship" (Fig. 14) between reduced track length r (= l/l_0 where l is the measured track length and l_0 is the initial length), log time (t), and inverse absolute temperature (T). They derived the following equation for constant temperature annealing that accounted for 98% of the variation in the observed data:

$$\left[\left\{\left(1-r^{2.7}\right)/2.7\right\}^{0.35}-1\right]/0.35=-4.87+0.000168T[\ln(t)+28.12] \quad (6)$$

Subsequent publications by the Melbourne research group illustrated the model's application to variable temperatures (Duddy et al. 1988) and then extrapolated the model to geological time-scales (Green et al. 1989). This collection of papers established this model as the pre-eminent forward model in apatite fission track analysis at the time. As a geological tool, however, its major weakness is that it does not allow for the effect of apatite composition on annealing, being developed entirely on the basis of a single apatite —Durango, a relatively low-chlorine apatite (~0.4 wt % Cl) (see Fig. 15). Hence, application to significantly different compositions will tend to give erroneous paleo-temperature estimates. Nevertheless, with the compositional caveat in mind, this model has proved invaluable in understanding low temperature thermal histories in many areas of the world. Laslett and Galbraith (1996) developed an improved version of the Laslett et al (1987) model with additional parameters. Further adaptations of the Laslett et al. (1987) model have been developed in the commercial environment to address the compositional issue, but remain as yet unpublished.

The Laslett et al. (1987) model was soon followed by additional research on the compositional dependence of annealing. Crowley et al. (1991) conducted over 100 heating experiments that included fluorapatite and Sr-apatite. Their observations suggested that track annealing was consistently anisotropic at all stages of fading, thus allowing tracks of varying orientations to be normalized to a mean track length. They concluded that their annealing results and the data of Green et al. (1986), based on Durango, could be fitted to the following equation:

$$\left[\left\{\left(1-r^{4.3}\right)/4.3\right\}^{0.76}-1\right]/0.76=-1.508+\left[\frac{2.076\times10^{-5}\ln(t)+2.143\times10^{-4}}{1/T-9.967\times10^{-4}}\right] \tag{7}$$

Annealing models with these equations lie on a fanning-linear Arrhenius-type plot similar to that of the Laslett et al. (1987) model. The Crowley et al. (1991) model appeared to accommodate the composition problem, in that a near end-member fluorapatite was observed to anneal more readily than Durango (0.4 wt % Cl). However, on extrapolation to geological time-scales this particular model actually predicts fluorine-rich apatite to be more resistant than chlorine-rich apatite (Gallagher 1995). Crowley et al. (1991) interpreted the observed increase in resistance to annealing that occurs during track fading as the prime cause of the fanning form of the "Arrhenius-type" plot (Fig. 14). The fanning results from steepening of the iso-annealing contours and implies an increase in "activation energy" (Gold et al. 1981), which has been interpreted as a type of anneal "hardening".

Only one serious attempt has been made to describe fission track annealing by a physical kinetic model based on atomic-scale mechanisms (Carlson 1990). In essence, Carlson suggested a mechanism where short-range atomic motion caused radial shrinkage of the fission track. Using a simplified track geometry, the change in radial defect distribution could be directly related to a concomitant reduction in track length. The model incorporated parameters for defect distribution, activation energy of the process, a rate constant as well as a parameter controlling the proportion of shortened tracks that suffer segmentation. The kinetic model established the following primary constitutive equation for defect elimination:

$$\frac{\mathrm{d}N}{\mathrm{d}t}=-c\left(\frac{\mathrm{k}T(t)}{\mathrm{h}}\right)\exp\left(\frac{-Q}{\mathrm{R}T(t)}\right) \tag{8a}$$

Here N is the number of defects, t is time, and c is an empirical rate constant. The constants k, h and R are Boltzmann's, Plank's and the gas constant respectively. The term Q is derived from published experimental data, as are the terms A and n in the following equation (Eqn. 8b) where τ is a dummy variable of integration over time and l_o is initial track length (Carlson 1990). When Equation 8a above is combined with the equations describing an approximately Gaussian radial defect distribution, and a function linking axial reduction to varying radius, the resulting equation is:

$$r=1-\frac{\mathrm{A}}{l_0}\left(\frac{\mathrm{k}}{\mathrm{h}}\right)^{n}\left[\int_0^{1}T(\tau)\exp\left(\frac{-Q}{\mathrm{R}T(\tau)}\right)d\tau\right]^{n} \tag{8b}$$

The model attracted considerable criticism and discussion (Crowley 1993a, Carlson 1993a, Green et al. 1993, Carlson 1993b) focusing on two key areas of the model. Firstly, both took issue with the validity of the physical model and its mechanisms. It was argued that the proposed structure was not based on available physical evidence (Green et al. 1993) and that the mechanism and kinetics of the defect elimination were implausible (Crowley 1993a). Secondly, both dismissed Carlson's (1990) model predictions as having an inadequate fit with the laboratory data sets.

Nevertheless, Carlson's essentially semi-empirical method has gained acceptance, and has been used as the basis for further developments such as the "multi-kinetic" model of Ketcham et al. (1999). This paper is one of three (Carlson et al. 1999, Donelick et al. 1999, Ketcham et al. 1999) that have addressed many of the earlier criticisms of the Carlson (1990) model. This research group has also produced a substantial annealing data-set of mixed-compositional apatites and established a model to deal with crystallographic effects, both of which have been incorporated into their full annealing model.

Modeling at an atomic level

Although the empirical models do not provide significant insight into the fundamental processes that occur during track formation and annealing, some indication of the types of mechanisms can be gleaned from basic observations for other materials. For example, one can infer from equations for energy loss phenomena of high-energy particles in solids that the number of displaced ions per unit of track length formed is a function of stopping power of the projectile in the solid and the ionization potential of the target atoms (see Vineyard 1976, Ziegler et al. 1985, Chadderton 1988, Spohr 1990). In a "displacement-spike" or "coulomb-explosion" model of track formation (Fleischer et al. 1965b) (Fig. 2), this implies that lattice atoms with a low ionization potential (e.g., chlorine) will be readily ionized and displaced into the surrounding lattice. In contrast, fluorine has a higher ionization potential and is therefore less readily ionized. The overall result predicted is a larger track diameter in high-chlorine apatite than in fluorine-rich specimens. This is consistent with the observations based on the bulk etch rates in routine fission track analysis (Carlson et al. 1999). Track diameters have been used as a proxy for halide content with the "etch figure" method where the mean etch pit diameter is measured parallel to the crystal *c*-axis (Burtner et al. 1994, Donelick et al. 1999, Stockli et al. 2001).

James and Durrani (1988) suggested that in addition to having a spatial distribution, defects, such as those in Figure 13, may also be represented by a "potential-energy-well distribution". This concept has a direct bearing on the energy subsequently required to mobilize the defect. If the bulk of these displaced ions cannot immediately be accommodated in their "normal" lattice sites, they will lodge in the matrix as "self-interstitials". Several processes are then likely to occur. Those within 5-10 lattice spacings may spontaneously recombine with their corresponding vacancy (Borg and Dienes 1988)—either in the track core or a lattice vacancy caused by the knock-on event (Itoh and Tanimura 1986). For interstitials beyond the strain field of such vacancies, the large size of anion interstitials makes them quite unstable. Anions tend to be the most mobile and have the lowest migration or activation energy. Cation interstitials, in contrast, are smaller and will have slightly higher activation energies. Interstitial ions typically have much lower activation energies than vacancies and are more mobile. Since in fission tracks, such processes involve significant movement through the lattice, the transport process (diffusion) is likely to be the rate controlling parameter (Mrowec 1980). Collectively, these diffusion processes contribute to the relative stability of the vacancy-dominated track in insulators when compared to metals (Chadderton 1988).

The kinetics of clusters further enhances the stability of tracks in insulators (Dartyge et al. 1981). Trapping of various defects by allovalent substitutional atoms is also highly likely and will result in an increase in apparent activation energies for participating ions due to sequential trapping and de-trapping reactions preceding a final recombination.

Given the abundance of trivalent, rare-earth elements commonly substituting for divalent calcium in the apatite (Roeder et al. 1987, Hughes et al. 1991a,b), the contribution of this process cannot be dismissed as insignificant. These reactions are further complicated by the crystal structure (Sudarsanan and Young 1978) with its characteristic anion channels parallel to the *c*-axis (Fig. 17). A less prominent cation channel is also present. Currently, atomic scale approaches remain theoretical, but there is little doubt that advances in this field could dramatically improve our understanding of the fundamental processes involved in track annealing in apatite.

Thermal history reconstruction and inversion modeling

Annealing models, however parameterized, are designed to simulate nature. In these models, time and temperature represent the input, and outputs are in the form of track length distributions and apparent age. Forward modeling enables one to predict the resulting fission track parameters from any hypothetical thermal history that can then be compared with the actual observations on geological samples. Forward modeling has been used initially as a means of checking annealing models against well-constrained geological examples (e.g., Green et al. 1989) and as a guide to interpretation (e.g., Willett 1992) of real samples. This approach, however, is a very inefficient means of finding solutions for unknown (or poorly constrained) samples. What is required is a method for extracting the actual, or at least a most probable, thermal histories directly from the observed data by inversion modeling.

One favored method involves the use of a 'genetic' algorithm (Gallagher and Sambridge 1994) to search time-temperature space for solutions that reproduce the observed data with the least amount of variance. Since the solutions matching the observed data are not unique, a large number (typically thousands) of possible histories must be tested and an assessment of likelihood or probability is made for each. Solutions can be more tightly constrained by incorporating additional geological and thermal information. Because track annealing is a unidirectional process, high temperatures have the effect of resetting (or at least significantly modifying) the track length distributions. As a result, inversion modeling only provides robust thermal histories subsequent to maximum heating (or burial) of the sample. Prior to such heating the solutions are only poorly constrained and certainly unreliable.

Because the annealing process is mathematically non-linear, solutions cannot be determined by routine least-squares optimization procedures, and Monte-Carlo approaches to sampling the time-temperature parameter space are used (Corrigan 1991, Lutz and Omar 1991, Gallagher 1995). Corrigan (1991) used a stochastic optimization, whereas Lutz and Omar (1991) applied an iterative method called the 'downhill simplex method.' Gallagher's (1995) inversion method relied on a stochastic search using a genetic algorithm to improve the number of good data-fitting solutions as illustrated in Figure 19. More recently, Willett (1997) used a controlled random search algorithm in which the solutions are similarly assessed on the quality of fit with the observed data. A wealth of software has been developed to automate the inversion modeling described here (Crowley 1993b, Gallagher 1995, Issler 1996, Ketcham et al. 2000). Typically the computer code written for inversion models allows the user the option of choosing the preferred annealing model (such as described above) as well as accommodating compositional variations.

GEOLOGICAL APPLICATIONS OF APATITE FISSION TRACK ANALYSIS

Apatite fission track analysis has been applied to a broad range of geological problems (e.g., Wagner and Van den haute 1992, Ravenhurst and Donelick 1992, Brown et al. 1994, Andriessen 1995, Gallagher et al. 1998, Gleadow and Brown 2000, Dumitru

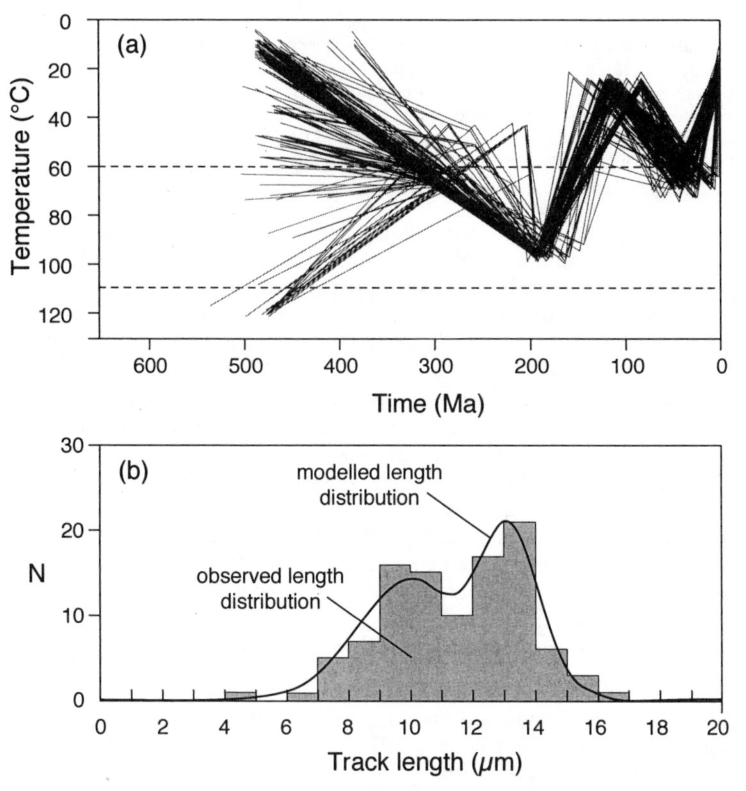

Figure 19. Output typical of an inversion model (searching time-temperature space using a Monte-Carlo approach). The upper frame shows model-generated histories that are each tested against the observed data (track length distribution and fission track age). Various algorithms are used to iteratively improve the model "fit" against the data. The lower frame illustrates a graphical output of the model "best fit" length-distribution compared to the measured track lengths. (generated using MonteTrax; based on Gallagher 1995).

2000, Gunnell 2000. Here, we present an overview of the mainstream applications and cite representative literature related to each topic. A flow chart showing the sequence of steps and inputs used to derive different geologically useful outputs from apatite fission track data and their linkages to various applications is summarized in Figure 20.

As indicated in the preceding discussion, apatite fission track data in most cases give rise to apparent ages, which reflect regional patterns of cooling, rather than the original formation ages of the rocks involved. Mostly, the apparent ages obtained are 'mixed' ages which reflect more than one component of the thermal history and only in a relatively limited number of circumstances do they directly date a particular episode or discrete geological event (Fig. 9).

Absolute dating

Absolute age dating using apatite is only possible in very restricted circumstances and has been applied to volcanic and shallow intrusive rocks, meteorite impact events and contact metamorphism. The fundamental requirement is that samples cooled rapidly and subsequently remained thermally 'undisturbed' and close to or at the surface. Due to its relatively low uranium content and restricted occurrence, apatite dating of Quaternary

Figure 20 (opposite page). Flow chart showing the sequence of steps and possible inputs which can be used to derive geologically useful output parameters from apatite fission track data. Outputs such as regional thermal history, denudation and paleotopography may also be displayed as images for different time slices or as movies. Sources of error involved are cumulative so that uncertainties increase with each step away from the primary apatite fission track data. Linkages of outputs to various geological applications are also shown (modified after Kohn et al. 2002b). → →

volcanic rocks is not commonly used, compared to the more ubiquitous zircon and/or glass. Zircon is most suitable in this time range, because of its relatively high uranium content and resistance to weathering. When used, apatite dating of volcanics is usually carried out together with fission track dating of coexisting zircon and/or glass or other radiometric dating methods (e.g., $^{40}Ar/^{39}Ar$). Apatite from three rapidly cooled igneous rocks, the Fish Canyon Tuff (Naeser et al. 1981, Hurford and Hammerschmidt 1985), the Cerro de Mercado Martite (the 'Durango' apatite, Naeser and Fleischer 1975), and the Mt Dromedary Igneous Complex (Green 1985), are commonly used as standards to calibrate fission track dating (Hurford 1990). Apatite has also been used in conjunction with other minerals to constrain ages of kimberlite and diatreme emplacement (e.g., Naeser 1971, Brookins and Naeser 1971). An early comparative study of apatite fission track ages and K/Ar whole rock and plagioclase ages from early Miocene seamounts in the Gulf of Alaska (Turner et al. 1973) indicated concordant ages (within errors quoted), suggesting little or no track annealing in the ocean floor environment. However, in this example the apatite ages are not robust, because their 2σ errors are more than three times those obtained by K-Ar and no track length measurements are reported.

The complete or partial resetting of apatite fission track clocks in target rocks by shock-wave heating has been used to constrain the timing of meteorite impacts events (e.g., Wagner 1977, Storzer and Wagner 1977, Hartung et al. 1986, Omar et al. 1987 and Kohn et al. 1995).

Calk and Naeser (1973) demonstrated the possibility of dating shallow intrusions indirectly. A late Cretaceous quartz monzonite body in Yosemite National Park, California, was intruded by a high level basaltic plug. Apatite and sphene fission track dates from the quartz monzonite within 5 feet of the contact with the basalt yield concordant late Miocene ages. These 'country rock' ages result from their total thermal resetting as a consequence of their proximity to the basalt intrusion, which they date by proxy. A further example of apatite dating of a discrete thermal event is the "Mottled Zone" in Israel and Jordan, formed during high-T, low-P metamorphism by near surface combustion of organic matter related to the development of the Dead Sea Transform system (Kolodny et al. 1971).

Apatite fission-track crustal profiles

The sensitivity of fission tracks in apatite to the relatively low temperatures characteristic of the upper few kilometers of the crust (~0-120°C), and the consequent patterns of track length distribution and apparent fission track age with depth that evolve within the crust, provide an unparalleled tool for studying the long-term thermal evolution of the Earth's crust. The detailed form of these fission track profiles reflects the distribution of temperature within the shallow crust and its variation through time. The profiles are therefore characteristic of the style of thermal history experienced by the crustal section. A sequence of samples representing a range of depths within the crust, such as a suite of samples collected from deep drill holes (e.g., the Otway Basin, Fig. 11), or from outcropping rocks collected over a range of surface elevations, can thus be used to document the vertical pattern of fission track parameters within the upper crust. These

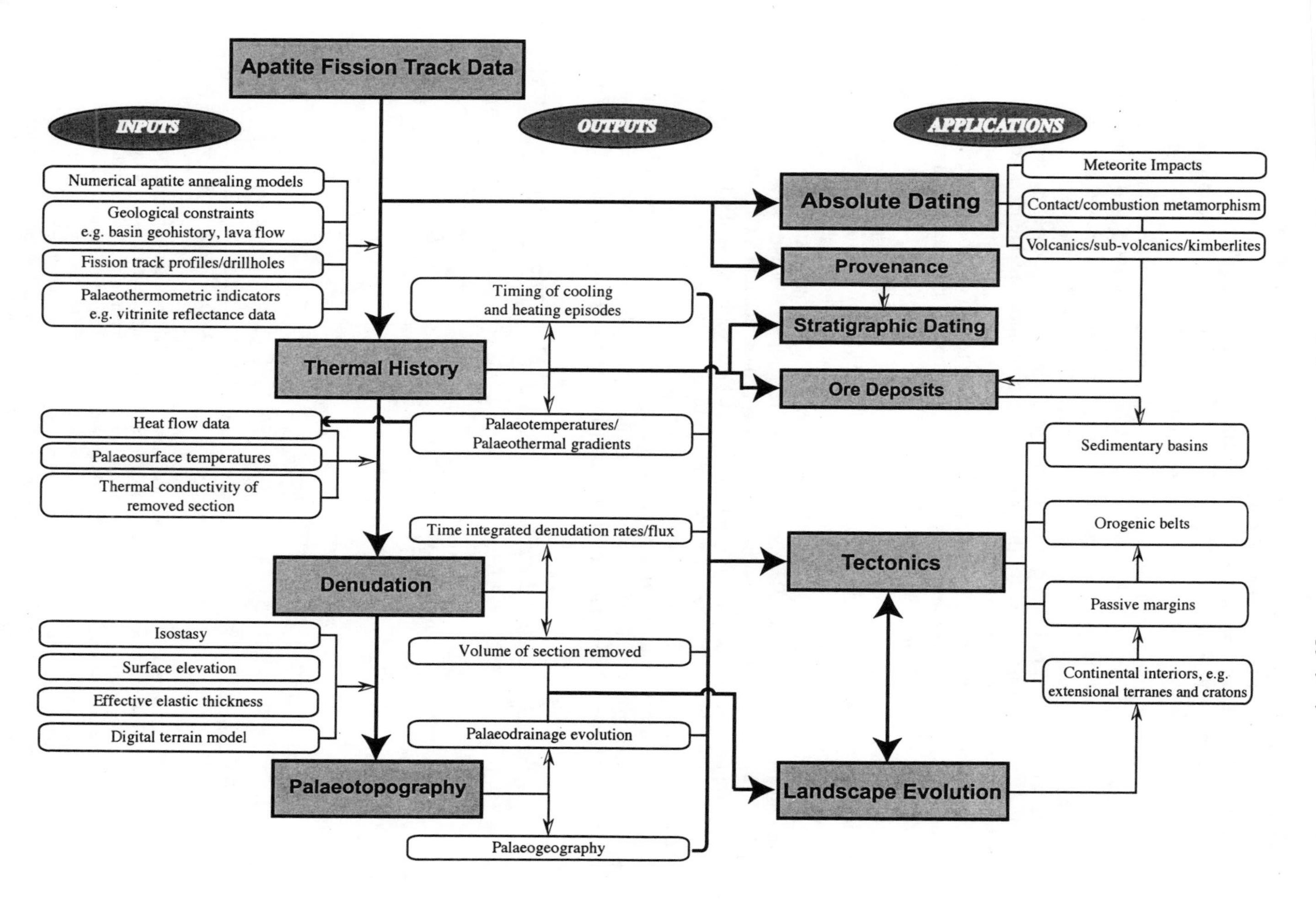

Apatite Fission Track Data
INPUTS
OUTPUTS
APPLICATIONS
Numerical apatite annealing models
Geological constraints e.g. basin geohistory, lava flow
Fission track profiles/drillholes
Palaeothermometric indicators e.g. vitrinite reflectance data
Thermal History
Heat flow data
Palaeosurface temperatures
Thermal conductivity of removed section
Denudation
Isostasy
Surface elevation
Effective elastic thickness
Digital terrain model
Palaeotopography
Timing of cooling and heating episodes
Palaeotemperatures/ Palaeothermal gradients
Time integrated denudation rates/flux
Volume of section removed
Palaeodrainage evolution
Palaeogeography
Absolute Dating
Provenance
Stratigraphic Dating
Ore Deposits
Tectonics
Landscape Evolution
Meteorite Impacts
Contact/combustion metamorphism
Volcanics/sub-volcanics/kimberlites
Sedimentary basins
Orogenic belts
Passive margins
Continental interiors, e.g. extensional terranes and cratons

fission track profiles can then be used to reconstruct the thermal history for the sampled section and thus place quantitative constraints on its thermal history. The potential of using apatite fission track profiles in this manner was first documented in the European Alps by Wagner and Reimer (1972), and Wagner et al. (1977), and subsequently in a profile from the Eielson drill-hole in Alaska (Naeser 1979b).

In regions of the crust that have not been disrupted by tectonic movements and where rates of erosion have been low (<~30 m Myr^{-1}) (Gleadow 1990, Brown et al. 1994) for a prolonged period (>10^6 years), such as cratonic interiors, or where progressive burial has been occurring, as in sedimentary basins, the typical form of the fission track crustal profile is controlled primarily by the progressive increase of temperature with depth. The pattern produced by this situation of continuous residence at constant or gradually increasing temperature is one of progressive reduction in mean track length and hence fission track age with depth (e.g., Fig. 11). The region of most rapid reduction in apparent fission track age and track length is described as the partial annealing zone (PAZ), typically between about 60° and 110°C.

More complicated thermal histories will produce correspondingly more complex apatite age profiles. If there is a rapid increase in the rate of erosion after a prolonged period of geomorphic stability, then the base of the existing concave-up apparent age profile will be shifted upwards towards the new mean topographic surface as denudation proceeds (e.g., Gleadow and Brown 2000). Apatite samples that were at temperatures > ~110°C prior to the acceleration in denudation will have accumulated no fission tracks and hence have zero apparent ages up to that point. On the initiation of cooling produced by the accelerated erosion these samples will begin to retain fission tracks below ~110°C and a new apparent age profile will begin to develop below the earlier profile as shown in Figure 21. If the amount of crust removed during this episode is of the order of a few kilometers then part of the earlier apparent age profile may be preserved within the upper sections of the new topographic relief. The transition from the earlier upper profile to the

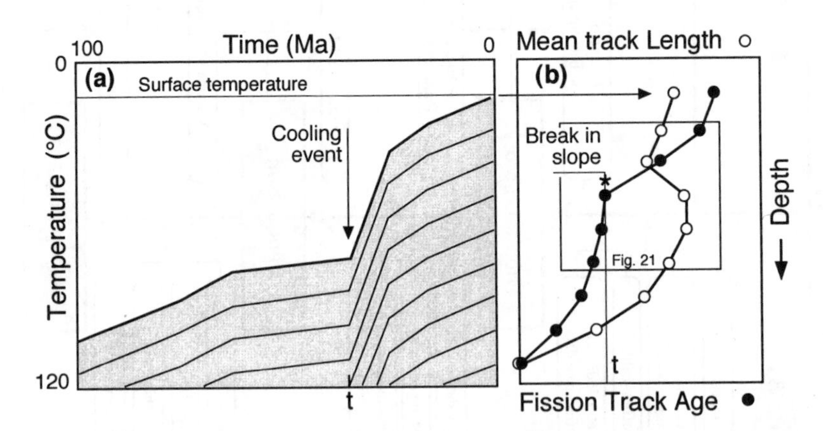

Figure 21. Modeled fission track age and mean track length profiles in apatite from a hypothetical two-stage cooling history incorporating a distinct cooling episode at time t. The cooling paths of samples at successively increasing depth in the final profile are shown in (a). The resulting profiles of both apatite age and track length (b) are distinctive and show a characteristic inflexion point or break in slope, the age of which closely approximates the time of onset of the cooling event. The break in slope represents the position which was at the base of the apatite fission track annealing zone prior to the onset of rapid cooling. The outlined rectangle in (b) represents the segment in the example shown in Figure 22. After Gleadow and Brown (2000).

new lower profile will be marked by a pronounced inflection (Fig. 21). The location of this inflection marks the depth at which the apparent apatite age was reduced to zero (i.e., the depth of the ~110°C paleo-isotherm) prior to the increase in erosion rate, and the corresponding age approximates the time at which the acceleration in erosion took place (Fig 21). Such inflection points have now been observed in vertical profiles sampled over significant topographic relief in various parts of the world (e.g., Gleadow and Fitzgerald 1987, Fitzgerald and Gleadow 1988, Fitzgerald et al. 1995, Foster and Gleadow 1996, Kohn et al. 1999) and are recognized as a feature of crustal blocks that have undergone episodic kilometer-scale denudation. An example from Denali in the Alaska Range is shown in Figure 22.

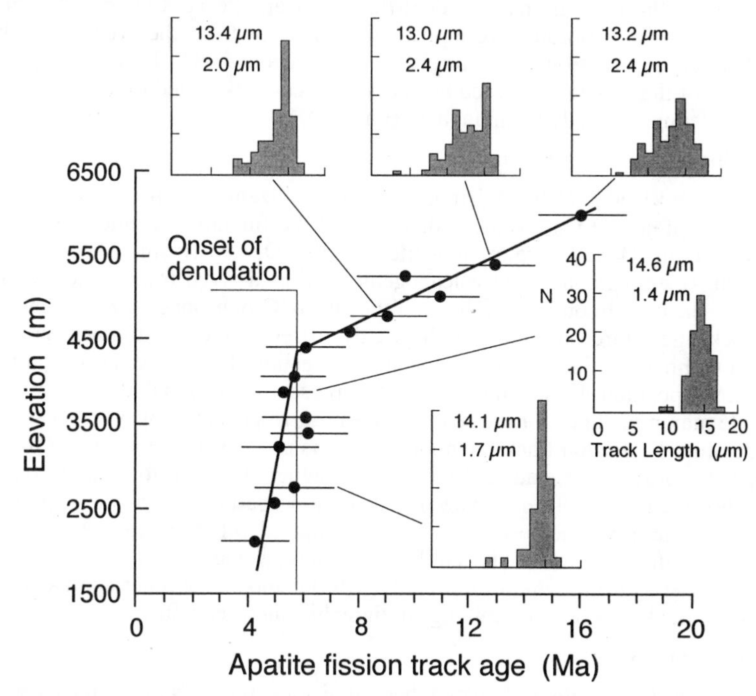

Figure 22. An apatite fission track age-elevation profile from Denali in the Alaska Range of North America. The profile shows the characteristic break in slope separating an upper zone of rapidly increasing age and mixed to bimodal length distributions from a lower steep zone where the age shows little variation and the length distributions are long and narrow. The inflection point at about 6 Ma represents the time of onset of rapid cooling in this profile, presumable related to rapid denudation consequent on uplift of the range. The two numbers on each length histogram represent the mean track length (top) and the standard distribution of the distribution (bottom). Modified after Fitzgerald et al. (1995).

The form of the new apparent age profile that develops below this inflection will depend on both the rate and duration of the accelerated period of denudation, and could be either a linear gradient in apparent age with depth if sufficient section were removed, or possibly a new concave-up PAZ profile. Clearly, it is possible to produce a composite profile with an inflection comprising two linear components with differing age-depth gradients representing a change from a moderate rate of erosion to a significantly higher rate. Interpreting an observed gradient in apparent apatite age therefore requires discriminating between prolonged annealing at constant temperature (very low rate or no erosion) and continuous erosion at some constant rate. Resolving these two interpretations of an age gradient is obviously a crucial step in arriving at a meaningful interpretation of any apatite fission track crustal profile and depends critically on the information represented by the distribution of individual track lengths of a sample.

Ore deposits

Several apatite studies have aided in deciphering the time-space relationships between igneous and/or hydrothermal events and their associated mineralization (Arne 1992). These studies, also employing other radiometric dating techniques, have been mainly applied to precious and base metal deposits environments in the western USA (e.g., Banks and Stuckless 1973, Ludwig et al. 1981, Cunningham et al. 1984, Lipman et al. 1976). Paleothermal anomalies, identified from apatite cooling patterns beneath mineralized rocks in Colorado, were postulated to result from the presence of buried stocks (Naeser et al. 1980, Cunningham and Barton 1984, Beaty et al. 1987). Identification of the anomaly reported by Naeser and co-workers led to the discovery of a major molybdenum ore body (Cunningham et al. 1987).

Stratigraphic dating and provenance

Gleadow and Duddy (1981a) demonstrated the utilization of apatite to constrain the maximum age of strata. They showed that most of the fluviatile sediments forming the Early Cretaceous Otway Group deposited in the Otway Basin, a rift graben in southeastern Australia, comprised volcanogenic detritus derived from contemporaneous volcanism to the east. In outcrop sections of the Otway Group, sphene, zircon and apatite fission track ages, three minerals with greatly different closure temperatures, were concordant within error, indicating that the outcrop sections have never been significantly heated since deposition. Hence, the sediments represent the original ages of the detritus and provide maximum ages for the sedimentary host. In general however, because of its low temperature for fission track retention, apatite is not as suitable as zircon for dating source ages for provenance studies (Storzer and Wagner 1982, Hurford and Carter 1991, Carter 1999). Carter (1999) emphasized that detrital apatite fission track applications should aim to improve the resolution of the temporal relationships between source evolution and sedimentation in adjacent basins. Brookins and Naeser (1971), Ross et al. (1976), Turner et al. (1983) and Carter et al. (1995) demonstrate other approaches using apatite fission track data for constraining stratigraphic interpretations.

Sedimentary basins

A widespread application of apatite thermal history studies is in sedimentary basins, particularly for the interpretation and quantitative modeling of thermal histories, and hydrocarbon resource evaluation. Naeser (1979b) described the expected apatite age versus depth trends for different basinal thermal histories, and presented the earliest discussion of the potential of this approach. Subsequent studies in the Otway Basin (e.g., Gleadow and Duddy 1981b, Gleadow et al. 1983, Green et al. 1989, Mitchell 1997) and other basins (e.g., Briggs et al. 1981, Storzer and Selo 1984, Dumitru 1988, Naeser et al. 1989, Seward 1989, Naeser et al. 1990, Duane and Brown 1991, Naeser 1993, Issler at al. 1990, Kamp and Green 1990, Crowley 1991, Steckler et al. 1993, Blackmer et al. 1994, Duddy et al. 1994, Gallagher et al. 1994a, Green et al. 1995, Hill et al. 1995, O'Sullivan 1996, Giles and Indrelid 1998) have demonstrated how apatite studies may provide unique constraints on paleo-geothermal gradients, estimates of maximum paleo-temperatures and their timing, basin inversion and erosion, fluid flow, and mechanisms of basin formation. The information may also be used to assess hydrocarbon potential that has led to increased commercial application of the fission track method. Several basin studies have also used a combined, multi-method approach for thermal history modeling using a variety of thermal indicators such as apatite fission track analysis, vitrinite reflectance, fluid inclusions, clay mineralogy and argon radiometric data (e.g., Feinstein et al. 1989, Kohn et al. 1990, Bray et al. 1992, Grist et al 1992, Arne and Zentilli 1994, Burtner et al. 1994, Kamp et al. 1996, Tseng et al. 1996, Zhao et al. 1996, Mitchell 1997, Pagel et al. 1997, Kohn et al. 1997, Parnell et al. 1999, Marshallsea et al. 2000, Mathiesen et al. 2000, Hu et al. 2001, Osadetz et al. 2002). Apatite studies related to carbonate-hosted Mississippi Valley-type (MVT) Pb-Zn ore deposits (Arne et al. 1990, Arne 1992) suggest that such mineralization is associated with broad regional heating rather than with discrete hot pulses of metalliferous brines migrating through sedimentary basins, as previously envisioned.

Orogenic belts

One of the earliest applications exploited by the temperature-sensitive nature of the apatite fission track system was for the reconstruction of thermal histories in mountain belts. Wagner (1968) and Wagner and Reimer (1972) first demonstrated the potential of the fission track method for studying tectonic problems, thus laying the foundation for expanding the scope from an 'age determination' approach alone to a unique thermo-tectonic tool. Quantitative information arising from fission track studies relate to the post-orogenic cooling history, and potentially provide estimates of the timing, magnitude and rate of erosional and tectonic denudation. Some early apatite studies in mountain belts used age versus elevation plots to directly derive denudation and uplift rates, but the relationship to the fission track parameters may be been oversimplified (e.g., Summerfield and Brown 1998, Gallagher et al. 1998). Some complicating factors are due to: (1) inclusion of samples containing tracks which accumulated prior to rapid cooling (i.e., from a paleo PAZ), (e.g., Roberts and Burbank 1993), (2) perturbation of crustal isotherms during denudation and the development of the topography (e.g., Parrish 1983, 1985, Kohn et al. 1984, Stüwe et al. 1994, Brown and Summerfield 1997), (3) post-cooling rotation or folding of the sampled rocks, complicating interpretation of data from vertical profiles (e.g., Johnson 1997, Rahn et al. 1997), and (4) lateral particle motion parallel to isotherms, rather than vertical transport perpendicular to isotherms (e.g., Willett et al. 1993, Batt and Brandon 2002). Apatite studies have been reported from most of the world's young orogenic belts including: the Alps (e.g., Schaer et al. 1975, Wagner et al. 1977, 1979; Hurford 1986, 1991; Schlunegger and Willett 1999), the Tatra Mountains (Burchart 1972), Pyrenees region (Yelland 1990, Fitzgerald et al. 1999, de Bruijne and Andriessen 2002), Calabrian Arc (Thomson 1994), Ural Mountains (Seward et al. 1997, Leech and Stockli 2000, Himalayas (Sharma et al. 1980, Zeitler et al. 1982, Zeitler 1985, Sorkhabi 1993, Foster et al. 1994, Sobel and Dumitru 1997), Rocky Mountains (Naeser 1979a, Bryant and Naeser 1980, Roberts and Burbank 1993), Taiwan (Liu 1982), Western North America (Harrison et al. 1979, Parrish 1983, Plafker et al. 1991, O'Sullivan et al. 1993, 1997; O'Sullivan and Parrish 1995, Fitzgerald et al. 1995, Brandon et al. 1998, Dumitru 1991), Andes (Nelson 1982, Crough 1983, Kohn et al. 1984, Shagam et al. 1984, Benjamin et al. 1987, Jordan et al. 1989, Coughlin et al. 1999, Spikings et al. 2000, Thomson et al. 2001); New Zealand Alps (White and Green 1986, Kamp et al. 1989, Tippett and Kamp 1993, Batt et al. 1999, 2002) and Papua-New Guinea (Hill and Gleadow 1989, Hill and Raza 1999).

Non-orogenic settings

Passive continental margins. Apatite fission track thermochronology also reveals the pattern and chronology of denudation in relation to the development of topography and landscape features in the on-shore region of rifted continental margins. Such data contribute to an improved understanding of the evolution of lithospheric thermal and strength distribution and the nature of rift-flank tectonic uplift associated with extension and surface processes (van der Beek et al. 1994, 1995). In the case of passive margins, the time-space relationship of onshore denudation is also important for calculating sediment volumes in adjacent offshore basins, which may have formed during extension. Apatite track studies on detrital grains and basement clasts in orogenic sediments (Wagner et al. 1979, Garver et al. 1999) and in offshore sediments (Clift et al. 1996, Gallagher and Brown 1999b) have been employed to reconstruct the exhumation histories of nearby mountainous areas and adjacent onshore passive margins, respectively. Onshore studies along rifted continental margins and rift flanks have been reported from southeastern Australia (Gleadow and Lovering 1978, Moore et al. 1986, Dumitru et al. 1991, Foster and Gleadow 1992a, O'Sullivan et al. 1995a, 1996, 1999, 2000b; Kohn et al. 1999), eastern Greenland (Gleadow and Brooks 1979, Thomson et al. 1999a, Hansen 2000, Johnson and Gallagher 2000, Mathiesen et al. 2000), Alaska (Dumitru et al. 1995), Rio Grande rift (May et al. 1994), northern United Kingdom (Green 1986, 2002; Lewis et al. 1992, Green et al. 1999, Thomson et al. 1999b), Baikal rift (van der Beek et al. 1996, van der Beek 1997), southern Norway (Rohrman et al. 1994, 1995), Red Sea-Gulf of Suez (Kohn and Eyal 1981, Bohannon et al. 1989, Omar et al. 1989, Omar and Steckler 1995, Kohn et al. 1997, Menzies et al. 1997), Transantarctic Mountains (Gleadow and Fitzgerald 1986, Fitzgerald et al. 1986, Fitzgerald 1994, Schäfer and Olesch 1998, Lisker 2002), southern Africa and southeast Brazil (Brown et al. 1990, Gallagher et al. 1994b, Gallagher and Brown 1997, 1999a,b; Brown et al. 2000, Cockburn et al. 2000, Raab et al 2002) and East Africa (Foster and Gleadow 1992b, 1993, 1996; Noble et al. 1997). Quantification of denudation history arising from apatite studies has been used to constrain models for the development of major escarpments formed during continental break-up (Gallagher and Brown 1999a, Cockburn et al. 2000). Blythe and Kleinspehn (1998) and Winkler et al. (1999) have also evaluated the possible role of climatically driven erosion as a component of the exhumation interpreted from apatite data. O'Sullivan and Brown (1998) explained Miocene cooling observed in well samples along the northern Alaska coastline as a response to a long-term decrease in the paleo-mean annual surface temperature.

Extensional terranes. In highly extended terranes, cooling recorded below large-scale detachment faults and crustal shear zones may result predominantly from normal faulting, i.e., tectonic denudation as opposed to erosional denudation. In extensional settings apatite fission track used together with other thermo-chronological methods may provide unique information on: (i) the timing of rapid crustal extension, (ii) rates of cooling, (iii) average slip rates on detachment faults, (iv) paleo-geothermal gradients prior to and following rapid extension, and (v) initial dips of detachment faults and tilted hanging wall blocks. Examples of apatite fission track studies in extensional settings include: Western USA (Foster et al. 1991, 1993; Fitzgerald et al. 1991, 1993; Dokka 1993, John and Foster 1993, Howard and Foster 1996, Foster and John 1999, Miller et al. 1999, Pease et al. 1999, Fayon et al. 2000, Stockli et al. 2001, Foster and Raza 2002), Gulf of California region (Axen et al. 2000, Fletcher et al. 2000), Greece (Thomson et al. 1998a,b; Hejl et al. 2002) and Papua-New Guinea (Baldwin et al. 1993).

Cratons. Ancient crystalline terranes, forming extensive cratons and large tracts of continental interiors, are generally viewed as tectonically and isostatically stable, and resistant to internal deformation. A growing body of evidence however, based on reconnaissance apatite studies of different cratons and Precambrian blocks, has revealed discrete, regional episodes of Phanerozoic km-scale crustal erosion of these ancient terrains (Crowley et al. 1986, 1991; Zeck et al. 1988, Brown et al. 1990, Noble et al. 1997, Spikings et al. 1997, Harman et al. 1998, Mitchell et al. 1998, Cederbom et al. 2000, Cederbom 2001, Gibson and Stüwe 2000, O'Sullivan et al. 2000a, Kohn et al. 2002b, Osadetz et al. 2002). These surprising results are difficult to reconcile with the widely held view of cratons as being tectonically and erosionally inert but open a promising avenue of research for further detailing the morpho-tectonic evolution of these ancient terranes.

Fault displacement and reactivation

In crystalline terranes traditional stratigraphic markers that might be used for reconstructing regional structure and tectonic evolution are usually not available. In such environments, apatite data from surface outcrops and deep drillhole samples may be used to characterize paleodepth thermo-chronological markers, in effect an 'invisible' stratigraphic tool. This approach has been used to place unique constraints on the magnitude and timing of relative fault displacement and regional structure (Tagami et al. 1988, Wagner et al. 1989, 1997; Fitzgerald and Gleadow 1990, Dumitru 1991, Foster and Gleadow 1992a,b, 1996; Kohn 1994, O'Sullivan et al. 1995b, 2000a; Coyle et al. 1997, Cox et al. 1998, Kohn et al. 1999), as well as the reactivation history of ancient, fundamental faults or lineaments (Harman et al. 1998, O'Sullivan et al. 1998, 1999; Raab et al. 2002).

Regional and continental-scale thermo-tectonic imaging

Apatite fission track ages rarely reflect the formation ages of the host rocks sampled, and on a regional or continental-scale often show broad spatial variations that reflect their thermal and denudation histories. Further, the form in which these data have been presented have often been difficult for non-specialists to interpret and their implications difficult to visualize. Recently, strategies have been developed for inverting such large data sets into time-temperature solutions, which can be visualized as a series of time slice images and movie sequences, depicting the cooling history of present day surface rocks during their passage through the upper crust (Gleadow et al. 1996, Gallagher and Brown 1999a,b; Gleadow et al. 2002, Kohn et al. 2002b). The data can be further extended and combined with other large-scale data sets, such as digital elevation and heat flow, to image other geologically useful parameters, such as denudation and paleotopography (see Fig. 20). Such images provide an important new perspective on the crustal processes and landscape evolution and permit tectonic and denudation events to be visualized in time and space.

CONCLUSIONS AND FUTURE DIRECTIONS

After some thirty years of investigation and development, fission track analysis of apatite has produced robust methodologies with a wide range application to geochronology and, particularly, to low-temperature thermochronology. The particular sensitivity of apatite fission tracks to the temperatures prevailing in the low-temperature environment of the upper crust has opened new opportunities for the chronological investigation of surface processes and tectonics not accessible to earlier approaches. Apatite fission track analysis has also proved to be of unparalleled utility for paleotemperature reconstruction in sedimentary basin environments with important implications for hydrocarbon exploration. These developments mean that the dominance of apatite as the mineral of choice in fission track analysis is likely to continue.

Despite the substantial progress achieved in techniques, interpretative strategies and modeling over the last twenty years, and an increasing standardization of approach in most laboratories, there is still room for considerable improvement. One limitation of current techniques is the necessarily limited precision of fission track analysis due to the relatively small numbers of tracks counted, and the reliability of track identification dependent to a large degree on the skill and experience of the analyst. While the generally good reproducibility achieved in inter-laboratory comparisons suggests these are not major concerns, there is no fundamental barrier to improved precision through an order of magnitude increase in the number of tracks counted. This situation is not likely to improve using existing manual track counting methods, but eventually, it should be possible to develop fully automated track counting which would remove the present limitations, determined as they are mostly by the endurance of the analyst. Previous attempts at automated track counting by image analysis have not as yet proved practical, but it can be anticipated that this is likely to change over the next five years.

A second area where a significant departure from traditional methods may be at hand is in direct analysis of uranium by, for example, laser-ablation inductively-coupled plasma mass spectrometry (LA-ICPMS). The possibility for the rapid determination of uranium at μg/g levels in individual apatite grains would be a radical departure from all previous approaches. Such an approach would make the present use of neutron irradiation unnecessary, and lead to improvements in analytical precision and sample turn-around time, as well as improved laboratory safety by eliminating the minor levels of radioactivity that must currently be managed. Such a method would also eliminate many of the calibration issues that have pre-occupied fission track workers for many years, although, of course, a new set of calibration problems would undoubtedly arise. Direct uranium determination would also open the way for a reinvestigation of monazite for fission track dating, which might have important applications in low-temperature thermochronology.

A third area where significant improvement can be anticipated is in closing the gap between the current generation of purely empirical annealing models and a fundamental understanding of the lattice processes involved in annealing. New models can be anticipated which will relate more closely to the atomic scale diffusion that controls the repair of fission damage in apatite and other minerals. Even without such developments, we can anticipate the emergence of a new generation of well-documented multi-compositional apatite annealing models that should significantly improve the reconstruction of thermal histories and enhance interpretation of apatite fission track data.

Many important insights in thermochronology have arisen from the application of apatite fission track analysis, either alone or in combination with other techniques, such as $^{40}Ar/^{39}Ar$ dating, which record higher temperature history information. Most fission track studies have produced only limited information on the very-low temperature history (<~60-70°C), characteristic of the shallowest crustal levels, due largely to the uncertainty of kinetic models in this range. This has limited the ability to close the gap between the deeper, subsurface evolution of the rock masses involved, which are reasonably well characterized, and processes acting at surface and near-surface levels. An important technical development in recent years, the advent of (U-Th)/He thermochronometry on apatites (see Farley and Stockli, this volume), provides an exciting opportunity to address this issue directly.

The temperature range of the He apatite partial retention zone (between ~40 and 80°C) records even lower temperature information, and often younger ages, than does apatite fission track thermochronology. Hence, in many cases shallower crustal depths and possibly less structural relief may be required to distinguish among various cooling histories using the apatite (U-Th)/He method. As the partial retention temperature ranges for the two methods partly overlap, apatite He ages will provide quantitative tests of track-length derived thermal models. Comparison of the two methods will thus increasingly be used to validate and crosscheck laboratory calibrations in both systems. Apatite (U-Th)/He thermochronometry is therefore an ideal complement to apatite fission track studies in that it is most sensitive over the temperature range where the apatite fission track system is least sensitive. Future challenges will involve the development of new protocols for the integrated modeling of low temperature histories obtained from both methods. Integration of these methods on both local and regional scales will lead to the development of new strategies for the time-space visualization and improved understanding of upper crustal and surface processes. The combined application of these two methods of thermochronology to the same apatites has great potential to provide significantly more information than either method alone. For this reason it can be foreseen that the predominance of apatite in low temperature thermochronology is likely to be sustained.

ACKNOWLEDGEMENTS

We are grateful to Günther Wagner, Tony Hurford, Matt Kohn and Cathy Skinner for their extremely helpful reviews that have substantially improved the original version of this manuscript. We also thank our numerous colleagues and students in the Melbourne Fission Track Research Group over many years from whom we have gained much and benefited in our understanding of the many issues covered in this chapter. This work has been supported by the Australian Research Council, the Australian Geodynamics Cooperative Research Centre and the Australian Institute of Nuclear Science and Engineering.

REFERENCES

Albrecht D, Armbruster P, Roth M, Spohr R (1982) Small angle neutron scattering observations from orientated latent nuclear tracks. Radiat Eff 65:145-148

Albrecht D, Armbruster P, Spohr R, Roth M, Schaupert K, Stuhrmann H (1985) Investigations of heavy ion produced defect structures in insulators by small angle scattering. Appl Phys A37:37-46

Andriessen PAM (1995) Fission-track analysis: principles, methodology and implications from tectono-thermal histories of sedimentary basins, orogenic belts and continental margins. Geologie Mijnbouw 74:1-12

Arne DC (1992) The application of fission track thermochronology to the study of ore deposits. *In* Zentilli M, Reynolds PM (eds) Short Course Handbook on Low Temperature Thermochronology, Mineral Assoc Can, Ottawa, p 75-96

Arne DC, Zentilli M (1994) Apatite fission track thermochronology integrated with vitrinite reflectance. *In* Mukhopadhyay PK, Dow WG (eds) Vitrinite Reflectance as a Maturity Parameter. Am Chem Soc Symp Ser 570:249-268

Arne DC, Green PF, Duddy IR (1990) Thermochronologic constraints on the timing of Mississippi Valley-type ore formation from apatite fission track analysis. Nucl Tracks Radiat Meas 17:319-323

Axen GJ, Grove M, Stockli D, Lovera O, Rothstein DA, Fletcher JM, Farley KA, Abbott PL (2000) Thermal evolution of Monte Blanco dome: Low-angle normal faulting during Gulf of California rifting and late Eocene denudation of the eastern Peninsular Ranges. Tectonics 197-212

Baldwin SL, Lister GS, Hill EJ, Foster DA, McDougall I (1993) Thermochronologic constraints on the tectonic evolution of an active metamorphic core complex, D'Entrecasteaux Islands, Papua New Guinea. Tectonics 12:611-628

Banks NG, Stuckless JS (1973) Chronology of intrusion and ore deposition at Ray, Arizona: Part II, Fission-track ages. Econ Geol 68:657-664

Batt GE, Brandon MT (2002) Lateral thinking: 2-D interpretation of thermochronology in convergent orogenic settings. Tectonophysics 349:185-201

Batt GE, Kohn BP, Braun J, McDougall I, Ireland TR (1999) New insight into the dynamic development of the Southern Alps, New Zealand, from detailed thermochronological investigation of the Mataketake Range pegmatites. *In* Ring U, Brandon MT, Lister GS, Willett SD (eds) Exhumation Processes: Normal Faulting, Ductile Flow, and Erosion, Geol Soc London Special Publ 154:261-282

Beaty DW, Naeser CW, Lynch WC (1987) The origin and significance of the strata-bound, carbonate-hosted gold deposits at Tennessee Pass, Colorado. Econ Geol 82:2158-2178

Benjamin MT, Johnson NM, Naeser CW (1987) Recent rapid uplift in the Bolivian Andes: Evidence from fission track dating. Geology 15:680-683

Bertagnolli E, Keil R, Pahl M (1983) Thermal history and length distribution of fission tracks in apatite: Part 1. Nucl Tracks 7:163-177

Bertel E, Märk TD (1983) Fission tracks in minerals: annealing kinetics, track structure and age correction. Phys Chem Mineral 9:197-204

Bertel E, Märk TD, Pahl M (1977) A new method for the measurement of the mean etchable track length and of extremely high fission track densities in minerals. Nucl Track Detection 1:123-126

Bhandari N, Bhat SC, Lal D, Rajagoplan G, Tamhane AS, Venkatavaradan VS (1971) Fission fragment tracks in apatite: Recordable track lengths. Earth Planet Sci Lett 13:191-199.
Blackmer GC, Omar GI, Gold DP (1994) Post-Alleghanian unroofing history of the Appalachian Basin, Pennsylvania, from apatite fission track analysis and thermal models. Tectonics 13:1259-1276
Blythe AE, Kleinspehn KL (1998) Tectonically versus climatically driven Cenozoic exhumation of the Eurasian Plate margin, Svalbard: fission track analyses. Tectonics. 17:621-639
Bohannon RG, Naeser CW, Schmidt DL, Zimmermann RA (1989) The timing of uplift, volcanism and rifting peripheral to the Red Sea: A case for passive rifting? J Geophys Res 94:1683-1702
Borg RJ, Dienes GJ (1988) An introduction to solid state diffusion. Academic Press, San Diego, p 360
Brandon MT, Roden-Tice MK, Garver JJ (1998) Late Cenozoic exhumation of the Cascadia wedge in the Olympic Mountains, northwest Washington State. Bull Geol Soc Am 110:985-1009
Bray RJ, Green PF, Duddy IR (1992) Thermal history reconstruction using apatite fission track analysis and vitrinite reflectance: a case study from the UK East Midlands and southern North Sea. *In* Hardman RFP (ed) Exploration Britain: Geological Insights for the Next Decade. Geol Soc London Special Publ 67:2-25
Briggs ND, Naeser CW, McCulloh TH (1981) Thermal history of sedimentary basins by fission-track dating. Nucl Tracks 5:235-237
Brookins DG, Naeser CW (1971) Age and emplacement of Riley County, Kansas, kimberlites and a possible minimum age for the Dakota Sandstone. Bull Geol Soc Am 82:1723-1726
Bros R, Carpéna J, Sère V, Beltritti A (1996) Occurrence of Pu and fissiogenic REE in hydrothermal apatites from the fossil nuclear reactor 16 of Oklo (Gabon). Radiochim Acta 74:277-282
Brown RW, Summerfield MA (1997) Some uncertainties in the derivation of rates of denudation from thermochronologic data. Earth Surf Proc Landforms 22:239-248
Brown RW, Rust DJ, Summerfield MA, Gleadow AJW, De Wit MCJ (1990) An accelerated phase of denudation on the south-western margin of Africa: Evidence from apatite fission track analysis and the offshore sedimentary record. Nucl Tracks Radiat Meas 17:339-350
Brown RW, Summerfield MA, Gleadow AJW (1994) Apatite fission track analysis: Its potential for the estimation of denudation rates and implications for models of long term landscape development. *In* Kirkby MJ (ed) Process Models and Theoretical Geomorphology. John Wiley & Sons Ltd, Chichester, UK, p 23-53
Brown RW, Gallagher K, Gleadow AJW, Summerfield MA (2000) Morphotectonic evolution of the South Atlantic margins of Africa and South America. *In* Summerfield MA (ed) Geomorphology and Global Tectonics. John Wiley and Sons Ltd, Chichester, UK, p 257-283
Bryant B, Naeser CW (1980) The significance of fission-track ages of apatite in relation to the tectonic history of the Front and Sawatch Ranges, Colorado. Bull Geol Soc Am 91:156-164
Burchart J (1972) Fission-track age determinations of accessory apatite from the Tatra Mountains. Earth Planet Sci Lett 15:418-422
Burnett D, Monnin M, Seitz M, Walker R, Yuhas D (1971) Lunar astrology—U-Th distributions and fission-track dating of lunar samples. Proc Second Lunar Sci Conf, Geochim Cosmochim Acta Suppl 2:1503-1519
Burtner RL, Nigrini A, Donelick RA (1994) Thermochronology of Lower Cretaceous source rocks in the Idaho-Wyoming Thrust Belt. Bull Am Assoc Petrol Geol 78:1613-1636
Calk LC, Naeser CW (1973) The thermal effect of a basalt intrusion on fission tracks in quartz monzonite. J Geol 81:189-198
Carlson WD (1990) Mechanisms and kinetics of apatite fission-track annealing. Am Mineral 75:1120-1139
Carlson WD (1993a) Mechanisms and kinetics of apatite fission-track annealing—reply to Crowley. Am Mineral 78:213-215
Carlson WD (1993b) Mechanisms and kinetics of apatite fission-track annealing—reply to Green et al. Am Mineral 78:446-449
Carlson WD, Donelick RA, Ketcham RA (1999) Variability of apatite fission-track annealing kinetics: I. Experimental results. Am Mineral 84:1213-1223
Carpéna J (1998) Uranium-235 fission track annealing in minerals of the apatite group. *In* Van den haute P, De Corte F (eds) Advances in Fission Track Geochronology. Kluwer Academic, Dordrecht, p 81-92
Carter A (1999) Present status and future avenues of source region discrimination and characterization using fission track analysis. Sed Geol 124:31-45
Carter A, Bristow C, Hurford AJ (1995) The application of fission track analysis to dating of barren sequences: examples from red beds in Scotland and Thailand. *In* Dunnay RE, Hailwood E (eds) Non Biostratigraphical Methods of Dating and Correlation. Geol Soc London Special Publ 89:57-68
Cederbom C (2001) Phanerozoic, pre-Cretaceous thermotectonic events in southern Sweden revealed by fission track thermochronology. Earth Planet Sci Lett 188:199-209

Cederbom C, Larson S-Å, Tullborg E-L, Stiberg J-P (2000) Fission track thermochronology applied to Phanerozoic thermotectonic events in central and southern Sweden. Tectonophysics 316:153-167

Chadderton LT (1988) On the anatomy of a fission track fragment track. Nucl Tracks Radiat Meas 15:11-29

Chambaudet A, Miellou JC, Igli H, Rebetez M, Grivet M (1993) Thermochronology by fission tracks—an exact inverse method associated with the resolution of a single ordinary differential equation (ODE). Nucl Tracks Radiat Meas 22:763-772

Clift PD, Carter A, Hurford AJ (1996) Constraints on the evolution of the East Greenland Margin: Evidence from detrital apatite in offshore sediments. Geology 24:1013-1016

Cockburn HAP, Brown RW, Summerfield MA, Seidl MA (2000) Quantifying passive margin denudation and landscape development using a combined fission-track thermochronology and cosmogenic isotope analysis approach. Earth Planet Sci Lett 179:429-435

Corrigan J (1991) Inversion of apatite fission track data for thermal history information. J Geophys Res 96:10374-10360

Corrigan J (1993) Apatite fission track analysis of Oligocene strata in South Texas, U.S.A.: Testing annealing models. Chem Geol 104:227-249

Coughlin TC, O'Sullivan PB, Kohn BP, Holcombe R (1998) Apatite fission-track thermochronology of the Sierras Pampeanas, central west Argentina: Implications for the mechanism of plateau-uplift in the Andes. Geology 106:999-1002

Cox R, Kosler J, Sylvester P, Hodych J (2000) Apatite fission track (FT) Dating by LAM-ICP-MS Analysis (Abstr) Goldschmidt 2000 Conf Abstr 5:322

Cox SJD, Kohn BP, Gleadow AJW (1998) From fission tracks to fault blocks: an approach to visualising tectonics in the Snowy Mountains. Austral Geol Surv Org Record 1998/2, p 44-47

Coyle DA, Wagner GA, Hejl E, Brown R, Van den haute P (1997) The Cretaceous and younger thermal history of the KTB site (Germany): Apatite fission-track data from the Vorbohrung. Geol Rundsch 86:203-209

Crough S T (1983) Apatite fission-track dating of erosion in the eastern Andes, Bolivia. Earth Planet Sci Lett 64:396-397

Crowley KD (1985) Thermal significance of fission-track length distributions. Nucl Tracks 10:311-322

Crowley KD (1991) Thermal history of the Michigan Basin and south Canadian Shield from apatite fission track analysis. J Geophys Res 96:697-711

Crowley KD (1993a) Mechanisms and kinetics of apatite fission-track annealing—Discussion. Am Mineral 78:210-212

Crowley KD (1993b) Lenmodel—a forward model for calculating length distributions and fission-track ages in apatite. Computer Geosci 19:619-626

Crowley KD, Young J (1988) Automated mirror-image stage (AMIS) for external-detector fission-track analysis. Nucl Tracks Radiat Meas 17:410

Crowley KD, Naeser CW, Babel C (1986) Tectonic significance of Precambrian apatite fission-track ages from the midcontinent United States. Earth Planet Sci Lett 79:329-336

Crowley KD, Cameron M, Schaffer LR (1991) Experimental studies of annealing of etched fission tracks in fluorapatite. Geochim Cosmochim Acta 55:1449-1465

Crozaz G, Tasker DR (1981) Thermal history of mesosiderites revisited. Geochim Cosmochim Acta 45:2037-2046

Cunningham CG, Barton PB Jr (1984) Recognition and use of paleothermal anomalies as a new exploration tool. Geol Soc Am Abstr Progr 16:481

Cunningham CG, Steven TA, Campbell DL, Naeser CW, Pitkin JA, Duval JS (1984) Multiple episodes of igneous activity, mineralization, and alteration in the western Tushar mountains, Utah. Igneous activity and related ore deposits in the western and southern Tushar Mountains, Maryvale Volcanic Field, West-Central Utah. U S Geol Surv Prof Paper 1299-A,B, p 1-22

Cunningham CG, Naeser CW, Cameron DE, Barrett LF, Wilson JC, Larson PB (1987) The Pliocene paleothermal anomaly at Rico, Colorado is related to a major molybdenum deposit. Geol Soc Am Abstr Progr 19:268-269

Dakowski M (1978) Length distributions of fission tracks in thick crystals. Nucl Track Det 2:181-189

Dartyge E, Duraud JP, Langevin Y, Maurette M (1981) New model of nuclear particle tracks in dielectric minerals. Phys Rev B23:5213-5229

De Bruijne CH, Andriessen PAM (2002) Far field effects of Alpine plate tectonism in the Iberian microplate recorded by fault related denudation in the Spanish Central system. Tectonophysics 349:161-184

Dokka RK (1993) Original dip and subsequent modification of a Cordilleran detachment fault, Mojave extensional belt, California. Geology 21:711-714

Donelick RA (1991) Crystallographic orientation dependence of mean etchable fission track length in apatite: An empirical model and experimental observations. Am Mineral 76:83-91

Donelick RA, Roden MK, Mooers JD, Carpenter BS, Miller DS (1990) Etchable length reduction of induced fission tracks in apatite at room temperature (~23°C): Crystallographic oreintation effects and "initial" mean lengths. Nucl Tracks Radiat Meas 17:261-265

Donelick RA, Ketcham RA, Carlson WD (1999) Variability of apatite fission-track annealing kinetics: II. Crystallographic orientation effects. Am Mineral 84:1224-1234

Dowty E (1977) Phosphate in Angra Dos Reis: structure and composition of the $Ca_3(PO_4)_2$ minerals. Earth Planet Sci Lett 35:347-351

Duane MJ, Brown RW (1991) Tectonic brines and sedimentary basins: Further applications of fission track analysis in understanding Karoo Basin evolution (South Africa). Basin Res 3:187-195

Duddy IR, Green PF, Laslett GM (1988) Thermal annealing of fission tracks in apatite 3. Variable temperature behaviour. Chem Geol 73:25-38

Duddy IR, Green PF, Bray RJ, Hegarty KA (1994) Recognition of the thermal effects of fluid flow in sedimentary basins. *In* Parnell J (ed) Geofluids: Origin. Migration and evolution of fluids in Sedimentary Basins. Geol Soc London Special Publ 78:325-345

Dumitru TA (1988) Subnormal geothermal gradients in the Great Valley forearc basin, California, during Franciscan subduction: A fission track study. Tectonics 7:1201-1221

Dumitru TA (1991) Major Quaternary uplift along the northernmost San Andreas fault, King Range, northwestern California. Geology 19:526-529

Dumitru TA (2000) Fission-Track Geochronology. *In* Noller JS, Sowers JM, Lettis, WR (eds) Quaternary Geochronology: Methods and Applications. Am Geophys Union Ref Shelf 4, Washington, DC, American Geophysical Union, p 131-155

Dumitru TA, Hill KC, Coyle DA, Duddy IR, Foster DA, Gleadow AJW, Green PF, Laslett GM, Kohn BP, O'Sullivan AB (1991) Fission track thermochronology: Application to continental rifting of southeastern Australia. Austral Petrol Exploration Assoc J 31:131-142

Dumitru TA, Miller EL, O'Sullivan PB, Amato JM, Hannula KA, Calvert AC, Gans PB (1995) Cretaceous to Recent extension in the Bering Strait region, Alaska. Tectonics 14:549-563

Fayon A, Peacock SM, Stump E, Reynolds SJ (2000) Fission track analysis of the footwall of the Catalina detachment fault, Arizona: Tectonic denudation, magmatism, and erosion. J Geophys Res 105: 11,047-11,062

Feinstein S, Kohn BP, Eyal M (1989) Significance of combined vitrinite reflectance and fission-track studies in evaluating thermal history of sedimentary basins: an example from southern Israel. *In* Naeser ND, McCulloh TH (eds) Thermal History of Sedimentary Basins: Methods and Case Histories. Springer-Verlag, Berlin, p 197-216

Fitzgerald PG (1994) Thermochronological constraints on the post-Paleozoic tectonic evolution of the central Transantarctic Mountains, Antarctica. Tectonics 13:818-836

Fitzgerald P, Gleadow AJW (1990) New approaches in fission track geochronology as a tectonic tool: Examples from the Transantarctic Mountains. Nucl Tracks Radiat Meas 17:351-357

Fitzgerald PG, Sandiford M, Barrett PJ, Gleadow AJW (1986) Asymmetric extension in the Transantarctic Mountains and Ross Embayment. Earth Planet Sci Lett 86:67-78

Fitzgerald PG, Fryxel JE, Wernicke BP (1991) Miocene crustal extension and uplift in southeastern Nevada: Constraints from fission track analysis. Geology 19:1013-1016

Fitzgerald PG, Reynolds SJ, Stump E, Foster DA, Gleadow AJW (1993) Thermochronologic evidence for timing of denudation and rate of crustal extension of the South Mountain metamorphic core complex and Sierra Estrella, Arizona. Nucl Tracks 21:555-563

Fitzgerald PG, Sorkhabi RB, Redfield TF, Stump E (1995) Uplift and denudation of the central Alaska Range: a case study in the use of apatite fission track thermochronology to determine absolute uplift parameters. J Geophys Res 100:20175-20191

Fitzgerald PG, Munoz JA, Coney PJ, Baldwin SL (1999) Asymmetric exhumation across the Pyrenean orogen: implications for the tectonic evolution of a collisional orogen. Earth Planet Sci Lett 173:157-170

Fleischer RL, Hart HR (1972) Fission track dating: Techniques and problems. *In* Calibration of hominid evolution.. Bishop WW, Miller DA, Cole S (eds) Scottish Academic Press, Edinburgh, p 135-170

Fleischer RL, Price PB (1964) Techniques for geological dating of minerals by chemical etching of fission fragment tracks. Geochim Cosmochim Acta 28:1705-1715

Fleischer RL, Price PB, Symes EM (1964) On the origin of anomalous etch figures in minerals. Am Mineral 49:794-800

Fleischer RL, Price PB, Walker RM (1965a) Effects of temperature, pressure and ionization of the formation and stability of fission tracks in minerals and glasses. J Geophys Res 70:1497-1502

Fleischer RL, Price PB, Walker RM (1965b) The ion explosion spike mechanism for formation of charged particle tracks in solids. J Appl Phys 36:3645-3652

Fleischer RL, Price PB, Walker RM (1975) Nuclear Tracks in Solids. University of California Press, Berkeley

Fletcher JM, Kohn BP, Foster DA, Gleadow AJW (2000) Heterogeneous Neogene cooling and uplift of the Los Cabos block, southern Baja California: Evidence from fission track thermochronology. Geology 28:107-110

Foster DA, Gleadow AJW (1992a) Reactivated tectonic boundaries and implications for the reconstruction of southeastern Australia and northern Victoria Land, Antarctica. Geology 20:267-270

Foster DA, Gleadow AJW (1992b) The morphotectonic evolution of rift-margin mountains in central Kenya: constraints from apatite fission track analyses. Earth Planet Sci Lett 113:157-171

Foster DA, Gleadow AJW (1993) Episodic denudation in East Africa - a legacy of intracontinental tectonism. Geophys Res Lett 20:2395-2398

Foster DA, Gleadow AJW (1996) Structural framework and denudation history of the flanks of the Kenya and Anza Rifts, East Africa. Tectonics 15:258-271

Foster DA, John BE (1999) Quantifying tectonic exhumation in an extensional orogen with thermochronology: Examples from the southern Basin and Range Province. *In* Ring U, Brandon MT, Lister G, Willett SD (eds) Exhumation Processes: Normal Faulting, Ductile Flow, and Erosion, Geol Soc London Special Publ 154:343-364

Foster DA, Raza A (2002) Low-temperature thermochronological record of exhumation of the Bitterroot metamorphic core complex, northern Cordilleran Orogen. Tectonophysics 349:23-36

Foster DA, Miller DS, Miller CF (1991) Tertiary extension in the Old Woman Mountains area, California: evidence from apatite fission track analysis. Tectonics 10:875-886

Foster DA, Gleadow AJW, Reynolds SJ, Fitzgerald PG (1993) The denudation of metamorphic core complexes and the reconstruction of the Transition Zone, west-central Arizona: constraints from apatite fission-track thermochronology. J Geophys Res 98:2167-2185

Foster DA, Gleadow AJW, Mortimer G (1994) Rapid Pliocene exhumation in the Karakoram, revealed by fission-track thermochronology of the K2 gneiss. Geology 22:19-22

Fuchs LH (1962) Occurrence of whitlockite in chondritic meteorites. Science 137:425.

Galbraith RF (1984) On statistical estimation in fission track dating. Math Geol 16:653-669

Galbraith RF (1990) The radial plot: graphical assessment of spread in ages. Nucl Tracks Radiat Meas 17: 207-214

Galbraith RF, Green PF (1990) Estimating the component ages in a finite mixture. Nucl Tracks Radiat Meas 17:197-206

Galbraith RF, Laslett GM (1993) Statistical models for mixed fission track ages. Nucl Tracks Radiat Meas 21:459-480

Gallagher K (1995) Evolving temperature histories from apatite fission-track data. Earth Planet Sci Lett 136:421-435

Gallagher K, Brown RW (1997) The onshore record of passive margin evolution. J Geol Soc London 154:451-457

Gallagher K, Brown RW (1999a) Denudation and uplift at passive margins: the record on the Atlantic Margin of southern Africa. Phil Trans Roy Soc London A 357:835-859

Gallagher K, Brown RW (1999b) The Mesozoic denudation history of the Atlantic margins of southern Africa and southeast Brazil and the relationship to offshore sedimentation. *In* Cameron N, Bate R, Clure V (eds) The Oil and Gas Habitats of the South Atlantic. Geol Soc London Special Publ No. 153: 41-53

Gallagher K, Sambridge M (1994) Genetic algorithms: a powerful method for large scale non-linear optimisation problems. Computer Geosci 20:1229-1236

Gallagher K, Dumitru TA, Gleadow AJW (1994a) Constraints on the vertical motion of eastern Australia during the Mesozoic. Basin Res 6:77-94

Gallagher K, Hawkesworth CJ, Mantovani MJM (1994b) The denudation history of the onshore continental margin of SE Brazil inferred from apatite fission track data. J Geophys Res 99:18,117-18,145

Gallagher K, Brown RW, Johnson C (1998). Fission track analysis and its applications to geological problems. Ann Rev Earth Planet Sci 26:519-572

Garver JI, Brandon MT, Roden-Tice M, Kamp PJJ (1999) Exhumation history of orogenic highlands determined by detrital fission-track thermochronology. *In* Ring U, Brandon MT, Lister GS, Willett SD (eds) Exhumation Processes: Normal Faulting, Ductile Flow, and Erosion, Geol Soc London Spec Publ 154:283-304

Gibson HJ, Stüwe K (2000) Multiphase cooling and exhumation of the southern Adelaide Fold Belt: Constraints from apatite fission track data. Basin Res 12:31-45

Giles MR, Indrelid SL (1998) Divining burial and thermal histories from indicator data: application and limitations: An example from the Irish Sea and Cheshire Basins. *In* Van den haute P, De Corte F (eds) Advances in Fission-Track Geochronology. Kluwer Academic Publishers, Dordrecht, p 115-150

Gleadow AJW (1981) Fission Track Dating Methods: what are the real alternatives? Nucl Tracks 5:3-14

Gleadow AJW (1990) Fission track thermochronology: reconstructing the thermal and tectonic evolution of the crust. *In* Pacific Rim Congress 90 Proceedings. Pacific Rim Congr 3:15-21.

Gleadow AJW, Brooks CK (1979) Fission track dating, thermal histories and tectonics of igneous intrusions of East Greenland. Contrib Mineral Petrol 71:45-60

Gleadow AJW, Brown RW (2000) Fission track thermochronology and the long-term denudational response to tectonics. *In* Summerfield MA (ed) Geomorphology and Global Tectonics. John Wiley & Sons Ltd, Chichester, UK, p 57-75

Gleadow AJW, Duddy IR (1981a) Early Cretaceous volcanism and early breakup history of southeastern Australia: Evidence from fission track dating of volcaniclastic sediments. *In* Cresswell MM, Vella P (eds) Gondwana Five - Proc Fifth Intl Gondwana Symp. Wellington, New Zealand, 11-16 February, 1980. Balkema, Rotterdam, p 295-300

Gleadow AJW, Duddy IR (1981b) A natural long-term track annealing experiment for apatite. Nucl Tracks 5:169-174

Gleadow AJW, Lovering JF (1974) The effect of weathering on fission track dating. Earth Planet Sci Lett 22:163-168

Gleadow AJW, Lovering JF (1977) Geometry factor for external detectors in fission track dating. Nucl Track Detection 1:99-106

Gleadow AJW, Lovering JF (1978) Fission track geochronology of King Island, Bass Strait, Australia: Relationship to continental rifting. Earth Planet Sci Lett 37: 429-437

Gleadow AJW, Fitzgerald PG (1987) Uplift history and structure of the Transantarctic Mountains: new evidence from fission track dating of basement apatites in the Dry Valleys area, southern Victoria Land. Earth Planet Sci Lett 82:1-14

Gleadow AJW, Leigh-Jones P, Duddy IR, Lovering JF (1982) An automated microscope stage system for fission track dating and particle track mapping. Workshop on Fission Track Dating. Fifth International Conference on Geochronology, Cosmochronology and Isotope Geology, Nikko Japan, Abstr, p 22-23.

Gleadow AJW, Duddy IR, Lovering JF (1983) Fission track analysis: A new tool for the evaluation of thermal histories and hydrocarbon potential. Austral Petrol Explor Assoc J 23:93-102

Gleadow AJW, Duddy IR, Green PF, Lovering JF (1986) Confined fission track lengths in apatite: a diagnostic tool for thermal history analysis. Contrib Mineral Petrol 94:405-415

Gleadow A, Kohn B, Gallagher K, Cox S (1996) Imaging the thermotectonic evolution of eastern Australia during the Mesozoic from fission track dating of apatites. Geol Soc Austral Abstr 43, (Mesozoic Geology of the Eastern Australia Plate Conference Brisbane, Sept 1996), p 195-204

Gleadow AJW, Kohn BP, Brown RW, O'Sullivan PB, Raza A (2002) Fission track thermotectonic imaging of the Australian continent. Tectonophysics 349:5-21

Gold R, Roberts JH, Ruddy F (1981) Annealing phenomena in solid state track recorders. Nucl Tracks 5:253-264

Goswami JN, Jha R, Lal D (1984) Quantitative treatment of annealing of charged particle tracks in common minerals. Earth Planet Sci Lett 71:120-128

Green PF (1980) On the cause of shortening of spontaneous fission tracks in certain minerals. Nucl Tracks 4:91-100

Green PF (1981) A new look at statistics in fission track dating. Nucl Tracks 5:77-86

Green PF (1985) Comparison of zeta calibration baselines for fission-track dating of apatite, zircon and sphene. Chem Geol 58:1-22

Green PF (1986) On the thermo-tectonic evolution of Northern England: Evidence from fission track analysis. Geol Mag 123:493-506

Green PF (1988) The relationship between track shortening and fission track age reduction in apatite: combined influences of inherent instability, annealing anisotropy, length bias and systems calibration. Earth Planet Sci Lett 89:335-352

Green PF, Durrani SA (1977) Annealing studies of tracks in crystals. Nucl Track Detection 1:33-39

Green PF, Hurford AJ (1984) Neutron dosimetry for fission track dating. Nucl Tracks 10:232-241

Green PF, Duddy IR, Gleadow AJW, Tingate PR, Laslett GM (1985) Fission track annealing in apatite: track length measurements and the form of the Arrhenius plot. Nucl Tracks 10:323-328

Green PF, Duddy IR, Gleadow AJW, Tingate PR, Laslett GM (1986) Thermal annealing of fission tracks in apatite, 1. A qualitative description. Chem Geol 59:237-253

Green PF, Duddy IR, Laslett GM (1988) Can fission track annealing in apatite be described by first order kinetics? Earth Planet Sci Lett 87:216-228.

Green PF, Duddy IR, Gleadow AJW, Lovering JF (1989a) Apatite fission track analysis as a paleotemperature indicator for hydrocarbon exploration. *In* Naeser ND, McCulloh, TH (eds) Thermal History of Sedimentary Basins: Methods and Case Histories. Springer-Verlag, Berlin, p 81-195

Green PF, Duddy IR, Laslett GM, Hegarty KA, Gleadow AJW, Lovering JF (1989b) Thermal annealing of fission tracks in apatite, 4. Quantitative modelling techniques and extension to geological timescales. Chem Geol 79:155-182

Green PF, Laslett GM, Duddy IR (1993) Mechanisms and kinetics of apatite fission-track annealing – Discussion. Am Mineral 78:414-445

Green PF, Duddy IR, Bray RJ (1995) Applications of thermal history reconstruction in inverted basins. *In* Buchanan JG, Buchanan PG (eds) Basin Inversion, Geol Soc London Spec Publ 88:149-165

Green PF, Duddy IR, Hegarty KA, Bray RJ (1999) Early Tertiary heat flow along the UK Atlantic margin and adjacent areas. *In* Fleet AJ, Boldy SAR (eds) Petroleum Geology of Northwest Europe: Proc 5th Conf, Geol Soc London, p 349-357

Grist AM, Li G, Reynolds PH, Zentilli M, Beaumont C (1992) Thermochronology of the Scotian Basin from apatite fission-track and $^{40}Ar/^{39}Ar$ data. *In* Zentilli M, Reynolds PM (eds) Short Course Handbook on Low-Temperature Thermochronology. Mineral Assoc Can, Ottawa, p 97-118

Gunnell Y (2000) Apatite fission track thermochronology: An overview of its potential and limitations in geomorphology. Basin Res 12:115-132

Haack U (1972) Systematics in the fission track annealing of minerals. Contrib Mineral Petrol 35:303-312

Haack U (1973) Suche nach überschweren Transuranelementen. Naturwissenschaften 60:65-70

Haack U (1977) The closing temperature for fission track retention in minerals. Am J Sci 277:459-464

Hansen K (2000) Tracking thermal history in East Greenland: An overview. Global Planet Change 24: 303-309

Harman R, Gallagher K, Brown R, Raza A, Bizzi L (1998) Accelerated denudation and tectonic/geomorphic reactivation of the cratons of northeastern Brazil during the Late Cretaceous. J Geophys Res 103:27,091-27,105

Harrison TM, Armstrong RL, Naeser CW, Harakal JE (1979) Geochronology and thermal history of the Coast plutonic complex, near Prince Rupert, British Columbia. Can J Earth Sci 16:400-410

Hartung JB, Izett GA, Naeser CW, Kunk MJ, Sutter JF (1986) The Manson, Iowa impact structure and the Cretaceous-Tertiary boundary event. Lunar Planet Sci 17:313-314

Hejl E (1995) Evidence for unetchable gaps in apatite fission tracks. Chem Geol 122:259-269

Hejl E, Reidl H, Weingartner H (2002) Post-plutonic unroofing and morphogenesis of the Attic-Cycladic complex (Aegea, Greece). Tectonophysics 349:37-56

Henk RP, Benton EV (1967) Charged particle tracks in polymers: No.5. A computer code for the computation of heavy ion range-energy relationships in any stopping material. U S Naval Radiological Defense Laboratory Techn Rep 67-122

Hill KC, Gleadow AJW (1989) Uplift and thermal history of the Papuan Fold Belt, Papua-New Guinea: Apatite fission track analysis. Austral J Earth Sci 36:515-539

Hill KC, Raza A (1999) Arc-continent collision in Papua Guinea: Constraints from fission track thermochronology. Tectonics 18:950-966

Hill KC, Hill KA, Cooper GT, O'Sullivan AJ, O'Sullivan PB, Richardson MJ (1995) Inversion around the Bass Basin, SE Australia. *In* Buchanan JG, Buchanan PG (eds) Basin Inversion. Geol Soc London Special Publ 88, p.525-547

Howard KA, Foster DA (1996) Thermal and unroofing history of a thick, tilted Basin and Range crustal section, Tortilla Mountains, Arizona. J Geophys Res 101:511-522

Hu S, O'Sullivan PB, Raza A, Kohn BP (2001) Thermal history and tectonic subsidence of the Bohai Basin, northern China: a Cenozoic rifted and local pull-apart basin. Phys Earth Planet Intl 126:231-245

Hughes JM, Cameron M, Crowley KD (1989) Structural variations in natural F, OH, Cl apatites. Am Mineral 74:870-876

Hughes JM, Cameron M, Crowley KD (1990) Crystal structures of natural ternary apatites: Solid solution in the $Ca_5(PO_4)3X$ (X = F, OH, Cl) system. Am Mineral 75:295-304

Hughes JM, Cameron M, Crowley KD (1991a) Ordering of divalent cations in the apatite structure: Crystal structure refinements of natural Mn- and Sr-bearing apatite. Am Mineral 76:1857-1862

Hughes JM, Cameron M, Mariano AN (1991b) Rare element ordering and structural variations in natural rare-earth-bearing apatites. Am Mineral 76:1165-1173

Hurford AJ (1986) Cooling and uplift patterns in the Lepontine Alps, south-cental Switzerland, and an age of vertical movement on the Insubric fault line. Contrib Mineral Petrol 92:413-42

Hurford AJ (1990) Standardization of fission track dating calibration: Recommendation by the Fission Track Working Group of the I.U.G.S. Subcommission on Geochronology. Chem Geol 80:171-178

Hurford AJ, Carter A (1991) The role of fission track dating in discrimination of provenance. *In* Morton AC, Todd SP, Haughton PDW (eds) Developments in Sedimentary Provenance Studies, Geol Soc London Spec Publ 57:67-78

Hurford AJ, Green PF (1982) A user's guide to fission track dating calibration. Earth Planet Sci Lett 59:343-354

Hurford AJ, Green PF (1983) The zeta calibration of fission track dating. Isotope Geosci 1:285-317

Hurford AJ, Hammerschmidt K (1985) $^{40}Ar/^{39}Ar$ and K/Ar dating of the Bishop and Fish Canyon tuffs: Calibration ages for fission-track dating standards. Chem Geol 58:23-32

Hurford AJ, Carter A, Barberand J, Walgenwwitz F (2000) Acid and chloro-tincles, rare earths and differing angles, just what might be important in understanding track annealing in apatite? Geol Soc Austral Abstr 58:175-176

Igli H, Miellou J-C, Chambaudet A, Rebetez M (1998) Mathematical convection methodology using Bertagnolli and Laslett fission track annealing laws. *In* Van den haute P, De Corte F (eds) Advances in Fission Track Geochronology. Kluwer Academic, Dordrecht, p 93-98

Issler DR (1996) Optimizing time-step size for apatite fission-track annealing models. Computer Geosci 22:67-74

Issler DR, Beaumont C, Willett SD, Donelick RA, Mooers J, Grist A (1990) Preliminary evidence from apatite fission track data concerning the Peace River Arch region, Western Canada Sedimentary Basin. Bull Can Petrol Soc 38A:250-269

Itoh N (1996) Self-trapped exciton model of heavy ion track registration. Nucl Instr Meth B116:33-36

Itoh N, Tanimura K (1986) Radiation effects in ionic solids. Radiat Effects 98:269-287

Iwano H, Kasuya M, Danhara T, Yamashita T, Tagami T (1993) Track counting efficiency and unetchable track range in apatite. Nucl Tracks Radiat Meas 21:513-517

James K, Durrani SA (1988) The registration-temperature dependence of heavy-ion track-etch rates and annealing sensitivity in crystals: implications for cosmic ray identification and fission track dating of meteorites. Earth Planet Sci Lett 87:229-236

John BE, Foster DA (1993) Structural and thermal constraints on the initiation angle of detachment faulting in the southern Basin and Range: The Chemehuevi Mountains case study. Bull Geol Soc Am 105:1091-1108

Johnson C (1997) Resolving denudational histories in orogenic belts with apatite fission track thermochronology and structural data: an example from northern Spain. Geology 25:623-626

Johnson C, Gallagher K (2000) A preliminary Mesozoic and Cenozoic denudation history of the North East Greenland onshore margin. Global Planet Change 24:261-274

Jordan TE, Zeitler P, Ramos V, Gleadow AJW (1989) Thermochronometric data on the development of the basement peneplain in the Sierra Pampeanas, Argentina. J South Am Earth Sci 2:207-222

Kamp PJJ, Green PF (1990) Thermal and tectonic history of selected Taranaki Basin (New Zealand) wells assessed by apatite fission track analysis. Bull Am Assoc Petrol Geol74:1401-1419

Kamp PJJ, Green PF, White SH (1989) Fission track analysis reveals character of collisional tectonics in New Zealand. Tectonics 8:169-195

Kamp PJJ, Webster KS, Nathan S (1996) Thermal history analysis by integrated modelling of apatite fission track and vitrinite reflectance data: Application of an inverted basin (Buller Coalfield, New Zealand). Basin Res 8:383-402

Kelly BT (1966) Irradiation Damage to Solids. Pergamon, Oxford

Ketcham RA, Donelick RA, Carlson WD (1999) Variability of apatite fission-track annealing kinetics: III. Extrapolation to geological time scales. Am Mineral 84:1235-1255

Ketcham RA, Donelick RA, Donelick MB (2000) AFTSolve: A program for multi-kinetic modeling of apatite fission-track data. Geol Mater Res 2

Kobetich EJ, Katz R (1968) Width of heavy ion tracks in emulsion. Phys Rev 170:405-411

Kohn BP, Eyal M (1981) History of uplift of the crystalline basement of Sinai and its relation to opening of the Red Sea as revealed by fission track dating of apatites. Earth Planet Sci Lett 52:129-141

Kohn BP, Shagam R, Banks PO, Burkley LA (1984) Mesozoic-Pleistocene fission-track ages on rocks of the Venezuelan Andes and their tectonic implications. Geol Soc Am Mem 162:365-384

Kohn BP, Feinstein S, Eyal M (1990) Cretaceous to present paleothermal gradients, central Negev, Israel: constraints from fission track dating. Nucl Tracks Radiat Meas 17:381-388

Kohn BP, Osadetz K, Bezys RK (1995) Apatite fission track dating of two crater structures in the Canadian Williston Basin: a preliminary report. Bull Can Petrol Geol 43:54-64

Kohn BP, Feinstein S, Foster DA, Steckler MS, Eyal M (1997) Thermal history of the eastern Gulf of Suez: II Reconstruction from apatite fission track and 40Ar/39Ar K-feldspar measurements. Tectonophysics 283:219-239

Kohn BP, Gleadow AJW, Cox SJD (1999) Denudation history of the Snowy Mountains: constraints from apatite fission track thermochronology. Austral J Earth Sci 46:181-198

Kohn BP, Foster DA, Farley KA (2002a) Low temperature thermochronology of apatites with exceptional compositional variations: the Stillwater Complex, Montana revisited. Workshop on Fission Track Analysis: Theory and Applications, El Puerto de Santa María, Spain, June 2002. Geotemas 4:103-105

Kohn BP, Gleadow AJW, Brown, RW, Gallagher K, O'Sullivan PB, Foster DA. (2002b) Shaping the Australian crust over the last 300 million years: Insights from fission track thermotectonic and denudation studies of key terranes. Austral J Earth Sci 49:697-717

Kolodny Y, Bar M, Sass E (1971) Fission track age of the 'Mottled Zone Event' in Israel. Earth Planet Sci Lett 11:269-272

Lal D, Rajan RS, Tamhane AS (1969) Chemical composition of nuclei of Z > 22 in cosmic rays using meteoric minerals as detectors. Nature 221:33-37

Laslett GM, Galbraith RF (1996) Statistical modelling of thermal annealing of fission tracks in apatite. Geochim Cosmochim Acta 60:5117-5131

Laslett GM, Kendall WS, Gleadow AJW, Duddy IR (1982) Bias in measurement of fission-track length distribution. Nucl Tracks 6:79-85

Laslett GM, Gleadow AJW, Duddy IR (1984) The relationship between fission track length and track density distributions. Nucl Tracks 9:29-38

Laslett GM, Green PF, Duddy IR, Gleadow A.J.W (1987) Thermal annealing of fission tracks in apatite, 2. A quantitative analysis. Chem Geol 65:1-13

Leech ML, Stockli DF (2000) The late exhumation history of the ultrahigh-pressure Maksyutov Complex, south Ural Mountains, from new apatite fission track data. Tectonics 19:153-167

Lewis CLE, Green PF, Carter A, Hurford AJ (1992) Elevated late Cretaceous to Early Teriary paleotemperatures throughout North-west England: Three kilometres of Teriary erosion? Earth Planet Sci Lett 112:131-145

Lipman PW, Fisher FS, Mehnert HM, Naeser CW, Luedke RG, Steven TA (1976) Multiple ages of mid-Tertiary mineralization and alteration in the western San Juan Mountains, Colorado. Econ Geol 71:571-588

Lisker F (2002) Review of fission track studies in northern Victoria Land, Antarctica—Passive-margin evolution versus uplift of the Transantarctic Mountains. Tectonophysics 349:57-73

Liu T-K (1982) Tectonic implication of fission track ages from the Central Range, Taiwan. Proc Geol Soc China 25:22-37

Ludwig KR, Nash JT, Naeser CW (1981) U-Pb isotope systematics and age of uranium mineralization Midnite Mine, Washington. Econ Geol 76:89-110

Lutz TM, Omar G (1991) An inverse method of modeling thermal histories from apatite fission-track data. Earth Planet Sci Lett 104:181-195

Marshallsea SJ, Green PF, Webb J (2000) Thermal history of the Hodgkinson Province and Laura Basin, Queensland: Multiple cooling episodes identified from apatite fission track analysis and vitrinite reflectance data. Austral J Earth Sci 47:779-797

Mathiesen A, Bidstrup T, Christiansen FG (2000) Denudation and uplift history of the Jameson Land basin – constrained from maturity and apatite fission track data. Global Planet Change 24:275-301

May SJ, Kelley SA, Russell LR (1994) Footwall unloading and rift shoulder uplifts in the Albuquerque Basin: Their relation to syn-rift fanglomerates and apatite fission-track ages. *In* Keller GR, Cather SM (eds) Basins of the Rio Grande: Structure, Statigraphy, and Tectonic Setting. Geol Soc Am Spec Paper 291:125-134

Menzies MA, Gallagher K. Hurford AJ, Yelland A (1997) Red Sea and Gulf of Aden rifted margins, Yemen: Denudational histories and margin evolution. Geochim Cosmochim Acta 61:2511-2527

Miellou JC, Igli H, Chambaudet A, Rebetez M, Grivet M (1997) Convection methods and functional equations in order to solve inverse thermochronology problems. Nucl Tracks Radiat Meas 28:549-554

Miller DS, Duddy IR, Green PF, Hurford AJ, Naeser CW (1985) Results of interlaboratory comparison of fission-track age standards: fission-track workshop 1984. Nucl Tracks 10:383-391

Miller DS, Eby N, McCorkell R, Rosenberg PE, Suzuki M (1990) Results of interlaboratory comparison of fission-track ages for the 1988 Fission Track Workshop 1984. Nucl Tracks Radiat Meas 17:237-245

Miller LE, Dumitru TA, Brown RW, Gans PB (1999) Rapid Miocene slip on the Snake Range–Deep Creek Range fault system, east-central Nevada. Bull Geol Soc Am 111:886–905

Mitchell MM (1997) Elevated mid-Cretaceous paleotemperatures in the western Otway basin: Consequences for hydrocarbon generation models. Austral Petrol Exploration Assoc J 37:505-523

Mitchell MM, Kohn BP, Foster DA (1998) Post-orogenic cooling history of Eastern South Australia from apatite FT thermochronology. *In* Van den haute P, De Corte F (eds) Advances in Fission-Track Geochronology, Kluwer Academic Publishers, Dordrecht, p 207-224

Mold P, Bull RK, Durrani SA (1984) Fission-track annealing characteristics of meteoritic phosphates. Nucl Tracks 9:119-128

Montel J, Foret S, Veschambre M, Nicollet C, Provost, A (1996) Electron microprobe dating of monazite. Chem Geol 131:37-53

Moore ME, Gleadow AJW, Lovering JF (1986) Thermal evolution of rifted continental margins: new evidence from fission tracks in basement apatites from southeastern Australia. Earth Planet Sci Lett 78:255-270

Morton A (1986) Dissolution of apatite in North Sea Jurassic Sandstones: implications for the generation of secondary porosity. Clay Minerals 21:711-733

Mrowec S (1980) Defects and diffusion in solids: an introduction. Elsevier Scientific Publishing, Amsterdam, p 466

Naeser CW (1967) The use of apatite and sphene for fission track age determinations. Bull Geol Soc Am 78:1523-1526

Naeser CW (1971) Geochronology of the Navajo-Hopi diatremes, Four Corners area. J Geophys Res 76:4978-4985

Naeser CW (1979a) Fission-track dating and geologic annealing of fission tracks. *In* Jäger E, Hunziker JC (eds) Lectures in Isotope Geology. Springer-Verlag, Berlin, p 154-169

Naeser CW (1979b) Thermal history of sedimentary basins: fission track dating of subsurface rocks. *In* Scholle PA, Schluger PR (eds) Aspects of Diagenesis, Soc Econ Paleo Mineral Spec Publ 26:109-112

Naeser CW (1981) The fading of fission tracks in the geologic environment—data from deep drill holes. Nucl Tracks 5:248-250

Naeser CW, Faul H (1969) Fission track annealing in apatite and sphene. J Geophys Res 74:705-710

Naeser CW, Fleischer RL (1975) The age of the apatite at Cerro de Mercado, Mexico: A problem for fission track annealing corrections. Geophys Res Lett 2:67-70

Naeser CW, Forbes (1976) Variation of fission track ages with depth in two deep drill holes. EOS Trans Am Geophys Union 57: 353

Naeser CW, Cunningham CG, Marvin RF, Obradovich JD (1980) Pliocene intrusive rocks and mineralization near Rico, Colorado. Econ Geol 75:122-133

Naeser CW, Zimmerman RA, Cebula GT (1981) Fission-track dating of apatite and zircon: an interlaboratory comparison. Nucl Tracks 5:65-72

Naeser ND (1993) Apatite fission-track analysis in sedimentary basins: A critical appraisal. *In* Dore AG, Augustson JH, Hermanrud C, Stewart DJ, Sylta O (eds) Basin modeling: Advances and applications. Norwegian Petrol Soc Spec Publ 3:147-160

Naeser ND, Naeser CW, McCulloh TH (1989) The application of fission track dating to the depositional and thermal history of rocks in sedimentary basins. *In* Naeser ND McCulloh TH (eds) Thermal History of Sedimentary Basins: Methods and Case Histories. Springer-Verlag, Berlin, p 157-180

Naeser ND, Naeser CW, McCulloh TH (1990) Thermal history of rocks in southern San Joaquin Valley, California: evidence from fission-track analysis. Bull Am Assoc Petrol Geol 74:13-29

Nagpaul KK, Metha PP, Gupta ML (1974) Annealing studies on radiation damages in biotite, apatite and sphene and corrections to fission track ages. Pure Appl Geophys 112:131-139

Nelson EP (1982) Post-tectonic uplift of the Cordillera Darwin orogenic core complex: evidence from fission track geochronology and closing temperature-time relationships. J Geol Soc London 139: 755-761

Noble WP, Foster DA, Gleadow AJW (1997) The post-Pan-African thermal and extensional history of crystalline basement rocks in eastern Tanzania. Tectonophysics 275:331-350

Omar GI, Steckler MS (1995) Fission-track evidence on the initial rifting of the Red Sea: two pulses, no propagation. Science 270:1341-1344

Omar GI, Johnson KR, Hickey LJ, Robertson PB, Dawson MR, Barnosky CW (1987) Fission-track dating of Haughton Astrobleme and included biota, Devon Island, Canada. Science 237:1603-1605

Omar GI, Steckler MS, Buck WR, Kohn BP (1989) Fission track analysis of basement apatites at the western margin of the Gulf of Suez rift, Egypt: Evidence for synchroneity of uplift and subsidence. Earth Planet Sci Lett 94:316-328

Osadetz KG, Kohn BP, Feinstein S, O'Sullivan PB (2002) Williston basin thermal history from apatite fission track thermochronology—implications for petroleum systems and geodynamic history. Tectonophysics 349:221-249

O'Sullivan PB (1996) Late Mesozoic and Cenozoic thermotectonic evolution of the Colville Basin, North Slope, Alaska. *In* Johnsson MJ, Howell DG (eds) Thermal Evolution of Sedimentary Basins in Alaska. U S Geol Surv Bull 2142:45-79

O'Sullivan PB, Brown RW (1998) Effects of surface cooling on apatite fission-track data: Evidence for Miocene climatic change, North Slope, Alaska. *In* Van den haute P, De Corte F (eds) Advances in Fission-Track Geochronology. Kluwer Academic Publishers, Dordrecht, p 255-267

O'Sullivan PB, Parrish RR (1995) The importance of apatite composition and single-grain ages when interpreting fission track data from plutonic rocks: A case study from the Coast Ranges, British Columbia. Earth Planet Sci Lett 132:213-224

O'Sullivan PB, Green PF, Bergman SC, Decker J, Duddy IR, Gleadow AJW, Turner DL (1993) Multiple phases of Tertiary uplift and erosion in the Arctic National Wildlife Refuge, Alaska, revealed by apatite fission track analysis. AAPG Bull 77:359-385

O'Sullivan PB, Foster DA, Kohn BP, Gleadow AJW, Raza A (1995a) Constraints on the dynamics of rifting and denudation on the eastern margin of Australia: Fission track evidence for two discrete causes of rock cooling. Austral Inst Min Metal Publ 9/95:441-446

O'Sullivan PB, Kohn BP, Foster DA, Gleadow AJW (1995b) Fission track data from the Bathurst batholith: Evidence for rapid mid-Cretaceous uplift and erosion within the eastern highlands of Australia. Austral J Earth Sci 42:597-607

O'Sullivan PB, Foster DA, Kohn BP, Gleadow AJW (1996) Multiple post orogenic denudation events: An example from the eastern Lachlan fold belt, Australia. Geology 24:563-566

O'Sullivan PB, Murphy JM, Blythe AE (1997) Late Mesozoic and Cenozoic thermotectonic evolution of the Central Brooks Range and adjacent North Slope foreland basin, Alaska: Including fission-track results from the Trans-Alaska Crustal Transect (TACT). J Geophys Res 102:20,821-20,845

O'Sullivan PB, Kohn BP, Mitchell MM (1998) Low temperature thermochronology of Phanerozoic reactivation along a fundamental Proterozoic crustal fault: The Darling River lineament, Australia. Earth Planet Sci Lett 164:451-465

O'Sullivan PB, Orr M, O'Sullivan AJ, Gleadow AJW (1999) Episodic Late Paleozoic to Recent denudation of the Eastern Highlands of Australia: evidence from the Bogong High Plains, Victoria. Austral J Earth Sci 46:199-216

O'Sullivan PB, Belton DX, Orr M (2000a) Post-orogenic thermotectonic history of the Mount Buffalo region, Lachlan fold belt, Australia: evidence for Mesozoic to Cenozoic. Tectonophysics. 317:1-26

O'Sullivan PB, Gibson DL, Kohn BP, Pillans B, Pain CF (2000b) Long-term landscape evolution of the Northparkes region of the Lachlan Fold Belt, Australia: Constraints from fission track and paleomagnetic data. J Geol 108:1-16

O'Sullivan PB, Mitchell MM, O'Sullivan AJ, Kohn BP, Gleadow AJW (2000c) Thermotectonic history of the Bassian Rise, Australia: Implications for the breakup of eastern Gondwana along Australia's southeastern margins. Earth Planet Sci Lett 182:31-47

Pagel M, Braun JJ, Disnar JR, Martinez L, Renac C, Vasseur G (1997) Thermal history constraints from studies of organic matter, clay minerals, fluid inclusions and apatite fission tracks at the Ardeche paleo-margin (BAI Drill Hole, CPF Program), France. J Sed Res 67:235-245

Paretzke HG (1986) Radiation track structure theory. *In* Freeman GR (ed) Kinetics of Homogeneous Processes. John Wiley, New York, p 89-170

Parnell J, Carey PF, Green PF, Duncan W (1999) Hydrocarbon migration history, West of Shetland: Integrated fluid inclusion and fission track studies. *In* Fleet AJ, Boldy SAR (eds) Petroleum Geology of Northwest Europe: Proc 5th Conf. Geol Soc London, p 613-625

Parrish RR (1983) Cenozoic thermal evolution and tectonics of the Coast Mountains of British Columbia 1, fission-track dating, apparent uplift rates, and patterns of uplift. Tectonics 2:601-632

Parrish RR (1985) Some cautions which should be exercised when interpreting fission-track and other dates with regard to uplift calculations. Nucl Tracks 10:425

Paul TA, Fitzgerald PG (1992) Transmission electron-microscopic investigation of fission tracks in fluorapatite. Am Mineral 77:336-344

Pease V, Foster D, Wooden J, O'Sullivan P, Argent J, Fanning C (1999) The Northern Sacramento Mountains, southwest United States. Part II: Exhumation history and detachment faulting. *In* Mac Niocaill C, Ryan PD (eds) Continental Tectonics. Geol Soc London Special Publ 164:199-237

Pellas P, Storzer D (1975) Uranium and Plutonium in chondritic phosphates. Meteoritics 10:471-473

Pellas P, Perron C, Crozaz G, Perelygin VP, Stetsenko SG (1983) Fission track age and cooling rate of the Marjalathi pallasite. Earth Planet Sci Lett 64:319-326

Pflaker G, Naeser CW, Zimmermann RA, Lull JS, Hudson T (1991) Cenozoic uplift history of the Mount McKinley area in the Central Alaska Range based on fission track dating. U S Geol Surv Bull 2041, p 202-212

Price PB, Walker RM (1962) Chemical etching of charged-particle tracks in solids. J Appl Phys 33: 3407-3412

Price PB, Walker RM (1963) Fossil tracks of charged particles in mica and the age of minerals. J Geophys Res 68:4847-4862

Raab MJ, Brown RW, Gallagher K, Carter A, Weber K (2002) Late Cretaceous reactivation of deep crustal shear zones in northern Namibia: Constraints from apatite fission track analysis. Tectonophysics 349:75-92

Rahn MK, Hurford AJ, Frey M (1997) Rotation and exhumation of a thrust plane: Apatite fission track data from the Glarus thrust, Switzerland. Geology 25: 599-602

Ravenhurst CE, Donelick RA (1992) Fission track thermochronology. *In* Zentilli M, Reynolds PM (eds) Short Course Handbook on Low Temperature Thermochronology, Mineral Assoc Can, Ottawa, p 21-42

Roberts SV, Burbank DW (1993) Uplift and thermal history of the Teton Range (northwestern Wyoming) defined by apatite fission-track dating. Earth Planet Sci Lett 118:295-308

Roeder PL, MacArthur D, Ma X-P, Palmer GR (1987) Cathodoluminescence and microprobe study of rare-earth elements in apatite. Am Mineral 72:801-811

Rohrman M, van der Beek P, Andriessen P (1994) Syn-rift thermal structure and post-rift evolution of the Oslo Rift (southeast Norway): New constraints from fission track thermochronology. Earth Planet Sci Lett 127:39-54

Rohrman M, van der Beek P, Andriessen P, Cloethingh S (1995) Meso-Cenozoic morphotectonic evolution of southern Norway: Neogene domal uplift inferred from apatite fission track thermochronology. Tectonics 14:704-718

Ross Jr RJ, Naeser CW, Izett GA (1976) Apatite fission-track dating of a sample from the type Caradoc (Middle Ordovician) Series in England. Geology 4:505-506

Schaer JP, Reimer GM, Wagner GA (1975) Actual and ancient uplift rate in the Gotthard region, Swiss Alps: a comparison between precise levelling and fission-track apatite age. Tectonophysics 29: 293-400.

Schäfer T, Olesch M (1998) Multiple thermal evolution of Oates Land (Northern Victoria Land, Antarctica): Evidence from apatite fission track analysis. *In* Van den haute P, De Corte F (eds) Advances in Fission Track Geochronology. Kluwer Academic, Dordrecht, p 241-253

Schlunegger F, Willett S (1999) Spatial and temporal variations in exhumation of the central Swiss Alps and implications for exhumation mechanisms. *In* Ring U, Brandon MT, Lister GS, Willett SD (eds) Exhumation Processes: Normal Faulting, Ductile Flow, and Erosion. Geol Soc London Special Publ 154:157-179

Seward D (1989) Cenozoic basin histories determined by fission-track dating of basement granites, South Island, New Zealand. Chem Geol 79:31-48

Seward D, Perez-Estaun A, Puchkov V (1997) Preliminary fission-track results from the southern Urals: Sterlitamak to Magnitogorsk. Tectonophysics 276:281-290

Shagam R., Kohn BP, Banks PO, Dasch LE. Vargas R, Rodriguez GI, Pimental N (1984) Tectonic implications of Cretaceous-Pliocene fission track ages from rocks of the circum-Maracaibo basin region of western Venezuela and eastern Colombia. Geol Soc Am Mem 162:385-412

Sharma KK, Bal KD, Parshad R, Lal N, Nagpaul K.K (1980) Palaeo-uplift and cooling rates from various orogenic belts in India, as revealed by radiometric ages. Tectonophysics 70:135-158

Shukoljukov JA, Komarov AN (1970) Tracks of uranium fission in monazite (Russian) Bulletin Commission Determination of Absolute Age of Geological Formations, Akad Nuak USSR, Moscow 9: 20-26

Sidall R, Hurford AJ (1998) Semi-quantitative determination of apatite anion composition for fission-track analysis using infra-red microspectroscopy. Chem Geol 150:181-190

Silk ECH, Barnes RS. (1959) Examination of fission fragment tracks with an electron microscope. Phil Mag 4:970-972

Smith MJ, Leigh-Jones P (1985) An automated microscope scanning stage for fission track dating. Nucl Tracks 10:395-400

Sobel ER, Dumitru TA (1997) Thrusting and exhumation around the margins of the western Tarim basin during the India-Asia collision. J Geophys Res 102:5043-5063

Sorkhabi RB (1993) Time-temperature pathways of Himalayan and Trans-Himalayan crystalline rocks: A comparison of fission-track ages. Nucl Tracks Radiat Meas 21:535-542

Spikings RA, Foster DA, Kohn BP (1997) Phanerozoic denudation history of the Mount Isa Inlier, Northern Australia: Response of a Proterozoic mobile belt to intraplate tectonics. Intl Geol Rev 39:107-124

Spikings RA, Seward D, Winkler W, Ruiz GM (2000) Low temperature thermochronology of the northern Cordillera Real, Ecuador: tectonic insights from zircon and apatite fission track analysis. Tectonics 19:649-668

Spohr R (1990) Ion Tracks and Microtechnology: Principles and Applications. Vieweg, Braunschweig

Steckler MS, Omar GI, Karner GD, Kohn BP (1993) Pattern of hydrothermal circulation within the Newark Basin from fission-track analysis. Geology 21:735-738

Stockli DF, Linn JK, Walker JD, Dumitru TA (2001) Miocene unroofing of the Canyon Range during extension along the Sevier Desert Detachment, west central Utah. Tectonics 20:289-307

Stormer JC, Pierson ML, Tacker RC (1993) Variation of F and Cl X-ray intensity due to anisotropic diffusion in apatite during electron microprobe analysis. Am Mineral 78:641-648
Storzer D, Pellas P (1977) Angra Dos Reis: Plutonium distribution and cooling history. Ear Planet Sci Lett 35:285-293
Storzer D, Selo M (1984) Toward a new tool in hydrocarbon resource evaluation: The potential of the apatite fission track chrono-thermometer. *In* Durand B (ed) Thermal Phenomena in Sedimentary Basins, International Colloquium Bordeaux, June 7-10, 1983, Collection Colloques et Séminaires 41, Editions Technip, Paris, p 89-110
Storzer D, Wagner GA (1977) Fission track dating of meteorite impacts, Meteoritics 12:368-369
Storzer D, Wagner GA (1982) The application of fission track dating in stratigraphy: A critical review. *In* Odin GS (ed) Numerical Dating in Stratigraphy, John Wiley and Sons Ltd, Chichester, UK, p 199-221
Stüwe K, White L, Brown RW (1994) The influence of eroding topography on steady-state isotherms. Application to fission track analysis. Earth Planet Sci Lett 124:63-74
Sudarsanan K, Young RA (1978) Structural interactions of F, Cl and OH in apatites. Acta Crystallogr B34:1401-1407
Summerfield MA, Brown RW (1998) Geomorphic factors in the interpretation of fission-track data. *In* Van den haute P, De Corte F (eds) Advances in Fission-Track Geochronology. Kluwer Academic Publishers, Dordrecht, p 269-284
Tagami T, Lal N, Sorkhabi RB, Nishimura S (1988) Fission track thermochronologic analysis of the Ryoke Belt and the Median Tectonic Line, Southwest Japan. J Geophys Res 93:705-713
Thomson K, Green PF, Whitham AG, Price SP, Underhill JR (1999a) New constraints on the thermal history of North-East Greenland from apatite fission-track analysis. Bull Geol Soc Am 111:1054-1068
Thomson K, Underhill JR, Green PF, Bray RJ, Gibson HJ (1999b) Evidence from apatite fission track analysis for the post-Devonian burial and exhumation history of the northern Highlands, Scotland. Marine Petrol Geol 16:27-39
Thomson SN (1994) Fission-track analysis of the crystalline basement rocks of the Calabrian Arc, southern Italy: evidence of Oligo-Miocene late-orogenic extension and erosion. Tectonophysics 238:331-352
Thomson SN, Rauche H, Brix MR (1998a) Apatite fission-track thermochronology of the uppermost tectonic unit of Crete, Greece: Implications for the post-Eocene tectonic evolution of the Hellenic subduction system. *In* Van den haute P, De Corte F (eds) Advances in Fission Track Geochronology. Kluwer Academic, Dordrecht, p 187-205
Thomson SN, Stöckhert B, Brix MR (1998b) Thermochronology of the high-pressure metamorphic rocks of Crete, Greece: Implications for the speed of tectonic processes. Geology 26:259-262
Thomson SN, Hervé E, Stöckhert B (2001) Mesozoic-Cenozoic denudation history of the Patagonian Andes (southern Chile) and its correlation to different subduction processes. Tectonics 20:693-711
Tippett JM, Kamp PJJ (1993) Fission track analysis of the late Cenozoic vertical kinematics of continental Pacific crust, South Island, New Zealand. J Geophys Res 98:16119-16148
Toulemonde M, Fuchs G, Nguyen N, Sturder F, Groult D (1987) Damage processes and magnetic field orientation in ferrimagnetic oxides Y3Fe5O12 and BaFe12O19 irradiated by high energy heavy ions: a Mossbauer study. Phys Rev B35:6560-6569
Toulemonde M, Constantini JM, Dufour Ch, Meftah A, Paumier E, Studer F (1996) Track creation in SiO2 and BaFe12O19 by swift heavy ions: a thermal spike description. Nucl Instr Meth B116:37-42
Tseng HY, Onstott TC, Burrus RC, Miller DS (1996) Constraints on the thermal history of the Taylorsville Basin, Virginia, U.S.A., from fluid-inclusion and fission-track analyses: implications for subsurface geomicrobiology experiments. Chem Geol 127:297-311
Turner DL, Forbes RB, Naeser CW (1973) Radiometric ages of Kodiak Seamount and Giacomini Guyot, Gulf of Alaska: implications for Circum-Pacific tectonics. Science 182:579-581.
Turner DL, Frizzell VA, Triplehorn DM, Naeser CW (1983) Radiometric dating for ash partings in coal of the Eocene Puget Group, Washington: Implications for paleobotanical stages. Geology 11:527-531
van der Beek PA (1997) Flank uplift and topography at the central Baikal Rift (SE Siberia): A test of kinematic models for continental extension. Tectonics 16:122-136
van der Beek PA, Cloetingh S, Andriessen PAM (1994) Mechanisms of extensional basin formation and vertical motions at rift flanks: Constraints from tectonic modelling and fission track thermochronology. Earth Planet Sci Lett 121:417-433
van der Beek PA, Andriessen PAM, Cloetingh S (1995) Morpho-tectonic evolution of rifted continental margins: inferences from a coupled tectonic-surface processes model and fission-track thermochronology. Tectonics 14:406-421
van der Beek PA, Delvaux D, Andriessen PAM, Levi KG (1996) Early Cretaceous denudation related to convergent tectonics in the Baikal region, SE Siberia. J Geol Soc London 153:515-523

Villa F, Grivet M, Rebetez M, Dubois C, Chambaudet A, Chevarier N, Blondiaux G, Sauvage T, Toulemonde M (2000) Damage morphology of Kr tracks in apatite: dependence on thermal annealing. Nucl Instr Meth B168:72-77

Vineyard GH (1976) Thermal spikes and activated processes. Radiat Effects 29:245-248

Wagner GA (1968) Fission track dating of apatites. Earth Planet Sci Lett 4:411-415

Wagner GA (1969) Spuren der spontanen Kernspaltung des ^{238}Urans als Mittel zur Datierung von Apatiten und ein Beitrag zur Geochronologie des Odenwaldes. N Jahrb Mineral Abh 110:252-286

Wagner GA (1977) Spaltspurendatierung an Apatite und Titanit aus dem Ries: Ein Beitrag zum Alter und zur Wärmegeschichte. Geol Bavarica 75:349-354

Wagner GA, Reimer GM (1972) Fission track tectonics: The tectonic interpretation of fission track apatite ages. Earth Planet Sci Lett 14:263-268

Wagner GA, Storzer D (1970) Die Interpretation von Spaltspurenaltern (fission track ages) am Beispiel von natürlichen Gläsern, Apatiten und Zirkonen. Eclogae Geol Helv 63:335-344

Wagner GA, Storzer D (1972) Fission track length reductions in minerals and the thermal history of rocks. Trans Am Nucl Soc 15:127-128

Wagner GA, Storzer D (1977) Fission track dating of meteorite impacts. Meteoritics 12:368-369

Wagner GA, Van den haute P (1992) Fission-Track Dating. Enke Verlag / Kluwer Academic Publishers, Dordrecht

Wagner GM, Reimer GM, Jäger E (1977) Cooling ages derived by apatite fission track, mica Rb-Sr, and K-Ar dating: the uplift and cooling history of the Central Alps. Mem Inst Geol Mineral Univ Padova 30:1-27

Wagner GM, Miller DS, Jäger E (1979) Fission track ages on apatite of Bergell rocks from Central Alps and Bergell boulders in Oligocene sediments. Earth Planet Sci Lett 45:355-360

Wagner GA, Gleadow AJW, Fitzgerald PG (1989) The significance of the partial annealing zone in apatite fission-track analysis: projected track length measurements and uplift chronology of the Transantarctic Mountains. Chem Geol . 79:295-305

Wagner GA, Coyle DA, Duyster F, Henjes-Kunst F, Peterek A, Schröder B Stoeckhert B, Wemmer K, Zulauf G, Ahrendt H, Bischoff R, Hejl E, Jacobs J, Menzel D, Lal N, Van den haute P, Vercoutere C, Welzel B (1997) Post-Variscan thermal and tectonic evolution of the KTB site and its surroundings. J Geophys Res 102:18,221-18,232

Weissbrod T, Perath I, Nachmias J (1987) Apatite as a Palaeoenvironmental indicator in the Precambrian-Mesozoic clastic sequence of the Middle East. J African Earth Sci 6:797-805

White SH, Green PF (1986) Tectonic development of the Alpine fault zone, New Zealand: A fission track study. Geology 14:124-127

Wien K (1989) Fast heavy ion induced desorption. Radiat Effects Defect Solid 109:137-167

Willet SD (1992) Modelling thermal annealing of fission tracks in apatite. *In* Zentilli M, Reynolds PH (eds) Short Course Handbook on low temperature thermochronology. Mineral Assoc Can, Ottawa, p 43-72

Willet SD (1997) Inverse modeling of annealing of fission tracks in apatite 1: A Controlled random search method. Am J Sci 297:939-969

Willett SD, Beaumont C, Fullsack P (1993) Mechanical model for the tectonics of doubly vergent compressional orogens. Geology 21:371-374

Williams ML, Jercinovic MJ, Terry M.P (1999) Age mapping and dating of monazite on the electron microprobe: deconvoluting multistage tectonic histories. Geology 27:1023-1026

Winkler JE, Kelley SA, Bergman SC (1999) Cenozoic denudation of the Witchita Mountains, Oklahoma, and southern mid-continent: apatite fission-track thermochronology constraints. Tectonophysics 305: 339-353

Yelland AJ (1990) Fission track thermotectonics in the Pyrenean orogen. Nucl Tracks Radiat Meas 17: 293-299

Young DA (1958) Etching of radiation damage in lithium fluoride. Nature 182:365-367

Zeck HP, Andriessen PAM, Hansen K, Jensen PK, Rasmussen BL (1988) Paleozoic paleo-cover of the southern part of the Fennoscandian Shield: fission track constraints. Tectonophysics 149:61-66

Zeitler PK (1985) Cooling history of the NW Himalaya, Pakistan. Tectonics 4:127-151

Zeitler PK Tahirkheli-Rashid AK, Naeser CW, Johnson NM (1982) Unroofing history of a suture zone in the Himalaya of Pakistan by means of fission-track annealing ages. Earth Planet Sci Lett. 57:227-240

Ziegler JF, Biersack JP, Littmark U (1985) The Stopping and Ranges of Ions in Solids. Pergamon, New York

Zhao M, Behr HJ, Ahrendt H, Wemmer K, Ren Z, Zhao Z (1996) Thermal and tectonic history of the Ordos Basin, China: evidence from apatite fission track analysis, vitrinite reflectance, and K-Ar dating. Bull Am Assoc Petrol Geol 80:1110-1134

17 Biomedical Application of Apatites

Karlis A. Gross

School of Physics and Materials Engineering
P O Box 69M, Monash University
Victoria 3800, Australia.

Christopher C. Berndt

Department of Materials Science and Engineering
State University of New York at Stony Brook
Stony Brook, New York 11794, USA.

INTRODUCTION

The field of biomedical materials has grown rapidly over the past 20 years and offers solutions to repair defects, correct deformities, replace damaged tissue and provide therapy. This has contributed to the increase in the average lifetime of individuals in developed countries. The market value for biomaterials is of the order of billions of dollars per annum worldwide and is growing as new products offer improved performance or provide new solutions health problems. Apatites are playing a key role in biomedical implants.

In developing materials used for implantation consideration must be given to both the influence of the implanted material on the body, and how the body affects the integrity of the material. The body will treat implants as inert, bioactive, or resorbable materials. Generally "inert" materials will evoke a physiological response to form a fibrous capsule, thus, isolating the material from the body. Calcium phosphates fall into the categories of bioactive and resorbable materials. A bioactive material will dissolve slightly, but promote the formation of an apatite layer before interfacing directly with the tissue at the atomic level. Such an implant will provide good stabilization for materials that are subject to mechanical loading. A bioresorbable material will, however, dissolve and allow tissue to grow into any surface irregularities but may not necessarily interface directly with the material (Neo et al. 1992).

The first use of calcium phosphate as an implanted biomaterial provided accelerated bone healing in surgically created defects in rabbits (Albee and Morrison 1920). Interest in apatite specifically started in the 1960s and initial studies principally involved the synthesis and analysis of hydroxylapatites in an attempt to better understand biological apatites (Le Geros 1965, McConnell 1965, Nancollas and Mohan 1970, Selvig et al. 1970). Hydroxylapatite is a specific form of apatite with a chemical composition of $Ca_{10}(PO_4)_6(OH)_2$. Because of its relationship to bone hydroxylapatites have been synthesized for the purpose of implantation (Aoki 1973, Jarcho et al. 1977, de Groot 1980, Winter et al. 1981, Frame et al. 1981). The early history of calcium phosphate implantation was discussed by Driskell (1994).

Calcium phosphate with an apatitic structure occurs naturally in the human body and can be described as a calcium deficient carbonate-hydroxylapatite. The chemical similarity of hydroxylapatite to the bone mineral suggests an intrinsic biocompatibility. Implantation of solid blocks of hydroxylapatite has revealed direct bonding to soft tissue (Jansen et al. 1985, Aoki et al. 1987), muscle tissue (Negami 1988) and bone tissue. This aspect of being able to create an artificial material, that provokes excellent tissue response, has provided the impetus for development of hydroxylapatite and other apatites for applications in the body.

1529-6466/00.0048-0017$05.00

Biomedical applications of hydroxylapatite are numerous. There are reviews addressing hydroxylapatite (Ben-Nissan et al. 1995, Suchanek 1998), hydroxylapatite coatings (Berndt et al. 1990, Dhert 1994, Thomas 1994, Jaffe and Scott 1996, Heimann et al. 1997, de Groot et al. 1998, Hlavac 1999, Ong and Chan 1999, Willmann 1999, Sun et al. 2001, Geesink 2002), and bioceramics (Le Geros 1993, Hench 1998a, Greenspan 1999, Blokhuis et al. 2000, Kim 2001). Several books also provide extensive discussion on apatite (Le Geros 1991, Elliott 1994, Driessens and Verbeeck 1990, and this volume). Hydroxylapatite is the OH end member of the calcium phosphate apatite group minerals, $Ca_{10}(PO_4)_6(F,OH,Cl)_2$. The most common abbreviations of hydroxylapatite are "OHA," "OHAp," "HA" or "Hap." Reference to hydroxylapatite within the present work will be abbreviated as HAp. It is worth noting that hydroxylapatite is often referred to in the literature as *hydroxyapatite*, by the medical community. The International Mineralogical Association, which represents the scientific community in regard to proper use of mineral nomen-clature, recommends use of *hydroxylapatite* for consistency (Smith 1994).

This chapter briefly reviews the formation of naturally occurring apatites in the body. For a more in-depth review see Elliott (this volume, p. 427-453). Powder synthesis will be discussed in light of its application for producing an implantable material. The manufacture, performance and applications of various forms of apatite will provide insight into its wide use in biomedical applications.

BIOLOGICAL APATITES

Several minerals are known to be essential by the human body for proper function. These include salts of calcium, magnesium, phosphorus, sodium, chlorine and potassium. The main functions of these minerals are as constituents of the skeleton, as soluble salts to maintain the composition of body fluids and as essential adjuncts to the action of many enzymes and other proteins.

About 20-30% of the calcium intake in a diet is absorbed into the body. The majority of the calcium is incorporated into calcium phosphate, confined to skeletal tissue and teeth. The skeleton provides shape to the body, supports the body weight, protects vital organs, anchors muscles and facilitates locomotion. The skeleton and teeth contain 99% of the total body calcium and 85% of the phosphorus that amounts to a combined mass of about 2 kg in an average person (Matkovic 1991, Power et al. 1999). The remaining 1% of calcium is used for physiological processes in the body. Unlike other nutrients in the body, calcium is stored in excess for short-term needs, but concurrently serves critical structural requirements. An imbalance within the metabolic system will preferentially sacrifice calcium from bones to maintain a balance in the physiological processes. Conversely, a higher serum calcium level may produce crystallization of calcium into kidney stones as calcium oxalate (Pineda et al. 1996). Although the conversion of calcium from the solution to a solid is an important aspect of calcium storage in the body, it may produce undesirable crystallization of apatite in urinary calculi (Konjiki et al. 1980) or on heart valves (Banas and Baier 2000, Deiwick et al. 2001).

Osteoblasts are responsible for the production of bone (a ceramic-polymer composite material filled with living cells). These mononuclear cells deposit an organic matrix that contains collagen, composed of a defined sequence of amino acids in a polypeptide chain. Three polypeptide chains are folded into rod-like triple-helical molecules about 300 nm long and 1.5 nm in diameter (Fig. 1). Collagen chains aggregate so that each molecule is longitudinally displaced by one quarter of the length relative to the nearest neighbor to form a fibril. Small apatite platelets fit into predetermined pockets of the collagen with the *c*-axis aligned with the fibral long axis (Fratzl et al. 1991, Landis et al. 1996). The fibrils are twisted around one another in the opposite direction to form a fiber. The fibers bundle together to build up the lamellae, 3 to 7 μm thick, located concentrically around a central canal. The numerous levels of order indicate the rich hierarchical structure of bone to assemble the osteon, the building block in bone. Osteons arrange themselves in a dense packing arrangement in compact bone, but take on more random arrangement in porous trabecular bone. Since osteons lie parallel to lines of stress in bone, the apatite crystals then have a set orientation with respect to stresses applied to bone (Martin and Burr 1989).

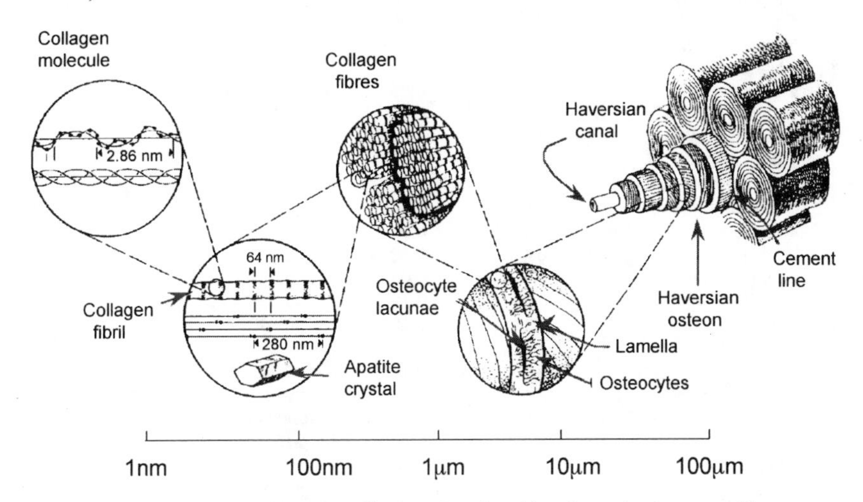

Figure 1. The structural location of hydroxylapatite with collagen in an osteon of bone.

In bone and teeth, apatites are based on a calcium hydroxylapatite composition modified with sodium, potassium, magnesium, zinc, and especially carbonate; see Table 1. Some of these substituent species, particularly carbonate, lead to large lattice strain and higher solubilities (Driessens 1988). Bioapatites with the highest degree of crystallinity and the highest concentration are found in the enamel of teeth. The formation of apatites is preceded by the differentiation of preameloblast cells that secrete enamel matrix proteins (amelogenins and enamelins). The high saturation of calcium and phosphate ions in the protein gel adjacent to the ameloblast causes the precipitation of carbonated apatite. Proteins dissolve or are resorbed to provide space for the growing apatite crystals. Eventually, ameloblasts withdraw, leaving apatite crystals which are stacked as rods or prisms, surrounded by a thin boundary film of enamelin (Ten Cate 1994). Upon completion of the tooth formation, the ameloblast departs and no repair is possible.

Apatite crystals in enamel are aligned perpendicular to the surface of the tooth (Johanssen 1964, Boyde 1997) and in an irregular fashion on the cusp tips and incisal edges, to impart strength in high stress areas. The underlying dentine acts as a support and and the inherent flexibility is thought to prevent fracture of the brittle enamel.

Teeth are subjected to a continuous demineralization/mineralization process and, therefore, the apatite in enamel is modified in response to the microenvironment around the teeth. Applications of fluoride through toothpaste and drinking water result in the exchange of F^- for OH^- in hydroxylapatite to form the more stable and acid resistant fluorapatite. Also, development of caries lesions is decelerated as dissolution removes carbonate and magnesium from the outer enamel layer, which further stabilizes the enamel apatite (Le Geros 1999).

Table 1. Chemical and crystallographic characteristics of natural and synthetic apatite (modified from Driessens and Verbeeck 1990 and Le Geros et al. 1995).

	Bone	Enamel	Dentine	Hydroxylapatite
Constituents (wt %)				
Calcium, Ca^{2+}	24.5	36.0	26.9	39.6
Phosphorus, P	11.5	17.7	13.2	18.5
(Ca/P ratio)	1.65	1.62	4.6	1.67
Carbonate, CO_3^{2-}	5.80	3.20	0.6	-
Sodium, Na^+	0.70	0.50	0.8	-
Magnesium, Mg^{2+}	0.55	0.44	0.06	-
Chloride, Cl^-	0.10	0.30	0.02	-
Potassium, K^+	0.03	0.08	0.1 (max)	-
Fluoride, F^-	0.02	0.01		-
Ash (total inorganic)	65.0	97.0		100
Total inorganic	25.0	1.0		-
Trace elements (max)				-
Strontium, Sr^{2+}	0.02	0.14	0.02	
Barium, Ba^{2+}	0.1	0.02	0.005	
Lead, Pb^{2+}	0.08	0.10	0.004	
Iron, Fe^{3+}	0.1	0.08	0.01	
Zinc, Zn^{2+}	0.04	0.12	0.07	
Copper, Cu^{2+}	0.1	0.008	0.005	
Aluminium, Al^{3+}		0.04	0.015	
Silicon, Si^{4+}	0.05	0.14	0.01	
Manganese, Mn^{2+}		0.006	present	
Selenium, Se^{2+}		0.002	present	
Tin, Sn^{2+}		0.009	present	
Lithium, Li^+		0.001	present	
Nickel, Ni^{2+}		0.001	present	
Silver, Ag^+		0.004	0.07	
Sulfur, S		0.005		
Cadmium, Cd^{2+}		0.007		
Lattice parameters				
a-axis	9.419	9.441		9.422
c-axis	6.880	6.882		6.88
Crystallite size, Å	250 x 25-50	1300 x 300		2000
Products after heating at 950°C	HAp + TCP	HAp + CaO		HAp

Because the natural mineral component of bones and teeth is carbonate-hydroxylapatite, the use of synthetic apatites as bone and tooth replacement materials has been extensively investigated. The first step in the manufacture of biomedical devices requires the ability to synthesize pure and reproducible apatite powders.

SYNTHESIS OF APATITE

Calcium phosphate apatite formulations that are available commercially from different suppliers exhibit high chemical variability. Thus, manufacturers of apatite for biomedical applications often produce their own powder. Furthermore, synthesis enables a range of chemical substitutions, crystal sizes, shapes and forms (separate crystals or cements, as discussed in the next section). Various reviews cover synthesis methods (Narasaraju and Phebe 1996, Le Geros et al. 1995, Orlovskii and Barinov 2001).

Precipitation from solution is the most common synthesis route and involves simultaneous addition of a calcium salt and a phosphate compound to water, or drop-wise addition of the phosphate into an aqueous solution of the calcium salt. Examples of calcium salts include calcium nitrate, calcium hydroxide, calcium chloride, or calcium acetate. The salts are reacted with a hydrogen phosphate or the phosphate ions are introduced in solution from di-ammonium hydrogen phosphate or orthophosphoric acid. Two reactions studied in detail involve (i) reaction of calcium nitrate with di-ammonium hydrogen phosphate, and (ii) the addition of orthophosphoric acid to calcium hydroxide (Tagai and Aoki 1980, Osaka et al. 1991).

The precipitation reaction is conducted with pure reactants at a pH greater than 9, with a controlled reactant addition rate, under stirred conditions and a temperature between 25 and 90°C. A yield of 87%, measured on the basis of the initial reactant stoichiometry, is achieved when orthophosphoric acid is reacted with calcium hydroxide at production rates of 50 g/hr. However, when calcium nitrate and diammonium hydrogen phosphate are selected as reactants the yield is 29%, for similar production rates. In reactions where ammonium is part of the precursor, dilute ammonium hydroxide is added continuously to restore the pH after a decrease caused by removal of a hydroxide from the solution to precipitate HAp. The slow incorporation of calcium into the apatitic structure must be accompanied by stirring and aging after the reaction. The Ca/P molar ratio of 1.67 is attained in as little as 5 hours after the completion of the reaction at 90°C (Rodríguez-Lorenzo and Vallet-Regí 2000). Post reaction maturation is an important step for ensuring the production of stoichiometric hydroxylapatite (Honda et al. 1990). During maturation, the crystal shape is modified and slender crystals become more "blocky" as the Ca/P molar ratio approaches 1.67 (Rodríguez-Lorenzo and Vallet-Regí 2000). After maturation, the precipitate is washed several times in double distilled water that may be adjusted for pH with ammonia. High water purity is essential at all times because the apatite lattice readily incorporates foreign elements into the structure. The precipitate is finally dried and calcined.

At lower pH values, a calcium deficient HAp can be formed (Silva et al. 2001, Raynaud et al. 2002a). Reactions conducted at pH $\cong$ 7.4 are primarily aimed at understanding the crystallization of apatites in the body (Okazaki et al. 1992).

During aqueous precipitation, other species such as NH_4^+ (Vignoles et al. 1987), H_2O (Le Geros et al. 1978, Young and Holcomb 1982), O^{2-} (Young and Holcomb 1982), CO_3^{2-} (Vignoles et al. 1987, Young and Holcomb 1982) and HPO_4^- (Young and Holcomb 1984) may be substituted in the structure or adsorbed onto the surface. Addition of more than one substituent element/group can lead to a combination of an expansion and contraction of the unit cell (Le Geros et al. 1977). For example, carbonate causes a decrease in the *a*-axis (Le Geros 1965) that could be counteracted by an increase from an acid phosphate group (Young and Holcomb 1984). This reaction sequence is complicated by the ability of carbonate to substitute phosphate or hydroxide, the former being the more common (Shimoda et al. 1990).

The incorporation of foreign ions during the crystallization in solution has inspired researchers to investigate the substitution of chemical groups found naturally in enamel or bone. Application of this knowledge can then be used to adjust properties such as solubility, mechanical behavior and bone bonding ability. Substituent elements and chemical groups can include fluoride (Jha et al. 1997, Rodríguez-Lorenzo et al. 2003), carbonate (Barralet et al. 1998, Nelson and Featherstone 1982), magnesium (Okazaki 1988, Mayer et al. 1997, Ben Abdelkader et al. 2001), zinc (Mayer et al. 1994, Bigi et al. 1995), silicon (Gibson et al. 1999), iron (Okazaki and Takahashi 1997) and strontium (Heijlijers et al. 1979, Leroux and Lacout 2001a, Marie et al. 2001). The addition of many of these chemical groups decreases the growth rate at low concentrations. Full substitution of fluoride for the hydroxyl ion removes lattice distortion, produces a more stable apatite and, thus, is able to drive precipitation to completion more easily (Rodríguez-Lorenzo et al. 2003). Carbonate replaces phosphate in reactions containing fluoride and at high pH (Shimoda et al. 1990).

Chemical elements not found in bone can be substituted for different effects. For example, addition of silver has been used for imparting antimicrobial properties (Kim et al. 1998).

Crystallinity, a term used to describe the crystal perfection and/or crystallite size, can vary depending upon the synthesis conditions. A high crystallinity is typically desired where an apatite is subjected to elevated processing temperatures for consolidation into dense forms (see sections on sintering, porous materials and coatings). However, low crystallinity can impart a higher resorbability in applications such as composites and cements. Synthesis at 90°C produces a more pure apatite with a higher degree of crystallinity, than at room temperature. Crystallinity also increases in the presence of strontium (Leroux and Lacout 2001a) and fluoride. Crystals can be plate-like, acicular or blocky and exhibit a surface area of 30-120 m^2/g (Shimoda et al. 1990, Rodríguez-Lorenzo and Vallet-Regí 2000, Senamaud et al. 1997) (Fig. 2). Crystal size in the *c*-axis direction can be as small as 50 nm when produced at 25°C to as large as 700 nm at 90°C. These crystallites agglomerate into clusters upon drying (Fig. 2c). For comparison, the unit cell parameters of biologic and synthetic hydroxylapatites are shown in Table 1. Details of the structure of apatite and the hydroxyl end member are given in Hughes and Rakovan (this volume).

High-temperature synthesis was the first method of apatite production reported and involved passing phosphorus trichloride vapor over red-hot lime (Daubrée 1851). This

Figure 2. Precipitated fluorapatite (a) at 65°C reaction temperature, (b) at 25°C reaction temperature as observed in transmission electron micrographs, and (c) a scanning electron micrograph of precipitate dried at 100°C showing agglomerated crystallites.

process involved the reaction of a gaseous and solid phase. Diffusion between two solid calcium phosphates also produces hydroxylapatite, at temperatures in excess of 1000°C, however, this method does not produce homogeneous apatites and leads to an increase in grain size through growth and reduction in surface area; two aspects that may be important in the further processing and application of calcium phosphates.

Other synthesis reactions include hydrothermal techniques, hydrolysis of other calcium phosphates (Monma and Kayima 1987) and sol-gel methods (Masuda et al. 1990). Hydrothermal synthesis is the second most common method and, in comparison to the wet chemical method, is able to produce well-crystallized, compositionally homogeneous apatite (Yoshimura and Suda 1994). In this process, a mixture of calcium carbonate and di-ammonium hydrogen phosphate is subjected to 12,000 psi and heated to 275°C (Roy and Linnehan 1974). A high crystallinity, carbonate substituted HAp is produced by this method. Calcium phosphates that have been hydrolysed to HAp include octacalcium phosphate (Graham and Brown 1996), tricalcium phosphate (Nakahira et al. 1999), and brushite (Monma and Kayima 1987, Fulmer and Brown 1998, Manjubala et al. 2001). The chemical formulas of these and other inorganic compounds are provided in Appendix 1.

APATITE CEMENTS

Traditionally, particles or blocks were used for reconstruction of defects in bone, but particles easily migrate or disperse into the surrounding tissue after implantation (Wittkampf 1988). In periodontal defects, calcium sulphate dental cement was used to prevent migration. Larger volumes (greater than about 5 ml) are more difficult to seal. Calcium phosphate cements take a special place in implantable ceramics. They are easily formed into bony defects of any geometry. Upon mixing of the reactants, they can be plastically formed into an osseous cavity to precisely fit the defect geometry.

Production of apatite cements

The initial work on calcium phosphate cements involved an equimolar mixture of tetracalcium phosphate and calcium hydrogen phosphate (Brown and Chow 1983). Finely ground and homogenized powders are mixed with water to form a paste. Initially, dicalcium phosphate dihydrate is formed with a plate-like morphology. Dissolution of this phase governs the initial reaction rate. The pH initially climbs to 10.6 where HAp crystallizes on the tetracalcium phosphate and then a decrease in pH yields a calcium deficient HAp (Walsh et al. 2001). The pH fluctuation can be minimized with a smaller particle size, presence of HAp seeds and a less than stoichiometric amount of tetracalcium phosphate (Liu et al. 1997, Matsuya et al. 2000). As HAp growth occurs according to Equation (1), the rate-limiting step is dictated by diffusion of ions through the acicular HAp layer to the tetracalcium phosphate. Unlike methacrylate bone cement, the reaction is isothermal and avoids cell and tissue damage that would normally occur from the heat of reaction.

$$Ca_4(PO_4)_2O + CaHPO_4 \rightarrow Ca_5(PO_4)_3OH \qquad (1)$$

Hardening of the cement occurs mostly within the first six hours, yielding an 80% conversion to HAp and a compressive strength of 50-60 MPa (Liu et al. 1997, Otsuka et al. 1995). Hardening can be accelerated with phosphate solution (Chow et al. 1999), sodium fluoride (Brown and Fulmer 1996), sodium hydrogen phosphate (Miyamoto et al 1995), and sodium alginate (Ishikawa et al. 1995). Alternatively, the use of alpha-tricalcium phosphate based bone cements has provided a means for fast curing times (Kon et al. 1998, Takagi et al. 1998, Fernandez et al. 1999). Thus, the setting time can be controlled to suit different sites within the body.

Inclusion of porosity, with the aim to improve the osteoconductivity, can be introduced by the addition of soluble inclusions such as sucrose, sodium hydrogen carbonate, sodium hydrogen phosphate (Takagi and Chow 2001), or calcium carbonate that reacts to evolve carbon dioxide (Walsh et al. 2001). These reactants produce ionic substitutions in the structure and further improve the resorbability. Another example is that of strontium, which has been shown to be incorporated at low concentrations, while higher concentrations require a heat treatment (Leroux and Lacout 2001b). Since the cement is designed to harden in the body, the composition is modified *in situ* by reaction with the physiological solutions.

The low temperature of formation and inherent porosity also permits the addition of antibiotics (Bohner et al. 1997, Takechi et al. 1998) or growth factors that stimulate the differentiation of preosteoblastic cells (Blom et al. 2000). However, the enhanced capability of the cement is somewhat offset by a longer hardening time (Ginebra et al. 2001).

Animal studies

Takagi et al. (2001) showed that a carbonate is not needed to seed the initial reaction. *In vivo* studies revealed that apatite incorporates 1 wt % carbonate from the available physiological fluids in as early as 12 hours.

In situ hardening in the body can result in particle release and a change in alkalinity in the surrounding environment. Implantation of a cement that is hardened prior to insertion in the body has revealed a less pronounced foreign body response (Frayssinet et al. 2000). This result may be attributed to the early particle release (Pioletti et al. 2000) or the increase in pH that has been known to trigger an inflammatory response and cell death (Silver et al. 2001).

Normal bone remodeling processes occur around calcium phosphate cement with osteoclastic resorption removing bone followed by deposition of new bone directly on the resorption line (Yuan et al. 2000). Bone growth on pre-hardened cement situated in muscle tissue suggests that bone cement is osteoinductive.

Clinical studies

The treatment for osteoporotic compression fracture of the vertebrae has been investigated on human cadaver vertebrae with calcium phosphate cement (Ikeuchi et al. 2001) and results indicate that cement provides an increased compressive strength where cancellous bone is replaced with cement.

An amorphous calcium phosphate cement retrieved from human biopsies, has indicated an absence of fibrous tissue and partial replacement by new bone. The surface of the cement was surrounded by cells indicative of a bone remodeling process leading to new bone with regular trabecular and osteonal patterns (Sarkar 2001). Patients with complex calcaneal fractures treated with calcium phosphate cement have indicated full weight-bearing as early as three weeks postoperatively (Schildhauer et al. 2000).

Other applications include reconstruction of an alveolar bone defect (Yoshikawa and Toda 2000), craniofacial reconstruction (Friedman et al. 1998), closure of cranial base and temporal bone defects following surgery (Kamerer et al. 1994), spinal surgery (Bohner 2001), filling of periodontal osseous defects (Brown et al. 1998), sealing of root canals (Macdonald et al 1994, Cherng et al. 2001) and, possibly, dental pulp-capping (Chaung et al. 1996). While filling of larger defects employs the use of a plastic mass, smaller voids can be filled by injecting a more fluid mass through a needle (Lim et al. 2002). This technique of injection may be adapted as a minimally invasive approach.

COMPOSITES

Matching the stiffness of the implant material to bone allows stress transfer from the implant to the surrounding bone in loading conditions. Known as Wolff's law (Wolff et al. 1986), this effect stimulates the surrounding bone for continued bone remodelling, a part of which includes bone deposition onto the biomaterial surface. The high elastic modulus (100 GPa for HAp) can be lowered to that of cortical bone (20 GPa) by blending with a polymer. In so doing, the low fracture toughness of HAp is also improved. Figure 3 shows a comparison of the component materials for elastic modulus vs. density (Gross and Ezerietis 2002).

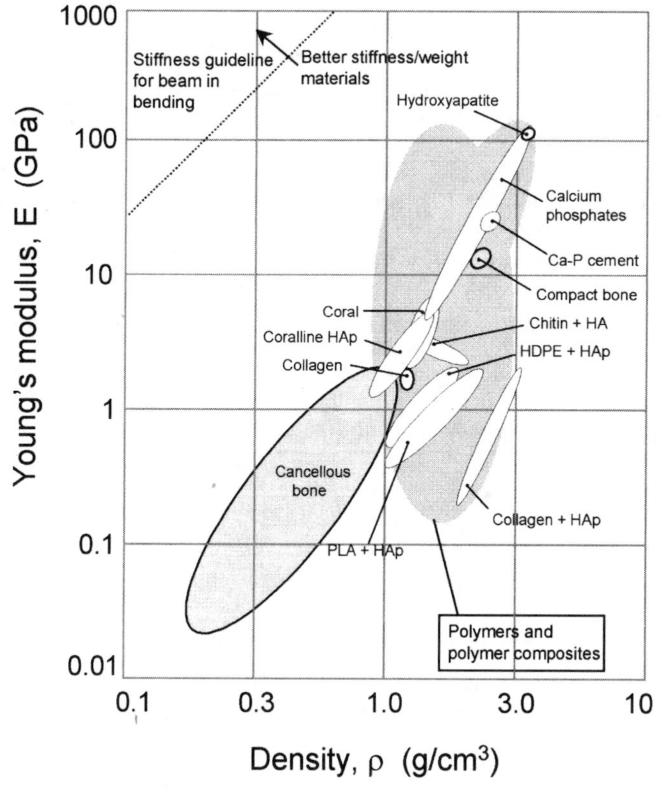

Figure 3. An Ashby diagram with the modulus of sintered hydroxylapatite, porous hydroxylapatite and polymer-apatite composites.

The first polymeric composite designed for implantation into bone was reported in 1981 (Bonfield et al. 1981). The limit of blending is dictated by the ductile/brittle transition at about 40 vol % HAp. At compositions lower than 40 vol %, the fracture toughness is higher than bone, but the elastic modulus is comparable to bone. Hydroxylapatite particles are blended in a twin screw extruder to promote mechanical bonding between the rough particle surface and high density polyethylene (Suwanprateeb et al. 1995, Wang et al. 1998). This bonding can be improved by chemically coupling the particles to the polymer (Deb et al. 1996). Use of fiber processing and compaction or compression molding together with extrusion has been shown to improve the ductility and tensile strength, respectively (Wang et al. 2000, Ladizesky et al. 1997). Other non-resorbable polymer composites that have been manufactured include polyetherketone

(Abu Bakar et al. 1999), polyhydroxybutyrate (Luklinska and Bonfield 1997) and polysulfone (Wang et al. 2001). These composites are easily formed and are used in maxillofacial augmentation. More specific uses include nasal reconstruction (Lovice et al. 1999), middle ear reconstruction (Geyer 1999, Meijer et al. 2002) and repair of orbital fractures (Tanner et al. 1994). Recently, a pilot study has been conducted to examine the feasibility of thin sheets for the outer ear canal (Zanetti et al. 2001). The bone grows up to the composite and establishes a bond with the low resorbable HAp particles.

The low elastic modulus of composites has been used to lower the stress level around implants. Such polymer-hydroxylapatite coatings have been successfully manufactured by thermal spraying (Sun et al. 2002). Finite element analysis has illustrated that a coating at the neck of a dental implant lowers the stress gradient at the coating-bone interface and the stress level in the surrounding bone (Abu-Hammad et al. 2000).

Resorbable polymer composites can be used when bone health is sufficient for remodeling to rebuild the bone. Bone can reclaim the empty space and adapt to the new loading conditions as the material resorbs. The polymer systems used for this approach include polylactic acid (Ignatovic et al. 2001), collagen (John et al. 2001), starch (Mano et al. 1999) , chitosan (Ito et al. 1999), and polyglycolic acid (Durucan and Brown 2000). A polylactic acid-hydroxylapatite composite has been successfully used for repair of the rib cage in a child (Watanabe et al. 1989). Alternatively, porous apatite containing bodies may be used for drug delivery (Yamashita et al. 1998).

Resorbable polymers are also used in the construction of porous scaffolds for tissue engineering (Laurencin et al. 1996). Stem cells are seeded onto the scaffold, which multiply and fill the pore volume as the material resorbs. The porous network serves to transport nutrients and remove the degradation products from the degrading scaffold. Inclusion of HAp as a filler in these resorbable polymers has provided a means of conducting tissue growth inside the pore system (Ma et al. 2001, Laurencin et al. 1999, Thomson et al. 1998, Devin et al. 1996). Furthermore, the dissolution of HAp minimizes the fall in pH associated with degradation of the polylactic/glycolic acid composites (Agrawal and Athanasiou 1997, Ignatius and Claes 1996). Porous scaffolds for stem cell growth are presently receiving immense interest from the commercial and scientific communities.

SINTERING OF DENSE CERAMICS

The sintering process involves calcination, and compaction at room temperature followed by heating at high temperatures. Calcination is performed at 600–900°C for apatites intended for high temperature processing. Adsorbed moisture, carbonates and chemicals remaining from the synthesis stage, such as ammonia and nitrates in some specific reactions, are removed as gaseous products. The removal of these gases facilitates the production of dense materials during sintering. These chemical changes are accompanied by a concurrent increase in crystal size and a decrease in the specific surface area. Apatites with a Ca/P molar ratio less than 1.67 will form beta tricalcium phosphate (as opposed to alpha tricalcium phopshate stable at high temperatures), but calcium rich compositions, with a Ca/P molar ratio greater than 1.67, will form calcium oxide. Thus, phase identification by X-ray diffraction after heating can be used to determine if the Ca/P ratio was above or below 1.67.

Apatite ceramics are consolidated by uniaxial or biaxial pressing (Rodríguez-Lorenzo et al. 2001a), cold isostatic pressing for a more homogeneous green density (Akimov et al. 1994), slip casting for complex shapes (Nordström and Karlsson 1990, Shareef et al. 1993, Toriyama et al. 1995, Rodríguez-Lorenzo et al. 2001b), or injection moulding (Cihlar and Trunec 1996). Some press-sintered ceramics are shown in Figure 4.

Figure 4. Sintered hydroxylapatite as powder, granules, pellets and other forms for implantation into bone and soft tissue.

Sintering of un-calcined powders produces lower shrinkage upon heating. The high surface area of the submicron crystallites leads to a lower fluidity and powder compaction is not very effective, but densification is greater on heating (Juang and Hon 1996, Landi et al. 2000). Densification occurs at the nanoscale between the individual crystallites within the particles and between the individual particles.

Chemically substituted apatites

Substituent elements and chemical groups play an important role in producing phase pure apatite ceramics. Sintering of HAp requires high temperature stability of the source powder and must avoid additions that decrease the stability at high temperatures. A desired powder will retain the apatite structure and not decompose to tricalcium phosphate and tetracalcium phosphate. The apatite structure can be destabilized by magnesium (Baravelli et al. 1984), carbonate (Ellies et al. 1988, Merry et al. 1998), and partial substitution with fluorine ions (Zhang et al. 2001). Manganese as a trace element, readily detected by a color change to blue by oxidation upon heating, is often present (Li et al. 1993b). The black color in teeth recovered from archeological sites suggests that manganese may also lead to black apatite (Stermer et al. 1996). The presence of silicon, replacing calcium ions, does not alter the thermal stability. Full substitution of hydroxyl groups by fluorine ions improves the thermal stability. More on the location of these substituent elements is available in Pan and Fleet (this volume).

A partial vapor pressure of water is needed to retain the structural OH^- at temperatures greater than 900°C (Riboud 1973); however this has also been noted to decrease the densification rate. Dehydroxylation in HAp produces an oxy-hydroxylapatite (Kijima and Tsutsumi 1979). The inclusion of peroxide ions (Zhao et al. 2000) is expected to decrease the diffusion and, hence, also slow densification at high temperatures.

Stoichiometry, described by the Ca/P molar ratio, has a major influence on the physical properties of the apatite and, hence, its application. For calcium-deficient hydroxylapatite, a decrease in surface area starting at 700°C leads to a lower shrinkage (8% less for a powder with Ca/P of 1.64 relative to stoichiometric HAp at 1200°C) (Raynaud et al. 2002b). Low shrinkage in the final sintered product is usually accompanied by porosity that leads to low mechanical strength. Tricalcium phosphate formed from the decomposition of the calcium deficient apatite decreases the sinterability. In Ca-rich hydroxylapatites, densification is again lower in comparison to stoichiometric hydroxyl-apatite (Slosarczyk et al. 1996).

Sintering additives

Densification is promoted by incorporating given chemical groups into the crystal lattice. Carbonate, when substituted for phosphate, promotes sintering to a higher density at lower temperatures (Ellies et al. 1988, Doi et al. 1993, Merry et al. 1998) achieving full density in an atmosphere of wet carbon dioxide (Barralet et al. 2000). No change in sintering is observed when carbonate substitutes for the hydroxyl group. Densification increases with sodium (Correia et al. 1996), lithium (Fanovich and Lopez 1998), fluoride (Senamaud et al. 1997), but is lowered with potassium and magnesium (Fanovich and Lopez 1998). These individual influences may be compounded. Use of sodium hydrogen carbonate in the synthesis of HAp has been shown to further accelerate the sintering process (Suchanek et al. 1997).

Chemical additives such as MnO_2 (Muralithran and Ramesh 2000a), lithium phosphate (Vaz et al. 1999) and sodium phosphate (Suchanek et al. 1997) may be used solely to improve densification and remain either at grain boundaries or included in the crystal structure. Particulate additives including zirconia, alumina, silicon nitride, silicon carbide (Ruys et al. 1993, Suchanek et al. 1997), stainless steel and titanium (Knepper et al. 1998) may be included to improve the mechanical strength. Suchanek et al. (1997) reviews processing aspects of HAp with an emphasis of improving strength. The affect of sintering additives, both chemical and particulate, after dissolution in the human body and the effect on normal cell function is presently unknown (Ballestri et al. 2001).

Another approach to enhance the sintering kinetics is to take advantage of the high surface area of ultrafine powders. Higher surface area powders can be sintered at 150°C lower than powders with a low surface area (Gibson et al. 2001). These particles can be incorporated into an emulsion to enable ease of movement between the crystallites in the particles and produce a high compacted and sintered density (Murray et al. 1995).

Sintering temperatures

Sintering of HAp is usually conducted at an average temperature of 1200 ± 100°C for periods up to three hours (Jarcho et al. 1976, Peelen et al. 1978, Akao et al. 1981, de With et al. 1981, Kondo et al. 1984, Wang and Chaki 1993, Puajindenetr et al. 1994, van Landuyt et al. 1995, Lu et al. 1998, Muralithran and Ramesh 2000b). The development of microstructure, as observed from the decrease in porosity in fractured ceramics, shows an increase in density upon sintering at high temperatures (Fig. 5). The densification rate depends on the atmosphere, decreasing as the environment is changed from vacuum to air to moist air for hydroxylapatite (Wang and Chaki 1993). Grain growth requires an activation energy of 235 kJ/K·mol (Jarcho et al. 1976). Grain growth can be minimized with other techniques such as microwave sintering (Fang et al. 1994), hot pressing (Halouani et al. 1994), or gel casting (Varma and Sivakumar 1996). At higher temperatures the decomposition to tetracalcium phosphate and tricalcium phosphate degrades the properties of the sintered body, the exact temperature depending on the synthesis technique employed and the impurities present.

The bending strength of HAp in three point bending is between 40 and 200 MPa (Jarcho et al. 1976, Akao et al. 1981) depending upon the surface finish and composition. It can be further noted that the three-point bend test intrinsically demonstrates a wide variability in results. Fracture strength reaches a maximum at a Ca/P ratio of 1.60-1.65 and decreases outside of this Ca/P range (Royer et al. 1993, Slósarczyk et al. 1996,

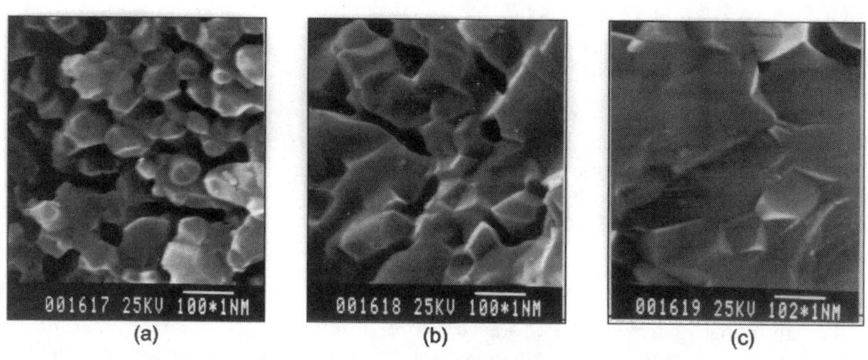

Figure 5. Fracture surface of hydroxylapatite sintered for one hour at: (a) 1150°C, (b) 1200°C, (c) 1250°C.

Raynaud et al. 2002c). Mechanical mixtures of tricalcium phosphate and HAp suggest that tricalcium phosphate is beneficial up to 30 mol % in increasing the composite strength (Toriyama et al. 1987). This may be attributed to the higher bending strength of tricalcium phosphate sintered at the same temperatures (Akao et al. 1984). At high sintering temperatures tricalcium phosphate assists densification by liquid phase sintering. Tricalcium phosphate can also be included as a submicron powder by the decomposition of a calcium deficient apatite into hydroxylapatite and tricalcium phosphate. This approach leads to a larger decrease in surface area of the powder at temperatures below 800°C and a resulting lower sinterability at higher temperatures (Asada et al. 1988). The reader is directed to Suchanek et al. (1997) for more information on mechanical properties.

In vitro and animal studies of sintered apatites

Materials implanted into bone are modified by degradation and the action of different cell types. Degradation occurs by dissolution and the action of osteoclasts. Osteoclasts are very effective in material removal and provide a material resorption mechanism at the surface of the ceramic and particulate removal by phagacytosis (Heymann et al. 2001). Degradation of calcium phosphates as obtained from *in vitro* studies employing cells or animal studies has been discussed in several reviews (Frayssinet et al. 1993, Le Geros 1993, Heymann et al. 1999).

Hydroxylapatite in the sintered form is less soluble than biologically formed HAp in bone and teeth, attributable to the fine crystal size and substituent elements. The solubility of sintered hydroxylapatite is increased with additions of strontium (Christoffersen et al. 1997, Okayama et al. 1991) and carbonate (Nelson 1981, Doi et al. 1998), but lowered with zinc (Mayer and Featherstone 2000) and fluorine. A discussion of dissolution mechanisms is available in a review by Dorozhkin (Dorozhkin 2002). The higher solubility of apatites may be linked to increased osteoclastic resorption, as found with carbonated apatite (Doi et al. 1999).

Bioactivity is determined by the ability of a material to invoke a crystallized carbonated apatite layer from a physiological fluid. Silica incorporation in HAp promotes early crystallization and a higher rate of osteoblastic cell proliferation on sintered materials. This response has not been isolated from the effects of different Ca/P ratio, surface area, and presence of other calcium phosphate phases (Best et al. 1997). Such comparative information on the influence of different substituent elements on the bioactivity still remains to be developed. In assessing the cell growth on apatites, it is important to separate the effects of the topology from the composition. Bodies fired at 1200°C to create a low surface porosity, have exhibited a higher cell growth (Frayssinet et al. 1997).

Sintered HAp particles are remodeled by the host tissue when implanted into bone. Implantation into the cortical bone of the femur of sheep has revealed that stoichiometrically pure HAp resorbs by several microns after 18 months with dissolution occurring mainly at the grain boundaries (Benhayoune et al. 2000). The tensile strength of bone from the tibia of a rabbit onto a HAp cylinder is 0.85 MPa after 3 months (Edwards et al. 1997).

Clinical studies

The ability of hydroxylapatite to accept foreign ions into its structure and onto the surface has been used to incorporate radionuclides. Radioactive samarium and rhenium have been incorporated into hydroxylapatite microspheres sized at 20-40 μm and injected into knee joints to treat rheumatoid joint synovitis (Chinol et al. 1993). Clinical studies have shown that hydroxylapatite particles are easily labeled with radionuclides, exhibit low leakage of radioactive species, and provide a reduced inflammation of the synovium and restored joint motion to the patient (Clunie et al. 1996).

Where hydroxylapatite is used in the powder form for orthopaedic or dental applications, the particles are normally immobilized by mixing with collagen (Sugaya et al. 1989), gelatin (Nagase et al. 1989), or fibrin glue (Wittkampf 1988). This limits the application to bone graft onlay applications. The mixture of collagen and HAp, known as Collagraft® (Collagen Corporation, Palo Alto, California, USA) has been shown to be an effective aid in fracture healing (Cornell 1992).

Fully densified bodies of hydroxylapatite have been employed for reconstruction of the middle ear (Shinohara et al. 2000), dental root implants (Ogiso 1998) and skull reconstruction (Koyama et al. 2000).

Large complex shapes require computer aided design and computer-aided manufacturing to fit the anatomical constraints. Mechanical strength has been optimized in terms of curvature, thickness, width, and porosity (Ono et al. 1998) and further employed for large complex cranial bone defects (Ono et al. 1999). The porosity serves the purpose for bone ingrowth, as explained in the next section.

Bioresorbable β-tricalcium phosphate is occasionally used in conjunction with hydroxylapatite to improve solubility (Klein et al. 1984, Yamada et al. 1997) and hence the osteoconductivity. Applications include nose reconstruction (Abe et al. 2001), fusion of the backbone (Ueda et al. 2001), and use as a bone graft (Fujibayashi et al. 2001).

POROUS APATITE BODIES

Implantation of "inert" ceramics with non-connecting cylindrical channels has shown that bone is capable of growing into pores larger than 100 μm. Bone growth occurs at 20 μm/week in a 100 μm pore and 70 μm/week in a 200 micrometer pore (Ravaglioli and Krajewski 1992). A porous surface provides mechanical fixation in addition to providing sites on the surface that allow chemical bonding between HAp and bone. The inclusion of pores increases the solubility at the expense of mechanical properties. Various types of pore geometries have been introduced into HAp, e.g., pore morphologies, which are both closed and open. An interconnecting pore network offers circulation of nutrients and facilitates deeper bone penetration. As bone grows into the porous network, the solubility of the filled pores decreases and the strength of the implant is improved by a mechanism of natural reinforcement.

Artificial porous structures

Pores can be created by a variety of techniques. In keeping with the sintering methodology, pores can be created by control of crystallite morphology (Nakahira et al. 2000) or sintering parameters (Liu 1996) to obtain a different degree of particle coalescence. These pores are small and cannot accommodate bone ingrowth. The process can be modified by including a foaming agent prior to heating (Dong et al. 2001), or by the evolution of gases from hydrogen peroxide (Peelen et al. 1978) or organic compounds such as napthalene (Monroe et al. 1971), polyvinylacrylate (Vaz et al. 1999) or starch (Rodríguez-Lorenzo et al. 2002b) during the heating cycle. Pore size and content can be further increased by adding the foaming step prior to removal of organics during the heating stage (Engin and Tas 1999). Gel-casting of HAp produces bodies with sufficient strength for shaping of porous bodies before the firing process (Sepulveda et al. 2000). Porosity leads to a decrease in elastic modulus and fracture toughness, i.e., to 100 GPa compared to 160 GPa and 1.1 $MPa{\cdot}m^{1/2}$ compared to 1.8 $MPa{\cdot}m^{1/2}$ for 100% dense materials (Rodríguez-Lorenzo et al. 2002).

Biologically architectured porous materials

The pore architecture of naturally occurring porous networks have been adapted for implantation. The exoskeleton of coral is a material with small crystallites of aragonite and pore connectivity (Fig. 6). *Porites* and *Goniopora* are coral species with a pore size range of 140 to 160 and 200 to 1000 μm, respectively. Coral skeletal material can be converted to carbonate-hydroxylapatite by hydrothermal exchange with di-ammonium hydrogen phosphate at 275°C and 82.7 MPa (Roy and Linnehan 1974). The pseudo-hexagonal structure of aragonite facilitates ease of conversion to the hexagonal unit cell of HAp. Conversion of calcite, another polymorph of calcium carbonate, under the same conditions produces tricalcium phosphate (Zaremba et al. 1998). This process preserves the interconnecting porosity and produces a carbonated, strontium enriched HAp along with magnesium substituted β-tricalcium phosphate (Le Geros et al. 1995). The carbonate and beta-tricalcium phosphate increases the material derived solubility in addition to the increase in surface area from the pores. The stimulation of bone growth by strontium shown in other studies improves the integration of converted corals in bone. Recent work has revealed that hydrothermal processing in the presence of a potassium dihydrogen phosphate can cause a complete transformation to an apatite (Xu et al. 2001).

Trabecular bone from a bovine source already possesses the desired interconnected porosity (Hing et al. 1999) and can be used as a suitable porous body after removal of the organic fraction by heating (Joschek et al. 2000). Large pores allow bone remodeling and trabecular bone formation within the pores (Chang et al. 2000).

Clinical applications of porous apatites

Porous HAp is used in a broad range of applications including filling bone defects (Yamamoto et al. 2000), facial reconstruction (Hobar et al. 2000), orbital implants in eyes (Jordan and Bawazeer 2001), hand surgery (Baer et al. 2002), correction of scoliosis (Delecrin et al. 2000) and drug delivery (Jain and Panchagnula 2000, Netz et al. 2001). The pore size and solubility are important aspects to promote osteoconduction (Kurioka et al. 1999). These porous bodies may be modified with tricalcium phosphate to enhance the solubility or may be enriched by the addition of biological species. The addition of a human osteogenic protein that adsorbs onto the surface of the porous body has been a key element in providing more complete bone growth inside a porous hydroxylapatite (Ripamonti et al. 2001).

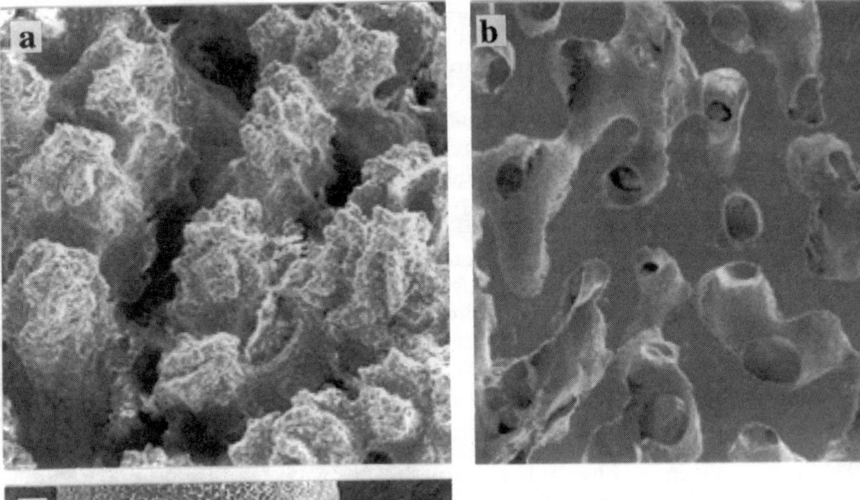

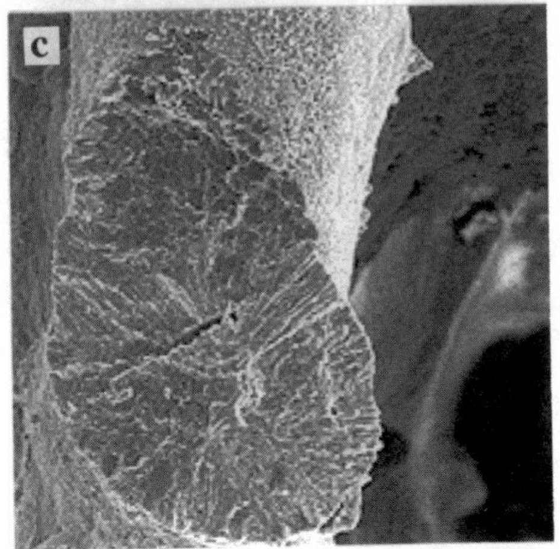

Figure 6. Coralline porous ceramic used for conversion to carbonated apatite by hydrothermal processing as viewed (a) from the surface, (b) in cross-section and (c) a fractured strut showing crystallite orientation.

COATINGS

The mechanical properties of sintered apatites has limited their application to low stress areas in the body. To overcome this difficulty, apatites are applied as coatings on the surface of metallic implants where high loads on the implant are expected. Various coating options are available including thermal spraying, sputter deposition, pulsed laser deposition, sol-gel deposition, electrophoretic coating, electrodeposition, and biomimetic deposition. These are discussed in turn.

Thermal spraying

Thermal spraying is the most widely used technology to manufacture a coating on implants (Fig. 7). Thermal spray was chosen as a candidate technology due to its high deposition rate. While conventional ceramics are delivered to the thermal heating zone at a rate of 5-10 kg/hr, hydroxylapatite powder is transported at 1kg/hr. The first clinical studies in the early 1980s revealed the greatly improved prosthetic bone performance of plasma sprayed coatings. Application of a coating minimizes the release of metallic ions from the underlying metal substrate (Sousa and Barbosa 1996, Browne and Gregson 2000, Finet et al. 2000, Ektessabi et al. 2001), provides a stimulus for bone growth, and a surface for bone to establish a strong bond. Thermal spraying involves concurrent heating and propulsion of feedstock though a thermokinetic medium to produce a melted particle traveling at high speeds. Hydroxylapatite has been sprayed with plasma spraying, vacuum plasma spraying (Chang et al. 1998, Heimann and Vu 1997, Ha et al. 1998, Cabrini et al. 1997, Bellemans 1999), high velocity oxy-fuel spraying (Oguchi et al. 1992, Wolke et al. 1992, Matsui et al. 1994, Haman et al. 1995, Li et al. 2000, Knowles et al. 1996, Brown et al. 1994), and the detonation gun process (Erkmen 1999, Gledhill et al. 1999). These processes all use powder as a feedstock material, sized between 10 and 150 μm, and subject the powder to different thermal and kinetic environments. Suspension plasma spraying is a unique technique in that an atomized suspension of HAp is fed into a radio-frequency plasma (Bouyer et al. 1997). The slow velocity allows the liquid to be vaporized and the remaining solid then melts. This technique avoids the high processing costs normally associated with drying, calcining, spheroidising, and particle sizing before injection into a plasma or a flame. Plasma spraying, also known as air plasma spraying, is the main method of choice, followed by vacuum plasma spraying.

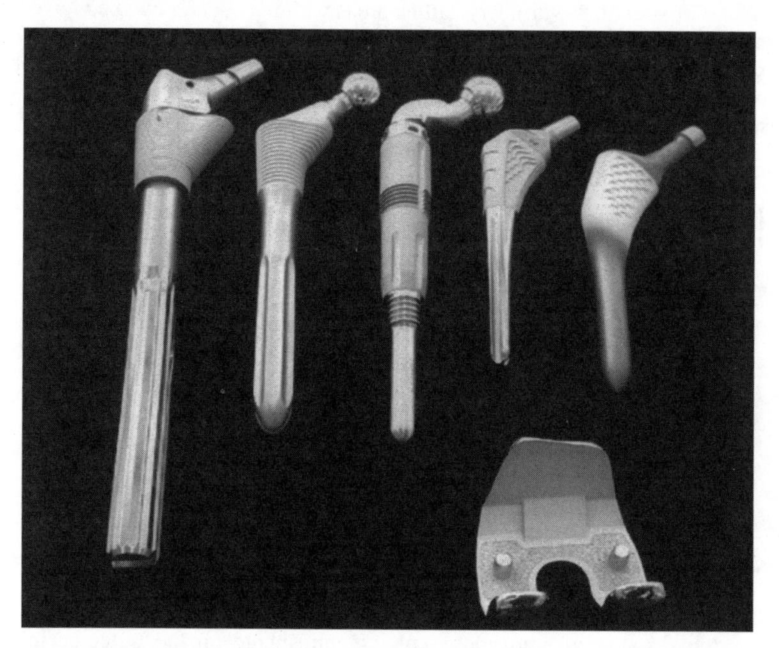

Figure 7. Hydroxylapatite plasma sprayed on various femoral stems (top) and a femoral component of the knee prosthesis showing hydroxylapatite on the internal surface (bottom).

Plasma spraying. The plasma spray process is controlled by a multitude of processing parameters, all of which influence the quality of the sprayed coating. The heat available for melting of the powder can be controlled by a selection of plasma gases. Plasma gas combinations used for depositing apatite coatings include Ar/H_2, N_2/H_2, Ar/N_2 and Ar/He. Powder injected into the heat source is rapidly heated over the melting temperature. Overheating can lead to decomposition into tetracalcium phosphate, tricalcium phosphate and calcium oxide (Lugscheider et al. 1991a, Radin and Ducheyne 1992, Palka et al. 1993, Yang et al. 1995, McPherson et al. 1995, Vogel et al. 1999, Tufekci et al. 1999). The particle characteristics such as particle morphology, particle size, inherent porosity, and crystallite size all influence the heat conduction and, hence, the thermal chemistry and particle melting at high temperatures.

Crystallinity of coatings. The molten droplets are impacted onto roughened substrates. The heat from the droplet is rapidly removed providing a quench rate of about 10^5 °C/sec (Gross et al. 1998d). The brief time at which the chemical species are mobile within the droplet during cooling is insufficient for crystallization. Crystallization of melted regions is further impeded due to the loss of structural water in the shell of the particles. Plasma spraying is thus programmed by a choice of spray conditions (Weng et al. 1995) to produce melting of the outer shell that upon deposition will predominantly form an amorphous calcium phosphate. The crystallinity of coatings is adjusted by controlling the amount of heat input into the powder. This is conducted by changing the spray conditions or the particle size; smaller size produces lower crystallinity (Klein et al. 1994, Tong et al. 1996, Gross et al. 1998d).

The relative location of the amorphous phase has been shown in coatings (Chen et al. 1994, Gross et al. 1997), and the location of the possible decomposition phases proposed in a model (Gross et al. 1998b). Crystalline areas within a coating can be viewed in terms of thermal changes within a droplet during particle traverse within the plasma, heat conduction within the coating as successive molten droplets release their heat to the already deposited material, and post-deposition heating operations. A typical microstructure of a coating showing crystalline regions is shown in Figure 8.

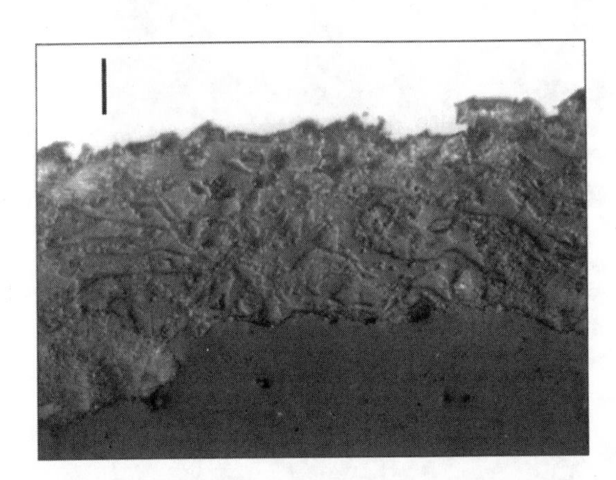

Figure 8. Cross-section of hydroxyl-apatite plasma-sprayed coating exhibiting small crystalline islands. The vertical bar represents 20 microns.

There may be three distinct regions identified within a HAp droplet. These arise from (1) dehydroxylated amorphous calcium phosphate on the outer periphery of a particle that may transform into oxyapatite (Gross et al. 1998c), (2) hydroxylated amorphous calcium phosphate inside the droplet that forms HAp where sufficient heat is conducted to promote crystallization (Zyman et al. 1994, Gross et al. 1998e), and (3) regions in the particle core which are insufficiently heated and remain in their original crystalline state. Crystalline regions can form during droplet deposition as successive droplets release their heat and crystallize underlying amorphous regions. Finally, crystallinity can be further increased after thermal spraying with an appropriate heat treatment between 500 and 700°C (Filiaggi et al. 1993, Ji and Marquis 1993, Brossa et al. 1994, Wang et al. 1995, Gross et al. 1998e, Burgess et al. 1999) but, as will be discussed in a latter section, post depositional processing influences the coating adhesion because the residual stresses of the material system are altered.

Powders for spraying. A range of powder compositions have been examined for use in plasma spraying. Amorphous calcium phosphate (Liu et al. 1994), biologically derived

(Joshi et al. 1993) and calcium deficient (Ellies et al. 1992) HAps have been shown to be unsuitable for plasma spray processing since they decompose. A biphasic calcium phosphate has been chosen as one coating preference for the improved dissolution characteristics imparted by the addition of tricalcium phosphate (Burr et al. 1993, Tisdel et al. 1994, Lee et al. 2001). Fluorapatite is more thermally stable than hydroxylapatite and accordingly produces higher crystallinity coatings when comparable plasma spray conditions are selected (Lugscheider et al. 1991b, Rocca et al. 1998, Overgaard et al. 1998). Regulatory bodies have proposed standards to limit the range in decomposition and loss in crystallinity, structural factors that can in part be linked to the purity of the spray material.

Powder agglomeration techniques such as spray drying (Lugscheider et al. 1992: Luo and Nieh 1996, Kweh et al. 1999), freeze drying (Hattori et al. 1987, Lu et al. 1998, Itatani et al. 2000), spray pyrolysis (Aizawa et al. 1996, Osaka et al. 1997, Vallet-Regí et al. 1994, Inoue and Ono 1987), or sintering and crushing can be used for developing a desired particle size range for thermal spray operations. Of these, spray drying is the most commonly used agglomeration process.

Implant materials for coating. Prosthetic materials coated with HAp include titanium, Ti-6Al-4V, stainless steel, Co-Cr-Mo, and alumina (Jiang and Shi 1998). These materials are roughened by grit blasting for a mechanical interlock between the melted component of the particle and the substrate. The Ti-6Al-4V and Cr-Co-Mo alloys are the most common. Ideally, the elastic modulus and co-efficient of thermal expansion of the substrate and the coating material will be matched to minimize any residual stresses at the interface. Hydroxylapatite (E = 100 GPa and $\alpha = 12 \times 10^{-6}$ °C^{-1} (Perdok et al. 1987)) is well matched to the titanium alloy (E = 110 GPa and $\alpha = 10 \times 10^{-6}$ °C^{-1}) and is coated directly onto the prepared surface.

Cr-Co alloys are known for their high strength and wear resistance. Despite these benefits, there is a concern about the adverse biocompatibility of chromium and cobalt ions, whereas other studies have indicated that chromium promotes the crystallization of HAp (Wakamura et al. 1997) and, therefore, might be an important contributing factor leading to good fixation in bone.

A bond coat, used to promote adhesion between the coating and the substrate, has been employed for a variety of purposes in biomedical applications. A rough titanium bond coat provides a porous surface for bone attachment after the calcium phosphate coating has dissolved. A titania (TiO_2) ceramic bond coat can decrease the quenching rate of deposited molten droplets and, thus, produce a crystalline coating (Heimann et al. 1999). An alumina (Al_2O_3) base layer between the Ti-6Al-4V substrate and the calcium phosphate coating has also been used (Labat et al. 1999). Ceramic bond coats have not been used commercially for apatite coatings.

Surface preparation and bonding. Diffusion has been proposed to occur between titanium and HAp to produce a calcium titanate (de Groot 1987). Sintering studies between titanium and HAp have confirmed this diffusion (Lacout et al. 1984, Chai et al. 1993, Knepper et al. 1998) with another study suggesting incorporation of titanium into the HAp structure (Weng et al. 1994). Titanium oxide (TiO_2) is an essential component of this diffusion reaction, and improved bond strength on oxidized substrates has been reported (Ueda et al. 2000).

The surface of the implant or prosthesis is grit blasted not only to establish a mechanical bond between the substrate and the coating, but also to establish a stronger mechanical bond with bone once the HAp has been resorbed. A surface roughness, R_a, as little as 1 μm can lead to a twofold increase in the removal torque of an implant from bone (Carlsson et al. 1988). The roughness for bonding of thermally sprayed HAp to the substrate is usually two to five fold higher (Yankee et al. 1991). Various types of macrotexturing or porosity have been designed to further improve bone attachment (Kienapfel et al. 1999). An implant surface may contain vacuum plasma sprayed porous titanium (Nakashima et al. 1997), sintered mesh (Wilke et al. 1993), or sintered beads (Moroni et al. 1994). A coating placed on top of the porous surface provides high bone to implant interfacial contact (Bloebaum et al. 1993). Grooves may also be placed on the surface of an implant to provide mechanical resistance against torsion of the hip prosthesis. Grooves 1 mm deep provide improved biological fixation and earlier fixation compared to porous coated implants (Hayashi et al. 1999). A comparison of the shear strength between the different macrostructured surfaces with bone, using the same implant site, is not av ilable.

Coating thickness and residual stress. A coating thickness of 50 μm has been determined to provide good fatigue resistance with good resorption and bone attachment characteristics in orthopedic applications (Geesink et al. 1987). In comparison, a 200-μm thick coating produced a 50% decrease in bonding strength (Wang et al. 1993a). Animal studies have shown that fracture occurs at the coating-bone interface for 50 μm thick coatings but within the coating for the 200-μm thick coatings (Wang et al. 1993b). The fracture mode can be described as "cohesive" in the latter case. At a low residual stress, failure occurs within the coating, but at higher values is shifted toward the coating-Ti alloy substrate (Yang et al. 2001). The lower bond strength can be related to the residual stress that reaches a maximum at the interface and exhibits larger values for thicker coatings (Yong et al. 2001). Real-time residual stress measurement during spraying reveals that the stress is tensile in nature and increases in value upon cooling (Tsui et al. 1998). Yang et al. (2000) has shown a higher residual stress with well-melted particles.

It can be assumed that coatings formed from lower crystallinity particles exhibit low residual stress because feedstock also has inherent porosity that would tend to relieve process-induced stresses. A comparison of the high velocity oxy-fuel and plasma spray processes has revealed that the residual stress is lower in the former process (Knowles et al. 1996). Heat treatment at temperatures of 800°C can minimize these process-induced stresses (Brown et al. 1994). Despite the large influence of thickness on coating strength, no difference in bone apposition has been found for 50 μm and 100 μm coatings.

Strength of coatings. Plasma sprayed coatings contain process-induced porosity, partially molten particles, and a range of grain sizes. The pancake structure of the flattened particles produces a higher population of pores parallel to the surface of the implant. Fracture toughness of these coatings is low and is attributed to the pore content in these coatings. Mancini et al (2001) have indicated that this pore level can vary between 2 and 10%. Attempts at improving the fracture mechanical properties of these coatings have involved incorporation of a second phase. Examples of the second phase include titanium (Zheng et al. 2000), titanium alloy (Khor et al. 2000), titania (Ramires et al. 2001), alumina (Morimoto et al. 1988), and zirconia (Chou and Chang 1999). The faster resorption time of plasma sprayed coatings in comparison to sintered apatites can potentially lead to the release of the non-resorbable particulate and possibly lead to implant loosening from an overload of particulate in bone. The inclusion of crack healing additives, such as glass, may improve the fracture toughness. Some initial work with calcium phosphate glass has indicated that other aspects such as the wetting, surface charge, and resorbability need to be considered to produce a coating that would integrate easily with bone (Ferraz et al. 2001). The mechanical performance of coatings is available in a review by Sun et al. (2001).

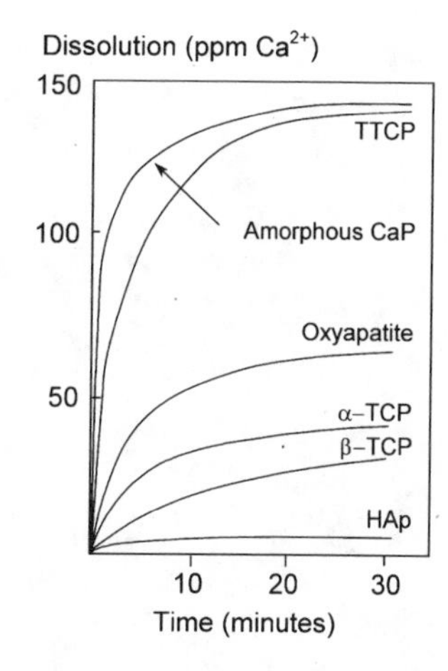

Figure 9. Solubility of hydroxylapatite compared to oxyapatite, amorphous calcium phosphate, tetracalcium phosphate (TTCP), α-tricalcium phosphate and β-tricalcium phosphate determined in 0.1M potassium acetate at pH 6 (modified from Le Geros et al. 1995).

Bioactivity of coatings. Immersion of HAp plasma sprayed coating into a buffered solution or a simulated body fluid leads to partial coating dissolution, the extent of which depends upon the phase composition, surface area and residual stress. The solubility of the calcium phosphate phases increases in the order HAp, tricalcium phosphate, oxyapatite, tetracalcium phosphate and amorphous calcium phosphate (Ducheyne et al. 1990, Le Geros et al. 1995) (Fig. 9). Dissolution of plasma sprayed coatings depends primarily on the coating crystallinity (Leali Tranquilli et al. 1994, Klein et al. 1994, Fazan and Marquis 2000). Coating loss occurs by (1) dissolution which is dominated by the amorphous calcium phosphate, and (2) particle release (Gross et al. 1997, Ogiso et al. 1998b). The loss of amorphous calcium phosphate is very clearly identified by a change in the surface morphology (Fig. 10). Despite the importance of crystallinity, other factors can contribute to a five-fold change in dissolution rate (Paschalis et al. 1995).The presence of protein in the testing solution can lead to faster initial calcium ion release (Bender et al. 2000). After the surrounding solution is saturated with respect to calcium and phosphate ions, reprecipitation occurs on the surface of the coating (Fig. 11). Fine crystallites form preferentially in recessed areas such as pores and cracks (Gross 1991, Weng et al. 1997). Low crystallinity coatings produce a higher concentration of dissolved ions in solution (Chou et al. 1999) and, hence, have a shorter induction time before precipitation of the carbonated apatite (Anselme et al. 1997). Cell proliferation on these surfaces has been difficult to compare, since cells are more sensitive to topographical variation preferentially depositing on smooth areas (Lumbikanonda and Sammons 2001).

A comparison in bone bonding to sintered HAp and HAp plasma sprayed coatings has revealed higher attachment strength of bone to the coating (Ogiso et al. 1998a). The higher bioactivity of the plasma sprayed coatings provides earlier fixation. The propensity for bone bonding is highlighted by the bonding that occurs during early loading of coated hip prostheses (Overgaard et al. 1998).

Clinical performance. The lifetime of these coatings extends up to several years depending upon the coating characteristics, prosthesis design, and the implantation site. Hydroxylapatite coatings placed in trabecular bone maintain bone contact and, thus, prevent high dissolution that could occur from the physiological fluids (Caulier et al. 1995, Lind et al. 1999). The design of the implant plays an important role. Areas of the implant receiving the most loading from the bone, such as the tip of dental root implants (Finet et al. 2000), the apical edges of blade implants (Baltag et al. 2000) and the thread tips of a collar on a femoral stem (Gross et al. 1998a) have been observed to produce the most coating resorption.

Figure 10. A surface view of a hydroxylapatite plasma sprayed coating revealing (a) smooth areas more commonly associated with the amorphous calcium phosphate and (b) a change in surface topography due to the loss of the amorphous phase after immersion in 0.1M potassium acetate at pH 6 for 2 days.

Figure 11. Precipitate formed on a plasma-sprayed coating exhibiting the small crystallite size associated with the newly deposited apatite layer.

The rate of coating loss is further dictated by the physical activity of the implant recipient. Higher loading on the prosthesis has been found to produce more active bone remodeling and result in more rapid coating loss (Tonino et al. 1999). Loading on the coating can, in part, be influenced by the amount of coating coverage on a femoral stem. Examination of radiographs has revealed that a coating shifts the load transfer distally for uncoated prostheses to the proximal region in a fully coated prosthesis (Abrahams and Crothers 1992). A coating placed on the upper half of the stem may further shift the loading to the surrounding bone in the proximal region and, thus, alter the applied stress distribution on the coating.

Hydroxylapatite plasma sprayed coatings have been used in total hip arthroplasties (Jaffe and Scott 1996), dental implants (Ong and Chan 1999), knee replacements, ankle arthroplasties (Zerahn et al. 2000), orthopaedic screws (Magyar et al. 1997), and spinal implants. The most widely used application is the femoral stem where 11 year clinical results show slightly better performance than cemented prostheses (Havelin et al. 2000).

Biomimetic processing

The biological response to metals can be improved by modifications to the surface composition of an implantable metal alloy. This follows from the work on bioactive glasses which, when implanted into the body, produce a modified surface that facilitates apatite precipitation (Hench 1998a). Various chemical enrichment treatments have been proposed to aid the precipitation of a carbonate apatite.

The most commonly used approach involves transforming the oxide layer of a metal surface that is rich in hydroxyl ions. Titanium soaked in sodium hydroxide and then heat-treated produces an amorphous sodium titanate layer. The minimum concentration found to create a layer of sodium titanate hydrogel is 0.5 M at 60°C for 24 hours (Kim et al. 1997). After immersion in simulated body fluid, sodium ions released from the layer are exchanged with hydronium ions (H_3O^+) to form a hydrated titanium layer. Formation of a hydrated layer incorporates calcium and phosphate ions from the fluid to form an amorphous calcium phosphate. It is thought that the gel takes on a positive charge that attracts phosphate ions into the structure resulting in a Ca/P ratio of 1.40. The release of sodium increases the pH and ionic activity in the solution, thus creating a supersaturated environment for apatite precipitation (Jonasova et al. 2002). Crystallization of the layer incorporates more calcium from solution to increase the Ca/P ratio to 1.65, similar to the value of bone mineral (Takadama et al. 2001a,b). The layer grows at 0.5 to 7 μm per day (Hata et al. 1995).

Li et al. (1993a) indicated that a silica gel induced an apatite layer on the surface. The precipitated crystal shape changed from a plate-like morphology to flakes by increasing the pH from 7.2 to 7.4. Crystals formed as rods in the presence of magnesium, and needles where fluorine ions were present. Studies on titania and alumina gels indicated that apatite precipitation can be induced on the surface of other gel compositions (Li et al. 1994). Noble metals such as tantalum and niobium have illustrated this capacity through an alkali treatment. Zirconia, alumina and silica can also be transformed to possess a bioactive surface. Hydroxylation of metals, zirconia (Uchida et al. 2001), tantalum (Miyazaki et al. 2001), zirconium (Uchida et al. 2002), and niobium (Kokubo et al. 2000), provides faster apatite formation on the surface. Such a treatment lowers the contact angle and facilitates spreading of neutrophils and osteoblasts (Lim et al. 2001). At four weeks, alkali modified titanium implanted into the femur of a rabbit exhibits a bond with bone that is eight times stronger (2.4 to 4.5 MPa) compared to untreated titanium (0.3 to 0.6 MPa). At 12 weeks, the superior bond strength is maintained, while the bonding between bone with titanium improves two-fold (Nishiguchi et al. 2001).

Precipitation has also been shown to occur on modified polymeric surfaces. An example of this includes Ca(II) containing hybrids of gelatin and 3-(glycidoxy-propyl)trimethoxysilane (Ren et al. 2001). Silanol groups on silicone (Oyane et al. 1999), poly (ethylene terephthalate), polyether sulfone and polyethylene (Kokubo 1998), polymethylmethacrylate (PMMA), polyamide 6, and polyethersulfone (PESF) (Tanahashi et al. 1994) can also provide sites for apatite formation.

An alternative method lies in ion implantation of the alkali ion into the metal surface (Pham et al. 2000). Fluorine (Ellingsen 1995) and calcium ion (Feng et al. 2002) enrichment of titanium alloys have also shown to promote apatite precipitation.

Sol-gel deposition

Sol-gel technology offers a chemically homogeneous and pure product and has been used for HAp production since 1988 (Masuda et al. 1990). A calcium alkoxide mixed with an organic phosphate undergoes various intermediate states before a stoichiometric composition is obtained. This chemical process requires an aging time (Chai et al. 1998). Use of an intermediate phosphate composition eliminates the aging step (Ben-Nissan et al. 2001). A coating can be applied by dipping or by spinning followed by a heat treatment to remove the organics (Weng and Baptista 1998). The small grain size permits sintering of dense apatite compositions at temperatures less than 900°C to avoid the phase transformation in the parent titanium alloy (Gross et al. 1998f, Lopatin et al. 1998, Cheng et al. 2001). Furthermore, this coating can be solidified during a rapid thermal processing schedule of 100°C/min (Russell et al. 1996).

Electro-deposition

Electrophoretic deposition involves movement of synthesized particles under the influence of an electrical field. The implant is placed as an electrode upon which the particles collect. Particle shape, composition, electric field and stirring conditions influence the deposit characteristics (Zhitomirsky and Galor 1997). After coating by this method the implant needs to be dried and sintered. Sintering requires temperatures greater than 1000°C to produce a dense coating layer. Cracking may evolve during the cooling stage due to the difference in thermal expansion substrate and the newly formed interlayer. This process has been useful in coating porous surfaces (Ducheyne et al. 1990). Electrocrystallization represents a low temperature process whereby calcium and phosphate in an electrolyte migrate to an electrode where crystals are formed. A heat treatment at 125°C followed by calcination at 425°C can densify an apatite. Carbonated and fluoridated apatites with a thickness of 50 μm have been produced in one hour (Shirkhanzadeh 1995).

Vacuum deposition

Sputtering was introduced as a possible coating technique alongside thermal spray processes in the late 1980s. This technique involves displacing atoms from a target material with high-energy ions and transfer to a flat substrate under vacuum conditions. Addition of another gas such as carbon dioxide to the plasma forming gas can modify the composition to a carbonated calcium phosphate (Yamashita et al. 1996). Use of a magnetron produces a deposition rate of 1 micrometer/hour. Such films produce an amorphous phase when deposited on a cold substrate and an oriented columnar HAp at higher temperatures. An increase in discharge power leads to a crystalline coating (van Dijk et al. 1995). The dense layer formed during magnetron sputtering creates a residual stress that, along with crystallinity (Wolke et al. 1998), influences the dissolution behavior due to the residual stress (Burke et al. 2001).

Pulsed laser deposition ablates a target material that results in transfer of a droplet onto the substrate to be coated. This process occurs under vacuum and like sputtering produces an amorphous phase at low substrate temperatures, and a crystalline phase at higher temperatures (Cotell et al. 1992).

In vitro cell culture studies on HAp coatings produced by plasma spraying, sol-gel and sputtering revealed that the sol-gel coating exhibited the highest cell growth (Massaro et al. 2001). The difference in purity, density, grain size, surface roughness, and contaminant phases of the various coatings makes it difficult to isolate the characteristic that enables a better *in vitro* response.

FUTURE WORK

While calcium phosphates are currently widely used as implants in the body, more detailed work on the role of chemically enriched apatites on the biological response, characterization of apatites and microstructural control will reveal the optimal characteristics of an apatite for clinical application.

The rich elemental substitution (Pan and Fleet this volume) of apatites can lead to a wide variety of biomaterials with differences in processing, resulting microstructure and tissue bonding behavior. This is an area of active research and the outcome can produce tissue-bonding maps similar to those established for bioactive calcium phosphate glasses developed by Hench (1998b). Such maps can indicate how substituent elements enhance the tissue bonding response and can serve as useful guides for selecting the appropriate apatite composition for implantation into different types of bone.

An important chemical aspect of apatites for biomedical use is the Ca/P molar ratio. Synthesized apatites may be calcium deficient or rich and yet display an X-ray diffraction pattern typical of apatite. Those apatites that are subjected to low temperature processing such as cements and biomimetically processed coatings require a wet chemical analysis for determination of the Ca/P molar ratio. With the new interest in chemically modified apatites, the Ca/P molar ratio, the impurity elements according to present standards (ASTM 1998) and the major substituent elements will need to be reported to provide a better understanding of the apatites being investigated for biomedical applications.

Hydroxylapatite is typically subjected to heating during processing. Heat treatment creates a stoichiometric apatite along with a secondary phase for non-stoichiometric compositions. For example, a calcium deficient apatite will lead to the formation of tricalcium phosphate upon heating, and quantitative phase analysis can be used to determine the Ca/P molar ratio (Toth et al. 1991; Ishikawa et al. 1993). Recently, Rietveld analysis has been proposed as a useful and fast alternative to time consuming solution techniques for the determination of Ca/P ratios (Raynaud et al. 2001). The X-ray diffraction can also provide information such as contaminant phase identification, crystallite size, and crystal orientation, which are very useful when dealing with complex material systems such as coatings produced by plasma spray techniques (Keller 1995; Keller et al. 2000). Further work will need to develop methods for ascertaining the location of substitutional elements within the apatite lattice.

The detection of different phases and their chemistry is presently limited. This, in part, is attributed to the small size, low concentration or the presence of numerous phases. X-ray diffraction is limited, since the very small grain size produces significant peak broadening, and the small quantity of phases is at the limit of detection. The best technique would provide viewing of the microstructure while an analysis is conducted. Raman microprobe data exhibit peak overlapping and, therefore, distinction between a tricalcium phosphate and an amorphous calcium phosphate is difficult (Tudor et al. 1993), however useful data can be obtained from interface studies with bone (Walters et al. 1991). Cathodoluminescence microscopy has been developed to distinguish between the amorphous calcium phosphate and HAp (Gross et al. 1998g). Transmission electron microscopy remains the only reliable technique for detection of chemical phases in very small quantities. For complicated microstructures, such as those produced in thermal spraying, the preparation of large areas will be necessary to view the location of the different chemical phases.

Recent studies reveal that microstructural surface features possibly influence the cell response to implanted synthetic apatites. The roughness is an important aspect of cell adhesion, but can influence the spreading of cells. A comparison of sputtered, sol-gel and plasma sprayed coatings has indicated that sol-gel coatings provide the best surface for cell multiplication and proliferation (Massaro et al. 2001). It is not clear, whether, this behavior can be attributed to the high chemical purity or low surface roughness of sol-gel coatings. Another study has shown that cells prefer to spread over smooth areas of plasma sprayed coatings that represent well melted areas (Lumbikanonda and Sammons 2001). HAp powder is not completely melted during the plasma spray process and thus rough areas on the coating surface are expected from unmolten particle segments. To improve the smoothness of plasma sprayed coatings, it would be necessary to increase the melting of the particle, that would produce a lower crystallinity, or improve the thermal stability of HAp powders, and thus retain a high crystallinity within the coating. Such modifications would provide good cell spreading in addition to retaining the osteoconductivity and good bone bonding achievable with present coatings.

The charge state of apatites has recently been found to be important in the integration into bone. Sintered apatite poled at high temperature to produce a negative charge has been shown to (i) induce earlier precipitation (Ohgaki et al. 2000), (ii) promote the formation of osteoblast-like cells (Ohgaki et al. 2001), and (iii) enhance bone bonding (Kobayashi et al. 2001). The long-term bone bonding ability remains to be assessed.

Apatites presently used in clinical applications are utilized solely for their ability to either bond to tissue or promote bone growth. The next stage of development will involve the optimization of the microstructure to allow incorporation of biological or chemical species that can be released for stimulation or therapy. Such work presently is being focused on microspheres (Paul et al. 2002; Sivakumar and Rao 2002).

The solubility of apatites is becoming more important as emphasis is being placed on biomaterials for regeneration of tissues (Hench 1998b). Where apatites are incorporated with resorbable polymers for tissue engineering applications, it will become necessary to match the solubility rate of the inorganic and organic components within the composite.

APPENDIX 1

CHEMICAL FORMULAS FOR INORGANIC COMPOUNDS

Alumina	Al_2O_3
Calcium acetate	$CH_3(COOCa)_2 \cdot H_2O$
Calcium carbonate	$CaCO_3$
Calcium chloride	$CaCl_2$
Calcium deficient hydroxylapatite	$Ca_{10-x}(HPO_4)_x(PO_4)_{6-x}(OH)_{2-x}$ where $0 \leq x \leq 2$
Calcium hydrogen phosphate	$CaHPO_4$ (also known as monetite)
Calcium hydroxide	$Ca(OH)_2$
Calcium sulphate	$CaSO_4 \cdot 2H_2O$
Calcium nitrate	$Ca(NO_3)_2 \cdot 4H_2O$
Carbonated hydroxylapatite	$Ca_{10-x}Na_x(PO_4)_{6-x}(CO_3)_x(OH)_2$ (B-type)
	$Ca_{10}(PO_4)_6(CO_3)_x(OH)_{2-2x}$ (A-type)
Diammonium hydrogen phosphate	$(NH_4)_2HPO_4$
Dicalcium phosphate dihydrate	$CaHPO_4 \cdot 2H_2O$ (also known as brushite)
Fluorapatite	$Ca_{10}(PO_4)_6F_2$
Hydrogen peroxide	H_2O_2
Hydroxylapatite	$Ca_{10}(PO_4)_6(OH)_2$

Lime	CaO	
Lithium phosphate	Li_3PO_4	
Manganese oxide	MnO_2	
Octacalcium phosphate	$Ca_8H_2(PO_4)_6 \cdot 5H_2O$	
Orthophosphoric acid	H_3PO_4	
Oxyapatite	$Ca_{10}(PO_4)_6O$	
Phosphorous trichloride	PCl_3	
Potassium dihydrogen phosphate	KH_2PO_4	
Silica	SiO_2	
Silicon carbide	SiC	
Silicon nitride	Si_3N_4	
Silica	SiO_2	
Sodium fluoride	NaF	
Sodium hydroxide	$NaOH$	
Sodium hydrogen carbonate	$NaHCO_3$	
Sodium hydrogen phosphate	Na_2HPO_4	
Sodium phosphate	Na_3PO_4	
Tetracalcium phosphate	$Ca_4P_2O_9$	
Titania	TiO_2	
Tricalcium phosphate	$Ca_3(PO_4)_2$	(exists in α, $\overline{\alpha}$ and β crystal forms)
Zirconia	ZrO_2	

ACKNOWLEDGMENTS

The authors are grateful for the fracture surfaces of sintered hydroxylapatite and various forms of sintered hydroxylapatite supplied by Mr. V. Gross, Commonwealth Scientific Industrial Research Organisation, Australia. We thank David Williams and Serena Best for their careful reviews of this chapter.

REFERENCES

Abe T, Matsumoto K, Kushima M (2001) Reconstruction of the sellar floor during transnasal pituitary surgery using ceramics composed of a combination of hydroxylapatite and tricalciumphosphate. Neurol Surg 29:511-505

Abrahams TG, Crothers OD (1992) Radiographic analysis of an investigational hydroxylapatite-coated total hip replacement. Investig Radiol 27:779-784

Abu Bakar MS, Cheang P, Khor KA (1999) Thermal processing of hydroxylapatite reinforced polyetherketone composites. J Mater Proces Technol 90:462-466

Abu-Hammad OA, Harrison A, Williams D (2000) The effect of a hydroxylapatite-reinforced polyethylene stress distributor in a dental implant on compressive stress levels in surrounding bone. Intl J Oral Maxillofac Implants 15:559-564

Agrawal CM, Athanasiou KA (1997) Technique to control pH in vicinity of biodegradable PLA-PGA implants. J Biomed Mater Res 38:105-114

Aizawa M, Itatani K, Howell FS, Kishioka A (1996) Effects of starting materials on properties of hydroxylapatite powders prepared by spray-pyrolysis technique. J Ceram Soc Japan 104:126-132

Akao M, Aoki H, Kato K (1981) Mechanical properties of sintered hydroxylapatite for prosthetic applications. J Mater Sci 16:809-812

Akao M, Miura N, Aoki H (1984) Fracture toughness of sintered hydroxylapatite and beta-tricalcium phosphate. Yogyo-Kyokai-Shi 92:672-4

Akimov GY, Timchenko VM, Arsenev PA (1994) Mechanical properties of hydroxylapatite samples produced by cold isostatic pressing. Ogneupory 5:19-21

Albee FH, Morrison HF (1920) Studies in bone growth—triple calcium phosphate as a stimulus to osteogenesis. Ann Surg 71:32-39

Anselme K, Sharrock P, Hardouin P, Dard M (1997) *In vitro* growth of human adult bone-derived cells on hydroxylapatite plasma-sprayed coatings. J Biomed Mater Res 34:247-259

Aoki H (1973) Synthetic apatite as an effective implant material. J Stomatol Soc 40:277 (in Japanese)

Aoki H, Akao M, Shin Y, Tsuji T, Togawa T (1987) Sintered hydroxylapatite for percutaneous devices and its clinical application. Med Progr Technol 12:213

Asada M, Oukami K, Nakamura S, Takahashi K (1988) Microstructure and mechanical properties of non-stoichiometric apatite ceramics and sinterability of raw powder. J Ceram Soc Jap 96:595-598

ASTM (1988). Standard specification for composition of ceramic hydroxylapatite for surgical implants. F1185-88, 415.

Baer W, Schaller P, Carl HD (2002). Spongy hydroxylapatite in hand surgery—A five-year follow-up. J Hand Surg 27B:101-103

Ballestri M, Baraldi A, Gatti AM, Furci L, Bagni A, Loria P, Rapana RM, Carulli N, Albertazzi A (2001) Liver and kidney foreign bodies granulomatosis in a patient with maloclussion, bruxism, and worn dental prostheses. Gastroenterology 121:1234-1238

Baltag I, Watanabe K, Kusakari H, Tagushi N, Miyakawa O, Kobayashi M, Ito N (2000) Long-term changes of hydroxylapatite-coated dental implants. J Biomed Mater Res-Appl Biomater 53:76-85

Banas MD, Baier RE (2000) Accelerated mineralization of prosthetic heart valves. Molec Cryst Liq Cryst 354:837-855

Baravelli S, Bigi A, Ripamonti A, Roveri N, Foresti E (1984) Thermal behavior of bone and synthetic hydroxylapatites submitted to magnesium interaction in aqueous medium. J Inorg Biochem 20:1-12

Barralet J, Best S, Bonfield W (1998) Carbonate substitution in precipitated hydroxylapatite: an investigation into the effects of reaction temperature and bicarbonate ion concentration. J Biomed Mater Res 41:79-86

Barralet JE, Best SM, Bonfield W (2000) Effect of sintering parameters on the density and microstructure of carbonated apatite. J Mater Sci: Mater in Med 11:719-724

Bellemans J (1999) Osseointegration in porous coated knee arthroplasty. The influence of component coating type in sheep. Acta Orthop Scand 288:1-35

Ben Abdelkader S, Khattech I, Rey C, Jemal M (2001) Synthesis, characterization and thermochemistry of calcium-magnesium hydroxylapatite and fluorapatite. Thermochim Acta 376:25-36

Bender SA, Bumgardner JD, Roach MD, Bessho K, Ong JL (2000) Effect of protein on the dissolution of HA coatings. Biomaterials 21:299-305

Benhayoune H, Jallot E, Laquerriere P, Balossier G, Bonhomme P, Fraysinet P (2000) Integration of dense HA rods into cortical bone. Biomaterials 21:235-242

Ben-Nissan B, Chai CS, Evans L (1995) Crystallographic and spectroscopic characterisation and morphology of biogenic and synthetic apatites. *In* Encyclopedic handbook of biomaterials and bioengineering. Vol 2. Wise DL, Tarantalo DJ, Altobelli DE, Yaszemski MJ, Gresser JD, Schwartz ER (eds) Marcel Dekker, p 191-222

Ben-Nissan B, Green DD, Kannangara GSK, Chai CS, Milev A (2001) P^{31} NMR studies of diethyl phosphite derived nanocrystalline hydroxylapatite. J Sol-Gel Sci Technol 21:27-37

Berndt CC, Haddad GN, Farmer AJD, Gross KA (1990) Thermal spraying for bioceramic applications—A review. Materials Forum 14:161-173

Best SM, Sim B, Kayser M, Downes S (1997) The dependence of osteoblastic response on variations in the chemical composition and physical properties of hydroxylapatite. J Mater Sci: Mater in Med 8:97-103

Bigi A, Foresti E, Gandolfi M, Gazzano M, Roveri N (1995) Inhibiting effect of zinc on hydroxylapatite crystallization. J Inorg Biochem 58:49-58

Bloebaum RD, Bachus KN, Rubman MH, Dorr LD (1993) Postmortem comparative analysis of titanium and hydroxylapatite porous-coated femoral implants retrieved from the same patient. A case study. J Arthroplasty 8:203-211

Blokhuis TJ, Termaat MF, den Boer FC, Patka P, Bakker FC, Haarman HJTM (2000) Properties of calcium phosphate ceramics in relation to their *in vivo* behavior. J Trauma 48:179-189

Blom EJ, Klein-Nulend J, Klein CPAT, Kurashina K, van Waas MAJ, Burger EH (2000) Transforming growth factor-beta1 incorporated during setting in calcium phosphate cement stimulates bone cell differentiation *in vitro*. J Biomed Mater Res 50:67-74

Bohner M (2001) Physical and chemical aspects of calcium phosphates used in spinal surgery. Eur Spine J S114-S121

Bohner M, Lemaitre J, van Landuyt P, Zambelli PY, Merkle HP, Gander B (1997) Gentamicin-loaded hydraulic calcium phosphate bone cement as antibiotic delivery system. J Pharmaceut Sci 86:565-572

Bonfield W, Grynpas MD, Tully AE, Bowman J, Abram J (1981) Hydroxylapatite reinforced polyethylene —a mechanically compatible implant material for bone replacement. Biomaterials 2:185-186

Bouyer E, Gitzhofer F, Boulos MI (1997) The suspension plasma spraying of bioceramics by induction plasma. J Metals 58-62

Boyde A (1997) Microstructure of enamel. *In:* Dental Enamel, Ciba Foundation Symposium. Vol 205. Wiley, New York, p 18-31

Brossa F, Cigada A, Chiesa R, Paracchini L, Consonni C (1994) Post-deposition treatment effects on hydroxylapatite vacuum plasma spray coatings. Biomaterials 5:855-857

Brown CD, Mealey BL, Nummikoski PV, Bifano SL, Waldrop TC (1998) Hydroxylapatite cement implant for regeneration of periodontal osseous defects in humans. J Periodontol 69:146-157

Brown PW, Fulmer M (1996) The effects of electrolytes on the rates of hydroxylapatite formation at 25 and 388°C. J Biomed Mater Res 31:395-400

Brown SR, Turner IG, Reiter H (1994) Residual stress measurement in thermal sprayed hydroxylapatite coatings. J Mater Sci: Mater in Med 5:756-759

Brown WE, Chow LC (1983) A new calcium phosphate setting cement. J Dent Res 63:672

Browne M, Gregson PJ (2000) Effect of mechanical surface pretreatment on metal ion release. Biomaterials 21:385-392

Burgess AV, Story BJ, La D, Wagner WR, Le Geros JP (1999) Highly crystalline Mp-1 ™ hydroxylapatite coating. Part I: *In vitro* characterization and comparison to other plasma-sprayed hydroxylapatite coatings. Clin Oral Implants Res 10:245-256

Burke EM, Haman JD, Weimer JJ, Cheney AB, Rigsbee JM, Lucas LC (2001) Influence of coating strain on calcium phosphate thin-film dissolution. J Biomed Mater Res 57:41-47

Burr DB, Mori S, Boyd RD, Sun TC, Blaha JD, Lane L, Parr J (1993) Histomorphometric assessment of the mechanisms for rapid ingrowth of bone to HA/TCP coated implants. J Biomed Mater Res 27: 645-653

Cabrini M, Cigada A, Rondelli G, Vicentini B (1997) Effect of different surface finishing and of hydroxylapatite coatings on passive and corrosion current of Ti6Al4V alloy in simulated physiological solution. Biomaterials 18:783-787

Carlsson L, Rostlund T, Albrektsson B, Albrektsson T (1988) Removal torques for polished and rough titanium implants. Intl J Oral Maxillofac Implants 3:21-24

Caulier H, van der Waerden JPCM, Paquay YCGJ, Wolke JGC, Kalk W, Naert I, Jansen JA (1995) Effect of calcium phosphate (Ca-P) coatings on trabecular bone response: A histological study. J Biomed Mater Res 29:1061-1069

Chai C, Ben-Nissan B (1993) Interfacial reactions between titanium and hydroxylapatite. J Aust Ceram Soc 29:71-80

Chai CS, Gross KA, Ben-Nissan B (1998) Critical ageing of hydroxylapatite sol-gel solutions. Biomaterials 19:2291-2296

Chang BS, Lee CK, Hong KS, Youn HJ, Ryu HS, Chung SS, Park KW (2000) Osteoconduction at porous hydroxylapatite with various pore configurations. Biomaterials 21:1291-1298

Chang C, Shi J, Huang J, Hu Z, Ding C (1998) Effect of power level on characteristics of vacuum plasma sprayed hydroxylapatite coating. J Thermal Spray Technol 7:484-488

Chaung HM, Hong CH, Chiang CP, Lin SK, Kuo YS, Lan WH, Hsieh CC (1996) Comparison of calcium phosphate cement mixture and pure hydroxide as direct pulp-capping agents. J Formoson Med Assoc 95:545-550

Chen J, Wolke JGC, de Groot K (1994) Microstructure and crystallinity in hydroxylapatite coatings. Biomaterials 15:396-399

Cheng K, Shen G, Weng WJ, Han GR, Ferreira JMF, Yang J (2001) Synthesis of hydroxylapatite/fluorapatite solid solution by a sol-gel method. Mater Lett 51:37-41

Cherng AM, Chow LC, Takagi S (2001) *In vitro* evaluation of a calcium phosphate cement root canal filler/sealer. J Endodontics 27:613-615

Chinol M, Vallabhajosula S, Goldsmith SJ, Klein MJ, Deutsch KF, Chinen LK, Brodack JW, Deutsch EA, Watson BA, Tofe AJ (1993) Chemistry and biological behavior of samarium-153 and rhenium-186-labeled hydroxylapatite particles: Potential radiopharmaceuticals for radiation synovectomy. J Nucl Med 34:1536-1542

Chou BY, Chang E (1999) Microstructural characterization of plasma-sprayed hydroxylapatite-10 wt % ZrO_2 composite coating on titanium. Biomaterials 20:1823-1832

Chou L, Marek B, Wagner WR (1999) Effects of hydroxylapatite coating crystallinity on biosolubility, cell attachment efficiency and proliferation *in vitro*. Biomaterials 20:977-985

Chow LC, Markovic M, Takagi S (1999) Formation of hydroxylapatite in cement systems: Effect of phosphate. Phosph Sulf Silicon and Rel Elements 146:129-132

Christoffersen J, Christoffersen MR, Kolthoff N, Barenholdt O (1997) Effects of strontium ions on growth and dissolution of hydroxylapatite and on bone mineral detection. Bone 20:47-54

Cihlar J, Trunec M (1996) Injection moulded hydroxylapatite ceramics. Biomaterials 17:1905-1911

Clunie G, Lui D, Cullum I, Ell PJ, Edwards JC (1996) Clinical outcome after one year following samarium-153 particulate hydroxylapatite radiation synovectomy. Scand J Rheumatol 25:360-366

Cornell CN (1992) Initial experience with the use of Collagraft as a bone graft substitute. Tech Orthop 7:55-62

Correia RN, Magalhaes MCF, Marques PAAP, Senos AMR (1996) Wet synthesis and characterisation of modified hydroxylapatite powders. J Mater Sci: Mater in Med 7:501-505

Cotell CM, Chrisey DB, Grabowski KS, Sprague JA, Gosset CR (1992) Pulsed laser deposition of hydroxylapatite thin films on Ti6Al4V. J Appl Biomater 3:87-93

Daubrée A (1851) Expériences sur la production artficielle de l'apatite, de la topza, et de quelques sutres métaux fruorifréres. Compt Rend Acad Sci Paris 32:625

de Groot K (1980) Bioceramics consisting of calcium phosphate salts. Biomaterials 1:47-50

de Groot K, Geesink R, Klein CPAT, Serekian P (1987) Plasma sprayed coatings of hydroxylapatite. J Biomed Mater Res 21:1375-1381

de Groot K, Wolke JGC, Jansen JA (1998) Calcium phosphate coatings for medical implants [Review]. Proc Inst Mech Eng Part H - J Eng in Med 212:137-147

de With G, van Dijk HJA, Hattu N, Prijs K (1981) Preparation, microstructure and mechanical properties of dense polycrystalline hydroxy apatite. J Mater Sci 16:1592-1598

Deb S, Wang M, Tanner KE, Bonfield W (1996) Hydroxylapatite polyethylene composites—Effect of grafting and surface treatment of hydroxylapatite. J Mater Sci: Mater in Med 7:191-193

Deiwick M, Glasmacher B, Pettenazzo E, Hammel D, Castellon W, Thiene G, Reul H, Berendes E, Scheld HH (2001) Primary tissue failure of bioprostheses: new evidence from *in vitro* tests. Thoracic Cardiovascular Surgeon 49:78-83

Delecrin J, Takahashi S, Gouin F, Passuti N (2000) A synthetic porous ceramic as a bone graft substitute in the surgical management of scoliosis: a prospective, randomized study. Spine 25:563-569

Devin JE, Attawia MA, Laurencin CT (1996) Three-dimensional degradable porous polymer-ceramic matrices for use in bone repair. J Biomater Sci Polymer Edn 7:661-669

Dhert WJA (1994) Retrieval studies on calcium phosphate-coated implants. Med Prog Technol 20:143-54

Doi Y, Koda T, Wakamatsu N, Goto T, Kamemizu H, Moriwaki Y, Adachi M, Suwa Y (1993) Influence of carbonate on sintering of apatites. J Dent Res 72:1279-1284

Doi Y, Shibutani T, Moriwaki Y, Kajimoto T, Iwayama Y (1998) Sintered carbonate apatites as bioresorbable bone substitutes. J Biomed Mater Res 39:603-610

Doi Y, Iwanaga H, Shibutani T, Moriwaki Y, Iwayama Y (1999) Osteoclastic responses to various calcium phosphates in cell culture. J Biomed Mater Res 47:424-433

Dong J, Kojima H, Uemara T, Tateishi T, Tanaka J (2001) *In vivo* evaluation of a novel porous hydroxylapatite to sustain osteogenesis of transplanted bone marrow-derived osteoblastic cells. J Biomed Mater Res 57:208-216

Dorozhkin SV (2002) A review on the dissolution models of calcium apatites [Review]. Progr Crystal Growth Charact Mater 44:45-61

Driessens FC (1988) Physiology of hard tissues in comparison with the solubility of synthetic calcium phosphates. [Review] Ann NY Acad Sci 523:131-136

Driessens FCM, Verbeeck RMH (1990) Biominerals. CRC Press, Boca Raton, Florida

Driskell TD (1994) Early history of calcium phosphate materials and coatings. Characterization and Performance of Calcium Phosphate Coatings for Implants. ASTM, Philadelphia, p 1-8

Ducheyne P, Radin S, Heughebaert M, Heughebaert JC (1990) Calcium phosphate ceramic coatings on porous titanium: effect of structure and composition on electrophoretic deposition, vacuum sintering and *in vitro* dissolution. Biomaterials 22:244-254

Ducheyne P, van Raemdonck W, Heughebaert JC, Heughebaert M (1986) Structural analysis of hydroxylapatite coatings on titanium. Biomaterials 7:97-103

Ducy P, Schinke T, Karsenty G (2000) The osteoblast: A sophisticated fibroblast under central surveillance. Science 289:1501-1504

Durucan C, Brown PW (2000) Low temperature formation of calcium-deficient hydroxylapatite-PLA/PLGA composites. J Biomed Mater Res 51:717-725

Edwards JT, Brunski JB, Higuchi HW (1997) Mechanical and morphologic investigation of the tensile strength of a bone-hydroxylapatite interface. J Biomed Mater Res 36:454-468

Ektessabi A, Shikine S, Kitamura N, Røkkum M, Johansson C (2001) Distribution and chemical states of iron and chromium released from orthopaedic implants into human tissues. X-ray Spectrometry 30: 44-48

Ellies LG, Nelson DGA, Featherstone JDB (1988) Crystallographic structure and surface morphology of sintered carbonated apatites. J Biomed Mater Res 22:541-553

Ellies LG, Nelson DGA, Featherstone JDB (1992) Crystallographic changes in calcium phosphates during plasma-spraying. Biomaterials 13:313-316

Elliot JC (1994) Structure and chemistry of the apatites and other calcium orthophosphates. Studies in Inorganic Chemistry 18. Elsevier Science, Amsterdam

Ellingsen JE (1995) Pre-treatment of titanium implants with fluoride improves their retention in bone. J Mater Sci: Mater in Med 6:749-753

Engin NO, Ta_ AC (1999) Manufacture of macroporous calcium hydroxylapatite bioceramics. J Eur Ceram Soc 19:1269-1272
Erkmen ZE (1999) The effect of heat treatment on the morphology of D-gun sprayed hydroxylapatite coatings. J Biomed Mater Res 48:861-868
Fang Y. Agrawal DK, Roy DM, Roy R (1994) Microwave sintering of hydroxylapatite ceramics. J Mater Res 9:180-187
Fanovich MA, Lopez JMP (1998) Influence of temperature and additives on the microstructure and sintering behavior of hydroxylapatites with different Ca/P ratios. J Mater Sci: Mater in Med 91:53-60
Fazan F, Marquis PM (2000) Dissolution behavior of plasma-sprayed hydroxylapatite coatings. J Mater Sci: Mater in Med 11:787-792
Feng B, Chen JY, Qi SK, He L, Zhao JZ, Zhang XD (2002) Carbonate apatite coating on titanium induced rapidly by precalcification. Biomaterials 23:173-179
Fernandez E, Planell JA, Best SM (1999) Precipitation of carbonated apatite in the cement system α-$Ca_3(PO_4)_2$-$Ca(H_2PO_4)_2$-$CaCO_3$. J Biomed Mater Res 47:466-471
Ferraz MP, Monteiro FJ, Serro AP, Saramago B, Gibson IR, Santos JD (2001) Effect of chemical composition on hydrophobicity and zeta potential of plasma sprayed HA/CaO-P_2O_5 glass coatings. Biomaterials 22:3105-3112
Filiaggi MJ, Pilliar RM, Coombs NA (1993) Post-plasma-spraying heat treatment of the HA coating/Ti-6Al-4V implant system. J Biomed Mater Res 27:191-198
Finet B, Weber G, Cloots R (2000) Titanium release from dental implants: an *in-vivo* study on sheep. Mater Lett 43:159-165
Frame JW, Browne RM, Brady CL (1981) Hydroxylapatite as a bone substitute in jaws. Biomaterials 2: 19-22
Fratzl P, Fratzl-Zelman N, Klaushofer K, Vogl G, Koller K (1991) Nucleation and growth of mineral crystals in bone studies by small-angle X-ray scattering. Calcif Tissue Intl 48:407-413
Frayssinet P, Rouquet N, Tourenne F, Fages J, Hardy D, Bonel G (1993) Cell-degradation of calcium phosphate ceramics [Review]. Cells Materi 3:383-394
Frayssinet P. Rouquet N. Fages J. Durand M. Vidalain PO. Bonel G (1997) The influence of sintering temperature on the proliferation of fibroblastic cells in contact with HA-bioceramics J Biomed Mater Res 35:337-347
Frayssinet P, Roudier M, Lerch A, Ceolin JL, Depres E, Rouquet N (2000) Tissue reaction against a self-setting calcium phosphate cement set in bone or outside the organism. J Mater Sci: Mater in Med 11:811-815
Friedman CD, Costantino PD, Takagi S, Chow LC (1998) BoneSource hydroxylapatite cement: A novel biomaterial for craniofacial skeletal tissue engineering and reconstruction. J Biomed Mater Res 53:428-432
Fujibayashi S, Shikata J, Tanaka C, Matsushita M, Nakamura T (2001) Lumbar posterolateral fusion with biphasic calcium phosphate ceramic. J Spinal Disord 14:214-221
Fulmer MT, Brown PW (1998) Hydrolysis of dicalcium phosphate dihydrate to hydroxylapatite. J Mater Sci: Mater in Med 9:197-202
Geesink RGT, de Groot K, Klein CPAT (1987) Chemical implant fixation using hydroxylapatite coatings. The development of a human total hip prosthesis for chemical fixation to bone using hydroxyl-apatite coatings on titanium substrates. Clin Orthop 225:147-170
Geesink RG (2002) Osteoconductive coatings for total joint arthroplasty [Review]. Clin Orthop Rel Res 395:53-65
Geyer C (1999) Materials for middle ear reconstruction. HNO 47:77-91
Gibson IR, Best SM, Bonfield W (1999) Chemical characterization of silicon-substituted hydroxylapatite J Biomed Mater Res 44:422-28
Gibson IR, Ke S, Best SM, Bonfield W (2001) Effect of powder characteristics on the sinterability of hydroxylapatite powders. J Mater Sci: Mater in Med 12:163-171
Ginebra MP, Rilliard A, Fernandez E, Elvira C, San Roman J, Planell JA (2001) Mechanical and rheological improvement of a calcium phosphate cement by the addition of a polymeric drug. J Biomed Mater Res 57:113-118
Gledhill HC, Turner IG, Doyle C (1999) Direct morphological comparison of vacuum plasma sprayed and detonation gun sprayed hydroxylapatite coatings for orthopedic applications. Biomaterials 20:315-322
Graham S, Brown PW (1996) Reactions of octacalcium phosphate to form hydroxylapatite. J Crystal Growth 165:106-115
Greenspan DC (1999) Bioactive ceramic implant materials. Curr Opin Solid State Mater Sci 4:389-393
Gross KA (1991) Surface modification of prostheses. M Eng Science thesis. Monash University.
Gross KA, Berndt CC, Goldschlag DD, Iacono VJ (1997) *In-vitro* changes of hydroxylapatite coatings. Intl J Oral Maxillofac Implants 12:589-597

Gross KA, Berndt CC (1998) Thermal processing of hydroxylapatite for coating production. J Biomed Mater Res 39:580-587

Gross KA, Phillips M (1998) Identification and mapping of the amorphous phase in plasma sprayed hydroxylapatite coatings using scanning cathodoluminescence microscopy. J Mater Sci: Mater in Med 9:797-802

Gross KA, Ben-Nissan B, Walsh WR, Swarts E (1998a) Analysis of retrieved hydroxylapatite coated orthopaedic implants. Thermal Spray: Meeting the Challenges of the 21st Century, ASM International, 1133-1138.

Gross KA, Berndt CC, Dinnebier R, Stephens P (1998c) Oxyapatite in hydroxylapatite coatings. J Mater Sci 33:3985-3991

Gross KA, Berndt CC, Herman H (1998d) Formation of the amorphous phase in hydroxylapatite coatings. J Biomed Mater Res 39:407-414

Gross KA, Gross V and Berndt CC (1998e) Thermal analysis of the amorphous phase in hydroxylapatite coatings. J Am Ceram Soc 81:106-112

Gross KA, Hanley L, Chai CS, Kannangara K, Ben-Nissan B (1998f) Thin hydroxylapatite coatings via sol gel synthesis. J Mater Sci: Mater in Med 9:839-843

Gross KA, Ezerietis E (2002) Juniper wood as an implant material. J Biomed Mater Res (in press)

Ha SW, Reber R, Eckert KL, Petitmermet M, Mayer J, Wintermantel E, Baerlocher C, Gruner H (1998) Chemical and morphological changes of vacuum plasma sprayed hydroxylapatite coatings during immersion in simulated physiological solutions. J Am Ceram Soc 81:81-88

Halouani R, Bernache-Assolant D, Champion E, Ababou A (1994) Microstructure and related mechanical properties of hot pressed hydroxylapatite ceramics. J Mater Sci: Mater in Med 5:563-568

Haman JD, Lucas LC, Crawmer D (1995) Characterisation of high velocity oxy-fuel combustion sprayed hydroxylapatite. Biomaterials 16:229-237

Hata K, Kokubo T, Nakamura T, Yamamuro T (1995) Growth of a bonelike apatite layer on a substrate by a biomimetic process. J Am Ceram Soc 78(4):1049-1053

Hattori T, Iwadate Y, Inai H, Sato K, Imai Y (1987) Preparation of hydroxylapatite powder using a freeze-drying method. J Ceram Soc Japan 95:825-827

Havelin LI, Engesaeter LB, Espehaug B, Furnes O, Lie SA, Vollset SE (2000) The Norwegian arhroplasty register—11 years and 73,000 arthroplasties. Acta Orthop Scand 71:337-353

Hayashi K, Mashima T, Uenoyama K (1999) The effect of hydroxylapatite coating on bony ingrowth into grooved titanium implants. Biomaterials 20:111-119

Heijlijers HJ, Driessens FC, Verbeeck RM (1979) Lattice parameters and cation distribution of solid solutions of calcium and strontium hydroxylapatite. Calcif Tiss Intl 29:127-131

Heimann RB, Vu TA (1997) Low-pressure plasma-sprayed (LPPS) bioceramic coatings with improved adhesion strength and resorption resistance. J Thermal Spray Technol 6:145-149

Heimann RB, Vu TA, Wayman ML (1997) Bioceramics coatings: State-of-the-art and recent development trends. Eur J Mineral 9:597-515

Heimann RB, Hemachandra K, Itiravivong P (1999) Material engineering approaches towards advanced bioceramic coatings on Ti6Al4V implants. J Met Mater Mineral 8:25-40

Hench LL (1998a) Bioceramics [Review]. J Am Ceram Soc 81:1705-1728

Hench LL (1998b) Biomaterials – A forecast for the future. Biomaterials 19:1419-23

Heymann D, Pradal G, Benahmed M (1999) Cellular mechanisms of calcium phosphate ceramic degradation [Review]. Histol Histopath 14:871-877

Heymann D, Guicheux J, Rouselle AV (2001) Ultrastructural evidience *in vitro* of osteoclast-induced degradation of calcium phosphate ceramic by simultaneous resorption and phagacytosis mechanisms. Histol Histopath 16:37-44

Hing KA, Best SM, Tanner KE, Bonfield W, Revell PA (1999) Quantification of bone ingrowth within bone-derived porous hydroxylapatite implants of varying density. J Mater Sci: Mater in Med 10: 663-670

Hlavac J (1999) Ceramic coatings on titanium for bone implants [Review]. Ceramics-Silikaty 43:133-139 (in Czech)

Hobar PC, Pantaloni M, Byrd HS (2000) Porous hydroxylapatite granules for alloplastic enhancement of the facial region. [Review] Clin Plast Surg 27:557-569

Honda T, Takagi M, Uchida N, Saito K, Uematsu K (1990) Post-composition control of hydroxylapatite in an aqeous medium. J Mater Sci: Mater in Med 1:114-117

Ignatius AA, Claes LE (1996) *In vitro* biocompatibility of bioresorbable polymers—poly(L,DL-lactide) and poly(L-lactide-co-glycolide). Biomaterials 17:831-839

Ignjatovic N, Savic V, Najman S, Plavsic M, Uskokovic D (2001) A study of HAp/PLLA composite as a substitute for bone powder, using FT-IR spectroscopy. Biomaterials 22:571-575

Ikeuchi M, Yamamoto H, Shibata T, Otani M (2001) Mechanical augmentation of the vertebral body by calcium phosphate cement injection. J Orthop Sci 6:39-45

Inoue S, Ono A (1987) Preparation of hydroxylapatite by spray-pyrolysis technique. J Ceram Soc Japan 95:759-763

Ishikawa K, Ducheyne P, Radin S (1993) Determination of the Ca/P ratio in calcium-deficient hydroxylapatite using X-ray diffraction analysis. J Mater Sci: Mater in Med 4:165-168

Ishikawa K, Miyamoto Y, Kon M, Nagayama M, Asaoka K (1995) Non-decay type fast-setting calcium phosphate cement: composite with sodium alginate. Biomaterials 16:527-532

Itatani K, Iwafune K, Howell FS, Aizawa M (2000) Preparation of various calcium-phosphate powders by ultrasonic spray freeze-drying technique. Mater Res Bull 35:574-85

Ito M, Hidaka Y, Nakajima M, Yagasaki H, Kafrawy AH (1999) Effect of hydroxylapatite content on physical properties and connective tissue reactions to a chitosan-hydroxylapatite composite membrane. J Biomed Mater Res 45:204-208

Jaffe WL, Scott DF (1996) Total hip arthroplasty with hydroxylapatite-coated prostheses [A review]. J Bone Joint Surg 78A:1918-1934

Jain AK, Panchagnula R (2000) Skeletal drug delivery systems. Intl J Pharmaceut 206:1-12

Jansen JA, de Wijn JR, Wolters-Lutgerhorst JML, van Mullern PJ (1985) Ultrastructural study of epithelial cell attachment to implant materials. J Dent Res 64:891-896

Jarcho M, Bolen CH, Thomas MB, Bobick J, Kay JF, Doremus RH (1976) Hydroxylapatite synthesis and characterization in dense polycrystalline form. J Mater Sci 11:2027-2035

Jarcho M, Kay JF, Gumaer KI, Doremus RH, Drobeck HP (1977) Tissue, cellular and subcellular events at a bone-ceramic hydroxylapatite interface. J Bioeng 1:79-92

Jha LJ, Best SM, Knowles JC, Rehman I, Santos JD, Bonfield W (1997) Preparation and characterization of fluoride-substituted apatites. J Mater Sci: Mater in Med 8:185-191

Ji H, Marquis PM (1993) Effect of heat treatment on the microstructure of plasma-sprayed hydroxylapatite coating. Biomaterials 14:64-68

Jiang G, Shi D (1998) Coating of hydroxylapatite on highly porous alumina substrate for bone substitutes. J Biomed Mater Res: Appl Biomater 43:77-81

Johanssen E (1964) Microstructure of enamel and dentin. J Dent Res 43:1007-1009

John A, Hong L, Ikada Y, Tabata Y (2001) A trial to prepare biodegradable collagen-hydroxylapatite composites for bone repair. J Biomater Sci Polym Ed 12:689-705

Jonasova L, Muller FA, Helebrant A, Strnad J, Greil P (2002) Hydroxylapatite formation on alkali-treated titanium with different content of Na+ in the surface layer. Biomaterials 24 (in press)

Jordan DR, Bawazeer A (2001) Experience with 120 synthetic hydroxyapatite implants. Ophthal Plast Reconstruct Surg 17:184-90

Joschek S, Nies B, Krotz R, Goepferich A (2000) Chemical and physicochemical characterization of porous hydroxylapatite ceramics made of natural bone. Biomaterials 21:1645-1658

Joshi SV, Srivasta MP, Pal A, Pal S (1993) Plasma spraying of biologically derived hydroxylapatite on implantable materials. J Mater Sci: Mater in Med 4:251-255

Juang HY, Hon MH (1996) Effect of calcination on sintering of hydroxylapatite. Biomaterials 17: 2059-2064

Kamerer DB, Hirsch BE, Snyderman CH, Costantino P, Friedman CD (1994) Hydroxylapatite cement: a new method for achieving watertight closure in transtemporal surgery. Am J Otology 15:47-49

Keller L (1995) X-ray powder diffraction patterns of calcium phosphates analyzed by the Rietveld method. J Biomed Mater Res 29:1403-1413

Keller L, Dollase WA (2000) X-ray determination of crystalline hydroxylapatite to amorphous calcium-phosphate ratio in plasma sprayed coatings. J Biomed Mater Res 49:244-249

Khor KA, Dong ZL, Quek CH, Cheang P (2000) Microstructure investigation of plasma sprayed HA/Ti6Al4V composites by TEM. Mater Sci Engineer A281:221-228

Kienapfel H, Sprey C, Wilke A, Griss P (1999) Implant fixation by bone ingrowth [A review]. J Arthroplasty 14:355-368

Kijima T and Tsutsumi M (1979) Preparation and thermal properties of dense polycrystalline oxyhydroxylapatite. J Am Ceram Soc 62:455-460

Kim HM (2001) Bioactive ceramics: challenges and perspectives. J Ceram Soc Japan 109:S49-S57

Kim HM, Miyaji F, Kokubo T, Nakamura T (1997) Apatite-forming ability of alkali-treated Ti metal in body environment. J Ceram Soc Japan 105:111-116

Kim TN, Feng QL, Kim JO, Wu J, Wang H, Chen GC, Cui FZ (1998) Antimicrobial effects of metal ions (Ag^+, Cu^{2+}, Zn^{2+}) in hydroxylapatite. J Mater Sci: Mater in Med 9:129-134

Klein CPAT, Driessen AA, de Groot K (1984) Relationship between the degradation behaviour of calcium phosphate ceramics and their physical chemical characteristics and ultrastructural geometry. Biomaterials 5:157-160

Klein CPAT, Wolke JGC, de Blieck-Hogervorst JMA, de Groot K (1994) Features of calcium phosphate plasma-sprayed coatings: An *in vitro* study. J Biomed Mater Res 28:961-967

Knepper M, Milthorpe BK, Moricca S (1998) Interdiffusion in short-fibre reinforced hydroxylapatite ceramics. J Mater Sci: Mater in Med 9:589-596

Knowles JC, Gross K, Berndt CC, Bonfield W (1996) Structural changes of thermally sprayed hydroxylapatite investigated by Rietveld Analysis. Biomaterials 17:639-645

Kobayashi T, Nakamura S, Yamashita K (2001) Enhanced osteobonding by negative surface charges of electrically polarized hydroxylapatite. J Biomed Mater Res 57:477-484

Kokubo T (1998) Apatite formation on surfaces of ceramics, metals and polymers in body environment. Acta Materialia 46:2519-27

Kokubo T, Kim HM, Kawashita M, Takadama H, Miyazaki T, Uchida M, Nakamura T (2000) Nucleation and growth of apatite an amorphous phases in simulated body fluid. Glass Sci Technol 73:247-254

Kon M, Miyamoto Y, Asaoka K, Ishikawa K, Lee HH (1998) Development of calcium phosphate cement for rapid crystallization to apatite. Dental Mater J 17:223-232

Kondo K, Okuyama M, Ogawa H, Shibata Y, Abe Y (1984) Preparation of high strength apatite ceramics. J Am Ceram Soc 67:222-223

Konjiki T, Sudo T, Kohyama N (1980) Mineralogical notes of apatite in urinary calculi. Calc Tiss Intl 30:101-107

Koyama J, Hongo K, Iwashita T, Kobayashi S (2000) A newly designed key-hole button. J Neurosurg. 93:506-508

Kurioka K, Umeda M, Teranobu O, Komori T (1999) Effect of various properties of hydroxylapatite ceramics on osteoconduction and stability. Kobe J Med Sci 45:149-163

Kweh SWK, Khor KA, Cheang P (1999) The production and characterization of hydroxylapatite (HA) powders. J Mater Proc Technol 9:373-377

Labat B, Demonet N, Rattner A, Aurelle JL, Rieu J, Frey J, Chamson A (1999) Interaction of a plasma-sprayed hydroxylapatite coating in contact with human osteoblasts and culture medium. J Biomed Mater Res 46:331-336

Lacout JL, Assarane J, Trombe JC (1984) Fixation of titanium by phosphate minerals. C R Acad Sci 298:173-175

Ladizesky NH, Ward IM, Bonfield W (1997) Hydroxylapatite high-performance polyethylene fiber composites for high-load-bearing bone replacement materials. J Appl Polymer Sci 65:1865-1882

Landi E, Tampieri A, Celotti G, Sprio S (2000) Densification behaviour and mechanisms of synthetic hydroxylapatites. J Europ Ceram Soc. 20:2377-2387

Landis WJ, Hodgens KJ, Arena J, Song MJ, McEwen BF (1996) Structural relations between collagen and mineral in bone as determined by high voltage electron microscopic tomography. Microscopy Res and Technique. 33:192-202

Laurencin CT, Attawia MA, Elgendy HE, Herbert KM (1996) Tissue engineered bone-regeneration using degradable polymers: The formation of mineralized matrices. Bone 19:S93-S99

Laurencin CT, Ambrosio AMA, Borden MD, Cooper JA (1999) Tissue engineering: Orthopaedic applications. Ann Rev Biomed Eng 1:19-46

Le Geros RZ (1965) Effect of carbonate on the lattice parameters of apatite. Nature 206:403-404

Le Geros RZ, Miravite MA, Quirolgico GB, Curzon ME (1977) The effect of some trace elements on the lattice parameters of human and synthetic bone. Calc Tiss Res 22:362-367

Le Geros RZ, Bonel B, Le Geros R (1978) Types of H_2O in human enamel and in precipitated apatites. Calcif Tiss Intl 26:111-118

Le Geros RZ (1991) Calcium phosphates in oral biology and medicine. Monographs in Oral Science Volume 15. Karger, Basel

Le Geros RZ (1993) Biodegradation and bioresorption of calcium phosphate ceramics [Review]. Clin Materials 14:65-88

Le Geros RZ, LeGeros JP, Daculsi G, Kijkowska R (1995) Calcium Phosphate Biomaterials: Preparation, Properties, and Biodegradation. *In:* Encyclopedic handbook of biomaterials and bioengineering. Vol 2. Wise DL, Tarantalo DJ, Altobelli DE, Yaszemski MJ, Gresser JD, Schwartz ER (eds) Marcel Dekker, p 1429-1463

Le Geros RZ (1999) Calcium phosphates in demineralization/remineralization processes J Clin Dent 10: 65-73

Leali Tranquilli P, Merolli A, Palmacci O, Gabbi C, Cacchiolo A, Gonizzi G (1994) Evaluation of different preparation of plasma-spray hydroxylapatite coating on titanium alloy and duplex stainless steel in the rabbit. J Mater Sci: Mater in Med 5:345-349

Lee TM, Wang BC, Yang YC, Chang E, Yang CY (2001) Comparison of plasma-sprayed hydroxylapatite coatings and hydroxylapatite/tricalcium phosphate composite coatings: An *in vivo* study. J Biomed Mater Res 55:360-367

Leroux L, Lacout JL (2001a) Synthesis of calcium-strontium phosphate fluor-hydroxylapatites by neutralisation. Phosph Sulf Sil Rel Elements 173:27-38
Leroux L, Lacout JL (2001b) Preparation of calcium strontium hydroxylapatites by a new route involving calcium phosphate cements. J Mater Res 16:171-178
Li H, Khor KA, Cheang P (2000) Effect of powder's melting state on the properties of HVOF sprayed hydroxylapatite coatings. Mater Sci Eng 293A:71-80
Li P, Nakanishi K, Kokubo T, de Groot K (1993a) Induction and morphology of hydroxylapatite, precipitated from metastable simulated body fluids on sol-gel prepared silica. Biomaterials 14:963-968
Li YB, Klein CPAT, Zhang XD, de Groot K (1993b) Relationship between the colour change of hydroxylapatite and the trace element manganese. Biomaterials 14:969-972
Li P, Ohtsuki Ch, Kokubo T, Nakanishi K, Soga N, de Groot K (1994) The role of hydrated silica, titania, and alumina in indicing apatite on implants. J Biomed Mater Res 28:7-15
Lim YJ, Oshida Y, Andres CJ, Barco MT (2001) Surface characterizations of variously treated titanium materials. Intl J Oral Maxillofac Implants 16:333-342
Lim TH, Brebach GT, Renner SM, Kim WJ, Kim JG, Lee RE, Andersson GBJ, An HS (2002) Biomechanical evaluation of an injectable calcium phosphate cement for vertebroplasty. Spine 27:1297-1302
Lind M, Overgaard S, Bunger C, Soballe K (1999) Improved bone anchorage of hydroxylapatite coated implants compared with tricalcium-phosphate coated implants in trabecular bone in dogs. Biomaterials 20:803-808
Liu C, Shen W, Gu Y, Hu L (1997) Mechanism of the hardening process for a hydroxylapatite cement. J Biomed Mater Res 35:75-80
Liu DM (1996) Control of pore geometry on influencing the mechanical property of porous hydroxylapatite bioceramic. J Mater Sci Lett 15:419-421
Liu DM, Chou HM, Wu JD, Tung MS (1994) Hydroxyl apatite coating via amorphous calcium phosphate. Mater Chem Phys 37:39-44
Lopatin CM, Pizziconi V, Alford TL, Laursen T (1998) Hydroxylapatite powders and thin films prepared by a sol-gel technique. Thin Solid Films 326:227-232
Lovice DB, Mingrone MD, Toriumi DM (1999) Grafts and implants in rhinoplasty and nasal reconstruction [Review]. Otolaryngol Clin North America 32:113ff
Lu H, Qu Z, Zhou YC (1998) Preparation and mechanical properties of dense polycrystalline hydroxylapatite through freeze-drying. J Mater Sci: Mater in Med 9:583-587
Lugscheider E, Weber Th, Knepper M (1991a) Production of biocompatible coatings by atmospheric plasma spraying. Mater Sci Eng 139A:45-48
Lugscheider E, Weber Th F, Knepper M (1991b) Processability of fluorapatite through atmospheric plasma spraying technology. Metalloberflache 45:129-132
Lugscheider E, Knepper M, Gross KA (1992) Production of spherical apatite powders—The first step for optimized thermal-sprayed apatite coatings. J Thermal Spray Techn 1:215-223
Luklinska ZB, Bonfield W (1997) Morphology and ultrastructure of the interface between hydroxylapatite-polyhydroxybutyrate composite implant and bone. J Mater Sci: Mater in Med 8:379-383
Lumbikanonda N, Sammons R (2001) Bone cell attachment to dental implants of different surface characteristics. Intl J Oral Maxillofac Implants 16:627-636
Luo P, Nieh TG (1996) Preparing hydroxylapatite powders with controlled morphology. Biomaterials 17:1959-1964
Ma PX, Zhang R, Xiao G, Franceschi R (2001) Engineering new bone tissue *in vitro* on highly porous poly (alpha-hydroxyl acids)/hydroxylapatite composite scaffolds. J Biomed Mater Res 54:284-293
Macdonald A, Moore BK, Newton CW, Brown CE (1994) Evaluation of an apatite cement as a root end filling material. J Endodontics 20:598-604
Magyar G, Toksvig-Larsen S, Moroni A (1997) Hydroxylapatite coating of threaded pins enhances fixation. J Bone Joint Surg 79B:487-489
Mancini CE, Berndt CC, Sun L, Kucuk A (2001) Porosity determinations in thermally sprayed hydroxylapatite coatings. J Mater Sci 36:3891-3896
Manjubala I, Sivakumar M, Najma Nikkath S (2001) Synthesis and characterisation of hydroxy/fluorapatite solid solutions. J Mater Sci: Mater in Med 36:5481-5486
Mano JF, Vaz CM, Mendes SC, Reis RL, Cunha AM (1999) Dynamic mechanical properties of hydroxylapatite-reinforced and porous starch-based degradable biomaterials. J Mater Sci: Mater in Med 10:857-862
Marie PJ, Ammann P, Boivin G, Rey C (2001) Mechanisms of action and therapeutic potential of strontium in bone. Calcif Tiss Intl 69:121-129
Martin RB, Burr DB (1989) Structure, function and adaptation of compact bone. Raven Press, 30 p

Massaro C, Baker MA, Cosentino F, Ramires PA, Klose S, Milella E (2001) Surface and biological evaluation of hydroxylapatite-based coatings on titanium deposited by different techniques. J Biomed Mater Res 58:651-657

Masuda Y, Matubara K, Sakka S (1990) Synthesis of hydroxylapatite from metal alkoxides through sol-gel technique. J Ceram Soc Japan 98:84-95

Matkovic V (1991) Calcium metabolism and calcium requirements during skeletal modeling and consolidation of bone mass. Am J Clin Nutr 54:245S-260S

Matsui Y, Ohno K, Michi K, Yamagata K (1994) Experimental study of high-velocity flame sprayed hydroxylapatite coated and noncoated titanium implants. Intl J Oral Maxillofac Implants 9:397-404

Matsuya S, Takagi S, Chow LC (2000) Effect of mixing ratio and pH on the reaction between $Ca_4(PO_4)_2O$ and $CaHPO_4$. J Mater Sci: Mater in Med 11:305-311

Mayer I, Apfelbaum F, Featherstone JDB (1994) Zinc ions in synthetic carbonated hydroxylapatites. Arch Oral Biol 39:87-90

Mayer I, Shlam R, Featherstone JDB (1997) Magnesium-containing carbonate apatites. J Inorg Biochem 66:1-6

Mayer I. Featherstone JDB (2000) Dissolution studies of Zn-containing carbonated hydroxylapatites. J Cryst Growth 219:98-101

McConnell D (1965) Crystal chemistry of hydroxylapatite. Its relation to bone mineral. Arch Oral Biol 10:421-431

McPherson R, Gane N, Bastow TJ (1995) Structural characterization of plasma-sprayed hydroxylapatite coatings. J Mater Sci: Mater in Med 6:327-334

Meijer AGW, Segenhout HM, Albers EWJ, van de Want HJL (2002) Histopathology and biocompatible hydroxylapatite-polyethylene composite in ossiculoplasty. J Oto-Rhino-Laryngol Rel Specialit 64: 173-179

Merry JC, Gibson IR, Best SM, Bonfield W (1998) Synthesis and characterisation of carbonate hydroxylapatite. J Mater Sci: Mater in Med 9:779-783

Miyamoto Y, Ishikawa K, Fukao H, Sawada M, Nagayama M, Kon M, Asaoka K (1995) *In vivo* setting behaviour of fast-setting calcium phosphate cement. Biomaterials 16:885-60

Miyazaki T, Kim HM, Kokubo T, Kato H, Nakamura T (2001) Induction and acceleration of bonelike apatite formation on tantalum oxide gel in simulated body fluid. J Sol-Gel Sci Technol 21:83-88

Monma H, Kayima T (1987) Preparation of hydroxylapatite by the hydrolysis of brushite. J Mater Sci 22:4247-4250

Monroe ZA, Votawa W, Bass DB, McMullen J (1971) New calcium phosphate ceramic material for bone and tooth implants. J Dent Res 50:860

Morimoto K, Kihara A, Takeshita F, Akedo H, Suetsugu T (1988) Differences between the bony interfaces of titanium and hydroxylapatite-alumina plasma-sprayed titanium blade implants. J Oral Implantol 14:314-324

Moroni A, Caja VL, Egger EL, Trinchese L, Chao EYS (1994) Histomorphometry of hydroxylapatite coated and uncoated porous titanium bone implants. Biomaterials 15:926-930

Muralithran G, Ramesh S (2000a) Effect of MnO_2 on the sintering behaviour of hydroxylapatite. Biomed Eng Appl Basis Comm 12:43-48

Muralithran G, Ramesh S (2000b) The effects of sintering temperature on the properties of hydroxylapatite. Ceram Intl 26:221-230

Murray MGS, Wang J, Ponton CB, Marquis PM (1995) An improvement in processing of hydroxylapatite ceramics. J Mater Sci 30:3061-3074

Nagase M, Chen R, Asada Y, Nakajima T (1989) Radiographic and microscopic evaluation of subperiosteally implanted blocks of hydroxylapatite-gelatin mixture in rabbits. J Oral Maxillofac Surg 47:40-45

Nakahira A, Sakamoto K, Yamaguchi S, Kijima K, Okazaki M (1999) Synthesis of hydroxylapatite by hydrolysis of alpha-TCP. J Ceram Soc Japan 107:89-91

Nakahira A, Tamai M, Sakamoto K, Yamaguchi S (2000) Sintering and microstructure of porous hydroxylapatite. J Ceram Soc Japan 108:99-104

Nakashima Y, Hayashi K, Inadome T, Uenoyama K, Hara T, Kanemaru T, Sugioka Y, Noda I (1997) Hydroxylapatite-coating on titanium arc sprayed titanium implants. J Biomed Mater Res 35:287-298

Nancollas GH, Mohan MS (1970) The growth of hydroxylapatite crystals. Arch Oral Biol 15:731-745

Narasaraju TSB, Phebe DE (1996) Some physico-chemical aspects of hydroxylapatite [Review]. J Mater Sci 31:1-21

Negami S (1988) Histological observations on muscle tissue reactions to porous hydroxyapatite sintered bodies in rats. J Japan Orthop Assoc 62:85-94

Nelson DGA (1981) The influence of carbonate on the atomic structure and reactivity of hydroxylapatite. J Dent Res 60:1621-1629

Nelson DGA, Featherstone JDB (1982) Preparation, analysis and characterization of carbonated apatites. Calcif Tiss Intl 34:S69-S81

Neo M, Kotani S, Fujita Y, Nakamura T, Yamamuro T, Bando Y, Ohtsuki C, Kokubo T (1992) Differences in ceramic-bone interface between surface-active ceramics and resorbable ceramics: a study by scanning and transmission electron microscopy. J Biomed Mater Res 26:255-267

Netz DJA, Sepulveda P, Padnolfelli VC, Sparado ACC, Alencastre JB, Bentley MVLB, Marchetti JM (2001) Potential use of gelcasting hydroxylapatite porous ceramic as an implantable drug delivery system. Intl J Pharmaceut 213:117-125

Nishiguchi S, Kato H, Fujita H, Oka M, Kim HM, Kokubo T, Nakamura T (2001) Titanium metals form direct bonding to bone after alkali and heat treatments. Biomaterials 22:2525-2533

Nordström EG, Karlsson KH (1990) Slip-cast apatite ceramics. Ceram Bull 69:824-827

Ogiso M (1998) Reassessment of long-term use of dense HA as dental implant: case report. J Biomed Mater Res 43:318-320

Ogiso M, Yamamura M, Kuo PT, Borgese D, Matsumoto T (1998a) Comparitive push-out test of dense HA implants and HA-coated implants: Findings in a canine study. J Biomed Mater Res 39:364-372

Ogiso M, Yamashita Y, Matsumoto T (1998b) The process of physical weakening and dissolution of the HA-coated implant in bone and soft tissue. J Dent Res 77:1426-1434

Oguchi H, Ishikawa K, Ojima S, Hirayama Y, Seto K, Eguchi G (1992) Evaluation of a high-velocity flame-spraying technique for hydroxylapatite. Biomaterials 13:471-477

Ohgaki M, Kizuki T, Katsura M, Yamashita K (2001) Manipulation of selective cell adhesion and growth by surface charges of electrically polarized hydroxylapatite. J Biomed Mater Res 57:366-373

Ohgaki M, Nakamura S, Okura T, Yamashita K (2000) Enhanced mineralization on electrically polarized hydroxylapatite ceramics in culture medium. J Ceram Soc Japan 108:1037-1040

Okayama S, Akao M, Nakamura S, Shin Y, Higashikata M, Aoki H (1991) The mechanical properties and solubility of strontium-substituted hydroxylapatite. Biomed Mater Eng 1:11-17

Okazaki M (1988) Magnesium-containing fluoridated apatites. J Fluor Chem 41:45-52

Okazaki M, Takahashi J (1997) Heterogeneous iron-containing fluoridated apatites. Biomaterials 18:11-14

Ong JL, Chan DCN (1999) Hydroxylapatite and their use as coatings in dental implants: A review. Crit Rev Biomed Engin 28:667-707

Ono I, Tateshita T, Nakajima T, Ogawa T (1998) Determinations of strength of synthetic hydroxylapatite ceramic implants. Plast Reconstruct Surg 102:807-813

Ono I, Tateshita T, Satou M, Sasaki T, Matsumoto M, Kodama N (1999) Treatment of large complex cranial bone defects by using hydroxylapatite ceramic implants. Plast Reconstruct Surg 104:339-349

Orlovskii VP, Barinov SM (2001) Hydroxylapatite and hydroxylapatite-matrix ceramics: A survey [Review]. Russ J Inorg Chem 46:S129-S149

Osaka A, Miura Y, Takeuchi K, Asada M, Takahashi K (1991) Calcium apatite prepared from calcium hydroxide and orthophosphoric acid. J Mater Sci: Mater in Med 2:51-55

Osaka A, Tsuru K, Iida H, Ohtsuki C, Hayakawa S, Miura Y (1997) Spray pyrolysis preparation of apatite-composite particles for biological application. J Sol-Gel Sci Technol 8:655-61

Otsuka M, Matsuda Y, Suwa Y, Fox JL, Higuchi WI (1995) Effect of particle size of metastable calcium phosphates on mechanical strength of a novel self-setting bioactive calcium phosphate cement. J Biomed Mater Res 29:25-32

Overgaard S, Lind M, Josephsen K, Maunsbach AB, Bunger C, Soballe K (1998) Resorption of hydroxylapatite and fluorapatite ceramic coatings on weight-bearing implants: A quantitative and morphological study in dogs. J Biomed Mater Res 39:141-152

Oyane A, Nakanishi K, Kim HM, Miyaji F, Kokubo T, Soga N, Nakamura T (1999) Sol-gel modification of silicone to induce apatite-forming ability. Biomaterials 20:79-84

Palka V, Brezovsky M, Zeman J, Cepera M (1993) Phase changes after plasma spraying hydroxylapatite. Met Mater 31:491-498

Paschalis EP, Zhao Q, Tucker BE, Mukhopadhayay S, Bearcroft JA, Beals NB, Spector M, Nancollas GH (1995) Degradation potential of plasma-sprayed hydroxylapatite-coated titanium implants. J Biomed Mater Res 29:1499-1505

Paul W, Nesamony J, Sharma CP (2002) Delivery of insulin from hydroxyapatite ceramic microspheres: Preliminary in vivo studies. J Biomed Mater Res 61:660-662

Peelen JGJ, Rejda BV, de Groot (1978) Preparation and properties of sintered hydroxylapatite. Ceram Intl 4:71-74

Perdok WG, Christoffersen J, Arends J (1987) The thermal lattice expansion of calcium hydroxylapatite. J Crystal Growth 80:149-154

Pham MT, Maitz MF, Matz W, Reuther H, Richter E, Steiner G (2000) Promoted hydroxylapatite nucleation on titanium ion-implanted with sodium. Thin Solid Films 379:50-56

Pineda CA, Rodgers AL, Prozecky VM, Przybylowicz WJ (1996) Microanalysis of calcium-rich human kidney stones at the NAC nuclear microprobe. Cellular and Mol Biol 42:119-126
Pioletti DP, Takei H, Lin T, van Landuyt P, Ma OJ, Kwon SY, Sung KL (2000) The effects of calcium phosphate cement particles on osteoblast functions. Biomaterials 21:1103-1114
Power ML, Heaney RP, Kalkwarf HJ, Pitkin RM, Repke JT, Tsang RC, Schulkin J (1999) The role of calcium in health and disease. Am J Obstetrics Gynaecol 181:1560-1569
Puajindenetr S, Best SM, Bonfield W (1994) Characterisation and sintering of precipitated hydroxylapatite. Brit Ceram Trans 93:96-99
Radin SR, Ducheyne P (1992) Plasma-spraying induced changes of calcium phosphate ceramic characteristics and the effect on *in vitro* stability. J Mater Sci: Mater in Med 3:33-42
Ramires PA, Romito A, Cosentino F, Milella E (2001) The influence of titania/hydroxylapatite composite coatings on *in vitro* osteoblasts behaviour. Biomaterials 22:1467-1474
Ravaglioli A, Krajewski A (1992) Bioceramics: Materials, Properties, Applications. Chapman and Hall, London
Raynaud S, Champion E, Bernache-Assolant D, Laval JP (2001) Determination of calcium/phosphorus atomic ratio of calcium phosphate apatites using X-ray diffractometry. J Am Ceram Soc 84:359-366
Raynaud S, Champion E, Bernache-Assolant D, Thomas P (2002a) Calcium phosphate apatites with variable Ca/P atomic ratio I. Synthesis, charcterisation and thermal stability of powders. Biomaterials 23:1065-1072
Raynaud S, Champion E, Bernache-Assolant D (2002b) Calcium phosphate apatites with variable Ca/P atomic ratio II. Calcination and Sintering. Biomaterials 23:1073-1080
Raynaud S, Champion E, Lafon JP, Bernache-Assolant D (2002c) Calcium phosphate apatites with variable Ca/P atomic ratio III. Mechanical properties and degradation in solution of hot pressed ceramics. Biomaterials 23:1081-1089
Ren L, Tsuru K, Hayakawa S, Osaka A (2001) Sol-gel preparation and *in vitro* deposition of apatite on porous gelatin-siloxane hybrids. J Non-Crystal Solids 285:116-122
Riboud PV (1973) Composition and stability of apatite phases in the system CaO-P_2O_5-iron oxide-H_2O at high temperature. Ann Chim Fr 8:381-390 (In French)
Ripamonti U, Crooks J, Rueger DC (2001) Induction of bone formation by recombinant human osteogenic protein-1 and sintered porous hydroxylapatite in adult primates. Plast Reconstruct Surg 107:977-988
Rocca M, Orienti L, Stea S, Moroni A, Fini M, Giardino R (1998) Comparison among three different biocoatings for orthopaedic prostheses – An experimental animal study. Intl J Artifical Organs 21: 553-558
Rodríguez-Lorenzo LM, Vallet-Regí M (2000) Controlled crystallization of calcium phosphate apatites. Chem Mater 12:2460-2465
Rodríguez-Lorenzo LM, Vallet-Regí M, Ferreira JMF (2001a) Fabrication of hydroxylapatite bodies by uniaxial pressing from a precipitated powder. Biomaterials 22:583-588
Rodríguez-Lorenzo LM, Vallet-Regí M, Ferreira JMF (2001b) Colloidal processing of hydroxylapatite. Biomaterials 22:1847-1852
Rodríguez-Lorenzo LM, Vallet-Regí M, Ferreira JMF, Ginebra MP, Aparicio C, Planell JA (2002) Hydroxylapatite ceramic bodies with tailored mechanical properties for different applications. J Biomed Mater Res 60:159-166
Rodríguez-Lorenzo LM, Hart J, Gross KA (2003) Influence of fluorine in the synthesis of apatites. Synthesis of solid solutions of hydroxyfluorapatite. Submitted to J Mater Chem
Roy D, Linnehan SK (1974) Hydroxylapatite formed from coral skeletal carbonate by hydrothermal exchange. Nature 247:220-222
Royer A, Viguie JC, Heughebaert M, Heughebaert JC (1993) Stoichiometry of hydroxylapatite: influence on flexural strength. J Mater Sci: Mater in Med 4:76-82
Russell SW, Luptak KA, Suchicital CTA, Alford TL, Pizziconi VB (1996) Chemical and structural evolution of sol-gel derived hydroxylapatite thin films under rapid thermal processing. J Am Ceram Soc 79:837-842
Ruys AJ, Milthorpe BK, Sorell CC (1993) Short-fibre reinforced hydroxylapatite: effects of processing on thermal stability. J Aust Ceram Soc 29:39-51
Sarkar MR, Wachter N, Patka P, Kinzl L (2001) First histological observations on the incorporation of a novel calcium phosphate bone substitute material in human cancellous bone. J Biomed Mater Res 58:329-334
Schildhauer TA, Bauer TW, Josten C, Muhr G (2000) Open reduction and augmentation of internal fixation with an injectable skeletal cement for the treatment of complex calcaneal fractures. J Orthop Trauma 14:309-317
Selvig KA (1970) Periodic lattice images of hydroxylapatite crystals in human bone and dental hard tissues. Calc Tiss Res 6:227-238

Senamaud N, Bernache-Assolant D, Champion E, Heughebaert M, Rey C (1997) Calcination and sintering of hydroxyfluorapatite powders. Solid State Ionics 101-103:1357-1362

Sepulveda P, Binner JGP, Rogero SO, Higa OZ, Bressiani JC (2000) Production of porous hydroxylapatite by the gel-casting of foams and cytotoxic evaluation. J Biomed Mater Res 50:27-34

Shareef MY, Messer PF, van Noort R (1993) Fabrication, characterisation and fracture study of machinable hydroxylapatite ceramic. Biomaterials 14:69-75

Shimoda S, Aoba T, Moreno EC, Miake Y (1990) Effect of solution composition on morphological and structural features of carbonated calcium apatites. J Dent Res 69:1731-1740

Shinohara T, Gyo K, Saiki T, Yanagihara N (2000) Ossiculoplasty using hydroxylapatite prostheses: long term results. Clin Otolaryngol 25:287-292

Shirkhanzadeh M (1995) Calcium phosphate coatings prepared by electrocrystallization from aqueous electrolytes. J Mater Sci: Mater in Med 6:90-93

Silva VV, Lameiras FS, Domingues RZ (2001) Evaluation of stoichiometry of hydroxylapatite powders prepared by coprecipitation method. Key Eng Mater 189-191:79-84

Silver IA, Deas J, Erecinska M (2001) Interactions of bioactive glasses with osteoblasts *in vitro*: Effects of 45S5 Bioglass, 58S and 77S bioactive glasses on metabolism, intracellular ion concentrations and cell viability. Biomaterials 22:175-185

Sivakumar M, Rao KP (2002) Preparation, characterization and in vitro release of gentamicin from coralline hydroxyapatite-gelatin composite microspheres. Biomaterials 23:3175-81

_lósarczyk A, Stobierska E, Paskiewicz Z, Gawlicki M (1996) Calcium phosphate materials prepared from precipitates with various calcium:phosphorus molar ratios. J Am Ceram Soc 79:2539-2544

Smith DK (1994) Calcium phosphate apatites in nature. In: Hydroxyapatite and Related Materials. Brown PW, Constantz B (eds) CRC Press, London, p 29-45

Sousa SR, Barbosa MA (1996) Effect of hydroxylapatite thickness on metal ion release from Ti6Al4V substrates. Biomaterials 17:397-404

Stermer EM, Risnes S, Fischer PM (1996) Trace element analysis of blackish staining on the crowns of human archeological teeth. Eur J Oral Sci 104:253-261

Suchanek W, Yoshimura M (1998) Processing and properties of hydroxylapatite-based biomaterials for use as hard tissue replacement implants [Review]. J Biomed Mater Res 13:94-117

Suchanek W, Yashima M, Kakihana M, Yoshimura M (1997) Hydroxylapatite ceramics with selected sintering additives. Biomaterials 18:923-933

Sugaya A, Minabe M, Tamura T, Hori T (1989) Effects on wound healing of hydroxylapatite-collagen complex implants in periodontal osseous defects in the dog. J Periodontal Res 24:284-288

Sun L, Kucuk A, Berndt CC, Gross KA (2001) Material fundamentals and clinical performance of plasma sprayed hydroxylapatite coatings: A Review. J Appl Biomater Res 58:570-592

Sun L, Berndt CC, Gross KA (2002) Hydroxylapatite-polymer composite flame sprayed coatings for bone bonding. J Biomed Mater Res [In press]

Suwanprateeb J, Tanner KE, Turner S, Bonfield W (1995) Creep in polyethylene and hydroxylapatite reinforced polyethylene composites. J Mater Sci: Mater in Med 6:804-807

Tagai H, Aoki H (1980) Preparation of synthetic hydroxylapatite and sintering of apatite ceramics. In: Mechanical Properties of Biomaterials. GW Hastings, DF Williams (eds) John Wiley and Sons, New York, p 477-488

Takadama H, Kim HM, Kokubo T, Nakamura T (2001a) An X-ray photoelectron spectroscopy study of the process of apatite formation on bioactive titanium metal. J Biomed Mater Res 55:185-193

Takadama H, Kim HM, Kokubo T, Nakamura T (2001b) TEM-EDX study of mechanism of bonelike apatite formation on bioactive titanium metal in simulated body fluid. J Biomed Mater Res 57:441-448

Takagi S, Chow LC (2001) Formation of macropores in calcium phosphate cement implants. J Biomed Mater Res 12:135-139

Takagi S, Chow LC, Ishikawa K (1998) Formation of hydroxylapatite in new calcium phosphate cements. Biomaterials 19:1593-1599

Takagi S, Chow LC, Markovic M, Friedman CD, Costantino PD (2001) Morphological and phase characterizations of retrieved calcium phosphate cement implants. J Biomed Mater Res 58:36-41

Takechi M, Miyamoto Y, Ishikawa K, Nagayama M, Kon M, Asaoka Kenzo, Suzuki K (1998) Effects of added antibiotics on the basic properties of anti-washout-type fast-setting calcium phosphate cement. J Biomed Mater Res 39:308-316

Tanahashi M, Yao T, Kokubo T, Minoda M, Miyamoto T, Nakamura T, Yamamuro T (1994) Apatite coated on organic polymers by biomimetic process—improvement in its adhesion to substrate by NaOH treatment. J Appl Biomater 5:339-347

Tanner KE, Downes RN, Bonfield W (1994) Clinical applications of hydroxylapatite reinforced materials. Brit Cer Trans 93:104-107

Ten Cate AR (1994) Oral Histology: Development, Structure and Function. 4th Edn. Mosby.

Thomas KA (1994) Hydroxylapatite coatings [Review]. Orthop 17:267-278

Thomson RC, Yaszemski MJ, Powers JM, Mikos AG (1998) Hydroxylapatite fiber reinforced poly (alpha-hydroxy ester) foams for bone regeneration. Biomaterials 19:1935-43

Tisdel CL, Goldberg VM, Parr JA, Bensusan JS, Staikoff LS, Stevenson S (1994) The influence of a hydroxylapatite and tricalcium-phosphate coating on bone growth into titanium fiber-metal implants. J Bone Joint Surg 76A:159-171

Tong WD, Chen JY, Li XD, Cao Y, Yang ZJ, Feng JM, Zhang XD (1996) Effect of particle size on molten states of starting powder and degradation of the relevant plasma-sprayed hydroxylapatite coatings. Biomaterials 17:1507-1513

Tonino AJ, Thèrin M, Doyle C (1999) Hydroxylapatite-coated femoral stems. Histology and histomorphometry around five components retrieved at post mortem. J Bone Joint Surg 81B:148-154

Toriyama M, Kawamura S, Shiba S (1987) Bending strength of hydroxylapatite ceramics containing α-tricalcium phosphate. Yogyo-Kyokai-Shi 95:92-94

Toriyama M, Ravaglioli A, Krajewski A, Galassi C, Roncari E, Piancastelli A (1995) Slip casting of mechanochemically synthesized hydroxylapatite. J Mater Sci 30:3216-3221

Toth JM, Hirthe WM, Hubbard WG, Brantley WA, Lynch KL (1991) Determination of the ratio of HA/TCP mixtures by X-ray diffraction. J Appl Biomater 2:37-40

Tsui YC, Doyle C, Clyne TW (1998) Plasma sprayed hydroxylapatite coatings on titanium substrates. Part 1: Mechanical properties and residual stress levels. Biomaterials 19:2015-2029

Tudor AM, Melia CD, Davies MC, Anderson D, Hastings G, Morrey S, Domingos-Sandos J, Barbosa M (1993) The analysis of biomedical hydroxylapatite powders and hydroxylapatite coatings on metallic implants by near-IR Fourier transform Raman spectroscopy. Spectrochim Acta 49A:675-680

Tufekci E, Brantley WA, Mitchell JC, Foreman DW, Georgette FS (1999) Crystallographic characteristics of plasma-sprayed calcium phosphate coatings on Ti-6Al-4V. Intl J Oral Maxillofac Implants 14: 661-672

Uchida M, Kim HM, Kokubo T, Miyaji F, Nakamura T (2001) Bonelike apatite formation induced on zirconia gel in a simulated body fluid and its modified solutions. J Am Ceram Soc 84:2041-2044

Uchida M, Kim HM, Miyaji F, Kokubo T, Nakamura T (2002) Apatite formation on zirconium metal treated with aqueous NaOH. Biomaterials 23:313-317

Ueda R, Imai Y, Motoe A, Uchida K, Aso N (2000) Adhesion of hydroxylapatite layer prepared by thermal plasma spraying to titanium or titanium (IV) oxide substrate. J Ceramic Soc Japan 108:865-868

Ueda K, Oba S, Omiya Y, Okada M (2001) Cranial-bone defects with depression deformity treated with ceramic implants and free-flap transfers. Brit J Plast Surg 54:403-408

Vallet-Regí M, Gutierrezrios MT, Alonso MP, Defrutos MI, Nicolopoulos S (1994) Hydroxylapatite particles synthesized by pyrolysis of an aerosol. J Solid State Chem 112:58-64

Van Dijk K, Schaekien HG, Wolke JC, Maree CH, Habraken FH, Verhoeven J, Jansen JA (1995) Influence of discharge power level on the properties of hydroxylapatite film deposited on Ti6Al4V with RF magnetron sputtering. J Biomed Mater Res 29:269-276

van Landuyt P, Li P, Keustermans JP, Streydio JM and Delannay F (1995) The influence of high sintering temperature on the mechanical properties of hydroxylapatite. J Mater Sci: Mater in Med 6:8-13

Varma HK, Sivakumar R (1996) Dense hydroxylapatite ceramics through gel casting technique. Mater Lett 29:57-61

Vaz L, Lopes AB, Almedia M (1999) Porosity control of hydroxylapatite implants. J Mater Sci: Mater in Med 10:239-242

Vignoles M, Bonel G, Young RA (1987) Occurrence of nitrogeneous species in precipitated B-type carbonated hydroxylapatites. Calcif Tiss Intl 40:64-70

Vogel J, Russel C, Hartmann P, Vizethum JF, Bergner N (1999) Structural changes in plasma sprayed hydroxylapatite. Cfi/Ber. DKG 76:28-32

Wakamura M, Kandori K, Ishikawa T (1997) Influence of chromium (III) on the formation of hydroxylapatite. Polyhedron 16:2047-2053

Walsh D, Tanaka J (2001) Preparation of a bone-like apatite foam cement. J Mater Sci: Mater in Med 12:339-344

Walters MA, Blumenthal NC, Leung Y, Wang Y, Ricci JL, Spivak JM (1991) Molecular structure at the bone-implant interface: A vibrational spectroscopic characterization. Calcif Tissue Intl 48:368-369

Wang BC, Chang E, Yang CY, Tu D, Tsai CH (1993a) Characteristics and osteoconductivity of three different plasma-sprayed hydroxylapatite coatings. Surf and Coatings Technol 58:107-117

Wang BC, Lee TM, Chang E, Yang CY (1993b) The shear strength and the failure mode of plasma-sprayed hydroxylapatite coating to bone: the effect of coating thickness. J Biomed Mater Res 27:1315-1327

Wang BC, Chang E, Lee TM, Yang CY (1995) Changes in phases and crystallinity of plasma-sprayed hydroxylapatite coatings under heat treatment: A quantitative study. J Biomed Mater Res 29: 1483-1492

Wang M, Joseph R, Bonfield W (1998) Hydroxylapatite-polyethylene composites forn bone substitution: effects of ceramic particle size and morphology Biomaterials 19:2357-2366

Wang M, Ladizesky NH, Tanner KE, Ward IM, Bonfield W (2000) Hydrostatically extruded HAPEX™. J Mater Sci 35:1023-1030

Wang M, Yue CY, Chua B (2001) Production and evaluation of hydroxylapatite reinforced polysulfone for tissue replacement. J Mater Sci: Mater in Med 12:821-826

Wang PE, Chaki TK (1993) Sintering behaviour and mechanical properties of hydroxylapatite and dicalcium phosphate. J Mater Sci: Mater in Med 4:150-158

Watanabe S, Nakamura T, Shimizu Y, Hitomi S, Ikada Y (1989) Traumatic sternal segment dislocation in a child. Chest. 96:684-686

Weng J, Liu X, Zhang X, Ji X (1994) Thermal deposition of hydroxylapatite structure induced by titanium and its dioxide. J Mater Sci Lett 13:159-161

Weng J, Liu XG, Li XD, Zhang XD (1995) Intrinsic factors of apatite influencing its amorphization during plasma-spray coating. Biomaterials 16:39-44

Weng J, Liu W, Wolke JGC, Zhang XD, de Groot K (1997) Formation and characteristics of the apatite layer on plasma-sprayed hydroxylapatite coatings in simulated body fluid. Biomaterials 18:1027-1035

Weng W, Baptista JL (1998) Alkoxide route for preparing hydroxylapatite and its coatings. Biomaterials 19:125-131

Wilke A, Orth J, Kraft M, Griss P (1993) Standardized infection model for examining the bony ingrowth dynamics of hydroxylapatite-coated and uncoated pure titanium mesh in the pig femur. Zeitschrift Orthop Grenzgeb 131:370-376

Willmann G (1999) Coating of implants with hydroxylapatite – Material connections between bone and metal. Adv Engin Materials 1:95-105

Winter M, Griss P, de Groot K, Tagai H, Heimke G, van Dijk HJ, Sawai K (1981) Comparative histocompatibility testing of seven calcium phosphate ceramics. Biomaterials 2:159-160

Wittkampf AR (1988) Augmentation of the maxillary alveolar ridge with hydroxylapatite and fibrin glue. J Oral Maxillofac Surg 17:1019-1021

Wolff J, Maquet P, Furlong R (1986) The Law of Bone Remodelling. Springer-Verlag, Berlin.

Wolke JGC, de Blieck-Hogervorst JMA, Dhert WJA, Klein CPAT, de Groot K (1992) Studies on the thermal spraying of apatite bioceramics. J Thermal Spray Technol 1:75-82

Wolke JGC, de Groot K, Jansen JA (1998) Dissolution and adhesion behaviour of radio-frequency magnetron-sputtered Ca-P coatings. J Mater Sci 33:3371-3376

Xu Y, Wang D, Yang L, Tang H (2001) Hydrothermal conversion of coral into hydroxylapatite. Mater Characterisation 47:83-87

Yamada S, Heymann D, Bouler JM, Daculsi G (1997) Osteoclastic resorption of calcium phosphate ceramics with different hydroxylapatite beta-tricalcium phosphate ratios. Biomaterials 18:1037-1041

Yamamoto T, Onga T, Marui T, Mizuno K (2000) Use of hydroxylapatite to fill cavities after excision of benign bone tumours. Clinical results. J Bone Joint Surg 82B:1117-1120

Yamashita K, Yagi T, Umegaki T (1996) Bonelike coatings onto ceramics by reactive magnetron sputtering. J Amer Ceram Soc 79:3313-3316

Yamashita Y, Uchida A, Yamakawa T, Shinto Y, Araki N, Kato K (1998) Treatment of chronic osteomyelitis using calcium hydroxylapatite ceramic implants impregnated with antibiotic. Intl Orthop 22:247-251

Yang CY, Wang BC, Chang E, Wu, JD (1995) The influence of plasma spraying parameters on the characteristics of hydroxylapatite coatings: a quantitative study. J Mater Sci: Mater in Med 6:249-257

Yang YC, Chang E (2001) Influence of residual stress on bonding strength and fracture of plasma-sprayed hydroxylapatite coatings on Ti-6Al-4V substrate. Biomaterials 22:1827-1836

Yang YC, Chang E, Hwang BH, Lee SY (2000) Biaxial residual stress states of plasma-sprayed hydroxylapatite coating on titanium alloy substrate. Biomaterials 21:1327-1337

Yankee SJ, Pletka BJ, Salsbury RL (1991) Quality control of hydroxylapatite coatings: The surface preparation stage. Proc. 4th National Thermal Spray Conference ASM International 475-479

Yong H, Kewei X, Jian L (2001) Dissolution response of hydroxylapatite coatings to residual stresses. J Biomed Mater Res 55:596-602

Yoshikawa M, Toda T (2000) Reconstruction of alveolar bone defect by calcium phosphate compounds. J Biomed Mater Res 53:430-437

Yoshimura M, Suda H (1994) Hydrothermal processing of hydroxylapatite: Past, present and future. *In*: Hydroxylapatite and related materials. Brown PW, Constantz B (eds.) CRC Press, Boca Raton, Florida, p 45-72

Young RA, Holcomb DW (1982) Variability of hydroxylapatite preparations. Calcif Tiss Intl 34:S17-S32

Young RA, Holcomb DW (1984) Role of acid phosphate in hydroxylapatite lattice expansion. Calcif Tiss Intl 26:60-63

Yuan H, Li Y, de Bruijn JD, de Groot K, Zhang X (2000) Tissue responses of calcium phosphate cement: A study in dogs. Biomaterials 21:1283-1290

Zanetti D, Nassif N, Antonelli AR (2001) Surgical repair of bone defects of the ear canal wall with flexible hydroxylapatite sheets: A pilot study. Otol Neurotol 22:745-753

Zaremba CM, Morse DE, Mann S, Hansma PK, Stucky GD (1998) Aragonite-hydroxylapatite conversion in gastropod (abalone) nacre. Chem Mater 10:3813-3824

Zerahn B, Kofoed H, Borgwardt A (2000) Increased bone mineral density adjacent to hydroxy-apatite-coated ankle arthroplasty. Foot & Ankle Intl 21:285-289

Zhang Y, Fu T, Xu K, An H (2001) Wet synthesis and characterization of fluoride-substituted hydroxylapatite. J Biomed Eng 18:173-176

Zhao HC, Li XD, Wang JX, Qu SX, Weng J, Zhang XD (2000) Characterization of peroxide ions in hydroxyapatite lattice. J Biomed Mater Res 52:157-63

Zheng X, Huang M, Ding C (2000) Bond strength of plasma-sprayed hydroxylapatite/Ti composite coatings. Biomaterials 21:841-849

Zhitomirsky I, Galor L (1997) Electrophoretic deposition of hydroxylapatite. J Mater Sci: Mater in Med 8:213-219

Zyman Z, CaO Y, Zhang Y (1994) Periodic crystallization effect in the surface layers of coatings during plasma spraying of hydroxylapatite. Biomaterials 14:1140-1144

18 Phosphates as Nuclear Waste Forms

Rodney C. Ewing[1,2] and LuMin Wang[1]

[1]Department of Nuclear Engineering and Radiological Sciences
[2]Department of Geological Sciences
University of Michigan
Ann Arbor, Michigan 48109

INTRODUCTION

The disposal of radioactive "waste" generated by the nuclear fuel cycle is among the most pressing and potentially costly environmental problems of the 21st century, a heritage from the Atomic Age of the 20th century. Proposed disposal strategies are complicated, not only because of the large volumes and activities of waste, but also because of the political and public-policy issues associated with the long times required for containment and disposal (10^4 to 10^6 years). The development and use of highly durable waste forms, materials that have a high chemical durability and resistance to radiation damage effects, can simplify the disposal strategy (Ewing 2001).

An interest in phosphate-based waste forms has developed because the high-level waste generated by reprocessing of spent nuclear fuel can contain substantial amounts of phosphates (up to 15 wt % P_2O_5) that result from processing technologies that utilized either a bismuth phosphate or tributylphosphate process (Bunker et al. 1995). In addition to the high phosphate content, other metal oxides may achieve significant proportions (up to 15 wt % Fe_2O_3; up to 30 wt % Bi_2O_3; up to 30 wt % UO_2) (Lambert and Kim 1994). These complex compositions have presented special challenges in developing crystalline ceramics that can accommodate the full compositional range of the waste streams.

The early work on phosphate glasses led to the idea that crystalline phosphates might make extremely durable waste forms, particularly for actinides. The earliest suggestion was for the use of monazite (Boatner 1978, Boatner et al. 1980, McCarthy et al. 1978, 1980). The attractive qualities of monazite as a nuclear waste form are: (1) a high solubility for actinides and rare earths (10 to 20 wt %); (2) evidence from natural occurrences of good chemical durability; (3) an apparent resistance to radiation damage, as natural monazites are seldom found in the metamict state, despite very high alpha-decay event doses (Boatner and Sales 1988). There have been extensive studies of monazite and apatite as potential waste form phases, and a considerable amount of work on a number of synthetic phosphate phases has been completed.

The crystal chemistry of monazite, apatite, and related phosphate minerals, has been discussed in detail (see this volume, Chapters 1, 2, and 4 by Hughes and Rakovan, Pan and Fleet, and Boatner, respectively) and will not be repeated here. Rather, we will summarize the work relevant to the consideration of these phosphate phases as nuclear waste forms.

TYPES OF NUCLEAR WASTE

The design and evaluation of nuclear waste forms requires some understanding of the sources, volumes, compositions and activities of the various waste streams generated by the nuclear fuel cycle. There are three primary sources of radioactive waste in the United States (DOE 1997): the high-level waste (HLW) from the reprocessing of spent nuclear fuel, the spent nuclear fuel itself, and plutonium reclaimed by reprocessing or obtained by the dismantlement of nuclear weapons.

1529-6466/00/0048-0018$0.500

High-level waste from reprocessing to reclaim fissile materials for weapons

Approximately 380,000 m^3 (100 million gallons) of HLW with a total radioactivity of 960 million Curies were generated by U.S. weapons programs over a period of nearly 40 years. The greatest volumes (340,000 m^3) are stored in tanks at Hanford, Washington, and Savannah River, South Carolina. Over 99% of the present radioactivity is from non-actinide radionuclides, with half-lives less than 50 years (reprocessing has removed much of the actinide content). After 500 years, the total activity will be substantially reduced, and the primary radionuclides in the inventory will be ^{238}Pu, ^{131}Sm and ^{241}Am. After 50,000 years, most of the activity will be associated with longer-lived radionuclides, such as ^{239}Pu and ^{240}Pu.

Also resulting from reprocessing are the lower-activity wastes that are contaminated with transuranic elements (TRU waste). These wastes are defined as containing 100 nanocuries of alpha-emitting transuranic isotopes (with half-lives >20 years) per gram of waste. Over 60,000 m^3 are stored retrievably at DOE sites and are destined for disposal at the Waste Isolation Pilot Plant in New Mexico (Ahearne 1997). The estimated cost of remediation and restoration programs in the DOE complex during the next few decades is on the order of 200 billion dollars (Crowley 1997).

Used or spent nuclear fuel resulting from commercial power generation

Just over 20% of the electricity generated in the United States is produced by nuclear power plants. In 1995, 32,200 metric tons of spent fuel, with a total activity of 30,200 MCi, was stored by the electric utilities at 70 sites (either in pools or in dry storage systems) (Ahearne 1997, Richardson 1997). By 2020, the projected inventory will be 77,100 metric tons of heavy metal (MTHM) with a total activity of 34,600 MCi. Although the volume of the spent fuel is only a few percent of the volume of HLW, over 95% of the total activity (defense-related plus commercially generated waste) is associated with the commercially generated spent nuclear fuel (Crowley 1997). At present in the United States, none of the spent fuel will be reprocessed; all is destined for direct disposal in a geological repository at Yucca Mountain, Nevada (Hanks et al. 1999).

Plutonium: Reprocessing of spent nuclear fuel and dismantled nuclear weapons

Since the creation of milligram quantities of plutonium by Glenn Seaborg and colleagues in 1941, the global inventory of plutonium has reached over 1,400 metric tons and continues to increase by approximately 70-100 MT/year (Ewing 2001). This commercially generated plutonium is mainly in two forms: (1) still incorporated in spent nuclear fuel destined for direct geological disposal (more than 600 metric tons of plutonium remains in the spent fuel in the U.S.); (2) plutonium separated by reprocessing of commercial fuel. The latter is estimated to have reached 300 metric tons, and this is greater than the amount of plutonium presently in nuclear weapons (Stoll 1998, Oversby et al. 1997). Considering that the bare critical mass for weapons-grade plutonium is 10 kg of metal (this number is substantially reduced in the presence of a neutron reflector), safeguarding this plutonium is essential. In fact, the need for safeguards to protect against the diversion of separated plutonium applies equally to all grades of plutonium (Mark 1993).

Under the first and second Strategic Arms Reduction Treaties, as well as unilateral pledges made by both the United States and Russia, thousands of nuclear weapons will be dismantled. Initially, this will result in between 70 and 100 metric tons of weapons-related plutonium that will require long-term disposition. The selected disposition strategy (stabilization and storage versus permanent disposal) should not only protect the public and the environment, but must also ensure that the plutonium is not readily recoverable for use in weapons (NRC 1994). The initial U.S. strategy called for "burning" the Pu as a mixed-oxide fuel in existing or modified reactors, followed by direct disposal with commercially generated spent fuel in a geological repository (NRC 1995, von Hippel 1998). A smaller portion of the Pu (tens of metric tons) was destined for immobilization in a durable solid followed by geological disposal. At present in the United States, the latter option has been abandoned, and the "scrap" plutonium will be processed and used in the fabrication of a mixed-oxide (MOX) fuel. The present U.S. program has an anticipated cost of 2 billion dollars. However, there are still active research programs in France (using apatite and monazite) and Russia (using zircon, pyrochlore, monazite and apatite) to develop crystalline ceramics for the immobilization and disposition of plutonium.

The interest in phosphate minerals is largely based on their ability to incorporate actinides (Ewing et al. 1995a). In addition to the ^{239}Pu that results from dismantling nuclear weapons and the plutonium separated during the reprocessing of spent nuclear fuel from nuclear power plants, substantial quantities of the "minor" actinides are generated annually: ^{237}Np (3.4 tons/yr), ^{241}Am (2.7 tons/yr), and ^{244}Cm (335 kg/yr). Minor actinides, such as ^{237}Np can have a substantial impact on repository performance because of its long half-life (2.1 Myr). The peaceful use of nuclear energy will inevitably require a strategy for the disposition of actinides, particularly fissile ^{239}Pu.

ROLE OF THE WASTE FORM

The fate of nuclear wastes generated by the nuclear fuel cycle had already been recognized as an important issue by the middle of the Twentieth Century, and at that time, the first proposal for the use of minerals, clays, as nuclear waste forms was made (Hatch 1953). In the 1970s innovative proposals were made by mineralogists, such as the tailored-ceramic by Rustum Roy and colleagues at Penn State and the Rockwell International Science Center (Roy 1975, 1977, 1979; McCarthy 1977) and Synroc, a polyphase, titanite assemblage, proposed by Ted Ringwood and colleagues at the Australian National University and the Australian Nuclear Science and Technology Organisation (Ringwood 1978, 1985; Ringwood et al. 1979, Reeve et al. 1984). During the late 1970s and early 1980s, there were extensive programs in the research and development of nuclear waste forms; in particular, a wide variety of single and polyphase ceramics, as well as new glass compositions, were developed. At this time the first work on a phosphate phase, monazite, as a waste form was completed (Boatner and Sales 1988). Subsequent research activity on alternative waste forms, however, was severely curtailed as a result of the decision in the United States to solidify defense nuclear waste in borosilicate glass and the subsequent construction of the Defense Waste Processing Facility (DWPF) at Savannah River Laboratory in South Carolina. Work on nuclear waste forms during this period was summarized by Lutze and Ewing (1988b).

Recently, there has been a resurgence of interest in crystalline nuclear waste forms due to the need to develop durable materials for the stabilization and disposal of "excess" plutonium that results from dismantling nuclear weapons or accumulates during the reprocessing of commercially generated nuclear fuels (Ewing, in press). This resurgence in interest has resulted in work on minerals, such as apatite, monazite, zirconolite, zircon and pyrochlore, as potential nuclear waste forms (Ewing et al. 1995a, Weber et al. 1998). Donald et al. (1997) and Trocellier (2001) have given recent overviews of some of the recent developments in nuclear waste form research.

Despite the secondary role of waste forms, as determined by total system performance assessments of nuclear waste repositories, there are logical and compelling reasons for emphasizing the importance of the material properties of the waste form (Ewing 1992):

1. The strategy of radionuclide containment and isolation should emphasize near-field containment. This is primarily a function of waste form or "waste package" performance. Strategies that rely solely on long travel times, mineral sorption, and dilution, implicitly presume release and movement of radionuclides into the environment.
2. Assessment of long-term performance of radionuclide containment requires the development of deterministic models of the future physical and chemical behavior of each barrier system. Although such an assessment is challenging, it is certainly easier to model the chemistry and physics of corrosion and alteration of waste forms than it is to develop coupled hydrological, geochemical and geophysical models of the movement of radionuclides through the far-field of a geological repository. Both near-field and far-field assessments are necessary, and both should provide important barrier functions. However, extrapolation of the corrosion behavior of the waste form over long periods can be modeled more convincingly than, as an example, hydrological systems that are site-specific and highly dependent on idealized boundary-conditions (Konikow 1986).
3. Natural phases (minerals and glasses) provide a means for testing the hypothesized long-term behavior of waste-form phases in specific geochemical environments (Ewing 1993). As an example, the mineralogical studies developed as part of the evaluation of radiometric dating techniques (e.g. Willigers et al. 2002) are immediately useful in the discussion of the release of radionuclides from actinide-bearing waste-form phases. Hence, there is a continuing interest in the use of phosphate phases, such as monazite and apatite, as nuclear waste forms.

PHOSPHATE WASTE FORMS: MINERALS

Apatite

Introduction. Apatite has been studied by mineralogists and petrologists because of its use as a geothermometer and as a measure of phosphorous and volatile element fugacities in igneous, metamorphic and hydrothermal processes (Hughes et al. 1990). However, because of its ability to incorporate rare earth elements and actinides (Hughes et al. 1991, Fleet and Pan 1995; Rakovan et al., in press), the apatite structure has also been investigated as a potential nuclear waste form (Weber 1982, Weber and Roberts 1983, Weber et al. 1985, 1991, 1997), as well as a host phase for toxic metals (Dong et al., in press, in preparation). Most recently, there has been considerable interest in the chemical durability of apatite because it may be an important source of calcium in base-poor ecosystems (Blum et al. 2002) and of the role of apatite in removing trace metals, such as Pb and Cd, from solution (E. Valsami-Jones et al. 1998).

Chemistry and structure. Because most nuclear waste glasses are silicates, silicate apatite has been proposed as a potential phase for the immobilization of actinides. There is a complete solid solution between hydroxylapatite, $Ca_{10}P_6O_{24}(OH)_2$, and abukumalite, $Ca_4Y_6Si_6O_{24}(OH)_2$ (Ito 1968). Rare-earth silicates with the apatite structure have been observed or proposed as actinide host phases in a borosilicate nuclear waste glass (Weber et al. 1979), a multiphase ceramic waste form (Turcotte et al. 1982), a glass ceramic waste form (Dé et al. 1976; Donald et al., in press a), a cement waste form (Jantzen and Glasser 1979), and as glass-ceramic for Pu (Zhao et al. 2001; Donald et al., in press b). These apatites generally have the compositions: $Ca_{4-x}REE_{6+x}(SiO_4)_{6-y}(PO_4)_y(F,OH,O)_2$ (where REE = La, Ce, Pr, Nd, Pm, Sm, Eu, and Gd) and are isostructural with apatite, $Ca_{10}(PO_4)_6(F,OH)_2$ (Hughes and Rakovan this volume). Plutonium and other actinides readily substitute for the rare earths in this hexagonal crystal structure. At least 6 wt % actinides (^{244}Cm and ^{240}Pu) have been incorporated into a silicate apatite phase in a devitrified waste glass (Weber et al. 1979), and a phase-pure apatite containing 1.8 wt % actinides (^{244}Cm and ^{240}Pu) has been prepared (Weber 1983). Much higher Pu-compositions are feasible in the rare-earth apatites by substitution for the rare earths. In addition, rare earths, such as Gd, are effective as neutron absorbers. The extensive data from fission track studies (Green et al. 1986, Ritter and Mark 1986, Crowley and Cameron 1987), particularly annealing kinetics, have been used to evaluate the radiation "resistance" of apatite as an actinide-bearing waste form (Weber et al. 1997).

There has been increased interest in britholite and fluorapatite compositions for the incorporation of actinides (Murray et al. 1983, Meis et al. 2000) and fission product elements, such as Cs (Meis et al. 2000; Chartier et al. 2001), Mo (Gaillard et al. 2000, 2001), Rh (Gaillard et al. 2001), and lanthanides (Moncoffre et al. 1998, Marin et al. 1999, Martin et al. 2000), as well as nuclides in the U- and Th-decay chains, such as Ra (Murray et al.1983). Fluorapatite-based ceramics are also proposed for the disposition of radioactively contaminated toxic fluoride salts that result from electro-refining processes used in the reprocessing of Al-matrix nuclear fuels (Lexa 1997). Apatite has been considered as a sorption-medium (e.g., as a back-fill around nuclear waste containers) to retard the mobility of radionuclides or toxic metals (Ohnuki et al 1997). Apatite is also considered as a stable disposal medium for toxic, heavy metals (Eighmy et al. 1998, Crannell et al. 2000; Dong et al., in press, in preparation).

Apatite with appreciable Th- and U-contents has been studied as natural analogues for the performance and properties of actinide-containing apatites over geologic periods. For example, natural apatites with significant amounts of rare earths and Th are reported to be partially metamict as a result of self-irradiation damage from the alpha-decay of Th (Lindberg and Ingram 1964). Fission track analysis in U-containing natural apatites is used for geologic age dating, and annealing studies provide data on fission track recovery kinetics (Koul 1979, Bertagnolli et al. 1983). Of more relevance is the recent discovery of natural apatites formed near the Oklo natural reactors in Gabon that indicate significant amounts of ^{239}Pu and U were incorporated into the apatites, leading to ^{235}U-enrichment (^{235}U is a decay product of ^{239}Pu) that has been retained for 2 Gyr (Bros et al. 1996).

Radiation effects. Apatite (Lindberg and Ingram 1964) and synthetic rare-earth silicate apatite containing actinides (Weber et al. 1979, Turcotte et al. 1982, Weber 1983, 1992, Weber et al. 1991) do undergo an alpha-decay-induced crystalline-to-amorphous transformation (Ewing et al. 1995b). The mechanism (Weber 1993) and temperature dependence (Weber and Wang 1994) of this transformation is well understood and modeled. The macroscopic swelling associated with the amorphization process is 9.5% and has also been modeled as a function of damage accumulation (Weber 1992, 1993). Property changes associated with this alpha-decay-induced transformation have been summarized by Ewing et al. (1995b). Fission track annealing studies of natural apatites (Koul 1979) and ion-beam irradiation studies of natural and synthetic apatites (Weber and Wang 1994, Wang et al. 1994) suggest that simultaneous thermal recovery processes will minimize or even prevent amorphization in Pu-containing apatites under deep borehole conditions ($T > 150$-200°C). In a study of a rare-earth silicate apatite containing ^{244}Cm and ^{240}Pu (Wald and Weber 1984), the dissolution in deionized water at 90°C occurred congruently, and the measured Pu dissolution rate for the undamaged crystalline phase was 0.035 g/m^{2}d (this dissolution rate does not account for the 22 percent porosity of this material). Alpha-decay-induced amorphization increased the average Pu dissolution rate in this apatite by a factor of 12. It is this general increase in release rate of radionuclides with increasing damage that has been the rationale for the extensive studies of radiation damage as a function of temperature in apatite structure types. For a detailed review of the techniques and principles of radiation damage studies, the reader is referred to Ewing et al. (2000).

The susceptibility of phosphate apatite [$Ca_5(PO_4)_3(OH)$] to ionizing irradiation damage has been noted by researchers in a variety of transmission electron microscopy (TEM) studies (Nelson et al. 1982, Bres et al. 1991). In these studies the effects of electron beam damage must be clearly delineated from primary structural features in order to interpret the TEM images. Cameron et al. (1992) have documented the damage evolution caused by electron irradiation in naturally occurring, hexagonal fluorapatite from Durango, Mexico, with *in situ* TEM using a 200 keV electron beam. Modification of the apatite microstructure developed very rapidly under the electron beam. The perfect (undamaged) high-resolution TEM (HRTEM) image shown in Figure 1a could only be obtained in the initial moments (within 30 seconds). Most noticeable in the initial stages of beam damage were discrete, evenly distributed lightly colored areas. Some of the areas were completely facetted with boundaries defined by {110} or {100} forms (Fig. 1b); others do not display the facetted forms. These lightly colored regions were interpreted as areas of lower mass, reflecting displacement by radiolytic processes and migration (or even sublimation) of weakly bound fluorine and perhaps calcium from their structural sites. With continued exposure to the electron beam, these areas coalesced by elongation—commonly in directions parallel to the {110} or {100} planes. Ultimately, the damaged areas formed a complex labyrinth of channels or voids, approximately 5 nm in diameter (Fig. 1c). Concomitant with the formation of this distinctive pattern was the development of Moiré patterns throughout the crystal between the complex of channels (Fig. 1d). These Moiré patterns are the crystallization sites of randomly oriented CaO crystals, which result from the radiation-induced chemical decomposition of the apatite (Wang et al. 2000a). With continued irradiation, some CaO crystals extended into the voids produced during the earlier stages of the damage process. Irradiation experiments have not revealed any significant instability of CaO under further electron irradiation. The microstructural modifications were accompanied by systematic changes in selected area electron diffraction patterns. Notably, the diffraction maxima associated with crystalline apatite decreased in intensity, and powder rings, which match the *d*-spacings of CaO form, increase in intensity as the irradiation dose increases (Fig. 1(e)). No new phosphorus-bearing or fluorine-bearing phases were identified, and no evidence of amorphization was observed by electron diffraction techniques.

Wang et al. (1994) have irradiated both a fluorapatite, $Ca_{10}(PO_4)_6F_2$, and a silicate apatite [$Ca_2La_8(SiO_4)_6O_2$] using 1.5 MeV Kr^+ and studied the damage evolution in the samples by *in situ* TEM observation over a wide temperature range (15-680 K) using the HVEM-Tandem Facility at Argonne National Laboratory. A detailed description of the experimental method can be found in Wang and Ewing (1992) and Wang (1998). Unlike the 200 keV electrons, the high-energy Kr^+ ions can directly displace the target atoms through ballistic interactions and cause displacement cascades through a branching chain of collisions. As a result, solid-state amorphization occurred in both apatite compositions. The critical dose for amorphization, given in a converted unit of displacements per atom (dpa, see Ewing et al. 2000 for a description of the method of calculation), was recorded when the diffraction maxima from the crystalline phases were completely replaced by the diffuse halo in the selected area electron diffraction pattern that is characteristic of amorphous material. The temperature dependences of the critical amorphization doses for the two apatite compositions are shown in Figure 2. At low temperatures, $Ca_{10}(PO_4)_6F_2$ amorphized at lower doses than $Ca_2La_8(SiO_4)_6O_2$. However, the critical amorphization dose increased much more rapidly for the phosphate phase above 350 K than the silicate composition due to a much lower activation energy for the recovery process in the phosphate structure. At 475 K, the critical amorphization dose

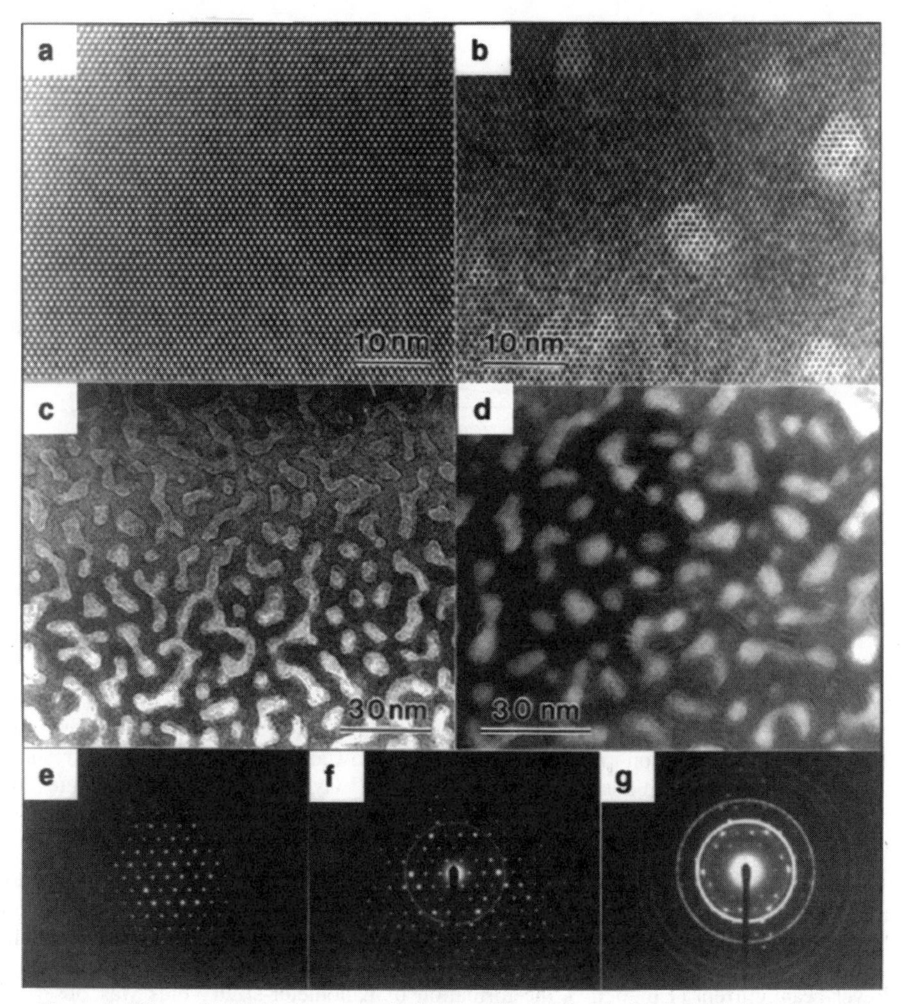

Figure 1. Effects of 200 keV electron irradiation on a fluorapatite, $Ca_5(PO_4)_3F$, (001) single crystal: (a) HRTEM image of undamaged structure; (b) early stage of the electron irradiation damage; (c) after five minutes of exposure to 200 keV electron irradiation showing a labyrinth of voids; (d) after ten minutes of exposure to 200 keV electron irradiation showing the formation of CaO nanocrystals (manifested by the Moiré fringes) in addition to the voids; (e, f, g) selected area electron diffraction patterns showing the progressive formation of CaO crystals in fluorapatite during electron irradiation. All of the rings in (f) and (g) have been indexed exactly with the cubic CaO structure. [Used by permission of Microscopy Society of America, from Cameron et al. (1992) *Proc 50th Annual Meeting Electron Microscopy Society of America* (edited by G.W. Bailey and J.A. Small, San Francisco Press), Figs. 1-5, p 379.]

for the phosphate apatite was five times higher than that for the silicate. The critical amorphization temperature, the temperature above which amorphization does not occur, was ~500 K for the phosphate vs. ~700 K for the silicate apatite. The lower critical amorphization dose at low temperatures and the lower activation energy for recovery of the phosphate apatite were attributed to the weaker P-O bond and the higher fluorine mobility in the phosphate (Wang et al. 1994).

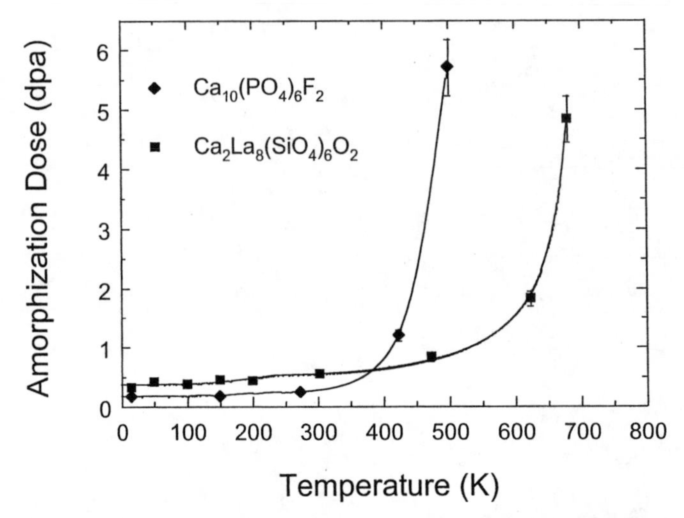

Figure 2. Comparison of temperature dependence of critical amorphization doses of a phosphate apatite [$Ca_{10}(PO_4)_6F_2$] and a silicate apatite [$Ca_2La_8(SiO_4)_6O_2$] under 1.5 MeV Kr^+ irradiation. [Used by permission of CRC Press, Inc., from Wang et al. (1994) in *Hydroxyapatite and Related Materials* (edited by P.W. Brown and B. Constantz), Fig. 5, p 247.]

Utsunomiya et al. (in press) have completed systematic studies of synthetic britholite irradiated by 1.0 MeV Kr^{2+} and 1.5 MeV Xe^+ over the temperature range of 50 to 973 K. The critical temperatures for amorphization were high, between 910 and 1010 K; however, electron irradiation experiments showed that ionizing radiation resulted in recrystallization at an absorbed dose of 6.2×10^{13} Gy. Thus, for the phosphates the competing processes of ballistic interactions with heavy particles and the ionization effects of lighter particles generally increase the "resistance" to radiation damage.

Meldrum et al. (1997c) have studied electron-irradiation-induced phase segregation in crystalline and amorphous phosphate fluorapatite by TEM over a range of electron beam energies and currents. Irradiation of crystalline apatite using a high current density (16 A/cm^2) caused the precipitation of cubic CaO in the crystalline apatite matrix. Using a lower beam current (1.6 A/cm^2), the formation of nanometer-sized voids was observed, but CaO did not crystallize even after prolonged irradiation. Amorphous apatite crystallized to a coarse-grained polycrystalline assemblage of apatite crystallites at 85-200 keV. Increasing the beam current caused the formation of fine-grained cubic CaO, and the crystallization of apatite was not observed, even at high doses. In each case, many beam-induced bubbles formed and were typically larger at the edge of the beam. Thermal annealing at 450°C resulted in epitaxial crystallization of the fluorapatite from the thicker, undamaged portions of the TEM foil, resulting in a single crystal with a high defect density. Electron-beam irradiations at 300°C confirmed that the different microstructures form as a function of variations in the current density and dose-rate. Temperature and dose-rate have competing effects in the precipitation of CaO from the fluorapatite.

Ionizing radiation can also cause radiation-enhanced diffusion that leads to defect annealing. Ouchani et al. (1997) demonstrated in a dual-beam irradiation (220 keV Pb and 0.3 to 3.2 MeV He) that the critical amorphization dose increases due to defect annealing caused by electron energy loss processes. Soulet et al. (2001) recently studied the influence of SiO_4/PO_4 and OH^-/F^- ratios on alpha-annealing efficiency in apatite-like structures by He^+ irradiation. In fluorapatite, the alpha-annealing is a recrystallization recovery process due to the electronic energy loss of alpha-particles emitted by radionuclides. This effect, which can be accompanied by thermal recovery, explains the persistence of the crystalline state of natural calcium phosphate apatites containing long-lived actinide isotopes. Using a TEM with an ion implanter, the main result obtained from this type of study is that annealing of alpha-particle damage is strongly dependent on chemical composition. This effect decreases when the SiO_4/PO_4 ratio increases and when F^- is substituted for OH^-. Chaumont et al. (2002) investigated defect recovery in $Ca_{10}(PO_4)_6F_2$ that has simultaneously been irradiated with He and Pb ions. The Pb-ion irradiation simulates the damage from the alpha-recoil nucleus, and the He-ion irradiation simulates the alpha-particle damage. The alpha-particle dissipates most of its energy by ionization, and this causes enhanced annealing of the recoil damage. Under ambient conditions, alpha-induced annealing is the dominant recovery process and will prevent amorphization.

Summary. Phosphate and silicate apatite offer a number of advantages as nuclear waste forms: (1) a high capacity for the incorporation of actinide elements, as well as selected fission products such as ^{90}Sr; (2) a reasonable chemical durability depending on the geochemical environment; and (3) a propensity for rapid annealing of radiation damage for the phosphate compositions. Considerable work remains to be done, mainly systematic studies, under relevant repository conditions, of the effects of composition on chemical durability.

Monazite

Introduction. Monazite has been proposed as a single-phase ceramic to incorporate a wide variety of nuclear wastes, particularly those rich in actinides (Boatner et al. 1980, Kelly et al. 1981, Davis et al. 1981, Boatner and Sales 1988). The mineral monazite is a mixed lanthanide orthophosphate, $LnPO_4$ (Ln = La, Ce, Nd, Gd, etc.), which often contains significant amounts of Th and U (up to 27 wt % combined). The structure of monazite is monoclinic ($P2_1/n$), and the relatively large LnO_9 coordination polyhedron can easily accommodate actinide elements (Ni et al. 1995, Boatner, this volume). Substitution of tetravalent actinides may be charged balanced by two mechanisms: $(Th,U)^{4+} + Ca^{2+} = 2\,Ln^{3+}$; $(Th,U)^{4+} + Si^{4+} = Ln^{3+} + P^{5+}$ (Van Emden et al. 1997). Because of their high U- and Th-contents, many natural monazites have been subjected to significant alpha-decay event damage over geologic time (some samples are over two billion years of age). In spite of the large radiation doses received by natural monazites, monazite is usually found in the crystalline state (Ewing 1975, Ewing and Haaker 1980). The apparent "resistance" of monazite to radiation-induced amorphization was an important factor in the initial proposal of monazite ceramics as potential candidates for the immobilization of nuclear waste. Further, the chemical durability of monazite has led to its increasing use in U/Pb dating of minerals in both igneous and metamorphic rocks (Parrish 1990, Harrison et al., this volume) and was an important aspect of the proposal of monazite as a waste form for actinides (Boatner 1980, Floran et al. 1981, Poitrasson et al. 2000).

Chemical durability. Leaching studies on synthetic monazite containing 20 wt % simulated Savannah River waste (MCC-1 leach test, 28 days at 90°C) showed release rates of uranium to be on the order of 0.001 g/m^2d (Sales et al. 1983). The leach rate of the host matrix of a synthetic monazite, $LaPO_4$, containing simulated waste remained low even after the material had been transformed to an amorphous state by irradiation with 250 keV Bi^+ ions (Sales et al. 1983). Perhaps, the most interesting aspect of the work by Sales et al. (1983) was the variety of techniques used to measure the leach rate. In addition to the standard MCC-1 leach test, they also used changes in the ionic conductivity of the solution to determine relative cation concentrations as a function of time. In another series of experiments, they used Rutherford backscattering to determine the depth of implanted Bi ions as the leaching experiment progressed. The decrease in depth is a measure of the loss of material at the surface. This latter technique should certainly find applications in measuring the dissolution rates of very durable materials. Using a different leaching experiment on natural monazite, Eyal and Kaufman (1982) showed preferential dissolution of the radionuclide daughter-products, ^{234}U, ^{230}Th, and ^{228}Th, by factors of between 1.1 and 10 relative to the structurally incorporated parent isotopes, ^{238}U and ^{232}Th. This isotopic fractionation is attributed to radiation damage in the tracks of the recoil nuclei emitted during alpha-decay of the parent isotopes. While there have been some concerns regarding these results (Boatner and Sales 1988), the increases in dissolution rates are similar to those observed in other actinide-host phases. In a series of detailed papers, Eyal and Olander (1990a,b,c) demonstrated that incongruent dissolution due to selective leaching of recoil nuclei depended on the degree of radiation damage and the chemistry of the precursors in the decay chains. This same phenomenon, differential release of recoil-nuclei, has been identified in Th-doped borosilicate glass (Eyal and Ewing 1993).

Radiation effects. As with apatite, the increased release rates of radionuclides as a function of radiation damage has lead to rather detailed studies of the behavior of monazite under a variety of irradiation conditions. Karioris et al. (1981) and Cartz et al. (1981) established that natural monazite can be readily transformed to an amorphous state by irradiation with 3 MeV Ar^+ ions at moderate doses. Robinson (1983) simulated the cascades that formed in monazite and found them to be similar in size and shape to simulated cascades in metals. The lack of observed radiation damage effects in monazite is related to the dominance of annealing processes. These early studies lead to detailed studies of damage accumulation as a function of temperature.

In natural monazite that was irradiated to a partially amorphous state (diffraction intensities were reduced by one-third), complete recovery of the fully crystalline state occurred after annealing at 300°C for 20 hours (Meldrum et al 1996). The apparent radiation "stability" of natural monazite is attributed to this relatively low temperature of recovery from radiation damage. Differential scanning calorimetry (DSC) measurements of natural (Gowda 1982) and synthetic (Gowda 1982, Ehlert et al. 1983) monazite irradiated with 3 MeV Ar^+ ions have been performed. In natural monazite, the stored energy released during recrystallization was 33 J/g, the peak in the release rate occurred at 450°C, and the activation energy associated with the recovery process was estimated to be 2.77 eV. For synthetic monazite, $CePO_4$, the stored energy release was 30.5 ± 1.3 J/g, the temperature for maximum release was slightly lower at 370°C, and the activation energy was estimated to be 2.7 ± 0.3 eV. Meldrum et al. (1996) irradiated a natural monazite with 1.5 MeV Kr^+ ions over the temperature range of 30 to 480 K and determined that the critical temperature above which the sample could not be amorphized was 428 K. This is much lower than the critical temperatures of related silicate phases, such as zircon (1100 K) (Meldrum 1997a), but similar to that of other phosphates, such as fluorapatite (475 K) (Wang et al. 1994).

In order to quantify the radiation damage process and the effect of thermal annealing, Meldrum et al. (1997b) conducted a systematic study on displacive-radiation effects in monazite- and zircon-structure orthophosphates. In their study, monazite-structure orthophosphates, including $LaPO_4$, $PrPO_4$, $NdPO_4$, $SmPO_4$, $EuPO_4$, $GdPO_4$, and natural monazite, and their zircon-structure analogs, including $ScPO_4$, YPO_4, $TbPO_4$, $TmPO_4$, $YbPO_4$, and $LuPO_4$, were irradiated by 800 keV Kr^{2+} ions in the temperature range of 20 to 600 K. The critical amorphization dose was determined *in situ* as a function of temperature using selected-area electron diffraction. Amorphization doses were determined to be in the range of 10^{14} to 10^{16} ions/cm^2, depending on the temperature. Materials with the zircon structure amorphized at higher temperatures than those with the monazite structure. The critical amorphization temperature ranged from 350 to 485 K for orthophosphates with the monazite structure and from 480 to 580 K for those with the zircon structure (Fig. 3). As a comparison, natural silicate zircon ($ZrSiO_4$) has been amorphized at over 1000 K (Meldrum et al. 2000). Within each structure type, the critical temperature of amorphization increased with the increasing atomic number of the lanthanide. Structural topology models of radiation "resistance" are consistent with the observed differences between the two structure types (Hobbs et al. 1994), but the topologically based model does not predict the relative amorphization doses for the different compositions of the same structure-type. The ratio of electronic-to-nuclear stopping correlated well with the observed susceptibility to amorphization within each structure type, consistent with previous results that electronic-energy loss enhances defect recombination in the orthophosphates.

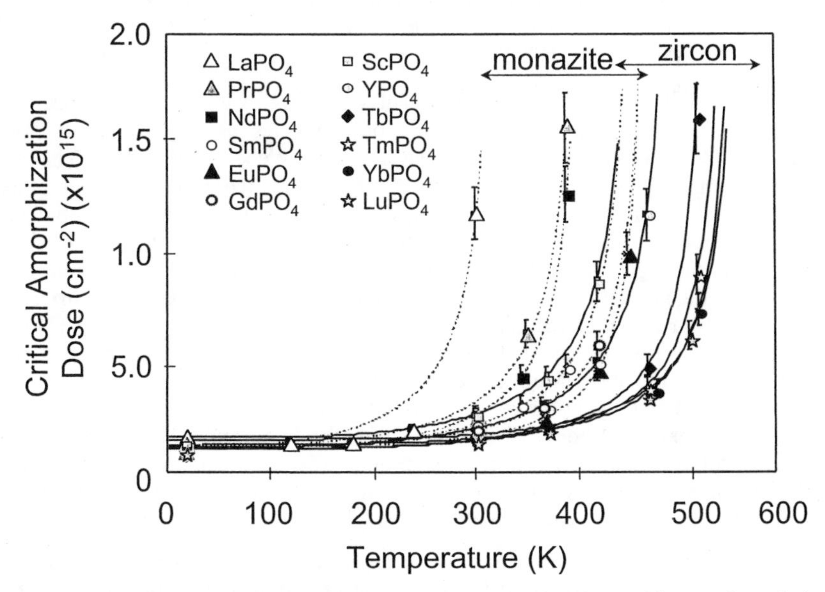

Figure 3. Amorphization dose as a function of temperature for phosphates with monazite and zircon structures under 800 keV Xe^+ irradiation. [Courtesy of A. Meldrum; data based on Meldrum et al. (1997a) *Phys Rev B*, Vol. 56, p 13805-13814.]

Meldrum et al. (2000) also completed a comparative study of radiation effects in crystalline ABO_4-type phosphates versus silicates. The effects of an 800 keV Kr^+ irradiation in the ABO_4-type compounds were compared by performing experiments on four materials ($ZrSiO_4$, monoclinic $ThSiO_4$, $LaPO_4$ and $ScPO_4$) that included both phosphates and silicates with the monazite and zircon structures. Again, radiation damage accumulation was monitored as a function of temperature *in situ* in a TEM. Based on the temperature dependent amorphization dose curves (Fig. 4), the activation energies for recrystallization during irradiation were calculated to be 3.1-3.3 eV for the orthosilicates, but only 1.0-1.5 eV for the isostructural orthophosphates, leading to lower critical amorphization temperatures for the phosphate phases. This result is qualitatively consistent with that observed by Wang et al. (1994) on phosphate and silicate apatite. For the ion-beam-irradiated samples, the critical temperature for amorphization is >700°C

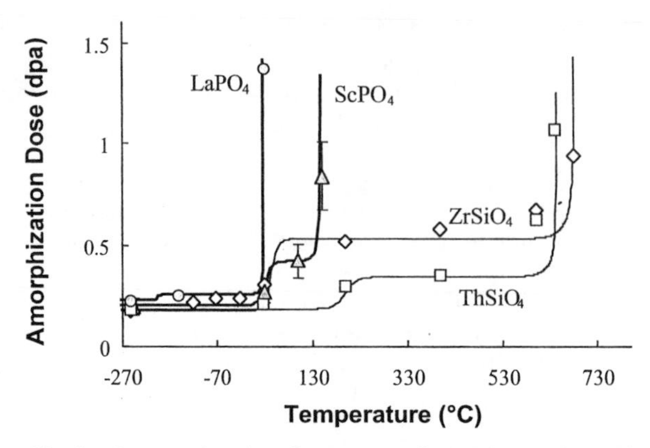

Figure 4. Amorphization dose as a function of temperature for $LaPO_4$, $ScPO_4$, $ZrSiO_4$ and $ThSiO_4$. The lines connecting the data points were calculated using an equation described by Meldrum et al. (2000). [Meldrum et al. (2000) from *Mineralogical Magazine*, Vol. 64, Fig. 2, p 185.]

for $ZrSiO_4$, but it is only 35°C for $LaPO_4$. The data were evaluated with respect to the proposed use of the orthophosphates and orthosilicates as host materials for the stabilization and disposal of high-level nuclear waste (Weber et al. 1997). The results show that zircon with 10 wt % Pu would have to be maintained at temperatures in excess of 300°C in order to prevent complete amorphization. In contrast, a similar analysis for the orthophosphates implies that monazite-based waste forms would not become amorphous or undergo phase decomposition. The large difference in the temperature dependence of the amorphization dose is considered to reflect a fundamental difference in the amorphization and recrystallization kinetics between the phosphates and silicates. This difference was attributed to differences in the structure of the amorphous phases produced by the irradiation. Meldrum et al. (2000) have argued that the PO_4 tetrahedra are probably less readily polymerized than their silicate counterparts, owing to the presence of a double P-O bond. If the SiO_4 tetrahedra in amorphous silicates are indeed more highly polymerized, then a significant amount of bond breaking would be required to form a structure based on isolated SiO_4 tetrahedra (i.e., zircon). Additionally, the rigid PO_4 units may be more easily rotated or realigned during recrystallization.

For both the orthophosphates and orthosilicates, the phases with the zircon structure ($ScPO_4$ and $ZrSiO_4$) have higher critical amorphization temperatures than the phases of a similar composition with the monazite structures. The monazite structure is thus more stable under irradiation at elevated temperatures; however, this difference is small as compared with the chemical effect discussed above. At low temperatures, the measured difference between the monazite- and zircon-structure phases was smaller, close to the experimental error. Nonetheless, the compositions with the monazite structure still required a higher radiation dose for amorphization. Additionally, the effects produced by energetic ions may be slightly different in the monazite structure owing to the lack of ion channeling and linear collision sequences in the lower symmetry structure. This may lead to more compact collision cascades and less defect survival than would be expected to occur in the higher symmetry structure of zircon (e.g., Robinson 1983).

Meldrum et al. (1997b) studied electron-irradiation-induced nucleation and growth of crystallites in amorphous domains of $LaPO_4$, $ScPO_4$, and zircon. In that study, synthetic $LaPO_4$, $ScPO_4$, and crystalline natural zircon ($ZrSiO_4$) from Mud Tanks, Australia, were irradiated by 1.5 MeV Kr^+ until complete amorphization was achieved. The resulting amorphous materials were subsequently irradiated by an 80 to 300 keV electron beam in the transmission electron microscope at temperatures between 130 and 800 K, and the resulting microstructural changes were monitored by *in situ* TEM. Thermal annealing in the range of 500 to 600 K was also conducted to compare the thermally-induced microstructure with that produced by the electron irradiation. Amorphous $LaPO_4$ and $ScPO_4$ annealed to form a randomly oriented polycrystalline assemblage of the same composition as the original material (Fig. 5), but zircon recrystallized to ZrO_2 and amorphous SiO_2 for high beam currents and at high temperatures. The rate of crystallization increased in the order: zircon, $ScPO_4$, $LaPO_4$ (Fig. 6). Submicron tracks of crystallites having a width equal to that of the electron beam could be "drawn" on the amorphous substrate (Fig. 7).

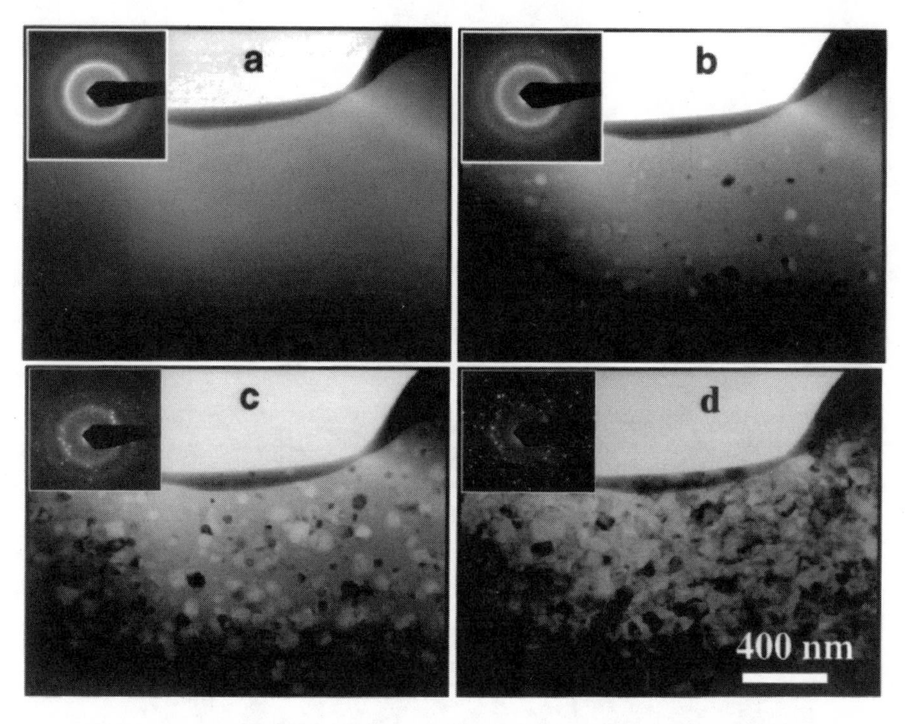

Figure 5. Electron beam (200 keV) induced crystallization sequence in $LaPO_4$ at room temperature using a beam current of 0.3 A/cm^2: (a) unirradiated, (b) 1.5 min, (c) 4 min and (d) 6 min of irradiation. The sample was completely amorphized using a 1.5 MeV Kr^+ irradiation prior to the electron beam irradiation. [Used by permission of the Materials Research Society, from Meldrum et al. (1997b) *Journal of Materials Research*, Vol. 12, Fig. 3, p 1820.]

In contrast, thermal annealing resulted in epitaxial recrystallization from the thick edges of the TEM samples. Electron-irradiation-induced nucleation and growth in these materials were explained by a combination of radiation-enhanced diffusion as a result of ionization processes and a strong thermodynamic driving force for crystallization. The structure of the amorphous orthophosphates was assumed to be less rigid than that of their silicate analogues because of the lower coordination across the PO_4 tetrahedron, and thus a lower energy is required for reorientation and recrystallization. The more highly constrained monazite structure-type recovers at a lower electron dose than the zircon structure-type, consistent with recent topological models used to predict the crystalline-to-amorphous transition as a result of ion irradiation (Hobbs et al. 1996).

Considerable work on the structures of rare-earth monazites (Boatner, Ch. 4, this

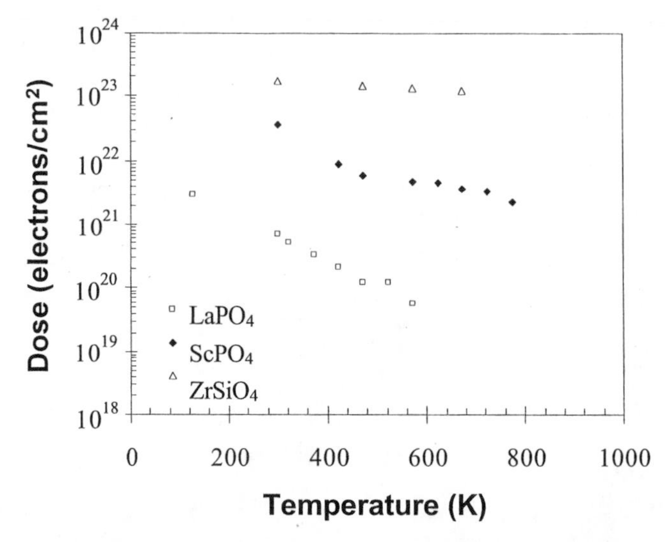

Figure 6. Crystallization dose of 200 keV electrons as a function of temperature for $LaPO_4$ (beam current = 1.45 A/cm^2), $ScPO_4$ (beam current = 1.52 A/cm^2) and $ZrSiO_4$ (beam current = 90 A/cm^2). [Used by permission of the Materials Research Society, from Meldrum et al. (1997b) *Journal of Materials Research*, Vol. 12, Fig. 7, p 1823.]

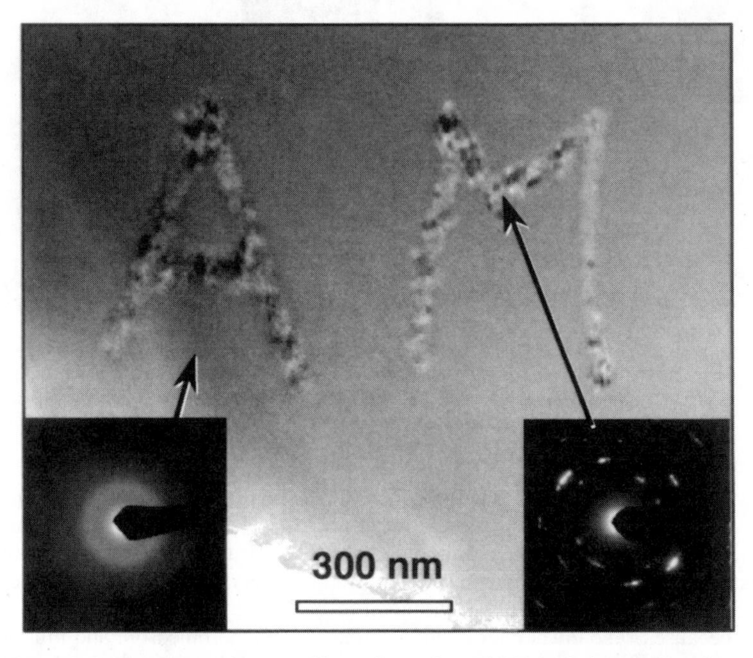

Figure 7. Letters drawn with crystallites using a focused 200 keV electron beam in an amorphous $ScPO_4$ matrix at the room temperature. The temperature rise due to the beam heating was estimated to be no more than 55°C. Crystallization was fast enough that the electron beam could be continuously moved. [Used by permission of the Materials Research Society, from Meldrum et al. (1997b) *Journal of Materials Research*, Vol. 12, Fig. 5, p 1822.]

volume) and their response to irradiation over a wide range of temperatures (Meldrum et al. 1997a) has now been complemented by systematic studies of the thermochemistry of rare-earth phosphates (Ushakov et al. 2001). The enthalpies of formation of the fourteen lanthanide orthophosphates show an almost linear dependence on the radii of the lanthanides; the enthalpies of formation become more negative with increasing ionic radii of the rare-earths. There is no significant discontinuity observed on the change in the structure-type from the monazite to zircon. This trend is consistent with rare-earth monazite solubility in aqueous solutions. These are exactly the types of data that are required to evaluate waste form performance as a function of composition and structure.

Summary. Monazite remains an important candidate material for the immobilization of actinides, as well as certain HLW stream compositions. Monazite exhibits the same tendency for low temperature recovery and recrystallization as apatite, and hence appears to be radiation "resistant" under most proposed repository conditions. Future work will require detailed studies of the effect of composition on chemical durability and radiation "resistance", but one may expect good performance of monazite as compared with other nuclear waste forms. In this area, those charged with developing monazite waste forms could take considerable advantage of the geologic literature on the durability of a variety of monazite compositions (see for example, Podor and Cuney 1997) and recent studies of thermal annealing (Seydoux-Guillaume et al. 2002).

PHOSPHATE WASTE FORMS: SYNTHETIC PHASES

Sodium zirconium phosphate (NZP)

Sodium zirconium phosphate, $NaZr_2P_3O_{12}$, was first proposed by Rustum Roy and colleagues at Penn State as a ceramic nuclear waste form (Roy et al. 1983, Ewing 1988). The main attraction to the use of NZP was that it is a single-phase ceramic that could accommodate the wide variety of radionuclides present in high-level waste streams. The advantages to a single-phase waste form are: (1) elimination of the need to make tedious determinations of radionuclide partition coefficients among phases in a polyphase waste form; (2) elimination of concern for anisotropic mechanical properties (e.g., cracking caused by variations in the coefficients of thermal expansion or anisotropic swelling that accompanies radiation damage); (3) simple models of corrosion; (4) potentially higher levels of waste loading, as compared with the dilute solid-solutions in Synroc; (5) a simple technology for processing and quality control. Recent work on NZP was summarized in detail by Scheetz et al. (1994). A natural analogue is kosnarite, $KZr_2(PO_4)_3$, a late-stage secondary phosphate mineral in pegmatites (Brownfield et al. 1993).

NZP has the structural formula of $[M']_1[M'']_2[A^{VI}]_2[B^{IV}]_3O_{12}$, where M′ and M″ are interstitial sites partially or fully-occupied by Na or other large cations, and the A- and B-sites are occupied by Zr and P, respectively. Extensive coupled substitutions are possible on all three sites. The structure of NZP, described by Hong (1976) and Goodenough et al. (1976), is a trigonal, open, three-dimensional framework with three types of crystallographic sites: (1) distorted octahedral sites, M′ and M″, which can contain large cations, such as Cs and Ca; (2) nearly regular octahedral sites, A, for Zr; (3) tetrahedral sites, B, that can contain P and Al. A remarkable feature of the NZP structure is its ability to accommodate a large number of cations through coupled substitutions. Indeed, nearly two-thirds of the periodic table may be found in NZP or related structure types. Large radionuclides, such as ^{137}Cs or ^{90}Sr, can substitute on the M′ site; coupled substitutions may occur between the M′ and M″ site; heterovalent substitutions on the M″ site can be accommodated by the incorporation of vacancies in order to maintain charge balance. Finally, a variety of coupled substitutions can occur between the A- and

B-sites and the larger M′ and M″ sites. Actinides can be incorporated in compositions such as $KZr_{2-x}U_x(PO_4)_3$, where $0 \leq x \leq 0.20$ (Hawkins et al. 1999). A full tabulation of documented compositions is provided in Scheetz et al. (1994). NZP compositions have been prepared using waste loadings of 10 to 30 wt % of PW-4b (this is a waste composition that simulates a waste stream from a pressurized water reactor, Scheetz et al. 1985, Hawkins et al. 1997) and 20 to 56 wt % of mixed high zirconia and alumina waste from the Idaho Chemical Processing Plant (ICPP) (Scheetz et al. 1985). For the 10 to 20 wt % loading of PW-4b waste, an essentially single-phase waste form was made with only trace amounts of a second phase, monazite. At higher loadings (e.g., 20 wt %), a two-phase ceramic formed consisting of NZP and monazite. For the ICPP wastes, NZP formed, but with an additional fluorite structure-type phase, and finally an apatite phase formed at higher waste loadings. One novel application of NZP is its use as a principal host for Cs in cement composites. NZP composites rank among the best cement-based nuclear waste forms (Scheetz et al. 1994). An analogous waste form phase (NTP) can be made by substituting Ti for Zr (Yang et al. 1984). This was first proposed because TiO_2 is less expensive than ZrO_2. The structure of NTP still preserves the compositional flexibility of NZP with three distinct sites: M′ and M″ (Cs, Ba, REE, actinides), A (Cr, Fe, REE, U, and Zr), B (P, Si, Mo, Te). Up to 60 wt % PW-4b waste compositions have been incorporated into NTP, although at waste loadings up to 43 wt % monazite is present as a second phase. At higher waste loadings, $CsZr_2(PO_4)_3$ is also present.

There are limited leach data on NZP and the closely related NTP for which Ti is substituted for Zr (Scheetz et al. 1994). Typically, the normalized leach rates based on Cs, Na or P release are on the order of 10^{-1} g/m^{2}d (at 100°C). These leach rates are comparable to those of borosilicate glass compositions under static conditions (Lutze 1988). In a study comparing the leach rate of NZP to Synroc-C (based on analysis of 19 elements in the leach solution using the MCC-1 procedure at 90°C), NZP-based waste ceramics were judged to have chemical durabilities within an order of magnitude of the Ti-based Synroc formulation (Zyryanov and Vance 1997). Under different experimental conditions (e.g., MCC-5 tests at 70°C), lower leach rates in the range of 10^{-4} to 10^{-6} g/m^{2}d have been obtained (Sugantha et al. 1998, Buvanesware and Varadaraju 2000). Systematic studies of leach rate as a function of temperature indicate that a substantial decrease in leach rate is attained by higher processing temperatures. Hence, the wide range of reported leach rates is not only a result of different experimental conditions, but also of differences in the samples prepared under different synthesis conditions. One may compare the leach rates of NZP to common minerals, and based on measured dissolution rates at 25°C at pH = 5, NZP crystals (10^{-3} m in diameter) have a calculated lifetime of 20,000 times greater than that of similar sized crystals of quartz (Scheetz et al. 1994). Certainly for short-lived fission products, such as ^{137}Cs and ^{90}Sr, NZP is a durable and attractive host material.

There are limited studies on radiation effects. Szirtes et al. (1999) studied the effect of ionizing radiation on various zirconium phosphates. They irradiated zirconium phosphate (alpha-type), and zirconium phosphate-phosphite, and their silica-containing forms, by gamma-rays (to 3×10^6 Gy). The samples were characterized with XRD before and after the irradiation. Comparison of the X-ray patterns revealed no detectable structural changes, probably due to the relatively low irradiation dose. For comparison, structural changes in zeolites, a phase very sensitive to ionizing irradiation, usually occur only after an absorbed dose greater than 10^9 Gy at room temperature under the electron beam irradiation (Wang et al. 2000b).

During this past decade, there have been a number of studies of NZP by Russian investigators (see summary by Kryukova et al. 1992). Orlova et al. (1994) have investigated $Cs_{1-x}Nd_xZr_{2-x}(PO_4)_3$ over a range of Cs concentrations where x = 0, 0.2, 0.4, 0.6 and 1.0, as well as $NaPu_2(PO_4)_3$ for different ratios of ^{238}Pu:^{239}Pu. The Cs-NZP compositions were irradiated by a ^{60}Co gamma-source to a dose of 3×10^8 Gy with no evidence of structural transformation. Because of the different half-lives of ^{238}Pu and ^{239}Pu (87.7 and 24,100 years, respectively), variations in the isotopic ratio of a Pu-doped phase give different dose rates. The $NaPu_2(PO_4)_3$, which was heat-treated at 1100°C to form the trigonal structure of NZP, became amorphous after a dose of 4.7×10^{15} alpha-decay events/g as determined by powder X-ray diffraction. These experiments are important because they are nearly the only systematic studies of radiation effects in NZP.

Thorium phosphate-diphosphate (TPD)

A variety of diphosphates, such as the mixed valence $U(UO_2)(PO_4)_2$ (Bénard et al. 1994, Brandel et al. 1996) and thorium phosphate-diphosphate (TPD), $Th_4(PO_4)_4P_2O_7$, (Bénard et al. 1996a, Dacheux et al. 1998a) have been investigated as potential hosts for actinide and fission product waste streams. One can expect a wide variety of actinide phosphate structures to be identified, such as $U_2O(PO_4)_2$, (Bénard 1996b) as a result of systematic investigations of the actinide–phosphorous systems (Dacheux et al. 1998b). Dacheux et al. (in press) have already reported the preparation of TPD containing U(IV), Np(IV) and Pu(IV) at concentrations of 47.6, 33.2 and 26.1 wt %, respectively.

Based on preliminary leaching results for TPD (Dacheux et al., in press), normalized dissolution rates are of the order of 10^{-6} g/m^{2}d for neutral solutions. These leaching data are comparable to those for monazite and better than those for apatite. Thomas et al. (2000, 2001) have completed detailed leaching studies of Th-TPD and Th-U-TPD solid solutions as a function of surface area, flow rate, temperature and pH of the solution. For the Th-TPD, remarkably low leach rates of 10^{-5} g/m^{2}d were measured. Th-concentrations in solution were controlled by the formation of a thorium phosphate phase, identified by HRTEM, with a solubility $<10^{-5}$ M in solutions in contact with the Th-TPD. With the substitution of uranium, the Th-U-TPD shows a slight increase in leach rate, 10^{-4} g/m^{2}d. The saturation concentrations in solution for U and Th were controlled by the formation of $(UO_2)_3(PO_4)_2 \cdot 5H_2O$ and $Th_2(PO_4)_2(HPO_4) \cdot H_2O$, respectively.

Pichot et al. (2001) conducted a preliminary study of irradiation effects on thorium phosphate-diphosphate. Powdered samples were irradiated with 1.5 Gy dose of gamma-rays. The formation of PO^{2-}_3 and POO′ free radicals were detected using electron spin resonance (ESR) and thermoluminescence (TL) methods. These free radicals do not modify the macroscopic properties of the TPD and disappear when the sample is heated at 400°C. The implantation of 1.6 MeV He^+ with a fluence of 10^{16} ions/cm^2 and 5 meV Au^{3+} with a fluence 4×10^{15} ions/cm^2 causes some surface damage to sintered samples. Amorphization and chemical decomposition of the matrix were observed for the dose of 10^{15} ions/cm^2 and higher when irradiated with Pb^{2+} (200 keV) and Au^{3+} (5 MeV). These effects were evidenced by means of X-ray diffraction (XRD) and X-ray photoelectron spectroscopy (XPS).

Although the work on TPD is relatively recent, there is already a substantial amount of data on its chemical durability, and leach rates are comparable to those of other phosphate waste forms. One can anticipate important advances in the use of TPD or mixed phase assemblages of TPD-monazite (Clavier et al., in press) in reprocessing spent nuclear fuel, processing HLW, and developing new nuclear waste forms for special applications.

PHOSPHATE GLASSES

Nuclear waste glasses are typically borosilicate glasses, and these glass compositions can experience phase separation at elevated concentrations of P_2O_5 (0.5 to 7 wt %, depending on the composition of the glass); thus, in borosilicate glasses, the maximum P_2O_5 concentrations must be limited to between 1 and 3 wt %. For some waste streams, this can require considerable dilution and a substantial increase in the volume of the waste glass produced. As an example, for the HLW in the tanks at the Hanford site, a 1% limit on the phosphate content can be met by dilution, but over 200,000 glass canisters (approximately 2 tons each) would be generated (Bunker et al. 1995); hence, there has been a continuing interest in developing phosphate glasses as waste forms. Further, typical borosilicate glasses are limited to no more than 5 wt % actinides (e.g., 2 wt % for Pu). In contrast iron phosphate glass with up to 15 wt% P_2O_5 can accommodate up to 40 wt % of simulated HLW or 10 to 20 wt % UO_2 (Day et al. 1998, Mesko and Day 1999).

Phosphate glasses have some advantages over borosilicate glasses, such as a lower melting temperature and higher solubility for problematic elements, such as sulfur, and were investigated as early as the 1960s (Clark et al. 1966, Tuthill et al. 1966). Later work on sodium-aluminum phosphate glass (van Geel 1976) and iron-aluminum phosphate glass (Grambow and Lutze 1980) showed that some of these glasses had comparable or better chemical durability than the borosilicate glasses. Present efforts are focused on the development of iron phosphate glasses (Sales and Boatner 1984, 1988; Day et al. 1998: Reis, in press). The main disadvantage of phosphate glass is that the melts are highly corrosive, and this limits the serviceable life of the melter. Still, a number of the engineering problems were overcome and in the 1980s at Mayak in the Urals, considerable amounts of waste, approximately 1,000 m^3, were immobilized in a phosphate glass (Lutze 1988, National Research Council, 1996). Vitrification using a Na-Al phosphate glass continues today at the Mayak Production Association in Chelyabinsk where 300 million Curies of activity of high-level waste have been immobilized in glass. In the United States, there have been studies to investigate the immobilization of Cs (Reis and Martinelli 1999), CsCl and SrF_2 (Mesko et al. 2000), mixed-waste sludge (Spence et al. 1999) and spent nuclear fuel (Mesko and Day 1999) in iron phosphate glass compositions. Systematic studies of the system K_2O-Na_2O-Fe_2O_3-P_2O_5 are underway to optimize glass compositions for higher waste loadings and increase chemical durability (Fang et al. 2000).

The most detailed work on the chemical durability of iron phosphate glass was done by Boatner and Sales (1984, 1988). They demonstrated that the leach rates (using the MCC-1 test, 90°C) for the iron phosphate glass were a factor of 100 to 1,000 lower for all elements than for a borosilicate glass (frit #131) under the same conditions. Jantzen (1986) was not able to duplicate these results; he found a factor of only 100 improvement over the borosilicate glass (SRL 131). Much of the difference between the results can be attributed to quality of the glass used in the experiments. Also, the corrosion rate for the phosphate glass is much more pH dependent than that of the borosilicate glass; thus, experimental conditions that effect the final pH have an important effect on the measured corrosion rate. The most important parameter in evaluating the long-term durability of a glass is its long-term corrosion rate. For the phosphate glass this is controlled by back-reactions that lead to the formation of phosphate alteration products on the surface of the glass. For the phosphate compounds formed during corrosion, the solubility limits depend on the metal-to-phosphorous ratio of the glass. The corrosion is suppressed by a high metal-to-phosphorous ratio. When this ratio is low, the corrosion products do not form, and the high concentration of phosphoric acid in the leaching solution lowers the pH and accelerates the direct hydrolysis of the P-O-P bonds at the glass surface (Schiewer et al. 1986). The most recent work (Day et al. 1998) has used the standard product consistency test (PCT) and compared the results to the environmental assessment borosilicate glass (standard used by the US DOE). The PCT is used to evaluate the quality of the glass produced, but it is not a measure of the long-term durability of the glass in a geologic repository. Elemental releases from the phosphate glass were generally 20 percent less than the environmental assessment (EA) glass standard under the PCT conditions.

A number of studies of radiation effects on phosphate glasses have been conducted because of the variety of technological applications for phosphate glasses. There are no studies of phosphate glasses that have been doped with actinides. Based on present studies, which mainly emphasize the effect of ionizing radiation on the structure of phosphate glass, one must anticipate important effects associated with ionization caused by the alpha-particle.

Griscom et al. (1998) studied the structure and radiation chemistry of iron phosphate glasses using electron spin resonance (ESR), Mössbauer, and evolved-gas mass spectroscopy methods. Several phosphate glass compositions for the immobilization of plutonium and/or HLW were investigated by (ESR) in order to detect evidence of radiolytic decomposition resulting from gamma doses of 30 MGy. While preliminary results were reported for Defense Waste Processing Facility (DWPF) borosilicate glass compositions and a lanthanum-silicate glass, the work focused primarily on glasses containing 40-75 mol% P_2O_5 and up to 40 mol% Fe_2O_3. Each of the six diverse compositions investigated displayed characteristic ESR signals (not resembling those of the Fe-containing, P-free glasses) comprising combinations of an extremely broad "X resonance" and a narrow "Z resonance," both centered near $g = 2.00$ and both displayed nearly perfect Lorentzian line shapes (peak-to-peak derivative widths similar to 300-600 mT and similar to 30 mT, respectively, at 300 K). The X-resonance intensities in the air-melted glasses varied linearly with Fe:P ratio up to approximately to 0.6. The X and Z ESR signals of the iron phosphate glasses do not resemble any other spectra in the literature except the correspondingly denoted signals in an Fe-free amorphous peroxyborate (APB). The X and Z resonances in the latter were judged to arise from superoxide ions ($O^{-2(-)}$) in the berate network and in a separated Na_2O_2 phase, respectively. An asymmetric Z resonance signal attributable to interstitial O species was a radiation-induced manifestation in a phosphate glass composition: $50P_2O_5$-$20Fe_2O_3$-$23Li_2O$-$7CeO_2$. Irradiated and unirradiated samples of this glass were studied by ESR isochronal annealing and differential thermal analysis, revealing a one-for-one conversion of X to Z resonances upon partial crystallization near 670°C and a Z to X reconversion upon partial remelting near 970°C. To explain these results, the authors suggested that air-melted iron phosphate glasses may contain macroscopic amounts of superoxide ions as an intrinsic chemical feature of their as-quenched structures. A specific four-connected phosphorus-oxygen glass network incorporating O^{-2} ions has been proposed.

Ezz-Eldin (1999) studied radiation effects on selected physical properties of the binary V_2O_5-P_2O_5 glass system over the composition range 50-85 mol % V_2O_5 before and after gamma-irradiation. He found that increasing P_2O_5 causes remarkable changes in the properties studied as shown by the increases in both the softening points and the Vickers microhardness values. The observed variations in the properties may be correlated with the changes in internal glass network that occur with changes in the chemical composition. Vanadium ions may be present in three possible valence states, V^{3+}, V^{4+} and V^{5+}, and the ratio of these states depends on glass composition. Observed decreases in electrical conductivity were assumed to be related to decreases in the mean cross-link densities, an increase in the number of non-bridging oxygens, and/or electron hopping between vanadate ions of different valence states. The changes obtained due to gamma-irradiation are correlated to several factors such as polarization and field strengths of the respective cations and to the number of defect centers created upon gamma-irradiation.

SUMMARY

Crystalline phosphates and phosphate glasses continue to receive attention as potential hosts for the immobilization and disposal of radionuclides, particularly actinides and waste streams with a high phosphorous content. The principal crystalline phases considered are apatite, silicates with the apatite-structure, and monazite. As has been discussed by Lutze and Ewing (1988a), there are a number of factors that have to be considered in selecting a nuclear waste form. The most important are:

(1) chemical durability as a function of waste loading;

(2) radiation damage effects;

(3) the thermal and chemical environment of the disposal site,

(4) the chemical constraints that result from the waste stream composition, and

(5) viable processing technologies that can be operated remotely.

The main advantages of phosphate phases as nuclear waste forms are the high capacity for actinides (up to 20 wt %), a relatively high chemical durability (Donald et al. 1997, Trocellier 2001), and the ability to anneal radiation damage effects at relatively low temperatures (300 to 400°C). Of particular importance for monazite and apatite is the fact that they occur naturally; thus, models that are used to extrapolate the physical and chemical behavior of the waste form over long periods can be confirmed by comparison to natural occurrences.

ACKNOWLEDGMENTS

During the past decade, both authors have benefited greatly from collaborations and discussions with Bill Weber of Pacific Northwest National Laboratory, Al Meldrum of the University of Alberta, Lynn Boatner of Oak Ridge National Laboratory and Werner Lutze of Catholic University. We thank Al Meldrum and John Rakovan for very detailed and useful review comments and suggestions. Our work on radiation effects in ceramics would not have been possible without the sustained funding of the Office of Basic Energy Sciences of the U.S. Department of Energy.

REFERENCES

Ahearne JF (1997) Radioactive waste: The size of the problem. Physics Today 50:24-29

Bénard P, Louër D (1994) $U(UO_2)(PO_4)_2$, a new mixed-valence uranium orthophosphate: *ab initio* structure determination from powder diffraction data and optical and X-ray photoelectron spectra. Chem Mater 6:1049-1058

Bénard P, Brandel V, Dacheux N, Jaulmes S, Launay S, Lindecker C, Genet M, Louër D, Quarton M (1996a) $Th_4(PO_4)_4P_2O_7$, a new thorium phosphate: Synthesis, characterization, and structure determination. Chem Mater 8:181-188

Bénard P, Louër D, Dacheux N, Brandel V, Genet M (1996b) Synthesis, *ab initio* structure determination from powder diffraction, and spectroscopic properties of a new diuranium oxide phosphate. Anales de Química Intl Ed 92:79-87

Bertagnolli E, Keil R, Pahl M (1983) Thermal history and length distribution of fission tracks in apatite: Part 1. Nuclear Tracks 7:163-177

Blum JD, Klaue A, Nezat CA, Driscoll CT, Johnson CE, Siccama TG, Eagar C, Fahey TJ, Likens GE (2002) Mycorrhizal weathering of apatite as an important calcium source in base-poor forest ecosystems. Nature 417:729-731

Boatner LA (1978) Letter to the U.S. Department of Energy, Office of Basic Energy Sciences, Division of Materials Sciences, Washington, DC, 28 April 1978

Boatner LA and Sales BC (1988) Monazite. *In* Radioactive Waste Forms for the Future. Lutze W and Ewing RC (eds) North-Holland, Amsterdam, 495-564

Boatner LA, Beall GW, Abraham MM, Finch CB, Huray PG, Rappaz M (1980) Monazite and other lanthanide orthophosphates as alternate actinide waste forms. *In* Scientific Basis for Nuclear Waste Management, vol. 2, Northrup, CJM Jr (ed) Plenum Press, New York, p 289-296

Brandel V, Dacheux N, Genet M (1996) Reexamination of uranium (IV) phosphate chemistry. J Solid State Chem 121:467-472

Bres EF, Barry JC, Hutchison JL (1984) A structural basis for the curious dissolution of the apatite crystals of human tooth enamel. Ultramicroscopy 12:367-372

Bres EF, Hutchison JL, Senger B, Voegel JC, Frank RM (1991) HREM study of irradiation damage in human dental enamel crystals. Ultramicroscopy 35:305-322

Bros R, Carpena J, Sere V, Beltritti A (1996) Occurrence of Pu and fissiogenic REE in hydrothermal apatites from the fossil nuclear reactor 16 at Oklo (Gabon). Radiochimica Acta 74: 277-282

Brownfield ME, Foord EE, Stuley SJ, Botinelly T (1993) Kosnarite, $KZr_2(PO_4)_3$, a new mineral from Mount Mica and Black Mountain, Oxford County, Maine. Am Mineral 78:653-656

Bunker B, Virden J, Kuhn B, Quinn R (1995) Nuclear materials, radioactive tank wastes. *In* Encyclopedia of Energy Technology and the Environment. John Wiley & Sons, New York, p 2023-2032

Buvaneswari G, Varadaraju UV (2000) Low leachability phosphate lattices for fixation of select metal ions. Mater Res Bull 35:1313-1323

Cameron M, Wang LM, Crowley KD and Ewing RC (1992) HRTEM observation on electron irradiation damage in F-apatite. Proceedings of the 50th Annual Meeting of the Electron Microscopy Society of America. Bailey GW, Small JA (eds) San Francisco Press, San Francisco, California, p 378-379

Cartz L, Karioris FG, Fournelle RA, Gowda KA, Ramasami K, Sarkar G (1981) Metamictization by heavy ion bombardment of alpha-quartz, zircon, monazite and nitride structures. *In* Scientific Basis for Nuclear Waste Management, vol. 3. Moore JG (ed) Plenum Press, New York, p 421-427

Chartier A, Meis C (2001) Computational study of Cs immobilization in the apatites $Ca_{10}(PO_4)_6F_2$, $Ca_4La_6(SiO_4)_6F_2$ and $Ca_2La_8(SiO_4)_6O_2$. Phys Rev B64:085110-1 – 085110-9

Chaumont J, Soulet S, Krupa JC, Carpena J (2002) Competition between disorder creation and annealing in fluoroapatite nuclear waste forms. J Nucl Mater 301:122-128

Clark WE, Godbee HW, Fitzgerald CL (1966) Laboratory development of pot processes for solidification of radioactive wastes. *In* Proc. Symp. Solidification and Long-Term Storage of Highly Radioactive Wastes. Regan WH (ed) The Atomic Energy Commission, CONF-660208, p 95-119

Clavier N, Dacheux N, Terra O, Le Coustumer P, Podor R (in press) Study of TPD-monazite systems as a ceramic for the nuclear waste storage. Proc 10th Intl Ceramics Congress. Vincenzini P (ed) Techna Publishers, Florence, Italy

Crannell BS, Eighmy TT, Krzanowski JE, Eusden JD Jr, Shaw EL, Francis CA (2000) Heavy metal stabilization in municipal solid waste combustion bottom ash using soluble phosphate. Waste Management 20:135-148

Crowley KD (1997) Nuclear waste disposal: The technical challenges. Physics Today 50:32-39

Crowley KD, Cameron M (1987) Annealing of etchable fission-track damage in apatite: effects of anion chemistry. Geol Soc Am Ann Meet, Progr Abstr 19:631-631

Dacheux N, Thomas AC, Brandel V, Genet M (1998a) Investigation of the system ThO_2-NpO_2-P_2O_5. Solid solutions of thorium-neptunium (IV) phosphate-diphosphate. J Nucl Mater 257:108-117

Dacheux N, Podor R, Brandel V, Genet M (1998b) Investigations of systems ThO_2-MO_2-P_2O_5 (M = U, Ce, Zr, Pu). Solid solutions of thorium-uranium(IV) and thorium-plutonium(IV) phosphate-diphosphates. J Nucl Mater 252:179-186

Dacheux N, Chassigneux B, Brandel V, Coustumer PL, Genet M, Cizeron G (in press) Reactive sintering of the thorium phosphate-diphosphate. Study of physical, thermal and thermomechanical properties. Chem Mater

Dacheux N, Clavier N, Le Coustumer P, Podor R (in press) Immobilization of tetravalent actinides in the TPD structure. Proc 10th Intl Ceramics Congress. Vincenzini P (ed) Techna Publishers, Florence, Italy

Davis DD, Vance ER, McCarthy GJ (1981) Crystal chemistry and phase relations in the synthetic minerals of ceramic waste forms. II. Studies of uranium-containing monazites. *In* Scientific Basis for Nuclear Waste Management, vol. 3. Moore JG (ed) Plenum Press, New York, p 197-200

Day DE, Wu Z, Ray CS, Hrma P (1998) Chemically durable iron phosphate glass waste forms. J Non-Crystalline Solids 241:1-12

Dé AK, Luckscheiter B, Lutze W, Malow G, Schiewer E (1976) Development of glass ceramics for the incorporation of fission products. Ceram Bull 55:500-503

Department of Energy (1997) Linking Legacies: Connecting the Cold War Nuclear Weapons Production Processes to Their Environmental Consequences. Office of Environmental Management 0319, Washington, DC

Donald IW, Metcalfe BL, Taylor RNJ (1997) Review: The immobilization of high level radioactive wastes using ceramics and glass. J Mater Sci 32:5851-5887

Donald IW, Metcalfe BL, Greedharee RS (in press, a) A glass-encapsulated ceramic wasteform for the immobilization of chloride-containing ILW: Formation of halite crystals by reaction between the glass encapsulant and ceramic host. *In* Scientific Basis for Nuclear Waste Management XXV, Materials Research Society Proceedings

Donald IW, Metcalfe BL, Scheele RD, Strachan DM (in press, b) A ceramic wasteform for the immobilization of chloride-containing radioactive wastes. Proc 10th Intl Ceramics Congress. Vincenzini P (ed) Techna Publishers, Florence, Italy

Dong Z, White TJ, Wei B, Laursen K (in press) Model apatite systems for the stabilization of toxic metals: I, calcium lead vanadate. J Am Ceram Soc

Dong Z, White TJ (in preparation) Model apatite systems for the stabilization of toxic metals: II, calcium phosphate vanadate. J Am Ceram Soc

Ehlert TC, Gowda KA, Karioris FG, Cartz L (1983) Differential scanning calorimetry of heavy ion bombarded synthetic monazite. Radiation Effects 70:173-181

Eighmy TT, Crannell BS, Krzanowski JE, Butler LG, Cartledge FK, Emery EF, Eusden JD Jr, Shaw EL, Francis CA (1998) Characterization and phosphate stabilization of dusts from the vitrification of MSW combustion residues. Waste Management 18:513-524

Ewing RC (1975) The crystal chemistry of complex niobium and tantalum oxides. IV. The metamict state. Am Mineral 60:728-733

Ewing RC (1988) Novel waste forms. *In* Radioactive Waste Forms for the Future. Lutze, W and Ewing RC (eds) North-Holland, Amsterdam, p 589-633

Ewing RC (1992) Preface—Thematic issue on nuclear waste. J Nucl Mater 190:vii-ix

Ewing RC (1993) The long-term performance of nuclear waste forms: Natural materials—three case studies. *In* Scientific Basis for Nuclear Waste Management XVI. Interrante CG, Pabalan RT (eds) Mater Res Soc Proc 294:559-568

Ewing RC (1999) Nuclear waste forms for actinides. Proc Nat Acad Sci 96:3432-3439

Ewing RC (2001) The design and evaluation of nuclear-waste forms: clues from mineralogy. Can Mineral 39:697-715

Ewing RC (in press) Materials research in nuclear waste management: reflections on twenty-five MRS symposia. *In* Scientific Basis for Nuclear Waste Management XXV, Materials Research Society Proceedings

Ewing RC, Haaker RF (1980) The metamict state: implications for radiation damage in crystalline waste forms. Nucl Chem Waste Management 1:51-57

Ewing RC, Meldrum A, Wang LM, Wang SX (2000) Radiation-induced amorphization. Rev Mineral Geochem 39:319-361

Ewing RC, Weber WJ, Lutze W (1995a) Crystalline ceramics: waste forms for the disposal of weapons plutonium. *In* Disposal of Weapon Plutonium Approaches and Prospects. Merz ER, Walter CE (eds) Kluwer Academic Publishers, Dordrecht, The Netherlands, p 65-83

Ewing RC, Weber WJ, Clinard FW, Jr (1995b) Radiation effects in nuclear waste forms for high-level radioactive waste. Progress Nucl Energy 29:63-112

Eyal Y, Ewing RC (1993) Impact of alpha-recoil damage on dissolution of thoriated glass. *In* Proceedings of the International Conference on Nuclear Waste Management and Environmental Remediation, Vol. 1: Low and Intermediate Level Radioactive Waste Management. Alexander D, Baker R, Kohout R, Marek J (eds) Am Soc Mechanical Engineers, p 191-196

Eyal Y, Kaufman A (1982) Alpha-recoil damage in monazite: preferential dissolution of radiogenic actinide isotopes. Nucl Techn 58:77-83

Eyal Y, Olander DR (1990a) Leaching of uranium and thorium from monazite: I. initial leaching. Geochim Cosmochim Acta 54:1867-1877

Eyal Y, Olander DR (1990b) Leaching of uranium and thorium from monazite: II. elemental leaching. Geochim Cosmochim Acta 54:1879-1887

Eyal Y, Olander DR (1990c) Leaching of uranium and thorium from monazite: III. Leaching of radiogenic daughters. Geochim Cosmochim Acta 54:1889-1896

Ezz-Eldin RM (1999) Radiation effects on some physical and thermal properties of V_2O_5-P_2O_5 glasses. Nucl Instr Meth Phys Res B159:166-175

Fang X, Chandra SR, Marasinghe GK, Day DE (2000) Properties of mixed Na_2O and K_2O iron phosphate glasses. J Non-Crystalline Solids 263 & 264:293-298

Fleet ME, Pan Y (1995) Site preference of rare earth elements in fluorapatite. Am Mineral 80:329-335

Floran RJ, Abraham MM, Boatner LA, Rappaz M (1981) Geologic stability of monazite and its bearing on the immobilization of actinide wastes. *In* Scientific Basis for Nuclear Waste Management, Vol. 3. Moore JG (ed) Plenum Press, New York, p 507-514

Gaillard C, Chevarier N, Millard-Pinard N, Delichère P, Sainsot Ph (2000) Thermal diffusion of molybdenum in apatite. Nucl Instr Meth Phys Res B161-163:646-650

Gaillard C, Chevarier N, Den Auwer C, Millard-Pinard N, Delichère P, Sainsot Ph (2001) Study of mechanisms involved in thermal migration of molybdenum and rhenium in apatites. J Nucl Mater 299:43-52

Goodenough JB, Hong HYP, Kafalas JA (1976) Fast Na^+-ion transport in skeleton structures. Mater Res Bull 11:203-220

Gowda KA (1982) Heavy ion bombardment of zircon, monazite and other crystal structures. PhD dissertation, Marquette University, Milwaukee, Wisconsin

Grambow B, Lutze W (1979) Chemical stability of phosphate glass under hydrothermal conditions. *In* Scientific Basis for Nuclear Waste Management, Vol. 2. Northrup, CJM, Jr (ed) Plenum Press, New York, p 109-116

Green PF, Duddy IR, Gleadow AJW, Tingate PR, Laslett GM (1986) Thermal annealing of fission tracks in apatite: 1. a qualitative description. Chemical Geology 59:237-253

Griscom DL, Merzbacher CI, Bibler NE, Imagawa H, Uchiyama S, Namiki A, Marasinghe GK, Mesko M, Karabullut M (1998) On the structure and radiation chemistry of iron phosphate glasses: new insights from electron spin resonance, Mössbauer, and evolved gas mass spectroscopy. Nucl Instr Meth Phys Res B141:600-615

Hanks TC, Winograd IJ, Anderson RE, Reilly TE and Weeks EP (1999) Yucca Mountain as a radioactive-waste repository. U S Geol Surv Circ 1184

Hatch LP (1953) Ultimate disposal of radioactive wastes. Am Scientist 41:97-13

Hawkins HT, Scheetz BE, Guthrie GD, Jr (1997) Preparation of monophasic (NZP) radiophases: potential host matrices for the immobilization of reprocessed commercial high-level wastes. *In* Scientific Basis for Nuclear Waste Management XX. Gray WJ and Triay IR (eds) Mater Res Soc Proc 465:387-394

Hawkins HT, Spearing DR, Veirs DK, Danis JA, Smith DM, Tait CD, Runde WH (1999) Synthesis and characterization of uranium(IV)-bearing members of the [NZP] structural family. Chem Mater 11: 2851-2857

von Hipple FN (1998) How to simplify the plutonium problem. Nature 394:415-416

Hobbs LW, Clinard FW Jr, Zinkle SJ, Ewing RC (1994) Radiation effects in ceramics. J Nucl Mater 216:291-321

Hobbs LW, Sreeram AN, Jesurum CE, Berger, BA (1996) Structural freedom, topological disorder, and the irradiation-induced amorphization of ceramic structures. Nucl Instr Meth Phys Res B116:18-25

Hong, HYP (1976) Crystal structures and crystal chemistry in the system $Na_{1+x}Zr_2Si_xP_{3-x}O_{12}$. Mater Res Bull 11:173-182

Hughes JM, Cameron M, Crowley KD (1990) Crystal structures of natural ternary apatites: Solid solution in $Ca_5(PO_4)_3X$ (X = F, OH, Cl) system. Am Mineral 75:295-304

Hughes JM, Cameron M, Mariano AN (1991) Rare-earth-element ordering and structural variations in natural rare-earth-bearing apatites. Am Mineral 76:1165-1173

Ito J (1968) Silicate apatites and oxyapatites. Am Mineral 53:890-907

Jantzen CM (1986) Investigation of lead-iron phosphate glass for SRP waste. *In* Nuclear Waste Management. Clark DE, White WB, Machiels AJ (eds) Adv Ceram 20:157-165

Jantzen CM, Glasser FP (1979) Stabilization of nuclear waste constituents in Portland cement. Ceramic Bull 58:459-466

Karioris FG, Gowda KA, Cartz L (1981) Heavy ion bombardment of monoclinic $ThSiO_4$, ThO_2, and monazite. Radiation Effects Lett 58:1-3

Kelly KL, Beall GW, Young JP, Boatner LA (1981) Valence states of actinides in synthetic monazites. *In* Scientific Basis for Nuclear Waste Management, Vol. 3. Moore JG (ed) Plenum Press, New York, p 189-195

Konikow LF (1986) Predictive accuracy of a groundwater model – lessons from a postaudit. Ground Water 24:173-184

Koul, SL (1979) On the fission track dating and annealing behavior of accessory minerals of eastern Ghats (Andhra Pradesh, India). Radiation Effects 40:187-192

Kryukova AI, Kulikov IA, Artem'eva GY (1992) Crystalline phosphates of the $NaZr_2(PO_4)_3$ family: radiation stability. Radiochem 34:82-89.

Lambert SL, Kim DS (1994) Tank Waste Remediation System High-Level Waste Feed Processability Assessment Report. Westinghouse Hanford Company, WHC-SP-1143, UC-811

Lexa D (1997) Development of a substituted-fluorapatite waste form for the disposition of radioactive and toxic fluoride salt materials. Argonne National Laboratory Report ANL-NT-52, 21 p

Lindberg ML, Ingram B (1964) Rare-earth silicate apatite from the Adirondack Mountains, New York. U S Geol Surv Prof Paper 501-B:64-65

Lutze W (1988) Silicate glasses. *In* Radioactive Waste Forms for the Future. Lutze W, Ewing RC (eds) North-Holland, Amsterdam, p 1-159

Lutze W, Ewing RC (1988a) Summary and evaluation of nuclear waste forms. *In* Nuclear Waste Forms for the Future. Lutze W, Ewing RC (eds) North-Holland, Amsterdam, p 699-740

Lutze W, Ewing RC (eds) (1988b) Radioactive Waste Forms for the Future. North-Holland, Amsterdam

Mark JC (1993) Explosive properties of reactor-grade plutonium. Science Global Security 4:111-128

Martin P, Chevarier A, Panczer G (2000) Diffusion under irradiation of rare elements in apatite. J Nucl Mater 278:202-206

McCarthy GJ (1977) High-level waste ceramics: materials considerations, process simulation and product characterization. Nuclear Techn 32:92-105

McCarthy GJ, White WB, Pfoetsch DE (1978) Synthesis of nuclear waste monazites, ideal actinide hosts for geologic disposal. Mater Res Bull 13:1239-1245

McCarthy GJ, Pepin JG, Davis DD (1980) Crystal chemistry and phase relations in the synthetic minerals of ceramic waste forms: I. fluorite and monazite structure phases. *In* Scientific Basis for Nuclear Waste Management, Vol. 2. Northrup CJM Jr (ed) Plenum Press, New York, p 289-296

Meis C, Gale JD, Boyer L, Carpena J, Gosset D (2000) Theoretical study of Pu and Cs incorporation in a mono-silicate neodymium fluoroapatite $Ca_9Nd(SiO_4)(PO_4)_5F_2$. J Phys Chem A 104:5380-5387

Meldrum A, Wang LM, Ewing RC (1996) Ion beam induced amorphization of monazite. Nucl Instr Meth Phys Rev B116:220-224

Meldrum A, Boatner LA, Ewing RC (1997a) Displacive radiation effects in the monazite- and zircon-structure orthophosphates. Phys Rev B 56:13805-13814

Meldrum A, Boatner LA, Ewing RC (1997b) Electron-irradiation-induced nucleation and growth in amorphous $LaPO_4$, $ScPO_4$, and zircon. J Mater Res 12:1816-1827

Meldrum A, Wang LM, Ewing RC (1997c) Electron-irradiation-induced phase segregation in crystalline and amorphous apatite: a TEM study. Am Mineral 82:858-869

Meldrum A, Boatner LA, Ewing RC (2000) A comparison of radiation effects in crystalline ABO_4-type phosphates and silicates. Mineral Mag 64:183-192

Mesko MG, Day DE (1999) Immobilization of spent nuclear fuel in iron phosphate glass. J Nucl Mater 273:27-36

Mesko MG, Day DE, Bunker BC (2000) Immobilization of CsCl and SrF_2 in iron phosphate glass. Waste Management 20:271-278

Moncoffre N, Barbier G, Leblond E, Martin Ph, Jaffrezic H (1998) Diffusion studies using ion beam analysis. Nucl Instr Meth Phys Res B140:402-408

Murray FH, Brown JR, Fyfe WS, Kronberg BI (1983) Immobilization of U-Th-Ra in mine wastes by phosphate mineralization. Can Mineral 21:607-610

National Research Council (1994) Management and Disposition of Excess Weapons Plutonium. National Academy Press, Washington, DC

National Research Council (1995) Management and Disposition of Excess Weapons Plutonium: Reactor-Related Options. National Academy Press, Washington, DC

National Research Council (1996) Glass as a Waste Form and Vitrification Technology: Summary of an International Workshop. RC Ewing (chair) National Academy Press, Washington, DC

Nelson DGA, McLean JD, Sanders JV (1982) High-resolution electron microscopy of electron-irradiation damage in apatite. Rad Effects Lett 68:51-56

Ni Y, Hughes JM, Mariano AN (1995) Crystal chemistry of the monazite and xenotime structures. Am Mineral 80:21-26

Ohnuki T, Kozai N, Isobe H, Murakami T, Yamamoto S, Aoki Y, Naramoto H (1997) Sorption mechanism of europium by apatite using Rutherford backscattering spectroscopy and resonant nuclear reaction analysis. J Nucl Sci Techn 34:58-62

Orlova AI, Volkov YF, Melkaya RF, Masterova LY, Kulikov IA, Alferov VA (1994) Synthesis and radiation stability of NZP phosphates containing f-elements. Radiochem 36:322-325

Ouchani S, Dran J-C, Chaumont J (1997) Nucl Instr Meth Phys Res B132:447-451

Oversby VM, McPheeters CC, Degueldre C, Paratte JM (1997) Control of civilian plutonium inventories using burning in a non-fertile fuel. J Nucl Mater 245:17-26

Parrish RR (1990) U-Pb dating of monazite and its application to geological problems. Can J Earth Sci 27:1431-1449

Pichot E, Dacheux N, Emery J, Chaumont J, Brandel V, Genet M (2001) Preliminary study of irradiation effects on thorium phosphate-diphosphate. J Nucl Mater 289:219-226

Podor R, Cuney M (1997) Experimental study of Th-bearing $LaPO_4$ (780°C, 200 MPa): implications for monazite and actinide orthophosphate stability. Am Mineral 82:765-771

Poitrasson F, Chenery S, Shepherd TJ (2000) Electron microprobe and LA-ICP-MS study of monazite hydrothermal alteration: Implications for U-Th-Pb geochronology and nuclear ceramics. Geochim Cosmochim Acta 64:3283-3297

Rakovan J, Reeder RJ, Elzinga EJ, Cherniak DJ, Tait CD, Morris DE (in press) Structural characterization of U(VI) in apatite by X-ray absorption spectroscopy. Environ Sci Techn

Robinson, MT (1983) Computer simulation of collision cascades in monazite. Phys Rev B 27:5347-5359

Reis ST, Martinelli JR (1999) Cs immobilization by sintered lead iron phosphate glasses. J Non-Crystalline Solids. 247:241-247

Reeve DD, Levins DM, Woolfrey JL, Ramm EJ (1984) Immobilisation of high-level radioactive waste in SYNROC. *In* Advances in Ceramics, Vol. 8. Wicks G, Ross WA (eds) American Ceramic Society, Columbus, Ohio, p 200-208

Richardson JA (1997) United States high-level radioactive waste management programme: Current status and plans. Proc Instn. Mech Engrs 211A:381-392

Ringwood AE (1978) Safe Disposal of High-level Nuclear Reactor Wastes: A New Strategy. Australian National University Press, Canberra, Australia

Ringwood AE (1985) Disposal of high-level nuclear wastes: a geological perspective. Mineral Mag 49: 159-176

Ringwood AE, Kesson, SE, Ware NG, Hibberson W, Major A (1979) Immobilisation of high level nuclear reactor wastes in SYNROC. Nature 278:219-223

Ritter W, Märk TD (1986) Radiation damage and its annealing in apatite. Nucl Instr Meth Phys Res B14:314-322

Robinson MT (1983) Computer simulation of collision cascades in monazite. Phys Rev B 27:5347-5359

Roy R (1975) Ceramic science for nuclear waste fixation. Am Ceram Soc Bull 54:459 (abstr)

Roy R (1977) Rational molecular engineering of ceramic materials. J Am Ceram Soc 60:358-359

Roy R (1979) Science underlying radioactive waste management: status and needs. *In* Scientific Basis for Nuclear Waste Management, Vol. 1. McCarthy GJ (ed) Plenum Press, New York, p 1-20

Roy R, Yang LJ, Alamo J, Vance ER (1983) A single phase ([NZP]) ceramic radioactive waste form. *In* Scientific Basis for Nuclear Waste Management VI. Brookins DG (ed) Mater Res Soc Proc 15:15-21

Sales BC, White CW, Boatner LA (1983) A comparison of the corrosion characteristics of synthetic monazite and borosilicate glass containing simulated nuclear waste glass. Nucl and Chem Waste Management 4:281-289

Sales BC, Boatner LA (1984) Lead-iron phosphate glass: a stable storage medium for high-level nuclear waste. Science 226:45-48

Sales BC, Boatner LA (1988) Lead-iron phosphate glass. *In* Radioactive Waste Forms for the Future. Lutze W, Ewing RC (eds) North-Holland, Amsterdam, 193-231

Scheetz BE, Komarneni S, Jajun W, Yang LJ, Ollinen M, Roy R (1985) Stability of NZP waste forms and their application to ICCP waste. *In* Scientific Basis for Nuclear Waste Management VIII. Jantzen CM, Stone JA, Ewing RC (eds) Mater Res Soc Proc 44:902-910

Scheetz BE, Agrawal DK, Breval E, Roy R (1994) Sodium zirconium phosphate (NZP) as a host structure for nuclear waste immobilization: a review. Waste Management 14:489-505

Schiewer E, Lutze W, Boatner LA, Sales BC (1986) Characterization of lead-iron phosphate nuclear waste glasses. *In* Scientific Basis for Nuclear Waste Management IX. Werme LO (ed) Mater Res Soc Proc 50:231-238

Seydoux-Guiollaume AM, Wirth R, Nasdala L, Gottschalk M, Montel JM, Heinrich W (2002) An XRD, TEM and Raman study of experimentally annealed natural monazite. Phys Chem Minerals 29:240-253

Soulet S, Carpena J, Chaumont J, Kaitasov O, Ruault MO, Krupa JC (2001) Simulation of the alpha-annealing effect in apatitic structures by He-ion irradiation: Influence of the silicate/phosphate ratio and of the OH^{-}/F^{-} substitution. Nucl Instr Meth Phys Res B184:383-390

Spence RD, Gilliam TM, Mattus CH, Mattus AJ (1999) Laboratory stabilization/solidification of surrogate and actual mixed-waste sludge in glass and grout. Waste Management 19:453-465

Stoll W (1998) What are the options for disposition of excess weapons plutonium? Mater Res Soc Bull 23:6-16

Sugantha M, Kumar NRS, Varadaraju UV (1998) Synthesis and leachability studies of NZP and eulytine phases. Waste Management 18:275-279

Szirtes L, Lázár K, Kuzmann E (1999) Effect of ionizing radiation on various zirconium phosphate derivatives. Rad Phys Chem 55:583-587

Thomas AC, Dacheux N, Le Coustumer P, Brandel V, Genet M (2000) Kinetic and thermodynamic study of the thorium phosphate-diphosphate dissolution. J Nucl Mater 281:91-105

Thomas AC, Dacheux N, Le Coustumer P, Brandel V, Genet M (2001) Kinetic and thermodynamic studies of the dissolution of thorium-uranium (IV) phosphate-diphosphate solid solutions. J Nucl Mater 295-264

Trocellier P (2001) Chemical durability of high level nuclear waste forms. Ann Chim Sci Mater 26: 113-130

Turcotte RP, Wald JW, Roberts FP, Rusin JM, Lutze W (1982) Radiation damage in nuclear waste ceramics. J Am Ceram Soc 65:589-593

Tuthill EJ, Weth, GG, Emma LC, Strickland G, Hatch LP (1966) Brookhaven National Laboratory process for the continuous conversion of high-level radioactive waste to phosphate glass. *In* Proc. Symp. Solidification and Long-Term Storage of Highly Radioactive Wastes. Regan WH (ed) The Atomic Energy Commission, CONF-660208, p 139-168

Ushakov SV, Helean KB, Navrotsky A, Boatner LA (2001) Thermochemistry of rare-earth orthophosphates. J Mater Res 16:2623-2633

Utsunomiya S, Wang LM, Yudintsev S, Ewing RC (in press) Ion irradiation experiments of synthetic britholite. J Nucl Mater

Valsami-Jones E, Ragnarsdottir KV, Putnis A, Bosbach D, Kemp AJ, Cressey G (1998) The dissolution of apatite in the presence of aqueous metal cations at pH 2-7. Chem Geol 151:215-233.

Van Emden B, Thornber MR, Graham J, Lincoln FJ (1997) The incorporation of actinides in monazite and xenotime from placer deposits in Western Australia. Can Mineral 35:95-104

Van Geel J, Eschrich H, Heimerl W, Grziwa P (1976) Management of Radioactive Wastes from the Nuclear Fuel Cycle, IAEA, Vienna, 22-26

Wald JW, Weber WJ (1984) Effects of self-radiation damage on the leachability of actinide-host phases. *In* Advances in Ceramics, Vol. 8. Wicks GG, Ross WA (eds) Am Ceram Soc, Columbus, Ohio, p 71-75

Wang LM (1998) Application of advanced transmission electron microscopy techniques in the study of radiation effects in insulators. Nucl Instr Meth Phys Res B 141:312-325

Wang LM and Ewing RC (1992) Ion beam induced amorphization of complex ceramic materials-minerals. Mater Res Soc Bull 17:38-44

Wang LM, Cameron M, Weber WJ, Ewing RC (1994) In situ TEM observation of radiation induced amorphization in crystals with the apatite structure. Brown PW, Constantz B (eds) CRC Press, Ann Arbor, Michigan p 243-249

Wang LM, Wang SX, Ewing RC, Meldrum A, Birtcher RC, Provencio PN, Weber WJ and Matzke Hj (2000a) Irradiation-induced nanostructures. Mater Sci Eng A286:72-80

Wang SX, Wang LM and Ewing RC (2000b) Electron and ion irradiation of zeolites. J Nucl Mater 278:233-241

Weber WJ (1982) Radiation damage in a rare-earth silicate with the apatite structure. J Am Ceram Soc 64:544-548

Weber WJ (1983) Radiation-induced swelling and amorphization of $Ca_2Nd_8(SiO_4)_6O_2$. Rad Effects 77: 295-308

Weber WJ (1992) Radiation-induced amorphization in complex silicates. Nucl Instr Meth Phys Res B65:88-92

Weber WJ (1993) Alpha-decay-induced amorphization in complex silicate structures. J Am Ceram Soc 76: 1729-1738

Weber WJ and Roberts FP (1983) A review of radiation effects in solid nuclear waste forms. Nucl Techn 60:178-198

Weber WJ and Wang LM (1994) Effect of temperature and recoil-energy spectra on irradiation-induced amorphization in $Ca_2La_8(SiO_4O_6)_2$. Nucl Instr Meth Phys Res B91:22-29

Weber WJ, Turcotte RP, Bunnell LR, Roberts FP, Westsik JH (1979) Radiation effects in vitreous and devitrified simulated waste glass. *In* Ceramics in Nuclear Waste Management, Chikalla TD, Mendel JE (eds) CONF-790420, National Technical Information Service, Springfield, Virginia, p 294-299

Weber WJ, Wald JW, Matzke Hj (1985) Self-radiation in actinide host phases of nuclear waste forms. Jantzen CM, Stone JA, Ewing RC (eds) Mater Res Soc Proc 44:679-686

Weber WJ, Eby RK, Ewing RC (1991) Accumulation of structural defects in ion-irradiated $Ca_2Nd_8(SiO_4)O_2$. J Mater Res 6:1334-1345

Weber WJ, Ewing RC, Meldrum A (1997) The kinetics of alpha-decay-induced amorphization in zircon and apatite containing weapons-grade plutonium or other actinides. J Nucl Mater 250:147-155

Weber WJ, Ewing RC, Catlow CRA, Diaz de la Rubia T, Hobbs LW, Kinoshita C, Matzke H, Motta AT, Nastasi M, Salje EKH, Vance ER, Zinkle SJ (1998) Radiation effects in crystalline ceramics for the immobilization of high-level nuclear waste and plutonium. J Mater Res 13:1434-1484

Willigers BJA, Baker JA, Krogstad EJ, Peate DW (2002) Precise and accurate *in situ* Pb-Pb dating of apatite, monazite, and sphene by laser ablation multiple-collector ICP-MS. Geochem Cosmochim Acta 66:1051-1066

Yang LJ, Komarneni S, Roy R (1984) Titanium phosphate (NTP) waste form. *In* Nuclear Waste Management. Wicks GG, Ross WA (eds) Adv Ceramics 8:255-262

Zhao DG, Li LY, Davis LL, Weber WJ, Ewing RC (2001) Gadolinium borosilicate glass-bonded Gd-silicate apatite: A glass-ceramic nuclear waste form for actinides. Hart KP, Lumpkin GR (eds) Mater Res Soc Proc 663:199-206

Zyryanov VN, Vance ER (1997) Sodium zirconium phosphate-structured HLW forms and synroc for high-level nuclear waste immobilization. *In* Scientific Basis for Nuclear Waste Management. Gray WJ, Triay IR (eds) Mater Res Soc Proc 465:409-415

19 Apatite Luminescence

Glenn A. Waychunas

E.O. Lawrence Berkeley National Laboratory
Earth Science Division MS 70-108B
1 Cyclotron Road
Berkeley, California 94720

INTRODUCTION

Apatite group minerals are among the most interesting of luminescent minerals due to their wonderful combinations of varying emission color, complex zoning, and intriguing associations. Several of the most famous mineral localities known for spectacularly fluorescent mineral associations involve apatite group minerals, calcite and ore minerals. For example, Långban, Sweden, with svabite, hedyphane and mimetite; and Franklin, New Jersey, USA, with fluorapatite, svabite and turneaurite (see e.g., Bostwick 1977, Robbins 1994). Apatite is also an extremely common mineral, formed under a wide range of conditions and in many types of host rock. The diversity of its luminescence is created in part by (1) the ability of the apatite structure to incorporate transition metal, REE and anion impurity activators and co-activators, often in combination; (2) the varying types of associations and formation conditions that promote luminescence activity; and (3) the nature of the structure of the apatite host itself. This favorable and flexible host structure has not been lost to commercial enterprises, as apatites have long been used as synthetic phosphors in industrial and consumer products, and more recently as laser matrix materials. Because the luminescence is often associated with rare earth elements (REE), apatite is frequently useful as a REE-indicator mineral. Indeed, apatite can act as a reservoir for REE, making fractionation and isotopic analyses feasible. Analysis of the REE content can also serve as an indicator of growth rate, growth conditions, local chemical mixing and redox conditions. As apatite is readily stimulated to luminescence by an electron beam, cathodoluminescence is frequently reported and contributes to trace contaminant REE analysis, and characterization of chemical zoning.

The thermo-luminescence of apatite has also been studied with an aim to extract details of the defect electronic structure. As bone and teeth materials have hydroxylapatite as their main mineral component, it would seem that thermoluminescence could be used to determine age and thermal history of fossil material. Even in meteorites, where apatite is rare, thermoluminescence has been used to investigate resetting and equilibration temperatures (McKeever 1985). Most detailed studies of the physics of apatite luminescence have been done on particular synthetic compositions, but a number of surveys are also available that describe the spectra of suites of natural samples (Gaft et al. 2001a, Gorobets 1968, 1981; Knutson et al. 1985, Lapraz and Baumer 1983, Kempe and Götze 2002, Marshall 1988, Mitchell et al. 1997, Panczer et al. 1998, Portnov and Gorobets 1969, Reisfeld et al. 1996, Remond et al. 1992, Roeder et al. 1987, Taraschan 1978). In the discussion herein the luminescence results from both natural and synthetic apatites have been combined to best summarize the current state of knowledge. Although the coverage of apatite group mineral compositions considered is not intended to be exhaustive, especially as the literature is heavily biased in favor of fluorapatite compositions, efforts have been made to include information of use to petrologists, geochemists, materials scientists, ceramists, and mineral collectors.

Studies of luminescence in apatite have not been widely used in geochemistry or mineralogy to date, a situation that the author believes is due (paradoxically) to the great

1529-6466/00/0048-0019$05.00

sensitivity of luminescence phenomena, which respond to any slight change in the nature of the valence electrons and their accessible electronic states. This has made spectral interpretation difficult for natural minerals, and requires that most studies utilize models with simplified chemistry. Luminescence processes can also involve several species in a structure and multiple energy transfer processes, so that amplitude information (and thus quantitative concentration or occupation information) can be difficult, if not impossible, to model. However such problems are mitigated by the potential information available on many types of electronic processes, and by the extremely low measurement thresholds afforded by emission spectroscopy techniques, laser excitation, and photon counting detection systems. Further, pulsed and gated spectroscopy techniques are enabling separation of electronic processes as never previously possible, thus opening the door to electronic characterization of natural materials that are heterogeneous on multiple levels (types of sites, surfaces, defects, substituents, traps, etc). Apatite group minerals are a perfect focus for luminescent studies, not only for their potential return on many aspects of geochemical processes, but also for their inherent richness of crystal chemical and petrological diversity.

DEFINITIONS

Types of luminescence

It is crucial to understand the nomenclature used for luminescence investigations, as there are important aspects not found in descriptions of optical spectroscopy, and numerous authors use terminology in somewhat different ways. The general approach here is to use terms as defined by Garlick 1958. *Fluorescence* is here taken to be the same as photoluminescence, i.e., luminescence excited by high energy visible or ultraviolet light, and created by a spin-allowed emission process. *Phosphorescence* is a term applied to spin-forbidden processes that are generally very slow compared to spin-allowed transitions. The term is usually associated with organic systems where the differences between spin-allowed and spin-forbidden transitions are very dramatic. In heavier atoms with more interacting electronic states and additional factors influencing transition probabilities, the distinction is not as clear (Blasse and Grabmaier 1994). *Thermoluminescence* (TL) is the result of the untrapping of excited electrons post-excitation. This is a separate process from phosphorescence although it also produces a delay in emission. A general term for all light emission once excitation has been removed is afterglow, which includes all emission mechanisms. *Cathodoluminescence* (CL) is the stimulation of luminescence via electron excitation, usually with an electron microprobe (EMP) or a scanning electron microscope (SEM), but other devices use low energy plasmas to produce wide-field electron and ion excited luminescence (e.g., Luminoscope). *Radioluminescence* is due to excitation from high energy photons (x-rays and gamma rays) or from high energy particles, with the latter sometimes called Ionoluminescence. *Laser-excited luminescence* is the same as photoluminescence, although the energy density applied to the sample can be orders of magnitude more than with ultraviolet sources. *Pulsed laser-excited time-resolved luminescence* involves extremely short pulses of laser light that excite luminescence whose decay rate can then be measured with high-speed spectroscopy, or alternatively, an entire spectrum can be collected after a fixed delay period or over a particular interval post-excitation. This allows separation of features in the emission spectrum on the basis of excited state lifetimes and energy transfer time (Gaft et al. 2001a,b, 1999, 1998a; Solomonov et al. 1993).

Types of luminescence spectra

The *emission spectrum* for a luminescent system is the spectrum of emitted light for a given excitation energy or type of excitation. A given system may have large variations in the emission spectrum depending on the excitation energy and internal electronic mechanisms. The great majority of luminescence spectra in the mineralogical literature are emission spectra. The *excitation spectrum* for a luminescent system is the spectrum of exciting energy (and especially photon energy) that gives rise to a particular emission feature (i.e., photon emission energy). In general, the excitation spectrum of a luminescent spectrum must be consistent with all other excitation spectra because the final state emission level and other internal energy levels are fixed in energy (see e.g., Blasse and Grabmaier 1994, Henderson and Imbusch 1989). The *absorption spectrum* consists of the energy of radiation that is absorbed by species in the luminescent system. For UV-Vis-IR light it is identical to optical absorption spectroscopy. As with the excitation spectrum the absorption spectrum must be related directly to the spacing of fixed energy levels in the material.

Luminescence terms

A detailed description of luminescence processes is beyond the scope of this article. Readers who wish to delve further into luminescence physics and theory can find much useful information in the extensive and approachable reviews by Blasse (1988), Henderson and Imbusch (1989), Marshall (1988), and Shionoya and Yen (1999). A more mineralogical introduction to the field is given by Waychunas (1989) and references therein. Here, the bare essentials are introduced. *Activators* are substituents or impurities in materials that help to produce or aid in luminescence by acting as absorption or emission centers and often as both. Co-activators or *sensitizers* are activators that operate in pairs or groups to produce luminescence, usually with one activator absorbing energy and transferring it to the other for emission. Sometimes the host lattice can function as the absorber in a luminescent system, in which case we refer to "lattice-sensitized" luminescence. Silicate host lattices often can function as lattice-sensitizers (Blasse and Grabmaier 1994). *Quenchers* are substituents or impurities that absorb excitation energy in some manner, usually dispersing the energy as lattice vibrations, and thereby reducing luminescence activity. *Energy bands* are quasi-continuous fields of *energy levels* that an electron can occupy in a solid. They are created by the action of the Pauli exclusion principle on the atomic energy levels of identical atoms on lattice sites within crystals. Forbidden energy bands or *band gaps* are related energy level fields that cannot be occupied by any of the electrons in the crystal structure, but may have within them levels produced by impurity atoms or defects. Apatite without a large concentration of impurities is an insulator, as it has a large band gap between the (filled) valence and (empty) conduction allowed bands. Most phosphors are insulators, with activators producing energy levels in the band gap.

A second class of semi-conducting phosphors and fluorescent minerals, including sphalerite, behave differently than insulator materials, but are not considered further here. *Trapping states* are energy levels associated with impurity atoms within the band gap. These states can hold onto or trap an excited electron which can be released later via thermal stimulation. This is one of the mechanisms contributing to afterglow, and is the basis for thermoluminescence spectroscopy. The *Stokes shift* is the difference between the absorption energy and the emission energy in a luminescent system. It is usually positive due to energy losses after absorption, and thus the emission is generally at longer wavelengths than the absorption. Energy absorbed that is not released in a luminescent (or *radiative*) emission process, is said to be lost to *non-radiative electronic transitions*. Transitions between energy levels, Stokes shifts and the width of transition bands (peak or line width) are exemplified with a configurational coordinate diagram. Many of the terms defined here are illustrated with configurational coordinate diagrams in Figures 1 and 2.

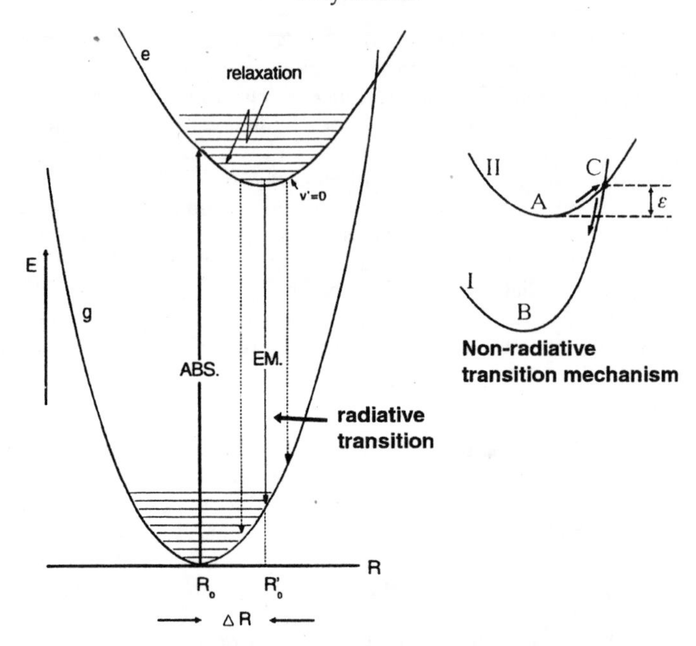

Figure 1. Configurational coordinate diagram. Each electronic state is represented as a harmonic oscillator system, with interatomic distance R on the abscissa, and energy on the ordinate. Vibrational states are represented by horizontal lines. Higher intensity vibrations result in larger R excursions in any given state. An absorption from the lowest energy state, g (for ground), takes place from the lowest vibrational level of the ground state into a higher vibrational level of the lowest excited state, e (for excited). Relaxation occurs by successive drops in the vibrational levels with loss of thermal energy as lattice phonons. Emission takes place as the electron transitions from the lowest vibrational levels of the excited state. The shift in energy between the absorption and emission transition is the Stokes shift. Differences in the curvature of the states are related to the symmetry of the state and several electrostatic factors. Transitions among similar symmetry states generally produce narrow bands, while transitions between different symmetry states give rise to large band widths. This is readily seen in divalent vs. trivalent REE spectra (see Fig. 14). The small diagram shows a non-radiative transition produced by crossing ground and excited states. If the vibrations in the excited state A are large enough to get the system to point C, no photon emission occurs. This is the main reason why increasing temperature quenches luminescence, and why vibrational states affect excited state lifetimes. Modified after Blasse and Grabmaier (1994) and Shionoya and Yen (1999).

LUMINESCENCE METHODS

Photoluminescence, cathodoluminescence and radioluminescence distinctions

Most amateur mineralogists and mineral collectors describe luminescence in minerals via response to ultraviolet (UV) light, often of several wavelengths. This characterization of the photoluminescence remains popular, and most species of luminescent minerals have been described via this method (e.g., Henkel 1989, Robbins 1994, Robbins 1983). Typical wavelengths of UV light from Hg-discharge lamps are 365, 312 and 253.7 nm, often referred to as “long wave” (LWUV), “middle wave” (MWUV) and “short wave” (SWUV), respectively. The actual emission wavelengths vary in the case of LWUV and MWUV as these are actually produced by different combinations of UV phosphors within the lamp envelopes. These wavelengths correspond to presently accepted divisions of UV called UVA, UVB and UVC,

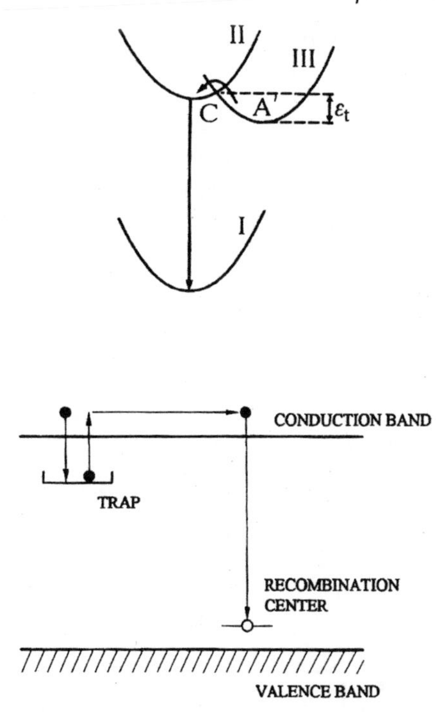

Figure 2. Thermoluminescence depicted from the configurational coordinate model (top) and from the band model (bottom). In the coordinate model the trapped electron occupies an excited state (III) slightly lower in energy than the main excited state (II). Heat can induce vibrations, which move the electron into the (II) state by direct crossover at C. A transition can then occur to the ground state (I), resulting in emission. In the band model heat moves the trapped electron to the conduction band where it can travel throughout the crystal until it recombines with a hole at the bottom of the band (forbidden energy) gap. [Used by permission of the CRC Press, from Shionoya and Yen (1999) Figs. 49 and 50, p. 90.]

respectively. Museums that display luminescent minerals often use one or more of these wavelengths in showcases. UV provides a convenient and relatively effective way of exciting visible luminescence, and the quantitative efficiency of the luminescence process can be readily measured. Laser-stimulated photoluminescence provides significantly higher energy density than Hg-discharge lamps. However, the energy density injected into a luminescent system by a Hg-UV source, and even a UV laser source is much lower than can be achieved by electron beam excitation. In cathodoluminescence (CL) excited with an SEM the electron beam is typically on the order of microns in diameter, penetrates to a few microns in depth, and can deposit up to a few mW of power into this region. A strong UV lamp, by comparison, will produce roughly a few tens of mW per centimeter diameter area with penetration of a few mm or so. Hence the CL excitation inputs some 10^9 times more energy per unit volume. This high energy density results in a cascade of x-ray emission and direct production of Auger electrons, both of which excite luminescence, plus the excitation of electron-hole pairs (Fig. 3). The net effect is greatly enhanced excitation of most luminescent processes including very inefficient ones that would not be visible or quantifiable with UV excitation. Additionally, the spatial resolution of the electron beam allows mapping of luminescent activity, which can be done in concert with chemical analysis by analysis of the x-ray emission. Such work is becoming increasingly important in geochemical studies of trace and minor elements. The only drawbacks to CL are the inability to study excitation spectra, saturation effects that produce a non-linear dependence of emission intensity with beam current, and degradation of the CL intensity with excitation time. Saturation effects can affect the apparent amplitude of peaks in the CL spectrum, and variations in the detector efficiency as a function of emission wavelength can also influence relative peak amplitudes (Barbarand and Pagel 2001) (Fig. 4). X-ray excited luminescence is intensely studied as many uses of phosphors are tied to x-ray measurements, medical fluoroscopy and

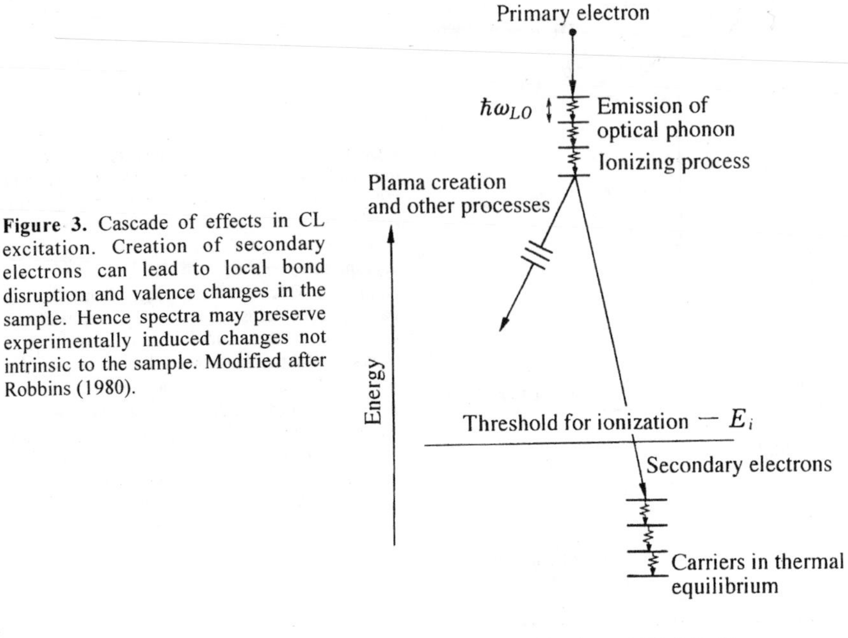

Figure 3. Cascade of effects in CL excitation. Creation of secondary electrons can lead to local bond disruption and valence changes in the sample. Hence spectra may preserve experimentally induced changes not intrinsic to the sample. Modified after Robbins (1980).

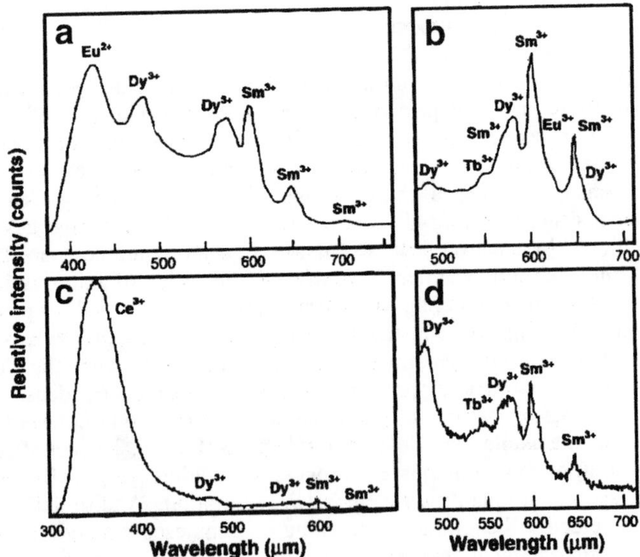

Figure 4. CL emission spectra from Durango Mexico fluorapatite reported in the literature. (a) Roeder et al. (1987). (b) Murray and Oreskes (1997). (c) and (d) Barbarand and Pagel (2001). (c) demonstrates the importance of having a detector that is sensitive to the shorter wavelength part of the spectrum. Modified after Barbarand and Pagel (2001).

radiology. However, x-ray sources suitable for the analysis of radioluminescence in mineral samples are mainly limited to synchrotrons, and hence this has been little explored. Particle-excited luminescence, generally as a byproduct of PIXE analysis, is performed only with proton and ion accelerators (Karali et al. 2001). The energy input into the sample in this case can be even higher than the CL case, and there is frequently massive local structural damage and electron transfer that can complicate spectral interpretation.

Emission, absorption and excitation spectroscopy

Luminescence emission spectroscopy is the spectral analysis of emitted photoluminescence, cathodoluminescence or other process. It has been described briefly in Waychunas (1989) and Marfunin (1979), and in considerable detail by Henderson and Imbusch (1989). This process is the modern day analog of Newton's analysis of sunlight with a wavelength-dispersing prism, and has been around for a century. Time-resolved emission spectroscopy is a relatively new technique usually paired with pulse laser excitation. If the laser pulse is brief enough (typically a few nanoseconds), the excitation is considerably shorter than most fluorescence excited state lifetimes, offering the possibility of separating emission from different electronic states. In practice, this technique is applied by the synchronous use of laser pulses and gated detectors. One timed signal will activate the laser, while the next, after a predetermined wait, will turn on the detector. Another signal will turn off the detector after a set dwell period. In this way, over many thousands of pulses, a complete emission spectrum can be acquired from electronic states with certain lifetimes. This is of particular value when trying to sort out the emission of activators whose emission spectra heavily overlap, but differ in lifetime (Gaft et al. 1998a).

Classic absorption spectroscopy has been described in detail by Burns (1993), Rossman (1988) and many others. The key aspect that relates to luminescence is that the absorption spectrum indicates the energy level separations, i.e., the electronic structure, of the absorbing species. Once these are known, comparison with the excitation spectrum can indicate type of activator and type of sensitizer. Host lattice sensitization can also be determined in this way.

Excitation spectroscopy requires instruments with monochromators both on the incident and emission side of the sample. The emission monochromator is set to a fixed energy consistent with an emission peak, and the excitation monochromator is scanned. This produces a spectrum of all absorption states that ultimately give rise to the particular emission peak.

Apatite occurrences

Apatite is commonly found in sedimentary, metamorphic, igneous and hydrothermal rocks, and its luminescent activity and emission spectrum seem to differ with paragenesis. For example, Mariano (1988) reported that apatite found in granitic rocks and pegmatites frequently has a strong yellow to yellow-orange fluorescence and CL, while CL from apatites in carbonatites is usually blue, and that from apatites in peralkaline syenite is pink-violet. John Hanchar (pers. comm.) also reports that granitic apatites emit a bright creamy yellow CL, based on 20-25 observations. Somewhat different variations can also be seen in the emission spectra (excited by UV light) collected by Gorobets (1981) reproduced in Figure 5. These spectra are dominated by a blue component created by Ce^{3+} and Eu^{2+}, and an orange component created mainly by Mn^{2+}. Roughly equal amounts of blue and orange emission lead to a pink luminescence, and hence the dominant emission colors range from orange to pink to blue. Notable occurrences of pegmatitic apatite with strong orange-yellow luminescence are those of the granite quarries of New Hampshire and Connecticut, the quarries and mines near Spruce Pine in North Carolina, and the famous Harding mine near Taos, New Mexico. Outside the USA, brightly yellow fluorescing apatite occurs in the pegmatites of the

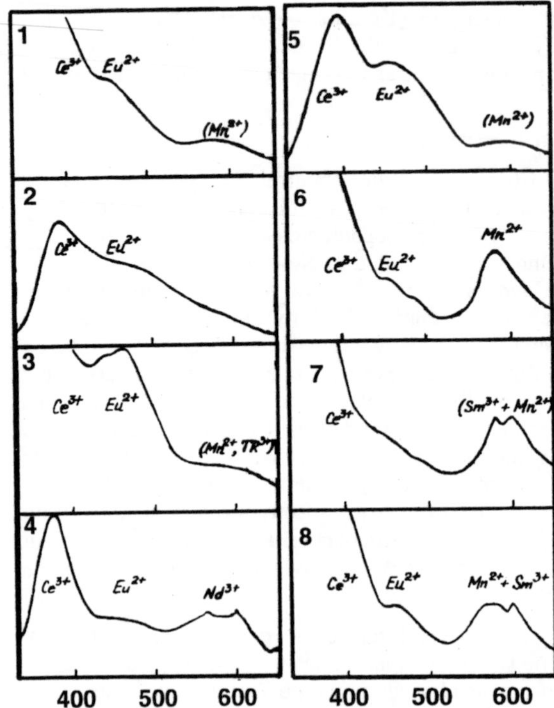

Figure 5. UV-excited emission spectra for apatite from different geological environments.

(Modified after Gorobets 1981.)

1-kimberlite.

2-carbonatite.

3-urtite (Khibiny).

4-nepheline syenite (Lovozero)

5-gabbro

6-diorite

7-granodiorite

8-granodiorite.

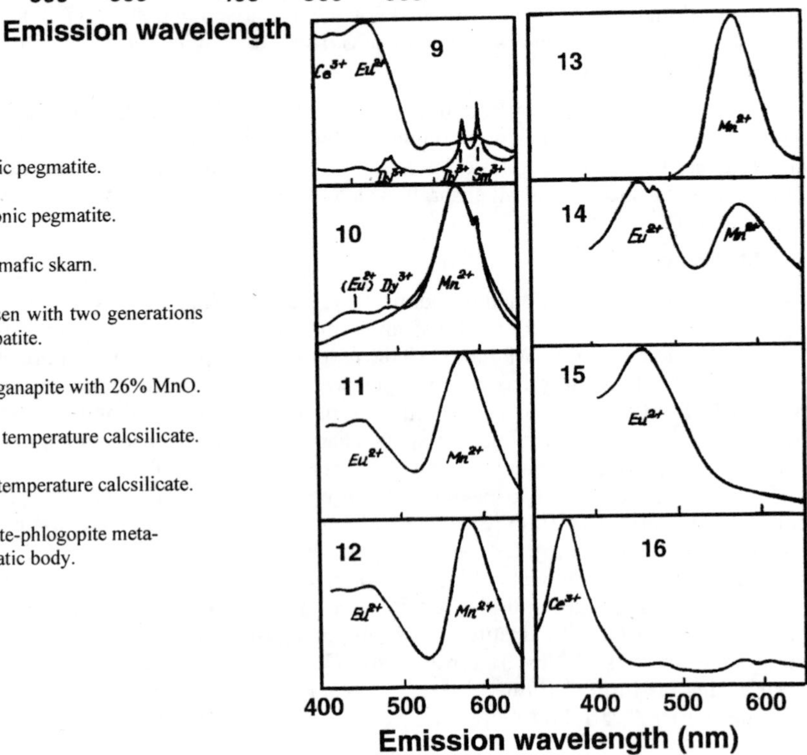

9-granitic pegmatite.

10-plutonic pegmatite.

11-ultramafic skarn.

12-greisen with two generations of apatite.

13-manganapite with 26% MnO.

14-high temperature calcsilicate.

15-low temperature calcsilicate.

16-calcite-phlogopite metasomatic body.

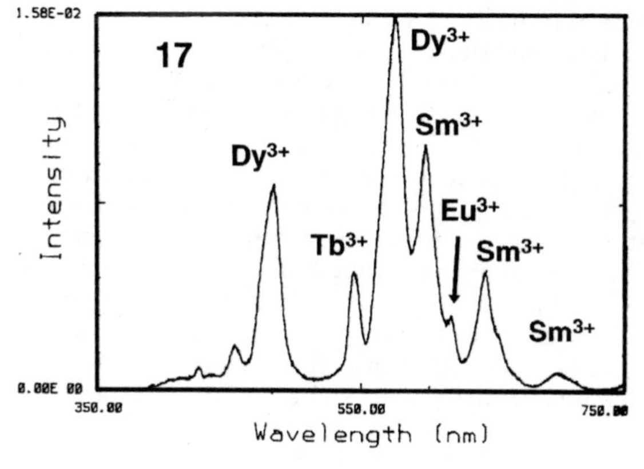

Figure 5, continued.

Cathodoluminescence.

17-supergene apatite from crusts on a botryooidal laterite derived from a carbonatite, Mt. Weld Western Australia.

Modified after Mariano (1989).

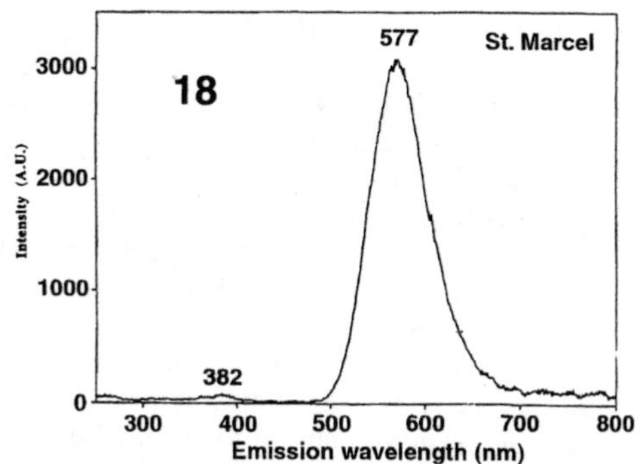

18-arsenic-containing fluorapatite from St. Marcel, Italy.

Modified after Perseil et al. (2000).

Minas Gerais region of Brazil, and in the gem pegmatites of the mountainous regions of Pakistan and Afghanistan (Robbins 1994). Occasionally, pockets in the pegmatites allow for the development of spectacular euhedral apatite crystals from colorless to light pink in color and with spectacular bright yellow photoluminescence. However, pegmatitic apatites do not always yield yellow-orange emission, and blue, pink-violet, and orange-red emission are also observed.

Magnetite ore apatites. Apatite is common in hydrothermal iron ore deposits such as at the Cerro de Mercado mine in Durango, Mexico, in southwestern Utah, in the Grenville formation at Mineville, New York, and in the Atlas mountains of Morocco. In all of these deposits, fluorapatite is found as substantial pods of intersecting crystals, and sometimes as the dominant rock-forming mineral. The iron ore acts to selectively partition Fe, Ni and other transition metals, leaving the rare earths to the fluorapatite. Accordingly, iron deposit fluorapatites display some of the most spectacular rare-earth-activated luminescence. In general these apatites are all rich in F and display blue-violet emission under SWUV, and a somewhat more violet-pink emission under MWUV. CL emission is generally bright blue-violet. As we shall see below, this emission is consistent with domination by Eu^{2+} and Ce^{3+} activation.

Other ore deposits. The classic fluorapatite specimens and fluorescent specimens are associated with these localities, notably Långban Sweden (ores of Mn, Fe, As), Franklin NJ (Zn), Panasqueria Portugal (Sn, W), and Llallagua, Bolivia (Sn, W). These ore bodies are mainly hydrothermal in origin, but in the case of Franklin multiple alteration events have created unusual species and chemistry. The Långban apatite group minerals are

Most of the Långban apatites fluoresce in shades of yellow to reddish-orange, and the author has not seen pink-violet or blue emission in any apatites from this mine in examining hundreds of specimens. This type of emission suggests mainly Mn^{2+} activation and a relative paucity of REE. However at the smaller Jacobsberg mine in Sweden, a similar ore body, johnbaumite has a distinctly pink-violet emission, which must be due to significant REE activation. At Franklin, fluorapatite is common but turneaurite, svabite, johnbaumite and other apatite group minerals are found occasionally, with most material showing strong luminescence. Most of the fluorescent emission is orange, with relatively slight variations in specimens seen by the author, suggesting Mn^{2+} activation with little REE contributions. At Panasqueria the fluorapatites can vary in luminescent intensity from nonluminescent to exceptionally strong. The crystals can also display spectacular luminescence zoning, typically with the bodies of the crystals fluorescing yellow, and the ends blue (Fig. 6, Color Plate 8), but other combinations and emission colors are also observed.

Figure 6. Figure 6 contains Color Plates 5-16. They are in the color signature in this chapter.

Marbles and carbonatites. Fluorapatites are commonly found in marble deposits such as the region around the Franklin ore body, and the marbles associated with skarns in the Bancroft and Wilberforce areas of Ontario, Canada. The Canadian fluorapatites usually have intense physical color which varies from a pale green to a dark green or a reddish brown. In general, the pale specimens have the strongest luminescence under UV, but most will show a blue to gray emission under SWUV or MWUV. CL response is generally a bright blue. Some of the most spectacular specimens came from the Silver Crater mine near Bancroft, where individual crystals weighing 10 kg were found by the author. Such crystals exhibit both Mn^{2+} and REE activation, including contributions from Eu^{2+}, Ce^{3+}, Dy^{3+}, Sm^{3+} and Eu^{3+}. Carbonatites, noted above, appear to be enriched in Eu^{2+} and Ce^{3+} with low Mn^{2+}. This yields an interesting difference in coloration between CL and UV excitation, as CL excites the Eu and Ce efficiently producing a strong blue emission, but the UV excitation energy is transferred to the Mn^{2+} so that the photoluminescence is usually a deep red or pinkish-red.

Sedimentary and biologic apatites. Diagenetically-produced sedimentary apatites show generally weak luminescence, though spectral analysis has been done to reveal the presence of REE and Mn. Fossil bone material, if luminescent, appears to be activated by substituents or defects that were incorporated during diagenetic alteration. In general the fluorescence of biological apatites is weak and has not been thoroughly explored.

APATITE STRUCTURE AND TYPES OF LUMINESCENT SUBSTITUENTS

Activator sites and occupations

Ca sites. The structure of fluorapatite is described elsewhere (Hughes and Rakovan, this volume, p. 1-12). The crystal structure of REE-rich apatite has been investigated by Kalsbeek et al. (1990). Five types of sites can be occupied by activator species: the Ca1 and Ca2 sites, the P site, the halogen site, and interstitial sites. The Ca1 site has trigonal-prism topology, with additional oxygens capping the main faces of the prism to yield 9-fold coordination. The six shorter Ca-O distances (3 × 2.40 Å, 3 × 2.46 Å) form the prism, with considerably longer Ca-O distances (2.81 Å) for the three caps. The overall site symmetry is C_3. The Ca2 site has six-fold coordination by oxygen (2.44 Å average distance, but range from 2.34-2.70 Å) and one halogen neighbor at (2.31 Å) with overall symmetry C_s. Because of their size, the Ca sites can be substituted with divalent Sr, Pb and to a lesser extent Ba, Y and REE. Na is the only major single-valent substituent

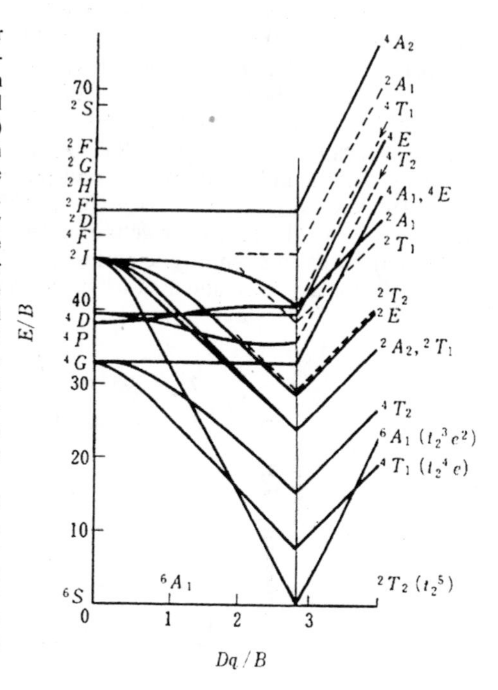

Figure 7. Tanabe-Sukano diagram for octahedral Mn^{2+} (d^5), constructed from all possible electronic states of five 3*d* electrons in an octahedral crystal field by solving the total energy as a function of the crystal (ligand) field. Pre-superscripts designate total spin state. Capital letters and subscripts designate group theory-based transformation symmetry. Such a diagram shows the relative energy placement and dependence of the possible electronic states on crystal field (Dq). For Mn^{2+} with a weak field as from oxygen, the ground state is $^6A_1$, first excited state $^4T_1$ and second excited state $^4T_2$. Increasing crystal field lowers $^4T_1$ and thus the transition energy from $^6A_1$. B is a scaling factor known as a Racah parameter. At high crystal field, $Dq/B \approx 2.8$, there is a ground state symmetry change due to electron pairing (high spin to low spin transition). Note that the symmetry of the Ca1 site in apatite is not octahedal, and the diagram is even less applicable for the Ca2 site. However the concepts are generally applicable. Note also that for some state symmetries transitions between states increase in energy with crystal field strength. For details on these diagrams see Cotton (1971) and Cotton and Wilkinson (1980).

found in natural samples. Pb^{2+} is a strong absorber of UV light, and can act as a powerful co-activator for luminescence (Botden and Kröger 1948). Pb, Sr and Ba substitution for Ca will expand the apatite lattice, and this would be expected to weaken the crystal field of activator ions in other Ca sites. Such expansion usually produces a shift in the luminescence emission toward higher energies because the weaker crystal field will result in a higher energy for the lowest excited state, and this increases the energy separation between the ground and the lowest excited state—($^6A_1$) and ($^4T_1$) states, respectively, for Mn^{2+} in an octahedrally coordinated site (Fig. 7). Such a change is observed in synthetic Sr and Ba apatite phosphors (Shionoya and Yen 1999). Sr preferentially occupies the Ca2 site (Rakovan and Hughes 2000), and it is probable that Pb^{2+} and Ba^{2+} also prefer the Ca2 site. Pb^{2+} is ordered into this site in hedyphane, but ordered Ba-Ca or Ba-Sr apatites have apparently been little studied. Other common substituents are Mn^{2+} and to a much lesser extent, Fe^{2+}. Interestingly, the apatite structure preferentially takes up Mn^{2+} over any of the other 3*d* transition metal ions. As Mn^{2+} is an activator of luminescence, and Cu^+, Ni^{2+} and Fe^{3+} strongly quench Mn^{2+} emission, this partially explains the frequent luminescence of natural apatites. The site partitioning of Mn^{2+} remains disputed, although most studies indicate that there is a preference for the Ca1 site. X-ray diffraction structural refinements indicate a Ca1 site preference that decreases with increasing Mn^{2+} (Hughes et al. 1991a, Suitch et al. 1985). Synthetic fluorapatites used in phosphors also have Mn^{2+} partitioned into Ca1 (Ryan et al. 1970, 1971). In contrast, EPR spectroscopy has indicated Mn^{2+} occupying both of the Ca sites (Warren 1970), with variations in partitioning in the Sr analogs (Ryan et al. 1972). Laser-excited photoluminescence study of the Mn^{2+} emission band in a natural fluorapatite shows that two Gaussian peaks are necessary to fit the band, and further that two relaxation times are needed to fit the emission decay profile (Gaft et al. 1997b). The larger band and relaxation component can be attributed to Mn^{2+} in the Ca1 site with a Ca1/Ca2 site occupation ratio of about 2.5:1

(Fig. 8). This component has a longer wavelength emission and a slower relaxation rate, both consistent with a stronger crystal field (Stefanos et al. 2000, Blasse and Grabmaier 1994). Trivalent lanthanides substituting for Ca require a coupled substitution for charge balance. This can be OH for O or F,Cl for O ions local to the substituted Ca site (Chen et al., 2002a). More common substitutions are SiO_4 for PO_4 and Na for a second Ca (Ronsbo 1989, Hughes et al. 1991). The trivalent lanthanides are selectively partitioned into the Ca2 site in fluorapatite, but with site ratio decreasing over the 4*f* series (Fleet and Pan 1995a,b; Hughes et al. 1991b, Mackie and Young 1973), and subject to the dependency of the apatite structure on the presence of multiple REE (Fleet and Pan 1997a,b). Sr has a stronger site preference for Ca2 than do the trivalent rare earths, and thus will displace them into the Ca1 site if present (Rakovan and Hughes 2000).

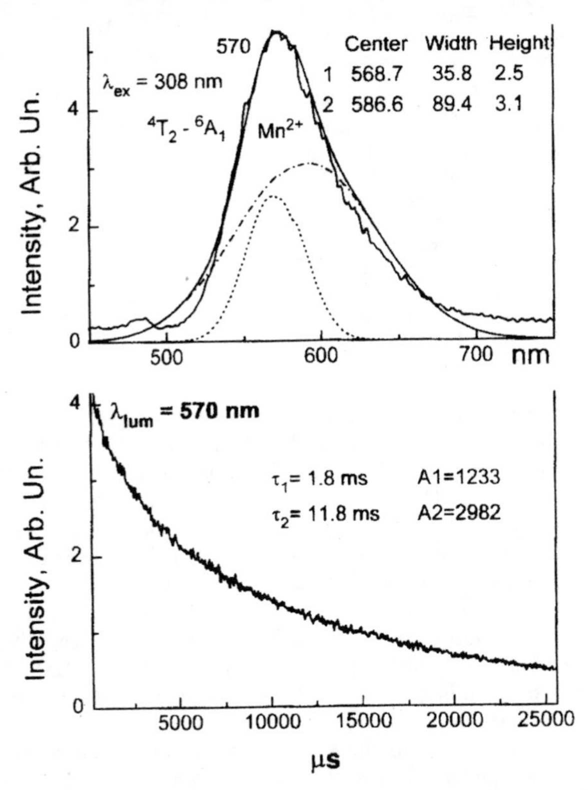

Figure 8. Mn^{2+} laser-excited emission spectrum from fluorapatite. Top: showing a possible fit for the two Ca site contributions. Bottom: Fitting to the excited state lifetime requires two components for a good fit. The components lifetimes are consistent with the symmetries of the two sites (and states). Modified after Gaft et al. (1997b).

The phosphate site. The phosphate site is highly regular and can be substituted by vanadate, chromate, arsenate, carbonate, silicate and sulfate, but only vanadate, arsenate and carbonate solid solution is widespread in natural apatites. Charge-coupled substitution schemes to aid incorporation involve the substitution of silicate for phosphate and REE^{3+} for Ca, and carbonate or sulfate for phosphate and Na for Ca. Numerous oxyanions may act as activators and sensitizers, with the most widespead being molybdate and tungstate, but vanadate and niobate can also activate luminescence (Blasse and Grabmaier 1994). Silicate is a strong absorber of UV radiation and can serve as a sensitizer for other activators. Unfortunately the activation of apatite group minerals via these oxyanions has been little explored. Substitution of arsenate for phosphate does however have a significant effect on the apatite group minerals lattice constants, and this should translate into changes in activator crystal fields. For example, svabite has cell constants of a = 9.750 Å c = 6.920 Å while fluorapatite has typical cell constants of a = 9.367 Å c = 6.884 Å, and synthetic $Ca_5[AsO_4]_3Cl$ has a = 10.076 Å and c = 6.807 Å compared to chlorapatite with a = 9.5979Å and c = 6.7762 Å. This should produce a weakened crystal field at the Ca sites as arsenate content is increased, shifting the emission spectrum to shorter wavelengths. Another possible occupant of the P site in apatite is Mn^{5+}. Gaft et al. (1997b) showed that some natural fluorapatites demonstrate the strong IR emission bands and diagnostic luminescence decay times characteristic of Mn^{5+} doped into the P site in synthetic apatite (Moncorge et al. 1994, Oetliker et al. 1994, Scott et al. 1997). Emission spectra at two temperatures are shown in Figure 9. Oetliker et al. (1994) showed that lattice expansion when arsenate and vanadate replaced phosphate in Sr and Ba synthetic apatites reduced the MnO_4 crystal field, resulted in both a shift of emission and excitation bands, and a change in luminescence lifetime. By comparing Mn^{5+} spectra in a variety of host lattices, Oetliker et al. (1994) also determined that distortions of the MnO_4 tetrahedron had greater effect on luminescence lifetime and temperature quenching than cell dimension changes. Mn^{5+} also produces absorption bands at 540 and 660 nm in tetrahedral coordination and is partially responsible for the coloring of blue and blue-green fluorapatites.

The halogen site. The halogen site may contain F, Cl, Br or OH though Br in significant levels is found only in synthetic apatites. None of these species act as activators or sensitizers, but they can contribute to cell dimension changes and thus

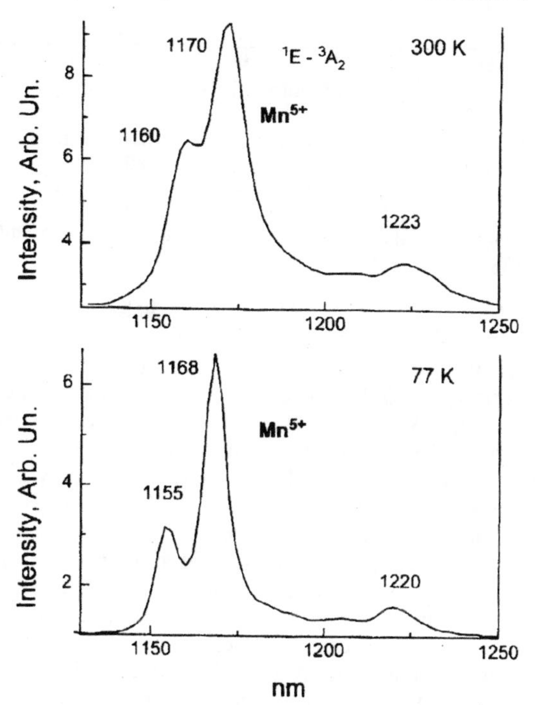

Figure 9. Mn^{5+} in P site emission spectra. Note the improved resolution at lower temperature. Modified after Gaft et al. (1997b).

activator crystal fields. By far the largest effect is due to substitution of Cl for F or OH. Compare fluorapatite with chlorapatite above, and also the synthetic varieties of pyromorphite:

$Pb_5[PO_4]_3Cl$	a = 10.976 Å	c = 7.351 Å
$Pb_5[PO_4]_3F$	a = 9.760 Å	c = 7.30 Å
$Pb_5[PO_4]_3OH$	a = 9.774 Å	c = 7.291 Å

These values indicate that Cl should have a definite effect on activator crystal fields, and one would expect a priori that the increased cell should result in a smaller field at the Mn site and an emission shift to shorter wavelengths. In fact, what is observed is a shift to longer wavelengths (Ryan et al. 1971; Fig. 10). The substitution of Cl^- for F^- has a large effect on the Stokes shift but little effect on the excitation spectrum of the Mn^{2+}. This suggests that the Cl^- is changing the energy configuration of the Mn^{2+} excited state possibly due to Ca2 site distortion (Ryan et al. 1971).

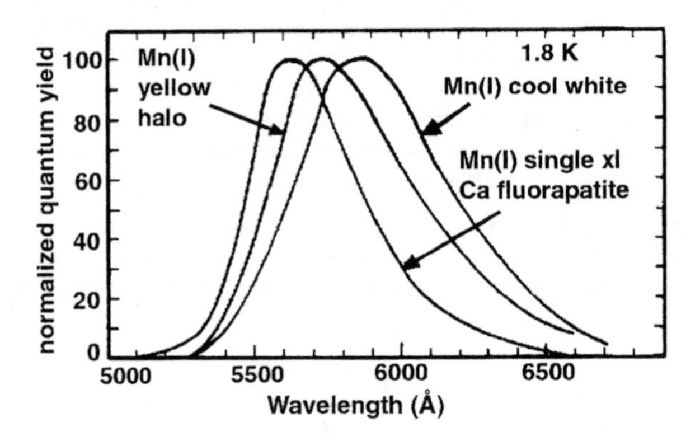

Figure 10. Changes in Mn emission (from Mn in Ca1 site) from fluorapatite with changes in other substituents. "Yellow Halo" synthetic fluorapatite has Sb and Cd added. "Cool white" has Sb, Cd and Cl added. The cool white composition should have increased cell dimensions, and thus a smaller crystal field. Ordinarily this should raise the energy of the emission. Modified from Ryan and Vodoklys (1971).

Defect sites. Besides lattice defects such as dislocations and stacking faults, vacancies can be created in the apatite structure by certain types of impurity substitutions. For example, two REE^{3+} can substitute for three Ca ions, creating a Ca vacancy (Fleet and Pan 1994), or a halogen can be lost in order to satisfy the charge of a silicate group substituting for phosphate. Anion vacancies can serve as sites for electron centers, such as F (from the German "farbe") centers. The pink color of pegmatitic fluorapatites (Marfunin 1979) is due to the F center formed from a fluorine vacancy with a trapped electron. Other defect sites are associated with molecular groups which do not readily fit into the apatite structure. One such species is the uranyl group, UO_2^{2+}, a linear ion with very strong luminescence, which has been observed in sedimentary apatite luminescence (Panczer et al. 1998). Other defect sites in apatite can act as traps for electrons or holes, and give rise to thermoluminescence-phosphorescence effects. One such trapping site can be produced by cation or anion vacancies. Chen et al. (2002b) suggested that two Gd^{3+} ions can substitute at adjacent apatite Ca1 sites with a neighboring Ca2 site vacancy, forming a trimer defect unit. This mechanism could possibly be used by other trivalent REE. The cation vacancy would have a net negative charge and could thus trap holes, leading to thermoluminescence behavior.

APATITE ELECTRONIC STRUCTURE

Band structure

The fluorapatite electronic structure is typical of a low conductivity insulator with a significant energy gap between the valence and conduction bands (Louis-Achille et al. 2000, Blasse and Grabmaier 1994). The energy gap of about 8 eV requires photons of wavelength near 1500 Å to create carriers, so that lattice UV absorption is unimportant for Ca apatites, i.e., the relevant absorption that excites any activators occurs at the activator or sensitizer sites themselves, which are located in the band gap. The case is entirely different for CL, where much greater energies are available for excitation, as well as much higher excitation densities. Hence, pure Mn^{2+} doped fluorapatites will hardly show any luminescence under ordinary UV excitation, but will emit brightly under the electron or ion beam. The situation changes for the Pb apatites, where strong absorption of short wave UV occurs, and the energy can be transferred to the Mn^{2+} to excite luminescence. This particular case has apparently not been studied in detail during phosphor research, as Pb-sensitized halophosphate phosphors are either not as efficient as compositions having other sensitizers, or because of deterioration with time in applications. Hence the action of Pb^{2+} as a sensitizer is drawn by analogy with the Pb^{2+}-Mn^{2+} system in calcite and wollastonite (Shionoya and Yen 1999, Leverenz 1968). Ce^{3+} and Sb^{3+} are other sensitizers that transfer excitation energy to Mn^{2+} and have been thoroughly explored due to phosphor research (Shinoya and Yen 1999).

Spectra of Mn^{2+} in the Ca sites. An approximate calculation for crystal field splitting of Mn^{2+} 3*d* states in the Ca1 site of fluorapatite has been done by Narita (1961, 1963). According to this model, a 2% isotropic lattice contraction is expected to produce a shift of 1200 cm^{-1}, equal to an emission wavelength change from 580 nm to 620 nm (yellow to reddish-orange). The calculation shows acceptable agreement with experimental observations for the replacement of about 10% of the Ca in Ca fluorapatite by Cd. This decreases the unit cell dimensions by 0.042% and 0.35% on the *a*- and *c*-axes, respectively, and creates a shift in the emission spectral peak to longer wavelengths of about 100 cm^{-1}. The electronic structure of Mn^{2+} in fluorapatite has been measured in detail by Ryan et al. (1970, 1971). Excitation spectra obtained at 1.8 K show fine structure that can be separated into contributions from both the Mn^{2+} in Ca1 and the Mn^{2+} in Ca2 (Fig. 11). Given slightly different excitation spectra for each site, it was then possible to excite the Mn^{2+} in each site separately (Fig. 12), demonstrating the effect of the less symmetric crystal field in the Ca2 site, namely, an emission shift of 10 nm from

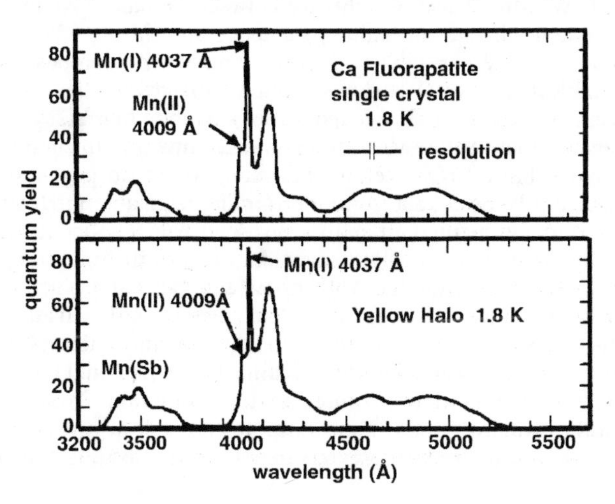

Figure 11. Excitation spectra for two of the synthetic samples depicted in Figure 10. At low temperature the Mn bands due to the two Ca sites can be separated. Modified after Ryan and Vodoklys (1971).

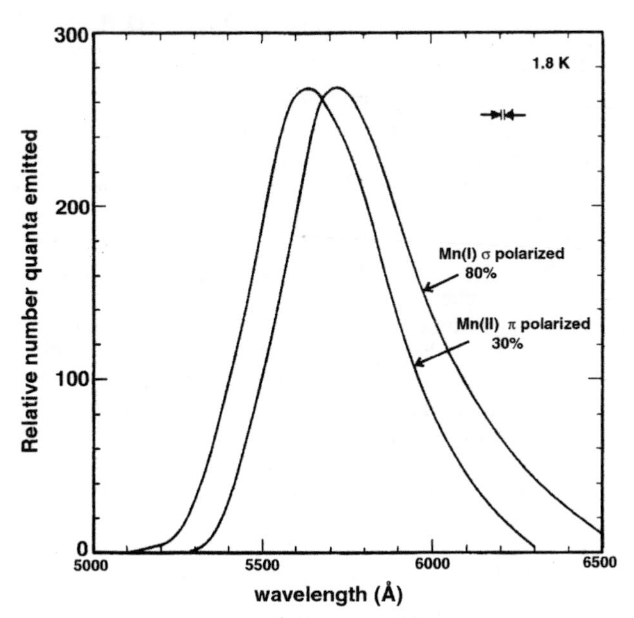

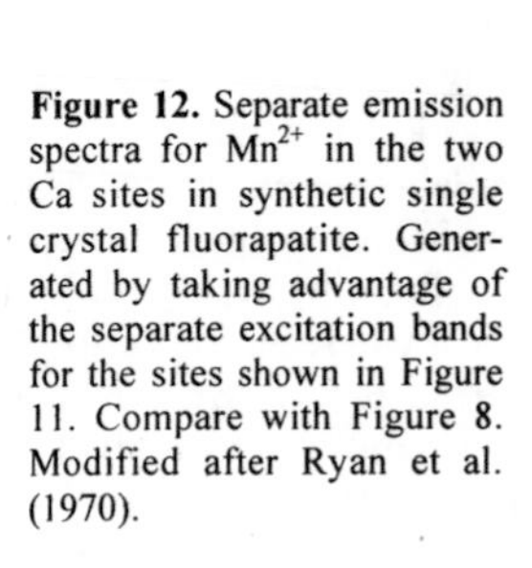
Figure 12. Separate emission spectra for Mn^{2+} in the two Ca sites in synthetic single crystal fluorapatite. Generated by taking advantage of the separate excitation bands for the sites shown in Figure 11. Compare with Figure 8. Modified after Ryan et al. (1970).

572 to 562 nm at 1.8 K. This means that the lowest excited state of the Mn^{2+} ion in the Ca2 site is about 300 cm^{-1} higher in energy than the analogous state for Mn^{2+} in the Ca1 site, consistent with a weaker crystal field at the Ca2 site.

The emission band for Mn^{2+} in apatite is always broad due to the different symmetries of the ground and excited states (Fig. 1). The band peaks at about 570 nm (yellow or yellow-orange) in fluorapatite (Gaft et al. 1997b) at room temperature excited by a UV laser. Cathodoluminescence studies (Perseil et al. 2000, Barbarand and Pagel 2001) of fluorapatites show a strong Mn^{2+} emission band at 577 nm. Asymmetry in the band has been attributed to contributions from both Ca sites which have slightly different crystal field parameters (Gaft et al. 1997b) , but the work of Ryan et al. (1970) shows that even the emission from only one of the Ca sites has asymmetry. Both studies agree on the sense of the crystal field and emission wavelength change between the two sites. The survey by Gorobets (1981) shows Mn^{2+} bands which vary between 560 and 590 nm in natural apatites, presumably due to changes in lattice parameters and Ca1 vs. Ca2 site occupation. Anecdotal evidence for the dependence of Mn^{2+} emission on composition comes from Henkel (1989) and Robbins (1994) and observations collected by the author for several dozen apatite specimens excited by UV (unpublished). Svabite from Långban, Sweden is a well characterized arsenate apatite with strong orange emission. In contrast, mimetite from Långban, Sweden has a bright yellow emission, similar to pegmatitic fluorapatites from Afghanistan and Pakistan, and from the Harding pegmatite in Dixon, New Mexico. The larger size of the mimetite unit cell is consistent with smaller crystal fields at the Mn^{2+} sites, and thus shorter wavelength emission, i.e., a shift from orange toward yellow. However the fluorapatites from the Afghanistan and Pakistan pegmatites appear to contradict this conclusion, unless the emission is dominated by REE activation. Interpretation of visual apatite emission colors are subject to large variations depending on several factors: filter effect of the physical color of the apatite, the amount and type of room lighting, and the nature of the excitation source in UV excitation. Thus, for example, physically green apatites with Mn^{2+} activation appear to emit yellow-green light due to filtering. In CL studies the spectral peaks observed are not subject to shifts due to these factors, but large differences in intensity, background and resolution are often observed for nominally very similar material.

Luminescence intensity and quenching. There is disagreement in the literature as to whether the luminescence intensity is related to the Mn^{2+} concentration, the presence of certain quenching agents, or the presence of particular sensitizers. For example, Perseil et al. (2000) found that in fluorapatites from one location the fluorescence emission intensity was correlated with Mn^{2+}, and partially quenched by higher concentrations of arsenate substituting for phosphate. This is remarkable, as pure arsenate apatites are known to be highly fluorescent (svabite and johnbaumite) due to Mn^{2+} emission. In contrast, Knutson et al. (1985) used statistical analysis to show that fluorapatites from Panasqueira, Portugal have fluorescence not strongly correlated with Mn^{2+} content, but more with coexistence of Mn^{2+} and sensitizing REE Eu^{2+} and Ce^{3+}. Ce^{3+} and other REE are well known to be sensitizers for Mn^{2+} in other minerals, notably calcite (Blasse and Aguilar 1984). Garcia and Sibley (1988) showed that Eu^{2+} acts as a coactivator for Mn^{2+} in synthetic fluoroperovskite. Further, Chenot et al. (1981) found that Ce^{3+} and Mn^{2+} can act as coactivators in synthetic fluorapatite phosphors.

Fe^{2+} in the Ca sites has been shown to quench the Mn^{2+} emission in apatite from studies of synthetic carbonate apatite (Filippeli and Delaney 1993). The manner in which Fe^{2+} affects fluorescence varies both with total Fe^{2+} concentration and with the Mn/Fe ratio. This is entirely consistent with local quenching effects of the Fe^{2+} impurity. At low Mn and Fe concentrations, most Mn will be unaffected by exchange of energy to a nearby Fe. However at high Fe content all Mn will be affected. The range of quenching effects depends on the type of interaction between quenching ion and activator. Magnetic and electric dipole interactions are relatively short range. However, some quenching ions operate by having their own absorption bands overlap the excitation bands of the activators. In this case quenching ions anywhere in a crystal can act to absorb the exciting photons before they reach an activator. A detailed study of the quenching effects of many impurities on the Mn^{2+} emission from a synthetic "warm white" calcium halophosphate phosphor ($Ca_{10}(PO_4)_6(F,Cl)_2$: Sb^{3+}, Mn^{2+}) was done by Wachtel (1958). In this study the phosphor was excited with SWUV at 2537 Å. At the 10,000 ppm level, Cu, V, Co, Cu and Fe removed essentially all luminescence, while Ni and Cr reduced luminescence intensity by over 90%. At the 1000 ppm level, these species reduced luminescence intensity by 40 to 60%.

Even if excited directly via band gap absorption or indirectly from sensitizers, Mn^{2+} can transfer some or all of its excitation energy to nearby REE ions. This is the primary reason why Mn^{2+} emission may not be observed in apatites that have significent Mn and REE impurities. Marfunin (1979) listed Nd^{3+}, Sm^{3+}, Eu^{3+} and Tm^{3+} as the main REE for sensitization by Mn^{2+}, with Sm^{3+} being the most significant recipient. However, this type of energy transfer process does not seem to have been explored in detail. A diagram that shows how a transition metal ion could transfer energy to a rare earth system is shown in Figure 13.

Spectra of rare-earths in the Ca sites. The rare earth ions have an incompletely filled $4f$ shell, which is strongly shielded from outside interactions by the surrounding filled $5s^2$ and $5p^6$ orbitals. This means that electronic transitions within the $4f$ orbitals are only weakly influenced by crystal fields and covalency effects, and thus $4f$ to $4f$ transitions for rare earths in apatite will not be distinguishable between the Ca sites. Further, the change in configurational coordinate (ΔR) between ground and excited state is small or zero, so that very narrow absorption and emission lines will be observed. $4f$ to $4f$ transitions are parity forbidden, and would be very weak if not for some mixing of other symmetry orbital character in either the ground or excited state. This is enhanced by lower

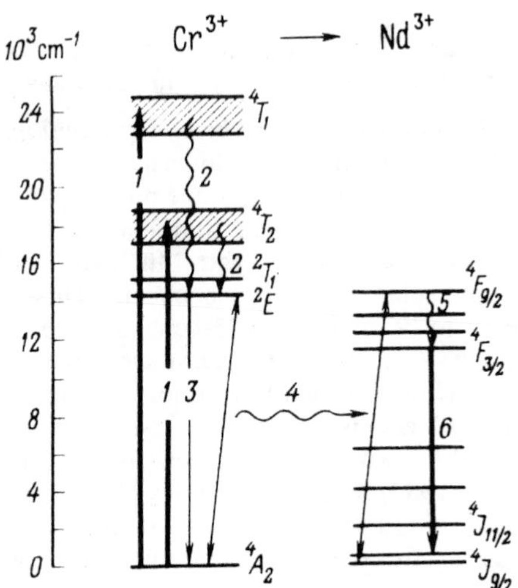

Figure 13. A scheme for energy transfer from excited Cr^{3+} to Nd^{3+}. Despite the varying transition symmetries, the energy of the two transitions shown by slanted arrows are similar enough so that there can be a coupled resonance (wiggly arrow 4) between them. This is sometimes called a "resonance radiationless transition." In this kind of diagram the curved states of the configurational coordinate plot (Fig. 1) are collapsed to show the band width in energy. Wiggly lines indicate non-radiative transitions due to the crossing of state levels and vibrational heat loss. Modified after Marfunin (1979).

symmetry crystal fields, and in this way 4*f* to 4*f* transitions will have intensity related to site symmetry. Stronger transitions that are parity allowed occur via charge transfer $4f^n$ to $4f^{n+1}L^{-1}$ or $4f^n$ to $4f^{n-1}$-5*d* transitions. The charge transfer transitions are strongest in rare earth ions that have a tendency for reduction, while the 4*f*-5*d* transitions are strongest in rare earths that have a tendency for oxidation. Hence the tetravalent rare earth ions (Ce^{4+}, Pr^{4+} and Tb^{4+}) have strong charge-transfer bands, while the divalent rare earth ions (Eu^{2+}, Sm^{2+} and Yb^{2+}) have strong 4*f*-5*d* transitions. Trivalent rare earth ions that can be reduced have charge-transfer bands in the ultraviolet, and trivalent ions that can be oxidized have 4*f*-5*d* transitions in the ultraviolet. Thus, in general, the strong transitions do not give rise to physical coloration, but have a strong channel for excitation luminescence because of UV absorption, either by their own internal transitions, or by energy transfer to nearby activators. Both charge transfer and 4*f*-5*d* transitions produce broad absorption and emission peaks due to $\Delta R \neq 0$ (Fig. 14).

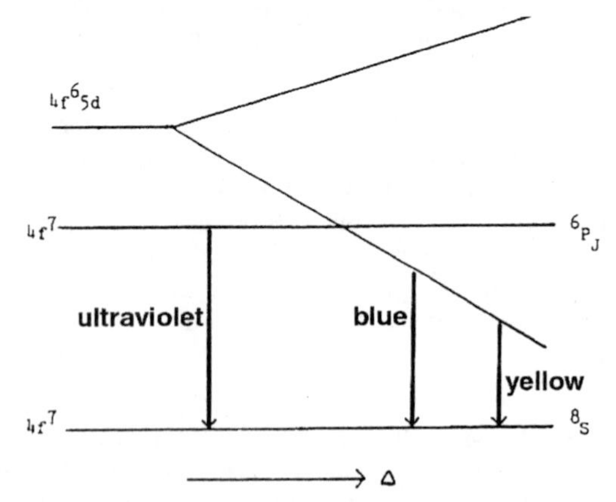

Figure 14. A simplified energy level diagram showing why there is significant difference in the line (band) widths for *f*-*f* vs. *d*-*f* transitions in Eu^{2+}. As the crystal field is increased (Δ) the spacing of the 4*f* states does not change due to their symmetry agreement. However, the 4*f*5*d* state drops appreciably in energy. For low crystal fields the *f*-*f* emission band will be quite narrow, while for high fields the $4f^65d$ to $4f^7$ transition will be very broad. See also Figures 1 and 7. Modified after Shionoya and Yen (1999).

***f-d* transitions: Eu^{2+}, Sm^{2+} and Yb^{2+}.** Divalent Eu is an important activator in many minerals (Mariano and Ring 1975, Roeder et al. 1987, Gaft et al. 2001b), and is the activator responsible for fluorite's frequent strong blue luminescence, a phenomenon that first attracted Stokes' attention in studying fluorescence (Stokes 1852). Eu^{2+} emission is dependent on the local crystal field strength and symmetry, and thus apatite structural

COLOR PLATES

Color Plates 1-4: From Chapter 3, Rakovan (this volume).

Color Plate 1. Figure 15 in Rakovan, this volume, page 72. Fluorapatite from the Siglo XX Mine, Llallagua, Bolivia. (*left*) In plane light. (*right*) In long wave ultraviolet light. Photoluminescence is activated by REEs. Differential luminescence between {001} faces (violet) and {100} faces (orange), indicates sectoral zoning of REEs.

Color Plate 2. Figure 20 in Rakovan, this volume, page 75. (*left*) Cathodoluminescence (REE-activated) photomicrograph of the growth hillock in the right image. Luminescence (orange) is homogenous between the two symmetrically equivalent vicinal faces. Differential luminescence exists between these and the symmetrically nonequivalent vicinal face (blue). *right*) DIC photomicrograph of a trigonal growth hillock on the {100} face of a fluorapatite from the Golconda Mine, Minas Gerais, Brazil. [Used by permission of the Mineralogical Society of America, from Rakovan and Reeder (1994) American Mineralogist, Vol. 79, Fig. 6 p. 897].

Color Plate 3. Figure 21 in Rakovan, this volume, page 75. (*left*) DIC photomicrograph of a trigonal growth hillock on the {100} face of a fluorapatite from the Siglo XX Mine, Llallagua, Bolivia. Steps of $[01\bar{1}]$ orientation comprise vicinal face **a**, and steps of [001] orientation comprise vicinal face **b**. (*right*) Cathodoluminescence photomicrograph of the hillock in left image. Luminescence is homogenous between the two symmetrically equivalent vicinal faces (yellow luminescence activated by Mn^{2+}). Differential luminescence exists between these and the symmetrically nonequivalent vicinal face (blue activated by REEs). [Modified after Rakovan and Reeder (1996)].

Color Plate 4. Figure 24 in Rakovan, this volume, page 77. (*left*) Cathodoluminescence photomicrograph of hexagonal growth hillocks on the {001} face of an apatite from the Siglo XX Mine, Llallagua, Bolivia. Luminescence (purple activated by REEs) is homogenous among the six symmetrically equivalent pyramidal vicinal faces. Yellow luminescence (Mn^{2+}-activated) dominates flat regions of the {001} face and the terminal surfaces of the hillocks. (*right*) DIC photomicrograph showing the microtopography of the {001} apatite face in the left image. [Used by permission of the Mineralogical Society of America, from Rakovan and Reeder (1994) American Mineralogist, Vol. 79, Fig. 7, p. 897].

Color Plates 5-16: Figure 6 in Waychunas (this chapter).

Photoluminescence (short-wave ultraviolet, SWUV, Mid-wave ultraviolet, MWUV, and long-wave ultraviolet, LWUV) of apatites from various localities with postulated major luminescent centers, based on various unpublished data. Photos taken with Canon 4 Mb digital camera equipped with UV filter and close-up capability. Exposures 0.5-10 seconds. Colors adjusted with Photoshop 6.0 to agree with observed emission in a dark room.

Color Plate 5. Fluorapatite, Cerro Grande Mine, La Paz, Bolivia. Sector zoned: blue {001} and {101} sectors and violet {100} sectors. Eu^{2+}/Ce^{3+}-activation. Largest crystal is 60 mm. SWUV.

Color Plate 6. Fluorapatite, Greenwood, Maine, USA. (pinkish-red) Eu^{3+}-activation. Matrix of luminescent albite (blue) activated by Eu^{2+}. Largest crystal is 10 mm. SWUV.

Color Plate 7. Fluorapatite, Nuristan, Afghanistan. Mixed areas of REE- and Mn^{2+}-activation. Specimen is 100 mm across. SWUV.

Color Plate 8. Fluorapatite, Panasqueira, Portugal. Center (yellow) Mn^{2+}-activation; ends (blue) Eu^{2+}-activation. Zoning is probably due to a combination of temporal variations in solution chemistry and face-specific incorporation. Largest crystal is 25 mm. SWUV.

Color Plate 9. Fluorapatite associated with (or altering to) autunite. Ritchfield Pegmatite, Connecticut, USA. Mn^{2+} (yellow) and uranyl (green) activation. 100 mm sample. SWUV.

Color Plate 10. Johnbaumite, Jacobsberg, Sweden. Eu^{3+} and Sm^{3+} activation (pink-violet). Matrix is fluorescent calcite (red) activated by Mn^{2+} and Pb^{2+}. 60 mm sample. SWUV.

Color Plate 11. Fluorapatite, Imichil, Anti Atlas Mtns., Morocco, exhibiting concentric zoning. Eu^{2+} activation. 300 mm specimen. MWUV.

Color Plate 12. Fluorapatite, Franklin, New Jersey, USA. Mn^{2+}-activation (red). Other luminescent minerals: (green) willemite, activated by tetrahedral Mn^{2+}; (pink-red) calcite, activated by Mn^{2+} and Pb^{2+}; non-luminescent (black) franklinite. 200 mm specimen. SWUV.

Color Plate 13. Fluorapatite, Durango, Mexico. See Figure 4 in (this chapter, page 706) for luminescence spectra. Purple fluorescence activated by Ce^{3+}, Eu^{2+}, Sm^{3+} and Dy^{3+}. Green fluorescence is hyalite opal with uranyl-activation. 200 mm specimen. MWUV.

Color Plate 14. Hydroxylapatite, Molina Mine, Potosi, Bolivia. Concentric zoning with cores Mn^{2+}-activated; exterior Eu^{2+}-activated. 50 mm crystals. MWUV.

Color Plate 15. Fluorapatite from a granite pegmatite pocket, Gilgit-Skardu Road, Northern areas, Pakistan. Mn^{2+}-activation with bright yellow zones (Dy^{3+}?). Specimen exhibits strong yellow phosphorescence. 75 mm wide. SWUV.

Color Plate 16. Fluorapatite, Silver Crater Mine, Bancroft, Ontario (blue). Ce^{3+}- and Sm^{3+}-activation. Largest crystal is 150 mm. Fluorescent matrix (pink-red) is calcite activated by Mn^{2+}, Pb^{2+} and REE. MWUV.

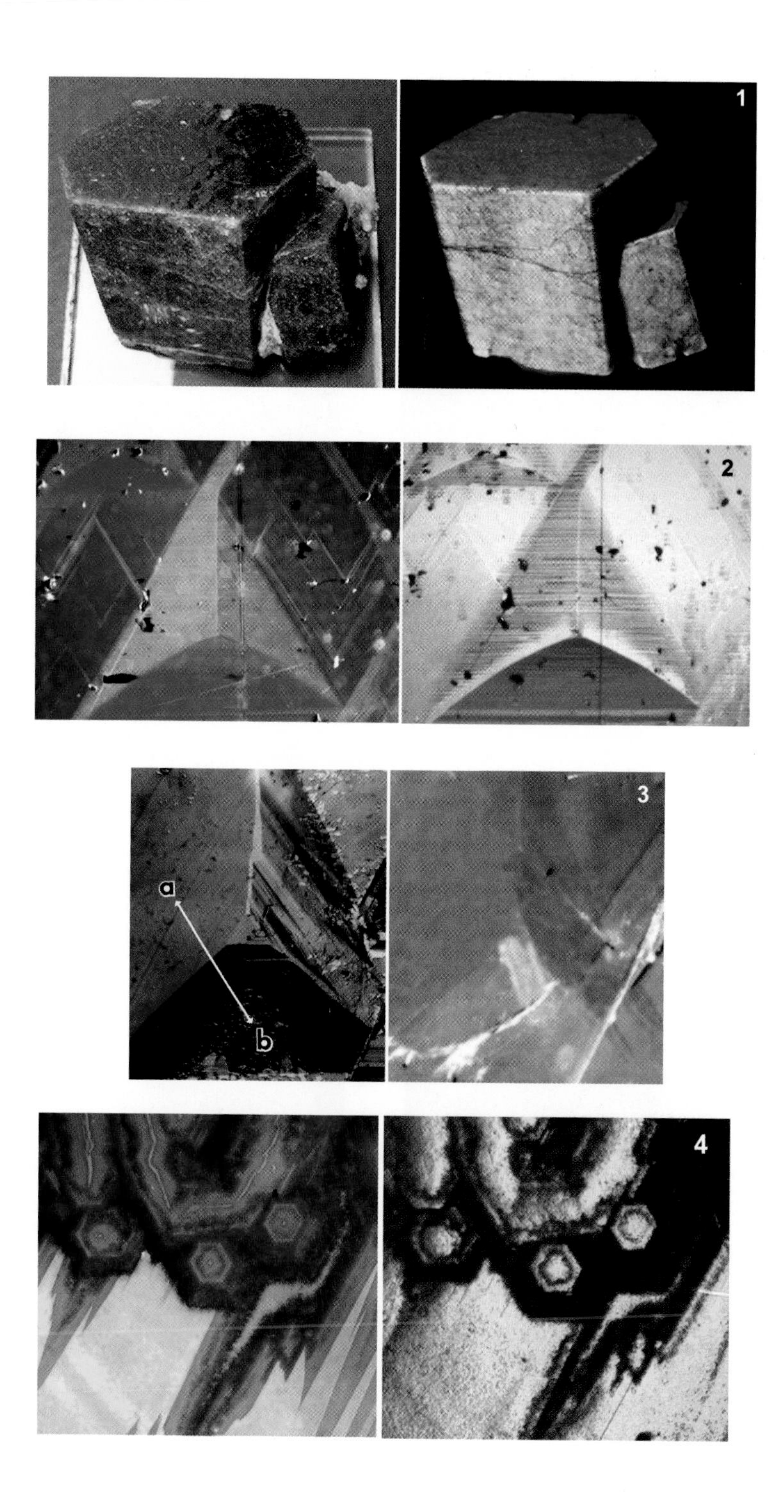
1
2
a
b
3
4

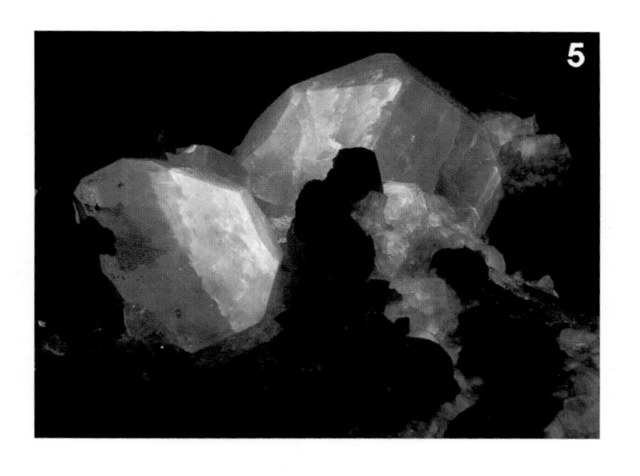

5

6

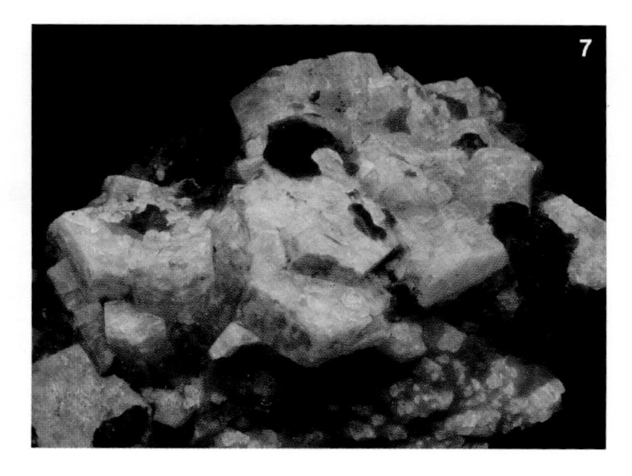

7

8

9

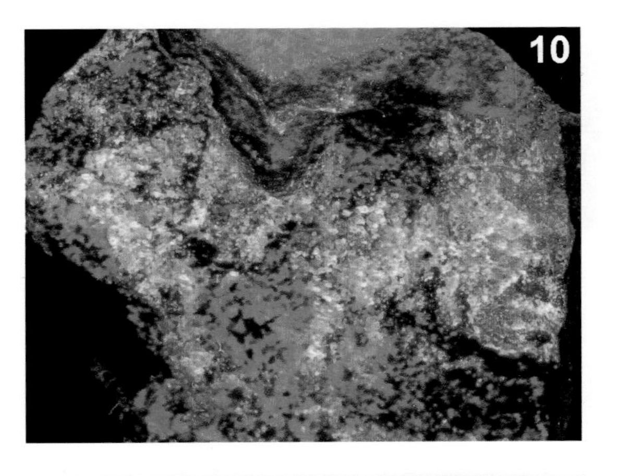
10

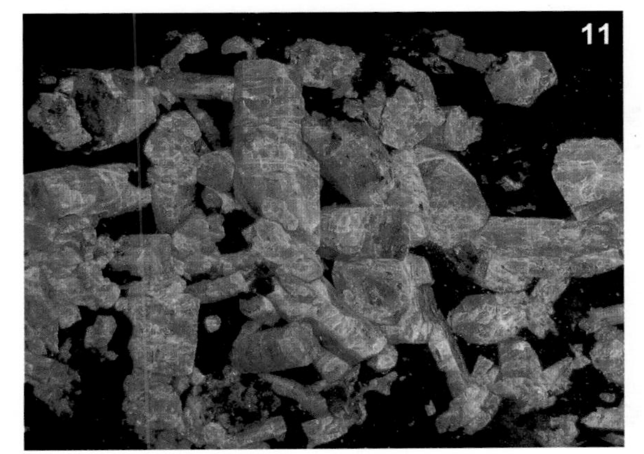
11

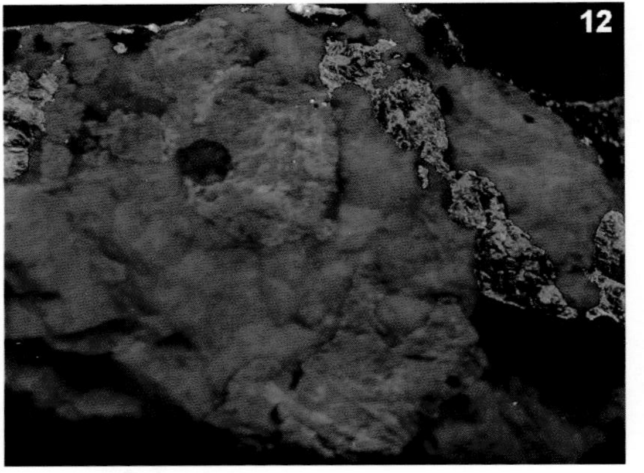
12

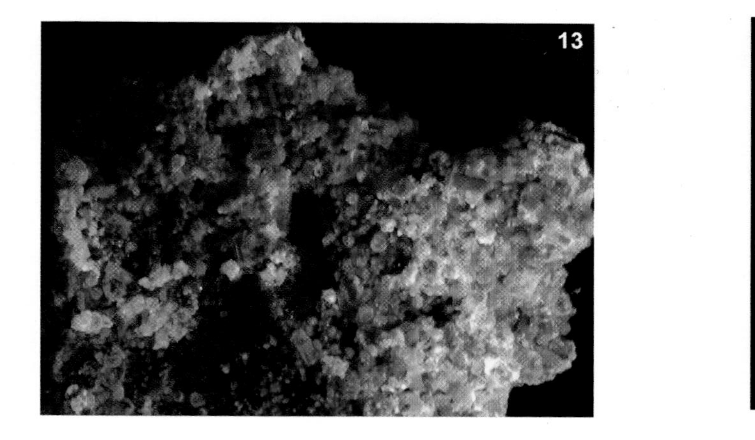

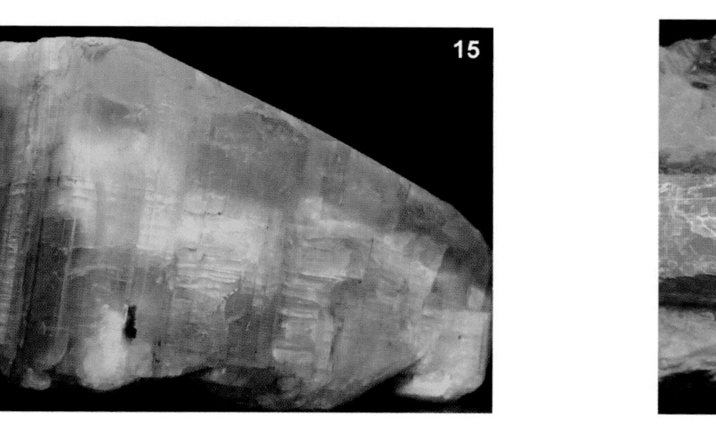

16

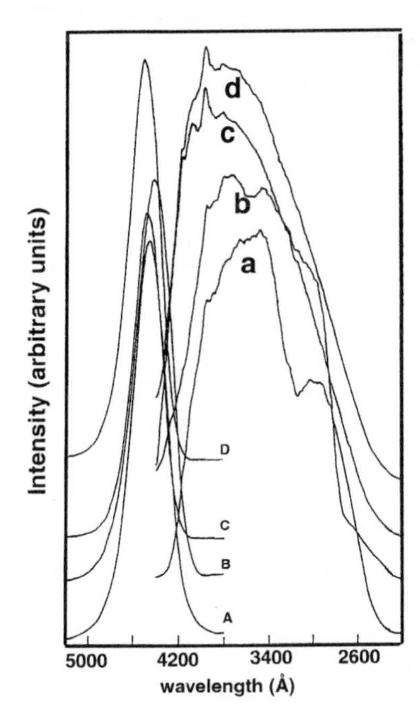

Figure 15. Shifts of both excitation (capital A-D) and emission bands (small a-d) in synthetic Sr apatites. (a) hydroxylapatite, (b) fluorapatite, (c) chlorapatite, (d) bromoapatite. Modified after Kottaisamy et al. (1994).

dimensions influence the luminescence spectrum. This effect is well shown in spectra collected by Kottaisamy et al. (1994) of doped synthetic apatites (Fig. 15) where Ca, Sr and Ba chlorapatites have Eu^{2+} emission bands at 450, 446 and 438 nm, respectively. In other minerals with 6-7 coordinated sites, Eu^{2+} emission can vary from 450 to 430 nm (Gaft et al. 2001b). CL spectra of apatites generally show Eu^{2+} bands at about 450 nm (Mitchell et al. 1997) although interference from Ce emission may overwhelm the Eu^{2+} signal. Gaft et al. (2001a) were able to separate the Eu^{2+} emission from a natural Ca-chlorapatite with considerable Ce and other rare earths using time-resolved spectroscopy (Fig. 16), also identifying a Eu^{2+} band at 450 nm. Other Eu^{2+} emission spectra have been reported by Jagannathan and Kutty (1997) and Gorobets (1981).

Eu^{2+}-Mn^{2+} energy transfer is an important process utilized in early halophosphate phosphors for fluorescent lamps. This is well known from excitation spectra of the Mn^{2+} emission, which show features of the Eu^{2+} absorption spectrum. In strontium chlorapatite doped with 2% Mn^{2+} and 2% Eu^{2+}, the Mn^{2+} emission increases in intensity and the Eu emission decreases in intensity with increase in temperature (Kottaisamy et al. 1994). This is due to increased transfer of energy from an excited Eu center to Mn^{2+} as temperature increases. Interestingly, the larger interatomic distances in barium chlorapatite result in significantly reduced Mn^{2+} emission, probably due to reduced energy transfer. Energy transfer of this type has a very strong dependence on intersite distance (R), on the order of R^{-6} for electric dipole interactions, and R^{-8} for electric dipole/electric quadrupole interactions (Blasse and Grabmaier 1994). Hence small changes in lattice dimensions can produce dramatic effects in luminescence emission, where several types of energy transfer can occur, as with REE doped minerals. Thus energy transfer can bleed energy away from an excited Eu^{2+} so that Eu emission may vanish or be very weakened in response to lattice expansion, REE concentration levels, and type of REE impurity. This explains in part why CL and UV studies of Eu^{2+} in apatites, except in pure doped samples, are hard pressed to correlate Eu^{2+} composition and luminescent intensity.

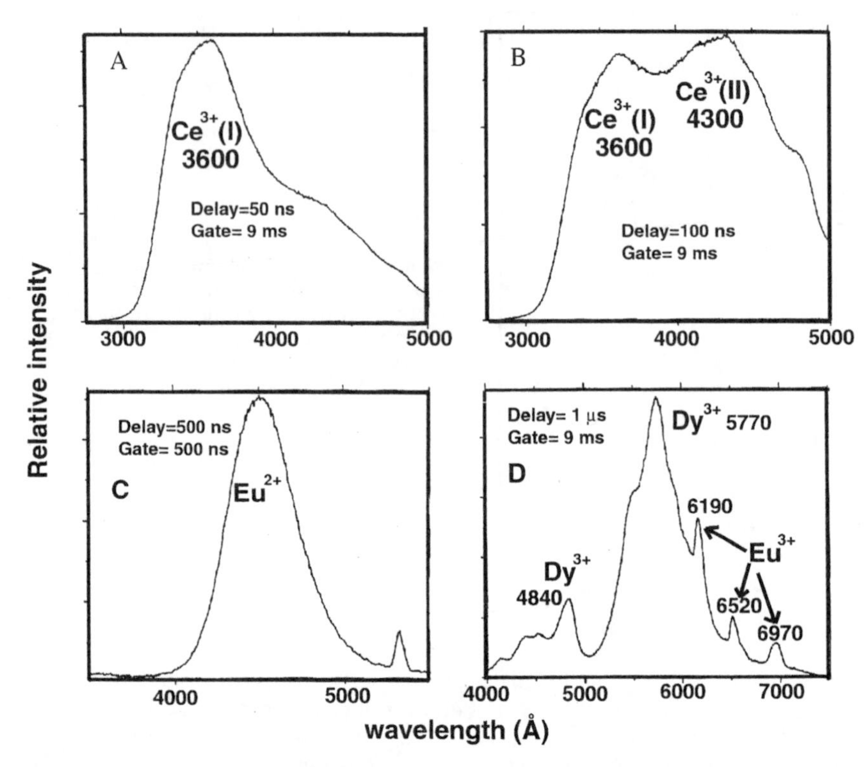

Figure 16. Time resolved laser-excited emission from green fluorapatite from Norway showing separation of REE emission due to lifetime differences. (a) Ce^{3+} emission with long gate and short delay. (b) slightly longer delay reveals both Ce centers. (c) shorter gate and longer delay separates out the Eu^{2+} emission. (d) longer delay and long gate show other REE emission. Modified after Gaft et al. (2001a).

In some cases, notably those with relatively short excited state lifetimes, energy transfer can be suppressed by excitation with laser pulses of shorter duration than the mean transfer time.

Sm^{2+} luminescence has been reported from natural anhydrite samples (Gaft et al. 1985, 2001a; Taraschan 1978) as a broad strong band at 630 nm. Other sharp bands are reported at 688, 700 and 734 nm. Sm^{2+} emission has not been reported for apatite, but the ion size and valence are amenable to the Ca sites, so ultimately it may be observed via pulsed laser techniques. $Sm^{2+}$is present in aqueous solution only under quite reducing conditions, so this may limit concentrations. Sm^{2+}does occur in fluorite, and is responsible for the strong green coloration and sensitized Eu^{2+} luminescence in that mineral (Robbins 1994). However, in fluorite the Sm^{2+} may be created by radiation effects which reduce bound Sm^{3+} ions. Yb^{2+} emission has not apparently been reported in any minerals, but has been studied in borates and oxides (Blasse and Grabmaier 1994). Its emission is in the UV at about the position of Eu^{2+} bands. Yb^{3+} emission is observed in apatite, hence Yb^{2+} might be created in naturally irradiated samples.

Trivalent rare earths. The trivalent rare earths may show *4f-5d* or *4f-4f* luminescence transitions. As with the divalent REE, the *4f-5d* transitions will be broad, while the *4f-4f* transitions will be sharp. Given the large number of possible states (depending on the crystal field symmetry) there are many possible transitions. Thus given the many REE present in a given structure, a good chance exists that many of the transitions will be close in energy and thus favor energy transfer. Accordingly, not only can single energy transfers occur, but multiple ones involving several REE ions can occur as well.

Cerium. Cerium is the most common of the rare earths, and can have relatively high concentrations in apatites, e.g., Gaft et al. (2001a) lists a blue apatite containing over 4000 ppm Ce. Despite such high concentrations of Ce, specific Ce luminescence is infrequently observed, due to efficient energy transfer to manganese and other REE, and identification of the Ce^{3+} emission bands is rather ambiguous. CL spectra of Ce^{3+}-doped synthetic apatites show broad Ce^{3+} emission bands at 440 nm (0.18 % Ce, Mitchell et al. 1997), and 350, 380 and 458 nm (1% Ce, Blanc et al. 1995). CL emission from a natural fluorapatite from Portugal gave a single, very broad band at 349 nm (Knutson et al. 1985). Using UV excitation, Morozov et al. (1970) found two different bands at 395 and 420 nm in Ce^{3+} doped fluorapatite crystals. Shionoya and Yen (1999) list a range of emission wavelengths for Ce^{3+} in phosphors of 300-420 nm. In general, Ce^{3+} emission appears to be structure and even sample specific (Blasse and Bril 1967). However, Ce^{3+} emission can be identified unambiguously via absorption and excitation spectroscopy, or by time-resolved laser spectroscopy. Figures 16 and 17 show several Ce^{3+} fluorapatite emission spectra obtained from time resolved measurements (Gaft et al. 2001a) that suggest the Ce^{3+} emission band may differ depending on which Ca site is occupied. Gaft et al. (2001a) obtained band positions of 360 nm for Ce^{3+} in the Ca1 site, and 430 nm for Ce^{3+} in the Ca2 site in two different natural samples. Other minerals showed similar Ce^{3+} bands, notably zircon (355 nm), anhydrite (320 and 340 nm), barite (330 and 360 nm), danburnite (346 and 367 nm) and datolite (335 and 360 nm), with Ce^{3+} presumably substituting at a Ca site in all cases except zircon. In these samples, the Eu^{2+} emission bands were always at longer wavelengths, and thus distinctive. As the Ce^{3+} luminescence transition is from the lowest level of the 5*d* manifold to a spin-orbit coupling split 4*f*

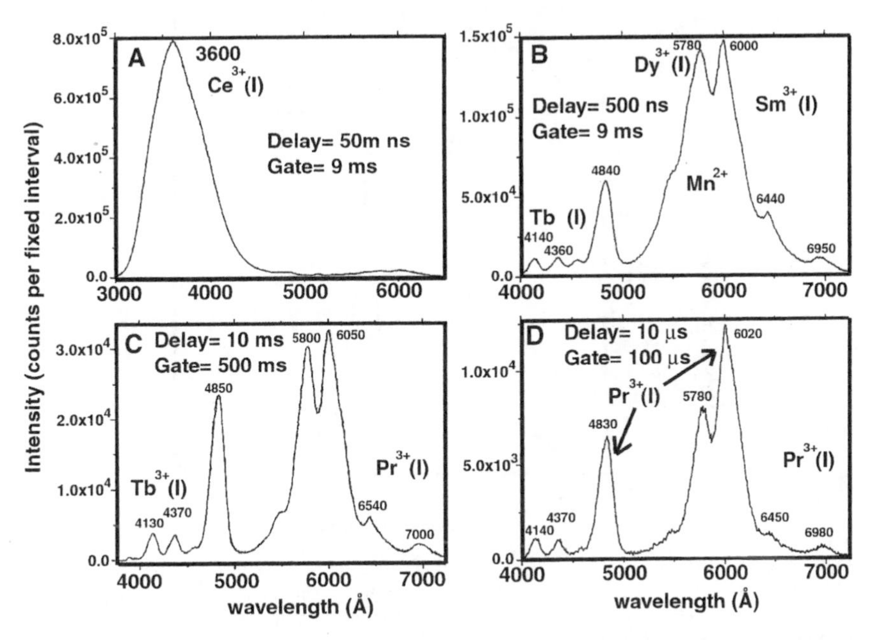

Figure 17. Emission from blue fluorapatite, Brazil, using a time resolved laser-excitation technique. Modified after Gaft et al. (2001a).

ground state, we expect two relatively broad bands for a given Ce^{3+} site (Blasse and Grabmaier 1994). Hence in all these minerals there are two bands for Ce^{3+} (except zircon), and in the Gaft et al. (2001a) apatite spectra the observed Ce^{3+} bands appear to each have two components that are incompletely resolved. In light of the result of Gaft et al. (2001a) the differing results of Mitchell et al. (1997) and Blanc et al. (1995) might conceivably be due to Ce^{3+} site partitioning as a function of dopant concentration.

Ce^{3+}-REE^{3+} energy transfer is strong for any REE ion that has transitions in the UV, such as Eu^{2+}, Pr^{3+}, Tm^{3+} and Tb^{3+}. Therefore in CL or photoluminescence spectra of apatites (and other minerals), Ce^{3+} emission may be quenched by energy transfer to these and other REE species, even with their concentration levels substantially lower than that of the Ce^{3+}. An example is given by the spectra of two fluorescent apatites from the Coldwell alkaline complex in Ontario (Mitchell et al. 1997). Both of these apatites have dominant Ce^{3+}, but only very weak emission in the general area of the Ce^{3+} bands. However, the Dy^{3+}, Sm^{3+} and Tb^{3+} emission bands are quite strong (Fig. 18). Contrariwise, Ce^{3+} emission might be expected to be strong in apatite group minerals with very low concentrations of other REE (and quenching ions), where the individual ions are too far apart on average for efficient energy transfer.

Praseodymium. Praseodymium is generally present at about 10% or less of the Ce^{3+} level in apatites. The CL spectrum of a synthetic doped sample shows two bands in the

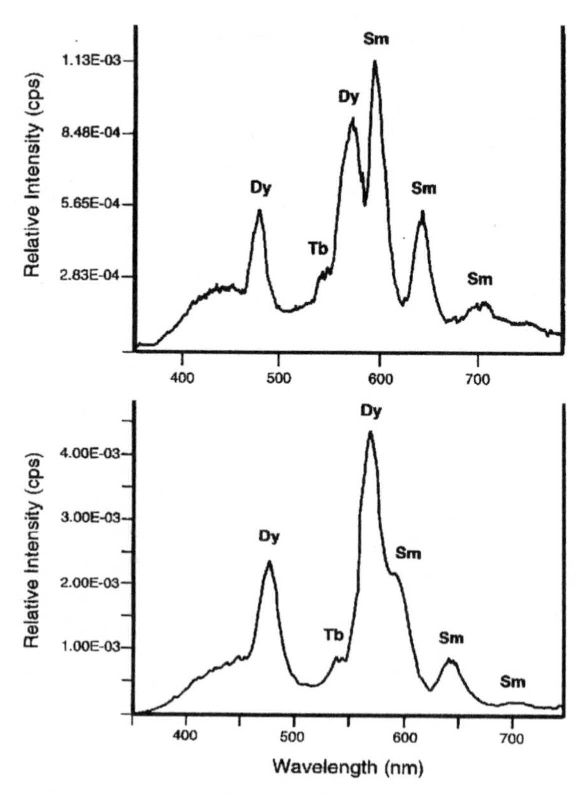

Figure 18. CL emission spectra of fluorapatite from the Coldwell complex, Ontario. Top: yellow-luminescent sample from a ferro-augite syenite in Center I. This type of emission may closely resemble Mn^{2+} broadband emission to the naked eye. Bottom: yellow-red luminescent sample from the same general region. Modified after Mitchell et al. (1997).

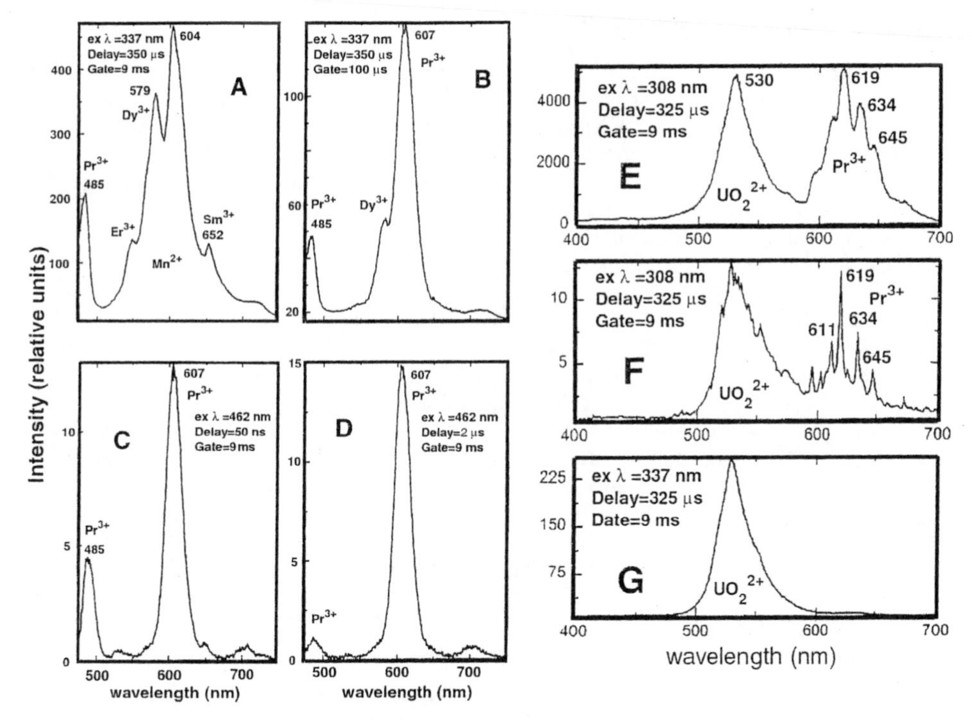

Figure 19. Pr^{3+} emission from a magmatic apatite revealed by time-resolved laser excitation. (A) and (B) 337 nm excitation. Shorter gate discriminates between Pr^{3+} and other REE emission. Delay stops quickly decaying Eu^{2+} and Ce^{3+} emission from obscuring the 485 nm band. (C) and (D) 462 nm excitation. Delay timing separation of Pr^{3+} emission. (E) Sedimentary apatite with uranyl and Pr^{3+} emission excited at 325 nm. (F) same as (E) but higher resolution spectral scan. (G) Pr^{3+} emission removed by excitation at 337 nm. Modified after Gaft et al. (1999).

UV at 248 and 278 nm, and sets of narrow bands centered at about 488 and 615 nm (Mitchell et al. 1997) yielding a brick-red emission color. Natural samples of hydroxylapatite and fluorapatite show similar CL bands (Gaft et al. 1999). Time-resolved emission spectra (Fig. 19) show well defined Pr^{3+} emission in natural magmatic apatite at 485 and 607 nm (Gaft et al. 1999). Studies using polarized emission measurements on this sample were interpreted as representative of only Ca1 site occupation by Pr^{3+} (Reisfeld et al. 1996). In contrast, a sedimentary apatite annealed in air showed a different Pr^{3+} spectrum, with a set of bands centered at about 630 nm (Fig. 19). This spectrum was interpreted to be due to Pr^{3+} in Ca2 (Reisfeld et al. 1996). Pr^{3+} appears to be an efficient sensitizer for Sm^{3+}, as many of its transition energies are almost identical to Sm^{3+}, and in general Pr^{3+} probably is more important as a sensitizer of other REE than for its own emission (Mitchell et al. 1997).

Neodymium. Neodymium can be present in relatively high concentrations in fluorapatites. Gaft et al. (2001a) lists Nd analyses for several natural apatites that are higher than any other REE except Ce, and at a concentration level of about 40% of the Ce value. Nd^{3+} emission is well into the IR, and it is not sensitized by most of the other REE. Hence, Nd^{3+} emission is expected to be relatively independent of other impurities, and will not contribute to visible luminescence. However, Nd^{3+}-doped synthetic apatites are excellent laser materials, due to several physical attributes of the Nd^{3+} electronic structure in the host lattice. Detailed evaluation of the optical properties of Nd^{3+} in Ba fluorapatite are given by Stefanos et al. (2000). The laser applications are considered below.

Samarium. Samarium is an important REE for apatite luminescence as it is frequently present in significant concentrations, is usually an efficient emission site, can receive energy from many REE sensitizers, and can have a large effect on the visible emission colors. The *f-f* transition spectra of Sm^{3+} in synthetic doped apatite has been reported by Taraschan (1978), Morozov et al. (1970), Mitchell et al. (1997) and Blanc et al. (1995) with reasonable agreement. The emission spectrum consists of 4 large bands, each composed of many transitions, at 558, 594, 639 and 701 nm (Mitchell et al. 1997). These bands produce an emission color that is reddish-orange both via CL and UV excitation. Sm^{3+} emission in natural apatites (Fig. 20) is consistent with the bands observed in the synthetics (Mitchell et al. 1997, Murray and Oreskes 1997, Roeder et al. 1987, Barbarand and Pagel 2001). Laser-excited time-resolved emission spectra show Sm^{3+} bands at 600, 644 and 695 nm in a high-REE blue fluorapatite (Gaft et al. 2001a), and 567, 598 and 646 in a colorless fluorapatite (Gaft et al. 1998a). Splitting of the Sm^{3+} bands in a yellow fluorapatite emission spectrum was interpreted as being due to Sm^{3+} in Ca1 and Ca2 sites (Reisfeld et al. 1996) (Fig. 21). Gruber et al. (1999) reported the absorption and emission spectra of Sm^{3+} in a synthetic Sr fluorapatite. The fluorescence intensities of the various bands were found to be similar to that for Sm^{3+} in doped glass laser material. These authors also have calculated the crystal field parameters for Sm^{3+} in the Ca2 site in Sr fluorapatite and the radiative lifetimes for the emission transitions.

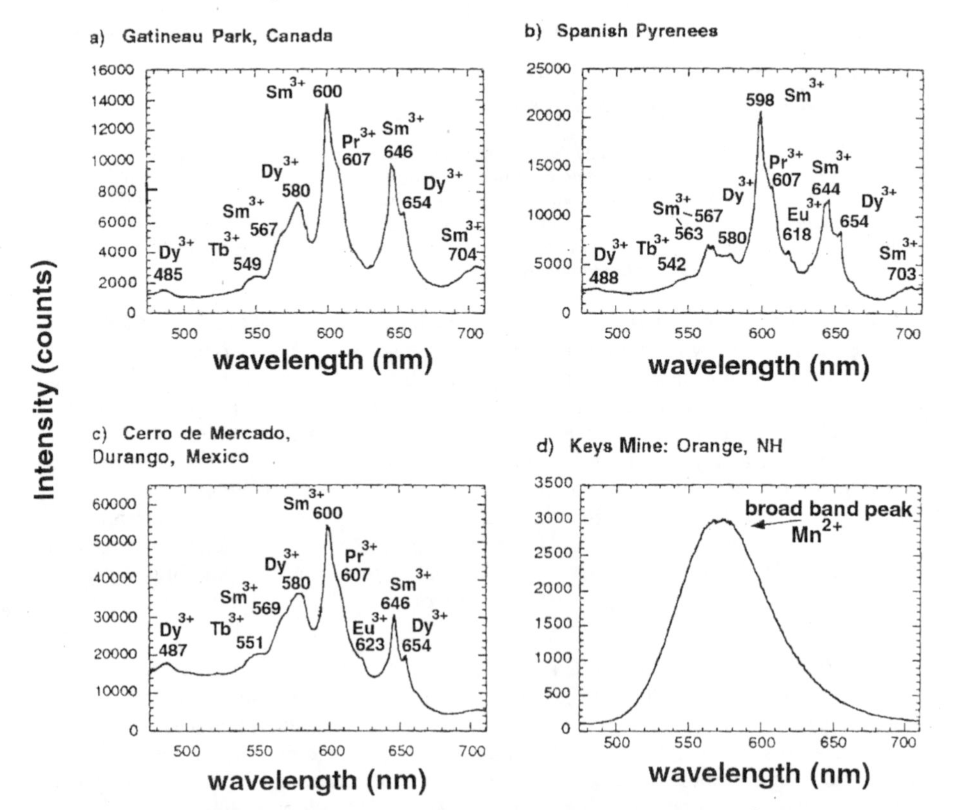

Figure 20. (a), (b) and (c) CL emission spectra for apatites with significant Sm^{3+} and other REE activators. Compare with the clean Mn^{2+} spectrum from the NH-apatite (d). Modified after Murray and Oreskes (1997).

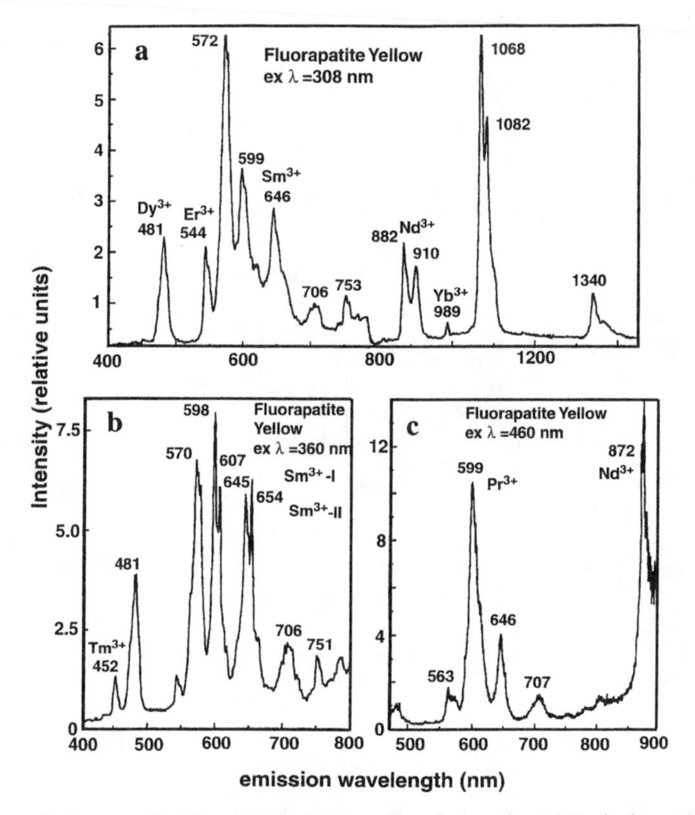

Figure 21. Laser excited fluorescence from a yellow fluorapatite. (a) Excitation at 308 nm. Note strong Nd^{3+} bands in the IR. (b) Sm^{3+} emission separated into contributions from Ca1 and Ca2 sites using 360 nm excitation. (c) Removal of Dy^{3+} emission using 460 nm excitation. Modified after Reisfeld et al. (1996).

Europium. Trivalent europium gives rise to red emission with CL bands at 587, 614, 645 and 694 nm (Mitchell et al. 1997, Mariano 1988, Mariano and Ring 1975, Roeder et al. 1987, Jagannathan and Kottaisamy 1995). All bands are due to *f-f* type transitions, and are composed of many such electronic transitions. As with other *f-f* emission spectra, doped synthetic apatite and natural apatite agree closely, but in natural samples there is considerable overlap of the 587 band with other REE emission features. Laser activated time-resolved spectra at 266 nm yield well defined bands at 579, 590, 618, 653 and 700 nm attributed to Eu^{3+} in the Ca1 site (Gaft et al. 1997a,b). Laser excitation at 337 and 355 nm allows detection of bands at 575, 628 and 712 nm that are attributed to Eu^{3+} in the Ca2 site because of shorter luminescence lifetime (Fig. 22). In general, a less symmetric site produces a shorter excited state lifetime by allowing increased lattice interactions. Eu^{3+} emission bands overlap with Sm^{3+} bands as well as bands from other REE. Hence assignment of Eu^{3+} features requires careful comparison with doped standards, and use of time-resolution measurement techniques if available. Piriou et al. (2001) obtained high resolution spectra of Eu^{3+} in a synthetic sodium lead apatite, and confirmed that Eu^{3+} spectra can be used to determine individual site occupations due to symmetry, splitting and lifetime effects. Eu^{3+} doped into synthetic Ca hydroxyapatite was also investigated by Ternane et al. (1999) with time-resolved techniques (see below).

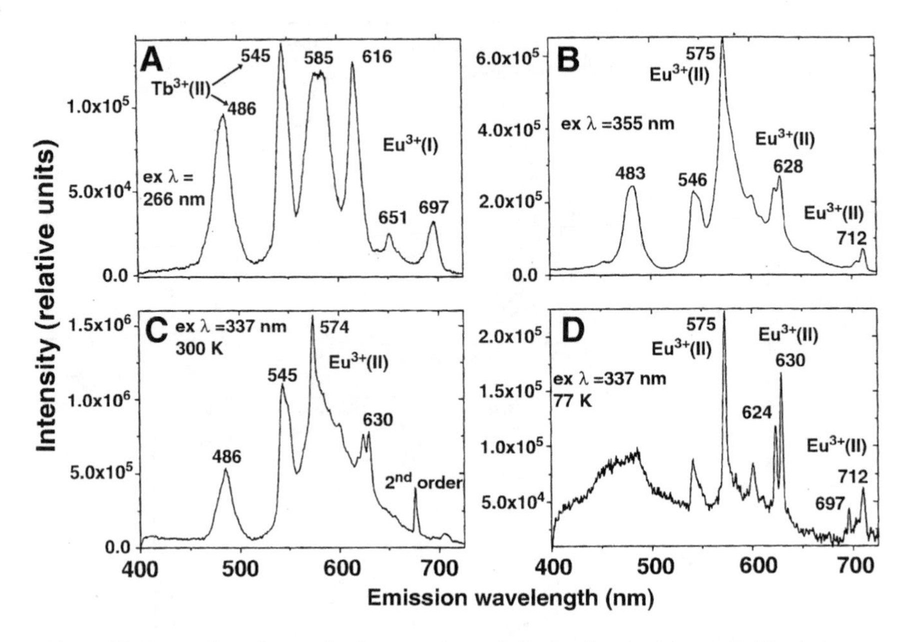

Figure 22. Comparison of strategies for separating emission bands using laser-excited luminescence on a Norway fluorapatite. (A-D) Effects of excitation wavelength and cooling.

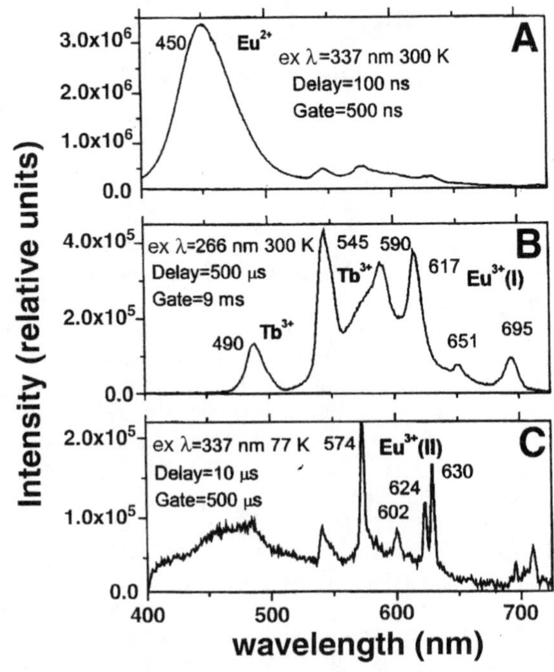

Figure 22, continued. Comparison of strategies for separating emission bands using laser-excited luminescence on a Norway fluorapatite.(A-C) Use of gate and delay timing to separate Eu^{2+} and multi-site Eu^{3+} emission bands. Modified after Gaft et al. (1997a,b; 2001a).

Gadolinium and terbium. Gd^{3+} *f-f* emission bands appear only in the mid-range UV (UVB), and do not contribute to visible luminescence. No spectra of Gd^{3+} in apatite or other Ca-containing minerals have been found in the literature. Terbium *f-f* emission spectra in synthetic fluorapatite have been measured by Blanc et al. (1995, 2000) via CL, and by Morozov et al. (1970) via photoluminescence with reasonable agreement. A large set of sharp bands is observed from 370 to 600 nm, with the main sets centered at about 375, 415, 490, 550 and 590 nm. The only band typically observed in natural apatites via CL or UV-excitation is at 545 nm (Roeder et al. 1987, Mitchell et al. 1997), and appears to have the longest lifetime of any of the emission states. Using laser-excitation and time-resolution Gaft et al. (2001a) found other Tb^{3+} bands at 380, 415 and 437 nm. These authors state that the 545 band is strongest in synthetic apatite and represents Tb^{3+} in the Ca2 site, while the 415 and 437 nm bands are due to Tb^{3+} in the Ca1 site and are strongest in natural apatites. This seems to be inconsistent with the other spectral observations for Tb^{3+} in natural apatite, including another spectrum reported by Gaft et al. (2001a) of a red natural apatite which shows only a strong 545 nm Tb^{3+} band (Fig. 22). Tb^{3+} can be a sensitizer for Eu^{3+} (Tachihante et al. 1996) so that its excited state energy may be resonantly siphoned off into excitation of Eu^{3+}.

Dysprosium. Dysprosium *f-f* emission bands produce a yellow luminescence. The CL emission spectrum in doped synthetic fluorapatite consists of bands at 475, 570, 657 and 746 nm, each composed of several sharp transitions visible in high resolution conditions (Mitchell et al. 1997). This spectrum agrees with CL and photoluminescence spectra collected by Blanc et al. (1995) and Morozov et al. (1970), respectively, also for synthetics. The Dy^{3+} bands in natural apatite group minerals are commonly observed, and often are some of the most intense REE contributions to the emission. We can infer from this that Dy^{3+} is probably the beneficiary of other REE acting as sensitizers. Mitchell et al. (1997) observed bands at 475 and 570 nm in all three of their natural apatite samples from the Coldwell alkaline complex. In another sample from Waldigee Hills, Australia with low REE content an unidentified band and another due to Mn could easily be reassigned to Dy^{3+}. Roeder et al. (1987) and Mariano (1988) observed similar bands in natural apatites from many additional areas. Barbarand and Pagel (2001) also detected a Dy^{3+} band at 665 nm in apatite from Arendal, Bamble, Norway. Laser excitation studies by Gaft et al. (2001a) in natural apatites reveals bands at the same positions, plus an additional band at 750 nm.

Holmium, erbium, thulium and ytterbium. Holmium appears to be a weak emission center compared to other divalent and trivalent REE, and will not contribute detectably to natural apatite emission spectra in the presence of Eu, Sm or Dy. A synthetic apatite CL spectra for Ho^{3+} has been collected by Blanc et al. (1995), and Gruber et al. (1997b) have investigated the emission spectrum in Sr fluorapatite. Holmium is a potential luminescence quencher due to non-radiative resonant energy transfer. Erbium, like Holmium, is a weak emission center, and not an important player in natural apatite luminescence. It has been explored as a dopant in Sr fluorapatite for use in laser applications (Gruber et al. 1997a). Er^{3+} spectra obtained from synthetic fluorapatite do not show agreement. The CL spectrum of a synthetic Er^{3+} apatite studied by Mitchell et al. (1997) shows only a broad band near 440 nm, resembling an Eu^{2+} band. In contrast, Blanc et al. (1995) obtained a CL spectrum with well defined bands at 319, 402, 471, 527 and 552 nm on a broad background. Although Er^{3+} bands are not assigned in any of their natural apatite emission spectra, several bands were observed in scheelite at 524 and 551 by Gaft et al. (2001a). Thulium is a weak luminescence activator (Mitchell et al. 1997) and not a significant contributor to apatite luminescence. A Tm^{3+} band has been observed by Gaft et al. (1998a) in a yellow fluorapatite from the Kola peninsula at 452 nm. This is in agreement with bands observed at 356 and 461 nm in synthetic doped fluorapatite using CL by Blanc et al. (1995). Reisfeld et al. (1996) and Gaft et al. (1997b) associate the Tm^{3+} emission with Ca2 occupation. Gruber et al. (1998) have obtained the crystal field parameters for Tm^{3+} in the Ca2 site from optical spectra and lattice-sum electrostatic calculations. Ytterbium has no luminescence emission in the visible, with bands near 1000 nm (DeLoach et al. 1994). It is not important as a sensitizer or activator in natural apatites. Reisfeld et al. (1996) assigns the emission bands as due to Yb^{3+} occupation of the Ca2 site. Gruber et al. (1998) have derived the crystal field parameters for Yb^{3+} in the Ca2 site in Sr fluorapatite. Ytterbium can be sensitized by other REE ions, an example of this is shown in a cartoon from Marfunin (1979) in Figure 23.

Uranium. Hexavalent uranium as the uranyl group, UO_2^{2+}, and the uranate ion U^6+, have characteristic luminescence spectra in model compounds (Blasse et al. 1979, Blasse 1988, Blasse and Grabmaier 1994, DeNeufville et al. 1981). The linear uranyl ion yields characteristic vibronic emission spectral structure that is readily identified at low temperatures and is quite distinct from uranate ion spectra. Uranyl ion is known to substitute in many mineral structures, sometimes on lattice sites, and sometimes as an interstitial or inclusional impurity. In synthetic powder fluorapatites, recent EXAFS analysis by Rakovan et al. (2002) showed that hexavalent uranium (2.6 wt %) was present mainly as the uranate species, with six approximately equal U-O distances of 2.06 Å. The uranate resided mainly in the Ca1 site and utilized the innermost six oxygen ligands. Luminescence spectra was consistent with this assignment. This result is interesting as most hexavalent uranium in mineral structures and in aqueous solutions is believed present as the uranyl specie. For example, Gaft et al. (1996b, 1999) examined fossil apatites (see below) and concluded that the uranium content was diagenetic in origin and only included uranyl ions. In sedimentary apatite Panczer et al. (1998) determined that uranyl was present both as sorbed complexes, and as films of uranyl-bearing minerals, based on both emission spectra and lifetime measurements. However, in magmatic apatites Panczer et al. (1998) did not observe uranium luminescence until after a 1100 K anneal. This led to the suggestion that the uranium was originally present in the tetravalent state. This is indeed permissible due to its similar ionic radius (0.97 Å) compared to Ca (0.99Å). Such a substitution could be charge balanced by Na substitution for Ca in two Ca sites. It was further conjectured that during the anneal uranium was oxidized to the hexavalent state, and formed a complex with OH and F ligands from the apatite.

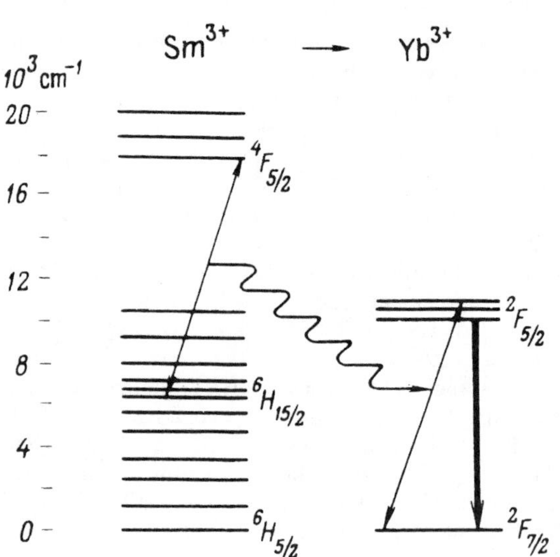

Figure 23. A possible mechanism for energy transfer from Sm^{3+} to Yb^{3+}. This type of transition is common among trivalent REE due to similar state symmetry and transition energies. The heavy line in the Yb^{3+} state diagram is the ultimate emission transition. Modified after Marfunin (1979).

Luminescence in other apatite group minerals

Hydroxylapatite $Ca_5(PO_4)_3(OH)$, carbonate apatites and biological apatite have been extensively examined due to their significance in human and animal skeletal and dental material (Elliott 1994, McConnel 1973, Grandjean-Lecuyer et al. 1993, Gotze et al. 2001). However, CL and other fluorescence emission have been less studied, in part due to the usually very low concentrations of Mn^{2+} (about 1 ppm) and REE activators (often well below 1 ppm) in the biological samples, and the inapplicability of hydroxylapatite and carbonate apatites for phosphor utilization. Habermann et al. (2000) studied apatite samples and fossil conodonts (francolite-carbonate apatite) with micro-PIXE, CL and ESR. They found that the concentration distribution of REE in the conodonts were inconsistent with REE concentrations in seawater, and also different from that in recent fish bone debris. In particular, Ce^{3+} and Eu^{3+} were enriched relative to both standards. They concluded that most of the REE were added during diagenetic alteration. CL spectra showed variations in REE composition with position in the conodonts, and strong contributions from trivalent Sm, Dy, and Tb (Fig. 24). Habermann et al. (2000) also refer to an "intrinsic" blue emission from the conodonts, which is not explained. Gaft et al. (1996a,b) examined the laser-excited emission from fossil fish teeth from phosphorite deposits in Israel. Untreated samples showed bands due to uranyl ion with a short lifetime and possible organic material with very long lifetimes, always occurring together in the

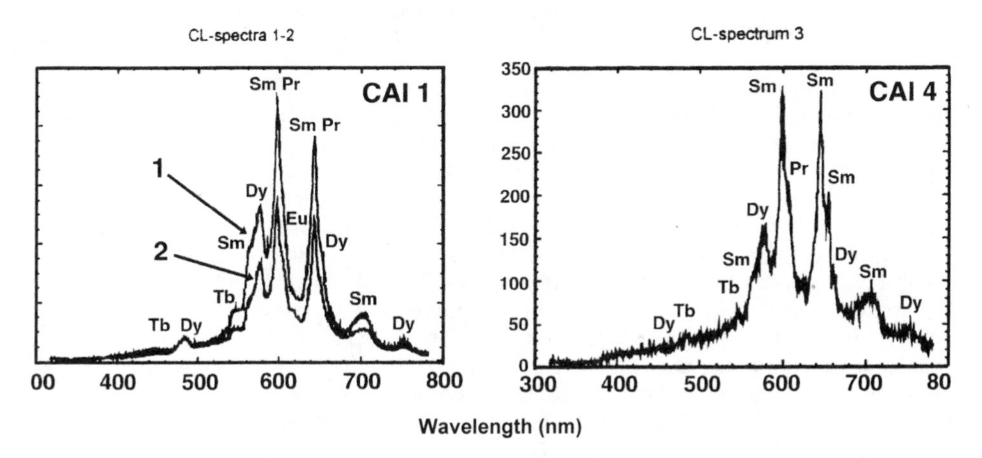

Figure 24. Cathodoluminescence spectra of Triassic conodonts (carbonate-fluorapatite) from Germany. 1 and 2 in the left-hand spectra refer to two points on the same fossil. Modified after Habermann et al. (2000).

same relative intensity, suggesting a mixture of uranyl with organic contaminants (Fig. 25). Annealing of the samples at 1000 K produced a better resolved and more intense uranyl emission, a reduced organic emission, and appearance of REE bands from several species. The authors speculated that dehydration and organic loss may have removed quenching mechanisms from the REE environments, e.g., coordination with water molecules. In work on synthetic hydroxylapatites, Ramesh and Jagannathan (2000) examined colloidally-formed nanocrystals of Sr chlorohydroxylapatite with Ce^{3+} dopant. They found that the luminescence was complex and markedly affected by changes in preparation conditions, possibly due to varying ratios of structural, surface and intergrain sites for the activator. Luminescence studies of Eu^{3+} in synthetic precipitated Ca hydroxylapatite (Ternane et al. 1999) showed less complexity, and a preference for the Ca2 site with two types of charge compensating neighbor arrangements.

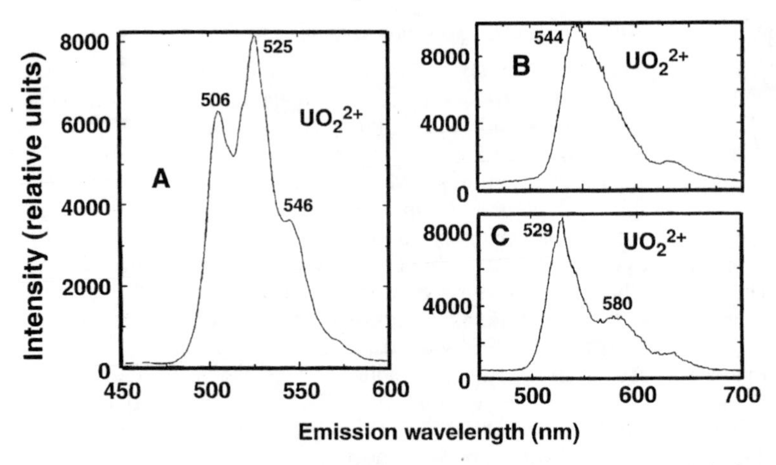

Figure 25. Laser-excited luminescence in fossil fish teeth from phosphorite beds in the Senonian Mishash formation, Israel. (A) Uranyl emission at 77 K indicating uranyl aqueous complexes. (B) after annealing at 1000 K the uranyl complexes are "dehydrated". (C) Same as in A but spectrum collected at 300 K. From these data it was concluded that the uranyl resides in organic material and not in the apatite structure. Modified after Gaft et al. (1996b).

Arsenate apatites Johnbaumite $Ca_5(AsO_4)_3(OH)$ and Svabite $Ca_5(AsO_4)_3F$ are noted above as having luminescence somewhat contradicting other observations. Some of the inconsistencies may be due to improper identification of samples characterized by UV light excitation. Svabite from Långban, Sweden has strong orange emission in samples studied by the author, but can be intimately admixed with other apatite group minerals with slightly different fluorescence. Svabite from Franklin, New Jersey has similar luminescence, perhaps slightly shifted to more yellow color. Johnbaumite from Jacobsberg, Sweden has interesting pink-violet UV-excited fluorescence that slightly varies over a specimen. It commonly occurs admixed with luminescent svabite and calcite, yielding samples with three color emission: deep red, orange, and pink-violet. Johnbaumite from Franklin, New Jersey, in contrast, seems to always show the orange emission. This suggests that the Jacobsberg samples have considerable REE activation (probably Eu^{2+} or Ce^{3+}), while the Franklin samples have mainly Mn^{2+} activation.

Lead arsenate apatites hedyphane $Pb_3Ca_2(AsO_4)_3Cl$, and mimetite $Pb_5(AsO_4)_3Cl$ are often luminescent, and the author has verified yellow-orange UV-excited emission in mimetite from Långban. Some of the mimetite samples have a golden emission that is particularly beautiful in association with luminescent calcite and svabite. Mimetite from Johanngeorgenstadt, Saxony, Germany show a yellow-orange emission under SWUV, but a pink emission under LWUV (Robbins 1994). This type of emission color change is not unusual in apatite group minerals. It is likely due to mainly Mn^{2+} emission from SWUV excitation, and combined Mn^{2+}, Ce^{3+}, and Eu^{2+} emission from LWUV excitation. As Pb^{2+} absorption is strong for SWUV but not LWUV, SWUV will excite $Pb^{2+} \rightarrow Mn^{2+}$ energy transfer and strong Mn^{2+} emission.

Lead apatites pyromorphite $Pb_5(PO_4)_3Cl$ and turneaurite $Pb_5(AsO_4,PO_4)_3Cl$ are also commonly fluorescent. Turneaurites from Franklin, New Jersey and Balmat, New York have an orange and orange-red emission, respectively, under UV excitation. Pyromorphite generally has strong physical coloration, which filters luminescence, but orange to yellow UV-excited luminescence has been observed in specimens from Broken Hill, NSW, Australia (Gait and Back 1992). Robbins (1994) mentions pyromorphite from

Granite Co., Montana that has a strong orange emission under LWUV, and pink-orange luminescence has also been observed in samples from several localities.

Vanadinite $Pb_5(VO_4)_3Cl$ is generally a strongly colored mineral that rarely shows fluorescence, but the weakest colored varieties, straw-yellow in physical color, can show yellow or greenish-yellow emission. Robbins (1994) notes that such vanadinite from Santa Eulalia, Chihuahua, Mexico is probably endlichite, with a large amount of arsenate substituting for vanadate.

LUMINESCENCE ZONING

Study of luminescence zoning can yield a wealth of information about the morphologic history of a crystal, changes in the environment during crystal growth, growth mechanism, structural differences at the crystal surface, and single crystal ages (Rakovan and Reeder 1994, 1996; Shore and Fowler 1996, Coulson and Chambers 1996, Rakovan et al. 1997, Rakovan et al. 2001). The most familiar type of luminescent zoning observed in apatite is concentric zoning; luminescent differences between concentric layers from the center of the crystal outward. Concentric zoning reflects temporal changes in the environment while a crystal is growing. Figure 6—Color Plates 11 and 14—show concentric zoning. The luminescence zoning seen in Figure 6—Color Plate 8—reflects a type of concentric zoning (temporal variation in luminescent activators) where the last stages of growth occurred at the ends of the crystals only. Oscillatory zoning, similar in appearance to concentric zoning, is thought in many cases to be due to quasiperiodic fluctuations in the chemistry at the mineral-solution or mineral-melt interface caused by diffusion-limited supply of growth units or substituent ions to the crystal surface (Shore and Fowler 1996). Oscillatory zoning is usually distinguished from normal concentric zoning by the scale of the zonation pattern, from tens of nanometers to tens of microns. Selective attachment and incorporation of substituent elements between structurally different regions of apatite surfaces give rise to two similar types of luminescent zoning, sectoral and intrasectoral zoning (Rakovan this volume). Sectoral and intrasectoral zoning of Eu^{2+}, Ce^{3+}, Eu^{3+}, Sm^{3+} as well as Mn^{2+} leads to spectacular cathodoluminescence zoning in apatites from Llallagua, Bolivia and the Golconda Mine, Minas Gerais, Brazil (see Color Plates 1-4; Figs. 15, 20, 21, 24-26 in Rakovan, this volume, p 51ff).

APATITE THERMOLUMINESCENCE

Thermoluminescence (TL) is a widely used technique for the determination of shallow energy states in solids due to trapped electrons and holes. The basic principle is that an applied temperature will release trapped electrons when kT (k = Boltzmann constant, and T = absolute temperature) is equivalent to the energy difference between the trapped state and the conduction band. Hence a type of spectrum is produced by slowly ramping the temperature of a material while its spectral emission is detected. Such a spectrum is called a glow curve, and usually plots all emitted photon intensity without discrimination of wavelength. After such a spectrum has been obtained, the full emission spectrum for every peak (i.e., temperature) in the glow curve can be obtained. In this way the ions associated with each trapping state can be identified (McKeever 1985).

TL is used for determination of equilibration temperatures in minerals, i.e., the highest temperature the mineral has experienced with sufficient time to empty traps corresponding to that or lower energies. In this case the threshold for the appearance of any emission is roughly equivalent to the equilibration temperature. Another use of TL is in dating materials and artifacts. In this case one must know the rate at which trapped electrons are generated in the material. As the electrons are normally trapped by the action of high energy radiation, the integrated radiation dose may potentially be estimated from the measured TL emission, and then the age of the material calculated. This technique is widely used in determination of the age of pottery after firing, and is as accurate as the estimation of the dose rate allows.

TL is used in meteorite geochemistry to determine thermal history, terrestrial age, petrologic type, cosmic ray exposure rate, and other characteristics of meteorites (Sears 1980, Guimon et al. 1984, McKeever and Sears 1979). Remarkably, information on the orbital evolution of selected meteorites can also be determined using TL (Benoit and Sears 1997, 1991). As apatite can occur in practically every type of rock type, one might think that TL studies of apatites could be of significant use in terrestrial paleothermometry, but TL work has been limited. Ratnam et al. (1980) found that TL of a Mn^{2+} doped synthetic fluorapatite irradiated by x-rays from a laboratory source yielded glow curves with emission peaks at temperatures of 145, 185, 260 and 395°C. Spectra collected at several of these peaks showed the broad Mn^{2+} emission band. The band appeared to have two components at 560 and 595 nm, which were attributed to Mn^{2+} in the two Ca sites. Lapraz and Baumer (1983) collected TL spectra from thirty natural and five synthetic fluorapatites at temperatures between 77 and 750 K, both without treatment, and after x-ray irradiation. They found a large number of TL glow curve peaks, most of which could be associated with Mn^{2+} , Ce^{3+} , Eu^{2+} and Eu^{3+}. Individual spectra taken at the glow curve peaks showed good agreement with luminescence spectra in the literature (Figs. 26 and 27). Emission from other REE, though know to occur in the natural samples, was not observed. This suggests that those REE are not associated with trapping sites. An interesting result of this study and subsequent ones (Baumer et al. 1987, Lapraz et al. 1985) was the identification of a "lattice emission" band at about 400 nm which was believed to be associated with defects at the phosphate site. Perhaps this is the band identified by Habermann et al. (2000) as intrinsic in conodont emission spectra. This band was found to be strongest in undoped synthetic materials, and was quenched with addition of activators to synthetics or in natural samples, presumably due to energy transfer to the activator sites. Another finding was that many of the TL glow curve peaks were present even in the absence of activators, and were intensified with doping, or in naturals. This suggested that the peaks are related to intrinsic lattice defects, rather than to defect states associated with the activators. Results on hydroxyapatites and chlorapatites followed those from fluorapatite.

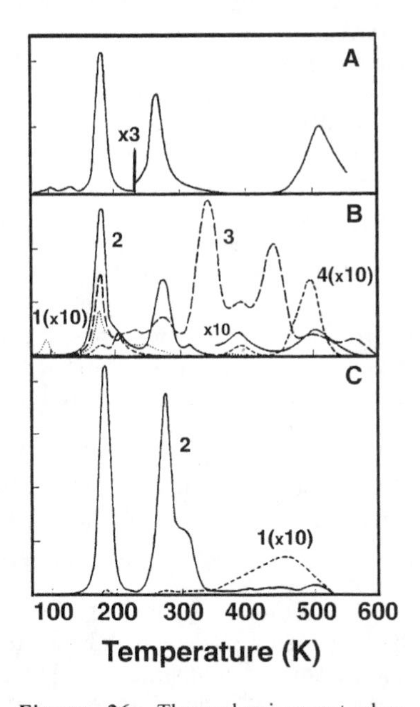

Figure 26. Thermoluminescent glow curves from synthetic fluorapatite after being irradiated with x-rays at 77 K: (A) doped with 0.2% Mn. (B) 1-undoped; 2-doped with 0.2% Mn; 3-doped with 0.2% Ce; 4-doped with 0.2% Eu. (C) 1- doped with 0.3% Mn; 2-doped with 0.3% Mn with hydrothermal anneal at 973 K. Each peak represents the emptying of a different set of trapped electrons. Modified after Lapraz and Baumer (1983).

A comparison of TL intensity with fission track density was done by Al-Khalifa et al. (1988) in a set of natural apatites. They found that increased densities of fission tracks were correlated with a decrease in TL sensitivity, or TL output per unit dose. The results with the naturals were supplemented with apatite samples subjected to varying doses of 30 MeV alpha particles from a cyclotron, which verified the TL sensitivity reduction. Chlorapatites were more prone to the TL sensitivity loss than fluorapatites. This work indicates that apatites may be difficult to employ as radiation flux detectors, and thus not practical for long-term age dating.

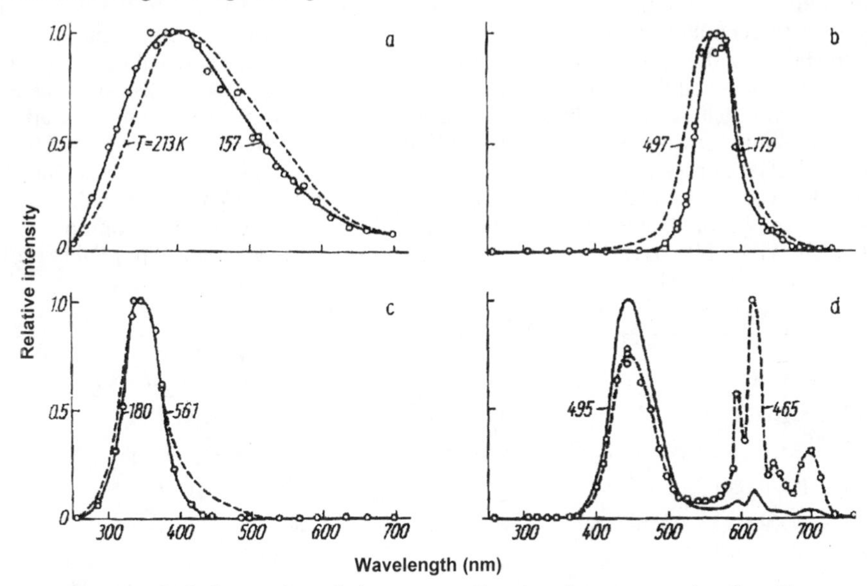

Figure 27. Synthetic fluorapatite emission spectra collected at glow curve maxima from Figure 26. (a) undoped. (b) 0.2% Mn. (c) 0.2% Ce. (d) 0.2% Eu. Modified after Lapraz and Baumer (1983).

SYNTHETIC APATITE USE IN FLUORESCENT LAMPS AND LASERS

Fluorescent lamp phosphors

Early fluorescent lamps (1938-1948) utilized mixtures of two phosphors, $MgWO_4$ (scheelite structure with broad band emission at 480 nm) and $(Zn,Be)_2SiO_4:Mn^{2+}$ (phenacite structure with two broad bands at 525 and 600 nm). These phosphors were initially replaced due to the poor lifetime of the phenacite phosphor, but the toxic aspects of the Be was also a consideration. Chloro- and fluor-apatites (called halophosphors) have been used since 1948 (Blasse and Grabmaier 1994, Jenkins et al. 1949). Numerous different activators and coactivators have been tested in apatite host matrices, but few have been found to be practical. For example, Ce^{3+} doped fluorapatite, either alone or coactivated with Mn^{2+}, develops parasitic emission bands that markedly affect emission color (Chenot et al. 1981). However, Eu^{2+}, Sb^{3+} and Mn^{2+}-Sb^{3+} doped halophosphors are efficient and resist degradation. Remarkably, the precise site occupation of the Sb^{3+} remains in question. Oomen et al. (1988) were able to explain the luminescence spectra and optical properties of Sb^{3+} by assuming a strong Jahn-Teller effect on the excited state. Their analysis was consistent with Sb^{3+} occupation of a Ca site. The structure of Sb^{3+} doped calcium fluorapatite was investigated by DeBoer et al. (1991) via Rietveld powder x-ray diffraction. They found that most of the Sb^{3+} appeared to enter the Ca2 site in a 2.2 wt % antimony sample (0.185 Sb atoms per 10 Ca atoms). However, a 3.1 wt % antimony sample gave conflicting results, indicating defect site occupation. In contrast, Moran et al. (1992) examined a series of Ca fluorapatites doped with 0-3.0 wt% Sb^{3+} via

^{19}F and ^{31}P magic-angle spinning NMR. They found evidence for Sb^{3+} occupation of the phosphate site. Table 1 (information from Shionoya and Yen 1999) shows a list of lamp phosphors in common use. Representative emission spectra are shown in Figure 28.

A white halophosphor has one major drawback, as it cannot be optimized for both high brightness and high color rendering. Color rendering refers to the ability of a light source to accurately match the output of a black body source with respect to color perception. If the rendering is poor, objects do not appear to the naked eye as having realistic colors and the color rendering index, or CRI, will be low. Blasse and Grabmaier (1994) note that at highest brightness for a white halophosphor (80 lumens/Watt), the CRI is about 60, while if the Mn and Sb levels are adjusted to obtain a CRI of 90, then the brightness drops to about 50 lumens/Watt. Using three phosphors (two activated by REE) solved this problem. The resulting "three color lamp" has phosphors with relatively narrow emission bands at 450, 550 and 610 nm. It can reach an output of 100 lumens/Watt while having a CRI of 85. Usually only the blue phosphor of this set is a

Table 1. Fluorescent lamp synthetic apatite phosphors.

Composition	**Emission color*	*Emission peak(s)*	*Application*
$3Ca_3(PO_4)_2\ Ca(F,Cl)_2{:}Sb^{3+}$	blue white	480	Fl lamps
§$3Ca_3(PO_4)_2\ Ca(F,Cl)_2{:}Sb^{3+},Mn^{2+}$	daylight	480, 575	Fl lamps
§$3Ca_3(PO_4)_2\ Ca(F,Cl)_2{:}Sb^{3+},Mn^{2+}$	white	480, 575	Fl lamps
§$3Ca_3(PO_4)_2\ Ca(F,Cl)_2{:}Sb^{3+},Mn^{2+}$	warm white	480, 580	Fl lamps
$Sr_{10}(PO_4)_6\ Cl_2{:}Eu^{2+}$	blue	447	Three-band and high-pressure lamps
$(Sr,Ca)_{10}(PO_4)_6\ Cl_2{:}Eu^{2+}$	blue	452	Three-band lamps
$(Sr,Ca)_{10}(PO_4)_6\ nB_2O_3{:}Eu^{2+}$	blue	452	Three-band lamps
$(Sr,Ca,Mg)_{10}(PO_4)_6\ Cl_2{:}Eu^{2+}$	blue-green	483	Color reproduction
$(Sr,Ca,Ba)_{10}(PO_4)_6\ Cl_2{:}Eu^{2+}$	blue	445	Color reproduction
$(Ba,Ca,Mg)_{10}(PO_4)_6\ Cl_2{:}Eu^{2+}$	varies with Ba/Ca ratio		Color reproduction

*Industry trade names

§Variable concentrations of Sb^{3+} and Mn^{2+} amongst the three formulas.

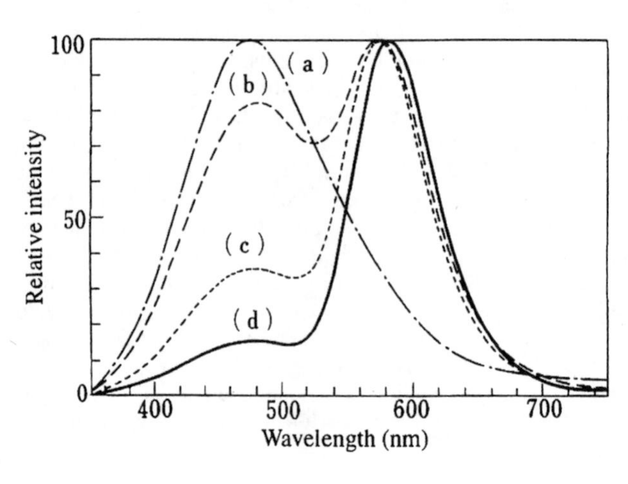

Figure 28. Comparison of halophosphate phosphor emission spectra. Modified after Shionoya and Yen (1999).

(a) Blue-white. Ratio of $Mn^{2+}{:}Sb^{3+}$ = 0:0.15,

(b) Daylight 0.08:0.08,

(c) Cool-white 0.17:0.08,

(d) Warm-white 0.24:0.08.

halophosphor. The white colors noted in Table 1 can be referred to temperatures for a black body emitter obeying Plank's law. In this scheme, white is equivalent to black body emission at 3500 K, while warm white is equivalent to emission at 3000 K. "Cooler" light richer in blue would be produced at higher black body temperatures. High-pressure Hg lamps require another type of lamp phosphor. This Hg discharge has increased emission at 365 nm relative to the low-pressure standard fluorescent lamp, so the phosphor must absorb both SWUV and LWUV (= UVC and UVA) to efficiently produce white light. In such high-pressure lamps, the operational temperature is in the range of 600 K, so the phosphor must have a very high threshold for thermal quenching. Under these conditions Sr chlorapatite doped with Eu^{2+} operates well, but other halophosphors do not perform as well as Eu^{3+} and Tb^{3+} activated oxide phosphors.

Laser applications

Fluorapatites have been examined fairly extensively as CW laser host crystals since the discovery by Ohlmann et al. (1968) that Nd^{3+} activated fluorapatite had much higher gain and efficiency than any other Nd^{3+} host. The excellent laser properties are due to the very sharp 1060 nm emission band and the broad absorption spectrum within the large band gap. Unfortunately, these advantages are not enough to overcome the low thermal conductivity of fluorapatite, which causes build up of temperature and crystal distortion even at low power outputs. Additionally, problems existed in the preparation of large optical grade crystals of fluorapatite. Thus other apatite compositions and other dopants have been examined to explore ways to improve thermal conductivity or reduce emission bandwidth. Steinbruegge et al. (1972) examined Nd^{3+}, Ho^{3+} and Tm^{3+} doped Ca, Mg, Sr and Ba fluorapatite, CaLa oxyapatite, and CaY, CaLa and SrLa silicate apatite. The silicate oxyapatites could accommodate more REE than the phosphate apatites, as well as REE of either trivalent or divalent charge, and had fluorescence line widths about ten times broader than in fluorapatite. They are also harder and more thermal shock resistant than Nd^{3+}doped fluorapatite, and could be grown into high quality large crystals without difficulty.

With the advent of diode laser pumping it became practical to use fluorapatite once again as a laser host for Nd^{3+} because the crystal size could be markedly reduced and thus the thermal load decreased. Zhang et al. (1994) prepared such small fluorapatite crystals and demonstrated diode pumped lasing action with high efficiency. Another set of apatite laser hosts were explored by Payne et al. (1994a) using Yb-doped Ca, Sr, and CaSr fluorapatites, and Sr vanadate apatite. Yb^{3+} is a useful laser dopant species in YAG, but is limited due to the paucity of available absorption transitions for pumping using flash lamps and other undirected excitation methods. As with Nd^{3+} doped fluorapatite, this difficulty was also solved by the use of semiconductor diode lasers as pumps, as they can be made to produce directed excitation beams with narrow emission band widths. One way to measure laser materials utility is via a plot of the absorption cross section versus the minimum pump saturation intensity, i.e., the minimum pump intensity needed to reach laser action. Such a plot is shown in Figure 29, and shows the favorable location of the Yb^{3+}-doped apatites prepared by Payne et al. (1994a).

Another approach in apatite structure laser hosts is that by Scott et al. (1997) using $3d^2$ ions substituted into the phosphate sites of Ca fluorapatite and Sr vanadate apatite. In Ca fluorapatite, Cr^{4+} substitution required that one of the tetrahedral ligand oxygens be replaced by fluorine for charge compensation, which causes inhomogeneous broadening of the spectral bands. There is no analogous problem for Mn^{5+} substitution, and $Sr_{10}(VO_4)_6F_2$:Mn^{5+} showed ideal laser characteristics with emission at 1160 nm. Mn^{5+} doped into Sr fluorapatite and other phosphates for laser applications has also been examined by Copobianco et al. (1992), and the luminescence properties of Mn^{5+} in a

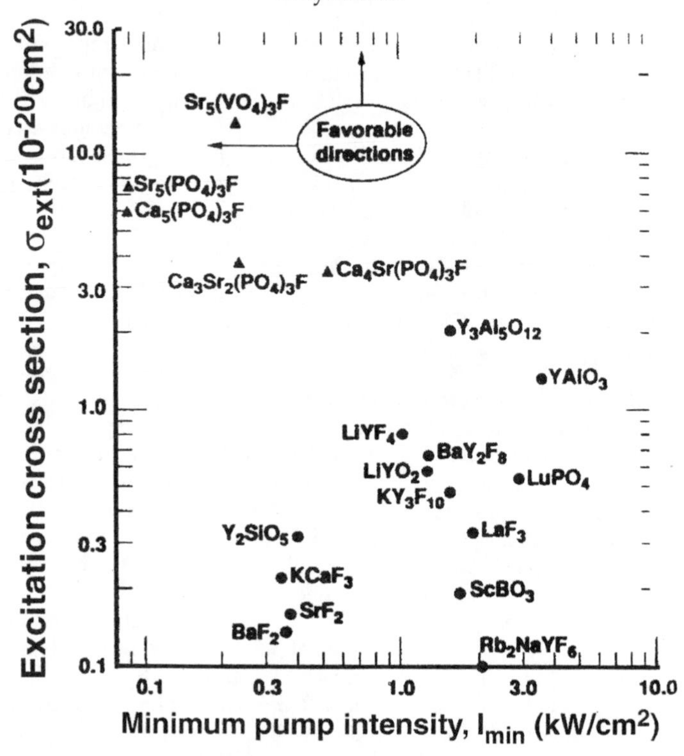

Figure 29. Laser performance parameter chart. Modified after Payne et al. (1994).

variety of apatite structures has been described by Oetliker et al. (1994). The latter work indicates very low thermal quenching at room temperature in the apatite structures compared to other Mn^{5+} host crystals, a distinct advantage for laser applications.

ISSUES CONCERNING APATITE LUMINESCENCE

Apatite luminescence is sufficiently complex that very many questions remain about activators, energy transfer, quenching effects, and site occupation. Although the crystal chemistry of fluorapatite is fairly well understood, the same cannot be said for many of the other apatite compositions with respect to activator placement, charge compensation, defect structures and lattice effects on luminescent properties. Outstanding questions also remain on the possible use of apatite luminescence for age dating or paleotemperature measurements, as the thermoluminescence behavior has had limited investigation. On the other hand, progress has been made in understanding the history of growth of apatites via luminescence mapping to yield REE selective uptake and redox information. With the continued development of micro CL methods (e.g., Kempe and Gotze 2002) a potentially powerful complementary tool is added to standard microprobe or SEM chemical analysis (Fig. 30). Quantitative measurement of REE concentration levels in apatite group minerals via luminescence spectroscopy, which is plagued by energy transfer issues in CL and UV-excited work, might be practically done using time-resolved laser-excitation techniques. This can potentially also be extended to microfocus situations and even raster-scanning of samples given adequate laser focusing and detector speed. Such capabilities depend on the ability to excite directly into the particular REE excitation bands, and measure emission before energy transfer has occurred. This may be possible for many of the REE, allowing improved feasibility for REE partitioning measurements at extremely low REE concentrations, and the use of REE relative abundance signatures to extract information on petrologic history. For the mineral collector who appreciates apatite luminescence, there seems to be almost no end to the diversity of natural materials which have been located, and are being newly discovered. With the advent of UV tunable lasers in the not so distant future, many of these enthusiasts will be turned into amateur spectroscopists by this single species.

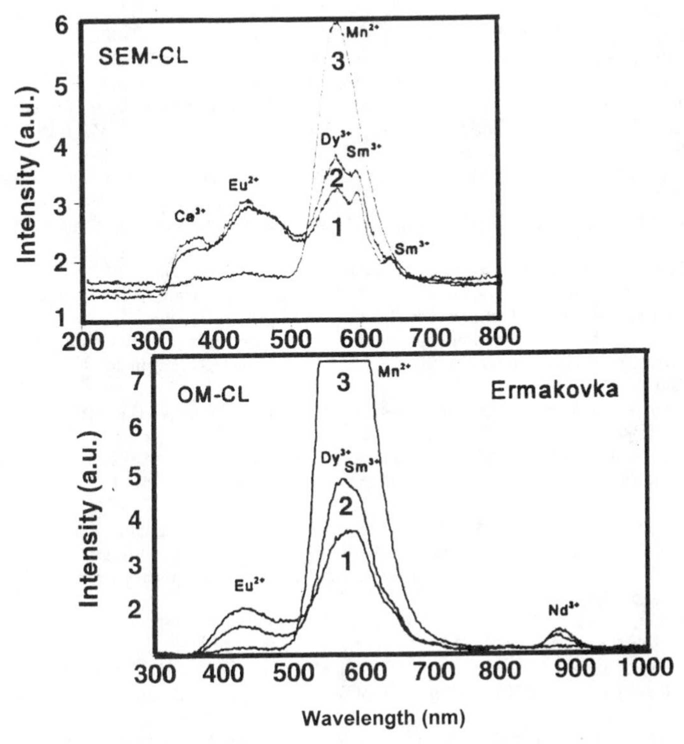

Figure 30. Micro (SEM) vs. bulk (optical microscope apparatus) CL from a sample of fluorapatite from a Be deposit at Ermakova, Russia. The numbers refer to positions on the crystal. 3 is near the crystal rim, while 1 and 2 are in the interior. Modified after Kempe and Götze (2002).

ACKNOWLEDGMENTS

Past discussions with Emmanuel Fritsche, Manuel Robbins, John Rakovan and many members of the Fluorescent Mineral Society (FMS) both contributed to and inspired this chapter. Thorough reviews by John Hanchar, John Rakovan, Will White and George Rossman contributed substantially to the final composition, and they are much appreciated. Manuscript and figure production was assisted by the staff of the Geochemistry department at the Lawrence Berkeley National Laboratory. Support for the this work came from the Office of Science, Office of Basic Energy Sciences, Division of Chemical Sciences, Geosciences Research Program of the U.S. Department of Energy under contract DE-AC03-76-SF00098 to LBNL.

REFERENCES

Al-Khalifa IJM, James K, Durrani SA, Khalifa MS (1988) Radiation damage studies of mineral apatite, using fission tracks and thermoluminescence techniques. Nucl Tracks Rad Measure 15:61-64

Barbarand J, Pagel M (2001) Cathodoluminescence study of apatite crystals. Am Mineral 86:473-484

Baumer A, Lapraz D, Klee WE (1987) Thermoluminescent properties of hydrothermally prepared apatites. Neues Jahrb Mineral Mon 1987:43-48

Benoit PH, Sears DWG (1997) The orbits of meteorites from natural thermoluminescence. Icarus 125: 281-287

Benoit PH, Sears DWG (1991) The natural thermoluminescence of meteorites. II. Meteorite orbits and orbital evolution. Icarus 94:311-325

Blanc P, Baumer A, Cesbron F, Ohnenstetter D (1995) Les activateurs de Cathodoluminescence dans les chlorapatites preparees par synthese hydrothermale. C R Acad Sci Paris Ser IIa 321:119-126

Blanc P, Baumer F, Cesbron F, Ohnenstetter D, Panczer G, Remond G (2000) Systematic cathodoluminescence spectral analysis of synthetic doped minerals: Anhydrite, apatite, calcite, fluorite, scheelite and zircon. *In* Cathodoluminescence in Geosciences. Pagel M, Barbin V, Blanc P, Ohnenstetter D (eds) Springer Verlag, Berlin

Blasse G, Bleijenberg KC, Krol DM (1979) The luminescence of hexavalent uranium in solids. J Lumin 18-19:57-62

Blasse G (1988) Luminescence of inorganic solids: From isolated centers to concentrated systems. Prog Sol State Chem 18:79-191

Blasse G, Aguilar M. (1984) Luminescence of natural calcite ($CaCO_3$). J Lumin 29:239-241

Blasse G, Bril A (1967) Investigation of Ce^{3+}-activated phosphors. J Chem Phys 47:5139-5145

Blasse G, Grabmaier BC (1994) Luminescent Materials. Springer-Verlag, Berlin

Bostwick RC (1977) The fluorescent minerals of Franklin and Sterling Hill. J Fluor Mineral Soc 6:7-40

Botden TPJ, Kröger FA (1948) Energy transfer in sensitized $Ca_3(PO_4)_2$:Ce:Mn and $CaSiO_3$:Pb:Mn. Physica 14:553-566

Burns RG (1993) Mineralogical Applications of Crystal Field Theory. Cambridge University Press, Cambridge, UK

Chen N, Pan Y, Weil JA (2002a) Electron paramagnetic resonance study of synthetic fluorapatite: Part 1. Local structural environment and substitution mechanism of Gd^{3+} at the Ca2 site. Am Mineral 87: 37-46

Chen N, Pan Y, Weil JA, Nilges MJ (2002b) Electron paramagnetic resonance study of synthetic fluorapatite: Part 2. Gd^{3+} at the Ca1 site, with a neighboring Ca1 vacancy. Am Mineral 87:47-55

Chenot CF, Kasenga AF, Poppalardo RE. (1981) Depreciation in cerium-activated fluorapatite phosphors. J Lumin 24-25:95-98

Copobianco JA, Cormier G, Moncourge R, Manaa H, Bertinelli M. (1992) Gain measurements of Mn^{5+} ($3d^2$) doped $Sr_5(PO_4)_3Cl$ and Ca_2PO_4Cl. Appl Phys Lett 60:163-165

Cotton FA (1971) Chemical Applications of Crystal Field Theory. John Wiley & Sons, New York

Cotton FA, Wilkinson G (1980) Advanced Inorganic Chemistry, 5th Edition. John Wiley & Sons, New York

DeBoer BG, Sakthivel A, Cagle JR, Young RA (1991) Determination of the antimony substitution site in calcium fluorapatite from powder x-ray diffraction data. Acta Crystallogr B 47:683-692

DeLoach LD, Payne SA, Kway WL, Tasssano JB, Dixit SN, Krupke WF (1994) Vibrational structure in the emission spectrum of Yb(3+) doped apatite crystals. J Lumin 62:85-94

DeNeufville JP, Kasdan A, Chimenti R (1981) Selective detection of uranium by laser-induced fluorescence: A potential remote sensing technique. I. Optical characteristics of uranyl geologic targets. Appl Optics 20:1279-1296

Diaz MA, Luff BJ, Townsned PD, Wirth KR (1991) Temperature dependence of luminescence from zircon, calcite, iceland spar and apatite. Nucl Tracks Rad Measure 18:45-51

Elliott JC (1994) Structure and Chemistry of the Apatites and Other Calcium Orthophosphates. Elsevier, Amsterdam, The Netherlands

Filippelli GM, Delaney ML (1993) The effect of manganese (II) and iron (II) on the cathodoluminescence signal in synthetic apatite. J Sed Petrol 63:167-173

Fleet ME, Pan Y (1994) Site preference of Nd in fluorapatite ($Ca_{10}(PO_4)_6F_2$). J Solid State Chem 112:78-81

Fleet ME, Pan Y (1995a) Crystal chemistry of rare earth elements in fluorapatite and some calc-silicates. Eur J Mineral 7:591-605

Fleet ME, Pan Y (1995b) Site preference of rare earth elements in fluorapatite. Am Mineral 80:329-335

Fleet ME, Pan Y (1997a) Site preference of rare earth elements in fluorapatite: Binary (LREE+HREE)-substituted crystals. Am Mineral 82:870-877

Fleet ME, Pan Y (1997b) rare earth elements in apatite: Uptake from H_2O-bearing phosphate-fluoride melts and the role of volatile components. Geochim Cosmochim Acta 61:4745-4760

Gaft M, Bershov L, Krasnaya A, Yaskolko V (1985) Luminescence centers in anhydrite, barite, celestite and their synthesized analogs. Phys Chem Minerals 11:255-260

Gaft M, Reisfeld R, Panczer G, Shoval S, Garapon C, Boulon G, Strek W (1996a) Luminescence of Eu(III), Pr(III) and Sm (III) in carbonate-fluor-apatite. Acta Phys Pol A 90:267-274

Gaft M, Shoval S, Panczer G, Nathan Y, Champagnon B, Garapon C (1996b) Luminescence of uranium and rare-earth elements in apatite of fossil fish teeth. Palaeogeogr Palaeoclimatol Palaeoecol 126: 187-193

Gaft M, Reisfeld R, Panczer G, Shoval S, Champagnon B, Boulon G (1997a) Eu^{3+} luminescence in high-symmetry sites of natural apatite. J Lumin 72:572-574

Gaft M, Reisfeld R, Panczer G, Boulon G, Shoval S, Champagnon B (1997b) Accommodation of rare-earths and manganese by apatite. Opt Materials 8:149-156

Gaft M, Reisfeld R, Panczer G, Blank P, Boulon G (1998a) Laser-induced time-resolved luminescence of minerals. Spectrochim Acta A 54:2163-2175

Gaft M, Reisfeld R, Panczer G, Uspensky E, Varrel B, Boulon G (1999) Luminescence of Pr^{3+} in minerals. Opt Materials 13:71-79

Gaft M, Panczer G, Reisfeld R, Shinno I, Champagnon B, Boulon G (2000) Laser-induced Eu^{3+} luminescence in zircon $ZrSiO_4$. J Lumin 87-89:1032-1035

Gaft M, Panczer G, Reisfeld R, Uspensky E (2001a) Laser-induced time-resolved luminescence as a tool for rare-earth element identification in minerals. Phys Chem Minerals 28:347-363

Gaft M, Reisfeld R, Panczer G, Ioffe O, Sigal I (2001b) Laser-induced time-resolved luminescence as a means for discrimination of oxidation states of Eu in minerals. J Alloys Compounds 323-324:842-846

Gait RI, Back ME (1992) Pyromorphite: A review. Rocks Minerals 67:22-36

Garcia J, Sibley WA (1988) Energy transfer between europium and manganese ions. J Lumin 42:109-116

Garlick GFJ (1958) *In* Handbook der Physik. Vol XXVI. Flugge S (ed) Springer, Berlin–Heidelberg–New York, p 1

Gorobets BS (1968) On the luminescence of fluorapatite activated by rare-earth elements. Optics Spectr 25:154-155

Gorobets BS (1981) Luminescence Spectra of Minerals. (in Russian) Moscow. 153 p

Gotze J, Heimann RB, Hildebrandt H, Gburek U (2001) Microstructural investigation into calcium phosphate biominerals by spatially resolved cathodoluminescence. Mater Wissen Werkstofftechnik 32:130-136

Grandjean-Lecuyer P, Feist R, Albarede, F (1993) Rare-earth elements in old biogenic apatites. Geochim Cosmochim Acta 57:2507-2514

Gruber JB, Wright AO, Seltzer MD, Zandi B, Merkle LD, Hutchinson JA, Morrison CA, Allik TH, Chai BHT (1997a) Site-selective excitation and polarized absorption and emission spectra of trivalent thulium and erbium in strontium fluorapatite. J Appl Phys 81:6585-6598

Gruber JB, Zandi B, Seltzer MD (1997b) Spectra and energy levels of trivalent holmium in strontium fluorapatite. J Appl Phys 81:7506-7513

Gruber JB, Zandi B, Merkle L (1998) Crystal-field splitting of energy levels of rare earth ions $Dy^{3+}(4f^9)$ and $Yb^{3+}(4f^{13})$ in M(II) sites in fluorapatite crystal $Sr_5(PO_4)_3F$. J Appl Phys 83:1009-1016

Gruber JB, Zandi B, Ferry M, Merkle L (1999) Spectra and energy levels of trivalent samarium in strontium fluorapatite. J Appl Phys 86:4377-4382

Guimon RK, Weeks KS, Keck BD, Sears DWG (1984) Thermoluminescence as a paleothermometer. Nature 311:363-365

Habermann D, Gotte T, Meijer J, Stephan A, Richter DK, Niklas JR (2000) High-resolution rare-earth element analyses of natural apatite and its applications in geo-sciences: Combined micro-PIXE, quantitative CL spectroscopy and electron spin resonance analyses. Nucl Inst Methods Phys Res B 161-163:846-851

Henderson B, Imbusch GF (1989) Optical Spectroscopy of Inorganic Solids. Clarendon Press, Oxford

Henkel G (1989) The Henkel glossary of fluorescent minerals. Verbeek E, Modreski P (eds) *In* J Fluor Mineral Soc, Vol. 15

Hughes JM, Cameron M, Crowley KD (1991a) Ordering of the divalent cations in the apatite structure: Crystal structure refinements of natural Mn- and Sr-bearing apatite. Am Mineral 76:1857-1862

Hughes JM, Cameron M, Mariano AN (1991b) Rare-earth-element ordering and structural variations in natural rare-earth-bearing apatites. Am Mineral 76:1165-1173

Jagannathan R, Kottaisamy M. (1995) Eu^{3+} luminescence: A spectral probe in $M_5(PO_4)_3X$ apatites (M = Ca or Sr, X = F^-, Cl^-, Br^- or OH^-. J Phys Condens Matter 7:8453-8466

Jagannathan R, Kutty TRN (1997) Anomalous fluorescence features of Eu^{2+} in apatite-pyromorphite type matrices. J Lumin 71:115-121

Jenkins HG, McKearg AH, Ranby PW (1949) Alkaline earth halophosphates and related phosphors. J Electrochem Soc 96:1-12

Kalsbeek N, Larsen S, Ronsbo JG (1990) Crystal structures of rare earth elements rich apatite analogues. Z Kristallogr 191:249-263

Kempe J, Gotze J (2002) Cathodoluminescence (CL) behavior and crystal chemistry of apatite from rare-metal deposits. Mineral Mag 66:151-172

Knutson C, Peacor DR, Kelly WC (1985) Luminescence, color and fission track zoning in apatite crystals of the Panasqueira tin-tungsten deposit, Beira-Baixa, Portugal. Am Mineral 70:829-837

Kottaisamy M, Jagannathan R, Jeyagopal P, Rao RP, Narayanan R (1994) Eu^{2+} luminescence in $M_5(PO_4)_3X$ apatites, where M is Ca^{2+}, Sr^{2+} and Ba^{2+}, and X is F^-, Cl^-, Br^- and OH^-. J Phys D 27:2210-2215

Lapraz D, Baumer A (1983) Thermoluminescent properties of synthetic and natural fluorapatite, $Ca_5(PO_4)3_F$. Phys Stat Solidi A 80:353-366

Lapraz D, Gaume F, Barland M (1985) On the thermoluminescent mechanism of a calcium fluorapatite single crystal doped with Mn^{2+}. Phys Stat Solidi A 89:249-253

Leckebusch R (1979) Comments on the luminescence of apatites from Panasqueria (Portugal). N Jahrb Mineral Monat 17-21

Leverenz HW (1968) An Introduction to Luminescence of Solids. Dover Publications, New York

Louis-Achille V, DeWindt L, Defranceschi M (2000) Electronic structure of minerals: The apatite group as a relevant example. Intl J Quantum Chem. 77:991-1006

Loutts GB, Hong P, Chai BHT (1994) Comparison of neodymium laser hosts based on a fluoro-apatite structure. Mater Res Soc 1994:45-49

Mackie PE, Young RA (1973) Location of Nd dopant in fluorapatite, $Ca_5(PO_4)_3F$:Nd. J Appl Crystallogr 6:26-31

Marfunin AS (1979) Spectroscopy, Luminescence and Radiation Centers in Minerals. Springer-Verlag, Berlin

Mariano AN (1988) Some further geological applications of cathodoluminescence. *In* Cathodoluminescence of Geological Materials. Marshall DJ (ed) Unwin Hyman, Boston

Mariano AN (1989) Cathodoluminescence emission spectra of REE activators in minerals. Rev Mineral 21:339-348

Mariano AN, Ring PJ (1975) Europium-activated cathodoluminescence in minerals. Geochim Cosmochim Acta 39:649-660

Marshall DJ (1988) Cathodoluminescence of Geologic Materials. Unwin Hyman, Boston, 146 p

Mayer I, Layani JD, Givan A, Gaft M, Blanc P (1999) La ions in precipitated hydroxyapatites. J Inorg Biochem 73:221-226

McConnel D (1973) Apatite: Its crystal chemistry, mineralogy, utilization, and geologic and biologic occurrences. Applied Mineralogy 5. Springer-Verlag, New York

McKeever SWS (1985) Thermoluminescence of Solids. Cambridge University Press, Cambridge, UK

McKeever SWS, Sears DW (1979) Meteorites and Thermoluminescence. Meteoritics 14:29-41

Mitchell RH, Xiong J, Mariano AN, Fleet ME (1997) Rare-earth-element-activated cathodoluminescence in apatite. Can Mineral 35:979-998

Moncorge R, Manaa H, Boulon G (1994) Cr^{4+} and Mn^{5+} active centers for new solid state laser materials. Opt Mater 4:139-151

Moran LB, Berkowitz JK, Yesinowski JP (1992) ^{19}F and ^{31}P magic-angle spinning nuclear magnetic resonance of antimony (III)-doped fluorapatite phosphors: Dopant sites and spin diffusion. Phys Rev B 45:5347-5360

Morozov A, Morozova L, Trefilov A, Feofilov P (1970) Spectral and luminescent characteristics of fluorapatite single crystals activated by rare earth ions. Opt Spectros 29:590-596

Murray JR, Oreskes N (1997) Uses and limitations of cathodoluminescence in the study of apatite paragenesis. Econ Geol 92:368-376

Narita K (1961) Energy spectrum of Mn^{++} ion in Calcium Fluorophosphate. I. J Phys Soc Japan 16:99-105

Narita K (1963) Energy spectrum of Mn^{++} ion in Calcium Fluorophosphate. II. J Phys Soc Japan 18:79-86

Nor AF, Amin YM, Mahat R, Kamaluddin B (1994) Thermoluminescence of Apatite. Solid State Sci Tech 2:294-300

Oetliker U, Herren M, Gugel HU, Kesper U, Albrecht C, Reinen D (1994) Luminescence properties of Mn^{5+} in a variety of host lattices: Effects of chemical and structural variation. J Chem Phys 100: 8656-8665

Ohlmann RC, Steinbruegge KB, Maczelsky R (1968) Spectroscopic and laser characteristics of neodymium-doped calcium fluorophosphate. Appl Opt 7:905-914

Oomen EWJL, Smit WMA, Blasse G (1988) Luminescence of the Sb^{3+} ion in calcium fluorapatite and other phosphates. Mater Chem Phys 19:357-368

Palilla FC, O'Reilly BE (1968) Alkaline-earth halophosphate phosphors activated by divalent europium. J Electrochem Soc 115:1076-1081

Suitch PR, LaCout JL, Hewat A, Young RA (1985) The structural location and role of Mn^{2+} partially substituted for Ca^{2+} in fluorapatite. Acta Crystallogr B41:173-179

Tachihante M, Zambon D, Cousseins JC (1996) optical study of the Tb^{3+} to Eu^{3+} energy transfer in calcium fluorapatite. Eur J Solid State Inorg Chem 33:713-725

Taraschan A. (1978) Luminescence of Minerals. Naukova Dumka, Kiev (in Russian)

Taraschan A, Waychunas G (1995) Luminescence of minerals. *In* Advanced Mineralogy 2. Methods and Instrumentation. Marfunin A (ed) Springer, Berlin, p 124-135

Ternane R, Trabelsi-Ayedi M, Kbir-Ariguib N, Piriou B (1999) Luminescent properties of Eu in calcium hydroxyapatite. J Lumin 81:153-236

Vasilenko VB, Sotnikov VI, Nikitina YI, Kholodova LD (1968) The possibility of using apatite luminescence in geologic thermometry (in Russian). Geologiya Geofizika 11:131-135

Voronko YK, Gorbachev AV, Zverev AA, Sobil AA, Morozov NN, Muravev EN, Niyazov SA, Orlovskii VP (1992) Raman scattering and luminescence spectra of compounds with the structure of apatite $Ca_5(PO_4)_3F$ and $Ca_5(PO_4)_3OH$ activated with Eu^{3+} ions. Inorg Mater 28:442-447

Voronko YK, Maksimova G, Sobol AA (1991) Anisotropic luminescence centers of TR^{3+} ions in fluorapatite crystals. Opt Spectros 70:203-206

Wachtel A (1958) The effect of impurities on the plaque brightness of a 3000 K calcium halophosphate phosphor. J Electrochem Soc 105:256-260

Warren RW (1970) EPR of Mn^{2+} in calcium fluorophosphates. I. The Ca (II) site. Phys Rev B 2:4383-4388

Waychunas G (1989) Luminescence, X-ray emission and new spectroscopies. Rev Mineral 18:638-698

Wright AO, Seltzer MD, Gruber JB, Zandi B, Merkle LD, Chai BHT (1996) Spectroscopic investigation of Pr^{3+} in fluorapatite crystals. J Phys Chem Solids 57:1337-1350

Xiong J (1995) Cathodoluminescence studies of feldspars and apatites from the Coldwell alkaline complex. MSc thesis, Lakehead Univ, Thunder Bay, Ontario

Zhang XX, Loutts GB, Bass M, Chai BHT (1994) Growth of laser-quality single crystals of Nd^{3+}-doped calcium fluorapatite and their efficient lasing performance. Appl Phys Lett 64:10-12

Panczer G, Gaft M, Reisfeld R, Shoval S, Boulon G, Champagnon B (1998) Luminescence of ura natural apatites. J. Alloys Compounds 275-277:269-272

Payne SA, DeLoach LD, Smith LK, Kway WL, Tassno JB, Krupke WF, Chai BHT, Loutts G Ytterbium-doped apatite-structure crystals-a new class of laser materials. J Appl Phys 76:497-50

Payne SA, Smith LK, DeLoach LD, Kway WL, Tassano JB, Krupke WF (1994b) Laser, optic thermomechanical properties of Yb-doped fluorapatite. IEEE J Quantum Elect 30:170-179

Perseil E-A, Blanc P, Ohnenstetter D (2000) As-bearing fluorapatite in manganiferous deposi St. Marcel-Praborna, Val D'Aosta, Italy. Can Mineral 38:101-117

Piriou B, Elfakir A, Quarton M (2001) Site-selective spectroscopy of Eu^{3+}-doped sodium lead pho apatite. J Lumin 93:17-26

Portnov AM, Gorobets BS (1969) Luminecence of apatite from different rock types. Dokl Akad SSSR 184:110-115 (in Russian)

Rakavon J, Hughes JM (2000) Strontium in the apatite structure: Strontian apatite and belovite-C Mineral 38:839-845

Rakovan J, Reeder RJ (1994) Differential incorporation of trace elements and dissymmetrization in a The role of surface structure during growth. Am Mineral 79:892-903

Rakovan J, Reeder RJ (1996) Intracrystalline rare earth element distributions in apatite:
Surface structural influences on zoning during growth. Geochem. Cosmochim. Acta 60:4435-4445

Rakovan J, Waychunas GA (1996) Luminescence in minerals. Mineral Record 27:7-19

Rakovan J, McDaniel DK, Reeder RJ (1997) Use of surface-controlled REE sectoral zoning in apatite Llallagua, Bolivia, to determine a single crystal Sm-Nd age. Earth Planet Sci Lett 146:329-336

Rakovan J, Newville M, Sutton S (2001) Evidence of Heterovalent Europium in Zoned Llallagua A using Wavelength Dispersive XANES. Am Mineral 86:697-700

Ramesh R, Jagannathan R (2000) Optical properties of Ce^{3+} in self-assembled strontium ch (hydroxyl)apatite nanocrystals. J Phys Chem B 104:8351-8360

Ratnam VV, Jayaprakish R, Daw NP (1980) Thermoluminescence and thermoluminescence spect synthetic fluorapatite. J Lumin 21:417-424

Reisfeld R, Gaft M, Boulon G, Panczer C, Jrgensen CK (1996) Laser-induced luminescence of rare-e elements in natural fluor-apatites. J Lumin 69:343-353

Remond G, Cesbron F, Chapoulie R, Ohnenstetter D, Roques-Carmes C, Schvoerer M (19 Cathodoluminescence applied to microcharacterization of mineral materials: The present statu experimentation and interpretation. Scanning Micros 6:23-68

Robbins DJ (1980) On predicting the maximum efficiency of phosphor systems excited by ioniz radiation. J Electrochem Soc 127:2694-2701

Robbins M (1983) The collectors book of fluorescent minerals. Van Nostrand Reinhold, New York

Robbins M (1994) Fluorescence: Gems and Minerals Under Ultraviolet Light. Geoscience Press, Phoenix

Roeder PL, MacArthur D, Ma X-P, Palmer GR, Mariano AN (1987) Cathodoluminescence and micropro study of rare-earth elements in apatite. Am Mineral 72:801-811

Ronsbo JG (1989) Coupled substitutions involving REEs and Na and Si in apatites in alkaline rocks fro the Ilimaussaq intrusion, South Greenland, and the petrological implications. Am Mineral 74:896-901

Ropp RC (1971) The emission colors of the strontium apatite phosphor system. J Electrochem So 118:1510-1512

Rossman GR (1988) Optical Spectroscopy. Rev Mineral 18:207-254

Ryan FM, Vodoklys FM (1971) The Optical Properties of Mn^{2+} in Calcium halophosphate phosphors. J Electrchem Soc 118:1814-1819

Ryan FM, Ohlmann RC, Murphy J, Mazelsky R, Wagner GR, Warren RW (1970) Optical properties of divalent manganese in calcium fluorophosphate. Phys Rev B 2:2341-2352

Scott MA, Han TPJ, Gallagher HG, Henderon B (1997) Near-infrared laser crystals based on $3d^2$ ions: Spectroscopic studies of $3d^2$ ions in oxide, melilite and apatite crystals. J Lumin 72-74:260-262

Sears DWG (1980) Thermoluminescence of meteorites: Relationships with their K-Ar age and their shock and reheating history. Icarus 44:190-206

Shionoya S, Yen WM (1999) Phosphor Handbook. CRC Press, Boca Raton, Florida

Smith JV, Stenstrom RC (1965) Electron-excited luminescence as a petrologic tool. J Geol 73:627-635

Stefanos SM, Bonner CE, Meegoda C, Rodriguez WJ, Loutts GB (2000) Energy levels and optical properties of neodymium-doped barium fluorapatite. J Appl Phys 88:1935-1942

Solomonov VI, Osipov VV, Mikhailov SG (1993) Pulse-periodic cathodoluminescence of apatite. Zh Prikladnoi Spectroskopii 59:107-113

Steinbruegge KB, Hennigsen T, Hopkins RH, Mazelsky R, Melamed NT, Riedel EP, Roland GW (1972) Laser properties of Nd^{3+} and Ho^{3+} doped crystals with the apatite structure. Appl Optics 11:999-1012

Stokes GG (1852) On the change of refrangibility of light. Phil Trans 142:463-562